LIBRARY
ZOOLOGICAL MUSEUM
UNIVERSITY OF WISCONSIN-MADISON

D1716363

DICTIONARY OF GENERIC NAMES

OF SEED PLANTS

DICTIONARY OF GENERIC NAMES OF

Seed Plants

Tatiana Wielgorskaya

Consulting Editor
ARMEN TAKHTAJAN

COLUMBIA UNIVERSITY PRESS
New York

Biology Library
University of Wisconsin - Madison
Birge Hall
430 Lincoln Drive
Madison, WI 53706-1381

Columbia University Press
New York Chichester, West Sussex
Copyright (c) 1995 Columbia University Press
All rights reserved
Library of Congress Cataloging-in-Publication Data

Wielgorskaya, Tatiana.
   Dictionary of generic names of seed plants / Tatiana Wielgorskaya.
     p. cm.
   ISBN 0–231–07892–7
   1. Botany—Nomenclature. I. Title.
QK11.W52 1995   95-6723
581′.014—dc20   CIP
Casebound editions of Columbia University Press books are printed
on permanent and durable acid-free paper.
Printed in the United States of America
c 10 9 8 7 6 5 4 3 2 1

CONTENTS

Foreword by Peter Raven	vii
Preface	ix
Acknowledgments	xi
List of Familes	1
Dictionary of Generic Names of Seed Plants	85

FOREWORD

The parents of a "new" dictionary of generic names about seed plants, a reference work to what is known about these genera, is an event of considerable importance for the botanical community. With its detailed information about geographical distribution, numbers of species in the different genera, and their dates of publication, this work will be an important step in our increasing knowledge of the genera of plants

The botanical community and those who have need for information about the nature and distribution of plants around the world will find this work to be of fundamental importance. The selection of names to be included has been carried out on a reasonable basis, with the result that the present book will be a very useful reference.

We live in a time when as many as a fifth—fully 50,000 species—of the estimated 250,000 of seed plants in the world may be facing extinction over the next three decades. Their wise use, proper management, study, and conservation ought to be a matter of concern for every human being. Since so much of our future will be based on our ability to carry out those activities well, understand what is known is the first priority in acting intelligently. For this reason, the botanical community owes a special debt of gratitude to Tatiana Wielgorskaya and especially to Academician Armen Takhtajan for his leadership in the realization of this important project.

PETER H. RAVEN
St. Louis, Missouri

have been of great help in the compilation of this book. I highly appreciate their untiring efforts in verification of my card indexes of the various plant families. I am grateful to L. V. Averyanov and R. L. Dressler (Orchidaceae), I. A. Grudzinskaya (Urticaceae), E. V. Mordak (Hyacinthaceae and Lilaceae s.str.), V. N. Shulkina (Campanulaceae), M. G. Pimenov and V. N. Tihkomirov (Apiaceae), made by D. Nicolson (Araceae), and H. van der Werff (Lauraceae). Useful conversations with W. G. D'Arcy (Solanaceae), I. A. Al-Shehbaz (Brassicaceae), Jeremy Bruhl (Cyperaceae), and Alwin Gentry (Bignoniaceae) during my work on dictionary in the Missouri Botanical Garden are greatly appreciated. For their critical comments on the manuscript I am grateful to Peter Goldblatt (Iridaceae) and Dale Johnson (Asteraceae).

I wish to express my sincere gratitude to Arthur Cronquist, who had constantly helped me in one way or other during my work on the Dictionary.

I am deeply indebted to the staff members of the Komarov Botanical Institute Library and the library of the Missouri Botanical Garden for their help and the steady warm support.

Reading the typescript of the dictionary needed truly stoical patience. Dr. M. E. Kirpicznikov has kindly read the whole typescript and it is difficult to overestimate the value of his help and advice. I am very thankful to him.

Special acknowledgment must be given to Dr. Peter H. Raven, Director of the Missouri Botanical Garden. Thanks to his help it was possible to computerize this dictionary. I would especially like to thank Dr. Nancy Morin, who has done so much to forward the publication of this book, and in arranging the printing out of hard copies at the Missouri Botanical Garden.

The idea of compilation of *Dictionary of Generic Names of Seed Plants* belongs to Armen Takhtajan and its realization was possible thanks to his consultation, every day help, continuous stimulation, and encouragement.

I pay tribute to the editors of the Columbia University Press for their work on the text of this book.

T. WIELGORSKAYA
St. Petersburg
March 1995

PREFACE

Dictionary of Generic Names of Seed Plants is a reference book containing brief information about names at the rank of genus. It has been written both for professional and amateur botanists, as well as for horticulturalists, foresters, ecologists, geographers, phytochemists, and all those who are interested in plants.

Two similar books have been published. In 1897 the first edition of *A Dictionary of the Flowering Plants and Ferns* by John Christopher Willis (1868–1958) appeared. Five subsequent editions were published during the life of the author. Subsequently two considerably revised editions (1966 and 1974) were prepared by H. K. Airy Shaw. In 1987 *The Plant-book: A Portable Dictionary of the Higher Plants* by D. J. Mabberley was published.

The present book differs from the other dictionaries in that it mentions more detailed geographical distribution, an updated number of species within the genera, as well as the dates of their publication. Authors of many genera and years of publications are corrected in many cases. The verification and updating of all these data has required careful study of all available botanical periodicals and monographs published in the last decades, e.g. *Genera Palmarum* by N. W. Uhl and J. Dransfield (1987), *The Genera of Chrysobalanaceae* by G. T. Prance and F. White (1988), *Legumes of Africa: A Checklist* by J. M. Lock (1989). *Legumes of West Asia: A Checklist* by J. M. Lock and K. Simpson (1991), *Australian Plant Name Index* by A. D. Chapman (1991), etc. a great number of the modern floras, including *Flora Reipublicae Popularis Sinicae, Flora Malesiana, Flora Iranica, Flora of Turkey and the East Aegean Islands, Flora of tropical East Africa, Flora of Southern Africa, Flore de la Nouvelle-Caledonie et Dependances, Flora Neotropica, Flora of the USSR,* and numerous regional floras. Both editions of *Die Naturlichen Pflanzenfamilien* were used as well as *Das Pflanzenreich, Index Nominum Genericorum, Index Kewensis, Kew Index, International Code of Botanical Nomenclature* (1988).

In contrast to Willis's Dictionary, which is based on the Englerian philosophy of the family rank, and D. J. Mabberley's book, which is based on the Cronquist's concept of family, the present book follows Takhtajan's system. There are also essential differences from the Airy Shaw's concept as it was expressed in his revised edition of Willis's Dictionary.

I have considerably reduced the number of generic synonyms (as compared with Willis's Dictionary) mentioning only the "living synonyms," that is only those which are found in modern botanical literature. Generic hybrid names are also excluded. On the other hand I have added many recently published generic names which are not to be found in other reference books. Each generic entry includes: name/ authority (in standard abbreviation)/the date (year)/ the family/ the number of species/geographical distribution. Family names (including synonyms) are listed alphabetically before the list of generic names. The family name entry consists of the order to which the family is assigned, the number of genera and species, and the distribution of the family. Subfamilies, and the principal genera (or all the genera for smaller families) are also listed.

Many botanists from various countries

ACKNOWLEDGMENTS

For their assistance in the preparation of this project, Columbia University Press is pleased to acknowledge Nancy Morin, Assistant Director and Anne L. Lehmann Curator of North American Botany at the Missouri Botanical Garden; and Dennis Stevenson, Dean of Graduate Studies at the New York Botanical Garden.

DICTIONARY OF GENERIC NAMES

OF SEED PLANTS

LIST OF FAMILIES

Abamaceae J. Agardh = Melianthaceae Batsch.

Abolbodaceae (Suess. et Beyerle) Nakai = Xyridaceae C. Agardh.

Abrophyllaceae Nakai = Escalloniaceae Dumort.

Acaciaceae Schimp. = Fabaceae Lindl.

Acalyphaceae J. Agardh = Euphorbiaceae Juss.

ACANTHACEAE Juss. 1789. Dicots. Scrophulariales. 322–323/3,910. Chiefly tropics, a few in temp.–warm regions. Genera: Acanthopale, Acanthopsis, Acanthostelma, Acanthura, Acanthus, Achyrocalyx, Adenacanthus, Aechmanthera, Afrofittonia, Ambongia, Ancylacanthus, Andrographis, Angkalanthus, Anisacanthus, Anisosepalum, Anisotes, Anomacanthus, Anthacanthus, Apassalus, Aphanoserma, Aphelandra, Aphelandrella, Ascotheca, Asystasia, Asystasiella, Ballochia, Baphicacanthus, Barleria, Barleriola, Benoicanthus, Bentia, Blechum, Blepharis, Borneacanthus, Boutonia, Brachystephanus, Bravaisia, Bremekampia, Brillantaisia, Brunoniella, Buceragenia, Buteraea, Calacanthus, Calophanoides, Calycacanthus, Camarotea, Cardiacanthus, Carlowrightia, Carvia, Celerina, Cephalacanthus, Chaetacanthus, Chaetochlamys, Chaetothylax, Chalarothyrsus, Chamaeranthemum, Championella, Chilranthemum, Chlamydocardia, Chlamydostachys, Chroesthes, Clinacanthus, Clistax, Codonacanthus, Conocalyx, Corymbostachys, Cosmianthemum, Crabbea, Crossandra, Crossandrella, Cryphiacanthus, Ctenopaepale, Cueotia, Cyclacanthus, Cyclocheilon, Cylindrsolenium, Cyphacanthus, Cystacanthus, Dactylostegium, Danguya, Dasytropis, Dichazothece, Dicladanthera, Dicliptera, Didyplosandra, Diflugossa, Distichocalyx, Ditrichospermum, Dolichostachys, Dossifluga, Drejera, Duosperma, Dyschoriste, Dyspemptemorion, Ecbolium, Echinacanthus, Echinopaepale, Elytraria, Encephalosphaera, Endosiphon, Epiclastopelma, Eranthemum, Eremomastax, Eusiphon, Filetia, Fittonia, Forcipella, Gantelbua, Gastranthus, Geissomeria, Gendarussa, Glossochilus, Golaea, Graphandra, Graptophyllum, Gutzlaffia, Gymapsis, Gymnophragma, Gymnostachyum, Gypsacanthus, Habracanthus, Haemacanthus, Hansteinia, Haplanthus, Harpochilus, Hemigraphis, Herpetacanthus, Heteradelphia, Holographis, Hoverdenia, Hulemacanthus, Hygrophila, Hymenochlaena, Hypoestes, Indoneesiella, Ionacanthus, Isoglossa, Isotheca, Jadunia, Juruasia, Justicia, Kalbreyeriella, Kjellbergia, Kolobochilus, Kosmosiphon, Kudoacanthus, Lankesteria, Larsenia, Lasiocladus, Leandriella, Lepidacanthus, Lepidagathis, Leptacanthus, Leptosiphonium, Leptostachya, Liberatia, Linariantha, Lindauea, Linocalix, Lissospermum, Listrobanthes, Lophostachys, Louteridium, Lychniothyrsus, Mackaya, Mackenziea, Mananthes, Marcania, Megalochlamys, Megalostoma, Megaskepasma, Melittacanthus, Mellera, Mendoncia, Metarungia, Mexacanthus, Meyenia, Mimulopsis, Mirandea, Monechma, Monothecium, Morsacanthus, Nelsonia, Heohallia, Neolindenia, Neriacanthus, Neuracanthus, Nilgirianthus, Odontonema, Ophiorrhiziphyllon, Oplonia, Oreacanthus, Orophochilus,

Pachystachys, Parachampionella, Paragoldfussia, Pararuellia, Parastrobilanthes, Parasympagis, Pelecostemon, Pentstemonacanthus, Perenideboles, Pericalypta, Periestes, Perilepta, Peristrophe, Petalidium, Phaulopsis, Phialacanthus, Phidiasia, Phlebophyllum, Phlogacanthus, Physacanthus, Pleocaulus, Podorungia, Poikilacanthus, Polylychnis, Populina, Pranceacanthus, Psacadopaepale, Pseudacanthopale, Pseudaechmanthera, Pseuderanthemum, Pseudocalyx, Pseudodicliptera, Pseudoruellia, Pseudostenosiphonium, Psilanthele, Psiloesthes, Pteracanthus, Pteroptychia, Ptyssiglottis, Pulchranthus, Pupilla, Pyrrothrix, Razisea, Rhaphidospora, Rhinacanthus, Rhombochlamys, Ritonia, Ruellia, Ruelliopsis, Rungia, Ruspolia, Ruttya, Saintpauliopsis, Salpinctium, Salpixantha, Samuelssonia, Sanchezia, Sapphoa, Sarojusticia, Satanocrater, Sautiera, Schaueria, Schwabea, Sciaphyllum, Sclerochiton, Sebastiano-Schaueria, Semnostachya, Semnothyrsus, Sericocalyx, Sericospora, Sinthroblastes, Siphonoglossa, Sooia, Spathacanthus, Sphacanthus, Sphinctacanthus, Spirostigma, Stachyacanthus, Standleyacanthus, Staurogyne, Steirosanchezia, Stenandrium, Stenosiphonium, Stenostephanus, Stenothyrsus, Stephanophysum, Streblacanthus, Streptosiphon, Strobilacanthus, Strobilantes, Strobilanthopsis, Styasasia, Suessenguthia, Supushpa, Sympagis, Synchoriste, Tacoanthus, Taeniandra, Talbotia, Tarphochlamys, Teliostachya, Tessmanniacanthus, Tetraglochidium, Tetragoga, Tetragompha, Tetramerium, Theileamea, Thelepaepale, Thomandersia, Thunbergia, Thysanostigma, Tremacanthus, Triaenacanthus, Trichanthera, Trichaulax, Trichocalyx, Trichosanchezia, Trybliocalyx, Ulleria, Vandasia, Vavara, Warpuria, Whitfieldia, Xantheranthemum, Xanthostachya, Xenacanthus, Xerothamnella, Yeatesia, Zygoruellia.

ACERACEAE Juss. 1789. Dicots. Sapindales. 2/120–160. Temp. Northern Hemisphere, East and Southeast Asia, Acer laurinum extending to the Southern Hemisphere (Timor I.). Genera: Acer, Dipteronia.

ACHARIACEAE Harms. 1897. Dicot. Violales. 3/3. South Africa. Genera: Acharia, Ceratiosicyos, Guthriea.

ACHATOCARPACEAE Heimerl. 1934. Dicots. Caryophyllales. 2/16. Trop. America. Genera: Achatocarpus, Phaulothamnus.

Achraceae Roberty = Sapotaceae Juss.

ACORACEAE Martinov. 1820. Monocot. Arales. 1/4. Temp. and subtrop. regions of the Northern Hemisphere. Genus Acorus.

Acristaceae O. F. Cook = Arecaceae Schultz Sch.

ACTINIDIACEAE Hutch. 1926. Dicots. Actinidiales. 3/375. Tropics and subtropics, a few species of Actinidia temp. East Asia. Genera: Actinidia, Clematoclethra, Saurauia.

Adenogrammataceae (Fenzl) Nakai = Molluginaceae Hutch.

ADOXACEAE Trautv. 1853. Dicots. Dipsacales. 3/5. Cold and temp. regions of the Northern Hemisphere south to Himalayas and central North America (Colorado and Illinois). Genera: Adoxa, Sinadoxa, Tetradoxa.

Aegialitidaceae Lincz. = Plumbaginaceae Juss.

Aegicerataceae Blume = Myrsinaceae R. Br.

Aeginetiaceae Livera = Orobanchaceae Vent.

Aegiphilaceae Raf. = Verbenaceae J. St.-Hil.

Aesculaceae Bercht. et J. Presl = Hippocastanaceae DC.

AEXTOXICACEAE Engl. et Gilg. 1920. Dicot. Euphorbiales. 1/1. Chile. Genus Aextoxicon.

Agapanthaceae Lotsy = Alliaceae J. Agardh.

Agathidaceae Baum.-Bodenh. = Araucariaceae Henkel et Hochst.

AGAVACEAE Endl. 1841. Monocots. Amaryllidales. 9/410. Trop. and subtrop. America, West Indies. Genera: Agave, Beschorneria, Furcraea, Hesperaloë, Manfreda, Polianthes, Prochnyanthes, Pseudobravoa, Yucca.

AGDESTIDACEAE Nakai. 1942. Dicots. Caryophyllales. 1/1. North and Central America, West Indies. Genus Agdestis.

Agrimoniaceae Gray = Rosaceae Juss.

Agrostidaceae Burnett = Poaceae Barnhart.

Ailanthaceae J. Agardh = Simaroubaceae DC.

Aitoniaceae Harv. = Meliaceae Juss.

AIZOACEAE F. Rudolphi. 1830. Dicots. Caryophyllales. 133–137/2,200. Chiefly arid and semiarid tropics and subtropics, esp. Southwest and South Africa, western and southern Australia. Genera: Acrodon, Aethephyllum, Aizoanthemum, Aizoon, Aloinopsis, Amoebophyllum, Amphibolia, Antegibbaeum, Antimima, Apatesia, Aptenia, Arenifera, Argyroderma, Aridaria, Aspazoma, Astridia, Bergeranthus, Bijlia, Braunsia, Brownanthus, Calamophyllum, Carpanthea, Carpobrotus, Carruanthus, Caryotophora, Cephalophyllum, Cerochlamys, Chasmatophyllum, Cheiridopsis, Circandra, Cleretum, Conicosia, Conophytum, Corpuscularia, Cylindrophyllum, Cypselea, Dactylopsis, Delosperma, Derenbergia, Dicrocaulon, Didymaotus, Dinteranthus, Diplosoma, Disphyma, Dorotheanthus, Dracophilus, Drosanthemum, Eberlanzia, Ebracteola, Enarganthe, Erepsia, Esterhuysenia, Faucaria, Fenestraria, Frithia, Galenia, Gibbaeum, Glottiphyllum, Gunniopsis, Hereroa, Herrea, Herreanthus, Hymenogyne, Ihlenfeldtia, Imitaria, Jacobsenia, Jensenobotrya, Jordaaniella, Juttadinteria, Kensitia, Khadia, Lampranthus, Lapidaria, Leipoldtia, Lithops, Machairophyllum, Malephora, Mesembryanthemum, Mestoklema, Meyerophytum, Mitrophyllum, Monilaria, Mossia, Muiria, Namaquanthus, Namibia, Nananthus, Nelia, Neohenricia, Octopoma, Odontophorus, Oophytum, Ophthalmophyllum, Opophytum, Orthopterum, Oscularia, Ottosonderia, Phyllobolus, Platythyra, Pleiospilos, Plinthus, Polymita, Prenia, Psammophora, Pseudobrownanthus, Psilocaulon, Rabiea, Rhinephyllum, Rhombophyllum, Ruschia, Ruschianthemum, Ruschianthus, Saphesia, Sarcozona, Schlechteranthus, Schwantesia, Scopelogena, Semnanthe, Sesuvium, Skiatophytum, Smicrostigma, Sphalmanthus, Stayneria, Stoeberia, Stomatium, Synaptophyllum, Tanquana, Tetragonia, Titanopsis, Trianthema, Tribulocarpus, Trichodiadema, Vanheerdia, Vanzijlia, Wooleya, Zaleya, Zeuktophyllum.

AKANIACEAE Stapf. 1912. Dicots. Sapindales. 1/1. Eastern Australia. Genus Akania.

ALANGIACEAE DC. 1828. Dicots. Cornales. 1/30 Tropics of the Old World. Genus Alangium.

Alchemillaceae J. Agardh = Rosaceae Juss.

Aldrovandaceae Nakai = Droseraceae Salisb.

ALISMATACEAE Vent. 1799. Monocots. Alismatales. 13/110. Cosmopolitan, but chiefly Northern Hemisphere, esp. North America. Genera: Alisma, Baldellia, Burnatia, Caldesia, Damasonium, Echinodorus, Helanthium, Limnophyton, Luronium, Ranalisma, Rautanenia, Sagittaria, Wiesneria.

ALLIACEAE J. Agardh. 1858. Monocots. Amaryllidales. 36/885. Cosmopolitan, excl. trop. Australia and New Zealand. Genera: Allium, Agapanthus, Ancrumia, Androstephium, Behria, Bessera, Bloomeria, Brodiaea, Caloscordum, Dandya, Dichelostemma, Diphalangium, Erinna, Garaventia, Gethyum, Gilliesia, Hesperoscordum, Ipheion, Leucocoryne, Miersia, Milla, Milula, Muilla, Nectaroscrodum, Nothoscordum, Pabellonia, Petronymphe, Schickendantziella, Speea, Solaria, Trichlora, Tristagma, Triteleia, Triteleiopsis, Tulbaghia, Zoellnerallium.

Allioniaceae Horan. = Nyctaginaceae Juss.

Aloaceae Batsch = Asphodelaceae Juss.

Alpiniaceae F. Rudolphi = Zingiberaceae Lindl.

ALSEUOSMIACEAE Airy Shaw. 1965. Dicots. Hydrangeales. 3/12–13. New Guinea, eastern Australia, New Zealand, New Caledonia. Genera: Alseuosmia, Crispiloba, Wittsteinia.

Alsinaceae (DC.) Bartl. = Caryophyllaceae Juss.

Alsodeiaceae J. Agardh = Violaceae Batsch.

ALSTROEMERIACEAE Dumort. 1829. Monocots. Alstroemeriales. 4/200. Trop. America. Genera: Alstroemeria, Bomarea, Leontochir, Schickendantzia.

Altheniaceae Lotsy = Zannichelliaceae Dumort.

ALTINGIACEAE Lindl. 1846. Dicots. Hamamelidales. 3/20. North and Central America, Mediterranean, trop. and subtrop. Asia from Eastern Himalayas to continental China, Taiwan, Indochina and Java. Genera: Altingia, Liquidambar, Sinoliquidambar.

ALZATEACEAE S. A. Graham. 1985. Dicots. Myrtales. 1/1–2 Trop. America. Genus Alzatea.

AMARANTHACEAE Juss. 1789. Dicots. Caryophyllales. 74/945. Trop. (esp America and Africa), subtrop. and temp. regions. Genera: Achyranthes, Achyropsis, Aerva, Allmania, Allmaniopsis, Alternanthera, Amaranthus, Apterantha, Arthraerua, Blutaparon, Bosea, Calicorema, Celosia, Centema, Centemopsis, Centrostachys, Chamissoa, Charpentiera, Chionothrix, Cyathula, Cyphocarpa, Dasysphaera, Deeringia, Digera, Eriostylos, Froelichia, Froelichiella, Goerziella, Gomphrena, Gossypianthus, Guilleminea, Henonia, Herbstia, Hermbstaedtia, Indobanalia, Iresine, Lagrezia, Leucosphaera, Lithophila, Lopriorea, Marcelliopsis, Mechowia, Nelsia, Neocentema, Nothosaerva, Nototrichium, Nyssanthes, Pandiaka, Pfaffia, Philoxerus, Pleuropetalum, Pleuropterantha, Polyrhabda, Pseudogomphrena, Pseudoplantago, Pseudosericocoma, Psilotrichopsis, Psilotrichum, Ptilotus, Pupalia, Quaternella, Rosifax, Saltia, Sericocoma, Sericocomopsis, Sericorema, Sericostachys, Siamosia, Stilbanthus, Tidestromia, Trichuriella, Volkensinia, Woehleria, Xerosiphon.

AMARYLLIDACEAE J. St.-Hil. 1805. Monocots. Amaryllidales. 60/760. Tropics and subtropics, esp. America, Mediterranean, and Africa. Genera: Amaryllis, Ammocharis, Androstephanos, Apodolirion, Bokkeveldia, Boophone, Brunsvigia, Caliphruria, Callithauma, Calostemma, Carpolyza, Castellanoa, Chapmanolirion, Chlidanthus, Clivia, Crinum, Cryptostephanus, Cybistetes, Cyrtanthus, Eucharis, Eucrosia, Eustephia, Famatina, Galanthus, Gemmaria, Gethyllis, Griffinia, Habranthus, Haemanthus, Hannonia, Haylockia, Hessea, Hieronymiella, Hippeastrum, Hyline, Hymenocallis, Kamiesbergia, Lapiedra, Leptochiton, Leucojum, Lycoris, Mathieau, Namaquanula, Narcissus, Nerine, Pamianthe, Pancratium, Paramongaia, Phaedranassa, Placea, Plagiolirion, Proiphys, Pseudostenomesson, Pucara, Pyrolirion, Rauhia, Scadoxus, Sprekelia, Stenomesson, Sternbergia, Strumaria, Tedingea, Traubia, Ungernia, Urceolina, Vagaria, Worsleya, Zephyranthella, Zephyranthes.

AMBORELLACEAE Pichon. 1948. Dicots. Laurales. 1/1. New Caledonia. Genus Amborella.

Ambrosiaceae Link = Asteraceae Dumort.

Ammentotaxaceae Kudo et Yamam. = Taxaceae Lindl.

Ammanniaceae Horan. = Lythraceae J. St.-Hil.

Ammiaceae (J. Presl et C. Presl) Barnhart = Apiaceae Lindl.

Amomaceae J. St. Hil. = Zingiberaceae Lindl.

Amygdalaceae (Juss.) D. Don = Rosaceae Juss.

Amyridaceae R. Br. = Rutaceae Juss.

ANACARDIACEAE Lindl. 1830. Dicots. Rutales. 78/1,010. Tropics and subtropics, a few species in warm–temp. regions. Genera: Actinocheita, Amphipterygium, Anacardium, Androtium, Antrocaryon, Apterokarpos, Astronium, Baronia, Blepharocarya, Bonetiella, Bouea, Buchanania, Campnosperma, Cardenasiodendron, Choerospondias, Comocladia, Cotinus, Cyrtocarpa, Dobinea, Dracontomelon, Drimycarpus, Ebandoua, Euleria, Euroschinus, Faguetia, Fegimanra, Gluta, Haematostaphis, Haplorhus, Harpephyllum, Heeria, Holigarna, Koordersiodendron, Lannea, Laurophyllus, Lithrea, Loxopterygium, Loxostylis, Mangifera, Mauria, Melanochyla, Metopium, Micronychia, Montagueia, Mosquitoxylum, Nothopegia, Ochoterenaea, Operculicarya, Orthopterygium, Ozoroa, Pachycormus, Parishia, Pegia, Pentaspadon, Pistacia, Plectomirtha, Pleiogynium, Poupartia, Protorhus, Pseudoprotorhus, Pseudosmodingium, Pseudospondias, Rhodosphaera, Rhus, Scassellatia, Schinopsis, Schinus, Sclerocarya, Semecarpus, Smodingium, Solenocarpus, Sorindeia, Spondias, Swintonia, Tapirira, Thyrsodium, Toxicodendron, Trichoscypha.

Anarthriaceae D. F. Cutler et Airy Shaw = Restionaceae R. Br.

ANCISTROCLADACEAE Planch. ex Walp. 1851. Dicots. Ancistrocladales. 1/19. Trop. Africa, trop. Asia. Genus Ancistrocladus.

Andrederaceae J. Agardh = Basellaceae Moq.

Andromedaceae (Endl.) Schnizl. = Ericaceae Juss.

Andropogonaceae Martinov = Poaceae Barnhart.

Androsacaceae Rchb. ex Barnhart = Primulaceae Vent.

Androstachydaceae Airy Shaw = Euphorbiaceae Juss.

Androsynaceae Salisb. = Tecophilaeaceae Leyb.

Anemonaceae von Vest = Ranunculaceae Juss.

Angelicaceae Martinov = Apiaceae Lindl.

ANISOPHYLLACEAE Ridl. 1922. Dicots. Rhizophorales. 4/36–37. Trop. South America, trop. Africa and Asia. Genera: Anisophyllea, Combretocarpus, Poga, Polygonanthus.

ANNONACEAE Juss. 1789. Dicots. Annonales. 127/2,230. Tropics, but Asimina extending to North America (to southern Ontario). Genera: Afroguatteria, Alphonsea, Ambavia, Anaxagorea, Ancana, Annickia, Annona, Anomianthus, Anonidium, Artabotrys, Asimina, Asteranthe, Balonga, Bocagea, Bocageopsis, Boutiquea, Cananga, Cardiopetalum, Cleistochlamys, Cleistopholis, Cremastosperma, Cyathocalyx, Cyathostemma, Cymbopetalum, Dasoclema, Dasymaschalon, Deeringothamnus, Dendrokingstonia, Dennettia, Desmopsis, Desmos, Diclinanona, Dielsiothamnus, Disepalum, Drapananthus, Duckeanthus, Duguetia, Ellipeia, Ellipeiopsis, Enicosanthum, Ephedranthus, Exellia, Fenerivia, Fissistigma, Fitzalania, Friesodielsia, Froesiodendron, Fusaea, Gilbertiella, Goniothalamus, Greenwayodendron, Guamia, Guatteria, Guatteriella, Guatteriopsis, Haplostichanthus, Heteropetalum, Hexalobus, Hornschuchia, Isolona, Letestudoxa, Lettowianthus, Malmea, Marsypopetalum, Meiocarpidium, Meiogyne, Melodorum, Mezzettia, Mezzettiopsis, Miliusa, Mischogyne, Mitrephora, Mkilua, Monanthotaxis, Monocarpia, Monocyclanthus, Monodora, Neostenanthera, Neo-uvaria, Oncodostigma, Onychopetalum, Ophrypetalum, Oreomitra, Orophea, Oxandra, Pachypodanthium, Papualthia, Pelticalyx, Petalolophus, Phaeanthus, Phoenicanthus, Piptostigma, Platymitra, Polyalthia, Polyaulax, Polyceratocarpus, Popowia, Porcelia, Pseudartabotrys, Pseudephedranthus, Pseudoxandra, Pseuduvaria, Raimondia, Rauwenhoffia, Reedrollinsia, Richella, Rollinia, Ruizodendron, Sageraea, Sapranthus, Schefferomitra, Sphaerocoryne, Stelechocarpus, Stenanona, Tetrameranthus, Tetrapetalum,

Toussaintia, Tridimeris, Trigynaea, Trivalvaria, Unonopsis, Uvaria, Uvariastrum, Uvariodendron, Uvariopsis, Woodiellantha, Xylopia.

Anomochloaceae Nakai = Poaceae Barnhart.

Anrederaceae J. Agardh = Basellaceae Moq.

Anthemidaceae Martinov = Asteraceae Dumort.

Anthericaceae J. Agardh = Asphodelaceae Juss.

Anthobolaceae Dumort. = Santalaceae R. Br.

Antidesmataceae Loudon = Euphorbiaceae Juss.

Antirrhinaceae Pers. = Scrophulariaceae Juss.

Antoniaceae J. Agardh = Loganiaceae R. Br. ex Mart.

Apamaceae A. Kern. = Aristolochiaceae Dumort.

Aparinaceae Hoffmanns. et Link = Rubiaceae Juss.

APHLOIACEAE **Takht.** 1985. Dicots. Violales. 1/6. Trop. and South Africa, Madagascar and islands of the western Indian Ocean. Genus Aphloia.

APHYLLANTHACEAE **Burnett.** 1835. Monocots. Amaryllidales. 1/1. Western Mediterranean. Genus Aphyllantes.

APIACEAE **Lindl.** 1836 or UMBELLIFERAE Juss. 1789 (nom. altern.). Dicots. Apiales. 455/3,245. Cosmopolitan. Genera: Aciphylla, Acronema, Actinanthus, Actinolema, Actinotus, Adenosciadium, Aegopodium, Aethusa, Aframmi, Afrocarum, Afroligusticum, Afrosison, Agasyllis, Ageomoron, Agrocharis, Albertia, Alepidea, Aletes, Alococarpum, Ammi, Ammiopsis, Ammodaucus, Ammoides, Ammoselinum, Anethum, Angelica, Anginon, Angoseseli, Anisopoda, Anisosciadium, Anisotome, Annesorrhiza, Anthriscus, Aphanopleura, Apiastrum, Apium, Apodocarpum, Arafoe, Arctopus, Arcuatopterus, Arracacia, Artedia, Asciadium, Asteriscium, Astomaea, Astrantia, Astrodaucus, Astydamia, Athamanta, Aulacospermum, Aulospermum, Austropeucedanum, Autumnalia, Azilia, Azorella, Berula, Bifora, Bilacunaria, Bolax, Bonannia, Bowlesia, Brachypium, Bunium, Bupleurum, Cachrys, Calyptrosciadium, Capnophyllum, Carlesia, Caropsis, Carum, Caucalis, Cenolophium, Centella, Cephalopodium, Cervaria, Chaerophyllopsis, Chaerophyllum, Chaetosciadium, Chámaele, Chamaesciadium, Chamaesium, Chamarea, Changium, Chlaenosciadium, Choritaenia, Chuanminshen, Chymsidia, Cicuta, Cnidiocarpa, Cnidium, Coaxana, Conioselinum, Conium, Conopodium, Coriandrum, Coristospermum, Cortia, Cortiella, Cotopaxia, Coulterophytum, Coxella, Crenosciadium, Crithmum, Cryptotaenia, Cuminum, Cyathoselinum, Cyclorhiza, Cyclospermum, Cymbocarpum, Cymopterus, Cynapium, Cynorhiza, Cynosciadium, Czernaevia, Dactylaea, Dasispermum, Daucosma, Daucus, Demavendia, Dethawia, Deverra, Dichosciadium, Dickinsia, Dicyclophora, Dimorphosciadium, Diplaspis, Diplolophium, Diplotaenia, Diposis, Distichoselinum, Domeykoa, Donnellsmithia, Dorema, Dracosciadium, Drusa, Ducrosia, Dystaenia, Echinophora, Elaeoselinum, Elaeosticta, Eleutherospermum, Enantiophylla, Endressia, Eremocharis, Eremodaucus, Ergocarpon, Erigenia, Eriocycla, Eriosciadium, Eriosynaphe, Eryngium, Erythroselinum, Fuernrohria, Eurytaenia, Exoa cantha, Ezosciadium, Falcaria, Fergania, Ferula, Ferulago, Foeniculum, Frommia, Froriepia, Fuernrohria, Galagania, Geocaryum, Gingidia, Claucosciadium, Glehnia, Glia, Clochidotheca, Gongylosciadium, Grafia, Grammosciadium, Guillonea, Gymnophyton, Gynophyge, Hacquetia, Halosciastrum, Hansenia, Haplosciadium, Phalosphaera, Harbouria, Harrysmithia, Haussknechtia, Hellenocarum, Heptaptera, Hieracleum, Hermas, Heteromorpha, Hladnikia, Hohenackeria, Homalocarpus, Homalosciadium, Horstrissea, Huanaca, Hyalolaena, Hydrocotyle, Hymenidium, Hymenolaena, Imperatoria, Itasina, Johrenia, Johreniopsis, Kadenia, Kafirnigania, Kalakia, Kandaharia, Karatavia, Karnataka, Kedarnatha, Kenopleurum, Keraymonia, Kitagawia, Klotzschia, Komarovia, Korshinskya, Kosopoljanskia, Kozlovia, Krasnovia, Krubera, Kundmannia, Ladyginia, Lagoecia, Lalldhwojia, Laretia, Laser, Laserpitium, Lecokia, Ledebouriella, Lefebvrea, Leutea, Levisticum, Lichtensteinia, Lignocarpa, Ligusticella, Ligusticopsis, Ligusticum, Lilaeopsis, Limnosciadium, Lipskya, Lisaea, Lithosciadium, Lomatium, Lomatocarpa, Macrosciadium, Macroselinum, Magadania, Magydaris, Malabaila, Mandenovia, Margotia, Marlothiella, Mastigosciadium, Mathiasella, Mediasia, Meeboldia, Melanosciadium, Melanoselinum, Meringogyne, Neum, Micropleura, Microsciadium, Mogoltavia, Molopospasma, Monizia, Mulinum, Musineon, Mutellina, Myrrhidendron, Myrrhis, Naufraga, Nematosciadium, Neoconopodium, Neogoezea, Neonelsonia, Noeparrya, Neosciadium, Niphogeton, Nirarathamnos, Nothosmyrnium, Notosciadium, Notopterygium, Oedibasis, Oenanthe, Oligocladus, Oliveria, Olymposciadium, Opoidia, Opopanax, Oreocome, Oreomyrrhis, Oreonana, Oreoschimperella, Oreoselinum, Oreoxis, Orlaya, Ormopterum, Ormosciadium, Ormosolenia, Orogenia, Osmorhiza, Ottoa, Oxypolis, Pachyctenium, Pachypleurum, Palimbia, Pancicia, Paraligusticum, Paraselinum, Parasilaus, Pastinaca, Pastinacopsis, Paulita, Pedinopetalum, Pentapeltis, Perderidia, Perissocoeleum, Petagnaea, Petroedmondia, Petorselinum, Peucedanum, Phellolophium, Phlojodicarpus, Phlyctidocarpa, Physocaulis, Physospermopsis, Physospermum, Physotrichia, Pilopleura, Pimpinella, Pinacantha, Pinda, Platysace, Pleurospermopsis, Pleurospermum, Podistera, Polemannia, Polemanniopsis, Polylophium, Polytaenia, Polyzygus, Portenschlagiella, Postiella, Pozoa, Prangos, Prionosciadium, Psammogeton, Pseudocarum, Pseudorlaya, Pseudoselinum, Pseudotaenidia, Pternopetalum, Pterocyclus, Pterygopleurum, Ptilimnium, Ptychotis, Pycnocycla, Pyramidoptera, Registaniella, Rhabdosciadium, Rhodosciadium, Rhopalosciadium, Rhysopterus, Ridolfia, Rouya, Rumia, Rupiphila, Rutheopsis, Sajanella, Sanicula, Saposhnikovia, Scaligeria, Scandia, Scandix, Schizeilema, Schoenolaena, Schrenkia, Schtschurowskia, Schulzia, Sclerochorton, Sclerotiaria, Scrithacola, Selenopsis, Selinum, Semenovia, Seseli, Seseelopsis, Shoshonea, Silaum, Sinocarum, Sinodielsia, Sinolimprichtia, Sison, Sium, Smyrniopsis, Smyrnium, Sonderina, Spananthe, Spermolepis, Sphaenolobium, Spaerosciadium Sphallerocarpus, Sphenosciadium, Spuriodaucus, Spuriopimpinella, Stefanoffia, Steganotaenia, Stenocoelium, Stenosemis, Stenotaenia, Stewartiella, Stoibrax, Symphyoloma, Szovitsia, Taenidia, Taeniopetalum, Tamamschjania, Tauschia, Tetrataenium, Thamnosciadium, Thapsia, Thaspium, Thecocarpus, Thyselium, Tilingia, Tinguarra, Todaroa, Tommasinia, Tongoloa, Tordyliopsis, Tordylium, Torilis, Tornabenea, Trachydium, Trachy mene, Trachyspermum, Trepocarpus, Tricholaser, Trigonosciadium, Trinia, Trochiscanthes, Turgenia, Uldinia, Vanasushava, Vicatia, Visnaga, Vvedenskia, Santhoselinum, Xanthosia, Xatardia, Yabea, Zeravschania, Zizia, Zosima.

APOCYNACEAE Juss. 1789. Dicots. Gentianales. 193/2,330. Trop., subtrop., and (a few species) temp. regions. Genera: Acokanthera, Adenium, Aganonerion, Aganosma, Alafia, Allamandia, Allomarkgrafia, Allowoodsonia, Alstonia, Alyxia, Amalocalyx, Ambelania, Amblyocalyx, Amsonia, Ancylobothrys, Anechites, Anagadenia, Anodendron, Anthoclitandra, Aphanostylis, Apocynum, Artia, Asketanthera, Aspidosperma, Baissea, Beaumontia, Bousigonia, Cabucala, Callichilia, Calocrater, Cameraria, Carissa, Carissophyllum, Carpodinopsis, Carpodinus, Carruthersia, Carvalhoa, Cascabela, Catharanthus, Ceratites, Cerbera, Cerberiopsis, Chamaeclitandra, Chavannesia, Chilocarpus, Chonemorpha, Chunechites, Cleghornia, Clitandra, Codonechites, Comularia, Condylocarpon, Couma, Craspidospermum, Crioceras, Cufodontia, Cyclocotyla, Cylindropsis, Cylindrosperma, Delphyodon, Dewevrella, Dictyophleba, Diplorhynchus, Discalyxia, Dyera, Ecdysanthera, Echites, Elytropus, Epigynum, Epitaberna, Farquharia, Fernaldia, Forsteronia, Funtumia, Galactophora, Geissospermum, Gonioma, Grisseea, Hancornia, Hanghomia, Haplophandra, Haplophyton, Holarrhena, Hunteria, Hymenolophus, Ichnocarpus, Isonema, Ixodonerium, Kamettia, Kibatalia, Kopsia, Lacmellea, Landolphia, Laubertia, Laxoplumeria, Lepinia, Lepiniopsis, Leuconotis, Macoubea, Macropharynx, Macrosiphonia, Malouetia, Mandevilla, Mascarenhasia, Melodinus, Mesechites, Micrechites, Microplumeria, Molongum, Mortoniella, Motandra, Muantum, Mucoa, Neisosperma, Neobracea, Neocouma, Nerium, Neurolobium, Notonerium, Nouettea, Ochrosia, Odontadenia, Oncinotis, Orthopichonia, Pachypodium, Papuechties, Parabarium, Parahancornia, Paralyxia, Parameria, Parepigynum, Parsonsia, Peltastes, Pentalinon, Petchia, Pezisicarpus, Picralima, Plectaneia, Pleiocarpa, Pleioceras, Plumeria, Poacynum, Pottsia, Prestonia, Pteralyxia, Pycnobotrya, Quiotania, Rauvolfia, Rhabdadenia, Rhazya, Rhigospira, Rhodocalyx, Rhynchodia, Saba, Salpinctes, Schizozygia, Secondatia, Sindechites, Skytanthus, Spirolobium, Spongiosperma, Stemmadenia, Stephanostegia, Stephanostema, Stipecoma, Strempeliopsis, Streptotrachelus, Strophanthus, Tabernaemontana, Tabernanthe, Temnadenia, Tetradoa, Thenardia, Thevetia, Tintinnabularia, Trachelospermum, Trachomitum, Urceola, Urnularia, Vahadenia, Vallariopsis, Vallaris, Vallesia, Vinca, Voacanga, Walidda, Willughbeia, Winchia, Woytkowskia, Wrightia, Xylinabaria, Xylinabariopsis, Zygodia.

APODANTHACEAE Tiegh. ex Takht. 1987. Dicots. Rafflesiales. 3/37. Southwest Asia, western Australia, America. Genera: Apodanthes, Berlinianche, Pilostyles.

APONOGETONACEAE J. Agardh. 1858. Monocots. Aponogetonales. 1/47. Tropics and subtropics of the Old World. Genus Aponogeton.

Aporosaceae Lindl. ex Miq. = Euphorbiaceae Juss.

Apostasiaceae Lindl. = Orchidaceae Juss.

Aptandraceae Miers = Olacaceae Mirb. ex DC.

AQUIFOLIACEAE Bartl. 1830. Dicots. Celastrales. 2/400. Trop., subtrop., and temp. regions. Genera: Ilex, Nemopanthus.

Aquilariaceae R. Br. = Thymelaeaceae Juss.

ARACEAE Juss. 1789. Monocots. Arales. 106/2,980. Tropics, a few species in temp. regions. Genera: Aglaodorum, Aglaonema, Alloschemone, Alocasia, Ambrosina, Amorphophallus, Amydrium, Anadendrum, Anaphyllopsis, Anaphyllum, Anchomanes, Anthurium, Anubias, Aridarum, Ariopsis, Arisaema, Arisarum, Arophyton, Arum, Asterostigma, Biarum, Bognera, Bucephalandra, Caladium, Calla, Callopsis, Carlephyton, Chlorospatha, Colletogyne, Colocasia, Cryptocoryne, Culcasia, Cyrtosperma, Dieffenbachia, Dracontium, Dracunculus, Eminium, Epipremnum, Filarum, Furtadoa, Gearum, Gonatanthus, Gonatopus, Gymnostachys, Hapaline, Heteroaridarum, Heteropsis, Holochlamys, Homalomena, Hottarum, Jasarum, Lagenandra, Lasia, Lasimorpha, Lazarum, Lysichiton, Mangonia, Monstera, Montrichardia, Nephthytis, Orontium, Pedicellarum, Peltandra, Philodendron, Phymatarum, Pinellia, Piptospatha, Pistia, Plesmonium, Podolasia, Pothoidium, Pothos, Protarum, Pseudodracontium, Pseudohydrosme, Pycnospatha, Remusatia, Rhaphidophora, Rhektophyllum, Rhodospatha, Sauromatum, Scaphispatha, Schismatoglottis, Scindapsus, Spathantheum, Spathicarpa, Spathiphyllum, Stenospermation, Steudnera, Stylochaeton, Symplocarpus, Synandrospasix, Syngonium, Taccarum, Theriophonum, Thomsonia, Typhonium, Typhonodorum, Ulearum, Urospatha, Urospathella, Xanthosoma, Zamioculcas, Zantedeschia, Zomicarpa, Zomicarpella.

Arachnitidaceae Muñoz = Corsiaceae Becc.

Aragoaceae D. Don = Scrophulariaceae Juss.

ARALIACEAE Juss. 1789. Dicots. Apiales. 56/1,290. Trop., subtrop. and warm–temp. regions. Genera: Anakasia, Anomopanax, Apiopetalum, Aralia, Arthrophyllum, Astrotricha, Boninofatsia, Brassaiopsis, Cephalaralia, Cheirodendron, Crepinella, Cromopanax, Cuphocarpus, Cussonia, Delarbrea, Dendropanax, Didymopanax, Eleutherococcus, Enochoria, Evodiopanax, Fatsia, Gamblea, Gastonia, Harmsiopanax, Hedera, Heteropanax, Hunaniopanax, Kalopanax, Mackinlaya, Macropanax, Megalopanax, Merilliopanax, Meryta, Motherwellia, Munroidendron, Myodocarpus, Oplopanax, Oreomyrrhis, Oreopanax, Osmoxylon, Panax, Pentapanax, Polyscias, Pseudopanax, Pseudosciadium, Reynoldsia, Schefflera, Schizomeryta, Sciadodendron, Seemannaralia, Sinopanax, Stilbocarpa, Tetrapanax, Tetraplasandra, Trevesia, Woodburnia.

ARALIDIACEAE Philipson et B. C. Stone. 1980. Dicots. Aralidiales. 1/1. Western Indochina, Malesia. Genus Aralidium.

ARAUCARIACEAE Henkel et W. Hochst. 1865. Pinopsida. Araucariales. 2/33. Southeast Asia, Malesia, warm–temp. regions of the Southern Hemisphere, excl. Africa. Genera: Agathis, Araucaria.

Arbutaceae J. Agardh = Ericaceae Juss.

Arceuthobiaceae Tiegh. ex Nakai = Viscaceae Miers.

Arctostaphylaceae J. Agardh = Ericaceae Juss.

Arctotidaceae Bessey = Asteraceae Dumort.

Ardisiaceae Juss. = Myrsinaceae R. Br.

ARECACEAE Schultz Sch. 1832 or **PALMAE** Juss.1789 (nom. altern.). Monocots. Arecales. 202/2,700. Tropics, subtropics and (very seldom) temp. regions. Genera: Acanthophoenix, Acoelorraphe, Acrocomia, Actinokentia, Actinorhytis, Aiphanes, Allagoptera, Alloschmidia, Alsmithia, Ammandra, Archontophoenix, Areca, Arenga, Asterogyne, Astrocaryum, Attalea, Bactris, Balaka, Barcella, Basselinia, Beccariophoenix, Bentinckia, Bismarckia, Borassodendron, Borassus, Brahea, Brassiophoenix, Brongniartikentia, Burretiokentia, Butia, Calamus, Calospatha, Calyptrocalyx, Calyptrogyne, Calyptronoma, Campecarpus, Carpentaria, Carpoxylon, Caryota, Catoblastus, Ceratolobus, Ceroxylon, Chamaedorea, Chamaerops, Chambeyronia, Chelyocarpus, Chrysalidocarpus, Chuniophoenix, Clinosperma, Clinostigma, Coccothrinax, Cocos, Colpothrinax, Copernicia, Corypha, Cryosophila, Cyphokentia, Cyphophoenix, Cyphosperma, Cyrtostachys, Daemonorops, Deckenia, Desmoncus, Dictyocaryum, Drymophloeus, Dypsis, Elaeis, Eleiodoxa, Eugeissona, Euterpe, Fremosphatha, Gastrococos, Gaussia, Geonoma, Goniocladus, Gronophyllum, Guihaia, Gulubia, Halmoorea, Hedyscepe, Heterospathe, Howea, Hydriastele, Hyophorbe, Hyospathe, Hyphaene, Iguanura, Irartella, Iriartea, Itaya, Jessenia, Johannesteijsmannia, Jubaea, Jubaeopsis, Kentiopsis, Kerriodoxa, Korthalsia, Kuania, Laccospadix, Laccosperma, Latana, Lavoixia, Leopoldinia, Lepidocaryum, Lepidorrhachis, Linospadix, Livistona, Lodoicea,

Louvelia, Loxococcus, Lucuala, Lytocaryum, Mackeea, Manicaria, Marojejya, Masoala, Mauritia, Mauritiella, Maxburretia, Maximiliana, Medemia, Metroxylon, Moratia, Myrialepis, Nannorrhops, Nenga, Neodypsis, Neonicholsonia, Neophloga, Neoveitchia, Nephrosperma, Normanbya, Nypa, Oenocarpus, Oncocalamus, Oncosperma, Orania, Oraniopsis, Orbignya, Palandra, Parajubaea, Pelagodoxa, Phloga, Phoeniocophorium, Phoenix, Pholidocarpus, Pholidostachys, Physokentia, Phytelephas, Pigafetta, Pinanga, Plectocomia, Plectocomiopsis, Podococcus, Pogonotium, Polyandrococos, Prestoea, Pritchardia, Pritchardiopsis, Pseudophoenix, Ptichosperma, Ptychococcus, Raphia, Ravenea, Reinhardtis, Retispatha, Rhapidophyllum, Rhapis, Rhopaloblaste, Rhopalostylis, Roscheria, Roystonea, Sabal, Salacca, Satakentia, Scheelea, Schippia, Sclerosperma, Serenoa, Siphokentia, Socratea, Sommieria, Syagrus, Synechanthus, Tectiphiala, Thrinax, Trachycarpus, Trithrinax, Veillonia, Veitchia, Verschaffeltia, Voaniola, Vonitra, Wallichia, Washingtonia, Welfia, Wendlandiella, Wettinia, Wodyetia, Zombia.

ARGOPHYLLACEAE Takht. 1987. Dicots. Hydrangeales. 2/15–17. Austrtalia, Lord Howe I., New Zealand, New Caledonia, Rapa I. Genera: Argophyllum, Corokia.

Arisaraceae Raf. = Araceae Juss.

ARISTOLOCHIACEAE Juss. 1789. Dicots. Aristolochiales. 7/c. 650–700. Tropics, subtropics, and temp. regions (excl. New Zealand). Genera: Aristolochia, Asarum, Euglypha, Holostyles, Pararistolochia, Saruma, Thottea.

Aristoteliaceae Dumort. = Elaeocarpaceae A. DC.

Arjonaceae Tiegh. = Santalaceae R. Br.

Armeriaceae Horan. = Plumbaginaceae Juss.

Artemisiaceae Martinov = Asteraceae Dumort.

Artocarpaceae R. Br. = Moraceae Link.

Arundinaceae (Dumort.) Herter = Poaceae Barnhart.

Arundinariaceae Baum.-Bodenh. = Poaceae Barnhart.

Arundinellaceae (Stapf) Herter = Poaceae Barnhart.

Asaraceae Vent. = Aristolochiaceae Juss.

ASCLEPIADACEAE R. Br. 1810. Dicots. Gentianales. 306/3,230. Tropics and subtropics, esp. Africa, a few species in warm–temp. regions. Genera: Absolmsia, Acomosperma, Adelostemma, Aidomene, Amblystigma, Anatropanthus, Anisopus, Anisotoma, Anomotassa, Araujia, Asclepias, Aspidoglossum, Aspidonepsis, Astephanus, Asterostemma, Atherandra, Atherolepis, Atherostemon, Baclea, Baeolepis, Barjonia, Baroniella, Baseonema, Basistelma, Belostemma, Bidaria, Biondia, Blepharodon, Blyttia, Brachystelma, Calotropis, Campestigma, Camptocarpus, Caralluma, Ceropegia, Chlorocyathus, Cibirhiza, Cionura, Clemensiella, Conomitra, Cordylogyne, Corollonema, Cosmostigma, Costantina, Cryptolepis, Cryptostegia, Curroria, Cyathostelma, Cynanchum, Cynoctonum, Cyprinia, Cystostemma, Dactylostelma, Dalzielia, Decalepis, Decanema, Decanemopsis, Decastelma, Dicarpophora, Diplolepis, Diplostigma, Dischidanthus, Dischidia, Ditassa, Dittoceras, Dolichoetalum, Dolichostegia, Dorystephania, Dregea, Drepanostemma, Duvalia, Duvaliandra, Echidnopsis, Ecliptostelma, Ectadium, Edithcolea, Elcomarhiza, Emicocarpus, Emplectanthus, Epistemma, Eustegia, Fanninia, Fimbristemma, Finlaysonia, Fischeria, Fockea, Frerea, Funastrum, Genianthus, Glaziostelma, Glossonema, Glossostelma, Gomphocarpus, Gongronema, Gongylosperma, Gonioanthela, Goniostemma, Gonocrypta, Gonolobus, Graphistemma, Gunnessia, Gymnanthera, Gymnema, Gymnemopsis, Harpanema, Hemidesmus, Hemipogon, Heterostemma, Heynella, Hickenia, Himantostemma, Holostemma, Hoodia,

Hoodiopsis, Hoya, Hoyella, Huernia, Huerniopsis, Hypolobus, Ibatia, Irmischia, Ischnolepsis, Ischnostemma, Jacaima, Janakia, Jobinia, Kanahia, Karimbolea, Kerbera, Kompitsia, Labidostelma, Lagoa, Lavrania, Leichhardtia, Leptadenia, Llotzkyella, Lobostephanus, Luckhoffia, Lugonia, Lygisma, Macropelma, Macropetalum, Macroscepis, Mahafalia, Mahawoa, Mangenotia, Margaretta, Marsdenia, Matelea, Meladerma, Melinia, Menabea, Meresaldia, Merrillanthus, Metaplexis, Micholitzia, Microdactylon, Microloma, Microstelma, Miraglossum, Mitolepis, Mitostigma, Mondia, Morrenia, Myriopteron, Nautonia, Nematostemma, Neoschumannia, Nephradenia, Notechidnopsis, Odontanthera, Odontostelma, Olympusa, Oncinema, Oncostemma, Ophionella, Orbea, Orbeanthus, Orbeopsis, Oreosparte, Orthanthera, Orthosia, Oxypetalum, Oxystelma, Pachycarpus, Pachycymbium, Papuastelma, Parapodium, Parquetina, Pattalias, Pectinaria, Pentabothra, Pentacyphus, Pentagonanthus, Pentanura, Pentarrhinum, Pentasacme, Pentastelma, Pentatropis, Pentopetia, Pentopetiopsis, Peplonia, Pergularia, Periglossum, Periploca, Petalostelma, Pherotrichis, Phyllanthera, Piaranthus, Platykeleba, Pleurostelma, Podandra, Podostelma, Prosopostelma, Pseudolithos, Ptycanthera, Pycnobregma, Pycnoneurum, Pycnorhachis, Quaqua, Quisumbingia, Raphionacme, Raphistemma, Rhynchostigma, Rhyssolobium, Rhyssostelma, Rhytidocaulon, Riocreuxia, Rojasia, Rothrockia, Sacleuxia, Sarcolobus, Sarcorrhiza, Sarcostemma, Schistogyne, Schistonema, Schizocorona, Schizoglossum, Schlechterella, Schubertia, Scyphostelma, Secamone, Secamonopsis, Seshagiria, Seutera, Sisyranthus, Socotora, Socotranthus, Solenostemma, Sphaerocodon, Spirella, Stapelia, Stapelianthus, Stapeliopsis, Stathmostelma, Steleostemma, Stelmacrypton, Stelmagonum, Stelmatocodon, Stenomeria, Stenostelma, Stephanotis, Stigmatorhynchus, Streptocaulon, Strobopetalum, Stuckertia, Swynnertonia, Symphytonema, Tacazzea, Tainionema, Tanulepis, Tassadia, Tavaresia, Telectadium, Telesilla, Telminostelma, Telosma, Tenaris, Tetracustelma, Tetraphysa, Thozetia, Toxocarpus, Treutlera, Trichocaulon, Trichosacme, Trichosandra, Trichostelma, Tridentea, Triodoglossum, Tromotriche, Tweedia, Tylophora, Tylophoropsis, Urostephanus, Utleria, Vailia, Vincetoxicopsis, Vincetoxicum, Voharanga, Vohemaria, Whitesloanea, Widgrenia, Woodia Xysmalobi um, Zacateza, Zygostelma.

Ascyraceae Martinov = Clusiaceae Lindl.

Aspalathaceae Martinov = Fabaceae Lindl.

ASPARAGACEAE Juss. 1789. Monocots. Asparagales. 3/210. Old World, esp. Africa. Genera: Asparagus, Myrsiphyllum, Protasparagus.

Asperifoliaceae Rchb. = Boraginaceae Juss.

Asperulaceae Spenn. = Rubiaceae Juss.

ASPHODELACEAE Juss. 1789. Monocots. Amaryllidales. 47/162. Trop., subtrop. and temp. regions. Genera: Alania, Aloe, Anemarrhena, Anthericum, Arnocrinum, Arthropodium, Asphodeline, Asphodelus, Astroloba, Borya, Bulbine, Bulbinella, Caesia, Chamaescilla, Chlorophytum, Chortolirion, Comospermum, Corynotheca, Diamena, Dichopogon, Diora, Diuranthera, Echeandia, Eremocrinum, Eremurus, Gasteria, Haworthia, Hensmania, Herpolirion, Hodgsoniola, Jodrellia, Johnsonia, Kniphofia, Laxmannia, Lomatophyllum, Murchisonia, Paradisea, Pasithea, Poellnitzia, Simethis, Sowerbaea, Stawellia, Terauchia, Thysanotus, Trachyandra, Trichopetalum, Tricoryne.

Aspidistraceae Endl. = Convallariaceae Horan.

ASTELIACEAE Dumort. 1829. Monocots. Asparagales. 6/53. Mascarene Is., New

Guinea, Australia, Tasmania, islands of Pacific Ocean, Falkland Is. Genera: Astelia, Cohnia, Collospermum, Cordyline, Milligania, Neoastelia.

ASTERACEAE Dumort. 1822 or COMPOSITAE Giseke.1792 (nom. altern.). Dicots. Asterales. 1,553–1,554/21,740. Cosmopolitan, chiefly temp. and subtrop. regions. Genera: Aaronsohnia, Abrotanella, Acamptopappus, Acanthocephalus, Acanthocladium, Acanthodesmos, Acantholepis, Acanthospermu, Acanthostyles, Achaenipodium, Achillea, Achnophora, Achnopogon, Achyrachaena, Achyrobaccharis, Achyrocline, Achyropappus, Achyrophorus, Achyrothalamus, Acilepidopsis, Acmella, Acomis, Acourtia, Acrisione, Acritopappus, Acroclinium, Acroptilon, Actinobole, Actinoseris, Adelostigma, Adenanthellum, Adenocaulon, Adenocritonia, Adenoglossa, Adenoon, Adenopappus, Adenophyllum, Adenostemma, Adenothamnus, Aedesia, Aegopordon, Aequatorium, Aetheolaena, Aetheopappus, Aetheorhiza, Ageratella, Ageratina, Ageratinastrum, Ageratum, Agoseris, Agrianthus, Ainsliaea, Ajania, Ajaniopsis, Alatoseta, Albertinia, Alcantara, Alciope, Aldama, Alepidocline, Alfredia, Aliella, Allagopappus, Allardia, Alloispermum, Allopterigeron, Alomia, Alomiella, Alvordia, Amauria, Amberboa, Amblyocarpum, Amblyolepis, Amblyopappus, Amblyopogon, Amboroa, Ambrosia, Amellus, Ammobium, Amolinia, Amphiachyris, Amphidoxa, Amphiglossa, Amphipappus, Amphoricarpos, Anacantha, Anacyclus, Anaphalioides, Anaphalis, Anaxeton, Ancathia, Ancistrocarphus, Ancistrophora, Andryala, Anemocarpa, Angelphytum, Angianthus, Anisochaeta, Anisocoma, Anisopappus, Anisothrix, Anomostephium, Antennaria, Anthemis, Antillia, Antiphiona, Antithrixia, Anvillea, Apalochlamys, Aphanactis, Aphanostephus, Aphyllocladus, Apodocephala, Aposeris, Apostates, Arbelaezaster, Archibaccharis, Arctanthemum, Arctium, Arctogeron, Arctotheca, Arctotis, Argentipallium, Argyranthemum, Argyroglottis, Argyrophanes, Argyrovernonia, Argyroxiphium, Aristeguietia, Arnaldoa, Arnica, Arnicastrum, Arnoglossum, Arnoseris, Arrhenechthites, Arrojadocharis, Arrowsmithia, Artemisia, Artemisiopsis, Asanthus, Ascidiogyne, Askellia, Aspilia, Asplundianthus, Aster, Asteridea, Asteriscus, Asterothamnus, Astranthium, Athanasia, Athrixia, Athroisma, Atractylis, Atractylodes, Atrichantha, Atrichoseris, Austrobrickellia, Austrocritonia, Austroeupatorium, Austrosynotis, Avellara, Axiniphyllum, Ayapana, Ayapanopsis, Aylacophora, Baccharidiopsis, Baccharis, Badilloa, Baeriopsis, Bafutia, Bahia, Bahianthus, Baileya, Balduina, Balsamorhiza, Baltimora, Barkleyanthus, Barnadesia, Barroetea, Barrosoa, Bartlettia, Bartlettina, Basedowia, Bebbia, Bedfordia, Bejaranoa, Bellida, Bellis, Bellium, Belloa, Benitoa, Berardia, Berkheya, Berlandiera, Berroa, Bidens, Bigelowia, Bishopalea, Bishopanthus, Bishopiella, Bishovia, Blainvillea, Blakeanthus, Blakiella, Blanchetia, Blennosperma, Blennospora, Blepharipappus, Blepharispermum, Blepharizonia, Blumea, Blumeopsis, Boeberastrum, Boeberoides, Bolanosa, Bolophyta, Boltonia, Bombycilaena, Borrichia, Bothriocline, Brachanthemum, Brachionostylum, Brachyactis, Brachychaeta, Brachyclados, Brachycome, Brachyglottis, Brachylaena, Brachythrix, Bracteantha, Bradburia, Brickellia, Brickelliastrum, Brintonia, Bryomorphe, Buphthalmum, Burkartia, Cabreriella, Cacalia, Cacaliopsis, Cacosmia, Cadiscus, Caesulia, Calea, Calendula, Calenduleae, Callicephalus, Callilepis, Callistephus, Calomeria, Calostephane, Calotesta, Calotis, Calycadenia, Calycocorsus, Calycoseris, Calyptocarpus, Camchaya, Campovassouria, Camptacra, Campuloclinium, Cancrinia, Cancriniella, Cardopatium, Carduncellus, Carduus, Carlina, Carminatia, Carpesium, Carphephorus, Carphochaete, Carramboa, Carterothamnus, Carthamus, Cassinia, Castalis, Castanedia, Catamixis, Catananche, Catatia, Cavalcantia, Cavea, Celmisia, Centaurea, Centaureodendron, Centauropsis, Centaurothamnus, Centipeda, Centratherum, Cephalipterum, Cephalopappus, Cephalorrhynchus, Cephalosorus, Ceratogyne, Ceruana, Chacoa, Chaenactis, Chaetadelpha, Chaetanthera, Chaetopappa, Chaetoseris, Chaetospira, Chaetymenia, Chamaechaenactis, Chamaegeron, Chamaemelum, Chamaepus, Chaptalia, Chardinia, Chartolepis, Cheirolepis, Cheirolophus, Chersodoma, Chevreulia, Chiliadenus, Chiliocephalum, Chiliophyllum, Chiliotrichiopsis, Chiliotrichum, Chimantaea, Chionolaena, Chionopappus, Chlamydophora, Chloracantha, Chondrilla, Chondropyxis, Chorisis, Chresta, Chromolaena, Chromolepis, Chronopappus, Chrisactinia, Chrysactinium, Chrysanthellum, Chrysanthemoides, Chrysanthemum, Chrysocephalum, Chrysocoma, Chrysogonum, Chrysopappus, Chrysophthalmum, Chrysopsis, Chrysothamnus, Chthonocephalus, Chucoa, Chuquiraga, Cicerbita, Ciceronia, Cichorium, Cineraria, Cirsium, Cissampelopsis, Cladanthus, Cladochaeta, Clappia, Clibadium, Cnicothamnus, Cnicus, Coespeletia, Coleocoma, Coleostephus, Colobanthera, Columbiadoria, Comaclinium, Comborhiza, Commidendrum, Complaya, Condylidium, Condylopodium, Conocliniopsis, Conoclinium, Conyza, Coreocarpus, Coreopsis, Corethamnium, Corethrogyne, Correllia, Corymbium, Cosmos, Cotula, Coulterella, Cousinia, Cousiniopsis, Craspedia, Crassocephalum, Cratystylis, Cremanthodium, Crepidiastrum, Crepidopsis, Crepis, Critonia, Critoniadelphus, Critoniella, Critoniopsis, Crocidium, Cronquistia, Cronquistianthus, Croptiolon, Crossostephium, Crossothamnus, Crupina, ? Cuatrecasanthus, Cuatrecasasiella, Cuchumatanea, Culcitium, Cullumia, Cuspidia, Cyanopsis, Cyanus, Cyathocline, Cyathomone, Cyclolepis, Cylindrocline, Cymbolaena, Cymbonotus, Cymbopappus, Cymophora, Cynara, Dacryo trichia, Dahlia, Damnamenia, Damnxanthodium, Darwiniothamnus, Dasycondylus, Dasyphyllum, Daveaua, Decachaeta, Decastylocarpus, Decazesia, Delairia, Delamerea, Delilia, Dendranthema, Dendrocacalia, Dendrophorbium, Dendrosenecio, Dendroseris, Denekia, Desmanthodium, Desynaphia, Dewildemania, Diacranthera, Dianthoseris, Diaphractanthus, Diaspananthus, Dicerococlados, Dichaetophora, Dichrocephala, Dichromochlamys, Dicoma, Dicoria, Dicranocarpus, Didelta, Dieltzia, Digitacalia, Dimeresia, Dimerostemma, Dimorphocoma, Dimorphotheca, Dinoseris, Diplazoptilon, Diplostephium, Dipterocome, Dipterocypsela, Disparago, Dissothrix, Distephanus, Disynaphia, Dithyrostegia, Dittrichia, Dolichlasium, Dolichoglottis, Dolichorrhiza, Dolichothrix, Dolomiaea, Doniophyton, Dorobaea, Doronicum, Dracopis, Dresslerothamnus, Dubautia, Dubyaea, Dugesia, Duhaldea, Duidaea, Duseniella, Dumondia, Dyscritogyne, Dyscritothamnus, Dysodiopsis, Dyssodia, Eastwoodia, Eatonella, Echinacea, Echinops, Eclipta, Edmondia, Egletes, Eitenia, Ekmania, Elachanthus, Elaphandra, Elephantopus, Eleutheranthera, Ellenbergia, Elytropappus, Emilia, Emiliella, Encelia, Enceliopsis, Endocellion, Endopappus, Engelmannia, Engleria, Enydra, Epaltes, Epilasia, Episcothamnus, Epitriche, Erato, Erechtites, Eremanthus, Eremothamnus, Eriachaenium, Ericameria, Ericentrodea, Erigeron, Erimocephala, Eriocephalus, Eriochlamys, Eriophyllum, Eriotrix, Erlangea, Erodiophyllum, Erymophyllum, Eryngiophyllum, Erythradenia, Erythrocephalum, Espejoa, Espeletia, Espeletiopsis, Ethulia, Euchiton, Eumorphia, Eupatoriastrum, Eupatorina, Eupatoriopsis, Euparotium, Euphrosyne, Eurybiopsis, Eurydochus, Euryops, Eutetras, Euthamia, Evacidium, Evax, Ewartia, Ewarthiothamnus, Exomiocarpon, Faberia, Facelis, Farfugium, Faujasia, Faujasiopsis, Faxonia, Feddea, Feldstonia, Felicia, Femeniasia, Fenixia, Ferreyranthus, Ferreyrella, Filago, Filifolium, Fitchia, Fitzwillia, Flaveria, Fleischmannia, Fleischmanniopsis, Florestina, Floscaldasia, Flosmutisia, Flourensia, Flyriella, Formania, Foveolina, Freya, Fulcaldea, Gaillardia, Galactites, Galeana, Galeomma, Galinsoga, Gamochaeta, Gamochaetopsis, Garberia, Garcibarrigoa, Garcilassa, Gardnerina, Garuleum, Gazania, Gazaniopsis, Geigeria, Geissolepis, Geraea, Gerbera, Geropogon, Gibbaria, Gilberta, Gilruthia, Gladiopappus, Glossarion, Glossocardia, Glossopappus, Glyptopleura, Gnaphaliothamnus, Gnaphalium, Gnephosis, Gochnatia, Goldmanella, Gongrostylus, Gongylolepis, Goniocaulon, Gorceixia, Gorteria, Gossweilera, Goyazianthus, Grangea, Grangeopsis, Graphistylis, Gratwickia, Crauanthus, Grazielia, Greenmaniella, Grindelia, Grisebachianthus, Grossheimia, Grosvenoria, Guardiola, Guayania, Guevaria, Guizotia, Gundelia, Gundlachia, Gutenbergia, Gutierrezia, Gymnarrhena, Gymnocondylus, Gymnocoronis, Gymnodiscus, Gymnolaena, Gymnolomia, Gymnopentzia, Gymnosperma, Gymnostephium, Gynoxys, Gynura, Gypothamnium, Gyptidium, Gyptis, Gyrodoma, Haastia, Haeckeria, Haegiela, Handelia, Haplocarpha, Haploesthes, Haplopappus, Haplostephium, Haptotrichion, Haradjania, Harleye, Harnackia, Hartwrightia, Hasteola, Hatschbachiella, Hazardia, Hebeclinium, Hecastocleis, Hedypnois, Helenium, Helianthella, Helianthus, Helichrysopsis, Helichrysum, Heliocauta, Heliopsis, Helminthotheca, Helogyne, Hemisteptia, Hemizonia, Henricksonia, Heptanthus, Herderia, Herodotia, Hesperevax, Hesperodoria, Hesperomannia, Heteracia, Heteranthemis, Heterochaeta, Heterocoma, Heterocondylus, Heterocypsela, Heteroderis, Heterolepis, Heteromera, Heteromma, Heteropappus, Heteroplexis, Heterorhachis, Heterosperma, Heterothalamus, Heterotheca, Hidalgoa, Hieracium, Hilliardia, Hinterhubera, Hippia, Hippolytia, Hirpicium, Hispidella, Hoehnephytum, Hoffmanniella, Hofmeisteria, Holocarpha, Holocheilus, Hololeion, Holozonia, Homognaphalium, Homogyne, Hoplophyllum, Huarpea, Hubertia, Hughesia, Hulsea, Hulteniella, Humbertacalia, Humeocline, Hyalea, Hyalis, Hyalochaete, Hyalochlamys, Hyaloseris, Hyalosperma, Hybridella, Hydrodyssodia, Hydroidea, Hydropectis, Hymenocephalus, Hymenoclea, Hymenolepis, Hymenonema, Hymenopappus, Hymenostemma, Hymenothrix, Hymenoxys, Hyoseris, Hypacanthium, Hypericophyllum, Hypochaeris, Hysterionica, Hystrichophora, Ichthyothere, Idiothamnus, Ifloga, Ighermia, Iltisia, Imeria, Inezia, Inula, Inulanthera, Iocenes, Iodocephalus, Iogeton, Iostephane, Iphiona, Iphionopsis, Iranecio, Ischnea, Isocarpha, Isoetopsis, Isostigma, Iva, Ixeridium, Ixeris, Ixiochlamys, Ixiolaena, Ixodia, Jacmaia, Jaegeria, Jalcophila, Jaliscoa, Jamesianthus, Jaramilloa, Jasonia, Jaumea, Jefea, Jeffreya, Joweanthus, Jungia, Jurinea, Kalimeris, Karelinia, Karvandarina, Kaschgaria, Kaunia, Keysseria, Kinghamia, Kingianthus, Kippistia, Kirkianella, Kleinia, Koanophyllon, Koelpinia, Krigia, Kyrsteniopsis, Lachanodes, Lachnophyllum, Lachnorhiza, Lachnospermum, Lactuca, Lactucella, Lactucosonchus, Laennecia, Laestadia, Lagascea, Lagenophora, Laggera, Lagophylla, Lamprachaenium, Laprocephalus, Lamyropappus, Langebergia, Lantanopsis, Lapsana, Lasianthaea, Lasiocephalus, Lasiolaena, Lasiopogon, Lasiospermum, Lasthenia, Launaea, Lawrencella, Layia, Lecocarpus, Leibnitzia, L eiboldia, Lembertia,

Lemooria, Leontodon, Leontopodium, Lepidesmia, Lepidolopha, Lepidolopsis, Lepidonia, Lepidophorum, Lepidophyllum, Lepidospartum, Lepidostephium, Leptinella, Leptocarpha, Leptoclinium, Leptorhynchos, Lescaillea, Lessingia, Leucactinia, Leucanthemella, Leucanthemopsis, Leucanthemum, Leucheria, Leucoblepharis, Leucochrysum, Leucocyclus, Leucogenes, Leucoglossum, Leucoptera, Leunisia, Leuzea, Leysera, Liabellum, Liabum, Liatris, Libanothamnus, Lidbeckia, Lifago, Ligularia, Limbarda, Lindheimera, Lipochaeta, Lipskyella, Litogyne, Litothamnus, Litrisa, Llerasia, Logfia, Lomatozona, Lonas, Lopholaena, Lophopappus, Lordhowea, Lorentzianthus, Loricaria, Lourteigia, Loxothysanus, Lucilia, Lucilliocline, Lugoa, Luina, Lulia, Lundellianthus, Lycapsus, Lychnophora, Lycoseris, Lygodesmia, Macdougalia, Machaeranthera, Macowania, Macrachaenium, Macraea, Macropodina, Macvaughiella, Madia, Mairia, Malacothrix, Malmeanthus, Malperia, Mantisalca, Marasmodes, Marshallia, Marshalljohnstonia, Matricorenia, Matricaria, Mattfeldanthus, Mattfeldia, Matudina, Mausolea, Mecomischus, Megalodonta, Melampodium, Melanodendron, Melanthera, Metalasia, Metastevia, Mexerion, Mexianthus, Micractis, Microcephala, Microchaete, Microglossa, Microgynella, Microliabum, Micropsis, Micropus, Microseris, Microspermum, Mikania, Mikaniopsis, Milleria, Millotia, Minuria, Miricacalia, Miyamayomena, Molpadia, Monactis, Monarrhenus, Monenteles, Monogereion, Monolopia, Monoptilon, Montanoa, Monticalia, Moonia, Moquinia, Morithamnus, Moscharia, Msuata, Mulgedium, Munnozia, Munzothamnus, Muschleria, Mutisia, Mycelis, Myopordon, Myriactis, Myriocephalus, Myripnois, Myxopappus, Nabalus, Nananthea, Nannoglottis, Nanothamnus, Nardophyllum, Narvalina, Nassauvia, Nauplius, Neblinaea, Nelsonianthus, Nemosenecio, Neocabreria, Neocaldasia, Neocuatrecasia, Neohintonia, Neojeffreya, Neomirandea, Neopallasia, Neotysonia, Nesomia, Neuractis, Neurolaena, Neurolakis, Nicolasia, Nicolletia, Nidorella, Nikitinia, Nipponanthemum, Nivellea, Nolletia, Nothobaccharis, Noticastrum, Notobasis, Notoseris, Nouelia, Novenia, Oaxacania, Oblivia, Ochrocephala, Odixia, Odontocline, Oedera, Oiospermum, Oldenburgia, Olearia, Olgaea, Oligactis, Oliganthes, Oligochaeta, Oligothrix, Olivaea, Omalotheca, Omphalopappus, Oncosiphon, Ondetia, Onopordum, Onoseris, Oonopsis, Oparanthus, Ophryosporus, Opisthopappus, Oreochrysum, Oreoleysera, Oreostemma, Oritrophium, Orochaenactis, Osbertia, Osmadenia, Osmiopsis, Osmitopsis, Osteospermum, Otanthus, Oteiza, Othonna, Otopappus, Otospermum, Outreya, Oxycarpha, Oxylaena, Oxylobus, Oxypappus, Oxyphyllum, Oyedaea, Ozothamnus, Pachylaena, Pachythamnus, Packera, Pacourina, Palaeocyanus, Palafoxia, Paleaepappus, Pamphalea, Pappobolus, Papuacalia, Paracalia, Parachionolaena, Parafaujasia, Paragynoxys, Paranephelius, Parantennaria, Parapiqueria, Paraprenanthes, Parasenecio, Parastrephia, Parthenice, Parthenium, Passacardoa, Pechuel-Loeschea, Pectis, Pegolettia, Pelucha, Pentacalia, Pentachaeta, Pentalepis, Pentanema, Pentatrichia, Pentzia, Perdicium, Perezia, Pericallis, Pericome, Perityle, Perralderia, Pertya, Perymeniopsis, Perymenium, Petalacte, Petasites, Peteravenia, Petradoria, Petrobium, Peucephyllum, Peyrousea, Phacellothrix, Phaenocoma, Phaeostigma, Phagnalon, Phalacrachena, Phalacraea, Phalacrocarpum, Phalacroseris, Phaneroglossa, Phanerostylis, Phania, Philactis, Philoglossa, Philyrophyllum, Phyllocephalum, Phymaspermum, Picnomon, Picris, Picrosia, Picrothamnus, Pilostemon, Pinaropappus, Pinillosia, Piora, Pippenalia, Piptocarpha, Piptocoma, Piptolepis, Piptothrix, Piqueria,

Piqueriella, Piqueriopsis, Pithecoseris, Pithocarpa, Pittocaulon, Pityopsis, Pladaroxylon, Plagiobasis, Plagiocheilus, Plagiolophus, Plagius, Planaltoa, Planea, Plateilema, Platycarpha, Platypodanthera, Platyschkuhria, Plazia, Plecostachys, Plectocephalus, Pleiotaxis, Pleocarphus, Pleurocarpaea, Pleurocoronis, Pleuropappus, Pleurophyllum, Pluchea, Plummera, Podachaenium, Podanthus, Podocoma, Podolepis, Podotheca, Poecilolepis, Pogonolepis, Pojarkovia, Pollalesta, Polyachyrus, Polyanthina, Polyarrhena, Polycalymma, Polychrysum, Polymnia, Polytaxis, Porophyllum, Porphyrostemma, Praxeliopsis, Praxelis, Prenanthella, Prenanthes, Prestelia, Printzia, Prionopsis, Prolobus, Prolongoa, Proteopsis, Proustia, Psacaliopsis, Psacalium, Psathyrotes, Psathyrotopsis, Psednotrichia, Pseudobahia, Pseudoblepharispermum, Pseudobrickellia, Pseudocadiscus, Pseudoclappia, Pseudoconyza, Pseudognaphalium, Pseudogynoxys, Pseudohandelia, Pseudokyrsteniopsis, Pseudonoseris, Pseudostifftia, Psiadia, Psilocarphus, Psilostrophe, Psychrophyton, Pterachaenia, Pterocaulon, Pterochaeta, Pterocypsela, Pteronia, Pteropogon, Pterothrix, Pterygopappus, Ptilostemon, Pulicaria, Pyrethropsis, Pyrrhopappus, Pyrrocoma, Quelchia, Quinetia, Quinqueremulus, Radlkoferotoma, Rafinesquia, Raillardella, Raillardiopsis, Rainiera, Raoulia, Raouliopsis, Rastrophyllum, Ratibida, Raulinoreitzia, Reichardia, Relhania, Remya, Rennera, Rensonia, Revealia, Rhagadiolus, Rhamphogyne, Rhanteriopsis, Rhanterium, Rhetinolepis, Rhinactinidia, Rhodanthe, Rhynchospermum, Rhysolepis, Richteria, Rien courtia, Rigiopappus, Robinsonia, Rochonia, Rojasianthe, Rolandra, Roldana, Rosenia, Rothmaleria, Rudbeckia, Rugelia, Ruilopezia, Rumfordia, Russowia, Rutidosis, Sabazia, Sabbata, Sachsia, Salmea, Santolina, Santosia, Sanvitalia, Sarcanthemum, Sartorina, Sartwellia, Saussurea, Scalesia, Scariola, Scherya, Schischkinia, Schistocarpha, Schistostephium, Schizogyne, Schizopsera, Schizotrichia, Schkuhria, Schlechtendalia, Schmalhausenia, Schoenia, Schumeria, Sciadocephala, Sclerocarpus, Sclerolepis, Sclerorhachis, Sclerostephane, Scolymus, Scorzonera, Scrobicaria, Scyphocoronis, Selloa, Senecio, Seriphidium, Serratula, Shafera, Sheareria, Shinnersia, Shinnersoseris, Siebera, Sigesbeckia, Siloxerus, Silphium, Silybum, Simsia, Sinacalia, Sinclairia, Sinoleontopodium, Sinosenecio, Sipolisia, Smallanthus, Soaresia, Solanecio, Solenogyne, Solidago, Soliva, Sommerfeltia, Sonchus, Sondottia, Soroseris, Spaniopappus, Sparganophorus, Sphaeranthus, Sphaereupatorium, Sphaeroclinium, Sphaeromeria, Sphaeromorphaea, Sphagneticola, Spilanthes, Spiracantha, Spiroseris, Squamopappus, Stachycephalum, Staehelina, Standleyanthus, Stanfieldia, Staurochlamys, Stebbinsia, Stebbinsoseris, Steiractinia, Steirodiscus, Stenachaenium, Stenocline, Stenopadus, Stenophalium, Stenops, Stenoseris, Stenotus, Stephanochilus, Stephanodoria, Stephanomeria, Steptorhamphus, Stevia, Steviopsis, Steyermarkina, Stifftia, Stilpnogyne, Stilpnolepis, Stilpnopappus, Stoebe, Stokesia, Stomatanthes, Stomatochaeta, Stramentopappus, Streptoglossa, Strotheria, Stuartina, Stuckertiella, Stuessya, Stylocline, Stylotrichium, Sventenia, Symphyllocarpus, Symphyopappus, Syncalathium, Syncarpha, Syncephalum, Syncretocarpus, Synedrella, Synedrellopsis, Syneilesis, Synotis, Syntrichopappus, Synurus, Syreitschikovia, Tagetes, Takhtajaniantha, Tamananthus, Tamania, Tamaulipa, Tanacetopsis, Tanacetum, Taplinia, Taraxacum, Tarchonanthus, Teixeiranthus, Telanthophora, Telekia, Telmatophila, Tenrhynea, Tephroseris, ? Terana, Tessaria, Tetrachyron, Tetradymia, Tetragonotheca, Tetramolopium, Tetraneuris, Tetranthus, Tetraperone, Thaminophyllum, Thamnoseris, Thelesperma, Thespidium, Thespis, Thevenotia, Thiseltonia, Thymophylla, Thymopsis, Tiarocarpus, Tietkensia, Tithonia, Tolbonia, Tolpis, Tomentaurum, Tonestus, Tourneuxia, Townsendia, Toxanthes, Tracyina, Tragopogon, Traversia, Trichanthemis, Trichanthodium, Trichocline, Trichocoronis, Trichocoryne, Trichogonia, Trichogoniopsis, Trichogyne, Tricholepis, Trichoiptilium, Trichospira, Tridactylina, Tridax, Trigonospermum, Trilisia, Trimorpha, Trioncinia, Tripleurospermum, Triplocephalum, Triptilion, Triptilodiscus, Trixis, Troglophyton, Tuberostylis, Tugarinovia, Turaniphytum, Tussilago, Tuxtla, Tyleropappus, Tyrimnus, Uechtritzia, Ugamia, Uleophytum, Unxia, Urbananthus, Urbinella, Urmenetea, Urolepis, Uropappus, Urospermum, Ursinia, Vanclevea, Varilla, Varthemia, Vellereophyton, Venegasia, Verbesina, Vernonanthura, Vernonia, Vernoniopsis, Viereckia, Vieria, Vigethia, Viguiera, Villanova, Vilobia, Vittadinia, Vittetia, Volutaria, Wagenitzia, Waitzia, Wamalchitamia, Warionia, Wedelia, Welwitschiella, Werneria, Westoniella, Whitneya, Wilkesia, Wollastonia, Wulffia, Wunderlichia, Wythia, Xanthisma, Xanthium, Xanthocephalum, Xanthopappus, Xeranthemum, Xerochrysum, Xerolakia, Xiphochaeta, Xylanthemum, Xylorhiza, Xylothamia, Yermo, Youngia, Zaluziana, Zandera, Zexmenia, Zinnia, Zoegea, Zyzyxia.

Asteranthaceae Knuth = Lecythidaceae Poit.

ASTEROPEIACEAE Takht. ex Reveal et Hoogland. 1990. Dicots. Theales. 1/7. Madagascar. Genus Asteropeia.

Astilbaceae Krach = Saxifragaceae Juss.

Astragalaceae Martinov = Fabaceae Lindl.

Astrocarpaceae A. Kern. = Resedaceae DC. ex Gray.

Athanasiaceae Martinov = Asteraceae Dumort.

Atherospermataceae R. Br. = Monimiaceae Juss.

Atriplicaceae Juss. = Chenopodiaceae Vent.

Atropaceae Martinov = Solanaceae Juss.

AUCUBACEAE J. Agardh. 1858. Dicots. Cornales. 1/6. Asia from eastern Himalayas to Japan and Russian Far East. Genus Aucuba.

Aurantiaceae Juss. = Rutaceae Juss.

AUSTROBAILEYACEAE Croizat. 1943. Dicots. Austrobaileyales. 1/1. Northeastern Australia. Genus Austrobaileya.

AUSTROTAXACEAE Nakai. 1938. Pinopsida. Taxales. 1/1. New Caledonia. Genus Austrotaxus.

Avenaceae Martinov = Poaceae Barnhart.

Averrhoaceae Hutch. = Oxalidaceae R. Br.

AVICENNIACEAE Endl. 1841. Dicots. Lamiales. 1/14. Mangroves of tropics and subtropics. Genus Avicennia.

Azaleaceae von Vest = Ericaceae Juss.

BALANITACEAE Endl. 1841. Dicots. Rutales. 1/25. North and trop. Africa, West and South Asia. Genus Balanites.

BALANOPACEAE Benth. 1880. Dicots. Balanopales. 2/10. Australia, New Caledonia, Vanuatu, Fiji. Genera: Balanops, Trilocularia.

BALANOPHORACEAE Rich. 1822. Dicots. Balanophorales. 3/30. Tropics of the Old World from Africa to Polynesia. Genera: Balanophora, Langsdorffia, Thonningia.

Baloghiaceae Baum.-Bodenh. = Euphorbiaceae Juss.

Balsameaceae Dumort. = Burseraceae Kunth.

BALSAMINACEAE A. Rich. 1822. Dicots. Balsaminales. 2/900. Trop. Africa and Asia, a few species in temp. Eurasia, temp. Africa and North America. Genera: Hydrocera, Impatiens.

Bambusaceae Burnett = Poaceae Barnhart.

BARBEUIACEAE Nakai. 1942. Dicots. Caryophyllales. 1/1. Madagascar. Genus Barbeuia.

Barbeyaceae Rendle. 1916. Dicots. Barbeyales. 1/1. Northeast Africa, Arabian Peninsula. Genus Barbeya.

Barclayaceae H. L. Li. 1955. Dicots. Nymphaeales. 1/2–3. Southeast Asia, New Guinea. Genus Barclaya.

Barringtoniaceae F. Rudolphi = Lecythidaceae Poit.

Basellaceae Moq. 1840. Dicots. Caryophyllales. 5/13–18. Tropics and subtropics, esp. America. Genera: Anredera, Basella, Boussingaultia, Tournonia, Ullucus.

Bataceae Mart. ex Meisn. 1842. Dicots. Batales. 1/2. Trop. and subtrop. America, New Guinea, Australia, Hawaiian Is. Genus Batis.

Baueraceae Lindl. 1830. Dicots. Cunoniales. 1/3. Australia and Tasmania. Genus Bauera.

Bauhiniaceae Martinov = Fabaceae Lindl.

Begoniaceae C. Agardh. 1825. Dicots. Begoniales. 3/1,000. Tropics and subtropics, excl. Australia and Polynesia. Genera: Begonia, Hillebrandia, Symbegonia.

Belangeraceae J. Agardh = Cunnoniaceae R. Br.

Belloniaceae Martinov = Gesneriaceae Dumort.

Bembiciaceae R. Keating et Takht. 1994. Dicots. Violales. 1/1. Madagascar. Genus Bembicia.

Berberidaceae Juss. 1789. Dicots. Ranunculales. 15/630–780. Chiefly temp. and subtrop. Northern Hemisphere, but a few species of Berberis extend to Strait of Magellan. Genera: Achlis, Berberis, Bongardia, Caulphyllum, Diphylleia, Dysosma, Epimedium, Gymnospermum, Jeffersonia, Leontica, Mahonia, Plagiorhegma, Podophyllum, Synopodophyllum, Vancouveria.

Berberidopsidaceae Takht. 1985. Dicots. Violales. 2/4. East Asia, northeastern Australia, Chile. Genera: Berberidopsis, Streptothamnus.

Bertyaceae J. Agardh = Euphorbiaceae Juss.

Berzeliaceae Nakai = Bruniaceae DC.

Besleriaceae Raf. = Gesneriaceae Dumort.

Betaceae Burnett = Chenopodiaceae Vent.

Betulaceae Gray. 1821. Dicots. Betulales. 6/200. Arctic and temp. regions of the Northern Hemisphere, a few species of Alnus extending to Chile and Argentina. Genera: Alnus, Betula, Carpinus, Corylus, Ostrya, Ostryopsis.

Biebersteiniaceae Endl. 1841. Dicots. Geraniales. 1/4–5. Southeast Europe eastward to West Siberia and western Tibet. Genus Biebersteinia.

Bifariaceae Nakai = Viscaceae Miers.

Bignoniaceae Juss. 1789. Dicots. Scrophulariales. 116/870. Chiefly tropics, a few species in subtropics and warm–temp. Asia. Genera: Adenocalymna, Amphilophium, Amphitecna, Anemopaegma, Argylia, Arrabidaea, Astianthus, Bignonia, Callichlamys, Campsidium, Campsis, Catalpa, Catophractes, Ceratophytum, Chilopsis, Clytostoma, Colea, Crescentia, Cuspidaria, Cybistax, Cydista, Delostoma, Deplanchea, Digomphia, Dinklageodoxa, Distictella, Distictis, Dolichandra, Dolichandrone, Eccremocarpus, Ekmanianthe, Exarata, Fernandoa, Fridericia, Gardnerodoxa, Glaziovia, Godmania, Haplolophium, Heterophragma, Hieris, Incarvillea, Jacaranda, Kigelia, Lamiodendron, Leucocalantha, Lundia, Macfadyena, Macranthisiphon, Manaosella, Mansoa, Markhamia, Martinella, Mayodendron, Melloa, Memora, Millingtonia, Mussatia, Neojobertia, Neosepicaea, Newbouldia, Niedzwedzkia, Nyctocalos, Onochualcoa, Ophiocolea, Oroxylum, Pajanelia, Pandorea, Parabignonia, Paracarpaea, Paragonia, Paratecoma, Parmentiera, Pauldopia, Paulownia, Perianthomega, Periarrabidaea, Perichlaena, Phryganocydia, Phyllarthron, Phylloctenium, Pithecoctenium, Pleonotoma, Podranea, Potamoganos, Pseudocatalpa, Pyrostegia, Radermachera, Rhigozum, Rhodocolea, Roentgenia, Romeroa, Santisukia, Saritaea, Schlegelia, Scobinaria, Setilobus, Sideropogon, Sparattosperma, Spathicalyx, Spathodea, Sphingiphila, Spirotecoma, Stereospermum, Stizophyllum, Synapsis, Tabebuia, Tanaecium, Tecoma, Tecomaria, Tecomella, Tourrettia, Tynnanthus, Urbanolophium, Wightia, Xylophragma, Zeyheria.

Bischofiaceae (Muell. Arg.) Airy Shaw = Euphorbiaceae Juss.

BIXACEAE Link. 1831. Dicots. Bixales. 1/1. Trop. America, West Indies. Genus Bixa.

Blakeaceae Rchb. ex Barnhart = Melastomataceae Juss.

BLANDFORDIACEAE R. Dahlgren et Clifford. 1985. Monocots. Amaryllydales. 1/4. Eastern Australia, Tasmania. Genus Blandfordia.

Blepharocaryaceae Airy Shaw = Anacardiaceae Lindl.

Blitaceae Adans. ex T. Post et Kuntze = Chenopodiaceae Vent.

Blyxaceae Nakai = Hydrocharitaceae Juss.

Boerlagellaceae H. J. Lam = Sapotaceae Juss.

Bolivariaceae Grieb. = Oleaceae Hoffmanns. et Link.

BOMBACACEAE Kunth. 1822. Dicots. Malvales. 32/320. Pantropics, but chiefly South America, esp. Brazil. Genera: Adansonia, Aguiaria, Bernoullia, Bombacopsis, Bombax, Camptostemon, Catostemma, Cavanillesia, Ceiba, Chorisia, Coelostegia, Cullenia, Durio, Eriotheca, Gyranthera, Huberodendron, Kostermansia, Matisia, Neesia, Neobuchia, Ochroma, Pachira, Paradombeya, Patinoa, Phragmotheca, Pseudobombax, Quararibea, Rhodognaphalopsis, Scleronema, Septotheca.

BONNETIACEAE L. Beauvis. ex Nakai. 1948. Dicots. Theales. 4/31–32. Trop. America, genus Ploiarium in Southeast Asia, Malesia. Genera: Acopanea, Archytaea, Bonnetia, Ploiarium.

Bontiaceae Horan. = Myoporaceae R. Br.

Boopidaceae Cass. = Calyceraceae Rich.

BORAGINACEAE Juss. 1789. Dicots. Boraginales. 141–142/2,120. Trop., subtrop. and temp. regions, esp. Mediterranean and Irano-Turanian Region. Genera: Actinocarya, Adelocaryum, Afrotysonia, Alkanna, Allocarya, Amblynotus, Amphibologyne, Amsinckia, Anchusa, Ancistrocarya, Antiotrema, Antiphytum, Argusia, Arnebia, Arnebiola, Asperugo, Auxemma, Beruniella, Bilegnum, Borago, Borrachinea, Bothriospermum, Bourreria, Brachybotrys, Brandellia, Brunnera, Buglossoides, Caccinia, Carmona, Cerinthe, Chionocharis, Choriantha, Coldenia, Cordia, Cortesia, Craniospermum, Crucicaryum, Cryptantha, Cynoglossopsis, Cynoglossum, Cynoglottis, Cystostemon, Dasynotus, Decalepidanthus, Echinoglochin, Echiochilon, Echiochilopsis, Echiostachys, Echium, Ehretia, Elizaldia, Embadium, Eritrichium, ? Galeata, Gastrocotyle, Gyrocaryum, Hackelia, Halascya, Halgania, Harpagonella, Heliotropium, Heterocaryum, Huynhia, Ivanjohnstonia, Ixorhea, Lacaitaea, Lappula, Lasiarrhenum, Lasiocaryum, Lepechiniella, Lepidocordia, Lindelofia, Lithodora, Lithospermum, Lobostemon, Lycopsis, Macromeria, Maharanga, Mairetis, Menais, Mertensia, Metaeritrichium, Microcaryum, Microparacaryum, Microula, Mimophytum, Moltkia, Moltkiopsis, Moritzia, Myosotidium, Myosotis, Neotostema, Nesocaryum, Nogalia, Nomosa, Nonea, Ogastemma, Omphalodes, Omphalolappula, Omphalotrigonotis, Onosma, Onosmodium, Oxyosmyles, Paracaryopsis, Paracynoglossum, Paramoltkia, Patagonula, Pectocarya, Pentaglottis, Perittostema, Phyllocara, Plagiobothrys, Pseudomertensia, Psilolaemus, Pteleocarpa, Pulmonaria, Rindera, Rochefortia, Rochelia, Rotula, Saccellium, Selkirkia, Sericostoma, Sinjohnstonia, Solenanthus, Stenosolenium, Stephanocaryum, Suchtelenia, Symphytum, Thaumatocaryon, Thyrocarpus, Tianschaniella, Tiquilia, Tournefortia, Trachelanthus, Trachystemon, Trichodesma, Trigonocaryum, Trigonotis, Ulugbekia, Valentiniella, Wellstedia.

Borassaceae Schultz Sch. = Arecaceae Schultz Sch.

Boroniaceae J. Agardh = Rutaceae Juss.

Bortyodendraceae J. Agardh = Araliaceae Juss.

Bougainvilleaceae J. Agardh = Nyctaginaceae Juss.

BOWENIACEAE D. W. Stev. 1981. Cycadopsida. Cycadales. 1/2. Northeastern Australia. Genus Bowenia.

BRASSICACEAE Burnett. 1835 or CRUCIFERAE Juss. 1789 (nom. altern.). Dicots. Capparales. 390/3,400. Cosmopolitan. Genera: Acachmena, Acanthocardamum, Achoriphragma, Aethionema, Agallis, Alliaria, Alyssoides, Alyssopsis, Alyssum, Ammosperma, Anastatica, Anchonium, Andrezeiowskia, Anelsonia, Aphragmus, Aplanodes, Arabidella, Arabidopsis, Arabis, Arcyosperma, Armoracia, Aschersoniodoxa, Asperuginoides, Asta, Ateixa, Atelanthera, Athysanus, Atropatenia, Aubrieta, Aurinia, Ballantinia, Barbarea, Berteroa, Berteroella, Biscutella, Bivonaea, Blennodia, Boleum, Boreava, Bornmuellera, Borodinia, Botschantzevia, Brachycarpaea, Brassica, Braya, Brayopsis, Brossardia, Bunias, Cakile, Calepina, Callothlaspi, Calymmatium, Camelina, Camelinopsis, Capsella, Cardamine, Cardaminopsis, Cardaria, Carinavalva, Carrichtera, Catadysia, Catenulina, Caulanthus, Caulostramina, Ceratocnemum, Ceriosperma, Chalcanthus, Chamira, Chartoloma, Cheesemania, Cheiranthus, Chlorocrambe, Chorispora, Christolea, Chrysobraya, Chrysochamela, Citharcloma, Clandestinaria, Clastopus, Clausia, Clypeola, Cochlearia, Cochleariella, Coelonema, Coincya, Coluteocarpus, Conringia, Cordylocarpus, Coronopus, Crambe, Crambella, Cremolobus, Cryptospora, Cuphonotus, Cusickiella, Cycloptychis, Cymatocarpus, Cyphocardamum, Dactylocardamum, Decaptera, Degenia, Delpinophytum, Dentaria, Descurainia, Desideria, Diceratella, Dichasianthus, Dictyophragmus, Didesmus, Didymophysa, Dielsiocharis, Dilophia, Dimorphocarpa, Dimorphostemon, Diplopilosa, Diplotaxis, Dipoma, Diptychocarpus, Dithyrea, Dolichorhynchus, Dontostemon, Douepea, Draba, Drabastrum, Drabopsis, Dryopetalon, Eigia, Elburzia, Enarthrocarpus, Englerocharis, Eremobium, Eremoblastus, Eremodraba, Eremophyton, Ermania, Ermaniopsis, Erophila, Eruca, Erucaria, Erucastrum, Erysimum, Euclidium, Eudema, Eutrema, Euzomodendron, Farsetia, Fibigia, Foleyola, Fortuynia, Galitzkya, Geococcus, Glaribraya, Glastaria, Glaucocarpum, Goldbachia, Gorodkovia, Graellsia, Grammosperma, Guiraoa, Gynophorea, Halimolobos, Harmsiodoxa, Hedinia, Heldreichia, Heliophila, Hemicrambe, Hemilophia, Henophyton, Hesperis, Heterodraba, Hexaptera, Hirschfeldia, Hollermayera, Hornungia, Horwoodia, Hugueninia, Hutchinsia, Hutchinsiella, Hymenolobus, Iberis, Idahoa, Iodanthus, Irenepharsus, Isatis, Ischnocarpus, Iskandera, Iti, Ivania, Jonopsidium, Kernera, Kremeriella, Lachnocapsa, Lachnoloma, Leavenworthia, Leiospora, Lemphoria, Lepidium, Lepidostemon, Leptaleum, Lesquerella, Lignariella, Lithodraba, Litwinowia, Lobularia, Loxoptera, Loxostemon, Lunaria, Lutzia, Lycocarpus, Lyrocarpa, Macropodium, Malcolmia, Mancoa, Maresia, Mathewsia, Matthiola, Megacarpaea, Megadenia, Menkea, Menonvillea, Micrantha, Microlepidium, Micromystria, Microsisymbrium, Microstigma, Morettia, Moricandia, Moriera, Morisia, Murbeckiella, Muricaria, Myagrum, Nasturtiopsis, Nasturtium, Neomartinella, Neotschihatchewia, Neotorularia, Nerisyrenia, Neslia, Neuontobotrys, Noccaea, Notoceras, Notothlaspi, Ochthodium, Octoceras, Onuris, Oreoblastus, Oreoloma, Oreophyton, Ornithocarpa, Orychophragmus, Otocarpus,

Oudneya, Pachycladon, Pachymitus, Pachyneurum, Pachyphragma, Pachypterygium, Papuzilla, Parlatoria, Parodiodoxa, Parolinia, Parrya, Parryodes, Parryopsis, Pegaeophyton, Peltaria, Peltariopsis, Pennellia, Petinio tia, Petrocallis, Phaeonychium, Phlebiophragmus, Phlebolobium, Phlegmatospermum, Phoenicaulis, Physaria, Physocardamum, Physoptychis, Physorhynchus, Platycraspedum, Polyctenium, Polypsecadium, Pringlea, Prionotrichon, Pritzelago, Pseudanastatica, Pseudarabidella, Pseuderucaria, Pseudocamelina, Pseudoclausia, Pseudofortuynia, Pseudovesicaria, Psychine, Pterygiosperma, Pterygostemon, Ptilotrichum, Puccionia, Pugionium, Pycnoplinthopsis, Pycnoplinthus, Quezelianthe, Quidproquo, Raffenaldia, Raphanorhyncha, Raphanus, Rapistrum, Redowskia, Rhizobotrya, Ricotia, Robeschia, Rollinsia, Romanschulzia, Roripella, Rorippa, Rytidocarpus, Sameraria, Sarcodraba, Savignya, Scambopus, Schimpera, Schivereckia, Schizopetalon, Schlechteria, Schoenocrambe, Schouwia, Scoliaxon, Selenia, Sibara, Silicularia, Sinapidendron, Sinapis, Sisymbrella, Sisymbriopsis, Sisymbrium, Smelowskia, Sobolewskia, Solms-Laubachia, Sophiopsis, Sphaerocardamum, Spirorhynchus, Spryginia, Staintoniella, Stanleya, Stenopetalum, Sterigmostemum, Stevenia, Straussiella, Streptanthella, Streptanthus, Streptoloma, Strigosella, Stroganowia, Stubendorffia, Subularia, Succowia, Synstemonanthus, Synthlipsis, Syrenia, Takhtajaniella, Taphrospermum, Tauscheria, Teesdalia, Teesdaliopsis, Tetracme, Tetracmidion, Thellungiella, Thelypodium, Thlaspeocarpa, Thlaspi, Thysanocarpus, Trachystoma, Trichochiton, Trichotolinum, Trochiscus, Tropidocarpum, Turritis, Vella, Veselskya, Vvedenskyella, Warea, Wasabia, Weberbauera, Werdermannia, Winklera, Xerodraba, Yinshania, Zerdana, Zilla, Zuvanda.

BRETSCHNEIDERACEAE **Engler et Gilg.** 1924. Dicots. Sapindales. 1/1. China, Thailand. Genus Bretschneidera.

BREXIACEAE **Loudon.** 1830. Dicots. Celastrales. 2/10. East Africa, Madagascar, Seycheles Is. Genera: Brexia, Ixerba.

Bromaceae K. Koch = Poaceae Barnhart.

BROMELIACEAE **Juss.** 1789. Monocots. Bromeliales. 61/2,340. Trop. and subtrop. America, 1 sp. Pitcairnia in West Africa. Genera: Abromeitiella, Acanthostachys, Aechmea, Ananas, Andrea, Androlepis, Araeococcus, Ayensua, Billbergia, Brewcaria, Brocchinia, Bromelia, Canistrum, Caraguata, Catopsis, Chevaliera, Cipuropsis, Connellia, Cottendorfia, Cryptanthus, Deuterocohnia, Disteganthus, Dyckia, Encholirium, Fascicularia, Fernseea, Fosterella, Glomeropitcairnia, Greigia, Guzmania, Hechtia, Hohenbergia, Hohenbergiopsis, Lamprococcus, Lymania, Mezobromelia, Navia,, Puya, Neoglaziovia, Neoregelia, Nidularium, Ochagavia, Ortgiesia, Orthophytum, Pepinia, Pitcairnia, Platyaechmea, Podaechmea, Portea, Pothuava, Pseudaechmea, Pseudananas, Puya, Quesnelia, Racinaea, Ronnbergia, Steyerbromelia, Streptocalyx, Tillandsia, Vriesea, Wittrockia.

BRUNELLIACEAE **Engl.** 1897. Dicots. Cunoniales. 1/52. Trop. America. Genus Brunellia.

BRUNIACEAE **R. Br. ex DC.** 1825. Dicots. Bruniales. 11/67–68. South Africa. Genera: Audouinia, Berzelia, Brunia, Linconia, Lonchostoma, Mniothamnea, Pseudobaeckea, Raspalia, Staavia, Thamnes, Tittmannia.

BRUNONIACEAE **Dumort.** 1829. Dicots. Goodeniales. 1/1. Australia, Tasmania. Genus Brunonia.

Brunsvigiaceae Horan. = Amarylldiaceae J. St.-Hil.

Bryoniaceae Adans. ex T. Post et Kuntze = Cucurbitaceae Juss.

Bucklandiaceae J. Agardh = Hamamelidaceae R. Br.

BUDDLEJACEAE **K. Wilh.** 1910. Dicots.

Scrophulariales. 8/120. Trop. and subtrop. America, Africa and Asia. Genera: Androya, Buddleja, Emorya, Gomphostigma, Nuxia, Peltanthera, Polypremum, Sanango.

Buglossaceae Hoffmanns. et Link = Boraginaceae Juss.

Bulbocodiaceae Salisb. = Melanthiaceae Batsch.

Bumeliaceae Barnhart = Sapotaceae Juss.

Bupleuraceae Martinov = Apiaceae Lindl.

BURMANNIACEAE **Blume.** 1827. Monocots. Burmanniales. 17/140. Tropics and subtropics. Genera: Afrothismia, Aptera, Burmannia, Campylosiphon, Cymbocarpa, Desmogymnosiphon, Dictyostega, Glaziocharis, Gymnosiphon, Haplothismia, Hexapterella, Mamorea, Marthella, Miersiella, Ophiomeris, Oxygyne, Thismia.

BURSERACEAE **Kunth.** 1824. Dicots. Rutales. 17/600. Pantropics, esp. trop. America, Northeast Africa and Malesia. Genera: Aucoumea, Beiselia, Boswellia, Bursera, Canarium, Commiphora, Crepidospermum, Dacryodes, Garuga, Haplolobus, Protium, Santiria, Scutinanthe, Tapirocarpus, Tetragastris, Trattinnickia, Triomma.

BUTOMACEAE **Rich.** 1816. Monocots. Butomales. 1/1. Temp. Eurasia. Genus Butomus.

BUXACEAE **Dumort.** 1822. Dicots. Buxales. 4/94. North and Central America, West Indies, Eurasia, Africa, Madagascar. Genera: Buxus, Notobuxus, Pachysandra, Sarcococca.

BYBLIDACEAE **Domin.** 1922. Dicots. Byblidales. 1/2. New Guinea, Australia. Genus Byblis.

Byttneriaceae R. Br. = Sterculiaceae (DC.) Bartl.

CABOMBACEAE **A. Rich.** 1828. Dicots. Nymphaeales. 2/8. Tropics, subtropics and a few species in temp. regions. Genera Brasenia, Cabomba.

CACTACEAE **Juss.** 1789. Dicots. Caryophyllales. 97/1,550. Chiefly arid and semiarid regions of America, extending to British Columbia and Patagonia, and high mountains in the Andes; Galapagos Is., species of Rhipsalis in trop. Africa, Madagascar, Seychelles, Mascarene Is., and Sri Lanka. Genera: Acanthocereus, Ariocarpus, Armatocereus, Arrojadoa, Arthrocereus, Astrophytum, Austrocactus, Aztekium, Bergerocactus, Blossfeldia, Brachycereus, Brasilicereus, Browningia, Calymmanthium, Carnegiea, Cephalocereus, Cereus, Cipocereus, Cleistocactus, Coleocephalocereus, Copiapoa, Corryocactus, Coryphantha, Denmoza, Discocactus, Disocactus, Echinocactus, Echinocereus, Echinopsis, Epiphyllum, Epithelantha, Eriosyce, Escobaria, Escontria, Espostoa, Espostoopsis, Eulychnia, Facheiroa, Ferocactus, Frailea, Geohintonia, Gymnocalycium, Haageocereus, Harrisia, Hatiora, Hylocereus, Jasminocereus, Leocereus, Lepismium, Leptocereus, Leuchtenbergia, Lophophora, Maihuenia, Mammillaria, Mammilloydia, Matucana, Melocactus, Micranthocereus, Mila, Myrtillocactus, Neobuxbaumia, Neolloydia, Neoporteria, Neoraimondia, Neowerdermannia, Obregonia, Opuntia, Oreocereus, Oroya, Ortegocactus, Pachycereus, Parodia, Pediocactus, Pelecyphora, Peniocereus, Pereskia, Pereskiopsis, Pilosocereus, Polaskia, Pseudorhipsalis, Pterocactus, Rathbunia, Rebutia, Rhipsalis, Samaipaticereus, Schlumbergera, Sclerocactus, Selenicereus, Stenocactus, Stenocereus, Stephanocereus, Stetsonia, Strombocactus, Tacinga, Thelocactus, Uebelmannia, Weberocereus.

Caesalpiniaceae R. Br. = Fabaceae Lindl.
Caladiaceae Salisb. = Araceae Juss.
Calectasiaceae Endl. = Dasypogonaceae Dumort.
Calendulaceae Link = Asteraceae Dumort.
Callaceae Rchb. ex Bartl. = Araceae Juss.
Callicomaceae J. Agardh = Cunoniaceae R. Br.
Calligonaceae Khalk. = Polygonaceae Juss.

CALLITRICHACEAE Link. 1821. Dicots. Lamiales. 1/17. Cosmopolitan. Genus Callitriche.

CALOCHORTACEAE Dumort. 1829. Monocots. Liliales. 1/60. North and Central America. Genus Calochortus.

Calophyllaceae J. Agardh = Clusiaceae Lindl.

Calthaceae Martinov = Ranunculaceae Juss.

CALYCANTHACEAE Lindl. 1819. Dicots. Laurales. 2/9. Continental China, North America. Genera: Calicanthus, Chimonanthus.

CALYCERACEAE R. Br. ex Rich. 1820. Dicots. Calycerales. 6/60. Central and South America. Genera: Acarpha, Acicarpha, Boopis, Calycera, Gamocarpha, Moschopsis.

Cambogiaceae Horan. = Clusiaceae Lindl.

Camelliaceae DC. = Theaceae D. Don.

CAMPANULACEAE Juss. 1789. Dicots. Campanulales. 56/1,000. Temp. Northern Hemisphere, montane areas of the tropics, South America, South Africa, Australia, New Zealand. Genera: Adenophora, Astrocodon, Asyneuma, Azorina, Berenice, Brachycodonia, Campanula, Campanumoea, Canarina, Cephalostigma, Codonopsis, Craterocapsa, Cryptocodon, Cyananthus, Cylindrocarpa, Echinocodon, Edraianthus, Feeria, Gadellia, Githopsis, Gunillaea, Hanabusaya, Heterochaenia, Heterocodon, Homocodon, Jasione, Legousia, Leptocodon, Merciera, Michauxia, Microcodon, Musschia, Mzymtella, Namacodon, Nesocodon, Numaeacampa, Ostrowskia, Peracarpa, Petkovia, Petromarula, Physoplexis, Phyteuma, Platycodon, Popoviocodonia, Prismatocarpus, Rhigiophyllum, Roella, Sergia, Siphocodon, Symphyandra, Theilera, Trachelium, Treichelia, Triodanis, Wahlenbergia, Zeugandra.

CAMPYNEMATACEAE Dumort. 1829. Monocots. Liliales. 2/4. Tasmania, New Caledonia. Genera: Campynema, Campynemanthe.

Canacomyricaceae Baum.-Bodenh. = Myricaceae Blume.

Candolleaceae F. Muell. = Stylidiaceae R. Br.

CANELLACEAE Mart. 1832. Dicots. Annonales. 5/16. Trop. America, West Indies, Africa, Madagascar. Genera: Canella, Cinnamodendron, Cinnamosma, Pleodendron, Warburgia.

CANNABACEAE Endl. 1837. Dicots. Urticales. 2/3–6. Temp. Northern Hemisphere. Genera: Cannabis, Humulus.

CANNACEAE Juss. 1789. Monocots. Zingiberales. 1/55. Trop. and subtrop. America. Genus Canna.

Canopodaceae C. Presl = Santalaceae R. Br.

Canotiaceae Airy Shaw = Celastraceae R. Br.

Cansjeraceae J. Agardh = Opiliaceae (Benth.) Valeton.

CAPPARACEAE Juss. 1789. Dicots. Capparales. 44/810. Trop., subtrop. and warm–temp. regions. Genera: Apophyllum, Atamisquea, Bachmannia, Belencita, Borthwickia, Boscia, Buchholzia, Buhsia, Cadaba, Calanthea, Capparidastrum, Capparis, Cladostemon, Cleome, Cleomella, Crateva, Dactylaena, Dhofaria, Dipterygium, Euadenia, Forchhammeria, Haptocarpum, Hypselandra, Koeberlinia, Linnaeobreynia, Maerua, Morisonia, Neocalyptrocalyx, Neothorelia, Oceanopapaver, Oxystylis, Pentadiplandra, Physostemon, Podandrogyne, Poilanedora, Polanisia, Ritchiea, Setchellanthus, Steriphoma, Stixis, Tetratelia, Thilachium, Tirania, Wislizenia.

Caprariaceae Martinov = Scrophulariaceae Juss.

CAPRIFOLIACEAE Juss. 1789. Dicots. Dipsacales. 12/274. Boreal and temp. regions, a few species in montane tropics. Genera: Abelia, Diervilla, Dipelta, Heptacodium, Kolkwitzia, Leycesteria, Linnaea, Lonicera, Symphoricarpos, Triosteum, Weigela, Zabelia.

Capusiaceae Gagnep. = Celastraceae R. Br.

CARDIOPTERIDACEAE Blume. 1847. Dicots. Celastrales. 1/2. India, Southeast Asia to New Guinea and Australia. Genus Cardiopteris.

Carduaceae Dumort. = Asteraceae Dumort.

CARICACEAE **Dumort.** 1829. Dicots. Violales. 4/32. Trop. and subtrop. America, West Indies, trop. West Africa. Genera: Carica, Cylicomorpha, Jacaratia, Jarilla.

CARLEMANNIACEAE **Airy Shaw.** 1965. Dicots. Gentianales. 2/6. Himalayas, southwestern China, Southeast Asia. Genera: Carlemannia, Silvianthus.

Carludovicaceae A. Kern. = Cyclanthaceae Poit. ex A. Rich.

Carpinaceae von Vest = Betulaceae Gray.

CARPODETACEAE **Fenzl.** 1841. Dicots. Hydrangeales. 1/2–10. Eastern Malesia, New Zealand. Genus Carpodetus.

Cartonemataceae Pichon = Commelinaceae R. Br.

CARYOCARACEAE **Szyszyl.** 1893. Dicots. Theales. 2/25. Trop. America. Genera: Anthodiscus, Caryocar.

CARYOPHYLLACEAE **Juss.** 1789. Dicots. Caryophyllales. 101/2,320. Cosmopolitan, but chiefly temp. Northern Hemisphere. Genera: Acanthophyllum, Achyronychia, Agrostemma, Allochrusa, Alsinidendron, Ankyropetalum, Arenaria, Bolanthus, Bolbosaponaria, Brachystemma, Bufonia, Calycotropis, Cardionema, Cerastium, Cerdia, Chaetonychia, Colobanthus, Cometes, Corrigiola, Cucubalus, Cyathophylla, Dadjoua, Dianthus, Diaphanoptera, Dicheranthus, Dichoglottis, Drymaria, Drypis, Gastrolychnis, Geocarpon, Gymnocarpos, Gypsophila, Habrosia, Haya, Herniaria, Holosteum, Honkenya, Illecebrum, Kabulia, Krauseola, Kuhitangia, Lepyrodiclis, Lochia, Loeflingia, Lychnis, Melandrium, Mesostemma, Microphyes, Minuartia, Moehringia, Moenchia, Myosoton, Ochotonophila, Ortegia, Ototes, Paronychia, Pentastemonodiscus, Petrocoma, Petrocoptis, Petrorhagia, Philippiella, Phrynella, Pinosia, Pirinia, Pleioneura, Plettkea, Pollichia, Polycarpaea, Polycarpon, Polytepalum, Psammosilene, Pseudostellaria, Pteranthus, Pycnophyllopsis, Pycnophyllum, Reicheella, Rhodalsine, Sagina, Sanctambrosia, Saponaria, Schiedea, Scleranthopsis, Scleranthus, Sclerocephalus, Scopulophila, Silene, Spanizium, Spergula, Spergularia, Sphaerocoma, Stellaria, Stipulicida, Telephium, Thurya, Thylacospermum, Uebelinia, Vaccaria, Velezia, Wangerinia, Wilhelmsia, Xerotia.

Caryotaceae O. F. Cook = Arecaceae Schultz Sch.

Cassiaceae von Vest = Fabaceae Lindl.

Cassiniaceae Schultz Sch. = Asteraceae Dumort.

Cassipureaceae J. Agardh = Rhizophoraceae R. Br.

Cassuviaceae R. Br. = Anacardiaceae Lindl.

Cassythaceae Bartl. ex Lindl. = Lauraceae Juss.

Castaneaceae Baill. = Fagaceae Dumort.

Castelaceae J. Agardh = Simaroubaceae A. DC.

CASUARINACEAE **R. Br.** 1814. Dicots. Casuarinales. 4/70–90. Tropics of the Old World, but chiefly in Australia and New Caledonia. Genera: Allocasuarina, Casuarina, Ceuthostoma, Gymnostoma.

Catesbaeaceae Martinov = Rubiaceae Juss.

CECROPIACEAE **C. C. Berg.** 1978. Dicots. Urticales. 6/230. Trop. America, Africa, and trop. Asia. Genera: Cecropia, Coussapoa, Musanga, Myrianthus, Poikilospermum, Puarouma.

Cedraceae von Vest = Pinaceae Lindl.

Cedrelaceae R. Br. = Meliaceae Juss.

CELASTRACEAE **R. Br.** 1814. Dicots. Celastrales. 94/1,360. Trop. subtrop. and warm–temp. regions. Genera: Acanthothamnus, Allocassine, Anthodon, Apatophyllum, Apodostigma, Arnicratea, Bhesa, Brassiantha, Brexiella, Campylostemon, Canotia, Cassine, Celastrus, Cheiloclinium, Crossopetalum, Cuervea, Denhamia, Dicarpellum, Elachyptera,

Empleuridium, Euonymus, Evonymopsis, Fraunhofera, Genitia, Glyptopetalum, Goniodiscus, Gyminda, Hartogiella, Hedraianthera, Herya, Hexaspora, Hippocratea, Hylenaea, Hypsophila, Kokoona, Lecardia, Loeseneriella, Lophopetalum, Lydenburgia, Maurocenia, Maytenus, Menepetalum, Microtropis, Monimopetalum, Monocelastrus, Mortonia, Moya, Myginda, Nicobariodendron, Orthosphenia, Paxistima, Peripterygia, Peritassa, Perrottetia, Platypterocarpus, Plenckia, Pleurostylia, Polycardia, Prionostemma, Pristimera, Psammomoya, Pseudosalacia, Ptelidium, Pterocelastrus, Putterlickia, Quadripterygium, Quetzalia, Reissantia, Rzedowskia, Salacia, Salacicratea, Salacighia, Salaciopsis, Salvadoropsis, Sarawakodendron, Schaefferia, Semialarium, Semicratea, Semirestis, Sinomerrillia, Siphonodon, Telemachia, Tetrasiphon, Thyrsosalacis, Tontelea, Torralbasia, Tricerma, Tripterygium, Tristemonanthus, Villaresia, Viposia, Wimmeria, Xylonymus, Zinowiewia.

Celosiaceae Martinov = Amaranthaceae Juss.

Celtidaceae Link = Ulmaceae Mirb.

Centaureaceae Martinov = Asteraceae Dumort.

CENTROLEPIDACEAE **Endl.** 1836. Monocots. Restionales. 3/36. Southeast Asia through Malesia to southern Australia, New Zealand, temp. South America, Falkland Is. Genera: Aphelia, Centrolepis, Gaimardia.

Cepaceae Salisb. = Alliaceae J. Agardh.

Cephalanthaceae Dumort. = Rubiaceae Juss.

CEPHALOTACEAE **Dumort.** 1829. Dicots. Saxifragales. 1/1. Australia. Genus Cephalotus.

CEPHALOTAXACEAE **Neger.** 1907. Pinopsida. Cephalotaxales. 2/11. Himalayas, East and Southeast Asia. Genera: Amentotaxus, Cephalotaxus.

Cerastiaceae von Vest = Caryophyllaceae Juss.

Ceratoniaceae Link = Fabaceae Lindl.

CERATOPHYLLACEAE **Gray.** 1821. Dicots. Ceratophyllales. 1/2–6-8. Cosmopolitan. Genus Ceratophyllum.

Cerberaceae Martinov = Apocynaceae Juss.

CERCIDIPHYLLACEAE **Engl.** 1909. Dicots. Cercidiphyllales. 1/1–2. East Asia. Genus Cercidiphyllum.

Cercocarpaceae J. Agardh = Rosaceae Juss.

Cercodiaceae Juss. = Haloragaceae R. Br.

Cereaceae Spreng. ex Jameson = Cactaceae Juss.

Cerinthaceae Martinov = Boraginaceae Juss.

Ceroxylaceae O. F. Cook = Arecaceae Schultz Sch.

Cestraceae Schltdl. = Solanaceae Juss.

Cevalliaceae Griseb. = Loasaceae Dumort.

Chailletiaceae R. Br. = Dichaetalaceae Baill.

Chamaedoraceae O. F. Cook = Arecaceae Schultz Sch.

Chamaelauciaceae Lindl. = Myrtaceae Juss.

Cheiranthodendraceae A. Gray = Sterculiaceae (DC.) Bartl.

Chelidoniaceae Martinov = Papaveraceae Juss.

Chelonaceae Martinov = Scrophulariaceae Juss.

CHENOPODIACEAE **Vent.** 1799. Dicots. Caryophyllales. 115/1,400. Cosmopolitan, but chiefly temp. and subtrop. regions. Genera: Acroglochin, Agatophora, Agriophyllum, Alexandra, Allenrolfea, Anabasis, Anthochlamys, Aphanisma, Archiatriplex, Arthrocnemum, Arthrophytum, Atriples, Axyris, Baolia, Bassia, Beta, Bienertia, Borsczowia, Camphorosma, Ceratocarpus, Ceratoides, Chenoleoides, Chenopodium, Climacoptera, Corispermum, Cornulaca, Cremnophyton, Cyathobasis, Cycloloma, Darniella, Didymanthus, Dissocarpus, Dysphania, Einadia, Enchylaena, Endolepis, Eremophea, Eremosemium, Eriochiton, Exomis, Fadenia, Fredolia, Gamanthus, Girgensohnia, Gyroptera, Hablitzia, Halanthium, Halarchon, Halimione, Halimocnemis, Halocharis, Halocnemum, Halogeton, Halopeplis, Halosarcia, Halostachys, Halothamnus, Halotis, Haloxylon, Hammada, Hemichora, Heterostachys, Holmbergia, Horaninovia, Iljinia, Kalidium, Kirilowia, Kochia, Lagenantha, Londesia, Maireana, Malacocera, Manochlamys, Microcnemum, Microgynoecium, Monolepis, Nanophyton, Neobassia, Nitrophila, Noaea, Nucularia, Ofaiston, Oreobliton, Osteocarpum, Pachycornia, Panderia, Patellifolia, Petrosimonia, Physandra, Piptoptera, Polycnemum, Rhagodia, Rhaphidophyton, Roycea, Salicornia, Salsola, Sarcobatus, Sarcocornia, Scleroblitum, Sclerochlamys, Sclerolaena, Sclerostegia, Seidlitzia, Sevada, Spinacia, Stelligera, Suaeda, Suckleya, Sympegma, Tecticornia, Tegicornia Threlkeldia, Traganopsis, Traganum, Zuckia.

Chingithamnaceae Hand.-Mazz. = Celastraceae R. Br.

CHIONOGRAPHIDACEAE Takht. 1994. Monocots. Melanthiales. 1/7. East Asia. Genus Chionographis.

Chironiaceae Horan. = Gentianaceae Juss.

CHLOANTHACEAE Hutch. 1959. Dicots. Lamiales. 11/110. Australia. Genera: Chloanthes, Cyanostegia, Dicrastylis, Hemiphora, Lachnostachys, Mallophora, Newcastelia, Physopsis, Pityrodia, Spartothamnella, Tectona.

CHLORANTHACEAE R. Br. ex Lindl. 1821. Dicots. Chlorantales. 4/70. Madagascar, trop. and subtrop. Asia, Oceania, trop. America. Genera: Ascarina, Chloranthus, Hedyosmum, Sarcandra.

Chloridaceae (Rchb.) Herter = Poaceae Barnhart.

CHRYSOBALANACEAE R. Br. 1818. Dicots. Rosales. 17/500. Trop. and subtrop. regions, esp. America. Genera: Acioa, Atuna, Bafodeya, Chrysobalanus, Couepia, Dactyladenia, Exellodendron, Grangeria, Hirtella, Hunga, Kostermanthus, Licania, Magnistipula, Maranthes, Neocarya, Parastemon, Parinari.

Cichoriaceae Juss. = Asteraceae Dumort.

Cinarocephalaceae Juss. = Asteraceae Dumort.

Cinchonaceae Batsch = Rubiaceae Juss.

Circaeaceae Lindl. = Onagraceae Juss.

CIRCAEASTERACEAE Hutch. 1926. Dicots. Ranunculales. 1/1. Himalayas, China. Genus Circaeaster.

Cissaceae Horan. = Vitaceae Juss.

CISTACEAE Juss. 1789. Dicots. Bixales (Cistales). 9/204. Temp. and warm-temp. regions, chiefly Mediterranean, a few species in the eastern U.S., West Indies, and South America. Genera: Atlanthemum, Cistus, Fumana, Halimium, Helianthemum, Hudsonia, Lechea, Therocistus, Tuberaria.

Citracea Drude = Rutaceae Juss.

Clematidaceae Martinov = Ranunculaceae Juss.

Cleomaceae Horan. = Capparaceae Juss.

CLETHRACEAE Klotzsch. 1851. Dicots. Ericales. 1/67. East and Southeast Asia, America, Madeira Is. Genus Clethra.

Cliffortiaceae Mart. = Rosaceae Juss.

CLUSIACEAE Lindl. 1836 or **GUTTIFERAE Juss.** 1789 (nom. altern.). Dicots. Theales. 35/920–940. Trop., subtrop., and temp. regions. Genera: Allanblackia, Balboa, Calophyllum, Caraipa, Clusia, Clusiella, Decaphalangium, Dystovomita, Endodesmia, Garcinia, Haploclathra, Havetia, Havetiopsis, Kayea, Kielmeyera, Lebrunia, Lorostemon, Mammea, Marila, Mesua, Montrouziera, Moronobea, Neotatea, Oedematopus, Pentadesma, Pilosperma, Platonia, Poeciloneuron, Quapoya, Renggeria, Rheedia, Symphonia, Thysanostemon, Tovomita, Tovomitopsis.

CNEORACEAE Link. 1831. Dicots. Rutales. 1/3. Cuba, Canary Is., Mediterranean. Genus Cneorum.

Cnicaceae von Vest = Asteraceae Dumort.

Cobaeaceae D. Don = Polemoniaceae Juss.

COCHLOSPERMACEAE Planch. 1847. Dicost. Bixales (Cistales). 2/24. Trop. Africa, Southeast Asia, northern Australia, trop. and subtrop. America. Genera: Amoreuxia, Cochlospermum.

Cocosaceae Schultz Sch. = Arecaceae Schultz Sch.

Coffeaceae Batsch = Rubiaceae Juss.

COLCHICACEAE **DC.** 1805. Dicots. Liliales. 16/190–195. Temp., subtrop., and trop. regions, chefly South Africa. Genera: Androcymbium, Baeometra, Bulbocodium, Camptorrhiza, Colchicum, Gloriosa, Hexacyrtis, Iphigenia, Iphigeniopsis, Littonia, Merendera, Beodregea, Onixotis, Ornithoglossum, Sandersonia, Wurmbea.

Coleogynaceae J. Agardh = Rosaceae Juss.

COLUMELLIACEAE **D. Don.** 1828. Dicots. Hydrangeales. 1/4. Andes. Genus Columellia.

COMBRETACEAE **R. Br.** 1810. Dicots. Myrtales. 18/560. Tropics, a few species in subtropics. Genera: Anogeissus, Buchenavia, Bucida, Calopyxis, Calicopteris, Combretum, Conocarpus, Dansiea, Guiera, Laguncularia, Lumnitzera, Macropteranthes, Meiostemon, Pteleopsis, Quisqualis, Strephonema, Terminalia, Thiloa.

COMMELINACEAE **R. Br.** 1810. Monocots. Commelinales. 41/640. Tropics and subtropics, a few species in temp. East Asia, Australia, and North America. Genera: Aetheolirion, Amischophacelus, Amischotolype, Aneilema, Anthericopsis, Ballya, Belosynapsis, Buforrestia, Callisia, Cartonema, Cochliostema, Coleotrype, Commelina, Cyanotis, Dichorisandra, Dictyospermum, Elasis, Floscopa, Geogenanthus, Gibasis, Gibasoides, Matudanthus, Murdannia, Palisota, Pollia, Polyspatha, Porandra, Pseudoparis, Rhopalephora, Sauvallea, Siderasis, Spatholirion, Stanfieldiella, Streptolirion, Thyrsanthemum, Tinantia, Tradescantia, Tricarpe lema, Triceratella, Tripogandra, Weldenia.

Comocladiaceae Martinov = Anacardiaceae Lindl.

COMPOSITAE **Giseke.** See ASTERACEAE Dumort.

CONNARACEAE **R. Br.** 1818. Dicots. Connarales. 20/406. Pantropics. Genera: Agelaea, Bernardinia, Burttia, Byrsocarpus, Cnestidium, Cnestis, Connarus, Ellipanthus, Hemandradenia, Jaundea, Jollydora, Manotes, Paxia, Pseudellipanthus, Pseudoconnarus, Rourea, Roureopsis, Schellenbergia, Spiropetalum, Vismianthus.

CONOSTYLIDACEAE **Takht.** 1987. Monocots. Haemodorales. 6/66. Western and southwestern Australia. Genera: Anigozanthos, Blanco, Conostylis, Macropodia, Phlebocarya, Tribonanthes.

CONVALLARIACEAE **Horan.** 1834. Monocots. Asparagales. 22/240. Chiefly Northern Hemisphere, esp. North America, Himalayas, and East Asia. Genera: Antherolophus, Aspidistra, Campylandra, Clintonia, Colania, Convallaria, Disporopsis, Disporum, Evrardiella, Gonioscypha, Liriope, Maianthemum, Ophiopogon, Peliosanthes, Polygonatum, Reineckea, Rohdea, Smilacina, Speirantha, Streptopus, Theropogon, Tupistra.

CONVOLVULACEAE **Juss.** 1789. Dicots. Convolvulales. 60/1,525. Trop., subtrop., and temp. regions, esp. America and Asia. Genera: Aniseia, Argyreia, Astripomoea, Blinkworthia, Bonamia, Calycobolus, Calystegia, Cardiochlamys, Cladostigma, Convolvulus, Cordisepalum, Cressa, Decalobanthus, Dichondra, Dichondropsis, Dicranostyles, Dinetus, Dipteropeltis, Ericybe, Evolvulus, Falkia, Hewittia, Hildebrandtia, Humbertia, Hyalocystis, Ipomoea, Iseia, Itzaea, Jacquemontia, Lepistemon, Lepistemonopsis, Lysiostyles, Maripa, Merremia, Metaporana, Mina, Nephrophyllum, Neuropeltis, Neuropeltopsis, Odoniellia, Operculina, Paralepistemon, Pentacrostigma, Pharbitis, Polymeria, Porana, Poranopsis, Quamoclit, Rapona, Remirema, Rivea, Sabaudiella, Seddera, Stictocardia, Stylisma, Tetralocularia, Tridynamia, Turbina, Wilsonia, Xenostegia.

Coptidaceae A. Löve et D. Löve = Ranunculaceae Juss.

Cordiaceae R. Br. ex Dumort. = Boraginaceae Juss.

Coreopsidaceae Link = Asteraceae Dumort.

Coriaceae J. Agardh = Primulaceae Vent.

Coriandraceae Burnett = Apiaceae Lindl.

CORIARIACEAE DC. 1824. Dicots. Coriariales. 1/20. Eurasia, New Zealand, Oceania, trop. America. Genus Coriaria.

Corispermaceae Link = Chenopodiaceae Vent.

CORNACEAE Dumort. 1829. Dicots. Cornales. 3/55–100. Arctic and temp. Northern Hemisphere, Himalayas, montane trop. Southeast Asia, Central and South America, trop. Africa. Genera: Afrocrania, Cornus, Swida.

Coronillaceae Martinov = Fabaceae Lindl.

Correaceae J. Agardh = Rutaceae Juss.

Corrigiolaceae (Dumort.) Dumort. = Caryophyllaceae Juss.

CORSIACEAE Becc. 1878. Monocots. Burmanniales. 2/27. New Guinea, Solomon Is., northern Australia, Chile. Genera: Arachnitis, Corsia.

Corylaceae Mirb. = Betulaceae Gray.

CORYNOCARPACEAE Engl. 1897. Dicots. Celastrales. 1/5–7. New Guinea, Solomon Is., Australia, New Caledonia, New Zealand. Genus Corynocarpus.

Coryphaceae Schultz Sch. = Arecaceae Schultz Sch.

COSTACEAE (Meisn.) Nakai. 1941. Monocots. Zingiberales. 4/85. Pantropics. Genera: Costus, Dimerocostus, Monocostus, Tapeinocheilos.

Cotyledonaceae Martinov = Crassulaceae DC.

Coulaceae Tiegh. ex Bullock = Olacaceae Mirb. ex DC.

Coutareaceae Martinov = Rubiaceae Juss.

Coutoubeaceae Martinov = Gentianaceae Juss.

CRASSULACEAE DC. 1805. Dicots. Saxifragales. 41/1,780. Almost cosmopolitan (excl. Australia and Polynesia), chiefly South Africa. Genera: Adromischus, Aeonium, Aichryson, Bryophyllum, Chiastophyllum, Clementsia, Cotyledon, Crassula, Diamorpha, Dudleya, Echeveria, Graptopetalum, Greenovia, Hylotelephium, Hypagophytum, Jovibarba, Kalanchoë, Kirpicznikovia, Lenophyllum, Macrosepalum, Meterostachys, Monanthes, Mucizonia, Orostachys, Pachyphytum, Perrierosedum, Pistorina, Prometheum, Pseudosedum, Rhodiola, Rosularia, Sedum, Sempervivella, Sempervivum, Sinocrassula, Telmissa, Thompsonella, Tillaea, Tylecodon, Umbilicus, Villadia.

Crescentiaceae Dumort. = Bignoniaceae Juss.

Cressaceae Raf. = Convolvulaceae Juss.

Crinaceae von Vest = Amaryllidaceae J. St.-Hil.

Crocaceae von Vest = Iridaceae Juss.

Croomiaceae Nakai = Stemonaceae Engl.

CROSSOSOMATACEAE Engl. 1897. Dicots. Crossosomatales. 3/8–10. Western North America. Genera: Apacheria, Crossosoma, Glossopetalon.

Crotonaceae J. Agardh = Euphorbiaceae Juss.

CRUCIFERAE Juss. See BRASSICACEAE Burnett.

CRYPTERONIACEAE A. DC. 1868. Dicots. Myrtales. 3/10. South and Southeast Asia. Genera: Axinandra, Crypteronia, Dactylocladus.

Cryptocarynaceae J. Agardh = Araceae Juss.

Cryptomeriaceae Hayata = Taxodiaceae Warm.

CTENOLOPHONACEAE (H. Winkl.) Exell et Mendonça. 1951. Dicots. Linales. 1/2. Trop. Africa, Malesia. Genus Ctenolophon.

CUCURBITACEAE Juss. 1789. Dicots. Cucurbitales. 122/960. Tropics, subtropics and a few species in temp. regions. Genera: Abobra, Acanthosicyos, Actinostemma, Alsomitra, Ampelosicyos, Anacaona, Apatringania, Apodanthera, Baijiania, Bambekea, Benincasa, Biswarea, Bolbostemma, Brandegea, Bryonia, Calycophysum, Cayaponia, Cephalopentandra, Ceratosanthes, Chalema, Cionosicyos, Citrullus, Coccinia, Cogniauxia, Corallocarpus, Cremastopus, Ctenolepis, Cucumella, Cucumeropsis, Cucumis, Cucurbita, Cucurbitella, Cyclanthera, Cyclantheropsis, Dactyliandra, Dendrosicyos, Dicoelospermum, Dieterlea, Diplocyclos, Doyerea, Ecballium, Echinocystis, Echinopepon, Edgaria, Elateriopsis, Eureiandra, Fevillea, Gerrardanthus, Gomphogyne, Gurania, Guraniopsis, Gymnopetalum, Gynostemma, Halosicyos, Hanburia, Helmontia, Hemsleya, Herpetospermum, Hodgsonia, Ibervillea, Indofevillea, Kedrostis, Lagenaria, Lemurosicyos, Luffa, Marah, Melancium, Melothria, Melothrianthus, Microlagenaria, Microsechium, Momordica, Muellerargia, Mukia, Myrmecosicyos, Neoalsomitra, Nothoalsomitra, Odosicyos, Oreosyce, Parasicyos, Penelopeia, Peponium, Peponopsis, Polyclathra, Posadaea, Praecitrullus, Pseudocyclanthera, Pseudosicydium, Psiguria, Pterope pon, Pterosicyos, Raphidiocystis, Ruthalicia, Rytidostylis, Schizocarpum, Schizopepon, Sechiopsis, Sechium, Selysia, Seyrigia, Sicana, Sicydium, Sicyos, Sicyosperma, Siolmatra, Siraitia, Solena, Tecunumania, Telfairia, Thladiantha, Trichosanthes, Tricyclandra, Trochomeria, Trochomeriopsis, Tumamoca, Vaseyanthus, Wilbrandia, Xerosicyos, Zanonia, Zehneria, Zombitsia, Zygosicyos.

Cunninghamiaceae Zucc. = Taxodiaceae Warm.

CUNONIACEAE R. Br. 1814. Dicots. Cunoniales. 22/c. 280. Temp. and subtrop. regions of the Southern Hemisphere, a few species of Weinmannia northward to southern Mexico, West Indies and Philippines. Genera: Acrophyllum, Acsmithia, Aistopetalum, Anodopetalum, Aphanopetalum, Caldcluvia, Callicoma, Ceratopetalum, Codia, Cunonia, Geissois, Gillbeea, Gumillea, Lamanonia, Pancheria, Platylophus, Pseudoweinmannia, Pullea, Schizomeria, Spiracanthemum, Vesselowskya, Weinmannia.

CUPRESSACEAE Rich. ex Bartl. 1830. Pinopsida. Cupressales. 20/115. Cosmopolitan. Genera: Actinostrobus, Austrocedrus, Callitris, Calocedrus, Chamaecyparis, Cupressus, Diselma, Fitzroya, Fokienia, Juniperus, Libocedrus, Microbiota, Neocallitropsis, Papuacedrus, Pilgerodendron, Platycladus, Tetraclinis, Thuja, Thujopsis, Widdringtonia.

Curcumaceae Dumort. = Zingiberaceae Lindl.

CURTISIACEAE Takht. 1987. Dicots. Cornales. 1/1. Southeast and South Africa. Genus Curtisia.

CUSCUTACEAE Dumort. 1829. Dicots. Convolvulales. 1/150–170. Cosmopolitan. Genus Cuscuta.

Cuspariaceae (DC.) Tratt. = Rutaceae Juss.

Cyananthaceae J. Agardh = Campanulaceae Juss.

CYANASTRACEAE Engl. 1900. Monocots. Liliales. 1/7. Trop. Africa. Genus Cyanastrum.

Cyanellaceae Salisb. = Tecophilaeaceae Leyb.

CYCADACEAE Pers. 1807. Cycadopsida. Cycadales. 1/20. Tropics of the Old World from East Africa to East Asia, Australia, and Polynesia. Genus Cycas.

CYCLANTHACEAE Dumort. 1829. Dicots. Cyclanthales. 12/200. Trop. America, West Indies. Genera: Asplundia, Carludovica, Chorigyne, Cyclanthus, Dianthoveus, Dicranopygium, Evodianthus, Ludovia, Schultesiophytum, Sphaeradenia, Stelestylis, Thoracocarpus.

CYCLOCHEILACEAE Marais. 1981. Dicots. Lamiales. 2/4. Northeast and East trop. Africa, southern Arabian Peninsula. Genera: Asepalum, Cyclocheilon.

Cymbanthaceae Salisb. = Melanthiaceae Batsch.

CYMODOCEACEAE N. Taylor. 1909. Monocots. Cymodoceales. 5/17. Trop. and subtrop. seas, a few species in warm seas. Genera: Amphibolis, Cymodocea,

Halodule, Syringodium, Thalassodendron.

Cynaraceae Juss. = Asteraceae Dumort.

CYNOMORIACEAE **Lindl.** 1833. Dicots. Cynomoriales. 1/1–2. Canary Is., Mediterranean, West and Central Asia. Genus Cynomorium.

CYPERACEAE **Juss.** 1789. Monocots. Cyperales. 125/5,300. Cosmopolitan, but chiefly cold and temp. regions. Genera: Abildgaar-dia, Acriulus, Actinoschoenus, Afrotrilepis, Alinula, Amphiscirpus, Androtrichum, Anosporum, Arthrostylis, Ascolepis, Ascopholis, Baumea, Becquerelia, Bisboeckelera, Blysmopsis, Blysmus, Boeckeleria, Bolboschoenus, Bulbostylis, ? Calisto, Calyptrocarya, Capitularina, Carex, Carpha, Caustis, Cephalocarpus, Chorizandra, Chrysitrix, Cladium, Coleochloa, Costularia, Courtoisina, Crosslandia, Cyathocoma, Cymbophyllus, Cyperus, Desmochoenus, Dichromena, Didymiandrum, Diplacrum, Diplasia, Dulichium, Duval-Jouvea, Egleria, Eleocharis, Epischenus, Eriophorella, Eriophorum, Evandra, Everardia, Exocarya, Exochogyne, Ficinia, Fimbristylis, Fuirena, Gahnia, Gymnoschoenus, Hellmuthia, Hemicarpha, Hypolytrum, Juncellus, Kobresia, Koyamaea, Kyllinga, Kyllingiella, Lagenocarpus, Lepidosperma, Lepironia, Lipocarpha, Machaerina, Macrochaetium, Mapania, Mapaniopsis, Mariscus, Mesomelaena, Microdracoides, Morelotia, Neesenbeckia, Nelmesia, Nemum, Neolophocarpus, Oreobolopsis, Oreobolus, Oreograstis, Oxycaryum, Paramapania, Pentasticha, Phylloscirpus, Pleurostachys, Principina, Pseudoschoenus, Ptilanthelium, Pycreus, Queenslandiella, Reedia, Remirea, Rhynchocladium, Rhynchospora, Rikliella, Schoenoides, Schoenoxiphium, Schoenus, Scirpidella, Scirpodendron, Scirpoides, Scripus, Scleria, Sorostachys, Sphaerocyperus, Sumatroscirpus, Syntrinema, Tetraria, Tetrariopsis, Thoracostachyum, Torulinium, Trachystylis, Trianoptiles, Trichoschoenus, Tricostularia, Trilepis, Tylocarya, Uncinia, Vincentia, Volkiella, Websteria.

CYPHIACEAE **A. DC.** 1839. Dicots. Campanulales. 1/60. Trop. and South Africa. Genus Cyphia.

CYPHOCARPACEAE **Miers.** 1848. Dicots. Campanulales. 1/2. Chile. Genus Cyphocarpus.

Cypripediaceae Lindl. = Orchidaceae Juss.

CYRILLACEAE **Endl.** 1841. Dicots. Ericales. 3/14. North, Central, and South America; West Indies. Genera: Cliftonia, Cyrilla, Purdiaea.

Cyrtandraceae Jack = Gesneriaceae Dumort.

Cyrtanthaceae Salisb. = Amaryllidaceae J. St.-Hil.

CYTINACEAE **A. Rich.** 1824. Dicots. Rafflesiales. 3/10. Europe, Mediterranean, trop., and South Africa, Madagascar, West Asia. Genera: Bdallophytum, Bortyocytinus, Cytinus.

DACTYLANTHACEAE **Takht.** 1987. Dicots. Balanophorales. 2/2. New Caledonia (Hachettea), New Zealand (Dactylanthus).

Damasoniaceae Nakai = Alismataceae Vent.

Daphnaceae Vent. = Thymelaeaceae Juss.

DAPHNIPHYLLACEAE **Muell. Arg.** 1869. Dicots. Daphniphyllales. 1/12. South, Southeast, and East Asia. Genus Daphniphyllum.

DASYPOGONACEAE **Dumort.** 1829. Monocots. Amaryllidales. 9/72. New Guinea, New Britain, Australia, Tasmania, New Caledonia. Genera: Acanthocarpus, Baxteria, Calectasia, Chamaexeros, Dasypogon, Kingia, Lomandra, Romnalda, Xerolirion.

DATISCACEAE **R. Br. ex Lindl.** 1830. Dicots. Begoniales. 3/4. Temp., subtrop., and trop. Asia, Malesia, trop. Australia, North America. Genera: Datisca, Octomeles, Tetrameles.

Daturaceae Raf. = Solanaceae Juss.

Daucaceae Martinov = Apiaceae Lindl.

DAVIDIACEAE **(Harms) H. L. Li.** 1955. Dicots. Cornales. 1/1. Western and central China. Genus Davidia.

DAVIDSONIACEAE G. G. J. Bange. 1952. Dicots. Cunoniales. 1/1. Eastern Australia. Genus Davidsonia.

Deeringiaceae J. Agardh = Amaranthaceae Juss.

DEGENERIACEAE I. W. Bailey et A. C. Sm. 1942. Dicots. Magnoliales. 1/2. Fiji. Genus Degeneria.

Delphiniaceae Baum.-Bodenh. = Ranunculaceae Juss.

Dendrophthoaceae Tiegh. ex Nakai = Loranthaceae Juss.

DESFONTAINIACEAE Endl. 1841. Dicots. Gentianales. 1/1. Central and South America. Genus Desfontainia.

Detariaceae (DC.) J. Hess = Fabaceae Lindl.

DIALYPETALANTHACEAE Rizzini et Occhioni. 1949. Dicots. Gentianales. 1/1. Eastern Brazil. Genus Dialypetalanthus.

Dianellaceae Salisb. = Phormiaceae J. Agardh.

Dianthaceae von Vest = Caryophyllaceae Juss.

DIAPENSIACEAE Lindl. 1836. Dicots. Diapensiales. 5/14. Cold and temp. Northern Hemisphere. Genera: Berneuxia, Diapensia, Galax, Pyxidanthera, Shortia.

DICHAPETALACEAE Baill. 1886. Dicots. Euphorbiales. 3/180. Tropics. Genera: Dichapetalum, Stephanopodium, Tapura.

Dichondraceae Dumort. = Convolvulaceae Juss.

Diclidantheraceae J. Agardh = Polygalaceae R. Br.

Dicrastylidaceae J. L. Drumm. ex Harv. = Verbenaceae J. St.- Hil.

Dictamnaceae von Vest = Rutaceae DC.

DIDIEREACEAE Drake. 1903. Dicots. Caryophyllales. 4/11. Madagascar. Genera: Alluaudia, Alluaudiopsis, Decaryia, Didierea.

DIDYMELACEAE Léandri. 1937. Dicots. Didymelales. 1/2. Madagascar. Genus Didymeles.

Didymocarpaceae D. Don = Gesneriaceae Dumort.

DIEGODENDRACEAE Capuron. 1964. Dicots. Ochnales. 1/1. Madagascar. Genus Diegodendron.

Digitalidaceae Martinov = Scrophulariaceae Juss.

DILLENIACEAE Salisb. 1807. Dicots. Dilleniales. 12/300. Tropics and subtropics, esp. Asia and Australia. Genera: Acrotrema, Curatella, Davilla, Didesmandra, Dillenia, Doliocarpus, Hibbertia, Neowormia, Pachynema, Pinzona, Schumacheria, Tetracera.

Dionaeaceae Raf. = Droseraceae Salisb.

DIONCOPHYLLACEAE Airy Shaw. 1952. Dicots. Dioncophyllales. 3/3. Trop. West Africa. Genera: Dioncophyllum, Habropetalum, Triphyophyllum.

DIOSCOREACEAE R. Br. 1810. Monocots. Dioscoreales. 5/c. 650. Tropics, a few species in subtropics and warm–temp. regions. Genera: Bornerea, Dioscorea, Epipetrum, Rajania, Tamus.

Diosmaceae R. Br. = Rutaceae Juss.

Diospyraceae von Vest = Ebenaceae Guerke.

DIPENTODONTACEAE Merr. 1941. Dicots. Violales. 1/1. Himalayas, India, Burma, China. Genus Dipentodon.

Diphylleiaceae Schultz Sch. = Berberidaceae Juss.

Diplantheraceae Baum.-Bodenh. = Cymodoceaceae N. Taylor.

Diplarchaceae Klotzsch = Ericaceae Juss.

Diplolaenaceae J. Agardh = Rutaceae Juss.

DIPSACACEAE Juss. 1789. Dicots. Dipsacales. 14/317. Eurasia, Africa, esp. Mediterranean and West Asia. Genera: Cephalaria, Dipsacus, Knautia, Lomelosia, Pseudoscabiosa, Pterocephalidium, Pterocephalus, Pycnocomon, Scabiosa, Scabiosiopsis, Simenia, Sixalix, Succisa, Succisella.

DIPTEROCARPACEAE Blume. 1825. Dicots. Malvales. 13/700. Tropics of the Old World, esp. rainforests of Malesia. Genera: Anisoptera, Cotylelobium, Dipterocarpus, Dryobalanops, Hopea, Neobalanocarpus, Parashorea, Shorea, Stemonoporus, Upuna, Vateria, Vateriopsis, Vatica.

Dirachmaceae Hutch. 1959. Dicots. Geraniales. 1/2. Somalia, Socotra Genus Dirachma.

Disanthaceae Nakai = Hamamelidaceae R. Br.

Dodonaeaceae Link = Sapindaceae Juss.

Dombeyaceae (DC) Bartl. = Sterculiaceae (DC.) Bartl.

Donatiaceae Hutch. 1959. Dicots. Stylidiales. 1/2. Tasmania, New Zealand, subantarctic South America. Genus Donatia.

Doryanthaceae R. Dahlgren et Clifford. 1985. Monocots. Amaryllidales. 1/2. Eastern Australia. Genus Doryanthes.

Drabaceae Martinov = Brassicaceae Burnett.

Dracaenaceae Salisb. 1866. Monocots. Asparagales. 3/210. Tropics of the Old World eastward to Hawaiian Is., Central America. Genera: Dracaena, Pleomele, Sansevieria.

Dracontiaceae Salisb. = Araceae Juss.

Drimeaceae Tiegh. = Winteraceae R. Br. ex Lindl.

Droseraceae Salisb. 1808. Dicots. Droserales. 4/90–100. Nearly cosmopolitan. Genera: Aldrovanda, Dionaea, Drosera, Drosophyllum.

Drosophyllaceae Chrtek, Slavíkova et Studicka = Droseraceae Salisb.

Drupaceae Gray = Rosaceae Juss.

Dryadaceae Gray = Rosaceae Juss.

Duabangaceae Takht. 1986. Dicots. Myrtales. 1/3. Himalayas, India, China, Indochina, Malesia. Genus Duabanga.

Duckeodendraceae Kuhlm. 1950. Dicots. Solanales. 1/1. Brazil. Genus Duckeodendron.

Dulongiaceae J. Agardh. 1858. Dicots. Hydrangeales. 1/4. Trop. America. Genus Phyllonoma.

Durantaceae J. Agardh = Verbenaceae J. St.-Hil.

Dysphaniaceae Pax = Chenopodiaceae Vent.

Ebenaceae Guerke. 1891. Dicots. Ebenales. 2/500. Tropics. Genera: Diospyros, Euclea.

Ecdeiocoleaceae D. F. Cutler et Airy Shaw. 1965. Monocots. Restionales. 1/1. Southwestern Australia. Genus Ecdeiocolea.

Echiaceae Raf. = Boraginaceae Juss.

Echinopaceae Dumort. = Asteraceae Dumort.

Ehretiaceae Mart. ex Lindl. = Boraginaceae Juss.

Elaeagnaceae Juss. 1789. Dicots. Elaeagnales. 3/57–87. Temp. and subtrop. Northern Hemisphere, trop. Asia. Elaeagnus extending to Australia. Genera: Elaeagnus, Hippophaë, Shepherdia.

Elaeocarpaceae Juss. ex DC. 1824. Dicots. Malvales. 10/290. Madagascar; Mauritius; Socotra; South, Southeast, and East Asia; Malesia; Australia; Tasmania; New Zealand; New Caledonia; Vanuatu; Fiji; Samoa; Tongo; trop. and temp. South America. Genera: Aceratium, Aristotelia, Crinodendron, Dubouzetia, Elaeocarpus, Peripentadenia, Petenaea, Sericolea, Sloanea, Vallea.

Elatinaceae Dumort. 1829. Dicots. Elatinales. 2/37. Temp., subtrop., and trop. regions. Genera: Bergia, Elatine.

Elegiaceae Raf. = Restionaceae R. Br.

Elismataceae Nakai = Alismataceae Vent.

Ellisiaceae Bercht. et J. Presl = Hydrophyllaceae R. Br.

Ellisiophyllaceae Honda = Scrophulariaceae Juss.

Elodeaceae Dumort. = Hydrocharitaceae Juss.

Elynaceae Rchb. ex Barnhart = Cyperaceae Juss.

Elytranthaceae Tiegh. ex Nakai = Loranthaceae Juss.

Embeliaceae J. Agardh = Myrsinaceae R. Br.

Emblingiaceae (Pax) Airy Shaw. 1965. Dicots. Sapindales. 1/1. Australia. Genus Emblingia.

Emmotaceae Tiegh. = Icacinaceae (Benth.) Miers.

Empetraceae Gray. 1821. Dicots. Ericales. 3/5–6 (11). Cold and temp. Northern Hemisphere, temp. South America. Genera: Ceratiola, Corema, Empetrum.

Enhalaceae Nakai = Hydrocharitaceae Juss.

EPACRIDACEAE R. Br. 1810. Dicots. Ericales. 32/390. Southeast Asia, Australia, Tasmania, New Zealand, New Caledonia, Hawaiian Is., temp. South America. Genera: Acrotriche, Andersonia, Archeria, Astroloma, Brachyloma, Budawangia, Choristemon, Coleanthera, Conostephium, Cosmelia, Cyathopsis, Daenikeranthus, Decatoca, Dracophyllum, Epacris, Lebetanthus, Leucopogon, Lissanthe, Lysinema, Melichrus, Monotoca, Needhamiella, Oligarrhena, Pentachondra, Prionotes, Richea, Rupicola, Sphenotoma, Sprengelia, Styphelia, Trochocarpa, Woollsia.

EPHEDRACEAE Dumort. 1829. Ephedropsida. Ephedrales. 1/40. North and South America, Eurasia. Genus Ephedra.

Ephemeraceae Batsch = Commelinaceae R. Br.

Epidendraceae A. Kern. = Orchidaceae Juss.

Epilobiaceae Vent. = Onagraceae Juss.

Epimediaceae F. Lestib. = Berberidaceae Juss.

Eragrostidaceae (Stapf) Herter = Poaceae Barnhart.

EREMOLEPIDACEAE Tiegh. ex Nakai. 1952. Dicots. Santalales. 3/12. West Indies, trop. South America, ? New Zealand (Tupeia). Genera: Antidaphne, Eubrachion, Lepidoceras.

EREMOSYNACEAE Dandy. 1959. Dicots. Saxifragales. 1/1. Australia. Genus Eremosyne.

ERICACEAE Juss. 1789. Dicots. Ericales. 114/3,550. Cold, temp., subtrop. regions and montane trop. areas, absent in steppes and deserts. Genera: Acrostemon, Agapetes, Agarista, Allotropa, Andromeda, Aniserica, Anomalanthus, Anthopteropsis, Anthopterus, Arachnocalyx, Arbutus, Arcterica, Arctostaphylos, Bejaria, Blaeria, Botryostege, Bruckenthalia, Bryanthus, Calluna, Calopteryx, Cassiope, Cavendishia, Ceratostema, Chamaedaphne, Cheilotheca, Chimaphila, Coccosperma, Coilostigma, Comarostaphylis, Costera, Craibiodendron, Daboecia, Demosthenesia, Didonica, Dimorphanthera, Diogenesia, Diplarche, Diplycosia, Disterigma, Elliotia, Enkianthus, Epigaea, Eremia, Eremiella, Erica, Eubotryoides, Findlaya, Gaultheria, Gaylussacia, Gonocalyx, Grisebachia, Hemitomes, Kalmia, Kalmiopsis, Killipiella, Lateropora, Ledothamnus, Ledum, Leiophyllum, Leucothoe, Loiseleuria, Lyonia, Macleania, Malea, Menziesia, Moneses, Monotropa, Monotropsis, Mycerinus, Nagelocarpus, Notopora, Oreanthes, Ornithostaphylos, Orthaea, Orthilia, Oxydendrum, Pellegrinia, Pernettya, Pernettyopsis, Philippia, Phyllodoce, Pieris, Pityopus, Platycalyx, Pleuricospora, Plutarchia, Polyclita, Psammisia, Pterospora, Pyrola, Rhododedndron, Rhodothamnus, Rusbya, Salaxis, Sarcodes, Satyria, Scyphogyne, Semiramisia, Simocheilus, Siphonandra, Sphyrospermum, Stokoeanthus, Sympieza, Syndesmanthus, Tepuia, Thamnus, Themistoclesia, Therorhodion, Thibaudia, Thoracosperma, Tsusiophyllum, Utleya, Vaccinium, Zenobia.

Erinaceae Pfeiff. = Scrophulariaceae Juss.

ERIOCAULACEAE P. Beauv. ex Desv. 1828. Monocots. Commelinales. 10/1,080. Tropics, esp. America, a few species of Eriocaulon extending to temp. East Asia and North America. Genera: Blastocaulon, Ericaulon, Lachnocaulon, Leiothrix, Mesanthemum, Paepalanthus, Philodoce, Rondonathus, Syngonanthus, Tonina.

Eriogonaceae (Dumort.) Meisn. = Polygonaceae Juss.

ERIOSPERMACEAE Endl. 1841. Monocots. Liliales. 1/90. Africa. Genus Eriospermum.

Erodiaceae Horan. = Geraniaceae Juss.

Erycibaceae Endl. ex Bullock = Convolvulaceae Juss.

Eringiaceae Raf. = Apiaceae Lindl.

Erodiaceae Horan. = Geraniaceae Juss.

Erycibaceae Endl. = Convolvulaceae Juss.

Erysimaceae Martinov = Brassicaceae Burnett.

Erythrinaceae Schimp. = Fabaceae Lindl.

Erythroniaceae Martinov = Liliaceae Vent.

Erythropalaceae Sleumer = Olacaceae Mirb. ex DC.

Erythrospermaceae Tiegh. ex Bullock = Flacourtiaceae DC.

ERYTHROXYLACEAE **Kunth.** 1822. Dicots. Linales. 4/260. Tropics and subtropics, esp. America and Madagascar. Genera: Aneulophus, Erythroxylum, Nectaropetalum, Pinacopodium.

ESCALLONIACEAE **R. Br. ex Dumort.** 1829. Dicots. Hydrangeales. 8/c. 150. South and Southeast Asia, Malesia, and Southern Hemisphere. Genera: Abrophyllum, Anopterus, Cuttsia, Escallonia, Forgesia, Polyosma, Quintinia, Valdivia.

Eschscholtziaceae Seringe = Papaveraceae Juss.

Eucomidaceae Salisb. = Hyacinthaceae Batsch.

EUCOMMIACEAE **Engl.** 1909. Dicots. Eucommiales. 1/1. China. Genus Eucommia.

EUCRYPHIACEAE **Endl.** 1841. Dicots. Cunoniales. 1/6. Australia, Tasmania, Chile. Genus Eucryphia.

Eupatoriaceae Martinov = Asteraceae Dumort.

EUPHORBIACEAE **Juss.** 1789. Dicots. Euphorbiales. 326/7,990. Cosmopolitan (excl. Arctic), chiefly trop. and subtrop. regions. Genera: Acalypha, Acidocroton, Acidoton, Actephila, Adelia, Adenocline, Adenopeltis, Adenophaedra, Adriana, Aerisilvaea, Afrotrewia, Agrostistachys, Alchornea, Alchorneopsis, Aleurites, Algernonia, Alphandia, Amanoa, Amperea, Amyrea, Andrachne, Androstachys, Angostyles, Annesijoa, Anomalocalyx, Anthostema, Antidesma, Apodiscus, Aporosa, Argomuellera, Argythamnia, Aristogeitonia, Ashtonia, Astrocasia, Astrococcus, Austrobuxus, Avellanita, Baccaurea, Baliospermum, Baloghia, Benoistia, Bernardia, Bertya, Beyeria, Bischofia, Blachia, Blotia, Blumeodendron, Bocquillonia, Bonania, Borneodendron, Bossera, Botryophora, Breynia, Bridelia, Burseranthe, Calycopeplus, Caperonia, Caryodendron, Casabitoa, Cavacoa, Celaenodendron, Celianella, Cephalocroton, Cephalomappa, Chaetocarpus, Chascotheca, Cheilosa, Chlamydojatropha, Chondrostylis, Chonocentrum, Choriceras, Chorisandrachne, Chrozophora, Cladogelonium, Cladogynos, Claoxylon, Claoxylopsis, Cleidiocarpon, Cleidion, Cleistanthus, Clutia, Cnesmosa, Cnidoscolus, Cocconerion, Codiaeum, Coelebogyne, Colliguaja, Conceveiba, Cordemoya, Croizatia, Croton, Crotonogyne, Crotonogynopsis, Crotonopsis, Ctenomeria, Cubacroton, Cubanthus, Cyathogyne, Cyrtogonone, Cyttaranthus, Dalechampia, Dalembertia, Danguyodrypetes, Deuteromallotus, Deutzianthus, Dichostemma, Didymocistus, Dimorphocalyx, Discocarpus, Discoclaoxylon, Discocleidon, Discoglypremna, Dissiliaria, Ditta, Dodecastigma, Domohinea, Doryxylon, Droceloncia, Drypetes, Duvigneaudia, Dysopsis, Elaeophorbia, Elateriospermum, Eleutherostigma, Endadenium, Endospermum, Enriquebeltrania, Epiprinus, Eremocarpus, Erismanthus, Erythrococca, Euphorbia, Excoecaria, Fahrenheitia, Flueggea, Fontainea, Fragariopsis, Garcia, Gavarretia, Gitara, Givotia, Glochidion, Glycydendron, Clyphostylus, Grimmeodendron, Grossera, Gymnanthes, Haematostemon, Hamilcoa, Hemicicca, Hevea, Heywoodia, Hippomane, Homonoia, Hura, Hyaenanche, Hyeronima, Hylandia, Hymenocardia, Jablonskia, Jatropha, Joannesia, Julocroton, Kairothamnus, Keayodendron, Klaineanthus, Koilodepas, Lachnostylis, Lasiococca, Lasiocroton, Lautembergia, Leeuwenbergia, Leidesia, Leptonema, Leptopus, Leucocroton, Liodendron, Lobanilia, Loerzingia, Longetia, Mabea, Macaranga, Maesobotrya, Mallotus, Manihot, Manihotoides, Manniophyton, Maprounea, Mareya, Mareyopsis, Margaritaria, Martretia, Megistostigma, Meineckia, Melanolepis, Mercurialis, Mettenia, Micrandra, Micrandropsis, Micrantheum, Micrococca, Mildbraedia, Mischodon, Moacroton, Monadenium, Monotaxis, Moultonianthus, Myladenia, Myricanthe, Nealchornea, Necepsia, Neoboutonia, Neoguillauminia, Neoholstia, Neoroepera, Neoscortechinia, Neotrewia, Octospermum, Oldfieldia, Oligoceras, Omalanthus, Omphalea, Ophellantha, Ophthalmoblapton, Oreoporanthera, Ostodes, Pachystroma, Pachystylidium, Pantadenia, Paradrypetes, Paranecepsia, Parapantadenia, Parodiodendron, Pausandra, Pedilanthus, Pera, Petalostigma, Philyra, Phyllanoa, Phyllanthodendron, Phyllanthus, Picrodendron, Pimelodendron, Piranhea, Plagiostyles, Plukenetia, Podadenia, Podocalyx, Pogonophora, Poilaniella, Poinsettia, Polyandra, Poranthera, Protomegabaria, Pseudagrostistachys, Pseudanthus, Pseudocroton, Pseudolachnostylis, Pseudosagotia, Pterococcus, Ptychopyxis, Pycnocoma, Reutealis, Reverchonia, Richeria, Richeriella, Ricinocarpos, Ricinodendron, Ricinus, Rockinghamia, Romanoa, Sagotia, Sampantaea, Sandwithia, Sapium, Sauropus, Savia, Scagea, Schinziophyton, Scortechinia, Sebastiania, Securinega, Seidelia, Senefeldera, Senefelderopsis, Sibangea, Spathiostemon, Speranskia, Sphaerostylis, Sphyranthera, Spirostachys, Spondianthus, Stachyandra, Stachystemon, Stillingia, Strophioblachia, Sumbaviopsis, Suregada, Synadenium, Syndyophyllum, Tacareuna, Tannodia, Tapoides, Tetracarpidium, Tetracoccus, Tetraplandra, Tetrorchidium, Thecacoris, Thyrsanthera, Tragia, Tragiella, Trewia, Trigonopleura, Trigonostemon, Uapaca, Vaupesia, Veconcibea, Vernicia, Victorinia, Voatamalo, Wetria, Whyanbeelia, Wielandia, Zimmermannia, Zimmermanniopsis.

Euphrasiaceae Martinov = Scrophulariaceae Juss.

EUPHRONIACEAE Marc.-Berti. 1989. Dicots. Polygalales. 1/3. Northern South America. Genus Euphronia.

EUPOMATIACEAE Endl. 1841. Dicots. Eupomatiales. 1/2. New Guinea, eastern and southeastern Australia. Genus Eupomatia.

EUPTELEACEAE K. Wilh. 1910. Dicots. Eupteleales. 1/2. South and East Asia. Genus Euptelea.

Euryalaceae J. Agardh = Nymphaeaceae Salisb.

Exocarpaceae J. Agardh = Santalaceae R. Br.

FABACEAE Lindl. 1836, or LEGUMINOSAE Juss.1789 (nom. altern.). Dicots. Fabales. 713/17,500. Cosmopolitan. Genera: Abarema, Abrus, Acacia, Acmispon, Acosmium, Acrocarpus, Adenanthera, Adenocarpus, Adenodolichos, Adenolobus, Adenopodia, Adesmia, Aenictophyton, Aeschynomene, Affonsea, Afgekia, Afzelia, Aganope, Airyantha, Albizia, Aldina, Alexa, Alhagi, Alistilus, Almaleea, Alysicarpus, Amblygonocarpus, Amburana, Amherstia, Amicia, Ammodendron, Ammopiptanthus, Ammothamnus, Amorpha, Amphicarpaea, Amphimas, Amphithalea, Anadenanthera, Anagyris, Anarthrophyllum, Andira, Androcalymma, Angylocalyx, Antheroporum, Anthonotha, Anthyllis, Antopetitia, Aotus, Aphanocalyx, Apios, Apoplanesia, Aprevalia, Apuleia, Apurimacia, Arachis, Arapatiella, Archidendron, Archidendropsis, Arcoa, Argyrocytisus, Argyrolobium, Arthrocarpum, Arthroclianthus, Aspalathus, Astracantha, Astragalus, Ateleia, Atylosia, Aubrevillea, Augouardia, Austrodolichos, Austrosteenisia, Baikiaea, Balisaea, Balsamocarpon, Baphia, Baphiastrum, Baphiopsis, Baptisia, Batesia, Bathiaea, Baudouinia, Bauhinia, Baukea, Behaimia, Belairia, Bembicidium, Benedictella, Bergeronia, Berlinia, Bisserula, Bituminaria, Bocoa, Bolusafra, Bolusanthus, Bolusia, Bossiaea, Bowdichia, Borwingia, Brachycylix, Brachypterum, Brachysema, Brachystegia, Brenierea, Brodriguesia, Brongniartia, Brownea, Browneopsis,

Brya, Bryaspis, Burgesia, Burkea, Burkilliodendron, Burtonia, Bussea, Butea, Cadia, Caesalpinia, Cajanus, Calis, Calicotome, Callerya, Calliandra, Calliandropsis, Calophaca, Calopogonium, Calpocalyx, Calpurnia, Camoensia, Campsiandra, Camptosema, Campylotropis, Canavalia, Candolleodendron, Caragana, Carmichaelia, Carrissoa, Cascaronia, Cassia, Castanospermum, Cedrelinga, Cenostigma, Centrolobium, Centrosema, Cephalostigmaton, Ceratonia, Cercidium, Cercis, Chadsia, Chaetocalyx, Chamaecrista, Chamaecytisus, Chapmannia, Chesneya, Chidlowia, Chordospartium, Chorizema, Christia, Chrysoscias, Cicer, Cladrastis, Clathrotropis, Cleobulia, Clianthus, Clitoria, Clitoriopsis, Cochlianthus, Codariocalyx, Coelidium, Cojoba, Collaea, Cologania, Colophospermum, Colutea, Colvillea, Conzattia, Copaifera, Corallospartium, Cordeauxia, Cordyla, Coronilla, Corynella, Coursetia, Cracca, Craibia, Cranocarpus, Craspedolobium, Cratylia, Crotalaria, Cruddasia, Crudia, Cryptosepalum, Cullen, Cupulanthus, Cyamopsis, Cyathostegia, Cyclocarpa, Cyclolobium, Cyclopia, Cyclodiscus, Cylindrokelupha, Cylista, Cymbosema, Cynometra, Cytisophyllum, Cytisopsis, Cytisus, Dahlstedtia, Dalbergia, Dalbergiella, Dalea, Dalhousiea, Daniellia, Daviesia, Decorsea, Deguelia, Delonix, Dendrolobium, Derris, Desmanthus, Desmodiastrum, Desmodium, Detarium, Dewevrea, Dialium, Dicerma, Dichilus, Dichrostachys, Dicorynia, Dicraeopetalum, Dicymbe, Didelotia, Dillwynia, Dimorphandra, Dinizia, Dioclea, Diphyllarium, Diphysa, Diplotropis, Dipogon, Dipteryx, Diptychandra, Discolobium, Distemonanthus, Disynstemon, Dolichopsis, Dolichos, Dorycnium, Dorycnopsis, Droogmansia, Dumasia, Dunbaria, Duparquetia, Dussia, Dysolobium, Ebenus, Echinospartum, Eleiotis, Elephantorrhiza, Eligmocarpus, Elizabetha, Eminia, Endertia, Endomallus, Endosamara, Englerodendron, Entada, Enterolobium, Eperua, Eremosparton, Erichsenia, Erinacea, Eriosema, Errazuri zia, Erythrina, Erythrophleum, Etaballia, Euchilopsis, Euchresta, Eurypetalum, Eutaxis, Eversmannia, Exostyles, Eysenhardtia, Factorovskya, Faidherbia, Fiebrigiella, Fillaeopsis, Fissicalyx, Flemingia, Fordia, Gagnebina, Galactia, Galega, Gastrolobium, Geissaspis, Genista, Genistidium, Geoffroea, Gigasiphon, Gilbertiodendron, Gilletiodendron, Gleditsia, Gliricidia, Glottidium, Glycine, Glycyrrhiza, Goldmania, Gompholobium, Goniorrhachis, Gonocytisus, Goodia, Gossweilerodendron, Grazielodendron, Gueldenstaedtia, Guibourtia, Gymnocladus, Haematoxylum, Halimodendron, Hallia, Hammatolobium, Hanslia, Haplormosia, Hardenbergia, Hardwickia, Harleyodendron, Harpalyce, Havardia, Hebestigma, Hedysarum, Herpyza, Hesperolaburnum, Heterostemon, Heylandia, Hippocrepis, Hoffmannseggia, Hoita, Holocalyx, Hovea, Humboldtia, Humboldtiella, Humularia, Hylodendron, Hymenaea, Hymenocarpos, Hymenolobium, Hymenostegia, Hypocalyptus, Indigastrum, Indigofera, Indopiptadenia, Inga, Inocarpus, Intsia, Isoberlinia, Isotropis, Jacksonia, Jacqueshuberia, Jansonia, Julbernardia, Kalappia, Kaoue, Kennedia, Kerstania, Kingiodendron, Klugiodendron, Koompassia, Kotschya, Kummerowia, Kunstleria, Labichea, Lablab, Laburnum, Lamprolobium, Lathyrus, Latrobea, Lebeckia, Lebruniodendron, Lecointea, Lembotropis, Lemurodendron, Lemuropisum, Lennea, Lens, Leonardoxa, Leptoderris, Leptodesmia, Leptosema, Lespedeza, Lessertia, Leucaena, Leucomphalos, Leucostegane, Librevillea, Liparia, Loddigesia, Loesenera, Lonchocarpus, Lotononis, Lotus, Lietzelburgia, Lupinus, Luzonia, Lysidice, Lysiloma, Maackia, Machaerium, Macrolobium, Macropsychanthus, Macroptilium, Macrotyloma, Macroule, Maniltoa, Margaritolobium, Marina, Marmaroxylon, Martiodendron, Mastersia, Mecopus, Medicago, Melanoxylum, Melilotoides, Melilotus, Melliniella, Melolobium, Mendoravia, Meristotropis, Michelsonia, Microberlinia, Microcharis, Mildbraediodendron, Millettia, Mimosa, Mimozyganthus, Minkelersia, Mirbelia, Moldenhawera, Monopetalanthus, Monopteryx, Mora, Moullava, Mucuna, Muellera, Muelleranthus, Mundulea, Myrocarpus, Myrospermum, Myroxylon, Neoapaloxylon, Neochevalierodendron, Neocollettia, Neocracca, Neodunnia, Neoharmsia, Neonotonia, Neorautanenia, Neorudolphia, Nephrodesmus, Neptunia, Nesphostylis, Newtonia, Nissolia, Nogra, Notodon, Notospartium, Obolinga, Oddoniodendron, Olneya, Onobrychis, Ononis, Ophrestia, Orbexilum, Oreophysa, Ormocarpopsis, Ormocarpum, Ormosia, Ornithopus, Orphanodendron, Ostryocarpus, Ostryoderris, Otholobium, Oxylobium, Oxyrhynchus, Oxystigma, Oxytropis, Pachecoa, Pachyelasma, Pachyrhizus, Padbruggea, Paloue, Paloveopsis, Panurea, Paracalyx, Paraderris, Paramachaerium, Paramacrolobium, Parapiptadenia, Pararchidendron, Paraserianthes, Paratephrosia, Parkia, Parkinsonia, Parochetus, Parryella, Pearsonia, Pediomelum, Pellegriniodendron, Peltogyne, Peltophorum, Pentaclethra, Periandra, Pericopsis, Petaladenium, Petalostylis, Peteria, Petteria, Phaseolus, Philenoptera, Phylacium, Phyllocarpus, Phyllodium, Phyllota, Phylloxylon, Physostigma, Pickeringia, Pictetia, Piliostigma, Piptadenia, Piptadeniastrum, Piptadeniopsis, Piptanthus, Piscidia, Pisum, Pithecolobium, Plagiocarpus, Plagiosiphon, Plathymenia, Platycelyphium, Platycyamus, Platylobium, Platymischium, Platypodium, Platysepalum, Podalyria, Podocytisus, Podolotus, Poecilanthe, Poeppigia, Poiretia, Poitea, Polhillia, Polystemonanthus, Pongamia, Pongamiopsis, Priestleya, Prioria, Prosopidastrum, Prosopis, Pseudarthria, Pseudeminia, Pseudoeriosema, Pseudoglycine, Pseudolotus, Pseudomacrolobium, Pseudopiptadenia, Pseudoprosopsis, Pseudovigna, Psophocarpus, Psoralea, Psoralidium, Psorothamnus, Pterocarpus, Pterodon, Pterogyne, Pterolobium, Ptycholobium, Ptychosema, Pueraria, Pultenaea, Punjuba, Pycnospora, Rafnia, Ramirezella, Ramorinoa, Recordoxylon, Requienia, Retama, Rhodopis, Rhynchosia, Rhynchotropis, Riedeliella, Robinia, Robynsiophyton, Rothia, Ruddia, Rupertia, Sabinea, Sakoanala, Salweenia, Saraca, Sarcodum, Sartoria, Schefflerodendron, Schizolobium, Schleinitzia, Schotia, Schranckiastrum, Schrankia, Sclerolobium, Scorodophloeus, Scorpiurus, Securigera, Sellocharis, Senna, Serianthes, Sesbania, Shuteria, Sindora, Sindoropsis, Singana, Sinodolichos, Smirnowia, Smithia, Soemmerignia, Sophora, Spartidium, Spartium, Spathionema, Spatholobus, Sphaerolobium, Sphaerophysa, Sphenostylis, Sphinctospermum, Spirotropis, Spongiocarpella, Stachyithyrsus, Stahlia, Stauracanthus, Stemonocoleus, Stenodrepanum, Storckiella, Stracheya, Streblorrhiza, Strongylodon, Strophostyles, Stryphnodendron, Stuhlmannia, Stylosanthes, Styphnolobium, Sutherlandia, Swainsonia, Swartzia, Sweetia, Sympetalandra, Tachigali, Tadehagi, Talbotiella, Tamarindus, Taralea, Taverniera, Templetonia, Tephrosia, Teramnus, Tessmannia, Tetraberlinia, Tetrapleura, Tetrapterocarpon, Teyleria, Thermopsis, Thylacanthus, Tibetia, Tipuana, Trachylobium, Trichocyamos, Trifidacanthus, Trifolium, Trigonella, Tripodion, Tylosema, Uleanthus, Ulex, Umtiza, Uraria, Urariopsis, Uribea, Vandasina, Vatairea, Vatovaea, Vavilovia, Vermifrux, Vexibia, Vicia, Vigna, Viminaria, Virgilia, Vouacapoua, Wajira, Wallaceodendron, Walpersia, Weberbauerella, Wiborgia, Wisteria, Xanthocercis, Xerocladia, Xeroderris, Xylia, Yucaratonia, Zapoteca, Zenia, Zenkerella, Zollernia, Zornia, Zuccagnia, Zygia.

FAGACEAE Dumort. 1829. Dicots. Fagales.

6/1,000. Temp., subtrop., and trop. regions (excl. trop. and South Africa and trop. South America). Genera: Castanea, Castanopsis, Fagus, Lithocarpus, Quercus Trigonobalanus.

Festucaceae (Dumort.) Herter = Poaceae Barnhard.

Fevilleaceae Augier = Cucurbitaceae Juss.

Ficaceae (Dumort.) Dumort. = Moraceae Link.

FLACOURTIACEAE DC. 1824. Dicots. Violales. 86/950. Tropics and subtropics, a few species in temp. East Asia, South Africa, North and South America. Genera: Abatia, Ahernia, Antinisa, Aphaerema, Azara, Baileyoxylon, Banara, Bartholomaea, Bennettiodendron, Bivinia, Buchnerodendron, Byrsanthus, Calantica, Caloncoba, Camptostylus, Carpotroche, Carrierea, Casearia, Chiangiodendron, Chlorocarpa, Dankia, Dasylepis, Dissomeria, Dovyalis, Eleutherandra, Erythrospermum, Euceraea, Flacourtia, Gerrardina, Grandidiera, Gynocardia, Hasseltia, Hasseltiopsis, Hecatostemon, Hemiscolopia, Homaliopsis, Homalium, Hydnocarpus, Idesia, Itoa, Kiggelaria, Laetia, Lasiochlamys, Lindackeria, Ludia, Lunania, Macrohasseltia, Mayna, Mocquerysia, Neopringlea, Neoptychocarpus, Neosprucea, Olmediella, Oncoba, Ophiobotrys, Osmelia, Pangium, Peterodendron, Phyllobotryon, Phylloclinium, Pineda, Pleuranthodendron, Poggea, Poliothyrsis, Priamosia, Prockia, Prockiopsis, Pseudoscolopia, Pseudosmelia, Rawsonia, Ryania, Ryparosa, Samyda, Scaphocalyx, Scolopia, Scottellia, Smeathmannia, Soyauxia, Tetrathylacium, Tisonia, Trichadenia, Trichostephanus, Trimeria, Xylosma, Xylotheca, Zuelania.

FLAGELLARIACEAE Dumort. 1829. Monocots. Restionales. 1/4. Tropics and subtropics of the Old World. Genus Flagellaria.

Flindersiaceae (Engl.) C. T. White ex Airy Shaw = Rutaceae Juss.

Foetidaceae (Nied.) Airy Shaw = Lecythidaceae Poit.

Forestieraceae Endl. = Oleaceae Hoffmanns. et Link.

Fothergillaceae Nutt. = Hamamelidaceae R. Br.

FOUQUIERIACEAE DC. 1828. Dicots. Fouquieriales. 2/11–12. Southern North America. Genera: Fouquieria and Idria.

Fragariaceae Rich. ex Nestl. = Rosaceae Juss.

FRANCOACEAE A. Juss. 1832. Dicots. Saxifragales. 2/2. Chile. Genera: Francoa, Titella.

Frangulaceae DC. = Rhamnaceae Juss.

FRANKENIACEAE Gray. 1821. Dicots. Tamaricales. 2/50 (85). Canary Is., Eurasia, West and Southwest Africa, Australia, California, South America. Genera: Frankenia, Hypericopsis.

Fraxinaceae Gray = Oleaceae Hoffmanns. et Link.

Fraxinellaceae Nees et Mart. = Rutaceae Juss.

Fremontiaceae J. Agardh = Sterculiaceae (DC.) Bartl.

Fritillariaceae Salisb. = Liliaceae Juss.

FUMARIACEAE DC. 1821. Dicots. Papaverales. 18/358–408. Temp. and warm–temp. regions, mountainous trop. East and South Africa. Genera: Adlumia, Capnoides, Ceratocapnos, Corydalis, Cryptocapnos, Cysticapnos, Dactylicapnos, Dicentra, Discocapnos, Dissosperma, Fumaria, Fumariola, Platycapnos, Pseudofumaria, Roborowskia, Rupicapnos, Sarcocapnos, Trigonocapnos.

Funkiaceae Horan. = Hostaceae Mathew.

Galacaceae D. Don = Diapensiaceae Lindl.

Galanthaceae Salisb. = Amaryllidaceae J. St.-Hil.

Galeaceae Bubani = Myricaceae Blume.

Galeniaceae Martinov = Aizoaceae F. Rudolphi.

Galiaceae Lindl. = Rubiaceae Juss.

Garciniaceae Bartl. = Clusiaceae Lindl.

Gardeniaceae Dumort. = Rubiaceae Juss.

Gardneriaceae J. Agardh = Loganiaceae R. Br. ex Mart.

GARRYACEAE Lindl. 1834. Dicots. Cornales. 1/13. North and Central America, West Indies. Genus Garrya.

GEISSOLOMATACEAE Endl. 1841. Dicots. Geissolomatales. 1/1. South Africa. Genus Geissoloma.

GENTIANACEAE Juss. 1789. Dicots. Gentianales. 82/1,230. Cosmopolitan, esp. temp. and subtrop. regions and mountainous tropics. Genera: Anagallidium, Anthocleista, Bartonia, Bisgoeppertia, Blackstonia, Canscora, Celiantha, Centaurium, Chironia, Chorisepalum, Cicendia, Comastoma, Congolanthus, Cotylanthera, Coutoubea, Cracosna, Curtia, Deianira, Djaloniella, Enicostema, Eustoma, Exaculum, Exacum, Fagraea, Faroa, Frasera, Geniostemon, Gentiana, Gentianella, Gentianopsis, Gentianothamnus, Halenia, Hockinia, Hoppea, Irlbachia, Ixanthus, Jaeschkea, Karina, Latouchea, Lehmanniella, Lisianthius, Lomatogoniopsis, Lomatogonium, Macrocarpaea, Megacodon, Microrphium, Monodiella, Neblinantha, Neurotheca, Obolaria, Ophelia, Oreonesion, Ornichia, Orphium, Parajaeschkea, Potalia, Prepusa, Pterygocalyx, Pycnosphaera, Qaisera, Rogersonanthus, Sabatia, Schinziella, Schultesia, Sebaea, Senaea, Sipapoantha, Swertia, Symbolanthus, Symphyllophyton, Tachia, Tachiadenus, Tapeinostemon, Tetrapollinia, Tripterospermum, Urogentias, Veratrilla, Voyria, Voyriella, Wurdackanthus, Zonanthus, Zygostigma.

Geonomaceae O. F. Cook = Arecaceae Schultz Sch.

GEOSIRIDACEAE Jonker. 1939. Monocots. Liliales. 1/1. Madagascar. Genus Geosiris.

GERANIACEAE Juss. 1789. Dicots. Geraniales. 5/680. Cosmopolitan, but chiefly temp. and subtrop. regions. Genera: Erodium, Geranium, Monsonia, Pelargonium, Sarcocaulon.

GESNERIACEAE Dumort. 1822. Dicots. Scrophulariales. 151/2,385. Tropics and subtropics, esp. Old World, genus Sarmienta in southern Chile. Genera: Acanthonema, Achimenes, Aeschynanthus, Agalmyla, Allocheilos, Alloplectus, Allostigma, Alsobia, Ancylostemon, Anetanthus, Anna, Anodiscus, Asteranthera, Beccarinda, Bellonia, Besleria, Boea, Boeica, Bournea, Briggsia, Briggsiopsis, Bucinellina, Calcareoboea, Capanea, Cathayanthe, Championia, Charadrophila, Chirita, Chiritopsis, Chrysothemis, Cobananthus, Codonanthe, Colpogyne, Columnea, Conandron, Corallodiscus, Coronanthera, Cremosperma, Cubitanthus, Cyrtandra, Dayaoshania, Deinocheilos, Deinostigma, Depanthus, Diastema, Dichrotrichum, Didissandra, Didymocarpus, Didymostigma, Diplolegnon, Dolicholoma, Drymonia, Episcia, Epithema, Eucodonia, Fieldia, Gasteranthus, Gesneria, Gloxinia, Goyazia, Gyrocheilos, Gyrogyne, Haberlea, Hemiboea, Hemiboeopsis, Heppiella, Hexatheca, Hygea, Isometrum, Jancaea, Jerdonia, Koellikeria, Kohleria, Kohlerianthus, Lagarosolen, Lembocarpus, Lenbrassia, Leptoboea, Lietzia, Linnaeopsis, Loxocarpus, Loxonia, Loxostigma, Lysionotus, Metabriggsia, Metapetrocosmea, Micraeschynanthus, Mitraria, Monophyllaea, Monopyle, Moussonia, Napeanthus, Nautilocalyx, Negria, Nematanthus, Neomortonia, Niphaea, Nodonema, Oerstedina, Opithandra, Orchadocarpa, Oreocharis, Ornithoboea, Oxychlamys, Paliavana, Paraboea, Paradrymonia, Parakohleria, Pearcea, Petrocodon, Petrocosmea, Pheidonocarpa, Phinaea, Phyllobaea, Platyadenia, Platystemma, Primulina, Protocyrtandra, Pseudochirita, Ramonda, Reldia, Resia, Rhabdothamnopsis, Rhabdothamnus, Rhoogeton, Rhynchoglossum, Rhynchotoechum, Rhytidophyllum, Rufodorsia, Saintpaulia, Sarmienta, Schistolobos, Schizoboea, Sepikea, Sinningia, Smithiantha, Solenophora, Stauranthera, Streptocarpus, Tengia, Tetraphyllum, Thamnocharis, Titanotrichum, Trachystigma, Tremacron, Trisepalum, Tylopsacas, Vanhouttea, Whytockia.

Gethyllidaceae J. Agardh = Amaryllidaceae J. St.-Hil.

Giadendraceae Tiegh. ex Nakai = Loranthaceae Juss.

Gilliesiaceae Lindl. = Alliaceae J. Agardh.

Ginalloaceae Tiegh. ex Nakai = Viscaceae Miers.

GINKGOACEAE Endl. 1897. Ginkgoopsida. Ginkgoales. 1/1. Eastern China. Genus Ginkgo.

GISEKIACEAE (Endl.) Nakai. 1942. Dicots. Caryophyllales. 1/2–5. Africa, Southwest, South and Southeast Asia. Genus Gisekia.

Gladiolaceae Raf. = Iridaceae Juss.

GLAUCIDIACEAE Tamura. 1972. Dicots. Glaucidiales. 1/1. Japan. Genus Glaucidium.

Glinaceae Link = Molluginaceae Hutch.

GLOBULARIACEAE DC. 1805. Dicots. Scrophulariales. 2/31. Macaronesia, Europe, Mediterranean, Somalia, West Asia, Socotra Genera: Globularia, Poskea.

Gnaphaliaceae F. Rudolphi = Asteraceae Dumort.

GNETACEAE Lindl. 1834. Gnetopsida. Gnetales. 1/30. Tropics. Genus Gnetum.

GOETZEACEAE Miers ex Airy Shaw. 1965. Dicots. Solanales. 4/7. West Indies. Genera: Coeloneurum, Espadaea, Goetzea, Henconia.

GOMORTEGACEAE Reiche. 1896. Dicots. Laurales. 1/1. Chile. Genus Gomortega.

Gomphiaceae DC ex Schnizl. = Ochnaceae DC.

Gomphrenaceae Raf. = Amaranthaceae Juss.

GONYSTYLACEAE Gilg. 1897 (nom. cons.). Dicots. Thymelaeales. 3/22. Nicobar Is., Malesia, Melanesia, eastward to Fiji. Genera: Aetoxylon, Amyxa, Gonystylus.

GOODENIACEAE R. Br. 1810. Dicots. Goodeniales. 11/460. Tropics and subtropics, esp. Australia, the genus Selliera in New Zealand and temp. South America. Genera: Anthotium, Coopernookia, Dampiera, Diaspasis, Goodenia, Lechenaultia, Pentaptilon, Scaevola, Selliera, Velleia, Verreauxia.

GOUPIACEAE Miers. 1862. Dicots. Celastrales. 1/3. Trop. South America. Genus Goupia.

GRAMINEAE Juss. See POACEAE Barnhart.

Gratiolaceae Martinov = Scrophulariaceae Juss.

GREYIACEAE Hutch. 1926. Dicots. Saxifragales. 1/3. South Africa. Genus Greyia.

Grielaceae Martinov = Neuradaceae Link.

GRISELINIACEAE Takht. 1987. Dicots. Hydrangeales. 1/6. New Zealand, South America. Genus Griselinia.

Gronoviaceae Endl. = Loasaceae Dumort.

GROSSULARIACEAE DC. 1805. Dicots. Saxifragales. 1/150. America from Canada through Andes to Tierra del Fuego, temp. Eurasia, Northwest Africa. Genus Ribes.

GRUBBIACEAE Endl. 1839. Dicots. Ericales. 1/3. South Africa. Genus Grubbia.

Guettardaceae Batsch = Rubiaceae Juss.

GUNNERACEAE Meisn. 1841. Dicots. Gunnerales. 1/40. Tropics (excl. trop. continental Asia) and warm–temp. regions of the Southern Hemisphere. Genus Gunnera.

Gustaviaceae Burnett = Lecythidaceae Poit.

GUTTIFERAE Juss. See CLUSIACEAE Lindl.

Gyrocarpaceae Dumort. = Hernandiaceae Blume.

GYROSTEMONACEAE Endl. 1841. Dicots. Batales. 5/17. Australia and Tasmania. Genera: Codonocarpus, Cypselocarpus, Gyrostemon, Tersonia, Walteranthus.

Haemanthaceae Salisb. = Amaryllidaceae J. St.-Hil.

HAEMODORACEAE R. Br. 1810. Monocots. Haemodorales. 9/40. North and South America, Africa, New Guinea, Australia. Genera: Barberetta, Dilatris, Haemodorum, Hagenbachia, Lachnanthes, Pyrrorhiza, Schiekia, Wachendorfia, Xiphidium.

Halesiaceae D. Don = Styracaceae Dumort.

Halleriaceae Link = Scrophulariaceae Juss.

HALOPHILACEAE J. Agardh. 1858. Monocots. Hydrocharitales. 1/9. Tropical coasts of the Indian and Pacific Oceans and Carribean. Genus Halophila.

HALOPHYTACEAE A. Soriano. 1984. Dicots. Caryophyllales. 1/1. Argentina. Genus Halophytum.

HALORAGACEAE R. Br. 1814. Dicots. Haloragales. 8/145. Cosmopolitan, esp. Southern Hemisphere (chiefly Australia). Genera: Glischrocaryon, Gonocarpus, Haloragis, Haloragodendron, Laurembergia, Meziella, Myriophyllum, Proserpinaca.

HAMAMELIDACEAE R. Br. 1818. Dicots. Hamamelidales. 27/115. Trop. and South Africa, southeastern Caucasus, northern Iran, Himalayas, Assam, East and Southeast Asia, New Guinea, Northeastern Australia, North and Central America. Genera: Chunia, Corylopsis, Dicoryphe, Disanthus, Distyliopsis, Distylium, Embolanthera, Eustigma, Exbucklandia, Fortunearia, Fothergilla, Hamamelis, Loropetalum, Maingaya, Matudaea, Molinodendron, Mytilaria, Neostrearia, Noahdendron, Ostrearia, Parrotia, Parrotiopsis, Shaniodendron, Sinowilsonia, Sycopsis, Tetrathyrium, Trichocladus.

HANGUANACEAE Airy Shaw. 1965. Monocots. Asparagales. 1/2. Sri Lanka, Indochina, Malesia, northern Australia. Genus Hanguana.

Harmandiaceae Tiegh. ex Bullock = Olacaceae Mirb. ex DC.

Haworthiaceae Horan. = Asphodelaceae Juss.

Hebenstreitiaceae Horan. = Scrophulariaceae Juss.

HECTORELLACEAE Philipson et Skipw. 1961. Dicots. Caryophyllales. 1/2. Kerguelen, New Zealand. Genus Lyallia.

Hederaceae Giseke = Araliaceae Juss.

Hedyotidaceae Dumort. = Rubiaceae Juss.

Hedysaraceae J. Agardh = Fabaceae Lindl.

Heisteriaceae Tiegh. ex Bullock = Olacaceae Mirb. ex DC.

Heleniaceae Bessey = Asteraceae Dumort.

Helianthaceae Dumort. = Asteraceae Dumort.

Helichrysaceae Link = Asteraceae Dumort.

HELICONIACEAE Nakai. 1941. Monocots. Zingiberales. 1/100. Trop. America, southwestern Oceania. Genus Heliconia.

Helicteraceae J. Agardh = Sterculiaceae (DC.) Bartl.

Heliotropiaceae Schrad. = Boraginaceae Juss.

Helleboraceae von Vest = Ranunculaceae Juss.

Heloniadaceae J. Agardh = Melanthiaceae Batsch.

HELOSEACEAE (Schott et Endl.) Tiegh. ex Reveal et Hoogland. 1990. Dicots. Balanophorales. 5/6. Trop. America, West Indies, Madagascar, South and Southeast Asia, Malesia. Genera: Corynaea, Ditepalanthus, Exorhopala, Helosis, Rhopalocnemis.

HELWINGIACEAE Decne. 1836. Dicots. Apiales (Araliales). 1/4–6. Asia from Himalayas to southern Japan. Genus Helwingia.

HEMEROCALLIDACEAE R. Br. 1810. Monocots. Amaryllidales. 2/16. Central Europe, Mediterranean, temp. Asia (Hemerocallis), North America (Leucocrinum).

Henriqueziaceae Bremek. = Rubiaceae Juss.

Henslowiaceae Lindl. = Crypteroniaceae DC.

Hermanniaceae Bercht. et J. Presl = Sterculiaceae (DC.) Bartl.

HERNANDIACEAE Blume. 1826. Dicots. Laurales. 5/60. Pantropics. Genera: Gyrocarpus, Hazomalania, Hernandia, Illigera, Sparattanthelium.

Herniariaceae Augier ex Martinov = Caryophyllaceae Juss.

HERRERIACEAE Endl. 1841. Monocots. Asparagales. 2/9. South America (Herreria), Madagascar (Herreriopsis).

HESPEROCALLIDACEAE Traub. 1972. Monocots. Amaryllidales. 1/1. Southwestern North America. Genus Hesperocallis.

Heterantheraceae J. Agardh = Pontederiaceae Kunth.

HETEROPYXIDACEAE Engl. et Gilg. 1920. Dicots. Myrtales. 1/3. Southeast Africa. Genus Heteropyxis.

Heterostylaceae Hutch. = Lilaeaceae Dumort.

Hewardiaceae Nakai = Iridaceae Juss.

Hibbertiaceae J. Agardh = Dilleniaceae Salisb.

Hibiscaceae J. Agardh = Malvaceae Juss.

Hilleriaceae Nakai = Petiveriaceae C. Agardh.

HIMANTANDRACEAE Diels. 1917. Dicots. Magnoliales. 1/2–3. Eastern Malesia, northeastern Australia. Genus Galbulimina.

HIPPOCASTANACEAE DC. 1824. Dicots. Sapindales. 2/18. Southeast Europe, Asia from western Himalayas to Japan and Indochina, America. Genera: Aesculus, Billia.

Hippocrateaceae Juss. = Celastraceae R. Br.

Hippomanaceae J. Agardh = Euphorbiaceae Juss.

HIPPURIDACEAE Link. 1821. Dicots. Hippuridales. 1/1. Cold and temp. Northern Hemisphere. Genus Hippuris.

Hirtellaceae Horan. = Chrysobalanaceae R. Br.

Homaliaceae R. Br. = Flacourtiaceae A. DC.

HOPLESTIGMATACEAE Gilg. 1924. Dicots. Boraginales. 1/2. Trop. West Africa. Genus Hoplestigma.

Hordeaceae Burnett = Poaceae Barnhart.

Hornschuchiaceae J. Agardh = Annonaceae Juss.

Hortensiaceae Bercht. et J. Presl = Hydrangeaceae Dumort.

Hortoniaceae (J. Perkins et Gilg) A. C. Sm. = Monimiaceae Juss.

HOSTACEAE Mathew. 1988. Monocots. Amaryllidales. 1/40. East Asia. Genus Hosta.

Houstoniaceae Raf. = Rubiaceae Juss.

HUACEAE A. Chev. 1947. Dicots. Malvales. 2/3. Trop. Africa. Genera: Afrostyrax, Hua.

HUGONIACEAE Arn. 1834. Dicots. Linales. 6/60–70. Trop. Old World from Africa to Fiji, trop. South America. Genera: Durandea, Hebepetalum, Hugonia, Indorouchera, Philbornea, Roucheria.

Humbertiaceae Pichon = Convolvulaceae Juss.

HUMIRIACEAE A. Juss. 1829. Dicots. Linales. 8/50. Trop. South America, 1 (Sacoglottis) trop. West Africa. Genera: Duckesia, Endopleura, Humiria, Humiriastrum, Hylocarpa, Sacoglottis, Schistostemon, Vantanea.

HYACINTHACEAE Batsch ex Borckh. 1797. Monocots. 44/650. Cosmopolitan, but chiefly Mediterranean, South Africa and Irano-Turanian floristic region. Genera: Albuca, Alrawia, Amphisiphon, Androsiphon, Battandiera, Bellevalia, Bowiea, Brimeura, Camassia, Chionodoxa, Chlorogalum, Daubenya, Dipcadi, Drimia, Drimiopsis, Eucomis, Galtonia, Hemiphylacus, Hyacinthella, Hyacinthoides, Hyacinthus, Kinia, Lachenalia, Ledebouria, Litanthus, Massonia, Muscari, Neopatersonia, Ornithogalum, Polyxena, Prospero, Pseudogaltonia, Puschkinia, Rhadamanthus, Rhodocodon, Schizobasis, Schoenolirion, Scilla, Tenicroa, Thuranthos, Tricalistra, Urginea, Veltheimia, Whiteheadia.

HYDATELLACEAE U. Hamann. 1976. Monocots. 2/8. Australia, Tasmania, New Zealand. Genera: Hydatella, Trithuria.

HYDNORACEAE C. Agardh. 1821. Dicots. Hydnorales. 2/17. South America, trop. and South Africa, Madagascar. Genera: Hydnora, Prosopanche.

HYDRANGEACEAE Dumort. 1829. Dicots. Hydrangeales. 17/200–250. Temp. and subtrop. Northern Hemisphere, a few species in Southeast Asia and Andes. Genera: Broussaisia, Cardiandra, Carpenteria, Decumaria, Deinanthe, Deutzia, Dichroa, Fendlera, Fendlerella, Hydrangea, Jamesia, Kirengeshoma, Philadelphus, Pileostegia, Platycrater, Schizophragma, Whipplea.

HYDRASTIDACEAE Martinov. 1820. Dicots.

Ranunculales. 1/2. Japan, North America. Genus Hydrastis.
Hydroceraceae Blume = Balsaminaceae A. Rich.
HYDROCHARITACEAE Juss. 1789. Monocots. Hydrocharitales. 16/c. 80. Cosmopolitan, excl. cold and arid regions. Genera: Apalanthe, Appertiella, Blyxia, Egeria, Elodea, Enhalus, Hydrilla, Hydrocharis, Hydromystria, Lagarosiphon, Limnobium, Maidenia, Nechamandra, Ottelia, Stratiotes, Vallisneria.
Hydrocotylaceae Hylander = Apiaceae Lindl.
Hydrogetonaceae Link = Potamogetonaceae Dumort.
Hydroleaceae Bercht. et J. Presl = Hydrophyllaceae R. Br.
Hydropeltidaceae Dumort. = Cabombaceae A. Rich.
Hydrophylacaceae Martinov = Rubiaceae Juss.
HYDROPHYLLACEAE R. Br. 1817. Dicots. Boraginales. 19/315. Almost cosmopolitan, but mostly America, a few species of Hydrolea in Africa, trop. Asia, one species of Nama on Hawaiian Is., not in Australia. Genera: Andropus, Codon, Decemium, Draperia, Ellisia, Emmenanthe, Eriodictyon, Eucrypta, Hesperochiron, Hydrolea, Hydrophyllum, Nama, Nemophila, Phacelia, Pholistoma, Romanzoffia, Tricardia, Turricula, Wigandia.
HYDROSTACHYACEAE Engl. 1898. Dicots. Hydrostachyales. 1/22. Trop. and South Africa, Madagascar. Genus Hydrostachys.
Hymenocardiaceae Airy Shaw = Euphorbiaceae Juss.
Hyoscyamaceae von Vest = Solanaceae Juss.
HYPECOACEAE Nakai ex Reveal et Hoogland. 1991. Dicots. Papaverales. 1/20. From western Mediterranean to Mongolia, northern China, and Himalayas. Genus Hypecoum.
Hyperantheraceae Link = Moringaceae R. Br.
HYPERICACEAE Juss. 1789. Dicots. Theales. 9/490–500. Trop., subtrop., and temp. regions. Genera: Cratoxylum, Eliea, Harungana, Hypericum, Psorospermum, Santomasia, Thornea, Triadenum, Vismia.
HYPOXIDACEAE R. Br. 1814. Monocots. Haemodorales. 9/218. America, West Indies, Africa, trop. and subtrop. Asia, Australia, New Zealand. Genera: Curculigo, Empodium, Hypoxidia, Hypoxis, Molineria, Pauridia, Rhodohypoxis, Saniella, Spiloxene.
HYPSEOCHARITACEAE Wedd. 1861. Dicost. Geraniales. 1/8. Andes of South America. Genus Hypseocharis.
ICACINACEAE (Benth.) Miers. 1851. Dicots. Celastrales. 56/340. Tropics, a few species in temp. regions. Genera: Acrocoelium, Alsodeiopsis, Apodytes, Calatola, Cantleya, Casimirella, Cassinopsis, Chlanydocarya, Citronella, Codiocarpus, Dendrobangia, Desmostachys, Discophora, Emmotum, Gastrolepis, Gomphandra, Gonocaryum, Grisollea, Hartleya, Hosiea, Icacina, Iodes, Irvingbaileya, Lasianthera, Lavigeria, Leptaulus, Leretia, Mappia, Mappianthus, Medusanthera, Merrilliodendron, Miquelia, Natsiatopsis, Natsiatum, Nothapodytes, Oecopetalum, Ottoschulzia, Pennantia, Phytocrene, Pittosporopsis, Platea, Pleurisanthes, Polycephalium, Polyporandra, Poraqueiba, Pseudobotrys, Pyrena cantha, Rhaphiostylis, Rhyticaryum, Sarcostigma, Stachyanthus, Stemonurus, Tridianisia, Urandra, Villaresiopsis, Whitmorea.
IDIOSPERMACEAE S. T. Blake. 1972. Dicots. Laurales. 1/1. Australia. Genus Idiospermum.
Ilicaceae Bercht. et J. Presl = Aquifoliaceae Bartl.
Illecebraceae R. Br. = Caryophyllaceae Juss.
ILLICIACEAE (DC.) A. C. Sm. 1947. Dicots. Illiciales. 1/42. South and Southeast Asia, North America, West Indies. Genus Illicium.
Illigeraceae Blume = Hernandiaceae Blume.

Impatientaceae Barnhart = Balsaminaceae Dumort.

Imperatoriaceae Martinov = Apiaceae Lindl.

Inulaceae Bessey = Asteraceae Dumort.

Iriarteaceae O. F. Cook = Arecaceae Schultz Sch.

IRIDACEAE Juss. 1789. Monocots. Liliales. 87/1,890. Trop., subtrop., and temp. regions, esp. eastern Mediterranean, South Africa, and trop. America. Genera: Ainea, Alophia, Anomalostylus, Anomatheca, Aristea, Babiana, Barnardiella, Belamcanda, Bobartia, Calydorea, Cardenanthus, Chasmanthe, Chlamydostylus, Cipura, Cobana, Crocosmia, Crocus, Cypella, Devia, Dierama, Dietes, Diplarrhena, Duthieastrum, Eleutherine, Ennealophus, Ferraria, Fosteria, Freesia, Galaxia, Geissorhiza, Gelasine, Gladiolus, Gynandriris, Herbertia, Hermodactylus, Hesperantha, Hesperoxiphion, Hexaglottis, Homeria, Homoglossum, Iridodictyum, Iris, Isophysis, Ixia, Kelissa, Klattia, Lapeirousia, Lethia, Libertia, Mastigostyla, Melasphaerula, Micranthus, Moraea, Nemastylis, Neomarica, Nivenia, Olsynium, Onira, Orthrosanthus, Pardanthopsis, Patersonia, Pillansia, Pseudotrimezia, Radinosiphon, Rheome, Rigidella, Roggeveldia, Romulea, Savannosiphon, Schizostylis, Sessilanthera, Siphonostylis, Sisyrinchium, Solenomelus, Sparaxis, Sympa, Syringodea, Tapeinia, Thereianthus, Tigridia, Trimezia, Tritonia, Tritonopsis, Tucma, Watsonia, Witsenia, Zygotritonia.

IRVINGIACEAE (Engl.) Exell et Mendonça. 1951. Dicots. Rutales. 4/13–24. Trop. Africa, Madagascar, South and Southeast Asia. Genera: Allanthospermum, Desbordesia, Irvingia, Klainedoxa.

Isnardiaceae Martinov = Onagraceae Juss.

ITEACEAE J. Agardh. 1858. Dicots. Saxifragales. 2/11–16. Trop. and South Africa, Himalayas, East and Southeast Asia, North America. Genera: Choristylis, Itea.

Ivaceae Rchb. = Asteraceae Dumort.

Ixerbaceae Griseb. = Brexiaceae Loudon.

Ixiaceae Horan. = Iridaceae Juss.

IXIOLIRIACEAE (Pax) Nakai. 1943. Monocots. Amaryllidales. 1/4–6. West, Middle, and Central Asia. Genus Ixiolirion.

IXONANTHACEAE (Benth.) Exell et Mendonça. 1951. Dicots. Linales. 4/16. Trop. Africa, Madagascar, India, Himalayas, southern China, Southeast Asia, New Guinea, trop. America. Genera: Cyrillopsis, Ixonanthes, Ochthocosmus, Phyllocosmus.

Jamboliferaceae Martinov = Rutaceae Juss.

JAPONOLIRIACEAE Takht. 1994. Monocots. Melanthiales. 1/1. Japan. Genus Japonolirion.

Jasionaceae Dumort. = Campanulaceae Juss.

Jasminaceae Juss. = Oleaceae Hoffmanns. et Link.

Johnsoniaceae Lotsy = Asphodelaceae Juss.

JOINVILLEACEAE A. C. Sm. et Toml. 1970. Monocots. Restionales. 1/2. Malesia, Oceania to Hawaiian Is. and Caroline atoll. Genus Joinvillea.

JUGLANDACEAE A. Rich. ex Kunth. 1824. Dicots. Juglandales. 9/70. Temp. and subtrop. Northern Hemisphere, a few species extending through Malesia to New Guinea and western South America. Genera: Alfaroa, Alfaropsis, Carya, Cyclocarya, Engelhardia, Juglans, Oreomunnea, Platycarya, Pterocarya.

Julianiaceae Hemsl. = Anacardiaceae Lindl.

JUNCACEAE Juss. 1789. Monocots. Juncales. 7–9/355. Genera Juncus and Luzula, chiefly in cold and temp. Northern Hemisphere; the rest in the Southern Hemisphere. Genera: Andesia (= Oxychloe), Distichia, Juncus, Luzula, Marsippospermum, Oxychloe, Patosia (= Oxychloe), Prionium, Rostkovia.

JUNCAGINACEAE Rich. 1808. Monocots. Juncaginales. 3/17. Cosmopolitan, esp. temp. and cold regions. Genera: Cycno-

geton, Lilaea, Maundia, Tetroncium, Triglochin.
Juniperaceae Bercht. et J. Presl = Cupressaceae Rich. ex Bartl.
Jussiaeaceae Martinov = Onagraceae Juss.
Justiciaceae Raf. = Acanthaceae Juss.
KALIPHORACEAE Takht. 1994. Dicots. Hydrangeales. 1/1. Madagascar. Genus Kaliphora.
Kaniaceae Nakai = Myrtaceae Juss.
Kiggelariaceae Link = Flacourtiaceae DC.
KINGDONIACEAE A. S. Foster ex Airy Shaw. 1965. Dicots. Ranunculales. 1/1. China. Genus Kingdonia.
Kingiaceae Endl. = Dasypogonaceae Dumort.
Kirengeshomaceae Nakai = Hydrangeaceae Dumort.
KIRKIACEAE (Engl.) Takht. 1967. Dicots. Rutales. 2/6. Trop. and South Africa. Genus Kirkia, Pleiokirkia.
Kobresiaceae Gilly = Cyperaceae Juss.
Koeberliniaceae Engl. = Capparaceae Juss.
Koelreuteriaceae J. Agardh = Sapindaceae Juss.
KRAMERIACEAE Dumort. 1829. Dicots. Polygalales. 1/17. Trop. and subtrop. America. Genus Krameria.
LABIATAE Juss. See **LAMIACEAE** Lindl.
Lacandoniaceae E. Martinez et C. H. Ramos = ? Triuridaceae Gardner.
Lachenaliaceae Salisb. = Hyacinthaceae Batsch ex Borckh.
LACISTEMATACEAE Mart. 1826. Dicots. Violales. 2/14. Trop. America. Genera: Lacistema, Lozania.
LACTORIDACEAE Engl. 1888. Dicots. Lactoridales. 1/1. Juan Fernández Is. Genus Lactoris.
Lactucaceae Drude = Asteraceae Dumort.
Lagerstroemiaceae J. Agardh = Lithraceae J. St.-Hil.
LAMIACEAE Lindl. 1836 or **LABIATAE** Juss. 1789 (nom. altern.). Dicots. Lamiales. 225/6,170. Cosmopolitan, chiefly Mediterranean and Irano-Turanian floristic region. Genera: Acanthomintha, Achyrospermum, Acinos, Acrocephalus, Acrotome, Acrymia, Aeollanthus, Agastache, Ajuga, Ajugoides, Alajja, Alvesia, Amethystea, Anisochilus, Anisomeles, Antonina, Ascocarydion, Asterohyptis, Ballota, Bancroftia, Basilicum, Becium, Benguellia, Blephilia, Bostrychanthera, Bovonia, Brazoria, Bystropogon, Calapodium, Capitanopsis, Capitanya, Catopheria, Cedronella, Ceratanthus, Ceratominthe, Chamaesphacos, Chaunostoma, Chelonopsis, Cleonia, Clerodendranthus, Clinopodium, Colebrookea, Collinsonia, Colquhounia, Comanthosphace, Conradina, Coridothymus, Craniotome, Cuminia, Cunila, Cyclotrichium, Cymaria, Dauphinea, Dicerandra, Dorystaechas, Dracocephalum, Drepanocaryum, Eichlerago, Elsholtzia, Endostemon, Englerastrum, Epimeredi, Eremostachys, Eriope, Eriophyton, Eriopidion, Erythrochlamys, Eurysolen, Eusteralis, Fuerstia, Galeopsis, Geniosporum, Glechoma, Glechon, Gomphostemma, Gontscharovia, Hanceola, Haplostachys, Haumaniastrum, Hedeoma, Hemiandra, Hemigenia, Hemizygia, Hesperozygis, Heterolamium, Hoehnea, Holocheila, Holostylon, Horminum, Hoslundia, Hymenocrater, Hypenia, Hypogomphia, Hyptidendron, Hyptis, Hyssopus, Isodictyophorus, Isoleucas, Keiskea, Kinostemon, Kudrjaschevia, Kurzamra, Lagochilus, Lagopsis, Lamiophlomis, Lamium, Lavandula, Leocus, Leonotis, Leonurus, Lepechinia, Leucas, Leucosceptrum, Limniboza, Lophanthus, Loxocalyx, Lycopus, Macbridea, Mahya, Marmoritis, Marrubium, Marsypianthes, Meehania, Melissa, Melittis, Mentha, Meriandra, Mesona, Metastachydium, Microcorys, Microtoena, Minthostachys, Moluccella, Monardia, Monardella, Mosla, Neoeplingia, Heohyptis, Nepeta, Nosema, Ocimum, Octomeron, Ombrocharis, Oreosphacus, Origanum, Orthosiphon, Otostegia, Panzerina, Paraeremostachys, Paralamium, Paraphlomis, Peltodon, Pentapleura, Perilla, Perillula, Perovskia, Perrierastrum, Phlomi-

doschema, Phlomis, Phlomoides, Phyllostegia, Physoleucas, Physostegia, Philoblephis, Pitardia, Platostoma, Plectranthastrum, Plectranthus, Pogogyne, Pogostemon, Poliomintha, Prasium, Prostanthera, Prunella, Pseuderemostachys, Pseudochamaespacos, Pseudomarrubium, Puntia, Pycnanthemum, Pycnostachys, Rabdosia, Rabdosiella, Renschia, Rhabdocaulon, Rhaphiodon, Rhododon, Rosmarinus, Rostrinucula, Roylea, Rubiteucris, Sabaudia, Saccocalyx, Salvia, Satureja, Schizonepeta, Scutellaria, Sideritis, Siphocranion, Skapanthus, Solenostemon, Sphacele, Stachyopsis, Stachys, Stenogyne, Stiptanthus, Sulaimania, Suzukia, Symphostemon, Synandra, Syncolostemon, Tetradenia, Teucrium, Thorncroftia, Thuspeinanta, Thymbra, Thymus, Tinnea, Trichostema, Wenchengia, Westringia, Wiedemannia, Wrixonia, Zataria, Zhumeria, Ziziphora.

LANARIACEAE H. Huber ex T. Dahlgren. 1988. Monocots. Liliales. 1/1. South Africa. Genus Lanaria.

Langsdorfiaceae Tiegh. ex Pilg. et K. Krause = Balanophoraceae Rich.

Lantanaceae Martinov = Verbenaceae J. St.-Hil.

Lapageriaceae Kunth = Philesiaceae Dumort.

LARDIZABALACEAE Decne. 1839. Dicots. Ranunculales. 8/40. Asia from western Himalayas to Korea, Japan and Taiwan, South America. Genera: Akebia, Boquila, Decaisnea, Holboellia, Lardizabala, Parvatia, Sinofranchetia, Stauntonia.

Lasiopetalaceae J. Agardh = Sterculiaceae (DC.) Bartl.

Lathyraceae Burnett = Fabaceae Lindl.

LAURACEAE Juss. 1789. Dicots. Laurales. 55/2,555. Tropics and subtropics, esp. Southeast Asia and trop. America, a few species in temp. regions. Genera: Actinodaphne, Aiouea, Alseodaphne, Anaueria, Aniba, Apollonias, Aspidostemon, Beilschmiedia, Caryodaphnopsis, Cassytha, Chlorocardium, Cinnadenia, Cinnamomum, Clinostemon, Cryptocarya, Dahlgrenodendron, Dehaasia, Dicypelli um, Dodecadenia, Endiandra, Endlicheria, Eusideroxylon, Gamanthera, Hexapora, Hypodaphnis, Laurus, Licaria, Lindera, Litsea, Machilus, Mezilaurus, Nectandra, Neocinnomomum, Neolitsea, Nothaphoebe Ocotea, Paraia, Persea, Phoebe, Phyllostemonodaphne, Pleurothyrium, Potameia, Potoxylon, Povedadaphne, Ravensara, Rhodostemonodaphne, Sassafras, Sinosassafras, Stemmatodaphne, Syndiclis, Systemonodaphne, Temmodaphne, Umbellularia, Urbanodendron, Williamodendron.

Lawsoniaceae J. Agardh = Lythraceae J. St.-Hil.

LECYTHIDACEAE Poit. 1825. Dicots. Lecythidales. 20/310. Tropics, esp. South America. Genera: Abdulmajidia, Allantoma, Asteranthos, Barringtonia, Bertholletia, Careya, Cariniana, Chydenanthus, Corythophora, Couratari, Couroupita, Crateranthus, Eschweilera, Foetidia, Grias, Gustavia, Lecythis, Napoleonaea, Petersianthus, Planchonia.

Ledaceae Link = Ericaceae Juss.

LEDOCARPACEAE Meyen. 1834. Dicots. Geraniales. 2/11. South America. Genera: Balbisia (Ledocarpon), Wendtia.

LEEACEAE Dumort. 1829. Dicots. Vitales. 1/34. Trop. Old World. Genus Leea.

Legnotidaceae Endl. = Rhizophoraceae R. Br.

LEGUMINOSAE Juss. See FABACEAE Lindl.

LEITNERIACEAE Benth. 1880. Dicots. Leitneriales. 1/1. Southeastern North America. Genus Leitneria.

LEMNACEAE Gray. 1821. Monocots. Arales. 6/30. Cosmopolitan. Genera: Lemna, Pseudowolffia, Spirodela, Wolffia, Wolffiella, Wolffiopsis.

LENNOACEAE Solms-Laub. 1870. Dicots. Boraginales. 2/4. America from southern U.S. to Colombia and Venezuela. Genera: Lennoa, Pholisma.

LENTIBULARIACEAE Rich. 1808. Dicots. Scrophulariales. 4/246. Cosmopolitan.

Genera: Genlisea, Pinguicula, Polypompholyx, Utricularia.

Leoniaceae DC. = Violaceae Batsch.

Leonticaceae Bercht. et J. Presl = Berberidaceae Juss.

Lepidariaceae Tiegh. = Loranthaceae Juss.

LEPIDOBOTRYACEAE J. Léonard. 1950. Dicots. Geraniales. 1/1. Trop. Africa. Genus Lepidobotrys.

Lepidocarpaceae Schultz Sch. = Proteaceae Juss.

Lepidocaryaceae O. F. Cook = Arecaceae Schultz Sch.

Lepidocerataceae Tiegh. ex Nakai = Eremolepidaceae Tiegh. ex Nakai.

Leptospermaceae F. Rudolphi = Myrtaceae Juss.

Lepturaceae (Holmberg) Herter = Poaceae Barnhart.

LEPUROPETALACEAE (Engl.) Nakai. 1943. Dicots. Saxifragales. 1/1. North and South America. Genus Lepuropetalon.

Leuchtenbergiaceae Salm-Dyck = Cactaceae Juss.

Leucojaceae Batsch ex Borckh. = Amaryllidaceae J. St-Hil.

Licaneaceae Martinov = Chrysobalanaceae R. Br.

Lilacaceae Vent. = Oleaceae Hoffmanns. et Link.

Lilaeaceae Durmort. = Juncaginaceae Rich.

LILIACEAE Juss. 1789. Monocots. Liliales. 13/540–580 (sensu. str.). Temp. and subtrop. Northern Hemisphere, esp. Asia. Genera: Amana, Cardiocrinum, Erythronium, Fritillaria, Gagea, Halongia, Korolkowia, Lilium, Lloydia, Nomocharis, Notholirion, Rhinopetalum, Tulipa.

LIMNANTHACEAE R. Br. 1833. Dicots. Limnanthales. 2/10. North America. Genera: Floerkea, Limnanthes.

LIMNOCHARITACEAE Takht. ex Cronquist. 1981. Monocots. Alismatales. 3/12. Tropics and subtropics. Genera: Butomopsis, Hydrocleys, Limnocharis.

Limodoraceae Horan. = Orchidaceae Juss.

Limoniaceae Seringe = Plumbaginaceae Juss.

Limosellaceae J. Agardh = Scrophulariaceae Juss.

LINACEAE DC. ex Gray. 1821. Dicots. Linales. 8/220–250. Cosmopolitan, esp. temp. and subtrop. regions. Genera: Anisadenia, Cliococca, Hesperolinon, Linum, Radiola, Reinwardtia, Sclerolinon, Tirpitzia.

Linariaceae Martinov = Scrophulariaceae Juss.

Linderniaceae Rchb. = Scrophulariaceae Juss.

Lindleyaceae J. Agardh = Rosaceae Juss.

Lippayaceae Meisn. = Rubiaceae Juss.

Liquidambaraceae Bromhead = Altingiaceae Lindl.

Liriodendraceae F. A. Barkley = Magnoliaceae Juss.

LISSOCARPACEAE Gilg. 1924. Dicots. Ebenales. 1/3. Trop. South America. Genus Lissocarpa.

Littorellaceae Gray = Plantaginaceae Juss.

LOASACEAE Dumort. 1822. Dicots. Loasales. 14/260. America, West Indies, Fissenia—in Africa and Arabian Peninsula, Plakothira—on the Marquesas Is. Genera: Blumenbachia, Caiophora, Cevallia, Eucnide, Fuertesia, Gronovia, Kissenia, Klaprothia, Loasa, Mentzelia, Petalonyx, Plakothira, Schismocarpus, Scyphanthus.

LOBELIACEAE R. Br. 1817. Dicots. Campanulales. 31/1,150. Tropics and subtropics, esp. America, a few in temp. regions. Genera: Apetahia, Brighamia, Burmeistera, Centropogon, Clermontia, Cyanea, Delissea, Dialypetalum, Diastatea, Dielsantha, Downingia, Grammatotheca, Heterotoma, Hippobroma, Howellia, Hypsela, Legenera, Lobelia, Lysipomia, Mezleria, Monopsis, Palmerella, Porterella, Rollandia, Ruthiella, Sclerotheca, Siphocampylus, Solenopsis, Trematolobelia, Trimeris, Unigenes.

LOGANIACEAE R. Br. ex Mart. 1827. Dicots. Gentianales. 12/313. Tropics and subtropics, a few species in warm–temp.

regions. Genera: Antonia, Bonyunia, Gardneria, Gelsemium, Geniostoma, Labordia, Logania, Mostuea, Neuburgia, Norrisia, Strychnos, Usteria.

Lomandraceae Lotsy = Dasypogonaceae Dumort.

Loniceraceae von Vest = Caprifoliaceae Juss.

Lophiolaceae Nakai = Melanthiaceae Batsch.

LOPHIRACEAE Loudon. 1830. Dicots. Ochnales. 1/2. Trop. Africa. Genus Lophira.

LOPHOPHYTACEAE Horan. 1847. Dicots. Balanophorales. 3/8. Trop. and subtrop. South America. Genera: Lathrophytum, Lophophytum, Ombrophytum.

LOPHOPYXIDACEAE (Engl.) H. Pfeiff. 1951. Dicots. Celastrales. 1/2. Malesia to Solomon Is. Genus Lophopyxis.

LORANTHACEAE Juss. 1808. Dicots. Santalales. 71/960. Tropics and subtropics. Genera: Actinanthella, Aetanthus, Alepis, Amyema, Amylotheca, Atkinsonia, Bakerella, Baratranthus, Benthamina, Berhautia, Botryoloranthus, Cecarria, Cladocolea, Cyne, Dactyliophora, Decaisnina, Dendrophthoe, Desmaria, Diplatia, Distrianthes, Elytranthe, Emelianthe, Erianthemum, Gaiadendron, Globimetula, Helicanthes, Helixanthera, Ileostylus, Ixocactus, Lampas, Lepeostegeres, Lepidaria, Ligaria, Loranthus, Loxanthera, Lysiana, Macrosolen, Moquiniella, Muellerina, Notanthera, Nuytsia, Odontella, Oncella, Oryctanthus, Oryctina, Panamanthus, Papuanthes, Pedistylis, Peraxilla, Phthirusa, Plicosepalus, Psittacanthus, Rhizomonanthes, Scurrula, Septulina, Socratina, Sogerianthe, Spragueanella, Struthanthus, Tapinanthus, Taxillus, Tetradyas, Thaumasianthes, Tieghemia, Tolypanthus, Trilepidea, Tripodanthus, Tristerix, Trithecanthera, Tupeia, Vanwykia.

Lotaceae Burnett = Fabaceae Lindl.

LOWIACEAE Ridl. 1924. Monocots. Zingiberales. 1/9. Southern China, Indochina, Malesia, Oceania. Genus Orchidantha.

Lupulaceae Link = Cannabaceae Endl.

Luxemburgiaceae Tiegh. ex Soler. = Sauvagesiaceae Dumort.

LUZURIAGACEAE Lotsy. 1911. Monocots. Smilacales. 5/9. Southeast Africa, Malesia, Australia, New Zealand, Oceania, temp. South America, Falkland Is. Genera: Behnia, Drymophila, Eustrephus, Geitonoplesium, Luzuriaga.

Lyciaceae Raf. = Solanaceae Juss.

Lygodisodeaceae Bartl. = Rubiaceae Juss.

Lysimachiaceae Juss. = Primulaceae Vent.

Lysinemaceae Rchb. = Epacridaceae R. Br.

LYTHRACEAE J. St.-Hil. 1805. Dicots. Myrtales. 29/585. Tropics and subtropics, esp. America; a few species in cold and temp. regions. Genera: Adenaria, Ammannia, Capuronia, Crenea, Cuphea, Decodon, Didiplis, Diplusodon, Galpinia, Ginoria, Haitia, Heimia, Hionanthera, Hydrolythrum, Koehneria, Lafoënsia, Lagerstroemia, Lawsonia, Lourtella, Lythrum, Nesaea, Pehria, Pemphis, Peplis, Physocalymma, Pleurophora, Rotala, Tetrataxis, Woodfordia.

Macarisiaceae J. Agardh = Rhizophoraceae R. Br.

MAGNOLIACEAE Juss. 1789. Dicots. Magnoliales. 13/200. Trop. and East Asia, Malesia, trop. and subtrop. America, West Indies. Genera: Alcimandra, Aromadendron, Elmerrillia, Kmeria, Liriodendron, Magnolia, Manglietia, Manglietiastrum, Michelia, Pachylarnax, Paramichelia, Talauma, Tsoongiodendron.

Malaceae Small ex Britton = Rosaceae Juss.

Malachodendraceae J. Agardh = Theaceae D. Don.

MALESHERBIACEAE D. Don. 1827. Dicots. Violales. 1/30. Western South America. Genus Malesherbia.

Malortieaceae O. F. Cook = Arecaceae Schultz Sch.

MALPIGHIACEAE Juss. 1789. Dicots. Polygalales. 71/1,120. Tropics and subtropics, esp. South America. Genera: Acmanthera, Acridocarpus, Adenoporces, Aspicarpa, Aspidopterys, Banisterioides, Banisteriopsis, Barnebya, Blepharandra, Brachylophon, Brachypterys, Brittonella, Bunchosia, Burdachia, Byrsonima, Callaeum, Calyptostylis, Camarea, Caucanthus, Clonodia, Coleostachys, Cordobia, Diacidia, Dicella, Digoniopterys, Dinemagonum, Dinemandra, Diplopterys, Echinopteris, Ectopopterys, Flabellaria, Flabellariopsis, Gallardoa, Galphimia, Gaudichaudia, Glandonia, Heladena, Hemsleyna, Henleophytum, Heteropterys, Hiptage, Hiraea, Janusia, Jubelina, Lasiocarpus, Lophanthera, Lophopterys, Malpighia, Mascagnia, Mcvaughia, Mezia, Microsteira, Mionandra, Peixotoa, Peregrina, Philgamia, Pterandra, Ptilochae ta, Rhynchophora, Rhyssopteris, Schwannia, Spachea, Sphedamnocarpus, Stigmaphyllon, Tetrapteris, Thryallis, Triaspis, Trichomariopsis, Tricomaria, Triopteris, Tristellateia, Verrucularina.

MALVACEAE Juss. 1789. Dicots. Malvales. 114/1,670. Cosmopolitan, but chiefly tropics. Genera: Abelmoschus, Abutilon, Abutilothamnus, Acaulimalva, Alcea, Allosidastrum, Allowissadula, Althaea, Alyogyne, Anisodontea, Anoda, Anotea, Arcynospermum, Asterotrichion, Bakeridesia, Bastardia, Bastardiastrum, Bastardiopsis, Batesimalva, Billieturnera, Blanchetiastrum, Bombycidendron, Briquetia, Brockmania, Callirhoe, Calyculogygas, Calyptraemalva, Cenocentrum, Cephalohibiscus, Cienfuegosia, Codonochlamys, Cristaria, Decaschistia, Dendrosida, Dicellostyles, Dirhamphis, Eremalche, Fryxellia, Gaya, Gossypioides, Gossypium, Gynatrix, Hampea, Helicteropsis, Herissantia, Hibiscadelphus, Hibiscus, Hochreutinera, Hoheria, Horsfordia, Howittia, Humbertianthus, Humbertiella, Iliamna, Julostylis, Jumelleanthus, Kearnemalvastrum, Kitaibelia, Kokia, Kosteletzkya, Krapovickasia, Kydia, Lagunaria, Lavatera, Lawrencia, Lebronnecia, Lopimia, Macrostelia, Malachra, Malacothamnus, Malope, Malva, Malvastrum, Malvella, Malvaviscus, Malvella, Megistostegium, Meximalva, Modiola, Modiolastrum, Monteiroa, Napaea, Nayariophyton, Neobaclea, Neobrittonia, Nototriche, Notoxylinon, Palaua, Pavonia, Peltaea, Periptera, Perrierophytum, Phragmocarpidium, Phymosia, Plagianthus, Pseudabutilon, Radyera, Rhynchosida, Robinsonella, Rojasimalva, Senra, Sida, Sidalcea, Sidastrum, Sphaeralcea, Symphyochlamys, Tarasa, Tetrasida, Thespesia, Uladendron, Urena, Urocarpidium, Wercklea, Wissadula.

Mamillariaceae (Rchb.) Dostál = Cactaceae Juss.

Manicariaceae O. F. Cook = Arecaceae Schultz Sch.

MARANTACEAE O. Petersen. 1888. Monocots. Zingiberales. 31/430. Tropics, esp. America. Genera: Afrocalathea, Calathea, Cominsia, Ctenanthe, Donax, Halopegia, Haumania, Hylaeanthe, Hypselodelphys, Ischnosiphon, Koernickanthe, Maranta, Marantochloa, Megaphrynium, Monophrynium, Monophyllanthe, Monotagma, Myrosma, Phacelophrynium, Phrynium, Pleiostachya, Sanblasia, Saranthe, Sarcophrynium, Schumannianthus, Stachyphrynium, Stromanthe, Thalia, Thaumatococcus, Thymocarpus, Trachyphrynium.

Marathraceae Dumort. = Podostemaceae Rich. ex C. Agardh.

MARCGRAVIACEAE Choisy. 1824. Dicots. Theales. 5/123. Trop. America. Genera: Caracasia, Marcgravia, Norantea, Ruyschia, Souroubea.

MARTYNIACEAE Stapf. 1895. Dicots. Scrophulariales. 4/20. Trop. and subtrop. America, West Indies. Genera: Craniolaria, Ibicella, Martynia, Proboscidea.

MASTIXIACEAE Calest. 1905. Dicots. Cornales. 2/14. Trop. Asia, Malesia to Solomon Is. Genera: Diplopanax, Mastixia.

Maundiaceae Nakai = Juncaginaceae Rich.

MAYACACEAE Kunth. 1842. Monocots.

Commelinales. 1/4. America, West Indies, trop. Africa. Genus Mayaca.

Meboreaceae Raf. = Euphorbiaceae Juss.

MEDEOLACEAE **Takht.** 1987. Monocots. Liliales. 1/2. North America. Genus Medeola.

MEDUSAGYNACEAE **Engl. et Gilg.** 1924. Dicots. Medusaginales. 1/1. Seychelles. Genus Medusagyne.

MEDUSANDRACEAE **Brenan.** 1952. Dicots. Santalales. 1/2. Trop. West Africa. Genus Medusandra.

Melaleucaceae von Vest = Myrtaceae Juss.

Melampyraceae Lindl. = Scrophulariaceae Juss.

MELANOPHYLLACEAE **Takht. ex Airy Shaw.** 1972. Dicots. Hydrangeales. 2/9. Madagascar. Genera: Kaliphora, Melanophylla.

MELANTHIACEAE **Batsch.** 1802. Monocots. Liliales. 27/160. Almost cosmopolitan. Genera: Aletris, Amianthium, Burchardia, Chamaelirium, Harperocallis, Helonias, Heloniopsis, Isidrogalvia, Kuntheria, Lophiola, Melanthium, Metanarthecium, Narthecium, Nietneria, Onixotis, Petrosavia, Schelhammera, Schoenocaulon, Scoliopus, Stenanthium, Tofieldia, Tricyrtis, Tripladenia, Uvularia, Veratrum, Ypsilandra, Zigadenus.

MELASTOMATACEAE **Juss.** 1789. Dicots. Myrtales. 202/5,025. Tropics and subtropics, esp. South America. Genera: Acanthella, Aciotis, Acisanthera, Adelobotrys, Allomaieta, Allomorphia, Alloneuron, Amphiblemma, Amphitoma, Amphorocalyx, Anaectocalyx, Anerincleistus, Antherotoma, Appendicularia, Arthrostemma, Aschistanthera, Astrocalyx, Astronia, Astronidium, Axinaea, Barthea, Beccarianthus, Behuria, Bellucia, Benevidesia, Bertolonia, Bisglaziovia, Blakea, Blastus, Boerlagea, Boyania, Brachyotum, Brachypremna, Bredia, Brittenia, Bucquetia, Cailliella, Calvoa, Calycogonium, Cambessedesia, Campimia, Carionia, Castratella, Catanthera, Catocoryne, Centradenia, Centradeniastrum, Centronia, Chaetolepis, Chaetostoma, Charianthus, Cincinnobotrys, Clidemia, Comolia, Comoliopsis, Conostegia, Copedesma, Creaghiella, Creochiton, Cyanandrium, Cyphostyla, Cyphotheca, Dalenia, Desmoscelis, Dicellandra, Dicerospermum, Dichaetanthera, Dinophora, Dionycha, Dionychastrum, Diplarpea, Diplectria, Dissochaeta, Dissotis, Dolichoura, Driessenia, Eisocreochiton, Enaulophyton, Eriocnema, Ernestia, Farringtonia, Feliciadamia, Fordiophyton, Fritzschia, Graffenrieda, Gravesia, Guyonia, Henriettea, Henriettella, Heterocentron, Heterotis, Heterotrichum, Huberia, Huilaea, Hylocharis, Hypenanthe, Kendrickia, Kerriothyrsus, Killipia, Kirkbridea, Lavoisiera, Leandra, Lijndenia, Lithobium, Llewelynia, Loreya, Loricalepis, Macairea, Macrocentrum, Macrolenes, Maguireanthus, Maieta, Mallophyton, Marcetia, Mecranium, Medinilla, Meiandra, Melastoma, Melastomastrum, Memecylon, Menendezia, Meriania, Merianthera, Miconia, Microlepis, Microlicia, Mommsenia, Monochaetum, Monolena, Mouriri, Myriaspora, Myrmidone, Neblinanthera, Necramium, Neodriessenia, Nepsera, Nerophila, Ochthephilus, Ochthocharis, Omphalopus, Opisthocentra, Oritrephes, Osbeckia, Ossaea, Otanthera, Oxyspora, Pachyanthus, Pachycentria, Pachyloma, Phainantha, Phyllagathis, Pilocosta, Plagiopetalum, Pleiochiton, Plethiandra, Podocaelia, Pogonanthera, Poikilogyne, Poilannammia, Poteranthera, Preussiella, Pseudodissochaeta, Pseudosbeckia, Pternandra, Pterogastra, Pterolepis, Rhexia, Rhynchanthera, Rousseauxia, Salpinga, Sandemania, Sarcopyramis, Schwackaea, Scorpiothyrsus, Siphanthera, Sonerila, Spathandra, Sporoxeia, Stapfiophyton, Stenodon, Stussenia, Styrophyton, Svitramia, Tateanthus, Tayloriophyton, Tessmannianthus, Tetraphyllaster, Tetrazygia, Tibouchina, Tibouchinopsis, Tigridiopalma, Tococa, Topobea, Trembleya,

Triolena, Tristemma, Tryginia, Tryssophyton, Tylanthera, Vietsenia, Votomita, Warneckea.

MELIACEAE Juss. 1789. Dicots. Rutales. 50/615. Tropics and subtropics, a few species in warm–temp. regions. Genera: Aglaia, Anthocarapa, Aphanamixis, Astrotrichilia, Azadirachta, Cabralea, Calodecaryia, Capuronianthus, Carapa, Cedrela, Chisocheton, Chukrasia, Cipadessa, Dysoxylum, Ekebergia, Entandrophragma, Guarea, Heckeldora, Humbertioturraea, Khaya, Lansium, Lepidotrichilia, Lovoa, Malleastrum, Melia, Munronia, Naregamia, Neobeguea, Nymania, Owenia, Pseudobersama, Pseudocarapa, Pseudocedrela, Pterorhachis, Quivisianthe, Reinwardtiodendron, Ruagea, Sandoricum, Schmardaea, Soymida, Sphaerosacme, Swietenia, Synoum, Toona, Trichelia, Turraea, Turraeanthus, Vavaea, Walsura, Xylocarpus.

MELIANTHACEAE Link. 1831. Dicots. Sapindales. 2/9. Trop. and South Africa. Genera: Bersama, Melianthus.

Melicaceae Martinov = Poaceae Barnhart.

Meliosmaceae Endl. = Sabiaceae Blume.

Melittaceae Martinov = Lamiaceae Lindl.

Melochiaceae J. Agardh = Sterculiaceae (DC.) Bartl.

Memecylaceae DC. = Melastomataceae Juss.

Mendonciaceae Bremek. = Acanthaceae Juss.

MENISPERMACEAE Juss. 1789. Dicots. Ranunculales. 79/540. Tropics and subtropics, a few species in temp. regions. Genera: Abuta, Adeliopsis, Albertisia, Anamirta, Anisocycla, Anomospermum, Antizoma, Arcangelisia, Aspidocarya, Beirnaertia, Borismene, Burasaia, Calycocarpum, Carronia, Caryomene, Chasmanthera, Chlaenandra, Chondrodendron, Cionomene, Cissampelos, Cocculus, Coscinium, Curarea, Cyclea, Dialytheca, Dioscoreophyllum, Diploclisia, Disciphania, Echinostephia, Elephantomene, Eleutharrhena, Elissarrhena, Fibraurea, Gamopoda, Haematocarpus, Hyperbaena, Hypserpa, Jateorhiza, Kolobopetalum, Legnephora, Leptoterantha, Limacia, Limaciopsis, Macrococculus, Menispermum, Odontocarya, Orthogynium, Orthomene, Pachygone, Parabaena, Paracyclea, Penianthus, Pericampylus, Platytinospora, Pleogyne, Pycnarrhena, Rameya, Rhaptonema, Rhigiocarya, Sarcolophium, Sarcopetalum, Sciadotenia, Sinomenium, Sphenocentrum, Spirospermum, Stephania, Strychnopsis, Synandropus, Synclisia, Syntriandrum, Syrrheonema, Taubertia, Telitoxicum, Tiliacora, Tinomiscium, Tinospora, Triclisia, Ungulipetalum, Znkerophytum.

Menthaceae Burnett = Lamiaceae Lindl.

MENYANTHACEAE Dumort. 1829. Dicots. Gentianales. 5/35. Cosmopolitan. Genera: Liparophyllum, Menyanthes, Nephrophyllidium, Nymphoides, Villarsia.

Menziesiaceae Klotzsch = Ericaceae Juss.

Mercurialaceae Martinov = Euphorbiaceae Juss.

Mesembryaceae Dumort. = Aizoaceae F. Rudolphi.

Mesembryanthemaceae Fenzl = Aizoaceae F. Rudolphi.

Mespilaceae Schultz Sch. = Rosaceae Juss.

Metasequoiaceae Hu et Cheng = Taxodiaceae Warm.

Methonicaceae Trautv. = Melanthiaceae Batsch.

METTENIUSACEAE Karst. ex Schnizl. 1860. Dicots. Celastrales. 1/7. Panama and northwestern South America. Genus Metteniusa.

Meyeniaceae Sreemadh. = Thunbergiaceae Bremek.

Miconiaceae K. Koch = Melastomataceae Juss.

Micrantheaceae J. Agardh = Euphorbiaceae Juss.

Microbiotaceae Nakai = Cupressaceae Rich. ex Bartl.

Miliaceae Burnett = Poaceae Barnhart.

Milulaceae Traub = Alliaceae J. Agardh.

Mimosaceae R. Br. = Fabaceae Lindl.

Mirabilidaceae W. R. B. Oliv. = Nyctaginaceae Juss.

MISODENDRACEAE J. Agardh. 1858. Dicots. Santalales. 1/8. Temp. South America. Genus Misodendrum.

MITRASTEMMATACEAE Makino. 1911. Dicots. Rafflesiales. 1/2. East and Southeast Asia, Malesia, Mexico, Central America. Genus Mitrastemma.

Miyoshiaceae Makino = Melanthiaceae Batsch.

Modeccaceae J. Agardh = Passifloraceae Juss. ex Kunth.

MOLLUGINACEAE Hutch. 1926. Dicots. Caryophyllales. 14/113. Tropics and subtropics, esp. Southern Hemisphere. Genera: Adenogramma, Coelanthum, Corbichonia, Glinus, Glischrothamnus, Hypertelis, Limeum, Macarthuria, Mollugo, Orygia, Pharnaceum, Polpoda, Psammotropha, Suessenguthiella.

Monieraceae Raf. = Scrophulariaceae Juss.

MONIMIACEAE Juss. 1809. Dicots. Laurales. 35/421. Trop., subtrop., and temp. Southern Hemisphere. Genera: Atherosperma, Austromatthaea, Bracteanthus, Daphnandra, Decarydendron, Doryphora, Dryadodaphne, Ephippiandra, Faika, Glossocalyx, Hedycarya, Hennecartia, Hortonia, Kairoa, Kibara, Kibaropsis, Laurelia, Laureliopsis, Lauterbachia, Levieria, Macropeplus, Macrotorus, Matthaea, Mollinedia, Monimia, Nemuaron, Palmeria, Parakibara, Peumus, Siparuna, Steganthera, Tambourissa, Tetrasynandra, Wilkiea, Xymalos.

Monodoraceae J. Agardh = Annonaceae Juss.

MONOTACEAE Maury ex Takht. 1987. Dicots. Malvales. 3/41. Northern South America, trop. Africa, Madagascar. Genera: Marquesia, Monotes, Pakaraimaea.

Monotropaeae Nutt. = Ericaceae Juss.

Montiaceae Dumort. = Portulacaceae Juss.

MONTINIACEAE Nakai. 1943. Dicots. Hydrangeales. 2/3. Trop. and South Africa, Madagascar. Genera: Grevea, Montinia.

MORACEAE Link. 1831. Dicots. Urticales. 46/1,460. Tropics and subtropics, a few species in warm–temp. regions. Genera: Antiaris, Antiaropsis, Artocarpus, Bagassa, Batocarpus, Bleekrodea, Bosqueiopsis, Brosimum, Broussonetia, Calpidochlamys, Castilla, Chlorophora, Clarisia, Cudrania, Dorstenia, Fatoua, Ficus, Helianthostylis, Helicostylis, Hullettia, Maclura, Maillarida, Maquira, Mesogyne, Metatrophis, Milicia, Morus, Naucleopsis, Parartocarpus, Perebea, Plecospermum, Poulsenia, Prainea, Pseudolmedia, Sahagunis, Scyphosyce, Sloetiopsis, Sorocea, Sparattosyce, Stenochasma, Streblus, Treculia, Trilepisium, Trophis, Trymatococcus, Utsetela.

MORINACEAE Raf. 1820. Dicots. Dipsacales. 3/13. Southeast Europe, West Asia to continental China and Himalayas. Genera: Acanthocalyx, Cryptothladia, Morina.

MORINGACEAE R. Br. ex Dumort. 1829. Dicots. Capparales. 1/14. Africa, Southwest and South Asia. Genus Moringa.

Moronobaceae Miers = Clusiaceae Lindl.

Mouririaceae Gardner = Melastomataceae Juss.

Moutabeaceae Endl. = Polygalaceae R. Br.

MUSACEAE Juss. 1789. Monocots. Zingiberales. 3/60. Trop. of the Old World and East Asia. Genera: Ensete, Musa, Musella.

Mutisiaceae Lindl. = Asteraceae Dumort.

MYOPORACEAE R. Br. 1810. Dicots. Scrophulariales. 3/213. Mauritius, East Asia, New Guinea, Australia, New Zealand, Oceania, West India, trop. South America. Genera: Bontia, Eremophila, Myoporum.

MYRICACEAE Blume. 1829. Dicots. Myricales. 3/52. Cold, temp., and subtrop. regions (excl. Australia), New Caledonia. Genera: Canacomyrica, Comptonia, Myrica.

Myriophyllaceae Schultz Sch. = Haloragaceae R. Br.

MYRISTICACEAE R. Br. 1810. Dicots. Annonales. 19/350. Tropics. Genera: Bicuiba, Brochoneura, Cephalosphaera, Coelocaryon, Compsoneura, Endocomia, Gymnacranthera, Haematodendron, Horsfieldia, Iryanthera, Knema, Mauloutchia, Myristica, Osteophloeum, Otoba, Pycnanthus, Scyphocephalium, Staudtia, Virola.

Myrobalanaceae Juss. ex Martinov = Combretaceae R. Br.

MYROTHAMNACEAE Nied. 1891. Dicots. Myrothamnales. 1/2. Trop. and South Africa, Madagascar. Genus Myrothamnus.

Myrrhiniaceae Arn. = Myrtaceae Juss.

MYRSINACEAE R. Br. 1810. Dicots. Primulales. 35/1,610. Tropics and subtropics. Genera: Aegiceras, Amblyanthopsis, Amblyanthus, Antistrophe, Ardisia, Badula, Conandrium, Correlliana, Ctenardisia, Cybianthus, Discocalyx, Elingamita, Embelia, Emblemantha, Fittingia, Geissanthus, Grenacheria, Heberdenia, Hymenandra, Labisia, Loheria, Maesa, Monoporus, Myrsine, Oncostemum, Parathesis, Pleiomeris, Rapanea, Sadiria, Solonia, Stylogyne, Tapeinosperma, Tetrardisia, Vegaea, Wallenia.

MYRTACEAE Juss. 1789. Dicots. Myrtales. 133/3,900. Tropics, subtropics, amd warm–temp. regions, mostly Australia and South America. Genera: Acca, Accara, Acicalyptus, Acmena, Acmenosperma, Acreugenia, Actinodium, Agonis, Allosyncarpia, Amomyrtella, Amomyrtus, Angasomyrtus, Angophora, Archirhodomyrtus, Arillastrum, Astartea, Asteromyrtus, Austromyrtus, Backhousia, Baeckea, Balaustion, Barongia, Basisperma, Beaufortia, Blepharocalyx, Callistemon, Calothamnus, Calycolpus, Calycorectes, Calyptranthes, Calyptorgenia, Calythropsis, Calytrix, Campomanesia, Carpolepis, Chamelaucium, Chamguava, Choricarpia, Cleistocalyx, Cloezia, Conothamnus, Corynanthera, Corynemyrtus, Cupheanthus, Darwinia, Decaspermum, Eremaea, Eucalyptopsis, Eucalyptus, Eugenia, Feijoa, Hexachlamys, Homalocalyx, Homalospermum, Homoranthus, Hottea, Hypocalymma, Kania, Kjellbergiodendron, Kunzea, Legrandia, Lencymmoea, Leptospermum, Lindsayomyrtus, Lophomyrtus, Lophostemon, Luma, Lysicarpus, Malleostemon, Marlierea, Melaleuca, Meteoromyrtus, Metrosideros, Micromyrtus, Mitranthes, Mitrantia, Monimiastrum, Mosiera, Mozartia, Myrceugenia, Myrcia, Myrcianthes, Myrciaria, Myrrhinium, Myrtastrum, Myrtella, Myrteola, Myrtus, Neofabricia, Neomitranthes, Neomyrtus, Ochrosperma, Octamyrtus, Osbornia, Paramyrciaria, Pericalymma, Phymatocarpus, Pileanthus, Pilidiostigma, Piliocalyx, Pimenta, Pleurocalyptus, Plinia, Pseudanamomis, Pseudeugenia, Psidium, Purpureostemon, Pyrenocarpa, Regelia, Rhodamnia, Rhodomyrtus, Rinzia, Ristantia, Rylstonea, Scaryomyrtus, Scholtzia, Siphoneugenia, Sphaerantia, Stereocaryum, Syncarpia, Syzygium, Tepualia, Thryptomene, Tristania, Tristaniopsis, Ugni, Uromyrtus, Verticordia, Waterhousea, Welchiodendron, Whiteodendron, Xanthomyrtus, Xanthostemon.

MYSTROPETALACEAE Hook. f. 1853. Dicots. Balanophorales. 1/2. South Africa. Genus Mystropetalon.

Myzodendraceae J. Agardh = Misodendraceae J. Agardh.

NAGEIACEAE D. Z. Fu. 1992. Pinopsida. Podocarpales. 1/5. South, Southeast, and East Asia; Malesia. Genus Nageia.

NAJADACEAE Juss. 1789. Monocots. Najadales. 1/40–50. Almost cosmopolitan, except Arctica and taiga zone of the Eurasia. Genus Najas.

NANDINACEAE Horan. 1858. Dicots. Ranunculales. 1/1. China, Japan. Genus Nandina.

Napoleonaeaceae A. Rich. = Lecythidaceae Poit.

Narcissaceae Juss. = Amaryllidaceae J. St.-Hil.
Nardaceae Martinov = Poaceae Barnhart.
Nartheciaceae Small = Melanthiaceae Batsch.
Nassauviaceae Burmeist. = Asteraceae Dumort.
Naucleaceae (DC.) Wernham = Rubiaceae Juss.
Nectaropetalaceae Exell et Mendonca = Erythroxylaceae Kunth.
Neilliaceae Miq. = Rosaceae Juss.
Nelsoniaceae Sreemadh. = Acanthaceae Juss.
NELUMBONACEAE **Dumort.** 1829. Dicots. Nelumbonales. 1/2. ? East Europe; South, East, and Southeast Asia; Malesia; northern Australia; America. Genus Nelumbo.
NEMACLADACEAE **Nutt.** 1843. Dicots. Campanulales. 3/12. Southwestern U.S. and Mexico. Genera: Nemacladus, Parishella, Pseudonemacladus.
Neottiaceae Horan. = Orchidaceae Juss.
NEPENTHACEAE **Dumort.** 1829. Dicots. Nepenthales. 1/72. Madagascar, Seychelles, South and Southeast Asia, Malesia, Australia, New Caledonia. Genus Nepenthes.
Nepetaceae Horan. = Lamiaceae Lindl.
NESOGENACEAE **Marais.** 1981. Dicots. Lamiales. 1/7. Trop. East Africa, Madagascar, Seychelles and Mascarene Is., Malesia, Pacific islands. Genus Nesogenes.
Neumanniaceae Tiegh. ex Bullock = Aphloiaceae Takht.
NEURADACEAE **Link.** 1831. Dicots. Rosales. 3/9. Arid Africa, Southwest and South Asia. Genera: Grielum, Neurada, Neuradopsis.
Neuwiediaceae (Burns-Balogh et Funk) R. Dahlgren ex Reveal et Hoogland = Orchidaceae Juss.
Nhandirobaceae T. Lestib. = Cucurbitaceae Juss.
Nicotianaceae Martinov = Solanaceae Juss.
Nigellaceae J. Agardh = Ranunculaceae Juss.
Nipaceae Brongn. ex Martinet = Arecaceae Schultz Sch.
NITRARIACEAE **Bercht. et J. Presl.** 1820. Dicots. Rutales. 1/10. From North Africa and Southeast Europe to East Siberia, Central Asia and southwestern Australia. Genus Nitraria.
NOLANACEAE **Dumort.** 1829. Dicots. Solanales. 1/18. Western South America and Galapagos Is. Genus Nolana.
NOLINACEAE **Nakai.** 1943. Monocots. Asparagales. 3/49. Southwestern U.S., Mexico. Genera: Calibanus, Dasylirion, Nolina.
Nonateliaceae Martinov = Rubiaceae Juss.
Nopaleaceae J. St.-Hil. = Cactaceae Juss.
Noranteaceae T. Post et Kuntze = Marcgraviaceae Choisy.
NOTHOFAGACEAE **Kuprian.** 1962. Dicots. Fagales. 1/35. Southern Hemisphere from New Guinea and temp. Australia to temp. South America. Genus Nothofagus.
Nucamentaceae Hoffmanns. et Link = Asteraceae Dumort.
Nupharaceae A. Kern. = Nymphaeaceae Salisb.
Nuytsiaceae Tiegh. ex Nakai = Loranthaceae Juss.
NYCTAGINACEAE **Juss.** 1789. Dicots. Caryophyllales. 34/350–360. Tropics and subtropics, esp. America, a few species in temp. regions. Genera: Abronia, Acleisanthes, Allionia, Allioniella, Ammocodon, Andradea, Anulocaulis, Belemia, Boerhavia, Boldoa, Bougainvillea, Caribea, Cephalotomandra, Colignonia, Commicarpus, Cryptocarpus, Cuscatlania, Cyphomeris, Grajalesia, Guapira, Leucaster, Mirabilis, Neea, Neeopsis, Nyctaginia, Okenia, Phaeoptilum, Pisonia, Pisoniella, Ramisia, Reichenbachia, Salpianthus, Selinocarpus, Tripterocalyx.
Nyctanthaceae J. Agardh = Oleaceae Hoffmanns. et Link.

NYMPHAEACEAE **Salisb.** 1805. Dicots. Nyphaeales. 5/60. Almost cosmopolitan, in fresh water. Genera: Euryale, Nuphar, Nymphaea, Odinea, Victoria.

Nypaceae Brongn. ex Le Maout et Decne. = Arecaceae Schultz Sch.

NYSSACEAE **Juss. ex Dumort.** 1829. Dicots. Cornales. 2/7. South and Southeast Asia, North and Central America. Genera: Camptotheca, Nyssa.

Obolariaceae Martinov = Gentianaceae Juss.

OCHNACEAE **DC.** 1811. Dicots. Ochnales. 8/370. Tropics and subtropics, esp. South America, West Africa, and Southeast Asia. Genera: Brackenridgea, Charidion, Elvasia, Gomphia, Ochna, Ouratea, Perissocarpa, Rhabdophyllum.

Ochranthaceae Lindl. ex Endl. = Staphyleaceae Lindl.

OCTOKNEMACEAE **Engl.** 1909. Dicots. Santalales. 1/6. Trop. Africa. Genus Octoknema.

Oenotheraceae von Vest = Onagraceae Juss.

OFTIACEAE **Takht. et Reveal.** 1993. Dicots. Scrophulariales. 2/4. South Africa and Madagascar. Genera: Oftia, Ranopisoa.

OLACACEAE **Mirb. ex DC.** 1824. Dicots. Santalales. 27/235. Tropics and subtropics. Genera: Anacolosa, Aptandra, Brachynema, Cathedra, Chaunochiton, Coula, Curupira, Diogoa, Douradoa, Dulacia, Erythropalum, Harmandia, Heisteria, Malania, Minquartia, Ochanostachys, Olax, Ongokea, Petalocaryum, Phanerodiscus, Ptychopetalum, Schoepfia, Scorodocarpus, Strombosia, Strombosiopsis, Tetrastylidium, Ximenia.

OLEACEAE **Hoffmanns. et Link.** 1813–1820. Dicots. Oleales. 26/960–970. Almost cosmopolitan, but chiefly tropics and subtropics. Genera: Abeliophyllum, Chionanthus, Comoranthus, Fontanesia, Forestiera, Forsythia, Fraxinus, Haenianthus, Hesperelaea, Jasminum, Ligustrum, Menodora, Myxopyrum, Nestegis, Noldeanthus, Noronhia, Notelaea, Nyctanthes, Olea, Osmanthus, Phillyrea, Picconia, Schrebera, Syringa, Tessarandra, Tetrapilus.

OLINIACEA **Harv. et Sond.** 1862. Dicots. Myrtales. 1/8. St. Helena, trop. East and South Africa. Genus Olinia.

ONAGRACEAE **Juss.** 1789. Dicots. Myrtales. 18/680. Trop., subtrop., and temp. regions, a few species in cold areas. Genera: Boisduvalia, Burragea, Calylophus, Camissonia, Circaea, Clarkia, Epilobium, Fuchsia, Gaura, Gayophytum, Gongylocarpus, Hauya, Heterogaura, Lopezia, Ludwigia, Oenothera, Stenosiphon, Xylonagra.

ONCOTHECACEAE **Kobuski ex Airy Shaw.** 1965. Dicots. Theales. 1/2. New Caledonia. Genus Oncotheca.

Onosmaceae Martinov = Boraginaceae Juss.

Onosmataceae Horan. = Boraginaceae Juss.

Operculariaceae Dumort. = Rubiaceae Juss.

Ophiopogonaceae Endl. = Convallariaceae Horan.

Ophiraceae Arn. = Grubbiaceae Endl.

Ophrydaceae Raf. = Orchidaceae Juss.

OPILIACEAE **(Benth.) Valeton.** 1886. Dicots. Santalales. 11/36. Tropics and subtropics, esp. Asia. Genera: Agonandra, Cansjera, Champereia, Gjellerupia, Izabalaea, Lepionurus, Melientha, Opilia, Pentarhopalopilia, Rhopalopilia, Urobotrya.

Oporanthaceae Salisb. = Amaryllidaceae J. St.-Hil.

Opuntiaceae Martinov = Cactaceae Juss.

ORCHIDACEAE **Juss.** 1789. Monocots. Orchidales. 867/20,600–21,400. Cosmopolitan, chiefly Southeast Asia and trop. America. Genera: Aa, Abdominea, Acampe, Acanthophippium, Aceras, Aceratorchis, Acianthus, Acineta, Ackermania, Acostaea, Acriopsis, Acrolophia, Acrorchis, Ada, Adactylus, Adenochilus, Adenoncos, Adrorhizon, Aerangis, Aeranthes, Aerides, Aganisia, Aglossorhyncha, Agrostophyllum, Alamania, Allochilus, Ambrella, Amerorchis, Amesiella,

Amitostigma, Amparoa, Amphigena, Amphorkis, Anacamptis, Ancistrochilus, Ancistrorhynchus, Androchilus, Androcorys, Angraecopsis, Angraecum, Anguloa, Anochilus, Anoectochilus, Ansellia, Anthogonium, Anthosiphon, Antillanorchis, Aorchis, Aphyllorchis, Aplectrum, Aporostylis, Apostasia, Appendicula, Aracamunia, Arachnis, Archineottia, Arethusa, Armodorum, Arpophyllum, Arthrochilus, Artorima, Arundina, Asarca, Ascidieria, Ascocentrum, Ascochilopsis, Ascochilus, Ascoglossum, Aspasia, Aspidogyne, Aulosepalum, Auxopus, Barbosella, Barbrodria, Barkeria, Barlia, Barombia, Bartholina, Basigyne, Basiphyllaea, Baskervillea, Batemannia, Beadlea, Beclardia, Beloglottis, Benthamia, Benzingia, Biermannia, Bifrenaria, Binotia, Bipinnula, Bletia, Bletilla, Bogoria, Bollea, Bolusiella, Bonatea, Bonniera, Bothriochilus, Brachionidium, Brachtia, Brachycorythis, Brachypeza, Brachystele, Bracisepalum, Braemia, Brassavola, Brassia, Brenesia, Bromheadia, Broughtonia, Brownleea, Buchtienia, Bulbophyllum, Bulleyia, Burnettia, Cadetia, Caladenia, Calanthe, Caleana, Calochilus, Calopogon, Caluera, Calymmanthera, Calypso, Calyptrochilum, Campylocentrum, Capanemia, Cardiochilos, Catasetum, Cattleya, Cattleyopsis, Caucaea, Caularthron, Centroglossa, Centropetalum, Centrostigma, Cephalanthera, Cephalantheropsis, Ceratandra, Ceratocentron, Ceratochilus, Ceratostylis, Chaenanthe, Chaetocephala, Chamaeangis, Chamaeanthus, Chamaegastrodia, Chamelophyton, Chamorchis, Changnienia, Chaseella, Chaubardia, Chaubardiella, Chauliodon, Cheiradenia, Cheirostylis, Chelonistele, Chiloglottis, Chilopogon, Chiloschista, Chitonanthera, Chitonochilus, Chloraea, Chondradenia, Chondrorhyncha, Chroniochilus, Chrysocycnis, Chrysoglossum, Chysis, Chytroglossa, Cirrhaea, Cischweinfia, Claderia, Cleisocentron, Cleisomeria, Cleisostoma, Cleisostomopsis, Cleistes, Clematepistephium, Clowesia, Coccineorchis, Cochleanthes, Cochlioda, Codonorchis, Codonosiphon, Coelia, Coeliopsis, Coeloglossum, Coelogyne, Cohniella, Coilochilus, Collabiopsis, Collabium, Comparettia, Comperia, Condylago, Constantia, Corallorrhiza, Cordanthera, Cordiglottis, Coryanthes, Corybas, Corycium, Corymborkis, Cottonia, Cranichis, Cremastra, Cribbia, Crocodeilanthe, Crybe, Cryptarrhena, Cryptocentrum, Cryptochilus, Cryptophoranthus, Cryptopus, Cryptopylos, Cryptostylis, Cyanaeorchis, Cybebus, Cyclopogon, Cycnoches, Cymbidiella, Cymbidium, Cynorkis, Cyphochilus, Cypholoron, Cypripedium, Cyrtidiorchis, Cyrtochilum, Cyrtopodium, Cyrtorchis, Cyrtosia, Cystorchis, Dactylorhiza, Dactylorhynchus, Dactylostalix, Deceptor, Degranvillea, Deiregyne, Dendrobium, Dendrochilum, Dendrophylax, Diadenium, Diaphananthe, Diceratostele, Dicerostylis, Dichaea, Dichromanthus, Dickasonia, Dictyophyllaria, Didiciea, Didymoplexiella, Didymoplexis, Diglyphosa, Dignathe, Dilochia, Dilomilis, Dinklageella, Diothonea, Diphylax, Diplandrorchis, Diplocaulobium, Diplocentrum, Diplolabellum, Diplomeris, Diploprora, Dipodium, Dipteranthus, Dipterostele, Disa, Discyphus, Disperis, Distylodon, Dithrix, Dithyridanthus, Diuris, Dodsonia, Domingoa, Doritis, Dossinia, Dracula, Drakaea, Dresslerella, Dressleria, Dryadella, Dryadorchis, Drymoanthus, Drymoda, Duckeella, Dunstervillea, Dyakia, Earina, Eggelingia, Eleorchis, Elleanthus, Eloyella, Eltroplectis, Elythranthera, Embreea, Encyclia, Endresiella, Entomophobia, Eparmatostigma, Ephippianthus, Epiblastus, Epiblema, Epidanthus, Epidendropsis, Epidendrum, Epigeneium, Epilyna, Epipactis, Epipogium, Epistephium, Eria, Eriaxis, Eriochilus, Eriodes, Eriopsis, Erycina, Erythrodes, Erythrorchis, Esmeralda, Euanthe, Eucosia, Eulophia, Eulophiella, Eurycentrum, Eurychone, Eurystyles, Evota, Evotella,

Evrardianthe, Fernándezia, Fimbriella, Flickingeria, Forficaria, Fregea, Frondaria, Fuertesiella, Funkiella, Galeandra, Galeola, Galeottia, Galeottiella, Gamosepalum, Gastorkis, Gastrochilus, Gastrodia, Gastrorchis, Gavilea, Geesinkorchis, Gennaria, Genoplesium, Genyorchis, Geodorum, Glomera, Glossoida, Glossorhyncha, Gomesa, Gomphichis, Gonatostylis, Gongora, Goniochilus, Goodyera, Govenia, Grammangis, Grammatophyllum, Graphorkis, Grobya, Grosourdya, Gunnarella, Gunnarorchis, Gymnadenia, Gymnochilus, Gynoglottis, Habenaria, Hagsatera, Hakoneaste, Hammarbya, Hancockia, Hapalochilus, Hapalorchis, Harrisella, Hederorkis, Helcia, Helleriella, Helonoma, Hemipilia, Hemiscleria, Herminium, Herpisma, Herschelia, Hetaeria, Hetaeria, Hexadesmia, Hexalectris, Hexisea, Himantoglossum, Hintonella, Hippeophyllum, Hirtzia, Hispaniella, Hoehneella, Hofmeisterella, Holcoglossum, Holothrix, Homalopetalum, Horichia, Hormidium, Horvatia, Houlletia, Huntleya, Huttonaea, Hybochilus, Hygrochilus, Hylophila, Hymenorchis, Imerinaea, Inobulbon, Ionopsis, Ipsea, Isabelia, Ischnogyne, Isochilus, Isotria, Jacquiniella, Jamaiciella, Jejosephia, Jumellea, Kalopternix, Kefersteinia, Kegeliella, Kingidium, Kitigorchis, Koellensteinia, Konantzia, Kraenzlinella, Kreodanthus, Kryptostoma, Kuhlhasseltia, Lacaena, Laelia, Laeliopsis, Lanium, Lankesterella, Leaoa, Lecanorchis, Lemboglossum, Lemurella, Lemurorchis, Leochilus, Lepanthes, Lepanthopsis, Lepidogyne, Leporella, Leptotes, Lesliea, Leucohyle, Ligeophila, Limodorum, Lindsayella, Liparis, Listera, Listrostachys, Lockhartia, Loefgrenianthus, Ludisia, Lueddemannia, Luisia, Lycaste, Lycomormium, Lyperanthus, Lyroglossa, Lysiella, Macodes, Macradenia, Macropodanthus, Malaxis, Malleola, Manniella, Margelliantha, Masdevallia, Maxillaria, Mediocalcar, Megalorchis, Megalotus, Megastylis, Meiracyllium, Mesadenella, Mesadenus, Mesoglossum, Mesospinidium, Mexicoa, Mexipedium, Microcoelia, Micropera, Microsaccus, Microtatorchis, Microterangis, Microthelys, Microtis, Miltonia, Miltoniopsis, Mischobulbon, Mobilabium, Monadenia, Monomeria, Monophyllorchis, Mormodes, Mormolyca, Myoxanthus, Myrmechis, Myrmecophila, Myrosmodes, Mystacidium, Nabaluia, Nageliella, Nanodes, Neobartlettia, Neobathiea, Neobenthamia, Neobolusia, Neoclemensia, Neocogniauxia, Neodryas, Neofinetia, Neogardneria, Neogyna, Neomoorea, Neotinea, Neottia, Neottianthe, Neo-urbania, Neowilliamsia, Nephelaphyllum, Nephrangis, Nervilia, Neuwiedia, Nidema, Nigritella, Nothodoritis, Nothostele, Notylia, Oakes-Amesia, Oberonia, Octarrhena, Octomeria, Odontochilus, Odontoglossum, Odontorrhynchus, Oeceoclades, Oeonia, Oeoniella, Oligophyton, Oliveriana, Omoea, Oncidium, Ophidion, Ophrys, Orchipedum, Orchis, Oreorchis, Orestias, Orleanesia, Ornithocephalus, Ornithochilus, Ornithophora, Orthoceras, Osmoglossum, Ossiculum, Otochilus, Otoglossum, Otostylis, Pabstia, Pabstiella, Pachites, Pachyphyllum, Pachyplectron, Pachystoma, Palmorchis, Palumbina, Panisea, Pantlingia, Paphinia, Paphiopedilum, Papilionanthe, Papillilabium, Papperitzia, Papuaea, Paracaleana, Paradisanthus, Paraphalaenopsis, Parapteroceras, Parhabenaria, Pecteilis, Pedilochilus, Pelatantheria, Pelexia, Pennilabium, Peristeranthus, Peristeria, Peristylus, Pescatorea, Phaius, Phalaenopsis, Pholidota, Phragmipedium, Phragmorchis, Phreatia, Phymatidium, Physoceras, Pilophyllum, Pinelianthe, Piperia, Pityphyllum, Platanthera, Platycoryne, Platyglottis, Platylepis, Platyrhiza, Platystele, Platythelys, Plectorhiza, Plectrelminthus, Plectrophora, Pleione, Pleurothallis, Plocoglottis, Poaephyllum, Podangis, Podochilus, Pogonia, Pogoniopsis, Polycycnis, Polyotidium, Polyradicion, Polystachya, Pomatocalpa, Ponera, Ponthieva, Porolabium, Porpax, Porphyrodesme, Porphyroglottis, Porphyrostachys, Porroglossum, Porrorhachis, Prasophyllum, Prescotia, Pristiglottis, Promenaea, Pseudacoridium, Pseuderia, Pseudocentrum, Pseudocranichis, Pseudogoodyera, Pseudolaelia, Pseudorchis, Pseudovanilla, Psilochilus, Psychopsiella, Psygmorchis, Pterichis, Pteroceras, Pteroglossa, Pteroglossaspis, Pterostemma, Pterostylis, Pterygodium, Pygmaeorchis, Quekettia, Quisqueya, Rangaeris, Rauhiella, Raycadenco, Reichenbachanthus, Renanthera, Renantherella, Renata, Restrepia, Restrepiella, Restrepiopsis, Rhaesteria, Rhamphorhynchus, Rhinerrhiza, Rhizanthella, Rhynchogyna, Rhyncholaelia, Rhynchophreatia, Rhynchostylis, Ridleyella, Rimacola, Risleya, Robiquetia, Rodriguezia, Rodrigueziella, Rodrigueziopsis, Roeperocharis, Roezliella, Rossioglossum, Rudolfiella, Rusbyella, Saccoglossum, Saccolabiopsis, Saccolabium, Sacoila, Salpistele, Sanderella, Sarcanthopsis, Sarcochilus, Sarcoglottis, Sarcoglyphis, Sarcophyton, Sarcostoma, Satyridium, Satyrium, Saundrsia, Scaphosepalum, Scaphyglottis, Scelochiloides, Scelochilus, Schaenomorphus, Schiedeella, Schistotylus, Schizochilus, Schizodium, Schlimmia, Schoenorchis, Schomburgkia, Schwartzkopffia, Scuticaria, Sedirea, Seidenfadenia, Selenipedium, Sepalosaccus, Sepalosiphon, Serapias, Sertifera, Sievekingia, Sigmatogyne, Sigmatostalix, Silvorchis, Sinorchis, Sirhookera, Smithorchis, Smithsonia, Smitinandia, Sobennikoffia, Sobralia, Solenangis, Solenidiopsis, Solenidium, Solenocentrum, Sophronitella, Sophronitis, Soterosanthus, Spathoglottis, Sphyrarhynchus, Sphyrastylis, Spiculaea, Spiranthes, Spongiola, Stanhopea, Staurochilus, Stalkya, Stanhopea, Staurochilus, Stauropsis, Stelis, Stellilabium, Stenia, Stenocoryne, Stenoglossum, Stenoglottis, Stenoptera, Stenorrhynchos, Stephanothelys, Stereochilus, Stereosandra, Steveniella, Stictophyllum, Stigmatodactylus, Stigmatorthos, Stigmatosema, Stolzia, Suarezia, Summerhayesia, Sunipia, Sutrina, Symphyglossum, Symphyosepalum, Synanthes, Systeloglossum, Taeniophyllum, Taeniorrhiza, Tainia, Tainiopsis, Tangtsinia, Teagueia, Telipogon, Tetramicra, Teuscheria, Thaia, Thecopus, Thecostele, Thelasis, Thelymitra, Thelyschista, Thrixspermum, Thulinia, Thunia, Thysanoglossa, Ticoglossum, Tipularia, Townsonia, Traunsteinera, Trevoria, Trias, Triceratorhynchus, Trichocentrum, Trichoceros, Trichoglottis, Trichopilia, Trichosalpinx, Trichotosia, Tridactyle, Trigonidium, Triphora, Trisetella, Trizeuxis, Tropidia, Tropilis, Trudelia, Tsaiorchis, Tuberolabium, Tubilabium, Tulotis, Tylostigma, Uleiorchis, Uncifera, Vanda, Vandopsis, Vanilla, Vargasiella, Vasqueziella, Ventricularia, Vexillabium, Vrydagzynea, Warmingia, Warrea, Warreella, Warreopsis, Wullschlaegelia, Xenikophyton, Xerorchis, Xylobium, Yoania, Yolanda, Ypsilopus, Zetagyne, Zeuxine, Zootrophion, Zygopetalon, Zygosepalum, Zygostates.

Orchidanthaceae Dostál = Lowiaceae Ridl.

Ornithogalaceae Salisb. = Hyacinthaceae Batsch.

Ornithropaceae Martinov = Sapindaceae Juss.

OROBANCHACEAE Vent. 1799. Dicots. Scrophulariales. 16/170–225. Temp., subtrop., and trop. regions of the Northern Hemisphere, south to Panama, Ethiopia, and Malesia. Genera: Aeginetia, Boschniakia, Christisonia, Cistanche, Conopholis, Epifagus, Gleadovia, Kopsiopsis, Lathraea, Mannagettaea, Myzorrhiza, Necranthus, Orobanche, Phacellanthus, Phelypaea, Platypholis.

Orontiaceae Bartl. = Araceae Juss.

Oryzaceae Burnett = Poaceae Barnhart.

Osyridaceae Juss. ex Martinov = Santalaceae R. Br.

OXALIDACEAE R. Br. 1817. Dicots. Geraniales. 5–6/700. Trop., subtrop., and temp. regions. Genera: Averrhoa, Biophytum, Dapania, Oxalis, Sarcotheca, ? Trifoliada.

Oxycladaceae (Miers) Schnizl. = Scrophulariaceae Juss.

Oxycoccaceae A. Kern. = Ericaceae Juss.

Oxystylidaceae Hutch. = Capparaceae Juss.

Pachysandraceae J. Agardh = Buxaceae Dumort.

PAEONIACEAE F. Rudolphi. 1830. Dicots. Paeoniales. 1/35–40. Temp. and subtrop. Eurasia, North America. Genus Paeonia.

Pagamaeaceae Martinov = Rubiaceae Juss.

Paivaeusaceae A. Meeuse = Euphorbiaceae Juss.

PALMAE Juss. See ARECACEAE Schultz Sch.

Pancratiaceae Horan. = Amaryllidaceae J. St.-Hil.

PANDACEAE Engl. et Gilg. 1913. Dicots. Euphorbiales. 5/21–22. Trop. Africa, Southeast Asia, Malesia. Genera: Centroplacus, Dicoelia, Galearia, Microdesmis, Panda.

PANDANACEAE R. Br. 1810. Monocots. Pandanales. 3/780. Trop. of the Old World, a few species in temp. regions (China, Japan, and New Zealand). Genera: Freycinetia, Pandanus, Sararanga.

Pangiaceae Endl. = Flacourtiaceae DC.

Panicaceae Voigt = Poaceae Barnhart.

PAPAVERACEAE Juss. 1789. Dicots. Papaverales. 23/325. Temp., warm–temp., and subtrop. Northern Hemisphere (chiefly southwestern U.S., Mediterranean, Asia), South Africa, Australia. Genera: Arctomecon, Agremone, Bocconia, Canbya, Chelidonium, Dendromecon, Dicranostigma, Eomecon, Eschscholzia, Glaucium, Hesperomecon, Hunnemannia, Hylomecon, Macleaya, Meconella, Meconopsis, Papaver, Platystemon, Roemeria, Romneya, Sanguinaria, Stylomecon, Stylophorum.

Papayaceae Blume = Caricaceae Dumort.

Papilionaceae Giseke = Fabaceae Lindl.

Pappophoraceae (Kunth) Herter = Poaceae Barnhart.

Papyraceae Burnett = Cyperaceae Juss.

PARACRYPHIACEAE Airy Shaw. 1965. Dicots. Paracryphiales. 1/1–2. New Caledonia. Genus Paracryphia.

Parianaceae Nakai = Poaceae Barnhart.

Paridaceae Dumort. = Trilliaceae Lindl.

PARNASSIACEAE Gray. 1821. Dicots. Saxifragales. 1/70. Temp. Northern Hemisphere. Genus Parnassia.

Paronychiaceae Juss. = Caryophyllaceae Juss.

Paropsiaceae Dumort. = Passifloraceae Juss. ex Kunth.

Parrotiaceae Horan. = Hamamelidaceae R. Br.

Partheniaceae Link = Asteraceae Dumort.

PASSIFLORACEAE Juss. ex Kunth. 1817. Dicots. Violales. 19/529. Trop., subtrop., and warm–temp. regions. Genera: Adenia, Ancistrothyrsus, Androsiphonia, Barteria, Basananthe, Crossostemma, Deidamia, Dilkea, Efulensia, Hollrungia, Mitostemma, Paropsia, Paropsiopsis, Passiflora, Schlechterina, Tetrapathaea, Tetrastylis, Triphostemma, Viridivia.

Pastinaceae Martinov = Apiaceae Lindl.

Patrisiaceae Mart. = Flacourtiaceae DC.

Paulowniaceae Nakai = Bignoniaceae Juss.

Paviaceae Horan. = Hippocastanaceae DC.

Pectiantiaceae Raf. = Saxifragaceae Juss.

PEDALIACEAE R. Br. 1840. Dicots. Scrophulariales. 13/85. Tropics of the Old World and Brazil (Rogeria). Genera: Ceratotheca, Dicerocaryum, Harpagophytum, Holubia, Josephinia, Linariopsis, Pedaliodiscus, Pedalium, Pterodiscus, Rogeria, Sesamothamnus, Sesamum, Uncarina.

Pediculariaceae Juss. = Scrophulariaceae Juss.

PEGANACEAE Tiegh. ex Takht. 1987. Dicots. Rutales. 2/6. South Europe, North Africa, arid and semiarid areas of Asia and North America. Genera: Malacocarpus, Peganum.

Peliosanthaceae Salisb. = Convallariaceae Horan.

PELLICIERACEAE **(Triana et Planch.) L. Beauvis. ex Bullock.** 1959. Dicots. Theales. 1/1. Central and South America. Genus Pelliciera.

PENAEACEAE **Sweet ex Guill.** 1828. Dicots. Myrtales. 7/21. South Africa. Genera: Brachysiphon, Endonema, Glischrocolla, Penaea, Saltera, Sonderothamnus, Stylapterus.

Pennantiaceae J. Agardh = Icacinaceae (Benth.) Miers.

Pentadiplandraceae Hutch. et Dalziel = Capparaceae Juss.

PENTAPHRAGMATACEAE **J. Agardh.** 1858. Dicots. Campanulales. 1/30. Southern China, Southeast Asia, Malesia. Genus Pentaphragma.

PENTAPHYLACACEAE **Engl.** 1897. Dicots. Theales. 1/1–2. Southern China, Indochina, western Malesia. Genus Pentaphylax.

PENTASTEMONACEAE **Duyfjes.** 1992. Monocots. Dioscoreales. 1/2. Sumatra. Genus Pentastemona.

PENTHORACEAE **Rydb. ex Britton.** 1910. Dicots. Saxifragales. 1/3. East and Southeast Asia, North America. Genus Penthorum.

PEPEROMIACEAE **A. C. Sm.** 1981. Dicots. Piperales. 3/Over 1,000. Tropics and subtropics. Genera: Manekia, Peperomia, Piperanthera.

Peraceae Klotzsch = Euphorbiaceae Juss.

Perdiciaceae Link = Asteraceae Dumort.

PERIDISCACEAE **Kuhlm.** 1950. Dicots. Violalaes. 2/2. Trop. South America. Genera: Peridiscus, Whittonia.

Periplocaceae Schltr. = Asclepiadaceae R. Br.

Peripterygiaceae F. N. Williams = Cardiopteridaceae Blume.

Perseaceae Horan. = Lauraceae Juss.

Persicariaceae Adans. ex T. Post et Kuntze = Polygonaceae Juss.

PETERMANNIACEAE **Hutch.** 1934. Monocots. Smilacales. 1/1. Eastern Australia. Genus Petermannia.

PETIVERIACEAE **C. Agardh.** 1825. Docots. Caryophyllales. 11/40. Tropics and subtropics, esp. Southern Hemisphere. Genera: Flueckigera, Gallesia, Hilleria, Lophiocarus, Microtea, Monococcus, Petiveria, Rivina, Schindleria, Seguieria, Trichostigma.

Petreaceae J. Agardh = Verbenaceae J. St.-Hil.

Petrosaviaceae Hutch. = Melanthiaceae Batsch.

Phalaridaceae Burnett = Poaceae Barnhart.

Phaleriaceae Meisn. = Thymelaeaceae Juss.

Pharaceae (Stapf) Herter = Poaceae Barnhart.

Pharnaceaceae Martinov = Aizoaceae F. Rudolphi.

Phaseolaceae Schnizl. = Fabaceae Lindl.

PHELLINACEAE **Takht.** 1987. Dicots. Celastrales. 1/10. New Caledonia. Genus Phelline.

Philadelphaceae Martinov = Hydrangeaceae Dumort.

PHILESIACEAE **Dumort.** 1829. Monocots. Smilacales. 2/2. Chile. Genera: Lapageria, Philesia.

Philippodendraceae Endl. = Malvaceae Juss.

Philocrenaceae Bong. = Podostemaceae Rich. ex C. Agardh.

PHILYDRACEAE **Link.** 1821. Monocots. Philydrales. 3/60. East and Southeast Asia, Malesia, Australia. Genera: Holmholtzia, Philydrella, Philydrum.

Phoenicaceae Schultz Sch. = Arecaceae Schultz Sch.

Phoradendraceae Karst. = Viscaceae Batsch.

PHORMIACEAE **J. Agardh.** 1858. Monocots. Amaryllidales. 8/40. Chiefly Southern Hemisphere from trop. East Africa to Hawaiian Is. and Fiji, South America. Genera: Agrostocrinum, Dianella, Excremis, Phormium, Rhuacophila, Stypandra, Thelionema, Xeronema.

Phrymaceae Schauer = Verbenaceae J. St.-Hil.

Phylicaceae J. Agardh = Rhamnaceae Juss.

Phyllanthaceae J. Agardh = Euphorbiaceae Juss.

Phyllocladaceae C. E. Bessey. 1907. Pinopsida. Podocarpales. 1/5. Malesia from Borneo to New Guinea, Tasmania, New Zealand. Genus Phyllocladus.

Phyllonomataceae Small = Dulongiaceae J. Agardh.

Physenaceae Takht. 1985. Dicots. Sapindales. 1/2. Madagascar. Genus Physena.

Phytelephantaceae Brongn. ex Martinet = Arecaceae Schultz Sch.

Phytocrenaceae Arn. ex R. Br. = Icacinaceae (Benth.) Miers.

Phytolaccaceae R. Br. 1818. Dicots. Caryophyllales. 4/32. Tropics and subtropics, esp. America. Genera: Anisomeria, Ercilla, Nowickea, Phytolacca.

Picridaceae Martinov = Asteraceae Dumort.

Picrodendraceae Small ex Britton et Millsp. = Euphorbiaceae Juss.

Pilocarpaceae J. Agardh = Rutaceae Juss.

Pinaceae Lindl. 1836. Pinopsida. Pinales. 13/200. Northern Hemisphere southward to Sumatra, Java, Central America and West Indies. Genera: Abies, Cathaya, Cedrus, Ducampopinus, Hesperopeuce, Keteleeria, Larix, Nothotsuga, Picea, Pinus, Pseudolarix, Pseudotsuga, Tsuga.

Pinguiculaceae Dumort. = Lentibulariaceae Rich.

Piperaceae C. Agardh. 1824. Dicots. Piperales. 7/over 2,000. Pantropics. Genera: Lindeniopiper, Macropiper, Ottonia, Piper, Sarcorhachis, Trianaeopiper, Zippelia.

Piriquetaceae Martinov = Theaceae D. Don.

Pisoniaceae J. Agardh = Nyctaginaceae Juss.

Pistaciaceae Mart. ex Caruel = Anacardiaceae Lindl.

Pistaceae Rich. ex C. Agardh = Araceae Juss.

Pittosporaceae R. Br. 1814. Dicots. Pittosporales. 11/240. Tropics and subtropics of the Old World, esp. Australia. Genera: Bentleya, Billardiera, Bursaria, Cheiranthera, Citriobatus, Hymenosporum, Marianthus, Pittosporum, Pronaya, Rhytidosporum, Sollya.

Plagianthaceae J. Agardh = Malvaceae Juss.

Plagiopteraceae Airy Shaw. 1965. Dicots. Malvales. 1/2. Southern India, China, Burma, Thailand. Genus Plagiopteron.

Plantaginaceae Juss. 1789. Dicots. Scrophulariales. 3/270. Cosmopolitan. Genera: Bougueria, Littorella, Plantago.

Platanaceae T. Lestib. ex Dumort. 1829. Dicots. Hamamelidales. 1/7–8. North America, Eurasia from South Europe to western Himalayas, Indochina. Genus Platanus.

Platycaryaceae Nakai = Juglandaceae A. Rich. ex Kunth.

Platymetraceae Salisb. = Convallariaceae Horan.

Platystemonaceae A. C. Sm. = Papaveraceae Juss.

Plocospermataceae Hutch. 1973. Dicots. Gentianales. 2/2. Southern Mexico and Guatemala. Genera: Lithophytum, Plocosperma.

Plumbaginaceae Juss. 1789. Dicots. Plumbaginales. 28/490. Almost all over, chiefly in arid and maritime habitats. Genera: Acantholimon, Aegialitis, Afrolimon, Armeria, Bakerolimon, Bamiania, Bukiniczia, Cephalorhizum, Ceratostigma, Chaetolimon, Dictyolimon, Dyerophytum, Eremolimon, Ghaznianthus, Gladiolimon, Goniolimon, Ikonnikovia, Limoniastrum, Limoniopsis, Limonium, Muellerolimon, Neogontschrovia, Plumbagella, Plumbago, Popoviolimon, Psylliostachys, ? Taxanthema, Vassilczenkoa.

Plumeriaceae Horan. = Apocynaceae Juss.

Poaceae Barnhart. 1895, or Gramineae Juss.1789. Monocots. Poales. 857/10,500–11,000. Cosmopolitan. Genera: Achnatherum, Aciachne, Acidosasa, Acostia, Acrachne, Acritochaete, Acroceras, Actinocladum, Aegilops, Aegopogon, Aeluropus, Afrotrichloris, Agenium, Agropyron, Agropyropsis, Agrostis, Aira, Airopsis, Alloeochaete, Allolepis, Alloteropsis, Alopecurus, Alvimia, Amblyopyrum, Ammochloa, Ammophila, Ampelodesmos, Amphibromus, Amphicarpum, Amphipogon, Anadelphia, Ancistrachne, Ancistragrostis, Ancistrochloa, Andropogon, Andropterum, Anemanthele, Ani sachne, Anisantha, Aniselytron, Anisopogon, Anomochloa, Anthaenantia, Anthaenantiopsis, Anthephora, Anthochloa, Anthoxanthum, Antinoria, Antonella, Apera, Aphanelytrum, Apluda, Apochiton, Apoclada, Apocopis, Arberella, Arcangelina, Arctagrostis, Arctophila, Arctopoa, Argopogon, Aristavena, Aristida, Arrhenatherum, Arthragrostis, Arthraxon, Arthropogon, Arthrostylidium, Arundinaria, Arundinella, Arundo, Arundoclaytonia, Asthenochloa, Astrebla, Athroostachys, Atractantha, Atractocarpa, Aulonemia, Australopyrum, Austrochloris, Austrofestuca, Avellinia, Avena, Avenula, Axonopus, Bambusa, Baptorhachis, Beckeropsis, Beckmannia, Bellardiochloa, Bewsia, Bhidea, Blakeochloa, Blepharidachne, Blepharoneuron, Boissiera, Boriskellera, Bothriochloa, Bouteloua, Brachiaria, Brachyachne, Brachychloa, Brachyelytrum, Brachypodium, Brachystachyum, Briza, Brizochloa, Bromidium, Bromopsis, Bromuniola, Bromus, Brylkinia, Buchloe, Buchlomimus, Buergersiochloa, Calamagrostis, Calamovilfa, Calderonella, Calyptochloa, Camusia, Camusiella, Capillipedium, Castellia, Catabrosa, Catabrosella, Catalepis, Catapodium, Cathestecum, Cenchrus, Centotheca, Centrochloa, Centropodia, Cephalostachyum, Ceratochloa, Chaboissaea, Chaetium, Chaetobromus, Chaetopoa, Chaetopogon, Chaetostichium, Chaetotropis, Chamaeraphis, Chandrasekharania, Chascolytrum, Chasechloa, Chasmanthium, Chasmopodium, Chevalierella, Chikusichloa, Chimonobambusa, Chionachne, Chionochloa, Chloachne, Chloris, Chlorocalymma, Chondrosum, Chrysochloa, Chrysopogon, Chumsriella, Chusquea, Cinna, Cinnagrostis, Cladoraphis, Clausospicula, Cleistachne, Cleistochloa, Cleistogenes, Cockaynea, Coelachne, Coelachyrum, Coelorachis, Coix, Colanthelia, Coleanthus, Colpodium, Commelinidium, Cornucopiae, Cortaderia, Corynephorus, Cottea, Craspedorhachis, Criciuma, Crinipes, Crithopsis, Crossotropis, Crypsis, Cryptochloa, Ctenium, Ctenopsis, Cutandia, Cyathopus, Cyclostachya, Cymbopogon, Cymbosetaria, Cynodon, Cynosurus, Cyperochloa, Cyphochlaena, Cypholepis, Cyrtococcum, Dactylis, Dactyloctenium, Daknopholis, Danthonia, Danthoniastrum, Danthonidium, Danthoniopsis, Dasypyrum, Davidsea, Decaryella, Decaryochloa, Dendrocalamus, Dendrochloa, Deschampsia, Desmazeria, Desmostachya, Diandranthus, Diandrochloa, Diandrolyra, Diandrostachya, Diarrhena, Dichaetaria, Dichanthium, Dichelachne, Dictyochloa, Diectomis, Dielsiochloa, Digastrium, Digitaria, Digitariella, Digitariopsis, Dignathia, Diheteropogon, Dilophotriche, Dimeria, Dimorphochloa, Dinebra, Dinochloa, Diplachne, Diplopogon, Disakisperma, Dissanthelium, Dissochondrus, Distichlis, Drake-Brockmania, Dryopoa, Dupontia, Duthiea, Dybowskia, Eccoptocarpha, Echinaria, Echinochloa, Echinolaena, Echinopogon, Ectrosia, Ectrosiopsis, Ehrharta, Ekmanochloa, Eleusine, Elionurus, Elthemidea, Elymandra, Elymus, Elytrigia, Elytrophorus, Elytrostachys, Enneapogon, Enteropogon, Entolasia, Entoplocamia, Epicampes, Eragrostiella, Eragrostis, Eremitis, Eremocaulon, Eremochloa, Eremopoa, Eremopogon, Eremopyrum, Eriachne, Erianthecium, Erianthus, Eriochloa, Eriochrysis, Eriocoma, Erioneuron, Erythranthera, Euclasta, Eulalia, Eulaliopsis, Eustachys, Euthryptochloa, Exotheca, Farrago, Ferrocalamus, Festuca, Festucella, Festucopsis, Fingerhuthia, Froesiochloa, Garnotia, Gastridium, Gaudinia,

Gaudinopsis, Germainia, Gerritea, Gigantochloa, Gilgiochloa, Glaziophyton, Glyceria, Glyphochloa, Gossweilerochloa, Gouinia, Gouldochloa, Graphephorum, Greslania, Griffithsochloa, Guadua, Guaduella, Gymnachne, Gymnopogon, Gynerium, Habrochloa, Hackelochloa, Hakonechloa, Halopyrum, Harpachne, Harpochloa, Helictotrichon, Hellerochloa, Hemarthria, Hemimunroa, Hemisorghum, Henrardia, Hesperochloa, Heterachne, Heteranthelium, Heteranthoecia, Heterocarpha, Heteropholis, Heteropogon, Hickelia, Hierochloe, Hilaria, Hitchcockella, Holcolemma, Holcus, Homoiachne, Homolepis, Homopholis, Homopogon, Homozeugos, Hookerochloa, Hordelymus, Hordeum, Hubbardia, Hubbardochloa, Humbertochloa, Hyalopoa, Hydrothauma, Hygrochloa, Hygroryza, Hylebates, Hymenachne, Hyparrhenia, Hyperthelia, Hypogynium, Hypseochloa, Hystrix, Ichnanthus, Imperata, Indocalamus, Indochloa, Indopoa, Indosasa, Isachne, Isalus, Ischaemum, Ischnochloa, Ischnurus, Iseilema, Ixophorus, Jacquesfelixia, Jansenella, Jardinea, Jouvea, Kampochloa, Kaokochloa, Karroochloa, Kengyilia, Kerriochloa, Klemachloa, Koeleria, Lachnagrostis, Lachnochloa, Lagurus, Lamarckia, Lamprothyrsus, Lasiacis, Lasiorrhachis, Lasiurus, Lecomtella, Leersia, Leleba, Lepargochloa, Leptagrostis, Leptaspis, Leptocanna, Leptocarydion, Leptochloa, Leptocoryphium, Leptoloma, Leptophyllochloa, Leptosaccharum, Leptothrium, Lepturella, Lepturidium, Lepturopetium, Lepturus, Lerchenfeldia, Leucophrys, Leucopoa, Leymus, Libyella, Limnas, Limnodea, Limnopoa, Lindbergella, Lingnania, Linkagrostis, Lintonia, Lithachne, Littledalea, Loliolum, Lolium, Lophacme, Lophatherum, Lophochlaena, Lopholepis, Lophopogon, Loudetia, Loudetiopsis, Louisiella, Loxodera, Luziola, Lycochloa, Lycurus, Lygeum, Maclurolyra, Macrobriza, Macrochloa, Maillea, Maltebrunia, Manisuris, Megalachne, Megaloprotachne, Megastachya, Melanocenchris, Melica, Melinis, Melocalamus, Melocanna, Menstruocalamus, Meringurus, Merostachys, Merxmuellera, Mesosetum, Metasasa, Metcalfia, Mezochloa, Mibora, Micraira, Microbriza, Microcalamus, Microchloa, Microlaena, Micropyropsis, Micropyrum, Microstegium, Mildbraediochloa, Milium, Miscanthidium, Miscanthus, Mitwabachloa, Mnesithea, Mniochloa, Molineriella, Molinia, Moliniopsis, Monachather, Monanthochloe, Monelytrum, Monerma, Monium, Monochadus, Monocymbium, Monodia, Monroa, Mosdenia, Muantijamvella, Muhlenbergia, Myriocladus, Myriostachya, Nabelekia, Narduroides, Nardurus, Nardus, Narenga, Nassella, Nastus, Neesiochloa, Nematopoa, Neobouteloua, Neohouzeaua, Neohusnotia, Neomolinia, Neoschischkinia, Neostapfia, Neostapfiella, Nephelochloa, Neurachne, Neurolepis, Neuropoa, Nevskiella, Neyraudia, Normanboria, Notochloe, Notodanthonia, Ochlandra, Ochtochloa, Odontelytrum, Odyssea, Olmeca, Olyra, Ophiuros, Opizia, Oplismenopsis, Oplismenus, Orcuttia, Oreiostachys, Oreobambos, Oreochloa, Orinus, Oropetium, Ortachne, Orthoclada, Orthoraphium, Oryza, Oryzidium, Oryzopsis, Otachyrium, Ottochloa, Oxychloris, Oxyrhachis, Oxytenanthera, Panicum, Pappagrostis, Pappophorum, Paracolpodium, Paractaenum, Parafestuca, Parahyparrhenia, Paraneurachne, Parapholis, Paratheria, Pariana, Parodiolyra, Parvotrisetum, Pascopyrum, Paspalidium, Paspalum, Patis, Pennisetum, Pentameris, Pentapogon, Pentarrhaphis, Pentaschistis, Pentatherum, Pereilema, Periballia, Perotis, Perrierbambus, Petriella, Petrina, Peyritschia, Phacelurus, Phaenanthoecium, Phaenosperma, Phalaris, Phalaroides, Phanopyrum, Pharus, Pheidochloa, Phippsia, Phleum, Pholiurus, Phragmites, Phyllorachis, Phyllostachys, Pilgerochloa, Piptatherum, Piptochaetium, Piptophyllum, Piptosta chya, Piresia, Plagiantha, Plagiochloa, Plagiosetum, Plectrachne, Pleiadelphia, Pleuropogon, Plinthanthesis, Poa, Poagrostis, Podophorus, Poecilostachys, Pogonachne, Pogonarthria, Pogonatherum, Pogonochloa, Pogononeura, Pohlidium, Polevansia, Polliniopsis, Polyanthus, Polypogon, Polytoca, Polytrias, Pommereulla, Porteresia, Potamophila, Pringleochloa, Prionanthium, Prosphytochloa, Psammagrostis, Psammochloa, Psathyrostachys, Pseudanthistiria, Pseudarrhenatherum, Pseudechinolaena, Pseudobromus, Pseudochaetochloa, Pseudocoix, Pseudodanthonia, Pseudodichanthium, Pseudolasiacis, Pseudopentameris, Pseudophleum, Pseudopogonatherum, Pseudoraphis, Pseudosasa, Pseudosorghum, Pseudostachyum, Pseudovossia, Pseudozoysia, Psilathera, Psilolemma, Psilurus, Pterochloris, Ptilagrostis, Puccinellia, Puelia, Pyrrhanthera, Racemobamnos, Raddia, Raddiella, Raimundochloa, Ramosia, Rattraya, Ratzeburgia, Redfieldia, Reederochloa, Rehia, Reimarochloa, Reitzia, Relchela, Rendlia, Reynaudia, Rhipidocladum, Rhizocephalus, Rhombolytrum, Rhynchelytrum, Rhynchoryza, Rhytachne, Richardsiella, Robynsiochloa, Rostraria, Rottboellia, Saccharum, Sacciolepis, Sartidia, Sasa, Sasaella, Sasamorpha, Saugetia, Schaffnerella, Schedonnardus, Schismus, Schizachne, Schizachyrium, Schizostachyum, Schmidtia, Schoenefeldia, Sclerachne, Sclerandrium, Sclerochloa, Sclerodactylon, Scleropoa, Scleropogon, Sclerostachya, Scolochloa, Scribneria, Scrotochloa, Scutachne, Secale, Sehima, Semiarundinaria, Senisetum, Sesleria, Setaria, Setariopsis, Setiacis, Shibataea, Sieglingia, Silentvalleya, Simplicia, Sinarundinaria, Sinobambusa, Sinocalamus, Sinochasea, Sitanion, Snowdenia, Soderstromia, Sohnsia, Solenachne, Sorghastrum, Sorghum, Spartina, Spartochloa, Spathia, Sphaerobambos, Sphaerocaryum, Spheneria, Sphenopholis, Sphenopus, Spinifex, Spodiopogon, Sporobolus, Steinchasma, Steirachne, Stenofestuca, Stenotaphrum, Stephanachne, Stereochlaena, Steyermarkochloa, Stipa, Stipagrostis, Streblochaete, Strephium, Streptochaeta, Streptogyna, Streptolophus, Streptostachys, Styppeiochloa, Sucrea, Suddia, Swallenia, Swallenochloa, Symplectrodia, Taeniatherum, Taeniorhachis, Tanzaniochloa, Tarigidia, Tatianyx, Teinostachyum, Tetrachaete, Tetrachne, Tetrapogon, Tetrarrhena, Thamnocalamus, Thaumastochloa, Thelepogon, Themeda, Thrasya, Thrasyopsis, Thuarea, Thyridachne, Thyridolepis, Thyrsia, Thyrsostachys, Thysanolaena, Torreyochloa, Tovarochloa, Trachynia, Trachypogon, Trachys, Tragus, Triavenopsis, Tribolium, Trichachne, Trichloris, Tricholaena, Trichoneura, Trichopteryx, Tridens, Trikeraia, Trilobachne, Triniochloa, Triodia, Triplachne, Triplasis, Triplopogon, Tripogon, Tripsacum, Triraphis, Triscenia, Trisetaria, Trisetum, Tristachya, Triticum, Tuctoria, Tzvelevia, Uniola, Uranthoecium, Urelytrum, Urochlaena, Urochloa, Urochondra, Vahlodea, Vaseyochloa, Ventenata, Veseyochloa, Vetiveria, Vietnamosasa, Viguierella, Vossia, Vulpia, Vulpiella, Wangenheimia, Whiteochloa, Willbleibia, Willkommia, Woodrowia, Xerochloa, Xerodanthia, Yakirra, Yvesia, Zea, Zenkeria, Zeugites, Zingeria, Zingeriopsis, Zizania, Zizaniopsis, Zonotriche, Zoysia, Zygochloa.

PODOCARPACEAE Endl. 1847. Pinopsida. Podocarpales. 16/172. Southern Hemisphere North to Central America, West Indies, northern Iran and Japan. Genera: Acmopyle, Afrocarpus, Dacrycarpus, Dacrydium, Falcatifolium, Holocarpus, Lagarostrobos, Lepidothamnus, Microcachrys, Microstrobos, Parasitaxus, Podocarpus, Prumnopitys, Retrophyllum, Saxegothaea, Sundacarpus.

PODOONACEAE Baill. ex Franch. 1889. Dicots. Rutales. 2/3. Himalayas, India,

China, Thailand. Genera: Campylopetalum, Dobinea.

Podophyllaceae DC. = Berberidaceae Juss.

PODOSTEMACEAE Rich. ex C. Agardh. 1822. Dicots. Podostemales. 54/260. Tropics. Genera: Angolaea, Apinagia, Blandowia, Butumia, Carajaea, Castelnavia, Ceratolacis, Cladopus, Crenias, Dalzellia, Devillea, Dicraeanthus, Diplobryum, Djinga, Endocaulos, Farmeria, Hanseniella, Heterotristicha, Hydrobryopsis, Hydrobryum, Indotristicha, Jenmaniella, Ledermanniella, Leiothylax, Letestuella, Lonchostephus, Lophogyne, Macarenia, Macropodiella, Maferria, Malaccotristicha, Marathrum, Mourera, Oserya, Paleodicraeia, Podostemum, Polypleurum, Rhyncholacis, Saxicolella, Sphaerothylax, Stonesia, Synstylis, Terniopsis, Thelethylax, Torrenticola, Tristicha, Tulasneantha, Vanroyenella, Weddellina, Wettsteiniola, Willisia, Winklerella, Zehnderia, Zeylanidium.

POLEMONIACEAE Juss. 1789. Dicots. Polemoniales. 21/330. Temp. regions of the north hemisphere, a few species in South America (Chile). Genera: Acanthogilia, Allophyllum, Bonplandia, Cantua, Cobaea, Collomia, Eriastrum, Gilia, Giliastrum, Gymnosteris, Huthia, Ipomopsis, Langloisia, Leptodactylon, Linanthus, Loeselia, Loeseliastrum, Microsteris, Navarretia, Phlox, Polemonium.

Polpodaceae Nakai = Molluginaceae Hutch.

POLYGALACEAE R. Br. 1814. Dicots. Polygalales. 22/1,040. Almost cosmopolitan. Genera: Ancylotropis, Atroxima, Badiera, Balgoya, Barnhartia, Bredemeyera, Carpolobia, Comesperma, Diclidanthera, Epirixanthes, Eriandra, Monnina, Monrosia, Moutabea, Muraltia, Nylandtia, Phlebotaenia, Polygala, Pteromonnina, Salomonia, Securidaca, Xanthophyllum.

POLYGONACEAE Juss. 1789. Dicots. Polygonales. 49/1,170. Cosmopolitan, chiefly northern temp. regions. Genera: Afrobrunnichia, Antigonon, Aristocapsa, Atraphaxis, Brunnichia, Calligonum, Centrostegia, Chorizanthe, Coccoloba, Dedeckera, Dodecahema, Emex, Eriogonella, Eriogonum, Fagopyrum, Fallopia, Gilmania, Goodmania, Gymnopodium, Harfordia, Harpagocarpus, Hollisteria, Homalocladium, Knorringia, Koenigia, Lastarriaea, Leptogonum, Mucronea, Muehlenbeckia, Nemacaulis, Neomillspaughia, Oxygonum, Oxyria, Oxytheca, Parapteropyrum, Persicaria, Podopterus, Polygonella, Polygonum, Pteropyrum, Pterostegia, Pteroxygonum, Rheum, Rumex, Ruprechtia, Stenogonum, Symmeria, Systenotheca, Triplaris.

Polygonanthaceae Croizat = Anisophyllaceae Ridl.

Polygonataceae Salisb. = Convallariaceae Horan.

Polyosmaceae Blume = Escalloniaceae R. Br. ex Dumort.

Pomaceae Gray = Rosaceae Juss.

PONTEDERIACEAE Kunth. 1816. Monocots. Pontederiales. 8/35. Trop., subtrop., and warm–temp. regions. Genera: Eichhornia, Eurystemon, Heteranthera, Hydrothrix, Monochoria, Pontederia, Scholleropsis, Zosterella.

Poranaceae J. Agardh = Convolvulaceae Juss.

Porantheraceae (Pax) Hurusawa = Euphorbiaceae Juss.

PORTULACACEAE Juss. 1789. Dicots. Caryophyllales. 31/400. Cosmopolitan, but chiefly in temp. and warm–temp. regions, esp. western North America. Genera: Amphipetalum, Anacampseros, Baitaria, Calandrinia, Calyptridium, Calyptrotheca, Ceraria, Cistanthe, Claytonia, Claytoniella, Erocallis, Grahamia, Lenzia, Lewisia, Limnalsine, Maxia, Mona, Monocosmia, Montia, Montiastrum, Montiopsis, Naiocrene, Neopaxia, Portulaca, Portulacaria, Rumicastrum, Schreiteria, Talinaria, Talinella, Talinopsis, Talinum.

POSIDONIACEAE Hutch. 1934. Monocots. Posidoniales. 1/3–5. Mediterranean, southern Australia, Tasmania. Genus Posidonia.

Potaliaceae Mart. = Gentianaceae Juss.

POTAMOGETONACEAE **Dumotr.** 1829. Monocots. Potamogetonales. 2/c. 100. Cosmopolitan. Genera Groenlandia, Potamogeton.

Potentillaceae Perleb. = Rosaceae Juss.

Poteriaceae Raf. = Rosaceae Juss.

Pothoaceae Raf. = Araceae Juss.

POTTINGERIACEAE **Takht.** 1987. Dicots. Hydrangeales. 1/1. Northeastern India, Burma, Thailand. Genus Pottingeria.

PRIMULACEAE **Vent.** 1799. Dicots. Primulales. 23/870. Cosmopolitan, but chiefly cold and temp. Northern Hemisphere. Genera: Anagallis, Androsace, Ardisiandra, Asterolinon, Bryocarpum, Coris, Cortusa, Cyclamen, Dionysia, Dodecatheon, Glaux, Hottonia, Kaufmannia, Lysimachia, Omphalogramma, Pelletiera, Pomatosace, Primula, Samolus, Soldanella, Sredinskya, Stimpsonia, Trientalis.

Prionotaceae Hutch. = Epacridaceae R. Br.

Prockiaceae Bertuch = Flacourtiaceae DC.

PROTEACEAE **Juss.** 1789. Dicots. Proteales. 77/1,500. Tropics and subtropics, chiefly Southern Hemisphere. Genera: Acidonia, Adenanthos, Agastachys, Alloxylon, Athertonia, Aulax, Austromuellera, Banksia, Beauprea, Beaupreopsis, Bellendena, Bleasdalea, Brabejum, Buckinghamia, Cardwellia, Carnarvonia, Cenarrhenes, Conospermum, Darlingia, Diastella, Dilobeia, Dryandra, Embothrium, Euplassa, Faurea, Finschia, Floydia, Franklandia, Garnieria, Gevuina, Grevillea, Hakea, Helicia, Heliciopsis, Hicksbeachia, Hollandaea, Isopogon, Kermadecia, Knightia, Lambertia, Leucadendron, Leucospermum, Lomatia, Macadamia, Malagasia, Mimetes, Musgravea, Neorites, Opisthiolepis, Oreocallis, Orites, Orothamnus, Panopsis, Paranomus, Persoonia, Petrophile, Placospermum, Protea, Pycnonia, Roupala, Serruria, Sleumerodendron, Sorocephalus, Spatalla, Sphalmium, Stenocarpus, Stirlingia, Strangea, Symphionema, Synaphea, Telopea, Toronia, Triunia, Turrillia, Vexatorella, Vitora, Xylomelum.

Protoliriaceae Makino = Melanthiaceae Batsch.

Prunaceae Bercht. et J. Presl = Rosaceae Juss.

Pseudanthaceae Endl. = Euphorbiaceae Juss.

Pseudodophoenicaceae O. F. Cook = Arecaceae Schultz Sch.

PSILOXYLACEAE **Croizat.** 1960. Dicots. Myrtales. 1/1. Mascarene Is. Genus Psiloxylon.

Psittacanthaceae Nakai = Loranthaceae Juss.

Psychotriaceae F. Rudolphi = Rubiaceae Juss.

PTAEROXYLACEAE **J.-F. Leroy.** 1960. Dicots. Rutales. 2/4–5. Trop. East and South Africa, Madagascar. Genera: Cedrelopsis, Ptaeroxylon.

Pteleaceae Kunth = Rutaceae Juss.

PTERIDOPHYLLACEAE **Sugiura ex Nakai.** 1943. Dicots. Papaverales. 1/1. Japan. Genus Pteridophyllum.

Pterisanthaceae J. Agardh = Vitaceae Juss.

Pterocaryaceae Nakai = Juglandaceae A. Rich. ex Kunth.

PTEROSTEMONACEAE **Small.** 1905. Dicots. Hydrangeales. 1/2. Mexico. Genus Pterostemon.

PUNICACEAE **Horan.** 1834. Dicots. Myrtales. 1/2. Southeast Europe, West Asia to western Himalayas, Socotra. Genus Pinica.

Putranjivaceae Endl. = Euphorbiaceae Juss.

Pyraceae Burnett = Rosaceae Juss.

Pyrenaceae Vent. = Verbenaceae J. St.-Hil.

Pyrolaceae Dumort. = Ericaceae Juss.

Quassiaceae Bertol. = Simaroubaceae DC.

Quercaceae Martinov = Fagaceae Dumort.

QUIINACEAE **Choisy ex Engl.** 1888. Dicots.

Ochnales. 4/44. Trop. America. Genera: Froesia, Lacunaria, Quiina, Touroulia.

Quillaiaceae D. Don = Rosaceae Juss.

RAFFLESIACEAE **Dumort.** 1829. Dicots. Rafflesiales (Cytinales). 3/15–18. Northeastern India, southern China, Indochina, Malesia. Genera: Rafflesia, Rhizanthes, Sapria.

Ramondaceae Godron et Grenier = Gesneriaceae Dumort.

Randiaceae Martinov = Rubiaceae Juss.

RANUNCULACEAE **Juss.** 1789. Dicots. Ranunculales. 63/c. 2,000. Cosmopolitan, but chiefly cold and temp. regions. Genera: Aconitum, Actaea, Adonis, Anemoclema, Anemone, Anemonopsis, Aphanostemma, Aquilegia, Archiclematis, Arcteranthis, Arsenjevia, Asteropyrum, Barneoudia, Beesia, Calathodes, Callianthemoides, Callianthemum, Caltha, Ceratocephala, Cimicifuga, Clematis, Clematopsis, Consolida, Coptis, Cyrtorhyncha, Delphinium, Dichocarpum, Enemion, Eranthis, Ficaria, Garidella, Halerpestes, Hamadryas, Helleborus, Hepatica, Isopyrum, Jurtsevia, Knowltonia, Komaroffia, Krapfia, Kumlienia, Laccopetalum, Leptopyrum, Megaleranthis, Metanemone, Miyakea, Myosurus, Naravelia, Nigella, Oreithales, Oxygraphis, Paraquilegia, Paropyrum, Paroxygrapis, Pulsatilla, Ranunculus, Semiaquilegia, Souliea, Thalictrum, Trautvetteria, Trollius, Urophysa, Xanthorhiza.

RANZANIACEAE **Takht.** 1994. Dicots. Ranunculales. 1/1. Japan. Genus Ranzania.

RAPATEACEAE **Dumort.** 1829. Monocots. Commelinales. 17/85. Tropics and subtropics. Genera: Amphiphyllum, Cephalostemon, Duckea, Epidryos, Guacamaya, Kunhardtia, Marahuacaea, Maschalocephalus, Monotrema, Phelpsiella, Potarophytum, Rapatea, Saxofridericia, Schoenocephalium, Spathanthus, Stegolepis, Windsorina.

Raphanaceae Horan. = Brassicaceae Burnett.

Reaumuriaceae Ehrenb. ex Lindl. = Tamaricaceae Link.

RESEDACEAE **DC. ex Gray.** 1821. Dicots. Capparales. 6/85. Southwestern North America, Macaronesia, Eurasia eastward to West Siberia, China and Himalayas, Africa, Socotra Genera: Caylusea, Ochradenus, Oligomeris, Randonia, Reseda, Sesamoides.

RESTIONACEAE **R. Br.** 1810. Monocots. Restionales. 38–39/380. South trop. and South Africa, Madagascar, Indochina, Malay Peninsula, New Guinea, Australia, Tasmania, New Zealand, temp. South America. Genera: Alexgeorgea, Anarthria, Anthochortus, Askidiosperma, Calopsis, Calorophus, Cannomois, Ceratocaryum, Chaetanthus, Chondropetalum, Coleocarya, Dielsia, Dovea, Elegia, Empodisma, Harperia, Hopkinsia, Hydrophilus, Hypodiscus, Hypolaena, Ischyrolepis, Lepidobolus, Leptocarpus, Lepyrodia, Loxocarya, Lyginia, Mastersiella, Meeboldina, Nevillea, Onychosepalum, Platycaulos, ? Pseudoloxocarya, Restio, Rhodocoma, Sporadanthus, Staberoha, Thamnochortus, Willdenowia, Winifredia.

RETZIACEAE **Bartl.** 1830. Dicots. Scrophulariales. 1/1. South Africa. Genus Retzia.

RHABDODENDRACEAE **(Huber) Prance.** 1968. Dicots. Rutales. 1/3. Trop. South America. Genus Rhabdodendron.

RHAMNACEAE **Juss.** 1789. Dicots. Rhamnales 52/925. Almost cosmopolitan, but chiefly tropics and subtropics. Genera: Adolphia, Alphitonia, Alvimiantha, Ampelozizyphus, Araliorhamnus, Auerodendron, Bathiorhamnus, Berchemia, Berchemiella, Blackallia, Ceanothus, Colletia, Colubrina, Condalia, Crumenaria, Cryptandra, Dallachya, Disaster, Discaria, Doerpfeldia, Emmenosperma, Gouania, Helinus, Hovenia, Karwinskia, Kentrothamnus, Krugiodendron, Lamellisepalum, Lasiodiscus, Maesopsis, Nesiota, Noltea, Paliurus, Phylica, Pomaderris, Reissekia, Retanilla, Reynosia, Rhamnella, Rhamnidium, Rhamnus, Sagertia, Schistocarpaea, Scutia, Siegfriedia, Smythea, Spyridium, Talguenea, Trevoa, Trymalium, Ventilago, Ziziphus.

Rhexiaceae Dumort. = Melastomataceae Juss.

Rhinanthaceae Vent. = Scrophulariaceae Juss.

Rʜɪᴢᴏᴘʜᴏʀᴀᴄᴇᴀᴇ **R. Br.** 1814. Dicots. Rhizophorales. 15/150. Pantropics. Genera: Anopyxis, Blepharistemma, Bruguiera, Carallia, Cassipourea, Ceriops, Comiphyton, Crossostylis, Dactylopetalum, Gynotroches, Kandelia, Macarisia, Pellacalyx, Rhizophora, Sterigmapetalum.

Rhodiolaceae Martinov = Crassulaceae DC.

Rhododendraceae Juss. = Ericaceae Juss.

Rhodolaenaceae Bullock = Sarcolaenaceae Caruel.

Rʜᴏᴅᴏʟᴇɪᴀᴄᴇᴀᴇ **Nakai.** 1943. Dicots. Hamamelidales. 1/1–9. Southeast Asia. Genus Rhodoleia.

Rhodotypaceae J. Agardh = Rosaceae Juss.

Rʜᴏɪᴘᴛᴇʟᴇᴀᴄᴇᴀᴇ **Hand.-Mazz.** 1932. Dicots. Rhoipteleales. 1/1. China, northern Vietnam. Genus Rhoiptelea.

Rhopalocarpaceae Hemsl. ex Takht. = Sphaerosepalaceae Tiegh. ex Bullock.

Rʜʏɴᴄʜᴏᴄᴀʟʏᴄᴀᴄᴇᴀᴇ **L. A. S. Johnson et B. G. Briggs.** 1985. Dicots. Myrtales. 1/1. South Africa. Genus Rhynchocalyx.

Rʜʏɴᴄʜᴏᴛʜᴇᴄᴀᴄᴇᴀᴇ **Endl.** 1841. Dicots. Geraniales. 1/1. South America. Genus Rhynchotheca.

Ribesiaceae Marquis = Grossulariaceae DC.

Ricinaceae Martinov = Euphorbiaceae Juss.

Ricinocarpaceae (Muell. Arg.) Hurusawa = Euphorbiaceae Juss.

Rɪᴘᴏɢᴏɴᴀᴄᴇᴀᴇ **Conran et Clifford.** 1985. Monocots. Smilacales. 1/6. New Guinea, northeastern Australia, New Zealand. Genus Ripogonum.

Rivinaceae C. Agardh = Petiveriaceae C. Agardh.

Robiniaceae von Vest = Fabaceae Lindl.

Rᴏʀɪᴅᴜʟᴀᴄᴇᴀᴇ **Engl. et Gilg.** 1924. Dicots. Hydrangeales. 1/2. South Africa. Genus Roridula.

Rᴏsᴀᴄᴇᴀᴇ **Juss.** 1789. Dicots. Rosales. 116–118/3,800. Cosmopolitan, but chiefly temp. and subtrop. regions of the Northern Hemisphere. Genera: Acaena, Adenostoma, Agrimonia, Alchemilla, Amelanchier, Amygdalus, Aphanes, Aremonia, Aria, Ariosorbus, Armeniaca, Aronia, Aruncus, ? Atomostigma, Bencomia, Brachycaulos, Cerasus, Cercocarpus, Chaenomeles, Chamaebatia, Chamaebatiaria, Chamaemeles, Chamaemespilus, Chamaerhodos, Cliffortia, Coleogyne, Coluria, Comarum, Cormus, Cotoneaster, Cowania, Crataegus, Cydonia, Dendriopoterium, Dichotomanthes, Docynia, Dryadanthe, Dryas, Duchesnea, Eriobotrya, Eriolobus, Exochorda, Fallugia, Filipendula, Fragaria, Geum, Guamatela, Hagenia, Hesperomeles, Heteromeles, Holodiscus, Horkelia, Horkeliella, Ivesia, Kageneckia, Kelseya, Kerria, Laurocerasus, Leucosidea, Lindleya, Luetkea, Lyonothamnus, Maddenia, Malacomeles, Malus, Marcetella, Margyricarpus, Mespilus, Neillia, Neviusua, Novosieversia, Oemleria, Oncostylus, Osteomeles, Padellus, Padus, Pentactina, Peraphyllum, Petrophyton, Photinia, Physocarpus, Polylepis, Porteranthus, Potaninia, Potentilla, Pourthiaea, Prinsepia, Prunus, Pseudocydonia, Purshia, Pygeum, Pyracantha, Pyrus, Quillaja, Raphiolepis, Rhodotypos, Rosa, Rubus, Sanguisorba, Sarcopoterium, Sibbaldia, Sibiraea, Sieversia, Sorbaria, Sorbus, Spenceria, Spiraea, Spiraeanthus, Stephanandra, Stranvaesia, ? Taihangia, Tetraglochin, Torminalis, Tylosperma, Vauquelinia, Waldsteinia, Woronowia, Xerospiraea.

Rᴏᴜssᴇᴀᴄᴇᴀᴇ **DC.** 1839. Dicots. Brexiales. 1/1. Mauritius. Genus Roussea.

Roxburghiaceae Wall. = Stemonaceae Engl.

Rᴜʙɪᴀᴄᴇᴀᴇ **Juss.** 1789. Dicots. Gentianales. 620/11,800. Cosmopolitan, but chiefly in trop., subtrop., and warm–

temp. regions, few species in Arctica and Antarctica. Genera: Abbottia, Acranthera, Acrobotrys, Acrosynanthus, Acunaeanthus, Adina, Adinauclea, Afroknoxia, Agatisanthemum, Aidia, Aidiopsis, Airosperma, Aitchisonia, Alberta, Aleisanthia, Alibertia, Allaeophania, Alleizettella, Allenanthus, Alseis, Amaioua, Amaracarpus, Amphiasma, Amphidasya, Ancylanthos, Anomanthodia, Anotis, Antherostele, Anthocephalus, Anthorrhiza, Anthospermum, Antirhea, Aoranthe, Aphaenandra, Aphanocarpus, Appunettia, Appunia, Arcytophyllum, Argocoffeopsis, Argostemma, Asemnantha, Asperula, Astiella, Atractocarpus, Atractogyne, Augusta, Aulacocalyx, Badusa, Balmea, Bathysa, Batopedina, Becheria, Belonophora, Benkara, Benzonia, Berghesia, Bertiera, Bikkia, Blandibractea, Blepharidium, Bobea, Boholia, Borojoa, Borreria, Bothriospora, Botryarrhena, Bouvardia, Brachytome, Bradea, Brenania, Breonadia, Breonia, Burchellia, Burttdavya, Byrsophyllum, Caelospermum, Calanda, Callipeltis, Calochone, Calycophyllum, Calycosia, Calycosiphonia, Canephora, Canthium, Capirona, Captaincookia, ? Carmenocania, Carpacoce, Carphalea, Carterella, Casasia, Catesbaea, Catunaregam, Cephaelis, Cephalanthus, Cephalodendron, Ceratopyxis, Ceriscoides, Ceuthocarpus, Chaetostachydium, Chalepophyllum, Chamaepentas, Chapelieria, Chassallia, Chazaliella, Chimarrhis, Chiococca, Chione, Chlorochorion, Cigarilla, Cinchona, Cladoceras, Clarkella, Coccochondra, Coccocypselum, Coddia, Coelopyrena, Coffea, Coleactina, Colletoecema, Commitheca, Condaminea, Conostomium, Coprosma, Coptophyllum, Coptosapelta, Corynanthe, Corynula, Coryphothamnus, Cosmibuena, Cosmocalyx, Coursiana, Coussarea, Coutaportla, Coutarea, Craterispermum, Cremaspora, Cremocarpon, Crobylanthe, Crocyllis, Crossopteryx, Crucianella, Cruciata, Cruckshanksia, Crusea, Cuatrecasasiodendron, Cubanola, Cuviera, Cyclophyllum, Damnacanthus, Deccania, Danais, Deccania, Declieuxia, Dendrosipanea, Dentella, Deppea, Diacrodon, Dibrachionostylus, Dichilanthe, Dictyandra, Didymaea, Didymchlamys, Didymopogon, Didymosalpinx, Diodella, Diodia, Dioecrescis, Dioicodendron, Diplospora, Diyaminauclea, Dolichodelphys, Dolicholobium, Dolichometra, Doricera, Duidiana, Dunnia, Duperrea, Duroia, Durringtonia, Ecpoma, Eionitis, Eizia, Elaeagia, Emmenopterys, Emmeorhiza, Eosanthe, Eriosemopsis, Erithalis, Ernodea, Etericius, Euclinia, Exallage, Exandra, Exostema, Fadogia, Fadogiella, Fagerlindia, Faramea, Ferdinandusa, Feretia, Fergusonia, Fernelia, Flagenium, Flexanthera, Gaertnera, Galiniera, Galium, Gallienia, Galopina, Gamotopea, Gardenia, Gardeniopsis, Genipa, Gentingia, Geophila, Gillespiea, Gleasonia, Glionnetia, Glossostipula, Gomphocalyx, Gonzalagunia, Greenea, Greeniopsis, Guettarda, Guettardella, Gynochtodes, Gynopachis, Gyrostipula, Habroneuron, Haldina, Hallea, Hamelia, Hayataella, Hedstromia, Hedyotis, Hedythyrsus, Heinsenia, Heinsia, Hekistocarpa, Henriquezia, Heterophyllaea, Hillia, Himalrandia, Hindsia, Hintonia, Hippotis, Hodgkinsonia, Hoffmannia, Holstianthus, Homollea, Homolliella, Hutchinsonia, Hydnophytum, Hydrophylax, Hymenocnemis, Hymenocoleus, Hymenodictyon, Hyperacanthus, Hypobathrum, Hyptianthera, Ibetralia, Indopolysolenia, Isertia, Isidorea, Ixora, Jackiopsis, Janotia, Jaubertia, Joosia, Jovetia, Kailarsenia, Kajewskiella, Keenania, Keetia, Kelloggia, Kerianthera, Khasiaclunea, Knoxia, Kochummenia, Kohautia, Kraussia, Kutchubaea, Ladenbergia, Lagynias, Lamprothamnus, Larsenaikia, Lasianthus, Lathraeocarpa, Lecananthus, Lecanosperma, Lecariocalyx, Lelya, Lemyrea, Lepidostoma, Leptactina, Leptodermis, Leptomischus,

Leptoscela, Leptunis, Lerchea, Leucocodon, Leucolophus, Limnosipanea, Lindenia, Litosanthes, Lucinaea, Luculia, Lucya, Ludekia, Macbrideina, Machaonia, Macronemum, Macrosphyra, Maguireocharis, Maguireothamnus, Malanea, Manettia, Manostachya, Mantalania, Margaritopsis, Maschalocorymbus, Maschalodesme, Massularia, Mastixiodendron, Mazaea, Melanopsidium, Mericarpaea, Merumea, Metadina, Meyna, Micrasepalum, Microphysa, Mitchella, Mitracarpus, Mitragyna, Mitrasacmopsis, Mitriostigma, Molopanthera, Monosalpinx, Montamans, Morelia, Morierina, Morinda, Morindopsis, Motleya, Mouretia, Multidentia, Mussaenda, Mussaendopsis, Mycetia, Myonima, Myrioneuron, Myrmecodia, Myrmeconauclea, Myrmephytum, Nargedia, Nauclea, Neanotis, Neblinathamnus, Nematostylis, Nenax, Neobertiera, Neoblakea, Neobreonia, Neofranciella, Neogaillonia, Neohymenopogon, Neolamarckia, Neolaugeria, Neonauclea, Neopentanisia, Nernstia, Nertera, Nesohedyotis, Neurocalyx, Nichallea, Nodocarpaea, Nonatelia, Normandia, Nostolachma, Ochreinauclea, Octotropis, Oldenlandia, Oldenlandiopsis, Oligocodon, Omiltemia, Opercularia, Ophiorrhiza, Ophryococcus, Oreopolus, Osa, Otiophora, Otocalyx, Otomeria, Ottoschmidtia, Oxyanthus, Oxyceros, Pachystigma, Pachystylus, Paederia, Paedicalyx, Pagamea, Pagameopsis, Palicourea, Pampletantha, Paracephaelis, Parachimarrhis, Paracorynanthe, Paragenipa, Paraknoxia, Parapentas, Paratriaina, Pauridiantha, Pausinystalia, Pavetta, Payera, Pelagodendron, Pentagonia, Pentaloncha, Pentanisia, Pentanopsis, Pentas, Pentodon, Peponidium, Perakanthus, Perama, Peratanthe, Peripeplus, Pertusadina, Petagomoa, Petitiocodon, Phellocalyx, Phialanthus, Phitopis, Phuopsis, Phyllacanthus, Phyllis, Phyllocrater, Phyllomelia, Phylohydrax, Picardaea, Pimentelia, Pinarophyllon, Pinckneya, Pittonniotis, Placocarpa, Placopoda, Platycarpum, Plectroniella, Pleiocarpidia, Pleiocoryne, Pleiocraterium, Pleuocoryne, Pleurocoffea, Plocama, Poecilocalyx, Pogonopus, Polysphaeria, Polyura, Pomax, Porterandia, Portlandia, Posoqueria, Pouchetia, Praravinia, Pravinaria, Preussiodora, Prismatomeris, Proscephaleium, Psathura, Pseudaidia, Pseudogaillonia, Pseudogardenia, Pseudohamelia, Pseudomantalania, Pseudomussaenda, Pseudonesohedyotis, Pseudopyxis, Pseudosabicea, Psilanthopsis, Psilanthus, Psychotria, Psydrax, Psyllocarpus, Pteridocalyx, Pterogaillonia, Pubistylus, Putoria, Pymgaeothamnus, Pyragra, Pyrostria, Ramosmania, Randia, Raritebe, Readea, Remijia, Rennellia, Retiniphyllum, Rhadinopus, Rhaphidura, Rhipidantha, Rhopalobrachium, Richardia, Riqueuria, Robynsia, Roigella, Rondeletia, Rothmannia, Rubia, Rudgea, Rustia, Rutidea, Rytigynia, Sabicea, Sacosperma, Saldinia, Salzmannia, Saprosma, Sarcocephalus, Sarcopygme, Schismatoclada, Schizenterospermum, Schizocalyx, Schizocolea, ? Schizomussaenda, Schizostigma, Schmidtottia, Schradera, Schumanniophyton, Schwendenera, Scolosanthus, Scyphiphora, Scyphochlamys, Scyphostachys, Sericanthe, Serissa, Shaferocharis, Sherardia, Sherbournia, Siderobombyx, Siemensia, Simira, Sinoadina, Sipanea, Sipaneopsis, Siphonandrium, Sommera, Spathichlamys, Spermacoce, Spermadictyon, Sphinctanthus, Spiradiclis, Squamellaria, Stachyarrhena, Stachyococcus, Staelia, Standleya, Steenisia, Stelechantha, Stephanococcus, Stevensia, Steyermarkia, Stichianthus, Stilpnophyllum, Stomandra, Streblosa, Streblosiopsis, Strempelia, Striolaria, Strumpfia, Stylosiphonia, Suberanthus, Sukunia, Sulitia, Syringantha, Tamilnadia, Tammsia, Tapiphyllum, Tarenna, Tarennoidea, Temnocalyx, Temnopteryx, Tennantia, Thecorchus, Thogsennia, Thyridocalyx,

Timonius, Tobagoa, Tocoyena, Tortuella, Trailliaedoxa, Tresanthera, Triainolepis, Tricalysia, Trichostachys, Trigonopyren, Trukia, Tsiangia, Uncaria, Urophyllum, Valantia, Vangueria, Vangueriella, Vangueriopsis, Versteegia, Villaria, Virectaria, Warburgina, Warszewiczia, Wendlandia, Wernhamia, Wittmackanthus, Xanthophytum, Xantonnea, Xantonneopsis, Xerococcus, Xeromphis, Yutajea, Zuccarinia.

Rumicaceae Martinov = Polygonaceae Juss.

RUPPIACEAE Horan. ex Hutch. 1934. Monocots. Potamogetonales. 1/2–10. Cosmopolitan. Genus Ruppia.

RUSCACEAE Spreng. ex Hutch. 1934. Monocots. Asparagales. 3/13. Macaronesia, West and Central Europe, Mediterranean, West Asia. Genera: Danae, Ruscus, Semele.

RUTACEAE Juss. 1789. Dicots. Rutales. 158–159/1,900. Trop., subtrop., warm–temp. regions, esp. South Africa and Australia. Genera: Achuaria, Acmadenia, Acradenia, Acronychia, Adenandra, Adiscanthus, Aegle, Aeglopsis, Afraegle, Afraurantium, Agathosma, Almeidea, Amyris, Angostura, Apocaulon, Araliopsis, Asterolasia, Atalantia, Balfourodendron, Balsamocitrus, Boenninghausenia, Boninia, Boronia, Bosistoa, Bouchardatia, Bouzetia, Brombya, Burkillanthus, Calodendrum, Casimiroa, Chloroxylon, Choisya, Chorilaena, Citropsis, Citrus, Clausena, Clymenia, Cneoridium, Coleonema, Comptonella, Correa, Crowea, Decagonocarpus, Decatropis, Decazyx, Dictamnus, Dictyoloma, Diosma, Diphasia, Diphasiopsis, Diplolaena, Drummondita, Dutaillyea, Empleurum, Eremocitrus, Eriostemon, Erythrochiton, Esenbeckia, Euchaetis, Euodia, Euxylophora, Evodiela, Fagaropsis, Feroniella, Flindersia, Fortunella, Galipea, Geijera, Geleznowia, Glycosmis, Halfordia, Halophyllum, Helietta, Hortia, Humblotiodendron, Ivodea, Kodalyodendron, Leptothyrsa, Limonia, Lubaria, Lunasia, Luvunga, Maclurodendron, Macrostylis, Medicosma, Megastigma, Melicope, Merope, Merrillia, Metrodorea, Microcitrus, Microcybe, Micromelum, Monanthocitrus, Monnieria, Muiriantha, Murraya, Nyrtopsis, Naringi, Naudinia, Nematolepis, Neobyrnesia, Neoraputia, Nycticalanthus, Oricia, Oriciopsis, Orixa, Oxanthera, Pamburus, Paramignya, Peltostigma, Pentaceras, Phebalium, Phellodendron, Philotheca, Phyllosma, Pilocarpus, Pitavia, Platydesma, Platyspermation, Pleiospermium, Plethadenia, Polyaster, Poncirus, Pseudiosma, Psilopeganum, Ptelea, Raputia, Raputiarana, Rauia, Raulinoa, Ravenia, Raveniopsis, Rhadinothamnus, Ruta, Rutaneblina, Sarcomelicope, Severinia, Sheilanthera, Skimmia, Spathelia, Spiranthera, Stauranthus, Swingle, Teclea, Tetractomia, ? Tetractys, Tetradium, Thamnosma, Ticorea, Toddalia, Toddaliopsis, Tractocopevodia, Triphasia, Vepris, Wenzelia, Zanthoxylum, Zieria, Zieridium.

Sabalaceae Schultz Sch. = Arecaceae Schultz Sch.

SABIACEAE Blume. 1851. Dicots. Sapindales. 3/90. Himalayas, South, Southeast and East Asia, Malesia eastward to Solomon Is., trop. America. Genera: Meliosma, Ophiocaryon, Sabia.

Sabiceaceae Martinov = Rubiaceae Juss.

Saccharaceae Martinov = Poaceae Barnhart.

SACCIFOLIACEAE Maguire et Pires. 1978. Dicots. Gentianales. 1/1. Northern South America. Genus Saccifolium.

Salaciaceae Raf. = Celastraceae R. Br.

Salaxidaceae J. Agardh = Ericaceae Juss.

Salazariaceae F. A. Barkley = Lamiaceae Lindl.

SALICACEAE Mirb. 1815. Dicots. Salicales. 3/335. Arctic, temp. and warm temp. regions of the Northern Hemisphere, few species in tropics; 2 in the Southern Hemisphere. Genera: Chosenia, Populus, Salix.

Salicorniaceae Martinov = Chenopodiaceae Vent.

Salpiglossidaceae (Benth.) Hutch. = Solanaceae Juss.

Salsolaceae Moq. = Chenopodiaceae Vent.

SALVADORACEAE Lindl. 1836. Dicots. Celastrales. 3/11. Africa, Madagascar, South and Southeast Asia. Genera: Azima, Dobera, Salvadora.

Salviaceae Raf. = Lamiaceae Lindl.

SAMBUCACEAE Batsch ex Borckh. 1797. Dicots. Dipsacales. 1/40. Temp. and subtrop. regions, trop. East Africa. Genus Sambucus.

Samolaceae Raf. = Primulaceae Vent.

Samydaceae Vent. = Flacourtiaceae DC.

Sanguisorbaceae Marquis. = Rosaceae Juss.

Saniculaceae (Drude) A. Löve et D. Löve = Apiaceae Lindl.

Sansevieriaceae Nakai = Dracaenaceae Salisb.

SANTALACEAE R. Br. 1810. Dicots. Santalales. 38/510. Almost cosmopolitan, but chiefly tropics and subtropics. Genera: Acanthosyris, Amphorogyne, Anthobolus, Arjona, Austroamericium, Buckleya, Cervantesia, Choretrum, Cladomyza, Colpoon, Comandra, Daenikera, Dendromyza, Dendrotrophe, Dufrenoya, Elaphanthera, Exocarpos, Geocaulon, Jodina, Kunkeliella, Leptomeria, Mida, Myoschilos, Nanodea, Nestronia, Okoubaka, Omphacomeria, Osyridicarpos, Osyris, Phacellaria, Pyrularia, Quinchamalium, Rhoiacarpos, Santalum, Scleropyrum, Spirogardnera, Thesidium, Thesium.

Santolinaceae Augier ex Martinov = Asteraceae Dumort.

SAPINDACEAE Juss. 1789. Dicots. Sapindales. 146/1,550. Tropics and subtropics, few species in warm–temp. regions. Genera: Alectryon, Allophylus, Allosanthus, Amesiodendron, Aporrhiza, Arfeuillea, Arytera, Atalaya, Athyana, Averrhoidium, Beguea, Bemarivea, Bizonula, Blighia, Blighiopsis, Blomia, Boniodendron, Bottegoa, Bridgesia, Camptolepis, Cardiophyllarium, Cardiospermum, Castanospora, Chimborazoa, Chonopetalum, Chouxia, Chytranthus, Conchopetalum, Cossinia, Cotylodiscus, Cubilia, Cupania, Cupaniopsis, Deinbollia, Delavaya, Diatenopteryx, Dictyoneura, Dilodendron, Dimocarpus, Diploglottis, Diplokeleba, Diplopeltis, Distichostemon, Dodonaea, Doratoxylon, Elattostachys, Eriandrostachys, Eriocoelum, Erythrophysa, Erythrophysopsis, Euchorium, Euphoria, Euphorianthus, Eurycorymbus, Exothea, Filicium, Ganophyllum, Glenniea, Gloeocarpus, Gongrodiscus, Gongrospermum, Guindilia, Guioa, Handeliodendron, Haplocoelum, Harpullia, Hippobromus, Hornea, Houssayanthus, Hypelate, Jagera, Koelreuteria, Laccodiscus, Lecaniodiscus, Lepiderema, Lepidopetalum, Lepisanthes, Litchi, Llagunoa, Lophostigma, Loxodiscus, Lychnodiscus, Macphersonia, Magonia, Majidea, Matayba, Melicoccus, Mischocarpus, Molinaea, Neotina, Nephelium, Omalocarpus, Otonephelium, Otophora, Pancovia, Pappea, Paranephelium, Paullinia, Pavieasia, Pentascyphus, Phyllotrichum, Placodiscus, Plagioscyphus, Podonephelium, Pometia, Porocystis, Pseudima, Pseudopancovia, Pseudopteris, Radlkofera, Rhysotoechia, Rutilia, Sapindus, Sarcopteryx, Sarcotoechia, Schleichera, Sciaplea, Scyphonychium, Serjania, Sisyrolepis, Smelophyllum, Stadmannia, Stocksia, Storthocalyx, Strophiodiscus, Synima, Talisia, Thinouia, Thouinia, Thouinidium, Tina, Tinopsis, Toechima, Toulicia, Trigonachras, Tripterodendron, Tristira, Tristiropsis, Tsingya, Ungnadia, Urvillea, Vouarana, Xanthoceras, Xerospermum, Zanha, Zollingeria.

Saponaceae Vent. = Sapindaceae Juss.

SAPOTACEAE Juss. 1789. Dicots. Styracales. 55/1,400. Tropics, few species in temp. regions. Genera: Argania, Aubregrinia, Aulandra, Autranella, Baillonella, Boerlagella, Breviea, Burckella, Capurodendron, Chromolucuma, Chrysophyllum, Delpydora, Diploknema, Diploon, Eberhardtia, Ecclinusa, Elaeoluma, Englerophytum, Faucherea,

Gluema, Inhambanella, Isonandra, Labourdonnaisia, Labramia, Lasersisia, Lecomtedoxa, Leptostylis, Letestua, Madhuca, Magodendron, Manilkara, Micropholis, Mimusops, Neohemsleya, Neolemonniera, Nesoluma, Niemeyera, Northea, Omphalocarpum, Palaquium, Payena, Pichonia, Pouteria, Pradosia, Pycnandra, Sarcaulus, Sarcosperma, Sideroxylon, Synsepalum, Tieghemella, Trisedmostemon, Tsebona, Vitellaria, Vitellariopsis, Xantolis.

SARCOLAENACEAE Caruel. 1881. Dicots. Malvales. 10/62. Madagascar. Genera: Eremolaena, Leptolaena, Mediusella, Pentachlaena, Perrierodendron, Rhodolaena, Sarcolaena, Schizolaena, Xerochlamys, Xyloolaena.

SARCOPHYTACEAE A. Kern. 1891. Dicots. Balanophorales. 2/3. Trop. West, East, and Southeast Africa. Genera: Chlamydophytum, Sarcophyte.

Sarcopodaceae Gagnep. = Santalaceae R. Br.

Sarcospermataceae Lam. = Sapotaceae Juss.

Sarcostigmataceae Tiegh. ex Bullock = Icacinaceae (Benth.) Miers.

SARGENTODOXACEAE Stapf ex Hutch. 1926. Dicots. Ranunculales. 1/2. China and Indochina. Genus Sargentodoxa.

SARRACENIACEAE Dumort. 1829. Dicots. Sarraceniales. 3/14. North and South America. Genera: Darlingtonia, Heliamphora, Sarracenia.

Sarumaceae Nakai = Aristolochiaceae Juss.

Saurauiaceae J. Agardh = Actinidiaceae Hutch.

SAURURACEAE Rich ex E. Mey. 1827. Dicots. Piperales. 4/7. South, East, and Southeast Asia; North America. Genera: Anemopsis, Gymnotheca, Houttuynia, Saururus.

SAUVAGESIACEAE Dumort. 1829. Dicots. Ochnales. 25/128. South America, trop. Africa, trop. Asia, Malesia. Genera: Adenanthe, Adenarake, Blastemanthus, Cespedesia, Euthemis, Fleurydora, Godoya, Indosinia, Indovethia, Krukoviella, Lauradia, Leitgebia, Luxemburgia, Neckia, Philacra, Poecilandra, Rhytidanthera, Sauvagesia, Schuurmansia, Schuurmansiella, Sinia, Testulea, Tyleria, Vausagesia, Wallacea.

SAXIFRAGACEAE Juss. 1789. Dicots. Saxifragales. 30/525. Cold, temp. regions and montane areas of tropics. Genera: Astilbe, Astilboides, Bensoniella, Bergenia, Bolandra, Boykinia, Chrysospleniella, Chrysosplenium, Conimitella, Darmera, Elmera, Heuchera, Jepsonia, Leptarrhena, Lithophragma, Mitella, Mukdenia, Oresitrophe, Peltoboykinia, Rodgersia, Saxifraga, Saxifragella, Saxifragodes, Saxifragopsis, Suksdorfia, Sullivantia, Tanakaea, Tellima, Tiarella, Tolmiea.

Scabiosaceae Adans. ex T. Post et Kuntze = Dipsacaceae Juss.

Scaevolaceae Lindl. = Goodeniaceae R. Br.

Scepaceae Lindl. = Euphorbiaceae Juss.

SCHEUCHZERIACEAE F. Rudolphi. 1830. Monocots. Scheuchzeriales. 1/1. Arctic and temp. regions of the Northern Hemisphere. Genus Scheuchzeria.

SCHISANDRACEAE Blume. 1830. Dicots. Illicales. 2/49. Asia from India to Japan and Malesia, southeastern North America. Genera: Kadsura, Schisandra.

Schizolaenaceae Barnhart = Sarcolaenaceae Caruel.

Schoepfiaceae Blume = Olacaceae Mirb. ex DC.

Schreberaceae (R. Wight) Schnizl. = Oleaceae Hoffmanns. et Link.

SCIADOPITYACEAE Luerss. 1877. Pinopsida. Sciadopityales. 1/1. Japan. Genus Sciadopitys.

Scillaceae von Vest = Hyacinthaceae Batsch.

Scirpaceae Batsch ex Borckh. = Cyperaceae Juss.

Scleranthaceae Bercht. et J. Presl = Caryophyllaceae Juss.

SCLEROPHYLACACEAE Miers. 1848. Dicots. Solanales. 1/12. South America. Genus Sclerophylax.

SCROPHULARIACEAE Juss. 1789. Dicots.

Scrophulariales. 270–274/5,000. Cosmopolitan. Genera: Acanthorrhinum, Achetaria, Adenosma, Agalinis, Albraunia, Alectra, Allocalyx, Alonsoa, Amalophyllon, Amphiolanthus, Anarrhinum, Anastrabe, Ancistrostylis, Angelonia, Antherothamnus, Anticharis, Antirrhinum, Aptosimum, Aragoa, Artanema, Asarina, Aureolaria, Bacopa, Bampsia, Bartsia, Basistemon, Baumia, Bellardia, Benjaminia, Besseya, Beyrichia, Bonnayodes, Bowkeria, Brachystigma, Brandisia, Brookea, Bryodes, Buchnera, Bungea, Buttonia, Bythophyton, Calceolaria, Camptoloma, Campylanthus, Capraria, Castilleja, Centranthera, Centrantheropsis, Chaenorrhinum, Cheilophyllum, Chelone, Chenopodiopsis, Chionohebe, Chionophila, Chodaphyton, Clevelandia, Cochlidiosperma, Collinsia, Colpias, Conobea, Cordylanthus, Craterostigma, Crepidorhopalon, Cycniopsis, Cycnium, Cymbalaria, Cymbaria, Cyrtandromoea, Darcya, Dasistoma, Decaryanthus, Deinostema, Dermatobotrys, Derwentia, Detzneria, Diascia, Diclis, Digitalis, Dintera, Dizygostemon, Dodartia, Dolichostemon, Dopatrium, Elacholoma, Ellisiophyllum, Encopella, Epixiphium, Eremogeton, Erinus, Escobedia, Esterhazya, Euphrasia, Eylesia, Faxonanthus, Fonkia, Freylinia, Galvezia, ? Gambelia, Gerardiina, Ghikaea, Gibsoniothamnus, Glekia, Glossostigma, Glumicalyx, Graderia, Gratiola, Gynocraterium, Halleria, Harveya, Hasslerella, Hebe, Hedbergia, Hemianthus, Hemichaena, Hemimeris, Hemiphragma, Hemisiphonia, Hiernia, Holmgrenanthe, Holzneria, Hueblia, Hydrotriche, Hyobanche, Ilysanthes, Isoplexis, Ixianthes, Jamesbrittenia, Jovellana, Kashmiria, Kechiella, Kickxia, Lafuentia, Lagotis, Lamourouxia, Lancea, Legazpia, Leptorhabdos, Lesquereuxia, Leucocarpus, Leucophyllum, Leucosalpa, Leucospora, Limnophila, Limosella, Linaria, Lindenbergia, Lindernia, Lophospermum, Mabrya, Macranthera, Maeviella, Magdalenaea, Manulea, Manuleopsis, Maurandella, Maurandya, Mazus, Mecardonia, Melampyrum, Melanospermum, Melasma, Melosperma, Micranthemum, Micrargeria, Micrargeriella, Microcarpaea, ? Mimetanthe, Mimulicalyx, Mimulus, Misopates, Mohavea, Monochasma, Monopera, Monttea, Mosheovia, Namation, Nathaliella, Nemesia, Neopicrorhiza, Nothochelone, Nothochilus, Nuttallanthus, ? Odicardis, Odontites, Omania, Omphalotrix, Ophiocephalus, Oreosolen, Orthanthella, Orthocarpus, Otacanthus, Ourisia, Ourisianthus, Paederota, Paederotella, Parahebe, Parastriga, Parentucellia, Pedicularis, Peliostomum, Pennellianthus, Penstemon, Peplidium, Petitmenginia, Phteirospermum, Phygelius, Phyllopodium, Physocalyx, Picria, Picrorhiza, Pierranthus, Poarium, Polycarena, Porodittia, Psammetes, Pseudobartsia, Pseudolysimachion, Pseudorontium, Pseudosopubia, Pseudostriga, Pterygiella, Radamaea, Rehmannia, Rhabdotosperma, Rhamphicarpa, Rhaphispermum, Rhinanthus, Rhodochiton, Rhynchocorys, Russelia, Sairocarpus, Schistophragma, Schizoseala, Schwalbea, Schweinfurthia, Scolophyllum, Scoparia, Scrofella, Scrophularia, Seymeria, Seymeriopsis, ? Shiuyinghua, Sibthorpia, Silviella, Sibobacopa, Siphonostegia, Sopubia, Spirostegia, Staurophragma, Stemodia, Stemodiopsis, Striga, Strobilopsis, Sutera, Synthyris, Teedia, Tetranema, Tetraspidium, Tetraulacium, Thunbergianthus, Tonella, Torenia, Tozzia, Triaenophora, Trieenea, Triphysaria, Trungbao, Tuerckheimocharis, Uroskinnera, Valeria, Vellosiella, Verbascum, Verena, Veronica, Veronicastrum, Wulfenia, Wulfeniopsis, Xizangia, Xylocalyx, Zaluzianskya.

Scutellariaceae Caruel = Lamiaceae Lindl.

SCYBALIACEAE A. Kern. 1891. Dicots. Balanophorales. 1/4. Trop. South America, West Indies. Genus Scybalium.

SCYPHOSTEGIACEAE Hutch. 1926. Dicots.

Violales. 1/1. Borneo. Genus Scyphostegia.
SCYTOPETALACEAE Engl. 1897. Dicots. Ochnales. 5/20. Trop. Africa. Genera: Brazzeia, Oubanguia, Pierrina, Rhaptopetalum, Scytpetalum.
Sedaceae von Vest = Crassulaceae DC.
Seguieriaceae Nakai = Petiveriaceae C. Agardh.
SELAGINACEAE Choisy. 1823 (nom. cons.). Dicots. Scrophulariales. 10/300. Trop. and South Africa, Genera: Agathelpis, Cromidon, Dischisma, Globulariopsis, Gosela, Hebenstretia, Microdon, Selago, Tetraselago, Walafrida.
Sempervivaceae Juss. = Crassulaceae DC.
Senecionaceae Spenn. = Asteraceae Dumort.
Sequoiaceae Arnoldi = Taxodiaceae Warm.
Serratulaceae Martinov = Asteraceae Dumort.
Sesamaceae R. Br. ex Bercht. = Pedaliaceae R. Br.
Sesuviaceae Horan. = Aizoaceae F. Rudolphi.
Sibthorpiaceae D. Don = Scrophulariaceae Juss.
Silenaceae (DC.) Bartl. = Caryophyllaceae Juss.
Simabaceae Horan. = Simaroubaceae DC.
SIMAROUBACEAE DC. 1811. Dicots. Rutales. 20/150. Tropics and subtropics, few in warm–temp. regions. Genera: Ailanthus, Alvaradoa, Amaroria, Brucea, Castela, Cedronia, Eurycoma, Gymnostemon, Harrisonia, Iridosma, Laumoniera, Nothospondias, Perriera, Picramnia, Picrasma, Picrolemma, Pleiokirkia, Quassia, Recchia, Soulamea.
SIMMONDSIACEAE (Muell. Arg.) Tiegh. ex Reveal et Hoogland. 1990. Dicots. Simmondsiales. 1/1. Southwestern North America. Genus Simmondsia.
Siparunaceae (A. DC.) Schodde = Monimiaceae Juss.
Siphonodontaceae Gagnep. et Tard. ex Tard. = Celastraceae R. Br.
Sisymbriaceae Martinov = Brassicaceae Burnett.

Sladeniaceae (Gilg et Werderm.) Airy Shaw = Theaceae D. Don.
SMILACACEAE Vent. 1799. Monocots. Smilacales. 2/320–370. Tropics and subtropics, few in warm–temp. regions. Genera: Heterosmilax, Smilax.
Smyrniaceae Burnett = Apiaceae Lindl.
SOLANACEAE Juss. 1789. Dicots. Solanales. 100/2,900. Almost cosmopolitan, but chiefly trop. South America. Genera: Acnistus, Anisodus, Anthocercis, Anthotroche, Archiphysalis, Athenaea, Atrichodendron, Atropa, Atropanthe, Aureliana, Benthamiella, Bouchetia, Brachistus, Browallia, Brugmansia, Brunfelsia, Calibrachoa, Capsicum, Cestrum, Chamaesaracha, Combera, Crenidium, Cuatresia, Cyphanthera, Cyphomandra, Datura, Deprea, Discopodium, Dittostigma, Duboisia, Dunalia, Dyssochroma, Ectozoma, Exodeconus, Fabiana, Grammosolen, Grabowskia, Hawkesiophyton, Hebecladus, Heteranthia, Hunzikeria, Hyoscyamus, Iochroma, Jaborosa, Jaltomata, Juanulloa, Latua, Leptoglossis, Leucophysalis, Lycianthes, Lycium, Lycopersicon, Mandragora, Margaranthus, Markea, Melananthus, Mellissia, Metternichia, Nectouxia, Nicandra, Nicotiana, Nierembergia, Nolana, Normania, Nothocestrum, Oryctes, Pantacantha, Parabouchetia, Pauia, Petunia, Phrodus, Physaliastrum, Physalis, Physochlaina, Plowmania, Protoschwenckia, Przewalskia, Quincula, Rahowardiana, Reyesia,- Salpichora, Salpiglossis, Saracha, Schizanthus, Schultesianthus, Schwenckia, Scopolia, Sessea, Solandra, Solanum, Streptosolen, Symonanthus, Trianae, Triguera, Tsoala, Tubocapsicum, Vassobia, Vestia, Withania, Witheringia.
SONNERATIACEAE Engl. et Gilg. 1924. Dicots. Myrtales. 1/7. Trop. coasts of the Old World. Genus Sonneratia.
Soulameaceae Endl. = Simaroubaceae DC.
SPARGANIACEAE F. Rudolphi. 1830. Monocots. Typhales. 1/15. Extratrop. regions of the Northern Hemisphere, 1–2

Southeast Asia, Australia and New Zealand. Genus Sparganium.

Sparmanniaceae J. Agardh = Tiliaceae Juss.

Spartinaceae Burnett = Poaceae Barnhart.

Spatheliaceae J. Agardh = Rutaceae Juss.

SPHAEROSEPALACEAE Tiegh. ex Bullock. 1959. Dicots. Malvales. 2/17. Madagascar. Genera: Dialyceras, Rhopalocarpus.

SPHENOCLEACEAE Mart. ex DC. 1839. Dicots. Campanulales. 1/2. Tropics of the Old World. Genus Sphenoclea.

SPHENOSTEMONACEAE P. Royen et Airy Shaw. 1972. Dicots. Celastrales. 1/8. Eastern Malesia, Australia, New Caledonia. Genus Sphenostemon.

Spielmanniaceae J. Agardh = Oftiaceae Takht. et Reveal.

SPIGELIACEAE Mart. 1827. Dicots. Gentianales. 3/ c. 95. Madagascar, South, Southeast and East Asia, Malesia, Australia, Tasmania, New Zealand, New Caledonia, America, West Indies. Genera: Mitrasacme, Mitreola, Spigelia.

Spiraeaceae Bartuch = Rosaceae Juss.

Spondiadaceae Martinov = Anacardiaceae Lindl.

Sporobolaceae Herter = Poaceae Barnhart.

Stachydaceae Salisb. = Lamiaceae Lindl.

STACHYURACEAE J. Agardh. 1858. Dicots. Theales. 1/16. Trop. Asia from eastern Himalayas to Japan. Genus Stachyurus.

STACKHOUSIACEAE R. Br. 1814. Dicots. Celastrales. 3/17. Malesia, Micronesia, Australia, Tasmania, New Zealand. Genera: Macgregoria, Stackhousia, Tripterococcus.

STANGERIACEAE (Pilg.) L. A. S. Johnson. 1959. Cycadopsida. Cycadales. 1/1. South Africa. Genus Stangeria.

Stanleyaceae Nutt. = Brassicaceae Burnett.

Stapeliaceae Horan. = Asclepiadaceae R. Br.

STAPHYLEACEAE Lindl. 1829. Dicots. Sapindales. 3/22–42. Temp. regions of the Northern Hemisphere, trop. Asia, Central and South America, West Indies. Genera: Euscaphis, Staphylea, Turpinia.

Staticaceae Hoffmanns. et Link ex Gray = Plumbaginaceae Juss.

STEGNOSPERMATACEAE (A. Rich.) Nakai. 1942. Dicots. Caryophyllales. 1/4. North and Central America, West Indies. Genus Stegnosperma.

Stellariaceae Dumort. = Caryophyllaceae Juss.

STEMONACEAE Engl. 1887. Monocots. Dioscoreales. 3/c. 30. South, Southeast, and East Asia; trop. Australia; North America. Genera: Croomia, Stemona, Stichoneuron.

STENOMERIDACEAE J. Agardh. 1858. Monocots. Dioscoreales. 2/6. Madagascar (Avetra), Malesia (Stenomeris).

STERCULIACEAE (DC.) Bartl. 1830. Dicots. Malvales. 67/1,755. Tropics, few species in subtrop. and warm–temp. regions. Genera: Acropogon, Ambroma, Astiria, Ayenia, Brachychiton, Byttneria, Cheirolaena, Chiranthodendron, Cola, Commersonia, Cotylonychia, Courtenia, Dicarpidium, Dombeya, Eriolaena, Firmiana, Franciscodendron, Fremontodendron, Gilesia, Glossostemon, Guazuma, Guichenotia, Hannafordia, Harmsia, Helicteres, Helmiopsiella, Helmiopsis, Heritiera, Hermannia, Herrania, Hildegardia, Keraudrenia, Kleinhovia, Lasiopetalum, Leptonychia, Lysiosepalum, Mansonia, Maxwellia, Megatritheca, Melhania, Melochia, Neoregnellia, Nesogordonia, Octolobus, Paramelhania, Pentapetes, Pimia, Pterocymbium, Pterospermum, Pterygota, Rayleya, Reevesia, Ruizia, Rulingia, Scaphium, Scaphopetalum, Seringia, Sterculia, Tetradia, Theobroma, Thomasia, Trichostephania, Triplochiton, Trochetia, Trochetiopsis, Ungeria, Waltheria.

Stilaginaceae C. Agardh = Euphorbiaceae Juss.

STILBACEAE Kunth. 1831. Dicots. Scrophulariales. 5/13. South Africa. Genera: Campylostachys, Eurylobium, Euthystachys, Stilbe, Xeroplana.

Stipaceae Burnett = Poaceae Barnhart.

STRASBURGERIACEAE Engl. et Gilg. 1924. Dicots. Ochnales. 1/1. New Caledonia. Genus Strasburgeria.

Stratiotaceae Link = Hydrocharitaceae Juss.

STRELITZIACEAE Hutch. 1934. Monocots. Zingiberales. 3/7. Trop. South America (Phenakospermum), South Africa (Strelitzia), Madagascar (Ravenala).

Strephonemataceae Venkat. et Prak. Rao = Combretaceae R. Br.

Streptochaetaceae Nakai = Poaceae Barnhart.

Strombosiaceae Tiegh. = Olacaceae Mirb. ex DC.

Strumariaceae Salisb. = Amaryllidaceae J. St.-Hil.

Strychnaceae Perleb = Loganiaceae R. Br. ex Mart.

STYLIDIACEAE R. Br. 1810. Dicots. Stylidiales. 5/170. South and Southeast Asia, Malesia, Australia, Tasmania, New Zealand, temp. South America. Genera: Forstera, Levenhookia, Oreostylidium, Phyllachne, Stylidium.

Stylobasiaceae J. Agardh = Surianaceae Arn.

STYLOCERATACEAE (Pax) Baill. ex Reveal et Hoogland. 1990. Buxales. 1/5. South America. Genus Styloceras.

Stypheliaceae Horan. = Epacridaceae R. Br.

STYRACACEAE Dumort. 1829. Dicots. Ebenales. 11/150. Eastern Mediterranean; South, Southeast, and East Asia; Malesia; America; West Indies. Genera: Alniphyllum, Bruinsmis, Halesia, Huodendron, Melliodendron, Pamphilia, Parastyrax, Pterostyrax, Rehderodendron, Sinojackia, Styrax.

SURIANACEAE Arn. 1834. Dicots. Rutales. 4/6 Coasts of tropical seas and coastal Australia. Genera: Cadellia, Guilfoylia, Stylobasium, Suriana.

Swartziaceae (DC.) Bartl. = Fabaceae Lindl.

Swieteniaceae Bercht. et J. Presl = Meliaceae Juss.

Symphoniaceae (C. Presl) Barnhart = Clusiaceae Lindl.

SYMPHOREMATACEAE Moldenke ex Reveal et Hoogland. 1991. Dicots. Lamiales. 3/35. South and Southeast Asia, Malesia. Genera: Congea, Sphenodesme, Symphorema.

SYMPLOCACEAE Desf. 1820. Dicots. Theales. 1/250. Tropics and subtropics, excl. Africa. Genus Symplocos.

Synechanthaceae O. F. Cook = Arecaceae Schultz Sch.

Syringaceae Horan. = Oleaceae Hoffmanns. et Link.

Tabernaemontanaceae Baum.-Bodenh. = Apocynaceae Juss.

TACCACEAE Dumort. 1829. Monocots. Taccales. 2/13. Trop. of the Old World, 1 sp. of Tacca in trop. South America. Genera: Schizocapsa, Tacca.

Taiwaniaceae Hayata = Taxodiaceae Warm.

Takhtajaniaceae (J.-F. Leroy) J.-F. Leroy = Winteraceae Lindl.

Tamaceae Martinov = Dioscoreaceae R. Br.

TAMARICACEAE Link. 1821. Dicots. Tamaricales. 3/75. Eurasia and Africa, chiefly Mediterranean and arid regions of Asia. Genera: Myricaria, Reaumuria, Tamarix.

Tamarindaceae Bercht. et J. Presl = Fabaceae Lindl.

Tanacetaceae von Vest = Asteraceae Dumort.

TAPISCIACEAE Takht. 1987. Dicots. Sapindales. 2/7. Southeastern China (Tapiscia), West Indies, and South America (Huertea).

TAXACEAE Gray. 1821. Pinopsida. Taxales. 3/15. Temp. Northern Hemisphere southward to southern Mexico and New Caledonia. Genera: Pseudotaxus, Taxus, Torreya.

TAXODIACEAE Warm. 1884. Pinopsida. Cupressales. 9/15. East and Southeast Asia, Tasmania, temp. North America. Genera: Athrotaxis, Cryptomeria, Cunninghamia, Glyptostrobus, Metasequoia, Sequoia, Sequoiadendron, Taiwania, Taxodium.

TECOPHILAEACEAE Leyb. 1862. Monocots.

Liliales. 6/25. Trop. and South Africa, Madagascar, California, Chile. Genera: Conanthera, Cyanella, Odontostomum, Tecophilaea, Walleria, Zephyra.

Telephiaceae Martinov = Caryophyllaceae Juss.

TEPUIANTHACEAE Maguire et Steyerm. 1981. Dicots. Sapindales. 1/6. Northern South America. Genus Tepuianthus.

Terebinthaceae Juss. = Anacardiaceae Lindl.

Terminaliaceae J. St.-Hil. = Combretaceae R. Br.

Ternstroemiaceae Mirb. ex DC. = Theaceae D. Don.

TETRACARPAEACEAE Nakai. 1943. Dicots. Hydrangeales. 1/1. Tasmania. Genus Tetracarpaea.

TETRACENTRACEAE A. C. Sm. 1945. Dicots. Trochodendrales. 1/1. Eastern Himalayas, Burma, central and southwestern China. Genus Tetracentron.

Tetraceraceae Baum.-Bodenh. = Dilleniaceae Salisb.

TETRACHONDRACEAE Skottsb. ex Wettst. 1924. Dicots. Lamiales. 1/2. New Zealand, temp. South America. Genus Tetrachondra.

Tetraclinaceae Hayata = Cupressaceae Rich. ex Bartl.

TETRADICLIDACEAE Takht. 1986. Dicots. Rutales. 1/1. East Europe, eastern Mediterranean, West Asia. Genus Tetradiclis.

Tetragoniaceae Nakai = Aizoaceae F. Rudolphi.

Tetramelaceae Airy Shaw = Datiscaceae Lindl.

TETRAMERISTACEAE Hutch. 1959. Dicots. Theales. 2/2. Western Malesia (Tetramerista), northern South America (Pentamerita).

Tetrastylidiaceae Calest. = Olacaceae Mirb. ex DC.

THALASSIACEAE Nakai. 1943. Monocots. Hydrocharitales. 1/2. Islands of the Indian and Pacific Oceans and Caribbean Sea. Genus Thalassia.

Thalictraceae Raf. = Ranunculaceae Juss.

THEACEAE D. Don. 1825. Dicots. Theales. 25/560. Tropics and subtropics, a few species in temp. East Asia and eastern North America. Genera: Adinandra, Anneslea, Apterosperma, Archboldiodendron, Balthasaria, Camellia, Cleyera, Eurya, Ficalhoa, Franklinia, Freziera, Glyptocarpa, Gordonia, Hartia, Killipiodendron, Paranneslea, Parapyrenaria, Pyrenaria, Schima, Sladenia, Stewartia, Symplococarpon, Ternstroemia, Tutcheria, Visnea.

THELIGONACEAE Dumort. 1829. Dicots. Gentianales. 1/3. Canary Is., Eurasia, Mediterranean. Genus Theligonum.

Themidaceae Salisb. = Alliaceae J. Agardh.

Theobromaceae J. Agardh = Sterculiaceae (DC.) Bartl.

THEOPHRASTACEAE Link. 1829. Dicots. Primulales. 5/90. Trop. America, West Indies. Genera: Clavija, Deherainia, Jacquinia, Neomezia, Theophrasta.

Thesiaceae von Vest = Santalaceae R. Br.

Thismiaceae J. Agardh = Burmanniaceae Blume.

Thlaspiaceae Martinov = Brassicaceae Burnett.

Thomandersiaceae Sreemadh. = Acanthaceae Juss.

Thujaceae Burnett = Cupressaceae Rich. ex Bartl.

Thunbergiaceae Bremek. = Acanthaceae Juss.

THURNIACEAE Engl. 1907. Monocots. Juncales. 1/2. Northern South America. Genus Thurnia.

THYMELAEACEAE Juss. 1789. Dicots. Thymelaeales. 49/760. Almost cosmopolitan, but mostly trop. Africa and Australia. Genera: Aquilaria, Arnhemia, Atemnosiphon, Craterosiphon, Cryptadenia, Dais, Daphne, Daphnimorpha, Daphnopsis, Deltaria, Diarthron, Dicranolepis, Dirca, Drapetes, Edgeworthia, Enkleia, Eriosolena, Funifera, Gnidia, Goodallia, Gyrinops, Jedda, Kelleria, Lachnaea, Lagetta, Lasiadenia, Lethedon, Linodendron, Linostoma, Lophostoma, Octolepis, Oreodendron, Ovidia, Passerina, Peddiea, Phaleria, Pimelea, Restella, Rahmnoneuron, Schoenobiblus, Solmsia, Stellera, Stephanodaphne, Struthiola, Synandrodaphne, Synaptolepis, Thecanthes, Thymelaea, Wikstroemia.

TICODENDRACEAE Gómez-Laurito et Gómez P. 1991. 1/1. Central America. Genus Ticodendron.

TILIACEAE Juss. 1789. Dicots. Malvales. 50/720. Tropics, the genus Tilia in temp. regions of the Northern Hemisphere. Genera: Ancistrocarpus, Apeiba, Asterophorum, Berrya, Brownlowia, Burretiodendron, Christiania, Clappertonia, Colona, Corchoropsis, Corchorus, Craigia, Desplatsia, Dicraspidia, Diplodiscus, Duboscia, Eleutherostylis, Endosteira, Entelea, Erinocarpus, Glyphaea, Goethalsia, Grewia, Hainania, Heliocarpus, Hydrogaster, Jarandersonia, Luehea, Lueheopsis, Microcos, Mollia, Mortoniodendron, Muntingia, Neotessmannia, Nettoa, Pentace, Pentaplaris, Pseudocorchorus, Schoutenia, Sicrea, Sparrmannia, Tahitia, Tetralix, Tilia, Trichospermum, Triumfetta, Triumfettoides, Vasivaea, Vinticena, Westphalina.

Tillandsiaceae A. Juss. = Bromeliaceae Juss.

Tithymalaceae Vent. = Euphorbiaceae Juss.

Toddaliaceae Baum.-Bodenh. = Rutaceae Juss.

TORICELLIACEAE Hu. 1934. Dicots. Toricelliales. 1/3. China, Himalayas, Burma. Genus Toricellia.

Torreyaceae Nakai = Taxaceae Gray.

TOVARIACEAE Pax. 1891. Dicots. Capparales. 1/2. Trop. America, West Indies. Genus Tovaria.

Tragiaceae Raf. = Euphorbiaceae Juss.

TRAPACEAE Dumort. 1828. Dicots. Myrtales. 1/1–5 (or 15–40). Tropics, subtropics, and temp. regions of the Old World, excl. Australia. Genus Trapa.

TRAPELLACEAE Honda et Sakis. 1930. Dicots. Scrophulariales. 1/1–2. East Asia. Genus Trapella.

TREMANDRACEAE R. Br. ex DC. 1824. Dicots. Polygalales. 3/43. Temp. Australia and Tasmania. Genera: Platytheca, Tetratheca, Tremandra.

Treviaceae Lindl. = Euphorbiaceae Juss.

TRIBELACEAE (Engl.) Airy Shaw. 1965. Dicots. Hydrangeales. 1/1. Temp. South America. Genus Tribeles.

Tribulaceae (Engl.) Hadidi = Zygophyllaceae R. Br.

Tribulaceae Trautv. = Zygophyllaceae R. Br.

TRICHOPODACEAE Hutch. 1934. Monocots. Dioscoreales. 1/1. South Asia. Genus Trichopus.

Triglochinaceae Dumort. = Juncaginaceae Rich.

TRIGONIACEAE Endl. 1841. Dicots. Polygalales. 4/27. Trop. America, Madagascar, Western Malesia. Genera: Humbertiodendron, Trigonia, Trigoniastrum, Trigoniodendron.

TRILLIACEAE Lindl. 1846. Monocots. Dioscoreales. 5/62. Europe, Mediterranean, Asia, North America. Genera: Daiswa, Kinugasa, Paris, Trillidium, Trillium.

TRIMENIACEAE (Perkins et Gilg) Gibbs. 1917. Dicots. Laurales. 1/5. Malesia, Australia, Melanesia eastward to Marquesas. Genus Trimenia.

Triplochitonaceae K. Schum. = Sterculiaceae (DC.) Bartl.

TRIPLOSTEGIACEAE Bobrov ex Airy Shaw. 1965. Dicots. Dipsacales. 1/2. China, Himalayas, India, Burma, Taiwan, Malesia. Genus Triplostegia.

Tripterygiaceae Huber = Celastraceae R. Br.

Tristichaceae J. C. Willis = Podostemaceae Rich ex C. Agardh.

TRIURIDACEAE Gardner. 1843. Monocots. Triuridales. 6/60. Trop. Africa, Asia, and America. Genera: Hyalisma, Peltophyllum, Sciaphila, Seychellaria, Soridium, Triuris.

TROCHODENDRACEAE Prantl. 1888. Dicots. Trochodendrales. 1/1. East Asia. Genus Trochodendron.

TROPAEOLACEAE Juss. ex DC. 1824. Dicots. Tropaeolales. 2–3/c. 90. Central and South America. Genera: Magellana, ? Tropaeastrum, Tropaeolum.

Tulbaghiaceae Salisb. = Alliaceae J. Agardh.

Tulipaceae Batsch ex Borckh. = Liliaceae Juss.

TURNERACEAE DC. 1828. Dicots. Violales. 10/120. Trop. and subtrop. America, trop. and South Africa, Madagascar, Rodriguez I. Genera: Adenoa, Erblichia, Hyalocalyx, Loewia, Mathurina, Piriqueta, Stapfiella, Streptopetalum, Turnera, Wormskioldia.

TYPHACEAE Juss. 1789. Monocots. Typhales. 1/15. Cosmopolitan. Genus Typha.

Uapacaceae (Muell. Arg.) Airy Shaw = Euphorbiaceae Juss.

Ullucaceae Nakai = Basellaceae Moq.

ULMACEAE Mirb. 1815. Urticales. 15/180. Trop., subtrop., and temp. regions. Genera: Ampelocera, Aphananthe, Celtis, Chaetachme, Gironniera, Hemiptelea, Holoptelea, Lozanella, Parasponia, Phyllostylon, Planera, Pteroceltis, Trema, Ulmus, Zelkova.

Ulmariaceae Gray = Rosaceae Juss.

UMBELIFERAE Juss. See APIACEAE Lindl.

URTICACEAE Juss. 1789. Dicots. Urticales. 54/1,300. Almost cosmopolitan, mostly tropics and subtropics. Genera: Aboriella, Achudemia, Archiboemeria, Astrothalamus, Australina, Boemeria, Boemeriopsis, Chamabainia, Cypholophus, Debregeasia, Dendrocnide, Didymodoxa, Discocnide, Droguetia, Elatostema, Forsskaolea, Gesnouinia, Gibbsia, Girardinia, Gonostegia, Gyrotaenia, Hemistylus, Hesperocnide, Hyrtanandra, Laportea, Lecanthus, Leycosyke, Maoutia, Meniscogyne, Myriocarpa, Nanocnide, Neodistemon, Neraudia, Nothocnide, Obetia, Oreocnide, Parietaria, Parsana, Pellionia, Petelotiella, Phenax, Pilea, Pipturus, Pouzolzia, Procris, Rousselia, Sarcochlamys, Sarcopilea, Sceptrocnide, Soleirolia, Touchardia, Urera, Urtica, Villebrunea.

Utriculariaceae Hoffmanns. et Link = Lentibulariaceae Rich.

Uvulariaceae A. Gray ex Kunth = Melanthiaceae Batsch.

Vacciniaceae DC. ex Gray = Ericaceae Juss.

VAHLIACEAE Dandy. 1959. Dicots. Saxifragales. 1/5. Africa, Madagascar, Southwest and South Asia. Genus Vahlia.

VALERIANACEAE Batsch. 1802. Dicots. Dipsacales. 8/430. Almost cosmopolitan, excl. trop. Africa, Australia, and New Zealand. Genera: Centranthus, Fedia, Nardostachys, Partinia, Plectritis, Pseudobetckea, Valeriana, Valerianella.

Vallesneriaceae Link = Hydrocharitaceae Juss.

Vanillaceae Lindl. = Orchidaceae Juss.

VELLOZIACEAE Endl. 1841. Monocots. Velloziales. 8/260. Trop. America, trop. Africa, Madagascar, Arabian Peninsula, southwestern China. Genera: Acanthochlamys, Barbacenia, Barbaceniopsis, Burlemarxis, Nanuza, Pleurostima, Vellozia, Xerophyta.

Veratraceae von Vest = Melanthiaceae Batsch.

Verbascaceae Nees = Scrophulariaceae Juss.

VERBENACEAE J. St.-Hil. 1805. Dicots. Lamiales. 75/c. 2,000. Tropics and subtropics, a few species in warm–temp. regions. Genera: Acantholippia, Adelosa, Aegiphila, Aloysia, Amasonia, Archboldia, Baillonia, Bouchea, Burroughsia, Callicarpa, Caryopteris, Casselia, Chascanum, Citharexylum, Clerodendrum, Coelocarpum, Cornutia, Diostea, Dipyrena, Duranta, Faradaya, Garrettia, Glandularia, Glossocarya, Gmelina, Hierobotana, Holmskioldia, Hosea, Huxleya, Hymenopyramis, Junellia, Kalaharia, Karomia, Lampayo, Lantana, Lippia, Monochilus, Nashia, Neorapinia, Neosparton, Oncinocalyx, Oxera, Paravitex, Parodianthus, Peronema, Petitia, Petraeovitex, Petrea, Phryma, Phyla, Pitraea, Premna, Priva, Pseudocarpidium, Pygmaeopremna, Recordia, Rehdera, Rhaphithamnus, Schnabelia, Stachytarpheta, Stylodon, Tamonea, Teijsmanniodendron, Tetraclea, Teucridium,

Tsoongia, Ubochea, Urbania, Verbena, Verbenoxylum, Vitex, Viticipremna, Xeroaloysia, Xerocarpa, Xolocotzia.

Vernoniaceae Burmeist. = Asteraceae Dumort.

Veronicaceae Raf. = Scrophulariaceae Juss.

VIBURNACEAE Raf. 1820. Dicots. Dipsacales. 1/225. Temp. and subtrop. regions, a few species in tropics. Genus Viburnum.

Viciaceae Bercht. et J. Presl = Fabaceae Lindl.

Vincaceae von Vest = Apocynaceae Juss.

VIOLACEAE Batsch. 1802. Dicots. Violales. 23–24/990. Cosmopolitan. Genera: Agatea, Allexis, Amphirrhox, Anchietea, Corynostylis, ? Cubelium, Decorsella, Fusispermum, Gloeospermum, Hybanthus, Hymenanthera, Isodendrion, Leonia, Mayanaea, Melicytus, Noisettia, Orthion, Paypayrola, Perissandra, Rinorea, Rinoreocarpus, Schweiggeria, Scyphellandra, Viola.

VISCACEAE Batsch. 1802. Dicots. Santalales. 8/250. Cosmopolitan, but chiefly tropics and subtropics. Genera: Arceuthobium, Dendrophthora, Distichella, Ginalloa, Korthalsella, Notothixos, Phoradendron, Viscum.

VITACEAE Juss. 1789. Dicots. Vitales. 14–15/840. Tropics, subtropics, and warm–temp. regions. Genera: Acareosperma, Ampelocissus, Ampelopsis, Cayratia, Cissus, Clematicissus, Cyphostemma, Parthenocissus, Pterisanthes, Pterocissus, Rhoicissus, ? Spondylantha, Tetrastigma, Vitis, Yua.

Viticaceae Juss. = Verbenaceae J. St.-Hil.

VIVIANIACEAE Klotzsch. 1836. Dicots. Geraniales. 4/6. South America. Genera: Araeoandra, Caesarea, Cissarobryon, Viviania.

VOCHYSIACEAE A. St.-Hil. 1820. Dicots. Polygalales. 7/220. Trop. America, West Indies, trop. West Africa. Genera: Callisthene, Erisma, Erismadelphus, Qualea, Ruizterania, Salvertia, Vochysia.

Wellstediaceae (Pilg.) Novák = Boraginaceae Juss.

WELWITSCHIACEAE Markgr. 1926. Class Welwitschiopsida. Welwitschiales. 1/1. South-West Africa. Genus Welwitschia.

Willughbeiaceae J. Agardh = Apocynaceae Juss.

WINTERACEAE R. Br. ex Lindl. 1830. Dicots. Winterales. 5/115. Madagascar, Malesia, eastern Australia, Pacific islands, South America. Genera: Drimys, Pseudowinter, Takhtajania, Tasmannia, Zygogonum.

Wolffiaceae Bubani = Lemnaceae Gray.

Xanthiaceae von Vest = Asteraceae Dumort.

Xanthophyllaceae (Chodat) Gagnep. ex Reveal et Hoogland = Polygalaceae R. Br.

XANTHORRHOEACEAE Dumort. 1829. Monocots. Amaryllidales. 1/28. Australia and Tasmania. Genus Xanthorrhoea.

XEROPHYLLACEAE Takht. 1994. Monocots. Melanthiales. 1/2. North America. Genus Xerophyllum.

Xerotaceae Endl. = Dasypogonaceae Dumort.

Ximeniaceae Martinet = Olacaceae Mirb. ex DC.

Xiphidiaceae Dumort. = Haemodoraceae R. Br.

XYRIDACEAE C. Agardh. 1823. Monocots. Commelinales. 5/270. Tropics and subtropics, a few species in warm–temp. regions. Genera: Abolboda, Achlyphila, Aratitiyopea, Orecanthe, Xyris.

Yuccaceae J. Agardh = Agavaceae Endl.

ZAMIACEAE Horan. 1834. Cycadopsida. Cycadales. 8/115. Trop. and subtrop. America, West Indies, trop. and South Africa, Australia. Genera: Ceratozamia, Chigua, Dioon, Encephalartos, Lepidozamia, Macrozamia, Microcycas, Zamia.

ZANNICHELLIACEAE Dumort. 1829. Monocots. Cymodoceales. 4/9–14. Cosmopolitan. Genera: Althenia, Lepilaena, Vleisia, Zannichellia.

Zanoniaceae Dumort. = Cucurbitaceae Juss.

Zanthoxylaceae Bercht. et J. Presl = Rutaceae Juss.

Zeaceae A. Kern. = Poaceae Barnhart.

Zephyranthaceae Salisb. = Amaryllidaceae J. St.-Hil.

Zingiberaceae **Lindl.** 1835. Monocots. Zingiberales. 52/1,350. Tropics. Genera: Aframomum, Alpinia, Amomum, Aulotandra, Boesenbergia, Burbidgea, Camptandra, Caulokaempferia, Cautleya, Cenolophon, Curcuma, Curcumorpha, Cyphostigma, Elettaria, Elettariopsis, Eltingera, Gagnepainia, Geocharis, Geostachys, Globba, Guillainia, Haniffia, Haplochorema, Hedychium, Hemiorchis, Hitchenia, Hornstedtia, Kaempferia, Kolowratia, Leptosolena, Mantisia, Nanochilus, Paracautleya, Parakaempferia, Paramomum, Plagiostachys, Pleuranthodium, Pommereschea, Pyrgophyllum, Re nealmia, Rhynchanthus, Riedelia, Roscoea, Scaphochlamys, Siliquamomum, Siphonochilus, Stadiochilus, Stahlianthus, Thylacophora, Vanoverberghia, Zerumbet, Zingiber.

Ziziphaceae Adans. ex T. Post et Kuntze = Rhamnaceae Juss.

Zosteraceae **Dumort.** 1829. Monocots. Zosterales. 3/23. Seas of temp. regions, a few species in trop. Africa, Southeast Asia, and Australia. Genera: Heterozostera, Phyllospadix, Zostera.

Zygophyllaceae **R. Br.** 1814. Dicots. Rutales. 22/235. Tropics and subtropics, a few species in warm–temp. regions. Genera: Augea, Bulnesia, Fagonia, Guaiacum, Kallstroemia, Kelleronia, Larrea, Metharme, Miltianthus, Morkillia, Neoluederitzia, Pintoa, Plectocarpa, Porlieria, Seetzenia, Sericodes, Sisyndite, Tetraena, Tribulopsis, Tribulus, Viscainoa, Zygophyllum.

DICTIONARY OF GENERIC NAMES OF SEED PLANTS

A

Aa Rchb. f. 1854. Orchidaceae. 30 Andes from Colombia to Argentina

Aalium Lam. ex Kuntze = Sauropus Blume. Euphorbiaceae

Aaronsohnia Warb. et Eig. 1927. Asteraceae. 1 North Africa, Southwest Asia

Abarema Pittier. 1927 (Archidendron F. Muell.). Fabaceae. 20 Central and South America

Abasola La Llave ex Lex. = Exlipta L. Asteraceae

Abatia Ruiz et Pav. 1794. Flacourtiaceae. 9 trop. America from southern Mexico to Bolivia, northwestern Argentina, and southern Brazil

Abauria Becc. = Koompassia Maingay. Fabaceae

Abbevillea O. Berg = Campomanesia Ruiz et Pav. Myrtaceae

Abbottia F. Muell. 1875 (Timonius DC.). Rubiaceae. 1 northern Australia

Abdominea J. J. Sm. 1914. Orchidaceae. 1 southern Thailand, Malay Peninsula, Java, Borneo, Philippines, Bali

Abdra Greene = Draba L. Brassicaceae

Abdulmajidia Whitemore. 1974. Lecythidaceae. 2 Malaysia

Abebaia Baehni = Manilkara Adans. Sapotaceae

Abelia R. Br. 1818. (Linnaea L.). Caprifoliaceae. 20–30 Middle Asia (1, A. corymbosa), Afghanistan, and from Himalayas to East Asia (incl. Russian Far East—A. koreana), Mexico

Abelicea Baill. = Zelkova Spach. Ulmaceae

Abeliophyllum Nakai. 1919. Oleaceae. 1 Korea

Abelmoschus Medik. 1787 (Hibiscus L.). Malvaceae. 15 trop. Africa, South and Southeast Asia, Indonesia, Philippines, and Australia (7)

Aberemoa Aubl. = Guatteria Ruiz et Pav. Annonaceae

Abies Hill. 1753. Pinaceae. c. 40 North and Central Europe, Russia (European part, the Urals, West and East Siberia, Far East); North Africa (Morocco, Algeria); temp. Asia: western Caucasus, Turkey, Tien Shan, Pamir, western and central China, Himalayas, Japan, Korea, Taiwan (1, A. kawakamii); America: Canada, U.S., northwestern Mexico, Guatemala

Abildgaardia Vahl. 1805 (Fimbristylis Vahl). Cyperaceae. 80 tropics

Abobra Naudin. 1862. Cucurbitaceae. 1 subtrop. South America

Abola Adans. = Cinna L. Poaceae

Abola Lindl. = Caucaea Schltr. Orchidaceae

Abolboda Bonpl. 1813. Xyridaceae. 17 trop. South America (excl. Andes)

Aboriella Bennet. 1981. Urticaceae. 1 eastern Himalayas, Tibet

Abortopetalum O. Deg. = Abutilon Hill. Malvaceae

Abramsia Gillespie = Airosperma K. Schum. et Lauterb. Rubiaceae

Abroma Jacq. = Ambroma L. f. Sterculiaceae

Abromeitia Mez = Fittingia Mez. Myrsinaceae

Abromeitiella Mez. 1927. Bromeliaceae. 3 trop. South America

Abronia Juss. 1789. Nyctaginaceae. c. 35 southwestern U.S. and northern Mexico

Abrophyllum Hook. f. ex Benth. 1864. Escalloniaceae. 2 eastern Australia

Abrotanella Cass. 1825. Asteraceae. c. 20 New Guinea, Australia, Tasmania, New

Zealand, Snares Is., South America (Tierra del Fuego), Falkland Is.

Abrus Adans. 1763. Fabaceae. 17 tropics

Absolmsia Kuntze. 1891. Asclepiadaceae. 2 southwestern China (1), Borneo (1)

Abryanthemum Rothm. = Carpobrotus N. E. Br. Aizoaceae

Abuta Aubl. 1775. Menispermaceae. 32 trop. South America

Abutilon Hill. 1753. Malvaceae. 150–200 tropics and subtropics, esp. America, a few species in temp. regions; 1 (A. theophrasti) extending to southern European part of Russia, Caucasus, and Middle Asia

Abutilothamnus Ulbr. 1915. Malvaceae. 1 trop. South America

Acachmena H. P. Fuchs. 1960. Brassicaceae. 2 Balkan Peninsula, Asia Minor, northern Iran, coasts of the Black Sea, Caucasus

Acacia Hill. 1753. Fabaceae. c. 1,200 tropics and subtropics, esp. of Africa and Australia

Acaciella Britton et Rose = Acacia Hill. Fabaceae

Acaciopsis Britton et Rose = Acacia Hill. Fabaceae

Acaena Mutis ex L. 1771. Rosaceae. 100–150 South Africa, New Guinea, Australia, Tasmania, New Zealand, subantarctic islands, Polynesia, Hawaiian Is., and from California to South America

Acalypha L. 1753. Euphorbiaceae. 450 tropics and subtropics; 1 (A. australis) extending to Caucasus and Russian Far East

Acalyphes Hassk. = Acalypha L. Euphorbiaceae

Acalyphopsis Pax et K. Hoffm. = Acalypha L. Euphorbiaceae

Acampe Lindl. 1853. Orchidaceae. 10 trop. and South Africa, Madagascar, Comoro Is., Seychelles Is., Aldabra I., trop. Asia from Himalayas (from Garhwal to Sikkim), India and Sri Lanka through Burma and Indochina to southern China (Yunnan to Kwangtun, Hong Kong and Hainan), Malay Peninsula, and Philippines

Acamptoclados Nash = Eragrostis Wolf. Poaceae

Acamptopappus (A. Gray) A. Gray. 1873 (Haplopappus Cass.). Asteraceae. 2 southwestern U.S.: Nevada, California, Utah, Arizona

Acanthambrosia Rydb. = Ambrosia L. Asteraceae

Acanthella Hook. f. 1867. Melastomataceae. 2 trop. South America

Acanthephippium Blume = Acanthophippium Blume. Orchidaceae

Acanthinophyllum Allemao = Clarisia Ruia et Pav. Moraceae

Acanthocalycium Backeb. = Echinopsis Zucc. Cactaceae

Acanthocalyx (DC.) M. Cannon. 1984 (Morina L.). Morinaceae. 3 Himalayas, southwestern China

Acanthocardamum Thell. 1906. Brassicaceae. 1 southern Iran

Acanthocarpus Lehm. 1848. Dasypogonaceae. 10 southwestern Australia

Acanthocaulon Klotzsch ex Endl. = Platygyna Mercier. Euphorbiaceae

Acanthocephala Backeb. = Parodia Speg. Cactaceae

Acanthocephalus Kar. et Kir. 1842. Asteraceae. 2 Middle Asia, Iran, Afghanistan

Acanthocereus (Engelm. ex A. Berger) Britton et Rose. 1909. Cactaceae. 10 America from southeastern U.S. to northeastern Brazil, West Indies

Acanthochiton Torr. = Amaranthus L. Amaranthaceae

Acanthochlamys P. C. Kao. 1980. Velloziaceae or Acanthochlamydaceae. 1 southwestern China

Acanthocladium F. Muell. 1861. Asteraceae. 1 southeastern Australia

Acanthocladus Klotzsch ex Hassk. = Polygala L. Polygalaceae

Acanthococos Barb. Rodr. = Acrocomia Mart. Arecaceae

Acanthodesmos C. D. Adams et du Quesnay. 1971. Asteraceae. 1 Jamaica

Acanthodium Delile = Blepharis Juss. Acanthaceae

Acanthogilia A. G. Day et Moran. 1986. Polemoniaceae. 1 North America

Acanthoglossum Blume = Pholidota Lindl. ex Hook. Orchidaceae

Acanthogonum Torr. = Chorizanthe R. Br. Polygonaceae

Acantholepis Less. 1831. Asteraceae. 1 eastern Mediterranean and from West Asia to western China, Caucasus, Middle Asia

Acantholimon Boiss. 1846. Plumbaginaceae. c. 170 Eurasia from Yugoslavia, Greece, and Crete through Asia Minor to Israel and Jordan and eastward to western Tibet; Caucasus and Middle Asia (c. 70)

Acantholippia Griseb. 1874 (Lippia L.). Verbenaceae. 6 subtrop. and temp. South America

Acantholobivia Backeb. = Echinopsis Zucc. Cactaceae

Acantholoma Gaudich. ex Baill. = Pachystroma Muell. Arg. Euphorbiaceae

Acanthomintha A. Gray ex Benth. et Hook. f. 1878. Lamiaceae. 4 California, Baja California, northern Mexico

Acanthonema Hook. f. 1862. Gesneriaceae. 2 West Africa

Acanthopale C. B. Clarke. 1899. Acanthaceae. 15 trop. Africa, Southeast Asia

Acanthopanax (Decne. et Planch.) Miq. = Eleutherococcus Maxim. Araliaceae

Acanthopetalus Y. Ito = Echinopsis Zucc. Cactaceae

Acanthophaca Nevski = Astragalus L. Fabaceae

Acanthophippium Blume. 1825. Orchidaceae. 15 trop. Asia from Himalayas, India, and Sri Lanka through Burma to southern China and eastern islands from Japan, Indochina, Malaysia from Java to New Guinea, Solomon Is., Vanuatu, New Caledonia, Fiji, Tonga, and the Horne Is.

Acanthophoenix H. A. Wendl. 1867. Arecaceae. 1 Mascarene Is.

Acanthophora Merr. = Aralia L. Araliaceae

Acanthophyllum C. A. Mey. 1831. Caryophyllaceae. c. 60 from Southwest and West Asia through Caucasus and Middle Asia eastward to Siberia, Mongolia, China, and northwestern India

Acanthopsis Harv. 1842. Acanthaceae. 8 South Africa

Acanthopteron Britton = Mimosa L. Fabaceae

Acanthopyxis Miq. ex Lanj. = Caperonia A. St.-Hil. Euphorbiaceae

Acanthorhipsalis (K. Schum.) Britton et Rose = Lepismium Pfeiff. Cactaceae

Acanthorhipsalis Kimnach = Lymanbensonia Kimnach. Cactaceae

Acanthorrhinum Rothm. 1943. Scrophulariaceae. 1 Northwest Africa.

Acanthorrhiza H. A. Wendl. = Cryosophila Blume. Arecaceae

Acanthosabal Prosch. = Acoelorrhaphe H. A. Wendl. Arecaceae

Acanthoscyphus Small = Oxytheca Nutt. Polygonaceae

Acanthosicyos Welw. ex Hook. f. 1867. Cucurbitaceae. 2 southern trop. Africa

Acanthospermum Schrank. 1820. Asteraceae. 6 southern U.S., Mexico, Central and trop. South America, West Indies, Galapagos Is.

Acanthosphaera Warb. = Naucleopsis Miq. Moraceae

Acanthostachys Klotzsch. 1840. Bromeliaceae. 1 southeastern Brazil, Paraguay, northeastern Argentina

Acanthostelma Bidgood et Brummitt. 1985. Acanthaceae. 1 northern Somalia

Acanthostemma (Blume) Blume = Hoya R. Br. Asclepiadaceae

Acanthostyles R. M. King et H. Rob. 1971 (Eupatorium L.). Asteraceae. 1 Bolivia, Argentina

Acanthosyris (Eichler) Griseb. 1879. Santalaceae. 3 temp. South America

Acanthothamnus Brandegee. 1909. Celastraceae. 1 Mexico

Acanthotreculia Engl. = Treculia Decne. ex Trécul. Moraceae

Acanthotrichilia (Urb.) O. F. Cook et G. N. Collins = Trichilia P. Browne. Meliaceae

Acanthoxanthium (DC.) Fourr. = Xanthium L. Asteraceae

Acanthura Lindau. 1901. Acanthaceae. 1 Brazil

Acanthus L. 1753. Acanthaceae. 30 Mediterranean, trop. and subtrop. Africa and Asia; A. ilicifolius extends to New Caledonia; A. dioscoridis extends to Armenia

Acanthyllis Pomell = Anthyllis L. Fabaceae

Acareosperma Gagnep. 1919. Vitaceae. 1 Southeast Asia

Acarpha Griseb. 1855. Calyceraceae. 8 Chile, Argentina, Falkland Is.

Acaulimalva Krapov. 1974. Malvaceae. 19 South America (Andes)

Acaulon N. E. Br. = Aloinopsis Schwantes. Aizoaceae

Acca O. Berg. 1856 (Psidium L.). Myrtaceae. 3 Peru, southeastern Brazil

Accara Landrum. 1990 (Acca O. Berg). Myrtaceae. 1 Brazil (Minas Gerais)

Accia A. St.-Hil. = Fragariopsis A. St.-Hil. Euphorbiaceae

Acelica Rizzini = Justicia L. Acanthaceae

Acelidanthus Trautv. et C. A. Mey. = Veratrum L. Melanthiaceae

Acer L. 1753. Aceraceae. c. 120–160 Northern Hemisphere from subarctic areas of Europe and Alaska to Central America and South Asia; A. laurinum extends to Southern Hemisphere (Timor I.)

Aceranthus C. Morren et Decne. = Epimedium L. Berberidaceae

Aceras R. Br. 1813. Orchidaceae. 1 Europe, North Africa, Asia Minor

Acerates Elliott = Asclepias L. Asclepiadaceae

Aceratium DC. 1824. Elaeocarpaceae. 20 Moluccas, New Guinea, Solomon Is., New Hebrides (Vanuatu), northeastern Australia

Aceratorchis Schltr. 1922. Orchidaceae. 2 Tibet

Aceriphyllum Engl. = Mukdenia Koidz. Saxifragaceae

Acetosa Hill = Rumex L. Polygonaceae

Acetosella Kuntze = Oxalis L. Oxalidaceae

Acetosella (Meisn.) Fourr. = Rumex L. Polygonaceae

Achaenipodium Brandegee. 1906. Asteraceae. 1 Mexico

Achaeta Fourn. = Calamagrostis Adans. Poaceae

Achaetogeron A. Gray = Erigeron L. Asteraceae

Acharia Thunb. 1794. Achariaceae. 1 South Africa (Natal, Cape)

Acharitea Benth. = Nesogenes A. DC. Verbenaceae

Achasma Griff. = Eltingera Giseke. Zingiberaceae

Achatocarpus Triana. 1858. Achatocarpaceae. 15 America from Texas and northwestern Mexico to Paraguay and Argentina

Achetaria Cham. et Schltdl. 1827. Scrophulariaceae. 5 trop. America

Achillea L. 1753. Asteraceae. 85 temp. Eurasia, esp. montane Mediterranean and Russia (c. 50)

Achimenes Pers. 1806. Gesneriaceae. 25. trop. America, West Indies

Achlaena Griseb. = Arthropogon Nees. Poaceae

Achlyphila Maguire et Wurdack. 1960. Xyridaceae. 1 Venezuela

Achlys DC. 1821. Berberidaceae. 2 East Asia and western North America

Achnatherum P. Beauv. 1812 (Stipa L.). Poaceae. c. 20 warm–temp. regions and montane areas of tropics

Achneria Benth. = Pentaschistis (Nees) Spach. Poaceae

Achnodon Link = Phleum L. Poaceae

Achnodonton P. Beauv. = Phleum L

Achnophora F. Muell. 1833. Asteraceae. 1 southern Australia

Achnopogon Maguire, Steyerm. et Wurdack. 1957. Asteraceae. 3 Venezuela, Guyana

Achoriphragma Soják. 1982. Brassicaceae. 51: c. 30 Middle Asia, Siberia, and Far East; Afghanistan, Pakistan, India, China, North America from Alaska and arctic Canada to Wyoming and Utah

Achradelpha O. F. Cook = Pouteria Aubl. Sapotaceae

Achradotypus Baill. = Pycnandra Benth. Sapotaceae

Achras L. = Manilkara Adans. Sapotaceae

Achroanthes Raf. = Malaxis Sol. ex Sw. Orchidaceae

Achrochloa B. D. Jacks = Koeleria Pers. Poaceae

Achroostachys Benth. = Athroostachys Benth. Poaceae

Achrouteria Eyma = Chrysophyllum L. Sapotaceae

Achuaria Gereau. 1990. Rutaceae. 1 (A. hirsuta) Peru (Amazonian)

Achudemia Blume. 1856 (Pilea Lindl.). Urticaceae. 2 East Asia, Java; Russia (1, A. japonica—Far East)

Achyrachaena Schauer. 1838. Asteraceae. 1 southwestern North America

Achyranthes L. 1753 [Pandiaka (Moq.) Hook. f.]. Amaranthaceae. 6 America, Mediterranean, trop. and subtrop. Africa, Asia, Australia, eastward to Polynesia

Achyrobaccharis Sch. Bip. ex Walp. 1843. (Baccharis L.). Asteraceae. 8 trop. America

Achyrocalyx Benoist. 1930. Acanthaceae. 4 Madagascar

Achyrocline (Less.) DC. 1832. Asteraceae. 22 trop. America and trop. Africa, Madagascar

Achyrodes Boehm. = Lamarckia Moench. Poaceae

Achyronychia Torr. et A. Gray. 1868. Caryophyllaceae. 2 southwestern U.S., Mexico

Achyropappus Kunth. 1818. Asteraceae. 1 Mexico

Achyrophorus Adans. 1763. Asteraceae. 20 Europe, Asia, South America; Russia (4 from European part to Far East)

Achyropsis (Moq.) Hook. f. 1880. Amaranthaceae. 6 trop. and South Africa

Achyrospermum Blume. 1826. Lamiaceae. 10 trop. Africa, Madagascar, India, Himalayas, southern China, Indochina, Malaysia

Achyrothalamus O. Hoffm. 1893. Asteraceae. 1 trop. East Africa

Aciachne Benth. 1881. Poaceae 1 trop. South America (Andes)

Acianthera Scheidw. = Pleurothallis R. Br. Orchidaceae

Acianthus R. Br. 1810. Orchidaceae. 22 New Guinea, Solomon Is. (1), Australia, Tasmania, ? New Zealand, New Caledonia (16)

Acicalyptus A. Gray. 1854 (Cleistocalyx Blume). Myrtaceae. 8 New Caledonia, Fiji

Acicarpa Raddi = Digitaria Haller. Poaceae

Acicarpha Juss. 1803. Calyceraceae. 5 Central and South America southward to Argentina, Paraguay, and Uruguay

Acidanthera Hochst. = Gladiolus L. Iridaceae

Acidocroton Griseb. 1859. Euphorbiaceae. 10 West Indies

Acidocroton P. Beauv. = Flueggea Willd. Eurphorbiaceae

Acidonia L. A. S. Johnson et B. Briggs. 1975. Proteaceae. 13 southwestern Australia

Acidosasa C. D. Chu et C. S. Chao. 1979 (Arundinaria Michx.). Poaceae. 6 southern China, Vietnam (1)

Acidoton P. Browne = Flueggea Willd. Euphorbiaceae

Acidoton Sw. 1788. Euphorbiaceae. 6 Central America, Venezuela, West Indies

Aciella Tiegh. = Amylotheca Tiegh. Loranthaceae

Acilepidopsis H. Rob. 1989 (Vernonia Schreb.). Asteraceae. 1 Brazil

Acineta Lindl. 1843. Orchidaceae. 15 trop. America from Mexico to Venezuela and Ecuador

Acinos Mill. 1754 (Satureja L.). Lamiaceae. 10 Europe, Mediterranean, West Asia to Caucasus, Middle Asia, and Iran

Acioa Aubl. 1775. Chrysobalanaceae. 4 trop. America

Aciotis D. Don. 1832. Melastomataceae. 30 Central and trop. South America, West Indies

Aciphylla J. R. Forst. et G. Forst. 1776. Apiaceae. 35–40 Australia (3), New Zealand

Acisanthera P. Browne. 1756. Melastomataceae. 35 trop. America, West Indies

Ackama A. Cunn. = Caldcluvia D. Don. Cunoniaceae

Ackermania Dodson et Escobar. 1993. Orchidaceae. 3 Andes of Colombia, Ecuador, and Peru

Acleisanthes A. Gray. 1853. Nyctaginaceae. 7 southwestern U.S. from California to Texas, Mexico

Aclinia Griff. = Dendrobium Sw. Orchidaceae

Aclisia E. Mey. ex C. Presl = Pollia Thunb. Commelinaceae

Acmadenia Bartl. et H. L. Wendl. 1824. Rutaceae. 33 South Africa

Acmanthera Griseb. 1858. Malpighiaceae. 6 South America

Acmella L. C. Rich. ex Pers. 1807. Asteraceae. 30 tropics

Acmena DC. 1828. Myrtaceae. 14 China, Southeast Asia, Malaysia, Australia

Acmenosperma Kausel. 1957. Myrtaceae. 2 Australia

Acmispon Raf. 1832 (Lotus L.). Fabaceae. 1 (A. roudairei) West Sahara, Tunisia, Algeria, Morocco

Acmopyle Pilg. 1903. Podocarpaceae. 2 New Caledonia (1), Fiji (1, Viti Levu)

Acmostemon Pilg. = Ipomoea L. Convolvulaceae

Acmostigma Raf. = Pavetta L. Rubiaceae

Acnida L. = Amaranthus L

Acnistus Schott. 1829. Solanaceae. 1 trop. America from Mexico to Peru and Brazil

Acoelorraphe H. A. Wendl. 1879. Arecaceae. 1 southern Florida, West Indies, Caribbean coasts of Central America

Acoidium Lindl. = Trichocentrum Poepp. et Endl. Orchidaceae

Acokanthera G. Don. 1837 (1838) (Carissa L.). Apocynaceae. 5 central (Zaire, Uganda), east (from Somalia to Tanzania), and South Africa, southern Arabian Peninsula

Acomastylis Greene = Geum L. Rosaceae

Acomis F. Muell. ex Benth. 1867. Asteraceae. 4 Australia

Acomosperma K. Schum. 1908. Asclepiadaceae. 1 trop. South America

Aconceveibum Miq. = Mallotus Lour. Euphorbiaceae

Aconitella Spach = Delphinium L. Ranunculaceae

Aconitum L. 1753. Ranunculaceae. c. 100 (or 300–350) temp. Northern Hemisphere; esp. Russia and adjacent territories (c. 100)

Aconogonon (Meisn.) Rchb. = Persicaria Hill. Polygonaceae

Aconogonum (Meisn.) Rchb. = Persicaria Hill. Polygonaceae

Acopanea Steyerm. 1984 (or = Bonnetia Mart.). Bonnetiaceae. 1 Venezuela

Acorellus Palla ex Kneuck. = Juncellus (Griseb.) C. B. Clarke. Cyperaceae Acoridium Nees et Meyen = Ceratostylis Blume. Orchidaceae

Acorus L. 1753. Acoraceae. 4 temp. and subtrop. Northern Hemisphere, montane trop. Asia

Acosmium Schott. 1827. Fabaceae. c. 20 trop. America, esp. Brazil

Acosmus Desv. = Aspicarpa Rich. Malpighiaceae

Acostaea Schltr. 1923. Orchidaceae. 2 Costa Rica, Panama

Acostia Swallen. 1968. Poaceae. 1 Ecuador

Acourtia D. Don. 1830. Asteraceae. 5 southern U.S., Mexico

Acrachne Wight et Arn. ex Chiov. 1907. Poaceae. 2 Africa from Senegal to Eritrea, Kenya and Tanzania, south to Namibia and Transvaal; South Asia (Afghanistan, western Pakistan, India), Australia

Acradenia Kippist. 1853. Rutaceae. 2 eastern Australia, Tasmania

Acraea Lindl. = Pterichis Lindl. Orchidaceae

Acrandra O. Berg = Campomanesia Ruiz et Pav. Myrtaceae

Acranthemum Tiegh. = Tapinanthus (Blume) Rchb. Loranthaceae

Acranthera Arn. ex Meisn. 1838. Rubiaceae. c. 40 Himalayas, India, Sri Lanka, southern China, Indochina, Malaysia, esp. Borneo

Acratherum Link = Arundinella Raddi. Poaceae

Acreugenia Kausel. 1956 (Myrcianthes O. Berg). Myrtaceae. 1 southern Brazil, northern Argentina

Acridocarpus Guill., Perr. et A. Rich. 1831. Malpighiaceae. 30 trop. Africa,

Madagascar (1) to Arabian Peninsula, New Caledonia (1)

Acrilia Griseb. = Trichilia P. Browne

Acriopsis Reinw. ex Blume. 1825. Orchidaceae. 6–12 Burma, Indochina, Malaysia (Malay Peninsula, Sumatra, Java, Philippines, Sulawesi, New Guinea) southward to Australia, Pacific Islands

Acrisione B. Nord. 1985. Asteraceae. 2 Chile

Acrista O. F. Cook = Prestoea Hook. f. Arecaceae

Acritochaete Pilg. 1902. Poaceae. 1 trop. Africa from Cameroun to Ethiopia and Tanzania

Acritopappus R. M. King et H. Rob. 1973. Asteraceae. 14 Brazil

Acriulus Ridl. 1883 (Scleria Bergius). Cyperaceae. 3 Central and East Africa, Madagascar

Acroanthes Raf. = Malaxis Sol. ex Sw. Orchidaceae

Acrobotrys K. Schum. et K. Krause. 1908. Rubiaceae. 1 Colombia

Acrocarpus Wight et Arn. 1836. Fabaceae. 2 trop. Asia from India to Indochina and Sumatra

Acrocephalus Benth. 1829. Lamiaceae. c. 130 trop. and South (1) Africa, trop. Asia

Acroceras Stapf. 1920. Poaceae. 21 tropics, chiefly Madagascar Acrochaena Lindl. = Monomeria Lindl. Orchidaceae

Acrochaeta Peter = Setaria P. Beauv. Poaceae

Acroclasia C. Presl = Mentzelia L. Loasaceae

Acroclinium A. Gray. 1852. Asteraceae. 3 Australia

Acrocoelium Baill. 1829. Icacinaceae. 1 Zaire

Acrocomia Mart. 1824. Arecaceae. 20 trop. America from Mexico southward to Argentina, West Indies

Acrodiclidium Nees et Mart. = Licaria Aubl. Lauraceae

Acrodon N. E. Br. 1927. Aizoaceae. 4 South Africa

Acrodryon Spreng. = Cephalanthus L. Rubiaceae

Acroelytrum Steud. = Lophatherum Brongn. Poaceae

Acroglochin Schrad. ex Schult. 1822. Chenopodiaceae. 1–2 Pakistan, India, Himalayas, western and central China

Acrolobus Klotzsch = Heisteria Jacq. Olacaceae

Acrolophia Pfitzer. 1887. Orchidaceae. 10 trop. and South Africa

Acronema Falc. ex Edgew. 1846. Apiaceae. 24 Himalayas from Kumaun to Bhutan, northeastern India, northern Burma, Tibet, southwestern China.

Acronodia Blume = Elaeocarpus L. Elaeocarpaceae

Acronychia J. R. Forst. et G. Forst. 1776. Rutaceae. 44 trop. Asia from India and Sri Lanka to southwestern China, Taiwan, Indochina, Malesia (esp. New Guinea—24), Solomon Is., northern, eastern, and southeastern Australia; Lord Howe I.; New Caledonia

Acropera Lindl. = Gongora Ruiz et Pav. Orchidaceae

Acrophyllum Benth. 1838. Cunoniaceae. 1 southeastern Australia

Acropogon Schltr. 1906. Sterculiaceae. 6 New Caledonia

Acroptilon Cass. 1827 (Centaurea L.). Asteraceae. 1 (A. repens) Asia Minor, Caucasus, Middle Asia, Iran to western China, Mongolia; Russia—European part, West Siberia

Acrorchis Dressler. 1990. Orchidaceae. 1 Costa Rica, Panama

Acrosanthes Eckl. et Zeyh. 1837. Molluginaceae. 6 South Africa (Cape)

Acroschizocarpus Gombocz = Christolea Cambess. ex Jacquem. Brassicaceae

Acrosepalum Pierre = Ancistrocarpus Oliv. Tiliaceae

Acrospelion Steud. = Trisetum Pers. Poaceae

Acrostachys (Benth.) Tiegh. = Helixanthera Lour. Loranthaceae

Acrostemon Klotzsch. 1838 (Eremia D. Don). Ericaceae. 9 South Africa (southwestern Cape)

Acrostephanus Tiegh. = Tapinanthus (Blume)

Acrostigma O. F. Cook et Doyle = Catoblastus H. A. Wendl. Arecaceae

Acrostoma Didr. = Remijia DC. Rubiaceae

Acrostylia Frapp. ex Cordem. = Gynorkis Thouars. Orchidaceae

Acrosynanthus Urb. 1913 (Remijia DC.). Rubiaceae. 7 Cuba

Acrotome Benth. ex Endl. 1838. Lamiaceae. 8 trop. (south of the equator) and South (6) Africa

Acrotrema Jack. 1820. Dilleniaceae. 10 southern India (1), Sri Lanka (8), A. costatum—Lower Burma, peninsular Thailand, Malay Peninsula

Acrotriche R. Br. 1810. Epacridaceae. 15 Australia (except Northern Territory), Tasmania

Acroxis Steud. = Muhlenbergia Schreb. Poaceae

Acrymia Prain. 1908. Lamiaceae. 1 Malay Peninsula

Acsmithia Hoogland. 1979 (Spiraeanthemum A. Gray). Cunoniaceae. 17 Moluccas, New Guinea, Australia, New Caledonia, Fiji

Actaea L. 1753. Ranunculaceae. 9 temp. Northern Hemisphere; Russia (3, from European part to Far East)

Actephila Blume. 1826. Euphorbiaceae. c. 40 India, Sri Lanka, Andaman Is., China, Southeast Asia, Malaysia, trop. Australia

Actephilopsis Ridl. = Trigonostemon Blume. Eurphorbiaceae

Actinanthella Balle. 1954. Loranthaceae. 2 Southeast and South (1, A. wyliei, Natal) Africa

Actinanthus Ehrenb. 1829 (or = Oenanthe L.) Apiaceae. 1 West Asia

Actinea Juss. = Helenium L. Asteraceae

Actinidia Lindl. 1836. Actinidiaceae. 54 China (52), East Asia north to Sakhalin and Kuril Is., south to Taiwan; Himalayas, northeastern India, Indochina, Malaysia

Actinobole Endl. 1843. Asteraceae. 4 Australia

Actinocarya Benth. 1876. Boraginaceae. 1 (A. tibetica) Pakistan, northwestern India, southwestern China

Actinocheita F. A. Barkley. 1937. Anacardiaceae. 1 Mexico

Actinochloa Roem. et Schult. = Chondrosum Desv. Poaceae

Actinochloris Steud. = Chloris Sw. Poaceae

Actinocladum McClure ex Soderstr. 1981. Poaceae. 1 Brazil

Actinodaphne Nees. 1831. Lauraceae. c. 100 India, Nepal, Burma, China, Japan, Indochina, Malaysia

Actinodium Schauer. 1836. Myrtaceae. 1 western Australia

Actinokentia Dammer. 1906. Arecaceae. 2 New Caledonia

Actinolema Fenzl. 1842. Apiaceae. 2 Caucasus (1, A. macrolema), Asia Minor, Syria, Iran, Iraq

Actinomeris Nutt. = Verbesina L. Asteraceae

Actinomorphe (Miq.) Miq. = Schefflera J. R. Forst. et G. Forst. Araliaceae

Actinophloeus (Becc.) Becc. = Ptychosperma Labill. Arecaceae

Actinophyllum Ruiz et Pav. = Schefflera J. R. Forst. et G. Forst. Araliaceae

Actinorhytis H. A. Wendl. et Drude. 1875. Arecaceae. 2 New Guinea (1), Solomon Is.

Actinoschoenus Benth. 1881. Cyperaceae. 4 Madagascar, India, Sri Lanka, southern China, Indochina, Malaysia

Actinoscirpus (Ohwi) R. W. Haines et Lye = Scirpus L. Cyperaceae

Actinoseris (Endl.) Cabrera. 1970. Asteraceae. 6 southeastern Brazil, Uruguay, northeastern Argentina

Actinospermum Elliott = Balduina Nutt. Asteraceae

Actinostemma Griff. 1841. Cucurbitaceae. 5 Asia from Himalayas to Japan; 1 (A. lobatum) extending to Russian Far East

Actinostemon Mart. ex Klotzsch = Gymnanthes Sw. Euphorbiaceae

Actinostigma Turcz. = Asterolasia F. Muell. Rutaceae

Actinostrobus Miq. 1845. Cupressaceae. 2 southwestern Australia

Actinotinus Oliv. = Viburnum L. Viburnaceae

Actinotus Labill. 1805. Apiaceae. 18. Australia, Tasmania, New Zealand
Actites Lander = Sonchus L. Asteraceae
Acuan Medik. = Desmanthus Willd. Fabaceae
Acunaeanthus Borhidi, Koml. et M. Moncada. 1981. Rubiaceae. 1 Cuba
Acustelma Baill. = Cryptolepis R. Br. Asclepiadaceae
Acyntha Medik. = Sansevieria Thunb. Dracaenaceae
Ada Lindl. 1854. Orchidaceae. 8 Andes of Colombia
Adactylus (Endl.) Rolfe. 1896 (Apostasia Blume). Orchidaceae. 3 India, Sri Lanka, Indochina, Malaysia, excl. Philippines
Adansonia L. 1753. Bombacaceae. 10 trop. Africa (1), Madagascar, Australia (2)
Adariantha Knoche = Pimpinella L. Apiaceae
Addisonia Rusby = Helogyne Nutt. Asteraceae
Adelaster Lindl. ex Veitch = Fittonia Coem. Acanthaceae
Adelbertia Meisn. = Merania Sw. Melastomataceae
Adelia L. 1759. Euphorbiaceae. 15 trop. and subtrop. America from Mexico to Brazil, West Indies
Adelia P. Browne = Forestiera Poir. Oleaceae
Adeliopsis Benth. 1862 (Hypserpa Miers). Menispermaceae. 1 northeastern Australia
Adelobotrys DC. 1828. Melastomataceae. 25 trop. America from Mexico and Jamaica to Peru, Bolivia, Brazil
Adelocaryum Brand. 1915 (Lindelofia Lehm.). Boraginaceae. 7 East Africa, western Himalayas
Adelodypsis Becc. = Dypsis Noronha ex Mart. Arecaceae
Adelonema Schott = Homalomena Schott. Araceae
Adelonenga Hook. f. = Hydriastele H. A. Wendl. et Drude. Arecaceae
Adelopetalum Fitzg. = Bulbophyllum Thouars. Orchidaceae
Adelosa Blume. 1850. Verbenaceae. 1 Madagascar
Adelostemma Hook. f. 1883. Asclepiadaceae. 3 Burma, China
Adelostigma Steetz. 1864. Asteraceae. 2 trop. Africa
Ademo Post et Kuntze = Euphorbia L. Euphorbiaceae
Adenacanthus Nees. 1832 (Strobilanthes Blume). Acanthaceae. 4 Burma, Southeast Asia, Malaysia
Adenandra Willd. 1809. Rutaceae. 18 South Africa (Cape)
Adenanthe Maguire, Steyerm. et Wurdack. 1961 (Tyleria Gleason). Sauvagesiaceae. 1 Venezuela
Adenanthellum B. Nord. 1979. Asteraceae. 1 South Africa
Adenanthemum B. Nord. = Adenanthellum B. Nord. Asteraceae
Adenanthera L. 1753. Fabaceae. 8 trop. and subtrop. Asia, Malaysia, Australia; widely cultivated and naturalized
Adenanthos Labill. 1805. Proteaceae. 32 southwestern (30) and southeastern (2) Australia
Adenarake Maguire et Wurdack. 1961. Sauvagesiaceae. 1 Venezuela
Adenaria Kunth. 1823. Lythraceae. 1 trop. America from southern Mexico to Argentina; West Indies
Adeneleuterophora Barb. Rodr. = Elleanthus C. Presl. Orchidaceae
Adenia Forssk. 1775. Passifloraceae. 93: 60—trop. and South (10) Africa, Madagascar (c. 20), Southeast Asia, Malaysia to Solomon Is., northern Australia (1, A. heterophylla)
Adenium Roem. et Schult. 1819. Apocynaceae. 5 trop. and subtrop. Africa, Arabian Peninsula, Socotra
Adenoa Arbo. 1977. Turneraceae. 1 Cuba
Adenocalymma Mart. ex Meisn. 1840. Bignoniaceae. 35 trop. America from Mexico to Argentina, West Indies
Adenocarpus DC. 1815. Fabaceae. 15 Canary Is., southern Europe, Morocco (9), Algeria; A. mannii—montane areas of trop. Africa south to Angola and Zambia
Adenocaulon Hook. 1829. Asteraceae. 5

western North America, Mexico, Guatemala, Chile, Argentina, Himalayas, East Asia; 1 (A. adhaerescens) extending to Russian Far East

Adenoceras Rchb. f. et Zoll. ex Baill. = Macaranga Thouars. Euphorbiaceae

Adenochilus Hook. f. 1853. Orchidaceae. 2 Australia, New Zealand

Adenochlaena Boivin ex Baill. = Cephalocroton Hochst. Euphorbiaceae

Adenocline Turcz. 1843. Euphorbiaceae. 8 South Africa

Adenocrepis Blume = Baccaurea Lour. Euphorbiaceae

Adenocritonia R. M. King et H. Rob. 1976. Asteraceae. 1 Jamaica

Adenodaphne S. Moore = Litsea Lam. Lauraceae

Adenodolichos Harms. 1902. Fabaceae. 22 trop. Africa, mainly Angola (9) and Zaire (13)

Adenoglossa B. Nord. 1976. Asteraceae. 1 trop. Africa

Adenogramma Rchb. 1828. Molluginaceae. 10 South Africa

Adenogynum Rchb. f. et Zoll. = Cladogynos Zipp. ex Spanoghe. Euphorbiaceae

Adenolinum Rchb. = Linum L. Linaceae

Adenolisianthus Gilg = Irlbachia Mart. Gentianaceae

Adenolobus (Harv. ex Benth.) Torre et Hillc. 1956 (Bauhinia L.). Fabaceae. 2 Angola, Namibia, Botswana, South Africa (northwestern Cape)

Adenoncos Blume. 1825. Orchidaceae. 10 Indochina, Malaysia

Adenoon Dalzell. 1850. Asteraceae. 1 India

Adenopappus Benth. 1840. Asteraceae. 1 Mexico

Adenopeltis Bertero ex A. Juss. 1832. Euphorbiaceae. 2 temp. South America

Adenopetalum Klotzsch et Garcke = Euphorbia L. Euphorbiaceae

Adenophaedra (Muell. Arg.) Muell. Arg. 1874. Euphorbiaceae. 4 Costa Rica, Panama, trop. South America

Adenophora Fisch. 1823. Campanulaceae. 50–60 temp. Eurasia, esp. China (c. 40); European part of Russia (1, A. liliifolia), the Crimea (A. taurica), Middle Asia, Siberia, Far East

Adenophyllum Pers. 1807 (Dyssodia Cav.). Asteraceae. 10 southwestern U.S., northern Mexico

Adenoplea Radlk. = Buddleja L. Buddlejaceae

Adenoplusia Radlk. = Buddleja L. Buddlejaceae

Adenopodia C. Presl. 1851. Fabaceae. 9 Mexico, Peru, Surinam, Brazil, trop. and South Africa (4)

Adenoporces Small. 1910 (Tetrapteris Cav.). Malpighiaceae. 1 West Indies

Adenopus Benth. = Lagenaria Ser. Cucurbitaceae

Adenorachis (DC.) Nieuwl. = Aronia Medik. Rosaceae

Adenorhopium Rchb. = Jatropha L. Euphorbiaceae

Adenorima Raf. = Euphorbia L. Euphorbiaceae

Adenoropium Pohl = Jatropha L. Euphorbiaceae

Adenosacme Wall. = Mycetia Reinw. Rubiaceae

Adenosciadium H. Wolff. 1927. Apiaceae. 1 southeastern Arabian Peninsula

Adenosma R. Br. 1810. Scrophulariaceae. 10 China, Indo-Malesia, Australia

Adenostachya Bremek. = Strobilanthes Blume. Acanthaceae

Adenostegia Benth. = Cordylanthus Nutt. ex Benth. Scrophulariaceae

Adenostemma J. R. Forst. et G. Forst. 1776. Asteraceae. 20 Central and trop. South America, West Indies, trop. Africa, trop. Asia, trop. Australia, Micronesia, and Polynesia

Adenostoma Hook. et Arn. 1832. Rosaceae. 2 California, Baja California

Adenostyles Cass. = Cacalia L. Asteraceae

Adenostylis Blume = Zeuxine Lindl. Orchidaceae

Adenothamnus D. Keck. 1935. Asteraceae. 1 Mexico (Baja California) Adenothola Lem. = Manettia L. Rubiaceae

Adenotrias Laub. et Spach = Hypericum L. Hypericaceae

Adesmia DC. 1825. Fabaceae. c. 230 South

America from Peru and southern Brazil to Tierra del Fuego

Adhatoda Hill = Justicia L. Acanthaceae

Adina Salisb. 1807. Rubiaceae. 8 Himalayas (from Nepal to Sikkim), India, Sri Lanka, Indochina, China, Korea, Japan

Adinandra Jack. 1822. Theaceae. 70 northeastern India, China, Indochina, Malaysia to New Guinea.

Adinandrella Exell = Ternstroemia Mutis ex L. f. Theaceae

Adinauclea Ridsdale. 1978. Rubiaceae. 1 Sulawesi, Moluccas

Adinobotrys Dunn = Callerya Endl. Fabaceae

Adisa Steud. = Sumbaviopsis J. J. Sm. Euphorbiaceae

Adisca Blume = Sumbaviopsis J. J. Sm. Euphorbiaceae

Adiscanthus Ducke. 1922. Rutaceae. 1 Brazil (Amazonia) Adlumia Raf. ex DC. 1821. Fumariaceae. 2 China (Manchuria), Korea, eastern North America; Russian Far East (1, A. asiatica).

Adnula Raf. = Pelexia Poit. ex Lindl. Orchidaceae

Adolphia Meisn. 1837. Rhamnaceae. 2 southwestern U.S., Mexico

Adonanthe Spach = Adonis L. Ranunculaceae

Adonis L. 1753. Ranunculaceae. 30 temp. Eurasia; esp. (17) European part of Russia, Caucasus, West and East Siberia

Adoxa L. 1753. Adoxaceae. 3 subarct. and temp. regions (incl. European part of Russia, Caucasus, Middle Asia, Siberia, Far East) south to Himalayas and central U.S. (Colorado and Illinois)

Adrastaea DC. = Hibbertia Andrews. Dilleniaceae

Adriana Gaudich. 1825. Euphorbiaceae. 5 Australia

Adromischus Lem. 1852. Crassulaceae. 27 Southwest (Namibia) and South (mostly Cape) Africa

Adrorhizon Hook. f. 1898. Orchidaceae. 1 India, Sri Lanka, Andaman Is.

Aechmanthera Nees. 1832. Acanthaceae. 3 Himalayas (from Kashmir to Bhutan), China (2)

Aechmea Ruiz et Pav. 1794. Bromeliaceae. 173 trop. America: Mexico, Guatemala, Belize, Honduras, Nicaragua, Costa Rica, Panama, Colombia, Venezuela, Trinidad, Guyana, Surinam, French Guyana, Ecuador, Peru, Brazil, Bolivia, Argentina; West Indies: Bahamas, Puerto Rico, Virgin Is., Leeward Is.

Aechmolepis Decne. = Tacazzea Decne. Asclepiadaceae

Aechmophora Steud. = Bromus L. Poaceae

Aedesia O. Hiffm. 1897. Asteraceae. 3 trop. West Africa

Aegialina Schult. = Rostraria Trin. Poaceae

Aegialitis Trin. = Rostraria Trin. Poaceae

Aegialitis R. Br. 1810. Plumbaginaceae. 2 coastal Bangladesh and Burma, Andaman and Nicobar Is., eastern Malaysia, trop. Australia

Aegiceras Gaertn. 1788. Myrsinaceae. 2 mangroves South and Southeast Asia, New Guinea, Bismark Arch., Solomon Is., northeastern Australia, Lord Howe I

Aegicon Adans. = Aegilops L. Poaceae

Aegilemma A. Löve = Aegilops L. Poaceae

Aegilonearum A. Löve = Aegilops L. Poaceae

Aegilopodes A. Löve = Aegilops L. Poaceae

Aegilops L. 1753. Poaceae. c. 25 Mediterranean, Southwest Asia eastward to Pakistan; southern European part of Russia, Caucasus, Middle Asia

Aeginetia L. 1753. Orobanchaceae. 3 Pakistan, China, Japan, Southeast Asia

Aegiphila Jacq. 1767. Verbenaceae. 175 trop. America, West Indies

Aegle Corrêa. 1800. Rutaceae. 1 (A. marmelos) Sri Lanka, Burma, Indochina

Aeglopsis Swingle. 1912. Rutaceae. 5 trop. Africa

Aegochloa Benth. = Navarretia Ruiz et Pav. Polemoniaceae

Aegokeras Raf. 1840. Apiaceae. 1 Turkey

Aegonychon Gray = Lithospermum L. Boraginaceae

Aegopicron Giseke = Maprounea Aubl. Euphorbiaceae

Aegopodium L. 1753. Apiaceae. 5–7 Eurasia, incl. (4) European part of Russia, Caucasus, Middle Asia, and Siberia to Enisei

Aegopogon Humb. et Bonpl. ex Willd. 1806. Poaceae. 3 southern U.S., Mexico, Central and South America

Aegopordon Boiss. 1846 (Jurinea Cass.). Asteraceae. 1 Southwest Asia

Aellenia Ulbr. = Halothamnus Jaub. et Spach. Chenopodiaceae

Aeluropus Trin. 1822. Poaceae. 6 Mediterranean, Russia (6, southern European part, West Siberia), Ethiopia, Arabian Peninsula, Iraq, Iran, Middle Asia, Afghanistan, Pakistan, China, Mongolia, India, Sri Lanka

Aenictophyton A. T. Lee. 1973. Fabaceae. 1 Australia

Aeollanthus Mart. ex Spreng. 1825. Lamiaceae. 40 trop. and subtrop. Africa

Aeoniopsis Rech. f. = Bukinezia Lincz. Plumbaginaceae

Aeonium Webb et Berthel. 1840. Crassulaceae. 38 Macaronesia (36), North Africa from Morocco to Ethiopia, Somalia, Uganda, Kenya, Tanzania, Arabian Peninsula

Aequatorium B. Nord. 1978. Asteraceae. 2 Colombia, Ecuador

Aerangis Rchb. f. 1865. Orchidaceae. 50–70 trop. and South Africa (26), Madagascar, Comoro and Mascarene Is., Sri Lanka (1, A. hologlottis)

Aeranthes Lindl. 1824. Orchidaceae. 46 Madagascar and Mascarene Is.

Aerides Lour. 1790. Orchidaceae. 20 trop. Asia from northwestern Himalayas and India to southern China and Indochina, Malaysia (excl. New Guinea)

Aeridostachya (Hook. f.) Brieger = Eria Lindl. Orchidaceae

Aerisilvaea Radcl. = Sm. 1990. Euphorbiaceae. 1 Tanzania

Aerva Forssk. 1775. Amaranthaceae. 10 North, trop., and South (rare) Africa, Seychelles Is., Chagos Arch., Asia from Palestine, Arabian Peninsula to India, Sri Lanka, Burma, southern China, Indochina, Malaysia to New Guinea

Aesandra Pierre = Diploknema Pierre. Sapotaceae

Aeschrion Vell. Conc. = Picrasma Blume. Simaroubaceae

Aeschynanthus Jack. 1823. Gesneriaceae. 100–140 India, Sri Lanka, Himalayas from Nepal to Bhutan, Burma, southwestern and southern China, Hainan I., Hong Kong, Taiwan, Indochina, ? Java, Philippines, Solomon Is.

Aeschynomene L. 1753. Fabaceae. c. 150 trop. and subtrop. Africa, Asia, and America

Aesculus L. 1753. Hippocastanaceae. 16 southeastern Europe (1, H. hippocastanum L.); Asia: from western Himalayas to Japan and Indochina (8); North America (7—temp. eastern U.S., California)

Aetanthus (Eichler) Engl. 1889. Loranthaceae. 10 Andes of Colombia, Ecuador, and Peru, Venezuela

Aetheocephalus Gagnep. = Athroisma DC. Asteraceae

Aetheolaena Cass. 1827 (Lasiocephalus Schltdl.). Asteraceae. 20 trop. South America (Andes)

Aetheolirion Forman. 1962. Commelinaceae. 1 Thailand

Aetheopappus Cass. 1827. Asteraceae. 3 Asia Minor, Caucasus

Aetheorhiza Cass. 1827. Asteraceae. 1 Mediterranean

Aethephyllum N. E. Br. 1928 (Cleretum N. E. Br.). Aizoaceae. 1 South Africa (Cape)

Aetheria Blume ex Endl. = Stenorrhynchos Rich. ex Spreng. Orchidaceae

Aethiocarpa Vollesen. 1986. Sterculiaceae. 1 Somalia

Aethionema R. Br. 1812. Brassicaceae. 46 Mediterranean, Southwest Asia to Middle Asia (13), Afghanistan, and Pakistan

Aethonopogon Kuntze = Polytrias Hack. Poaceae

Aethusa L. 1753. Apiaceae. 1 Europe, North Africa, West Asia, incl. Caucasus

Aetia Adans. = Combretum Loefl. Combretaceae

Aetoxylon (Airy Shaw) Airy Shaw. 1950. Gonystylaceae. 1 Borneo

Aextoxicon Ruiz et Pav. 1794. Aextoxicaceae. 1 Chile

Affonsea A. St.-Hil. 1833. Fabaceae. 7 eastern Brazil

Afgekia Craib. 1927. Fabaceae. 3 southern China, Burma, Thailand

Afrachneria Sprague. 1922 [Pentaschistis (Nees) Spach]. Poaceae. 10 South Africa

Afraegle (Swingle) Engl. 1915. Rutaceae. 4 West Africa

Afrafzelia Pierre = Afzelia Sm. Fabaceae

Aframmi C. Norman. 1929. Apiaceae. 1 Angola

Aframomum K. Schum. 1904. Zingiberaceae. 50 trop. Africa from Senegal and Ethiopia to Angola and Mozambique, Madagascar, Mascarene Is.

Afrardisia Mez = Ardisia Sw. Myrsinaceae

Afraurantium A. Chev. 1949. Rutaceae. 1 West Africa

Afridia Duthie = Nepeta L. Lamiaceae

Afrobrunnichia Hutch. et Dalziel. 1927. Polygonaceae. 2 trop. West Africa from Liberia to Congo (A. erecta) and Angola (A. africana)

Afrocalathea K. Schum. 1902. Maranthaceae. 1 West Africa

Afrocarpus (Buchholz et E. Gray) C. N. Page. 1988. Podocarpaceae. 3–6 Ethiopia, Congo, Uganda, Zaire southward to Cape and Southeast Africa

Afrocarum Rauschert. 1982. Apiaceae. 1 trop. Africa (Central African Rep., Zaire, Zambia, Zimbabwe, Malawi, Tanzania, Mozambique)

Afrocrania (Harms) Hutch. 1942. Cornaceae. 1 trop. Africa

Afrodaphne Stapf = Beilschmiedia Nees. Lauraceae

Afrofittonia Lindau. 1913. Acanthaceae. 1 trop. West Africa

Afroguatteria Boutique. 1951. Annonaceae. 1 trop. Africa

Afrohamelia Wernham = Atractogyne Pierre. Rubiaceae

Afroknoxia Verdc. 1981 (Knoxia L.). Rubiaceae. 1 Zaire

Afrolicania Mildbr. = Licania Aubl. Chrysobalanaceae

Afroligusticum C. Norman. 1927. Apiaceae. 1 trop. Africa (Uganda, Burundi, Zaire, Tanzania)

Afrolimon Lincz. 1979. Plumbaginaceae. 7 South Africa (Cape)

Afromendoncia Gilg ex Lindau = Mendoncia Vell. ex Vand. Acanthaceae

Afroraphidophora Engl. = Rhaphidophora Hassk. Araceae

Afrormosia Harms = Pericopsis Thwaites. Fabaceae

Afrosersalisia A. Chev. = Synsepalum (A. DC.) Daniell. Sapotaceae

Afrosison H. Wolff. 1912. Apiaceae. 3 trop. Africa

Afrostyrax Perkins et Gilg. 1909. Huaceae. 2 trop. Africa

Afrothismia (Engl.) Schltr. 1906. Burmanniaceae. 2 trop. West Africa

Afrotrewia Pax et K. Hoffm. 1914. Euphorbiaceae. 1 trop. West Africa

Afrotrichloris Chiov. 1915. Poaceae. 2 trop. East Africa

Afrotrilepis (Gilly) J. Raynal. 1963. Cyperaceae. 2 trop. West Africa

Afrotysonia Rauschert. 1982. Boraginaceae. 3 trop. and South Africa

Afrovivella A. Berger = Rosularia (DC.) Stapf. Crassulaceae

Afzelia Sm. 1798. Fabaceae. 13 trop. and South (1, A. quanzensis) Africa (7), trop. Asia

Afzeliella Gilg = Guyonia Naudin. Melastomataceae

Agalinis Raf. 1837. Scrophulariaceae. 5 c. 60 North (c. 20), Central and South America

Agallis Phil. 1864. Brassicaceae. 1 Chile

Agalma Miq. = Schefflera J. R. Forst. et G. Forst. Araliaceae

Agalmanthus (Endl.) Hombr. et Jacquinot = Metrosideros Banks ex Gaertn. Myrtaceae

Agalmyla Blume. 1826. Gesneriaceae. 6 Malaysia

Agaloma Raf. = Euphorbia L. Euphorbiaceae

Aganippea Moçino et Sessé ex DC. = Jaegeria Kunth. Asteraceae

Aganisia Lindl. 1839. Orchidaceae. 3 Colombia, Venezuela, Trinidad, Guyana, Surinam, Brazil, Peru

Aganonerion Pierre ex Spire. 1894. Apocynaceae. 1 Indochina

Aganope Miq. 1855. Fabaceae. 6 trop. Africa (4, from Sierra Leone to Angola and Zaire), trop. Asia

Aganosma (Blume) G. Don. 1837. Apocynaceae. 16 India, China, Indochina, Philippines

Agaosizia Spach = Camissonia Link. Onagraceae

Agapanthus L'Hér. 1789. Alliaceae. 10 (variable) South Africa (Transvaal, Natal, Orange Prov., Cape)

Agapetes D. Don ex G. Don. 1834. Ericaceae. c. 100 Sikkim, southeastern Tibet, northeastern India, Burma, southern China (Yunnan), northern Vietnam, Thailand, Malaysia, esp. New Guinea, northeastern Australia (2), New Caledonia (1), Fiji (1)

Agarista D. Don ex G. Don. 1834. Ericaceae. 30 trop. South America (28: 3 Andes, c. 25 southeastern Brazil), 1 (A. salicifolia) central Africa, Madagascar, Mascarene Is.

Agastache Clayton ex Gronov. 1762. Lamiaceae. 30 North America, East Asia (1, A. rugosa from Tibet to Ussuri, Korea and Japan, south Taiwan)

Agastachys R. Br. 1810. Proteaceae. 1 Tasmania

Agasyllis Spreng. 1813. Apiaceae. 1 Great Caucasus

Agatea A. Gray. 1852. Violaceae. 1 (very polymorph.) New Guinea, Solomon Is., New Caledonia, Fiji, Tonga

Agathelpis Choisy. 1824. Selaginaceae. 3 South Africa (Cape)

Agathis Salisb. 1807. Araucariaceae. 21 Malay Peninsula, Sumatra, Borneo, Philippines, Sulawesi, Moluccas, New Guinea, New Britain, coastal Queensland, northern New Zealand, New Caledonia, Vanuatu (New Hebrides), Fiji

Agathisanthemum Klotzsch. 1861 (Hedyotis L.). Rubiaceae. 5–6 trop. and South (Transvaal) Africa, Madagascar, Comoro Is.

Agathomeris de Launay = Calomeria Vent. Asteraceae

Agathophora (Fenzl) Bunge. 1863. Chenopodiaceae. 5 Morocco, Algeria, Egypt, Israel, Jordan, Arabian Peninsula, Syria, Iraq, Iran?, Pakistan

Agathosma Willd. 1809. Rutaceae. 135 South Africa (esp. Cape)

Agauria (DC.) Hook. f. = Agarista D. Don ex G. Don. Ericaceae

Agave L. 1753. Agavaceae. 150–300 America from southern U.S. to trop. South America

Agdestis Moçino et Sessé ex DC. 1817. Agdestidaceae. 1 southern U.S. (Texas and southern Florida), Mexico to Nicaragua, Cuba

Agelaea Sol. ex Planch. 1850. Connaraceae. 50 trop. Africa, Madagascar, Burma, southern China, Indochina, Malaysia (4, Malay Peninsula, Sumatra, Java, Bali, Borneo, Philippines, Moluccas)

Agelanthus Tiegh. = Tapinanthus (Blume) Rchb. Loranthaceae

Agenium Nees. 1836. Poaceae. 4 trop. America

Ageomoron Raf. 1840. Apiaceae. 4 from eastern Mediterranean to Iran

Ageratella A. Gray ex S. Watson. 1887. Asteraceae. 1 Mexico

Ageratina Spach. 1841. Asteraceae. 230 America, West Indies

Ageratinastrum Mattf. 1932. Asteraceae. 5 trop. East Africa

Ageratum L. 1753. Asteraceae. 45 U.S. (Florida), Mexico, Guatemala, El Salvador, Honduras, Belize, Nicaragua, Costa Rica, Panama, Venezuela, Guyana, Brazil; West Indies

Agiabampoa Rose ex O. Hoffm. = Alvordia Brandegee. Asteraceae

Agianthus Greene = Streptanthus Nutt. Brassicaceae

Aglaia Lour. 1790. Meliaceae. 130 South Asia from India to southern China and Taiwan, Malaysia, Australia, Melanesia, Polynesia

Aglaodorum Schott. 1849. Araceae. 1 Sumatra, Borneo

Aglaonema Schott. 1829. Araceae. 50 northeastern India, Burma, southern China, Indochina, Malaysia

Aglossorhyncha Schltr. 1905. Orchidaceae. 14 eastern Malaysia, Palau I., Bismarck Arch., Bougainville and Solomon Is., Vanuatu, Fiji

Agnirictus Schwantes = Stomatium Schwantes. Aizoaceae

Agnostus A. Cunn. ex Loudon = Stenocarpus R. Br. Proteaceae

Agonandra Miers ex Benth. 1862. Opiliaceae. 12 America from Mexico to trop. South America

Agonis (DC.) Sweet. 1830. Myrtaceae. 12 western Australia

Agoseris Raf. 1819 (Troximon Nutt.). Asteraceae. 10 western North (9 from British Columbia to western Mexico and New Mexico) and temp. South (1) America

Agrianthus Mart. ex DC. 1836. Asteraceae. 6 Brazil

Agrimonia L. 1753. Rosaceae. c. 20 temp. Northern Hemisphere, esp. Russia (8, from European part to Far East), montane trop. and South Africa (1, A. odorata)

Agriophyllum M. Bieb. 1819. Chenopodiaceae. 5 southern European part of Russia, northern Caucasus, Armenia, Azerbaijan, West Siberia, Middle Asia; Arabian Peninsula, Iran, Afghanistan, Pakistan, western and northwestern China, Mongolia

Agriphyllum Juss. = Berkheya Ehrh. Asteraceae

Agrocharis Hochst. 1844. Apiaceae. 4 Northeast, trop., and South Africa

Agropyron Gaertn. 1770. Poaceae. 15 temp. Eurasia (esp. Russia, c. 10–15), North America

Agropyropsis (Batt. et Trab.) A. Camus. 1935. Poaceae. 2 North Africa

Agrostemma L. 1753. Caryophyllaceae. 3 Eurasia, incl. (3) European part of Russia, Caucasus, western and eastern Siberia, Far East, Middle Asia

Agrostis L. 1753. Poaceae. c. 150 temp. and warm regions, montane tropics

Agrostistachys Dalzell. 1850. Euphorbiaceae. 10 India, Sri Lanka to western Malaysia

Agrostocrinum F. Muell. 1860. Phormiaceae. 1 southwestern Australia

Agrostophyllum Blume. 1825. Orchidaceae. 40–50 Seychelles Is. (1, A. occidentale), trop. Asia northward to Himalayas, Caroline Is. (Palau), Malaysia from Malay Peninsula to Philippines (13) and New Guinea, Bougainville and Solomon Is., Vanuatu, Fiji, Samoa Is.

Aguiaria Ducke. 1935. Bombacaceae. 1 Brazil

Ahernia Merr. 1909. Flacourtiaceae. 1 Hainan I., Philippines

Ahouai Hill = Thevertia L. Apocynaceae

Ahzolia Standl. et Steyerm. = Sechium P. Browne. Cucurbitaceae

Aichryson Webb et Berthel. 1840. Crassulaceae. 15 Micronesia, Northwest Africa

Aidia Lour. 1790. Rubiaceae. 24 trop. Africa (8, 3 of which endemic to S. Tomé), trop. Asia, Malaysia to eastern Australia, Pacific islands (Marianas, Caroline, Marshall, Nauru, Banaba, and Gilbert islands)

Aidiopsis Tirveng. 1986 (Randia L.). Rubiaceae. 1 Malaysia

Aidomene Stopp. 1967. Asclepiadaceae. 1 Angola

Ailanthus Desf. 1788. Simaroubaceae. 5 tropics and subtropics, 1 (A. altissima) from Middle Asia through India, southern China, Indochina, and Malaysia to Solomon Is. and Australia

Ainea Ravenna. 1979. Iridaceae. 1 southern Mexico

Ainsliaea DC. 1838. Asteraceae. 40 East and Southeast Asia south to western Malaysia

Ainsworthia Boiss. = Tordylium L. Apiaceae

Aiolon Lunell. 1916. Ranunculaceae. 1 North America

Aiolotheca DC. = Zaluzania Pers. Asteraceae

Aiouea Aubl. 1775. Lauraceae. c. 20 Guatemala (2), Costa Rica (1), Colombia, Venezuela, Trinidad and Tobago, Guyana,

Surinam, French Guyana, Peru, Brazil, Bolivia, Paraguay (1)

Aiphanes Willd. 1807. Arecaceae. 41 West Indies, northern South America, esp. Colombia and Ecuador

Aipyanthus Steven = Nonea Medik. Boraginaceae

Aira L. 1753. Poaceae. 12 West Europe, Mediterranean, Northwest and South Africa, Southwest Asia

Airopsis Desv. 1809. Poaceae. 1 South Europe, Northwest Africa, Crimea, Caucasus, Middle Asia

Airosperma Lauterb. et K. Schum. 1900. Rubiaceae. 6 New Guinea (4), Fiji

Airyantha Brummitt. 1968. Fabaceae. 2 trop. Africa (1, A. schweinfurthii, from Ivory Coast to Central African Rep. and Zaire), Borneo (1)

Aisandra Pierre ex Airy Shaw = Diploknema Pierre. Sapotaceae

Aistocaulon Poelln. ex H. J. Jacobson = Nananthus N. E. Br. Aizoaceae

Aistopetalum Schltr. 1914. Cunoniaceae. 2 New Guinea

Aitchisonia Hemsl. ex Aitch. 1882. Rubiaceae. 1 Afghanistan

Aitonia Thunb. = Nymania Lindb. Meliaceae

Aizoanthemum Dinter ex Friedrich. 1957. Aizoaceae. 4 southern Angola, Namibia

Aizoon L. 1753. Aizoaceae. 25 Europe (1), Mediterranean, Africa, South Asia, Australia

Aizopsis Grulich = Sedum L. Crassulaceae

Ajania Poljakov. 1955 [Dendranthema (DC.). Des Moul.]. Asteraceae. 22 West Siberia, Middle Asia, Afghanistan, northern India, China, Mongolia, Korea, Japan, ? Alaska

Ajaniopsis C. Shih. 1978. Asteraceae. 1 Tibet

Ajuga L. 1753. Lamiaceae. c. 100 temp. and subtrop. Eurasia, montane trop. East and South (1) Africa, Madagascar, temp. and subtrop. Australia; Russia (16, from southern European part and Caucasus to Far East)

Ajugoides Makino. 1915 (Ajuga L.). Lamiaceae. 1 Japan

Akania Hook. f. 1862. Akaniaceae. 1 eastern Australia

Akebia Decne. 1837. Lardizabalaceae. 5 subtrop. East Asia to southern China (Yunnan)

Akersia Buining = Cleistocactus Lem. Cactaceae

Aladenia Pichon = Farquharia Stapf. Apocynaceae

Alafia Thouars. 1806. Apocynaceae. 26 trop. Africa, Madagascar

Alajja Ikonn. 1971(Lamium L.). Lamiaceae. 3 Middle Asia; Afghanistan, western China

Alamania La Llave et Lex. 1824. Orchidaceae. 1 Mexico

Alangium Lam. 1783. Alangiaceae. 30 trop. Africa, Madagascar, Comoro Is., trop. and East Asia, Malaysia to New Guinea and adjacent islands, Australia (Queensland, New South Wales), New Caledonia, Fiji

Alania Endl. 1836. Asphodelaceae. 1 southeastern Australia

Alatoseta Compton. 1931. Asteraceae. 1 South Africa

Alberta E. Mey. 1838. Rubiaceae. 5 South Africa, Madagascar

Albertia Regel et Schmalh. 1878. Apiaceae. 1 southern Middle Asia, Afghanistan

Albertinia Spreng. 1820. Asteraceae. 1 Brazil

Albertisia Becc. 1877. Menispermaceae. 18 trop. and subtrop. Africa (12), India (Assam), southern China, Indochina, Malaysia (3, Malay Peninsula, Sumatra, Java, Banka I., Borneo, southern Sulawesi, Moluccas, Aru Is., New Guinea)

Albertisiella Pierre ex Aubrév. = Pouteria Aubl. Sapotaceae

Albidella Pichon = Echinodorus Rich. ex Engelm. Alismataceae

Albizia Durazz. 1772. Fabaceae. c. 150 tropics, a few species in warm–temp. regions

Albizzia Benth. = Albizia Durazz. Fabaceae

Albovia Schischk. = Pimpinella L. Apiaceae

Albraunia Speta. 1982(Antirrhinum L.). Scrophulariaceae. 3 Southwest Asia

Albuca L. 1762. Hyacinthaceae. c. 30 trop. and (chiefly) South Africa, Arabian Peninsula

Alcantara Glaz. ex G. M. Barroso. 1909. Asteraceae. 1 Brazil

Alcea L. 1753. Malvaceae. 60 Europe, Mediterranean, West and Southwest Asia eastward to western China and Afghanistan; esp. (c. 40) in European part of Russia, Caucasus, West Siberia, and Middle Asia; A. rosea cosmopolitan in cultivation

Alchemilla L. 1753. Rosaceae. c. 300 temp. Northern Hemisphere, esp. (c. 260) Russia, Caucasus, and Middle Asia; montane tropics

Alchornea Sw. 1788. Euphorbiaceae. 50–70 tropics

Alchorneopsis Muell. Arg. 1865. Euphorbiaceae. 3 trop. America, West Indies

Alcimandra Dandy. 1927(Magnolia L.). Magnoliaceae. 1 eastern Himalayas, Assam, southern China

Alcineanthus Merr. = Neoscortechinia Pax. Euphorbiaceae

Alciope DC. ex Lindl. 1836. Asteraceae. 2 South Africa (Cape)

Alcmene Urb. = Duguetia A. St. = Hil. Annonaceae

Alcoceratothrix Nied. = Byrsonima Rich. ex Kunth. Malpighiaceae

Aldama La Llave. 1824. Asteraceae. 2 trop. America from Mexico to Colombia and Venezuela

Aldina Endl. 1840. Fabaceae. 15 trop. South America

Aldinia Raf. = Croton L. Euphorbiaceae

Aldrovanda L. 1753. Droseraceae. 1 (A. vesiculosa) Central Europe, Caucasus, Middle Asia, Far East of Russia, trop. Africa, Asia (except subarctic regions), Timor I., northern Australia.

Alectorurus Makino = Comospermum Rauschert. Asphodelaceae

Alectra Thunb. 1784. Scrophulariaceae. 40 trop. and South Africa, trop. Asia, trop. America and West Indies (2)

Alectryon Gaertn. 1788. Sapindaceae. c. 30 eastern Malaysia, Solomon Is., Australia, New Zealand, New Caledonia, Vanuatu Fiji, Samoa Is., Hawaiian Is. (1)

Aleisanthia Ridl. 1920. Rubiaceae. 2 Malay Peninsula

Alepidea F. Delaroche. 1808. Apiaceae. c. 30 trop. and South (27) Africa

Alepidocalyx Piper = Phaseolus L. Fabaceae

Alepidocline S. F. Blake. 1934. Asteraceae. 1 Guatemala

Alepis Tiegh. 1894 (Elyrtanthe Blume). Loranthaceae. 1 New Zealand

Aletes J. M. Coult. et Rose. 1888. Apiaceae. 15–20 North and Central America

Aletris L. 1753. Melanthiaceae. 15 East Asia, Malaysia, North America

Aleurites J. R. Forst. et G. Forst. 1776. Euphorbiaceae. 2 trop. Asia, Malaysia, western Oceania

Aleuritia (Duby) Opez = Primula L. Primulaceae

Alexa Moq. 1849. Fabaceae. 10 trop. South America

Alexandra Bunge. 1843. Chenopodiaceae. 1 Middle Asia (Kazakhstan, Uzbekistan)

Alexeya Pakhomova = Paraquilegia J. R. Drumm. et Hutch. Ranunculaceae

Alexgeorgea Carlquist. 1976 (Restio Rottb.). Restionaceae. 2 southwestern Australia

Alexitoxicon St.-Lag. = Vincetoxicum Wolf. Asclepiadaceae

Alfaroa Standl. 1927. Juglandaceae. 8 trop. America from Mexico to Colombia

Alfaropsis Iljinsk. 1993. Juglandaceae. 1 Southeast Asia extending to western India and Central China

Alfonsia Kunth = Elaeis Jacq. Arecaceae

Alfredia Cass. 1816(Carduus L.). Asteraceae. 5 Middle Asia, West Siberia, western China

Algernonia Baill. 1858. Euphorbiaceae. 3 Brazil

Alhagi Hill. 1753. Fabaceae. 4 arid and semiarid regions of the Mediterranean, Chad, Sudan, Arabian Peninsula, Caucasus and Middle Asia (1, A. maurorum), Iraq, Iran, Afghanistan, Pakistan, India, Nepal

Alibertia A. Rich. ex DC. 1830. Rubiaceae. 35 trop. America, West Indies

Alibrexia Miers = Nolana L. f. Solanaceae

Alicabon Raf. = Withania Pauquy. Solanaceae

Alicastrum P. Browne = Brosimum Sw. Moraceae

Aliciella Brand = Gilia Ruiz et Pav. Polemoniaceae

Aliella Qaiser et Lack. 1986. Asteraceae. 3 Morocco

Alifana Raf. = Brachyotum (DC.) Triana. Melastomataceae

Alifanus Adans. = Rhexia L. Melastomataceae

Aligera Suksd. = Plectritis (Lindl.) DC. Valerianaceae

Aliniella J. Raynal = Alinula J. Raynal. Cyperaceae

Alinula J. Raynal. 1977. Cyperaceae. 4 Ethiopia, Kenya, Uganda, Tanzania, Mozambique, Namibia, Zambia, Malawi, South Africa, Madagascar

Alisma L. 1753. Alismataceae. 10 temp. Northern Hemisphere, esp. (7) Russia, Caucasus, Middle Asia; Australia

Alistilus N. E. Br. 1921. Fabaceae. 2 southern Africa (A. bechuanicus—Namibiam Botswana, South Africa), Madagascar (1)

Alkanna Tausch. 1824. Boraginaceae. c. 50 southern Europe, Mediterranean, Southwest Asia to Caucasus (3) and Iran

Alkibias Raf. = Vernonia Schreb. Asteraceae

Allaeophania Thwaites. 1859. Rubiaceae. 3 Sri Lanka, Malay Peninsula

Allagopappus Cass. 1828. Asteraceae. 2 Canary Is.

Allagoptera Nees. 1821 (Polyandrococos Barb. Rodr.). Arecaceae. 4 Brazil, Bolivia, Argentina, Paraguay

Allamanda L. 1771. Apocynaceae. 12 West Indies and trop. South America

Allanblackia Oliv. ex Benth. 1867. Clusiaceae. 10 trop. Africa

Allantoma Miers. 1874. Lecythidaceae. 1 (A. lineata) Amazonian Venezuela and Brazil

Allantospermum Forman. 1965. Irvingiaceae. 2 Madagascar (1), Borneo (1)

Allardia Decne. 1836. Asteraceae. 8 Afghanistan, Pakistan, Tibet, Himalayas, Pamir-Alai, Tian Shan, Altai, Mongolia

Alleizettea Dubard et Dop = Danais Comm. ex Vent. Rubiaceae

Alleizettella Pit. 1923. Rubiaceae. 1 Indochina

Allenanthus Standl. 1940. Rubiaceae. 2 Central America

Allendea La Llave et Lex. = Liabum Adans. Asteraceae

Allenrolfea Kuntze. 1891. Chenopodiaceae. 3 North (southwestern U.S.), Central, and South America

Allexis Pierre. 1898. Violaceae. 3 trop. West Africa

Alliaria Heist. ex Fabr. 1759. Brassicaceae. 5 Europe, Northwest Africa, West Asia to Caucasus, Middle Asia, and eastward to China and Nepal

Allionia L. 1759. Nyctaginaceae. 3 trop. America from southwestern and central U.S. and Mexico to Chile and Argentina; West Indies

Allioniella Rydb. 1902 (Mirabilis L.). Nyctaginaceae. 25 chiefly South America, Himalayas (1)

Allium L. 1753. Alliaceae. c. 700 Northern Hemisphere, most concentrated in Asia Minor, Iran, Middle Asia, and Russia (from European part to Far East); 1 (A. dregeanum) South Africa

Allmania R. Br. ex Wight. 1834. Amaranthaceae. 1 trop. Asia from India and Sri Lanka eastward to southern China, Indochina, Indonesia, and Philippines

Allmaniopsis Suess. 1950. Amaranthaceae. 1 East Africa (Kenya)

Alloburkillia Whitmore = Burkilliodendron Sastry. Fabaceae

Allocalyx Cordem. 1895 (Monocardia Pennell). Scrophulariaceae. 1 Reunion I.

Allocarya Greene. 1887 (Plagiobothrys Fisch. et C. A. Mey.). Boraginaceae. 83 Pacific America (80), Australia (3), 1 (A. orientalis) Kamchatka

Allocaryastrum A. Brand = Plagiobothrys Fisch. et C. A. Mey. Boraginaceae

Allocassine N. Robson. 1965. Celastraceae. 1–2 trop. Southeast and South Africa

Allocasuarina L. A. S. Johnson. 1982. Casuarinaceae. c. 60 Australia

Allocheilos W. T. Wang. 1983. Gesneriaceae. 1 southwestern China

Allochilus Gagnep. 1932 (Goodyera R. Br.). Orchidaceae. 1 Southeast Asia

Allochrusa Bunge ex Boiss. 1867 (Acanthophyllum C. A. Mey.). Caryophyllaceae. 7 Caucasus (2), Middle Asia (4); Turkey, Iran

Alloeochaete C. E. Hubb. 1940. Poaceae. 6 trop. Africa: Malawi, Angola, Tanzania, Mozambique

Alloispermum Willd. 1807. Asteraceae. 12 Central and western South America

Allolepis Soderstr. et H. F. Decker. 1965. Poaceae. 1 southern U.S., Mexico

Allomaieta Gleason. 1929. Melastomataceae. 1 Colombia

Allomarkgrafia Woodson. 1932. Apocynaceae. 6 Central and trop. South America

Allomorphia Blume. 1831 (or = Oxyspora DC.). Melastomataceae. c. 30 China, Indochina, Malaysia

Alloneuron Pilg. 1905 (or = Loreya DC.). Melastomataceae. 6 Colombia and Peru

Allophyllum (Nutt.) A. D. Grant et V. E. Grant. 1955. Polemoniaceae. 5 western U.S.: Oregon, California, Nevada, Arizona; Baja California

Allophylus L. 1753. Sapindaceae. 1 (A. cobbe—very polymorph.) trop. and subtrop. Africa, Asia, New Guinea, and Australia

Alloplectus Mart. 1829. Gesneriaceae. 65 trop. America, West Indies

Allopterigeron Dunlop. 1981. Asteraceae. 1 Australia (Northern Territory, Queensland)

Allosanthus Radlk. 1933. Sapindaceae. 1 Peru

Alloschemone Schott. 1858 (Scindapsus Schott). Araceae. 1 Brazil

Alloschmidia H. E. Moore. 1978. Arecaceae. 1 northeastern New Caledonia

Allosidastrum (Hochr.) Krapov., Fryxell et D. M. Bates. 1988. Malvaceae. 4 Mexico, Belize, Guatemala, Salvador, Nicaragua, Costa Rica, Panama, Colombia, Bolivia, Venezuela, Surinam, Cuba, Puerto Rico, Haiti, Jamaica

Allostigma W. T. Wang. 1984. Gesneriaceae. 1 southern China

Allosyncarpia S. T. Blake. 1977. Myrtaceae. 1 northern Australia

Alloteropsis J. Presl. 1830. Poaceae. 5 tropics of the Old World

Allotropa Torr. et A. Gray ex A. Gray. 1858. Ericaceae. 1 western U.S.

Allowissadula D. M. Bates. 1978. Malvaceae. 11 North America from Texas to Mexico

Allowoodsonia Markgr. 1967. Apocynaceae. 1 Solomon Is.

Alloxylon P. H. Weston et Crisp. 1991. Proteaceae. 4 Australia (3, Queensland, New South Wales), Aru Is., and southern Guinea (1)

Allozygia Naudin = Oxyspora DC. Melastomataceae

Alluaudia (Drake) Drake. 1903. Didiereaceae. 6 Madagascar

Alluaudiopsis Humbert et Choux. 1934. Didiereaceae. 2 Madagascar

Almaleea Crisp et P. H. Weston. 1991. Fabaceae. 5 southeastern Australia from northern New South Wales to Victoria, and Tasmania (1)

Almana Raf. = Sinningia Nees. Gesneriaceae

Almeidea A. St.-Hil. 1823. Rutaceae. 7 northeastern South America

Almutaster A. Löve et D. Löve = Aster L. Asteraceae

Alniphyllum Matsum. 1901. Styracaceae. 3 southwestern China, Taiwan, Japan, Indochina

Alnus Hill. 1753. Betulaceae. 35–40 temp. Northern Hemisphere (Russia 9, all areas), south to Assam, Southeast Asia, and Andes of Chile and Argentina

Alocasia (Schott) G. Don. 1839. Araceae. 70 India, Sri Lanka to Indochina, southern China, Japan, Malaysia, Caroline Is., Australia

Alococarpum H. Riedl ex Kuber. 1964. Apiaceae. 1 Iran

Aloe L. 1753. Asphodelaceae. 365 Micronesia, trop. and South Africa, Madagascar, Arabian Peninsula, Socotra

Aloinopsis Schwantes. 1926 (Nananthus N. E. Br.). Aizoaceae. 12 western and central South Africa

Aloitis Raf. = Gentiana L. Gentianaceae

Alomia Kunth. 1820. Asteraceae. 5 trop. America from Mexico to Chile and southern Brazil

Alomiella R. M. King et H. Rob. 1972. Asteraceae. 2 Brazil

Alona Lindl. 1844 (or = Nolana L.). Nolanaceae. 6 South America

Alonsoa Ruiz et Pav. 1798. Scrophulariaceae. 16 trop. America from Mexico through Central America to Chile and Bolivia; South Africa (2)

Alopecurus L. 1753. Poaceae. 50 temp., warm–temp. and subtrop. regions and montane tropics

Alophia Herb. 1840. Iridaceae. 5 trop. and subtrop. America

Alophotropsis (Jaub. et Spach) Grossh. = Vavilovia Fedorov. Fabaceae

Aloysia Palau. 1784. Verbenaceae. 58 America from southwestern U.S. to Chile and Argentina

Alpaminia O. E. Schulz = Weberbauera Gilg et Muschler. Brassicaceae

Alphandia Baill. 1873. Euphorbiaceae. 3 New Guinea, New Caledonia, Vanuatu

Alphitonia Ressek ex Endl. 1840. Rhamnaceae. 15 southern China, Malaysia (Philippines and Borneo), Bismarck Arch., northern and northeastern Australia, New Caledonia, Fiji, Samoa Is., Polynesia to Hawaiian Is.

Alphonsea Hook. f. et Thomson. 1855 (Mitrephora Hook. f. et Thomson). Annonaceae. 20 India, Sri Lanka, Indochina, Malaysia from Malay Peninsula to Phillipines, ? New Guinea

Alphonseopsis E. G. Baker = Polyceratocarpus Engl. et Diels. Annonaceae

Alpinia Roxb. 1810. Zingiberaceae. 230 southern and northern India, Sri Lanka, Andaman Is., Burma, southern China, Taiwan, Japan south to Bonin Is., Indochina, Malaysia, eastern Australia, Caroline Is., New Britain, Bismarch Arch., Solomon Is., Vanuatu, New Caledonia, Fiji, Samoa Is., Hawaiian Is.

Alposelinum Pimenov = Lomatocarpa Pimenov. Apiaceae

Alrawia (Wendelbo) K. Persson et Wendelbo. 1979 (Bellevalia Lapeyr.). Hyacinthaceae. 2 Iran, Iraq

Alseis Schott. 1827. Rubiaceae. 20 Mexico, Central and trop. South America southward to Peru and Brazil

Alseodaphne Nees. 1831 (Persea Mill.). Lauraceae. 50 India, southwestern China, Indochina, Malaysia

Alseuosmia A. Cunn. 1838. Alseuosmiaceae. 8 New Zealand

Alsinanthe (Fenzl ex Endl.) Rchb. = Arenaria L. Caryophyllaceae

Alsine Gaertn. = Minuartia L. Caryophyllaceae

Alsine L. = Spergularia (Pers.) J. Presl et C. Presl. Caryophyllaceae

Alsinidendron H. Mann. 1866. Caryophyllaceae. 4 Hawaiian Is.

Alsinodendron H. Mann = Alsinidendron H. Mann. Caryophyllaceae

Alsmithia H. E. Moore. 1982. Arecaceae. 1 Fiji (Taveuni)

Alsobia Hanst. 1854 (Episcia Mart.) Gesneriaceae. 2 trop. America

Alsocydia Mart. ex J. C. Gomes f. = Cuspidaria DC. Bignoniaceae

Alsodeia Thouars = Rinorea Aubl. Violaceae

Alsodeidium Engl. = Alsodeiopsis Oliv. Icacinaceae

Alsodeiopsis Oliv. 1867. Icacinaceae. 11 trop. Africa

Alsomitra (Blume) M. Roem. 1846 (Neoalsomitra Hutch.). Cucurbitaceae. 2: A. macrocarpa—Indochina, Malaysia to Philippines and New Guinea; A. clarkei (King) Hutch.—Malay Peninsula

Alstonia R. Br. 1810. Apocynaceae. 40–45 trop. Africa, trop. Asia, Malaysia from Malay Peninsula to New Guinea, Australia, Solomon Is., Vanuatu, New Caledonia (14), Fiji (3), Samoa (2), eastward to Society and Marquesas Is., Central America (2, Guatemala)

Alstroemeria L. 1762. Alstroemeriaceae. 50–70 Central and South America

Altamiranoa Rose = Villadia Rose. Crassulaceae

Altensteinia Kunth. 1816. Orchidaceae. 8 Andes or Colombia, Venezuela, Ecuador, Peru, and Bolivia

Alternanthera Forssk. 1775. Amaranthaceae. 80 temp., subtrop., and trop. regions; 1 (A. sessilis) extending to Caucasus

Althaea L. 1753. Malvaceae. 40 from Western Europe to Northeast Asia, China, and northwestern India; Russia (8, from southern European part to East Siberia)

Althenia F. Petit. 1829. Zannichelliaceae (1, A. filiformis) southern and southeastern Europe, Asia Minor, South Africa; Russia—from Rostov region (Manychskoe Lake) to Altai

Althoffia K. Schum. = Trichospermum Blume. Tiliaceae

Altingia Noronha. 1790. Altingiaceae. 12 eastern Himalayas (Bhutan and Assam), continental China, ? Taiwan, Indochina to Malay Peninsula, Sumatra, Java

Alvaradoa Liebm. 1854. Simaroubaceae. 6 trop. America from Florida and Mexico to Argentina, West Indies

Alvesia Welw. 1869. Lamiaceae. 3 trop. Africa

Alvimia Calderón ex Soderstr. et Londono. 1988. Poaceae. 3 Brazil

Alvimiantha Grey = Wilson. 1978. Rhamnaceae. 1 Brazil

Alvordia Brandegee. 1899. Asteraceae. 3 Mexico (Baja California)

Alyogyne Alef. 1863. Malvaceae. 4 Australia

Alysicarpus Desv. 1813. Fabaceae. 25–30 trop. and South (3) Africa, Madagascar, trop. Asia, Malaysia, trop. Australia

Alyssoides Hill. 1753. Brassicaceae. 3 Eurasia, incl. Caucasus (1, A. utriculata)

Alyssopsis Boiss. 1842. Brassicaceae. 2 Caucasus, Middle Asia (Turkmenistan), Iran

Alyssum L. 1753. Brassicaceae. c. 170 Eurasia from Mediterranean through Russia and Caucasus to Russian Far East, China, and Kashmir

Alyxia Banks ex R. Br. 1810. Apocynaceae. 120–130 Madagascar, India, Sri Lanka, southern China, Indochina, Malaysia eastward to Hawaiian Is., northern Australia, New Caledonia (31), Fiji (5), Tonga, Samoa, Horne and Wallis Is., Society Is., Tuamotu Archipelago

Alzatea Ruiz et Pav. 1794. Alzateaceae. 1–2 Central America (Costa Rica, Panama), Andes of Colombia, Peru, and Bolivia

Amaioua Aubl. 1775. Rubiaceae. 25 Central and trop. South America

Amalocalyx Pierre. 1898. Apocynaceae. 2 Southeast Asia

Amalophyllon Brandegee. 1914. Scrophulariaceae. 1 Mexico

Amana Honda. 1935 (Tulipa L.). Liliaceae. 3 Japan, ? China

Amanoa Aubl. 1775. Euphorbiaceae. 10 west India, Panama, trop. South America, trop. Africa, Madagascar

Amaraboya Linden ex Mast. = Blakea P. Browne. Melastomataceae

Amaracarpus Blume. 1827. Rubiaceae. 60 Malay Peninsula, Sumatra, Java, Philippines, New Guinea, Caroline Is., Solomon Is., Fiji

Amaracus Gled. = Origanum L. Lamiaceae

Amaranthus L. 1753. Amaranthaceae. 60: 16 Russia (from European part to Far East) and Caucasus; subtrop. and temp. regions, a few species in tropics

Amarolea Small = Osmanthus Lour. Oleaceae

Amaroria A. Gray. 1853 (Soulamea Lam.). Simaroubaceae. 1 Fiji

Amaryllis L. 1753. Amaryllidaceae. 1 (A. belladonna) South Africa

Amasonia L. f. 1782. Verbenaceae. 8 South America, Trinidad

Amatlania Lundell = Ardisia Sw. Myrsinaceae

Amauria Benth. 1844. Asteraceae. 3 California, Mexico

Amauriella Rendle = Anubias Schott. Araceae

Amauriopsis Rydb. = Bahia Lag. Asteraceae.

Ambavia Le Thomas. 1972. Annonaceae. 2 Madagascar

Ambelania Aubl. 1775. Apocynaceae. 3 trop. South America

Amberboa (Pers.) Less. 1832. Asteraceae. 7 southeastern European part of Russia, Caucasus, West Siberia, Middle Asia; and from Southwest Asia to China

Amblostoma Scheidw. = Epidendrum L. Orchidaceae

Amblyanthe Rauschert = Dendrobium Sw. Orchidaceae

Amblyanthopsis Mez. 1902. Myrsinaceae. 4 Indo-Malesia

Amblyanthus A. DC. 1841. Myrsinaceae. 4 northeastern India, New Guinea

Amblyanthus (Schlechter) Brieger = Dendrobium Sw. Orchidaceae

Amblygonocarpus Harms. 1897. Fabaceae. 1 Africa

Amblygonum (Meisn.) Rchb. = Polygonum L. Polygonaceae

Amblynotopsis J. F. Macbr. = Antiphytum DC. ex Meisn. Boraginaceae

Amblynotus (A. DC.) I. M. Johnst. 1924. Boraginaceae. 1 West and East Siberia, Mongolia, China (Manchuria)

Amblyocalyx Benth. 1876 (Alstonia R. Br.). Apocynaceae. 1 Borneo

Amblyocarpum Fisch. et C. A. Mey. 1837. Asteraceae. 1 Azerbaijan, Iran

Amblyolepis DC. 1836 (Helenium L.). Asteraceae. 1 (A. setigera) southern U.S. (Texas), northeastern Mexico

Amblyopappus Hook. et Arn. 1841. Asteraceae. 1 California, northeastern Mexico, Peru, Chile

Amblyopetalum (Griseb.) Malme. 1927 (Oxypetalum R. Br.). Asclepiadaceae. 2 Argentina

Amblyopogon (DC.) Fisch. et C. A. Mey. ex Jaub. et Spach = Centaurea L. Asteraceae

Amblyopyrum (Jaub. et Spach) Eig. 1929 (Aegilops L.). Poaceae. 1 eastern Mediterranean, Southwest Asia to southern Turkmenistan

Amblystigma Benth. 1876. Asclepiadaceae. 7 Bolivia, Argentina

Amblytropis Kitag. = Gueldenstaedtia Fisch. Fabaceae

Ambongia Benoist. 1939. Acanthaceae. 1 Madagascar

Amborella Baill. 1869. Amborellaceae. 1 New Caledonia

Amboroa Cabrera. 1956. Asteraceae. 2 Peru, Bolivia

Ambrella H. Perrier. 1934. Orchidaceae. 1 Madagascar

Ambroma L. f. 1782. Sterculiaceae. 2 trop. Asia, Malaysia, Australia

Ambrosia L. 1753. Asteraceae. 43. North (from Canada to Mexico) and South America, West Indies, Mediterranean, trop. Africa

Ambrosina Bassi. 1763. Araceae. 1 Corsica and Sardinia Is., Italy, Algeria

Amburana Schwacke et Taub. 1894. Fabaceae. 1–2 South America from Peru to northeastern Brazil and northern Argentina

Ameghinoa Speg. = Trixis R. Br. Asteraceae

Amelanchier Medik. 1789. Rosaceae. 33 temp. Northern Hemisphere, chiefly North America (26); Crimea and Caucasus—A. ovalis, ? Middle Asia

Amellus L. 1753. Asteraceae. 18 South Africa

Amentotaxus Pilg. 1916. Cephalotaxaceae. 5 northeastern and southeastern India, Burma, continental China, Taiwan, Vietnam

Amerorchis Hulten. 1968. Orchidaceae. 1 North America

Amerosedum A. Löve et D. Löve = Sedum L. Crassulaceae

Amesia A. Nelson et J. F. Macbr. = Epipactis Zinn. Orchidaceae

Amesiella Schltr. ex Garay. 1972. Orchidaceae. 1 Philippines

Amesiodendron Hu. 1936. Sapindaceae. 3 China

Amethysanthus Nakai = Rabdosia (Blume) Hassk. Lamiaceae

Amethystea L. 1753. Lamiaceae. 1 Middle

Asia, West and East Siberia, Russian Far East; Asia Minor, Iran, China, Mongolia, Korea, Japan

Amherstia Wall. 1826. Fabaceae. 1 Burma

Amianthium A. Gray. 1837. Melanthiaceae. 1 eastern and southeastern U.S.

Amicia Kunth. 1824. Fabaceae. 7 Mexico, Ecuador, Peru, Bolivia, Argentina

Amischophacelus Rao et Kammathy. 1966 (or = Cyanotis D. Don). Commelinaceae. 2 Himalayas, India, China, Malaysia, Australia

Amischotolype Hassk. 1863. Commelinaceae. 15 trop. Africa, eastern Himalayas, Tibet, Burma, Indochina, Malaysia.

Amitostigma Schltr. 1919. Orchidaceae. 15–20 India, Sikkim, Burma, Thailand, Indochina, China, Taiwan, ? Korea, Japan

Ammandra O. F. Cook. 1927. Arecaceae. 2 northwestern South America (Colombia, Ecuador)

Ammannia L. 1753. Lythraceae. 25–30 trop., subtrop., and warm–temp. regions; 5 in southeastern European part of Russia, Caucasus, and Middle Asia

Ammanthus Boiss. et Heldr. ex Boiss. = Anthemis L. Asteraceae

Ammi L. 1753. Apiaceae. 10 Macaronesia, Mediterranean, Ethiopia, Southwest Asia (incl. Caucasus—A. majus).

Ammiopsis Boiss. 1856 (or = Daucus L.). Apiaceae. 1 Northwest Africa

Ammobium R. Br. 1824. Asteraceae. 3–4 southeastern Australia, Tasmania

Ammobroma Torr. ex A. Gray = Pholisma Nutt. ex Hook. Lennoaceae

Ammocharis Herb. 1821. Amaryllidaceae. 3 trop. (Sudan, Ethiopia, Somalia, Chad, Central African Rep., Rwanda, Uganda, Kenya, Tanzania, Burundi, Zaire, Zambia, Malawi, Mozambique, Zimbabwe, Angola, Namibia and Botwana) and South (1) Africa

Ammochloa Boiss. 1854. Poaceae. 3 Mediterranean, Southwest Asia; Caucasus (Apsheron Peninsula, 1—A. palaestina)

Ammocodon Standl. 1916 (Selinocarpus A. Gray). Nyctaginaceae. 1 southern U.S. ? and adjacent Mexico

Ammocyanus (Boiss.) Dostal = Centaurea L. Asteraceae

Ammodaucus Coss. et Durieu ex Coss. 1859. Apiaceae. 1 Canary Is., Algeria

Ammodendron Fisch. ex DC. 1825. Fabaceae. 3 deserts of Middle Asia, Iran, Afghanistan

Ammodenia J. G. Gmel. ex S. G. Gmel. = Honckenya Ehrh. Caryophyllaceae

Ammoides Adans. 1763. Apiaceae. 2 South Europe, Mediterranean

Ammophila Host. 1809. Poaceae. 4 eastern North America, Europe, North Africa

Ammopipthanthus S. H. Cheng. 1959. Fabaceae. 2 Tien Shan, Mongolia, China

Ammopurus Small = Liatris Gaertn. ex Schreb. Asteraceae

Ammoselinum Torr. et A. Gray. 1855. Apiaceae. 3 North and temp. South America

Ammosperma Hook. f. 1862. Brassicaceae. 2 North Africa: Algeria, Tunisia, Libya

Ammothamnus Bunge. 1848 (Sophora L.). Fabaceae. 3 Syria, Iraq; Middle Asia (2)

Amoebophyllum N. E. Br. 1925 (Phyllobolus N. E. Br.). Aizoaceae. 1 (A. angustum N. E. Br.) South Africa

Amolinia R. M. King et H. Rob. 1972. Asteraceae. 1 Mexico, Guatemala

Amomis O. Berg = Pimenta Lindl. Myrtaceae

Amomum Roxb. 1820. Zingiberaceae. c. 100 trop. Asia from India and Sri Lanka eastward to China and Indochina, and throughout Malaysia to northern Australia

Amomyrtella Kausel. 1956. Myrtaceae. 1 northern Argentina

Amomyrtus (Burret) D. Legrand et Kausel. 1948. Myrtaceae. 2 temp. South America (Chile)

Amoora Roxb. = Aglaia Lour. Meliaceae

Amoreuxia Moçino et Sessé. ex DC. 1825. Cochlospermaceae. 4 trop. southern U.S. (Arizona, Texas), Mexico, Guatemala, El Salvador, Nicaragua, Colombia, Peru, West Indies (Curaçao I.)

Amoria C. Presl = Trifolium L. Fabaceae

Amorpha L. 1753. Fabaceae. 15 North America from southern Canada to northern Mexico, esp. California and Texas

Amorphophallus Blume ex Decne. 1834. Araceae. c. 100 trop. Old World, a few species in temp. China and Japan

Amorphospermum F. Muell. = Niemeyera F. Muell. Sapotaceae

Ampalis Bojer ex Bur. = Streblus Lour. Moraceae

Amparoa Schltr. 1923. Orchidaceae. 2 Mexico, Costa Rica

Ampelamus Raf. = Gonolobus Michx. Asclepiadaceae

Ampelocalamus S. L. Chen, T. H. Wen et G. Y. Sheng = Sinarundinaria Nakai. Poaceae

Ampelocera Klotzsch. 1847. Ulmaceae. 9 Honduras, Nicaragua, Costa Rica, Panama, Colombia, Venezuela, Ecuador, Peru, Bolivia, Brazil, West Indies

Ampelocissus Planch. 1884. Vitaceae. 60 tropics

Ampelodesmos Link. 1827. Poaceae. 1 western Mediterranean

Ampelopsis Michx. 1803. Vitaceae. 25 temp. and subtrop. Asia, incl. Middle Asia (A. vitifolia), and Russian Far East (3), America

Ampelosicios Thouars. 1807. Cucurbitaceae. 3 Madagascar

Ampelothamnus Small = Pieris D. Don. Ericaceae

Ampelozizyphus Ducke. 1935. Rhamnaceae. 1 Brazil

Amperea A. Juss. 1824. Euphorbiaceae. 8 Australia, esp. western; A. xiphoclada— Queensland, New South Wales, Victoria, and Tasmania

Amphiachyris (DC.) Nutt. 1841. Asteraceae. 2 California

Amphianthus Torr. = Bacopa Aubl. Scrophulariaceae

Amphiasma Bremek. 1952. Rubiaceae. 6 trop. and Southwest Africa

Amphiblemma Naudin. 1850. Melastomataceae. 13 trop. West Africa

Amphibolia L. Bolus ex A. G. J. Herre. 1971 (Eberlanzia Schwantes). Aizoaceae. 1 coastal Namibia, southwestern South Africa

Amphibolis C. Agardh. 1823 (Cymodocea K. König). Cymodoceaceae. 3 coastal western and southern Australia, Tasmania

Amphibologyne Brand. 1931. Boraginaceae. 1 Mexico

Amphibromus Nees. 1843. Poaceae. 9 Australia, New Zealand, South America

Amphicarpa Nutt. = Amphicarpaea Elliott ex Nutt. Fabaceae

Amphicarpaea Elliott ex Nutt. 1818. Fabaceae. 3 North America, trop. Africa (1, A. africana), temp. Asia

Amphicarpon Raf. = Amphicarpum Kunth. Poaceae

Amphicarpum Kunth. 1829. Poaceae. 2 southeastern U.S.

Amphicome Royle = Incarvillea Juss. Bignoniaceae

Amphidasya Standl. 1936. Rubiaceae. 7 Panama, Colombia, Venezuela, Brazil

Amphidetes Fourn. = Matelea Aubl. Asclepiadaceae

Amphidoxa DC. 1838 (Gnaphalium L.). Asteraceae. 16 trop. and South Africa, Madagascar

Amphigena Rolfe. 1913. Orchidaceae. 2 South Africa

Amphiglossa DC. 1838. Asteraceae. 4 South Africa

Amphilophium Kunth. 1818. Bignoniaceae. 5 trop. America from Mexico to Argentina, West Indies

Amphimas Pierre ex Harms. 1906. Fabaceae. 4 trop. Africa from Senegal to Sudan south to Angola and Zaire

Amphineurion (A. DC.) Pichon = Aganosma G. Don. Apocynaceae

Amphinomia DC. = Lotononis (DC.) Eckl. et Zeyh. Fabaceae

Amphiodon Huber = Poecilanthe Benth. Fabaceae

Amphiolanthus Griseb. 1866 (Micranthemum Michx). Scrophulariaceae. 3 Cuba

Amphipappus Torr. et A. Gray. 1844 (Amphiachyris Nutt.). Asteraceae. 1 south-

western U.S.: eastern California, Utah, Arizona
Amphipetalum Bacigalupo. 1988. Portulacaceae. 1 Paraguay
Amphiphyllum Gleason. 1931. Rapateaceae. 2 Venezuela
Amphipogon R. Br. 1810. Poaceae. 7 Australia
Amphipterygium Schiede ex Standl. 1923. Anacardiaceae. 4 Mexico, Peru
Amphirrhox Spreng. 1827. Violaceae. 6 trop. America
Amphiscirpus Oteng = Yeb. 1974. Cyperaceae. 1 Canada, western U.S., Argentina
Amphiscopia Nees = Justicia L. Acanthaceae
Amphisiphon W. F. Barker. 1836. Hyacinthaceae. 1 South Africa (Cape)
Amphitecna Miers. 1868. Bignoniaceae. 20 Florida, Mexico, Central (esp.) and trop. South America, West Indies
Amphithalea Eckl. et Zeyh. 1836. Fabaceae. 20 South Africa
Amphitoma Gleason. 1925 (Miconia Ruiz et Pav.). Melastomataceae. 1 Colombia
Amphoricarpos Vis. 1847. Asteraceae. 4 Southeast Europe (Balkan Peninsula), Asia Minor, Caucasus (1, A. elegans)
Amphorkis Thouars. 1809. Orchidaceae. 4 Mascarene Is.
Amphorocalyx Baker. 1887. Melastomataceae. 5 Madagascar
Amphorogyne Stauffer et Huerl. 1957. Santalaceae. 2 New Caledonia
Amsinckia Lehm. 1831. Boraginaceae. 50 Pacific coastal North and temp. South America
Amsonia Walter. 1788. Apocynaceae. 30 China, Japan, North America, 1 (A. orientalis) Greece and Asia Minor
Amydrium Schott. 1863. Araceae. 4–6 southern China, Thailand, Malaysia
Amyema Tiegh. 1894. Loranthaceae. c. 90 Malaysia, Australia, Micronesia, New Caledonia
Amygdalus L. 1753 (Prunus L.). Rosaceae. 40 Mediterranean (Algeria, Balkan Peninsula, Asia Minor, Syria, Palestine) east to Mongolia and continental China; c. 20 Caucasus, Middle Asia, western and southern Siberia; 1—A. nana, from southern European part of Russia to southern East Siberia
Amylotheca Tiegh. 1894 (Loranthus Jacq.) Loranthaceae. 10 Malaysia (incl. New Guinea, 4), Australia (2, Queensland, New South Wales), Melanesia (1)
Amyrea Leandri. 1940. Euphorbiaceae. 2 Madagascar
Amyris P. Browne. 1756. Rutaceae. 31 trop. America, West Indies
Amyrsia Raf. = Myrteola O. Berg. Myrtaceae
Amyxa Tiegh. 1893. Gonystylaceae. 1 Borneo
Anabaena A. Juss. = Romanoa Trevis. Euphorbiaceae
Anabasis L. 1753. Chenopodiaceae. 27 Spain, Mauritania, Morocco, Algeria, Tunisia, Libya, Egypt, Syria, Israel, Jordan, Iraq, Iran, Afghanistan, Pakistan, China, Mongolia; c. 20 Ukraine, southern European part of Russia, Caucasus (Dagestan, Armenia, Azerbaijan), West Siberia, Middle Asia
Anacampseros L. 1758. Portulacaceae. c. 30–70 South Africa, Australia (2)
Anacampta Miers = Tabernaemontana L. Apocynaceae
Anacamptis Rich. 1817. Orchidaceae. 1 Europe, incl. Crimea, North Africa, Caucasus, Asia Minor, northern Iran
Anacantha (Iljin) Soják. 1982. Asteraceae. 3 Middle Asia (Pamir-Alai)
Anacaona A. H. Liogier. 1980. Cucurbitaceae. 1 Haiti
Anacardium L. 1753. Anacardiaceae. 8 trop. America
Anacharis Rich. = Elodea Michx. Hydrocharitaceae
Anachortus Jirásek et Chrtek = Corynophorus P. Beauv. Poaceae
Anaclanthe N. E. Br. = Babiana Ker-Gawl. ex Sims. Iridaceae
Anacolosa (Blume) Blume. 1851. Olacaceae. 22 trop. Africa (1), Madagascar (1), India, Assam, Andaman and Nicobar Is., Burma, Indochina, Malaysia (excl. Lesser Sunda Is.), Australia (1, Queens-

land), Solomon Is. (from Bougainville to San Cristobal), Fiji, Tonga Is.

Anacyclia Hoffmanns. = Billbergia Thunb. Bromeliaceae

Anacyclus L. 1753. Asteraceae. 12 Mediterranean, Southwest Asia, Caucasus (1, A. ciliatus)

Anadelphia Hack. 1885. Poaceae. 14 trop. Africa

Anadenanthera Speg. 1923. Fabaceae. 2 Central and northern South America

Anadendrum Schott. 1857. Araceae. 9 India, southern China, Indochina

Anaectocalyx Triana ex Hook. f. 1867. Melastomataceae. 3 Venezuela

Anagallidium Griseb. 1838 (Swertia L.). Gentianaceae. 2 Middle Asia, West and East Siberia; Mongolia, China

Anagallis L. 1753. Primulaceae. 28–30 Europe, Russia (2, from European part to Far East), North and South Africa, Madagascar, West Asia (incl. Caucasus), Himalayas from Kashmir to Bhutan, South America; A. arvensis—cosmopolitan, A. pumila—in pantropics

Anaglypha DC. = Gibbaria Cass. Asteraceae

Anagyris L. 1753. Fabaceae. 1–2 Canary Is., Mediterranean, Arabian Peninsula, Iraq, Iran

Anakasia Philipson. 1973. Araliaceae. 1 New Guinea

Anamirta Colebr. 1821. Menispermaceae. 1 India, Sri Lanka, Thailand, Vietnam, Malaysia (Sumatra, Java, Lesser Sunda Is., Moluccas, Philippines, Aru Is., New Guinea)

Anamomis Griseb. = Myrcianthes O. Berg. Myrtaceae

Ananas Hill. 1753. Bromeliaceae. 8 trop. and subtrop. South America

Anapalina N. E. Br. = Tritoniopsis L. Bolus. Iridaceae

Anaphalioides (Benth.) Kirp. 1950 (Gnaphalium L.). Asteraceae. 6 Australia, New Zealand

Anaphalis DC. 1838. Asteraceae. 60–100 Eurasia, Malaysia, America

Anaphyllopsis A. Hay. 1988. Araceae. 3 Venezuela, Brazil

Anaphyllum Schott. 1857. Araceae. 2 southern India

Anarrhinum Desf. 1798. Scrophulariaceae. 12 Mediterranean, extending to Ethiopia and Yemen

Anarthria R. Br. 1810. Restionaceae. 10 southwestern Australia

Anarthrophyllum Benth. 1865. Fabaceae. 15 Andes of Chile and Argentina

Anartia Miers = Tabernaemontana L. Apocynaceae

Anaspis Rech. f. = Scutellaria L. Lamiaceae

Anastatica L. 1753. Brassicaceae. 1 North Africa, Arabian Peninsula, Israel, Jordan, southern Iran, Iraq, Pakistan

Anastrabe E. Mey. ex Benth. 1836. Scrophulariaceae. 1 South Africa

Anastrophea Wedd. = Sphaerothylax Bisch. ex Krauss. Podostemaceae

Anatherum Nabelek, non-P. Beauv. = Nabelekia Roshev. Poaceae

Anatropanthus Schltr. 1908. Asclepiadaceae. 1 Borneo

Anatropostylia (Plitmann) Kupicha = Vicia L. Fabaceae

Anaueria Kosterm. 1938 (Beilschmiedia Nees). Lauraceae. 1 Amazonian of Brazil, Peru

Anax Ravenna = Stenomesson Herb. Amaryllidaceae

Anaxagorea A. St.-Hil. 1825. Annonaceae. 27 trop. America, a few species in Sri Lanka and Southeast Asia

Anaxeton Gaertn. 1791. Asteraceae. 9 southwestern South Africa

Ancana F. Muell. 1865 (Fissistigma Griff.). Annonaceae. ssp ? trop. Africa, Southeast Asia, northeastern Australia

Ancathia DC. 1833. Asteraceae. 1 Caucasus, West and East Siberia, Middle Asia; Mongolia, western China

Anchietea A. St.-Hil. 1824. Violaceae. 8 trop. South America

Anchomanes Schott. 1853. Araceae. 5 trop. Africa

Anchonium DC. 1821. Brassicaceae. 2 Armenia (1, A. elichrysifolium), Iran, Syria, Lebanon

Anchusa L. 1753. Boraginaceae. 35–50 Europe, central and southern European part of Russia, Caucasus, Middle Asia, Mediterranean, South Africa, Southwest Asia eastward to northwestern India, Nepal, western Tibet

Ancistrachne S. T. Blake. 1941. Poaceae. 4 Philippines, Australia, New Caledonia

Ancistragrostis S. T. Blake. 1946. Poaceae. 1 New Guinea

Ancistranthus Lindau. 1900. Acanthaceae. 1 Cuba

Ancistrocactus Britton et Rose = Sclerocactus Britton et Rose. Cactaceae

Ancistrocarphus A. Gray. 1859. Asteraceae. 1 southwestern North America from Oregon to Baja California

Ancistrocarpus Oliv. 1865. Tiliaceae. 5 trop. Africa

Ancistrocarya Maxim. 1872. Boraginaceae. 1 Japan

Ancistrochilus Rolfe. 1897. Orchidaceae. 3 trop. West (from Sierra Leone to Cameroun) and East (Uganda) Africa

Ancistrochloa Honda. 1936 (Calamagrostis Adans.) Poaceae. 1 Japan

Ancistrocladus Wall. 1829. Ancistrocladaceae. 19 trop. Africa (9), trop. Asia (10) from India, Sri Lanka and eastern Himalayas to southern China, Indochina, Sumatra, and Borneo

Ancistrophora A. Gray. 1859 (Verbesina L.). Asteraceae. 1 Cuba

Ancistrophyllum (G. Mann et H. A. Wendl.) H. A. Wendl. (1878, non-Goeppert. 1841) = Laccosperma (G. Mann et H. A. Wendl.) Drude. Arecaceae

Ancistrorhynchus Finet. 1907. Orchidaceae. 16 trop. Africa

Ancistrostylis Yamaz. 1980. Scrophulariaceae. 1 Laos

Ancistrothyrsus Harms. 1931. Passifloraceae. 1–2 Andes of South America

Ancrumia Harv. ex Baker. 1877. Alliaceae. 1 Chile

Ancylacanthus Lindau. 1913 (Ptyssiglottis T. Anderson). Acanthaceae. 1 New Guinea

Ancylanthos Desf. 1818. Rubiaceae. 5 trop. Africa eastward to Tanzania (1, A. rogersii)

Ancylobothrys Pierre. 1898 (Landolphia P. Beauv.). Apocynaceae. 10 trop. Africa, Comoro Is., Madagascar

Ancylostemon Craib. 1919. Gesneriaceae. 12 western and southwestern China

Ancylotropis Eriksen. 1993 (Monnina Ruiz et Pav.). Polygalaceae. 2 trop. America

Andersonia R. Br. 1810. Epacridaceae. 22 southwestern Australia

Anderssoniopiper Trel. = Piper L. Piperaceae

Andesia Hauman. 1915 (Oxychloe Phil.). Juncaceae. 1 Andes of South America

Andira Juss. 1789. Fabaceae. 20 trop. America, 1 (A. inermis) Cape Verde Is., trop. Africa from Senegal to Sudan south to Uganda

Andrachne L. 1753. Euphorbiaceae. 25 trop. America, West Indies, Mediterranean, Crimea (1), Caucasus and Middle Asia, Southwest Asia eastward to western Himalayas

Andradea Allemao. 1845. Nyctaginaceae. 1 eastern Brazil

Andrea Mez. 1896. Bromeliaceae. 1 central Brazil

Andreettaea Luer = Pleurothallis R. Br. Orchidaceae

Andresia Sleumer = Cheilotheca Hook. f. Ericaceae

Androcalymma Dwyer. 1957. Fabaceae. 1 Brazil (Amazonia)

Androcentrum Lem. = Bravaisia DC. Acanthaceae

Androcera Nutt. = Solanum L. Solanaceae

Androchilus Liebm. ex Hartman. 1844. Orchidaceae. 1 Mexico

Androcorys Schltr. 1919. Orchidaceae. 4 from India to East Asia

Androcymbium Willd. 1808. Melanthiaceae. 30 Europe (2), Mediterranean south to South Africa

Andrographis Wall. ex Nees. 1832. Acanthaceae. 20 trop. Asia, Australia (1, A. paniculata)

Androlepis Brongn. ex Houllet. 1870. Bromeliaceae. 1 Guatemala, Belize, Honduras, Costa Rica

Andromeda L. 1753. Ericaceae. 1–2 arctic and temp. Northern Hemisphere

Andromycia A. Rich. = Asterostigma Fisch. et C. A. Mey. Araceae

Andropogon L. 1753. Poaceae. c. 100 tropics and subtropics

Andropterum Stapf. 1917. Poaceae. 1 trop. Africa: Zambia, Malawi, Tanzania, Zimbabwe, Mozambique

Andropus Brand. 1912 (Nama L.). Hydrophyllaceae. 1 southwestern U.S.

Androsace L. 1753. Primulaceae. c. 100 temp. Northern Hemisphere, esp. China (71), a few species in temp. South America (Tierra del Fuego)

Androsiphon Schltr. 1924. Hyacinthaceae. 1 South Africa

Androsiphonia Stapf. 1905 (Paropsia Noronha ex Thouars). Passifloraceae. 1 trop. West Africa

Androstachys Prain. 1908. Euphorbiaceae. 5 trop. Madagascar

Androstephanos J. Fernandez Casas. 1983. Amaryllidaceae. 1 Bolivia

Androstephium Torr. 1859. Alliaceae. 2 North America

Androstylanthus Ducke = Helianthostylis Baill. Moraceae

Androtium Stapf. 1903. Anacardiaceae. 1 Malaysia.

Androtrichum (Brongn.) Brongn. 1834. Cyperaceae. 3 eastern temp. South America

Androya H. Perrier. 1952. Buddlejaceae. 1 Madagascar

Andruris Schltr = Sciaphila Blume. Triuridaceae

Andryala L. 1753. Asteraceae. 25 Macaronesia, Mediterranean

Andrzeiowskia Rchb. 1824. Brassicaceae. 1 Greece, Crete I., Asia Minor, Syria, western Caucasus

Anechites Griseb. 1861. Apocynaceae. 1 Costa Rica, Panama, Colombia, Venezuela, Ecuador, southern Peru; West Indies: Cuba, Jamaica, Haiti, Puerto Rico

Aneilema R. Br. 1810. Commelinaceae. 62: 57 trop. and subtrop. Africa from Senegal to Ethiopia south to northern Namibia, Zimbabwe, Mozambique, Transvaal, and northeastern Cape; 5 New Guinea, Solomon Is., New Caledonia, northeastern and eastern Australia; 1 Venezuela and Brazil

Anelsonia J. F. Macbr. et Payson. 1917. Brassicaceae. 1 Pacific coastal U.S.

Anelytrum Hack. = Avena L. Poaceae

Anemanthele Veldkamp. 1985. Poaceae. 1 New Zealand

Anemarrhena Bunge. 1833. Asphodelaceae. 1 northern and northeastern China, southern Korea

Anemocarpa Paul G. Wilson. 1992. Asteraceae. 3 Australia

Anemoclema (Franch.) W. T. Wang. 1964 (Anemone L.). Ranunculaceae. 1 southwestern China

Anemonastrum Holub = Anemone L. Ranunculaceae

Anemone L. 1753. Ranunculaceae. 120–150 cosmopolitan, esp. temp. Northern Hemisphere

Anemonella Spach = Thalictrum L. Ranunculaceae

Anemonidium (Spach) Holub = Anemone L. Ranunculaceae

Anemonoides Mill. = Anemone L. Ranunculaceae

Anemonopsis Siebold et Zucc. 1845. Ranunculaceae. 1 Japan

Anemopaegma Mart. ex Meisn. 1840. Bignoniaceae. 43 trop. America from Mexico to Brazil and Argentina

Anemopsis Hook. et Arn. 1840. Saururaceae. 1 southwestern U.S., Mexico

Anepsias Schott = Rhodospatha Poepp. Araceae

Anerincleistus Korth. 1844. Melastomataceae. 35 Indochina, western Malaysia

Anetanthus Hiern ex Benth. 1876. Gesneriaceae. 1 Brazil

Anethum L. 1753. Apiaceae. 1–4 Eurasia from Western Europe to European part of Russia, and throughout Mediterranean to Caucasus, Middle Asia, and India

Anetilla Galushko = Pulsatilla Hill. Ranunculaceae

Aneulophus Benth. 1862. Erythroxylaceae. 2 trop. West Africa

Aneurolepidium Nevski = Leymus Hochst. Poaceae

Angadenia Miers. 1878. Apocynaceae. 2 Florida, West Indies

Angasomyrtus Trudgen et Keighery. 1983. Myrtaceae. 1 southwestern Australia

Angelianthus H. Rob. et Brettell = Microliabum Cabrera. Asteraceae

Angelica L. 1753. Apiaceae. c. 110 North America, Eurasia, North Africa

Angelocarpa Rupr. = Angelica L. Apiaceae

Angelonia Bonpl. 1812. Scrophulariaceae. 25 trop. America, West Indies

Angelphytum G. M. Barroso. 1980. Asteraceae. 14 Brazil, Argentina

Angianthus J. C. Wendl. 1810. Asteraceae. 38 temp. Australia

Anginon Raf. 1840. Apiaceae. 7 South Africa

Angkalanthus Balf. f. 1883. Acanthaceae. 2 South Africa (Transvaal), Socotra (1)

Angolaea Wedd. 1873. Podostemaceae. 1 Angola

Angolluma R. Munster (1990) = Pachycymbium Leach. Asclepiadaceae

Angophora Cav. 1797 (Eucalyptus L'Hér.). Myrtaceae. 11–17 eastern and southeastern Australia

Angoseseli Chiov. 1924. Apiaceae. 1 Angola

Angostura Roem. et Schult. 1819. Rutaceae. 30 trop. South America

Angostyles Benth. 1854. Euphorbiaceae. 2 trop. South America

Angostylidium (Muell. Arg.) Pax et K. Hoffm. = Tetracarpidium Pax. Euphorbiaceae

Angraecopsis Kraenzl. 1900. Orchidaceae. 14–20 trop. Africa, Madagascar, Mascarene Is.

Angraecum Bory. 1804. Orchidaceae. 225–250 West Indies, Brazil, trop. and South Africa, Madagascar, Mascarene Is., Sri Lanka, Philippines

Anguillaria R. Br. = Wurmbea Thunb. Melanthiaceae

Anguloa Ruiz et Pav. 1794. Orchidaceae. 10 Colombia, Ecuador, Peru, and Venezuela

Anguria Jacq. = Psiguria Neck. ex Arn. Cucurbitaceae

Anguriopsis J. R. Johnst. = Doyerea Grosourdy. Cucurbitaceae

Angylocalyx Taub. 1896. Fabaceae. 7 trop. Africa south to Zaire

Ania Lindl. = Tainia Blume. Orchidaceae

Aniba Aubl. 1775. Lauraceae. 40–41 Colombia, Venezuela, Guyana, Surinam, French Guyana, Peru, Brazil, Bolivia; West Indies

Anigozanthos Labill. 1800. Conostylidaceae. 11 southwestern Australia

Aningueria Aubrev. et Pellegr. = Pouteria Aubl. Sapotaceae

Anisacanthus Nees. 1842 (Idanthisa Raf.). Acanthaceae. 7–15 southeastern U.S., Mexico, Central and trop. South America

Anisachne Keng. 1958 (Calamagrostis Adans.). Poaceae. 1 southwestern China

Anisadenia Wall. ex Meisn. 1838. Linaceae. 2 Asia from central China to Himalayas and northeastern India

Anisantha K. Koch. 1848 (Bromus L.). Poaceae. 10 Europe, North Africa, Asia Minor, Transcaucasia, Iran, Middle Asia, Himalayas, western China, ? Japan; Russia (southern European part, Caucasus, West Siberia)

Anisantherina Pennell ex Britton = Agalinis Raf. Scrophulariaceae

Aniseia Choisy. 1834. Convolvulaceae. 5 pantropics

Aniselytron Merr. 1910 (Calamagrostis Adans.). Poaceae. 1 Philippines

Aniserica N. E. Br. 1905. Ericaceae. 2 South Africa (Cape)

Anisocalyx L. Bolus = Drosanthemum Schwantes. Aizoaceae

Anisochaeta DC. 1836. Asteraceae. 1 South Africa

Anisochilus Wall. ex Benth. 1830. Lamiaceae. 20 trop. Africa, trop. continental Asia

Anisocoma Torr. et A. Gray. 1845. Asteraceae. 1 southwestern U.S.: Nevada, eastern California, Arizona

Anisocycla Baill. 1887. Menispermaceae. 7 trop. and South Africa, Madagascar

Anisodontea C. Presl. 1845. Malvaceae. 20 South Africa (esp. Cape)

Anisodus Link et Spreng. 1824 (Scopolia Jacq.). Solanaceae. 4 China, Nepal, Bhutan, Sikkim, eastern India

Anisomallon Baill. = Apodytes E. Mey. ex Arn. Icacinaceae

Anisomeles R. Br. 1810. Lamiaceae. 7 trop. and subtrop. Asia, Malaysia, trop. Australia

Anisomeria D. Don. 1832. Petiveriaceae. 3 Chile

Anisomeris C. Presl = Chomelia Jacq. Rubiaceae

Anisopappus Hook. et Arn. 1837. Asteraceae. 30 trop. Africa, Madagascar, trop. Asia

Anisophyllea R. Br. ex Sabine. 1824. Anisophyllaceae. 32 trop. Africa, trop. Asia and Malaysia (Malay Peninsula, Sumatra, Borneo), trop. South America (1, A. guyanensis)

Anisopoda Baker. 1890. Apiaceae. 1 Madagascar

Anisopogon R. Br. 1810. Poaceae. 1 southeastern Australia

Anisoptera Korth. 1841. Dipterocarpaceae. 11–13 India (Assam), Burma, Southeast Asia, Malaysia from Malay Peninsula to New Guinea

Anisopus N. E. Br. 1895. Asclepiadaceae. 4 trop. West Africa

Anisosciadium DC. 1829. Apiaceae. 3 arid and semiarid Southwest Asia eastward to Iran and Pakistan

Anisosepalum E. Hossain. 1972. Acanthaceae. 2 Central Africa

Anisosperma Silva Manso = Fevillea L. Cucurbitaceae

Anisotes Nees. 1847. Acanthaceae. 19 trop. Africa, Madagascar, Arabian Peninsula

Anisothrix O. Hoffm. ex Kuntze. 1898. Asteraceae. 2 South Africa (Cape)

Anisotoma Fenzl. 1844. Asclepiadaceae. 2 South Africa

Anisotome Hook. f. 1844 (Aciphylla J. R. et G. Forster). Apiaceae. 15 New Zealand and subantarctic islands

Anisum Hill = Pimpinella L. Apiaceae

Ankyropetalum Fenzl. 1843. Caryophyllaceae. 4 Asia Minor, Iran, Syria, Lebanon, Israel, Sinai, Jordan, northern Iraq

Anna Pellegr. 1930. Gesneriaceae. 3 southern China, Indochina

Annaea Kolak. = Campanula L. Campanulaceae

Annamocarya A. Chev. = Carya Nutt. Juglandaceae

Anneliesea Brieger et Lueckel = Miltonia Lindl. Orchidaceae

Annesijoa Pax et K. Hoffm. 1919. Euphorbiaceae. 1 New Guinea

Anneslea Wall. 1829. Theaceae. 6 China, Taiwan, Indochina, Malaysia

Annesorrhiza Cham. et Schltdl. 1826. Apiaceae. 15 South Africa

Annickia Setten et Maas. 1990. Annonaceae. 10 trop. East (1) and South (9) Africa

Annona L. 1753. Annonaceae. 130 tropics and subtropics, esp. America and Africa

Anochilus (Schltr.) Rolfe. 1913 (Pterygodium Sw.). Orchidaceae. 3 South Africa

Anoda Cav. 1785. Malvaceae. 23 America from southwestern and southern U.S. through Central America to Bolivia, Chile, and Argentina

Anodendron A. DC. 1844. Apocynaceae. 17 India, Sri Lanka east to Japan and Taiwan, Malaysia, Solomon Is.

Anodiscus Benth. 1876. Gesneriaceae. 1 Peru

Anodopetalum A. Cunn. ex Endl. 1839. Cunoniaceae. 1 Tasmania

Anoectochilus Blume. 1828. Orchidaceae. 40–60 trop. Asia from India and Sri Lanka to Japan, Malaysia, Australia, Pacific Islands eastward to New Caledonia, Fiji, and Hawaii

Anogeissus (DC.) Guillemin, Perrottet et A. Rich. 1832. Combretaceae. 8 trop. Africa, Arabian Peninsula, India, Sri Lanka, Nepal, China, Indochina

Anoiganthus Baker = Cyrtanthus Aiton. Amaryllidaceae

Anomacanthus R. D. Good. 1923. Acanthaceae. 1 western Zaire and Cabinda

Anomalanthus Klotzsch. 1838 (Scyphogyne Brongn.). Ericaceae. 10 South Africa (southwestern and southern Cape)

Anomalesia N. E. Br. = Gladiolus L. Iridaceae

Anomalocalyx Ducke. 1932. Euphorbiaceae. 1 trop. South America

Anomalostylus R. C. Foster. 1947 (Trimezia Salisb. ex Herb.). Iridaceae. 3 trop. South America

Anomanthodia Hook. f. 1873 (Randia L.). Rubiaceae. 10 South and Southeast Asia, Malaysia.

Anomatheca Ker-Gawl. 1804 (Freesia Eckl. ex Klatt). Iridaceae. 5 trop. and South Africa

Anomianthus Zoll. 1858. Annonaceae. 1 northeastern India, Thailand, Java

Anomochloa Brongn. 1851. Poaceae. 1 Brazil

Anomoctenium Pichon = Distictis Mart. ex Meisn. Bignoniaceae

Anomopanax Harms. 1904 (Mackinlaya F. Muell.). Araliaceae. 9 eastern Malaysia.

Anomospermum Miers. 1851. Menispermaceae. 8 trop. America from Panama to Brazil

Anomostephium DC. 1836 (Aspilia Thouars). Asteraceae. 50 trop. America

Anomotassa K. Schum. 1898. Asclepiadaceae. 1 Ecuador

Anonidium Engl. et Diels. 1900. Annonaceae. 5 trop. Africa

Anoplocaryum Ledeb. = Microula Benth. Boraginaceae

Anopterus Labill. 1805. Escalloniaceae. 2 southeastern Australia, Tasmania

Anopyxis Pierre ex Engl. 1900. Rhizophoraceae. 3 trop. Africa Anosporum Nees. 1834 (Cyperus L.). Cyperaceae. 3 trop. Africa, Madagascar, trop. Asia from India to southern China, trop. Australia, ? West Indies

Anota Schltr. = Rhynchostylis Blume. Orchidaceae

Anotea (DC.) Kunth. 1846 (Pavonia Cav.). Malvaceae. 1 (A. flavida) Mexico

Anotis DC. 1830. Rubiaceae. c. 30 China, Indo-Malaysia, Australia, South America (1)

Anplectrella Furtado = Creochiton Blume. Melastomataceae

Anplectrum A. Gray = Diplectria (Blume) Rchb. Melastomataceae

Anredera Juss. 1789. Basellaceae. 5–10 trop. and subtrop. America, West Indies, Galapagos Is., ? Tasmania

Ansellia Lindl. 1844. Orchidaceae. 1–2 trop. and South (Natal) Africa

Antegibbaeum Schwantes ex C. Weber. 1968 (Gibbaeum Haw.). Aizoaceae. 1 South Africa

Antennaria Gaertn. 1791. Asteraceae. 40–50 arctic and temp. regions, Andes of South America; Russia (10, from European part to Far East)

Antenonoron Raf. = Persicaria Hill. Polygonaceae

Anteriorchis E. Klein et Strack = Orchis L. Orchidaceae

Anthacanthus Nees. 1847. Acanthaceae. 6 West Indies

Anthaenantia P. Beauv. 1812. Poaceae. 2 southeastern U.S.

Anthaenantiopsis Mez ex Pilg. 1931. Poaceae. 2 Brazil, Argentina

Anthagathis Harms = Jollydora Pierre ex Gilg. Connaraceae

Antheliacanthus Ridl. = Pseuderanthemum Radlk. Acanthaceae

Anthemis L. 1753. Asteraceae. 130 Europe, Mediterranean, Southwest Asia to Iran, 1 (A. subtinctoria) from Southeast Europe to Siberia

Anthephora Schreb. 1772–1779. Poaceae. 12 trop. America, trop. Africa, Arabian Peninsula

Anthericopsis Engl. 1895. Commelinaceae. 1 trop. East Africa

Anthericum L. 1753. Asphodelaceae. 50 Central America ?, Europe (3), Caucasus, trop. and subtrop. Africa

Antherolophus Gagnep. 1934 (Aspidistra Ker-Gawl.). Convallariaceae. 1 Indochina

Antheropeas Rydb. = Eriophyllum Lag. Asteraceae

Antheroporum Gagnep. 1915. Fabaceae. 2 China, Thailand, Indochina

Antherostele Bremek. 1940. Rubiaceae. 4 Philippines

Antherostylis C. A. Gardner = Velleia Sm. Goodeniaceae

Antherothamnus N. E. Br. 1915. Scrophulariaceae. 1 Namibia, Botswana, Zimbabwe, South Africa

Antherotoma (Naudin) Hook. f. 1867. Melastomataceae. 2 trop. Africa from Guinea to Ethiopia, Uganda, Kenya, Tanzania, south to Angola and Transvaal, Madagascar

Anthobembix Perkins = Steganthera Perkins. Monimiaceae

Anthobolus R. Br. 1810. Santalaceae. 3 Australia, except southern part

Anthobryum Phil. = Frankenia L. Frankeniaceae

Anthocarapa Pierre. 1897. Meliaceae. 1–2 Philippines, Sulawesi, Lesser Sunda Is., New Guinea, Solomon Is., eastern Australia, New Caledonia, Vanuatu

Anthocephalus auct. non A. Rich. = Neolamarckia Bosser. Rubiaceae

Anthocephalus A. Rich. 1830 (Breonia A. Rich. ex DC.). Rubiaceae. 2 India, Sri Lanka, Nepal, northern Burma, southern China, Indochina, Malesia

Anthocercis Labill. 1806. Solanaceae. 9 temp. southwestern and southern Australia

Anthochlamys Fenzl. 1837. Chenopodiaceae. 6 Azerbaijan, Middle Asia (Turkmenistan, Uzbekistan, Tajikistan, Kirgisia), Iran, Afghanistan, Pakistan

Anthochloa Nees et Meyen. 1834. Poaceae. 1 Andes of Peru and Chile

Anthochortus Nees. 1836. Restionaceae. 7 South Africa (Cape)

Anthocleista Afzel. ex R. Br. 1818. Gentianaceae. 14 trop. Africa, Madagascar, Comoro Is. and Mascarene Is.

Anthoclitandra (Pierre) Pichon. 1853 (Landolphia P. Beauv.). Apocynaceae. 2 trop. West Africa

Anthodiscus G. Mey. 1818. Caryocaraceae. 9 Colombia, Venezuela, Peru, Bolivia, Brazil (western Amazonia)

Anthodon Ruiz et Pav. 1798. Celastraceae. 2 Central and trop. South America

Anthogonium Wall. ex Lindl. 1840. Orchidaceae. 1 (A. gracile) Nepal, Sikkim, Bhutan, Tibet, northeastern India, Burma, southern China (Yunnan, Szechuan), Thailand, Laos, Vietnam

Antholoma Labill. = Sloanea L. Elaeocarpaceae

Antholyza L. = Babiana Ker-Gawl. Iridaceae

Anthonotha P. Beauv. 1806 (Macrolobium Schreb.). Fabaceae. 30 trop. Africa

Anthopteropsis A. C. Sm. 1941. Ericaceae. 1 Central America

Anthopterus Hook. 1839. Ericaceae. 6 Andes

Anthorrhiza C. Huxley et Jebb. 1991. Rubiaceae. 9 southeastern New Guinea and adjacent islands

Anthosachne Steud. = Elymus L. Poaceae

Anthosiphon Schltr. 1920. Orchidaceae. 1 Colombia

Anthospermum L. 1753. Rubiaceae. 40 trop. and South Africa, Madagascar, Arabian Peninsula (1, A. herbaceum)

Anthostema A. Juss. 1824. Euphorbiaceae. 3 trop. Africa, Madagascar

Anthotium R. Br. 1810. Goodeniaceae. 2–4 southwestern Australia

Anthotroche Endl. 1839. Solanaceae. 3 arid and temp. western Australia

Anthoxanthum L. 1753. Poaceae. 20 temp. and warm Eurasia (incl. 4 European part of Russia, Caucasus, Siberia, except northern areas, montane Middle Asia), North Africa, montane tropics

Anthriscus Pers. 1805. Apiaceae. 6–12 Eurasia, North, trop. and South Africa

Anthurium Schott. 1829. Araceae. 700 trop. America, West Indies

Anthyllis L. 1753. Fabaceae. 20–25 (A. vulneraria is very various) Europe, Caucasus, Mediterranean, Ethiopia (1, A. vulneraria), eastward to Iran

Antiaris Lesch. 1811. Moraceae. 1–3 trop. Africa, Madagascar, trop. Asia, Australia, Melanesia to Fiji

Antiaropsis K. Schum. 1889. Moraceae. 1–2 New Guinea

Anticharis Endl. 1839. Scrophulariaceae.

14 Africa from Namibia to Ethiopia, Somalia and north to Asia Minor, Arabian Peninsula, India, Malesia

Antidaphne Poepp. et Endl. 1838. Eremolepidaceae. 8: A. viscoideae—southern Mexico, Guatemala, Costa Rica, Panama, Colombia (2), Venezuela (2), Ecuador, Peru, Bolivia; Brazil (2); A. punctulata—southern Chile; A. wrightii—Cuba, southern Haiti, Puerto Rico

Antidesma L. 1753. Euphorbiaceae. 160 trop. and subtrop. Old World, esp. Asia

Antigonon Endl. 1837. Polygonaceae. 5 Mexico, Central America

Antillanorchis Garay. 1974. Orchidaceae. 1 Cuba

Antillia R. M. King et H. Rob. 1971. Asteraceae. 1 Cuba

Antimima N. E. Br. 1930 (Ruschia Schwantes). Aizoaceae. c. 60 Southwest Africa

Antinisa (Tul.) Hutch. 1941 (Homalium Jacq.). Flacourtiaceae. 3 Madagascar

Antinoria Parl. 1845. Poaceae. 2 Western Europe, Mediterranean

Antiotrema Hand. = Mazz. 1920. Boraginaceae. 1 southwestern China

Antiphiona Merxm. 1954. Asteraceae. 2 trop. and Namibia (Namibia)

Antiphytum DC. ex Meisn. 1840. Boraginaceae. 8 trop. America from Mexico to Brazil

Antirhea Comm. ex Juss. 1789. Rubiaceae. 50–100 West Indies, Panama (1, A. trichanta), Madagascar, Mascarene Is., trop. Asia, Malesia, Australia, Oceania eastward to Samoa Is.

Antirrhinum L. 1753. Scrophulariaceae. 42 Pacific coastal North America, Europe (17), Mediterranean, esp. western

Antistrophe A. DC. 1841. Myrsinaceae. 5 Indo-Malesia

Antithrixia DC. 1838. Asteraceae. 1 (A. flavicoma DC.) South Africa

Antitoxicum Pobed. = Vincetoxicum Wolf. Asclepiadaceae

Antizoma Miers. 1851. Menispermaceae. 3–4 South Africa

Antonella Caro. 1981 (Tridens Roem. et Schult.). Poaceae. 1 Argentina

Antongilia Jum. = Neodypsis Baill. Arecaceae

Antonia Pohl. 1829. Loganiaceae. 1 trop. South America

Antonina Vved. 1961 (Satureja L.). Lamiaceae. 1 Caucasus, Middle Asia

Antopetitia A. Rich. 1840 (Ornithopus L.). Fabaceae. 1 trop. Africa from Nigeria to Ethiopia south to Zambia and Zimbabwe

Antoschmidtia Boiss. = Schmidtia Steud. ex J. A. Schmidt. Poaceae

Antrocaryon Pierre. 1898. Anacardiaceae. 3 trop. Africa from Sierra Leone to Zaire, Uganda (1, A. micraster A. Chev.)

Antrophora I. M. Johnst. = Lepidocordia Ducke. Boraginaceae

Antunesia O. Hoffm. = Distephanos (Cass.) Cass. Asteraceae

Anubias Schott. 1857. Araceae. 8 trop. West Africa

Anulocaulis Standl. 1909 (Boerhavia L.). Nyctaginaceae. 4–5 southern U.S., Mexico

Anura (Juz.) Tscherneva = Coussinia Cass. Asteraceae

Anurosperma (Hook. f.) Hallier = Nepenthes L. Nepenthaceae

Anvillea DC. 1836. Asteraceae. 2 North Africa from Morocco to Egypt, Southwest Asia east to Iran

Anvilleina Maire = Anvillea DC. Asteraceae

Aoranthe Somers. 1988 (Porterandia Ridl.). Rubiaceae. 5 trop. Africa: Guinea, Gabon, Congo, Zaire, Angola, Nigeria, Cameroun, Central African Rep., Tanzania (1, A. penduliflora)

Aorchis Vermuelen. 1972. (Habenaria Willd.). Orchidaceae. 1 Himalayas

Aostea Buscalioni et Muschler = Vernonia Schreb. Asteraceae

Aotus Sm. 1805. Fabaceae. 15 Australia, Tasmania

Apacheria C. T. Mason. 1975. Crossosomataceae. 1 U.S. (Arizona)

Apalanthe Planch. 1848 (Elodea Michx.). Hydrocharitaceae. 1 trop. South America

Apalochlamys Cass. 1828. Asteraceae. 1 southeastern Australia

Apaloxylon Drake = Neoapaloxylon Rauschert. Fabaceae

Apama Lam. = Thottea Rottb. Aristolochiaceae

Apargidium Torr. et A. Gray = Microseris D. Don. Asteraceae

Aparisthmium Endl. = Conceveibum Aubl. Euphorbiaceae

Apassalus Kobuski. 1928. Acanthaceae. 3 southeastern U.S., West Indies

Apatesia N. E. Br. 1927. Aizoaceae. 3 southwestern South Africa

Apatophyllum McGill. 1971. Celastraceae. 2 Australia (Queensland and New South Wales)

Apatostelis Garay = Stelis Sw. Orchidaceae

Apatringania Dieterle. 1974. Cucurbitaceae. 1 Mexico

Apeiba Aubl. 1775. Tiliaceae. 10 trop. South America

Apera Adans. 1763. Poaceae. 4 Europe, North Africa, Asia Minor, Transcaucasia, Iran, Iraq, Middle Asia, Afghanistan; Russia (all areas except Arctica, northern Siberia, and Far East)

Apetahia Baill. 1882. Lobeliaceae. 3 Society Is., Rapa I., Marquesas Is.

Aphaca Adans. = Lathyrus L. Fabaceae

Aphaenandra Miq. 1857. Rubiaceae. 2 Indochina, Sumatra, Java

Aphaerema Miers. 1863. Flacourtiaceae. 1 southern Brazil

Aphanactis Wedd. 1856. Asteraceae. 4 trop. America

Aphanamixis Blume. 1825 (Aglaia Lour.). Meliaceae. 3 China, India, Malesia, Solomon Is.

Aphanandrium Lindau = Neriacanthus Benth. Acanthaceae

Aphananthe Planch. 1848. Ulmaceae. 5 Madagascar (1), Sri Lanka, Andaman Is., Indochina, China, Japan, Malesia, Solomon Is., eastern Australia (3), Mexico (1)

Aphandra Barfod. 1991. Arecaceae. sp. Distr. ?

Aphanelytrum Hack. 1902. Poaceae. 1 Andes of Colombia, Ecuador, and Bolivia

Aphanes L. 1753. Rosaceae. 20 America, Europe (5), Mediterranean, Ethiopia, Australia; 1 (A. arvensis) Crimea, Caucasus

Aphania Blume = Lepisanthes Blume. Sapindaceae

Aphanisma Nutt. ex Moq. 1849. Chenopodiaceae. 1 U.S. (California), Baja California

Aphanocalyx Oliv. 1810. Fabaceae. 3 trop. Africa from Sierra Leone to Zaire

Aphanocarpus Steyerm. 1965. Rubiaceae. 1 Venezuela

Aphanococcus Radlk. = Lepisanthes Blume. Sapindaceae

Aphanopetalum Endl. 1839. Cunoniaceae. 2 southwestern and southeastern Australia

Aphanopleura Boiss. 1872. Apiaceae. 4 Caucasus, Middle Asia; Iran, Afghanistan, western China

Aphanosperma (Leonard et Gentry) Daniel. 1988. Acanthaceae. 1 northwestern Mexico

Aphanostelma Schltr. = Metaplexia R. Br. Asclepiadaceae

Aphanostemma A. St.-Hil. 1925. Ranunculaceae.—1 southern Brazil, northern central Argentina

Aphanostephus DC. 1836. Asteraceae. 5 southern U.S., Mexico

Aphanostylis Pierre. 1898 (Landolphia P. Beauv.). Apocynaceae. 3 trop. Africa

Aphelandra R. Br. 1810. Acanthaceae. 175 trop. America

Aphelandrella Mildbr. 1926. Acanthaceae. 1 Peru

Aphelexis D. Don = Edmondia Cass. Asteraceae

Aphelia R. Br. 1810. Centrolepidaceae. 6 southern Australia, Tasmania

Aphloia (DC.) Benn. 1840. Aphloiaceae. 6 trop. East (Kenya, Tanzania, Malawi, Mozambique, Simbabwe) and South (Transvaal, Natal) Africa, Madagascar, Seycheles, Comoro and Mascarene Is., and Rodriguez I.

Aphoma Raf. = Iphigenia Kunth. Melanthiaceae

Aphragmia Nees = Ruellia L. Acanthaceae

Aphragmus Andrz. ex DC. 1824. Brassicaceae. 6 West and northeastern Siberia, Aleutian Is. (Unalashke I.), Mongolia, China (2), India (3, Kashmir)

Aphyllanthes L. 1753. Aphyllanthaceae. 1 Iberian Peninsula, Balearic Is., southern France, northern Italy, Sardinia I., Morocco, Algeria

Aphyllarum S. Moore = Caladium Vent. Araceae

Aphyllocladus Wedd. 1855. Asteraceae. 5 Bolivia, China, Argentina

Aphyllodium (DC.) Gagnep. = Hedysarum L. Fabaceae

Aphyllon Mitch. = Orobanche L. Orobanchaceae

Aphyllorchis Blume. 1825. Orchidaceae. 10–20 India, Sri Lanka, Sikkim, southern China, Taiwan, Japan, Indochina, Malesia from Malay Peninsula to New Guinea, Australia

Apiastrum Nutt. ex Torr. et A. Gray. 1840. Apiaceae. 2 North America

Apinagia Tul. 1849. Podostemaceae. 50 trop. South America

Apiopetalum Baill. 1878. Araliaceae. 4 New Caledonia

Apios Fabr. 1759. Fabaceae. 10 East Asia, North America

Apium L. 1753. Apiaceae. c. 30 Europe, Mediterranean, North, Northwest, trop. and South Africa, Madagascar, West (incl. Caucasus and Middle Asia), Southwest, South, East and Southeast Asia, Malesia, Australia, Oceania, North, Central, and South America

Aplanodes Marais. 1966. Brassicaceae. 2 South Africa (Natal, Lesotho, eastern Cape)

Aplectrum (Nutt.) Torr. 1826. Orchidaceae. 2 Japan and temp. North America

Apleura Phil. = Azorella Lam. Apiaceae

Aploleia Raf. = Callisia Loefl. Commelinaceae

Apluda L. 1753. Poaceae. 3 Mauritius, Socotra, trop. Asia, New Guinea, New Caledonia; A. mutica—extending to Middle Asia

Apocaulon R. S. Cowan. 1953. Rutaceae. 1 Venezuela

Apochaete (C. E. Hubb.) J. B. Phipps = Tristachya Nees. Poaceae

Apochiton C. E. Hubb. 1936. Poaceae. 1 Tanzania

Apoclada McClure. 1967. Poaceae. 4 Brazil

Apocopis Nees. 1841. Poaceae. 15 trop. Asia

Apocynum L. 1753. Apocynaceae. 7 North America

Apodandra Pax et K. Hoffm. = Plukenetia L. Euphorbiaceae

Apodanthera Arn. 1841. Cucurbitaceae. 15 trop. and subtrop. America

Apodanthes Poit. 1824. Apodanthaceae. 7 trop. South America; A. caseariae Poit.—extending to Panama and Honduras

Apodicarpum Makino. 1891 (Apium L.). Apiaceae. 1 Japan (Honshu)

Apodiscus Hutch. 1912. Euphorbiaceae. 1 West Africa

Apodocephala (Baker) Humbert. 1885. Asteraceae. 8 Madagascar

Apodolirion Baker. 1878. Amaryllidaceae. 6 South Africa from Transvaal to Cape

Apodostigma R. Wilczek. 1956. Celastraceae. 1 trop. Africa

Apodytes E. Mey. ex Bernh. 1838. Icacinaceae. 3 trop. and South Africa, trop. Asia, Australia (1, Queensland), New Caledonia (1)

Apoia Merr. = Sarcosperma Hook. f. Sapotaceae

Apollonias Nees. 1833. Lauraceae. 2 Madeira and Canary Is. (1), India (western Ghats—A. arnottii)

Apomuria Bremek. = Psychotria L. Rubiaceae

Aponogeton L. f. 1782. Aponogetonaceae. 47 Africa, Madagascar, Comoro Is., India, Sri Lanka, China, Malesia, Australia

Apophyllum F. Muell. 1857. Capparaceae. 1 Australia (Queensland, New South Wales)

Apoplanesia C. Presl. 1832. Fabaceae. 1 trop. America from southern Mexico to Venezuela

Aporocactus Lem. = Disocactus Lindl. Cactaceae

Aporosa Blume. 1826. Euphorbiaceae. 75 Himalayas, India, Burma, southern China, Indochina, Malesia, Solomon Is., Australia

Aporosella Chodat et Hassler = Phyllanthus L. Euphorbiaceae

Aporostylis Rupp et Hatch. 1946. Orchidaceae. 1 New Zealand

Aporrhiza Radlk. 1878. Sapindaceae. 6 trop. Africa

Aporum Blume = Dendrobium Sw. Orchidaceae

Aporusa Blume = Aporosa Blume. Euphorbiaceae

Aposeris Neck. ex Cass. 1827. Asteraceae. 1 from central Europe to Carpathians

Apostasia Blume. 1825. Orchidaceae. 10 trop. Asia, Malesia, Australia

Apostates Lander. 1990. Asteraceae. 1 Rapa I.

Apotaenium Kozo-Polj. = Neoconopodium (Kozo-Polj.) Pimenov et Kljuykov. Apiaceae

Appendicula Blume. 1825. Orchidaceae. 60–100 trop. Asia from India to China and Indochina, Malesia from Malay Peninsula to New Ireland I., Caroline Is. (Palau), Bougainville and Solomon Is., Vanuatu, the Horn Is., New Caledonia, Fiji, Samoa

Appendicularia DC. 1903. Melastomataceae. 1 northeastern South America

Appertiella C. D. K. Cook et Triest. 1982. Hydrocharitaceae. 1 Madagascar

Appunettia R. D. Good. 1926 (Morinda L.). Rubiaceae. 1 Angola

Appunia Hook. f. 1873 (Morinda L.). Rubiaceae. c. 10 trop. America

Aprevalia Baill. 1884 (Delonix Raf.). Fabaceae. 2 Madagascar

Aptandra Miers. 1851. Olacaceae. 3: 2 Amazonian Peru, Brazil, and Bolivia, Venezuela, Guyana; 1 (A. zenkeri) trop. West Africa

Aptandropsis Ducke = Heisteria Jacq. Olacaceae

Aptenia N. E. Br. 1925. Aizoaceae. 4 Namibia, South Africa

Apterantha C. H. Wright. 1918 (Lagrezia Moq.). Amaranthaceae. 1 Seychelles (Aldabra I.)

Apteria Nutt. 1836. Burmanniaceae. 1 trop. America, West Indies

Apterigia Galushko = Atropatenia F. K. Mey., p.p. = Noccaea Moench, p.p. Brassicaceae

Apterokarpos Rizzini. 1975. Anacardiaceae. 1 trop. South America

Apterosperma H. T. Chang. 1976. Theaceae. 1 China

Apterygia Baehni = Sideroxylon L. Sapotaceae

Aptosimum Burch. ex Benth. 1836. Scrophulariaceae. c. 40 trop. and South Africa

Apuleia Mart. 1837. Fabaceae. 1 northeastern Peru, Brazil, northern Argentina

Apurimacia Harms. 1923. Fabaceae. 4 trop. and subtrop. South America

Aquifolium Hill = Ilex L. Aquifoliaceae

Aquilaria Lam. 1783. Thymelaeaceae. 15 northeastern India, Bhutan, Burma, southern China (incl. Hainan), Hong Kong, Indochina, Malesia (except Java and Lesser Sunda Is.)

Aquilegia L. 1753. Ranunculaceae. 70–100 temp. Northern Hemisphere, incl. (c. 30) southern European part of Russia, Caucasus, Middle Asia, Siberia, and Far East

Arabidella (F. Muell.) O. E. Schulz. 1924. Brassicaceae. 6 Australia

Arabidopsis Heynh. 1842 (Arabis L.). Brassicaceae. 25 temp. Northern Hemisphere, Mediterranean, a few species extending trop. East Africa

Arabis L. 1753. Brassicaceae. 100–120 North America, temp. Eurasia (incl. c. 40 in European part of Russia, Caucasus, Middle Asia, Siberia, Far East), Mediterranean, montane trop. Africa

Aracamunia Carnevali et I. Ramirez. 1989. Orchidaceae. 1 Venezuela

Arachis L. 1753. Fabaceae. 22. trop.

America; A. hypogaea—very widely cultivated

Arachne Neck. = Andrachne L. Euphorbiaceae

Arachnis Blume. 1825. Orchidaceae. 11 Himalayas, southeastern India, southern China, ? Ryukyu Is., Thailand, Vietnam, Malesia from Malay Peninsula to New Guinea, Bougainville and Solomon Is.

Arachnitis Phil. 1864. Corsiaceae. 1 Chile

Arachnocalyx Compton. 1935 (Eremia D. Don). Ericaceae. 1–2 South Africa (southwestern Cape)

Arachnothryx Planch. = Rondeletia L. Rubiaceae

Araeoandra (Barneoud) Lefor. 1975. Vivianiaceae. 1 Chile

Araeococcus Brongn. 1841. Bromeliaceae. 5 Costa Rica, Colombia, Venezuela, Trinidad, Tobago, Surinam, Guyana, Amazonian and southeastern (Bahia) Brazil

Arafoe Pimenov et Lavrova. 1989. Apiaceae. 1 western Caucasus

Aragoa Kunth. 1819. Scrophulariaceae. 5 Andes

Aralia L. 1753. Araliaceae. c. 50 India, Himalayas, China, Russian Far East (3), Japan, Indochina, Malesia, North America

Aralidium Miq. 1856. Aralidiaceae. 1 southern Thailand, Malay Peninsula, Singapore, Sumatra, Anambas I., Borneo

Araliopsis Engl. 1896. Rutaceae. 3 trop. West Africa

Araliorhamnus Perrier. 1943. Rhamnaceae. 2 Madagascar

Arapatiella Rizzinni et A. Mattos. 1972. Fabaceae. 2 southeastern Brazil

Aratitiyopea Steyerm. 1984. Xyridaceae. 1 southeastern Colombia, Amazonas of Venezuela and northwestern Brazil

Araucaria Juss. 1789. Araucariaceae. 19 New Guinea, coastal Queensland, New Caledonia, Norfolk I., central and southern Chile, southeastern Brazil, ? northeastern Argentina

Araujia Brot. 1817. Asclepiadaceae. 5 South America, A. hortorum naturalized in Australia

Arbelaezaster Cuatrec. 1986. Asteraceae. 1 Colombia

Arberella Soderstr. et Calderón. 1979. Poaceae. 1 Central and trop. South America

Arbulocarpus Tennant = Borreria G. Mey. Rubiaceae

Arbutus L. 1753. Ericaceae. 14–20 North and Central America, West Europe, Mediterranean, Southwest Asia; A. andrachne—extending to Crimea and western Transcaucasia

Arcangelina Kuntze. 1891 (Tripogon Roem. et Schult.). Poaceae. 1 North Africa

Arcangelisia Becc. 1877. Menispermaceae. 2 southern China (Hainan), southern peninsular Thailand, Malesia (Sumatra, Malay Peninsula, Java, Borneo, Sulawesi, Philippines, Moluccas, New Guinea)

Arceuthobium M. Bieb. 1819. Viscaceae. 36 North America, West Indies, Mediterranean, West Asia eastward to Himalayas and from China to western Malesia; A. oxycedri—extending to Crimea, Caucasus

Arceuthos Antoine et Kotschy = Juniperus L. Cupressaceae

Archangelica N. M. Wolf = Angelica L. Apiaceae

Archboldia E. Beer et H. J. Lam. 1936. Verbenaceae. 1 New Guinea

Archboldiodendron Kobuski. 1940. Theaceae. 1 New Guinea

Archeria Hook. f. 1857. Epacridaceae. 4 Australia, Tasmania, New Zealand

Archiatriplex G. L. Chu. 1987. Chenopodiaceae. 1 China

Archibaccharis Heering. 1904. Asteraceae. 21 Mexico, Guatemala, El Salvador, Honduras, Nicaragua, Costa Rica, Panama

Archiboehmeria C. J. Chen. 1980. Urticaceae. 1 China, North Vietnam

Archiclematis Tamura. 1968 (Clematis L.). Ranunculaceae. 1 Himalayas, Tibet

Archidendron F. Muell. 1865. Fabaceae. c. 100 trop. Asia, Malesia, eastward to Solomon Is., trop. Australia

Archidendropsis I. C. Nielsen. 1983. Fabaceae. 14 Malesia, New Caledonia

Archineottia S. C. Chen. 1979. Orchidaceae. 5 northwestern India, southern and southwestern China, Japan (Honshu)

Archiphysalis Kuang. 1966 (Leucophysalis Rydb.). Solanaceae. 3 China, Japan

Archirhodomyrtus (Nied.) Burret. 1941. Myrtaceae. 4 New Caledonia, Australia (1, New South Wales)

Archontophoenix H. A. Wendl. et Drude. 1875. Arecaceae. 2 eastern Australia from northern Queensland to southern New South Wales

Archytaea Mart. 1826. Bonnetiaceae. 3 Colombia, Venezuela, Bolivia, Brazil

Arcoa Urb. 1923. Fabaceae. 1 Haiti

Arctagrostis Griseb. 1852. Poaceae. 3 Arctica, Siberia, Russian Far East, montane Asia, and North America

Arctanthemum (Tsvelev) Tsvelev. 1985. Asteraceae. 4 Arctica

Arcteranthis Greene. 1897 (Oxygraphis Bunge). Ranunculaceae. 1 northwestern North America

Arcterica Coville. 1901. Ericaceae. 1 northeastern Siberia, Kamchatka, Sakhalin I. Kuril Is., Japan (Hokkaido and Honshu Is.)

Arctium L. 1753. Asteraceae. 10 temp. Eurasia; A. tomentosum from Europe to Russian Far East

Arctogeron DC. 1836. Asteraceae. 1 West and East Siberia

Arctomecon Torr. et Frém. 1845. Papaveraceae. 3 southwestern U.S.

Arctophila (Rupr.) N. J. Andersson. 1852. Poaceae. 1 Arctica, montane Asia, and North America

Arctopoa (Griseb.) Probatova. 1974 (Poa L.). Poaceae. 5 temp. Asia, western North America

Arctopus L. 1753. Apiaceae. 3 South Africa (Cape)

Arctostaphylos Adans. 1763. Ericaceae. c. 50 Pacific coastal North and Central America; circumpolar and temp. regions; Russia (2, Arctica, European part, Caucasus, West and East Siberia, Far East)

Arctotheca J. C. Wendl. 1798. Asteraceae. 4 South Africa

Arctotis L. 1753. Asteraceae. 50 Southwest Africa (Angola, Namibia)

Arctottonia Trel. = Piper L. Piperaceae

Arctous (A. Gray) Nied. = Arctostaphylos Adans. Ericaceae

Arcuatopterus Shen et Shan. 1986. Apiaceae. 3 China, Tibet

Arcynospermum Turcz. 1858. Malvaceae. 1 Mexico

Arcyosperma O. E. Schulz. 1924. Brassicaceae. 1 Pakistan, northwestern India, Nepal, Bhutan

Arcytophyllum Willd. ex Schult. et Schult. f. 1827. Rubiaceae. 15 Costa Rica, Colombia, Venezuela, Ecuador, Peru, Bolivia

Ardisia Sw. 1788. Myrsinaceae. 300–400 tropics and subtropics, esp. America and Asia

Ardisiandra Hook. f. 1864. Primulaceae. 3 montane trop. Africa

Areca L. 1753. Arecaceae. c. 60 trop. Asia from India and southern China to Indochina, Malesia to New Guinea and Solomon Is., ? Australia (1)

Arecastrum (Drude) Becc. = Syagrus Mart. Arecaceae

Arechavaleteia Speg. = Azara Ruiz et Pav. Flacourtiaceae

Aregelia Kuntze = Neoregelia L. B. Sm. Bromeliaceae

Aremonia Neck. ex Nestl. 1816. Rosaceae. 1 Southeast Europe

Arenaria L. 1753. Caryophyllaceae. c. 160 cold and temp. Northern Hemisphere

Arenga Labill. 1800. Arecaceae. 17 from India, southern China, Ryuku Is. and Taiwan throughout Southeast Asia, Malesia, to northern Australia, Christmas I. (Indian Ocean)

Arenifera A. G. J. Herre. 1948 (Psammophora Dinter et Schwantes). Aizoaceae. 1 southwestern South Africa

Arethusa L. 1753. Orchidaceae. 1 North America

Aretiastrum (DC.) **Spach** = Valeriana L. Valerianaceae

Arfeuillea Pierre ex Radlk. 1895. Sapindaceae. 1 Southeast Asia

Argania Roem. et Schult. 1819. Sapotaceae. 1 Morocco

Argemone L. 1753. Papaveraceae. 32 Hawaiian Is. (1, A. glauca), North and South America, West Indies

Argentipallium Paul G. Wilson. 1992. Asteraceae. 6 Australia and Tasmania

Argeta N. E. Br. = Gibbaeum Haw. Aizoaceae

Argillochloa W. A. Weber. 1984 (Festuca L.). Poaceae. 1 North America

Argocoffeopsis Lebrun. 1941. Rubiaceae. 8 trop. Africa

Argomuellera Pax. 1894. Euphorbiaceae. 10 trop. Africa (4), Comoro Is., Madagascar

Argophyllum J. R. Forst. et G. Forst. 1775. Argophyllaceae. 11 trop. Australia, New Caledonia

Argopogon Mimeur. 1951 (Ischaenum L.). Poaceae. 1 West Africa

Argostemma Wall. 1824. Rubiaceae. 100 trop. Africa, Asia

Argostemmella Ridl. = Argostemma Wall. Rubiaceae

Argusia Boehm. 1760. (Tournefortia L.). Boraginaceae. 4: A. sibirica—temp. Eurasia from Romania through Russia and Siberia to China, Korea, Japan, southeastern U.S., ? Central America; A. sogdiana—Middle Asia; A. argentea—coastal East Africa, Madagascar, Indian Ocean islands, coastal Vietnam, Hainan I., Taiwan, Ryukyu Is., Philippines, northern Australia, southern Pacific islands; A. gnaphalodes—West Indies

Argylia D. Don. 1823. Bignoniaceae. 12 southern Peru, north-central Chile and adjacent Argentina; A. uspallatensis—extending to southern Bolivia

Argyranthemum Webb. 1839. Asteraceae. 23 Madeira Is., Canary Is., Salvage Is.

Argyreia Lour. 1790. Convolvulaceae. 90 trop. continental Asia, Malesia, northeastern Australia, New Caledonia (1)

Argyrocytisus (Maire) Frodin et Heywood ex Raynaud. 1974 (Cytisus Desf.). Fabaceae. 1 Morocco (Atlas Mts.)

Argyroderma N. E. Br. 1922. Aizoaceae. 10 South Africa

Argyroglottis Turcz. 1851. Asteraceae. 1 northwestern Australia

Argyrolobium Eckl. et Zeyh. 1836. Fabaceae. c. 90 Mediterranean (incl. Europe—2), Crimea, Caucasus, montane trop. and South (68) Africa, Madagascar, West and Southwest Asia eastward to India

Argyronerium Pit. = Epigynum Wight. Apocynaceae

Argyrophanes Schltdl. 1847. Asteraceae. 1 Australia

Argyrovernonia MacLeish. 1984 (Chresta Vell. ex DC.). Asteraceae. 1 Brazil

Argyroxiphium DC. 1836 (Dubautia Gaudich.). Asteraceae. 5 Hawaiian Is.

Argythamnia P. Browne. 1756. Euphorbiaceae. 19 South America, West Indies

Aria (Pers.) **Host.** 1831. Rosaceae. c. 15 East and Southeast Asia

Ariadne Urb. 1922. Rubiaceae. 2 Cuba

Ariaria Cuervo = Bauhinia L. Fabaceae

Aridaria N. E. Br. 1925. Aizoaceae. 35 South Africa

Aridarum Ridl. 1913. Araceae. 7 northern Borneo

Arikury Becc. = Syagrus Mart. Arecaceae

Arikuryroba Barb. Rodr. = Syagrus Mart. Arecaceae

Arillastrum Panch. ex Baill. 1877. Myrtaceae. 1 New Caledonia

Ariocarpus Scheidw. 1838. Cactaceae. 6 southern U.S. (Texas), eastern Mexico

Ariopsis Nimmo. 1839. Araceae. 1 coastal western India, Himalayas from Nepal to Assam, Burma

Ariosorbus Koidz. 1934 (Sorbus L.). Rosaceae. 2 Japan

Arisaema Mart. 1831. Araceae. 160 coastal North America to Mexico (6); montane trop. East Africa; South, Southeast, and East Asia; 4 sp. extending to Russian Far East

Arisarum Hill. 1753. Araceae. 3 Azores, Northwest Africa, Mediterranean

Arischrada Pobed. = Salvia L. Lamiaceae

Aristavena F. Albers et Butzin. 1977 (Deschampsia P. Beauv.). Poaceae. 1 West Europe

Aristea Sol. ex Aiton. 1789. Iridaceae. 50 trop. (6) and South Africa, Madagascar (6)

Aristeguietia R. M. King et H. Rob. 1975. Asteraceae. 20 Andes

Aristella Bertol. = Achnatherum P. Beauv. Poaceae

Aristeyera H. E. Moore = Asterogyne H. A. Wendl. ex Hook. f. Arecaceae

Aristida L. 1753. Poaceae. 280 North America, Mediterranean, temp. Asia (incl. 1, A. adsensionis: Caucasus, Middle Asia), tropics and subtropics

Aristocapsa Reveal et Hardham. 1989. Polygonaceae. 1 southwestern U.S. (California)

Aristogeitonia Prain. 1908. Euphorbiaceae. 3 trop. (Angola, Kenya, Tanzania) Africa, Madagascar

Aristolochia L. 1753. Aristolochiaceae. c. 550 (or 300–350) trop., subtrop., and temp. regions; 8 extending to European part of Russia, Caucasus, and Russian Far East

Aristopsis Catasus = Aristida L. Poaceae

Aristotelia L'Hér. 1786. Elaeocarpaceae. 5 eastern Australia, Tasmania, New Zealand, South America from Peru to Chile

Arjona Cav. 1798. Santalaceae. 10 temp. South America

Armatocereus Backeb. 1938. Cactaceae. c. 10 Colombia, Ecuador, Peru

Armeniaca Hill. 1753 (Prunus L.). Rosaceae. 9 temp. Asia, incl. (6) Middle Asia, southern East Siberia, Far East of Russia

Armeria Willd. 1809. Plumbaginaceae. 80–100 temp. Northern Hemisphere, South America from Chilean Andes to Tierra del Fuego; Russia (5, Arctica, central European part, East Siberia, Far East)

Armodorum Breda. 1829. Orchidaceae. 3 India, China, Thailand, Sumatra, Java

Armoracia P. Gaertn., B. Mey. et Scherb. 1800. Brassicaceae. 4 North America from central U.S. to Quebec and Florida, Europe, Russia from European part to Siberia and Far East, China (1)

Armouria Lewton = Thespesia Sol. ex Corrêa. Malvaceae

Arnaldoa Cabrera. 1962. Asteraceae. 3 Peruvian Andes

Arnanthus Baehni = Pichonia Pierre. Sapotaceae

Arnebia Forssk. 1775. Boraginaceae. 25 Mediterranean, southeastern Europe (1), Caucasus, West Siberia, Middle Asia; trop. Africa, Southwest Asia from Arabian Peninsula to Himalayas, China, Mongolia

Arnebiola Chiov. 1929 (Arnebia Forssk.). Boraginaceae. 1 Somalia

Arnhemia Airy Shaw. 1978. Thymelaeaceae. 1 northern Australia

Arnica L. 1753. Asteraceae. 30 arctic and temp.; Russia (10, Arctica, northern European part, Siberia, Far East)

Arnicastrum Greenm. 1903. Asteraceae. 2 Mexico

Arnicratea N. Hallé. 1984. Celastraceae. 3 Malabar coast of India, Andaman Is., Burma, Thailand, Cambodia, southern Vietnam, Sumatra, Java, Borneo, Philippines, Moluccas, New Guinea

Arnocrinum Endl. et Lehm. 1846. Asphodelaceae. 3 southwestern Australia

Arnoglossum Raf. 1817. Asteraceae. 7–8 eastern and southeastern U.S.

Arnoseris Gaertn. 1791. Asteraceae. 1 Europe, Mediterranean

Arnottia A. Rich. = Amphorkis Thouars. Orchidaceae

Aromadendron Blume. 1825 (Magnolia L.). Magnoliaceae. 4 Malay Peninsula, Java, Borneo

Aronia Medik. 1789 (Amelanchier Medik.). Rosaceae. 3 eastern U.S.

Arophyton Jum. 1928. Araceae. 7 Madagascar

Aropsis Rojas = Spathicarpa Hook. Araceae

Arpophyllum La Llave et Lex. 1826. Orchidaceae. 6 trop. America, West Indies

Arrabidaea DC. 1838. Bignoniaceae. 70 trop. America from Mexico to Argentina, West Indies

Arracacia Bancr. 1828. Apiaceae. c. 55 trop. America from Mexico to Peru

Arrhenatherum P. Beauv. 1812. Poaceae. 6 Europe, North Africa, Southwest Asia to Caucasus and Turkmenistan

Arrhenechthites Mattf. 1939. Asteraceae. 7 New Guinea, southeastern Australia (1, A. mixta)

Arrhostoxylum Nees = Ruellia L. Acanthaceae

Arrojadoa Britton et Rose. 1920. Cactaceae. 4 eastern Brazil

Arrojadocharis Mattf. 1930. Asteraceae. 1 eastern Brazil

Arrowsmithia DC. = Macowania Oliv. Asteraceae

Arrudaria Macedo = Corypha L. Arecaceae

Arsenjevia Starod. 1989 (or = Anemone L.). Ranunculaceae. 5 Siberia, Russian Far East

Artabotrys R. Br. ex Ker-Gawl. 1820. Annonaceae. 85 trop. Old World

Artanacetum (Rzazade) Rzazade = Artemisia L. Asteraceae

Artanema D. Don. 1834. Scrophulariaceae. 4 India, Sri Lanka, Malesia, Australia

Artedia L. 1753. Apiaceae. 1 Cyprus I., Asia Minor, Syria, Lebanon, Israel, Iran, Iraq

Artemisia L. 1753. Asteraceae. 300 temp. Northern Hemisphere, esp. Russia and adjacent territories (c. 220), trop. and South Africa, South America

Artemisiastrum Rydb. = Artemisia L. Asteraceae

Artemisiopsis S. Moore. 1902. Asteraceae. 1 trop. and South Africa

Arthraerua (Kuntze) Schinz. 1893. Amaranthaceae. 1 Southwest Africa

Arthragrostis Lazarides. 1984. Poaceae. 2 northeastern Australia

Arthraxon P. Beauv. 1812. Poaceae. 20 trop. and temp. Old World; A. hispidus—extending to Caucasus, Middle Asia, and southern Far East of Russia

Arthrocarpum Balf. f. 1882. Fabaceae. 2 Somalia, Socotra

Arthrocereus (A. Berger) A. Berger. 1929. Cactaceae. 5 western and southeastern Brazil

Arthrochilus F. Muell. 1858 (Spiculaea Lindl.). Orchidaceae. 3 Australia

Arthrochloa Lorch, non R. Br. = Normanboria Butzin. Poaceae

Arthroclianthus Baill. 1870. Fabaceae. 10 New Caledonia

Arthrocnemum Moq. 1840. Chenopodiaceae. 2 Canary Is., Mediterranean, Somalia, Kenya, Tanzania, West Asia, coastal India from Bombay to Bengal, Sri Lanka; North America (California, U.S.)

Arthrophyllum Blume. 1826. Araliaceae. 21 Andaman and Nicobar Is., Indochina, Malesia

Arthrophytum Schrenk. 1845. Chenopodiaceae. 9 Middle Asia (Turkmenistan, Kazakhstan, Uzbekistan), northwestern China

Arthropodium R. Br. 1810. Asphodelaceae. 8 Madagascar (1), Australia (4), New Zealand (2), New Caledonia (2)

Arthropogon Nees. 1829. Poaceae. 6 Antilles Is., Brazil

Arthrosamanea Britton et Rose = Albizia Durazz. Fabaceae

Arthrosolen C. A. Mey. = Gnidia L. Thymelaeaceae

Arthrostemma Pav. ex D. Don. 1823. Melastomataceae. 4 trop. America from Mexico to Bolivia, West Indies

Arthrostylidium Rupr. 1840. Poaceae. 25 Antilles Is., trop. South America

Arthrostylis R. Br. 1810. Cyperaceae. 2 Australia

Artia Guillaumin. 1941. Apocynaceae. 7: 1 Thailand and Malay Peninsula, 2 southern China [Kuangsi (1), Hainan (1)], 4 New Caledonia

Artocarpus J. R. Forst. et G. Forst. 1775. Moraceae. c. 50 India, Sri Lanka, southern China through Malesia to Solomon Is. and Australia

Artorima Dressler et G. E. Pollard. 1971. Orchidaceae. 1 Mexico

Artrolobium Desv. = Coronilla L. Fabaceae

Arum L. 1753. Araceae. 16 Europe, Mediterranean, Caucasus, northern Iraq, Iran, Middle Asia

Aruncus Hill. 1753. Rosaceae. 6 arctic, temp., and montane regions of the Northern Hemisphere

Arundina Blume. 1825. Orchidaceae. 6 India, Sri Lanka, Himalayas, Burma, southern China, Taiwan, Ryukyu Is., Indochina, Malesia (except Philippines and New Guinea), Tahiti Is.

Arundinaria Michx. 1803. Poaceae. c. 50 mainly temp. Asia, esp. China and Japan; 2 South Asia (Sikkim, Bhutan, Vietnam), 1 (A. gigantea) southeastern U.S.

Arundinella Raddi. 1823. Poaceae. c. 50 East Asia and tropics and subtropics; A. hirta—extending to southern East Siberia and southern Far East of Russia

Arundo L. 1753. Poaceae. 5 Mediterranean, Southwest and South Asia; A. donax—in Caucasus, and Middle Asia

Arundoclaytonia Davidse et R. P. Ellis. 1988. Poaceae. 1 Brazil

Arytera Blume. 1849. Sapindaceae. 20–25 China, Indochina, Malesia to New Guinea, Australia, New Caledonia, Vanuatu, Fiji, Tonga, Samoa

Asaemia (Harv.) Harv. ex Benth. = Athanasia L. Asteraceae

Asanthus R. King et H. Rob. 1972. Asteraceae. 3 southwestern U.S., Mexico

Asarca Lindl. 1827. Orchidaceae. 20 temp. South America

Asarina Hill. 1753. Scrophulariaceae. 16 North America, Europe (1)

Asarum L. 1753. Aristolochiaceae. c. 90 temp. Northern Hemisphere; 4 sp. in European part of Russia, Caucasus, West Siberia, and Far East

Ascarina J. R. Forst. et G. Forst. 1776. Chloranthaceae. 12 Madagascar (1, A. coursii), Malesia (4: Borneo, Sulawesi, Philippines, New Guinea, New Britain, Manus, and Bougainville Is.), Solomon Is., New Zealand, New Caledonia (2), Vanuatu eastward to Marquesas Is.

Ascarinopsis Humbert et Capuron = Ascarina J. R. Forst. et G. Forst. Chloranthaceae

Aschenbornia S. Schuer = Calea L. Asteraceae

Aschersoniodoxa Gilg et Muschler. 1909. Brassicaceae. 3 Andes of Peru, western Bolivia and northwestern Argentina

Aschistanthera C. Hansen. 1987. Melastomataceae. 1 Vietnam

Asciadium Griseb. 1866. Apiaceae. 1 Cuba

Ascidieria Seidenf. 1984. Orchidaceae. 1 Thailand, Malay Peninsula, Sumatra, Borneo

Ascidiogyne Cuatrec. 1965. Asteraceae. 2 Peru

Asclepias L. 1753. Asclepiadaceae. 150 America (esp. U.S.), trop. and South Africa, Arabian Peninsula

Ascocarydion G. Taylor. 1931 (Plectranthus L'Hér). Lamiaceae. 1 Zaire, Malawi, Zambia, Angola, Namibia

Ascocentrum Schltr. ex J. J. Sm. 1914. Orchidaceae. 8 Himalayas, northeastern India, Sri Lanka, Andaman Is., Burma, southern China, Taiwan, Thailand, Vietnam, Laos, Malay Peninsula, Indonesia, Philippines

Ascochilopsis C. E. Carr. 1929. Orchidaceae. 1 Malay Peninsula, Sumatra

Ascochilus Ridl. 1896. Orchidaceae. 5 Burma, Thailand, Malay Peninsula, Sumatra, Java, ? Borneo, Philippines (3)

Ascoglossum Schltr. 1913 (Sarcochilus R. Br.). Orchidaceae. 2 Malesia from Malay Archipelago to New Guinea, Bougainville, and Solomon Is.

Ascolabium S. S. Ying = Ascocentrum Schltr. ex J. J. Sm. Orchidaceae

Ascolepis Nees ex Steud. 1855. Cyperaceae. 22 South America, trop. and South Africa, Madagascar, East Asia

Ascopholis C. E. C. Fisch. 1931 (Cyperus L.). Cyperaceae. 1 southern India

Ascotainia Ridl. = Taihia Blume. Orchidaceae

Ascotheca Heine. 1966. Acanthaceae. 1 trop. West Africa

Ascyrum Hill = Hypericum L. Hypericaceae

Asemanthia (Stapf) Ridl. 1940 (Mussaenda L.). Rubiaceae. 5 Malay Peninsula, Borneo

Asemnantha Hook. f. 1873. Rubiaceae. 1 Mexico

Asepalum Marais. 1981. Acanthaceae. 1 Ethiopia, Somalia, Uganda, Kenya, Tanzania, southwestern Arabian Peninsula

Ashtonia Airy Shaw. 1968. Euphorbiaceae. 2 Malay Peninsula, Borneo

Asiasarum F. Maek. = Asarum L. Aristolochiaceae

Asimina Adans. 1763. Annonaceae. 8 eastern North America

Asiphonia Griff. = Thottea Rottb. Aristolochiaceae

Askellia W. A. Weber. 1984 (or = Crepis L.). Asteraceae. 9 North America, North and Central Asia, northwestern Himalayas

Asketanthera Woodson. 1932. Apocynaceae. 4 trop. America, West Indies

Askidiosperma Steud. 1855. Restionaceae. 11 South Africa (Cape)

Aspalathus L. 1753. Fabaceae. c. 280 South Africa from Natal to southwestern Cape

Asparagopsis (Kunth) Kunth = Protasparagus Oberm. Asparagaceae

Asparagus L. 1753. Asparagaceae. c. 100 arid and semiarid regions of the Old World; incl. (c. 30), Russia (from Europeana part to Far East), Caucasus, and Middle Asia

Aspasia Lindl. 1832. Orchidaceae. 10 Central (Guatemala, Nicaragua, Costa Rica, Panama) and South (Colombia, Ecuador, Venezuela, Trinidad, Guyana, Surinam, Brazil) America

Aspazoma N. E. Br. 1925. Aizoaceae. 1 western South Africa

Asperuginoides Rauschert. 1982. Brassicaceae. 1 Caucasus, Middle Asia; Iran, Afghanistan, Pakistan

Asperugo L. 1753. Boraginaceae. 1 (A. procumbens) Europe, Mediterranean, Southwest Asia through Caucasus and Middle Asia to West Siberia, Mongolia, China, and Himalayas (from Kashmir to Nepal)

Asperula L. 1753. Rubiaceae. 90 (or c. 200) Eurasia, chiefly Mediterranean, and Russia (c. 50 from European part to Far East), eastern Australia and Tasmania

Asphodeline Rchb. 1830. Asphodelaceae. 15–16 Central Europe, Sicily, eastern Mediterranean, Asia Minor, northern Iraq, Iran; Crimea and Caucasus (6)

Asphodelus L. 1753. Asphodelaceae. 12 northern Mexico, Azores and Canary Is., Mediterranean, Southwest Asia eastward to India

Aspicarpa Rich. 1815. Malpighiaceae. 12 trop. America

Aspidistra Ker-Gawl. 1822. Convallariaceae. c. 40 eastern Himalayas, India (Assam), southwestern and southern China, Taiwan, northern Vietnam

Aspidocarya Hook. f. et Thomson. 1855. Menispermaceae. 1 eastern Himalayas, northeastern India, southwestern and southern China

Aspidogenia Burret = Myrcianthes O. Berg. Myrtaceae

Aspidoglossum E. Mey. 1836. Asclepiadaceae. 37 trop. and South Africa

Aspidogyne Garay. 1977. Orchidaceae. 26 trop. America

Aspidonepsis A. Nicholas et D. J. Goyder. 1992. Asclepiadaceae. 5 South Africa

Aspidophyllum Ulbr. 1922. Ranunculaceae. 1 Peru

Aspidopterys A. Juss. 1840. Malpighiaceae. 21 India, Nepal, Sikkim, Andaman Is., Bangladesh, Burma, southern China, incl. Hainan, Indochina, Malesia: Malay Peninsula, Sumatra, Java, Borneo, Philippines, Sulawesi

Aspidosperma Mart. et Zucc. 1824. Apocynaceae. 50–56 trop. America, Antilles Is.

Aspidostemon Rohwer et H. G. Richt. 1987. Lauraceae. 15 Madagascar

Aspilia Thouars. 1806. Asteraceae. 25 trop. Africa, Madagascar

Aspiliopsis Greenm. = Podachaenium Benth. ex Oerst. Asteraceae

Asplundia Harling. 1954. Cyclanthaceae. 90 trop. America from Mexico to Bolivia, Lesser Antilles Is., Trinidad, and Tobago

Asplundianthus R. M. King et H. Rob.

1975. Asteraceae. 10 Colombia (5), Ecuador (2), Peru (3)

Asraoa Jozeph. = Wallichia Roxb. Arecaceae

Asta Klotzsch ex O. E. Schulz. 1933. Brassicaceae. 2 Mexico

Astartea DC. 1828. Myrtaceae. 7 Australia

Astelia Banks et Sol. ex R. Br. 1810. Asteliaceae. 25 Mascarene Is., New Guinea, Australia, Tasmania, New Zealand, New Caledonia, Polynesia to Hawaiian and Marquesas Is., Falkland Is.

Astemma Less. = Monactis Kunth. Asteraceae

Astenolobium Nevski = Astragalus L. Fabaceae

Astephania Oliv. = Anisopappus Hook. f. et Arn. Asteraceae

Astephanus R. Br. 1811. Asclepiadaceae. 2 South Africa

Aster L. 1753. Asteraceae. 250 America, Hawaiian Is., temp. Eurasia, trop. and South Africa, Madagascar

Asteranthe Engl. et Diels. 1901. Annonaceae. 3 trop. East Africa

Asteranthera Klotzsch et Hanst. 1854. Gesneriaceae. 1 Chile

Asteranthos Desf. 1820. Lecythidaceae. 1 Rio Negro region of Colombia, Venezuela, and Brazil

Asteridea Lindl. 1839. Asteraceae. 7 Australia

Asterigeron Rydb. = Aster L. Asteraceae

Asteriscium Cham. et Schltdl. 1826. Apiaceae. 8 America from Mexico to Patagonia

Asteriscus Hill. 1753. Asteraceae. 3 Mediterranean

Asterochaete Nees = Carpha Banks et Sol. ex R. Br. Cyperaceae

Asterogyne H. A. Wendl. ex Hook. f. 1883. Arecaceae. 3–5 Central and northern South America

Asterohyptis Epling. 1932. Lamiaceae. 3 Mexico

Asterolasia F. Muell. 1855. Rutaceae. 11 Australia

Asterolinon Hoffmanns. et Link. 1813–1820 (Lysimachia L.). Primulaceae. 2 Mediterranean (except Malta), Northeast (Sudan, Ethiopia) and East (Uganda, Kenya, Tanzania) Africa (Ethiopia), Iran; A. linum stellatum—extending to Crimea and Caucasus

Asteromoea Blume = Kalimeris (Cass.) Cass. Asteraceae

Asteromyrtus Schauer (Melaleuca L.) Myrtaceae

Asteropeia Thouars. 1806. Asteropeiaceae. 7 Madagascar

Asterophorum Sprague. 1908. Tiliaceae. 2 Ecuador, Surinam

Asteropsis Less. = Podocoma Cass. Asteraceae

Asteropterus Adans. = Leysera L. Asteraceae

Asteropyrum J. R. Drumm. et Hutch. 1920. Ranunculaceae. 2 China (Guangxi, Guizhou, Hubei, Hunan, Sichuan, Yunnan)

Asterosedum Grulich = Sedum L. Crassulaceae

Asterostemma Decne. 1838. Asclepiadaceae. 1 Java

Asterostigma Fisch. et C. A. Mey. 1845. Araceae. 6 southern Brazil, northern Argentina

Asterothamnus Novopokr. 1950. Asteraceae. 7 West and East Siberia, Middle Asia; western China, Mongolia

Asterotricha V. V. Boczantzeva = Pterygostemon V. V. Boczantzeva. Brassicaceae

Asterotrichion Klotzsch. 1840 (Plagianthus J. R. Forst. et G. Forst.). Malvaceae. 1 (A. discolor) Tasmania

Asthenatherum Nevski = Centropodia Reichenb. Poaceae

Asthenochloa Büse. 1854. Poaceae. 1 Southeast Asia

Astianthus D. Don. 1823. Bignoniaceae. 1 Mexico, Guatemala, Honduras, El Salvador, Nicaragua

Astiella Jovet. 1941. Rubiaceae. 1 Madagascar

Astilbe Buch. = Ham. ex D. Don. 1825. Saxifragaceae. 18 Himalayas (from Kashmir to Bhutan), China (15) eastward to Russian Far East (2–3), Japan, and Indochina, North America (2)

Astilboides (Hemsl.) Engl. 1930. Saxifragaceae. 1 northern China

Astiria Lindl. 1844. Sterculiaceae. 1 Mascarene Is.

Astoma DC. = Astomaea Rchb. Apiaceae

Astomaea Rchb. 1837. Apiaceae. 2 Southwest and Middle (1, A. galiocarpa) Asia

Astematopsis Korovin = Astomaea Rchb. Apiaceae

Astracantha Podlich. 1983 (Astragalus L.). Fabaceae. c. 120–135 Spain (2), Sicily (3), Italy (1), northeastern Mediterranean from Yugoslavia to Israel and Sinai, Morocco (1, A. granatensis), Arabian Peninsula, Crimea (1, A. arnacantha), Iraq, Iran (esp., c. 75), Afghanistan, Pakistan, India

Astragalus L. 1753. Fabaceae. c. 2,200 North America, Andes, temp. Eurasia (esp. Russia, Caucasus, and Middle Asia—c. 1,100), montane trop. Africa

Astranthium Nutt. 1840. Asteraceae. 12 southern U.S., Mexico

Astrantia L. 1753. Apiaceae. 10 Central, South, and East Europe; Asia Minor; Caucasus; Iran

Astrebla F. Muell. ex Benth. 1878. Poaceae. 4 Australia

Astrephia Durf. = Valeriana L. Valerianaceae

Astridia Dinter et Schwantes. 1926. Aizoaceae. 7 southern Namibia, western South Africa

Astripomoea A. D. J. Meeuse. 1958. Convolvulaceae. 12 trop. Africa

Astrocalyx Merr. 1910. Melastomataceae. 1 (A. calycina) Philippines

Astrocarpa Neck. ex Dumort. = Sesamoides Hill. Resedaceae

Astrocarpus Neck. = Sesamoides Hill. Resedaceae

Astrocaryum G. Mey. 1818. Arecaceae. c. 40 trop. America from Mexico southward to Bolivia and Brazil, absent from the West Indies, except Trinidad

Astrocasia B. L. Rob. et Millsp. 1905. Euphorbiaceae. 5 Mexico (3), Guatemala (2); A. tremula—also in Belize, Panama, Venezuela, West Indies (Cuba, Jamaica, Cayman Is.); A. jacobinensis—Bolivia, Brazil

Astrococcus Benth. 1854. Euphorbiaceae. 2 Venezuela, Brazil

Astrocodon Fedorov. 1957 (Campanula L.). Campanulaceae. 1 Russian Far East

Astrodaucus Drude. 1898. Apiaceae. 2–3 southern Europe, Crimea, Caucasus, Asia Minor, Syria, Iran, Iraq

Astroloba Uitew. 1947. Asphodelaceae. 7 South Africa

Astroloma R. Br. 1810. (Styphelia Sm.). Epacridaceae. 18 Australia

Astronia Blume. 1827. Melastomataceae. 59 Andaman and Nicobar Is., Burma, Taiwan, Malesia (excl. Lesser Sunda Is.)

Astronidium A. Gray. 1854. Melastomataceae. 67 Borneo, Philippines, Caroline Is., New Guinea, New Ireland, New Britain, Bougainville I., Solomon Is., Vanuatu, Fiji, Samoa, Society Is.

Astronium Jacq. 1760. Anacardiaceae. 15 trop. America, West Indies

Astrophytum Lem. 1839. Cactaceae. 5 southern U.S. (Texas), eastern Mexico

Astrostemma Benth. = Absolmsia Kuntze. Asclepiadaceae

Astrothalamus C. B. Rob. 1911. Urticaceae. 1 Indonesia, Philippines

Astrotricha DC. 1829. Araliaceae. 18 mainly southeastern Australia; A. hamptonii—western Australia

Astrotrichilia (Harms) J.-F. Leroy. 1975. Meliaceae. 14 Madagascar

Astydamia DC. 1829. Apiaceae. 2 Canary Is., Northwest Africa

Astyposanthes Hertev = Stylosanthes Sw. Fabaceae

Asyneuma Griseb. et Schenck. 1852. Campanulaceae. 50 Europe; Mediterranean; Caucasus; Middle Asia; Southwest Asia eastward to Himalayas; India; Sri Lanka; Burma; western, southern, and northeastern China; Korea; Japan; A. japonicum—extending to Russian Far East

Asystasia Blume. 1826. Acanthaceae. c. 70 trop. and subtrop. Old World

Asystasiella Lindau. 1895. Acanthaceae. 3 trop. Africa, trop. Asia

Ataenidia Gagnep. = Phrynium Willd. Marantaceae

Atalanthus D. Don = Sonchus L. Asteraceae

Atalantia Corrêa. 1805. Rutaceae. 11 India, Sri Lanka, southern China, Indochina, Malesia to Australia

Atalaya Blume. 1847. Sapindaceae. 11 South Africa, Lesser Sunda Is., New Guinea, Solomon Is., northern, central, and eastern Australia and Tasmania (9)

Atamisquea Miers ex Hook. et Arn. 1833. Capparaceae. 1 southern California, southern Arizona, Baja California, and Argentina

Ateixa Ravenna. 1972 (Sarcodraba Gilg et Muschler). Brassicaceae. 1 Argentina

Atelanthera Hook. f. et Thomson. 1861. Brassicaceae. 1 (A. perpusilla) Middle Asia, Afghanistan, Tibet, western Himalayas

Ateleia (DC.) Benth. 1837. Fabaceae. 16–18 Central and northern South America

Atemnosiphon Leandri. 1947 (Gnidia L.). Thymelaeaceae. 1 Madagascar

Ateramnus P. Browne = Sapium P. Browne. Euphorbiaceae

Athamanta L. 1753. Apiaceae. 5–6 Europe, Mediterranean

Athanasia L. 1763. Asteraceae. 39. South Africa (esp. Cape)

Athenaea Sendtn. 1846. Solanaceae. 10 Brazil

Atherandra Decne. 1844. Asclepiadaceae. 1 Southeast Asia, western Malesia

Atheranthera Mast. = Gerrardanthus Harv. ex Hook. f. Brassicaceae

Atherolepis Hook. f. 1883. Asclepiadaceae. 3 Burma, Thailand

Atherosperma Labill. 1806. Monimiaceae. 1–2 southeastern Australia, Tasmania

Atherostemon Blume. 1850. Asclepiadaceae. 1 Burma, Malesia

Athertonia L. A. S. Johnson et B. Briggs. 1975 (Helicia Lour.). Proteaceae. 1 northeastern Australia

Athrixia Ker-Gawl. 1823. Asteraceae. 14 trop. and South Africa

Athroisma DC. 1833. Asteraceae. 8 trop. Africa, Indo-Malesia

Athroostachys Benth. 1883. Poaceae. 1 Brazil

Athrotaxis D. Don. 1838. Taxodiaceae. 2–3 Tasmania

Athyana (Griseb.) Radlk. 1887. Sapindaceae. 1 Paraguay, Argentina

Athysanus Greene. 1885. Brassicaceae. 1 western North America from British Columbia to California and Arizona

Atkinsia R. A. Howard = Thespesia Sol. ex Corrêa. Malvaceae

Atkinsonia F. Muell. 1865. Loranthaceae. 1 Australia (New South Wales)

Atlanthemum Raynaud. 1988. Cistaceae. 1 Mediterranean, except Italy and Crete I.

Atomostigma Kuntze. 1898. Rosaceae. 1 Brazil

Atopostema Boutique = Monanthotaxis Baill. Annonaceae

Atractantha McClure. 1973. Poaceae. 5 Venezuela (1, Amazonas), Brazil (Amazonas and Bahia)

Atractocarpa Franch. 1887 (Puelia Franch.). Poaceae. 1 trop. Africa

Atractocarpus Schltr. et K. Krause. 1908. Rubiaceae. 10 New Caledonia

Atractogyne Pierre. 1896. Rubiaceae. 3 West Africa

Atractylis L. 1753. Asteraceae. c. 30 Macaronesia, Mediterranean, Southwest Asia

Atractylodes DC. 1838. Asteraceae. 7–8 northeastern China, Korea, Japan; A. ovata—extending to Russian Far East

Atragene L. = Clematis L. Ranunculaceae

Atraphaxis L. 1753. Polygonaceae. 26. Southeast Europe, Caucasus, North Africa, Southwest Asia through Middle Asia eastward to West and East Siberia, China, and Mongolia

Atrichantha Hilliard et B. L. Burtt. 1981. Asteraceae. 2 South Africa Atrichodendron Gagnep. 1950 (Lycium L.). Solanaceae. 1 Indochina

Atrichoseris A. Gray. 1884. Asteraceae. 1 southwestern U.S. (California, Utah, Arizona)

Atriplex L. 1753. Chenopodiaceae. Over 200 cosmopolitan; Russia (c. 40, from European part to Far East)

Atropa L. 1753. Solanaceae. 5 Europe,

Mediterranean, Crimea, Caucasus, Southwest Asia eastward to Middle Asia (1, A. komarovii) and western Himalayas (Kashmir)

Atropanthe Pascher. 1909 (Scopolia Jacq.). Solanaceae. 1 southern China (Yunnan)

Atropatenia F. K. Mey. 1973 (Thlaspi L.). Brassicaceae. 2 Caucasus

Atroxima Stapf. 1905. Polygalaceae. 2 West and Central Africa

Atrutegia Bedd. = Goniothalamus (Blume) Hook. f. et Thomson. Annonaceae

Attalea Kunth. 1816. Arecaceae. 25 trop. America from Panama southward to Peru and Brazil

Atuna Raf. 1838. Chrysobalanaceae. 11–13 southern India, Thailand, Malesia (5, excl. Lesser Sunda Is.), Admiralty, Caroline, and Solomon Is. to Fiji, Tonga, and Samoa Is.

Atylosia Wight et Arn. 1834 (Cajanus DC.). Fabaceae. 35 trop. Asia, Malesia, Australia

Aubregrinia Heine. 1960. Sapotaceae. 1 trop. West Africa

Aubrevillea Pellegr. 1933. Fabaceae. 2 trop. Africa from Sierra Leone to Zaire

Aubrieta Adans. 1763. Brassicaceae. 15 Southeast Europe, Asia Minor, Syria, Iran, Iraq

Aucoumea Pierre. 1896. Burseraceae. 1 trop. West Africa

Aucuba Thunb. 1784. Aucubaceae. 6 eastern Himalayas, northern Burma, continental China, Taiwan, Korea, Japan, Ryukyu Is.; A. japonica—extending to Russian Far East

Audouinia Brongn. 1826. Bruniaceae. 1 South Africa

Auerodendron Urb. 1924. Rhamnaceae. 7 West Indies

Auganthus Link = Primula L. Primulaceae

Augea Thunb. 1794. Zygophyllaceae. 1 South Africa

Augouardia Pellegr. 1924. Fabaceae. 1 trop. Africa (Gabon)

Augusta Pohl. 1829. Rubiaceae. 1 eastern Brazil

Augustinea Karsten, non St.-Hil. et Naudin = Bactris Jacq. ex Scop. Arecaceae

Aulacocalyx Hook. f. 1873. Rubiaceae. 8 trop. Africa

Aulacocarpus O. Berg = Mouriri Aubl. Melastomataceae

Aulacolepis Hack., non Ettingsh. = Neoaulacolepis Rauschert. Poaceae

Aulacophyllum Regel = Zamia L. Zamiaceae

Aulacospermum Ledeb. 1833. Apiaceae. 12–15 eastern European part of Russia, Middle Asia, southern Siberia, Iran, Afghanistan, Pakistan, ? western China

Aulandra H. J. Lam. 1927. Sapotaceae. 3 Borneo

Aulax Bergius. 1767. Proteaceae. 3 South Africa

Aulojusticia Lindau = Justicia L. Acanthaceae

Aulonemia Goudot. 1846. Poaceae. 24 trop. America from Mexico to Bolivia and Brazil

Aulosepalum Garay. 1982. Orchidaceae. 4 Mexico, Guatemala

Aulosolena Kozo = Polj. = Sanicula L. Apiaceae

Aulospermum J. M. Coult. et Rose = Cymopterus Raf. Apiaceae

Aulostylis Schltr. 1912 (Calanthe R. Br.). Orchidaceae. 1 New Guinea

Aulotandra Gagnep. 1901. Zingiberaceae. 6 trop. West Africa (1), Madagascar (5)

Aureliana Sendtn. 1846. Solanaceae. 1 eastern South America

Aureolaria Raf. 1837. Scrophulariaceae. 11 eastern U.S. (10), Mexico (1)

Auriculardisia Lundell = Ardisia Sw. Myrsinaceae

Auricula = ursi Adans. = Primula L. Primulaceae

Aurinia Desv. 1815 (Alyssum L.). Brassicaceae. 11 Central and Southeast Europe, Asia Minor, Caucasus

Australina Gaudich. 1830. Urticaceae. 2 Ethiopia and Kenya (1, A. flaccida), Australia, Tasmania, New Zealand

Australopyrum (Tsvelev) A. Löve. 1984 (Agropyron Gaertner). Poaceae. 3 Australia (2), New Guinea (1)

Australorchis Brieger = Dendrobium Sw. Orchidaceae

Austroamericium Hendrych. 1963 (or = Thesium L.). Santalaceae. 3 Venezuela, southeastern Brazil

Austrobaileya C. T. White. 1933. Austrobaileyaceae. 1–2 Australia (Queensland)

Austrobassia Ulbr. p.p. = Sclerochlamys F. Muell.; p.p. = Sclepolaena R. Br., p.p. = Stelligera A. J. Scott. Chenopodiaceae

Austrobrickellia R. M. King et H. Rob. 1972. Asteraceae. 3 Brazil, Bolivia, Argentina, Paraguay

Austrobuxus Miq. 1861. Euphorbiaceae. 20 Malesia (except Java and Lesser Sunda Is.), eastern Australia, New Caledonia (15) to Fiji

Austrocactus Britton et Rose. 1922. Cactaceae. 5 central and southern Chile, western and southern Argentina

Austrocedrus Florin et Boutelje. 1954 (Libocedrus Endl.). Cupressaceae. 1 southern Chile, southern Argentina

Austrocephalocereus Backeb. = Espostoopsis F. Buxb. Cactaceae

Austrochloris Lazarides. 1972. Poaceae. 1 Australia (Queensland)

Austrocritonia R. M. King et H. Rob. 1975. Asteraceae. 3 Brazil

Austrocylindropuntia Backeb. = Opuntia Hill. Cactaceae

Austrocynoglossum Popov ex R. R. Mill = Cynoglossum L. Boraginaceae

Austrodolichos Verdc. 1970. Fabaceae. 1 northern Australia

Austroeupatorium R. M. King et H. Rob. 1970. Asteraceae. 12 trop. South America southward to Uruguay

Austrofestuca (Tzvelev) E. B. Alexeev. 1976. Poaceae. 4 Australia, New Zealand

Austrogambeya Aubrév. et Pellegr. = Chrysophyllum L. Sapotaceae

Austroliabum H. Rob. et Brettell. = Microliabum Cabrera. Asteraceae

Austromatthaea L. S. Sm. 1969. Monimiaceae. 1 northeastern Australia

Austromimusops A. D. J. Meeuse = Vitellariopsis Baill. ex Dubard. Sapotaceae

Austromuellera C. T. White. 1930. Proteaceae. 1 northeastern Australia

Austomyrtus (Nied.) Burret. 1941. Myrtaceae. 37 New Guinea (3), eastern Australia (13), New Caledonia, Vanuatu

Austropeucedanum Mathias et Constance. 1952. Apiaceae. 1 Argentina

Austrosteenisia Geesink. 1986 (Lonchocarpus Kunth). Fabaceae. 3 New Guinea, northern Australia

Austrosynotis C. Jeffrey. 1986. Asteraceae. 1 trop. East Africa

Austrotaxus Compton. 1922. Austrotaxaceae. 1 New Caledonia

Autranella A. Chev. 1917 (Mimusops L.). Sapotaceae. 1 trop. Africa

Autrania C. Winkl. et Barbey = Jurinea Cass. Asteraceae

Autumnalia Pimenov. 1989. Apiaceae. 2 Middle Asia

Auxemma Miers. 1875. Cordiaceae. 2 Brazil

Auxopus Schltr. 1900. Orchidaceae. 3 trop. Africa (2), Madagascar (1)

Avellanita Phil. 1864. Euphorbiaceae. 1 Chile

Avellara Blanca et Diaz Guard. 1985. Asteraceae. 1 Portugal, Spain

Avellinia Parl. 1842 (Trisetaria Forssk.). Poaceae. 1 Mediterranean

Avena L. 1753. Poaceae. 25. Europe, North Africa, Arabian Peninsula; Crimea, Caucasus, Transcaucasia, Asia Minor, Kurdistan, Iraq, Iran, Middle Asia, Afghanistan, Pakistan, India, Mongolia; Russia (15, all areas, except northern)

Avenastrum Opiz = Helictotrichon Besser. Poaceae

Avenella (Bluff et Fingerh.) Drejer = Aira L. Poaceae

Avenochloa Holub = Avenula (Dumort.) Dumort. Poaceae

Avenula (Dumort.) Dumort. 1868 (Helictotrichon Besser ex Schult. et Schult. f.). Poaceae. 30 temp. Eurasia, North Africa

Averia Leonard = Tetramerium Nees. 1940. Acanthaceae

Averrhoa L. 1753. Oxalidaceae. 2 tropics

Averrhoidium Baill. 1874. Sapindaceae. 2 Brazil, Paraguay

Avetra H. Perrier. 1924. Stenomeridaceae. 1 eastern Madagascar

Avicennia L. 1753. Avicenniaceae. 14 mangroves of tropics and subtropics

Aviceps Lindl. = Satyrium Sw. Orchidaceae

Axinaea Ruiz et Pav. 1794. Melastomataceae. 20 trop. America

Axinandra Thwaites. 1854. Crypteroniaceae. 5 Sri Lanka (1), Southeast Asia (4)

Axiniphyllum Benth. 1872. Asteraceae. 4 southern Mexico

Axonopus P. Beauv. 1812. Poaceae. c. 110 trop. and subtrop. America, trop. Africa (1, A. flexuosus—from Guinea to Angola, Zambia, Uganda, Kenya, Tanzania

Axyris L. 1753. Chenopodiaceae. 5 Russia (5: southeastern European part, northern Caucasus, West and East Siberia, Far East), Middle Asia, Afghanistan, Pakistan, China, Mongolia

Ayapana Spach. 1841 (Eupatorium L.). Asteraceae. 15 trop. America, West Indies

Ayapanopsis R. M. King et H. Rob. 1972. Asteraceae. 16 Andes of the South America

Ayenia L. 1756. Sterculiaceae. c. 70 trop. and subtrop. America

Ayensua L. B. Sm. 1969. Bromeliaceae. 1 Venezuela

Aylacophora Cabrera. 1953. Asteraceae. 1 Argentina

Aylostera Speg. = Rebutia K. Schum. Cactaceae

Aylthonia N. L. Menezes = Barbacenia Vand. Velloziaceae

Aynia H. Rob. = Vernonia Schreb. Asteraceae

Azadehdelia Braem = Cribbia Senghas. Orchidaceae

Azadirachta A. Juss. 1830. Meliaceae. 2 Indo-Malesia

Azalea L. = Rhododendron L. Ericaceae

Azanza (DC.) Alef. = Thespesia Sol. ex Corrêa. Malvaceae

Azara Ruiz et Pav. 1794. Flacourtiaceae. 10: 8 temp. Chile (incl. Juan Fernandez Is.) and adjacent Argentina; 2 subtrop. and trop. northwestern Argentina, southern Bolivia, southeastern Brazil, Uruguay

Azedarach Hill = Melia L. Meliaceae

Azilia Hedge et Lamond. 1987. Apiaceae. 1 Iran

Azima Lam. 1783. Salvadoraceae. 4 trop. and South Africa, Madagascar, Aldabra and Comoro Is., Arbabian Peninsula, trop. Asia from India eastward to Philippines and Lesser Sunda Is.

Azorella Lam. 1783. Apiaceae. c. 70 New Zealand, sub-Antarctic islands, South America, Falkland Is.

Azorina (Watson) Feer. 1890 (Campanula L.). Campanulaceae. 1 Azores Is.

Aztekium Boed. 1929. Cactaceae. 2 northeastern Mexico

Azukia Takah. ex Ohwi = Vigna Savi. Fabaceae

Azureocereus Akers et H. Johnson = Browningia Britton et Rose. Cactaceae

B

Babbagia F. Muell. = Osteocarpum F.Muell. Chenopodiaceae

Babcockia Boulos = Sonchus L. Asteraceae

Babiana Ker-Gawl. ex Sims. 1801. Iridaceae. 64 trop. and South (63) Africa, Socotra (1)

Babingtonia Lindl. = Baeckea L. Myrtaceae

Bacbockia Boulos = Sonchus L. Asteraceae

Baccaurea Lour. 1790. Euphorbiaceae. 80 Himalayas, India, Andaman Is., Burma, China, Indochina, Malesia, Polynesia

Baccharidastrum Cabrera = Baccharis L. Asteraceae

Baccharidiopsis G. M. Barroso. 1975 (Aster L.). Asteraceae. 1 Brazil

Baccharis L. 1753. Asteraceae. c. 510 America, esp. South America

Bachmannia Pax. 1897. Capparaceae. 2 Mozambique and South Africa (Natal, Cape)

Backhousia Hook. et Harv. 1845. Myrtaceae. 8 New Guinea (1) eastern Australia

Baclea E. Fourn. 1877. Asclepiadaceae. 2 Brazil

Bacopa Aubl. 1775. Scrophulariaceae. 60 tropics and subtropics, esp. America

Bactris Jacq. ex Scop. 1777. Arecaceae. c. 240 trop America from Mexico to Paraguay, West Indies

Bactyrilobium Willden. = Cassia L. Fabaceae

Bacularia F. Muell. ex Hook. f. = Linospadix H. A. Wendl. Arecaceae

Badiera DC. 1824 (Polygala L.). Polygalaceae. 15 trop. America, West Indies

Badilloa R. M. King et H. Rob. 1975. (Eupatorium L.). Asteraceae. 9 northern Andes

Badula Juss. 1789. Myrsinaceae. 13 Mascarene Is.

Badusa A. Gray. 1859. Rubiaceae. 3 Philippines (Palawan I.), ? Moluccas, New Guinea, Palau Is., Caroline Is., Solomon Is., Vanuatu, Fiji, Tonga, Society Is.

Baeckea L. 1753. Myrtaceae. 69: 1 China, Indochina, Malesia (except Java and Lesser Sunda Is.), Borneo (1, endemic), Australia (66), New Caledonia (1)

Baeolepis Decne. ex Moq. 1849. Asclepiadaceae. 1 southern India

Baeometra Salisb. 1813. Melanthiaceae. 1 South Africa

Baeothryon A. Dietr. = Eleocharis R. Br. Cyperaceae

Baeria Fisch. et C. A. Mey. = Lasthenia Cass. Asteraceae

Baeriopsis J. T. Howell. 1942. Asteraceae. 1 Baja California

Bafodeya Prance ex F. White. 1976 (Parinari Aubl.). Chrysobalanaceae. 1 West Africa (Guinea)

Bafutia C. D. Adams. 1962. Asteraceae. 1 trop. West Africa

Bagassa Aubl. 1775. Moraceae. 2 northeastern South America

Bahia Lag. 1816. Asteraceae. 15 southwestern U.S., Mexico, Chile

Bahianthus R. M. King et H. Rob. 1972. Asteraceae. 1 northeastern Brazil

Baijiania A. M. Lu et J. Q. Li. 1993. Cucurbitaceae. 3 southern China (Yunnan), Taiwan, Thailand, Malesia from Malay Peninsula to Borneo

Baikiaea Benth. 1865. Fabaceae. 7 trop. Africa from Nigeria to Tanzania, trop. Asia (1, B. insignis)

Baileya Harv. et A. Gray ex A. Gray. 1849. Asteraceae. 4 southwestern U.S. (California, Nevada, Utah, Arizona, Texas), Mexico

Baileyoxylon C. T. White. 1941. Flacourtiaceae. 1 Australia (Queensland)

Baillonella Pierre. 1890. Sapotaceae. 1 trop. West Africa

Baillonia Bocquillon. 1862. Verbenaceae. 1 South America

Baissea A. DC. 1844. Apocynaceae. 24 trop. and South (1) Africa, trop. Asia

Baitaria Ruiz et Pav. 1794 (Calandrinia Kunth). Portulacaceae. c. 40 South America: Peru, Bolivia, Chile, Argentina; 1 sp. Mexico

Bakerantha L. B. Sm. = Hechtia Klotzsch. Bromeliaceae

Bakerella Tiegh. 1895 (Taxillus Tiegh.). Loranthaceae. 16 Madagascar

Bakeridesia Hochr. 1913 (Abutilon Hill). Malvaceae. 13 Mexico, Central America; B. integerrima extending to Venezuela

Bakerolimon Lincz. 1968 (Limonium Mill.). Plumbaginaceae. 2 Peru, northern Chile

Bakerophyton (J. Léonard) Hutch. = Aeschynomene L. Fabaceae

Balaka Becc. 1885. Arecaceae. 7 Fiji (5), Samoa (2)

Balanites Delile. 1813. Balanitaceae. 9 North and trop. Africa, West Asia eastward to India and Burma

Balanocarpus Bedd. = Hopea Roxb. Dipterocarpaceae

Balanophora J. R. Forst. et G. Forst. 1776. Balanophoraceae. 16 trop. and subtrop. Old World from Africa and Madagascar through South Asia to Japan, Australia (1, northeastern Queensland), Micronesia, Polynesia

Balanops Baill. 1871. Balanopaceae. 9 northeastern Australia, New Caledonia, Vanuatu, Fiji

Balanostreblus Kurz = Sorocea A. St.-Hil. Moraceae

Balansaea Boiss. et Reut. = Conopodium Koch. Apiaceae

Balaustion Hook. 1851. Myrtaceae. 2 southwestern Australia

Balbisia Cav. 1804. Ledocarpaceae. 8 South America

Balboa Liebm. ex Ditrich = Cracca Medik. Fabaceae

Balboa Planch. et Triana. 1860 (Tovomitopsis Planch. et Triana). Clusiaceae. 1 Colombia

Baldellia Parl. 1854 (Echinodorus Rich.). Alismataceae. 2 West and South Europe, North Africa

Baldomiria Herter = Leptochloa P. Beauv. Poaceae

Balduina Nutt. 1818. Asteraceae. 3 southern U.S.

Balfourodendron Corr. Mello ex Oliv. 1877. Rutaceae. 1 southern Brazil, Paraguay, northern Argentina

Balgoya Marat et Meijden. 1991. Polygalaceae. 1 New Caledonia

Balisaea Taub. 1895. Fabaceae. 1 Brazil

Baliospermum Blume. 1826. Euphorbiaceae. 10 Himalayas from Kashmir to Assam, India, Burma, southwestern China, Indochina, Malesia

Ballantinia Hook. f. ex E. A. Shaw. 1974. Brassicaceae. 1 Australia, Tasmania

Ballardia Montrouz. = Carpolepis (J. W. Dawson) J. W. Dawson. Myrtaceae

Ballochia Balf. f. 1884. Acanthaceae. 2 Socotra

Ballota L. 1753. Lamiaceae. 33 Europe (7), Mediterranean (27), Northeast (Ethiopia, Somalia) and South Africa (1), West Asia, incl. Caucasus

Ballya Brenan. 1964 (Aneilema R. Br.). Commelinaceae. 1 trop. East Africa

Balmea Martinez. 1942. Rubiaceae. 1 Mexico

Baloghia Endl. 1833. Euphorbiaceae. 15 eastern Australia, Lord Howe I., Norfolk I., New Caledonia (13)

Balonga Le Thomas. 1968. Annonaceae. 1 trop. West Africa (Cameroun, Gabon)

Balsamita Hill = Tanacetum L. Asteraceae

Balsamocarpon Clos. 1847 (Caesalpinia L.). Fabaceae. 1 Chile

Balsamocarpon C. Gay = Balsamocarpon Clos. Fabaceae

Balsamocitrus Stapf. 1906. Rutaceae. 2 trop. West and East (Uganda, B. dawei) Africa

Balsamodendrum Kunth = Commiphora Jacq. Burseraceae

Balsamorhiza Hook. 1833. Asteraceae. 14 western North America, Mexico

Balthasaria Verdc. 1969. Theaceae. 2–3 trop. Africa

Baltimora L. 1771. Asteraceae. 2 Mexico, Central and northern South America, West Indies

Bambekea Cogn. 1916. Cucurbitaceae. 1 trop. West Africa

Bambusa Schreb. 1789. Poaceae. c. 70 South and Southeast Asia, New Guinea, northern Australia

Bamiania Lincz. 1971. Plumbaginaceae. 1 Afghanistan

Bamlera K. Schum. et Lauterb. = Beccarianthus Cogn. Melastomataceae

Bampsia Lisowski et Mielcarek. 1983. Scrophulariaceae. 2 Zaire

Banalia Moq. = Indobanalia A. N. Henry et Roy. Amaranthaceae

Banara Aubl. 1775. Flacouritaceae. 31 trop. and subtrop. America from southern Mexico to Uruguay, Paraguay and northern Argentina, West Indies

Bancroftia Porter. 1838. Lamiaceae. 1 Venezuela

Bandeiuraea Welw. ex Benth. = Griffonia Baill. Fabaceae

Banisterioides Dubard et Dop. 1908. Malpighiaceae. 1 Madagascar

Banisteriopsis C. B. Rob. ex Small. 1910. Malpighiaceae. 92 trop. America: Mexico, Guatemala, El Salvador, Honduras, Nicaragua, Costa Rica, Panama, Colombia, Venezuela, the Guyanas, Ecuador, Peru, Bolivia, Brazil (c. 55), Paraguay, Argentina; West Indies (1, B. pauciflora—Cuba)

Banjolea Bowdich. = Nelsonia R. Br. Acanthaceae

Banksia L. f. 1782. Proteaceae. 71 New Guinea (1), Australia (esp. southwestern—57 endemics)

Baolia H. W. Kung et G. L. Chu. 1978 (Chenopodium L.). Chenopodiaceae. 1 China (Gansu and Sichuan)

Baphia Afzel. ex Lodd. 1820. Fabaceae. 45 trop. and South (2) Africa, Madagascar (1)

Baphiastrum Harms. 1913. Fabaceae. 2 trop. Africa (Cameroun, Gabon, Zaire)

Baphicacanthus Bremek. 1944 (Strobilanthes Blume). Acanthaceae. 1 northeastern India, southern China, Indochina

Baphiopsis Benth. ex Baker. 1871. Fabaceae. 1 Gabon, Angola, Zaire, Uganda, Tanzania

Baptisia Vent. 1808. Fabaceae. 17 eastern U.S.

Baptorhachis Clayton et Renvoize. 1986. Poaceae. 2 Mozambique

Baratranthus (Korth.) Miq. 1856. Loranthaceae. 4 Sri Lanka, western Malesia

Barbacenia Vand. 1788. Velloziaceae. 104 South America

Barbaceniopsis L. B. Sm. 1962. Velloziaceae. 3 Andes

Barbarea W. T. Aiton. 1812. Brassicaceae. c. 20 temp. Northern Hemisphere, Mediterranean, North and trop. East Africa, Himalayas, Tibet

Barberetta Harv. 1868. Haemodoraceae. 1 southern Africa

Barbeuia Thouars. 1806. Barbeuiaceae. 1 Madagascar

Barbeya Schweinf. 1891. Barbeyaceae. 1 Ethiopia, Somalia, western Arabian Peninsula

Barbieria DC. = Clitoria L. Fabaceae

Barbosa Becc. = Syagrus Mart. Arecaceae

Barbosella Schltr. 1918. Orchidaceae. 10—20 trop. and temp. South America

Barbrodria Luer. 1981. Orchidacea. 1 Brazil

Barcella (Trail) Trail ex Drude. 1881. Arecaceae. 1 Bolivia, Brazil (the banks of the Rio Negro)

Barclaya Wall. 1827. Barclayaceae. 2–3 Burma, Thailand, Malay Peninsula, Singapore, Sumatra, Borneo, New Guinea

Barjonia Decne. 1844. Asclepiadaceae. 12 Brazil

Barkeria Knowles et Westc. 1838. Orchidaceae. 10 Central America

Barkerwebbia Becc. = Heterospathe Scheff. Arecaceae

Barkleyanthus H. Rob. et Brettell. 1974. Asteraceae. 1 southern U.S., Mexico, Central America

Barkleya F. Muell. = Bauhinia L. Fabaceae

Barleria L. 1753. Acanthaceae. 230–250 tropics and subtropics

Barleriola Oerst. 1854. Acanthaceae. 6 West Indies

Barlia Parl. 1860. Orchidaceae. 1 Canary Is., Mediterranean

Barnadesia Mutis ex L. f. 1782. Asteraceae. 22 trop. Andes, eastern Brazil

Barnardiella Goldblatt. 1976. Iridaceae. 1 South Africa (Namaqualand)

Barnebya W. R. Anderson et B. Gates. 1981. Malpighiaceae. 2 eastern and southeastern Brazil

Barneoudia Gay. 1845. Ranunculaceae. 3 Chile, Argentina

Barnettia Santisuk = Santisukia Brummitt. Bignoniaceae

Barnhartia Gleason. 1926. Polygalaceae. 1 Venezuela, Guyana, Surinam, Brazil

Barombia Schltr. 1914. Orchidaceae. 1 trop. West Africa

Barongia Peter G. Wilson et B. Hyland. 1988. Myrtaceae. 1 Australia (Queensland)

Baronia Baker. 1882. Anacardiaceae. 1 Madagascar

Baroniella Costantin et Gallaud. 1907. Asclepiadaceae. 4 Madagascar

Barosma Willd. = Agathosma Willd. Rutaceae

Barringtonia J. R. Forst. et G. Forst. 1775. Lecythidaceae. 40 trop. East Africa (2), Madagascar, Seychelles, Comores and Mauritius Is., South and Southeast Asia eastward to Taiwan, Indonesia, Aru Is., New Guinea, northern Australia, western Pacific

Barroetea A. Gray. 1880. Asteraceae. 6 Mexico

Barrosoa R. M. King et H. Rob. 1971 (Eupatorium L.). Asteraceae. 11 trop. South America

Barteria Hook. f. 1860. Passifloraceae. 1 trop. Africa from southern Nigeria to Angola, Uganda, and western Tanzania

Barthea Hook. f. 1867. Melastomataceae. 1 montane northern Vietnam, southeastern China, Taiwan

Bartholina R. Br. ex W. Aiton et W. T. Aiton. 1913. Orchidaceae. 2–3 South Africa (Cape)

Bartholomaea Standl. et Steyerm. 1940. Flacourtiaceae. 2 Central America: southern Mexico, Guatemala, Belize

Bartlettia A. Gray. 1855. Asteraceae. 1 southern U.S., Mexico

Bartlettina R. M. King et H. Rob. 1971 (Eupatorium L.). Asteraceae. 23 trop. America

Bartonia Mühl. ex Willd. 1801. Gentianaceae. 4 eastern North America

Bartonia Pursh ex Sims = Mentzelia L. Loasaceae

Bartschella Britton et Rose = Mammillaria Haw. Cactaceae

Bartsia L. 1753. Scrophulariaceae. 60 temp. Northern Hemisphere and montane trop. areas; B. alpina northern European part of Russia

Barylucuma Ducke = Pouteria Aubl. Sapotaceae

Basanacantha Hook. f. = Randia L. Rubiaceae

Basananthe Peyr. 1859 (Tryphostemma Harv.). Passifloraceae. 25 Central, trop. East and South (5) Africa

Basedowia E. Pritz. 1918. Asteraceae. 1 southern Australia

Basella L. 1753. Basellaceae. 5: B. alba—West Indies, the Guyanas, Brazil, Madeira Is., trop. Africa, trop. Asia eastward to China and Japan, Philippines, Borneo, Fiji, and Hawaiian Is.; B. paniculata—Kenya, Tanzania, Mozambique, and South Africa (Transvaal); Madagascar (3)

Baseonema Schltr. et Rendle. 1896. Asclepiadaceae. 1 trop. East Africa

Bashania Keng f. et Yi = Arundinaria Michx. Poaceae

Basigyne J. J. Sm. 1917 (Dendrochilum Blume). Orchidaceae. 1 Sulawesi

Basilicum Moench. 1802 (Ocimum L.). Lamiaceae. 6–7 trop. and South (1, B. polystachyon) Africa, trop. and East Asia, Malesia, trop. Australia

Basiphyllaea Schltr. 1921. Orchidaceae. 3 Florida, West Indies

Basisperma C. T. White. 1942. Myrtaceae. 1 New Guinea

Basistelma Bartlett. 1909. Asclepiadaceae. 2 Mexico, Central America

Basistemon Turcz. 1863. Scrophulariaceae. 8 eastern Andes from Venezuela to Argentina

Baskervillea Lindl. 1840. Orchidaceae. 3 Colombia, Venezuela, Peru, and Brazil

Basselinia Vieil. 1873. Arecaceae. 11 New Caledonia

Bassia All. 1766. Chenopodiaceae. 7 Europe, Northwest Africa, Arabian Peninsula, West Asia (incl. Caucasus), Middle Asia and eastward to West and East Siberia, Mongolia, China, and India

Bassovia Aubl. = Solanum L. Solanaceae

Bastardia Kunth. 1822. Malvaceae. 2 trop. and subtrop. America from Texas to Peru and Venezuela, West Indies

Bastardiastrum (Rose) D. M. Bates. 1978. Malvaceae. 6 Pacific coastal Mexico from Sonora to Chiapas

Bastardiopsis (K. Schum.) Hassl. 1910. Malvaceae. 1 South America

Basutica E. Phillips = Gnidia L. Thymelaeaceae

Batania Hatus = Pycnarrhena Miers ex Hook. f. ex Thomson. Menispermaceae

Batemannia Lindl. 1834. Orchidaceae. 5

Colombia, Venezuela, the Guyanas, Peru, Bolivia, Brazil

Batesanthus N. E. Br. = Cryptolepis R. Br. Asclepiadaceae

Batesia Spruce ex Benth. 1865. Fabaceae. 1 Peru and Brazil (Amazonian)

Batesimalva Fryxell. 1975. Malvaceae. 3 southern U.S. (Texas), northeastern Mexico, western Venezuela (1)

Bathiaea Drake. 1902. Fabaceae. 1 Madagascar

Bathiorhamnus Capuron. 1966. Rhamnaceae. 2 Madagascar

Bathysa C. Presl. 1845. Rubiaceae. 10 trop. America from Panama to Peru and Brazil

Batidaea (Dumort.) Greene = Rubus L. Rosaceae

Batis P. Browne. 1756. Bataceae. 2 New Guinea, northern and northeastern Australia, Hawaiian and Galapagos Is., southwestern U.S., West Indies, Atlantic coastal South America

Batocarpus Karst. 1863. Moraceae. 4 trop. America

Batodendron Nutt. = Vaccinium L. Ericaceae

Batopedina Verdc. 1953. Rubiaceae. 2 West and trop. South Africa

Batrachium (DC.) Gray = Ranunculus L. Ranunculaceae

Battandiera Maire. 1926 (Ornithogalum L.). Hyacinthaceae. 1 North Africa

Baudouinia Baill. 1866. Fabaceae. 4 Madagascar

Bauera Banks ex Andr. 1801. Baueraceae. 3 temp. eastern Australia, Tasmania

Bauerella Borzi = Sarcomelicope Endl. Rutaceae

Baueropsis Hutch. = Cullen Medik. Fabaceae

Bauhinia L. 1753. Fabaceae. c. 250 pantropics

Baukea Vatke. 1881. Fabaceae. 1 Madagascar

Baumea Gaudich. 1829. Cyperaceae. 30 Madagascar, Mascarene Is., India, southern China, Malesia, Australia, New Caledonia, Oceania to Hawaiian Is.

Baumia Engl. et Gilg. 1903. Scrophulariaceae. 1 trop. Africa

Baumiella H. Wolff = Afrocarum Rauschert. Apiaceae

Baxteria R. Br. ex Hook. 1843. Dasypogonaceae. 1 southwestern Australia

Bdallophyton Eichler. 1872. Cytinaceae. 4 Mexico, El Salvador

Beadlea Small. 1903 (Cyclopogon C. Presl). Orchidaceae. 54 trop. and subtrop. America

Beata O. F. Cook = Coccothrinax Sargent. Arecaceae

Beaucarnea Lem. = Nolina Michx. Nolinaceae

Beaufortia R. Br. 1812. Myrtaceae. 16 southwestern Australia

Beaumontia Wall. 1824. Apocynaceae. 10 trop. Asia, Malesia

Beauprea Brongn. et Gris. 1871. Proteaceae. 12 New Caledonia

Beaupreopsis Virot. 1968. Proteaceae. 1 New Caledonia

Beauverdia Herter = Ipheion Raf. Alliaceae

Bebbia Greene. 1885. Asteraceae. 2 southwestern U.S. form Nevada and California to western Texas, Mexico

Beccarianthus Cogn. 1890 (Astronidium A. Gray). Melastomataceae. 20 Malesia: Philippines, Borneo, and eastward to New Guinea

Beccariella Pierre = Planchonella Pierre. Sapotaceae

Beccarinda Kuntze. 1891. Gesneriaceae. 7 Burma, southwestern China, Hainan I., Indochina,

Beccariodendron Warb. = Goniothalamus (Blume) Hook. f. et Thomson. Annonaceae

Beccariophoenix Jum. et H. Perrier. 1915. Arecaceae. 1 eastern Madagascar

Becheria Ridl. 1912 (Psychotrya L.). Rubiaceae. 1 Malay Peninsula

Becium Lindl. 1842 (Ocium L.). Lamiaceae. 14 trop. and South Africa, trop. Asia from southern Arabian Peninsula eastward to India

Beckeropsis Figari et De Not. 1853 (Pennisetum Rich.). Poaceae. 6 trop. Africa

Beckmannia Host. 1805. Poaceae. 4 North America, temp. and subtrop. Eurasia

Beckwithia Jeps. = Ranunculus L. Ranunculaceae

Beclardia A. Rich. 1828. Orchidaceae. 2 Mascarene Is.

Becquerelia Brongn. 1829. Cyperaceae. 10 trop. South America

Bedfordia DC. 1833. Asteraceae. 3 southeastern Australia, Tasmania

Beesia Balf. f. et W. W. Sm. 1915. Ranunculaceae. 2 northern Burma, western and southwestern China

Befaria Mutis ex L. = Bejaria Mutis ex L. Ericaceae

Begonia L. 1753. Begoniaceae. c. 1,000 tropics and subtropics, excl. Australia and Polynesia

Begoniella Oliv. = Begonia L. Begoniaceae

Beguea Capuron. 1969. Sapindaceae. 1 Madagascar

Behaimia Griseb. 1866. Fabaceae. 1 Cuba

Behnia Didr. 1855. Luzuriagaceae. 1 Southeast Africa

Behria Greene. 1886 (Bessera Schult.). Alliaceae. 1 southern California

Behuria Cham. 1834. Melastomataceae. 10 southern Bazil

Beilschmiedia Nees. 1831. Lauraceae. c. 200 tropics (America from Mexico to southern Brazil), Australia, New Zealand

Beirnaertia Louis ex Troupin. 1949. Menispermaceae. 1 trop. Africa

Beiselia Forman. 1987. Burseraceae. 1 Mexico

Bejaranoa R. M. King et H. Rob. 1978. (Eupatorium L.). Asteraceae. 2 Bolivia, Brazil, Paraguay

Bejaria Mutis ex L. 1771. Ericaceae. 25–30 southeastern U.S. (Georgia and Florida), West Indies, trop. America from Mexico to Bolivia

Bejaria Vent. = Bejaria Mutis ex L. Ericaceae

Bejaudia Gagnep. = Myrialepis Becc. Arecaceae

Belairia A. Rich. 1846. Fabaceae. 5–6 Cuba

Belamcanda Adans. 1763. Iridaceae. 1–2 East Asia

Belandra S. F. Blake = Prestonia R. Br. Apocynaceae

Belangera Cambess. = Lamononia Vell. Cunnoniaceae

Belemia Pires. 1981. Nyctaginaceae. 1 eastern Brazil

Belencita Karst. 1857. Capparaceae. 1 Colombia

Beliceodendron Lundell = Lecointea Ducke. Fabaceae

Belicia Lundell = Morinda L. Rubiaceae

Bellardia All. 1785 (Bartsia L.). Scrophulariaceae. 1 (B. trixago) Mediterranean, Ethiopia, South Africa, Caucasus, Asia Minor, northwestern Iran, northern Iraq

Bellardiochloa Chiov. 1929 (Poa L.). Poaceae. 2 montane Europe to the Carpathians, Caucasus, Southwest Asia to Iraq

Bellendena R. Br. 1810. Proteaceae. 1 Tasmania

Bellevalia Lapeyr. 1808 (Muscari Hill). Hyacinthaceae. c. 50 Mediterranean, Southeast Europe, West and Southwest Asia eastward to Middle Asia and Pakistan

Bellida Ewart. 1907. Asteraceae. 1 southwestern Australia

Bellidastrum Scop. = Aster L. Asteraceae

Belliolum Tiegh. = Zygogynum Baill. Winteraceae

Bellis L. 1753. Asteraceae. 7 Europe, Mediterranean, Caucasus, Middle Asia

Bellium L. 1771. Asteraceae. 3 Mediterranean

Belloa J. Remy. 1848 (Lucilia Cass.). Asteraceae. 9 Andes from Venezuela to central Chile

Bellonia L. 1753. Gesneriaceae. 2 West Indies

Bellucia Neck. ex Raf. 1838. Melastomataceae. 7 trop. America: Mexico, Guatemala, Belize, Honduras, Nicaragua, Costa Rica, Panama, Guadeloupe, Colombia, Venezuela, the Guyanas, Ecuador, Peru, Bolivia, and Brazil

Bellynkxia Muell. Arg. = Morinda L. Rubiaceae

Belmontia E. Mey. = Sebaea Sol. ex R. Br. Gentianaceae

Beloglottis Schltr. 1920 (Spioranthes Rich.). Orchidaceae. 8 trop. and subtrop. America

Belonanthus Graebn. = Valeriana L. Valerianaceae

Belonophora Hook. f. 1873. Rubiaceae. 8 trop. Africa from Sierra Leone to Sudan and Tanzania

Beloperone Nees = Justicia L. Acanthaceae

Belostemma Wall. 1834 (Tylophora R. Br.). Asclepiadaceae. 2 Himalayas, India, China

Belosynapsis Hassk. 1871 (Cyanotis D. Don). Commelinaceae. 4 trop. Asia

Belotia A. Rich. = Trichospermum Blume. Tiliaceae

Beltrania Miranda = Enriquebeltrania Rzed. Euphorbiaceae

Bemarivea Choux. 1927 (Tinopsis Radkl.). Sapindaceae. 1 Madagascar

Bembicia Oliv. 1883. Bembiciaceae. 1 Madagascar

Bembicidium Rydb. 1920. Fabaceae. 1 Cuba

Bembiciopsis H. Perrier = Camelia L. Theaceae

Bencomia Webb et Berthel. 1842. Rosaceae. 5 Canary Is.

Benedictella Maire. 1924 (Lotus L.). Fabaceae. 1 Morocco

Beneditaea Toledo = Ottelia Pers. Hydrocharitaceae

Benevidesia Saldanha et Cogn. 1888. Melastomataceae. 2 southeastern Brazil

Benguellia G. Taylor. 1931. Lamiaceae. 1 Angola

Benincasa Savi. 1818. Cucurbitaceae. 1 (B. hispida) trop. Asia, Malesia, Australia

Benitoa D. Keck. 1956. Asteraceae. 1 California

Benjaminia Mart. ex Benj. 1847. Scrophulariaceae. 1 trop. America

Benkara Adans. 1763 (Randia L.). Rubiaceae. 1 southern India, Sri Lanka

Bennetia Miq. = Bennettiodendron Merr. Flacourtiaceae

Bennettiodendron Merr. 1927. Falcourtiaceae. 4 India, Burma, southern China, Indochina, Malesia (Sumatra and Java)

Benoicanthus Heine et A. Raynal. 1968. Acanthaceae. 2 Madagascar

Benoistia H. Perrier et Leandri. 1938. Euphorbiaceae. 2 Madagascar

Bensonia Abrams et Bacigalupo = Bensoniella C. Morton. Saxifragaceae

Bensoniella C. Morton. 1965. Saxifragaceae. 1 western U.S. (southwestern Oregon and northwestern California)

Benthamantha Alef. = Cracca Benth. Fabaceae

Benthamia A. Rich. 1828. Orchidaceae. 25 Mascarene Is.

Benthamidia Spach = Cornus L. Cornaceae

Benthamiella Speg. 1883. Solanaceae. 12 southern Chile (4), southern Argentina (Patagonia)

Benthamina Tiegh. 1895 (Amyema Tiegh.). Loranthaceae. 1 eastern Australia

Benthamistella Kuntze = Buchnera L. Scrophulariaceae

Bentia Rolfe. 1894 (Justicia L.). Acanthaceae. 1 southern Arabian Peninsula

Bentinckia Berry ex Roxb. 1832. Arecaceae. 2 southern India (Tranvancore, 1), Nicobar Is. (1)

Bentinckiopsis Becc. = Clinostigma H. A. Wendl. Arecaceae

Bentleya E. M. Benn. 1986. Pittosporaceae. 1 western Australia

Benzingia Dodson. 1989. Orchidaceae. ? 2 western trop. South America

Benzonia Schumach. 1827. Rubiaceae. 1 West Africa

Bequaertia R. Wilczek = Campylostemon Welw. Celastraceae

Bequaertiodendron De Wild. = Englerophytum K. Krause. Sapotaceae

Berardia Vill. 1779. Asteraceae. 1 western Alps

Berberidopsis Hook. f. 1862. Berberidopsidaceae. 2 Australia (B. beckleri—Queensland and New South Wales), Chile (B. corallina)

Berberis L. 1753. Berberidaceae. 450 Eurasia, North Africa, North and South America along the Andes to Strait of Magellan

Berchemia DC. 1825. Rhamnaceae. 31 trop. Old World from trop. East Africa to East Asia and northern Australia (1, B. ecorollata); western North America

Berchemiella Nakai. 1923. Rhamnaceae. 3 China, southern Korea, Japan

Berendtiella Wettst. et Harms = Hemichaena Benth. Scrophulariaceae

Berenice Tul. 1857. Campanulaceae. 1 Reunion I.

Bergenia Moench. 1794. Saxifragaceae. 14: 6 Middle Asia, West and East Siberia, Far East; Afghanistan, Himalayas from Kashmir to Bhutan, northeastern India, Burma, China, Mongolia, Japan

Bergeranthus Schwantes. 1926. Aizoaceae. 12 South Africa

Bergerocactus Britton et Rose. 1909. Cactaceae. 1 southwestern U.S. (California), Mexico

Bergeronia Micheli. 1883. Fabaceae. 1 Paraguay, Argentina

Berghesia Nees. 1847. Rubiaceae. 1 Mexico

Bergia L. 1771. Elatinaceae. 25 trop., subtrop. and warm–temp. regions; 2 sp. in Caucasus and Middle Asia

Berginia Harv. = Holographis Nees. Acanthaceae

Berhautia Balle. 1956. Loranthaceae. 1 West Africa

Berkheya Ehrh. 1788. Asteraceae. 75 trop. and South Africa

Berkheyopsis O. Hoffm. = Hirpicium Cass. Asteraceae

Berlandiera DC. 1836. Asteraceae. 4 U.S. from California to Kansas and Arkansas, Mexico

Berlinia Sol. ex Hook. f. 1849. Fabaceae. 18 trop. Africa

Berlinianche (Harms) Vattimo. 1955 (Pilostyles Guill.). Apodanthaceae. 2 trop. Africa

Bernardia Mill. 1754. Euphorbiaceae. 50 trop. and subtrop. America, West Indies

Bernardinia Planch. 1850 (Rourea Aubl.). Connaraceae. 1 eastern Brazil

Berneuxia Decne. 1873. Diapensiaceae. 2 eastern Tibet, southwestern China, northern Burma

Bernieria Baill. = Beilschmiedia Nees. Lauraceae

Bernoullia Oliv. 1873. Bombacaceae. 2 trop. America

Berresfordia L. Bolus = Conophytum N. E. Br. Aizoaceae

Berroa Beauverd. 1913 (Lucilia Cass.). Asteraceae. 1 southern Brazil, Uruguay, central and northeastern Argentina

Berrya Roxb. 1820. Tiliaceae. 6 India, Sri Lanka, southern China (1), Taiwan, Indochina, Malesia, Polynesia

Bersama Fres. 1837. Melianthaceae. 2 (1—very polymorphic) trop. and South Africa

Berteroa DC. 1821. Brassicaceae. 5 temp. Eurasia, incl. (2) European part of Russia, Caucasus, West and East Siberia, Middle Asia

Berteroella O. E. Schulz. 1919. Brassicaceae. 1 northern and northeastern China, Korea, Japan (Honshu, Kyushu)

Bertholletia Bonpl. 1807. Lecythidaceae. 1 (B. excelsa Humboldt et Bonpl.) Colombia, Venezuela, Guyana, Surinam, French Guyana, Peru, Brazil, Bolivia

Bertiera Aubl. 1775. Rubiaceae. 55 trop. America, trop. Africa (41), Madagascar

Bertolonia Raddi. 1820. Melastomataceae. 14 Brazil

Bertya Planch. 1845. Euphorbiaceae. 23 Australia, Tasmania

Berula Besser ex W. D. J. Koch. 1826. Apiaceae. 2 North and Central America, Europe, trop. and South Africa, Caucasus, Middle Asia, Southwest and South Asia, Australia

Beruniella Zak. et Nabiev. 1986 (Heliotropium L.). Boraginaceae. 1 Middle Asia

Berzelia Brongn. 1826. Bruniaceae. 12 South Africa (Cape)

Beschorneria Kunth. 1850. Agavaceae. 7 Mexico

Besleria L. 1753. Gesneriaceae. 150 trop. and subtrop. America, West Indies

Bessera Schult. 1829. Alliaceae. 2 southern U.S., Mexico

Besseya Rydb. 1903. Scrophulariaceae. 9 North America

Beta L. 1753. Chenopodiaceae. 12 West and Southeast Europe, North Africa, West Asia eastward to Caucasus, Turkmenistan, and Iran

Betchea Schltr. = Caldeluvia D. Don. Cunoniaceae

Betonica L. = Stachys L. Lamiaceae

Betula L. 1753. Betulaceae. c. 110 arct. and temp. Northern Hemisphere, esp. Russia (c. 110)

Bewsia Goossens. 1941. Poaceae. 1 Central and Southern Africa

Beyeria Miq. 1844. Euphorbiaceae. 12 Australia

Beyrichia Cham. et Schltdl. 1828 (Achetaria Cham. et Schltdl.). Scrophulariaceae. 3 West Indies, Brazil

Bhesa Buch.-Ham. ex Arn. 1834. Celastraceae. 5 southern and northeastern India, Sri Lanka, Bangladesh, Andaman Is., Burma, peninsular Thailand, Indochina, Malesia (Sumatra, Malay Peninsula, Borneo, Philippines, New Guinea)

Bhidea Stapf ex Bor. 1949. Poaceae. 2 India

Biarum Schott. 1832. Araceae. 15 Mediterranean, eastward to Iran

Biasolettia W. D. J. Koch = Geocaryum Coss. Apiaceae

Bicuiba W. J. de Wilde. 1991. Myristicaceae. 1 southeastern Brazil

Bidaria (Endl.) Decne. 1844 (Gymnema R. Br.). Asclepiadaceae. 10 India, Sri Lanka, Bangladesh, southern Burma, Thailand, Malesia, northern Australia

Bidens L. 1753. Asteraceae. c. 230 cosmopolitan

Bidwillia Herb. = ? Caesia R. Br. Asphodelaceae

Biebersteinia Stephan. 1811. Biebersteiniaceae. 4–5 Eurasia from Southeast Europe (Greece) through Caucasus, Middle Asia to West Siberia and western Tibet

Bienertia Bunge ex Boiss. 1879. Chenopodiaceae. 1 Southeast Europe, Transcaucasia, Kazakhstan, Turkmenistan, Iran, Iraq

Biermannia King et Pantl. 1897. Orchidaceae. 9 India, China, Thailand, Vietnam, Malaya, Sumatra, Java, Bali

Bifora Hoffm. 1816. Apiaceae. 3 North America, Central and South Europe, North Africa, Southwest and West Asia, incl. Caucasus), Middle Asia

Bifrenaria Lindl. 1832. Orchidaceae. 10 trop. America from Panama to Peru and Brazil, West Indies

Bigelowia DC. 1836. Asteraceae. 2 North America

Bignonia L. 1753. Bignoniaceae. 1 southeastern U.S.

Bijlia N. E. Br. 1928. Aizoaceae. 1 (B. cana) or 12 ? South Africa (Cape)

Bikkia Reinw. 1825. Rubiaceae. 20 eastern Malesia eastward to Marianas and Caroline Is., Solomon Is., Vanuatu, New Caledonia (c. 10), Fiji (1), the Horne Is., Tonga and Niue Is.

Bilacunaria Pimenov et V. N. Tichom. 1983. Apiaceae. 4 eastern Mediterranean, Caucasus (2)

Bilderdykia Dumort. = Fallopia Adans. Polygonaceae

Bilegnum Brand. 1915 (Rindera Pall.). Boraginaceae. 1 northern Iran, Middle Asia

Billardiera Sm. 1793. Pittosporaceae. 8 Australia and Tasmania

Billbergia Thunb. 1821. Bromeliaceae. 56 Mexico, Guatemala, Belize, El Salvador, Nicaragua, Costa Rica, Panama, Colombia, Venezuela, Trinidad, Surinam, French Guyana, Brazil (c. 45), Peru, Bolivia, Paraguay, Uruguay, Argentina; West Indies: Cuba, Leeward Is., Windward Is.

Billia Peyr. 1858. Hippocastanaceae. 2 trop. America from Mexico to northwestern South America

Billieturnera Fryxell. 1982. Malvaceae. 1 southern U.S. (Texas) and northeastern Mexico

Billya Cass. = Petalacte D. Don. Asteraceae

Binotia Rolfe. 1905. Orchidaceae. 1 Brazil

Biolettia Greene = Trichocoronis A. Gray. Asteraceae

Biondia Schltr. 1905. Asclepiadaceae. 7 China

Biophytum DC. 1824. Oxalidaceae. 75 pantropics

Biota (D. Don) Endl. = Platycladus Spach. Cupressaceae

Biovularia Kamienski = Utricularia L. Lentibulariaceae

Bipinnula Comm. ex Juss. 1789. Orchidaceae. 8 temp. South America

Bipontia S. F. Blake = Soaresia Sch. Bip. Asteraceae

Bisboeckelera Kuntze. 1891. Cyperaceae. 8 South America

Bischofia Blume. 1827. Euphorbiaceae. 2 central and southeastern China, Taiwan, India, Indochina, Malesia, Polynesia

Biscutella L. 1753. Brassicaceae. 40 Central and South Europe, eastern Mediterranean, West Asia; B. laevigata extending to southern European part of Russia

Bisedmondia Hutch. = Calycophysum Karst. et Triana. Cucurbitaceae

Biserrula L. 1753 (or = Astragalus L.). Fabaceae. 1 Mediterranean, Northeast and East Africa

Bisglaziovia Cogn. 1891. Melastomataceae. 1 Brazil

Bisgoeppertia Kuntze. 1891. Gentianaceae. 4 West Indies

Bishopalea H. Rob. 1981. Asteraceae. 1 Brazil

Bishopanthus H. Rob. 1983. Asteraceae. 1 Peru

Bishopiella R. M. King et H. Rob. 1981. Asteraceae. 1 Brazil

Bishovia R. M. King et H. Rob. 1978. Asteraceae. 2 Bolivia and Argentina

Bismarckia Hildebrandt et H. A. Wendl. 1881. Arecaceae. 1 drier parts of Madagascar

Bisquamaria Pichon = Laxoplumeria Markgr. Apocynaceae

Bistella Adans. = Vahlia Thunb. Vahliaceae

Bistorta Hill = Polygonum L. Polygonaceae

Biswarea Cogn. 1882. Cucurbitaceae. 1 Nepal, Sikkim, southwestern China

Bituminaria Heist. ex Fabr. 1759. Fabaceae. 2–4 Mediterranean

Biventraria Small = Asclepias L. Asclepiadaceae

Bivinia Jaub. ex Tul. 1857. Flacourtiaceae. 1 Kenya, Tanzania, Mozambique, Zimbabwe, Madagascar

Bivonaea DC. 1821. Brassicaceae. 1 Sardinia, Sicily, Algeria, Tunisia

Bixa L. 1753. Bixaceae. 1 (B. orellana) trop. America, West Indies

Bizonula Pellegr. 1924. Sapindaceae. 1 trop. Africa

Blabeia Baehni = Pouteria Aubl. Sapotaceae

Blaberopus A. DC. = Alstonia R. Br. Apocynaceae

Blachia Baill. 1858. Euphorbiaceae. 10 India, Andaman Is., Sri Lanka, Bangladesh, Burma, China, Indochina, Malesia to Sulawesi

Blackallia C. A. Gardner. 1942. Rhamnaceae. 2 western Australia

Blackiella Aellen = Atriplex L. Chenopodiaceae

Blackstonia Huds. 1762. Gentianaceae. 4 Europe, Mediterranean; 1 (B. perfoliata) Crimea, Caucasus

Blaeria L. 1753 (or = Erica L.). Ericaceae. 34 trop. and South Africa, Madagascar

Blainvillea Cass. 1823. Asteraceae. 10 pantropics

Blakea P. Browne. 1756. Melastomataceae. c. 100 trop. America, West Indies

Blakeanthus R. M. King et H. Rob. 1972. Asteraceae. 1 Guatemala

Blakeochloa Veldkamp. 1981 (Plinthanthesis Steud.). Poaceae. 3 Australia

Blakiella Cuatrec. 1968. Asteraceae. 1 Venezuela

Blanchetia DC. 1836. Asteraceae. 1 northeastern Brazil

Blanchetiastrum Hassl. 1910. Malvaceae. 1 Brazil

Blancoa Lindl. 1840. Conostylidaceae. 1 southwestern Australia

Blandfordia Sm. 1804. Blandfordiaceae. 4 eastern Australia and Tasmania

Blandibractea Wernham. 1917. Rubiaceae. 1 Brazil

Blandowia Willd. 1809 (Apinagia Tul.). Podostemaceae. 1 South America

Blastemanthus Planch. 1846. Sauvagesiaceae. 5 northeastern South America

Blastocaulon Ruhland. 1903. Eriocaulaceae. 4 Brazil

Blastus Lour. 1790. Melastomataceae. 18 northeastern India, China, Indochina, Malesia

Bleasdalea F. Muell. [1865] ex Domin. 1921 (Grevillea R. Br. ex Knight). Proteaceae. 3–5 northeastern Australia, western Oceania

Blechum P. Browne. 1756. Acanthaceae. 6–10 trop. America, West Indies

Bleekeria Hassk. = Ochrosia Juss. Apocynaceae

Bleekrodea Blume. 1865 (Streblus Lour.). Moraceae. 3 Madagascar, Indochina, Borneo

Blennodia R. Br. 1849. Brassicaceae. 2 Australia

Blennosperma Less. 1832. Asteraceae. 3 California, Chile (1)

Blennospora A. Gray. 1851. Asteraceae. 2 southern Australia

Blepharandra Griseb. 1849. Malpighiaceae. 6 southern Venezuela, Guyana, Brazil

Blepharidachne Hack. 1888. Poaceae. 4 U.S. (2 California, Nevada, Utah, Texas), Mexico (1), Argentina (2)

Blepharidium Standl. 1918. Rubiaceae. 2 Mexico, Central America

Blephariglottis Raf. = Habenaria Willd. Orchidaceae

Blepharipappus Hook. 1833. Asteraceae. 1 western U.S.

Blepharis Juss. 1789. Acanthaceae. c. 100 eastern Mediterranean, South Africa and trop. Old World

Blepharispermum DC. ex Wight. 1834. Asteraceae. 10 trop. Africa, Madagascar, trop. Asia from Arabian Peninsula to India

Blepharistemma Wall. ex Benth. 1858. Rhizophoraceae. 1 India

Blepharitheca Pichon = Cuspidaria DC. Bignoniaceae

Blepharizonia (A. Gray) Greene. 1855. Asteraceae. 1 (B. plumosa) California

Blepharocalyx O. Berg. 1856. Myrtaceae. 3 trop. and subtrop. South America, West Indies

Blepharocarya F. Muell. 1878. Anacardiaceae. 2 northern and eastern Australia

Blepharodon Decne. 1844. Asclepiadaceae. 45 trop. America from Mexico to Chile

Blepharoneuron Nash. 1898. Poaceae. 2 North America from Utah and Colorado to southern Mexico

Blephilia Raf. 1819. Lamiaceae. 2 North America

Bletia Ruiz et Pav. 1794. Orchidaceae. 26 trop. America from Mexico and Florida to Peru, Bolivia and Brazil, West Indies

Bletilla Rchb. f. 1852–53. Orchidaceae. 9 temp. East (southward to Taiwan) and Southeast Asia

Blighia K. König. 1806. Sapindaceae. 4 trop. Africa

Blighiopsis Veken. 1960. Sapindaceae. 1 trop. Africa

Blinkworthia Choisy. 1834. Convolvulaceae. 2 Burma, southern China

Blomia Miranda. 1953. Sapindaceae. 1 Mexico

Bloomeria Kellogg. 1863. Alliaceae. 2 California

Blossfeldia Werderm. 1937 (Frailea Britton et Rose). Cactaceae. 1 southern Bolivia and northwestern and northern Argentina

Blotia Leandri. 1957. Euphorbiaceae. 5 Madagascar

Blumea DC. 1833. Asteraceae. c. 75 trop. Africa, Madagascar, Himalayas from Kumaon to Sikkim, India, Sri Lanka, Bangladesh, Burma, China, Indochina,, Malesia to New Guinea, northern Australia, Micronesia; 1 Caribbean region

Blumenbachia Schrad. 1825. Loasaceae. 6 South America

Blumeodendron (Muell. Arg.) Kurz. 1873. Euphorbiaceae. 6 Andaman Is., Malesia

Blumeopsis Gagnep. 1920. Asteraceae. 1 India, Nepal, Southeast Asia

Blutaparon Raf. 1838 (Philoxerus R. Br.). Amaranthaceae. 4 America, West Africa, Ryukyu Is.

Blysmocarex N. Ivanova = Kobresia Willd. Cyperaceae

Blysmopsis Oteng-Yeb. 1974 (Blysmus Panz. ex Schult.). Cyperaceae. 1 temp. and subtrop. Eurasia, eastern North America

Blysmoschoenus Palla. 1910. Cyperaceae. 1 Bolivia

Blysmus Panz. ex Schult. 1824. Cyperaceae. 3 temp. and suptrop. Eurasia, North Africa

Blyttia Arn. 1838 (Gynanchum L.). Asclepiadaceae. 2 East Africa, southern Arabian Peninsula

Blyxa Noronha et Thouars. 1806. Hydrocharitaceae. 9 trop. Old World

Bobartia L. 1753. Iridaceae. 12 South Africa (Cape)

Bobea Gaudich. 1830 (Timonius DC.). Rubiaceae. 4 Hawaiian Is.

Bocagea A. St.-Hil. 1825. Annonaceae. 3 trop. America

Bocageopsis R. E. Fr. 1931. Annonaceae. 4 trop. America

Bocconia L. 1753. Papaveraceae. 10 trop. and subtrop. America, West Indies

Bocoa Aubl. 1775. Fabaceae. 7 Guyana, Surinam, Amazonian

Bocquillonia Baill. 1862. Euphorbiaceae. 14 New Caledonia

Boea Comm. ex Lam. 1785. Gesneriaceae. 20 trop. Asia, Australia

Boeberastrum (A. Gray) Rydb. 1915. Asteraceae. 2 Mexico

Boeberoides (DC.) Strother. 1987. Asteraceae. 1 Mexico

Boechera A. Löve et D. Löve = Arabis L. Brassicaceae

Boecherarctica A. Löve = Saxifraga L. Saxifragaceae

Boeckeleria T. Durand. 1888 (Tetraria P. Beauv.). Cyperaceae. 1 South Africa

Boehmeria Jacq. 1760. Urticaceae. 50 trop. and subtrop. Northern Hemisphere

Boehmeriopsis Kom. 1901 (Fatoua Gaudich.). Urticaceae. 1 Korea

Boeica T. Anderson ex C. B. Clarke. 1874. Gesneriaceae. 12 southern China, Hong Kong, Indochina

Boeicopsis H. W. Li = Boecia T. Anderson ex C. B. Clarke. Gesneriaceae

Boenninghausenia Rchb. ex Meisn. 1837. Rutaceae. 2 Himalayas from Kashmir to Bhutan, India, China, Taiwan, Japan, Malesia

Boerhavia L. 1753. Nyctaginaceae. 20 tropics and subtropics

Boerlagea Cogn. 1890. Melastomataceae. 1 Borneo

Boerlagella Cogn. 1891. Sapotaceae. 1 Sumatra

Boerlagiodendron Harms = Osmoxylon Miq. Araliaceae

Boesenbergia Kuntze. 1891. Zingiberaceae. 43 India, China, Indochina, Malesia

Bognera Mayo et Nicolson. 1984. Araceae. 1 Brazil

Bogoria J. J. Sm. 1905. Orchidaceae. 3 Java, Philippines, New Guinea

Boholia Merr. 1926. Rubiaceae. 1 Philippines

Boisduvalia Spach. 1835. Onagraceae. 8 western North and temp. South America

Boissiaea Lem. = Bossiaea Vent. Fabaceae

Boissiera Hochst. et Steud. 1854. Poaceae. 1 eastern Mediterranean, Caucasus, Kurdistan, Iraq, Iran, Middle Asia, Afghanistan, Pakistan, northwestern India

Bojeria DC. = Inula L. Asteraceae

Bokkeveldia D. Müll.-Doblies et U. Müll.-Doblies. 1985. Amaryllidaceae. 4 South Africa (Cape)

Bolandra A. Gray. 1868. Saxifragaceae. 2 Pacific coastal North America (Washington, Oregon, Califronia)

Bolanosa A. Gray. 1852. Asteraceae. 1 Mexico

Bolanthus (Ser.) Rchb. 1841. Caryophyllaceae. 14–16 South Europe Asia Minor, Lebanon, ? Syria, Israel and Sinai Peninsula

Bolax Comm. ex A. Juss. 1789 (Azorella Lam.). Apiaceae. 4–5 temp. South America

Bolbidium Lindl. = Dendrobium Sw. Orchidaceae

Bolbosaponaria Bondar. 1971. Caryophyllaceae. 6 Middle Asia

Bolboschoenus (Asch.) Palla. 1907. Cyperaceae. 8 temp., subtrop. and trop. regions

Bolbostemma Franquet. 1930. Cucurbitaceae. 2 China

Boldoa Cav. ex Lag. 1816. Nyctaginaceae. 1 Mexico, Central America, Colombia, Venezuela; West Indies

Boleum Desv. 1815. Brassicaceae. 1 Spain

Bollea Rchb. f. 1852. Orchidaceae. 3 northern South America

Bolocephalus Hand.-Mazz. = Dolomiaea DC. Asteraceae

Bolophyta Nutt. 1840 (Parthenium L.). Asteraceae. 3 western U.S.

Boltonia L'Hér. 1789. Asteraceae. 2 North America

Bolusafra Kuntze. 1891. Fabaceae. 1 South Africa

Bolusanthus Harms. 1906. Fabaceae. 1 Zambia, Malawi, Botswana, Zimbabwe, Mozambique, South Africa

Bolusia Benth. 1873. Fabaceae. 6 Zaire (1), Tanzania (1), Angola (1), Namibia, Zambia, Zimbabwe, South Africa (1)

Bolusiella Schltr. 1918. Orchidaceae. 6 West and Southeast Africa

Bomarea Mirb. 1802. Alstroemeriaceae. c. 150 Mexico, West Indies, trop. America

Bombacopsis Pittier. 1916 (Pachira Aubl.). Bombacaceae. 30 Central and trop. South America, West Indies, West Africa (1)

Bombax L. 1753. Bombacaceae. c. 20 pantropics

Bombycidendron Zoll. et Moritzi. 1845 (Hibiscus L.). Malvaceae. 4 western Malesia

Bombycilaena (DC.) Smoljan. 1955. Asteraceae. 3 Central, South and Southeast Europe, North Africa, West Asia eastward to Afghanistan, Pacific coastal North America

Bonafousia A. DC. = Tabernaemontana L. Apocynaceae

Bonamia Thouars. 1804. Convolvulaceae. 45 pantropics

Bonania A. Rich. 1853. Euphorbiaceae. 10 West Indies

Bonannia Guss. 1843. Apiaceae. 1 Sicily I., Italy, Greece

Bonatea Willd. 1805. Orchidaceae. c. 20 trop. and South (11) Africa, Arabian Peninsula

Bonatia Schltr. et K. Krause = Tarrena Gaertn. Rubiaceae

Bonaveria Scop. = Securigera DC. Fabaceae

Bonetiella Rzed. 1957 (Pseudosmodingium Endl.). Anacardiaceae. 1 Mexico

Bongardia C. A. Mey. 1831. Berberidaceae. 1 North Africa (Libya), Aegean Isles, Cyprus, Turkey, Caucasus, Syria, Lebanon, Israel, Jordan, Iraq, Middle Asia from Turkmenistan to Pamir Alai, Iran, Afghanistan, western Pakistan, northwestern India

Boninia Planch. 1872. Rutaceae. 2 Ogasawara (Bonin) Is.

Boninofatsia Nakai. 1924. Araliaceae. 2 Ogasawara (Bonin) Is.

Boniodendron Gagnep. 1946. Sapindaceae. 2 China, Indochina

Bonnaya Link et Otto = Lindernia All. Scrophulariaceae

Bonnayodes Blatter et Hallberg. 1921 (Limnophila R. Br.). Scrophulariaceae. 1 India

Bonnetia Mart. 1826. Bonnetiaceae. 25 Colombia, Venezuela, Peru, Brazil, West Indies (Cuba, 1)

Bonniera Cordem. 1899. Orchidaceae. 2 Reunion I.

Bonnierella R. Vig. = Polyscias J. R. Forst. et G. Forst. Araliaceae

Bonplandia Cav. 1800. Polemoniaceae. 2 Mexico

Bontia L. 1753. Myoporaceae. 1 West Indies, trop. South America

Bonyunia Schomb. ex Progel. 1868. Loganiaceae. 4 Colombia, Venezuela, Guyana, Brazil

Boophone Herb. 1821. Amaryllidaceae. 6 trop. and South Africa

Boopis Juss. 1803. Calyceraceae. 13 Andes, south to Chile and Argentina (incl. Patagonia), southern Brazil

Boottia Wall. = Ottelia Pers. Hydrocharitaceae

Boquila Decne. 1837. Lardizabalaceae. 1 central and southern Chile and adjacent Argentina

Borago L. 1753. Boraginaceae. 3 Eurasia, Mediterranean; B. officinalis from European part of Russia and Caucasus to West Siberia and Middle Asia

Borassodendron Becc. 1914. Arecaceae. 2 southern Thailand and northern Peninsular Malaysia (B. machadonis), Borneo (B. borneense)

Borassus L. 1753. Arecaceae. 3–7 Africa, Madagascar, northeastern Arabian Peninsula, through India and Southeast Asia to New Guinea and Australia

Borderea Miègev. 1866 (Dioscorea L.). Dioscroeaceae. 2 Pyrenees

Boreava Jaub. et Spach. 1841. Brassicaceae. 2 Turkey, Syria

Boriskellera Terekhov. 1938 (Eragrostis Wolf). Poaceae. 1 Southeast Europe, Southwest Asia, West Siberia, Kazakhstan

Borismene Barneby. 1972. Menispermaceae. 1 Colombia, Venezuela, Peru, Brazil

Borissa Raf. = Lysimachia L. Primulaceae

Borkonstia Ignatov = Rhinactinidia Novopokr. Asteraceae

Borneacanthus Bremek. 1960. Acanthaceae. 6 Borneo

Borneodendron Airy Shaw. 1963. Euphorbiaceae. 1 Borneo

Bornmuellera Hausskn. 1897. Brassicaceae. 6 Yugoslavia, Albania, Greece, Asia Minor

Borodinia N. Busch. 1921. Brassicaceae. 1 East Siberia

Borojoa Cuatrec. 1950. Rubiaceae. 8 Panama, Colombia, Venezuela, Brazil

Boronella Baill. = Boronia Sm. Rutaceae

Boronia Sm. 1789. Rutaceae. 96 Australia, New Caledonia

Borrachinea Lavy. 1830 (Borago L.). Boraginaceae. 1 northwestern Italy

Borreria G. Mey. 1818. Rubiaceae. 150 tropics and subtropics

Borrichia Adans. 1763. Asteraceae. 2 southeastern U.S., Central America, West Indies

Borszczowia Bunge. 1877. Chenopodiaceae. 1 Kazakhstan, Middle Asia (Uzbekistan), northwestern China

Borthwickia W. W. Sm. 1912. Capparaceae. 1 Burma, southern China (Yunnan)

Borya Labill. 1805. Asphodelaceae. 10 northern (3), southwestern (6) and southeastern (1) Australia

Borzicactella F. Ritter = Cleistocactus Lem. Cactaceae

Borzicactus Riccob. = Cleistocactus Lem. Cactaceae

Boschia Korth. = Durio Adans. Bombacaceae

Boschniakia C. A. Mey. ex Bong. 1833. Orobanchaceae. 2: B. rossica—subarct. Russian Eurasia south to Altai, northern China, Kuril Is., Japan, northwestern North America; B. himalaica—Himalayas, China, Taiwan

Boscia Lam. 1793. Capparaceae. 37 North, trop. and South Africa, Arabian Peninsula (c. 20)

Bosea L. 1753. Amaranthaceae. 3 Canary Is., Cyprus I., northwestern India

Bosistoa F. Muell. 1863. Rutaceae. 7 northeastern Australia

Bosleria A. Nelson = Solanum L. Solanaceae

Bosqueia Thouars ex Baill. = Trilepisium Thouars. Moraceae

Bosqueiopsis De Willd. et T. Durand. 1901. Moraceae. 1 trop. Africa: Congo, Zaire, Tanzania and Mozambique

Bossera Leandri. 1962 (Alchornea Sw.). Euphorbiaceae. 1 Madagascar

Bossiaea Vent. 1800. Fabaceae. c. 40 Australia, Tasmania

Bostrychanthera Benth. 1876. Lamiaceae. 2 China, Taiwan

Boswellia Roxb. ex Colebr. 1807. Burseraceae. 25 trop. Africa, Madagascar, trop. Asia

Botelua Lag. = Bouteloua Lag. Poaceae

Bothriochilus Lem. 1856. (Coelia Lindl.). Orchidaceae. 4 Central America

Bothriochloa Kuntze. 1891. Poaceae. 35 temp., subtrop. and trop. regions

Bothriocline Oliv. ex Benth. 1873. Asteraceae. c. 40 trop. Africa, Madagascar

Bothriospermum Bunge. 1833. Boraginaceae. 5 trop. and East Asia; B. tenellum—extending to Middle Asia, Russian Far East

Bothriospora Hook. f. 1870. Rubiaceae. 1 trop. America

Bothrocaryum (Koehne) Pojark. = Cornus L. Cornaceae

Botryarrhena Ducke. 1932. Rubiaceae. 2 Venezuela, Brazil

Botryocytinus Watanabe. 1936 (Cytinus L.). Cytinaceae. 1 Madagascar

Botryoloranthus (Engl. et K. Krause) Balle. 1954 (Oncella Tiegh.). Loranthaceae. 1 trop. East Africa

Botryomeryta R. Vig. = Meryta J. R. Forst. et G. Forst. Araliaceae

Botryophora Hook. f. 1888. Euphorbiaceae. 1 southern Burma, Thailand, Malay Peninsula, Sumatra, Java, Borneo

Botryostege Stapf. 1934. Ericaceae. 1 Kuril Is., Japan (Hokkaido, northern Honshu)

Botschantzevia Nabiev. 1972. Brassicaceae. 1 Kazakhstan

Bottegoa Chiov. 1916. Sapindaceae. 1 Somalia

Bottionea Colla = Trichopetalum Lindl. Asphodelaceae

Bouchardatia Baill. 1867. Rutaceae. 2 New Guinea, eastern Australia

Bouchea Cham. 1831. Verbenaceae. 10 trop. America (9), Northeast Africa (Ethiopia, 1)

Bouchetia Dunal. 1852 (Salpiglossis Ruiz et Pav.). Solanaceae. 3 Mexico, southern Brazil, Uruguay, northeastern Argentina

Bouea Meisn. 1837. Anacardiaceae. 3 China, Indochina, Malesia, except Philippines

Bougainvillea Comm. ex Juss. 1789. Nyctaginaceae. 18 Central and trop. South America

Bougueria Decne. 1836. Plantaginaceae. 1 Andes (southern Peru, Bolivia, and northern Argentina)

Bourdaria A. Chev. = Cincinnobotrys Gilg. Melastomataceae

Bournea Oliv. 1893. Gesneriaceae. 2 China

Bourreria P. Browne. 1756. Ehretiaceae. 55 trop. America, West Indies, trop. East Africa (5) from eastern Ethiopia to Mozambique, Madagascar, Mauritius

Bousigonia Pierre. 1898. Apocynaceae. 2 China, Indochina

Boussingaultia Kunth. 1825 (Anredera Juss.). Basellaceae. 1 trop. America

Bouteloua Lag. 1816. Poaceae. 40 America from Canada to Argentina, esp. Mexico, and West Indies

Boutiquea Le Thomas. 1966. Annonaceae. 1 trop. West Africa

Boutonia DC. 1838. Acanthaceae. 1 Madagascar

Bouvardia Salisb. 1805. Rubiaceae. 20 southern U.S. (Arizona, New Mexico, Texas), Mexico (20), Guatemala, Belize, El Salvador, Honduras, Nicaragua, Costa Rica

Bouzetia Montrouz. 1860. Rutaceae. 1 New Caledonia

Bovonia Chiov. 1923. Lamiaceae. 1 trop. Africa

Bowdichia Kunth. 1824. Fabaceae. 4–5 South America from Venezuela to central Brazil

Bowenia Hook. ex Hook. f. 1863. Boweniaceae. 2 northeastern Australia

Bowiea Harv. ex Hook. f. 1867. Hyacinthaceae. 3 trop. East and South Africa

Bowkeria Harv. 1859. Scrophulariaceae. 3 South Africa

Bowlesia Ruiz et Pav. 1794. Apiaceae. 14 southern U.S. (1, B. incana), Central and South America

Bowringia Champ. et Benth. 1852. Fabaceae. 4 West Africa (2) from Ivory Coast to Zaire and Angola, Madagascar (1), Southeast Asia (1)

Boyania Wurdack. 1964. Melastomataceae. 1 northeastern South America

Boykinia Nutt. 1834. Saxifragaceae. 9 North America

Braasiella Braem, Lueckel et Ruessmann = Oncidium Sw. Orchidaceae

Brabejum L. 1753. Proteaceae. 1 South Africa (Cape)

Brachanthemum DC. 1838 (Chrysanthemum L.). Asteraceae. 8 Middle Asia, West Siberia, Mongolia, western China

Brachiaria (Trin.) Griseb. 1853. Poaceae. c. 100 trop. and warm–temp. regions; B. eruciformis—extending to Caucasus and Middle Asia

Brachionidium Lindl. 1859. Orchidaceae. 12 West Indies, trop. America from Costa Rica to Peru, Bolivia, Brazil, and Guyana

Brachionostylum Mattf. 1932. Asteraceae. 1 New Guinea

Brachistus Miers. 1849. Solanaceae. 3 Central America

Brachtia Rchb. f. 1850. Orchidaceae. 4 Venezuela, Colombia, Ecuador

Brachyachenium Baker = Dicoma Cass. Asteraceae

Brachyachne (Benth.) Stapf. 1922. Poaceae. 10 trop. Africa, Malesia, Australia

Brachyactis Ledeb. 1845 (Aster L.). Asteraceae. 5 Russia (1, B. ciliata—West and East Siberia, Far East), Middle Asia, Afghanistan, northeastern Pakistan, Tibet, China, Mongolia, western Himalayas, Sikkim; North America

Brachyandra Phil. = Helogyne Nutt. Asteraceae

Brachyapium (Baill.) Maire. 1932 (or = Stoibrax Raf.). Apiaceae. 4 Spain (1), North Africa (3)

Brachybotrys Maxim. ex Oliv. 1878. Boraginaceae. 1 Russian Far East, northeastern China (Manchuria)

Brachycarpaea DC. 1821. Brassicaceae. 1 (variable) South Africa (Cape)

Brachycaulos Dixit et Panigrahi. 1981 (Chamaerhodos Bunge). Rosaceae. 1 India (Sikkim)

Brachycereus Britton et Rose. 1920. Cactaceae. 1 Galapagos

Brachychaeta Torr. et A. Gray. 1842 (Solidago L.). Asteraceae. 1 southeastern U.S.

Brachychilum (R. Br. ex Wall.) Petersen = Hedychium J. Koenig. Zingiberaceae

Brachychiton Schott et Endl. 1832. Sterculiaceae. 31 Australia (29), eastern New Guinea (2)

Brachychloa S. M. Phillips. 1982. Poaceae. 2 Mozambique, South Africa

Brachyclados Gillies ex D. Don. 1832. Asteraceae. 3 temp. South America (Chile and Argentina)

Brachycodon Fedorov (1957, non Progel. 1805) = Brachycodonia Fedorov. Campanulaceae

Brachycodonia Fedorov. 1961 (or = Campanula L.). Campanulaceae. 1 Spain, Morocco, Algeria, Asia Minor, Lebanon, Syria, eastern Transcaucasia, Middle Asia (Pamir-Alai)

Brachycome Cass. 1825 (Brachyscome Cass.). Asteraceae. 66 New Guinea, Australia, New Zealand

Brachycorythis Lindl. 1838. Orchidaceae. 33 trop. and South (21) Africa, trop. Asia

Brachycylix (Harms) R. S. Cowan. 1975. Fabaceae. 1 western Colombia

Brachycyrtis Koidz. = Tricyrtis Wall. Melanthiaceae

Brachyelytrum P. Beauv. 1812. Poaceae. 2 East Asia, eastern U.S.

Brachyglottis J. R. Forst. et G. Forst. 1775. Asteraceae. 30 Tasmania, New Zealand

Brachyhelus (Benth.) Post et Kuntze = Schwenckia Royen ex L. Solanaceae

Brachylaena R. Br. 1817. Asteraceae. 17 trop. and South Africa, Madagascar, Mascarene Is.

Brachylepis C. A. Mey. = Anabasis L. Chenopodiaceae

Brachyloma Sond. 1845. Epacridaceae. 7 Australia

Brachylophon Oliv. 1887. Malpighiaceae. 2 Thailand (1), Malesia (1, Malay Peninsula and Sumatra)

Brachymeris DC. = Phymaspermum Less. Asteraceae

Brachynema Benth. 1857. Olacaceae. 1 northern Brazil (Amazonia)

Brachyotum (DC.) Triana. 1867. Melastomataceae. 50 Andes of Colombia, Ecuador, Peru and Bolivia

Brachypeza Schltr. ex Garay. 1972. Orchidaceae. 7 Thailand and Laos (1, B. laotica), Malesia eastward to New Guinea, western Pacific islands

Brachypodium P. Beauv. 1812. Poaceae. 15 temp. Eurasia, montane tropics

Brachypremna Gleason. 1935 (Ernestia DC.). Melastomataceae. 1 northeastern South America

Brachypterum (Wight et Arn.) Benth. 1837. Fabaceae. 6 Indo-Malesia

Brachypterys A. Juss. 1838. Malpighiaceae. 3 West Indies, trop. South America

Brachysema R. Br. 1811. Fabaceae. 14–15 Australia

Brachysiphon A. Juss. 1846. Penaeaceae. 4 South Africa

Brachystachyum Keng. 1940 (Semiarundinaria Nakai). Poaceae. 1 southern China

Brachystegia Benth. 1865. Fabaceae. 30 trop. Africa

Brachystele Schltr. 1920. Orchidaceae. 28 trop. and subtrop. America, Trinidad

Brachystelma R. Br. 1822. Asclepiadaceae. c. 100 trop. and South Africa, Madagascar, Southwest Asia eastward to India, Burma, Thailand, Philippines and New Guinea, northern Australia

Brachystemma D. Don. 1825. Caryophyllaceae. 1 eastern Himalayas (Nepal, Bhutan), Assam, southwestern China, Indochina

Brachystephanus Nees. 1847. Acanthaceae. 10 trop. Africa, Madagascar

Brachystigma Pennell. 1928 (Agalinis Raf.). Scrophulariaceae. 1 southwestern U.S. (Arizona, New Mexico), northern Mexico

Brachythalamus Gilg = Gyrinops Gaertn. Thymelaeaceae

Brachythrix Wild et G. V. Pope. 1978. Asteraceae. 5 trop. Southeast Africa

Brachytome Hook. f. 1871. Rubiaceae. 5 India, Burma, southern China, Malay Peninsula

Bracisepalum J. J. Sm. 1933. Orchidaceae. 2 Sulawesi I.

Brackenridgea A. Gray. 1854. Ochnaceae. 12 trop. and subtrop. Africa, Madagascar, South and Southeast Asia, Malesia (Malay Peninsula, Sumatra, Borneo, Philippines, Sulawesi, New Guinea), Australia, Fiji

Bracteantha Anderb. et Haegi. 1991. Asteraceae. 5 Australia

Bracteanthus Ducke. 1930. Monimiaceae. 1 eastern Brazil

Bracteolanthus De Wit = Bauhinia L. Fabaceae

Bradburia Torr. et A. Gray. 1842. Asteraceae. 1 southern U.S., Mexico

Bradea Standl. ex Brade. 1932. Rubiaceae. 5 Brazil

Braemia Jenny. 1985 (Houlletia Brongn.). Orchidaceae. 1 trop. South America

Brahea Mart. ex Endl. 1837. Arecaceae. 15 Baja California, Guadalupe I., Mexico, Guatemala

Brandegea Cogn. 1890. Cucurbitaceae. 1 southwestern U.S., Mexico

Brandellia R. R. Mill. 1986. Boraginaceae. 1 Egypt, Sudan, Ethiopia, Arabian Peninsula

Brandisia Hook. f. et Thomson. 1865. Scrophulariaceae. 11 Burma, China (8)

Brandzeia Baill. = Gagnebina Neck. ex DC. + Bathiaea Drake. Fabaceae

Brasenia Schreb. 1789. Cabombaceae. 1 (B. schreberi) North and Central America, trop. Africa, temp. East Asia, India, eastern Australia

Brasilia G. M. Barroso = Calea L. Asteraceae

Brasilicereus Backeb. 1938 (Cereus Mill.). Cactaceae. 2 eastern Brazil

Brasiliopuntia (K. Schum.) A. Berger = Opuntia Hill. Cactaceae

Brasiliparodia F. Ritter = Parodia Speg. Cactaceae

Brasilocalamus Nakai = Merostachys Spreng. Poaceae

Brassaia Endl. = Schefflera J. R. Forst. et G. Forst. Araliaceae

Brassaiopsis Decne. et Planch. 1854. Araliaceae. 36 India, Nepal, southern and southeastern China, Indochina, Malesia

Brassavola R. Br. 1813. Orchidaceae. 15 trop. America from Mexico to Argentina, West Indies

Brassia R. Br. 1813. Orchidaceae. 50

southeastern U.S. (Florida), Mexico, Central and South (southward to Brazil) America, West Indies

Brassiantha A. C. Sm. 1941. Celastraceae. 1 New Guinea

Brassica L. 1753. Brassicaceae. c. 40 Eurasia, Mediterranean

Brassiophoenix Burret. 1935. Arecaceae. 2 New Guinea

Braunblanquetia Eskuche = Fonkia Phil. Scrophulariaceae

Braunsia Schwantes. 1928. Aizoaceae. 4–6 South Africa

Bravaisia DC. 1838. Acanthaceae. 3 trop. America from northeastern Mexico to Colombia and Venezuela, Cuba

Bravoa La Llave ex Lex. = Polianthes L. Agavaceae

Braxireon Raf. = Narcissus L.

Braya Sternb. et Hoppe. 1815. Brassicaceae. 20 circumpolar regions of the Northern Hemisphere, Alps (Italy), western China, Tibet, Himalayas

Brayopsis Gilg et Muschler. 1909 (Eudema Humb. et Bonpl.). Brassicaceae. 5 Andes of Colombia, Ecuador, Peru, Bolivia and northern Argentina

Brayulinea Small = Guilleminea Kunth. Amaranthaceae

Brazoria Engelm. et A. Gray. 1845. Lamiaceae. 4 southern U.S.

Brazzeia Baill. 1886. Scytopetalaceae. 3 trop. Africa

Bredemeyera Willd. 1801. Polygalaceae. 20 West Indies and South America

Bredia Blume. 1849. Melastomataceae. 30 East and Southeast Asia

Breitungia A. Löve et D. Löve = Sedum L. Crassulaceae

Bremekampia Sreem. 1965 (Haplanthodes Kuntze). Acanthaceae. 3 India

Bremontiera DC. = Indigofera L. Fabaceae

Brenania Keay. 1958. Rubiaceae. 1 trop. West Africa

Brenesia Schltr. 1923 (or = Pleurothallis R. Br.). Orchidaceae. 1 Costa Rica

Brenierea Humbert. 1959. Fabaceae. 1 Madagascar

Breonadia Ridsdale. 1975 (Adina Salisb.). Rubiaceae. 1 trop. and South (Transvaal, Swaziland) Africa, Madagascar, Yemen

Breonia A. Rich. ex DC. 1830. Rubiaceae. 2 Madagascar, Mauritius

Bretschneidera Hemsl. 1901. Bretschneideraceae. 1 southwestern and southern China, northern Thailand, Taiwan

Breviea Aubrév. et Pellegr. 1934. Sapotaceae. 1 trop. West Africa

Brevipodium A. Löve et D. Löve = Brachypodium P. Beauv. Poaceae

Brevoortia A. Wood = Dichelostemma Kunth. Alliaceae

Brewcaria L. B. Sm., Steyerm., et H. E. Rob. 1984. Bromeliaceae. 1 Venezuela

Brewerina A. Gray = Arenaria L. Caryophyllaceae

Brexia Noronha ex Thouars. 1806. Brexiaceae. 9 trop. East Africa, Madagascar, Seychelles Is.

Brexiella H. Perrier. 1933. Celastraceae. 2 Madagascar

Breynia J. R. Forst. et G. Forst. 1775. Euphorbiaceae. 25 India, Burma, China, Taiwan, Ryukyu Is., Indochina, Malesia, Australia, New Caledonia

Brezia Moq. = Suaeda Forssk. ex Scop. Chenopodiaceae

Brickellia Elliott. 1823. Asteraceae. c. 100 Canada (1, British Columbia), U.S. (c. 40), Mexico (76), Guatemala, El Salvador, Honduras, Nicaragua, Costa Rica; 1 (B. diffusa) Mexico, Central America, Colombia, Venezuela, Ecuador, Galapagos Is., Peru, Bolivia, Brazil, Argentina, Greater Antilles

Brickelliastrum R. M. King et H. Rob. 1972. Asteraceae. 1 southwestern U.S.

Bridelia Willd. 1806. Euphorbiaceae. c. 60 trop. Old World

Bridgesia Bertero ex Cambess. 1834. Sapindaceae. 1 Chile

Briegeria Senghas. = Epidendrum L. Orchidaceae

Brieya De Wild. = Piptostigma Oliv. Annonaceae

Briggsia Craib. 1919. Gesneriaceae. 22–26 eastern Himalayas, Burma, southern China

Briggsiopsis K. Y. Pan. 1985. Gesneriaceae. 1 southwestern China

Brighamia A. Gray. 1866. Lobeliaceae. 2 Hawaiian Is.

Brillantaisia P. Beauv. 1818. Acanthaceae. c. 40 trop. Africa, Madagascar, Sri Lanka (1)

Brimeura Salisb. 1866. Hyacinthaceae. 2 South Europe

Brintonia Greene. 1895 (Solidago L.). Asteraceae. 1 southeastern U.S.

Briquetastrum Robyns et Lebrun = Leocus A. Chev. Lamiaceae

Briquetia Hochr. 1902. Malvaceae. 4 northern Mexico (2), Brazil and Paraguay (1), B. spicata—from Mexico and Cuba through Central America to Peru, Bolivia, Venezuela, Guyana, and Brazil

Briquetina J. F. Macbr. = Citronella D. Don. Icacinaceae

Brittenia Cogn. ex Boerl. 1890. Melastomataceae. 1 Borneo

Brittonastrum Briq. 1896 (Agastache Gronov.). Lamiaceae. 5–6 Central Asia

Brittonella Rusby. 1893 (Mionandra Griseb.). Malpighiaceae. 1 Bolivia

Briza L. 1753. Poaceae. 10 Europe, Southwest Asia, North Africa; Russia (4, European part, northern Caucasus)

Brizochloa Jirasek et Chrtek. 1966 (Briza L.). Poaceae. 1 Mediterranean, Southwest Asia, Crimea, Caucasus

Brizopyrum Stapf, non Link = Plagiochloa Adamson et Sprague. Poaceae

Brizula Hieron. = Aphelia R. Br. Centrolepidaceae

Brocchinia Schult. et Schult. f. 1830. Bromeliaceae. 18 Colombia, Venezuela, Guyana

Brochoneura Warb. 1896. Myristicaceae. 10 trop. Africa, Madagascar

Brockmania W. Fitzg. 1918 (Hibiscus L.). Malvaceae. 1 western Australia

Brodiaea Sm. 1810. Alliaceae. 15 western North America

Brodriguesia R. S. Cowan. 1981. Fabaceae. 1 eastern Brazil

Brombya F. Muell. 1865 (Melicope J. R. et G. Forster). Rutaceae. 1–2 northeastern Australia

Bromelia L. 1753. Bromeliaceae. 47 trop. America: Mexico, Guatemala, Belize, El Salvador, Honduras, Nicaragua, Costa Rica, Panama, Colombia, Ecuador, Venezuela, Trinidad, Tobago, the Guyanas, Brazil, Bolivia, Paraguay, northern Argentina, Uruguay; West Indies: Jamaica, Hispaniola, Puerto Rico, Leeward Is., Windward Is.; B. grandiflora—Cameroun

Bromelica (Thurb.) Farw. = Melica L. Poaceae

Bromheadia Lindl. 1841. Orchidaceae. 17 Burma, Indochina, Malesia from Malay Peninsula to New Guinea, Australia

Bromidium Nees et Meyen. 1843 (Agrostis L.). Poaceae. 5 southern U.S. and subtrop. South America

Bromopsis (Dumort.) Fourr. 1869 (Bromus L.). Poaceae. 50 temp., subtrop. and montane trop. regions

Bromuniola Stapf et C. E. Hubb. 1926. Poaceae. 1 trop. Africa

Bromus L. 1753. Poaceae. 25 temp. Eurasia, North and South Africa, montane trop. Asia

Brongniartia Kunth. 1824. Fabaceae. 56 southern U.S., Mexico, Central and South America

Brongniartikentia Becc. 1920. Arecaceae. 2 New Caledonia

Brookea Benth. 1876. Scrophulariaceae. 3 Borneo

Brosimopsis S. Moore = Brosimum Sw. Moraceae

Brosimum Sw. 1788. Moraceae. 13 trop. America, West Indies

Brossardia Boiss. 1841. Brassicaceae. 1 Iran, Iraq

Broughtonia R. Br. 1813. Orchidaceae. 1–2 West Indies

Brousemichea Bal. 1890 (Zoysia Willd.). Poaceae. 1 Indochina

Broussaisia Gaudich. 1830. Hydrangeaceae. 1 (B. arguta) Hawaiian Is.

Broussonetia L'Hér. ex Vent. 1799. Moraceae. 7–8 trop. Asia from Sri Lanka to Japan and New Guinea, Polynesia (1)

Browallia L. 1753. Solanaceae. 3 trop.

America from Mexico to Bolivia, West Indies

Brownanthus Schwantes. 1927 (Psilocaulon N. E. Br.). Aizoaceae. 12 southern Angola, western Namibia, South Africa

Brownea Jacq. 1760. Fabaceae. c. 30 trop. America, West Indies

Browneopsis Huber. 1906. Fabaceae. 1 Central America, Brazil (Amazonia)

Brownetera Rich. ex Tratt. 1825 (Phyllocladus Rich.). Phyllocladaceae. 1 Taiwan

Browningia Britton et Rose. 1920. Cactaceae. 7 Peru, Bolivia, Chile

Brownleea Harv. ex Lindl. 1842. Orchidaceae. 7 trop. and South (7) Africa, Madagascar

Brownlowia Roxb. 1820. Tiliaceae. 25 Southeast Asia, Malesia (excl. Java and Lesser Sunda Is.) eastward to Solomon Is.

Brucea J. F. Mill. 1780. Simaroubaceae. 6 trop. Old World from Africa to Australia

Bruckenthalia Rchb. 1831. Ericaceae. 1 Central and Southeast Europe, Asia Minor

Brugmansia Pers. 1805 (Datura L.). Solanaceae. 5–6 Andes of Colombia, Ecuador, Peru, and Bolivia

Bruguiera Lam. 1793. Rhizophoraceae. 6–7 trop. and South Africa, Madagascar, Seychelles Is., trop. Asia eastward to Ryukyu Is., Malesia (6, from Malay Peninsula to New Guinea and New Britain), northern and northeastern Australia (5), Micronesia (Palau, Yap, Marshall Is., etc.), Melanesia (Solomon Is., Vanuatu), Polynesia (Samoa, Fiji)

Bruinsmia Boerl. et Koord. 1893. Styracaceae. 2: 1 (B. polysperma)—Nepal, northeastern India, northern Burma Thailand, Malesia (1)

Brunellia Ruiz et Pav. 1794. Brunelliaceae. 52 southern Mexico, Central America; Andes of Colombia, Peru and Bolivia, West Indies

Brunfelsia L. 1753. Solanaceae. 40–45 trop. America, West Indies

Brunfelsiopsis (Urb.) Post et Kuntze = Brunfelsia L. Solanaceae

Brunia Lam. 1785. Bruniaceae. 7 South Africa

Brunnera Steven. 1851. Boraginaceae. 3 Asia Minor, Iran, Lebanon, Israel, Jordan, northern Iraq, Caucasus, Transcaucasia, Russia (2, West and East Siberia)

Brunnichia Banks ex Gaertn. 1788. Polygonaceae. 1 central to southern U.S.

Brunonia Sm. ex R. Br. 1810. Brunoniaceae. 1 Australia, Tasmania

Brunoniella Bremek. 1964. Acanthaceae. 4 Australia

Brunsvigia Heist. 1755. Amaryllidaceae. 20 Africa

Brya P. Browne. 1756. Fabaceae. 4–5 West Indies

Bryantea Raf. = Neolitsea (Benth.) Merr. Lauraceae

Bryanthus J. G. Gmel. 1769. Ericaceae. 1 Kamchatka, Komandorskiye, and Kuril Is., Japan

Bryaspis P. A. Duvign. 1954. Fabaceae. 2 trop. West Africa from Senegal to Guinea Bissau

Brylkinia F. Schmidt. 1868. Poaceae. 1 East Asia, incl. Sakhalin and Kuril Is.

Bryocarpum Hook. f. et Thomson. 1857. Primulaceae. 1 eastern Himalayas (Nepal, Bhutan), Assam, southern Tibet

Bryodes Benth. 1846. Scrophulariaceae. 3 Mascarene Is.

Bryomorphe Harv. 1863. Asteraceae. 1 South Africa

Bryonia L. 1753. Cucurbitaceae. 12 Europe, Canary Is., Mediterranean, West Asia, incl. Caucasus and Middle Asia

Bryophyllum Salisb. 1805 (Kalanchoë Adans.). Crassulaceae. c. 30 trop. and South (1, B. delagoense) Africa, Madagascar (c. 30), B. pinnatum widely distributed in trop. areas eastward to China and Japan

Bubalina Raf. = Burchellia R. Br. Rubiaceae

Bubania Girard = Limoniastrum Fabr. Plumbaginaceae

Bubbia Tiegh. = Zygogynum Baill. Winteraceae

Bucephalandra Schott. 1858. Araceae. 2 Borneo

Bucephalophora Pau = Rumex L. Polygonaceae

Buceragenia Greenm. 1897. Acanthaceae. 5 Mexico and Central America (to Costa Rica)

Buchanania Spreng. 1802. Anacardiaceae. 25 Himalayas, India, Burma, Thailand, China, Taiwan, Malesia, trop. Australia, Oceania

Buchenava Eichler. 1866. Combretaceae. 20 West Indies, trop. South America

Buchenroedera Eckl. et Zeyh. = Lotononis (DC.) Eckl. et Zeyh.

Buchholzia Engl. 1886. Capparaceae. 3 trop. West Africa

Buchingera Boiss. et Hohen = Asperuginoides Rauschert. Brassicaceae

Buchloë Engelm. 1859. Poaceae. 1 U.S., Mexico, ? China

Buchlomimus J. R. Reeder, C. F. Reeder et Rzed. 1965. Poaceae. 1 Mexico

Buchnera L. 1753. Scrophulariaceae. c. 100 tropics and subtropics, esp. Old World (c. 60)

Buchnerodendron Gürke. 1893. Flacourtiaceae. 2 trop. Africa from Cameroun to Angola and Mozambique

Buchtienia Schltr. 1929. Orchidaceae. 3 Ecuador, Peru, Bolivia

Bucida L. 1759. Combretaceae. 4 southeastern U.S. (Florida), Central America, West Indies

Bucinella Wiehler = Bucinellina Wiehler. Gesneriaceae

Bucinellina Wiehler. 1981. Gesneriaceae. 2 Colombia

Buckinghamia F. Muell. 1868. Proteaceae. 1 northeastern Australia

Bucklandia R. Br. ex Griff. = Exbucklandia R. W. Br. Hamamelidaceae

Buckleya Torr. 1843. Santalaceae. 4 continental China, Japan, southern U.S.

Bucquetia DC. 1828. Melastomataceae. 3 trop. South America (Andes)

Budawangia Telford. 1992. Epacridaceae. 1 Australia (New South Wales)

Buddleja L. 1753. Buddlejaceae. c. 100 trop. and subtrop. America, Africa and (esp.) Southeast Asia

Buergeria Miq. = Maackia Rupr. et Maxim. Fabaceae

Buergersiochloa Pilg. 1914. Poaceae. 1 New Guinea

Bufonia L. 1753. Caryophyllaceae. 20 Canary Is., Mediterranean, West Asia to Iran; coastal Black Sea, Crimea, Caucasus, Middle Asia

Buforrestia C. B. Clarke. 1881. Commelinaceae. 3 northeastern South America, trop. West Africa

Buglossoides Moench. 1794 (Lithospermum L.). Boraginaceae. 15 temp. Eurasia

Buhsia Bunge. 1861. Capparaceae. 2 Turkmenistan, Iran

Buiningia Buxb. = Coleocephalocereus Backeb. Cactaceae

Bukiniczia Lincz. 1971. Plumbaginaceae. 1 Afghanistan, Pakistan

Bulbine Wolf. 1776. Asphodelaceae. 35 trop. and South Africa, Australia, Tasmania

Bulbinella Kunth. 1843. Asphodelaceae. 15 South Africa, New Zealand, Auckland and Campbell Is.

Bulbinopsis Borzi = Bulbine Wolf. Asphodelaceae

Bulbocapnos Bernh. = Corydalis Vent. Fumariaceae

Bulbocodium L. 1753 (Colchicum L.). Colchicaceae. 2 Europe (except northern) and Mediterranean; Russia (1, B. versicolor—European part)

Bulbophyllum Thouars. 1822. Orchidaceae. c. 1,200 tropics and subtropics

Bulbostylis Kunth. 1837. Cyperaceae. 80 tropics and subtropics;

Bulbulus Swallen = Rehia Fijten. Poaceae

Bulleyia Schltr. 1912. Orchidaceae. 1 from eastern Himalayas to southwestern China

Bulnesia Gay. 1846. Zygophyllaceae. 9 South America from Venezuela to Chile and Argentina

Bumelia Sw. = Sideroxylon L. Sapotaceae

Bunchosia Rich. ex Kunth. 1822. Malpighiaceae. 55 trop. America from Mexico to Bolivia and Paraguay; West Indies

Bungea C. A. Mey. 1831. Scrophulariaceae. 2 Asia Minor, Iran, Caucasus (B. trifida), Middle Asia (B. resiculifera)

Bunias L. 1753. Brassicaceae. 5–6 Europe, Mediterranean, Asia

Buniella Schischk. = Bunium L. Apiaceae

Buniotrinia Stapf et Wettst. ex Stapf = Ferula L. Apiaceae

Bunium L. 1753. Apiaceae. 45–50 West and South Europe, Mediterranean, Northwest Africa; Caucasus and Middle Asia

Buphthalmum L. 1753. Asteraceae. 2 Europe

Bupleurum L. 1753. Apiaceae. c. 150 temp. Eurasia (esp. Russia, Caucasus, and Middle Asia—c. 60), southward to Himalayas, Tibet, northeastern India and northern Burma; Canary Is., Mediterranean, South Africa (1), arctic North America (1)

Buraeavia Baill. (1873) = Baloghia Endl. Euphorbiaceae

Burasaia Thouars. 1806. Menispermaceae. 4 Madagascar

Burbidgea Hook. f. 1879. Zingiberaceae. 6 Borneo

Burchardia R. Br. 1810. Melanthiaceae. 5 Australia, Tasmania

Burchellia R. Br. 1820. Rubiaceae. 1 South Africa

Burckella Pierre. 1890. Sapotaceae. 12 Moluccas, New Guinea, Solomon Is., eastward to Fiji, Samoa Is., and Tonga

Burdachia Mart. ex A. Juss. 1840. Malpighiaceae. 4 trop. South America

Bureavella Pierre = Lucuma Molina. Sapotaceae

Burgesia F. Muell. 1859 (Brachysema R. Br.). Fabaceae. 3 Australia

Burkartia Crisci. 1976. Asteraceae. 1 Argentina

Burkea Benth. 1843. Fabaceae. 1 trop. and South Africa

Burkillanthus Swingle. 1939. Rutaceae. 1 Malay Peninsula, Sumatra

Burkillia Ridl. = Burkilliodendron Sastry. Fabaceae

Burkilliodendron Sastry. 1969. Fabaceae. 1 Malay Peninsula

Burlemarxia N. L. Menezes et Semir. 1991. Velloziaceae. 3 Brazil

Burmabambus Keng f. = Sinarundinaria Nakai. Poaceae

Burmannia L. 1753. Burmanniaceae. 60 America (19), trop. Africa (11), trop. and subtrop. Asia (30)

Burmeistera Karst. et Triana. 1857. Lobeliaceae. c. 70 trop. America from Guatemala to Peru

Burnatastrum Briq. = Plectranthus L'Hér. Lamiaceae

Burnatia Micheli. 1818. Alismataceae. 3 trop. Africa from Ghana to the Sudan, Uganda, Kenya, Tanzania, and southward to South Africa

Burnettia Lindl. 1840. Orchidaceae. 1 southeastern Australia, Tasmania

Burragea Donn. Sm. et Rose. 1913. Onagraceae. 1 Baja California

Burretiodendron Rehder. 1936. Tiliaceae. 3 China (2), Southeast Asia

Burretiokentia Pic. Serm. 1955. Arecaceae. 2 New Caledonia

Burrielia DC. = Lasthenia Cass. Asteraceae

Burroughsia Moldenke. 1940 (Lippia L.). Verbenaceae. 2 southwestern North America

Bursaria Cav. 1797. Pittosporaceae. 3 Australia and Tasmania

Bursera Jacq. ex L. 1762. Burseraceae. 80 trop. America

Burseranthe Rizzini. 1974. Euphorbiaceae. 1 eastern Brazil

Burtonia R. Br. (non Salisb. 1807). 1811. Fabaceae. 12 Australia

Burttdavya Hoyle. 1936. Rubiaceae. 1 Africa: Tanzania, Malawi, Mozambique

Burttia Baker f. et Exell. 1931. Connaraceae. 1 trop. East Africa

Buschia? Ovczinnikov = Ranunculus L. Ranunculaceae

Buseria T. Durand = Coffea L. Rubiaceae

Bussea Harms. 1902. Fabaceae. 6: 1 (B. occidentalis) Sierra Leone, Liberia, Ivory Coast, Ghana; 4 Angola, Zaire, Zambia, Tanzania; 1 Madagascar

Bustelma Fourn. = Oxypetalum R. Br. Asclepiadaceae

Bustillosia Clos = Astiriscium Cham. et Schltdl. Apiaceae

Butania Keng f. = Sinarundinaria Nakai. Poaceae

Butea Roxb. ex Willd. 1802. Fabaceae. 4 India, Sri Lanka, Indochina, Malay Peninsula, Java

Buteraea Nees. 1832 (Strobilanthes Blume). Acanthaceae. 3 Burma

Butia (Becc.) Becc. 1916. Arecaceae. 8 southern Brazil, Paraguay, Uruguay, and Argentina

Butomopsis Kunth. 1841. Limnocharitaceae. 1 trop. Old World

Butomus L. 1753. Butomaceae. 1 temp. Eurasia

Buttonia McKen ex Benth. 1871. Scrophulariaceae. 3 trop. and South Africa

Butumia G. Taylor. 1953. Podostemaceae. 1 Nigeria

Butyrospermum Kotschy = Vitellaria C. F. Gaertn. Sapotaceae

Buxiphyllum W. T. Wang et C. Z. Gao = Paraboea (C. B. Clarke) Ridl. Gesneriaceae

Buxus L. 1753. Buxaceae. 70–80 North and Central America, West Indies, temp. Eurasia (except Russia and Middle Asia), trop. and South Africa, Madagascar eastward to Malay Peninsula, Borneo, and Philippines

Byblis Salisb. 1808. Byblidaceae. 2 southern Guinea, northern and western Australia

Byrsanthus Guill. 1839. Flacourtiaceae. 1 West Africa

Byrsocarpus Schumach. et Thonn. 1827 (Rourea Aubl.). Connaraceae. 20 trop. Africa, Madagascar

Byrsonima Rich. ex Kunth. 1822. Malpighiaceae. 150 trop. America from Mexico to Argentina; West Indies

Byrsophyllum Hook. f. 1873. Rubiaceae. 2 southern India, Sri Lanka

Bystropogon L'Hér. 1789. Lamiaceae. 11 Madeira and Canary Is.

Bythophyton Hook. f. 1884. Scrophulariaceae. 1 Indo-Malesia

Byttneria Loefl. 1758. Sterculiaceae. c 130 pantropics

Byurlingtonia Lindl. = Rodriguezia Ruiz et Pav. Orchidaceae

C

Caballeroa Font Quer = Limoniastrum Fabr. Plumbaginaceae

Cabi Ducke = Callaeum Small. Malpighiaceae

Cabomba Aubl. 1775. Cabombaceae. 5: 4 Mexico, Belize, Guatemala, Honduras, Nicaragua, El Salvador, Costa Rica, Panama, Colombia, Ecuador, Bolivia, Venezuela, the Guyanas, Brazil, West Indies (Cuba, Jamaica, Haiti, Puerto Rico, Trinidad), 1 (C. caroliniana) in the U.S. from Kansas and Massachusetts to Texas and Florida, and in South America (southern Brazil, Paraguay, Uruguay, and northeastern Argentina)

Cabralea A. Juss. 1830. Meliaceae. 1–6 Costa Rica, Guyana, Peru, Brazil, Bolivia, Paraguay, northeastern Argentina

Cabreriella Cuatrec. 1980. Asteraceae. 2 Colombia

Cabucala Pichon. 1948. Apocynaceae. 18 Madagascar, Comoro Is.

Cacabus Bernh. = Exodeconus Raf. Solanaceae

Cacalia DC. = Arnoglossum Raf. (American species) and = Parasenecio W. W. Sm. at Small (species of the Old World). Asteraceae

Cacalia L. 1753. Asteraceae. 3 montane Central and South Europe

Cacaliopsis A. Gray. 1883. Asteraceae. 1 Pacific coastal U.S.

Caccinia Savi. 1832. Boraginaceae. 6 West and Southwest Asia to Caucasus, Middle Asia, and Iran

Cachrys L. 1753. Apiaceae. 3–4 Mediterranean

Cacosmanthus De Vriese = Madhuca Ham. ex J. F. Gmel. Sapotaceae

Cacosmia Kunth. 1820. Asteraceae. 3 Andes of Ecuador and Peru

Cadaba Forssk. 1775. Capparaceae. c. 30 trop. Africa, Madagascar, Southwest Asia eastward to India and Sri Lanka, 1 (C. capparoides) Java, Lesser Sunda Is., and coastal northern Australia

Cadellia F. Muell. 1860. Surianaceae. 2 subtrop. Australia

Cadetia Gaudich. 1829. Orchidaceae. 55 Indonesia, Philippines (1), New Guinea, Australia (Queensland), Bougainville and Solomon Is. to Santa Cruz Is., New Caledonia, Vanuatu, Fiji

Cadia Forssk. 1775. Fabaceae. 7 Madagascar (6), C. purpurea—dry highlands of Somalia, Ethiopia, Kenya, and southern Arabian Peninsula

Cadiscus E. Mey. ex DC. 1838. Asteraceae. 1 South Africa

Caelebogyne J. Sm. = Coelebogyne J. Sm. Euphorbiaceae

Caelospermum Blume. 1826. Rubiaceae. 7 southern China (Hainan), Thailand, Vietnam, Malesia (Malay Peninsula, Sumatra, Java, Borneo, Philippines, Moluccas, New Guinea), Caroline and Solomon Is., Australia (Queensland and New South Wales), New Caledonia (2)

Caesalpinia L. 1753. Fabaceae. c. 100 (or c. 200) pantropics, subtrop. U.S., Argentina, and South Africa (Namibia)

Caesarea Cambess. 1829 (or = Viviania Cav.). Vivianiaceae. 1 Brazil

Caesia R. Br. 1810. Asphodelaceae. 11–12 South Africa, Madagascar, New Guinea, Australia, Tasmania

Caesulia Roxb. 1789. Asteraceae. 1 northeastern India

Cailliella Jacq.-Fel. 1939. Melastomataceae. 1 trop. West Africa

Caiophora C. Presl. 1831. Loasaceae. 65 South America

Cajalbania Urb. = Poitea Vent. Fabaceae

Cajanus DC. 1813. (Atylosia Wight et Arn.). Fabaceae. 3 trop. Africa (1, C. kerstingii), Madagascar, trop. Asia, Malesia

Cakile Mill. 1754. Brassicaceae. 7 North America, coastal Europe, incl. Russia, Mediterranean, Southwest and West Asia (incl. Caucasus), eastward to Australia

Calacanthus T. Anderson ex Benth. 1876. Acanthaceae. 1 western India

Caladenia R. Br. 1810. Orchidaceae. 80 Malesia, Australia, Tasmania, New Zealand, New Caledonia (1)

Caladiopsis Engl. = Chlorospatha Engl. Araceae

Caladium Vent. 1801. Araceae. 9 trop. South America

Calamagrostis Adans. 1763. Poaceae. c. 200 temp. and subtrop. regions and montane tropics

Calamintha Hill = Satureja L. Lamiaceae

Calamophyllum Schwantes. 1927 (Cylindrophyllum Schwantes). Aizoaceae. 3 South Africa

Calamovilfa (A. Gray) Hack. ex Scribn. et Southworth. 1890. Poaceae. 4 temp. North America

Calamus L. 1753. Arecaceae. c. 370 humid trop. Africa (1 variable or several closely related species), India, Burma, southern China, Taiwan, through the Malay Archipelago to Australia (Queensland) and eastward to Fiji (esp. Borneo and New Guinea)

Calanda K. Schum. 1903. Rubiaceae. 1 trop. Africa

Calandrinia Kunth. 1823. Portulacaceae. c. 10–15 America from western Canada to Chile

Calandriniopsis Franz = Baitaria Ruiz et Pav. Portulacaceae

Calanthe R. Br. 1821. Orchidaceae. 200–250 chiefly warm–temp., subtrop., and trop. Asia; Malesia; Madagascar; Australia; Pacific Is.; a few species in Central and South America; West Indies; in trop. and South Africa, on Mascarene Is.

Calanthea (DC.) Miers. 1865 (Capparis L.).

Capparaceae. 10 trop. America from Mexico to Peru, West Indies

Calantica Jaub. ex Tul. 1857. Flacourtiaceae. 8 East Africa, Madagascar

Calapodium Holub. 1976. Lamiaceae. 1 Hungary

Calathea G. Mey. 1818. Maranthaceae. 300 trop. America, West Indies

Calathiana Delarb. = Gentiana L. Gentianaceae

Calathodes Hook. f. et Thomson. 1855. Ranunculaceae. 5 Himalayas (Nepal, Bhutan), northeastern India, China, Taiwan

Calathostelma Fourn. = Ditassa R. Br. Asclepiadaceae

Calatola Standl. 1932. Icacinaceae. 7 trop. America from Mexico to Ecuador

Calcareoboea C. Y. Wu et H. W. Li. 1982 (Platyadenia B. L. Burtt). Gesneriaceae. 1 southern China (Yunnan)

Calceolaria L. 1770. Scrophulariaceae. 240–270 America from Mexico to Tierra del Fuego

Caldcluvia D. Don. 1830. Cunoniaceae. 11 Philippines, Sulawesi, Moluccas, New Guinea (7), Bismarck Arch., New Britain I., Solomon Is., trop. Australia (3), New Zealand (Northern I.), southern Chile (1)

Calderonella Soderstr. et H. F. Decker. 1974. Poaceae. 1 Panama

Calderonia Standl. = Simira Aubl. Rubiaceae

Caldesia Parl. 1860. Alismataceae. 4 temp. Eurasia (2), trop. and North Africa, Madagascar, trop. Asia from India to China, Malesia, south to Australia

Calea L. 1763. Asteraceae. 80 trop. America

Caleana R. Br. 1810. Orchidaceae. 5 temp. Australia, Tasmania (2), New Zealand

Calectasia R. Br. 1810. Dasypogonaceae. 3 western and southern Australia

Calendula L. 1753. Asteraceae. 12 Macaronesia, Mediterranean, West Asia eastward to Caucasus, Iran, Middle Asia, and Afghanistan

Caleopsis Fedde = Goldmanella Greenm. Asteraceae

Calepina Adans. 1763. Brassicaceae. 1 Europe, Mediterranean, Southwest Asia to Iran; Crimea, Caucasus, and Middle Asia

Calestania Kozo-Polj. = Thyselium Raf. Apiaceae

Calia Teran et Berland. 1832. (Sophora L.). Fabaceae. 5 southern U.S., Mexico

Calibanus Rose. 1906. Nolinaceae. 1 Mexico

Calibrachoa La Llave et Lex. 1825. Solanaceae. 37 South America

Calicorema Hook. f. 1880. Amaranthaceae. 2 trop. and South Africa

Calicotome Link. 1808. Fabaceae. 4 Mediterranean (except Malta, Egypt, and Sinai Peninsula)

Caliphruria Herb. 1844. Amaryllidaceae. 4 western Colombia (3), Peru (1)

Calispepla Vved. = Argyrolobium Eckl. et Zeyh. Fabaceae

Calisto Gaudich. 1826. Cyperaceae. 1 Mauritius

Calla L. 1753. Araceae. 1 subarctic and temp. Northern Hemisphere

Callaeum Small. 1910. Malpighiaceae. 10 trop. America from southern U.S. and Mexico to northeastern Argentina, southern Brazil, and Uruguay

Callerya Endl. 1843 (Millettia Wight et Arn.). Fabaceae. 16 Southeast and East Asia, Australia

Calliandra Benth. 1840. Fabaceae. c. 200 Central and South America, Africa (2), Madagascar, trop. Asia (11)

Calliandropsis H. M. Hern. et Guinet. 1990. Fabaceae. 1 (C. nervosus) central and southwestern Mexico

Callianthemoides Tamura. 1992. Ranunculaceae. 1 southern South America

Callianthemum C. A. Mey. 1830. Ranunculaceae. 14 montane Europe (3), West and East Siberia, Middle Asia, Afghanistan, Himalayas from Kashmir to Bhutan, western and southwestern China, Mongolia

Callicarpa L. 1753. Verbenaceae. 180 tropics and subtropics

Callicephalus C. A. Mey. 1831. Asteraceae. 1 Caucasus, Middle Asia, Iran

Callichilia Stapf. 1902. Apocynaceae. 7 trop. Africa

Callichlamys Miq. 1845. Bignoniaceae. 1 trop. America from Mexico to Brazil

Callicoma Andrews. 1809. Cunoniaceae. 1 eastern Australia

Calligonum L. 1753. Polygonaceae. 85 North Africa, eastern Mediterranean (except Cyprus), West Asia eastward to Middle Asia, Mongolia, and China; C. aphyllum from southeastern European part of Russia and Caucasus to West Siberia

Callilepis DC. 1836. Asteraceae. 3 South Africa

Callipeltis Steven. 1829. Rubiaceae. 3 Spain (1), Northeast Africa, West Asia to Pakistan, Middle Asia (2)

Calliphyllon Bubani = Epipactis Zinn. Orchidaceae

Callipsyche Herb. = Eucrosia Ker-Gawl. Amaryllidaceae

Callirhoe Nutt. 1821. Malvaceae. 8–10 temp. North America from Wisconsin to northeastern Mexico

Callisia Loefl. 1758. Commelinaceae. 22 trop. and subtrop. America

Callistachys Vent. = Oxylobium Andrews. Fabaceae

Callistemon R. Br. 1814. Myrtaceae. 20 Australia, Tasmania (2), New Caledonia

Callistephus Cass. 1825. Asteraceae. 1 China, Japan

Callisthene Mart. 1826. Vochysiaceae. 12 central and southern Brazil, eastern Bolivia, northern Paraguay

Callistigma Dinter et Schwantes = Mesembryanthemum L. Aizoaceae

Callistylon Pittier = Coursetia DC. Fabaceae

Callithauma Herb. 1837 (Paramongaia Velarde). Amaryllidaceae. 1 Andes of Peru

Callitriche L. 1753. Callitrichaceae. 17 cosmopolitan, excl. Arctica; Russia (8, all areas)

Callitris Vent. 1808. Cupressaceae. 14 arid Australia, eastern Tasmania, Flinders and Cape Barren Is., New Caledonia, Loyalty Is.

Callopsis Engl. 1895. Araceae. 1 Kenya, Tanzania, ? Cameroun

Calloscrodum Herb. 1844. Alliaceae. 1 (C. nerinifolium) East Siberia, Mongolia, China, Japan

Callostylis Blume = Eria Lindl. Orchidaceae

Callothlaspi F. K. Mey. 1973 (Thlaspi L.). Brassicaceae. 1 Caucasus

Calluna Salisb. 1802. Ericaceae. 1 Atlantic coastal North America, Azores, Europe from Scandinavia to Mediterranean, Morocco, Turkey, West and East Siberia

Calocarpum Pierre et Urb. = Pouteria Aubl. Sapotaceae

Calocedrus Kurz. 1873. Cupressaceae. 3 northern Burma, northeastern Thailand, southwestern China, Taiwan, western North America from Oregon to Baja California

Calocephalus R. Br. 1817. Asteraceae. 18 temp. Australia

Calochilus R. Br. 1810. Orchidaceae. 11 New Guinea, Australia, Tasmania, New Caledonia (1), New Zealand

Calochone Keay. 1958. Rubiaceae. 2 trop. West Africa

Calochortus Pursh. 1814. Calochortaceae. 60 America from British Columbia to Guatemala, esp. California

Calocrater K. Schum. 1895. Apocynaceae. 1 trop. West Africa

Calodecaryia J.-F. Leroy. 1960. Meliaceae. 1–2 Madagascar

Calodendrum Thunb. 1782. Rutaceae. 1 (C. capense) trop. (Uganda, Kenya, Tanzania, Malawi, Zimbabwe) and South Africa

Calogyne R. Br. = Goodenia Sm. Goodeniaceae

Calolisianthus Gilg = Irlbachia Mart. Gentianaceae

Calomeria Vent. 1804. Asteraceae. 1 Australia

Caloncoba Gilg. 1908. Flacourtiaceae. 10 trop. Africa

Calonyction Choisy = Ipomoea L. Convolvulaceae

Calophaca Fisch. ex DC. 1825. Fabaceae. 4–6 southeastern European part of Russia, Middle Asia (Tien Shan, Pamir-Alai), western China

Calophanoides (C. B. Clarke) Ridl. 1923 (*Justicia* L.). Acanthaceae. 13 Indo-Malesia, southern China, Australia (1, *C. hygrophiloides*)

Calophyllum L. 1754. Clusiaceae. 190 Madagascar, Mascarene Is., trop. Asia, Malesia, trop. Australia, Oceania, trop. America, West Indies

Calopogon R. Br. 1813. Orchidaceae. 4 North America

Calopogonium Desv. 1826. Fabaceae. 8 trop. America from Mexico to Argentina, West Indies

Calopsis P. Beauv. ex Desv. 1828 (*Restio* Rottb.). Restionaceae. 24 South Africa

Calopteryx A. C. Sm. 1946. Ericaceae. 2 western trop. South America

Calopyxis Tul. 1856. Combretaceae. 23 Madagascar

Calorhabdos Benth. = *Veronicastrum* Moench. Scrophulariaceae

Calorophus Labill. 1806. Restionaceae. 1 northeastern and eastern Australia, Tasmania, New Zealand

Calospatha Becc. 1911. Arecaceae. 1 Malay Peninsula

Calosteca Desv. = *Briza* L. Poaceae

Calostemma R. Br. 1810. Amaryllidaceae. 2 eastern Australia

Calostephane Benth. 1872. Asteraceae. 3 southern trop. Africa

Calostigma Decne. = *Oxypetalum* R. Br. Asclepiadaceae

Calotesta Karis. 1990. Asteraceae. 1 South Africa (Cape)

Calothamnus Labill. 1806. Myrtaceae. 24 western Australia

Calotis R. Br. 1820. Asteraceae. 24 Southeast Asia, Australia

Calotropis R. Br. 1811. Asclepiadaceae. 3–6 North Africa, Arabian Peninsula, trop. Asia from Pakistan to southern China and Malesia, northern Australia

Calpidochlamys Diels. 1935 (*Trophis* P. Browne). Moraceae. 2 New Guinea

Calpocalyx Harms. 1897. Fabaceae. 11 trop. West Africa from Sierra Leone to Angola

Calpurnia E. Mey. 1836. Fabaceae. 8 South Africa (7), *C. capensis* also in Botswana; *C. aurea*—extending through montane areas to Angola, Zaire, Sudan, Ethiopia, and eastward to India

Caltha L. 1753. Ranunculaceae. 10 (or c. 30) arctic and temp. Northern Hemisphere, New Zealand, temp. South America

Caluera Dodson et Determann. 1983. Orchidaceae. 2 Ecuador (1), Surinam, French Guyana

Calvaria C. F. Gaertn. = *Sideroxylon* L. Sapotaceae

Calvelia Moq. = *Suaeda* Forssk. ex Scop. Chenopodiaceae

Calvoa Hook. f. 1867. Melastomataceae. 18 trop. Africa

Calycacanthus K. Schum. 1889. Acanthaceae, 1 New Guinea

Calycadenia DC. 1836. Asteraceae. 11 western U.S.

Calycampe O. Berg = *Myrcia* DC. ex Guill. Myrtaceae

Calycanthus L. 1759. Calycanthaceae. 3 southwestern and southeastern U.S.

Calycera Cav. 1797. Calyceraceae. 15–20 temp. South America

Calycobolus Willd. ex J. A. Schult. 1819. Convolvulaceae. 8 trop. America (5) and trop. Africa

Calycocarpum Nutt. ex Spach. 1839. Menispermaceae. 1 eastern North America

Calycocorsus F. W. Schmidt. 1795. Asteraceae. 2 Europe, Southwest Asia, Caucasus (1, *C. tuberosus*)

Calycogonium DC. 1828. Melastomataceae. 40 West Indies, esp. Cuba

Calycolpus O. Berg. 1856. Myrtaceae. 13 trop. America, West Indies

Calycomis D. Don = *Acrophyllum* Benth. Cunoniaceae

Calycopeplus Planch. 1861. Euphorbiaceae. 4 western Australia

Calycophyllum DC. 1830. Rubiaceae. 7 trop. America, West Indies

Calycophysum Karst. et Triana. 1854. Cucurbitaceae. 5 trop. South America

Calycopteris Lam. 1793 (*Getonia* Roxb.). Combretaceae. 1 China, Indo-Malesia

Calycorectes O. Berg. 1856. Myrtaceae. 15 trop. America from Mexico to Brazil, West Indies

Calycoseris A. Gray. 1853. Asteraceae. 2 southwestern U.S., Mexico

Calycosia A. Gray. 1858. Rubiaceae. 6 Solomon Is. (1), Vanuatu, Fiji (4), Samoa Is.

Calycosiphonia Pierre ex Robbr. 1981. Rubiaceae. 2 trop. Africa

Calycotropis Turcz. 1862. Caryophyllaceae. 1 Mexico

Calyculogygas Krapov. 1960. Malvaceae. 1 Uruguay

Calydorea Herb. 1843. Iridaceae. 18 trop. and subtrop. America

Calylophus Spach. 1835. Onagraceae. 6 central U.S., northern Mexico

Calymmanthera Schltr. 1913. Orchidaceae. 5 New Guinea, Solomon Is., Fiji

Calymmanthium F. Ritter. 1962. Cactaceae. 1 northern Peru

Calymmatium O. E. Schulz. 1933. Brassicaceae. 2 Middle Asia (Tajikistan), Afghanistan

Calymmostachya Bremek. = Justicia L. Acanthaceae

Calypso Salisb. 1807. Orchidaceae. 1 cold and temp. regions; Russia—northwestern and central European part, West and East Siberia, Far East

Calyptocarpus Less. 1832. Asteraceae. 3 southern U.S., Mexico, West Indies

Calyptochloa C. E. Hubb. 1933. Poaceae. 1 northeastern Australia

Calyptostylis Arènes. 1946. Malpighiaceae. 1 Madagascar

Calyptranthes Sw. 1788. Myrtaceae. 100 trop. America from Florida and Mexico to Brazil; West Indies

Calyptrella Naudin = Graffenrieda DC. Melastomataceae

Calyptridium Nutt. ex Torr. et A. Gray. 1838 (Cistanthe Spach). Portulacaceae. c. 15 western North America

Calyptraemalva Krapov. 1965. Malvaceae. 1 Brazil

Calyptrocalyx Blume. 1838. Arecaceae. 38 Moluccas (1), New Guinea (37)

Calyptrocarya Nees. 1834. Cyperaceae. 7 trop. America

Calyptrochilum Kraenzl. 1895. Orchidaceae. 2 trop. Africa

Calyptrogenia Burret. 1941. Myrtaceae. 5 trop. America, West Indies

Calyptrogyne H. A. Wendl. 1859. Arecaceae. 5 trop. America from Mexico to Guatemala and Colombia

Calyptronoma Griseb. 1864 (Calyptrogyne H. Wendl.). Arecaceae. 3 Greater Antilles

Calyptrosciadium Rech. f. et Kuber ex Kuber, Rech. f. et Riedle. 1964. Apiaceae. 1 (C. polycladum) Iran, Afghanistan

Calyptrotheca Gilg. 1897. Portulacaceae. 2 trop. Northeast Africa

Calystegia R. Br. 1810 (Convolvulus L.). Convolvulaceae. 25 temp. and trop. both hemispheres

Calythropsis C. A. Gardner. 1942. Myrtaceae. 1 western Australia

Calytrix Labill. 1806. Myrtaceae. 60 Australia, Tasmania (1)

Camarea A. St.-Hil. 1823. Malpighiaceae. 8 eastern South America

Camarotea Scott-Elliot. 1891. Acanthaceae. 1 Madagascar

Camarotis Lindl. = Micropera Lindl. Orchidaceae

Camassia Lindl. 1832. Hyacinthaceae. 6 North (5) and South (1) America

Cambessedesia DC. 1828. Melastomataceae. 18 southern Brazil

Camchaya Gagnep. 1920. Asteraceae. 5 Southeast Asia

Camelina Crantz. 1762. Brassicaceae. 10 Europe, Mediterranean, West Asia eastward to China and Mongolia, Middle Asia, West and East Siberia, Far East

Camelinopsis A. G. Mill. 1978. Brassicaceae. 1 Iran, Iraq

Camellia L. 1753. Theaceae. 190 trop. and subtrop. Asia

Camelostalix Pfitzer et Kraenzl. = Pholidota Lindl. ex Hook. Orchidaceae

Cameraria L. 1753. Apocynaceae. 4 Central America, West Indies

Camerunia (Pichon) Boiteau = Tabernaemontana L. Apocynaceae

Camissonia Link. 1818. Onagraceae. 62 western North America (61 from British Columbia to California, Arizona, Texas, and Mexico) and temp. South America (1)

Camoensia Welw. ex Benth. 1865. Fabaceae. 2 trop. West Africa

Campanocalyx Valeton = Keenania Hook. f. Rubiaceae

Campanolea Gilg et Schellenb. = Chionanthus L. Oleaceae

Campanula L. 1753. Campanulaceae. 350 temp. Northern Hemisphere, Mediterranean, montane tropics; Russia (c. 150, from European part to Far East)

Campanulastrum Small = Campanula L. Campanulaceae

Campanulorchis Brieger = Eria Lindl. Orchidaceae

Campanumoea Blume. 1826 (Codonopsis Wall.). Campanulaceae. 5 Himalayas, India, Burma, southern China, Taiwan, Japan, Ryukyu Is., Malesia

Campbellia Wight = Christisonia Gardner. Orobanchaceae

Campecarpus H. A. Wendl. ex Becc. 1920. Arecaeae. 1 New Caledonia

Campeiostachys Drobov = Elymus L. Poaceae

Campelia Rich. = Tradescantia L. Commelinaceae

Campestigma Pierre ex Costantin. 1912. Asclepiadaceae. 1 Indochina

Camphorosma L. 1753. Chenopodiaceae. c. 10 South Europe, North Africa, Caucasus, West Siberia, Middle Asia, Asia Minor, Iran, Afghanistan, Pakistan, northwestern China, Mongolia

Campimia Ridl. 1910. Melastomataceae. 1 Malay Peninsula, ? Borneo

Campnosperma Thwaites. 1854. Anacardiaceae. 10 Madagascar, Seychelles, trop. Asia, Malesia, western Oceania, trop. America (1)

Campomanesia Ruiz et Pav. 1794. Myrtaceae. 30 trop. South America

Campovassouria R. M. King et H. Rob. 1971 (Eupatorium L.). Asteraceae. 1 Brazil to northern Argentina

Campsiandra Benth. 1840. Fabaceae. 3 trop. South America

Campsidium Seem. 1862. Bignoniaceae. 1 Chile, southern Argentina

Campsis Lour. 1790. Bignoniaceae. 2: C. grandiflora—Asia from western Pakistan to Japan and Indochina; C. radicans—eastern U.S.

Camptacra N. T. Burb. 1982. Asteraceae. 2 New Guinea, Australia (Northern Territory, Queensland, New South Wales)

Camptandra Ridl. 1899. Zingiberaceae. 6 western Malesia

Camptocarpus Decne. 1844. Asclepiadaceae. 5 Madagascar, Mauritius Is.

Camptolepis Radlk. 1907. Sapindaceae. 4 trop. East Africa, Madagascar

Camptoloma Benth. 1846. Scrophulariaceae. 2 trop. Africa

Camptopus Hook. f. = Psychotria L. Rubiaceae

Camptorrhiza Hutch. 1934. Melanthiaceae. 4 South Africa

Camptosema Hook. et Arn. 1833. Fabaceae. 12 South America

Camptostemon Mast. 1872. Bombacaceae. 3 Philippines, Aru Is., northern Australia

Camptostylus Gilg. 1898. Flacourtiaceae. 4 trop. West Africa

Camptotheca Decne. 1873. Nyssaceae. 1 China, Tibet

Campuloclinium DC. 1836. Asteraceae. 20 Central and South America southward to Brazil

Campylandra Baker. 1875 (Tupistra Ker-Gawl.). Convallariaceae. 10 India, eastern Himalayas, northern Burma, southwestern China, Indochina

Campylanthus Roth. 1821. Scrophulariaceae. 9 Canary Is. and Cape Verde Is. (1, C. salsoloides), 8 Somalia, Arabian Peninsula, Socotra, southeastern Pakistan

Campylocentrum Benth. 1881. Orchidaceae. 35–45 trop. America from Florida and Mexico to Brazil, West Indies

Campylopetalum Forman. 1954. Podoonaceae. 1 Thailand

Campyloptera Boiss. = Aethionema R. Br. Brassicaceae

Campylosiphon Benth. 1882. Burmanniaceae. 1 trop. South America

Campylospermum Tiegh. = Ouratea Aubl. Ochnaceae

Campylostachys Kunth. 1832. Stilbaceae. 2 South Africa

Campylostemon Welw. 1867. Celastraceae. 12 trop. West Africa

Campylotheca Cass. = Bidens L. Asteraceae

Campylotropis Bunge. 1835. Fabaceae. 65 trop. and subtrop. Asia

Campynema Labill. 1805 (Campynemanthe Baill.). Campynemataceae. 1 (C. lineare Labill.) Tasmania

Campynemanthe Baill. 1893. Campynemataceae. 3 New Caledonia

Camusia Lorch. 1961 (Acrachne Wight et Arn. ex Chiov.). Poaceae. 1 Madagascar

Camusiella Bosser. 1966 (Setaria P. Beauv.). Poaceae. 2 Madagascar

Canacomyrica Guillaumin. 1940. Myricaceae. 1 New Caledonia

Canacorchis Guillaumin = Bulbophyllum Thouars. Orchidaceae

Cananga (DC.) Hook. f. et Thomson. 1855 (or = Guatteria Ruiz et Pav.). Annonaceae. 2–4 Burma, Indochina, Malesia from Malay Peninsula to Philippines, Australia (Queensland)

Canarina L. 1771. Campanulaceae. 3 Canary Is. (1, C. canariensis), trop. Africa (2 southern Sudan, Ethiopia, Uganda, Kenya, Tanzania, eastern Zaire, Rwanda, Burundi, and Malawi)

Canarium L. 1759. Burseraceae. c. 80 trop. Africa, Madagascar, Indian Ocean islands, trop. Asia, Malesia, northern Australia, Oceania

Canavalia DC. 1825. Fabaceae. 50 pantropics, esp. trop. America; a few species extending to subtropics (South Africa, 2)

Canbya Parry ex A. Gray. 1877. Papaveraceae. 2 western U.S. (Oregon, California, Nevada), Mexico

Cancrinia Kar. et Kir. 1842. Asteraceae. 30 Middle Asia and West Siberia (4), Iran, Afghanistan, China, Mongolia

Cancriniella Tzvelev. 1961. Asteraceae. 1 Middle Asia

Candolleodendron R. S. Cowan. 1966. Fabaceae. 1 northeastern South America

Canella P. Browne. 1756. Canellaceae. 1–2 trop. America, West Indies

Canephora Juss. 1789. Rubiaceae. 5 Madagascar

Canistrum E. Morren. 1873. Bromeliaceae. 7 eastern Brazil

Canizaresia Britton = Piscidia L. Fabaceae

Canna L. 1753. Cannaceae. 55 America from South Carolina and Florida to northern Chile and Argentina

Cannabis L. 1753. Cannabaceae. 1–3 temp. Asia from Caucasus to Mongolia and from West Siberia to Gindukush and Himalayas

Cannomois P. Beauv. ex Desv. 1828. Restionaceae. 6–7 South Africa (Cape)

Canotia Torr. 1857. Celastraceae. 2 southwestern U.S. (California, Arizona, Utah), Mexico (Sonora)

Canscora Lam. 1785. Gentianaceae. 30 trop. Old World from Africa and Madagascar to China, Malesia, and Australia, esp. Southeast Asia

Cansjera Juss. 1789. Opiliaceae. 3 India, Sri Lanka, Nepal, Burma, Andaman and Nicobar Is., southern China (incl. Hainan), Indochina, Malesia (excl. Sulawesi), coastal northern Australia

Canthium Lam. 1785. Rubiaceae. 50 trop. Old World from eastern and southern trop. Africa, Seychelles and Madagascar to trop. Asia and Malesia, Australia, Micronesia, Melanesia eastward to eastern Polynesia

Cantleya Ridl. 1922. Icacinaceae. 1 Malesia: Sumatra, Malay Peninsula, Riouw and Lingga Arch., Bangka, Borneo

Cantua Juss. ex Lam. 1785. Polemoniaceae. 6 Andes of Ecuador, Peru, and Bolivia

Capanea Decne. ex Planch. 1849. Gesneriaceae. 6 Central and South America

Capanemia Barb. Rodr. 1877. Orchidaceae. 15 Brazil

Capassa Klotzsch = Lonchocarpus Kunth. Fabaceae

Caperonia A. St.-Hil. 1826. Euphorbiaceae. 40–60 trop. America, Africa, Madagascar

Capillipedium Stapf. 1917. Poaceae. 15 trop. East Africa, trop. Asia, Malesia, Australia

Capirona Spruce. 1859. Rubiaceae. 2 South America

Capitanopsis S. Moore. 1916. Lamiaceae. 2 Madagascar

Capitanya Schweinf. ex Gürke. 1895. Lamiaceae. 1 East Africa

Capitularia J. V. Suringar = Capitularina Kern. Cyperaceae

Capitularina Kern. 1912. Cyperaceae. 2 New Guinea, Solomon Is.

Capnoides Mill. 1754 (Dicentra Bernh.). Fumariaceae. 1 northern North America

Capnophyllum Gaertn. 1792. Apiaceae. 4 Canary Is., Mediterranean from Portugal to Turkey and Israel, Africa south to Cape

Capparidastrum Hutch. 1967 (Capparis L.). Capparaceae. 12 trop. America, West Indies

Capparis L. 1753. Capparaceae. 250 trop., subtrop. and warm–temp. regions, esp. America and Africa; 2 Crimea, Caucasus, and Middle Asia

Capraria L. 1753. Scrophulariaceae. 4 trop. and subtrop. America, West Indies

Caprifolium Hill = Lonicera L. Caprifoliaceae

Capsella Medik. 1792. Brassicaceae. 5 temp. and subtrop. regions; C. bursa-pastoris—nearly all over

Capsicodendron Hoehne = Cinnamodendron Endl. Canellaceae

Capsicum L. 1753. Solanaceae. 10 southern U.S., Mexico, Central and trop. South America (esp. Bolivia, eastern Brazil, Paraguay, northern Argentina), West Indies; widely cultivated

Captaincoockia N. Hallé. 1973. Rubiaceae. 1 New Caledonia

Captosperma Hook. f. = Tarenna Gaertn. Rubiaceae

Capurodendron Aubrév. 1962. Sapotaceae. 23 Madagascar

Capuronetta Markgr. = Tabernaemontana L. Apocynaceae

Capuronia Lourteig. 1960. Lythraceae. 1 Madagascar

Capuronianthus J.-F. Leroy. 1958. Meliaceae. 2 Madagascar

Capusia Lecomte = Siphonodon Griff. Celastraceae

Caquepira J. F. Gmel. = Gardenia Ellis. Rubiaceae

Caracasia Szyszyl. 1894. Marcgraviaceae. 2 Venezuela

Caragana Fabr. 1763. Fabaceae. c. 80 East Europe, southern Siberia, Far East of Russia, Asia Minor (1), Caucasus, Middle Asia, Afghanistan, Pakistan, China, Himalayas (India, Nepal)

Caraguata Adans. 1763 (Tillandsia L.). Bromeliaceae. 4 West Indies, South America

Caraipa Aubl. 1775. Clusiaceae. 36 Colombia, Venezuela, Guyana, Brazil (Amazonia)

Carajaea (Tul.) Wedd. 1873. Podostemaceae. 1 trop. South America

Carallia Roxb. 1811. Rhizophoraceae. 10 Madagascar, India, Sri Lanka, Nepal, southern China, Indochina, Malesia (Malay Peninsula, Sumatra, Borneo, Philippines, Moluccas, New Guinea), New Britain, Solomon Is., trop. Australia (1, C. brachiata)

Caralluma R. Br. 1811. Asclepiadaceae. 110 Mediterranean, trop. East and South Africa, Southwest Asia eastward to India and Burma

Caramuri Aubrév. et Pellegr. = Pouteria Aubl. Sapotaceae

Carania Chiov. = Basananthe Peyr. Passifloraceae

Carapa Aubl. 1775. Meliaceae. 2 Guatemala, Belize, Honduras, Nicaragua,

Costa Rica, Panama, Colombia, Venezuela, the Guyanas, Ecuador, Peru, Brazil; West Indies; C. procera also in West and Central Africa

Cardamine L. 1753. Brassicaceae. 130 cosmopolitan, but chiefly temp. and cold Northern Hemisphere and montane areas of the tropics

Cardaminopsis (C. A. Mey.) Hayek. 1908. Brassicaceae. 13 arctic and temp. Northern Hemisphere, esp. Russia (12)

Cardaria Desv. 1815. Brassicaceae. 6 Mediterranean, West Asia to western Mongolia, China, and western India; Caucasus; Middle Asia; West and East Siberia

Cardenanthus R. C. Foster. 1945. Iridaceae. 8 trop. America

Cardenasiodendron F. A. Barkley. 1954. Anacardiaceae. 1 Bolivia

Cardiacanthus Nees et Schauer. 1847 (or = Carlowrightia A. Gray). Acanthaceae. 2 Mexico

Cardiandra Siebold et Zucc. 1839. Hydrangeaceae. 5 China, Taiwan, Japan

Cardiochilus P. J. Cribb. 1977. Orchidaceae. 1 Malawi and adjacent Tanzania

Cardiochlamys Oliv. 1883. Convolvulaceae. 2 Madagascar

Cardiocrinum (Endl.) Lindl. 1847. Liliaceae. 3 Himalayas, northern and northeastern India, northern Burma, southwestern and central China, Japan; C. glehnii—southern Sakhalin, Kuril Is.

Cardionema DC. 1828. Caryophyllaceae. 6 Pacific coastal America from U.S. to Chile

Cardiopetalum Schltdl. 1834. Annonaceae. 1 trop. South America

Cardiophyllarium Choux. 1926. Sapindaceae. 1 Madagascar

Cardiopteris Blume. 1847. Cardiopteridaceae. 2 northeastern India (Assam), Bangladesh, Burma, Thailand, Indochina, southern China (Yunnan, Hainan), Malesia (Sumatra, Malay Peninsula, Borneo, Java, Lesser Sunda Is., Sulawesi, Moluccas, New Guinea, and New Britain), eastward to Solomon Is. (Bougainville I.), Australia (northern Queensland)

Cardiospermum L. 1753. Sapindaceae. 14 tropics, esp. America

Cardiostigma Baker = Calydorea Herb. Iridaceae

Cardioteucris C. Y. Wu = Caryopteris Bunge. Verbenaceae

Cardonaea Aresteg., Maguire et Steyerm. = Gongylolepis R. H. Schomb. Asteraceae

Cardopatium Juss. 1805. Asteraceae. 2–3 Mediterranean

Carduncellus Adans. 1763. Asteraceae. 30 Mediterranean, Southwest Asia

Carduus L. 1753. Asteraceae. 80 Eurasia, Mediterranean, North and trop. Africa; C. crispis—from European part of Russia to East Siberia and Far East

Cardwellia F. Muell. 1865. Proteaceae. 1 Australia (Queensland)

Carex L. 1753. Cyperaceae. c. 2,000 cosmopolitan

Careya Roxb. 1811. Lecythidaceae. 4 trop. Asia

Caribea Alain. 1960. Nyctaginaceae. 1 Cuba

Carica L. 1753. Caricaceae. 23 trop. and subtrop. America

Carinavalva Ising. 1955. Brassicaceae. 1 southern Australia

Cariniana Casar. 1842. Lecythidaceae. 15: 1 (C. pyriformis) eastern Panama, northern Colombia and northwestern Venezuela, 14 Brazil south to Santa Catarina, but mostly Amazonia

Carionia Naudin. 1851. Melastomataceae. 1 Philippines

Carissa L. 1767. Apocynaceae. 37 trop. and South Africa, Madagascar, Mascarene Is., Arabian Peninsula, trop. Asia, Australia, New Caledonia (1, C. ovata)

Carissophyllum Pichon. 1950 (Tachiadenus Griseb.). Apocynaceae. 1 Madagascar

Carlemannia Benth. 1853. Carlemanniaceae. 4 eastern Himalayas, Assam, northern Burma, southwestern China, Sumatra I.

Carlephyton Jum. 1919. Araceae. 3 Madagascar

Carlesia Dunn. 1902. Apiaceae. 1 (C. sinensis) China

Carlina L. 1753. Asteraceae. 28 Canary Is., temp. Eurasia, Mediterranean; Russia (7, European part, West and East Siberia), Caucasus, Middle Asia (1),

Carlowrightia A. Gray. 1878. Acanthaceae. 20–23 southwestern U.S. (southern Arizona, New Mexico, Texas), Mexico, El Salvador; C. arizonica—from Arizona to Costa Rica

Carludovica Ruiz et Pav. 1794. Cyclanthaceae. 3 trop. America from southern Mexico to central Bolivia

Carmenocania Wernham. 1912 (Pogonopus Klotzsch). Rubiaceae. 1 trop. America

Carmichaelia R. Br. 1825. Fabaceae. 40 Lord Howe I. (1, C. exsul), New Zealand

Carminatia Moçiño ex DC. 1838. Asteraceae. 2 southern U.S. (Arizona), Mexico, Guatemala, El Salvador

Carmona Cav. 1799 (Ehretia L.). Ehretiaceae. 1 China

Carnarvonia F. Muell. 1867. Proteaceae. 1 Australia (Queensland)

Carnegiea Britton et Rose. 1908. Cactaceae. 1 southwestern U.S. and northwestern Mexico

Carnegieadoxa Perkins = Hedycarya J. R. Forst. et G. Forst. Monimiaceae

Carolofritschia Engl. = Acanthonema Hook. Gesneriaceae

Caropodium Stapf et Wettst. = Grammosciadium DC. Apiaceae

Caropsis (Rouy et Camus) Rauschert. 1982. Apiaceae. 1 West Europe

Carpacoce Sond. 1865. Rubiaceae. 7 South Africa (southern Cape)

Carpanthea N. E. Br. 1925. Aizoaceae. 1 (C. pomeridiana) South Africa

Carpentaria Becc. 1885. Arecaceae. 1 Australia (Northern Territory)

Carpenteria Torr. 1853. Hydrangeaceae. 1 western U.S. (California, Nevada)

Carpesium L. 1753. Asteraceae. 20 South and Southeast Europe, temp. Asia (incl. Caucasus, Middle Asia, southern Far East or Russia), Malesia, trop. Australia

Carpha Banks et Sol. ex R. Br. 1810. Cyperaceae. 15 trop. and South Africa, Madagascar, Mascarene Is., ? Japan, New Guinea, Australia, Tasmania, New Zealand, temp. South America

Carphalea Juss. 1789. Rubiaceae. 10 trop. Africa (3), Madagascar (6), Socotra (1)

Carphephorus Cass. 1816. Asteraceae. 5 southeastern U.S.

Carphochaete A. Gray. 1849. Asteraceae. 7 southern U.S. (Arizona, New Mexico, Texas—C. bigelovii), Mexico (5)

Carpinus L. 1753. Betulaceae. 40 temp. Northern Hemisphere, chiefly East Asia

Carpobrotus N. E. Br. 1925 (Mesembryanthemum L.). Aizoaceae. 30 South Africa, Australia and Tasmania, New Zealand, Oceania, California (1), Chile (1, C. aequilaterus)

Carpoceras (DC.) Link = Thlaspi L. Brassicaceae

Carpodetus J. R. Forst. et G. Forst. 1776. Carpodetaceae. 2–10 New Guinea, Solomon Is., New Zealand, Stewart I.

Carpodinopsis Pichon. 1953 (Pleiocarpa Benth.). Apocynaceae. 4 trop. West Africa

Carpodinus R. Br. ex G. Don. 1837 (Landolphia P. Beauv.). Apocynaceae. 50 trop. Africa

Carpohypogaea Gibbs et Belli = Trifolium L. Fabaceae

Carpolepis (J. W. Dawson) J. W. Dawson. 1984. Myrtaceae. 3 New Caledonia

Carpolobia G. Don. 1831. Polygalaceae. 4 trop. Africa, Madagascar

Carpolyza Salisb. 1807. Amaryllidaceae. 1 South Africa

Carponema Eckl. et Zeyh. = Heliophila Burm. f. ex L. Brassicaceae

Carpothalis E. Mey. = Tricalysia A. Rich. ex DC. Rubiaceae

Carpotheca Tamamsch. = Echinophora L. Apiaceae

Carpotroche Endl. 1839. Flacourtiaceae. 11 (or 1, C. brasiliensis) trop. America from Guatemala to Brazil (south to Sao Paulo)

Carpoxylon H. A. Wendl. et Drude. 1875. Arecaceae. 1 Vanuatu

Carptotepala Moldenke = Syngonanthus Ruhland. Eriocaulaceae

Carramboa Cuatrec. 1976. Asteraceae. 5 Venezuela

Carrichtera DC. 1821. Brassicaceae. 1 Canary Is., Mediterranean, West Asia to Iran

Carrierea Franch. 1896. Flacourtiaceae. 3 western and southern China, Indochina

Carrissoa Baker f. 1933 [Eriosema (DC.) G. Don]. Fabaceae. 1 Angola

Carronia F. Muell. 1875. Menispermaceae. 3 New Guinea (1, C. thyrsiflora), Australia (Queensland, New South Wales)

Carruanthus (Schwantes) Schwantes. 1927. Aizoaceae. 2 South Africa

Carruthersia Seem. 1866. Apocynaceae. 10–12 eastern Malesia from Philippines to Solomon Is., Vanuatu, Fiji

Carsonia Greene = Cleome L. Capparaceae

Carterella Terrell. 1987 (Bouvardia Salisb.). Rubiaceae. 1 Baja California

Carterothamnus R. M. King et H. Rob. 1967. Asteraceae. 1 Baja California

Carthamus L. 1753. Asteraceae. 20 Madeira and Canary Is.; Europe; Mediterranean North Africa to Ethiopia; Southwest, West (incl. Caucasus), and Middle Asia; China (2)

Cartiera Greene = Streptanthus Nutt. Brassicaceae

Cartonema R. Br. 1810. Commelinaceae. 6–8 Aru Is., New Guinea, trop. Australia

Carum L. 1753. Apiaceae. c. 30 North America; Europe; temp. and subtrop. Asia; North, trop., and South Africa; Madagascar

Carvalhoa K. Schum. 1895. Apocynaceae. 1 trop. East and Southeast Africa

Carvia Bremek. 1944 (Strobilanthes Blume). Acanthaceae. 1 India

Carya Nutt. 1818. Juglandaceae. 17 China, few spp. extending to eastern Pakistan and northern Vietnam; eastern North America southward to southern Mexico

Caryocar L. 1771. Caryocaraceae. 15 trop. America from Costa Rica and Colombia to Brazil (Parana) and Paraguay

Caryodaphnopsis Airy Shaw. 1940. Lauraceae. 15 Sri Lanka, southern China, Indochina, Borneo, Philippines, New Guinea; trop. America (8—from Costa Rica to Peru and Brazil)

Caryodendron Karst. 1860. Euphorbiaceae. 3 Panama (1, C. angustifolium), trop. South America

Caryolobis Gaertn. = Shorea Roxb. ex C. F. Gaertn. Dipterocarpaceae

Caryomene Barneby et Krukoff. 1971. Menispermaceae. 5 Peru, Bolivia, Brazil, Guyana

Caryophyllus Hill = Dianthus L. Caryophyllaceae

Caryopteris Bunge. 1835. Verbenaceae. 23 Pakistan, Tibet, Himalayas from northwestern India to Bhutan, Burma, Thailand, China, Mongolia, Korea, Japan; C. mongolica—southern East Siberia (Burjatia)

Caryota L. 1753. Arecaceae. 12 India, Sri Lanka, southern China, southward through Indochina and Malesia to Solomon Is. and northern Australia

Caryotophora Leistner. 1958. Aizoaceae. 1 coastal southern South Africa

Casabitoa Alain. 1980. Euphorbiaceae. 1 Haiti (Dominican Republic)

Casasia A. Rich. 1853. Rubiaceae. 10 southern North and Central America, West Indies

Cascabela Raf. 1838. Apocynaceae. 7–8 trop. America from Mexico to Paraguay, West Indies

Cascadia A. M. Johnson = Saxifraga L. Saxifragaceae

Cascarilla (Endl.) Wedd. = Ladenbergia Klotzsch. Rubiaceae

Cascaronia Griseb. 1879. Fabaceae. Bolivia, Argentina

Casearia Jacq. 1760. Flacourtiaceae. c. 200 tropics and subtropics, mostly in rainforests

Cashalia Standl. = Dussia Krug et Urb. ex Taub. Fabaceae

Casimirella Hassl. 1913. Icacinaceae. 4 Colombia, Brazil, Paraguay

Casimiroa La Llave et Lex. 1825. Rutaceae. 7 Mexico, Central America

Caspia Galushko = Anabasis L. Chenopodiaceae

Cassandra D. Don = Chamaedaphne Moench. Ericaceae

Casselia Nees et Mart. 1823. Verbenaceae. 12 trop. America

Cassia L. 1753. Fabaceae. 30 pantropics

Cassidispermum Hemsl. = Burchella Pierre. Sapotaceae

Cassine L. 1753. Celastraceae. 80 South Africa, Madagascar, trop. Asia, Australia, Oceania

Cassinia R. Br. 1817. Asteraceae. 20 Australia, New Zealand

Cassinopsis Sond. 1860. Icacinaceae. 4 South Africa, Madagascar

Cassiope D. Don. 1834. Ericaceae. 12 circumpolar regions, Himalayas from Kashmir to Bhutan, northern Burma, China

Cassipourea Aubl. 1775. Rhizophoraceae. 82 trop. America, West Indies, trop. and South Africa, Madagascar, India, Sri Lanka

Cassupa Bonpl. = Isertia Schreb. Rubiaceae

Cassytha L. 1753. Lauraceae. 20 South Africa, Australia, New Zealand; 1 (C. filiformis) pantropics

Castalis Cass. 1824. Asteraceae. 3 South Africa

Castanea Hill. 1754. Fagaceae. 12 Atlantic North America, Mediterranean, Caucasus (1, C. sativa), Himalayas, East Asia

Castanedia R. M. King et H. Rob. 1978 (Eupatorium L.). Asteraceae. 1 Colombia

Castanella Spruce ex Hook. f. = Paullinia L. Sapindaceae

Castanopsis (D. Don) Spach. 1841. Fagaceae. 130 trop. and subtrop. Asia from Himalayas to Japan and New Guinea, California (2)

Castanospermum A. Cunn. ex Hook. 1830. Fabaceae. 1 northeastern Australia, New Caledonia, Vanuatu

Castanospora F. Muell. 1875. Sapindaceae. 1 northeastern Australia

Castela Turpin. 1806. Simaroubaceae. 15 Galapagos Is., trop. and subtrop. America, West Indies

Castelia Cav. = Pitraea Turcz. Verbenaceae

Castellanoa Traub. 1953 (Chlidanthus Herb.). Amaryllidaceae. 1 Argentina

Castellanosia Cárdenas = Browningia Britton et Rose. Cactaceae

Castellia Tineo. 1846. Poaceae. 1 Canary Is., Mediterranean, Sudan, Somalia, Pakistan

Castelnavia Tul. et Wedd. 1849. Podostemaceae. 9 Brazil

Castilla Cerv. 1794. Moraceae. 3 trop. America from southern Mexico to Bolivia, Cuba

Castilleja Mutis ex L. f. 1782. Scrophulariaceae. c. 200 arctic and temp. Eurasia, North (western part—c. 180, eastern 3), Central (8) and South (5, Andes) America

Castratella Naudin. 1850. Melastomataceae. 1 Colombia, Venezuela

Casuarina L. 1759. Casuarinaceae. 65–70 Australia, Tasmania, New Caledonia, Oceania, a few species in continental Southeast Asia and Malesia

Catabrosa P. Beauv. 1812. Poaceae. 4 temp. Eurasia, North America, Chile; Russia (3, European part, Caucasus, southern Siberia)

Catabrosella (Tzvelev) Tzvelev. 1965 (Poa L.). Poaceae. 9 Southeast Europe, Caucasus, southern West Siberia, Middle Asia, Southwest Asia eastward to western China and Himalayas

Catadysia O. E. Schulz. 1929. Brassicaceae. 1 Peru

Catalepis Stapf et Stent. 1929. Poaceae. 1 South Africa

Catalpa Scop. 1777. Bignoniaceae. 10 Central (Tibet), East and Southeast Asia, southeastern U.S. (2), West Indies (4 Bahamas, Cuba, Haiti, Jamaica, Windward Is.)

Catamixis Thomson. 1867. Asteraceae. 1 northwestern Himalayas

Catananche L. 1753. Asteraceae. 5 Mediterranean

Catanthera F. Muell. 1866 (Hederella Stapf). Melastomataceae. 5 Borneo, New Guinea

Catapodium Link. 1833. Poaceae. 2 Mediterranean, Crimea, Caucasus, Transcaucasia, Iraq, Iran

Catasetum Rich. ex Kunth. 1822. Orchidaceae. 70 trop. America from Mexico to Peru and Brazil

Catatia Humbert. 1923. Asteraceae. 2 Madagascar

Catenularia Botsch. = Catenulina Sojak. Brassicaceae

Catenulina Soják. 1980. Brassicaceae. 1 Middle Asia (Tajikistan)

Catesbaea L. 1753. Rubiaceae. 10 Florida Keys, West Indies

Catha Forssk. = Maytenus Molina. Celastraceae

Catharanthus G. Don. 1837. Apocynaceae. 6–8 Madagascar, India, Sri Lanka

Cathaya Chun et Kuang. 1958 (Pinus L.). Pinaceae. 1 (C. argyrophylla) western and southern China (Guangxi, Hunan ?, Sichuan)

Cathayanthe Chun. 1974. Gesneriaceae. 1 Hainan I.

Cathayeia Ohwi = Idesia Maxim. Flacourtiaceae

Cathedra Miers. 1852. Olacaceae. 5 Brazil (5), C. acuminata also in Colombia, Peru, Venezuela, and Guyana

Cathestecum C. Presl. 1830. Poaceae. 6 southern U.S., Mexico, Central America

Cathormion Hassk. = Albizia Durazz. Fabaceae

Catila Ravenna = Calydorea Herb. Iridaceae

Catoblastus H. A. Wendl. 1860. Arecaceae. 12–17 Panama, Colombia, Venezuela, Ecuador, Peru, Brazil

Catocoryne Hook. f. 1867. Melastomataceae. 1 Peru

Catopheria (Benth.) Benth. 1876. Lamiaceae. 4 trop. America

Catophractes D. Don. 1839. Bignoniaceae. 1 trop. and South Africa

Catopsis Griseb. 1864. Bromeliaceae. 20 trop. America from Florida and Mexico to Peru and Brazil, West Indies

Catospermum Benth. = Goodenia Sm. Goodeniaceae

Catostemma Benth. 1843. Bombacaceae. 9 Colombia, Venezuela, Guyana, Brazil

Catostigma O. F. Cook et Doyle = Catoblastus H. A. Wendl. Arecaceae

Cattleya Lindl. 1821. Orchidaceae. 60 trop. America from Mexico to Peru and Brazil; West Indies

Cattleyopsis Lem. 1853. Orchidaceae. 3 West Indies

Catunaregam Wolf. 1776. Rubiaceae. 7 trop. Africa, Madagascar, trop. Asia

Caucaea Schltr. 1920. Orchidaceae. 1 Colombia

Caucaliopsis H. Wolff = Agrocharis Hochst. Apiaceae

Caucalis L. 1753. Apiaceae. 1 (C. platycarpos) Europe, Mediterranean, Caucasus, Middle Asia, Iraq, Iran, Afghanistan

Caucanthus Forssk. 1775. Malpighiaceae. 5 Northeast and East Africa from Ethiopia and Somalia to Zimbabwe and Mozambique, Arabian Peninsula

Caulanthus S. Watson. 1871. Brassicaceae. 15 western U.S.

Caularthron Raf. 1837. Orchidaceae. 2 Guatemala, El Salvador, Honduras, Costa Rica, Panama, Colombia, Venezuela, Trinidad, Tobago, Guyana, Brazil

Caulinia Willd. = Najas L. Najadaceae

Caulocarpus Baker f. = Tephrosia Pers. Fabaceae

Caulokaempferia K. Larsen. 1964. Zingiberaceae. 10 from Himalayas to Southeast Asia

Caulophyllum Michx. 1803. Berberidaceae. 3 eastern North America (2); East Asia, incl. Russian Far East (1, C. robustum)

Caulostramina Rollins. 1973. Brassicaceae. 1 southwestern U.S.

Caustis R. Br. 1810. Cyperaceae. 10 Australia

Cautleya (Benth.) Royle ex Hook. f. 1888. Zingiberaceae. 5 Himalayas from Kashmir to Bhutan, India (Assam, Manipur), southwestern China

Cavacoa J. Léonard. 1955. Euphorbiaceae. 3 trop. and subtrop. Africa

Cavalcantia R. M. King et H. Rob. 1980 (Eupatorium L.). Asteraceae. 2 Brazil

Cavanillesia Ruiz et Pav. 1794. Bombacaceae. 3 trop. America

Cavea W. W. Sm. et Small. 1917. Asteraceae. 1 Tibet, eastern Himalayas

Cavendishia Lindl. 1835. Ericaceae. c. 100 trop. America from southern Mexico through Central America to northern Bolivia and east to Guyana highland to Pará (Brazil)

Cayaponia Silva Manso. 1836. Cucurbitaceae. 60 trop. and subtrop. America, West Indies, trop. West Africa, Madagascar (1)

Caylusea A. St.-Hil. 1838. Resedaceae. 3 Cape Verde Is., North Africa (Morocco, Central Sahara, Egypt), trop. East Africa (Sudan, Ethiopia, Uganda, Kenya, Tanzania), Crete, Israel, Jordan, Arabian Peninsula, Socotra, Iraq, Iran, Afghanistan, Pakistan, northwestern India

Cayratia Juss. 1818. Vitaceae. 45 trop. Africa, Madagascar, India, Himalayas, China, Indochina, Malesia, Australia, New Caledonia, Polynesia

Ceanothus L. 1753. Rhamnaceae. 55 Canada, U.S. (esp. California), Mexico, Guatemala

Cecarria Barlow. 1973. Loranthaceae. 1 Philippines, New Guinea, New Britain, Solomon Is., Australia (northern Queensland)

Cecropia Loefl. 1758. Cecropiaceae. c. 100 trop. America

Cedrela P. Browne. 1756. Meliaceae. 8 Mexico, Guatemala, Belize, Honduras, El Salvador, Nicaragua, Costa Rica, Panama, Colombia, Venezuela, the Guyanas, Ecuador, Peru, Brazil, Bolivia, Paraguay, Argentina, West Indies

Cedrelinga Ducke. 1922. Fabaceae. 1 Brazil

Cedrelopsis Baill. 1893. Ptaeroxylaceae. 7 Madagascar

Cedronella Moench. 1794. Lamiaceae. 1 Madeira and Canary Is.

Cedronia Cuatrec. 1951. Simaroubaceae. 1 Colombia

Cedrus Trew. 1757. Pinaceae. 4 montane areas of Morocco and Algeria, Cyprus, montane Asia Minor, Syria, Lebanon and Israel, northern Afghanistan, Pakistan, Himalayas from Kashmir to western Nepal; C. deodora—extending to Middle Asia

Ceiba Mill. 1754. Bombacaceae. 10 trop. America, trop. Africa (1, C. pentandra), Malesia

Celaenodendron Standl. 1927. Euphorbiaceae. 1 Mexico

Celastrus L. 1753. Celastraceae. c. 50 trop. and subtrop. America, Madagascar, Asia (esp. East), Malesia, eastern Australia, New Caledonia; Far East of Russia (3)

Celerina Benoist. 1964. Acanthaceae. 1 Madagascar

Celianella Jabl. 1965. Euphorbiaceae. 1 Venezuela, the Guyanas

Celiantha Maguire. 1981. Gentianaceae. 3 Venezuela, Guyana

Celmisia Cass. 1825. Asteraceae. 61 Australia, Tasmania, New Zealand

Celome Greene = Cleome L. Capparaceae

Celosia L. 1753. Amaranthaceae. 60 trop., subtrop. and temp. regions

Celsia L. = Verbascum L. Scrophulariaceae

Celtis L. 1753. Ulmaceae. c. 80–100 trop. and warm–temp. regions (from 40 degrees of north latitude to 35 degrees south latitude); 4 spp. extending to southern European part of Russia, Caucasus, Middle Asia

Cenarrhenes Labill. 1805. Proteaceae. 1 Tasmania

Cenchrus L. 1753. Poaceae. 25 trop. and warm regions; C. pauciflorus—southern Ukraine

Cenocentrum Gagnep. 1909. Malvaceae. 1 China, Indochina

Cenolophium W. D. J. Koch. 1824. Apiaceae. Europe, Siberia, Middle Asia, China

Cenolophon Blume. 1830 (Alpinia Roxb.). Zingiberaceae. 16 Indochina, Malesia

Cenostigma Tul. 1843. Fabaceae. 6 Brazil, Paraguay

Centaurea L. 1753. Asteraceae. c. 500 Eurasia from West Europe to Far East of Russia, northern China and northern India, North and trop. Africa, Australia, New Zealand, North America

Centaurium Hill. 1756. Gentianaceae. c. 30–40 cosmopolitan, except trop. and South Africa

Centaurodendron Johow. 1896. Asteraceae. 2 Juan Fernández Is.

Centauropsis Bojer ex DC. 1836. Asteraceae. 8 Madagascar

Centaurothamnus Wagenitz et Dittrich. 1982. Asteraceae. 1 southwestern Arabian Peninsula

Centella L. 1763 (Hydrocotyle L.). Apiaceae. c. 40 trop. and subtrop. America, trop. and South Africa; Madagascar; Southwest, South, East, and Southeast Asia; Malesia; Australia; Oceania (Hawaii); Antarctic Arch.; C. asiatica—extending to Caucasus

Centema Hook. f. 1880. Amaranthaceae. 2 trop. Africa

Centemopsis Schinz. 1911. Amaranthaceae. 11 trop. Africa

Centipeda Lour. 1790. Asteraceae. 6 Madagascar, Mascarene Is., Afghanistan, India, China, Mongolia, Russian Far East (1, C. orbicularis), Korea, Japan, Taiwan, Indochina, Malesia, northern Australia, New Zealand, New Caledonia, Tahiti, Chile

Centotheca Desv. 1810. Poaceae. 4 trop. Africa from Guinea to Angola, China, Japan, Southeast Asia, Philippines, northeastern Australia, Polynesia

Centradenia G. Don. 1832. Melastomataceae. 4 Mexico, Central America

Centradeniastrum Cogn. 1908. Melastomataceae. 2 trop. South America (Andes)

Centranthera R. Br. 1810. Scrophulariaceae. 7 India, Sri Lanka, Nepal, Bhutan, Burma, China, Indochina, Malesia, Australia

Centrantheropsis Bonati. 1914 (Phtheirospermum Bunge ex Fisch. et C. A. Mey.). Scrophulariaceae. 1 China

Centranthus Neck. ex Lam. et DC. 1805. Valerianaceae. 9 Europe, Mediterranean, Crimea and Caucasus (3)

Centratherum Cass. 1817. Asteraceae. 2 Philippines, Australia, trop. America

Centrilla Lindau = Justicia L. Acanthaceae

Centrochloa Swallen. 1935. Poaceae. 1 Brazil

Centrogenium Schltr. = Eltroplectris Raf. Orchidaceae

Centroglossa Barb. Rodr. 1882. Orchidaceae. 7 Peru, Brazil, Paraguay

Centrolepis Labill. 1804. Centrolepidaceae. 28 Hainan I., Indochina, Malesia (Sumatra, Borneo, Philippines, Sulawesi, New Guinea), Australia, Tasmania, New Zealand

Centrolobium Mart. ex Benth. 1837. Fabaceae. 6 Panama, trop. South America southward to Ecuador, Bolivia, and Brazil

Centronia D. Don. 1823. Melastomataceae. 20 trop. America

Centropetalum Lindl. 1838 (or = Fernandezia Ruiz et Pav.). Orchidaceae. 10 trop. South America (Andes)

Centroplacus Pierre. 1899. Pandaceae. 1 trop. West Africa

Centropodia Rchb. 1828. Poaceae. 4 North, East, and South Africa; Southwest Asia to Pakistan and northern India; C. forskalii—Middle Asia

Centropogon C. Presl. 1836. Lobeliaceae. 230 trop. America from southern Mexico throughout much of Andean South America, esp. Colombia; West Indies

Centrosema (DC.) Benth. 1837. Fabaceae. 45 trop. and subtrop. America

Centrosolenia Benth. = Episcia Mart. Gesneriaceae

Centrostachys Wall. ex Roxb. 1824. Amaranthaceae. 1 trop. Africa from Nigeria, the Sudan, Ethiopia south to Zimbabwe; India, Sri Lanka, Nepal, Burma, China, Indochina, Java, Norfolk I.

Centrostegia A. Gray ex Benth. 1856

(Chorizanthe R. Br.). Polygonaceae. 3 southwestern U.S. (California, Nevada, Arizona), Mexico (northern Baja California)

Centrostemma Decne. = Hoya R. Br. Asclepiadaceae

Centrostigma Schltr. 1915. Orchidaceae. 3 trop. and South (3) Africa

Centunculus L. = Anagallis L. Primulaceae

Ceodes J. R. Forst. et G. Forst. = Pisonia L. Nyctaginaceae

Cephaelis Sw. 1788 (Psychotria L.). Rubiaceae. c. 100 trop. Asia and trop. America

Cephalacanthus Lindau. 1905 Acanthaceae. 1 Peru

Cephalanthera Rich. 1817. Orchidaceae. 12 temp. Eurasia, incl. (6) European part of Russia, Caucasus, Middle Asia, Far East

Cephalantheropsis Guillaumin. 1960. Orchidaceae. 6 India, upper Burma, southern China, Taiwan, Ryukyu Is., Indochina, Malay Peninsula, Philippines (2)

Cephalanthus L. 1853. Rubiaceae. 6 trop. Asia, North (1) and trop. America

Cephalaralia Harms. 1896. Araliaceae. 1 Australia

Cephalaria Schrad. ex Roem. et Schult. 1818. Dipsacaceae. c. 70 Europe, Mediterranean, South Africa, West (incl. Caucasus), Southwest and Middle Asia eastward to West Siberia, Afghanistan and Pakistan

Cephalipterum A. Gray. 1852. Asteraceae. 1 western and southern Australia

Cephalobembix Rydb. = Schkuhria Roth. Asteraceae

Cephalocarpus Nees. 1842. Cyperaceae. 7 trop. South America

Cephalocereus Pfeiff. 1838. Cactaceae. 3 Mexico, ? Chile

Cephalocleistocactus F. Ritter = Cleistocactus Lem. Cactaceae

Cephalocroton Hochst. 1841. Euphorbiaceae. 5 trop. and South (Natal) Africa, Madagascar, Comoro Is. (1), Socotra (1), Sri Lanka (1)

Cephalocrotonopsis Pax = Cephalocroton Hochst. Euphorbiaceae

Cephalodendron Steyerm. 1972. Rubiaceae. 1 northern South America

Cephalohibiscus Ulbr. 1935 (Thespesia Sol. ex Corrêa.). Malvaceae. 1 New Guinea, Solomon Is.

Cephalomappa Baill. 1874. Euphorbiaceae. 5 southern China, western Malesia

Cephalonema K. Schum. = Clappertonia Meisn. Tiliaceae

Cephalopappus Nees et Mart. 1824. Asteraceae. 1 southeastern Brazil

Cephalopentandra Chiov. 1929. Cucurbitaceae. 1 Ethiopia, Somalia, Uganda, Kenya

Cephalophyllum (Haw.) N. R. Br. 1925. Aizoaceae. 30 southern Namibia, South Africa

Cephalopodum Korovin. 1973. Apiaceae. 2 Middle Asia

Cephalorhizum Popov et Korovin. 1923. Plumbaginaceae. 3 Turkmenistan, Iran, northern Afghanistan

Cephalorrhynchus Boiss. 1844. Asteraceae. 10 Southeast Europe, Southwest Asia through Caucasus and Middle Asia to China and Afghanistan

Cephaloschefflera (Harms) Merr. = Schefflera J. R. Forst. et G. Forst. Araliaceae

Cephalosorus A. Gray. 1851. Asteraceae. 1 western Australia

Cephalosphaera Warb. 1903. Myristicaceae. 1 trop. Africa

Cephalostachyum Munro. 1868 (Schizostachyum Nees). Poaceae. 12 Madagascar, South Asia

Cephalostemon R. H. Schomb. 1845. Rapateaceae. 5 trop. South America

Cephalostigma A. DC. 1830 (Wahlenbergia Schrader ex Roth). Campanulaceae. 1 pantropics

Cephalostigmaton (Yakovlev) Yakovlev. 1970 (Sophora L.). Fabaceae. 1 southwestern and southern China, northern Vietnam

Cephalotaxus Siebold et Zucc. 1842. Cephalotaxaceae. 9 eastern Himalayas, northeastern India, Burma, Thailand, Vietnam, China, Taiwan, Korea, Japan

Cephalotomandra Karst. et Triana. 1854. Nyctaginaceae. 1–2 Panama, Colombia

Cephalotus Labill. 1806. Cephalotaceae. 1 southwestern Australia

Ceranthera Elliott = Dicerandra Benth. Lamiaceae

Ceraria H. Pearson et Stephens. 1912. Portulacaceae. 6 trop. and South Africa

Cerastium L. 1753. Caryophyllaceae. 90 cosmopolitan

Cerasus Hill. 1753 (or = Prunus L.). Rosaceae. c. 100 temp. Northern Hemisphere

Ceratandra Eckl. ex F. A. Bauer. 1837. Orchidaceae. 6 South Africa (southwestern Cape)

Ceratanthus F. Muell. ex G. Taylor. 1936. Lamiaceae. 10 China, Indochina, New Guinea, Australia (Queensland)

Ceratephorus De Vries = Payena A. DC. Sapotaceae

Ceratiola Michx. 1803. Empetraceae. 1 southeastern U.S.

Ceratiosicyos Nees. 1836. Achariaceae. 1 South Africa (Transvaal, Swaziland, Natal, and Cape)

Ceratites Sol. ex Miers. 1878. Apocynaceae. 1 southeastern Brazil

Ceratocapnos Durieu. 1844. Fumariaceae. 3 West Europe from British Isles and southwestern Norway to central Iberian Peninsula (C. claviculata), southern Spain, Morocco and northwestern Algeria (C. heterocarpa), Cyprus, Syria, Lebanon, northern Israel, and Jordan (C. turbinata)

Ceratocarpus L. 1753. Chenopodiaceae. 2 Southeast Europe (Bulgaria, Romania, Russia), Caucasus, West and East Siberia, Middle Asia, Turkey, Iran, Afghanistan, Pakistan, western China, Mongolia

Ceratocaryum Nees. 1836 (Willdenowia Thunb.). Restionaceae. 5 South Africa

Ceratocentron Senghas. 1989. Orchidaceae. 1 Philippines

Ceratocephala Moench. 1794 (or = Ranunculus L.). Ranunculaceae. 3 Central and Southeast Europe, Mediterranean eastward to Caucasus, West Siberia, Middle Asia, Iran, western China, New Zealand (1)

Ceratochilus Blume. 1825. Orchidaceae. 2 Java (C. biglandulosus), ? Sumatra, Borneo (1)

Ceratochloa P. Beauv. 1812 (Bromus L.). Poaceae. 15 North and South America

Ceratocnemum Coss. et Balansa. 1873. Brassicaceae. 1 Morocco

Ceratogyne Turcz. 1851. Asteraceae. 1 temp. western Australia

Ceratoides Gagnebin. 1755 (Ceratocarpus L.). Chenopodiaceae. 5–9 Europe (Spain, Austria, Hungary, Romania, Ukraine, Russia), Caucasus, West and East Siberia, Middle Asia, Egypt, Turkey, Syria, Iran, Afghanistan, Pakistan, India, China, Mongolia, North America

Ceratolacis (Tul.) Wedd. 1873. Podostemaceae. 1 Brazil

Ceratolobus Blume. 1830. Arecaceae. 6 Malay Peninsula (2), Sumatra (4), Java (2), Borneo (4)

Ceratominthe Briq. 1896 (Satureja L.). Lamiaceae. 10 Andes

Ceratonia L. 1753. Fabaceae. 2: C. siliqua—Mediterranean, Northeast Africa, Turkey, Caucasus; C. oreothauma—Somalia, Oman

Ceratopetalum Sm. 1793. Cunoniaceae. 6 New Guinea, eastern Australia

Ceratophyllum L. 1753. Ceratophyllaceae. 2–6–8 (very variable species, sometimes = 30) cosmopolitan

Ceratophytum Pittier. 1928. Bignoniaceae. 1 trop. America from Mexico to Guyana, Surinam and Amazonian Peru, ? Bolivia, Grenada I., Trinidad

Ceratopyxis Hook. f. 1872. Rubiaceae. 1 Cuba

Ceratosanthes Burm. ex Adans. 1763. Cucurbitaceae. 5 west Indies, trop. South America

Ceratosepalum Oliv. = Triumfettoides Rauschert. Tiliaceae

Ceratospermum Pers. = Ceratoides Gagnebin. Chenopodiaceae

Ceratostema Juss. 1789. Ericaceae. 17 trop. South America (Andes)

Ceratostigma Bunge. 1835 (Plumbago L.). Plumbaginaceae. 8 trop. East Africa (Sudan, Ethiopia, Somalia, Kenya), Himalayas, Tibet, China, Burma, Thailand

Ceratostylis Blume. 1825. Orchidaceae. 60 northern India, Nepal, Bhutan, Sikkim, Burma, southern China, Indochina, Malesia from Malay Peninsula to Philippines (13) and (esp.) New Guinea, Oceania from Solomon Is. through New Caledonia (1) and Vanuatu eastward to Fiji

Ceratotheca Endl. 1832. Pedaliaceae. 5 trop. and South Africa

Ceratozamia Brongn. 1846. Zamiaceae. 10 Mexico, Guatemala, Belize

Cerbera L. 1753. Apocynaceae. 8–9 Madagascar, coasts of tropical Indian Ocean islands, Sri Lanka, southern China, Indochina, Malesia, New Caledonia, Fiji (1), Tonga, Samoa, eastward to Tuamotu Is.

Cerberiopsis Vieill. ex Pancher et Sebert. 1874. Apocynaceae. 3 New Caledonia

Cercestis Schott. 1857. Araceae. 11 West Africa from Guinea to Gabon

Cercidiopsis Britton et Rose = Parkinsonia L. Fabaceae

Cercidiphyllum Siebold et Zucc. 1846. Cercidiphyllaceae. 1–2 continental China, Kunashiri I., Japan

Cercidium Tul. 1844 (Parkinsonia L.). Fabaceae. 10 trop. and subtrop. America

Cercis L. 1753. Fabaceae. 6 temp. Northern Hemisphere; C. siliquastrum in Middle Asia

Cercocarpus Kunth. 1824. Rosaceae. 8 western and southwestern U.S., Mexico

Cerdia Moçino et Sessé ex DC. 1828. Caryophyllaceae. 4 Mexico

Cereus Mill. 1754. Cactaceae. 35–50 trop. America, West Indies

Cerinthe L. 1753. Boraginaceae. 10 Europe, Mediterranean, Caucasus, West Siberia

Ceriops Arn. 1838. Rhizophoraceae. 2 coastal East Africa from Somalia to Tanzania, Madagascar, Seychelles, coastal Pakistan, India, Sri Lanka, Andaman Is., Indochina, Taiwan, Malesia from Malay Peninsula to New Guinea, northern Australia, Micronesia (Carolines: Palau, Yap), Melanesia (New Ireland, Solomon Is., New Caledonia)

Ceriosperma (O. E. Schulz) Greuter et Burdet. 1983. Brassicaceae. 1 Syria

Ceriscoides (Hook. f.) Tirveng. 1978. Rubiaceae. 4–6 continental trop. Asia, Sri Lanka, Java, Philippines

Cerochlamys N. E. Br. 1928. Aizoaceae. 1 South Africa

Ceropegia L. 1753. Asclepiadaceae. 170 Canary Is., trop. and South Africa, Madagascar, trop. and subtrop. Asia from Arabian Peninsula to New Guinea, Australia (Queensland)

Ceroxylon Bonpl. ex DC. 1804. Arecaceae. c. 20 high elevations in the Andes from Venezuela through Colombia, Ecuador, and Peru to Bolivia

Ceruana Forssk. 1775. Asteraceae. 1 Egypt and northern trop. Africa

Cervantesia Ruiz et Pav. 1794. Santalaceae. 5 South America (Andes)

Cervaria M. M. Wolf. 1776. Apiaceae. 4 Europe, Southwest Asia, Caucasus, Middle Asia

Cervia Rodr. ex Lag. = Rochelia Rchb. Boraginaceae

Cespedesia Goudot. 1844. Sauvagesiaceae. 6 trop. and South Africa

Cestrum L. 1753. Solanaceae. c. 200 trop. America with major concentrations in Brazil and the Andean region, West Indies

Ceuthocarpus Aiello. 1979 (Portlandia P. Browne). Rubiaceae. 1 eastern Cuba

Ceuthostoma L. A. S. Johnson. 1988. Casuarinaceae. 2 Borneo, Philippines, New Guinea, Caroline Is.

Cevallia Lag. 1805. Loasaceae. 1 southwestern U.S., Mexico

Chaboissaea Fourn. 1881 (Muhlenbergia Schreb.). Poaceae. 4: 3 Mexico, 1 Argentina (Prov. de Jujuy and Salta)

Chabrea Raf. = Peucedanum L. Apiaceae

Chachasanum E. Mey. = Plexipus Raf. Verbenaceae

Chacaya Escal = Discaria Hook. Rhamnaceae

Chacoa R. M. King et H. Rob. 1975. Asteraceae. 2 Paraguay, Argentina

Chadsia Bojer. 1842. Fabaceae. 17 Madagascar

Chaenactis DC. 1836. Asteraceae. 40 western U.S., northwestern Mexico

Chaenanthe Lindl. 1838. Orchidaceae. 1 Brazil

Chaenocephalus Griseb. = Verbesina L. Asteraceae

Chaenomeles Lindl. 1821. Rosaceae. 4 Tibet, Bhutan, Burma, China, Japan (1)

Chaenorrhinum (Duby) Rchb. 1828. Scrophulariaceae. 20 Europe, Mediterranean, West Asia to Afghanistan, Caucasus, Middle Asia

Chaenostoma Benth. = Sutera Roth. Scrophulariaceae

Chaerophyllopsis Boissieu. 1909. Apiaceae. 1 western China

Chaerophyllum L. 1753. Apiaceae. 35 temp. Northern Hemisphere; Russia (European part; C. prescottii—to East Siberia), Caucasus (esp.), Middle Asia

Chaetacanthus Nees. 1836. Acanthaceae. 4 South Africa

Chaetachme Planch. 1848. Ulmaceae. 4 trop. and South Africa, Madagascar

Chaetacme Planch. = Chaetachme Planch. Ulmaceae

Chaetadelpha A. Gray ex S. Watson. 1873. Asteraceae. 1 southwestern U.S.

Chaetanthera Ruiz et Pav. 1794. Asteraceae. 41 southern Peru, Bolivia, Chile, Argentina

Chaetanthus R. Br. 1810. Restionaceae. 1 southwestern Australia

Chaetium Nees. 1829. Poaceae. 3 trop. America from Mexico to Brazil, West Indies

Chaetobromus Nees. 1836. Poaceae. 4 South Africa

Chaetocalyx DC. 1825. Fabaceae. 12 trop. America

Chaetocarpus Thwaites. 1854. Euphorbiaceae. 11 tropics (excl. eastern Malesia, Australia, and Oceania)

Chaetocephala Barb. Rodr. 1882 (or = Myoxanthus Poepp. et Endl.). Orchidaceae. 2 Brazil

Chaetochlamys Lindau. 1895 (or = Justicia L.). Acanthaceae. 14 trop. America

Chaetolepis (DC.) Miq. 1840. Melastomataceae. 17 trop. America, West Indies

Chaetolimon (Bunge) Lincz. 1940. Plumbaginaceae. 2 Middle Asia, northern Afghanistan

Chaetonychia (DC.) Sweet. 1839 (Paronychia Hill). Caryopyllaceae. 1 Portugal, Spain, southern France, Corsica, Sardinia, Morocco, Tunisia

Chaetopappa DC. 1836. Asteraceae. 11 southwestern U.S., Mexico

Chaetopoa C. E. Hubb. 1967. Poaceae. 2 Tanzania

Chaetopogon Janchen. 1913. Poaceae. 2 South Europe

Chaetoptelea Liebm. = Ulmus L. Ulmaceae

Chaetosciadium Boiss. 1872. Apiaceae. 1 eastern Mediterranean

Chaetoseris C. Shih. 1991. Asteraceae. 18 Tibet, China (Sichuan, Yunnan)

Chaetospira S. F. Blake. 1935. Asteraceae. 1 Colombia

Chaetostachydium Airy Shaw. 1965. Rubiaceae. 1 New Guinea

Chaetostachys Valeton = Chaetostachydium Airy Shaw. Rubiaceae

Chaetostichium C. E. Hubb. 1937 (Oropetium Trin.). Poaceae. 1 montane East Africa

Chaetostoma DC. 1828. Melastomataceae. 20 Brazil

Chaetosus Benth. = Parsonia R. Br. Apocynaceae

Chaetothylax Nees. 1847 (Justicia L.). Acanthaceae. 8 trop. America

Chaetotropis Kunth. 1829 (Polypogon Desf.). Poaceae. 4 temp. South America

Chaeturus Link, non Chaiturus Willd. = Chaetopogon Janchen. Poaceae

Chaetymenia Hook. et Arn. 1838. Asteraceae. 1 Mexico

Chaffeyopuntia Fric et Schelle = Opuntia Hill. Cactaceae

Chailletia DC. = Dichapetalum Thouars. Dichapetalaceae

Chaiturus Willd. = Leonurus L. Lamiaceae

Chalarothyrsus Lindau. 1904. Acanthaceae. 1 Mexico

Chalcanthus Boiss. 1867. Brassicaceae. 1 (C. renifolius) Middle Asia (Kopetdag, western Tien Shan, Pamir-Alai), Iran, Afghanistan

Chalema Dieterle. 1980. Cucurbitaceae. 1 Mexico

Chalepophyllum Hook. f. 1873. Rubiaceae. 5 Venezuela, Guyana

Chalybea Naudin = Pachyanthus A. Rich. Melastomataceae

Chamabainia Wight. 1853. Urticaceae. 2 India, Sri Lanka, Nepal, Bhutan, Burma, China, Taiwan, Malesia

Chamaeacanthus Chiov. = Campylanthus Roth. Scrophulariaceae

Cahamaealoe A. Berger = Aloë L. Asphodelaceae

Chamaeangis Schltr. 1915. Orchidaceae. 14–16 trop. Africa, Madagascar, Comoro and Mascarene Is.

Chamaeanthus Schltr. 1905. Orchidaceae. 10 southern China, Malesia, western Oceania to New Caledonia (4), and Vanuatu

Chamaeanthus Ule = Goegenanthus Ule. Commelinaceae

Chamaebatia Benth. 1849. Rosaceae. 2 California, Baja California

Chamaebatiaria (Porter) Maxim. 1879. Rosaceae. 1 western North America

Chamaecereus Britton et Rose = Echinopsis Zucc. Cactaceae

Chamaechaenactis Rydb. 1906. Asteraceae. 1 southwestern U.S.: Wyoming, Colorado, Utah, northeastern Arizona

Chamaeclitandra (Stapf) Pichon. 1953. Apocynaceae. 1 trop. Africa: Zaire, Angola, Zambia

Chamaecrista (L.) Moench. 1794 (Cassia L.). Fabaceae. 265 tropics, chiefly trop. America

Chamaecrypta Schltdl. et Diels = Diascia Link et Otto. Scrophulariaceae

Chamaecyparis Spach. 1841 (Cupressus L.). Cupressaceae. 8 Sino-Himalayan region, Taiwan, Japan, western and eastern North America

Chamaecytisus Link. 1831. Fabaceae. 25–30 Canary Is., Morocco (1, C. pulvinatus), Europe, Caucasus, Asia Minor, Lebanon, and Syria

Chamaedaphne Moench. 1794. Ericaceae. 1 (C. calyculata) temp. Northern Hemisphere; Russia—arctic, northern, and central European part, West and East Siberia, Far East

Chamaedorea Willd. 1806. Arecaceae. c. 100 trop. America from central Mexico to Bolivia and Brazil

Chamaegastrodia Makino et F. Maek. 1935. Orchidaceae. 1 Japan

Chamaegeron Schrenk. 1845. Asteraceae. 4 Middle Asia (2), Iran, Afghanistan, Pakistan

Chamaegigas Dinter ex Heil = Lindernia All. Scrophulariaceae

Chamaegyne Suess. = Eleocharis R. Br. Cyperaceae

Chamaelaucium DC. = Chamelaucium Desf. Myrtaceae

Chamaele Miq. 1867. Apiaceae. 1 Japan

Chamaeleon Cass. = Atractylis L. Asteraceae

Chamaelirium Willd. 1808. Melanthiaceae. 1 eastern North America

Chamaelobivia Y. Ito = Echinopsis Zucc. Cactaceae

Chamaemeles Lindl. 1821. Rosaceae. 1 Madeira Is.

Chamaemelum Hill. 1753 (Anthemis L.). Asteraceae. 4 West, Central, and Southeast Europe; Mediterranean; C. nobile—from Baltic Sea to Crimea

Chamaemespilus Medik. 1789 (Sorbus L.). Rosaceae. 1 Central and southern Europe, esp. Alps and western Carpathians

Chamaenerium Hill = Epilobium L. Onagraceae

Chamaepentas Bremek. 1952. Rubiaceae. 1 trop. East Africa (Tanzania)

Chamaepericlymenum Hill = Cornus L. Cornaceae

Chamaepus Wagenitz. 1980. Asteraceae. 1 Afghanistan

Chamaeranthemum Nees. 1836. Acanthaceae. 4 trop. America

Chamaeraphis R. Br. 1810. Poaceae. 1 Australia

Chamaerhodiola Nakai = Rhodiola L. Crassulaceae

Chamaerhodos Bunge. 1829. Rosaceae. 7: 6—Middle Asia, West and East Siberia, Far East of Russia; Mongolia, northern and western China, temp. North America (1)

Chamaerops L. 1753. Arecaceae. 1 (C. humulis) western Mediterranean, eastward to Malta I.

Chamaesaracha (A. Gray) Benth. 1876. Solanaceae. 10 southwestern U.S., northern Mexico

Chamaesciadium C. A. Mey. 1831 (Trachydium Lindl.). Apiaceae. 1 (C. acaule) Caucasus, Asia Minor, Iran, western China

Chamaescilla F. Muell. ex Benth. 1878. Asphodelaceae. 2 temp. southern Australia, Tasmania

Chamaesium H. Wolff. 1925. Apiaceae. 5 western China, Tibet, Himalayas

Chamaespartium Adans. = Genista L. Fabaceae

Chamaesphacos Fisch. et C. A. Mey. 1841. Lamiaceae. 1 Middle Asia, Iran, Afghanistan, China

Chamaesyce Gray. 1821. Euphorbiaceae. c. 250 cosmopolitan

Chamaethrinax H. A. Wendl. ex R. Pfister. 1892. Arecaceae. Dubious application

Chamaexeros Benth. 1878. Dasypogonaceae. 3 southwestern Australia

Chamarea Eckl. et Zeyh. 1837. Apiaceae. 5 South Africa

Chambeyronia Vieill. 1873. Arecaceae. 2 New Caledonia

Chamelaucium Desf. 1819. Myrtaceae. 12 western Australia

Chamelophyton Garay. 1974. Orchidaceae. 1 Venezuela, Guyana

Chamelum Phil. = Olsynium Raf. Iridaceae

Chamerion (Raf.) Raf. = Epilobium L. Onagraceae

Chamguava Landrum. 1991 (Pimenta Lindl. and Blepharocalyx O. Berg). Myrtaceae. 3 southern Mexico, Guatemala, Belize, Honduras, Panama

Chamira Thunb. 1782. Brassicaceae. 1 South Africa (Cape)

Chamissoa Kunth. 1818. Amaranthaceae. 7 trop. and subtrop. America

Chamissoniophila Brand = Antiphytum DC. ex Meisn. Boraginaceae

Chamomilla Gray = Matricaria L. Asteraceae

Chamorchis Rich. 1817. Orchidaceae. 1 Europe

Champereia Griff. 1844. Opiliaceae. 2 Andaman and Nicobar Is., Burma, Thailand, Vietnam, southern China, Taiwan, Malesia

Championella Bremek. 1944 (Strobilanthes Blume). Acanthaceae. 6 Indochina, China, Japan

Championia Gardner. 1846. Gesneriaceae. 1 Sri Lanka

Chandrasekharania V. J. Nair, W. S. Ramach. et Sreek. 1982. Poaceae. 1 southern India

Changium H. Wolff. 1924 (Tongoloa H. Wolff). Apiaceae. 1 eastern China

Changnienia S. S. Chien. 1935. Orchidaceae. 1 China

Chapelieria A. Rich. ex DC. 1830. Rubiaceae. 2 Madagascar

Chapmannia Torr. et A. Gray. 1838. Fabaceae. 1 southeastern U.S.

Chapmanolirion Dinter. 1909 (or = Pancratium L.). Amaryllidaceae. 1 Namibia

Chaptalia Vent. 1802. Asteraceae. 50 trop. and subtrop. America, West Indies

Charadrophila Marloth. 1899. Gesneriaceae. 1 South Africa

Chardinia Desf. 1817. Asteraceae. 2 Mediterranean, West Asia to Iran, Caucasus, and Middle Asia (2)

Charia C. DC. = Ekebergia Sparrm. Meliaceae

Charianthus D. Don. 1823. Melastomataceae. 9 West Indies

Charidion Bong. 1836 (Luxemburgia A. St.-Hil.). Ochnaceae. 2 Brazil

Charieis Cass. = Felicia Cass. Asteraceae

Chariessa Miq. = Citronella D. Don. Icacinaceae

Charpentiera Gaudich. 1829. Amaranthaceae. 6 Hawaii, Australia, and Cook Is.

Chartolepis Cass. 1826. (or = Centaurea L.). Asteraceae. 7 Southeast Europe, Caucasus, Middle Asia, eastern Asia Minor, Iran, Afghanistan, China

Chartoloma Bunge. 1844. Brassicaceae. 1 Middle Asia

Chascanum E. Mey. 1838. Verbenaceae. 27 trop. and South Africa (c. 25), Arabian Peninsula, Pakistan, India

Chascolytrum Desv. 1810 (Briza L.). Poaceae. 6 Central and South America

Chascotheca Urb. 1904 (Securinega Comm. ex Juss.). Euphorbiaceae. 2 Cuba (1), Haiti (1)

Chasechloa A. Camus. 1949 (Echinolaena Desv.). Poaceae. 3 Madagascar

Chaseëlla Summerh. 1961. Orchidaceae. 1 Kenya, Zimbabwe

Chasmanthe N. E. Br. 1932 (Gladiolus L.). Iridaceae. 3 trop. and South Africa

Chasmanthera Hochst. 1844. Menispermaceae. 2 trop. Africa

Chasmanthium Link. 1827. Poaceae. 5 eastern U.S., Mexico

Chasmatocallis R. C. Foster = Lapeirousia Pourr. Iridaceae

Chasmatophyllum (Schwantes) Dinter et Schwantes. 1927. Aizoaceae. 6 southeastern Namibia, central South Africa

Chasmopodium Stapf. 1917. Poaceae. 2 trop. West Africa from Guinea to Zaire

Chassalia Comm. ex Poir. 1812 (Psychotria L.). Rubiaceae. 40–50 Africa, Madagascar, Mascarene Is., a few species in India, Sri Lanka, Burma, southern China, Malesia east to Philippines

Chaubardia Rchb. f. 1852. Orchidaceae. 3 trop. South America

Chaubardiella Garay. 1969. Orchidaceae. 3 trop. America

Chauliodon Summerh. 1943. Orchidaceae. 1 trop. West Africa

Chaunochiton Benth. 1867. Olacaceae. 3 Costa Rica, Colombia, Venezuela, Guyana, Surinam, French Guyana, Brazil

Chaunostoma Donn. Sm. 1895. Lamiaceae. 1 Central America

Chavannesia A. DC. 1844 (Urceola Roxb.). Apocynaceae. 8 trop. Asia, Malesia

Chaydaia Pit. = Rhamnella Miq. Rhamnaceae

Chazaliella E. Petit et Verdc. 1975 (Chassalia Comm. ex Poir.). Rubiaceae. 25 trop. Africa: Sierra Leone, Cameroun, Fernando Po, Zaire, Angola, Uganda, Kenya, Tanzania

Cheesemania O. E. Schulz. 1929. Brassicaceae. 5 Tasmania, New Zealand

Cheiloclinium Miers. 1872. Celastraceae. 23 Central and trop. South America

Cheilophyllum Pennell ex Britton. 1920. Scrophulariaceae. 8 West Indies

Cheilosa Blume. 1826. Euphorbiaceae. 2 western Malesia

Cheilotheca Hook. f. 1876. Ericaceae. 8 Himalayas, Tibet, northern Burma, Indochina, China, Taiwan, Korea, Japan; Russia (1, Far East: Ussuri, Sakhalin I., Kuril Is.)

Cheiradenia Lindl. 1853. Orchidaceae. 2 northeastern South America

Cheiranthera A. Cunn. ex Brongn. 1834. Pittosporaceae. 4 southwestern (2) and southeastern (2) Australia

Cheiranthus L. 1753 (Erysimum L.). Brassicaceae. 10 Macaronesia, Europe, West and South Asia eastward to Himalayas; C. cheiri in the Crimea

Cheiridopsis N. E. Br. 1925. Aizoaceae. 23 Namibia, western South Africa

Cheirodendron Nutt. ex Seem. 1867. Araliaceae. 6 Hawaiian (5) and Marquesas Is. (1)

Cheirolaena Benth. 1862. Sterculiaceae (? Malvaceae). 1 Mauritius

Cheirolepis Boiss. 1849. (or = Centaurea L.). Asteraceae. 9 Caucasus (1, C. persica), Turkey, Syria, Iran

Cheirolophus Cass. 1827. Asteraceae. 12 Canary Is., Europe, North Africa, temp. South America

Cheirorchis C. E. Carr = Cordiglottis J. J. Sm. Orchidaceae

Cheirostemon Bonpl. = Chiranthodendron Larreat. Sterculiaceae

Cheirostylis Blume. 1825. Orchidaceae. c. 25 trop. and South Africa (Natal), Madagascar, Comoro Is., trop. Asia, trop. Australia, New Caledonia (1)

Chelidonium L. 1753. Papaveraceae. 1 (C. majus) subarctic and temp. Eurasia, incl. European part of Russia, Caucasus, Siberia, and Middle Asia

Chelonanthera Blume = Coelogyne Lindl. Orchidaceae

Chelonanthus Gilg = Irlbachia Mart. Gentianaceae

Chelone L. 1753 (Penstemon Schmidel). Scrophulariaceae. 5–6 North America

Chelonespermum Hemsl. = Burckella Pierre. Sapotaceae

Chelonistele Pfitzer et Kraenzl. 1907. Orchidaceae. 20 Burma, Malay Peninsula, Sumatra (1), Borneo (10)

Chelonopsis Miq. 1865. Lamiaceae. 16 Himalayas (Kashmir), eastern Tibet, China, Japan

Chelyocarpus Dammer. 1920. Arecaceae. 4 western Colombia, Amazonian Peru, Bolivia, and Brazil

Chennapyrum A. Löve = Aegilops L. Poaceae

Chenolea Thunb. = Bassia All. Chenopodiaceae

Chenoleoides (Ulbr.) Botsch. 1976 (Bassia All.). Chenopodiaceae. 3 Madeira and Canary Is., North (Morocco, Libya, Egypt), Northwest and Southwest Africa, Syria, Lebanon, Israel, Jordan, Arabian Peninsula

Chenopodiopsis Hilliard et B. L. Burtt. 1990. Scrophulariaceae. 3 South Africa

Chenopodium L. 1753. Chenopodiaceae. c. 150 cosmopolitan

Cherleria L. = Minuartia L. Caryophyllaceae

Chersodoma Phil. 1891. Asteraceae. 9 southern Peru, Bolivia, Chile, Argentina

Chesneya Lindl. ex Endl. 1840. (Astragalus L.). Fabaceae. 20 Caucasus and Middle Asia (15), Turkey, Syria, Iraq, Iran, Afghanistan, Pakistan, northwestern China, Mongolia, Himalayas

Chesniella Boriss. = Chesneya Lindl. ex Endl. Fabaceae

Chevaliera Gaudich. ex Beer. 1856 (Aechmea Ruiz et Pav.). Bromeliaceae. 24 trop. America

Chevalierella A. Camus. 1933. Poaceae. 1 Zaire

Chevreulia Cass. 1817. Asteraceae. 5 southern Brazil, Bolivia, Paraguay, Uruguay, Chile, northern and central Argentina, Falkland Is.

Chiangiodendron Wendt. 1988. Flacourtiaceae. 1 southeastern Mexico

Chiarinia Chiov. = Lecaniodiscus Planch. ex Benth. Sapindaceae

Chiazospermum Bernh. = Hypecoum L. Hypecoaceae

Chiastophyllum Stapf ex A. Berger. 1930. Crassulaceae. 1 Caucasus

Chiclea Lundell = Manilkara Adans. Sapotaceae

Chidlowia Hoyle. 1932. Fabaceae. 1 trop. West Africa from Sierra Leone to Ghana

Chienia W. T. Wang = Delphinium L. Ranunculaceae

Chieniodendron Tsiang et P. T. Li = Oncodostigma Diels. Annonaceae

Chigua D. W. Stev. 1990. Zamiaceae. 2 Colombia

Chikusichloa Koidz. 1925. Poaceae. 3 East and Southeast Asia from Japan to Malesia

Chileranthemum Oerst. 1854. Acanthaceae. 2 Mexico

Chiliadenus Cass. 1825. Asteraceae. 9 Mediterranean

Chiliocephalum Benth. 1873 (Helichrysum Miller). Asteraceae. 1 Ethiopia

Chiliophyllum Phil. 1864. Asteraceae. 3 southern Andes

Chiliotrichiopsis Cabrera. 1937. Asteraceae. 3 Argentina

Chiliotrichum Cass. 1817. Asteraceae. 2 temp. South America

Chillania Roiv. = Eleocharis R. Br. Cyperaceae

Chilocardamum O. E. Schulz = Sisymbrium L. Brassicaceae

Chilocarpus Blume. 1823. Apocynaceae. 16 India, Andaman and Nicobar Is., Malesia, northeastern Australia

Chiloglottis R. Br. 1810. Orchidaceae. 8 Australia, Tasmania, New Zealand

Chilopogon Schltr. 1912 (Appendicula Blume). Orchidaceae. 5 New Guinea, New Britain, Bougainville and Solomon Is.

Chilopsis D. Don. 1823. Bignoniaceae. 1 southwestern U.S. (California, Utah, Arizona, New Mexico, Texas), Mexico (northern Baja California, Chihuahuan and Sonoran deserts)

Chiloschista Lindl. 1832. Orchidaceae. 10–15 trop. Asia, Malesia, northern Australia, Palau, Fiji (1, C. godeffroyana)

Chimantaea Maguire, Steyerm. et Wurdack. 1957. Asteraceae. 8 Venezuela, Guyana

Chimaphila Pursh. 1814. Ericaceae. 4–8 North and Central America, West Indies, Eurasia; Russia (2, northern European part, Siberia, Far East)

Chimarrhis Jacq. 1763. Rubiaceae. 15 West Indies, Costa Rica, Panama, trop. South America

Chimborazoa H. Beck. 1992. Sapindaceae. 1 Ecuador

Chimonanthus Lindl. 1819. Calycanthaceae. 6 China

Chimonobambusa Makino. 1914. Poaceae. 10 Himalayas (2), East and Southeast Asia

Chimonocalamus Hsueh et Yi = Sinarundinaria Nakai. Poaceae

Chingiacanthus Hand.-Mazz. = Isoglossa Oerst. Acanthaceae

Chiococca P. Browne. 1756. Rubiaceae. 20 southern Florida, West Indies, trop. America

Chiogenes Salisb. ex Torr. = Gaultheria L. Ericaceae

Chionachne R. Br. 1838. Poaceae. 7 India, Indochina, Malesia, Australia, Oceania

Chionanthus L. 1753. Oleaceae. c. 120 trop. Africa (9), Madagascar, trop., and subtrop. Asia; a few species in temp. China and eastern North America

Chione DC. 1830. Rubiaceae. 10 Central America, West Indies

Chionocharis I. M. Johnst. 1924 (Eritrichium Schrad.). Boraginaceae. 1 Himalayas (Nepal, Bhutan), Tibet, western China

Chionochloa Zotov. 1963. Poaceae. 22 southeastern Australia (1), New Zealand (21), Campbell and Auckland Is., Stewart I.

Chionodoxa Boiss. 1844. Hyacinthaceae. 5 eastern Mediterranean

Chionographis Maxim. 1867. Chionographidaceae. 7 East Asia

Chionohebe B. S. Briggs et Ehrend. 1976. Scrophulariaceae. 7 Tasmania (1), New Zealand (6)

Chionolaena DC. 1836. Asteraceae. 8 Mexico, South America

Chionopappus Benth. 1873. Asteraceae. 1 Peru

Chionophila Benth. 1846. Scrophulariaceae. 2 North America (Rocky Mountains)

Chionothrix Hook. f. 1880. Amaranthaceae. 3 Somalia

Chiranthodendron Larréat. 1795. Sterculiaceae. 1 Mexico, Guatemala

Chirita Buch.-Ham. ex D. Don. 1822. Gesneriaceae. c. 130 Himalayas, India, Burma, southwestern and southern China (c. 80), Indochina, Malesia

Chiritopsis W. T. Wang. 1981 (Chirita Buch.-Ham. ex D. Don). Gesneriaceae. 7 China

Chironia L. 1753. Gentianaceae. 15 Africa, Madagascar

Chiropetalum A. Juss. = Argithamnia Sw. Euphorbiaceae

Chirripoa Suess. = Guzmania Ruiz et Pav. Bromeliaceae

Chisocheton Blume. 1825. Meliaceae. 51 southern China, Indochina, New Guinea, Solomon Is., eastward to Vanuatu, Australia

Chitonanthera Schltr. 1905. Orchidaceae. 11 New Guinea

Chitonochilus Schltr. 1905. Orchidaceae. 1 New Guinea

Chlaenandra Miq. 1868. Menispermaceae. 1 New Guinea

Chlaenosciadium C. Norman. 1838. Apiaceae. 1 western Australia

Chalmydacanthus Lindau = Theileamea Baill. Acanthaceae

Chlamydites J. R. Drumm. = Aster L. Asteraceae

Chlamydoboea Stapf = Paraboea (C. B. Clarke) Ridl. Gesneriaceae

Chlamydocardia Lindau. 1894. Acanthaceae. 4 trop. West Africa

Chlamydocarya Baill. 1872. Icacinaceae. 6 trop. West Africa

Chlamydocola (K. Schum.) Bodard = Cola Schott et Endl. Sterculiaceae

Chlamydojatropha Pax et K. Hoffm. 1912. Euphorbiaceae. 1 trop. West Africa

Chlamydophora Ehrenb. ex Less. 1832. Asteraceae. 3 North Africa, Crete and Rhodes

Chlamydophytum Midlbr. 1925. Sarcophytaceae. 1 trop. West Africa

Chlamydostachya Midlbr. 1934. Acanthaceae. 1 trop. East Africa

Chlamydostylus Baker. 1876 (Nemastylus Nutt.). Iridaceae. 1 Bolivia

Chlidanthus Herb. 1821. Amaryllidaceae. 2 trop. America

Chloachne Stapf. 1916 (Poecilostachys Hack.). Poaceae. 1 trop. Africa from Nigeria and Ethiopia to Zimbabwe and Mozambique

Chloanthes R. Br. 1810. Chloanthaceae. 4 Australia

Chloothamnus Büse = Nastus Juss. Poaceae

Chloracantha G. L. Nesom, Y. B. Suh, D. R. Morgan, S. D. Sundb. et B. B. Simpson. 1991. Asteraceae. 1 North and Central America from southwestern U.S. to Panama

Chloraea Lindl. 1827. Orchidaceae. 50 temp. South America

Chloranthus Sw. 1787. Chloranthaceae. 20 India, Sri Lanka eastward to China, Japan, Indochina, Malesia (Sumatra, Borneo, Philippines, New Guinea, New Britain); Russia (2, Far East: Primorsk, Sakhalin I., southern Kuril Is.)

Chloris Sw. 1788. Poaceae. c. 70 trop. and temp. regions; 1 (C. virgata)—eastern Caucasus, Middle Asia, southern Far East of Russia

Chlorocalymma W. D. Clayton. 1970. Poaceae. 1 Tanzania

Chlorocardium Rohwer, H. G. Richt. et van der Werff. 1991 (Ocotea Aubl.). Lauraceae. 2 Colombia and Ecuador (C. vednenosum), Guyana, Surinam (C. rodiei)

Chlorocarpa Alston. 1931. Flacourtiaceae. 1 Sri Lanka

Chlorochorion Puff et Robbr. 1989 (Pentanisia Harv.). Rubiaceae. 2 Kenya, Tanzania, Burundi, Zaire, Malawi, Zambia

Chlorocrambe Rydb. 1907. Brassicaceae. 1 western North America

Chlorocyathus Oliv. 1887. Asclepiadaceae. 1 trop. East Africa (Mozambique)

Chlorogalum (Lindl.) Kunth. 1843. Hyacinthaceae. 5 western North America (California)

Chloroleucon (Benth.) Britton et Rose = Pithecellobium Mart.

Chloroluma Baill. = Chrysophyllum L. Sapotaceae

Chloromyrtus Pierre = Eugenia L. Myrtaceae

Chloropatane Engl. = Erythrococca Benth. Euphorbiaceae

Chlorophora Gaudich. 1830 (Maclura Nutt.). Moraceae. 5 trop. America, trop. Africa, Madagascar

Chlorophytum Ker-Gawl. 1807. Asphodelaceae. c. 300 trop. and South Africa, Madagascar, Arabian Peninsula, Socotra, India, southern China, Indochina, Malay Peninsula, Australia, South America

Chlorosa Blume = Cryptostylis R. Br. Orchidaceae

Chlorospatha Engl. 1878. Araceae. 10 trop. America

Chloroxylon DC. 1824. Rutaceae. 1 Madagascar, southwestern India, Sri Lanka

Choananthus Rendle = Scadoxus Raf. Amaryllidaceae

Chodaphyton Minod. 1918 (Stemodia L.). Scrophulariaceae. 1 Paraguay

Chodsha = Kasiana Rauschert = Catenulina Sojak. Brassicaceae

Choeradoplectron Schauer = Habenaria Willd. Orchidaceae

Choerospondias B. L. Burtt et A. W. Hill. 1937. Anacardiaceae. 1 Himalayas from Nepal to Assam, Thailand, continental and southern China, southern Japan

Choisya Kunth. 1823. Rutaceae. 7 southern U.S., Mexico

Chomelia Jacq. = Tarenna Gaertner. Rubiaceae

Chondradenia Maxim. ex Maek. 1971. Orchidaceae. 1 Japan

Chondrachyrum Nees = Briza L. Poaceae

Chondrilla L. 1753. Asteraceae. 25 temp. Eurasia

Chondrococcus Steyerm. = Coccochondra Rauschert. Rubiaceae

Chondrodendron Ruiz et Pav. 1794. Menispermaceae. 3 Panama, Peru, Bolivia, Brazil

Chondropetalum Rottb. 1772. Restionaceae. 12 South Africa (Cape)

Chondropyxis D. A. Cooke. 1986. Asteraceae. 1 western and southern Australia

Chondrorhyncha Lindl. 1846. Orchidaceae. 11–15 Mexico, Central and trop. South America

Chondrostylis Boerl. 1897. Euphorbiaceae. 2 Indochina, western Malesia

Chondrosum Desv. 1810 (Bouteloua Kunth). Poaceae. 14 America from Canada to Argentina

Chonemorpha G. Don. 1837. Apocynaceae. 20 Himalayas, India, Sri Lanka, Andaman Is., Burma, southern China, Indochina, Malesia

Chonocentrum Pierre ex Pax et K. Hoffm. 1922. Euphorbiaceae. 1 Brazil (Amazonia)

Chonopetalum Radlk. 1920. Sapindaceae. 1 trop. West Africa

Chontalesia Lundell = Ardisia Sw. Myrsinaceae

Chordospartium Cheeseman. 1910. Fabaceae. 1 New Zealand

Choretrum R. Br. 1810. Santalaceae. 6 southern Australia

Choriantha Riedl. 1961. Boraginaceae. 1 Iraq

Choricarpia Domin. 1928. Myrtaceae. 2 eastern Australia

Choriceras Baill. 1873 (Dissiliaria F. Muell.). Euphorbiaceae. 2 New Guinea, northeastern Australia

Chorigyne R. Erikss. 1989. Cyclanthaceae. 7 Costa Rica, Panama

Chorilaena Endl. 1837. Rutaceae. 1 southwestern Australia

Chorioluma Baill. = Pycnandra Benth. Sapotaceae

Choriptera Botsch. 1967. Chenopodiaceae. 1 Abd-El-Kuri and Samkha Is. (near Socotra)

Chorisandrachne Airy Shaw. 1969. Euphorbiaceae. 1 Thailand

Chorisepalum Gleason et Wodehouse. 1931. Gentianaceae. 5 Venezuela, Guyana

Chorisia Kunth. 1822. Bombacaceae. 5 trop. South America

Chorisis DC. 1838. Asteraceae. 1 East Asia from Russian Far East to Vietnam

Chorispora R. Br. ex DC. 1821. Brassicaceae. 12 Southeast Europe, Caucasus, West and East Siberia, Middle Asia, Turkey, Syria, Israel, Jordan, Iran, Afghanistan, Pakistan, western Himalayas, China, Mongolia

Choristemon H. B. Williamson. 1924. Epacridaceae. 1 Australia (Victoria)

Choristylis Harv. 1842. Iteaceae. 1 trop. (Burundi, Zaire, Uganda, Tanzania, Malawi, Mozambique, Zimbabwe) and South Africa (Transvaal, Natal, Cape)

Choritaenia Benth. 1867. Apiaceae. 1 South Africa (Cape)

Chorizandra R. Br. 1810. Cyperaceae. 5 Australia, Tasmania

Chorizanthe R. Br. ex Benth. 1836. Polygonaceae. 50 arid and semiarid western America

Chorizema Labill. 1800. Fabaceae. 18 western and southwestern Australia

Chortolirion A. Berger. 1908 (Haworthia Duval). Asphodelaceae. 1 (C. angolense) southern trop. and South Africa

Chosenia Nakai. 1920. Salicaceae. 1 subarctic and temp. Northeast and East Asia

Chouxia Capuron. 1969. Sapindaceae. 1 Madagascar

Chresta Vell. ex DC. 1836. Asteraceae. 11 Brazil

Christia Moench. 1802. Fabaceae. 10 subtrop. and trop. Asia, Malesia, and northern Australia

Christiana DC. 1824. Tiliaceae. 2 trop. South America, trop. Africa, Madagascar

Christisonia Gardner. 1847. Orobanchaceae. 16 India, southwestern China (1, C. hookeri), Indochina, Malesia

Christolea Cambess. ex Jacquem. 1835. Brassicaceae. 2 Middle Asia, Afghanistan, Pakistan, western China, Tibet, India, Nepal

Chroesthes Benoist. 1927. Acanthaceae. 3 southwestern China, Burma, Thailand, Laos, Malay Peninsula

Chroilema Bernh. = Haplopappus Cass. Asteraceae

Chromolaena DC. 1836 (Eupatorium L.). Asteraceae. c. 150 southern U.S., trop. America, West Indies

Chromolepis Benth. 1840. Asteraceae. 1 Mexico

Chromolucuma Ducke. 1925. Sapotaceae. 2 Venezuela, Guyana, Amazonian Brazil

Chroniochilus J. J. Sm. 1918. Orchidaceae. 4 Thailand, Malay Peninsula, Sumatra, Java, Borneo

Chronopappus DC. 1836. Asteraceae. 1 Brazil

Chrozophora A. Juss. 1824. Euphorbiaceae. 12 Eurasia from western Mediterranean to India, Burma and Indochina, trop. and South Africa; Crimea, Caucasus, Middle Asia

Chrysactinia A. Gray. 1849. Asteraceae. 5 southwestern U.S., Mexico

Chrysactinium (Kunth) Wedd. 1857 (Liabum Adans.). Asteraceae. 6 trop. South America

Chrysalidocarpus H. A. Wendl. 1878. Arecaceae. c. 20 Madagascar (18), Comoro Is., Pemba I. (off the coast of Tanzania)

Chrysallidosperma H. Moore = Syagrus Mart. Arecaceae

Chrysanthellum Rich. ex Pers. 1807. Asteraceae. 8 tropics

Chrysanthemoides Fabr. 1759. Asteraceae. 2 trop. and South Africa

Chrysanthemum L. 1753. Asteraceae. 2 Europe, North Africa, Cyprus, Asia Minor, Caucasus, Syria, northwestern Iran

Chrysaspis Desv. = Trifolium L. Fabaceae

Chrysithrix L. 1771. Cyperaceae. 4 South Africa, western Australia

Chrysobalanus L. 1753. Chrysobalanaceae. 3 America from Florida to southern Brazil, West Indies, Namibia (1, C. icaco) from Liberia to Angola and Zambia

Chrysobraya Hara. 1974 (Braya Sternb. et Hoppe). Brassicaceae. 1 eastern Nepal, Bhutan

Chrysocephalum Walp. 1841. Asteraceae. 7 Australia

Chrysochamela (Fenzl) Boiss. 1867. Brassicaceae. 4 East Europe, Armenia (1, C. elliptica), Turkey, Syria, Iraq

Chrysochlamys Poepp. = Tovomitopsis Planch. et Triana. Clusiaceae

Chrysochloa Swallen. 1941. Poaceae. 4 trop. Africa

Chrysocoma L. 1753. Asteraceae. 18 southern Africa

Chrysocoryne Endl. = Gnephosis Cass. Asteraceae

Chrysocoryne Zoellner = Pabellonia Quezada et Martic. Alliaceae

Chrysocycnis Linden et Rchb. f. 1854. Orchidaceae. 5 Colombia, Ecuador

Chrysoglossum Blume. 1825. Orchidaceae. 25 trop. Asia from Nepal, Bhutan, northeastern India and Sri Lanka through Indochina to China, Taiwan, Indochina, Malesia (Sumatra, Java, Borneo, Sulawesi) eastward to New Caledonia (1) and Fiji

Chrysogonum L. 1753. Asteraceae. 4 eastern U.S. (1, C. virginianum), Australia (3)

Chrysolaena H. Rob. = Vernonia Schreb. Asteraceae

Chrysolepis Hjelmq. = Castanopsis (D. Don) Spach. Fagaceae

Chrysoma Nutt. = Solidago L. Asteraceae

Chrysopappus Takht. 1938 (Centaurea L.). Asteraceae. 1 Kurdistan (southeastern Turkey, northern Iraq)

Chrysophae Kozo-Polj. = Chaerophyllum L. Apiaceae

Chrysophtalmum Sch. Bip. ex Walp. 1843. Asteraceae. 3 Southwest Asia

Chrysophyllum L. 1753. Sapotaceae. 70 tropics: 43 trop. America and West Indies; 15 Africa; 10 Madagascar; 2–3 trop. Asia, Malesia to trop. Australia

Chrysopogon Trin. 1822. Poaceae. 26 trop., subtrop. and temp. regions, esp. Asia and Australia; C. gryllus in Caucasus

Chrysopsis (Nutt.) Elliott. 1823. Asteraceae. 10 U.S., Mexico, Bahama Is.

Chrysosciadium Tamamsch. = Echinophora L. Apiaceae

Chrysoscias E. Mey. 1836. Fabaceae. 6 South Africa

Chrysospleniella B. Sparre. 1956. Saxifragaceae. 1 Chile

Chrysosplenium L. 1753. Saxifragaceae. 64 North (5) and temp. South America (2), Europe (4), North Africa (1), arctic and temp. Asia (56, esp. China—42); Russia (28, from European part and Caucasus to Far East)

Chrysothamnus Nutt. 1840. Asteraceae. 16 Canada (British Columbia to Saskatchewan), U.S. (Washington, North Dakota south to California, Nevada, Utah, Arizona, Colorado, Kansas, New Mexico, Texas), Mexico

Chrysothemis Decne. 1849. Gesneriaceae. 7 trop. America, West Indies

Chthonocephalus Steetz. 1845. Asteraceae. 6 southern Australia from southwestern Australia to New South Wales and Victoria north to the southern part of Northern Territory

Chuanminshen M. L. Sheh et R. H. Shan. 1980. Apiaceae. 1 China

Chucoa Cabrera. 1955. Asteraceae. 1 Andes of Peru

Chukrasia A. Juss. 1830. Meliaceae. 1 India, Nepal to southwestern and southern China, Malesia

Chumsriella Bor. 1968 (Germainia Balansa et Poitrasson). Poaceae. 1 Thailand

Chunechites Tsiang. 1937. Apocynaceae. 1 Southeast China

Chunia H. T. Chang. 1948. Hamamelidaceae. 1 Hainan I.

Chuniodendron Hu = Aphanamixis Blume. Meliaceae

Chuniophoenix Burret. 1937. Arecaceae. 3 southern China, incl. Hainan, Vietnam

Chuquiraga Juss. 1789. Asteraceae. 25 Andes of South America, Patagonia

Chusquea Kunth. 1822. Poaceae. c. 100 Central and South America, West Indies

Chusua Nevski = Orchis L. Orchidaceae

Chydenanthus Miers. 1875. Lecythidaceae. 2 Burma, Andaman Is., Malesia (Sumatra, Borneo to New Guinea)

Chymsidia Albov. 1895. Apiaceae. 1 (C. agasylloides) western Transcaucasia

Chysis Lindl. 1837. Orchidaceae. 6 trop. America from Mexico to Peru and Venezuela

Chytranthus Hook. f. 1862. Sapindaceae. 30 trop. Africa

Chytroglossa Rchb. f. 1863. Orchidaceae. 3 Brazil

Chytroma Miers = Lecythis Loefl. + Eschweilera Mart. Lecythidaceae

Chytropsia Bremek. = Psychotria L. Rubiaceae

Cibirhiza Bruyns. 1988 (Anisopus N. E. Br.). Asclepiadaceae. 1 Oman

Cibotarium O. E. Schulz = Sphaerocardamum Schauer. Brassicaceae

Cicca L. = Phyllanthus L. Euphorbiaceae

Cicendia Adans. 1763. Gentianaceae. 2: C. filiformis—South Europe from Iberian Peninsula to Creece, Morocco, Algeria, Tunisia, Turkey, Syria, Lebanon; C. quadrangularis—California, western South America

Cicer L. 1753. Fabaceae. 40 Canary Is.,

Morocco (1), Egypt and Ethiopia (1), Southeast Europe, Caucasus and Middle Asia (14), Arabian Peninsula, Turkey, Syria, Lebanon, Iraq, Iran, Afghanistan, Pakistan, China, Tibet

Cicerbita Wallr. 1822. Asteraceae. 18 Europe, North Africa, eastern Mediterranean, incl. Asia Minor (7), Caucasus, Iraq, Iran, Middle Asia, West and East Siberia

Ciceronia Urb. 1925. Asteraceae. 1 Cuba

Cichorium L. 1753. Asteraceae. 10–12 Europe, Mediterranean, Ethiopia, Caucasus, West and East Siberia, Middle Asia; 1 (C. intybus) from West Asia to western China and Mongolia

Ciclospermum Lag. = Cyclospermum Lag. Apiaceae

Cicuta L. 1753. Apiaceae. 8 cold and temp. Northern Hemisphere; C. virosa—European part of Russia, Caucasus (Abkhasia), West and East Siberia, Far East

Cienfuegosia Cav. 1786. Malvaceae. 26 trop. and subtrop. America, from Texas and Florida to Argentina, Africa, Madagascar, Arabian Peninsula, Australia

Cienkowskia Schweinf. = Siphonochilus J. M. Wood et Franks. Zingiberaceae

Cienkowskiella Y. K. Kam = Siphonochilus J. M. Wood et Franks. Zingiberaceae

Cigarilla Aiello. 1979. Rubiaceae. 1 Mexico

Cimicifuga Wernisch. 1763 (Actaea L.). Ranunculaceae. 18 temp. Northern Hemisphere; Russia (5, European part, West (1) and East Siberia, Far East)

Cinchona L. 1753. Rubiaceae. 40 Costa Rica, Panama, Andes

Cinchonaceae Lindl. = Rubiaceae Juss.

Cincinnobotrys Gilg. 1897. Melastomataceae. 6 trop. Africa

Cineraria L. 1763. Asteraceae. 50 trop. and South Africa, Madagascar, southern Arabian Peninsula

Cinna L. 1753. Poaceae. 4: C. arundinacea—eastern North America; C. bolanderi—California; C. poaeformis—Mexico, Guatemala, Costa Rica, Panama, Colombia, northwestern Venezuela, Peru, northern Bolivia; C. latifolia—Canada, U.S., North Europe, Russia (European part, northern Caucasus, Siberia, Far East), northeastern China, Japan

Cinnabarinea Fric et F. Ritter = Echinopsis Zucc. Cactaceae

Cinnadenia Kosterm. 1973. Lauraceae. 2 Nepal, Bhutan, Burma, Malay Peninsula

Cinnagrostis Griseb. 1874 (Calamagrostis Adans.). Poaceae. 1 Argentina

Cinnamodendron Endl. 1840. Canellaceae. 7 trop. South America, West Indies

Cinnamomum Schaeffer. 1760. Lauraceae. 250 trop. America (50) from Mexico to southern Brazil and Paraguay, West Indies, trop. and subtrop. Asia, Australia, Polynesia

Cinnamosma Baill. 1867. Canellaceae. 3 Madagascar

Cionomene Krukoff. 1979. Menispermaceae. 1 Ecuador, Brazil (Amazonia)

Cionosicyos Griseb. 1860. Cucurbitaceae. 3 Central America, West Indies

Cionura Griseb. 1844 (Marsdenia R. Br.). Asclepiadaceae. 1 Yugoslavia, Greece, Asia Minor, Iran, western Afghanistan

Cipadessa Blume. 1825. Meliaceae. 1 (C. baccifera) India, Sri Lanka, Himalayas, western China, Indochina, Malesia

Cipocereus F. Ritter. 1979 (Pilosocereus Byles et G. Rowley). Cactaceae. 5 Brazil (Minas Gerais)

Cipura Aubl. 1775. Iridaceae. 6 trop. America

Cipuropsis Ule. 1907 (or = Vriesea Lindl.). Bromeliaceae. 1 Peru

Circaea L. 1753. Onagraceae. 8 arctic and temp. Northern Hemisphere; Russia (6, from European part to Far East)

Circaeaster Maxim. 1882. Circaeasteraceae. 1 Himalayas from Garhwal to Bhutan, western China, southeastern Tibet

Circaeocarpus C. Y. We = Zippelia Blume ex Schult. et Schult. f. Piperaceae

Circanda N. E. Br. 1930 (Erepsia N. E. Br.). Aizoaceae. 1 southwestern South Africa

Cirrhaea Lindl. 1832. Orchidaceae. 7 Brazil

Cirrhopetalum Lindl. = Bulbophyllum Thouars. Orchidaceae

Cirsium Hill. 1753. Asteraceae. 250–300 North and Central America, temp. Eurasia, North and trop. Africa

Cischweinfia Dressler et N. H. Williams. 1970. Orchidaceae. 6 Central and South America, southward to Ecuador and Bolivia

Cissampelopsis (DC.) Miq. 1856. Asteraceae. 20 trop. Old World

Cissampelos L. 1753. Menispermaceae. 20–25 pantropics

Cissarobryon Poepp. 1833 (or = Viviania Cav.). Vivianiaceae. 1 southern South America

Cissus L. 1753. Vitaceae. 350 tropics, rarely subtropics

Cistanche Hoffmanns. et Link. 1809. Orobanchaceae. 20 Mediterranean, Northwest Africa, Ethiopia, Southeast Europe (1, C. salsa), Caucasus, West and Middle Asia eastward to Mongolia, northwestern China and western India

Cistanthe Spach. 1836. Portulacaceae. c. 25 western arid and semiarid regions of North (3, California, Arizona) and South (Peru, Bolivia, Chile, Argentina) America

Cistus L. 1753. Cistaceae. 20 Canary Is., Mediterranean, southern Crimea, West Asia to Caucasus and Iran

Citharelonia Bunge. 1845. Brassicaceae. 3 Middle Asia (2), Iran, Afghanistan, Pakistan, western China

Citharexylum L. 1753. Verbenaceae. 70 (or 145) trop. America from southern U.S. to Argentina and Uruguay, Bermuda Is., West Indies

Citriobatus A. Cunn. et Putterl. 1839. Pittosporaceae. 5 Philippines, Java, Lesser Sunda Is. (Flores I.), New Guinea, northeastern Australia (4)

Citronella D. Don. 1832. Icacinaceae. 21 Malesia to Solomon Is., eastern Australia (2), New Caledonia, Loyalty Is., Vanuatu, Fiji, Samoa Is. and Tonga Is., Central and trop. South America (10)

Citropsis (Engl.) Swingle et M. Kellerm. 1914. Rutaceae. 8 trop. and South Africa

Citrullus Schrad. ex Eckl. et Zeyh. 1836. Cucurbitaceae. 9 South Europe, Africa, eastern Mediterranean, trop. Asia

Citrus L. 1753. Rutaceae. 16 trop. Asia from India and southern China through Indochina and Malesia easward to Fiji and Samoa; C. medica having been known in Babylon by. 4000 B.C.

Cladanthus Cass. 1816. Asteraceae. 1 southern Spain, Northwest Africa

Claderia Hook. f. 1890. Orchidaceae. 2 Thailand, Malesia, except Philippines

Cladium P. Browne. 1756. Cyperaceae. 5 temp., subtrop., and trop. regions

Cladocarpa (St. John) St. John. = Sicyos L. Cucurbitaceae

Cladoceras Bremek. 1940. Rubiaceae. 1 coastal Kenya and Tanzania

Cladochaeta DC. 1838 (Helichrysum Miller). Asteraceae. 2 Caucasus

Cladocolea Tiegh. 1895. Loranthaceae. c. 25 Mexico (mostly), Guatemala, Honduras, Panama, Colombia, Ecuador, Peru, Venezuela, Brazil

Cladogelonium Leandri. 1939. Euphorbiaceae. 1 Madagascar

Cladogynos Zipp. ex Spanoghe. 1841. Euphorbiaceae. 1 China, Indochina, Malesia

Cladomyza Danser. 1940. Santalaceae. 17 Borneo, Moluccas, New Guinea, Solomon Is.

Cladopus H. Moeller. 1899. Podostemaceae. 5 China, Japan, Thailand, Java, Sulawesi

Cladoraphis Franch. 1887. Poaceae. 2 South Africa, Namibia

Cladostachys D. Don = Deeringia R. Br. Amaranthaceae

Cladostemon A. Braun et Vatke. 1877. Capparaceae. 1 Kenya, Tanzania, Malawi, Zambia, Zimbabwe, Mozambique, South Africa (Natal, Swaziland)

Cladostigma Radlk. 1883. Convolvulaceae. 2 trop. Northeast Africa

Cladothamnus Bong. = Elliottia Mühl. ex Elliott. Ericaceae

Cladrastis Raf. 1824. Fabaceae. 6 China, Japan, North America (1)

Clandestinaria Spach. 1838 (Rorippa Scop.). Brassicaceae. 1 Egypt

Claoxylon A. Juss. 1824. Euphorbiaceae. 125 from Northeast Africa and Madagascar to Southeast Asia, Malesia, Australia, Oceania eastward to Hawaiian and Society Is.

Claoxylopsis Leandri. 1939. Euphorbiaceae. 1 Madagascar

Clappertonia Meisn. 1837. Tiliaceae. 2 trop. West Africa

Clappia A. Gray. 1859. Asteraceae. 2 southern U.S., Mexico

Clara Kunth = Herreria Ruiz et Pav. Herreriacea

Clarisia Ruiz et Pav. 1794. Moraceae. 2 trop. America

Clarkella Hook. f. 1880. Rubiaceae. 2 Himalayas, China, Thailand

Clarkia Pursh. 1813–1814. Onagraceae. c. 40 western North America from British Columbia to Baja California and Arizona, but chiefly California (c. 37); Chile (1)

Clastopus Bunge ex Boiss. 1867. Brassicaceae. 2 northern Iran, Iraq

Clathrotropis (Benth.) Harms. 1901. Fabaceae. 6 trop. South America

Clausena Burm. f. 1768. Rutaceae. 23 trop. and South Africa, subtrop. and trop. Asia, Malesia, eastern Australia

Clausenellia A. Löve et D. Löve. 1985 (Sedum L.). Crassulaceae. 1 U.S. (North Carolina)

Clausenopsis (Engl.) Engl. = Fagaropsis Mildbr. Rutaceae

Clausia Trotzky ex Hayek. 1911 (Hesperis L.). Brassicaceae. 6 East Europe, Caucasus, West and East Siberia, West and Middle Asia to Mongolia, China, and Afghanistan

Clavija Ruiz et Pav. 1794. Theophrastaceae. 55 Central (c. 5) and trop. South America

Clavinodum T. N. Wen = Arundinaria Michx. Poaceae

Claytonia L. 1753. Portulacaceae. 15–20 Northern Hemisphere from East Siberia and Russian Far East (8) to western coastal North America

Claytoniella Jurtzev. 1972 (Montia L.). Portulacaceae. 1 arcto-alpine Northeast Asia, northwestern North America

Cleanthe Salisb. ex Benth. = Aristea Sol. ex Aiton. Iridaceae

Cleanthes D. Don = Trixis P. Browne. Asteraceae

Cleghornia Wight. 1848. Apocynaceae. 2–3 Sri Lanka, Malesia

Cleidiocarpon Airy Shaw. 1965. Euphorbiaceae. 2 Burma, western China

Cleidion Blume. 1826. Euphorbiaceae. 25 pantropics, esp. New Caledonia (12)

Cleisocentron Brühl. 1926. Orchidaceae. 2–3 Himalayas, Indochina

Cleisomeria Lindl. ex G. Don. 1855. Orchidaceae. 2 Burma, Thailand, Laos, Cambodia, Vietnam, Malay Peninsula

Cleisostoma Blume. 1825 (Sarcanthus Lindl.). Orchidaceae. c. 100–160 trop. Asia from Nepal and eastern India to southern China, Indochina, Malesia from Malay Peninsula to New Guinea and New Ireland, Solomon Is., Australia, New Caledonia (1), Fiji (1, C. longipaniculatum)

Cleisostomopsis Seidenf. 1992. Orchidaceae. 1 Vietnam

Cleistachne Benth. 1882. Poaceae. 4 trop. and South Africa, India

Cleistanthopsis Capuron = Allantospermum Forman. Irvingiaceae

Cleistanthus Hook. f. ex Planch. 1848. Euphorbiaceae. 140 trop. Old World

Cleistes Rich. ex Lindl. 1840. Orchidaceae. 25–50 temp. North (1) and South America

Cleistocactus Lem. 1861. Cactaceae. 40–50 southern Ecuador, Peru, Bolivia, western Brazil, northern Argentina, Paraguay, Uruguay

Cleistocalyx Blume. 1850 (Syzygium P. Browne ex Gaertn.). Myrtaceae. c. 20 trop. Asia from Burma and southern China through Indochina to Malesia (from Malay Peninsula to New Guinea), and northern Australia, Lord Howe I., New Caledonia, Fiji (7)

Cleistochlamys Oliv. 1865. Annonaceae. 1

Tanzania, Malawi, Zambia, Zimbabwe, Mozambique

Cleistochloa C. E. Hubb. 1933. Poaceae. 3 New Guinea, Australia

Cleistogenes Keng. 1934 (or = Kengia Packer). Poaceae. 20 temp. Eurasia

Cleistopholis Pierre ex Engl. 1897. Annonaceae. 5 trop. West and East (1, C. patens—Uganda) Africa

Clelandia J. M. Black = Hybanthus Jacq. Violaceae

Clematepistephium N. Hallé. 1977 (Epistephium Kunth). Orchidaceae. 1 New Caledonia

Clematicissus Planch. 1887. Vitaceae. 1 southwestern Australia

Clematis L. 1753. Ranunculaceae. c. 250–300 temp. regions of both hemispheres, esp. China (c. 110); a few spp. in trop. Africa; Russia (20, southern European part, northern Caucasus, southern Siberia, Far East)

Clematoclethra (Franch.) Maxim. 1890. Actinidaceae. 1 (C. scandens) western and central China from western Henan to northern Guangxi

Clematopsis Bojer ex Hutch. 1920 (Clematis L.). Ranunculaceae. 18 trop. and South Africa, Madagascar

Clemensiella Schltr. 1915. Asclepiadaceae. 1 Philippines

Clementsia Rose. 1903. Crassulaceae. 2 U.S. (C. rhodantha—Rocky Mountains: Arizona and Utah to Montana), Middle Asia (C. semenovii)

Cleobulia Mart. ex Benth. 1837. Fabaceae. 3 Brazil

Cleome L. 1753. Capparaceae. c. 200 tropics and subtropics, esp. America

Cleomella DC. 1824. Capparaceae. 10 southwestern U.S., Mexico

Cleonia L. 1763. Lamiaceae. 1 Portugal, Spain, Morocco, Algeria, Tunisia

Cleophora Gaertn. = Latania Comm. ex Juss. Arecaceae

Cleretum N. E. Br. 1925. Aizoaceae. 3 southwestern South Africa

Clermontia Gaudich. 1829. Lobeliaceae. 22 Hawaiian Is.

Clerodendranthus Kudo. 1929 (or = Orthosiphon Benth.). Lamiaceae. 5 East Asia, Indo-Malesia

Clerodendrum L. 1753. Verbenaceae. 400 tropics and subtropics; C. bungei—extending to Caucasus

Clethra L. 1753. Clethraceae. 67 East and Southeast Asia, southeastern U.S., Mexico, Central and trop. South America; 1 (C. arborea) Madeira Is.

Clevelandia Greene. 1886. Scrophulariaceae. 1 Baja California

Cleyera Thunb. 1783. Theaceae. 9 Nepal, northern India, China, Taiwan, Indochina, Korea, Japan, America from Mexico to Panama, West Indies

Clianthus Sol. ex Lindl. 1835. Fabaceae. 3 Australia (2), New Zealand (1)

Clibadium L. 1771. Asteraceae. 40 trop. America, West Indies

Clidemia D. Don. 1823. Melastomataceae. 165 trop. America from Mexico to Argentina, West Indies

Cliffortia L. 1753. Rosaceae. 115 trop. and South Africa, chiefly montane Cape

Cliftonia Banks ex Gaertn. f. 1807. Cyrillaceae. 1 southeastern U.S., except Florida

Climacoptera Botsch. 1956. Chenopodiaceae. 42 Southeast Europe, Caucasus, West Siberia, Middle Asia, Turkey, Syria, Iran, Iraq, Afghanistan, Pakistan, China, Mongolia

Clinacanthus Nees. 1847. Acanthaceae. 2 southern China, Indochina, Malesia

Clinelymus (Griseb.) Nevski = Elymus L. Poaceae

Clinogyne Salisb. ex Benth. = Donax Lour. + Schumannianthus Gagnepain + Marantochloa Brongn. et Gris. Maranthaceae

Clinopodium L. 1753 (Satureja L.). Lamiaceae. 20 temp. Eurasia, southward to India, Burma, Indochina, Malaya, and Taiwan; North America; Russia (all areas)

Clinosperma Becc. 1921. Arecaceae. 1 New Caledonia

Clinostemon Kuhlm. et Samp. 1928 (Licaria Aubl.). Lauraceae. 2 trop. America

Clinostigma H. A. Wendl. 1862. Arecaceae. 13 from Ogasawara (Bonin) and Caroline Is. to Solomon Is., New Ireland, Vanuatu, Fiji, and Samoa Is.

Clinostigmopsis Becc. = Clinostigma H. A. Wendl. Arecaceae

Clintonia Raf. 1818. Convallariaceae. 6 Himalayas and temp. East Asia, North America (4); C. udenis—extending to Far East of Russia

Cliococca Bab. 1841 (Linum L.). Linaceae. 1 temp. South America

Clistax Mart. 1829. Acanthaceae. 2 Brazil

Clistoyucca (Engelm.) Trel. = Yucca L. Agavaceae

Clitandra Benth. 1849. Apocynaceae. 1 trop. Africa: Sierra Leone, Liberia, Ivory Coast, Guinea, Ghana, Nigeria, Cameroun, Central African Rep., Gabon, Congo, Zaire, Angola, Uganda, and Tanzania

Clitandropsis S. Moore = Melodinus J. R. Forst. et G. Forst. Apocynaceae

Clitoria L. 1753. Fabaceae. 70 pantropics, esp. America

Clitoriopsis R. Wilczek. 1954. Fabaceae. 1 Zaire

Clivia Lindl. 1828. Amaryllidaceae. 3 South Africa

Cloezia Brongn. et Gris. 1864. Myrtaceae. 8 New Caledonia

Clonodia Griseb. 1858. Malpighiaceae. 2–3 Colombia, Venezuela, Guyana, Brazil

Clonostylis S. Moore = Spathiostemon Blume. Euphorbiaceae

Closia Remy = Peristyle Benth. Asteraceae

Clowesia Lindl. 1843 (Catasetum Rich. ex Kunth). Orchidaceae. 5 Central America

Clusia L. 1753. Clusiaceae. 145 trop. and subtrop. America; a few species in Madagascar and New Caledonia

Clusiella Planch. et Triana. 1860. Clusiaceae. 1–4 Panama, Colombia

Clutia L. 1753. Euphorbiaceae. 70 Africa, Arabian Peninsula

Clybatis Phil. = Leucheria Lag. Asteraceae

Clymenia Swingle. 1939. Rutaceae. 1 Bismarck Arch.

Clypeola L. 1753. Brassicaceae. 9 Europe to Crimea, Mediterranean, Caucasus, West and Middle Asia to Pakistan

Clytostoma Miers ex Bur. 1868. Bignoniaceae. 10–12 trop. America from Mexico to Uruguay, Argentina and Brazil

Clytostomanthus Pichon = Cydista Miers. Bignoniaceae

Cnemidiscus Pierre = Glenniea Hook. f. Sapindaceae

Cneoridium Hook. f. 1862. Rutaceae. 1 southern California and Baja California

Cneorum L. 1753. Cneoraceae. 3 Cuba (1), Mediterranean (Spain, France, western Italy, Balearic Is., Sardinia)

Cnesmone Blume = Cnesmosa Blume. Euphorbiaceae

Cnesmosa Blume. 1826. Euphorbiaceae. 10 India (Assam), southern China, Indochina, western Malesia

Cnestidium Planch. 1850. Connaraceae. 2 trop. America from Mexico to Ecuador and Guyanas, Cuba

Cnestis Juss. 1789. Connaraceae. 40 trop. Africa, Madagascar; 2 southern China, Burma, Andaman Is., Indochina, Malesia (Malay Peninsula, Sumatra, Borneo, Philippines, Sulawesi)

Cnicothamnus Griseb. 1874. Asteraceae. 2 Bolivia, Argentina

Cnicus L. 1753. Asteraceae. 1 (C. benedictus) Europe, Mediterranean, Caucasus, Southwest and Middle Asia eastward to western China and western Himalayas

Cnidiocarpa Pimenov. 1983. Apiaceae. 2 Caucasus, Middle Asia

Cnidium Cuss. ex Juss. 1787. Apiaceae. 4–5 Siberia and Russian Far East (3), China, Korea, Japan, North America

Cnidoscolus Pohl. 1827. Euphorbiaceae. 75 trop. America

Coaxana J. M. Coult. et Rose. 1895. Apiaceae. 2 Mexico

Cobaea Cav. 1791. Cobaeaceae. 12 trop. America from Mexico to Venezuela and Andes from Colombia to northern Chile

Cobana Ravenna. 1974. Iridaceae. 1 Guatemala, Honduras

Cobananthus Wiehler. 1977 (Alloplectus Mart.). Gesneriaceae. 1 Guatemala

Coccineorchis Schltr. 1920. Orchidaceae. 4 trop. America

Coccinia Wight et Arn. 1834. Cucurbitaceae. 30 trop. and South Africa; C. grandis—trop. India, Malesia, northern Australia

Coccochondra Rauschert. 1982. Rubiaceae. 1 northern South America

Coccocypselum P. Browne. 1756. Rubiaceae. 20 trop. America, West Indies

Coccoloba P. Browne. 1756. Polygonaceae. c. 120 Central and trop. and subtrop. South America, West Indies

Cocconerion Baill. 1873. Euphorbiaceae. 2 New Caledonia

Coccosperma Klotzsch. 1838. Ericaceae. 5 South Africa (southwestern and southern Cape)

Coccothrinax Sarg. 1899. Arecaceae. 30–50 West Indies, esp. Cuba (c. 30), southern Florida, southern Mexico, Belize

Cocculus DC. 1817. Menispermaceae. 8–10 Africa, Israel, Jordan, Sinai and Arabian peninsulas, South and Southeast Asia (from India to Japan and Taiwan), Malesia (Sumatra, Malay Peninsula, Java, Philippines), northern Australia, Oceania eastward to Hawaiian Is. and Polynesia, North and Central America

Cochemiea (K. Brandegee) Walton = Mammillaria Haw. Cactaceae

Cochiseia Earle = Escobaria Britton et Rose. Cactaceae

Cochleanthes Raf. 1838. Orchidaceae. 15 Central and trop. South America, southward to Peru and Brazil; West Indies

Cochlearia L. 1753. Brassicaceae. 35 temp. Northern Hemisphere southward to Himalayas; Russia (c. 10, arctic and temp. European part, Siberia, Far East)

Cochleariella Y. H. Zhang et R. Vogt. 1991. Brassicaceae. 1 southern China (Yunnan)

Cochleariopsis Y. H. Zhang = Cochleariella Y. H. Zhang. Brassicaceae

Cochlianthus Benth. 1852 (Apios Fabr.). Fabaceae. 2 Himalayas, western China

Cochlidiosperma (Rchb.) Rchb. 1837. Scrophulariaceae. 8–10 China, Indochina

Cochlioda Lindl. 1853. Orchidaceae. 4–6 Andes of South America

Cochliostema Lem. 1859. Commelinaceae. 2 South America from Colombia to Bolivia

Cochlospermum Kunth. 1822. Cochlospermaceae. c. 20 trop. Africa, Southeast Asia through Malesia to northern Australia, trop. and subtrop. America

Cockaynea Zotov. 1943 (Hystrix Moench). Poaceae. 2 New Caledonia

Cockerellia (Clausen et Uhl) A. Löve et D. Löve. 1985 (or = Sedum L.). Crassulaceae. 2 U.S. (California and New Mexico)

Cocos L. 1753. Arecaceae. 1 (C. nucifera), origin uncertain, but probably western Pacific, widely cultivated

Codariocalyx Hassk. 1842. Fabaceae. 2 Southeast Asia, Malesia to trop. Australia

Coddia Verdc. 1981. Rubiaceae. 1 Mozambique

Coddingtonia S. Bowd. = Psychotria L. Rubiaceae

Codia J. R. Forst. et G. Forst. 1775. Cunoniaceae. 11 New Caledonia

Codiaeum A. Juss. 1824. Euphorbiaceae. 16 Malesia, northern Australia, New Caledonia (2), Melanesia, Polynesia

Codiocarpus R. A. Howard. 1943. Icacinaceae. 2 Andaman Is., Malesia

Codon L. 1767. Hydrophyllaceae. 2 South Africa

Codonacanthus Nees. 1847. Acanthaceae. 2 northeastern India (Assam), southern China

Codonanthe (Mart.) Hanst. 1854. Gesneriaceae. 15 trop. America from Mexico to Brazil, West Indies

Codonanthopsis Mansf. = Codonanthe (Mart.) Hanst. Gesneriaceae

Codonechites Markgr. 1924. Apocynaceae. 2 Bolivia, western Brazil

Codonemma Miers = Tabernaemontana L.

Codonoboea Ridl. = Didymocarpus Wall. Gesneriaceae

Codonocarpus A. Cunn. ex Endl. 1837. Gyrostemonaceae. 2 Australia (mostly arid areas)

Codonocephalum Fenzl = Inula L. Asteraceae

Codonochlamys Ulbr. 1915. Malvaceae. 2 Brazil

Codonopsis Wall. 1824. Campanulaceae. c. 40 Middle Asia (1, C. clematidea), Iran, Afghanistan, India, Himalayas, China, Mongolia, Korea, Japan, Soviet Far East (3), Burma, Indochina, Taiwan, Malesia (except Lesser Sunda Is.)

Codonorchis Lindl. 1840. Orchidaceae. 3 southern trop. and temp. South America

Codonosiphon Schltr. 1912. Orchidaceae. 3 New Guinea

Codonura K. Schum. = Baissea A. DC. Apocynaceae

Coelachne R. Br. 1810. Poaceae. 5 trop. Old World

Coelachyropsis Bor = Coelachyrum Hochst. et Nees. Poaceae

Coelachyrum Hochst. et Nees. 1842. Poaceae. 6 Northeast, East (1, C. longiglume—Kenya) and South Africa, Southwest Asia eastward to Pakistan

Coelanthum E. Mey. ex Fenzl. 1840. Molluginaceae. 3 South Africa

Coelebogyne J. Sm. 1839 (Alchornea Sw.). Euphorbiaceae. 2 eastern Australia

Coelia Lindl. 1830. Orchidaceae. 1–2 Central America, West Indies

Coelidium Vogel ex Walp. 1840 (Amphithalea Ecklon et Zeyher). Fabaceae. 20 South Africa

Coeliopsis Rchb. f. 1872. Orchidaceae. 1 Costa Rica, Panama, Colombia

Coelocarpum Balf. f. 1883. Verbenaceae. 6 Somalia (1), Socotra (1), Madagascar (4)

Coelocaryon Warb. 1895. Myristicaceae. 3 trop. West and Central Africa

Coelococcus H. A. Wendl. = Metroxylon Rottb. Arecaceae

Coelodiscus Baill. = Mallotus Lour. Euphorbiaceae

Coeloglossum Hartm. 1820 (Habenaria Wild.). Orchidaceae. 2 North America, temp. Eurasia

Coelogyne Lindl. 1821. Orchidaceae. c. 300 trop. and subtrop. Asia, Malesia from Malay Peninsula to Philippines (c. 25) and New Guinea, Micronesia, Bougainville and Solomon Is., New Caledonia (1), Vanuatu, Fiji, Samoa

Coelonema Maxim. 1880. Brassicaceae. 1 southwestern China

Coeloneurum Radlk. 1890. Goetzeaceae. 1 Haiti

Coelophragmus O. E. Schulz = Sisymbrium L. Brassicaceae

Coelopleurum Ledeb. = Angelica L. Apiaceae

Coelopyrena Valeton. 1909. Rubiaceae. 1 Moluccas, New Guinea

Coelorachis Brongn. 1831. Poaceae. 20 pantropics

Coelospermum Blume = Caelospermum Blume. Rubiaceae

Coelostegia Benth. 1862. Bombacaceae. 5 Western Malesia

Coelostelma Fourn. = Matelea Aubl. Asclepiadaceae

Coespeletia Cuatrec. 1976 (Espeletia Mutis). Asteraceae. 8 Venezuela

Coffea L. 1753. Rubiaceae. c. 90 trop. Africa, Madagascar, Mascarene Is.; C. arabica and C. liberica—widely cultivated throughout the tropics

Cogniauxia Baill. 1884. Cucurbitaceae. 2 trop. Africa

Cogniauxiocharis (Schltr.) Hoehne = Teroglossa Schltr. Orchidaceae

Cohnia Kunth. 1850 (Cordyline Comm. ex R. Br.). Asteliaceae. 3 Mascarene Is., New Caledonia

Cohniella Pfitzer. 1889 (or = Oncidium Sw.). Orchidaceae. 1 Central America

Coilocarpus F. Muell. ex Domin = Sclerolaena R. Br. Chenopodiaceae

Coilochilus Schltr. 1906. Orchidaceae. 1 New Caledonia

Coilostigma Klotzsch. 1838. Ericaceae. 2 South Africa (southeastern Cape)

Coilostylis Raf. = Epidendrum L. Orchidaceae

Coincya Porta et Rigo ex Rouy. 1891. Brassicaceae. 6 Great Britain, France, Germany, Iberian Peninsula, Corsica, Italy, Morocco (1)

Coinochlamys T. Anderson ex Benth. et Hook. f. = Mostuea Didr. Loganiaceae

Coix L. 1753. Poaceae. 6 trop. Asia; C.

lacryma-jobi—extending to Caucasus, Middle Asia

Cojoba Britton et Rose. 1928 (Archidendron F. Muell.). Fabaceae. 13 Mexico, Belize, Nicaragua, Costa Rica, Panama, western Colombia, Peru

Cola Schott et Endl. 1832. Sterculiaceae. 125 trop. Africa

Colania Gagnep. 1934 (Aspidistra Ker-Gawl.). Convallariaceae. 1 Indochina

Colanthelia McClure et E. W. Sm. 1973. Poaceae. 7 southeastern Brazil

Colax Lindl. = Pabstia Garay. Orchidaceae

Colchicum L. 1753. Melanthiaceae. 65 Europe (except northern), Mediterranean, Caucasus, West and Middle Asia eastward to Tibet and northern India

Coldenia L. 1753. Boraginaceae. 1 (C. procumbens), or 20 ? trop. and subtrop. Old World

Colea Bojer ex Meisn. 1840. Bignoniaceae. 20 Madagascar, Mascarene Is.

Coleactina N. Hallé. 1970. Rubitaceae. 1 trop. West Africa

Coleanthera Stschegl. 1859. Epacridaceae. 3 western Australia

Coleanthus J. Seidel. 1817. Poaceae. 1 temp. Eurasia, western U.S.

Colebrookea Sm. 1806. Lamiaceae. 1 India, Nepal, Bhutan, Burma, southwestern China, Indochina

Coleocarya S. T. Blake. 1943. Restionaceae. 1 Australia (Queensland)

Coleocephalocereus Backeb. 1938. Cactaceae. 6–10 eastern Brazil

Coleochloa Gilly. 1943. Cyperaceae. 7 trop. and subtrop. Africa, Madagascar

Coleocoma F. Muell. 1857. Asteraceae. 1 northwestern Australia

Coleogyne Torr. 1853. Rosaceae. 1 southwestern U.S.

Coleonema Bartl. et H. L. Wendl. 1824. Rutaceae. 1 South Africa

Coleophora Miers = Daphnopsis Mart. et Zucc. Thymelaeaceae

Coleostachys A. Juss. 1840. Malpighiaceae. 1 northern South America

Coleostephus Cass. 1826 (Chrysanthemum L.). Asteraceae. 2 West Europe, North Africa

Coleotrype C. B. Clarke. 1881. Commelinaceae. 9 Southeast Africa, Madagascar

Coleus Lour. = Plectranthus L'Hér. Lamiaceae

Colignonia Endl. 1837. Nyctaginaceae. 6 Andes of Colombia, Ecuador, Peru, Bolivia, and Argentina

Collabiopsis S. S. Ying. 1977 (or = Collabium Blume). Orchidaceae. 2 Taiwan

Collabium Blume. 1825. Orchidaceae. 8 northeastern India, China, Hainan, Taiwan, Indochina, Malesia, Polynesia

Colladonia DC. = Heptaptera Margot et Reut. Apiaceae

Collaea DC. 1825. Fabaceae. 3 South America

Colletia Comm. ex Juss. 1789. Rhamnaceae. 17 subtrop. and temp. South America

Colletoecema E. Petit. 1963. Rubiaceae. 1 trop. Africa

Colletogyne Buchet. 1939. Araceae. 1 Madagascar

Colliguaja Molina. 1781. Euphorbiaceae. 5 temp. South America

Collinia (Liebm.) Liebm. ex Oerst. = Chamaedorea Willd. Arecaceae

Collinsia Nutt. 1817. Scrophulariaceae. 20 North America, esp. Pacific coasts

Collinsonia L. 1753. Lamiaceae. 4 eastern North America

Collomia Nutt. 1818. Polemoniaceae. 15 western North (from British Columbia to Colorado, Arizona, New Mexico, and Mexico) and South America (from Bolivia to Patagonia)

Collospermum Skottsb. 1934. Asteliaceae. 5 New Zealand, Fiji, and Samoa Is.

Colmeiroa Reut. = Flueggea Willd. Euphorbiaceae

Colobanthera Humbert. 1923. Asteraceae. 1 Madagascar

Colobanthium (Rchb.) G. Taylor = Spenopholis Scribn. Poaceae

Colobanthus Bartl. 1830. Caryophyllaceae. 20 Kerguelen I., Australia, New Zealand, Auckland and Campbell Is., temp. South America, Falkland Is.

Colobanthus (Trin.) Spach = Sphenopholis Scribn. Poaceae

Colobogyne Gagnep. = Acmella Rich. ex Pers. Asteraceae

Colocasia Schott. 1832. Araceae. 13 western India, Sri Lanka, China (8), Indochina, Malesia, Polynesia

Cologania Kunth. 1824. Fabaceae. 10 trop. America from Mexico to Argentina

Colombiana Ospina = Pleurothallis R. Br. Orchidaceae

Colombobalanus Nixon et Crepet = Trigonobalanus Forman. Fagaceae

Colona Cav. 1797. Tiliaceae. 20–30 India, southern China (2), Indochina, Malesia

Colophospermum Kirk ex J. Léonard. 1949. Fabaceae. 1 Angola, Zambia, Malawi, Mozambique, Namibia, Botswana, Zimbabwe, South Africa

Coloradoa Boissev. et Davidson = Sclerocactus Britton et Rose. Cactaceae

Colpias E. Mey. ex Benth. 1836. Scrophulariaceae. 1 South Africa

Colpodium Trin. 1822. Poaceae. 5 montane East Africa, Southeast Asia; C. versicolor—extending to Caucasus

Colpogyne B. L. Burtt. 1971. Gesneraceae. 1 Madagascar

Colpoon Bergius. 1767. Santalaceae. 1–2 South Africa

Colpothrinax Griseb. et H. A. Wendl. 1879. Arecaceae. 2 Guatemala and Panama (1), Cuba (1)

Colquhounia Wall. 1822. Lamiaceae. 6 Himalayas from Kashmir to Assam, Burma, southwestern China, Indochina

Colubrina Rich. ex Brongn. 1827. Rhamnaceae. 31 southern U.S., Mexico, Central and trop. South America, West Indies; East Africa (1, C. asiatica—Kenya, Tanzania, Mozambique), islands of the Indian (from Madagascar and Seychelles to New Guinea and Queensland) and Pacific Oceans eastward to Hawaii; also Burma, Thailand, Laos, Cambodia, Vietnam, southern China (Yunnan)

Columellia Ruiz et Pav. 1794. Columelliaceae. 4 Andes from Colombia to Bolivia

Columnea L. 1753. Gesneriaceae. 160 trop. America, West Indies

Coluria R. Br. 1823. Rosaceae. 4–7 south of West and East Siberia (1, C. geoides), Mongolia, China

Colutea L. 1753. Fabaceae. 26–30 Mediterranean, Southeast Europe, Caucasus, Middle Asia, Northeast and East Africa (C. abyssinica), Arabian Peninsula, Iraq, Iran, Afghanistan, Pakistan, China, India, Nepal

Coluteocarpus Boiss. 1841. Brassicaceae. 1 (C. vesicaria) Caucasus, Asia Minor, western Iran, Syria, Lebanon, Iraq, Iran

Colvillea Bojer ex Hook. 1834. Fabaceae. 1 Madagascar

Comaclinium Schweidw. et Planch. 1852. Asteraceae. 1 Central America

Comandra Nutt. 1818. Santalaceae. 1 (C. umbellata) Southeast Europe, Asia Minor, North America ?

Comanthera L. B. Sm. = Syngonanthus Ruhland. Eriocaulaceae

Comanthosphacea S. Moore. 1877. Lamiaceae. 5 southeastern China, Taiwan, Japan

Comarostaphylis Zucc. 1837 (Arctostaphylos Adans.). Ericaceae. 11 America from southwestern U.S. and Mexico to Panama

Comarum L. 1753. (Potentilla L.). Rosaceae. 5 temp. Northern Hemisphere

Comastoma (Wettst.) Toyok. 1961 (Gentiana L.). Gentianaceae. 15 temp. Asia, Tibet, Himalayas

Combera Sandwith. 1936. Solanaceae. 2 Andes of Chile and Argentina

Combretocarpus Hook. f. 1865. Anisophylleaceae. 1 Malesia: Sumatra, Karimun Is., Banka and Billiton Is., Borneo

Combretodendron A. Chev. ex Exell = Petersianthus Merr. Lecythidaceae

Combretum Loefl. 1758. Combretaceae. c. 250 pantropics, incl. Australia (1)

Comesperma Labill. 1806 (Bredemeyera Willd.). Polygalaceae. 24 Australia

Cometes L. 1767. Caryophyllaceae. 2 deserts of Egypt, Ethiopia, Israel, Jordan and Arabian Peninsula eastward to northwestern India

Cometia Thouars ex Baill. = Drypetes Vahl. Euphorbiaceae

Cominsia Hemsl. 1891. Marantaceae. 5 Moluccas, New Guinea, Solomon Is.

Comiphyton Floret. 1974. Rhizophoraceae. 1 Gabon, Congo, eastern Zaire

Commelina L. 1753. Commelinaceae. 170–250 tropics and subtropics

Commelinantia Tharp = Tinantia Scheidw. Commelinaceae

Commelinidium Stapf. 1920 (Acroceras Stapf). Poaceae. 3 trop. West Africa

Commelinopsis Pichon = Commelina L. Commelinaceae

Commersonia J. R. Forst. et G. Forst. 1775. Sterculiaceae. 10 Southeast Asia, southern China, Malesia to Pacific Is., Australia

Commicarpus Standl. 1909 (Boerhavia L.). Nyctaginaceae. 25–30 tropics and subtropics, chiefly arid trop. Northeast Africa and trop. Arabian Peninsula

Commidendrum Burchell ex DC. 1833. Asteraceae. 4 St. Helena

Commiphora Jacq. 1797. Burseraceae. 200 South America, trop. Africa, Madagascar, Mascarene Is., Arabian Peninsula eastward to western India

Commitheca Bremek. 1940. Rubiaceae. 1 trop. West Africa

Comocladia P. Browne. 1756. Anacardiaceae. 20 Central America, West Indies

Comolia DC. 1828. Melastomataceae. 30 trop. South America

Comoliopsis Wurdack. 1984. Melastomataceae. 1 Venezuela

Comopyrum (Jaub. et Spach) A. Löve = Aegilops L. Poaceae

Comoranthus Knobl. 1934. Oleaceae. 3 Madagascar, Comoro Is.

Comospermum Rauschert. 1982 (Anthericum L.). Asphodelaceae. 2 Japan

Comparettia Poepp. et Endl. 1836. Orchidaceae. 7 trop. America from Mexico to Brazil, West Indies

Comperia K. Koch. 1849 (Orchis L.). Orchidaceae. 1 eastern Mediterranean, Southwest Asia to Iran

Complaya Strother. 1991. Asteraceae. 3–4 America: U.S. (Hawaii, Florida), Mexico, Guatemala, Belize, Honduras, Nicaragua, Costa Rica, Panama, Colombia, Venezuela, Brazil, West Indies (2); trop. Old World (1, C. chinensis)

Compsoneura (DC.) Warb. 1896. Myristicaceae. 11 trop. America

Comptonanthus B. Nord. = Ifloga Cass. Asteraceae

Comptonella Baker f. 1921. Rutaceae. 8 New Caledonia

Comptonia L'Hér. ex Aiton. 1789 (Myrica L.). Myricaceae. 1 eastern North America

Comularia Pichon. 1953. Apocynaceae. 1 trop. West Africa

Conamomum Ridl. = Amomum Roxb. Zingiberaceae

Conandrium (K. Schum.) Mez. 1902. Myrsinaceae. 9 Moluccas, Aru Is., New Guinea, Admiralty Is., New Britain, New Ireland

Conandron Siebold et Zucc. 1843. Gesneriaceae. 1 (C. ramondioides) China, Taiwan, Japan, Indochina

Conanthera Ruiz et Pav. 1802. Tecophilaeacea. 8 Chile

Conceveiba Aubl. 1775. Euphorbiaceae. 7 trop. South America

Conceveibastrum (F. Muell.) Pax et Hoffm. = Conceveiba Aubl. Euphorbiaceae

Conceveibastrum Steyerm. = Conceveiba Aubl. Euphorbiaceae

Conceveibum A. Rich. ex A. Juss. = Conceveiba Aubl. Euphorbiaceae

Conchidium Griff. = Eria Lindl. Orchidaceae

Conchopetalum Radlk. 1887. Sapindaceae. 1 Madagascar

Conchophyllum Blume = Dischidia R. Br. Asclepiadaceae

Condalia Cav. 1799. Rhamnaceae. 18 trop. and subtrop. America

Condaliopsis (Weberb.) Suess. = Ziziphus Mill. Rhamnaceae

Condaminea DC. 1830. Rubiaceae. 3 Andes from Costa Rica to Bolivia

Condylago Luer. 1982. Orchidaceae. 1 Colombia

Condylidium R. M. King et H. Rob. 1972

(Eupatorium L.). Asteraceae. 2 trop. America

Condylocarpon Desf. 1822. Apocynaceae. 7 trop. America from Nicaragua to Brazil

Condylocarpus Hoffm. = Tordylium L. Apiaceae

Condylopodium R. M. King et H. Rob. 1972 (Eupatorium L.). Asteraceae. 4 Colombia

Condylostylis Piper = Vigna Savi. Fabaceae

Congdonia Muell. Arg. = Declieuxia Kunth. Rubiaceae

Congea Roxb. 1820. Symphoremataceae. 7–10 Bangladesh, northeastern India, Burma, southern China, Indochina, Java, Sumatra, Borneo

Congolanthus A. Raynal. 1968. Gentianaceae. 1 trop. Africa

Conicosia N. E. Br. 1925. Aizoaceae. 3 western South Africa

Conimitella Rydb. 1905 (Heuchera L.). Saxifragaceae. 1 U.S. (Montana, Idaho, Wyoming, and Colorado)

Conioselinum Hoffm. 1814. Apiaceae. 10–12 temp. Northern Hemisphere; Russia (6—forest areas)

Conium L. 1753. Apiaceae. 7 North America, Mediterranean, trop. and South Africa, Eurasia; C. maculatum—European part of Russia, Caucasus, West Siberia, Middle Asia

Connarus L. 1754. Connaraceae. c. 120: c. 50 trop. America (Mexico, Guatemala, Honduras, Belize, Nicaragua, Costa Rica, Panama, Colombia, Ecuador, Peru, Venezuela, the Guyanas, Bolivia, Brazil) and West Indies; trop. Africa, trop. Asia, Malesia (from Malay Peninsula to New Guinea and New Britain), Australia, Micronesia, Melanesia

Connellia N. E. Br. 1910. Bromeliaceae. 4 Venezuela, Guyana, Brazil

Conobea Aubl. 1775. Scrophulariaceae. 7 America, West Indies

Conocalyx Benoist. 1967. Acanthaceae. 1 Madagascar

Conocarpus L. 1753. Combretaceae. 2: C. erectus—Florida, trop. America, West Indies; C. lancifolius—trop. West, Northeast, and East Africa; Arabian Peninsula

Conocephalus Blume = Poikilospermum Zipp. ex Miq. Cecropiaceae

Conocliniopsis R. M. King et H. Rob. 1972 (Eupatorium L.). Asteraceae. 1 Colombia, Venezuela, Brazil

Conoclinium DC. 1836. Asteraceae. 3 eastern U.S., Mexico

Conomitra Fenzl. 1839 (Glossonema Decne.). Asclepiadaceae. 1 Sudan

Conomorpha A. DC. = Cybianthus Mart. Myrsinaceae

Conopharyngia G. Don = Tabernaemontana L. Apocynaceae

Conopholis Wallr. 1825. Orobanchaceae. 2 America from southeastern U.S. to Panama

Conophyllum Schwantes = Mitrophyllum Schwantes. Aizoaceae

Conophytum N. E. Br. 1922. Aizoaceae. 290 southern Namibia, southwestern South Africa

Conopodium W. D. J. Koch. 1824 (Bunium L.). Apiaceae. 20 Southwest and South Europe, Northwest and North Africa

Conospermum Sm. 1798. Proteaceae. 38 Australia and Tasmania

Conostalix (Schltr.) Brieger = Dendrobium Sw. Orchidaceae

Conostegia D. Don. 1823. Melastomataceae. 50 trop. America from Central America southward to Peru, West Indies

Conostephium Benth. 1837. Epacridaceae. 5 western and southern Australia

Conostomium (Stapf) Cufod. 1948. Rubiaceae. 9 Sudan, Ethiopia, Kenya, Uganda and South Africa

Conostylis R. Br. 1810. Conostylidaceae. 45 southwestern Australia

Conothamnus Lindl. 1839. Myrtaceae. 3 southwestern Australia

Conradina A. Gray. 1870. Lamiaceae. 5 southeastern U.S.

Conringia Heist. ex Fabr. 1759. Brassicaceae. 8 Europe, Mediterranean, Caucasus, West Siberia, Middle Asia, Asia Minor, Iran, Afghanistan, Pakistan, China, Mongolia

Consolea Lem. = Opuntia Hill. Cactaceae

Consolida Brunsfel ex Gray. 1821 (Delphinium L.). Ranunculaceae. c. 43 Europe (12, except northern areas), Mediterranean, West Asia eastward to West Siberia, Middle Asia, and China

Constantia Barb. Rodr. 1877. Orchidaceae. 3 Brazil

Convallaria L. 1753. Convallariaceae. 1 (C. majalis) cold and temp. Northern Hemisphere from Atlantic Europe and western Mediterranean to Japan, southeastern U.S.

Convolvulus L. 1753. Convolvulaceae. c. 250 warm–temp. and arid regions of Eurasia, Mediterranean, western North and Andean South America, tropics and subtropics

Conyza Less. 1832. Asteraceae. 50 tropics and subtropics

Conyzanthus Tamamsch. = Aster L. Asteraceae

Conzattia Rose. 1909. Fabaceae. 3 Mexico

Coombea R. Royen = Medicosma Hook. f. Rutaceae

Cooperia Herb. = Zephyranthes Herb. Amaryllidaceae

Coopernookia Carolin. 1968. Goodeniaceae. 6 southwestern and southeastern Australia, Tasmania

Copaiba Adans. = Copaifera L. Fabaceae

Copaifera L. 1762. Fabaceae. 5 trop. Africa from Sierra Leone to Angola and Zaire

Copedesma Gleason. 1925 (Miconia Ruiz et Pav.). Melastomataceae. 1 northeastern South America

Copernicia Mart. ex Endl. 1837. Arecaceae. 25 Cuba (20), Haiti (2), South America (3) from Colombia and Venezuela to Argentina

Copiapoa Britton et Rose. 1922. Cactaceae. 1,520 coastal northern Chile

Coprosma J. R. Forst. et G. Forst. 1775. Rubiaceae. 90 Malesia, Australia, Tasmania, Lord Howe I., New Zealand, Norfolk and Kermadec Is. eastward to Hawaiian Is. and Juan Fernández Is., ? Chile

Coptis Salisb. 1807. Ranunculaceae. 16 arctic and temp. Northern Hemisphere; C. trifolia—East Siberia, Far East of Russia

Coptocheile Hoffmanns. = Gesneria L. Gesneriaceae

Coptophyllum Korth. 1850. Rubiaceae. 4 Nicobar Is. (1, C. nicobaricum), western Malesia

Coptosapelta Korth. 1851. Rubiaceae. 13 southeastern China, Taiwan, Ryukyu Is., Indochina, Malesia

Coptosperma Hook. f. = Tarrena Gaertn. Rubiaceae

Corallocarpus Welw. ex Benth. et Hook. f. 1867. Cucurbitaceae. 17 trop. and South Africa, Madagascar, Southwest Asia eastward to India

Corallodiscus Batalin. 1892. Gesneriaceae. 11 Asia from northwestern China to Himalayas, Indochina

Corallorhiza Châtel. 1760. Orchidaceae. 15 temp. Eurasia and America from Canada to Honduras; Russia (1, C. trifida)

Corallospartium J. B. Armstr. 1881. Fabaceae. 1 New Zealand

Corbassona Aubrév. = Niemeyera F. Muell. Sapotaceae

Corbichonia Scop. 1777. Molluginaceae. 2 southwest and trop. Africa (1), trop. Asia (1)

Corchoropsis Siebold et Zucc. 1843. Tiliaceae. 3–4 East Asia

Corchorus L. 1753. Tiliaceae. c. 40 tropics and subtropics

Cordanthera L. O. Williams. 1941 (or = Stellilabium Schltr.). Orchidaceae. 1 Colombia

Cordeauxia Hemsl. 1907. Fabaceae. 1 (C. edulis) Ethiopia, Somalia

Cordemoya Baill. 1861. Euphorbiaceae. 1 Mascarene Is.

Cordia L. 1753. Boraginaceae. c. 300 tropics and subtropics, esp. America

Cordiglottis J. J. Sm. 1922. Orchidaceae. 6 Thailand, Malay Peninsula, Sumatra

Cordisepalum Verdc. 1971. Convolvulaceae. 1 Indochina

Cordobia Nied. 1912. Malpighiaceae. 2 South America

Cordyla Lour. 1790. Fabaceae. 9 trop. Africa (9), Madagascar

Cordylanthus Nutt. ex Benth. 1846. Scrophulariaceae. 18 western U.S. (Washington, Montana, Wyoming, California, Oregon, Nevada, Idaho, Utah, Arizona, New Mexico) and adjacent Mexico (mainly Baja California)

Cordyline Comm. ex R. Br. 1810. Asteliaceae. 20 trop. and subtrop. Africa; South, Southeast, and East Asia; Malesia; eastern Australia; New Zealand; trop. South America

Cordyloblaste Henschel ex Moritzi = Symplocos Jacq. Symplocaceae

Cordylocarpus Desf. 1798. Brassicaceae. 1 Morocco, Algeria, Arabian Peninsula, southern Israel (coastal Negev)

Cordylogyne E. Mey. 1838. Asclepiadaceae. 1 South Africa

Corema D. Don. 1826. Empetraceae. 2 eastern North America, Azores and Canary Is., Iberian Peninsula

Coreocarpus Benth. 1844. Asteraceae. 9 southwestern U.S. (Arizona), northwestern Mexico and Baja California

Coreopsis L. 1753. Asteraceae. c. 100 North and South America, Africa

Corethamnium R. M. King et H. Rob. 1978. Asteraceae. 1 Colombia

Corethrodendron Fisch. et Basiner = Hedysarum L. Fabaceae

Corethrogyne DC. 1836. Asteraceae. 3 California

Coriandropsis H. Wolff = Coriandrum L. Apiaceae

Coriandrum L. 1753. Apiaceae. 2–3 Mediterranean, Southwest Asia; C. sativum—European part of Russia, Caucasus, Middle Asia, and southern Far East of Russia

Coriaria L. 1753. Coriariaceae. 20 (or 5 ?) western Mediterranean, Asia from temp. and subtrop. Himalayas to Japan and New Guinea, New Zealand, southern Oceania, America from Mexico to Chile

Coridothymus Rchb. f. 1857 (Thymus L.). Lamiaceae. 1 Mediterranean from Iberian Peninsula and Morocco eastward to Israel

Coris L. 1753. Primulaceae. 1–2 Mediterranean, except northeastern part (1, C. monspeliensis), Somalia

Corispermum L. 1753. Chenopodiaceae. c. 70 North America (Canada, U.S.), Central, South, and East Europe; Caucasus; Middle Asia; West and East Siberia; Far East of Russia; Asia Minor; Iran; Afghanistan; Pakistan; China; Mongolia

Coristospermum Bertol. 1837. Apiaceae. 3 South Europe

Cormonema Reissek ex Endl. = Colubrina Rich. ex Brongn. Rhamnaceae

Cormus Spach. 1834 (Sorbus L.). Rosaceae. 1 southern Europe eastward to Asia Minor

Cornera Furtado = Calamus L. Arecaceae

Cornucopiae L. 1753. Poaceae. 2 Eastern Mediterranean to Iraq

Cornuella Pierre = Chrysophyllum L. Sapotaceae

Cornulaca Delile. 1813–1814. Chenopodiaceae. 9 Algeria, Tunisia, Libya, Egypt, Sudan, Ethiopia, Somalia, Arabian Peninsula, Socotra, Syria, Iraq, Iran, Afghanistan, China; Middle Asia (1, C. koprshinskyi)

Cornus L. 1753. Cornaceae. 45 temp. Northern Hemisphere, a few species in South America and Africa; 1 (C. mas) extending to Ukraine, Crimea, and Caucasus

Cornutia L. 1753. Verbenaceae. 10–15 trop. America, West Indies

Corokia A. Cunn. 1839. Argophyllaceae. 4–6 Australia, New Zealand, Polynesia, Rapa I.

Corollonema Schltr. 1914. Asclepiadaceae. 1 Bolivia

Coronanthera Vieill. ex C. B. Clarke. 1883. Gesneriaceae. 10 northeastern Australia (1), New Caledonia

Coronaria Hill = Lychnis L. Caryophyllaceae

Coronilla L. 1753. Fabaceae. 20 Europe, Mediterranean, Caucasus, Middle Asia, Iraq, Iran

Coronopus Zinn. 1757. Brassicaceae. 10 almost cosmopolitan

Corothamnus Presl = Cytisus L. Fabaceae

Coroya Pierre = Dalbergia L. Fabaceae

Corozo Jacq. ex Giseke = Elaeis Jacq. Arecaceae

Corpuscularia Schwantes. 1926 (or = Delosperma N. E. Br.). Aizoaceae. 3–5 southeastern South Africa

Correa Andrews. 1798. Rutaceae. 7 temp. Australia

Correllia A. M. Powell. 1973. Asteraceae. 1 Mexico

Correlliana D'Arcy. 1973 (Cybianthus Mart.). Myrsinaceae. 4 trop. America

Corrigiola L. 1753. Caryophyllaceae. 13 South America (Andes of Colombia, Chile), Central Europe, Mediterranean, Africa, Madagascar

Corryocactus Britton et Rose. 1920. Cactaceae. 15–20 Peru, Bolivia, Chile

Corsia Becc. 1878. Corsiaceae. 26 New Guinea, Solomon Is. (1), northern Australia (1)

Cortaderia Stapf. 1897. Poaceae. 25 New Guinea (1), New Zealand (4), South America

Cortesia Cav. 1798. Ehretiaceae. 2 temp. South America

Cortia DC. 1830. Apiaceae. 10 eastern Afghanistan, western China, Himalayas from Kumaun to Bhutan, Tibet

Cortiella C. Norman. 1937. Apiaceae. 5 Himalayas (Nepal, Bhutan), Tibet

Cortusa L. 1753. Primulaceae. 8–10 Eurasia from montane Central Europe to West and East Siberia, Far East of Russia, northwestern and northern China (1, C. matthioli), Mongolia, Korea, Japan

Corunastylis Fitzg. = Prasophyllum R. Br. Orchidaceae

Coryanthes Hook. 1831. Orchidaceae. 17 trop. America from Guatemala to Peru, Bolivia, and Brazil, also in Trinidad and Tobago

Corybas Salisb. 1807. Orchidaceae. c. 160 India, southern China, Taiwan, Indochina, Malesia, Solomon Is., Vanuatu, Australia, New Caledonia (1), New Zealand, southern Oceania

Corycium Sw. 1800. Orchidaceae. 14 trop. and South (14) Africa, esp. Cape (10)

Corydalis Vent. 1805. Fumariaceae. 250–300 temp. Eurasia (esp. Irano-Turanian region, China, Himalayas), Algeria, montane trop. East Africa, North America

Corylopsis Siebold et Zucc. 1836–1838. Hamamelidaceae. 29 East Asia from Tibet and Bhutan to Japan

Corylus L. 1753. Betulaceae. 20 temp. Northern Hemisphere, esp. (c. 10) European part of Russia, Caucasus, southern East Siberia, and Far East of Russia

Corymbium L. 1753. Asteraceae. 9 South Africa (southwestern Cape)

Corymborkis Thouars. 1809. Orchidaceae. 6–7 (or 32) trop. and South Africa, Madagascar, Mascarene Is., trop. Asia from India and Sri Lanka to southern China and Indochina, Malesia, northern Australia, Oceania eastward to Fiji, trop. America from Mexico to Colombia and souteastern Brazil, West Indies

Corymbostachys Lindau. 1897 (Justicia L.). Acanthaceae. 1 Madagascar

Corynabutilon (K. Schum.) Kearney = Abutilon Hill. Malvaceae

Corynaea Hook. f. 1856. Helosiaceae. 1 trop. America

Corynanthe Welw. 1869. Rubiaceae. 8 trop. Africa

Corynanthera J. W. Green. 1979. Myrtaceae. 1 western Australia

Corynella DC. 1825. Fabaceae. 2 Haiti and Puerto Rico

Corynemyrtus (Kiaersk.) Mattos. 1963 (or = Myrtus L.). Myrtaceae. 1 Brazil

Corynephorus P. Beauv. 1812. Poaceae. 6 Europe, Mediterranean, Southwest Asia to eastern Caucasus

Corynephyllum Rose = Sedum L. Crassulaceae

Corynocarpus J. R. Forst. et G. Forst. 1776. Corynocarpaceae. 7 Aru Is., New Guinea, Solomon Is., Australia (Queensland, New South Wales), New Zealand, New Caledonia, Vanuatu

Corynopuntia F. Knuth = Opuntia Hill. Cactaceae

Corynostylis Mart. 1823. Violaceae. 4 trop. America

Corynotheca F. Muell. ex Benth. 1878 (Caesia R. Br.). Asphodelaceae. 6 trop. and temp. (esp. western) Australia

Corynula Hook. f. 1872 (Leptostigma Arn.). Rubiaceae. 2 Colombia

Corypha L. 1753. Arecaceae. 8 from southern India, Sri Lanka to the Bay of Bengal, and Indochina through Malesia to northern Australia

Coryphantha (Engelm.) Lem. 1868. Cactaceae. c. 45 southwestern U.S., Mexico, Cuba

Coryphomia Rojas Acosta = Copernicia Mart. ex Endl. Arecaceae

Coryphothamnus Steyerm. 1965. Rubiaceae. 1 southeastern Venezuela

Corythea S. Watson = Acalypha L. Euphorbiaceae

Corytholoma (Benth.) Decne. = Sinningia Nees. Gesneriaceae

Corythophora R. Knuth. 1939 Lecythidaceae. 4 Surinam, French Guyana, Brazil (Amazonia)

Corytoplectus Oerst. = Alloplectus Mart. Gesneriaceae

Coscinium Colebr. 1821. Menispermaceae. 2 southern India, Sri Lanka, Cambodia, Vietnam, Peninsular Thailand, Malay Peninsula, Banka I., Sumatra, Java, Borneo

Cosmea Willd. = Cosmos Cav. Asteraceae

Cosmelia R. Br. 1810. Epacridaceae. 1 southwestern Australia

Cosmianthemum Bremek. 1960. Acanthaceae. 9 Thailand, Malay Peninsula, Borneo

Cosmibuena Ruiz et Pav. 1802. Rubiaceae. 12 Mexico, Central and trop. South America

Cosmocalyx Standl. 1930. Rubiaceae. 1 Mexico

Cosmos Cav. 1791. Asteraceae. c. 30 trop. and subtrop. America, esp. Mexico, West Indies

Cosmostigma Wight. 1834. Asclepiadaceae. 3 India, Sri Lanka, Burma, southern China, Hainan I.

Cossignia Comm. ex Juss. = Cossinia Comm. ex Lam. Sapindaceae

Cossinia Comm. ex Lam. 1786. Sapindaceae. 4 Mascarene Is., eastern Australia, New Caledonia, Fiji

Costaea A. Rich. = Purdiaea Planch. Cyrillaceae

Costantina Bullock. 1965. Asclepiadaceae. 1 Indochina

Costarica L. D. Gomez = Sicyos L. Cucurbitaceae

Costera J. J. Sm. 1910 (Vaccinium L.). Ericaceae. 9 Sumatra, Borneo, Philippines

Costularia C. B. Clarke. 1898. Cyperaceae. 20 South Africa, Madagascar, Seychelles and Mascarene Is., Borneo, New Guinea, ? Australia, New Caledonia

Costus L. 1753. Costaceae. c. 70 trop. America (c. 40), Africa (25), trop. Asia, Australia (1, C. potierae—Queensland)

Cotinus Hill. 1753 (Rhus L.). Anacardiaceae. 5 southeastern U.S. (1, C. obovatus), Eurasia from South Europe, West Asia eastward to western Himalayas (1, C. coggygria), China (3)

Cotoneaster Medik. 1789. Rosaceae. 264 temp. Northern Hemisphere, esp. China (c. 135), montane trop. Asia

Cotopaxia Mathias et Constance. 1952. Apiaceae. 2 Colombia, Ecuador

Cottea Kunth. 1829. Poaceae. 1 southern U.S., Central and South America

Cottendorfia Schult. f. 1830. Bromeliaceae. 25 Venezuela, northeastern Brazil

Cottonia Wight. 1851. Orchidaceae. 2 southwestern India, Sri Lanka

Cotula L. 1753. Asteraceae. c. 80 chiefly South Africa and New Zealand, some in North Africa, Asia, New Guinea, Australia, South America and Falkland Is.

Cotylanthera Blume. 1826. Gentianaceae. 4 Himalayas (Nepal, Bhutan), Assam, southwestern China, Burma, Thailand, Java, West Pacific Islands

Cotyledon L. 1753. Crassulaceae. 9 South Africa; C. barbeyi—extending along East

Africa to Ethiopia and southwestern Arabian Peninsula

Cotylelobium Pierre. 1890. Dipterocarpaceae. 6 Sri Lanka, southern Thailand, Malay Peninsula, Borneo

Cotylodiscus Radlk. 1878 (Plagioscyphus Radlk.). Sapindaceae. 1 Madagascar

Cotylolabium Garay = Stenorrhynchos Rich. ex Spreng. Orchidaceae Cotylonychia Stapf. 1908. Sterculiaceae. 1 trop. Africa

Couepia Aubl. 1775. Chrysobalanaceae. 68 trop. America from Mexico to southern Brazil

Coula Baill. 1862. Olacaceae. 1 trop. West Africa

Coulterella Vasey et Rose. 1890. Asteraceae. 1 Mexico (California)

Coulterophytum B. L. Rob. 1892. Apiaceae. 5 Mexico

Couma Aubl. 1775. Apocynaceae. 15 Guyana, Brazil

Coumarouna Aubl. = Dipteryx Schreb. Fabaceae

Couratari Aubl. 1775. Lecythidaceae. 19 Costa Rica and Panama (2), Colombia, Venezuela, Guyana, Surinam, French Guyana, Peru, Brazil, and Bolivia

Couroupita Aubl. 1775. Lecythidaceae. 3 trop. America from El Salvador, Nicaragua, and Panama through Colombia to Ecuador, Peru, and Amazonian Brazil

Coursetia DC. 1825. Fabaceae. 20 trop. and subtrop. America from southwestern U.S. and Mexico to Peru and Brazil

Coursiana Homolle. 1942 (Schismatoclada Baker). Rubiaceae. 1–2 Madagascar

Courtenia Bennett et R. Br. 1844. Sterculiaceae. 3 trop. Africa

Courtoisia Nees = Courtoisina Soják. Cyperaceae

Courtoisina Soják. 1980. Cyperaceae. 2 Africa, Madagascar, India, southern China

Cousinia Cass. 1827. Asteraceae. 580 from eastern Mediterranean to Mongolia, China and western Himalayas, north to Caucasus and Middle Asia

Cousiniopsis Nevski. 1937. Asteraceae. 1 Middle Asia, ? northern Iran and northern Afghanistan

Coussapoa Aubl. 1775. Cecropiaceae. 46 trop. America from Mexico to Peru, Bolivia, and Brazil; not in West Indies

Coussarea Aubl. 1775. Rubiaceae. 100 trop. America, West Indies

Coutaportla Urb. 1923 (Portlandia P. Browne). Rubiaceae. 2 Mexico

Coutarea Aubl. 1775. Rubiaceae. 7 trop. America from Mexico to Argentina, West Indies

Coutinia Vell. Conc. = Aspidosperma Mart. et Zucc. Apocynaceae

Coutoubea Aubl. 1775. Gentianaceae. 5 Mexico, through Central America to Venezuela, Guyana, Brazil, Lesser Antilles

Covillea Vail = Larrea Cav. Zygophyllaceae

Cowania D. Don ex Tilloch et Taylor. 1825. Rosaceae. 5 southwestern North America

Cowiea Wernham = Petunga DC. Rubiaceae

Coxella Cheesman et Hemsl. 1911 (Aciphylla J. R. Forst. et G. Forst.). Apiaceae. 1 Chatham Is.

Crabbea Harv. 1842. Acanthaceae. 12 trop. and South Africa

Cracca Benth. 1853. Fabaceae. 15–20 trop. and subtrop. America from southwestern U.S. to Argentina

Cracosna Gagnep. 1929. Gentianaceae. 1 Indochina

Craibia Dunn. 1911. Fabaceae. 10 trop. Africa

Craibiodendron W. W. Sm. 1911. Ericaceae. 7 northeastern India, China, Indochina

Craigia W. W. Sm. et W. E. Evans. 1921. Tiliaceae. 2 southwestern China (Kwangtung and Yunnan)

Crambe L. 1753. Brassicaceae. 28 Europe, Mediterranean, Macaronesia, montane areas of trop. Africa, West Asia to Caucasus, West Siberia and Middle Asia, eastward to western Himalayas

Crambella Maire. 1924. Brassicaceae. 1 Morocco

Cranichis Sw. 1788. Orchidaceae. 34 trop. America from Mexico and Florida to Argentina, West Indies

Craniolaria L. 1753. Martyniaceae. 3 South America

Craniospermum Lehm. 1818. Boraginaceae. 4–5 West and East Siberia (4), Mongolia, China

Craniotome Rchb. 1825. Lamiaceae. 1 western Pakistan, Himalayas, northern India, Burma, southwestern China, Indochina

Cranocarpus Benth. 1859. Fabaceae. 3 Brazil

Craspedia G. Forst. 1786. Asteraceae. c. 15 temp. Australia, Tasmania, New Zealand

Craspedolobium Harms. 1921. Fabaceae. 1 western China

Craspedorhachis Benth. 1882. Poaceae. 2 trop. South Africa

Craspedospermum Airy Shaw = Craspidospermum Bojer ex A. DC. Apocynaceae

Craspedostoma Domke = Gnidia L. Thymelaeaceae

Craspidospermum Bojer ex A. DC. 1844. Apocynaceae. 1 Madagascar

Crassocephalum Moench. 1794. Asteraceae. 30 trop. and South Africa, Madagascar

Crassula L. 1753. Crassulaceae. c. 200 Europe, trop. East and South Africa (c. 150), Madagascar, Arabian Peninsula, Australia, America

Crataegus L. 1753. Rosaceae. c. 265 temp. Northern Hemisphere, esp. North America (c. 200); Russia (c. 80, from European part to Far East)

Crateranthus Baker f. 1913. Lecythidaceae. 3 trop. West Africa

Craterispermum Benth. 1849. Rubiaceae. 17 trop. Africa, Madagascar, Seychelles

Craterocapsa Hillard et B. L. Burtt. 1973 (Wahlenbergia Roth). Campanulaceae. 4 trop. and South Africa

Craterogyne Lanj. = Dorstenia L. Moraceae

Craterosiphon Engl. et Gilg. 1894 (Synaptolepis Oliv.). Thymelaeaceae. 9 trop. Africa

Craterostemma K. Schum. = Brachystelma R. Br. Asclepiadaceae

Craterostigma Hochst. 1841. Scrophulariaceae. c. 20 trop. and South Africa, Madagascar, Arabian Peninsula, India

Crateva L. 1753. Capparaceae. 8 tropics, except New Caledonia

Cratoxylum Blume. 1823. Hypericaceae. 6 eastern India, Burma, southern China, Indochina, western Malesia

Cratylia Mart. ex Benth. 1837. Fabaceae. 5–8 South America from northern Brazil to northern Argentina

Cratystylis S. Moore. 1905. Asteraceae. 3 Australia

Crawfurdia Wall. (1826) = Tripterospermum Blume (1826). Gentianaceae

Creaghiella Stapf. 1896 (Anerincleistus Korth.). Melastomataceae. 3 Borneo, Philippines

Creatantha Standl. = Isertia Schreb. Rubiaceae

Cremanthodium Benth. 1873. Asteraceae. 64 China (64), Taiwan, Himalayas

Cremaspora Benth. 1849. Rubiaceae. 3–4 trop. Africa, Madagascar, Comoro Is.

Cremastogyne (H. Winkler) Czerep. = Betula L. Betulaceae

Cremastopus Paul G. Wilson. 1962. Cucurbitaceae. 3 Mexico

Cremastosperma R. E. Fr. 1931. Annonaceae. 18 trop. South America

Cremastra Lindl. 1833. Orchidaceae. 3 Nepal, northeastern India, Thailand, China, Taiwan, Japan

Cremastus Miers = Arrabidaea DC. Bignoniaceae

Crematosciadium Rech. f. Eriocycla Lindl. Apiaceae

Cremnophila Rose = Sedum L. Crassulaceae

Cremnophyton Brullo et Pavone. 1987 (Atriplex L.). Chenopodiaceae. 1 Malta I.

Cremocarpon Baill. 1879. Rubiaceae. 1 Comoro Is.

Cremolobus DC. 1821. Brassicaceae. 7 Andes of South America

Cremosperma Benth. 1846. Gesneriaceae.

23 trop. America from Costa Rica and Panama to Ecuador and Peru

Crenea Aubl. 1775. Lythraceae. 2 Colombia, Venezuela, Trinidad, Guyana, northeastern Brazil

Crenias Spreng. 1827 (Mniopsis Mart. et Zucc.). Podostemaceae. 5 southeastern Brazil

Crenidium Haegi. 1981. Solanaceae. 1 southwestern Australia

Crenosciadium Boiss. et Heldr. ex Boiss. 1849 (or = Opopanax W. D. J. Koch). Apiaceae. 1 Turkey

Creochiton Blume. 1821. Melastomataceae. 6 Malesia

Crepidiastrum Nakai. 1920. Asteraceae. 7 East Asia

Crepidopsis Arv. = Touv. 1897 (Hieracium L.). Asteraceae. 1 Mexico

Crepidorhopalon Eb. Fisch. 1989 (Craterostigma Hochst.). Scrophulariaceae. 2 Zaire (1), Zambia (1)

Crepidospermum Hook. f. 1862. Burseraceae. 6 trop. South America

Crepinella Marchal. 1887 (or = Schefflera J. R. Forst. et G. Forst.). Araliaceae. 1 northeastern South America

Crepis L. 1753. Asteraceae. c. 200 Northern Hemisphere, trop. and South Africa, ? southern Australia

Crescentia L. 1753. Bignoniaceae. 6 Mexico, Guatemala, Belize, El Salvador, Honduras, Nicaragua, Costa Rica, Panama, Colombia, Venezuela, Peru, Brazil (Amazonia), West Indies from Bahamas to Windward Is.

Cressa L. 1753. Convolvulaceae. 1–5 tropics and subtropics; C. cretica—extending to Caucasus and Middle Asia

Cribbia Senghas. 1986. Orchidaceae. 1–2 trop. Africa from Liberia to Zaire, Zambia, Kenya, and Uganda

Criciuma Soderstr. et Londoño. 1987. Poaceae. 1 Brazil

Crimaea Vassilcz. = Medicago L. or Trigonella L. Fabaceae

Crinipes Hochst. 1855. Poaceae. 2 Sudan, Ethiopia, Uganda

Crinitaria Cass. = Galatella Cass. Asteraceae

Crinodendron Molina. 1782. Elaeocarpaceae. 4 Bolivia, northern Argentina, central Chile, Brazil (Prov. Santa Catarina)

Crinonia Blume = Pholidota Lindl. ex Hook. Orchidaceae

Crinum L. 1753. Amaryllidaceae. 115 tropics and subtropics, esp. on sea coasts

Crioceras Pierre. 1897 (Tabernaemontana L.). Apocynaceae. 2 trop. Africa

Criosanthes Raf. = Cypripedium L. Orchidaceae

Crispiloba Steenis. 1984. Alseuosmiaceae. 1 Australia (Queensland)

Cristaria Cav. 1799. Malvaceae. c. 80 temp. South America

Cristatella Nutt. = Polanisia Raf. Capparaceae

Critesion Raf. = Hordeum L. Poaceae

Crithmum L. 1753. Apiaceae. 1 (C. maritimum) on coasts of Atlantic islands, West Europe, Mediterranean and coasts of Black Sea, incl. coastal Crimea and Caucasus

Crithopsis Jaub. et Spach. 1851. Poaceae. 1 North Africa, Crete I., Southwest Asia

Critonia P. Browne. 1756 (Eupatorium L.). Asteraceae. 33 trop. America from Mexico to Paraguay, West Indies

Critoniadelphus R. M. King et H. Rob. 1971 (Eupatorium L.). Asteraceae. 2 Mexico, Central America

Critoniella R. M. King et H. Rob. 1975 (Eupatorium L.). Asteraceae. 5 Colombia, Venezuela

Critoniopsis Sch. Bip. 1863 (Vernonia Schreb.). Asteraceae. 26 South America (Andes)

Crobylanthe Bremek. 1940. Rubiaceae. 1 Borneo

Crocanthemum Spach = Helianthemum Hill. Cistaceae

Crocidium Hook. 1833. Asteraceae. 2 northwestern North America

Crockeria Greene ex A. Gray = Lasthenia Cass. Asteraceae

Crocodeilanthe Rchb. f. et Warsc. 1854 (or = Pleurothallis R. Br.). Orchidaceae. 1 Peru

Crocopsis Pax = Stenomesson Herbert. Amaryllidaceae

Crocosmia Planch. 1851–1852 (Tritonia Ker-Gawl.). Iridaceae. 9–11 trop. and South Africa, Madagascar (1)

Crocoxylon Eckl. et Zeyh. = Cassine L. Celastraceae

Crocus L. 1753. Iridaceae. c. 80 North America, Mediterranean, Eurasia (incl. 20 spp. in Crimea, Caucasus, and Middle Asia), eastward to China and western Pakistan

Crocyllis E. Mey. ex Hook. f. 1873. Rubiaceae. 1 South Africa

Croftia Small = Carlowrightia A. Gray. Acanthaceae

Croizatia Steyerm. 1952. Euphorbiaceae. 3 Panama and Colombia (1, C. panamensis), Venezuela (2)

Cromidon Compton. 1931. Selaginaceae. 12 Namibia and South Africa (western and central Cape to Orange Free State)

Cromopanax Grierson. 1991 (Macropanax Miq.). Araliaceae. 1 Bhutan

Cronquistia R. M. King. 1968. Asteraceae. 1 Mexico

Cronquistianthus R. M. King et H. Rob. 1972 (Eupatorium L.). Asteraceae. 16 Andes from Colombia to Peru

Croomia Torr. ex Torr. et A. Gray. 1840. Stemonaceae. 3 Japan (Honshu I.), eastern U.S. (1, C. pauciflora)

Croptilon Raf. 1837 (Haplopappus Cass.). Asteraceae. 3 southern U.S., Mexico

Crossandra Salisb. 1806. Acanthaceae. 50 trop. Africa, Madagascar, Arabian Peninsula, India, Sri Lanka, southern China

Crossandrella C. B. Clarke. 1906. Acanthaceae. 2 trop. Africa

Crosslandia W. Fitzg. 1918. Cyperaceae. 1 western Australia

Crossonephelis Baill. = Glenniea Hook. f. Sapindaceae

Crossopetalum P. Browne. 1756. Celastraceae. c. 40 trop. America, West Indies

Crossopteryx Fenzl. 1839. Rubiaceae. 1 Africa from Senegal to Sudan and Ethiopia south to northern Namibia and Transvaal

Crossosoma Nutt. 1848. Crossosomataceae. 3–4 southwestern U.S., northwestern Mexico

Crossostemma Planch. ex Hook. 1849. Passifloraceae. 2 trop. Africa

Crossostephium Less. 1831 (Artemisia L.). Asteraceae. 1 (C. artemisioides) East Asia

Crossostylis J. R. Forst. et G. Forst. 1775. Rhizophoraceae. 10 Polynesia

Crossothamnus R. M. King et H. Rob. 1972. Asteraceae. 1 Peru

Crossotropis Stapf. 1898 (Trichoneura Andersson). Poaceae. 4 trop. and South Africa

Crotalaria L. 1753. Fabaceae. c. 600 tropics and subtropics, chiefly Africa and Madagascar (c. 500)

Croton L. 1753. Euphorbiaceae. 750–800 tropics and subtropics

Crotonogyne Muell. Arg. 1864. Euphorbiaceae. 1 trop. (excl. East) Africa

Crotonogynopsis Pax. 1899. Euphorbiaceae. 1 trop. Africa: Ivory Coast, Cameroun and Zaire, Uganda, Tanzania

Crotonopsis Michx. 1803. Euphorbiaceae. 2 eastern North America

Crowea Sm. 1798. Rutaceae. 4 Australia

Crucianella L. 1753. Rubiaceae. c. 30 Europe, Mediterranean, West Asia to Iran; Crimea, Caucasus, and Middle Asia

Cruciata Hill. 1753. Rubiaceae. 20 Europe (5), Mediterranean eastward to Iran, Caucasus, Middle Asia, West and East Siberia (1, C. krylovii)

Crucicaryum Brand. 1929. Boraginaceae. 1 New Guinea

Cruckshanksia Hook. et Arn. 1833. Rubiaceae. 7 Chile

Cruddasia Prain. 1898. Fabaceae. 1 Burma, Thailand

Crudia Schreb. 1789. Fabaceae. 55 trop. America (esp. Amazonia), trop. Africa, Southeast Asia, New Guinea

Crumenaria Mart. 1826. Rhamnaceae. 6 Central and South America southward to Argentina

Crunocallis Rydb. = Montia L. Portulacaceae

Crupina (Pers.) DC. 1810. Asteraceae. 3 Central Europe, Mediterranean, Crimea to Caucasus, Middle Asia, Iraq, Iran

Crusea Cham. et Schltdl. 1830. Rubiaceae. 13 southern U.S. (Arizona and New Mexico), Mexico, Central America southward to northern Panama

Cruzia Phil. 1895 (or = Scutellaria L.). Lamiaceae. 1 Patagonia

Crybe Lindl. 1836 (or = Bletia Ruiz et Pav.). Orchidaceae. 1 Central America

Cryophytum N. E. Br. = Mesembryanthemum L. Aizoaceae

Cryosophila Blume. 1838. Arecaceae. 8 trop. America from western Mexico to northern Colombia

Cryphiacanthus Nees. 1841 (Ruellia L.). Acanthaceae. 10 South America

Crypsis Aiton. 1789. Poaceae. 10 Mediterranean, South and East Europe, trop. Africa, West, Southwest to Central Asia

Cryptadenia Meisn. 1841. Thymelaeaceae. 5 South Africa

Cryptandra Sm. 1798. Rhamnaceae. 40 temp. Australia

Cryptangium Schrad. ex Nees = Lagenocarpus Nees. Cyperaceae

Cryptanopsis Ule = Orthophytum Beer. Bromeliaceae

Cryptantha Lehm. ex G. Don. 1837. Boraginaceae. c. 100 western North America, C. albida—extending to Veracruz

Cryptanthemis Rupp = Rhizanthella R. S. Rogers. Orchidaceae

Cryptanthus Otto et A. Dietr. 1836. Bromeliaceae. 21 eastern and southeastern Brazil

Cryptarrhena R. Br. 1816. Orchidaceae. 3–4 trop. America

Crypteronia Blume. 1827. Crypteroniaceae. 4 India (Assam), Burma, southern China, Indochina, Malesia

Cryptocapnos Rech. f. 1968. Fumariaceae. 1 central–southern Afghanistan

Cryptocarpus Kunth. 1817. Nyctaginaceae. 1 Galapagos Is., Andes of Ecuador, Peru, and Bolivia

Cryptocarya R. Br. 1810. Lauraceae. 200–250 tropics (excl. Central Africa) and subtropics

Cryptocentrum Benth. 1883. Orchidaceae. 14 trop. America from Costa Rica to Peru

Cryptocereus Alexander = Selenicereus (A. Berger) Britton et Rose. Cactaceae

Cryptochilus Wall. 1824. Orchidaceae. 2–6 southwestern China, Himalayas, northeastern India, Indochina

Cryptochloa Swallen. 1924. Poaceae. 15 trop. America from Mexico to Brazil

Cryptocodon Fedorov. 1957. Campanulaceae. 1 Middle Asia

Cryptocoryne Fisch. ex Wydl. 1830. Araceae. c. 50 India, Sri Lanka, China, Indochina, Malesia

Cryptodiscus Schrenk ex Fisch. et C. A. Mey. = Prangos Lindl. Apiaceae

Cryptogyne Hook. f. = Sideroxylon L. Sapotaceae

Cryptolepis R. Br. 1810. Asclepiadaceae. 12 trop. Old World

Cryptomeria D. Don. 1838. Taxodiaceae. 2 Japan, southeastern China

Cryptophoranthus Barb. Rodr. 1881 (Pleurothallis R. Br.). Orchidaceae. 3–4 Costa Rica, Panama

Cryptophragmium Nees = Gymnostachyum Nees. Acanthaceae

Cryptophysa Standl. et J. F. Macbr. = Conostegia D. Don. Melastomataceae

Cryptopus Lindl. 1824. Orchidaceae. 3 Madagascar, Mascarene Is.

Cryptopylos Garay. 1972. Orchidaceae. 1 Thailand, Laos, Vietnam, Sumatra

Cryptorhiza Urb. = Pimenta Lindl. Myrtaceae

Cryptorrhynchus Nevski = Astragalus L. Fabaceae

Cryptosepalum Benth. 1865. Fabaceae. 11 trop. Africa

Cryptospora Kar. et Kir. 1842. Brassicaceae. 3 Kazakhstan, Middle Asia, Iran, Afghanistan, western China

Cryptostegia R. Br. 1820. Asclepiadaceae. 2 Africa, Madagascar, India

Cryptostephanus Welw. ex Baker. 1878. Amaryllidaceae. 5 trop. Africa

Cryptostylis R. Br. 1810. Orchidaceae. 12–15 trop. Asia from India and Sri Lanka to Indochina, Malesia from Malay Peninsula to New Guinea, Australia, Solomon Is., Vanuatu, New Caledonia, to Fiji (1, C. arachnites) and Samoa

Cryptotaenia DC. 1829. Apiaceae. 4–7 North America, Macaronesia, Europe, Mediterranean, Transcaucasia (2), trop. Africa, East Asia, Hawaiian Is.

Cryptotaeniopsis Dunn. = Pternopetalum Franch. Apiaceae

Cryptothladia (Bunge) M. J. Cannon. 1984 (Morina L.). Morinaceae. 6 Himalayas, Tibet

Ctenanthe Eichler. 1884 (Myrosma L. f.). Marantaceae. 20 Costa Rica (1), Brazil

Ctenardisia Ducke. 1930. Myrsinaceae. 1 northeastern Brazil

Ctenium Panz. 1813. Poaceae. 20 trop. and subtrop. America, Africa, Madagascar

Ctenocladium Airy Shaw = Dorstenia L. Moraceae

Ctenolepis Hook. f. 1867. Cucurbitaceae. 2 trop. Africa, India

Ctenolophon Oliv. 1873. Ctenolophonaceae. 2: 1 (C. engleriana) trop. West Africa (Gabon, Nigeria, Zaire, Angola); 1 (C. parvifolius) Malay Peninsula, Borneo, Philippines, New Guinea

Ctenomeria Harv. 1842 (Tragia L.). Euphorbiaceae. 2 South Africa

Ctenopaepale Bremek. 1944. Acanthaceae. 1 Java

Ctenophrynium K. Schum. = Saranthe (Regel et Koern.) Eichl. Marantaceae

Ctenopsis De Not. 1847 (Vulpia C. C. Gmel.). Poaceae. 4 South Europe, North Africa, West Asia

Cuatrecasanthus H. Rob. 1989. Asteraceae. 3 trop. America

Cuatrecasasia Standl. = Cuatrecasasiodendron Standl. et Steyerm. Rubiaceae

Cuatrecasasiella H. Rob. 1985. Asteraceae. 2 Ecuador, Peru, Chile, Argentina

Cuatrecasasiodendron Standl. et Steyerm. 1964. Rubiaceae. 2 Colombia

Cuatrecasea Dugand = Iriartella H. A. Wendl. Arecaceae

Cuatresia Hunz. 1977. Solanaceae. 9 Guatemala, Costa Rica, Colombia, Bolivia

Cubacroton Alain. 1960. Euphorbiaceae. 1 Cuba

Cubanola Aiello. 1979. Rubiaceae. 2 Cuba, Haiti

Cubanthus (Boiss.) Millsp. 1913 (Pedilanthus Neck. ex Poit.). Euphorbiaceae. 3 Cuba, Haiti

Cubelium Raf. 1824 (or = Hybanthus Jacq.). Violaceae. 1 eastern U.S.

Cubilia Blume. 1849. Sapindaceae. 1 Philippines, Sulawesi, Moluccas

Cubitanthus Barringer. 1984. Gesneriaceae. 1 Brazil

Cubospermum Lour. = Ludwigia L. Onagraceae

Cuchumatanea Seid. et Beaman. 1966. Asteraceae. 1 Guatemala

Cucubalus L. 1753. Caryophyllaceae. 2 temp. Northern Hemisphere, incl. European part of Russia, Caucasus, Middle Asia, West Siberia, and Russian Far East

Cucumella Chiov. 1929. Cucurbitaceae. 7 trop. Africa, Madagascar, India

Cucumeropsis Naudin. 1866. Cucurbitaceae. 1 trop. West Africa

Cucumis L. 1753. Cucurbitaceae. 31 trop. and South Africa, trop. Asia, Malesia, Australia; widely cultivated in trop. America

Cucurbita L. 1753. Cucurbitaceae. c. 30 trop. and subtrop. America

Cucurbitella Walp. 1846. Cucurbitaceae. 3 South America

Cudrania Trécul. 1847 (Maclura Nutt.). Moraceae. 8 northeastern India, China, Korea, Japan, Indochina, Malesia, Australia, New Caledonia

Cuenotia Rizzini. 1956. Acanthaceae. 1 northeastern Brazil

Cuervea Triana ex Miers. 1872 (Hippocratea L.). Celastraceae. 3 trop. America, West Indies

Cufodontia Woodson. 1934 (Aspidosperma Mart. et Zucc.). Apocynaceae. 4 Mexico, Central America

Cuitlauzina La Lllave ex Lex. = Odontoglossum Kunth. Orchidaceae

Culcasia P. Beauv. 1805. Araceae. 15 trop. Africa

Culcitium Humb. et Bonpl. 1808 (Senecio L.). Asteraceae. 15 Andes

Cullen Medik. 1787 (Psoralea L.). Fabaceae. 35 trop. Old World, a few species in North Africa and Palestine

Cullenia Wight. 1851 (Durio Adans.). Bombacaceae. 3 southern India, Sri Lanka

Cullumia R. Br. 1813. Asteraceae. 15 South Africa

Cumarinia (Knuth) Buxb. = Coryphantha (Engelm.) Lem. Cactaceae

Cumingia Vidal = Camptostemon Mast. Bombacaceae

Cuminia Colla. 1835. Lamiaceae. 1 Juan Fernández Is.

Cuminum L. 1753. Apiaceae. 4 North Africa, Middle Asia (3), Iran, Afghanistan, Pakistan, India, China

Cumulopuntia F. Ritter = Opuntia Hill. Cactaceae

Cunila L. 1759 (Hedyosmos Mithc.). Lamiaceae. 15 America from eastern U.S. to Uruguay

Cunninghamia R. Br. ex Rich. 1826. Taxodiaceae. 2 Central and southern China, Taiwan, Vietnam

Cunonia L. 1759. Cunoniaceae. 17 South Africa (1, C. capensis), New Caledonia (16)

Cunuria Baill. = Micrandra Benth. Euphorbiaceae

Cupania L. 1753. Sapindaceae. 44 trop. and subtrop. America

Cupaniopsis Radlk. 1879. Sapindaceae. c. 70 Sulawesi, Moluccas, New Guinea, Australia, New Caledonia, Vanuatu, Fiji, Samoa

Cuphea P. Browne. 1756. Lythraceae. 250–300 trop. America: southern U.S., Mexico, Guatemala, Belize, El Salvador, Honduras, Nicaragua, Costa Rica, Panama, Colombia, Venezuela, Ecuador, Peru, Bolivia, Brazil, Argentina and Paraguay; West Indies

Cupheanthus Seem. 1865. Myrtaceae. 5 New Caledonia

Cuphocarpus Decne. et Planch. 1854. Araliaceae. 1 Madagascar

Cuphonotus O. E. Schulz. 1933. Brassicaceae. 2–3 southern and eastern Australia, Tasmania

Cupressus L. 1753. Cupressaceae. 14 Mediterranean, Sakhara, Crimea and Caucasus (1, C. sempervirens), northern Iran, Himalayas, China, America: Pacific coasts from Oregon to Guatemala

Cupulanthus Hutch. 1964. Fabaceae. 1 western Australia

Cupularia Godron et Gren. = Dittrichia Greuter. Asteraceae

Curanga Juss. = Picria Lour. Scrophulariaceae

Curarea Barneby et Krukoff. 1971. Menispermaceae. 4 Panama, Colombia, Ecuador, Peru, Brazil

Curatella Loefl. 1758. Dilleniaceae. 2 trop. America, West Indies

Curculigo Gaertn. 1788. Hypoxidaceae. 14–15 trop. Southern Hemisphere

Curcuma L. 1753. Zingiberaceae. 52 Himalayas, India, Sri Lanka, Nicobar and Andaman Is., China, Malesia, Australia

Curcumorpha A. S. Rao et D. M. Verma. 1974. Zingiberaceae. 1 Himalayas

Curinila Schult. = Leptadenia R. Br. Asclepiadaceae

Curroria Planch. ex Benth. 1849. Asclepiadaceae. 5 trop. and subtrop. Africa, southern Arabian Peninsula, Socotra

Curtia Cham. et Schltdl. 1826. Gentianaceae. 10 trop. South America

Curtisia Aiton. 1789. Curtisiaceae. 1 trop. Southeast and South Africa

Curtonus N. E. Br. = Crocosmia Planch. Iridaceae

Curupira G. A. Black. 1948. Olacaceae. 1 Brazil (Amazonia)

Cuscatlania Standl. 1923. Nyctaginaceae. 1 Central America

Cuscuta L. 1753. Cuscutaceae. 150–170 temp., subtrop. and trop. regions, esp. America, Mediterranean, trop. Africa

Cusickia M. E. Jones = Lomatium Raf. Apiaceae

Cusickiella Rollins. 1988 (Draba L.). Brassicaceae. 2 western U.S. (Washington, Idaho, Utah, California, Nevada)

Cusparia Humb. ex R. Br. = Angostura Roem. et Schult. Rutaceae

Cuspidaria DC. 1838. Bignoniaceae. 8 trop. America from eastern Panama to Argentina, esp. Brazil

Cuspidia Gaertn. 1791. Asteraceae. 1 South Africa

Cussonia Thunb. 1780. Araliaceae. 25 trop. and South Africa, Madagascar, Mascarene Is.

Cutandia Willk. 1860. Poaceae. 6 Mediterranean, Caucasus, Southwest, West, and Middle Asia

Cuthbertia Small = Callisia Loefl. Commelinaceae

Cutsis Burns-Bal., E. W. Greenwood et Gonzales = Dichromantus Garay. Orchidaceae

Cuttsia F. Muell. 1865. Escalloniaceae. 1 eastern Australia

Cuviera DC. 1807. Rubiaceae. 16 trop. Africa

Cwangayana Rauschert = Aralia L. Araliaceae

Cyamopsis DC. 1826. Fabaceae. 3 deserts of Africa and Arabian Peninsula, Pakistan, India

Cyanaeorchis Barb. Rodr. 1877. Orchidaceae. 2 Brazil

Cyanandrium Stapf. 1895. Melastomataceae. 4 Borneo

Cyananthus Wall. ex Benth. 1830. Campanulaceae. 24 Himalayas from Kashmir and Punjab to Bhutan and Assam, Tibet, southwestern China (Gansu, western Sichuan, Qinghai, and northwestern Yunnan), Burma

Cyanastrum Oliv. 1891. Cyanastraceae. 7 trop. Africa

Cyanea Gaudich. 1829. Lobeliaceae. 52 Hawaiian Is.

Cyanella Royen ex L. 1754. Tecophilaeaceae. 7 Namibia, southwestern South Africa

Cyanopsis Cass. 1816 (Volutaria Cass.). Asteraceae. 1 Spain, North Africa

Cyanorkis Thouars = Phaius Lour. Orchidaceae

Cyanoseris (W. D. J. Koch) Schur = Lactuca L. Asteraceae

Cyanostegia Turcz. 1849. Chloanthaceae. 5 western and central Australia

Cyanothamnus Lindl. = Boronia Sm. Rutaceae

Cyanothyrsus Harms = Daniellia Benn. Fabaceae

Cyanotis D. Don. 1825. Commelinaceae. 50 trop. and South (7) Africa, trop. Asia, northern Australia

Cyanthillium Blume = Vernonia Schreb. Asteraceae

Cyanus Hill. 1753 (Centaurea L.). Asteraceae. 22 Europe, Mediterranean, Southwest Asia

Cyathanthus Engl. = Scyphosyce Baill. Moraceae

Cyathobasis Aellen. 1949. Chenopodiaceae. 1 Turkey

Cyathocalyx Champ. ex Hook. f. et Thomson. 1855. Annonaceae. 15 India, Sri Lanka, Indochina, Malesia eastward to Fiji (Viti-Lewu Is.)

Cyathochaeta Nees = Cyathocoma Nees. Cyperaceae

Cyathocline Cass. 1829. Asteraceae. 3 Himalayas, India, Burma, China, Southeast Asia

Cyathocoma Nees. 1834 (Tetraria P. Beauv.). Cyperaceae. 5 Australia

Cyathodes Labill. = Styphelia Sm. Epacridaceae

Cyathogyne Muell. Arg. 1864. Euphorbiaceae. 4–5 trop. Africa, Madagascar

Cyathomone S. F. Blake. 1923. Asteraceae. 1 Ecuador

Cyathophylla Bocquet et Strid. 1986. Caryophyllaceae. 1 Greece and Turkey

Cyathopsis Brongh. et Gris. 1864 (Styphelia Sm.). Epacridaceae. 1 New Caledonia

Cyathopus Stapf. 1895. Poaceae. 1 eastern Himalayas

Cyathoselinum Benth. 1867 (or = Seseli L.). Apiaceae. 1 Southeast Europe

Cyathostegia (Benth.) Schery. 1950. Fabaceae. 2 Ecuador, Peru

Cyathostelma Fourn. 1885. Asclepiadaceae. 2 Brazil

Cyathostemma Griff. 1854. Annonaceae. 8 Burma, southern China (Yunnan), Thailand, Indochina, Malesia: Malay Peninsula, Sumatra (1), Borneo (1)

Cyathula Blume. 1825. Amaranthaceae. 25–30 trop. America, trop. and South Africa, Madagascar, Himalayas, India, Sri Lanka, southern China, Taiwan, Indochina, Malesia, Australia, Polynesia

Cybebus Garay. 1978. Orchidaceae. 1 Colombia

Cybianthopsis (Mez) Lundell = Cybianthus Mart. Myrsinaceae

Cybianthus Mart. 1831 (Wallenia Sw.). Myrsinaceae. c. 150 trop. America, West Indies

Cybistax Mart. ex Meisn. 1840. Bignoniaceae. 1 trop. America: Surinam, Ecuador, Peru, Brazil, Bolivia, Paraguay, Argentina

Cybistctcs Milne-Redh. et Schweick. 1939. Amaryllidaceae. 1 South Africa

Cycas L. 1753. Cycadaceae. 20 trop. East Africa, Madagascar, Mascarene Is., Sri Lanka, southern, eastern and northeastern India, southern China, Taiwan (1, C. taiwaniana), Japan (Kyushu and Ryukyu Is.), Southeast Asia, Malesia to New Guinea, Micronesia, trop. Australia, New Caledonia, eastward to Fiji, Samoa, and Tonga

Cyclacanthus S. Moore. 1921. Acanthaceae. 2 Indochina

Cyclachaena Fres. = Iva L. Asteraceae

Cycladenia Benth. 1849. Apocynaceae. 1 southwestern U.S.

Cyclamen L. 1753. Primulaceae. 20 montane Europe (8), Mediterranean (except Egypt and Sinai Peninsula), northeastern Somalia (1, C. somalense), coasts of Black Sea and southern Caspian Sea; C. persicum extends to China; Crimea (C. coum), Caucasus (8)

Cyclanthera Schrad. 1831. Cucurbitaceae. 30 trop. America

Cyclantheropsis Harms. 1896. Cucurbitaceae. 3 trop. Africa, Madagascar

Cyclanthus Poit. 1822. Cyclanthaceae. 2 Central and South America (southward to Peru and Brazil), Lesser Antilles, and Trinidad

Cyclea Arn. ex Wight. 1840. Menispermaceae. c. 30 trop. and subtrop. Asia from India to central and southern China, Japan, Indochina, Malesia (Malay Peninsula, Sumatra, Java, Borneo, Philippines)

Cyclobalanopsis (Endl.) Oerst. = Quercus L. Fagaceae

Cyclocarpa Afzel. ex Baker. 1871. Fabaceae. 1 trop. Africa, Malesia, Australia (northern Queensland)

Cyclocarya Iljinsk. 1953 (Pterocarya Kunth). Juglandaceae. 1 continental China

Cyclocheilon Oliv. 1895. Acanthaceae. 3 eastern Ethiopia, Somalia

Cyclocotyla Stapf. 1908. Apocynaceae. 1 trop. West Africa

Cyclolepis Gillies ex D. Don. 1832. Asteraceae. 1 temp. South America

Cyclolobium Benth. 1837. Fabaceae. 4 trop. South America

Cycloloma Moq. 1840. Chenopodiaceae. 1 U.S.

Cyclophyllum Hook. f. 1873. Rubiaceae. 30 Moluccas, New Guinea, Caroline and Solomon Is., Australia, New Caledonia (14), Vanuatu, Fiji (3), Tonga, Tuamotu Arch.

Cyclopia Vent. 1808. Fabaceae. 19 South Africa

Cyclopogon C. Presl. 1827. Orchidaceae. c. 70 America from southern U.S. to Argentina, West Indies (or 1 Andes ?)

Cycloptychis E. Mey. ex Sond. 1860. Brassicaceae. 2 South Africa (southwestern Cape)

Cyclorhiza M. L. Shen et R. H. Shan. 1980 (Nicatia DC.). Apiaceae. 1 Tibet

Cyclospermum Lag. 1821. Apiaceae. 4 trop. and subtrop. America

Cyclostachya J. R. Reeder et C. G. Reeder. 1963. Poaceae. 1 Mexico

Cyclotrichium (Boiss.) Manden. et Scheng. 1953. Lamiaceae. 6 Asia Minor, southwestern Iran

Cycniopsis Engl. 1905. Scrophulariaceae. 2 Uganda, southeastern Ethiopia, Kenya, Yemen

Cycnium E. Mey. ex Benth. 1836 (Escobedia Ruiz et Pav.). Scrophulariaceae. 15 trop. and subtrop. Africa

Cycnoches Lindl. 1832. Orchidaceae. 12 trop. America from Mexico to Peru and Brazil

Cycnogeton Endl. 1838 (Triglochin L.). Juncaginaceae. 1 Australia, Tasmania

Cydista Miers. 1863. Bignoniaceae. 4 trop.

America from Mexico to Brazil, West Indies

Cydonia Hill. 1753. Rosaceae. 2: C. oblonga—Asia Minor, northern Iran, Caucasus, Middle Asia; C. sinensis—eastern China

Cylicodiscus Harms. 1897. Fabaceae. 1 trop. West and Central Africa

Cylicomorpha Urb. 1901. Caricaceae. 2 trop. West Africa

Cylindrocarpa Regel. 1877. Campanulaceae. 1 Middle Asia

Cylindrocline Cass. 1817. Asteraceae. 2 Mauritius

Cylindrokelupha Kosterm. 1954 (Archidendron F. Muell.). Fabaceae. 10 southern China, Indochina, western Malesia

Cylindrolobus (Blume) Brieger = Eria Lindl. Orchidaceae

Cylindrophyllum Schwantes. 1927. Aizoaceae. 5 South Africa

Cylindropsis Pierre. 1898. Apocynaceae. 1 trop. West Africa

Cylindropuntia (Engelm.) F. Knuth = Opuntia Hill. Cactaceae

Cylindorpyrum (Jaub. et Spach) A. Löve = Aegilops L. Poaceae

Cylindrosolenium Lindau. 1897. Acanthaceae. 1 Peru

Cylindrosperma Ducke. 1930 (Microplumeria Baill.). Apocynaceae. 1 Brazil (Amazonia)

Cylista Aiton. 1789 (or = Paracalyx Ali). Fabaceae. 7 trop. and South Africa, Socotra, Mauritius, eastern India

Cymaria Benth. 1830. Lamiaceae. 3 China, Burma, Java, Timor, Philippines

Cymatocarpus O. E. Schulz. 1924. Brassicaceae. 3 Caucasus, Middle Asia, Iran, Afghanistan

Cymbalaria Hill. 1756 (Linaria Hill). Scrophulariaceae. 10 Europe, Mediterranean; C. muralis—Crimea, Abkhazia

Cymbaria L. 1753. Scrophulariaceae. 4 southern European part of Russia (C. borystenica), East Siberia (C. daurica); Mongolia, China, Japan

Cymbidiella Rolfe. 1918. Orchidaceae. 5 Madagascar

Cymbidium Sw. 1799. Orchidaceae. 50–60 Madagascar, trop. Asia from the Himalayas through China to Korea and Japan, Taiwan, Indochina, Malesia from Malay Peninsula to New Guinea, trop. Australia, Oceania

Cymbiglossum Halbinger = Odontoglossum Kunth. Orchidaceae

Cymbispatha Pichon = Tradescantia L. Commelinaceae

Cymbocarpa Miers. 1840. Burmanniaceae. 2 trop. America, West Indies

Cymbocarpum DC. ex C. A. Mey. 1830. Apiaceae. 3–4 Asia Minor, Caucasus (1, C. anethoides), northwestern Iran

Cymbochasma (Endl.) Klok. et Zoz. = Cymbaria L. Scrophulariaceae

Cymbolaena Smoljan. 1955. Asteraceae. 1 Balkan Peninsula, Transcaucasia, Asia Minor, Israel, Syria, Iran, Middle Asia, Afghanistan, Pakistan

Cymbonotus Cass. 1825. Asteraceae. 1 temp. Australia

Cymbopappus R. Nord. 1976. Asteraceae. 3 South Africa

Cymbopetalum Benth. 1860. Annonaceae. 12 trop. America

Cymbopogon Spreng. 1815. Poaceae. 60 trop. of the Old World

Cymbosema Benth. 1840. Fabaceae. 1 southern Central and northern South America

Cymbosetaria Schweick. 1936 (Setaria P. Beauv.). Poaceae. 1 Sudan, Kenya, Tanzania, Malawi, Zambia, Zimbabwe, Mozambique, Botswana, South Africa, southern Arabian Peninsula (Yemen)

Cymodocea K. König. 1805. Cymodoceaceae. 4 Atlantic coasts from southwestern France to Senegal, Canary Is., Mediterranean, trop. coasts of the Indian and the Pacific Oceans

Cymophora B. L. Rob. 1907 (Tridax L.). Asteraceae. 4 Mexico, Venezuela

Cymophyllus Mack. ex Britton et A. Br. 1913 (Carex L.). Cyperaceae. 1 southeastern U.S.

Cymopterus Raf. 1819. Apiaceae. 32 western North America

Cynanchica Fourr. = Asperula L. Rubiaceae

Cynanchum L. 1753. Asclepiadaceae. c. 100 trop. and temp. regions;

Cynapium Nutt. ex Torr. et A. Gray. 1840 (Ligusticum L.). Apiaceae. 1 North America

Cynara L. 1753. Asteraceae. 8 Canary Is., Mauritania, Mediterranean (Iberian Peninsula, France, Italy, Greece, Morocco, Algeria, Libya, Egypt, Cyprus, Crete, islands of Turkey, Asia Minor, Syria, Lebanon, Israel), northern Iraq, western Iran

Cynaroides (Boiss. ex Walp.) Dostal = Centaurea L. Asteraceae

Cyne Danser. 1929. Loranthaceae. 4 Philippines

Cynoctonum E. Mey. 1838. Asclepiadaceae. 1 Mongolia, China, Korea; Russia—East Siberia, Far East

Cynodendron Baehni = Chrysophyllum L. Sapotaceae

Cynodon Rich. 1805. Poaceae. 10 temp., subtrop. and trop. regions; C. dactylon in southern European part of Russia, Caucasus, Altai, Middle Asia

Cynoglossopsis Brand. 1931. Boraginaceae. 2 Ethiopia, Somalia

Cynoglossum L. 1753. Boraginaceae. 60 temp. and subtrop. regions, rarely montane tropics

Cynoglottis (Gusul.) Vural et Kit Tan. 1983 (Anchusa L.). Boraginaceae. 2 Southeast Europe, Asia Minor, Syria, Lebanon, Israel

Cynometra L. 1753. Fabaceae. c. 70 tropics

Cynomorium L. 1753. Cynomoriaceae. 1–2 Canary Is., Mediterranean, West Asia eastward to China and Mongolia; C. songaricum—Middle Asia

Cynorchis Thouars = Cynorkis Thouars. Orchidaceae

Cynorhiza Eckl. et Zeyh. 1837 (Peucedanum L.). Apiaceae. 4 South Africa

Cynorkis Thouars. 1809. Orchidaceae. 125–143 trop. and South (13) Africa, Madagascar, Comoro and Mascarene Is., Seychelles

Cynosciadium DC. 1829. Apiaceae. 1–2 central and southern U.S.

Cynosurus L. 1753. Poaceae. 10 Europe, North Africa, Southwest and West and Middle (Turkmenistan) Asia

Cynoxylon Raf. = Cornus L. Cornaceae

Cypella Herb. 1826. Iridaceae. 20 trop. America from Mexico to Argentina

Cyperochloa Lazarides et L. Watson. 1987. Poaceae. 1 western Australia

Cyperorchis Blume = Cymbidium Sw. Orchidaceae

Cyperus L. 1753. Cyperaceae. 300 trop., subtrop. and temp. regions

Cyphacanthus Leonard. 1953. Acanthaceae. 1 Colombia

Cyphanthera Miers. 1853 (7E Anthocercis Labill.). Solanaceae. 9 temp. southern Australia, Tasmania (1)

Cyphia Bergius. 1767. Cyphiaceae. 60 Africa from Ethiopia to Cape

Cyphisia Rizzini = Justicia L. Acanthaceae

Cyphocalyx Gagnep. = Trungbao Rauschert. Scrophulariaceae

Cyphocardamum Hedge. 1968. Brassicaceae. 1 Afghanistan

Cyphocarpa (Fenzl) Lopr. 1899. Amaranthaceae. 5 trop. and South Africa

Cyphocarpus Miers. 1848. Cyphocarpaceae. 2 Chile

Cyphochilus Schltr. 1912. Orchidaceae. 1–7 New Guinea

Cyphochlaena Hack. 1901. Poaceae. 2 Madagascar

Cyphokentia Brongn. 1873. Arecaceae. 1 (C. macrostachya) New Caledonia

Cypholepis Chiov. 1907 (Coelachyrum Hochst. et Nees). Poaceae. 1 Ethiopia, Somalia, Kenya, Tanzania, and South Africa (Transvaal and adjacent parts of Cape Province); Arabian Peninsula (Yemen)

Cypholophus Wedd. 1854. Urticaceae. 35 southern China, Malesia, Macaronesia, Fiji and Samoa, Hawaiian Is.

Cypholoron Dodson et Dressler. 1972. Orchidaceae. 1 Ecuador

Cyphomandra Mart. ex Sendtn. 1845. Solanaceae. c. 50 Central and trop. montane South America, West Indies; widely cultivated

Cyphomeris Standl. 1909 (Boerhavia L.). Nyctaginaceae. 2 Texas, Mexico

Cyphophoenix H. A. Wendl. ex Hook. f. 1883. Arecaceae. 2 New Caledonia (1), Loyalty Is. (1)

Cyphosperma H. A. Wendl. ex Hook. f. 1883. Arecaceae. 3 New Caledonia (1), Fiji (2)

Cyphostemma (Planch.) Alston. 1931. Vitaceae. 160 tropics and subtropics

Cyphostigma Benth. 1882. Zingiberaceae. 1 Sri Lanka

Cyphostyla Gleason. 1929. Melastomataceae. 3 Colombia

Cyphotheca Diels. 1932 (Phyllagathis Blume). Melastomataceae. 1 (C. montana) southern China (Yunnan)

Cyprinia Browicz. 1966. Asclepiadaceae. 1 Cyprus, southeastern Turkey

Cypripedium L. 1753. Orchidaceae. 35–50 temp. Eurasia, North America; Russia (5, northern and central European part, Siberia, Far East)

Cypselea Turpin. 1806. Aizoaceae. 2 America from southern Florida to Venezuela, West Indies

Cypselocarpus F. Muell. 1873. Gyrostemonaceae. 1 southwestern Australia

Cypselodontia DC. = Dicoma Cass. Asteraceae

Cyrilla Garden ex L. 1767. Cyrillaceae. 1 America from southeastern U.S. to northern South America, West Indies

Cyrillopsis Kuhlm. 1925. Ixonanthaceae. 1 northeastern Brazil

Cyrilwhitea Ising = Sclerolaena R. Br. Chenopodiaceae

Cyrtandra J. R. Forst. et G. Forst. 1775. Gesneriaceae. c. 90 Australia (1), Santa Cruz Is., Vanuatu, Loyalty Is., Rotuma I., Horn Is., Fiji, Tonga, Samoa, Cook Is., Society Is., Marquesas Is.

Cyrtandroidea F. B. H. Br. = Cyrtandra J. R. Forst. et G. Forst. Gesneriaceae

Cyrtandromoea Zoll. 1855. Scrophulariaceae. 11 Burma, Andaman Is., Indochina

Cyrtanthera Nees = Jusiticia L. Acanthaceae

Cyrtanthus Aiton. 1789. Amaryllidaceae. 47 trop. and (mostly) South Africa

Cyrtidiorchis Rauschert. 1982. Orchidaceae. 5 trop. America

Cyrtidium Schltr. = Cyrtidiorchis Rauschert. Orchidaceae

Cyrtocarpa Kunth. 1824. Anacardiaceae. 4: 2 western and southern Mexico, incl. Baja California; 1 northern Colombia, Venezuela, Guyana; 1 Brazil (Roraima and Bahia)

Cyrtochilum Kunth. 1816 (or = Oncidium Sw.). Orchidaceae. 115 Andes

Cyrtococcum Stapf. 1917. Poaceae. 12 trop. Old World

Cyrtocymura H. Rob. = Vernonia Schreb. Asteraceae

Cyrtogonone Prain. 1911. Euphorbiaceae. 1 trop. Africa

Cyrtopera Lindl. = Eulophia R. Br. ex Lindl. Orchidaceae

Cyrtopodium R. Br. 1813. Orchidaceae. 10 trop. America from Mexico and Florida to Argentina

Cyrtorchis Schltr. 1914. Orchidaceae. 16 trop. and South Africa

Cyrtorhyncha Nutt. ex Torr. et A. Gray. 1838 (Ranunculus L.). Ranunculaceae. 1 western North America

Cyrtosia Blume. 1825 (Galeola Lour.). Orchidaceae. 2 Sri Lanka, India, Thailand, Vietnam, Malay Peninsula, Sumatra, Java, Borneo to Sulawesi

Cyrtosperma Griff. 1851. Araceae. 11 Malesia (Malay Peninsula, Sumatra, Java, Borneo, Philippines, Aru Is., New Guinea), C. merkusii—Palau Is., Caroline and Mariana Is., Solomon Is., Samoa, Marshall Is., Vanuatu, Washington I., Gilbert Is., Cook Is., Fiji, Tahiti, Society Is., Marquesas Is.

Cyrtostachys Blume. 1838. Arecaceae. 8 southern Thailand, Malay Peninsula, Sumatra, and Borneo (1, C. renda), New Guinea, and Melanesia

Cyrtostylis R. Br. = Acianthus R. Br. Orchidaceae

Cyrtoxiphus Harms = Cylicodiscus Harms. Fabaceae

Cystacanthus T. Anderson. 1867. Acanthaceae. 8 southwestern China, Burma, Indochina

Cysticapnos Mill. 1754. Fumariaceae. 5 South Africa

Cysticorydalis Fedde ex Ikonn. = Corydalis DC. Fumariaceae

Cystopus Blume (1858, not Levl. 1847—Fungi) = Pristiglottis Cretz. et J. J. Sm. Orchidaceae

Cystorchis Blume. 1858. Orchidaceae. 21 trop. Asia from India to China, Indochina, Malesia (Malay Peninsula, Sumatra, Java, Borneo, Philippines, New Guinea)

Cystostemma E. Fourn. 1885 (Oxypetalum R. Br.). Asclepiadaceae. 2 Brazil

Cystostemon Balf. f. 1883. Boraginaceae. 15 Northeast and trop. Africa, southwestern Arabian Peninsula, Socotra

Cytinus L. 1764. Cytinaceae. 7–8: 2 Canary Is., and Mediterranean (C. hypocistis) to western Caucasus (C. rubra), 5–6 trop. and South Africa, Madagascar

Cytisanthus O. Lang = Genista L. Fabaceae

Cytisophyllum O. Lang. 1843. Fabaceae. 1 montane Spain, France, and Italy

Cytisopsis Jaub. et Spach. 1844 (Anthyllis L.). Fabaceae. 2 Morocco, eastern Mediterranean from Asia Minor to Israel

Cytisus L. 1753. Fabaceae. 33 Europe, Canary Is., Morocco (9), Algeria, Tunisia (1), West Asia

Cyttaranthus J. Léonard. 1955. Euphorbiaceae. 1 trop. Africa

Czernaevia Turcz. ex Ledeb. 1844 (or = Angelica L.). Apiaceae. 1 Siberia, Russian Far East, China, Korea, Japan

D

Daboecia D. Don. 1834. Ericaceae. 2 Azores, Atlantic coastal Europe from Ireland to Spain

Dacrycarpus (Endl.) de Laub. 1969. Podocarpaceae. 9 Burma, southern China, abundant in Malesia (7) eastward to New Caledonia and Fiji, New Zealand

Dacrydium Sol. ex Lamb. 1806. Podocarpaceae. 20 subalpine forests of southern China, Indochina, Malesia (excl. Java and Lesser Sunda Is.), New Guinea, Tasmania, New Zealand, Stewart I., New Caledonia, Vanuatu, Fiji

Dacryodes Vahl. 1810. Burseraceae. 40 tropics

Dacryotrichia Wild. 1973. Asteraceae. 1 Zambia

Dactyladenia Welw. 1859 (Acioa Aubl.). Chrysobalanaceae. 27 trop. and Namibia

Dactylaea Fedde ex H. Wolff. 1930. Apiaceae. 2 southwestern and southern China

Dactylaena Schrad. ex Schult. et Schult. f. 1829. Capparaceae. 6 West Indies, Brazil

Dactylanthus Hook. f. 1859. Dactylanthaceae. 1 New Zealand

Dactyliandra (Hook. f.) Hook. f. 1871. Cucurbitaceae. 3 South-West and East Africa, India

Dactylicapnos Wall. 1826 (Dicentra Bernh.) Fumariaceae. 10 Himalayas from Kumaon eastward to southwestern China (Yunnan), Indochina, ? Sri Lanka

Dactyliophora Tiegh. 1895 (Amyema Tiegh.). Loranthaceae. 3 New Guinea, New Ireland I., Solomon Is., Australia (Queensland)

Dactylis L. 1753. Poaceae. 5 temp. Eurasia, North America; D. glomerata—European part of Russia, Caucasus, Siberia, Far East, Middle Asia

Dactylocardamum Al-Shehbaz. 1989. Brassicaceae. 1 Peru

Dactylocladus Oliv. 1895. Crypteroniaceae. 1 Borneo

Dactyloctenium Willd. 1809. Poaceae. 13 tropics, mainly Africa (south to Transvaal, Natal, and eastern Cape), Madagascar, Indian Ocean islands south to Mauritius, north to Arabian Peninsula and Pakistan, eastward to northwestern India; D. aegyptiacum in trop. and warm–temp. regions of the Old World, introduced into America

Dactylopetalum Benth. = Cassipourea Aubl. Rhizophoraceae

Dactylopsis N. E. Br. 1925. Aizoaceae. 2 South Africa

Dactylorhiza Neck. ex Nevski. 1937. Orchidaceae. 75 Macaronesia, temp. Eurasia, Mediterranean, North America (Alaska); Russia (c. 25, from European part to Far East)

Dactylorhynchus Schltr. 1912. Orchidaceae. 1 New Guinea

Dactylostalix Rchb. f. 1878. Orchidaceae. 1 southern Kuril Is. and Japan

Dactylostegium Nees. 1847 (Dicliptera Juss.). Acanthaceae. 2 trop. America, West Indies

Dactylostelma Schltr. 1895. Asclepiadaceae. 1 Bolivia

Dactylostigma D. F. Austin = Hildebrandtia Vatke ex A. Braun. Convolvulaceae

Dactyophyllaria Garay. 1986 (Vanilla Mill.). Orchidaceae. 1 Brazil

Dadjoua A. Parsa. 1960. Caryophyllaceae. 1 Iran

Daedalacanthus T. Anderson = Eranthemum L. Acanthaceae

Daemonorops Blume. 1830. Areaceae. 115 trop. Asia from India and southern China through the Malay Archipelago to New Guinea, esp. Malay Peninsula, Sumatra, and Borneo

Daenikera Hürl. et Stauffer. 1957. Santalaceae. 1 New Caledonia

Daenikeranthus Baumann-Bodenheim. 1989 (Dracophyllum Labill.) Epacridaceae. 1 New Caledonia

Dahlgrenia Steyerm. = Dictyocaryum H. A. Wendl. Arecaceae

Dahlgrenodendron J. M. van der Merwe, A. E. van Wyk et P. D. F. Kok. 1988 (Cryptocarya R. Br.). Lauraceae. 1 South Africa (Natal)

Dahlia Cav. 1791. Asteraceae. 28 trop. America from Mexico to Colombia

Dahlstedtia Malme. 1905. Fabaceae. 1 southern Brazil

Dais L. 1762. Thymelaeaceae. 2 trop. (Tanzania, Malawi, Zimbabwe) and Southeast (from Transvaal to eastern Cape) Africa (D. cotinifolia), Madagascar (1)

Daiswa Raf. 1838. Trilliaceae. 15 Himalayas, northeastern India, northern Burma, continental China, Hainan, Taiwan, Indochina

Daknopholis Clayton. 1967. Poaceae. 1 East Africa (Kenya, Tanzania), Madagascar, islands of the Indian Ocean (Aldabra I., Astove, Cosmoledo)

Dalanum J. Dostál = Galeopsis L. Lamiaceae

Dalbergia L. f. 1782. Fabaceae. c. 100 pantropics, esp. Africa (c. 70), a few species in South Africa

Dalbergiella Baker f. 1928. Fabaceae. 3 trop. Africa from Gabon to Zimbabwe

Dalea L. 1758. Fabaceae. c. 160 America from Canada to Argentina (chiefly Mexico)

Dalechampia L. 1753. Euphorbiaceae. 110 tropics and subtropics, esp. America

Dalembertia Baill. 1858. Euphorbiaceae. 4 Mexico

Dalenia Korth. 1844. Melastomataceae. 3 Borneo

Dalhousiea Wall. ex Benth. 1837. Fabaceae. 3 trop. West Africa (1), northeastern India, Bangladesh

Dalibarda L. = Rubus L. Rosaceae.

Dallachya F. Muell. 1875 (Rhamnella Miq.). Rhamnaceae. 1 eastern Australia

Dalzellia Wight. 1852. Podostemaceae. 1 southern India, Thailand, Vietnam

Dalzielia Turrill. 1916. Asclepiadaceae. 1 West Africa

Damasonium Hill. 1753. Alismataceae. 5 Europe, Caucasus, Middle Asia, West Siberia, Mediterranean, West Asia, southern Australia, North America (California)

Dammaropsis Warb. = Ficus L. Moraceae
Dammera Lauterb. et K. Schum. = Licuala Thunb. Arecaceae
Damnacanthus C. F. Gaertn. 1805. Rubiaceae. 6 Assam, Bangladesh, China, Korea, Japan
Damnamenia Given. 1973 (Celmisia Cass.). Asteraceae. 1 Auckland and Campbell Is.
Damnxanthodium Strother. 1987. Asteraceae. 1 Mexico
Dampiera R. Br. 1810. Goodeniaceae. 66 Australia
Damrongia Kerr ex Craib = Chirita Buch.-Ham. ex D. Don. Gesneriaceae
Danae Medik. 1787. Ruscaceae. 1 (D. racemosa) Caucasus, southeastern Asia Minor, Syria, northern Iran
Danais Comm. ex Vent. 1799. Rubiaceae. 40 Tanzania, Madagascar, Comoro and Mascarene Is.
Dandya H. E. Moore. 1953. Alliaceae. 3 Mexico
Danguya Benoist. 1930. Acanthaceae. 1 Madagascar
Danguyodrypetes Leandri. 1939. Euphorbiaceae. 1 Madagascar
Daniellia Benn. 1854. Fabaceae. 9 trop. Africa
Dankia Gagnep. 1939. Flacourtiaceae. 1 Southeast Asia
Dansera Steenis = Dialium L. Fabaceae
Danserella Balle = Odontella Tiegh. Loranthaceae
Dansiea Byrnes. 1981 (Macropteranthes F. Muell. ex Benth.). Combretaceae. 2 Australia (Queensland)
Danthonia DC. 1805. Poaceae. 20 North and South America, montane Europe, Caucasus, Camchatka (1)
Danthoniastrum (Holub) Holub. 1970. Poaceae. 2 Balkan Peninsula, Caucasus (1, D. compactum—Abkhazia)
Danthonidium C. E. Hubb. 1937. Poaceae. 1 India
Danthoniopsis Stapf. 1916. Poaceae. 15 Africa, Arabian Peninsula, Iran, western Pakistan
Danthorhiza Ten. = Helictotrichon Besser ex Schult. Poaceae

Dapania Korth. 1855. Oxalidaceae. 3 Madagascar (1), Malesia (2 Sumatra, Malay Peninsula, Borneo)
Daphnandra Benth. 1870. Monimiaceae. 6 New Guinea, eastern Australia
Daphne L. 1753. Thylemaeaceae. c. 90 Eurasia, incl. (c. 20) European part of Russia, Caucasus, Middle Asia, Far East; D. mesereum—from arctic areas to Caucasus and West Siberia; North America, Australia, Oceania
Daphnimorpha Nakai. 1937. Thymelaeaceae. 2 Japan
Daphniphyllum Blume. 1827. Daphniphyllaceae. 12 southwestern India, Sri Lanka, Eastern Himalayas, Assam, continental China, Taiwan, Korea, Japan, Malesia; D. humile—Kuril Is.
Daphnopsis Mart. et Zucc. 1824 (Schoenobiblus Mart.). Thymelaeaceae. 55 America from Mexico to eastern Argentina, West Indies
Darbya A. Gray = Nestronia Raf. Santalaceae
Darcya B. L. Turner et C. Cowan. 1993 (Stemodia L.). Scrophulariaceae. 3 Costa Rica, Panama
Darlingia F. Muell. 1866. Proteaceae. 2 Australia (Queensland)
Darlingtonia Torr. 1853. Sarraceniaceae. 1 western U.S. (Oregon and California)
Darmera Voss. 1899. Saxifragaceae. 1 western U.S. (Oregon and California)
Darniella Maire et Weiller. 1939 (Salsola L.). Chenopodiaceae. 11 Malta I. (1), North Africa (Sakhara), West Asia
Darwinia Rudge. 1815. Myrtaceae. 23 Australia
Darwiniella Braas et Lückel = Stellilabium Schltr. Orchidaceae
Darwiniera Braas et Lückel = Stellilabium Schltr. Orchidaceae
Darwiniothamnus Harling. 1962. Asteraceae. 2 Galapagos Is.
Dasiola Raf. = Vulpia C. C. Gmelin. Poaceae
Dasiphora Raf. = Potentilla L. Rosaceae
Dasispermum Neck. ex Raf. 1840. Apiaceae. 1 South Africa

Dasistoma Raf. 1819 (Seymeria Pursh). Scrophulariaceae. 1 southeastern U.S.

Dasoclema J. Sinclair. 1955. Annonaceae. 1 Thailand, western Malesia

Dasycephala (DC.) Hook. f. = Diodia L. Rubiaceae

Dasycondylus R. M. King et H. Rob. 1972. Asteraceae. 7 Peru, Brazil

Dasydesmus Craib = Oreocharis Benth. Gesneriaceae

Dasylepis Oliv. 1865. Flacourtiaceae. 6 trop. Africa

Dasylirion Zucc. 1838. Nolinaceae. 18 southwestern U.S., Mexico

Dasymaschalon (Hook. f. et Thomson) Dalla Torre et Harms. 1901 (Desmos Lour.). Annonaceae. 16 southwestern and southern China, Indochina, Philippines, Sunda Is.

Dasynotus I. M. Johnst. 1948. Boraginaceae. 1 northwestern U.S.

Dasyochloa Rydb. = Erioneuron Nash. Poaceae

Dasyphyllum Kunth. 1818. Asteraceae. 36 South America

Dasypoa Pilg. = Poa L. Poaceae

Dasypogon R. Br. 1810. Dasypogonaceae. 3 southwestern Australia

Dasypyrum (Coss. et Durieu) T. Durand. 1888. Poaceae. 2 South Europe from southern France to southern Romania and Turkey, Crimea, Caspian Sea (Krasnovodsk), islands of Mediterranean Sea, North Africa (Morocco, Algeria)

Dasysphaera Volkens ex Gilg. 1897. Amaranthaceae. 4 Northeast and East Africa

Dasystachys Baker = Chlorophytum Ker-Gawl. Asphodelaceae

Dasystachys Oerst. = Chamaedorea Willd. Arecaceae

Dasystephana Adans. = Gentiana L. Gentianaceae

Dasytropis Urb. 1924. Acanthaceae. 1 Cuba

Datisca L. 1753. Datiscaceae. 2: D. glomerata—southwestern North America; D. cannabina—Crete and Cyprus Is., Asia Minor, Caucasus eastward to Middle Asia, India, and Nepal

Datura L. 1753 (excl. Brugmansia Pers.). Solanaceae. 1 (D. stramonium) Mexico [or 11 warm–temp. and trop. regions, esp. India, East Asia, Australia, southwestern North America; southern European part of Russia, Caucasus, West Siberia, and Middle Asia); or 8 America from southwestern U.S. to northern South America], widely introduced and naturalized

Daturicarpa Stapf = Tabernanthe Baill. Apocynaceae

Daubentonia DC. = Sesbania Scop. Fabaceae

Daubentoniopsis Rydb. = Aeschynomene L. Fabaceae

Daubenya Lindl. 1835. Hyacinthaceae. 1 South Africa

Daucosma Engelm. et A. Gray ex A. Gray. 1850. Apiaceae. 1 North America

Daucus L. 1753. Apiaceae. 22 Europe, Mediterranean, trop. Africa, Southwest Asia eastward to western China, Australia, New Zealand, ? America; D. carota—southern European part of Russia, Caucasus, Middle Asia

Daumailia Arènes = Urospermum Scop. Asteraceae

Daumalia Airy Shaw = Urospermum Scop. Asteraceae

Dauphinea Hedge. 1983. Lamiaceae. 1 Madagascar

Daveaua Willk. ex Mariz. 1891. Asteraceae. 2 Portugal, North Africa

Davidia Baill. 1871. Davidiaceae. 1 western and central China

Davidsea Soderstr. et R. P. EllIs. 1988 (Schizostachyum Nees). Poaceae. 1 Sri Lanka

Davidsonia F. Muell. 1867. Davidsoniaceae. 1 Australia (Queensland and New South Wales)

Daviesia Sm. 1798. Fabaceae. c. 110 Australia

Davilla Vand. 1788. Dilleniaceae. 20 trop. America, West Indies

Dayaoshania W. T. Wang. 1983. Gesneriaceae. 1 southern China

Deamia Britton et Rose = Selenicereus Britton et Rose. Cactaceae

Deanea J. M. Coult. et Rose = Rhodosciadium S. Watson. Apiaceae

Debesia Kuntze = Anthericum L. Asphodelaceae

Debregeasia Gaudich. 1844. Urticaceae. 4 Ethiopia, Arabian Peninsula, Iran, Afghanistan, Himalayas from Kashmir to Bhutan, India, Sri Lanka, Burma, China, Indochina, Malesia

Decabelone Decne. = Tavaresia Welw. ex N. E. Br. Asclepiadaceae

Decachaena Torr. et A. Gray ex A. Gray = Gaylussacia Kunth. Ericaceae

Decachaeta DC. 1836. Asteraceae. 7 Mexico, Central America

Decadianthe Rchb. ? . Rutaceae. 1 Australia

Decagonocarpus Engl. 1874. Rutaceae. 1 Brazil

Decaisnea Hook. f. et Thomson. 1855. Lardizabalaceae. 2 Himalayas (Nepal, Bhutan), Assam, Tibet, southern China

Decaisnina Tiegh. 1895. Loranthaceae. c. 30 Philippines, Moluccas, New Guinea (6), northern Australia (6), Oceania from Bismarck Arch. and Solomon Is. eastward to Tahiti

Decalepidanthus Riedl. 1963. Boraginaceae. 1 western Pakistan

Decalepis Wight et Arn. 1834. Asclepiadaceae. 1 India

Decalobanthus Ooststr. 1936. Convolvulaceae. 1 Sumatra

Decamerium Nutt. = Gaylussacia Kunth. Ericaceae

Decanema Decne. 1838. Asclepiadaceae. 3 Madagascar

Decanemopsis Costantin et Gallaud. 1906 (Sarcostemma R. Br.). Asclepiadaceae. 1 Madagascar

Decaphalangium Melch. 1930 (Clusia L.). Clusiaceae. 1 Peru

Decaptera Turcz. 1846. Brassicaceae. 1 Chile

Decaryanthus Bonati. 1927. Scrophulariaceae. 1 Madagascar

Decarydendron Danguy. 1928. Monimiaceae. 3 Madagascar

Decaryella A. Camus. 1931. Poaceae. 1 Madagascar

Decaryia Choux. 1929. Didieraceae. 1 Madagascar

Decaryochloa A. Camus. 1946. Poaceae. 1 Madagascar

Decaschistia Wight et Arn. 1834. Malvaceae. 16 India, Burma, Indochina, Hainan, Malesia, Australia

Decaspermum J. R. Forst. et G. Forst. 1775. Myrtaceae. 40 Burma, southern China, Indochina, Malesia, Caroline Is., Australia, Vanuatu, Fiji, Samoa, Society Is.

Decastelma Schltr. 1899. Asclepiadaceae. 2 West Indies

Decastylocarpus Humbert. 1923. Asteraceae. 1 Madagascar

Decatoca F. Muell. 1889. Epacridaceae. 1 New Guinea

Decatropis Hook. f. 1862. Rutaceae. 2–3 southern Mexico, Guatemala

Decazesia F. Muell. 1879. Asteraceae. 1 western Australia

Decazyx Pittier et S. F. Blake. 1922. Rutaceae. 2 Mexico, Central America

Deccania Tirveng. 1983 (Randia L.). Rubiaceae. 1 South and Southeast Asia

Decemium Raf. 1817 (Hydrophyllum L.). Hydrophyllaceae. 1 North America

Deceptor Seidenf. 1992. Orchidaceae. 1 Thailand, Vietnam

Deckenia H. A. Wendl. ex Seem. 1870. Arecaceae. 1 Seychelles Is.

Declieuxia Kunth. 1819. Rubiaceae. 27 trop. America, West Indies

Decodon J. F. Gmel. 1791. Lythraceae. 1 eastern North America

Decorsea R. Vig. 1951. Fabaceae. 4: 3 Tanzania, Malawi, Namibia, Botswana, Zambia, Zimbabwe, Mozambique, South Africa; 1 Madagascar

Decorsella A. Chev. 1917. Violaceae. 1 trop. West Africa

Decumaria L. 1763. Hydrangeaceae. 2 East Asia, southeastern North America

Decussocarpus de Laub. = Retrophyllum C. N. Page and Nageia Gaertn. Podocarpaceae

Dedea Baill. = Quintinia A. DC. Escalloniaceae

Dedeckera Reveal et J. T. Howell. 1976

(Erigonum Michx.). Polygonaceae. 1 western North America (California)

Deeringia R. Br. 1810. Amaranthaceae. 12 Madagascar, trop. Asia from India through Indochina to southern China and Taiwan, Malesia, Australia, Oceania

Deeringothamnus Small. 1924. Annonaceae. 2 Florida peninsula

Degeneria I. W. Bailey et A. C. Sm. 1942. Degeneriaceae. 2 Fiji

Degenia Hayek. 1910. Brassicaceae. 1 Yugoslavia

Degranvillea Determann. 1985. Orchidaceae. 1 French Guyana

Deguelia Aubl. 1775 (Derris Lour.). Fabaceae. 16 northern South America

Dehaasia Blume. 1837. Lauraceae. 35 southern China, Taiwan, Indochina, Malesia eastward to New Guinea

Deherainia Decne. 1876. Theophrastaceae. 2 Mexico, Guatemala to Costa Rica

Deianira Cham. et Schltdl. 1826. Gentianaceae. 7 trop. America

Deidamia Noronha ex Thouars. 1805. Passifloraceae. 5 Madagascar

Deilanthe N. E. Br. = Nananthus N. E. Br. Aizoaceae

Deinacanthon Mez = Bromelia L. Bromeliaceae

Deinanthe Maxim. 1867. Hydrangeaceae. 2 Central China, Japan

Deinbollia Schumach. et Thonn. 1827. Sapindaceae. 40 trop. and subtrop. Africa, Madagascar

Deinocheilos W. T. Wang. 1986. Gesneriaceae. 2 Central and southern China

Deinostema Yamaz. 1953 (Gratiola L.). Scrophulariaceae. 1 northeastern China, Korea, Japan

Deinostigma W. T. Wang et Z. Y. Li. 1992 (Hemiboea C. B. Clarke). Gesneriaceae. 1 southern Vietnam

Deiregyne Schltr. 1920. Orchidaceae. 8 Mexico, Costa Rica (1)

Deiregynopsis Rauschert = Aulosepalum Garay. Orchidaceae

Dekindtia Gilg = Chionanthus L. Oleaceae

Dekinia M. Martens et Gal. = Agastache Clayton ex Gronov. Lamiaceae

Delabechea Mitchell ex Lindl. = Brachychiton Schott et Endl. Sterculiaceae

Delaetia Backeb. = Neoporteria Britton et Rose. Cactaceae

Delairia Lem. 1844 (Senecio L.). Asteraceae. 1 South Africa

Delamerea S. Moore. 1900. Asteraceae. 1 trop. East Africa

Delaportea Thorel ex Gagnep. = Acacia Hill. Fabaceae

Delarbrea Vieill. 1865. Araliaceae. 4–6 Lesser Sunda Is. and Moluccas, New Guinea, Australia (1, Queensland), Solomon Is., New Caledonia, Vanuatu (New Hebrides)

Delavaya Franch. 1866. Sapindaceae. 1 southwestern China

Delilia Spreng. 1823. Asteraceae. 2 Galapagos Is., trop. America

Delissea Gaudich. 1829. Lobeliaceae. 9 Hawaiian Is.

Delonix Raf. 1837. Fabaceae. 10 Madagascar (8), Northeast and East Africa (3, from Egypt to Zaire and Tanzania); D. elata—extending to Arabian Peninsula

Delopyrum Small = Polygonella Michx. Polygonaceae

Delosperma N. E. Br. 1925. Aizoaceae. 155 Ethiopia, Kenya, Zimbabwe, eastern Botswana, western Mozambique, South Africa, Madagascar, southwestern Arabian Peninsula

Delostoma D. Don. 1823. Bignoniaceae. 4 trop. Andes of Colombia, Venezuela, Ecuador, Peru, and Bolivia

Delphinacanthus Benoist = Pseudodicliptera Benoist. Acanthaceae

Delphinium L. 1753. Ranunculaceae. 250-320 temp. and warm–temp. Northern Hemisphere, esp. China (c. 110), Russia, and adjacent territories (c. 110), and Mediterranean (42)

Delphyodon K. Schum. 1898. Apocynaceae. 1 New Guinea

Delpinophytum Speg. 1903. Brassicaceae. 1 Patagonia

Delpya Pierre ex Radkl. = Sisyrolepis Radkl. Sapindaceae

Delpydora Pierre. 1897. Sapotaceae. 2 trop. West Africa

Deltaria Steenis. 1959. Thymelaeaceae. 1 New Caledonia

Deltocheilos W. T. Wang = Chirita Buch.-Ham. ex D. Don. Gesneriaceae

Demavendia Pimenov. 1987. Apiaceae. 1 Iran, Middle Asia

Demidium DC. = Gnaphalium L. Asteraceae

Demosthenesia A. C. Sm. 1936. Ericaceae. 12 Andes

Dendragrostis B. D. Jacks. = Chusquea Kunth. Poaceae

Dendranthema (DC.) Des Moulins. 1860. Asteraceae. c. 50 West and Southeast Europe (1), temp. Asia; Russia (17, northern European part and Siberia, Far East)

Dendriopoterium Svent. 1948 (Sanguisorba L.). Rosaceae. 2 Canary Is. (Grand Canary I.)

Dendrobangia Rusby. 1896. Icacinaceae. 3 Panama, Colombia, Ecuador, Venezuela, Guyana, French Guyana, Surinam, Brazil (Amazonas)

Dendrobenthamia Hutch. = Cornus L. Cornaceae

Dendrobium Sw. 1799. Orchidaceae. c. 1,400 trop. and subtrop. Asia, Malesia (esp. Philippines—over 85), Bougainville and Solomon Is. (55), Australia, New Zealand, Oceania

Dendrobrychis Galushko = Onobrychis Hill. Fabaceae

Dendrocacalia Nakai ex Tuyama. 1936. Asteraceae. 1 Ogasawara (Bonin) Is.

Dencrocalamopsis (L. C. Chia et H. L. Fung) Keng f. = Bambusa Schreb. Poaceae

Dendrocalamus Nees. 1835. Poaceae. 20 South and Southeast Asia

Dendrocereus Britton et Rose = Acanthocereus (A. Berger) Britton et Rose. Cactaceae

Dendrochilum Blume. 1825. Orchidaceae. c. 100–150 Burma, Indochina, Malesia, esp. Philippines (c. 90)

Dendrochloa C. E. Parkinson. 1933 (Schizostachyum Nees). Poaceae. 1 Burma

Dendrocnide Miq. 1851. Urticaceae. 36 Himalayas, India, Sri Lanka, Indochina, Taiwan, Malesia, trop. Australia, Oceania

Dendrocoryne (Lindl. et Paxton) Brieger = Dendrobium Sw. Orchidaceae

Dendrocousinsia Millsp. = Sebastiania Spreng. Euphorbiaceae

Dendrokingstonia Rauschert. 1982. Annonaceae. 1 western Malesia: Malay Peninsula, Java

Dendroleandria Arènes = Helmiopsiella Arènes. Sterculiaceae

Dendrolobium (Wight et Arn.) Benth. 1852. Fabaceae. 12 trop. Asia, Malesia, Australia

Dendromecon Benth. 1835. Papaveraceae. 1–3 California, Channel Is., and northern Baja California

Dendromyza Danser. 1940. Santalaceae. 5 Malesia, New Guinea, Solomon Is., Australia (Queensland)

Dendropanax Decne. et Planch. 1854. Araliaceae. c. 60 tropics and subtropics

Dendropemon (Blume) Rchb. = Loranthus L. Loranthaceae

Dendrophorbium (Cuatrec.) C. Jeffrey. 1992. Asteraceae. c. 50 Andes of Colombia, Ecuador, Peru, Bolivia, Argentina, Paraguay, Brazil; West Indies

Dendrophthoe Mart. 1830. Loranthaceae. 30–35 trop. Africa, trop. Asia, Malesia, and trop. Australia (6)

Dendrophthora Eichler. 1868. Viscaceae. c. 70 trop. America from southern Mexico to Peru and Bolivia, West Indies

Dendrophylax Rchb. f. 1864. Orchidaceae. 4 West Indies

Dendrosenecio (Hauman ex Humbert) B. Nord. 1978 (Senecio L.). Asteraceae. 5 montane trop. East Africa

Dendroseris D. Don. 1832. Asteraceae. 10 Juan Fernández Is.

Dendrosicus Raf. = Amphitecna Miers. Bignoniaceae

Dendrosicyos Balf. f. 1882. Cucurbitaceae. 1 Socotra

Dendrosida Fryxell. 1971. Malvaceae. 6 Mexico (4), Colombia, and Venezuela (2)

Dendrosipanea Ducke. 1935. Rubiaceae. 3 northern South America

Dendrosma Pancher et Sebert = Geijera Schott. Rutaceae

Dendrostellera (C. A. Mey.) Tiegh. = Diarthron Turcz. Thymelaeaceae

Dendrostigma Gleason = Mayna Aubl. Flacourtiaceae

Dendrotrophe Miq. 1856. Santalaceae. 10 northern India, southern China (6), Indochina, Malesia, trop. Australia (1, D. varians)

Denea O. F. Cook = Howea Becc. Arecaceae

Denekia Thunb. 1800. Asteraceae. 1 (D. capensis) southern trop. and South Africa

Denhamia Meisn. 1837. Celastraceae. 9 northern and eastern (Queensland and New South Wales) Australia

Denisonia F. Muell. = Pityrodia R. Br. Chloanthaceae

Denisophytum R. Vig. = Caesalpinia L. Fabaceae

Denmoza Britton et Rose. 1922 [Oreocereus (A. Berger) Riccob.]. Cactaceae. 1 northwestern and western Argentina

Dennettia Baker f. 1913. Annonaceae. 1 Ivory Coast, Nigeria

Dentaria L. 1753 (Cardamine L.). Brassicaceae. 20 Eurasia, eastern North America

Dentella J. R. Forst. et G. Forst. 1775. Rubiaceae. 10 southern China, Indochina, Malesia, Australia, New Caledonia

Dentoceras Small = Polygonella Michx. Polygonaceae

Depacarpus N. E. Br. = Meyerophytum Schwantes. Aizoaceae

Depanthus S. Moore. 1921. Gesneriaceae. 1 New Caledonia

Deplanchea Vieill. 1863. Bignoniaceae. 5 Malay Peninsula, Sumatra, Borneo, Sulawesi, New Guinea, northern Australia, New Caledonia

Deppea Cham. et Schltdl. 1830. Rubiaceae. 25 trop. America from Mexico to Brazil

Deprea Raf. 1838. Solanaceae. 15 Mexico, Costa Rica, Andes from Colombia and Venezuela to Bolivia

Derenbergia Schwantes. 1825 (Conophytum N. E. Br.). Aizoaceae. 15 South Africa

Derenbergiella Schwantes = Mesembryanthemum L. Aizoaceae

Dermatobotrys Bolus. 1890. Scrophulariaceae. 1 South Africa

Dermatocalyx Oerst. = Schlegelia Miq. Bignoniaceae

Derris Lour. 1790. Fabaceae. c. 65 trop. South America, trop. and South Africa (1, D. trifoliata), South and Southeast Asia, trop. Australia; 1 mangroves of the Indian and Pacific Oceans

Derwentia Raf. 1836 (Parahebe W. R. B. Oliv.). Scrophulariaceae. 8 Australia (Queensland, New South Wales, south Australia, and Victoria) and Tasmania

Desbordesia Pierre ex Tiegh. 1905. Irvingiaceae. 4 trop. West Africa

Deschampsia P. Beauv. 1812. Poaceae. c. 30 temp. and subtrop. regions and montane tropics

Descurainia Webb et Berthel. 1836. Brassicaceae. c. 50 arctic and temp. Northern Hemisphere, South Africa; Russia (2–3 all areas)

Desdemona S. Moore = Basistemon Turcz. Scrophulariaceae

Desfontainia Ruiz et Pav. 1794. Desfontainiaceae. 1 Andes from Costa Rica southward to Horn Cape

Desideria Pamp. 1926 (Christolea Cambess. ex Jacquem.). Brassicaceae. 3 Himalayas (India, Nepal); D. pamirica—Tajikistan

Desmanthodium Benth. 1872. Asteraceae. 10 trop. America from Mexico to Venezuela

Desmanthus Willd. 1806. Fabaceae. 25 trop. and subtrop. America from U.S. to Argentina, West Indies

Desmaria Tiegh. 1895 (Loranthus Jacq.). Loranthaceae. 1 southern Chile

Desmazeria Dumort. 1822. Poaceae. 4 Mediterranean

Desmidorchis Ehrenb. = Caralluma R. Br. Asclepiadaceae

Desmodiastrum (Prain) A. Pramanik et Thoth. 1986 (Alysicarpus Neck. ex Desv.). Fabaceae. 2 South Asia

Desmodium Desv. 1813. Fabaceae. c. 300 tropics and subtropics, esp. East Asia, Mexico, and Brazil

Desmogymnosiphon Guinea. 1946. Burmanniaceae. 1 trop. West Africa

Desmoncus Mart. 1824. Arecaceae. 20 trop. America from Mexico to Bolivia and Brazil, Trinidad

Desmopsis Saff. 1916. Annonaceae. 15 Mexico, Central America, Cuba (1)

Desmos Lour. 1790. Annonaceae. 25–30 India, Sri Lanka, Southeast Asia, Malesia to northern Australia and eastward to New Caledonia

Desmoscelis Naudin. 1849. Melastomataceae. 2 trop. South America

Desmoschoenus Hook. f. 1853. Cyperaceae. 1 New Zealand, Stewart and Chatham Is.

Desmostachya (Hook. f.) Stapf. 1898. Poaceae. 1 North Africa, Arabian Peninsula, Palaestine, Iraq, Iran, Afghanistan, western Pakistan, India, Indochina, southern China

Desmostachys Miers. 1852. Icacinaceae. 7 trop. Africa, Madagascar

Desmothamnus Small = Lyonia Nutt. Ericaceae

Desmotrichum Blume = Flickingeria A. D. Hawkes. Orchidaceae

Desplatsia Bocquillon. 1866. Tiliaceae. 7 trop. West Africa

Desynaphia Hook. et Arn. ex DC. 1938. Asteraceae. 16 Brazil (12), Argentina (1), Paraguay, and Uruguay

Detarium Juss. 1789. Fabaceae. 3–4 trop. Africa from Senegal to the Sudan south to Zaire

Dethawia Endl. 1839. Apiaceae. 1 Spain, southern France

Detzneria Schltr. ex Diels. 1929. Scrophulariaceae. 1 New Guinea

Deuterocohnia Mez. 1894. Bromeliaceae. 7 Peru, Bolivia, southwestern Brazil, northern Chile, northern Argentina

Deuteromallotus Pax et K. Hoffm. 1914. Euphorbiaceae. 1 Madagascar

Deutzia Thunb. 1781. Hydrangeaceae. c. 50 from Himalayas eastward to East Asia and Philippines, southern U.S., Mexico, Central America

Deutzianthus Gagnep. 1924. Euphorbiaceae. 1 Indochina

Deverra DC. 1829 (Pituranthos Viv.). Apiaceae. 7 North and South Africa, Southwest Asia

Devia Goldblatt et J. C. Manning. 1990 (Crocosmia Planch.). Iridaceae. 1 South Africa (Cape)

Devillea Tul. et Wedd. 1849. Podostemaceae. 1 Brazil

Dewevrea Micheli. 1898. Fabaceae. 2 Cameroun, Central African Rep., Zaire, Angola

Dewevrella De Wild. 1907. Apocynaceae. 1 trop. Africa

Dewildemania O. Hoffm. 1903. Asteraceae. 3 southern trop. Africa

Dewindtia De Wild. = Cryptosepalum Benth. Fabaceae

Deyeuxia Clarion ex P. Beauv. = Calamagrostis Adans. Poaceae

Dhofaria A. G. Mill. 1988 (Apophylla F. Muell.). Capparaceae. 1 Oman

Diacarpa T. R. Sim = Atalaya Blume. Sapindaceae

Diachyrium Griseb. = Sporobolus R. Br. Poaceae

Diacidia Griseb. 1858. Malpighiaceae. 12 Colombia, Venezuela, Brazil

Diacranthera R. M. King et H. Rob. 1972 (Eupatorium L.). Asteraceae. 2 Brazil

Diacrium Lindl. ex Benth. = Caularthron Raf. Orchidaceae

Diacrodon Sprague. 1928. Rubiaceae. 1 Brazil

Diadenium Peopp. et Endl. 1836. Orchidaceae. 2 Andes of South America

Dialium L. 1767. Fabaceae. c. 40 pantropics, but mainly trop. Africa (28, esp. Zaire—15); D. schlechteri in South Africa

Dialyanthera Warb. = Otoba (DC.) Karst. Myristicaceae

Dialyceras Capuron. 1962. Sphaerosepalaceae. 3 Madagascar

Dialypetalanthus Kuhlm. 1925. Dyalypetalanthaceae. 1 eastern Brazil

Dialypetalum Benth. 1873. Lobeliaceae. 5 Madagascar

Dialytheca Exell et Mendonça. 1935. Menispermaceae. 1 trop. Africa

Diamena Ravenna. 1987 (Anthericum L.). Asphodelaceae. 1 Peru

Diamorpha Nutt. 1818. Crassulaceae. 1 (D. smallii) eastern U.S.

Diandranthus Liou. 1987 (Miscanthus Andersson). Poaceae. 8 China, Tibet, Himalayas

Diandriella Engl. = Homalomena Schott. Araceae

Diandrochloa de Winter. 1960 (Eragrostis Wolf). Poaceae. 7 trop. and temp. regions; D. diarrhena—southeastern European part of Russia

Diandrolyra Stapf. 1906. Poaceae. 7 Brazil

Diandrostachya (C. E. Hubb.) Jacq.-Fél. 1960 (Loudetiopsis Conert). Poaceae. 4 trop. Africa

Dianella Juss. 1789. Phormiaceae. 25–30 trop. East Africa, Madagascar, Mascarene Is., India, Sri Lanka, southern China, Indochina, Malesia, Australia, Tasmania, Norfolk I., New Zealand, New Caledonia, Fiji, Polynesia to Hawaiian Is., western and northern South America

Dianthoseris Sch. Bip. ex A. Rich. 1842. Asteraceae. 1 Ethiopia

Dianthoveus Hammel et Wilder. 1989. Cyclanthaceae. 1 Andes of southwestern Colombia, northern Ecuador

Dianthus L. 1753. Caryophyllaceae. c. 300 Eurasia, Africa; Russia (c. 100, from European part to Far East)

Diapensia L. 1753. Diapensiaceae. 4 arctic, subarctic and montane areas of western China, Tibet, eastern Himalayas (Nepal, Bhutan), northern Burma; Russia (2, arctic areas of European part, East Siberia, Far East)

Diaphananthe Schltr. 1914. Orchidaceae. c. 50 trop. and South Africa

Diaphanoptera Rech. f. 1940 (Acanthophyllum C. A. Mey.). Caryophyllaceae. 3 Turkmenistan, northeastern Iran, Afghanistan

Diaphractanthus Humbert. 1923. Asteraceae. 1 Madagascar

Diaphycarpus Calest. = Bunium L. Apiaceae

Diarrhena P. Beauv. 1812. Poaceae. 1 eastern U.S.

Diarthron Turcz. 1832. Thymelaeaceae. 19 Europe (1), Caucasus, Middle Asia, Siberia, Far East of Russia; Asia Minor, Iran, northwestern India, China, Mongolia, Korea

Diascia Link et Otto. 1820. Scrophulariaceae. 50 South Africa

Diaspananthus Miq. 1866 (Ainsliaea DC.). Asteraceae. 1 Japan

Diaspasis R. Br. 1810. Goodeniaceae. 1 southwestern Australia

Diastatea Scheidw. 1841. Lobeliaceae. 7 trop. America from Mexico to Peru and Bolivia

Diastella Salisb. 1809 (Leucospermum R. Br.). Proteaceae. 7 South Africa

Diastema Benth. 1845. Gesneriaceae. 40 trop. America

Diateinacanthus Lindau = Odontonema Nees. Acanthaceae

Diatenopteryx Radlk. 1878. Sapindaceae. 2 South America

Dibrachionostylus Bremek. 1952 (Agathisanthemum Klotzsch). Rubiaceae. 1 trop. East Africa (Kenya)

Dicarpellum (Loes.) A. C. Sm. 1941 (Salacia L.). Celastraceae. 5 New Caledonia

Dicarpidium F. Muell. 1857. Sterculiaceae. 1 Australia

Dicarpophora Speg. 1926. Asclepiadaceae. 1 Bolivia

Dicella Griseb. 1839. Malpighiaceae. 6 trop. South America: Colombia, Peru, Bolivia, northern Argentina, and mostly Brazil

Dicellandra Hook. f. 1867. Melastomataceae. 3 trop. Africa

Dicellostyles Benth. 1862. Malvaceae. 1 Sri Lanka

Dicentra Borkh. ex Bernh. 1833. Fumariaceae. 16 East Siberia (1, D. peregrina), China, Himalayas, Burma, Korea, Japan, North America

Dicerandra Benth. 1830. Lamiaceae. 6 southeastern U.S.

Diceratella Boiss. 1844. Brassicaceae. 8 North (Algeria), Northeast (Sudan, Ethiopia, Somalia) and trop. East (Kenya) Africa, Socotra, Iran (2)

Diceratostele Summerh. 1938. Orchidaceae. 1 trop. Africa

Dicercoclados C. Jeffrey et Y. L. Chen. 1984. Asteraceae. 1 China

Dicerma DC. 1825. Fabaceae. 1 trop. Asia, Australia

Dicerocaryum Bojer. 1835. Pedaliaceae. 3 southern trop. Africa

Dicerospermum Bakh. f. 1943 (Poikilogyne Baker f.). Melastomataceae. 1 New Guinea

Dicerostylis Blume. 1859. Orchidaceae. 5 southern China, Taiwan, Philippines, Indonesia

Dichaea Lindl. 1833. Orchidaceae. 45 trop. America from Mexico to Peru and Brazil, West Indies

Dichaelia Harv. = Brachystelma R. Br. Asclepiadaceae

Dichaetanthera Endl. 1840. Melastomataceae. 34 trop. Africa (7), Madagascar (27)

Dichaetaria Nees ex Steud. 1854. Poaceae. 1 southern India, Sri Lanka

Dichaetophora A. Gray. 1849. Asteraceae. 1 southern U.S. northern Mexico

Dichanthelium (Hitchc. et Chase) Gould = Panicum L. Poaceae

Dichanthium Willemet. 1796. Poaceae. 20 trop. Old World

Dichapetalum Thouars. 1806. Dichapetalaceae. c. 150 tropics, except Micronesia and Polynesia

Dichasianthus Ovcz. et Junusov. 1978. Brassicaceae. 1 (or 13) Middle Asia (Kazakhstan, Uzbekistan, Tajikistan)

Dichazothece Lindau. 1898. Acanthaceae. 1 eastern Brazil

Dichelachne Endl. 1833. Poaceae. 8 New Guinea, Timor I., Australia, Tasmania, New Zealand

Dichelostemma Kunth. 1843 (Brodiaea Sm.). Alliaceae. 6 North America

Dicheranthus Webb. 1846. Caryophyllaceae. 1 Canary Is.

Dichilanthe Thwaites. 1856. Rubiaceae. 2 Sri Lanka, Borneo

Dichiloboea Stapf = Trisepalum C. B. Clarke. Gesneriaceae

Dichilus DC. 1825. Fabaceae. 5 Namibia, Zimbabwe, South Africa

Dichocarpum W. T. Wang et Hsiao. 1964 (Isopyrum L.). Ranunculaceae. 20 Asia from Himalayas and northern Burma through China to Japan

Dichodon (Bartl.) Rchb. = Cerastium L. Caryophyllaceae

Dichoglottis Fisch. et C. A. Mey. 1835. Caryophyllaceae. 2 eastern European part of Russia, West Siberia, Middle Asia, Iran

Dichondra J. R. Forst. et G. Forst. 1775. Convolvulaceae. 5–8 tropics and subtropics

Dichondropsis Brandegee. 1909. Convolvulaceae. 1 Mexico

Dichopogon Kunth. 1843. Asphodelaceae. 5 New Guinea (1), southern Australia, Tasmania (1)

Dichorisandra J. C. Mikan. 1823. Commelinaceae. 25 trop. America, West Indies (1)

Dichosciadium Domin. 1908. Apiaceae. 1 Australia

Dichostemma Pierre. 1896. Euphorbiaceae. 3 trop. Africa

Dichostylis P. Beauv. ex Lestib. = Cyperus L. Cyperaceae

Dichotomanthes Kurz. 1873. Rosaceae. 1 southwestern China (Yunnan, Sichuan)

Dichroa Lour. 1790. Hydrangeaceae. 12 China, Indo-Malesia to New Guinea

Dichrocephala L'Hér. ex DC. 1833. Asteraceae. 10 trop. Africa, Madagascar, trop. Asia through Malesia eastward to Japan, Hawaii, Fiji, Society and Austral Is.; D. integrifolia—extending to Caucasus

Dichromanthus Garay. 1980. Orchidaceae. 1 Mexico, Guatemala

Dichromena Michx. 1803 (or = Rhynchospora Vahl). Cyperaceae. 60 trop. and subtrop. America, West Indies

Dichromochlamys Dunlop. 1980. Asteraceae. 1 central Australia

Dichrospermum Bremek. = Spermacoce L. Rubiaceae

Dichrostachys (DC.) Wight et Arn. 1834. Fabaceae. 12 North (Egypt) and trop. Africa (4), Madagascar, Arabian Peninsula, trop. Asia, Malesia, and trop. Australia

Dichrotrichum Reinw. ex Vriese. 1856 (Agalmyla Blume). Gesneriaceae. 35 Philippines, Borneo, Sulawesi, Moluccas, New Guinea

Dickasonia L. O. Williams. 1941. Orchidaceae. 2 India, Burma

Dickinsia Franch. 1886. Apiaceae. 1 western China

Dicladanthera F. Muell. 1882. Acanthaceae. 1 western Australia

Diclidanthera Mart. 1827. Polygalaceae. 8 trop. South America

Diclinanona Diels. 1927. Annonaceae. 2 eastern Peru, western Brazil

Dicliptera Juss. 1805. Acanthaceae. 250 tropics and subtropics

Diclis Benth. 1836. Scrophulariaceae. 10 trop. East and South Africa, Madagascar (1)

Dicoelospermum C. B. Clarke. 1879. Cucurbitaceae. 1 western India

Dicoelia Benth. 1879. Pandaceae. 2–3 Malay Peninsula, Sumatra, Banda I., Borneo

Dicoma Cass. 1817. Asteraceae. 35 trop. and South Africa, Madagascar, Arabian Peninsula, Socotra, India

Dicoria Torr. et A. Gray. 1859. Asteraceae. 4 southwestern U.S., Mexico

Dicorynia Benth. 1840. Fabaceae. 1–2 trop. South America

Dicoryphe Thouars. 1804. Hamamelidaceae. 15 Madagascar, Comoro Is.

Dicraeanthus Engl. 1905. Podostemaceae. 4 trop. West Africa

Dicraeia Thouars = Podostemum Michx. Podostemaceae

Dicraeopetalum Harms. 1902. Fabaceae. 3–4 Ethiopia, Kenya, Somalia, Madagascar (2)

Dicranocarpus A. Gray. 1854. Asteraceae. 1 southern U.S., Mexico

Dicranolepis Planch. 1848. Thymelaeaceae. 16 trop. Africa

Dicranopygium Harling. 1954. Cyclanthaceae. 48 trop. America from southern Mexico to central Peru and Surinam, Tobago I.

Dicranostegia (A. Gray) Pennell = Cordylanthus Nutt. ex Benth. Scrophulariaceae

Dicranostigma Hook. f. et Thomson. 1855. Papaveraceae. 5 Himalayas (India, Nepal), western and southern China

Dicranostyles Benth. 1846. Convolvulaceae. 15 trop. South America: Colombia, Venezuela, Guyana, French Guyana, Surinam, Peru, Brazil

Dicraspidia Standl. 1929. Tiliaceae. 1 Central America

Dicrastylis J. L. Drumm. ex Harv. 1855. Chloanthaceae. 20 Australia

Dicraurus Hook. f. = Iresine P. Browne. Amaranthaceae

Dicrocaulon N. E. Br. 1928. Aizoaceae. 12 western South Africa

Dictamnus L. 1753. Rutaceae. 6 Eurasia from Central and South Europe through Caucasus, Middle Asia, and southern Siberia eastward to Far East of Russia, China, and Korea

Dictyandra Welw. ex Benth. et Hook. f. 1873. Rubiaceae. 2 trop. Africa from Sierra Leone to Angola, east to Uganda and Zaire

Dictyanthus Decne. = Matelea Aubl. Asclepiadaceae

Dictyocaryum H. A. Wendl. 1860. Arecaceae. 3 Panama, Colombia, Ecuador, Peru, Bolivia, Venezuela, Guyana

Dictyochloa (Murb.) E. G. Camus. 1901 (Ammochloa Boiss.). Poaceae. 1 Morocco

Dictyolimon Rech. f. 1974 (Limonium Mill.). Plumbaginaceae. 4 Afghanistan, Pakistan, India

Dictyoloma A. Juss. 1825. Rutaceae. 2 Peru (1), Brazil (1)

Dictyoneura Blume. 1847. Sapindaceae. 3 Philippines, eastern Borneo, Sulawesi, Moluccas, New Guinea, New Britain and New Ireland Is., Australia

Dictyophleba Pierre. 1898. Apocynaceae. 5 trop. Africa: Sierra Leone, Liberia, Guinea, Ivory Coast, Ghana, Nigeria, Cameroun, Central African Rep., Congo, Zaire, Angola, Zambia, Kenya, Tanzania, Malawi, Zimbabwe, Mozambique, Comoro Is.

Dictyophragmus O. E. Schulz. 1933. Brassicaceae. 1 Peru

Dictyophyllaria Garay. 1986 (Vanilla Mill.). Orchidaceae. 1 Brazil

Dictyosperma H. A. Wendl. et Drude. 1875. Arecaceae. 1 Mascarene Is.

Dictyospermum Wight. 1853. Commelinaceae. 5–10 Himalayas (Nepal, Bhutan), India, Sri Lanka, China, Taiwan, Indochina, Malesia

Dictyostega Miers. 1840. Burmanniaceae. 1 trop. America from Mexico to Bolivia and southern Brazil

Dicyclophora Boiss. 1844. Apiaceae. 1 Iran

Dicymanthes Danser = Amyema Tiegh. Loranhtaceae

Dicymbe Spruce ex Benth. 1865. Fabaceae. 13 trop. South America

Dicymbopsis Ducke = Dicymbe Spruce ex Benth. Fabaceae

Dicypellium Nees et Mart. 1833. Lauraceae. 2 Amazonian Brazil

Dicyrta Regel = Achimenes Pers. Gesneriaceae

Didelotia Baill. 1865. Fabaceae. 12 trop. Africa from Sierra Leone to Chad and Congo

Didelta L'Hér. 1786. Asteraceae. 2 Namibia

Didesmandra Stapf. 1900. Dilleniaceae. 1 Borneo

Didesmus Desv. 1915 (Rapistrum Crantz). Brassicaceae. 2 eastern Mediterranean, North Africa

Didiciea King et Prain. 1896. Orchidaceae. 2 eastern Himalayas (1), Japan (1)

Didierea Baill. 1880. Didiereaceae. 2 Madagascar

Didiplis Raf. 1833 (Lythrum L.). Lythraceae. 1 North America

Didissandra C. B. Clarke. 1883. Gesneriaceae. 31 India, China (5), Borneo

Didonica Luteyn et Wilbur. 1977. Ericaceae. 1 Panama

Didymaea Hook. f. 1873. Rubiaceae. 8–10 Mexico, Central America south to Panama

Didymanthus Endl. 1839. Chenopodiaceae. 1 southwestern Australia

Didymaotus N. E. Br. 1925. Aizoaceae. 1 western South Africa

Didymeles Thouars. 1804. Didymelaceae. 2 Madagascar

Didymia Phil. 1886. Cyperaceae. 1 Chile

Didymiandrum Gilly. 1941. Cyperaceae. 3 trop. South America

Didymocarpus Wall. 1819. Gesneriaceae. c. 180 trop. Africa, Madagascar, South and Southeast Asia, Malesia, trop. Australia

Didymocheton Blume = Dysoxylum Blume. Meliaceae

Didymochlamys Hook. f. 1872. Rubiaceae. 2 Panama, Colombia, Venezuela

Didymocistus Kuhlm. 1938. Euphorbiaceae. 1 trop. South America

Didymocolpus S. C. Chen = Acanthochlamys P. C. Kao. Acanthochlamidaceae

Didymodoxa Wedd. (non E. Mey.). 1857. Urticaceae. 2 northern Ethiopia, Kenya, Tanzania, ? Zaire, Zambia, Namibia, South Africa from Natal to Cape

Didymoecium Bremek. = Rennellia Korth. Rubiaceae

Didymopanax Decne. et Planch. 1854 (Schefflera J. R. Forst. et G. Forst.). Araliaceae. 20 trop. America, mostly Brazil; West Indies

Didymopelta Regel et Schmalh. = Astragalus L. Fabaceae

Didymophysa Boiss. 1841. Brassicaceae. 2 Turkey, Caucasus, Middle Asia, Iran, Afghanistan, western Pakistan

Didymoplexiella Garay. 1954. Orchidaceae. 6 Thailand, Malesia

Didymoplexis Griff. 1843. Orchidaceae. 23 trop. East and South Africa, Madagascar, trop. Asia from Himalayas and India to Indochina, Malesia, Oceania east to Fiji

Didymopogon Bremek. 1940. Rubiaceae. 1 Sumatra

Didymosalpinx Keay. 1958. Rubiaceae. 5 trop. Africa

Didymosperma H. A. Wendl. et Drude ex Hookf. = Arenga Labill. Arecaceae

Didymostigma W. T. Wang. 1984 (Chirita Buch. = Ham. ex D. Don). Gesneriaceae. 1 southern China

Didymotheca Hook. f. = Gyrostemon Desf. Gyrostemonaceae

Didyplosandra Wight ex Bremek. 1944. Acanthaceae. 3–7 India, Sri Lanka

Diectomis Kunth. 1815 (Andropogon L.). Poaceae. 3 tropics

Diedropetala Galushko = Delphinium L. Ranunculaceae

Dieffenbachia Schott. 1829. Araceae. c. 30 trop. America, West Indies

Diegodendron Capuron. 1964. Diegodendraceae. 1 Madagascar

Dieltzia P. S. Short. 1989. Asteraceae. 1 western Australia

Dielsantha E. Wimm. 1948. Lobeliaceae. 1 trop. Africa

Dielsia Gilg. 1904. Restionaceae. 1 Australia

Dielsiocharis O. E. Schulz. 1924. Brassicaceae. 1 Iran, Middle Asia (Turkmenistan)

Dielsiochloa Pilg. 1943. Poaceae. 1 Peru, Bolivia, and Chile

Dielsiothamnus R. E. Fr. 1955. Annonaceae. 1 trop. East Africa (Tanzania, Malawi, Mozambique)

Dielytra Cham. et Schlechtd. = Dicentra Benth. Fumariaceae

Dierama K. Koch et C. Bouché. 1855 (Ixia L.). Iridaceae. 44 trop. East and South (39) Africa

Diervilla Mill. 1754. Caprifoliaceae. 3 North America

Dieterlea Lott. 1986. Cucurbitaceae. 1 western Mexico

Dietes Salisb. ex Klatt. 1866 (Iris L.). Iridaceae. 6 South Africa (5), Lord Howe I. (1, D. robinsoniana)

Dieudonnaea Cogn. = Gurania (Schltdl.) Cogn. Cucurbitaceae

Diflugossa Bremek. 1944. Acanthaceae. 16 Himalayas (Nepal, Bhutan), Assam, southwestern China, Malesia

Digastrium (Hackel) A. Camus. 1921 (Ischaemum L.). Poaceae. 2 Australia (Queensland)

Digera Forssk. 1775. Amaranthaceae. 1 (D. muricata) trop. Africa, Madagascar, Socotra, Arabian Peninsula eastward to Afghanistan, Pakistan, India, Sri Lanka, and Malesia

Digitacalia Pippen. 1968. Asteraceae. 4 Mexico

Digitalis L. 1753. Scrophulariaceae. c. 20 Canary Is., Europe, Mediterranean, Asia Minor, Iran, Caucasus; D. grandiflora—eastward to West Siberia

Digitaria Haller. 1768. Poaceae. c. 300 trop. and temp. regions; Russia (6, all areas, except northern)

Digitariella de Winter. 1961 (Digitaria Haller). Poaceae. 1 Namibia

Digitariopsis C. E. Hubb. 1940 (Digitaria Haller). Poaceae. 2 trop. Africa

Diglyphosa Blume. 1825. Orchidaceae. 5 Southeast Asia, Malesia: Malay Peninsula, Sumatra, Philippines (1, D. elmeri)

Dignathe Lindl. 1849. Orchidaceae. 1 Mexico

Dignathia Stapf. 1911. Poaceae. 4 trop. East Africa, northwestern India

Digomphia Benth. 1846. Bignoniaceae. 3 Colombia, Venezuela, Guyana, Brazil

Digoniopterys Arènes. 1946. Malpighiaceae. 1 Madagascar

Digraphis Trin. = Phalaroides Wolf. Poaceae

Diheteropogon (Hack.) Stapf. 1922. Poaceae. 5 trop. and South Africa

Dilatris Bergius. 1767. Haemodoraceae. 4 South Africa (Cape)

Dilepis Suess. et Merxm. = Flaveria Juss. Asteraceae

Dilkea Mast. 1871. Passifloraceae. 5 trop. South America

Dillenia L. 1753. Dilleniaceae. c. 60 Madagascar, Seychelles and Mascarene Is., South and Southeast Asia, north to Himalayas and southern China (Yunnan, Kwangsi, and Kwangtung), Malesia (c. 40 from Sumatra to New Guinea and adjacent islands), trop. eastern Australia (1, D. alata), Melanesia easward to Fiji

Dillwynia Sm. 1805. Fabaceae. 12–16 Australia

Dilobeia Thouars. 1806. Proteaceae. 1 Madagascar

Dilochia Lindl. 1830. Orchidaceae. 7 Thailand (1, D. wallichii), Malesia: 5 Borneo, 2 Malay Peninsula, Sumatra and Java, Philippines (1, D. elmeri)

Dilodendron Radlk. 1878. Sapindaceae. 3 trop. America

Dilomilis Raf. 1838. Orchidaceae. 5 West Indies, Brazil

Dilophia Thomson. 1853. Brassicaceae. 5 Himalayas from Kashmir to Nepal, Tibet; D. salsa—Middle Asia

Dilophotriche (C. E. Hubb.) Jacq.-Fél. 1960. Poaceae. 5 trop. and South Africa

Dimerandra Schltr. = Epidendrum L. Orchidaceae

Dimeresia A. Gray. 1886. Asteraceae. 1 western U.S.

Dimeria R. Br. 1810. Poaceae. c. 40 Madagascar, Mascarene Is., trop. and East Asia, Australia, Oceania; D. neglecta—to southern Far East of Russia

Dimerocostus Kuntze. 1891. Costaceae. 2 Panama and trop. South America

Dimerostemma Cass. 1817. Asteraceae. 6 Brazil and Bolivia

Dimetra Kerr. 1838. Verbenaceae. 1 Thailand

Dimitria Ravenna = Sisymbrium L. Brassicaceae

Dimocarpus Lour. 1790 (Litchi Sonner). Sapindaceae. 6–20 India, southern China, Indochina, Malesia, Australia (2)

Dimorphandra Schott. 1827. Fabaceae. c. 25 trop. America

Dimorphanthera (Drude) F. Muell. ex J. J. Sm. 1886. Ericaceae. 70 Philippines, Moluccas, New Guinea, New Ireland I.

Dimorphocalyx Thwaites. 1861. Euphorbiaceae. 12 India, China, Malesia, Australia

Dimorphocarpa Rollins. 1979. Brassicaceae. 4 southwestern U.S., Mexico

Dimorphochloa S. T. Blake. 1941 (Cleistochloa C. E. Hubb.). Poaceae. 1 Australia (Queensland)

Dimorphocoma F. Muell. et Tate. 1883. Asteraceae. 1 central Australia

Dimorpholepis A. Gray = Triptilodiscus Turcz. Asteraceae

Dimorpholepis (G. M. Barroso) R. M. King et H. Rob. = Grazielia R. M. King et H. Rob. Asteraceae

Dimorphorchis Rolfe = Arachnis Blume. Orchidaceae

Dimorphosciadium Pimenov. 1975. Apiaceae. 1 Middle Asia

Dimorphostemon Kitag. 1939. Brassicaceae. 3 Middle Asia, Siberia, Far East of Russia; Pakistan, India, China, Mongolia

Dimorphotheca Moench. 1794. Asteraceae. 7 South Africa

Dinacria Harv. ex Sond. = Crassula L. Crassulaceae

Dinebra Jacq. 1809. Poaceae. 3 trop. Africa, Madagascar, trop. Asia

Dinema Lindl. = Epidendrum L. Orchidaceae

Dinemagonum A. Juss. 1843. Malpighiaceae. 3 Chile

Dinemandra A. Juss. 1840. Malpighiaceae. 6 Peru, Chile

Dinetopsis Roberty = Porana Burm. f. Convolvulaceae

Dinetus Buch.-Ham. ex Sweet. 1825 (Porana Burm. f.). Convolvulaceae. 6 Himalayas, Assam, northern Burma, southwestern China, Malaysia

Dinizia Ducke. 1922. Fabaceae. 1 Guyana, Brazil

Dinklageanthus Melch. ex Mildbr. = Dinklageodoxa Heine et Sandwith. Bignoniaceae

Dinklageella Mansf. 1934. Orchidaceae. 2 trop. West Africa

Dinklageodoxa Heine et Sandwith. 1962. Bignoniaceae. 1 trop. West Africa (Liberia)

Dinocanthium Bremek. = Pyrostria Comm. ex Juss. Rubiaceae

Dinochloa Büse. 1854. Poaceae. 25 Southeast Asia from Burma to Philippines

Dinophora Benth. 1849. Melastomataceae. 2 trop. West Africa

Dinoseris Griseb. 1879. Asteraceae. 1 southern Bolivia, northwestern Argentina

Dintera Stapf ex Schinz. 1900. Scrophulariaceae. 1 trop. Africa

Dinteracanthus C. B. Clarke ex Schinz = Ruellia L. Acanthaceae

Dinteranthus Schwantes. 1926. Aizoaceae. 6 southeastern Namibia, northwestern South Africa

Dioclea Kunth. 1824. Fabaceae. 30 tropics, esp. America

Diodella (Torr. et A. Gray) Small. 1913 (Diodia L.). Rubiaceae. 1 Florida, West Indies

Diodia L. 1753. Rubiaceae. c. 50 mostly America from southern U.S. and Mexico to trop. South America, West Indies, trop. Africa, Madagascar, Mascarene Is., a few species on Socotra and in trop. Asia

Dioecrescis Tirveng. 1983 (Gardenia Ellis). Rubiaceae. 1 South and Southeast Asia

Diogenesia Sleumer. 1934. Ericaceae. 13 Andes

Diogoa Exell ex Mendonça. 1951. Olacaceae. 1 trop. Africa

Dioicodendron Steyerm. 1963. Rubiaceae. 2 northwestern South America

Dion Lindl. = Dioon Lindl. Zamiaceae

Dionaea Ellis. 1773. Droseraceae. 1 southeastern U.S. (North Carolina)

Dioncophyllum Baill. 1890. Dioncophyllaceae. 1 trop. West Africa

Dionycha Naudin. 1850. Melastomataceae. 2 Madagascar

Dionychastrum A. Fern. et R. Fern. 1956 (Dionycha Naudin). Melastomataceae. 1 trop. East Africa

Dionysia Fenzl. 1843. Primulaceae. 42 Asia Minor (3), Iran (25, 22 are endemics). northern Iraq, Afghanistan, Middle Asia (4)

Dioön Lindl. 1843. Zamiaceae. 10 Mexico, Honduras

Diopogon A. Jord. et Fourr. = Sempervivum L. Crassulaceae

Diora Ravenna. 1987. Asphodelaceae. 1 Peru

Dioscorea L. 1753. Dioscoreaceae. c. 600–650 trop., subtrop. and warm–temp. regions (except Sakhara and Arabian Peninsula); 3 spp. extending to Caucasus and Russian Far East

Dioscoreophyllum Engl. 1895. Menispermaceae. 10 trop. Africa

Diosma L. 1753. Rutaceae. 28 South Africa

Diosphaera Buser = Campanula L. Campanulaceae

Diospyros L. 1753. Ebenaceae. c. 500 tropics, esp. Malesia, a few species in West Asia, Japan, and southeastern U.S.

Diostea Miers. 1870 (Baillonia Bocquillon). Verbenaceae. 4 South America

Diotacanthus Benth. = Phlogacanthus Nees. Acanthaceae

Diothonea Lindl. 1834. Orchidaceae. 7 western trop. South America

Diotis Schreb. = Ceratoides Gagnebin. Chenopodiaceae

Diotocranus Bremek. = Mitrasacmopsis Jovet. Rubiaceae

Dipanax Seem. = Tetraplasandra A. Gray. Araliaceae

Dipcadi Medik. 1790. Hyacinthaceae. 30 Mediterranean, Africa, Madagascar, Socotra, Iraq, Iran, Afghanistan, Pakistan, India

Dipelta Maxim. 1877. Caprifoliaceae. 4 China

Dipentodon Dunn. 1911. Dipentodontaceae. 1 eastern Himalayas, northeastern India, Burma, Tibet, southwestern China

Diphalangium Schauer. 1847. Alliaceae. 1 Mexico

Diphasia Pierre. 1898. Rutaceae. 6 trop. Africa, Madagascar

Diphasiopsis Mendonça. 1961. Rutaceae. 2: 1 trop. Africa; D. fadenii—Kenya

Diphelypaea Nicolson = Phelypaea L. Orobanchaceae

Dipholis A. DC. = Sideroxylon L. Sapotaceae

Diphylax Hook. f. 1889. Orchidaceae. 3 northeastern India, China (1)

Diphyllarium Gagnep. 1915. Fabaceae. 1 Indochina

Diphylleia Michx. 1803. Berberidaceae. 3 western China, Japan, eastern North America; D. grayi also—Sakhalin, Kuril Is.

Diphysa Jacq. 1760. Fabaceae. 15–20 trop. America from Mexico to Colombia and Venezuela

Dipidax Lawson ex Salisb. = Onixotis Raf. Melanthiaceae

Diplachne P. Beauv. 1812 (Leptochloa P. Beauv.). Poaceae. 15 tropics and subtropics

Diplacrum R. Br. 1810. Cyperaceae. 6 tropics and subtropics (Japan, Taiwan)

Dipladenia A. DC. = Mandevilla Lindl. Apocynaceae

Diplandra Hook. et Arn. = Lopezia Cav. Onagraceae

Diplandrorchis S. C. Chen. 1979. Orchidaceae. 1 China

Diplanthera Thouars = Halodule Endl. Cymodoceaceae

Diplarche Hook. f. et Thomson. 1854. Ericaceae. 2 eastern Himalayas, southwestern China

Diplarpea Triana. 1867. Melastomataceae. 1 Colombia

Diplarrhena Labill. 1800. Iridaceae. 2 eastern and southeastern Australia, Tasmania

Diplasia Pers. 1805. Cyperaceae. 3 Indochina (1), West Indies, and trop. South America (2)

Diplaspis Hook. f. 1847. Apiaceae. 1 Southeast Asia, ? Tasmania0

Diplatia Tiegh. 1894. Loranthaceae. 3 northwestern (1), northern, central, and eastern Australia

Diplazoptilon Y. Ling. 1965. Asteraceae. 2 southwestern China

Diplectria (Blume) Rchb. 1841. Melastomataceae. 11 Burma, southern China, Indochina, Malesia, except Lesser Sunda Is.

Diplobryum C. Cusset. 1972. Podostemaceae. 1 southern Vietnam

Diplocarex Hayata = Carex L. Cyperaceae

Diplocaulobium (Rchb. f.) Kraenzl. 1910 (Dendrobium Sw.). Orchidaceae. 70 Malesia, Australia, Oceania

Diplocentrum Lindl. 1832. Orchidaceae. 2 India, Sri Lanka

Diploclisia Miers. 1851. Menispermaceae. 3 India, Sri Lanka, Burma, southern China, Indochina, Malesia (excl. Lesser Sunda Is.)

Diplocyatha N. E. Br. = Orbea Haw. Asclepiadaceae

Diplocyclos (Endl.) Post et Kuntze. 1903. Cucurbitaceae. 4 trop. Africa (3); D. palmatus—trop. Asia, Malesia, Australia, western Oceania

Diplodiscus Turcz. 1858. Tiliaceae. 7 Sri Lanka (1, D. verrucosus), Malay Peninsula, Borneo, Philippines

Diplofatsia Nakai = Fatsia Decne. et Planch. Araliaceae

Diploglottis Hook. f. 1862. Sapindaceae. 10 eastern Malesia, eastern Australia (8), New Caledonia, Vanuatu

Diplokeleba N. E. Br. 1894. Sapindaceae. 2 trop. South America

Diploknema Pierre. 1884. Sapotaceae. 9–10 Himalayas (Nepal, Bhutan), northern India, Burma, southwestern and southern China, Indochina (Thailand, Cambodia, Vietnam), Malesia (Borneo, Philippines, and Amboina)

Diplolabellum F. Maek. 1935. Orchidaceae. 1 Korea

Diplolaena R. Br. 1814. Rutaceae. 6–8 western Austrlia

Diplolegnon Rusby. 1900 (Corytoplectus Oerst.). Gesneriaceae. 1 South America

Diplolepis R. Br. 1810. Asclepiadaceae. 2 China

Diplolophium Turcz. 1847. Apiaceae. 5–7 trop. and South Africa, Madagascar

Diplomeris D. Don. 1825. Orchidaceae. 2–4 Himalayas (Sikkim, Buthan), Burma, China, Indochina (1, D. pulchella)

Diploön Cronquist. 1946. Sapotaceae. 1 Venezuela, Guyana, Peru, Bolivia, Brazil

Diplopanax Hand.-Mazz. 1933. Mastixiaceae. 1 China, Indochina

Diplopeltis Endl. 1837. Sapindaceae. 5 northeastern Australia

Diplophractum Desf. = Colona Cav. Tiliaceae

Diplopilosa Dvorák. 1967 (or = Hesperis L.). Brassicaceae. 1 Asia Minor

Diplopogon R. Br. 1810. Poaceae. 1 western Australia

Diploprora Hook. f. 1890. Orchidaceae. 5 India, Sri Lanka, Burma, Thailand, China (Yunnan to Hongkong, Hainan), Taiwan

Diplopterys A. Juss. 1838 (Banisteriopsis C. B. Rob. et Small). Malpighiaceae. 4 trop. America: Mexico (1, D. mexicana), Colombia, Venezuela, Guyana, French Guyana, Ecuador, Peru, Brazil

Diplorhynchus Welw. ex Ficalho et Hiern. 1881. Apocynaceae. 1 trop. and South Africa

Diplosoma Schwantes. 1926. Aizoaceae. 2 South Africa (Cape)

Diplospora DC. 1830 (Tricalysia A. Rich. ex DC.). Rubiaceae. 25 India, Sri Lanka, southern China, Indochina, Malesia, Australia

Diplostephium Kunth. 1820. Asteraceae. 90 trop. Andes from Colombia and Venezuela to Bolivia and northern Chile

Diplostigma K. Schum. 1895. Asclepiadaceae. 1 East Africa

Diplotaenia Boiss. 1844. Apiaceae. 2 Turkey, Iran

Diplotaxis DC. 1821. Brassicaceae. 27 Europe, Caucasus, Mediterranean, West Asia eastward to Himalayas (India, Nepal)

Diplothemium Mart. = Allogoptera Nees. Arecaceae

Diplotropis Benth. 1837. Fabaceae. 7–8 Colombia, Venezuela, the Guyanas, and Brazil

Diplusodon Pohl. 1827. Lythraceae. 57 Brazil

Diplycosia Blume. 1826. Ericaceae. c. 100 East and Southeast Asia, Malesia, esp. New Guinea

Dipodium R. Br. 1810. Orchidaceae. 22 Southeast Asia, Malesia to New Guinea and New Ireland, Palau Is., Australia, Tasmania, Solomon Is. (1, D. picutm), New Caledonia, Vanuatu

Dipogon Liebm. 1854. Fabaceae. 1 South Africa

Dipoma Franch. 1886. Brassicaceae. 1 Tibet

Diposis DC. 1829. Apiaceae. 3 temp. South America

Dipsacus L. 1753. Dipsacaceae. 16 temp. Eurasia, Mediterranean, trop. Africa, Asia (Turkey, Caucasus, Iran, Iraq, Turkmenistan, Kazakhstan, Kirgizstan, Tajikistan, Afghanistan, Pakistan, West Siberia, China, India, Nepal)

Dipteracanthus Nees = Ruellia L. Acanthaceae

Dipteranthemum F. Muell. = Ptilotus P. Browne. Amaranthacae

Dipteranthus Barb. Rodr. 1882. Orchidaceae. 8 trop. South America

Dipterella Moggi, nom. illeg. Brassicaceae

Dipterocarpus C. F. Gaertn. 1805. Dipterocarpaceae. 70 India, Sri Lanka, eastward to southern China (3), Indochina, Malesia

Dipterocome Fisch. et C. A. Mey. 1835. Asteraceae. 1 eastern Mediterranean, Asia Minor, Iraq, Caucasus, Middle Asia, Iran, Afghanistanh

Dipterocypsela S. F. Blake. 1945. Asteraceae. 1 Colombia

Dipterodendron Radlk. = Dilodendron Radlk. Sapindaceae

Dipteronia Oliv. 1889. Aceraceae. 2 central and southern China

Dipteropeltis Hallier f. 1899. Convolvulaceae. 2 trop. West Africa

Dipterostele Schltr. 1921. Orchidaceae. 1 Ecuador

Dipterygia C. Presl ex DC. = Gymnophyton Clos. Apiaceae

Dipterygium Decne. 1835. Capparaceae. 1 Egypt, West and Southwest Asia to western Pakistan

Dipteryx Schreb. 1791. Fabaceae. 10 Central and trop. South America

Diptychandra Tul. 1843. Fabaceae. 3 Bolivia, Brazil, Paraguay

Diptychocarpus Trautv. 1860. Brassicaceae. 1 southeastern Europe part of Russia, Caucasus, Middle Asia, Iran, Afghanistan, western Pakistan, western China

Dipyrena Hook. 1830. Verbenaceae. 1 temp. South America

Dirachma Schweinf. ex Balf. f. 1884. Dirachmaceae. 2 central Somalia (1), Socotra (1)

Dirca L. 1753. Thymelaeaceae. 2 North America

Dirhamphis Krapov. 1970. Malvaceae. 2 Mexico (D. mexicana). Bolivia and Paraguay (M. balansae)

Dirichletia Klotzsch = Carphalia Juss. Rubiaceae

Disa Bergius. 1767. Orchidaceae. 120–130 trop. and South (116) Africa, Madagascar, Mascarene Is., Arabian Peninsula

Disaccanthus Greene = Streptanthus Nutt. Brassicaceae

Disakisperma Steud. 1854 (Leptochloa P. Beauv.). Poaceae. 1 Mexico

Disanthus Maxim. 1866. Hamamelidaceae. 1 continental China, Japan

Disaster Gilli = Trymalium Fenzl. Rahmnaceae

Discadia Raf. = Oxypolis Raf. Apiaceae

Discalyxia Markgr. 1927. Apocynaceae. 3 New Guinea

Discanthus Spruce = Cyclanthus Poit. Cyclanthaceae

Discaria Hook. 1829. Rhamnaceae. 15 Australia, New Zealand, South America

Dischidanthus Tsiang. 1936. Asclepiadaceae. 1 southern China, Indochina

Dischidia R. Br. 1810. Asclepiadaceae. 23 Sri Lanka, northeastern India (Assam), China, Taiwan, Indochina, Malesia, Australia

Dischidiopsis Schltr. = Dischidia R. Br. Asclepiadaceae

Dischisma Choisy. 1824. Selaginaceae. 11 South Africa (Cape)

Dischistocalyx T. Anderson ex Benth. = Distichocalyx T. Anderson ex Benth. Acanthaceae

Disciphania Eichler. 1864. Menispermaceae. c. 25 trop. America, esp. Amazonia, West Indies

Discipiper Trel. et Stehle = Piper L. Piperaceae

Discocactus Pfeiff. 1837. Cactaceae. 8 Brazil, eastern Bolivia, Paraguay

Discocalyx (A. DC.) Mez. 1902. Myrsinaceae. c. 50 Borneo (?), Philippines, Sulawesi, Moluccas, New Guinea (12), New Britain, New Ireland, Melanesia, Polynesia: Fiji (3), Tonga (1)

Discocapnos Cham. et Schltdl. 1826. Fumariaceae. 1 (D. mundtii) South Africa (Cape)

Discocarpus Klotzsch. 1841. Euphorbiaceae. 5 Guyana, Brazil

Discoclaoxylon (Muell. Arg.) Pax et K. Hoffm. 1914 (Claoxylon A. Juss.). Euphorbiaceae. 4 trop. Africa from Sierra Leone and the islands in the Gulf of Guinea eastward to Uganda (1, D. hexandrum)

Discocleidion (Muell. Arg.) Pax et K. Hoffm. 1914. Eurphobiaceae. 3 China, Ryukyu Is.

Discocnide Chew. 1965. Urticaceae. 1 Mexico, Guatemala

Discocrania (Harms) Kral = Cornus L. Cornaceae

Discoglypremna Prain. 1911. Euphorbiaceae. 1 trop. West Africa

Discolobium Benth. 1837. Fabaceae. 8 Brazil, Paraguay, Argentina

Discophora Miers. 1852. Icacinaceae. 2 Panama, Colombia, Peru, Guyana, Brazil

Discopleura DC. = Ptilimnium Raf. Apiaceae

Discopodium Hochst. 1844. Solanaceae. 2 montane trop. Africa: Fernando Po, Cameroun, Zaire, Rwanda, Burundi, Sudan, Ethiopia, Uganda, Kenya, Tanzania, Malawi

Discospermum Dalzell = Diplospora DC. Rubiaceae

Discovium Raf. = Lesquerella S. Watson. Brassicaceae

Discyphus Schltr. 1919. Orchidaceae. 1 Venezuela, Trinidad

Diselma Hook. f. 1857. Cupressaceae. 1 Tasmania

Disepalum Hook. f. 1860. Annonaceae. 6 China, Indochina, western Malesia

Disocactus Lindl. 1845. Cactaceae. 16 Central America, 1 extending to northern South America; 2 West Indies

Disparago Gaertn. 1791. Asteraceae. 7 South Africa

Disperis Sw. 1800. Orchidaceae. 75–90: trop. and South (c. 50) Africa; Mascarene Is., India, Sri Lanka, Thailand, Java, Philippines (1, D. philippinensis), New Guinea, Australia (1, D. alata)

Disphyma N. E. Br. 1925 (Mesembryanthemum L.). Aizoaceae. 3–4 South Africa, 1 (D. crassifolium) temp. southern and southeastern Australia, Tasmania, and New Zealand

Disporopsis Hance. 1883. Convallariaceae. 4 Thailand, southeastern China, Taiwan, Philippines

Disporum Salisb. 1825. Convallariaceae. 20 Himalayas, East and Southeast Asia, North America

Dissanthelium Trin. 1836. Poaceae. 16 western America from California to Bolivia

Dissiliaria F. Muell. ex Baill. 1867. Euphorbiaceae. 1–2 northeastern Australia

Dissocarpus F. Muell. 1858. Chenopodiaceae. 4 Australia

Dissochaeta Blume. 1831. Melastomataceae. 20 Indo-Malesia

Dissochondrus (W. F. Hillebr.) Kuntze. 1891. Poaceae. 1 Hawaiian Is.

Dissomeria Hook. f. ex Benth. 1849. Flacourtiaceae. 2 trop. West Africa

Dissosperma Soják. 1986 (Fumaria L.). Fumariaceae. 1 Europe

Dissothrix A. Gray. 1851. Asteraceae. 1 northeastern Brazil

Dissotis Benth. 1849. Melastomataceae. 140 trop. and South Africa

Disteganthus Lem. 1847. Bromeliaceae. 2 Guyana, Brazil

Distemon Wedd. = Neodistemon Babu et A. N. Henry. Urticaceae

Distemonanthus Benth. 1865. Fabaceae. 1 trop. West Africa from Sierra Leone to Gabon

Distephanus Cass. 1817 (or = Vernonia Schreb.). Asteraceae. 40 trop. Africa, Madagascar, Mauritius, Southeast Asia

Disterigma (Klotzsch) Nied. 1889. Ericaceae. 25 trop. America from Guatemala to Bolivia and Guyana

Distichella Tiegh. 1896 (Dendrophthora Eichler). Viscaceae. 3 West Indies

Distichia Nees et Meyen. 1843. Juncaceae. 4 South America (Andes)

Distichlis Raf. 1819. Poaceae. 5–12 Australia, America from Canada to Argentina

Distichocalyx T. Anderson ex Benth. 1876. Acanthaceae. 20 trop. Africa

Distichoselinum Garc. Mart. et Silvestre. 1983 (Elaeoselinum W. D. J. Koch ex DC.). Apiaceae. 1 Portugal, Spain

Distichostemon F. Muell. 1857. Sapindaceae. 6 trop. Australia

Distictella Kuntze. 1903. Bignoniaceae. 11 trop. America from Costa Rica to Brazil, Tobago I.

Distictis Mart. ex Meisn. 1840. Bignoniaceae. 10 trop. America from Mexico to Brazil, West Indies

Distoecha Phil. = Hypochaeris L. Asteraceae

Distomocarpus O. E. Schulz = Rytidocarpus Coss. Brassicaceae

Distrianthes Danser. 1929. Loranthaceae. 1 New Guinea

Distyliopsis P. K. Endress. 1970. Hamamelidaceae. 3–4 Southeast Asia from Burma to southern China and Taiwan, Malesia eastward to New Guinea

Distylium Siebold et Zucc. 1841. Hamamelidaceae. 18 Assam, China, Japan, Indochina, Malesia, Central America (2)

Distylodon Summerh. 1966. Orchidaceae. 1 trop. East Africa (Uganda)

Disynaphia Hook. et Arn. ex DC. 1838 (Eupatorium L.). Asteraceae. 12 Brazil, Uruguay

Disynstemon R. Vig. 1951. Fabaceae. 1 Madagascar

Ditassa R. Br. 1810. Asclepiadaceae. 75 South America

Ditaxis Vahl ex A. Juss. = Argythamnia P. Browne. Euphorbiaceae

Ditepalanthus Fagerl. 1938. Helosiaceae. 1 Madagascar

Dithrix Schltr. 1905. Orchidaceae. 1 Afghanistan, Pakistan

Dithyrea Harv. 1845. Brassicaceae. 5 southwestern U.S., Mexico

Dithyridanthus Garay. 1980 (or = Schiedeella Schltr.). Orchidaceae. 1 Mexico

Dithyrostegia A. Gray. 1851 (Angianthus Wendl.). Asteraceae. 3 Namibia, western Australia

Ditrichospermum Bremek. 1944 (or = Strobilanthes Blume). Acanthaceae. 1 Assam, Burma

Ditta Griseb. 1861. Euphorbiaceae. 1 West Indies

Dittoceras Hook. f. 1883. Asclepiadaceae. 3 eastern Himalayas, Thailand

Dittostigma Phil. 1870 (Nicotiana L.). Solanaceae. 1 Argentina

Dittrichia Greuter. 1973 (Inula L.). Asteraceae. 2 Europe, Mediterranean

Diuranthera Hemsl. 1902 (Chlorophytum Ker=Gawl.). Asphodelaceae. 2 western China

Diuris Sm. 1798. Orchidaceae. 38 Java, Australia, and Tasmania (37)

Diyaminauclea Ridsd. 1978. Rubiaceae. 1 Sri Lanka

Dizygostemon (Benth.) Radlk. ex Wettst. 1891. Scrophulariaceae. 1 Brazil

Dizygotheca N. E. Br. = Schefflera J. R. Forst. et G. Forst. Araliaceae

Djaloniella P. Taylor. 1963. Gentianaceae. 1 trop. West Africa

Djinga C. Cusset. 1987. Podostemaceae. 1 Cameroun

Dobera Juss. 1789. Salvadoraceae. 2 trop. East Africa from Sudan, Ephiopia and Somalia through Uganda, Kenya, and Tanzania to Mozamibique; southern Arabian Peninsula, Pakistan, northwestern India

Dobinea Buch.-Ham. ex D. Don. 1825. Anacardiaceae. 2 northwestern India, southwestern China, central and eastern Himalayas

Docynia Decne. 1874. Rosaceae. 2 Himalayas (Nepal, Bhutan, Sikkim), northeastern India, Bangladesh, Burma, China, Indochina (northern Thailand and northern Vietnam), northeastern India, Burma, China, Indochina (Vietnam)

Dodartia L. 1753. Scrophulariaceae. 1 (D. orientalis) southern European part of Russia, Caucasus, Middle Asia, West Siberia; Asia Minor, northwestern Iran, Afghanistan, northwestern China, Mongolia

Dodecadenia Nees. 1831 (Litsea Lam.). Lauraceae. 2 eastern Himalayas, southern Tibet, Burma

Dodecahema Reveal et Hardham. 1989 (Centrostegia A. Gray). Polygonaceae. 1 southern California

Dodecastigma Ducke. 1932. Euphorbiaceae. 3 trop. South America

Dodecatheon L. 1753. Primulaceae. 16 arctic and temp. North America, Chukotski (1, D. frigidum)

Dodonaea Hill. 1753. Sapindaceae. 68 tropics and subtropics, esp. Australia (59), absent on Galapagos Is.

Dodsonia Ackerman. 1979. Orchidaceae. 2 Ecuador

Doellingeria Nees = Aster L. Asteraceae

Doerpfeldia Urb. 1924. Rhamnaceae. 1 Cuba

Doga (Baill.) Baill. ex Nakai = Storckiella Seem. Fabaceae

Dolichandra Cham. 1832. Bignoniaceae. 1 southern Brazil, Paraguay, Argentina

Dolichandrone (Fenzl) Seem. 1862. Bignoniaceae. 13 trop. East Africa, Madagascar, trop. Asia from India to southern China and Indochina, Malesia from Malay Peninsula to New Guinea, Solomon Is., northern Australia, New Caledonia

Dolichlasium Lag. 1811 (Trixis P. Browne). Asteraceae. 1 Argentina

Dolichocentrum (Schltr.) Brieger = Dendrobium Sw. Orchidaceae

Dolichochaete (C. E. Hubb.) J. B. Phipps = Tristachya Nees. Poaceae

Dolichodelphys K. Shum. et Krause. 1908. Rubiaceae. 1 Peru

Dolichoglottis B. Nord. 1978. Asteraceae. 2 New Zealand

Dolichokentia Becc. = Cyphokentia Brongn. Arecaceae

Dolicholobium A. Gray. 1859. Rubiaceae. 30 Philippines, Moluccas, New Guinea, New Hebrides, Solomon Is., Vanuatu, Fiji (3)

Dolicholoma D. Fang et W. T. Wang. 1983 (Didymocarpus Wall.). Gesneriaceae. 1 China

Dolichometra K. Schum. 1904. Rubiaceae. 1 trop. East Africa (Tanzania)

Dolichopetalum Tsiang. 1973. Asclepiadaceae. 1 southern China

Dolichopsis Hassl. 1907. Fabaceae. 2 South America

Dolichopsis (Hook. f.) Brieger = Eria Lindl. Orchidaceae

Dolichopterys Kosterm. = Lophopterys A. Juss. Malpighiaceae

Dolichorhynchus Hedge et Kit Tan. 1987. Brassicaceae. 1 Arabian Peninsula

Dolichorrhiza (Pojark.) Galushko. 1970 (or = Ligularia Cass.) Asteraceae. 4 Caucasus, Iran

Dolichos L. 1753. Fabaceae. c. 70 trop. and subtrop. Old World from Africa (63) to East Asia and Australia

Dolichostachys Benoist. 1962. Acanthaceae. 1 Madagascar

Dolichostegia Schltr. 1915. Asclepiadaceae. 1 Phillipines

Dolichostemon Bonati. 1924. Scrophulariaceae. 1 Indochina

Dolichothrix Hilliard et B. L. Burtt. 1981. Asteraceae. 1 South Africa

Dolichoura Brade. 1959. Melastomataceae. 1 Brazil

Dolichovigna Hayata = Vigna Savi. Fabaceae

Doliocarpus Rol. 1756. Dilleniaceae. 40 Central and trop. South America, West Indies

Dollinera Endl. = Desmodium Desv. Fabaceae

Dolomiaea DC. 1833. Asteraceae. 5 Himalayas, Tibet

Dombeya Cav. 1787. Sterculiaceae (? or Malvaceae). 200 Africa, Madagascar, Mascarene Is.

Domeykoa Phil. 1860. Apiaceae. 4 Andes of Peru and Chile

Dominella E. Wimm. = Lysipomia Kunth. Lobeliaceae

Domingoa Schltr. 1913. Orchidaceae. 3 West Indies, Mexico (1, D. kienastii)

Dominia Fedde = Uldinia J. M. Black. Apiaceae

Domkeocarpa Markgr. = Tabernaemontana L. Apocynaceae

Domohinea Leandri. 1941. Euphorbiaceae. 1 Madagascar

Donatia J. R. Forst. et G. Forst. 1775. Donatiaceae. 2 Tasmania, New Zealand, subantarctic South America

Donax Lour. 1790. Maranthaceae. 6 India, Burma, southwestern China, Indochina, Malesia, Oceania

Donella Pierre ex Baill. = Chrysophyllum L. Sapotaceae

Donepea Airy Shaw = Douepea Cambess. Brassicaceae

Doniophyton Wedd. 1855. Asteraceae. 2 Chile, Argentina

Donnellsmithia J. M. Coult. et Rose. 1890. Apiaceae. 15–20 Mexico and Central America

Dontostemon Andrz. ex C. A. Mey. 1831. Brassicaceae. 10 Siberia and Russian Far East (8), Himalayas (northwestern India and Nepal), China, Mongolia, Korea, Japan

Doona Thwaites = Shorea Roxb. ex C. F. Gaertn. Dipterocarpaceae

Dopatrium Buch.-Ham. ex Benth. 1835. Scrophulariaceae. 12 trop. Africa, Asia, Australia; D. junceum—extending to Middle Asia

Doratoxylon Thouars ex Benth. et Hook. f. 1862. Sapindaceae. 1 Mascarene Is.

Dorema D. Don. 1831. Apiaceae. 12 Southwest (incl. Caucasus) and Middle Asia

Doricera Verdc. 1983. Rubiaceae. 1 Mascarene Is.

Doritis Lindl. 1833. Orchidaceae. 2 northeastern India, Burma, Thailand, Laos, Cambodia, Vietnam, China (Hainan), Malesia (Malay Peninsula, Sumatra, Borneo)

Dorobaea Cass. 1827 (Senecio L.). Asteraceae. 1 Andes

Doronicum L. 1753. Asteraceae. 35 temp. Eurasia, esp. (14) southern European part of Russia, Caucasus, Middle Asia, and southern Siberia; North Africa

Dorothea Wernham = Aulacocalyx Hook. f. Rubiaceae

Dorotheanthus Schwantes. 1927. Aizoaceae. 5 southwestern South Africa

Dorstenia L. 1753. Moraceae. 105 trop. America (45), trop. Africa and Madagascar (58), Arabian Peninsula (1), southern India and Sri Lanka (1, D. indica)

Doryalis Warb. = Dovyalis E. Mey. ex Arn. Flacourtiaceae

Doryanthes Corrêa. 1802. Doryanthaceae. 2 Australia (southeastern Queensland and coastal New South Wales)

Dorycnium Mill. 1754 (Lotus L.). Fabaceae. 10 Central Europe, Mediterranean, Crimea, Caucasus, Iraq, Iran

Dorycnopsis Boiss. 1840 (Anthyllis Adans.). Fabaceae. 1 Iberian Peninsula, southern France, Corsica and Sardinia Is., Italy, Morocco

Doryphora Endl. 1837. Monimiaceae. 2 northeastern Australia

Dorystaechas Boiss. et Heldr. ex Benth. 1848. Lamiaceae. 1 Asia Minor

Dorystephania Warb. 1904. Asclepiadaceae. 1 Philippines

Doryxylon Zoll. 1857. Euphorbiaceae. 1 Java, Philippines

Dossifluga Bremek. 1944. Acanthaceae. 3 Nepal, Thailand

Dossinia C. Morren. 1848. Orchidaceae. 1 Borneo

Douepea Cambess. 1835. Brassicaceae. 1 western Pakistan, northwestern India

Douglasia Lindl. = Androsace L. Primulaceae

Douradoa Sleumer. 1984. Olacaceae. 1 Brazil

Dovea Kunth. 1841 (Chondropetalum Rottb.). Restionaceae. 1 D. macrocarpa) South Africa (Cape)

Dovyalis E. Mey. ex Arn. 1841. Flacourtiaceae. c. 20 trop. (from Guinea Bissau to Somalia and south to Angola and Mozambique) and South Africa; 1—extending from Socotra to Sri Lanka and New Guinea

Downingia Torr. 1857. Lobeliaceae. 11 western North America, 1 extends to Chile

Doxantha Miers = Macfadyena A. DC. Bignoniaceae

Doxanthemum D. Hunt. 1864. Bignoniaceae. 1 North America

Doyerea Grosourdy. 1864. Cucurbitaceae. 1 Mexico, Central America, Colombia, northern Venezuela, West Indies

Draba L. 1753. Brassicaceae. 300 arctic and temp. Northern Hemisphere, esp. (c. 110) Russia, Caucasus, Middle Asia; montane Central and South America

Drabastrum (F. Muell.) O. E. Schulz. 1924. Brassicaceae. 1 southeastern Australia

Drabopsis K. Koch. 1841. Brassicaceae. 1 Caucasus, Middle Asia, Turkey, Iraq, Iran, Afghanistan, western Pakistan, western Himalayas

Dracaena Vand. ex L. 1767. Dracaenaceae. c. 50 Canary Is., trop. Old World, Central America (1)

Dracocephalum L. 1753. Lamiaceae. c. 70 temp. Eurasia, Mediterranean, North America (1)

Dracomonticola Linder et Kurzweil. 1993 (Neobolusia Schltr.). Orchidaceae. 1 South Africa (Drakensberg)

Dracontioides Engl. 1911. Araceae. 1 southern Brazil

Dracontium L. 1753. Araceae. 15 trop. America

Dracontomelon Blume. 1850–1851. Anacardiaceae. 8 China, Indochina, Malesia, Solomon Is., Vanuatu, Fiji, Tongo, Samoa

Dracophilus (Schwantes) Dinter et Schwantes. 1927 (or = Juttadinteris Schwantes). Aizoaceae. 3–4 southwestern Namibia, South Africa

Dracophyllum Labill. 1800. Epacridaceae. 48 Australia, Tasmania, Lord Howe I., New Caledonia (7), New Zealand (c. 35)

Dracopis Cass. 1825 (Rudbeckia L.).

Asteraceae. 1 southern U.S., northern Mexico
Dracosciadium Hilliard et B. L. Burtt. 1986. Apiaceae. 2 South Africa (Natal)
Dracula Luer. 1978. Orchidaceae. c. 50 trop. America: Guatemala, Honduras, Nicaragua, Costa Rica, Panama, Colombia, Ecuador, Peru
Dracunculus Hill. 1753. Araceae. 3 Mediterranean
Drakaea Lindl. 1840. Orchidaceae. 4 southwestern Australia
Drake-Brockmania Stapf. 1912. Poaceae. 2 East Africa: Sudan, Ethiopia, Somalia, Kenya, Tanzania
Draperia Torr. 1868. Hydrophyllaceae. 1 California
Drapetes Banks ex Lam. 1792. Thymelaeaceae. 4 antarctic South America (Tierra del Fuego), Falkland Is.
Dregea E. Mey. 1838. Asclepiadaceae. 12 trop. and South Africa, trop. Asia, China, Taiwan, Malesia
Dregeochloa Conert = Danthonia DC. Poaceae
Drejera Nees. 1847. Acanthaceae. 4 trop. America
Drejerella Lindau = Justicia L. Acanthaceae
Drepanocarpus G. Mey. = Machaerium Pers. Fabaceae
Drepanocaryum Pojark. 1954. Lamiaceae. 1 Middle Asia
Drepanostachyum Keng f. = Sinarundinaria Nakai. Poaceae
Drepanostemma Jum. et H. Perrier. 1911. Asclepiadaceae. 1 Madagascar
Dresslerella Luer. 1976. Orchidaceae. 8 Nicaragua, Costa Rica, Panama, Colombia, Ecuador, Peru
Dressleria Dodson. 1975. Orchidaceae. 4 Central America
Dressleriopsis Dwyer = Lasianthus Jack. Rubiaceae
Dresslerothamnus H. Rob. 1978. Asteraceae. 4 Costa Rica, Panama, Colombia
Driessenia Korth. 1844. Melastomataceae. 14: 3 Sumatra, Java, Sulawesi; 11 Borneo

Drimia Jacq. ex Willd. 1799. Hyacinthaceae. 15 trop. East and South Africa
Drimiopsis Lindl. et Paxton. 1851. Hyacinthaceae. 15 trop. and South Africa
Drimycarpus Hook. f. 1862. Anacardiaceae. 2 Himalayas (Nepal, Bhutan), northeastern India, Thailand, southern China, Malay Peninsula, Sumatra, Borneo
Drimys J. R. Forst. et G. Forst. 1775. Winteraceae. 70 Philippines, Borneo, Lesser Sunda Is., New Guinea, Australia, Tasmania, New Zealand, New Caledonia, South America
Droceloncia J. Léonard. 1959. Euphorbiaceae. 1 trop. Africa
Droguetia Gaudich. 1830. Urticaceae. 7 Africa from Cameroun, Fernando Po, and Zaire to northern Ethiopia, through highland of eastern Africa southward to Cape, Madagascar (2), Comoro Is., Reunion I. (2), Yemen, southern India, China, Java
Droogmansia De Wild. 1902. Fabaceae. 24 trop. Africa from Guinea to Zaire, Tanzania (1), Angola, Zambia, Zimbabwe (1)
Drosanthemopsis Rauschert = Drosanthemum Schwantes. Aizoaceae
Drosanthemum Schwantes. 1927. Aizoaceae. 95–100 southern Namibia, South Africa
Drosera L. 1753. Droseraceae. 87 cosmopolitan, but mostly Australia and Tasmania (54), and New Zealand, absent in Middle Asia
Drosophyllum Link. 1805. Droseraceae. 1 Portugal, Spain, Morocco
Drudeophytum J. M. Coult. et Rose = Tauschia Schltdl. Apiaceae
Drummondita Harv. 1855 (Philotheca Rudge). Rutaceae. 4 Australia
Drusa DC. 1807. Apiaceae. 1 Canary Is., Somalia
Dryadanthe Endl. 1840 (Potentilla L.). Rosaceae. 1 (D. pentaphylla) western Tibet, Himalayas
Dryadella Luer. 1978. Orchidaceae. 24 trop. America: Guatemala, El Salvador,

Nicaragua, Honduras, Costa Rica, Panama, Colombia, Ecuador, Peru, Brazil

Dryadodaphne S. Moore. 1923. Monimiaceae. 3 New Guinea (2), Australia (Queensland)

Dryadorchis Schltr. 1913 (Thrixspermum Lour.). Orchidaceae. 2 New Guinea

Dryandra R. Br. 1810. Proteaceae. 55 Australia

Dryas L. 1753. Rosaceae. 3–13 arctic–alpine and temp. Northern Hemisphere; Russia (13 arctic Eurasian part, Siberia, Far East, and Caucasus)

Drymaria Willd ex Roem. et Schult. 1819–1820. Caryophyllaceae. 50 trop. and South Africa, India, Himalayas, southwestern China, southern Japan, Malesia, Australia, Oceania, America from southern U.S. to Patagonia, West Indies

Drymoanthus Nicholls. 1943. Orchidaceae. 3 northeastern Australia (1), New Zealand (1), New Caledonia (1)

Drymocallis Fourr. ex Rydb. = Potentilla L. Rosaceae

Drymochloa Holub. 1984 (Festuca L.). Poaceae. 2 Europe, West and Middle Asia

Drymoda Lindl. 1838. Orchidaceae. 2 Burma (D. picta), Thailand and Laos (D. siamensis)

Drymonia Mart. 1829. Gesneriaceae. c. 100 trop. America, West Indies

Drymophila R. Br. 1810. Luzuriagaceae. 2 eastern Australia, Tasmania

Drymophloeus Zipp. 1829. Arecaceae. 15 Halmahera I., Ceram and Ambon Is. to New Guinea, Solomon Is. eastward to Fiji and Samoa Is.

Dryobalanops C. F. Gaertn. 1805. Dipterocarpaceae. 7 Malay Peninsula, Sumatra, Borneo

Dryopetalon A. Gray. 1853. Brassicaceae. 4 southwestern U.S., Mexico

Dryopoa Vickery. 1963. Poaceae. 1 southeastern Australia, Tasmania

Drypetes Vahl. 1807. Euphorbiaceae. 200 trop. America, trop. and South Africa, South, Southeast and East Asia, Malesia, Australia, Lord Howe I.

Drypis L. 1753. Caryophyllaceae. 1 Italy, Yugoslavia, Albania, Greece, Lebanon

Duabanga Buch.-Ham. 1837. Duabangaceae. 3 Himalayas (Nepal, Bhutan), Assam, Burma, Tibet, southern China (Yunnan), Indochina, Malesia

Dubardella H. J. Lam. = Pyrenaria Blume. Theaceae

Dubautia Gaudich. 1830. Asteraceae. 21 Hawaiian Is.

Duboisia R. Br. 1810. Solanaceae. 3 Australia, New Caledonia (1, D. myoporoides)

Duboscia Bocquillon. 1866. Tiliaceae. 3 trop. West Africa

Dubouzetia Pancher ex Brongn. et Gris. 1861. Elaeocarpaceae. 10 New Guinea (4), northern Australia, New Caledonia (6)

Dubyaea DC. 1838. Asteraceae. 10 Himalayas, western China

Ducampopinus A. Chev. 1944 (Pinus L.). Pinaceae. 1 Vietnam and Cambodia

Duchesnea Sm. 1811 (Fragaria L.). Rosaceae. 2 (or 5–6) Afghanistan, Himalayas, Tibet, India, Sri Lanka, Indochina, Japan, Malesia

Duckea Maguire. 1958. Rapateaceae. 4 Venezuela, Brazil

Duckeanthus R. E. Fr. 1934. Annonaceae. 1 trop. South America (northern Brazil)

Duckeella Porto et Brade. 1940. Orchidaceae. 3 Colombia, Venezuela, Brazil

Duckeodendron Kuhlm. 1925. Duckeodendraceae. 1 Brazil (Amazonia)

Duckera F. A. Barkley = Rhus L. Anacardiaceae

Duckesia Cuatrec. 1961. Humiriaceae. 1 Brazil (Amazonia)

Ducrosia Boiss. 1844. Apiaceae. 3 Syria, Arabian Peninsula, Pakistan

Dudleya Britton et Rose. 1903. Crassulaceae. c. 45 southwestern U.S., northwestern Mexico

Dufrenoya Chatin. 1860. Santalaceae. 14 Indo-Malesia

Dugaldia (Cass.) Cass. = Helenium L. Asteraceae

Dugandia Britton et Kullip = Acacia Hill. Fabaceae

Dugandiodendron Loz-Contr. 1975 (or = Magnolia L.). Magnoliaceae. 10 Venezuela, Colombia, Ecuador

Dugesia A. Gray. 1882. Asteraceae. 1 Mexico

Duggena Vahl = Gonzalagunia Ruiz et Pav. Rubiaceae

Duguetia A. St.-Hil. 1824. Annonaceae. 65 West Indies, trop. South America

Duhaldea DC. 1836 (Inula L.). Asteraceae. 13 East Africa, trop. Asia

Duidaea S. F. Blake. 1931. Asteraceae. 3 Venezuela

Duidania Standl. 1931. Rubiaceae. 1 Venezuela

Dukea Dwyer = Raritebe Wernham. Rubiaceae

Dulacia Vell. 1825. Olacaceae. 13 Colombia, Ecuador, Peru, Bolivia, Venezuela, the Guyanas, Brazil

Dulichium Pers. 1805. Cyperaceae. 1 North America

Dulongia Kunth = Phyllonoma Willd ex Schult. Dulongiaceae

Dumasia DC. 1825. Fabaceae. 8 trop. East and South Africa (1, D. villosa) to trop. Asia, Malesia, trop. Australia, ? Oceania

Dunalia Kunth. 1818 (Acnistus Schott). Solanaceae. 6–7 Andes from Colombia to Argentina

Dunbaria Wight et Arn. 1834. Fabaceae. 15 trop. Asia, Australia

Dunnia Tutcher. 1905. Rubiaceae. 2 India (Assam), southern China

Dunniella Rauschert = Aboriella Bennet. Urticaceae

Dunstervillea Garay. 1972. Orchidaceae. 1 Venezuela

Duosperma Dayton. 1945. Acanthaceae. 15 trop. and South Africa

Duparquetia Baill. 1865. Fabaceae. 1 trop. West Africa from Liberia to Angola, Zaire

Duperrea Pierre ex Pit. 1924. Rubiaceae. 2 India, China, Indochina

Dupontia R. Br. 1823. Poaceae. 1 arctic regions

Durandea Planch. 1847 (nom. cons.) (Hugonia L.). Hugoniaceae. 14–15 eastern Malesia, eastern Australia, New Caledonia, and eastward to Fiji (1, D. vitensis)

Duranta L. 1753. Verbenaceae. 17 suptrop. and trop. America from Texas and Florida to Argentina, West Indies

Duriala (R. H. Anderson) Ulbr. = Maireana Moq. Chenopodiaceae

Durio Adans. 1763. Bombacaceae. 27 Burma, western Malesia

Duroia L. f. 1782. Rubiaceae. 21 Central (from Guatemala to Costa Rica) and trop. South America

Durringtonia R. J. F. Hend. et Guymer. 1985. Rubiaceae. 1 Australia (Queensland and New South Wales)

Duschekia Opiz = Alnus Hill. Betulaceae

Duseniella K. Schum. 1902. Asteraceae. 1 Argentina (Patagonia)

Dussia Krug et Urb. ex Taub. 1892. Fabaceae. 10 trop. America from Mexico to Peru, West Indies

Dutaillyea Baill. 1872–1873. Rutaceae. 2 New Caledonia

Duthieastrum M. P. de Vos. 1975. Iridaceae. 1 South Africa

Duthiea Hack. 1896. Poaceae. 2 Afghanistan, Pakistan, Himalayas

Duthiella M. P. de Vos = Duthieastrum M. P. de Vos. Iridaceae

Duvalia Haw. 1812. Asclepiadaceae. 19 trop. and South Africa

Duvaliandra M. G. Gilbert. 1980 (Caralluma R. Br.). Asclepiadaceae. 1 Socotra

Duval-Jouvea Palla. 1905. Cyperaceae. 7 Eurasia, South America

Duvaucellia Bowdich = Kohautia Cham. et Schltdl. Rubiaceae

Duvernoia E. Mey. ex Nees = Justicia L. Acanthaceae

Duvigneaudia J. Léonard. 1959. Euphorbiaceae. 1 trop. Africa

Dyakia Christenson. 1986. Orchidaceae. 1 Borneo

Dybowskia Stapf. 1919 (Hyperrhenia Fourn.). Poaceae. 1 trop. West Africa

Dyckia Schult. f. 1830. Bromeliaceae. 106 Brazil, Bolivia, Paraguay, northern Argentina, Uruguay

Dyera Hook. f. 1882. Apocynaceae. 2–3 western Malesia: Malay Peninsula, Sumatra, Borneo

Dyerophytum Kuntze. 1891 (Plumbago L.). Plumbaginaceae. 3 trop. and South Africa, Arabian Peninsula, Socotra, India

Dymondia Compton. 1953. Asteraceae. 1 South Africa

Dypsidium Baill. = Neophloga Baill. Arecaceae

Dypsis Noronha ex Mart. 1837. Arecaceae. 21 Madagascar

Dyschoriste Nees. 1832. Acanthaceae. c. 65 tropics and subtropics

Dyscritogyne R. M. King et H. Rob. 1971 (Eupatorium L.). Asteraceae. 2 Mexico

Dyscritothamnus B. L. Rob. 1922. Asteraceae. 2 Mexico

Dysodiopsis (A. Gray) Rydb. 1915 (or = Dyssodia Cav.). Asteraceae. 1 southern U.S.

Dysolobium (Benth.) Prain. 1897 (Psophocarpus DC.). Fabaceae. 4 Indo-Malesia

Dysophylla Blume = Eusteralis Raf. Lamiaceae

Dysopsis Baill. 1858. Euphorbiaceae. 1 Juan Fernández Is., Andes

Dysosma Woodson. 1928 (Podophyllum L.). Berberidaceae. 7 temp. Asia (esp. China and Taiwan), North America

Dysoxylum Blume. 1825. Meliaceae. c. 75 Himalayas, India, western and southern China, Taiwan, Indochina, Malesia, Australia, New Zealand (1), Melanesia, Polynesia

Dyspemptemorion Bremek. 1948. Acanthaceae. 1 trop. South America

Dysphania R. Br. 1810. Chenopodiaceae. 17 Australia, New Zealand

Dyssochroma Miers. 1849 (Markea Rich. and Schultesianthus Hunz.). Solanaceae. 2 Brazil

Dyssodia Cav. 1803. Asteraceae. 32 (sensu lato: trop. America from southwestern U.S. to Colombia, Argentina); 8 (sensu str.: southwestern U.S., Mexico)

Dystaenia Kitag. 1937 (or = Ligusticum L.). Apiaceae. 2 Japan (Honshu I.), Korea

Dystovomita (Engl.) D'Arcy. 1978. Clusiaceae. 3 trop. America

Eadesia F. Muell. = Anthocercis Labill. Solanaceae

Earina Lindl. 1834. Orchidaceae. 7 New Zealand, New Caledonia, Vanuatu, Tonga, Fiji, Samoa, Tahiti

Eastwoodia Brandegee. 1894. Asteraceae. 1 southwestern U.S., Mexico

Eatonella A. Gray. 1883. Asteraceae. 1 (E. nivea) southern U.S.

Ebandoua Pellegr. 1955. Anacardiaceae. 1 trop. West Africa

Ebenus L. 1753. Fabaceae. 20 Greece, Crete I., east Aegean Is., Asia Minor, North Africa from Morocco to Egypt, and from Arabian Peninsula, to Iran, Afghanistan, and Pakistan

Eberhardtia Lecomte. 1920. Sapotaceae. 3 China, Indochina

Eberlanzia Schwantes. 1926. Aizoaceae. 12 southwestern Namibia, western South Africa

Ebingeria Chrtek et Krisa = Luzula DC. Juncaceae

Ebracteola Dinter et Schwantes. 1927. Aizoaceae. 5 southern Namibia, northwestern South Africa

Eburopetalum Becc. = Anaxagorea A. St.-Hil. Annonaceae

Ecastaphyllum P. Browne = Dalbergia L. f. Fabaceae

Ecballium A. Rich. 1824. Cucurbitaceae. 1 Mediterranean, West Asia

Ecbolium Kurz. 1871 (Justicia L.). Acanthaceae. 8 trop. East and South Africa, Madagascar, southern Arabian Peninsula, ? Socotra, India, Sri Lanka

Ecclinusa Mart. 1839. Sapotaceae. 11 Panama, Colombia, Venezuela, Guyana, Surinam, French Guiana, Trinidad, Brazil, Peru

Eccoilopus Steud. 1854 (Spodiopogon Trin.). Poaceae. 4 South and East Asia

Eccoptocarpha Launert. 1965. Poaceae. 1 Tanzania, Zambia

Eccremisl Baker = Excremis Willd. ex Baker. Phormiaceae

Eccremocactus Britton et Rose = Weberocereus Britton et Rose. Cactaceae

Eccremocarpus Ruiz et Pav. 1794. Bignoniaceae. 5 Peru, Chile

Ecdeiocolea F. Muell. 1874. Ecdeiocoleaceae. 1 southwestern Australia

Ecdysanthera Hook. et Arn. 1837. Apocynaceae. 15 China, Indo-Malesia

Echeandia Ortega. 1800. Asphodelaceae. 13 trop. America from Mexico to Ecuador and Venezuela

Echetrosis Phil. = Parthenium L. Asteraceae

Echeveria DC. 1828. Crassulaceae. c. 160 trop. America from southern U.S. to northern Chile and Argentina

Echidnium Schott = Dracontium L. Araceae

Echidnopsis Hook. f. 1871. Asclepiadaceae. 29 East Africa, Arabian Peninsula

Echinacanthus Nees. 1832. Acanthaceae. 10 Himalayas, southern China, Thailand, Java

Echinacea Moench. 1794. Asteraceae. 9 eastern North America

Echinaria Desf. 1799. Poaceae. 2 Mediterranean, Southwest Asia; E. capitata—also in Crimea, Caucasus, and Middle Asia

Echinocactus Link et Otto. 1827. Cactaceae. 5 southern U.S., Mexico

Echinocarpus Blume = Sloanea L. Elaeocarpaceae

Echinocaulon (Meisn.) Spach = Polygonum L. Polygonaceae

Echinocephalum Gardner = Melanthera Rohr. Asteraceae

Echinocereus Engelm. 1848. Cactaceae. 44 southern U.S., Mexico

Echinochloa P. Beauv. 1812. Poaceae. 30–40 trop. and warm–temp. regions

Echinocitrus Tanaka = Triphasia Lour. Rutaceae

Echinocodon D. Y. Hong. 1984. Campanulaceae. 1 eastern China (Hubei)

Echinocodon Kolak. = Campanula L. Campanulaceae

Echinocoryne H. Rob. = Vernonia Schreb. Asteraceae

Echinocystis Torr. et A. Gray. 1840. Cucurbitaceae. 1 North America

Echinodorus Rich. ex Engelm. 1848. Alismataceae. c. 50 trop. and subtrop. America, Africa

Echinofossulocactus Lawr. = Stenocactus (K. Schum.) A. W. Hill. Cactaceae

Echinoglochin (A. Gray) A. Brand. 1925 (Plagiobothrys Fisch. et C. A. Mey.). Boraginaceae. 8 western North America

Echinolaena Desv. 1813. Poaceae. 7 Central and South America, Madagascar (1)

Echinolobivia Y. Ito = Echinopsis Zucc. Cactaceae

Echinomastus Britton et Rose = Sclerocactus Britton et Rose. Cactaceae

Echinopaepale Bremek. 1944. Acanthaceae. 1 Java

Echinopanax Decne. et Planch. ex Harms = Oplopanax Miq. Araliaceae

Echinopepon Naudin. 1866. Cucurbitaceae. 15 America

Echinophora L. 1753. Apiaceae. 10 Mediterranean, West Asia eastward to Afghanistan; 2 Crimea, Caucasus, Middle Asia

Echinopogon P. Beauv. 1812. Poaceae. 7 New Guinea, Australia, New Zealand

Echinops L. 1753. Asteraceae. c. 120 Pyrenees, South Europe, North and Northeast Africa, steppe and semiarid Asia, esp. (c. 70) Caucasus and Middle Asia

Echinopsis Zucc. 1837. Cactaceae. 50–100 central and southern South America

Echinopteris A. Juss. 1843. Malpighiaceae. 3 Mexico

Echinosophora Nakai = Sophora L. Fabaceae

Echinospartum (Spach) Fourr. 1868. Fabaceae. 3 Iberian Peninsula, southern France

Echinostephia (Diels) Domin. 1926. Menispermaceae. 1 Australia (Queensland)

Echiochilon Desf. 1798. Boraginaceae. 6 North Africa, Arabian Peninsula

Echiochilopsis Caball. 1935. Boraginaceae. 1 Northwest Africa

Echioides Ortega = Arnebia Forssk. Boraginaceae

Echiostachys Levyns. 1934 (Lebostemon Lehm.). Boraginaceae. 3 South Africa

Echitella Pichon = Mascarenhasia A. DC. Apocynaceae

Echites P. Browne. 1756. Apocynaceae. 6 Florida, Mexico, Central America, Colombia, West Indies

Echium L. 1753. Boraginaceae. 40 Azores and Canary Is., Europe, North and South Africa, West (incl. Caucasus) and Middle Asia eastward to Siberia and China

Echyrospermum Schott = Plathymenia Benth. Fabaceae

Ecklonea Steud. = Trianoptiles Fenzl. Cyperaceae

Eclipta L. 1771. Asteraceae. 3–4 tropics and subtropics; Caucasus (2)? naturalized

Ecliptostelma Brandegee. 1917. Asclepiadaceae. 1 Mexico

Ecpoma K. Schum. 1896. Rubiaceae. 1 trop. Africa

Ectadiopsis Benth. = Cryptolepis R. Br. Asclepiadaceae

Ectadium E. Mey. 1838. Asclepiadaceae. 3 western Namibia, South Africa (northwestern Cape)

Ectinocladus Benth. = Alafia Thouars. Apocynaceae

Ectopopterys W. R. Anderson. 1980. Malpighiaceae. 1 Colombia, Peru

Ectosperma Swallen = Swallenia Soderstr. et H. F. Decker. Poaceae

Ectotropis N. E. Br. = Delosperma N. E. Br. Aizoaceae

Ectozoma Miers. 1849 (Juanulloa Ruiz et Pav.). Solanaceae. 1 Ecuador, Peru

Ectrosia R. Br. 1810. Poaceae. 12 New Guinea (1), trop. Australia

Ectrosiopsis (Ohwi) Jansen. 1952. Poaceae. 5 Indonesia, Philippines, New Guinea, Australia, Caroline Is.

Edbakaria R. Vig. = Pearsonia Dümmer. Fabaceae

Edgaria C. B. Clarke. 1876. Cucurbitaceae. 1 eastern Himalayas, Tibet

Edgeworthia Meisn. 1841. Thymelaeaceae. 4–5 Himalayas (Nepal, Bhutan), India (Assam), northern Burma, southwestern China, Japan

Edisonia Small. 1933 (Gonolobus Michx.). Asclepiadaceae. 1 southeastern U.S.

Edithcolea N. E. Br. 1895. Asclepiadaceae. 2 Northeast and East Africa, Socotra

Edithea Standl. = Omiltemia Standl. Rubiaceae

Edmondia Cass. 1818. Asteraceae. 3 South Africa (Cape)

Edosmia Nutt. ex Torr. et A. Gray = Perideridia Rchb. Apiaceae

Edraianthus (A. DC.) DC. 1839 (Wahlenbergia Roth). Campanulaceae. 13 South and Southeast Europe; E. owerinianus—Caucasus

Eduardoregelia Popov = Tulipa L. Liliaceae

Edwardsia Salisb. = Sophora L. Fabaceae

Eenia Hiern et S. Moore = Anisopappus Hook. et Arn. Asteraceae

Efulensia C. H. Wright. 1895 (Deidamia Noronha ex Thouars). Passifloraceae. 2 equatorial Africa from Nigeria to western Uganda

Eganthus Tiegh. = Minquartia Aubl. Olacaceae

Egeria Planch. 1849 (Elodea Michx.). Hydrocharitaceae. 2 southern Brazil, Uruguay, northern Argentina, Paraguay, Chile

Eggelingia Summerh. 1951. Orchidaceae. 3 trop. Africa from Ghana to Uganda and Malawi

Egleria G. Eiten. 1964. Cyperaceae. 1 Brazil (Amazonia)

Eglerodendron Aubrév. et Pellegr. = Pouteria Aubl. Sapotaceae

Egletes Cass. 1817. Asteraceae. 12 trop. America, West Indies

Ehretia P. Browne. 1756. Ehretiaceae. 50 tropics and subtropics

Ehrharta Thunb. 1779. Poaceae. 25 trop. and South Africa

Eichhornia Kunth. 1842. Pontederiaceae. 7 trop. America, West Indies, West and South Africa, Madagascar (1)

Eichlerago Carrick. 1977. Lamiaceae. 1 western Australia

Eichleria Progel = Rourea Aubl. Connaraceae

Eichlerodendron Briq. = Xylosma G. Forst. Flacourtiaceae

Eigia Soják. 1979. Brassicaceae. 1 Syria, Lebanon, Israel

Einadia Raf. 1838. Chenopodiaceae. 6 central, eastern, and southeastern Australia and Tasmania (4), New Zealand (2)

Einomeia Raf. = Aristolochia L. Aristolochiaceae

Eionitis Bremek. = Oldenlandia L. Rubiaceae

Eirmocephala H. Rob. = Vernonia Schreb. Asteraceae

Eisocreochiton Quisumb. et Merrill. 1928 (or = Creochinton Blume). Melastomataceae. 3 Borneo, Philippines

Eitenia R. M. King et H. Rob. 1974. Asteraceae. 2 Brazil

Eizia Standl. 1940. Rubiaceae. 1 Mexico

Ekebergia Sparrm. 1779. Meliaceae. 5 trop. (from Senegal to Ethiopia and Kenya, south to Angola, Zimbabwe, and Mozambique) and South Africa, Madagascar

Ekmania Gleason. 1919. Asteraceae. 1 Cuba

Ekmanianthe Urb. 1924. Bignoniaceae. 2 Cuba, Haiti

Ekmaniocharis Urb. = Mecranium Hook. f. Melastomataceae

Ekmanochloa Hitchc. 1936. Poaceae. 2 Cuba

Elachanthemum Y. Ling et Y. R. Ling = Stilpnolepis I. M. Kraschen. Asteraceae

Elachanthera F. Muell. = Myrsiphyllum (L.) Willd. Asparagaceae

Elachanthus F. Muell. 1853. Asteraceae. 2 temp. Australia

Elacholoma F. Muell. 1895. Scrophulariaceae. 1 central Australia

Elachyptera A. C. Sm. 1940. Celastraceae. 3 Central and trop. South America

Elaeagia Wedd. 1849. Rubiaceae. 15 trop. America, Cuba

Elaeagnus L. 1753. Elaeagnaceae. c. 50 (or 80). South Europe, Asia from eastern Mediterranean to Japan and Malesia, Australia (1, Queensland), North America

Elaeis Jacq. 1763. Arecaceae. 2: E. oleifera—northern and central South America; E. guineensis—trop. Africa

Elaeocarpus L. 1753. Elaeocarpaceae. 135 or c. 200 Madagascar, Mauritius, India, Sri Lanka, Burma, China (38), Taiwan, Japan, Indochina, Malesia, Australia, Tasmania, New Zealand, New Caledonia (29), Vanuatu eastward to Fiji, Hawaiian Is.

Elaeodendron J. F. Jacq. ex Jacq. = Cassine L. Celastraceae

Elaeoluma Baill. 1891. Sapotaceae. 4 Panama, Venezuela, Guyana, Surinam, Colombia, Peru, Brazil

Elaeophora Ducke = Plukentia L. Euphorbiaceae

Elaeophorbia Stapf. 1906 (Euphorbia L.). Euphorbiaceae. 4 trop. and South Africa

Elaeopleurum Korovin = Seseli L. Apiaceae

Elaeoselinum W. D. J. Koch ex DC. 1830. Apiaceae. 4 Mediterranean

Elaeosticta Fenzl. 1843 (Scaligeria DC.). Apiaceae. 24–26 Southwest and South Asia; c. 20—Caucasus and (chiefly) Middle Asia; E. lutea—southern European part of Russia

Elaphandra Strother. 1991. Asteraceae. 1 Panama

Elaphanthera N. Hallé. 1988. Santalaceae. 1 New Caledonia

Elasis D. R. Hunt. 1978 (Tradescantia L.). Commelinaceae. 1 Ecuador

Elateriopsis Ernst. 1873. Cucurbitaceae. 5 Central and trop. South America

Elateriospermum Blume. 1826. Euphorbiaceae. 1 southern Thailand, western Malesia

Elatine L. 1753. Elatinaceae. 15 temp., subtrop., and trop. regions

Elatostema J. R. Forst. et G. Forst. 1775. Urticaceae. 200 (or 350) tropics of the Old World eastward to Polynesia

Elatostematoides C. B. Rob. = Elatostema J. R. Forst. et G. Forst. Urticaceae

Elattospermum Soler. = Breonia A. Rich. ex DC. Rubiaceae

Elattostachys (Blume) Radlk. 1879. Sapindaceae. c. 20 Java, Borneo, Sulawesi, Lesser Sunda and Moluccas, Philippines, New Guinea, Solomon Is., Australia, New Caledonia, Fiji, Samoa, and Tongo

Elburzia Hedge. 1969. Brassicaceae. 1 northwestern Iran

Elcomarhiza Barb. Rodr. 1891. Asclepiadaceae. 2 Brazil

Elegia L. 1771. Restionaceae. 35 South Africa

Eleogiton Link, nom. illeg. = Scirpidella Rauschert. Cyperaceae

Eleiodoxa (Becc.) Burret. 1942. Arecaceae. 1 (E. conferta) southern Thailand, Malay Peninsula, Sumatra, Borneo

Eleiosine Raf. = Sibiraea Maxim. Rosaceae

Eleiotis DC. 1825. Fabaceae. 2 India, Sri Lanka

Eleocharis R. Br. 1810. Cyperaceae. c. 200 cosmopolitan

Eleogiton Link = Scirpus L. Cyperaceae

Eleorchis F. Maek. 1935. Orchidaceae. 2 Japan

Elephantomene Barneby et Krukoff. 1974. Menispermaceae. 1 French Guiana

Elephantopus L. 1753. Asteraceae. c. 30 North (9) and trop. America; few sp. in trop. Old World

Elephantorrhiza Benth. 1841. Fabaceae. 10 southern trop. and South Africa

Elettaria Maton. 1811. Zingiberaceae. 9 India, Sri Lanka (1, E. cardamomum), Burma, Indochina, western Malesia

Elettariopsis Baker. 1892. Zingiberaceae. 11 Indo-Malesia

Eleusine Gaertn. 1788. Poaceae. 9 Africa (Ethiopia, Somalia, Uganda, Kenya, Tanzania, Rwanda, Burundi to Cape), Arabian Peninsula, Iran, Afghanistan, Pakistan; 2 sp. extending to Caucasus and Middle Asia; E. indica (L.) Gaertner—tropics and subtropics

Eleutharrhena Forman. 1975. Menispermaceae. 1 India (Assam), China

Eleutherandra Slooten. 1925. Flacourtiaceae. 1 Sumatra, Borneo

Eleutheranthera Poit. ex Bosc. 1803. Asteraceae. 1 trop. America

Eleutherine Herb. 1843. Iridaceae. 2 Central and trop. South America, West Indies

Eleutherococcus Maxim. 1859. Araliaceae. c. 40 northern and northeastern India, Tibet, Nepal (2), Bhutan (1, A. cissifolius), China (c. 30), Indochina, Malay Peninsula; A. trifoliatus extending to Taiwan and Philippines; Russia (2, Far East)

Eleutheropetalum (H. A. Wendl.) H. A. Wendl. ex Oersted = Chamaedorea Willd. Arecaceae

Eleutherospermum K. Koch. 1842. Apiaceae. 1 (E. cicutarium) Caucasus, Turkey, Iran

Eleutherostigma Pax et K. Hoffm. 1919. Euphorbiaceae. 1 Colombia

Eleutherostylis Burret. 1926. Tiliaceae. 1 New Guinea

Eleuthranthes F. Muell. ex Benth. = Opercularia Gaertn. Rubiaceae

Eliea Cambess. 1830. Hypericaceae. 1 Madagascar

Eligmocarpus Capuron. 1968. Fabaceae. 1 southeastern Madagascar

Elingamita G. T. S. Baylis. 1951. Myrsinaceae. 1 New Zealand (Three Kings Is.)

Elionurus Kunth ex Willd. 1806. Poaceae. 15 tropics and subtropics

Elissarrhena Miers. 1864 (Anomospermum Miers). Menispermaceae. 1 Brazil

Elizabetha Schomb. ex Benth. 1840. Fabaceae. 10 trop. South America

Elizaldia Willk. 1852. Boraginaceae. 2 southwestern Spain, Morocco, Algeria, Tunisia, Libya

Elleanthus C. Presl. 1827. Orchidaceae. 50–60 trop. America from Mexico to Peru, West Indies

Elleimataenia Kozo-Polj. = Osmorhiza Raf. Apiaceae

Ellenbergia Cuatrec. 1964. Asteraceae. 1 Peru

Ellertonia Wight = Kamettia Kostel. Apocynaceae

Elliottia Mühl. ex Elliott. 1817. Ericaceae. 4 Japan, North America

Ellipanthus Hook. f. 1862. Connaraceae. 13 East Africa (3), Madagascar (2), India, Sri Lanka (1, E. unifoliatus), Burma, Andaman Is., Indochina, Hainan I., Malesia (Malay Peninsula, Java, Borneo, Philippines, Sulawesi)

Ellipeia Hook. f. et Thomson. 1855. Annonaceae. 5 Malesia: Malay Peninsula, Sumatra, Borneo

Ellipeiopsis R. E. Fr. 1955. Annonaceae. 2 eastern India, Indochina

Ellisia L. 1763. Hydrophyllaceae. 1 North America

Ellisiohyllum Maxim. 1871. Scrophulariaceae. 1 (E. pinnatum) Nepal, Bhutan, northeastern India, western China, Taiwan, Japan, Philippines, eastern New Guinea

Elmera Rydb. 1905. Saxifragaceae. 1 western North America (Washington)

Elmerrillia Dandy. 1927. Magnoliaceae. 5 Sumatra, Borneo, Sulawesi, Philippines, Moluccas, New Guinea, New Britain

Elodea Michx. 1803. Hydrocharitaceae. 12 America, esp. temp. regions

Eloyella Ortiz Vald. 1979 (or = Phymatidium Lindl.). Orchidaceae. 2 Colombia

Elsholtzia Willd. 1790. Lamiaceae. 40 Eurasia (esp. China—33 and Himalayas—10), Ethiopia

Elsiea F. M. Leighton = Ornithogalum L. Hyacinthaceae

Elsneria Walp. = Homalocarpus Hook. et Arn. Apiaceae

Elthemidea Zheng. 1992. Poaceae. 2 China (Sichuan Prov.)

Eltroplectis Raf. 1837 (Stenorrhynchos Rich. ex Spreng.). Orchidaceae. 10 trop. America southward to Paraguay, West Indies

Elvasia DC. 1811. Ochnaceae. 11 trop. South America

Elvira Cass. = Delilia Spreng. Asteraceae

Elwendia Boiss. = Bunium L. Apiaceae

Elymandra Stapf. 1919. Poaceae. 6 Brazil (1, E. grallata), trop. Africa

Elymus L. 1753. Poaceae. c. 150 temp. and subtrop. regions, montane areas of tropics

Elynanthus P. Beauv. ex T. Lestib. = Tetraria P. Beauv. Cyperaceae

Elythranthera (Endl.) A. S. George. 1963. Orchidaceae. 2 western Australia

Elytranthe (Blume) Blume. 1830. Loranthaceae. 10 southern China, Indochina, western Malesia

Elytraria Michx. 1803. Acanthaceae. 7 tropics and subtropics

Elytrigia Desv. 1810 (Agropyron Gaertner and Elymus L.). Poaceae. c. 50 temp. regions

Elytropappus Cass. 1816. Asteraceae. 8 South Africa

Elytrophorus P. Beauv. 1812. Poaceae. 2 trop. Old World

Elytropus Muell. Arg. 1860. Apocynaceae. 1 Chile

Elytrostachys McClure. 1942. Poaceae. 2 Central America, Colombia, Ecuador

Embadium J. M. Black. 1931. Boraginaceae. 3 southern Australia

Embelia Burm. f. 1768. Myrsinaceae. c. 100 trop. and subtrop. Africa; Madagascar; Mascarene Is.; South, Southeast, and East Asia; Oceania (or: = 10 New Guinea, New Britain, Solomon Is., eastern Australia [Queensland, New South Wales])

Embergeria Boulos = Sonchus L. Asteraceae

Emblemantha B. C. Stone. 1988. Myrsinaceae. 1 Sumatra

Emblica Gaertn. = Phyllanthus L. Euphorbiaceae

Emblingia F. Muell. 1860

Emblingiaceae. 1 western and southwestern Australia

Embolanthera Merr. 1909. Hamamelidaceae. 2 Indochina, Philippines

Embothrium J. R. Forst. et G. Forst. 1775. Proteaceae. 8 central and southern Andes

Embreea Dodson. 1980 (Stanhopea Frost ex Hook.). Orchidaceae. 1 Colombia, Ecuador

Emelianthe Danser. 1933. Loranthaceae. 1 trop. East Africa

Emerus Hill = Hippocrepis L. Fabaceae

Emex Campderá. 1819. Polygonaceae. 2 Mediterranean (E. spinosa), South Africa, Australia

Emicocarpus K. Schum. et Schltr. 1900. Asclepiadaceae. 1 Southeast Africa

Emilia Cass. 1817. Asteraceae. c. 100 trop. Old World

Emiliella S. Moore. 1918. Asteraceae. 5 trop. South Africa

Emiliomarcetia T. Durand et H. Durand = Trichoscypha Hook. f. Anacardiaceae

Eminia Taub. 1891. Fabaceae. 4 Angola, Zaire, Botswana, Zambia, Zimbabwe, Tanzania, Malawi, Mozambique

Eminium (Blume) Schott. 1855. Araceae. 7 eastern Mediterranean eastward to Middle Asia (2) and Afghanistan

Emmenanthe Benth. 1835. Hydrophyllaceae. 1 southwestern North America (from California, Nevada to Utah)

Emmenopterys Oliv. 1889. Rubiaceae. 2 Burma, Thailand, China

Emmenosperma F. Muell. 1862. Rhamnaceae. 3 Australia (2), New Caledonia, Fiji

Emmeorhiza Pohl ex Endl. 1838. Rubiaceae. 1 trop. South America

Emmotum Desv. ex Ham. 1825. Icacinaceae. 12 trop. South America

Emorya Torr. 1859. Buddlejaceae. 1 southern U.S. (Texas), Mexico

Empedoclesia Sleumer = Orthaea Klotzsch. Ericaceae

Empetrum L. 1753. Empetraceae. 2–3 (8) arctic and temp. hemisphere, temp. South America, Falkland Is., Is. of Tristan da Cunha; Russia (8, northern areas)

Emplectanthus N. E. Br. 1908. Asclepiadaceae. 2 South Africa

Emplectocladus Torr. = Prunus L. Rosaceae

Empleuridium Sond. et Harv. 1859. Celastraceae. 1 South Africa (Cape)

Empleurum Aiton. 1789. Rutaceae. 2 South Africa

Empodisma L. A. S. Johnson et D. F. Cutler. 1974. Restionaceae. 2 Australia, Tasmania, New Zealand

Empodium Salisb. 1866 (Curculigo Gaertner). Hypoxydaceae. 10 South Africa

Empogona Hook. f. = Tricalysia A. Rich. ex DC. Rubiaceae

Enallagma (Miers) Baill. = Amphitecna Miers. Bignoniaceae

Enantia Oliv. = Annickia Van Setten et Maas. Annonaceae

Enantiophylla J. M. Coult. et Rose. 1893. Apiaceae. 1 Mexico and Central America

Enarganthe N. E. Br. 1930 (or = Ruschia Schwantes). Aizoaceae. 1 western South Africa

Enarthrocarpus Labill. 1812. Brassicaceae. 5 eastern Mediterranean, North Africa

Enaulophyton Steenis. 1932. Melastomataceae. 2 Bunguran (Natuna) Is., Borneo

Encelia Adans. 1763. Asteraceae. 15 America from western U.S. to Chile, Galapagos Is.

Enceliopsis (A. Gray) A. Nelson. 1909. Asteraceae. 4 western U.S. from Idaho to California and Arizona

Encephalartos Lehm. 1834. Zamiaceae. 46 Africa: Ghana, Togo, Nigeria, Central African Rep., southern Sudan, Kongo, Zaire, Ugand, Kenya, Tanzania, northern Angola, Zambia, Zimbabwe, Malawi, Mozambique, but esp. (c. 30)—eastern South Africa from Transvaal to Cape

Encephalosphaera Lindau. 1904. Acanthaceae. 2 trop. South America

Encheiridion Summerh. = Microcoelia Lindl. Orchidaceae

Encholirium Mart. ex Schult. et Schult. f. 1830. Bromeliaceae. 14 northeastern and eastern Brazil

Enchosanthera King et Stapf ex Guillaumin = Creochiton Blume. Melastomataceae

Enchylaena R. Br. 1810. Chenopodiaceae. 2 Australia

Encopella Pennell. 1920. Scrophulariaceae. 1 Cuba

Encyclia Hook. 1828. Orchidaceae. 150 trop. America

Endadenium L. C. Leach. 1973 (Monadenium Pax). Euphorbiaceae. 1 Angola

Endertia Steenis et de Wit. 1947. Fabaceae. 1 Borneo

Endiandra R. Br. 1810. Lauraceae. c. 80–100 India (Assam), Burma, China, Taiwan, Indochina, Malesia from Malay Peninsula to New Guinea, Australia, and eastward to Polynesia

Endlicheria Nees. 1833. Lauraceae. 40 trop. America from Costa Rica to Paraguay and southern Brazil, the Lesser Antilles

Endocaulos C. Cusset. 1972 (1973). Podostemaceae. 1 Madagascar

Endocellion Turcz. ex Herder. 1865. Asteraceae. 2 Arctic and East Siberia, Chukotka, Russian Far East, northern Mongolia

Endocomia W. J. de Wilde. 1984 (Horsfieldia Willd.). Myristicaceae. 4 Andaman Is., Burma, southern China (Yunnan), Thailand, Laos, Malesia: Malay Peninsula, Sumatra, Java, Borneo, Sulawesi, Moluccas, Philippines, New Guinea

Endodeca Raf. = Aristolochia L. Aristolochiaceae

Endodesmia Benth. 1862. Clusiaceae. 1 trop. West Africa

Endolepis Torr. ex A. Gray. 1860. Chenopodiaceae. 3 U.S., northern Mexico

Endomallus Gagnep. 1915 (Dunbaria Wight et Arn.). Fabaceae. 2 Indochina

Endonema A. Juss. 1846. Penaeaceae. 2 South Africa

Endopappus Sch. Bip. 1860 (Chrysanthemum L.). Asteraceae. 1 North Africa

Endopleura Cuatrec. 1961. Humiriaceae. 1 Brazil (Amazonia)

Endosamara Geesink. 1984 (Millettia Wight et Arn.). Fabaceae. 1–2 Indo-Malesia

Endosiphon T. Anderson ex Benth. et Hook. f. 1876 (Ruellia L.). Acanthaceae. 1 trop. Africa

Endospermum Benth. 1861. Euphorbiaceae. 12–13 India, China, Indochina, Malesia, Australia (Queensland), Melanesia to Fiji

Endosteira Turcz. 1863. Tiliaceae. 1 West Indies

Endostemon N. E. Br. 1910. Lamiaceae. 17: 16 trop. and South (3) Africa; E. tenuiflorus reaches southern Arabian Peninsula and Socotra; E. viscosus—Pakistan, India

Endresiella Schltr. 1921 (or = Trevoria Lehm.). Orchidaceae. 1 Costa Rica

Endressia J. Gay. 1832. Apiaceae. 2 northern Spain, southern France

Endusa Miers ex Benth. = Minquartia Aubl. Olacaceae

Endymion Dumort. = Hyacinthoides Heister ex Fabr. Hyacinthaceae

Enemion Raf. 1820 (Isopyrum L.). Ranunculaceae. 2: 1 China, Korea, Japan, Russian Far East; 1 western North America

Engelhardia Lesch. ex Blume. 1826. Juglandaceae. 10 Himalayas from Kashmir to Assam, Burma, southwestern and southern China, Taiwan, Indochina, Malesia

Engelhardtia Blume = Engelhardia Lesch. ex Blume. Juglandaceae

Engelmannia A. Gray ex Nutt. 1840. Asteraceae. 1 U.S. from Colorado and Kansas to Arizona, New Mexico, Texas, Louisiana; northern Mexico

Englerastrum Briq. 1894. Lamiaceae. 5 trop. and South (1, E. schweinfurthii) Africa, Sri Lanka (1)

Englerella Pierre = Pouteria Aubl. Sapotaceae

Engleria O. Hoffm. 1888. Asteraceae. 2 trop. and South Africa Englerina Tiegh. = Tapinanthus (Blume) Rchb. Loranthaceae

Englerocharis Muschler. 1908. Brassicaceae. 2 Andes of Peru, Bolivia, northern Argentina

Englerodaphne Gilg = Gnidia L. Thymelaeaceae

Englerodendron Harms. 1907. Fabaceae. 1 Tanzania

Englerophytum K. Krause. 1914. Sapotaceae. 5–10 trop. Africa

Engysiphon G. J. Lewis = Geissorhiza Ker-Gawl. Iridaceae

Enhalus Rich. 1814. Hydrocharitaceae. 1 coasts bordering the Indian and western Pacific Oceans

Enhydra DC. = Enydra Lour. Asteraceae

Enicosanthellum Tien Ban = Disepalum Hook. f. Annonaceae

Enicosanthum Becc. 1871. Annonaceae. 16 Sri Lanka, Burma, Thailand, Malay Peninsula, Borneo, Philippines

Enicostema Blume. 1826. Gentianaceae. 3–4 Central America, West Indies, trop. and South Africa, Madagascar (1), India, Sri Lanka, Java, Lesser Sunda Is.

Enkianthus Lour. 1790. Ericaceae. 14 Himalayas, China (10), Japan

Enkleia Griff. 1844. Thymelaeaceae. 3 Andaman Is., Burma, Indochina, Malesia (Sumatra, Malay Peninsula, Borneo, Philippines, New Guinea)

Ennealophus N. E. Br. 1909 (Trimezia Salisb.). Iridaceae. 5 Brazil (Amazonia)

Enneapogon Desv. ex P. Beauv. 1812. Poaceae. 30–40 temp. Asia and America, tropics and subtropics

Enneastemon Exell = Monanthotaxis Baill. Annonaceae

Enneatypus Herz. = Ruprechtia C. A. Mey. Polygonaceae

Enochoria Baker f. 1921. Araliaceae. 1 New Caledonia

Enriquebeltrania Rzed. 1980. Euphorbiaceae. 1 Mexico

Ensete Horan. 1862. Musaceae. 20 trop. Africa, Madagascar, southern China, Indo-Malesia

Enslenia Nutt. = Cynanchum L. Asclepiadaceae

Entada Adans. 1763. Fabaceae. c. 30 tropics, esp. Africa (18)

Entadopsis Britton = Entada Adans. Fabaceae

Entandrophragma C. DC. 1894. Meliaceae. 11 trop. Africa

Entelea R. Br. ex Sims. 1824. Tiliaceae. 1 New Zealand

Enterolobium Mart. 1837. Fabaceae. 5 Central and South America south to Argentina, West Indies

Enteropogon Nees. 1836. Poaceae. 15 trop. Old World

Enterospermum Hiern = Tarenna Gaertn. Rubiaceae

Entolasia Stapf. 1920. Poaceae. 5 trop. Africa, eastern Australia

Entomophobia de Vogel. 1984 (Pholidota Lindl. ex Hook.). Orchidaceae. 1 Borneo

Entoplocamia Stapf. 1898. Poaceae. 3 trop. and South Africa

Enydra Lour. 1790. Asteraceae. 10 tropics and subtropics

Eomatucana F. Ritter = Oreocereus (A. Berger) Riccob. Cactaceae

Eomecon Hance. 1884. Papaveraceae. 1 eastern and southeastern China

Eosanthe Urb. 1923. Rubiaceae. 1 Cuba

Epacris Cav. 1797. Epacridaceae. 35 southeastern Australia, Tasmania, New Zealand, New Caledonia (1, E. pauciflora)

Epallage DC. = Anisopappus Hook. et Arn. Asteraceae

Epaltes Cass. 1818. Asteraceae. 15 tropics

Eparmatostigma Garay. 1972. Orchidaceae. 1 India (?), Vietnam

Eperua Aubl. 1775. Fabaceae. 14 northeastern South America

Ephedra L. 1753. Ephedraceae. 40 North (14 western U.S. and northern Mexico) and South (13 Andes from Ecuador to Patagonia) America, Canary Is., Eurasia from Mediterranean through arid subtrop. regions of Inner Asia to Lena and Amur rivers

Ephedranthus S. Moore. 1895. Annonaceae. 4 trop. South America

Ephemerantha P. F. Hunt et Summerh. = Flickingeria A. D. Hawkes. Orchidaceae

Ephippiandra Decne. 1858. Monimiaceae. 6 Madagascar

Ephippianthus Rchb. f. 1868. Orchidaceae. 1 Russian Far East (Sakhalin I.), Korea, Japan

Ephippiocarpa Markgr. = Callichilia Stapf. Apocynaceae

Epiblastus Schltr. 1905. Orchidaceae. 15–20 Philippines (1, E. merrillii), Sulawesi I., New Guinea, Solomon Is., Vanuatu, Fiji (1), and Samoa Is. (2)

Epiblema R. Br. 1810. Orchidaceae. 1 southwestern Australia

Epicampes C. Presl. 1830 (Muhlenbergia Schreb.). Poaceae. 10 America from California to Argentina

Epicharis Blume = Dysoxylum Blume. Meliaceae

Epicion Small = Cynanchum L. Asclepiadaceae

Epiclastopelma Lindau. 1895. Acanthaceae. 2 trop. East Africa

Epidanthus L. O. Williams. 1940. Orchidaceae. 3 Mexico, Central America

Epidendropsis Garay et Dunst. 1976 (Epidendrum L.). Orchida ceae. 2 Venezuela

Epidendrum L. 1763. Orchidaceae. c. 500 trop. and subtrop. America from southern U.S. to Argentina

Epidryos Maguire. 1962. Rapataceae. 3 Panama, Colombia

Epifagus Nutt. 1818. Orobanchaceae. 1 North America

Epigaea L. 1753. Ericaceae. 3: E. repens—eastern North America, E. gaultherioides—Caucasus, Asia Minor; E. asiatica—Japan

Epigeneium Gagnep. 1932. Orchidaceae. 35 trop. Asia from India to souhtern China, Taiwan and Indochina, Malesia (Indonesia, Philippines, New Guinea)

Epigynum Wight. 1848. Apocynaceae. 14 South and Southeast Asia, Malesia

Epilasia (Bunge) Benth. 1873. Asteraceae. 4 Asia from eastern Mediterranean, Caucasus through Middle Asia to western China, Afghanistan, Pakistan, and northwestern India

Epilobium L. 1753. Onagraceae. 165–200 cold and temp. Northern Hemisphere, montane trop. areas; Russia (c. 60, all areas)

Epilyna Schltr. 1918. Orchidaceae. 1 Costa Rica

Epimedium L. 1753. Berberidaceae. 22 northern Italy, Southeast Europe, Caucasus (4), North Africa, West Asia to Iran, western Himalayas, China, Japan; E. koreanum—extending to Russian Far East

Epimeredi Adans. 1763. Lamiaceae. 7–8 Madagascar, Mascarene Is., Australia

Epinetrum Hiern = Albertsia Becc. Menispermaceae

Epipactis Zinn. 1757. Orchidaceae. 25 temp. Northern Hemisphere southward to Mexico, trop. Africa, Thailand; Russia (c. 10, all areas)

Epipetrum Phil. 1864. Dioscoreaceae. 3 Chile

Epiphyllanthus A. Berger = Schlumbergera Lem. Cactaceae

Epiphyllum Haw. 1812. Cactaceae. 13 trop. America, West Indies

Epipogium J. F. Gmel. ex Borkh. 1792. Orchidaceae. 3–5 temp. Eurasia; E. roseum—trop. Africa, trop. Asia, Malesia, trop. Australia, Melanesia eastward to New Caledonia and Vanuatu; E. aphyllum—European part of Russia, Caucasus, Siberia, Far East

Epipremnopsis Engl. = Amydrium Schott. Araceae

Epipremnum Schott. 1857. Araceae. c. 10 India, Burma, southern China, Taiwan, Japan, Indochina, Malesia, Marshall Is., Australia

Epiprinus Griff. 1854. Euphorbiaceae. 5–6 India (Assam), southern China, Indochina, Malay Peninsula

Epirixanthes Blume. 1823 (Salomonia Lour.). Polygalaceae. 5 eastern India, Burma, southern China, northern Vietnam, Malesia (excl. Lesser Sunda Is.), Solomon Is. (San Cristobal I.)

Epischoenus C. B. Clarke ex Thiselton = Dyer. 1898. Cyperaceae. 8 South Africa

Episcia Mart. 1829. Gesneriaceae. 40 trop. America, West Indies

Episcothamnus H. Rob. 1981. Asteraceae. 1 Brazil

Epistemma D. V. Field et J. Hall. 1982. Asclepiadaceae. 3 trop. West Africa (montane forests of Ghana, Ivory Coast, and Cameroun)

Epistephium Kunth. 1822. Orchidaceae.

24 West Indies, trop. South America: Colombia, Venezuela, Trinidad, the Guianas, Peru, Bolivia, Brazil

Epitaberna K. Schum. 1903. Apocynaceae. 2 trop. West Africa

Epithelantha F. A. C. Weber ex Britton et Rose. 1922. Cactaceae. 1–2 southern U.S., northeastern Mexico

Epithema Blume. 1826. Gesneriaceae. 10 trop. Africa, Asia, and Malesia

Epitriche Turcz. 1851. Asteraceae. 1 southwestern Australia

Epixiphium (Engelm. ex A. Gray) Munz. 1926. Scrophulariaceae. 1 Mexico

Eplingia L. O. Williams = Trichostema L. Lamiaceae

Eragrostiella Bor. 1940. Poaceae. 9 East Africa, South Asia, northern Australia

Eragrostis Wolf. 1776. Poaceae. c. 300 trop. and temp. regions

Eranthemum L. 1753. Acanthaceae. 30 trop. and subtrop. Asia from India and Sri Lanka eastward to China and Lesser Sunda Is.

Eranthis Salisb. 1807. Ranunculaceae. 8 temp. Eurasia; Russia (3, Siberia, Far East)

Erato DC. 1836. Asteraceae. 4 trop. America from Costa Rica to Bolivia

Erblichia Seem. 1854 (Piriqueta Aubl.). Turneraceae. 5: E. odorata—Mexico, Guatemala, Belize, El Salvador, Honduras, Costa Rica, and Panama; South Africa (1), Madagascar (3)

Ercilla A. Juss. 1832. Phytolaccaceae. 2 Peru, Chile

Erdisia Britton et Rose = Corryocactus Britton et Rose. Cactaceae

Erechtites Raf. 1817. Asteraceae. 5 North (3) and South America, West Indies

Eremaea Lindl. 1839. Myrtaceae. 8 western Australia

Eremaeopsis Kuntze = Eremaea Lindl. Myrtaceae

Eremalche Greene. 1906 (Malvastrum A. Gray). Malvaceae. 4 western U.S. (southern California, Arizona), Mexico (northern Baja California and Sonora)

Eremanthus Less. 1829. Asteraceae. 25 Brazil

Eremia D. Don. 1834. Ericaceae. 7 South Africa (southwestern Cape)

Eremiastrum A. Gray = Monoptilon Torr. et A. Gray. Asteraceae

Eremiella Compton. 1953. Ericaceae. 1 South Africa (southern Cape)

Eremiopsis N. E. Br. = Eremia D. Don. Ericaceae

Eremitis Doell. 1877. Poaceae. 7 eastern Brazil

Eremobium Boiss. 1867. Brassicaceae. 3–4 North Africa (Morocco, Algeria, Libya, Egypt), Sinai Peninsula, Syria, Israel, Jordan, Arabian Peninsula, Iraq, Kuwait, southern Iran, western Pakistan

Eremoblastus Botsch. 1980. Brassicaceae. 1 southeastern Soviet European part, western Kazakhstan

Eremocarpus Benth. 1844. Euphorbiaceae. 1 western North America

Eremocarya Greene = Cryptantha Lehm. ex G. Don. Boraginaceae

Eremocaulon Soderstr. et Londoño. 1987. Poaceae. 1 Brazil

Eremocharis Phil. 1860. Apiaceae. 9 Andes of Peru and Chile

Eremochion Gilli = Horaninovia Fisch. et C. A. Meyer. Chenopodiaceae

Eremochloa Büse. 1854. Poaceae. 9 South and Southeast Asia, Malesia

Eremocitrus Swingle. 1914. Rutaceae. 1 northeastern Australia

Eremocrinum M. E. Jones. 1893. Asphodelaceae. 1 U.S. (Utah and Arizona)

Eremodaucus Bunge. 1843. Apiaceae. 1 (E. lehmannii) Caucasus, Middle Asia, Iran, Afghanistan

Eremodraba O. E. Schulz. 1924. Brassicaceae. 2 southern Peru, northern Chile

Eremogeton Standl. et L. O. Williams. 1953. Scrophulariaceae. 1 Mexico, Guatemala

Eremogone Fenzl = Arenaria L. Caryophyllaceae

Eremohylema A. Nelson = Pluchea Cass. Asteraceae

Eremolaena Baill. 1884. Sarcolaenaceae. 2 eastern Madagascar

Eremolepis Griseb. = Antidaphne Poepp. et Endl. Eremolepidaceae

Eremolimon Lincz. 1985. Plumbaginaceae. 7 Middle Asia

Eremomastax Lindau. 1894. Acanthaceae. 1 trop. Africa, ? Madagascar

Eremonanus I. M. Johnston = Eriophyllum Lag. Asteraceae

Eremopanax Baill. = Arthrophyllum Blume. Araliaceae

Eremopappus Takht. = Hyalea (DC.) Jaub. et Spach. Asteraceae

Eremophea Paul G. Wilson. 1984. Chenopodiaceae. 2 western and central Australia

Eremophila R. Br. 1810. Myoporaceae. c. 180 Australia

Eremophyton Bég. 1913. Brassicaceae. 1 Algeria, Morocco, Libya

Eremopoa Roshev. 1934. Poaceae. 7 Southeast Europe; North Africa; Southwest, West, Middle, and central Asia, Himalayas; Russia (1, E. altaica—southern West Siberia)

Eremopogon Stapf. 1917 (Dichanthium Willem.). Poaceae. 4 trop. Old World

Eremopyrum (Ledeb.) Jaub. et Spach. 1851. Poaceae. 10 South and Southeast Europe; North Africa; Southwest, West (incl. Caucasus) and Middle Asia; West Siberia, Himalayas

Eremosemium Greene. 1900. Chenopodiaceae. 2 western North America

Eremosis (DC.) Gleason = Vernonia Schreb. Asteraceae

Eremosparton Fisch. et C. A. Mey. 1841. Fabaceae. 3 southeastern Russian European part, northern Caucasus, Middle Asia, northern China

Eremospatha (G. Mann et H. A. Wendl.) H. A. Wendl. 1878. Arecaceae. 12 Africa from western Congo Basin eastward to Tanzania

Eremostachys Ledeb. 1830. Lamiaceae. 70 (or 10 ?) Europe (1), West (incl. Caucasus) and Middle Asia eastward to China and western Himalayas

Eremosyne Endl. 1837. Eremosynaceae. 1 southwestern Australia

Eremothamnus O. Hoffm. 1888. Asteraceae. 1 Namibia

Eremotropa Andres = Pyrola L. Ericaceae

Eremurus M. Bieb. 1810. Asphodelaceae. 50 montane Europe, Crimea, Caucasus, West Siberia, West and Middle Asia, eastward to Mongolia, western China, and western Himalayas

Erepsia N. E. Br. 1925. Aizoaceae. 27 southwestern and central South Africa

Ergocarpon C. C. Townsend. 1964. Apiaceae. 1 Iraq, Iran

Eria Lindl. 1825. Orchidaceae. c. 375 India, Sri Lanka, Burma, China, Japan, Indochina, Malesia, Australia, New Zealand, New Caledonia

Eriachaenium Sch. Bip. 1855. Asteraceae. 1 Tierra del Fuego

Eriachne R. Br. 1810. Poaceae. c. 40 South and Southeast Asia, Malesia, Australia

Eriadenia Miers = Mandevilla Lindl. Apocynaceae

Eriandra P. Royen et Steenis. 1952. Polygalaceae. 1 New Guinea, Solomon Is.

Eriandrostachys Baill. 1874. Sapindaceae. 1 Madagascar

Erianthecium Parodi. 1943. Poaceae. 1 southern Brazil, Uruguay

Erianthemum Tiegh. 1895. Loranthaceae. 15 trop. and South (2) Africa

Erianthera Benth. = Alajja S. Ikonn. Lamiaceae

Erianthus Michx. 1803 (Saccharum L.). Poaceae. c. 30 trop. and temp. regions; E. ravennae—in Caucasus and Middle Asia

Eriastrum Woot. et Standl. 1913. Polemoniaceae. 14 southwestern U.S. from California, Nevada and Utah to western Texas; Mexico

Eriaxis Rchb. f. 1876. Orchidaceae. 3 New Caledonia

Erica L. 1753. Ericaceae. c. 735 Atlantic islands from Faroes to Canary Is., Europe, Mediterranean, West Asia, montane trop. and South (650, 520—endemic of Cape) Africa, Madagascar, Mascarene Is.; E. tetralix—the Baltic Sea area; E. arborea—Caucasus (near Pitzunda)

Ericameria Nutt. 1841 (Haplopappus Cass.). Asteraceae. 22 southern U.S., Mexico

Ericentrodea S. F. Blake et Sherff. 1923. Asteraceae. 4 Andes

Erichsenia Hemsl. 1905. Fabaceae. 1 western Australia

Ericinella Klotzsch. 1838 (7E or = Erica L.). Ericaceae. 7 trop. and South Africa, Madagascar

Erigenia Nutt. 1818. Apiaceae. 1 eastern U.S.

Erigeron L. 1753. Asteraceae. Over 200 cosmopolitan, esp. North America

Erimocephala H. Rob. 1987. Asteraceae. 3 ? trop. America

Erinacea Hill. 1753. Fabaceae. 1 Spain, southern France, Tunisia, Algeria, Morocco

Erinna Phil. 1864. Alliaceae. 1 Chile

Erinocarpus Nimmo ex J. Graham. 1839. Tiliaceae. 1 southwestern India

Erinus L. 1753. Scrophulariaceae. 2 South Europe (Pyrenees and Alps), North Africa

Erioblastus Honda = Deschampsia P. Beauv. Poaceae

Eriobotrya Lindl. 1821. Rosaceae. 26 Himalayas, India, Bangladesh, Burma, central and southern China, Taiwan, southern Japan, Indochina, Malesia (Sumatra and Borneo); E. japonica—largely cultivated.

Eriocapitella Nakai = Anemone L. Ranunculaceae

Eriocaulon L. 1753. Eriocaulaceae. c. 400 tropics and subtropics, a few species in temp. Eurasia and North America

Eriocephalus L. 1753. Asteraceae. c. 30 South Africa

Eriocereus (A. Berger) Riccob. = Harrisia Britton. Cactaceae

Eriochilus R. Br. 1810. Orchidaceae. 6 Australia, Tasmania

Eriochiton (R. H. Anderson) A. J. Scott. 1978. Chenopodiaceae. 1 temp. southern Australia

Eriochlamys Sond. et F. Muell. ex Sond. 1853. Asteraceae. 2 southern Australia

Eriochloa Kunth. 1816. Poaceae. c. 30 trop. and temp. regions

Eriochrysis P. Beauv. 1812. Poaceae. 8 trop. America, Africa, India

Eriocnema Naudin. 1844. Melastomataceae. 2 Brazil

Eriocoelum Hook. f. 1862. Sapindaceae. 10 trop. Africa

Eriocoma Nutt. 1818 (Oryzopsis Michx.). Poaceae. 3 warm–temp. North America

Eriocycla Lindl. 1835. Apiaceae. 8 Asia—montane areas from Iran to China Eriodes Rolfe.1915. Orchidaceae. 1 northeastern India, Burma, southern China, Thailand, Vietnam

Eriodictyon Benth. 1844. Hydrophyllaceae. 9 southwestern U.S., Mexico

Erioglossum Blume = Lepisanthes Blume. Sapindaceae

Eriogonella Goodman. 1934. Polygonaceae. 2 California, Chile

Eriogonum Michx. 1803. Polygonaceae. 150–240 North America from central Alaska south to central Mexico and eastward to the Appalachian Mountains and Florida

Eriolaena DC. 1823. Sterculiaceae. 17 India, Nepal, Bhutan, China, Indochina, Malesia

Eriolobus (DC.) M. Roem. 1847 (Malus Hill). Rosaceae. 1 Eastern Mediterranean

Eriolopha Ridl. = Alpinia Roxb. Zingiberaceae

Erioneuron Nash. 1903. Poaceae. 5 southern U.S., Mexico, Bolivia and Argentina

Eriope Humb. et Bonpl. ex Benth. 1833. Lamiaceae. 28 trop. and subtrop. South America

Eriopexis (Schltr.) Brieger = Dendrobium Sw. Orchidaceae

Eriophorella Holub. 1984 (Baeothryon A. Dietr.). Cyperaceae. 2 cold and temp. Northern Hemisphere

Eriophorum L. 1753. Cyperaceae. 20 cold and temp. Northern Hemisphere, South Africa (1); Russia (13)

Eriophyllum Lag. 1816. Asteraceae. 15 western North America

Eriophyton Benth. 1829. Lamiaceae. 1 Himalayas (Nepal, Bhutan), Tibet, southern China (Yunnan) Eriopidion Harley.1976 (Eriope Humb. et Bonpl. ex

Benth.). Lamiaceae. 1 South America (Brazil)

Eriopsis Lindl. 1847. Orchidaceae. 6 trop. America from Costa Rica to Peru and Brazil

Erioscirpus Palla = Eriophorum L. Cyperaceae

Eriosema (DC.) Rchb. 1828. Fabaceae. 130 tropics and subtropics, esp. Africa (108) Eriosemopsis Robyns.1928. Rubiaceae. 1 South Africa

Eriosolena Blume. 1826 (Daphne L.). Thymelaeaceae. 5 eastern Himalayas, India, Burma, southwestern China, Indochina, Malesia (Sumatra, Malay Peninsula, Java, Borneo)

Eriospermum Jacq. ex Willd. 1799. Eriospermaceae. c. 90 Africa, abundant in South Africa (60)

Eriosphaera Less. = Galeomma Rauschert. Asteraceae

Eriospora Hochst. ex A. Rich. = Coleochloa Gilly. Cyperaceae

Eriostemon Sm. 1798. Rutaceae. 33 Australia, New Caledonia

Eriostrobilus Bremek. = Strobilanthes Blume. Acanthaceae

Eriostylos C. Towns. 1979. Amaranthaceae. 1 Somalia

Eriosyce Phil. 1872. Cactaceae. 2–4 Chile

Eriosynaphe DC. 1829 (Ferula L.). Apiaceae. 1 (E. longifolia) southeastern European part of Russia, northern Caucasus, Kazakhstan

Eriotheca Schott et Endl. 1832. Bombacaceae. 21 trop. America Eriothymus J. A. Schmidt = Hedeoma Pers. Lamiaceae

Eriotrix Cass. 1817. Asteraceae. 2 Reunion I.

Erioxylum Rose et Standl. = Gossypium L. Malvaceae

Eriphilema Herb. = Olsynium Raf. Iridaceae

Erisma Rudge. 1805. Vochysiaceae. 20 trop. South America

Erismadelphus Mildbr. 1913. Vochysiaceae. 3 trop. West Africa

Erismanthus Wall. ex Muell. Arg. 1866. Euphorbiaceae. 2 Hainan I., Indochina, Malay Peninsula, Sumatra, Borneo

Erithalis P. Browne. 1756. Rubiaceae. 10 Florida, West Indies, trop. America

Eritrichium Schrad. ex Gaudin. 1828. Boraginaceae. c. 90 temp Northern Hemisphere

Erlangea Sch. Bip. 1853. Asteraceae. 8 trop. Africa

Ermania Cham. 1831. Brassicaceae. 1 Russia—eastern Arctica, East Siberia, Far East

Ermaniopsis Hara. 1974. Brassicaceae. 1 western Nepal

Ernestia DC. 1828. Melastomataceae. 10 trop. South America

Ernestimeyera Kuntze = Alberta E. Mey. Rubiaceae

Ernodea Sw. 1788. Rubiaceae. 6 Florida, West Indies Erocallis Rydb.1906. Portulacaceae. 1 western North America

Erodiophyllum F. Muell. 1875. Asteraceae. 2 western and southern Australia

Erodium L'Hér. ex Aiton. 1789. Geraniaceae. c. 60 Europe (34); Mediterranean, West Asia eastward to Mongolia, China, and India; temp. Australia; southern trop. South America

Erophila DC. 1821. Brassicaceae. 10 Europe, Mediterranean, Caucasus, Middle Asia, northern Iraq, Iran, northwestern India (Kashmir)

Erpetion Sweet = Viola L. Violaceae

Errazurizia Phil. 1872. Fabaceae. 4 southwestern U.S. and Mexico (3), Chile (1)

Ertela Adans. = Moniera Loefl. Rutaceae

Eruca Hill. 1753. Brassicaceae. 5 Mediterranean, Northeast Africa, Southwest Asia; E. sativa also in southern European part of Russia, Caucasus, southern West Siberia, Middle Asia; Iran, Afghanistan, Mongolia, China, northwestern India

Erucaria Gaertn. 1791. Brassicaceae. 9 North Africa, eastern Mediterranean, Arabian Peninsula, Turkey, Iraq, Iran

Erucastrum (DC.) C. Presl. 1826. Brassicaceae. 18 Central and South Europe, Canary Is., Mediterranean, trop. (Uganda and northern Tanzania) and 3 South Africa, Arabian Peninsula

Ervatamia (A. DC.) Stapf = Tabernaemontana L. Apocynaceae

Erxlebenia Opiz = Pyrola L. Ericaceae

Erycibe Roxb. 1802. Convolvulaceae. c. 70 southern China, Taiwan, Japan, Indo-Malesia, Australia (northern Queensland)

Erycina Lindl. 1853. Orchidaceae. 2 Mexico

Erymophyllum Paul G. Wilson. 1989. Asteraceae. 5 southwestern Australia

Eryngiophyllum Greenm. 1903. Asteraceae. 1 Mexico

Eryngium L. 1753. Apiaceae. 230–250 America (esp. trop. regions), Eurasia, Mediterranean, trop. Africa, Malesia, Australia

Erysimum L. 1753. Brassicaceae. c. 100 Madeira and Canary Is., Cape Verde Is., Mediterranean, Europe, Asia

Erythea S. Watson = Brahea Mart. ex Endl. Arecaceae

Erythradenia (B. L. Rob.) R. M. King et H. Rob. 1969. Asteraceae. 1 Mexico

Erythraea Borkh. = Centaurium Hill. Gentianaceae

Erythranthera Zotov. 1963 (Notodanthonia Zotov). Poaceae. 2 Australia, New Zealand Erythranthus Oerst. ex Hanst. = Drymonia Mart. Gesneriaceae

Erythrina L. 1753. Fabaceae. 115 tropics and subtropics, esp. Africa (50)

Erythrocephalum Benth. 1873. Asteraceae. 12 trop. East Africa

Erythrochiton Nees et Mart. 1823. Rutaceae. 4–5 trop. America

Erythrochlamys Gürke. 1894. Lamiaceae. 5 trop. East Africa

Erythrococca Benth. 1849. Euphorbiaceae. c. 50 trop. and South Africa, southern Arabian Peninsula (1) Erythrocoma Greene = Geum L. Rosaceae

Erythrodes Blume. 1825. Orchidaceae. 60–100 trop. America (c. 80); trop. and subtrop. Asia from northern India and China south to Malesia; New Caledonia, Vanuatu, Fiji, Samoa, Tonga

Erythronium L. 1753. Liliaceae. 24 western (15) and eastern (5) North America, temp. Eurasia (4)

Erythropalum Blume. 1826. Olacaceae. 1 (E. scandens) southern and northeastern (Assam) India, eastern Himalayas, Bangladesh, Burma, Andaman Is., Indochina, southwestern China (incl. Hainan), Malesia (excl. New Guinea)

Erythrophleum Afzel. ex R. Br. 1826. Fabaceae. 9 trop. and South (1) Africa (5), Madagascar, East and Southeast Asia, trop. Australia

Erythrophysa E. Mey ex Arn. 1841. Sapindaceae. 3 Somalia and South Africa

Erythrophysopsis Verdc. 1962 (Erythrophysa E. Mey. ex Arn.). Sapindaceae. 1 Madagascar

Erythropsis Lindl. ex Schott et Endl. 1832 (Firmiana Marsigli). Sterculiaceae. 8 trop. Africa, Madagascar, India, China, Indochina, western Malesia

Erythrorhipsalis A. Berger = Rhipsalis Gaertn. Cactaceae

Erythrorchis Blume. 1837. Orchidaceae. 3: 1 (E. ochobiensis) trop. Asia eastward to Japan and Taiwan; 1 (E. altissima) Assam, Indochina, Taiwan, Ryukyu Is., Malay Peninsula, Java, Borneo, Philippines; 1—Australia

Erythroselinum Chiov. 1911. Apiaceae. 1 (E. atropurpureum) Ethiopia, Uganda, Kenya

Erythrospermum Lam. 1792. Flacourtiaceae. 5 Madagascar, Mauritius, India, Sri Lanka, Burma, China, Malesia (1, E. candidum Malay Peninsula, Borneo, Sulawesi, Moluccas, New Guinea), Melanesia from Solomon Is. eastward to Fiji and Samoa Is.

Erythroxylum P. Browne. 1756. Erythroxylaceae. c. 250 tropics and subtropics, chiefly America and Madagascar

Escallonia Mutis ex L. f. 1782. Escalloniaceae. c. 40 South America, esp. Andes Eschscholzia Cham.1820. Papaveraceae. 13 western North America

Eschweilera Mart. ex DC. 1828. Lecythidaceae. c. 100 trop. America from Mexico to northern Ecuador, Venezuela, the Guianas, Brazil south to Planalto

Esclerona Raf. = Xylia Benth. Fabaceae

Escobaria Britton et Rose. 1923. Cactaceae. 16 southern Canada, western U.S., northern Mexico, Cuba

Escobedia Ruiz et Pav. 1794. Scrophulariaceae. 6 trop. America

Escontria Britton et Rose. 1906 (Myrtillocactus Console). Cactaceae. 1 southern Mexico

Esenbeckia Kunth. 1825. Rutaceae. 26 trop. America: U.S. (1, E. runyonii—Texas, New Mexico), Mexico, Guatemala, Honduras, Belize, Nicaragua, Costa Rica, Panama, Colombia, Venezuela, the Guyanas, Ecuador, Peru, Bolivia, Brazil, Argentina, Paraguay; West Indies

Esfandiaria Charif et Aellen = Anabasis L. Chenopodiaceae

Eskemukerjea Malick et Sengupta = Fagopyrum Hill. Polygonaceae

Esmeralda Rchb. f. 1862. Orchidaceae. 2–3 Himalayas from Nepal to Bhutan, northeastern India, Burma, southern China (Yunnan and Hainan), Thailand

Espadaea A. Rich. 1850. Goetzeaceae. 1 Cuba

Espejoa DC. 1836. Asteraceae. 1 Mexico

Espeletia Mutis ex Humb. et Bonpl. 1808. Asteraceae. c. 50 Andes of Costa Rica, Panama, Venezuela, Colombia, Ecuador

Espeletiopsis Cuatrec. 1976. Asteraceae. 24 Andes of Colombia and Venezuela

Espostoa Britton et Rose. 1920. Cactaceae. c. 10 southern Ecuador, Peru, Bolivia

Espostoopsis F. Buxb. 1968. Cactaceae. 1 eastern Brazil

Esterhazya J. C. Mikan. 1821. Scrophulariaceae. 4 Brazil, Bolivia

Esterhuysenia L. Bolus. 1967 (or = Lampranthus N. E. Br.). Aizoaceae. 1 western South Africa

Esula Morandi = Euphorbia L. Euphorbiaceae

Etaballia Benth. 1840. Fabaceae. 1 Venezuela, Guyana, Brazil

Etericius Desv. 1825. Rubiaceae. 1 Guyana

Ethulia L. f. 1762. Asteraceae. 20 trop. Old World

Etiosedum A. Löve et D. Löve = Sedum L. Crassulaceae

Etlingera Giseke. 1792. Zingiberaceae. c. 60 Himalayas, southern China, Indochina, Sumatra, Borneo, Sulawesi, Philippines, Moluccas, New Guinea, Bismarck Arch., Solomon Is., Australia (1, Queensland), ? New Caledonia, Fiji, Samoa Is.

Euadenia Oliv. 1867. Capparaceae. 3 trop. Africa

Euanthe Schltr. 1914. Orchidaceae. 1 Philippines

Euaraliopsis Hutch. = Brassaiopsis Decne. et Planch. Araliaceae

Eubotryoides (Nakai) Hara. 1935 (Leucothoe D. Don). Ericaceae. 2 Sakhalin, Japan (Hokkaido, Honshu, Sikoku Is.)

Eubotrys Nutt. = Leucothoe D. Don. Ericaceae

Eubrachion Hook. f. 1846. Eremolepidaceae. 2: E. arnottii—Jamaica, Hispaniola, Puerto Rico, Brazil, Argentina, Uruguay; E. gracile-Venezuela

Eucalyptopsis C. T. Wight. 1951. Myrtaceae. 2 Moluccas, New Guinea

Eucalyptus L'Hér. 1792. Myrtaceae. c. 450 Australia and Tasmania, a few species in Malesia (Lesser Sunda Is., Moluccas, Sulawesi, Philippines, New Guinea), Hawaiian Is.

Eucarpha Spach = Knightia R. Br. Proteaceae

Eucarya T. L. Mitch. ex Sprague et Summerh. = Santalum L. Santalaceae

Eucephalus Nutt. = Aster L. Asteraceae

Euceraea Mart. 1831. Flacourtiaceae. 2 trop. South America: Amazonian Colombia, Venezuela, Guyana, Surinam, northern Brazil

Euchaetis Bartl. et H. L. Wendl. 1824. Rutaceae. 23 South Africa

Eucharis Planch. et Linden. 1852–1853. Amaryllidaceae. 17 trop. America from Guatemala to Bolivia

Euchilopsis F. Muell. 1882. Fabaceae. 1 western and southwestern Australia

Euchiton Cass. 1828. Asteraceae. 17 East and Southeast Asia, Malesia to Australia, New Zealand

Euchlaena Schrad. = Zea L. Poaceae

Euchlora Eckl. et Zeyh. = Lotononis (DC.) Eckl. et Zeyh. Fabaceae

Euchorium Ekman et Radlk. 1925. Sapindaceae. 1 Cuba

Euchresta Benn. 1840. Fabaceae. 4–5 northeastern India, China, southern Japan, Java

Euclasta Franch. 1895. Poaceae. 1 trop. America, Africa, India

Euclea L. ex Murr. 1784. Ebenaceae. 20 East and South Africa, Comoro Is., Arabian Peninsula

Euclidium R. Br. 1812. Brassicaceae. 2 Southeast Europe, Turkey, Iraq, Syria, Lebanon, Iran, Afghanistan, Pakistan, India, western and southwestern China; E. syriacum—also in Caucasus, West Siberia, and Middle Asia

Euclinia Salisb. 1808 (Randia L.). Rubiaceae. 3 trop. West and Central Africa to Uganda, Madagascar

Eucnide Zucc. 1844. Loasaceae. 8 southwestern North America

Eucodonia Hanst. 1854 (Achimenes Pers.). Gesneriaceae. 2 Central America

Eucomis L'Hér. 1789. Hyacinthaceae. 10 trop. and South Africa

Eucommia Oliv. 1890. Eucommiaceae. 1 continental China

Eucorymbia Stapf. 1903. Apocynaceae. 1 Borneo

Eucosia Blume. 1825. Orchidaceae. 2 Java, New Guinea

Eucrosia Ker-Gawl. 1817. Amaryllidaceae. 7 Andes of Ecuador and Peru

Eucryphia Cav. 1798. Eucryphiaceae. 6 southeastern Australia, Tasmania, Chile

Eucrypta Nutt. 1848. Hydrophyllaceae. 2 southwestern U.S. (California, Utah, Nevada, Arizona, Texas) and northwestern Mexico

Eudema Humb. et Bonpl. 1813. Brassicaceae. 6 Andes from Ecuador to Argentina

Eudianthe (Rchb.) Rchb. = Silene L. Caryophyllaceae

Eufournia J. R. Reeder = Sohnsia Airy Shaw. Poaceae

Eugeissona Griff. 1844. Arecaceae. 6 Malay Peninsula (2), Borneo (4)

Eugenia L. 1753. Myrtaceae. c. 1,000 tropics and subtropics, esp. America

Euglypha Chodat et Hassl. 1906. Aristolochiaceae. 1 northern Argentina, Paraguay

Euhesperida Brullo et Furnari = Satureja L. Lamiaceae

Euklisia Rydb. = Streptanthus Nutt. Brassicaceae

Eulalia Kunth. 1829. Poaceae. c. 30 trop. and subtrop. Africa and Asia

Eulaliopsis Honda. 1923. Poaceae. 2 South Asia, Taiwan, Philippines

Euleria Urb. 1925. Anacardiaceae. 1 Cuba

Eulophia R. Br. ex Lindl. 1823. Orchidaceae. 200–250 trop. America, trop. and South Africa, Madagascar, Mascarene Is., trop. and subtrop. Asia eastward to Japan, Malesia from Malay Peninsula to New Guinea, Solomon Is. (1, E. nuda), trop. Australia, Mariana Is., Vanuatu, Fiji, Tonga

Eulophidum Pfitz. = Oeceoclades Lindl. Orchidaceae

Eulophiella Rolfe. 1892. Orchidaceae. 4 Madagascar

Eulophus Nutt. ex DC. = Perideridia Rchb. Apiaceae

Eulychnia Phil. 1860. Cactaceae. 6–8 Peru and coastal Chile

Eumorphia DC. 1838. Asteraceae. 4 South Africa

Eunomia DC. = Aethionema R. Br. Brassicaceae

Euodia J. R. Forst. et G. Forst. 1775. Rutaceae. 130 trop. Africa, Madagascar, Mauritius and Réunion Is., trop. Asia from India to southeastern China, Taiwan, Ogasawara (Bonin) Is., Indochina, Malesia, northeastern Australia, Micronesia, Fiji, Samoa Is., Tongo

Euonymus L. 1753. Celastraceae. c. 180 mainly temp. and trop. Asia, Europe (4), North and Central America, a few in Africa, Australia (2)

Eupatoriadelphus R. M. King et H. Rob. = Eupatorium L. Asteraceae

Eupatoriastrum Greenm. 1904. Asteraceae. 4 Mexico, Central America

Eupatorina R. M. King et H. Rob. 1971. Asteraceae. 1 Haiti

Eupatoriopsis Hieron. 1893. Asteraceae. 1 Brazil

Eupatorium L. 1753. Asteraceae. 45 Europe, Caucasus, Middle and East Asia, eastern North America

Euphlebium (Kranzel) Brieger = Dendrobium Sw. Orchidaceae

Euphorbia L. 1753. Euphorbiaceae. 1,600–2,000 cosmopolitan

Euphoria Comm. ex Juss. 1789. Sapindaceae. 15 India, southern China, Indochina, western Malesia

Euphorianthus Radlk. 1933 (Diploglottis Hook. f. Sapindaceae. 1 Sulawesi, New Guinea

Euphrasia L. 1753. Scrophulariaceae. c. 450 temp. Northern Hemisphere, montane trop. Asia and Malesia, Australia, New Zealand, temp. South America

Euphronia Mart. et Zucc. 1825. Euphroniaceae. 3 northern South America

Euphrosyne DC. 1836. Asteraceae. 1 Mexico

Euplassa Salisb. ex Knight. 1809. Proteaceae. 25 trop. America

Euploca Nutt. = Cryptantha Lehm. ex G. Don. Boraginaceae

Eupomatia R. Br. 1814. Eupomatiaceae. 2 New Guinea, coastal eastern and southeastern Australia

Euptelea Siebold et Zucc. 1840–1841. Eupteleaceae. 2 eastern Himalayas, northeastern India, southwestern and central China, Japan

Eureiandra Hook. f. 1867. Cucurbitaceae. 8 trop. Africa, Socotra

Euroschinus Hook. f. 1862. Anacardiaceae. 6 New Guinea, New Britain, northeastern Australia, New Caledonia (4)

Eurotia Adans. = Ceratoides Gagnebin. Chenopodiaceae

Eurya Thunb. 1783. Theaceae. 70 Himalayas, India, Sri Lanka, Indochina, China (incl. Hainan), Malesia, Caroline and Mariana Is. eastward to Hawaiian and Samoa Is.

Euryale Salisb. 1805. Nymphaeaceae. 1 India, Bangladesh, continental China, Taiwan, Japan; Russian Far East

Eurybiopsis DC. 1836. Asteraceae. 1 trop. Australia

Eurycarpus Botsch. = Christolea Cambess. Brassicaceae

Eurycentrum Schltr. 1905. Orchidaceae. 5 New Guinea, Solomon Is.

Eurychone Schltr. 1918. Orchidaceae. 2 trop. Africa

Eurycles Salisb. ex Schult. et Schult. f. = Proiphys Herb. Amaryllidaceae

Eurycoma Jack. 1822. Simaroubaceae. 4 Andaman Is., Indochina, Malesia

Eurycorymbus Hand. = Mazz.1922. Sapindaceae. 1 southern China, Taiwan

Eurydochus Maguire et Wurdack. 1958. Asteraceae. 1 Venezuela, Guyana

Eurylobium Hochst. 1842. Stilbaceae. 2 South Africa (Cape) Eurynotia R. C. Froster = Ennealophus N. E. Br. Iridaceae

Euryops (Cass.) Cass. 1818. Asteraceae. 100 trop. and South Africa, Arabian Peninsula, Socotra

Eurypetalum Harms. 1910. Fabaceae. 3 Ghana, Cameroun, Gabon

Eurysolen Prain. 1898. Lamiaceae. 2 China, Indochina, Malesia

Eurystemon Alexander. 1937. Pontederiaceae. 1 Mexico

Eurystigma L. Bolus = Mesembryanthemum L. Aizoaceae

Eurystyles Wawra. 1863. Orchidaceae. 11 trop. America, West Indies

Eurytaenia Torr. et A. Gray. 1840. Apiaceae. 2 U.S. (Texas)

Euscaphis Siebold et Zucc. 1840. Staphyleaceae. 1 (E. japonica) southern China, Indochina, Taiwan, Japan

Eusideroxylon Teijsm. et Binn. 1863. Lauraceae. 2 Sumatra, Borneo

Eusiphon Benoist. 1939. Acanthaceae. 3 Madagascar

Eustachys Desv. 1810. Poaceae. 12 tropics and subtropics, esp. America

Eustegia R. Br. 1810. Asclepiadaceae. 5 South Africa

Eustephia Cav. 1795. Amaryllidaceae. 6 Peru, Argentina

Eusteralis Raf. 1837. Lamiaceae. 27 temp. and trop. Asia, Australia; E. jatabeanus—extending to Russian Far East

Eustigma Gardner et Champ. 1849. Hamamelidaceae. 3 southern China, Indochina, Taiwan

Eustoma Salisb. 1806. Gentianaceae. 3 U.S. from Wyoming, Colorado, Nebraska, Kansas, and Oklahoma to California, New Mexico, Texas, and Florida; Mexico, Central and northern South America; West Indies from Bahamas to Great Antilles

Eustrephus R. Br. ex Ker-Gawl. 1809 (Luzuriaga Ruiz et Pav.). Luzuriagaceae. 1 Malesia, New Guinea, eastern Australia, New Caledonia, Loyalty Is.

Eustylis Engelm. et A. Gray = Alophia Herb. Iridaceae

Eutaxis R. Br. 1811. Fabaceae. 8–9 western, southwestern, and eastern Australia

Euteline Raf. = Genista L. Fabaceae

Euterpe Mart. 1837. Arecaceae. 10 Central and trop. South America, southward to Peru and Brazil, Lesser Antilles Is. (1, E. dominicana)

Eutetras A. Gray. 1879. Asteraceae. 2 Mexico

Euthamia Nutt. 1818. Asteraceae. 8 North America

Euthemis Jack. 1820. Sauvagesiaceae. 2 Indochina (1, E. leucocarpa), Malesia (2—Sumatra, Riouw and Lingga Is., Banka, Malay Peninsula, Anambas Is., Borneo)

Eutheta Standl. = Melasma P. Bergius. Scrophulariaceae

Euthryptochloa Cope. 1987. Poaceae. 1 China (Sichuan)

Euthyra Salisb. = Daiswa Raf. Trilliaceae

Euthystachys A. DC. 1848. Stilbaceae. 1 South Africa

Eutrema R. Br. 1823. Brassicaceae. 16 arctic Europe (1, E. edwardsii), arcto-alpine and temp. Asia (incl. Russia, 7 arctic European part, Siberia, Far East), montane Middle Asia, southward to Himalayas, southwestern North America

Euxylophora Huber. 1910. Rutaceae. 1 Brazil (Amazonia)

Euzomodendron Coss. 1852. Brassicaceae. 1 southern Spain

Evacidium Pomel. 1875. Asteraceae. 1 Northwest Africa, Sicily I.

Evandra R. Br. 1810. Cyperaceae. 2 southwestern Australia

Evax Gaertn. = Filago L. Asteraceae

Evea Aubl. = Cephaelis Sw. Rubiaceae

Everardia Ridl. 1886. Cyperaceae. 12 Venezuela, Guyana

Eversmannia Bunge. 1838. Fabaceae. 1 southeastern European part of Russia, Middle Asia, Iran, Afghanistan

Evodia Lam. = Tetradium Lour. Rutaceae

Evodianthus Oerst. 1857. Cyclanthaceae. 1 trop. America from Costa Rica to Peru and central and eastern Brazil, Trinidad, and Tobago Is.

Evodiella van der Linden. 1959. Rutaceae. 2 New Guinea, Australia (Queensland)

Evodiopanax (Harms) Nakai. 1924 (Acanthopanax Miq.). Araliaceae. 4 western and southwestern China, Burma, Vietnam, Japan

Evolvulus L. 1762. Convolvulaceae. 100 tropics and subtropics, esp. America

Evonymopsis H. Perrier. 1942. Celastraceae. 4 Madagascar

Evota (Lindl.) Rolfe. 1913 (Ceratandra Echl. ex Bauer). Orchidaceae. 3 South Africa (Cape)

Evotella Kurzweil et Linder. 1991. Orchidaceae. 1 South Africa (southwestern and southern Cape)

Evrardia Gagnep. = Ervardianthe Rauschert. Orchidaceae

Evrardiana Aver. = Evrardianthe Rauschert. Orchidaceae

Evrardianthe Rauschert. 1983. Orchidaceae. 3 Tibet and Southeast Asia

Evrardiella Gagnep. 1934 (Aspidistra Ker-Gawl.). Convallaria ceae. 1 Vietnam

Ewartia Beauverd. 1910. Asteraceae. 5 southeastern Australia, Tasmania, New Zealand

Ewartiothamnus Anderb. 1991. Asteraceae. 1 New Zealand

Exaculum Caruel. 1886. Gentianaceae. 1 Iberian Peninsula, southern France, Corsica and Sardinia, Italy, Morocco, Algeria, Tunisia

Exacum L. 1753. Gentianaceae. 65 Africa from Senegal to Kenya and south to northern Angola and Transvaal, Madagascar and Comoro Is. (38), Mauritius (1), coastal Oman, Socotra, India, Sri Lanka, Himalayas, Burma, southern China, Indochina, Malesia to New Guinea, northern Australia

Exallage Bremek. 1952. Rubiaceae. 24 Indo-Malesia

Exandra Standl. 1923 (Simira Aubl.). Rubiaceae. 1 Mexico, Central America

Exarata Gentry. 1992. Bignoniaceae. 1 Colombia, Ecuador

Exarrhena R. Br. = Myosotis L. Boraginaceae

Exbucklandia R. W. Br. 1946. Hamamelidaceae. 4 Himalayas (Nepal, Bhutan), northeastern India, Burma, southern China, Indochina, Malay Peninsula, Sumatra, Java

Excavatia Markgr. = Ochrosia Juss. Apocynaceae

Excentrodendron H. T. Chang et R. H. Miau = Burretiodendron Rehder. Tiliaceae

Excoecaria L. 1759. Euphorbiaceae. c. 40 trop. Old World from Africa throughout trop. Asia and Malesia to Tongo and Niue I.

Excremis Willd. ex Baker. 1830. Phormiaceae. 1 Andes of Colombia, Venezuela, Peru, and Bolivia

Exechostylus K. Schum. = Pavetta L. Rubiaceae

Exellia Boutique. 1951. Annonaceae. 1 trop. Africa (Gabon, Congo, Angola)

Exellodendron Prance. 1972. Chrysobalanaceae. 5 trop. South America from Venezuela to central Brazil

Exoacantha Labill. 1791. Apiaceae. 2 Turkey, Syria, Israel, Iran

Exocarpos Labill. 1800. Santalaceae. 25 Indochina, Java, Borneo, Lesser Sunda and Moluccas, Philippines, Aru Is., New Guinea, Australia and Tasmania (11), New Zealand, New Caledonia (7), Fiji, Hawaiian Is.

Exocarya Benth. 1877. Cyperaceae. 2 New Guinea, eastern Australia

Exochogyne C. B. Clarke. 1905. Cyperaceae. 4 trop. South America

Exochorda Lindl. 1858. Rosaceae. 5 Middle Asia (2), China

Exodeconus Raf. 1838. Solanaceae. 8–10 Galapagos Is., Ecuador, northern Peru

Exogonium Choisy = Ipomoea L. Convolvulaceae

Exohebea R. C. Foster = Tritoniopsis L. Bolus. Iridaceae

Exolobus Fourn. = Gonolobus Michx. Asclepiadaceae

Exomiocarpon Lawalrée. 1943. Asteraceae. 1 Madagascar

Exomis Fenzl ex Moq. 1840. Chenopodiaceae. 2–3 St. Helena, Angola, Namibia, and South Africa

Exorhopala Steenis. 1931. Helosiaceae. 1 Malay Peninsula

Exorrhiza Becc. = Clinostigma H. A. Wendl. Arecaceae

Exospermum Tiegh. = Zygogynum Baill. Winteraceae

Exostema (Pers) Rich. ex Humb. et Bonpl. 1807. Rubiaceae. 50 trop. America from Mexico to Peru and Brazil, West Indies

Exostyles Schott. 1827. Fabaceae. 3 Brazil

Exothea Macfad. 1837. Sapindaceae. 3 Florida, Mexico, Central America, West Indies

Exotheca Andersson. 1856. Poaceae. 1 Sudan, Ethiopia, Uganda, Kenya, Tanzania, Zambia, Malawi, Vietnam

Eylesia S. Moore. 1908 (Buchnera L.). Scrophulariaceae. 1 trop. East Africa

Eysenhardtia Kunth. 1824. Fabaceae. 10 southwestern U.S., Mexico, Guatemala

Ezosciadium Burtt. 1991. Apiaceae. 1 South Africa (Cape)

Faba Hill = Vicia L. Fabaceae

Faberia Hemsl. 1888. Asteraceae. 4 southwestern China

Fabiana Ruiz et Pav. 1794. Solanaceae. 15 southern Peru, Bolivia, Chile (7), and Argentina

Fabrisinapis C. C. Towns. = Hemicrambe Webb. Brassicaceae

Facelis Cass. 1819. Asteraceae. 3 southern Brazil, Peru, Bolivia, Paraguay, Chile, Uruguay, and Argentina

Facheiroa Britton et Rose. 1920. Cactaceae. 3 northeastern Brazil

Factorovskya Eig. 1927. Fabaceae. 1 eastern Mediterranean eastward to Iran

Fadenia Aellen et C. C. Towns. 1972. Chenopodiaceae. 1 Ethiopia, Kenya

Fadogia Schweinf. 1868. Rubiaceae. c. 50 trop. (eastward to Kenya, Tanzania, and Mozambique) and South Africa

Fadogiella Robyns. 1928. Rubiaceae. 2–4 trop. Africa from Zaire to Tanzania, Malawi, Zambia

Fagara L. = Zanthoxylum L. Rutaceae

Fagaropsis Mildbr. ex Siebenl. 1914. Rutaceae. 2 trop. Africa (Ethiopia, Uganda, Kenya, Tanzania, Zaire, Angola, Malawi, Zimbabwe)

Fagelia Neck. ex DC. = Bolusafra Kuntze. Fabaceae

Fagerlindia Tirveng. 1983 (Randia L.). Rubiaceae. 8 South and Southeast Asia, Philippines

Fagonia L. 1753. Zygophyllaceae. 30 coastal South Europe, islands of Mediterranean Sea, arid Africa, Southwest Asia eastward to India

Fagopyrum Mill. 1754. Polygonaceae. 6 temp. Eurasia; Russia (3, from European part to Far East)

Fagraea Thunb. 1782. Gentianaceae. 36 trop. Asia, Malesia, Australia (Queensland), New Caledonia, Oceania from Mariana Is. to Marquesas Is. and Tubuai Is.

Faguetia Marchand. 1869. Anacardiaceae. 1 Madagascar

Fagus L. 1753. Fagaceae. 14 Europe, Caucasus, Asia Minor, northern Iran, continental China, Taiwan, Japan, eastern U.S.

Fahrenheitia Rchb. f. et Zoll. ex Muell. Arg. 1866 (Ostodes Blume). Euphorbiaceae. 4 southern India, Sri Lanka, western Malesia to Bali

Faidherbia A. Chev. 1934. Fabaceae. 1 trop. and subtrop. Africa, Arabian Peninsula, Syria, Lebanon, Israel, Jordan, Iran

Faika Philipson. 1985. Monimiaceae. 1 New Guinea

Falcaria Fabr. 1759. Apiaceae. 1 (F. vulgaris) Central, South, and East Europe; Northwest Africa; Southwest, West (incl. Caucasus), and Central Asia (incl. Middle Asia); West Siberia

Falcata J. F. Gmel. = Amphicarpaea Elliott ex Nutt. Fabaceae

Falcatifolium de Laub. 1969. Podocarpaceae. 5 Malay Peninsula, Borneo, Philippines, Moluccas, Sulawesi, Philippines, Riouw-Lingga Arch., New Guinea, New Caledonia (1)

Falconeria Hook. f. = Kashmiria D. Y. Hong. Scrophulariaceae

Falkia L. f. 1782. Convolvulaceae. 3 Africa from Ethiopia to Cape

Fallopia Adans. 1763 (Polygonum L.). Polygonaceae. 9 temp. Northern Hemisphere

Fallugia Endl. 1840. Rosaceae. 1 southwestern U.S. from California to Colorado and Texas, Mexico

Falya Descoings = Dichapetalum Thouars. Dichapetalaceae

Famatina Ravenna. 1972. Amaryllidaceae. 4 Andes of Chile and Argentina

Fanninia Harv. 1868. Asclepiadaceae. 1 South Africa

Faradaya F. Muell. 1865. Verbenaceae. 20 Eastern Malesia to New Guinea, Solomon Is. and northeastern Australia eastward to Vanuatu, Fiji (5), Samoa Is., and Tonga

Faramea Aubl. 1775. Rubiaceae. 120 trop. America, West Indies

Farfugium Lindl. 1857. Asteraceae. 2 East Asia

Fargesia Franch. = Thamnocalamus Munro. Poaceae

Farinopsis Chrtek et Soják = Potentilla L. Rosaceae

Farmeria Willis ex Trimen. 1900. Podostemaceae. 2 southwestern India, Sri Lanka

Faroa Welw. 1869. Gentianaceae. 19 trop. and South Africa

Farquharia Stapf. 1912. Apocynaceae. 1 Ivory Coast, Ghana, southern Nigeria, extending to Congo

Farrago Clayton. 1967. Poaceae. 1 Tanzania

Farringtonia Gleason. 1952 (Siphanthera Pohl). Melastomataceae. 1 Venezuela

Farsetia Turra. 1765. Brassicaceae. 20 North Africa (from Morocco to Egypt) and trop. Africa (Nigeria, Niger, Chad, Sudan, Ethiopia, Somalia, Kenya, Tanzania), Southwest and South Asia from Arabian Peninsula eastward to Afghanistan, Pakistan, and northwestern India

Fascicularia Mez. 1894. Bromeliaceae. 5 Chile

Fatoua Gaudich. 1830. Moraceae. 1 China, Japan, Southeast Asia, Malesia, northern Australia, New Caledonia

Fatsia Decne. et Planch. 1854. Araliaceae. 2 Japan (1), southern China, Taiwan

Faucaria Schwantes. 1926. Aizoaceae. 33 South Africa

Faucherea Lecomte. 1920. Sapotaceae. 11 Madagascar

Faujasia Cass. 1819. Asteraceae. 9 Réunion I.

Faujasiopsis C. Jeffrey. 1992. Asteraceae. 3 Réunion I. and Mauritius

Faurea Harv. 1847. Proteaceae. 18 trop. and South Africa, Madagascar

Fauria Franch. = Nephrophyllidium Gilg. Menyanthaceae

Favargera A. Löve et D. Löve = Gentiana L. Gentianaceae

Fawcettia F. Muell. = Tinospora Miers. Menispermaceae

Faxonanthus Greenm. 1902. Scrophulariaceae. 1 Mexico

Faxonia Brandegee. 1894. Asteraceae. 1 Baja California

Feddea Urb. 1925. Asteraceae. 1 Cuba

Fedia Gaertn. 1790. Valerianaceae. 3 Mediterranean

Fedorovia Kolak. = Campanula L. Campanulaceae

Fedorovia Yakovlev = Ormosia G. Jacks. Fabaceae

Fedtschenkiella Kudr. = Dracocephalum L. Lamiaceae

Feeria Buser. 1894 (Trachelium L.). Campanulaceae. 1 Morocco

Fegimanra Pierre. 1892. Anacardiaceae. 2 trop. Africa

Feijoa O. Berg. 1858 (or = Acca O. Berg). Myrtaceae. 2 Brazil

Feldstonia P. S. Short. 1989. Asteraceae. 1 western Australia

Felicia Cass. 1818. Asteraceae. 83 trop. and South Africa, Arabian Peninsula

Feliciadamia Bullock. 1962 (Miconia Ruiz et Pav.). Melastomataceae. 1 West Africa

Felipponia Hichen = Mangonia Schott. Araceae

Felipponiella Hicken = Mangonia Schott. Araceae

Femeniasia Susanna de la Serna. 1987 (Centaurea L.). Asteraceae. 1 western Mediterranean

Fendlera Engelm. et A. Gray. 1852. Hydrangeaceae. 1 southwestern U.S.

Fendlerella A. Heller. 1898. Hydrangeaceae. 1 U.S. from Utah and Colorado to California and New Mexico; northern Mexico

Feneriva Airy Shaw = Fenerivia Diels. Annonaceae

Fenerivia Diels. 1925 (or = Polyalthia Blume). Annonaceae. 1 Madagascar

Fenestraria N. E. Br. 1925 (Jordaaniella H. Hartmann). Aizoaceae. 1 coastal southwestern Namibia, South Africa

Fenixia Merr. 1917. Asteraceae. 1 Philippines

Fenzlia Endl. = Myrtella F. Muell. Myrtaceae

Ferdinandusa Pohl. 1829. Rubiaceae. 20 Panama, trop. South America, West Indies

Feretia Delile. 1843. Rubiaceae. 2 trop. Africa from Mauritania to Somalia and south to Botswana

Fergania Pimenov. 1982. Apiaceae. 1 (F. polyantha) Middle Asia (northern Pamir-Alai and western Tien Shan)

Fergusonia Hook. f. 1873. Rubiaceae. 1 southern India, Sri Lanka

Fernaldia Woodson. 1932. Apocynaceae. 4 Mexico, Central America

Fernandezia Ruiz et Pav. 1794 (Centropetalum Lindl. et Dichaea Lindl.). Orchidaceae. 7 Peru

Fernandoa Welw. ex Seem. 1865. Bignoniaceae. 14 trop. Africa (4), Madagascar (3), southern China (1), Indochina, Sumatra

Fernelia Comm. ex Lam. 1788. Rubiaceae. 4 Mascarene Is.

Fernseea Baker. 1889. Bromeliaceae. 1 southeastern Brazil

Ferocactus Britton et Rose. 1922. Cactaceae. 23 southwestern U.S., Mexico

Feronia Corrêa = Limonia L. Rutaceae

Feroniella Swingle. 1913. Rutaceae. 3 Southeast Asia, Java

Ferraria Burm. ex Mill. 1759. Iridaceae. 10 trop. and South Africa

Ferreirea Allemao = Sweetia Spreng. Fabaceae

Ferreyanthus H. Rob. et Brettell = Ferreyranthus H. Rob. et Brettell. Asteraceae

Ferreyranthus H. Rob. et Brettell. 1974. Asteraceae. 7 Andes of southern Ecuador and Peru

Ferreyrella S. F. Blake. 1958. Asteraceae. 2 Peru

Ferrocalamus C. J. Hsueh et Keng f. 1982. Poaceae. 2 China

Ferula L. 1753. Apiaceae. c. 175 Eurasia, chiefly arid areas from Southwest Europe to Mongolia and China; esp. (c. 115) European part of Russia, Caucasus, West Siberia, and Middle Asia; Northwest, North, Northeast, and trop. Africa

Ferulago W. D. J. Koch. 1824. Apiaceae. 45 Europe, Mediterranean, West (incl. Caucasus) and Middle Asia

Ferulopsis Kitag. = Phlojodicarpus Turcz. ex Ledeb. Apiaceae

Festuca L. 1753. Poaceae. c. 500 cosmopolitan

Festucella E. B. Alexeev. 1985 [Austrofestuca (Tzvelev) E. B. Alexeev]. Poaceae. 1 Australia

Festucopsis (C. E. Hubb.) Meld. 1978 (Elymus L.). Poaceae. 1 Balkan Peninsula

Fevillea L. 1753. Cucurbitaceae. 5–10 trop. America, West Indies

Fezia Pit. 1931. Brassicaceae. 1 Morocco

Fibigia Medik. 1792. Brassicaceae. 14 eastern Mediterranean, Crimea, Caucasus, Middle Asia, Iraq, Iran, Afghanistan

Fibraurea Lour. 1790. Menispermaceae. 2 northeastern India (Manipur), Nicobar Is., Burma, southern China, Indochina, Malesia (1, F. tinctoria—Sumatra, Malay Peninsula, Java, Borneo, Sulawesi, Philippines)

Ficalhoa Hiern. 1898. Theaceae. 1 trop. Africa

Ficaria Guett. 1754 (Ranunculus L.). Ranunculaceae. 8 temp. Eurasia; esp. (8) European part of Russia, Caucasus, West Siberia, and Middle Asia

Ficinia Schrad. 1832. Cyperaceae. 75 trop. and South Africa, Madagascar

Ficus L. 1753. Moraceae. c. 750 tropics and subtropics: America (c. 150), Africa (c. 100), Asia and Australasia (c. 500)

Fiebrigia Fritsch = Gloxinia L'Hér. Gesneriaceae

Fiebrigiella Harms. 1908. Fabaceae. 1 Peru, Bolivia

Fieldia A. Cunn. 1825. Gesneriaceae. 1 southeastern Australia

Filaginella Opiz = Gnaphalium L. Asteraceae

Filago L. 1753. Asteraceae. 35 Eurasia, North Africa, southwestern and eastern North America

Filarum Nicolson. 1967. Araceae. 1 Peru

Filetia Miq. 1858. Acanthaceae. 8 Malay Peninsula, Sumatra

Filicium Thwaites ex Hook. f. 1862. Sapindaceae. 3 trop. Africa, trop. Asia

Filifolium Kitam. 1940 (Artemisia L.). Asteraceae. 1 East Siberia, Russian Far East, Mongolia, northeastern China, Korea

Filipedium Raizada et S. K. Jain = Capillipedium Stapf. Poaceae

Filipendula Mill. 1754. Rosaceae. 15 temp. Northern Hemisphere; esp. (11) Russia; F. ulmaria—from Arctica to Caucasus, Middle Asia, and Russian Far East

Fillaeopsis Harms. 1899. Fabaceae. 1 Nigeria, Cameroun, Gabon, Zaire, Angola

Fimbriella Farw. ex Butzin. 1981 (Platanthera Rich.). Orchidaceae. 4 North America

Fimbripetalum (Turcz.) Ikonn. = Stellaria L. Caryophyllaceae

Fimbristemma Turcz. 1852. Asclepiadaceae. 5 Central and trop. South America

Fimbristylis Vahl. 1805–1806. Cyperaceae. 300 tropics and subtropics

Fimbrolina Raf. = Sinningia Nees. Gesneriaceae

Findlaya Hook. f. 1876 (Orthaea Klotzsch). Ericaceae. 1 Trinidad

Finetia Gagnep. = Anogeissus (DC.) Guillemin, Perrottet et A. Rich. Combretaceae

Fingerhuthia Nees. 1834. Poaceae. 2 South Africa, Arabian Peninsula, Afghanistan, Pakistan

Finlaysonia Wall. 1831. Asclepiadaceae. 1 Indo-Malesia

Finschia Warb. 1891. Proteaceae. 4 Palau Is., Aru Is., New Guinea, New Britain, Solomon Is., Vanuatu

Fintelmannia Kunth = Trilepis Nees. Cyperaceae

Firmiana Marsigli. 1786. Sterculiaceae. 12 trop. East Africa (1), trop. Asia

Fischeria DC. 1813. Asclepiadaceae. 16 Central and trop. South America, West Indies

Fissendocarpa (Haines) Bennet = Ludwigia L. Onagraceae

Fissenia R. Br. ex Endl. 1842 (Kissenia R. Br. ex Endl.). Loasaceae. 2 southwest (1) and East Africa, southern Arabian Peninsula, Socotra

Fissicalyx Benth. 1860. Fabaceae. 1 Panama, Venezuela, Guyana

Fissistigma Griff. 1854. Annonaceae. 75 trop. Africa, India, southwestern China, Taiwan, Malesia, northeastern Australia (2)

Fitchia Hook. f. 1845. Asteraceae. 6 Polynesia

Fittingia Mez. 1922. Myrsinaceae. 5 New Guinea, New Britain

Fittonia Coem. 1865. Acanthaceae. 3 Colombia, Peru

Fitzalania F. Muell. 1863. Annonaceae. 1 trop. eastern Australia

Fitzroya Hook. f. ex Lindl. 1851. Cupressaceae. 1 southern Chile, Chiloé I., southern Argentina

Fitzwillia P. S. Short. 1989. Asteraceae. 1 western Australia

Flabellaria Cav. 1790. Malpighiaceae. 1 trop. Africa from Senegal to Sudan, Uganda, and south to Angola and Tanzania

Flabellariopsis R. Wilczek. 1955. Malpighiaceae. 1 trop. Africa: Congo, Zaire, Uganda, Tanzania

Flacourtia Comm. ex L'Hér. 1786. Flacourtiaceae. 17 trop. and South Africa, Madagascar, Mascarene Is., South and Southeast Asia, Malesia, northern Australia, Melanesia to Fiji

Flagellaria L. 1753. Flagellariaceae. 4 trop. and subtrop. Old World

Flagellarisaema Nakai = Arisaema Mart. Araceae

Flagenium Baill. 1880. Rubiaceae. 6 trop. West Africa (2), Madagascar (4)

Flanagania Schltr. = Cynanchum L. Asclepiadaceae

Flaveria Juss. 1789. Asteraceae. 21 U.S. (Colorado, Kansas, Missouri, Oklahoma, Alabama, Arizona, New Mexico, Texas, Florida), Mexico (18), Colombia, Venezuela, Ecuador, Peru, Bolivia, Brazil, Chile, Argentina, Paraguay; West Indies (2)—extending to Africa and India; Australia (1, F. australasica)

Fleischmannia Sch. Bip. 1850. Asteraceae. 79 southern and eastern U.S., Mexico, Guatemala, El Salvador, Honduras, Nicaragua, Costa Rica, Panama, Colombia, Venezuela, Ecuador, Peru, Bolivia, Brazil, Argentina, Paraguay; West Indies; adventive in Africa

Fleischmanniana Sch. Bip. = Fleischmanniopsis R. M. King et H. Rob. Asteraceae

Fleischmanniopsis R. M. King et H. Rob. 1971. Asteraceae. 5 Mexico, Central America

Flemingia Roxb. ex W. T. Aiton. 1812. Fabaceae. 30 trop. Old World; F. grahamiana—extending to South Africa

Fleurya Gaudich. = Laportea Gaudich. Urticaceae

Fleurydora A. Chev. 1933. Sauvagesiaceae. 1 trop. West Africa

Flexanthera Rusby. 1927. Rubiaceae. 2 Colombia, Bolivia

Flexularia Raf. = Muhlenbergia Schreb. Poaceae

Flickingeria A. D. Hawkes. 1961. Orchidaceae. 65–70 trop. Asia, north to Taiwan, south to Christmas I. (Indian Ocean), Malesia, Caroline Is., Australia (2), Bougainville and Solomon Is., Vanuatu, New Caledonia, Fiji, Samoa

Flindersia R. Br. 1814. Rutaceae. 17 Salajar I., Moluccas, New Guinea, eastern Australia, New Caledonia (1)

Floerkea Willd. 1801. Limnanthaceae. 1 North America

Florestina Cass. 1817. Asteraceae. 8 Mexico

Floribunda F. Ritter = Cipocereus F. Ritter. Cactaceae

Floscaldasia Cuatrec. 1968. Asteraceae. 1 Colombia

Floscopa Lour. 1790. Commelinaceae. 20 tropics and subtropics

Flosmutisia Cuatrec. 1985. Asteraceae. 1 Colombia

Flourensia DC. 1836. Asteraceae. 30 America from southwestern U.S. to Argentina

Floydia L. A. S. Johnson et B. S. Briggs. 1975 (Helicia Lour.). Proteaceae. 1 northeastern Australia

Flueckigera Kuntze. 1891. Petiveriaceae. 3 trop. America from Mexico to Venezuela, West Indies

Flueggea Willd. 1809 (Securinega Juss.). Euphorbiaceae. 14 tropics; F. ussuriensis—southern Far East of Russia

Flyriella R. M. King et H. Rob. 1972. Asteraceae. 6 Mexico

Fockea Endl. 1839. Asclepiadaceae. 10 trop. and South Africa

Foeniculum Hill. 1753. Apiaceae. 1 (F. vulgare, very polymorphic) Europe, Caucasus, Middle Asia, North and trop. Africa, West Asia eastward to China, India, and Nepal

Foetidia Comm. ex Lam. 1788. Lecythidaceae. 17 East Africa (1), Madagascar (14), Mascarene Is. (2)

Fokienia A. Henry et H. H. Thomas. 1911. Cupressaceae. 1 southwestern and southeastern China, Laos, Vietnam

Foleyola Maire. 1925. Brassicaceae. 1 Morocco, Algeria

Fonkia Phil. 1859–1860. Scrophulariaceae. 1 temp. South America

Fontainea Heckel. 1870. Euphorbiaceae. 6 New Guinea, northeastern Australia, New Caledonia (1), Vanuatu

Fontanesia Labill. 1791. Oleaceae. 1–2 Sicily, Asia Minor, Syria, Lebanon, China

Fontquera Maire = Perralderia Coss. Asteraceae

Forbesina Ridl. = Eria Lindl. Orchidaceae

Forchhammeria Liebm. 1854. Capparaceae. 10 southwestern (from California) North and Central America, West Indies

Forcipella Baill. 1891. Acanthaceae. 5 Madagascar

Fordia Hemsl. 1886. Fabaceae. 18 southeastern China, southern peninsular

Thailand, Malesia (Malay Peninsula, Sumatra, Borneo, Anambas Is., Sulu Arch.)

Fordiophyton Stapf. 1892. Melastomataceae. 8 southern China, Indochina

Forestiera Poir. 1812. Oleaceae. 15 southeastern U.S., Mexico, Central and South America, West Indies

Forficaria Lindl. 1838. Orchidaceae. 16 trop. and South Africa

Forgesia Comm. ex Juss. 1789. Escalloniaceae. 1 Réunion I.

Formania W. W. Sm. et Small. 1922. Asteraceae. 1 southwestern China

Formanodendron Nixon et Crepet = Quercus L. Fagaceae

Formosia Pichon = Anodendron A. DC. Apocynaceae

Forrestia A. Rich. = Amischotolype Hassk. Commelinaceae

Forsellesia (A. Gray) Greene = Glossopetalon A. Gray. Crossosomataceae

Forsskaolea L. 1764. Urticaceae. 6 Canary Is. (F. angustifolia), Cape Verde Is. (2), southeastern Spain, North (from Morocco through Sahara to Egypt), trop. (Angola, Namibia, Sudan, Ethiopia, Kenya, Tanzania) and South Africa, Israel, Arabian Peninsula, Iran, Afghanistan, Pakistan, western India

Forstera L. f. 1780. Stylidiaceae. 5 Tasmania (1), New Zealand

Forsteronia G. Mey. 1818. Apocynaceae. c. 50 Central and trop. South America, West Indies

Forsythia Vahl. 1804. Oleaceae. 7 Yugoslavia, Albania (1, F. europaea), East Asia (6)

Forsythiopsis Baker = Oplonia Raf. Acanthaceae

Fortunatia J. F. Macbr. = Camassia Lindl. Hyacinthaceae

Fortunearia Rehder et E. H. Wilson. 1913. Hamamelidaceae. 1 central and eastern China

Fortunella Swingle. 1915. Rutaceae. 6 East and Southeast Asia south to Malay Peninsula; F. japonica—extending to Russian Far East

Fortuynia Shuttlew. ex Boiss. 1841. Brassicaceae. 2 Iran, Afghanistan, Pakistan

Fosterelia Airy Shaw = Fosterella L. B. Sm. Bromeliaceae

Fosterella L. B. Sm. 1960. Bromeliaceae. 14 trop. America (Mexico, Peru, Bolivia [c. 10], western Brazil, northern Argentina, Paraguay [1, F. rojasii])

Fosteria Molseed. 1968. Iridaceae. 1 Mexico

Fothergilla L. 1774. Hamamelidaceae. 2 eastern North America

Fouquieria Kunth. 1823. Fouquieriaceae. 11 southwestern U.S., Mexico

Fourraea Greuter et Burdet = Arabis L. Brassicaceae

Foveolaria Ruiz et Pav. = Styrax L. Styracaceae

Foveolina Källersjö. 1988. Asteraceae. 5 South Africa

Fragaria L. 1753. Rosaceae. 20 North and South (Brazil, Chile) America, temp. Eurasia south to India and Sri Lanka

Fragariopsis A. St.-Hil. 1840 (Plukenetia L.). Euphorbiaceae. 2 trop. South America

Fragosa Ruiz et Pav. = Azorella Lam. Apiaceae

Frailea Britton et Rose. 1922. Cactaceae. 15 eastern Bolivia, southern Brazil, Paraguay, northeastern Argentina, Uruguay

Franciscodendron B. Hyland et Steenis. 1987 (Sterculia L.). Sterculiaceae. 1 Australia

Francoa Cav. 1801. Francoaceae. 1 (F. sonchifolia, polymorphic) Chile

Francoevria Cass. = Pulicaria Gaertn. Asteraceae

Frangula Hill = Rhamnus L. Rhamnaceae

Frankenia L. 1753. Frankeniaceae. 25–50 (or 80?) temp. and subtrop. coastal regions in arid, saline habitats, margins of salt lakes, salt deserts (esp. Mediterranean; Africa; St. Helena; Australia; c. 45, North and South America)

Franklandia R. Br. 1810. Proteaceae. 2 western Australia

Franklinia W. Bartr. ex Marshall. 1785. Theaceae. 1 southeastern U.S.

Franseria Cav. = Ambrosia L. Asteraceae
Frantzia Pittier = Sechium P. Browne. Cucurbitaceae
Frasera Walter. 1788 (Swertia L.). Gentianaceae. 15 North America
Fraunhofera Mart. 1831. Celastraceae. 1 Brazil
Fraxinus L. 1753. Oleaceae. 65 temp. Northern Hemisphere, esp. Mediterranean, East Asia, and North America (a few species extending to tropics)
Fredolia (Cass. et Durieu) Ulbr. 1934. Chenopodiaceae. 1 Morocco, Algeria
Freesia Eckl. ex Klatt. 1866. Iridaceae. 11 South Africa
Fregea Rchb. f. 1852. (or = Sobralia Ruiz et Pav.). Orchidaceae. 2 Costa Rica, Panama
Fremontia Torr. = Fremontodendron Coville. Sterculiaceae
Fremontodendron Coville. 1893. Sterculiaceae. 3 southwestern U.S. (California, Arizona), northwestern Mexico (Baja California)
Frerea Dalzell. 1864. Asclepiadaceae.
Fresenia DC. = Felicia Cass. Asteraceae
Freya Badillo. 1985. Asteraceae. 1 Venezuela
Freycinetia Gaudich. 1824. Pandanaceae. 180 Assam, Sri Lanka, Andaman Is., Southeast Asia eastward to Japan, Malesia, Australia (Queensland), New Zealand, Norfolk Is., Melanesia, eastward to Polynesia
Freyera Rchb. = Geocaryum Coss. Apiaceae
Freylinia Colla. 1824. Scrophulariaceae. 7 trop. and South Africa
Freyliniopsis Engl. = Manuleopsis Thell. ex Schinz. Scrophulariaceae
Freziera Willd. 1799–1800. Theaceae. 414 trop. America, West Indies
Fridericia Mart. 1827. Bignoniaceae. 1 southern Brazil
Friesodielsia Steenis. 1948. Annonaceae. 55 trop. West Africa, northeastern India and Andaman Is. (1, F. fornicata), Southeast Asia, Malesia
Frithia N. E. Br. 1925. Aizoaceae. 1 South Africa
Fritillaria L. 1753. Liliaceae. c. 75 temp. Northern Hemisphere, esp. Russia (European part, northern Caucasus, southern Siberia, Far East); Mediterranean, Iraq, Iran, Middle Asia, Himalayas from Kashmir to Nepal
Fritzschia Cham. 1834. Melastomataceae. 4 Brazil
Froelichia Moench. 1794. Amaranthaceae. 20 trop. and subtrop. America, Galapagos Is.
Froelichiella R. E. Fr. 1920. Amaranthaceae. 1 Brazil
Froesia Pires. 1948. Quiinaceae. 3 Colombia, Venezuela, Brazil
Froesiochloa G. A. Black. 1950. Poaceae. 3 trop. South America
Froësiodendron R. E. Fr. 1956. Annonaceae. 2 trop. South America
Frolovia (Ledeb. ex DC.) Lipschitz = Saussurea DC. Asteraceae
Frommia H. Wolff. 1912. Apiaceae. 1 Tanzania, Zambia, Malawi
Frondaria Luer. 1986. Orchidaceae. 1 Peru
Froriepia K. Koch. 1842. Apiaceae. 2 Caucasus (1, F. subpinnata), Turkey, Iran
Fryxellia D. M. Bates. 1974 (Anoda Cav.). Malvaceae. 1 northern Mexico and adjacent Texas
Fuchsia L. 1753. Onagraceae. 100 Mexico, Central and South America (esp. Andes), New Zealand, Tahiti
Fuernrohria K. Koch. 1842. Apiaceae. 1 (F. seritolia) Caucasus, Turkey, Iran
Fuerstia T. C. E. Fr. 1929. Lamiaceae. 6 trop. Africa
Fuertesia Urb. 1911. Loasaceae. 1 West Indies
Fuertesiella Schltr. 1913. Orchidaceae. 2 West Indies
Fuirena Rottb. 1773. Cyperaceae. 30 tropics and subtropics
Fulcaldea Poir. ex Less. 1817. Asteraceae. 1 western trop. South America
Fumana (Dunal) Spach. 1836. Cistaceae. 15 Mediterranean, Southeast Europe, Southwest Asia; Crimea and Caucasus (3)
Fumaria L. 1753. Fumariaceae. 50 Europe, Macaronesia, Mediterranean (c. 40); 1

(F. abyssinica) East Africa from Ethiopia to Lake Nyasa, and southwestern Arabian Peninsula, West and South Asia eastward to Himalayas; Caucasus, Middle Asia, southern Siberia

Fumariola Korsh. 1898. Fumariaceae. 1 Middle Asia (Alai Mts.)

Funastrum Fourn. 1882. Asclepiadaceae. 10–15 trop. America

Funifera Leandro ex C. A. Mey. 1843. Thymelaeaceae. 3–4 Brazil

Funkia Spreng. = Hosta Tratt. Hostaceae

Funkiella Schltr. 1920. Orchidaceae. 3 Mexico, Guatemala

Funtumia Stapf. 1901. Apocynaceae. 2 trop. Africa

Furarium Rizzini = Phthirusa Mart. Loranthaceae

Furcaria (DC.) Kostel. = Hibiscus L. Malvaceae

Furcatella Baum.-Bod. 1989 (or = Psychotria L.). Rubiaceae. 1 New Caledonia

Furcraea Vent. 1793. Agavaceae. 21 trop. America

Furtadoa M. Hotta. 1981. Araceae. 2 Sumatra

Fusaea (Baill.) Saff. 1914. Annonaceae. 1 Guyana, eastern Peru, and Brazil (Amazonia)

Fusispermum Cuatrec. 1950. Violaceae. 3 Colombia, Peru

G

Gabunia K. Schum. ex Stapf = Tabernaemontana L. Apocynaceae

Gadellia Schulkina. 1979 (Campanula L.). Campanulaceae. 1 Caucasus

Gaertnera Lam. 1792. Rubiaceae. 32 trop. Africa, Madagascar, Mascarene Is., Sri Lanka, Assam, southern China, Indochina, Malay Peninsula, Borneo, trop. Australia

Gagea Salisb. 1806. Liliaceae. 200–250 temp. Eurasia, esp. (c. 130–140) Crimea, southern European part of Russia, Caucasus, but chiefly Middle Asia (c. 110); Mediterranean, Arabian Peninsula, Iran, Afghanistan, Himalayas

Gagnebina Neck. ex DC. 1825. Fabaceae. 4–5 Madagascar and the Indian Ocean islands

Gagnepainia K. Schum. 1904. Zingiberaceae. 3 Indochina

Gagria M. Kral = Pachyphragma (DC.) Rchb. Brassicaceae

Gahnia J. R. Forst. et G. Forst. 1775. Cyperaceae. 40 southern China, Ryukyu and Ogasawara Is., Indochina, Malesia, Australia, New Zealand, Polynesia eastward to Hawaiian and Marquesas Is.

Gaiadendron G. Don. 1834 (Phrygilanthus Eichler). Loranthaceae. 1 Andes from Costa Rica to Bolivia

Gaillardia Foug. 1786. Asteraceae. 28 North (26) and temp. South (2) America

Gaimardia Gaudich. 1825. Centrolepidaceae. 2 New Guinea, New Zealand, Tasmania, Tierra del Fuego, Falkland Is.

Galactia P. Browne. 1756. Fabaceae. 50 tropics and subtropics, esp. America

Galactites Moench. 1794. Asteraceae. 3 Canary Is., Mediterranean

Galactophora Woodson. 1932. Apocynaceae. 7 trop. South America

Galagania Lipsky. 1900. Apiaceae. 5–7 Middle Asia (2), Iran, Afghanistan

Galanthus L. 1753. Amaryllidaceae. 17 Central, South and Southeast Europe, West Asia, incl. Caucasus

Galatella Cass. = Aster L. Asteraceae

Galax Sims. 1804. Diapensiaceae. 1 southeastern U.S.

Galaxia Thunb. 1782. Iridaceae. 14 South Africa

Galbulimima F. M. Bailey. 1894. Himantandraceae. 2–3 Moluccas, New Guinea, New Britain I., Australia (Queensland)

Gale Duhamel = *Myrica* L. Myricaceae

Galeana La Llave ex Lex. 1824. Asteraceae. 3 Mexico, Central America

Galeandra Lindl. et Bauer. 1832. Orchidaceae. 20 Mexico, Central and South America, West Indies

Galearia Zoll. et Moritzi. 1846. Pandaceae. 6 Southeast Asia, Malesia, Solomon Is.

Galearis Raf. = *Orchis* L. Orchidaceae

Galeata Wendl. 1798. ? Boraginaceae or Convolvulaceae. 1 trop. Asia

Galeatella (E. Wimmer) O. Deg. et I. Deg. = *Lobelia* L. Lobeliaceae

Galega L. 1753. Fabaceae. 6 Central and East Europe, Caucasus, Mediterranean, Ethiopia (1), Kenya, Uganda, Iran, Afghanistan, Pakistan

Galenia L. 1753. Aizoaceae. c. 15–20 Namibia and western South Africa

Galeobdolon Adans. = *Lamium* L. Lamiaceae

Galeola Lour. 1790. Orchidaceae. 15–25 (or 80?) Madagascar, Mascarene Is., Seychelles, Comoro Is., eastern Himalayas, Tibet, India, Burma, Indochina, Taiwan, Ryukyu Is., Malesia from Malay Peninsula to Philippines (3) and New Guinea, trop. Australia, New Caledonia

Galeomma Rauschert. 1982. Asteraceae. 2 South Africa

Galeopsis L. 1753. Lamiaceae. 10 temp. Eurasia

Galeorchis Ridb. = *Galearis* Raf. = *Orchis* L. Orchidaceae

Galeottia A. Rich. 1845. Orchidaceae. 11 trop. America from Mexico to Peru, the Guianas, and Brazil

Galeottiella Schltr. 1920. Orchidaceae. 1 Mexico, Guatemala

Galiniera Delile. 1843. Rubiaceae. 2 (G. saxifraga) trop. Africa from Sudan and Ethiopia south to Zaire and Tanzania; Madagascar (1)

Galinsoga Ruiz et Pav. 1794. Asteraceae. 16 America from southern U.S. and Mexico to Argentina

Galipea Aubl. 1775. Rutaceae. 8–10 trop. America

Galitzkya V. V. Boczantzeva. 1979. Brassicaceae. 3 northern Kazakhstan (1, G. spathulata), Mongolia, western China

Galium L. 1753. Rubiaceae. 400 cosmopolitan, but chiefly temp. regions

Gallardoa Hicken. 1916. Malpighiaceae. 1 Argentina

Gallesia Casar. 1843. Petiveriaceae. 1 Ecuador, Peru, Bolivia, Brazil

Gallienia Dubard et Dop. 1925. Rubiaceae. 1 Madagascar

Galopina Thunb. 1781. Rubiaceae. 4 southern trop. and South Africa

Galphimia Cav. 1799. Malpighiaceae. 10 America from Texas to Argentina

Galpinia N. E. Br. 1894. Lythraceae. 1 South and Southeast Africa

Galtonia Decne. 1880. Hyacinthaceae. 4 South Africa (Natal, Lesoto, Transkei)

Galumpita Blume = *Gironniera* Gaudich. Ulmaceae

Galvezia Dombey ex Juss. 1789. Scrophulariaceae. 6 California, Mexico, Ecuador, Peru

Gamanthera van der Werff. 1991. Lauraceae. 1 Costa Rica

Gamanthus Bunge. 1862. Chenopodiaceae. 4 Middle Asia (4), Iran, Afghanistan

Gambelia Nutt. = *Antirrhinum* L. Scrophulariaceae

Gambeya Pierre = *Chrysophyllum* L. Sapotaceae

Gambeyobotrys Aubrév. = *Chrysophyllum* L. Sapotaceae

Gamblea C. B. Clarke. 1879. Araliaceae. 2 southeastern Tibet, eastern Himalayas (Nepal, Bhutan), Burma

Gamocarpha DC. 1836 (Boopis Juss.). Calyceraceae. 6 temp. South America

Gamochaeta Wedd. 1856. Asteraceae. 40 trop. America, West Indies

Gamochaetopsis Anderb. et Freire. 1991. Asteraceae. 1 southern Andes of Chile and Argentina

Gamolepis Less. = *Steirodiscus* Less. Asteraceae

Gamoplexis Falc. ex Lindl. = *Gastrodia* R. Br. Orchidaceae

Gamopoda Baker. 1887 (Rhaptonema

Miers). Menispermaceae. 1 Madagascar

Gamosepalum Hausskn. = Alyssum L. Brassicaceae

Gamosepalum Schltr. 1920 (Spiranthes Rich.). Orchidaceae. 1 Mexico

Gamotopea Bremek. 1934. Rubiaceae. 5 trop. South America

Gamwellia E. G. Baker = Pearsonia Duemmer. Fabaceae

Ganophyllum Blume. 1850. Sapindaceae. 1 trop. West Africa, Andaman Is., Sumatra, Java, Philippines, New Guinea, northeastern Australia

Gantelbua Bremek. 1944 (Hemigraphis Nees). Acanthaceae. 1 India

Ganua Pierre ex Dubard = Madhuga Ham. ex J. F. Gmel. Sapotaceae

Garaventia Looser. 1945. Alliaceae. 1 Chile

Garberia A. Gray. 1880. Asteraceae. 1 Florida Peninsula

Garcia Rohr. 1792. Euphorbiaceae. 2 eastern Mexico, Central America, Colombia

Garcibarrigoa Cuatrec. 1986. Asteraceae. 1 Colombia, Ecuador

Garcilassa Poepp. et Endl. 1843. Asteraceae. 1 trop. America from Costa Rica to Peru and Bolivia

Garcinia L. 1753. Clusiaceae. c. 450 tropics (esp. Asia), South Africa (2)

Gardenia Ellis. 1761. Rubiaceae. 250 trop. and subtrop. Old World, extending to New Caledonia

Gardeniopsis Miq. 1868. Rubiaceae. 1 Sumatra, Borneo

Gardneria Wall. 1820. Loganiaceae. 5 India, Himalayas, southern China, Japan

Gardnerina R. M. King et H. Rob. 1981. Asteraceae. 1 Brazil

Gardnerodoxa Sandwith. 1955. Bignoniaceae. 1 Brazil

Garhadiolus Jaub. et Spach = Rhagadiolus Juss. Asteraceae

Garidella L. 1753 (or = Nigella L.). Ranunculaceae. 2 Spain, southern France, Crete, east Aegean islands, Asia Minor, Cyprus, Syria, Lebanon, Israel, Jordan, Caucasus (2)

Garnieria Brongn. et Gris. 1871. Proteaceae. 1 New Caledonia

Garnotia Brongn. 1832. Poaceae. c. 30 trop. Asia, Australia (Queensland), and Oceania

Garnotiella Stapf = Asthenochloa Büse. Poaceae

Garrettia H. R. Fletcher. 1937. Verbenaceae. 2 southern China, Thailand, Java

Garrya Douglas ex Lindl. 1834. Garryaceae. 13 western North and Central America from Washington state to Panama, West Indies (1)

Garuga Roxb. 1811. Burseraceae. 5 India, Himalayas, Burma, southern China, Indochina, Malesia (Java, Lesser Sunda Is., Borneo, Philippines, Sulawesi, Moluccas, New Guinea), northeastern Australia, Solomon Is., Vanuatu, Samoa, Tonga

Garuleum Cass. 1819. Asteraceae. 8 South Africa

Gasoul Adans. = Mesembryanthemum L. Aizoaceae

Gasparinia Bertol. = Seseli L. Apiaceae

Gasteranthus Benth. 1846 (Bessleria L.). Gesneriaceae. 25 trop. America from Guatemala to Peru

Gasteria Duval. 1809. Asphodelaceae. 13 South Africa

Gastonia Comm. ex Lam. 1788. Araliaceae. 12 trop. East Africa, Madagascar, Seychelles, Mascarene Is., Borneo, New Guinea

Gastorkis Thouars. 1809 (Phaius Lour.). Orchidaceae. 9 Madagascar

Gastranthus Moritz ex Benth. et Hook. f. 1876 (Stenostephanus Nees). Acanthaceae. 1 Venezuela

Gastridium P. Beauv. 1812. Poaceae. 3 Canary Is., West and South Europe, North Africa southward to Sudan, Ethiopia, and Kenya; West and Southwest Asia; Caucasus (2)

Gastrocalyx Schischk. = Silene L. Caryophyllaceae

Gastrochilus D. Don. 1825. Orchidaceae. 15–20 India, Sri Lanka, Himalayas, Burma, China, Japan, Indochina, western Malesia (Java, Sumatra—G. sororius), Philippines (1, G. calceolaris)

Gastrococos Morales. 1865. Arecaceae. 1 Cuba

Gastrocotyle Bunge. 1854 (Anchusa L.). Boraginaceae. 2 eastern Mediterranean, West Asia eastward to Middle Asia (1, G. hispida), China, and northwestern India

Gastrodia R. Br. 1810. Orchidaceae. 20 trop. West Africa, trop. and East Asia, Malesia, Australia, New Zealand; G. elata—extending to Russian Far East

Gastrolepis Tiegh. 1897. Icacinaceae. 1 New Caledonia

Gastrolobium R. Br. 1811. Fabaceae. c. 50 Australia

Gastrolychnis (Fenzl) Rchb. 1841. Caryophyllaceae. 17 cold and temp. Eurasia, esp. Russia (c. 17, all areas)

Gastronychia Small = Paronychia Adans. Caryophyllaceae

Gastropyrum (Jaub. et Spach) A. Löve = Aegilops L. Poaceae

Gastrorchis Schltr. 1924 (Phaius Lour.). Orchidaceae. 8–10 Madagascar, Seychelles and Mascarene Is., trop. and subtrop. Asia, Philippines

Gaudichaudia Kunth. 1821. Malpighiaceae. 10 trop. America from Mexico to Bolivia

Gaudinia P. Beauv. 1812. Poaceae. 4 Azores, Mediterranean, Crimea (1, G. fragilis)

Gaudinopsis (Boiss.) Eig. 1929 (Ventenata Koeler). Poaceae. 5 Mediterranean, Southwest Asia; G. macra—the Crimea, Caucasus, Turkmenistan

Gaultheria L. 1753. Ericaceae. c. 150 East (32) and Southeast Asia, Malesia (24), Australia and New Zealand (10), North (6), Central and South America, West Indies; G. miqueliana—extending to Sakhalin

Gaura L. 1753. Onagraceae. 21 North, esp. Texas, and Central America

Gaurella Small = Oenothera L. Onagraceae

Gauropsis C. Presl = Clarkia Pursh. Onagraceae

Gaussia H. A. Wendl. 1865. Arecaceae. 4 Mexico (2), Guatemala, Belize, Cuba, and Puerto Rico

Gavarretia Baill. 1861. Euphorbiaceae. 1 Venezuela, Brazil (Amazonas), Peru

Gavilea Poepp. 1833. Orchidaceae. 17 temp. South America

Gaya Kunth. 1822. Malvaceae. 23 New Zealand (3), trop. America, and West Indies

Gayella Pierre = Pouteria Aubl. Sapotaceae

Gaylussacia Kunth. 1819. Ericaceae. 49 North (9) and South America

Gayoides Small = Herissantia Medik. Malvaceae

Gayophytum A. Juss. 1832. Onagraceae. 9 temp. North America and western temp. South America (2)

Gazachloa J. B. Phipps = Danthoniopsis Stapf. Poaceae

Gazania Gaertn. 1791. Asteraceae. 16 trop. and South Africa

Gazaniopsis C. Huber. 1880. Asteraceae. 1 southwestern U.S. (California), Mexico

Geanthemum (R. E. Fr.) Saff. = Duguetia A. St.-Hil. Annonaceae

Geanthus Valeton = Etlingera Giseke. Zingiberaceae

Gearum N. E. Br. 1882. Araceae. 1 southern Brazil

Geesinkorchis de Vogel. 1984. Orchidaceae. 2 Borneo

Geigeria Griess. 1830. Asteraceae. 26 trop. and South Africa

Geijera Schott. 1834. Rutaceae. 7 New Guinea, eastern Australia, New Caledonia, Loyalty Is.

Geissanthus Hook. f. 1876. Myrsinaceae. 35 western trop. South America

Geissaspis Wight et Arn. 1834. Fabaceae. 3 trop. and subtrop. Asia

Geissois Labill. 1825. Cunoniaceae. 20 eastern Australia, Solomon Is., New Caledonia, Vanuatu, Fiji

Geissolepis B. L. Rob. 1892. Asteraceae. 1 Mexico

Geissoloma Lindl. ex Kunth. 1830. Geissolomataceae. 1 South Africa (southwestern Cape)

Geissomeria Lindl. 1827. Acanthaceae. 15 trop. America

Geissopappus Benth. = Calea L. Asteraceae

Geissorhiza Ker-Gawl. 1803. Iridaceae. 82 South Africa, Madagascar

Geissospermum Allemao. 1846. Apocynaceae. 7 Peru and Brazil

Geitonoplesium A. Cunn. ex R. Br. 1832 (Luzuriaga Ruiz et Pav.). Luzuriagaceae. 1 eastern Malesia (except Sulawesi), Solomon Is., eastern Australia, Melanesia eastward to Fiji

Gelasine Herb. 1840. Iridaceae. 4 trop. and subtrop. America

Geleznowia Turcz. 1849. Rutaceae. 1 western Australia

Gelibia Hutch. = Polyscias J. R. Forst. et G. Forst. Araliaceae

Gelidocalamus T. H. Wen = Indocalamus Nakai. Poaceae

Gelonium Roxb. = Suregada Roxb. ex Rottler. Euphorbiaceae

Gelsemium Juss. 1891. Loganiaceae. 3: 1 southern China, Indochina, Sumatra, Borneo; 2 America from southern U.S. to Guatemala

Gemmaria Salisb. 1866. Amaryllidaceae. 10 South Africa

Gendarussa Nees. 1832 (Justicia L.). Acanthaceae. 2 Indo-Malesia from India to southern China, Taiwan, Philippines, and Java

Genianthus Hook. f. 1883. Asclepiadaceae. 10 India, China, Indochina, Malesia

Geniosporum Wall. ex Benth. 1830. Lamiaceae. 25 trop. and South Africa, Madagascar, India, Himalayas, Burma, southern China, Indochina

Geniostemon Engelm. et A. Gray. 1881. Gentianaceae. 2 Mexico

Geniostoma J. R. Forst. et G. Forst. 1775. Loganiaceae. 52 Mascarene Is.; Malesia (except Malay Peninsula); Indochina; southern China; Kyushu and Ogasawara Is.; Mariana, Caroline, and Solomon Is.; northeastern Australia; Lord Howe I.; New Zealand (North I.); Vanuatu; Polynesia

Genipa L. 1759. Rubiaceae. 7 tropical Florida, Mexico, Central and trop. South America southward to Paraguay, West Indies

Genista L. 1753. Fabaceae. 87–95 Canary Is., Europe, Mediterranean, Caucasus, West Siberia, West Asia

Genistella Ortega = Genista L. Fabaceae

Genistidium I. M. Johnst. 1941. Fabaceae. 1 Mexico, extending to Texas

Genitia Nakai. 1943. Celastraceae. 2 Japan, Taiwan

Genlisea A. St.-Hil. 1833. Lentibulariaceae. 16 trop. and South Africa, Madagascar, trop. America, West Indies

Gennaria Parl. 1860. Orchidaceae. 1 Macaronesia, western Mediterranean

Genoplesium R. Br. 1810 (Prasophyllum R. Br.). Orchidaceae. 1 Australia (New South Wales)

Gentiana L. 1753. Gentianaceae. 361 Europe, Northwest Africa (Morocco), Asia (esp. China [c. 250]), eastern Australia, America

Gentianella Moench. 1794 (Gentiana L.). Gentianaceae. 120 temp. regions of both hemispheres, excl. Africa

Gentianodes A. Löve et D. Löve = Gentiana L. Gentianaceae

Gentianopsis Ma. 1951. Gentianaceae. 24 cold and temp. Northern Hemisphere

Gentianothamnus Humbert. 1937. Gentianaceae. 1 Madagascar

Gentingia J. T. Johanss. et K. M. Wong. 1988 (Prismatomeris Thwaites). Rubiaceae. 1 Malay Peninsula

Gentlea Lundell = Ardisia Sw. Myrsinaceae

Gentrya Breedlove et Heckard = Castilleja Mutis ex L. f. Scrophulariaceae

Genyorchis Schltr. 1905. Orchidaceae. 6 trop. Africa

Geoblasta Barb. Rodr. = Chloraea Lindl. Orchidaceae

Geocarpon Mack. 1914. Caryophyllaceae. 1 U.S. (Missouri, Arkansas)

Geocaryum Coss. 1851. Apiaceae. 13–15 Mediterranean, West Asia

Geocaulon Fernald. 1928. Santalaceae. 1

North America: Alaska, Canada, northeastern U.S.

Geocharis (K. Schum.) Ridl. 1908. Zingiberaceae. 4 western Malesia

Geochorda Cham. et Schltdl. = Bacopa Aubl. Scrophulariaceae

Geococcus J. L. Drumm. ex Harv. 1855. Brassicaceae. 1 northwestern Australia, Tasmania

Geodorum G. Jacks. 1810. Orchidaceae. 10 trop. Africa, Madagascar, trop. Asia from India to northern China and Indochina, Malesia from Malay Peninsula to New Guinea and New Britain, and northern Australia, Solomon Is., Vanuatu, New Caledonia, Fiji, Samoa, Tonga, Niue Is.

Geoffraya Bonati = Lindernia All. Scrophulariaceae

Geoffroea Jacq. 1760. Fabaceae. 3 South America from Colombia and Venezuela to southern Argentina

Geogenanthus Ule. 1913. Commelinaceae. 4 trop. South America

Geohintonia Glass et Fitz Maurice. 1991. Cactaceae. 1 northeastern Mexico

Geomitra Becc. = Thismia Griff. Burmanniaceae

Geonoma Willd. 1805. Arecaceae. 80 trop. America from Mexico to Bolivia and Brazil, West Indies

Geopanax Hemsl. = Schefflera J. R. Forst. et G. Forst. Araliaceae

Geophila D. Don. 1825. Rubiaceae. c. 30 trop. America, trop. Africa, trop. Asia and Malesia, Mariana and Caroline Is.; 1 extends eastward to the Society Is.

Geosiris Baill. 1890. Geosiridaceae. 1 Madagascar, Saint Mary I. (near eastern Madagascar)

Geostachys (Baker) Ridl. 1899. Zingiberaceae. 13 Thailand, Malay Peninsula, Sumatra

Geraea Torr. et A. Gray. 1847 (Encelia Adans.). Asteraceae. 2 southwestern U.S. (California, Utah, Arizona), northwestern Mexico

Geraniopsis Chrtek. 1968 (Geranium L.). Geraniaceae. 2 coasts of the Red Sea, Arabian Peninsula, southern Iran

Geranium L. 1753. Geraniaceae. c. 300 cosmopolitan, esp. temp. regions and montane areas of tropics

Gerardia Benth. = Agalinis Raf. Scrophulariaceae

Gerardiina Engl. 1897. Scrophulariaceae. 2 trop. (Zaire, Angola, Burundi, Malawi, Tanzania, Zambia, Zimbabwe, Mozambique) and South Africa

Gerardiopsis Engl. = Anticharis Endl. Scrophulariaceae

Gerbera L. 1758. Asteraceae. 27 Andes, trop. and South Africa, Madagascar, trop. Asia

Germainia Balansa et Poitrasson. 1873. Poaceae. 4 Southeast Asia, northern Australia

Gerocephalus F. Ritter = Espostoopsis F. Buxb. Cactaceae

Geropogon L. 1763. Asteraceae. 1 Madeira and Canary Is., Europe, Mediterranean, Asia Minor; the Crimea, Caucasus

Gerrardanthus Harv. ex Hook. f. 1867. Cucurbitaceae. 4 trop. and South Africa

Gerrardina Oliv. 1870. Flacourtiaceae. 2 (G. eylesiana) Tanzania, Malawi, Zimbabwe, Mozambique; G. foliosa—South Africa from Transvaal to Cape

Gerritea Zuloaga, Morrone et Killeen. 1993 (Panicum L.). Poaceae. 1 Bolivia

Geruma Forssk. 1775. ? Aizoaceae. 1 Arabian Peninsula

Gesneria L. 1753. Gesneriaceae. 50 trop. America, West Indies

Gesnouinia Gaudich. 1830. Urticaceae. 2 Canary Is.

Gethyllis L. 1753. Amaryllidaceae. 32 South Africa (Cape)

Gethyum Phil. 1873. Alliaceae. 1 Chile

Getonia Roxb. = Calycopteris Lam. Combretaceae

Geum L. 1753. Rosaceae. c. 70 arctic and temp. regions

Geunsia Blume = Callicarpa L. Verbenaceae

Gevuina Molina. 1781. Proteaceae. 3 New Guinea, Australia (Queensland), Chile, Argentina

Ghaznianthus Lincz. 1979. Plumbaginaceae. 1 Afghanistan

Ghikaea Volkens et Schweinf. 1898. Scrophulariaceae. 1 trop. Africa

Ghinia Schreb. = Tamonea Aubl. Verbenaceae

Giadotrum Pichon = Cleghornia Wight. Apocynaceae

Gibasis Raf. 1837. Commelinaceae. 11 trop. America, West Indies

Gibasoides D. R. Hunt. 1978 (Tradescantia L.). Commelinaceae. 1 Mexico

Gibbaeum Haw. 1821. Aizoaceae. 15–20 South Africa

Gibbaria Cass. 1817. Asteraceae. 2 South Africa

Gibbesia Small = Paronychia Adans. Caryophyllaceae

Gibbsia Rendle. 1917. Urticaceae. 2 New Guinea

Gibsoniothamnus L. O. Williams. 1970. Scrophulariaceae. 7 southern Mexico, Central America

Gigantochloa Kurz ex Munro. 1868. Poaceae. 20 South and Southeast Asia

Gigasiphon Drake. 1902 (Bauhinia L.). Fabaceae. 5 Africa (2 Gabon, Angola, Zaire, Kenya, Tanzania), Madagascar (1), Malesia

Gigliolia Becc. = Areca L. Arecaceae

Gilberta Turcz. 1851. Asteraceae. 1 Australia

Gilbertiella Boutique. 1951. Annonaceae. 1 trop. Africa

Gilbertiodendron J. Léonard. 1952. Fabaceae. 27 trop. West Africa

Gilesia F. Muell. 1875 (Hermannia L.). Sterculiaceae. 1 central Australia

Gilgiochloa Pilg. 1914. Poaceae. 2 trop. East Africa

Gilia Ruiz et Pav. 1794. Polemoniaceae. 25 western North America

Giliastrum (A. Brand) Rydb. 1917 (Gilia Ruiz et Pav.). Polemoniaceae. 2 U.S. from Colorado and Kansas to Arizona and Texas; Mexico

Gilibertia Ruiz et Pav. = Dendropanax Decne. et Planch. Araliaceae

Gilipus Raf. = Myrica L. Myricaceae

Gillbeea F. Muell. 1865. Cunoniaceae. 2 New Guinea, Australia (northeastern Queensland)

Gillenia Moench = Porteranthus Britton ex Small. Rosaceae

Gillespiea A. C. Sm. 1936. Rubiaceae. 1 Fiji

Gilletiella De Wild. et T. Durand = Anomacanthus R. D. Good. Acanthaceae

Gilletiodendron Vermoesen. 1923. Fabaceae. 5 trop. Africa from Mali to Zaire and Angola

Gilliesia Lindl. 1826. Alliaceae. 3 Andes of Chile

Gilmania Coville. 1963. Polygonaceae. 1 southwestern U.S. (California)

Gilruthia Ewart. 1909. Asteraceae. 1 western Australia

Ginalloa Korth. 1839. Viscaceae. 8 South and Southeast Asia

Gingidia J. W. Dawson. 1974. Apiaceae. 10 New Zealand, ? Australia

Gingidium J. R. Forst. et G. Forst. = Gingidia J. W. Dawson. Apiaceae

Ginkgo L. 1771. Ginkgoaceae. 1 eastern China (Tan-Mu-Shan Mts.)

Ginoria Jacq. 1760. Lythraceae. 15 Mexico, West Indies

Giraldiella Dammer = Lloydia Salisb. ex Rchb. Liliaceae

Girardinia Gaudich. 1830. Urticaceae. 2 (or 11?) mountains of Africa from Senegal and Sudan to Angola, Zimbabwe and Transvaal, Madagascar, Yemen, India, Sri Lanka eastward to southern China, Taiwan, and western Malesia; 1 extending to southern Far East of Russia

Girgensohnia Bunge. 1851. Chenopodiaceae. 4 southeastern European part of Russia, southeastern Caucasus, Middle Asia, Syria, Palestine, Iraq, Iran, Afghanistan, Pakistan, northwestern China

Gironniera Gaudich. 1844. Ulmaceae. 6–15 India, Sri Lanka, Burma, Thailand, China, Indochina, Malesia, Micronesia, Melanesia, and Polynesia

Gisekia L. 1771. Gisekiaceae. 2–5 North, trop., and South Africa, Arabian Peninsula, India, Sri Lanka, Indochina

Gitara Pax et K. Hoffm. 1924. Euphorbiaceae. 2 trop. America

Githopsis Nutt. 1843. Campanulaceae. 4 western North America

Giulianettia Rolfe = Glossorhuncha Ridl. Orchidaceae

Givotia Griff. 1843. Euphorbiaceae. 4 southeastern Ethiopia, Somalia, Kenya, Madagascar (2), India, Sri Lanka

Gjellerupia Lauterb. 1912. Opiliaceae. 1 New Guinea

Glabraria L. = Brownlowia Roxb. Tiliaceae

Gladiolimon Mob. 1964 (Acantholimon Boiss.). Plumbaginaceae. 1 Afghanistan.

Gladiolus L. 1753. Iridaceae. c. 180 Madeira and Canary Is.; West and Central Europe; Mediterranean; Crimea; Caucasus; Southwest, West, and Middle Asia; trop. and South Africa

Gladiopappus Humbert. 1948. Asteraceae. 1 Madagascar

Glandonia Griseb. 1858. Malpighiaceae. 3 Venezuela, Guyana, Brazil

Glandularia J. F. Gmel. 1792 (Verbena L.). Verbenaceae. 28 America

Glaribraya Hara. 1978. Brassicaceae. 1 Nepal

Glastaria Boiss. 1841. Brassicaceae. 1 Turkey, Syria, Iraq, ? Iran

Glaucidium Siebold et Zucc. 1846. Glaucidiaceae. 1 Japan (Hokkaido and Honshu Is.)

Glaucium Hill. 1753. Papaveraceae. 25 Europe, Mediterranean, Southwest and West Asia; Crimea, Caucasus, Middle Asia

Glaucocarpum Rollins. 1938. Brassicaceae. 1 southwestern U.S.

Glaucocochlearia (O. E. Schulz) Pobed. = Cochlearia L. Brassicaceae

Glaucosciadium B. L. Burtt et P. H. Devis. 1949. Apiaceae. 1 eastern Mediterranean, incl. Asia Minor and Cyprus

Glaux L. 1753. Primulaceae. 1 subarctic and temp. Northern Hemisphere; Russia—all areas

Glaziocharis Taub. ex Warm. 1902 (or = Thismia Griff.). Burmanniaceae. 2 Japan (1), southeastern Brazil (1)

Glaziophyton Franch. 1889. Poaceae. 1 Brazil

Glaziostelma E. Fourn. 1885 (Tassadia Decne.). Asclepiadaceae. 1 Brazil

Glaziova Bureau. 1868. Bignoniaceae. 1 Brazil

Glaziova Mart. ex Drude = Lytocaryum Toledo. Arecaceae

Glaziovanthus G. M. Barroso = Chresta Vell. ex DC. Asteraceae

Gleadovia Gamble et Prain. 1901. Orobanchaceae. 6 China, western Himalayas

Gleasonia Standl. 1931. Rubiaceae. 4–5 trop. South America

Glechoma L. 1753. Lamiaceae. 4 temp. Eurasia, incl. Russia (3, all areas)

Glechon Spreng. 1827. Lamiaceae. 6 Brazil, Paraguay

Gleditschia Scop. = Gleditsia L. Fabaceae

Gleditsia L. 1753. Fabaceae. 14 eastern Caucasus and northern Iran (1, E. caspia), trop. Asia from India to Japan and Philippines, eastern North and South America

Glehnia F. Schmidt ex Miq. 1867. Apiaceae. 1 (G. litoralis) Soviet Far East, China, Japan, North America

Glekia Hilliard. 1989 (Phyllopodium Benth.). Scrophulariaceae. 1 South Africa

Glenniea Hook. f. 1862. Sapindaceae. 8 trop. Africa, Madagascar, southern India, Sri Lanka, Malesia (Malay Peninsula, Sumatra, Philippines, Borneo, New Guinea)

Glia Sond. 1862 (Annesorhiza Cham. et Schltdl.). Apiaceae. 3 Canary Is., St. Helena, South Africa (Cape)

Glinus L. 1753. Molluginaceae. 6 tropics and subtropics; G. lotoides—Caucasus

Glionnetia Tirveng. 1984. Rubiaceae. 1 Seychelles

Gliopsis Rauschert = Rutheopsis A. Hansen et Kunkel. Apiaceae

Gliricidia Kunth. 1824. Fabaceae. 4 Central and northern South America, West Indies

Glischrocaryon Endl. 1838. Haloragaceae. 4 southwestern and southern Australia

Glischrocolla (Endl.) A. DC. 1856. Penaeaceae. 1 South Africa

Glischrothamnus Pilg. 1908. Aizoaceae. 1 northeastern Brazil

Globba L. 1771. Zingiberaceae. c. 100 India, eastern Himalayas, Burma, southern China, Indochina, Malesia, Australia (1)

Globimetula Tiegh. 1895. Loranthaceae. 14 trop. Africa

Globularia L. 1753. Globulariaceae. 29 Canary Is., Cape Verde Is., Europe, Mediterranean, West Asia; 2 Crimea, Caucasus

Globulariopsis Compton. 1931. Selaginaceae. 1 South Africa

Globulostylis Wernham = Cuviera DC. Rubiaceae

Glochidion J. R. Forst. et G. Forst. 1775. Euphorbiaceae. c. 300 Madagascar, trop. Asia, Malesia, trop. Australia, Oceania, trop. America

Glochidocaryum W. T. Wang = Actinocarya Benth. Boraginaceae

Glochidotheca Fenzl. 1843 (Caucalis L.). Apiaceae. 1 (G. foeniculacea Fenzl) Bulgaria, Southwest Asia

Glockeria Nees = Hansteinia Oersted. Acanthaceae

Gloeocarpus Radlk. 1914. Sapindaceae. 1 Philippines

Gloeospermum Triana et Planch. 1862. Violaceae. 12 trop. America

Glomera Blume. 1825. Orchidaceae. 50 Java, Sumatra, Philippines (1 [G. merrillii] Mindanao), Caroline Is., New Guinea (c. 40), New Ireland I., Bougainville I., Solomon Is., New Caledonia, Vanuatu, Fiji, Samoa

Glomeropitcairnia Mez. 1905. Bromeliaceae. 2 Lesser Antilles Is., Trinidad

Gloriosa L. 1753. Colchicaceae. 1 (G. superba) trop. Old World

Glossanthis P. Poljakov = Trichanthemis Regel et Schmalh. Asteraceae

Glossarion Maguire et Wurdack. 1957. Asteraceae. 1 Venezuela, Guyana

Glossocalyx Benth. 1880. Monimiaceae. 4 trop. West Africa

Glossocardia Cass. 1817. Asteraceae. 1 northern trop. Africa (Niger, Chad), western and central India, Bangladesh

Glossocarya Wall. ex Griff. 1843. Verbenaceae. 9 India, Sri Lanka, Burma, Indochina, Malesia, Australia

Glossochilus Nees. 1847. Acanthaceae. 2 South Africa

Glossodia R. Br. 1810. Orchidaceae. 2–5 Australia, Tasmania

Glossogyne Cass. = Glossocardia Cass. Asteraceae

Glossolepis Gilg = Chytranthus Hook. f. Sapindaceae

Glossonema Decne. 1838. Asclepiadaceae. 4–5 North and trop. Africa, trop. Asia

Glossopappus Kunze. 1846. Asteraceae. 1 Southwest Europe, North Africa

Glossopetalon A. Gray. 1853. Crossosomataceae. 4–5 southwestern U.S., Mexico

Glossopholis Pierre = Tiliacora Colebr. Menispermaceae

Glossorhyncha Ridl. 1891. Orchidaceae. 60–100 Amboina, Kai I., Sulawesi, New Guinea, Bougainville and Solomon Is., Vanuatu, New Caledonia, Fiji

Glossostelma Schltr. 1895. Asclepiadaceae. 2–3 trop. and South Africa

Glossostemon Desf. 1817. Sterculiaceae. 1 Iran

Glossostigma Wight et Arn. 1836. Scrophulariaceae. 5 India (1, G. diandrum), Australia and Tasmania (5), New Zealand

Glossostipula Lorence. 1986. Rubiaceae. 2 Mexico, Guatemala

Glottidium Desv. 1813. Fabaceae. 1 southeastern U.S. from southeastern North Carolina to Florida and west into Texas

Glottiphyllum Haw. 1921. Aizoaceae. 16 South Africa (Cape)

Gloxinia L'Hér. 1789. Gesneriaceae. 15 trop. South America from Colombia and Venezuela to Bolivia and Brazil

Gluema Aubrév. et Pellegr. 1934. Sapotaceae. 2 trop. West Africa

Glumicalyx Hiern. 1903. Scrophulariaceae. 6 South and Southeast Africa

Gluta L. 1771. Anacardiaceae. c. 30 Madagascar (1), India, Burma, China, Indochina, Malesia, except Philippines

Glutago Comm. ex Raf. = Oryctanthus (Griseb.) Eichl. Loranthaceae

Glyceria R. Br. Poaceae. c. 50 temp. and subtrop. regions and montane areas of tropics

Glycine Wendl. 1798. Fabaceae. 9 Asia, Malesia, Micronesia, Australia; G. soja—extending to the Russia Far East

Glycosmis Corrêa. 1805. Rutaceae. 43 India, Sri Lanka, southern China, Taiwan, Indochina, Malesia, Australia (Queensland)

Glycoxylon Ducke = Pradosia Liais. Sapotaceae

Glycydendron Ducke. 1922. Euphorbiaceae. 1 Brazil (Amazonia)

Glycyrrhiza L. 1753. Fabaceae. 20 North and temp. South America, North Africa (1 [G. foetida] Morocco, Algeria, Tunisia), Eurasia (incl. [7] southern European part of Russia, Caucasus, Middle Asia, southern Siberia, Russian Far East), Australia

Glycyrrhizopsis Boiss. et Bal. = Glycyrrhiza L. Fabaceae

Glyphaea Hook. f. 1848. Tiliaceae. 2 trop. Africa

Glyphochloa Clayton. 1981. Poaceae. 8 India

Glyphosperma S. Watson = Asphodelus L. Asphodelaceae

Glyphostylus Gagnep. 1925. Euphorbiaceae. 1 Indochina

Glyptocarpa Hu. 1965 (Pyrenaria Blume). Theaceae. 1 southwestern China

Glyptocaryopsis Brand = Plagiobothrys Fisch. et C. A. Mey. Boraginaceae

Glyptopetalum Thwaites. 1856. Celastraceae. 27 India, Sri Lanka, Burma, Thailand, Indochina, Hainan, Malesia (Malay Peninsula, Borneo, Philippines, and Sulawesi)

Glyptopleura Eaton. 1871. Asteraceae. 2 western U.S. (Oregon, California, Utah, Nevada)

Glyptostrobus Endl. 1847. Taxodiaceae. 1 southeastern China, Vietnam

Gmelina L. 1753. Verbenaceae. 35 trop. Africa, Mascarene Is. (2), Pakistan, India, Himalayas (Nepal, Bhutan), southern China, Indochina, Malesia, northern Australia, Micronesia, and Melanesia

Gnaphaliothamnus Kirp. 1950. Asteraceae. 10 Mexico, Central America

Gnaphalium L. 1753. Asteraceae. c. 150 cosmopolitan

Gnaphalopdes A. Gray = Actinobole Fenzl ex Endl. Asteraceae

Gnephosis Cass. 1820. Asteraceae. 6 temp. Australia

Gnetum L. 1767. Gnetaceae. 30 northern South America (7, esp. Amazonia [6]; G. leybodlii—extends to Panama and Costa Rica), trop. West Africa (2: Cameroun, Angola), northeastern India (2), Southeast Asia, New Guinea, Caroline and Solomon Is.; 1 extends to Fiji

Gnidia L. 1753. Thymelaeaceae. 160 trop. and South Africa, Madagascar (c. 20), southwestern Arabian Peninsula (1), western India, and Sri Lanka (2)

Goadbyella R. S. Rogers = Microtis R. Br. Orchidaceae

Gochnatia Kunth. 1818. Asteraceae. 66 southern U.S., Mexico, Central and South America, montane Southeast Asia

Godetia Spach = Clarkia Pursh. Onagraceae

Godmania Hemsl. 1879. Bignoniaceae. 2 Mexico, Guatemala, Belize, El Salvador, Honduras, Nicaragua, Costa Rica, Panama, Colombia, Venezuela, Guyana, Ecuador, Brazil (2), Bolivia

Godoya Ruiz et Pav. 1794. Sauvagesiaceae. 5 western trop. South America

Godwinia Seem. = Dracontium L. Araceae

Goebelia Bunge ex Boiss. = Vexibia Raf. Fabaceae

Goeldinia Huber = Allantoma Miers. Lecythidaceae

Goerziella Urb. 1924 (Amaranthus L.). Amaranthaceae. 1 Cuba

Goethalsia Pittier. 1914. Tiliaceae. 1 Central America, Colombia

Goethartia Herz. = Pouzolzia Gaudich. Urticaceae

Goethea Nees = Pavonia Cav. Malvaceae

Goetzea Wydl. 1830. Goetzeaceae. 2 Haiti, Puerto Rico, Hispaniola.

Golaea Chiov. 1929. Acanthaceae. 1 Somalia

Goldbachia DC. 1821. Brassicaceae. 6 Europe, Caucasus, West Siberia, Asia Minor, Iran, Middle Asia, Afghanistan, Pakistan, northwestern India, China, Mongolia

Goldfussia Nees = Strobilanthes Blume. Acanthaceae

Goldmanella Greenm. 1908. Asteraceae. 1 Mexico

Goldmania Rose ex Micheli. 1903. Fabaceae. 2 Mexico, Central America, Argentina, Paraguay

Golionema S. Watson = Olivaea Sch. Bip. ex Benth. Asteraceae

Gomara Ruiz et Pav. = Sanango Bunting et Duke. Loganiaceae

Gomaranthus Rauschert = Sanango Bunting et Duke. Loganiaceae

Gomesa R. Br. 1815. Orchidaceae. 12 Brazil

Gomidesia O. Berg = Myrcia DC. ex Guillemin. Myrtaceae

Gomortega Ruiz et Pav. 1794. Gomortegaceae. 1 central Chile

Gomphandra Wall. ex Lindl. 1836 (Stemonurus Blume). Icacinaceae. 33 China, Indochina, Malesia eastward to Solomon Is., Santa Cruz I., Australia (1, G. australiana)

Gomphia Schreb. 1789. Ochnaceae. 30–35 trop. and South Africa, Madagascar, southern India, Sri Lanka, southern China (incl. Hainan), Indochina, Malesia; 1 (G. serrata) Sumatra, Banka, Malay Peninsula, Borneo, Philippines, Sulawesi

Gomphichis Lindl. 1840. Orchidaceae. 25 Costa Rica, Andes, Brazil

Gomphocalyx Baker. 1887. Rubiaceae. 1 Madagascar

Gomphocarpus R. Br. 1810 (Asclepias L.). Asclepiadaceae. 5 trop. and South Africa; G. sinaicus—extending to Egypt, Sinai, and southern Israel

Gomphogyne Griff. 1841. Cucurbitaceae. 2–6 from eastern Himalayas to central China, Indochina, and Malesia

Gompholobium Sm. 1798. Fabaceae. 25 New Guinea (1), Australia, Tasmania

Gomphostemma Benth. 1830. Lamiaceae. 40 India, Himalayas, Burma, China, Taiwan, Indochina, Malesia

Gomphostigma Turcz. 1843. Buddlejaceae. 2 trop. and South Africa from southern Angola to southern Zaire, Zimbabwe south to Cape

Gomphotis Raf. = Thryptomene Endl. Myrtaceae

Gomphrena L. 1753. Amaranthaceae. 90–100 trop. Africa, Indochina, Lesser Sunda Is. and Moluccas, Australia, Central and South America

Gonatanthus Klotzsch. 1841. Araceae. 2 Himalayas, Assam, southwestern China, Indochina

Gonatopus Hook. f. ex Engl. 1879. Araceae. 6 trop. Africa (Kenya, Tanzania, Zaire, Malawi, Zambia, Zimbabwe, Mozambique) and South Africa (Transvaal and Natal)

Gonatostylis Schltr. 1906. Orchidaceae. 2 New Caledonia

Gongora Ruiz et Pav. 1794. Orchidaceae. 25 trop. America from Mexico to Peru and Brazil; West Indies

Gongrodiscus Radlk. 1879. Sapindaceae. 2 New Caledonia

Gongronema (Endl.) Decne. 1844. Asclepiadaceae. 16 trop. Old World

Gongrospermum Radlk. 1914. Sapindaceae. 1 Philippines

Gongrostylus R. M. King et H. Rob. 1972. Asteraceae. 1 trop. South America from Costa Rica to Ecuador

Gongrothamnus Steetz. = Distephanus (Cass.) Cass. Asteraceae

Gongylocarpus Schltdl. et Cham. 1830. Onagraceae. 2 Mexico, Central America

Gongylolepis R. H. Schomb. 1847. Asteraceae. 12 trop. South America

Gongylosciadium Rech. f. 1988 (Pimpinella L.). Apiaceae. 1 Caucasus, Turkey, Iran

Gongylosperma King et Gamble. 1908. Asclepiadaceae. 2 Malay Peninsula

Gonioanthela Malme. 1927. Asclepiadaceae. 6 southern Brazil

Goniocaulon Cass. 1817. Asteraceae. 1 Sudan, India

Goniochilus M. W. Chase. 1987 (Rodriguezia Ruiz et Pav.). Orchidaceae. 1 trop. America

Goniocladus Burret. 1940. Arecaceae. 1 Fiji (Viti Levu I.)

Goniodiscus Kuhlm. 1933. Celastraceae. 1 Brazil

Goniolimon Boiss. 1848. Plumbaginaceae. 20 South (Italy) and Southeast Europe, Caucasus, Middle Asia, southern Siberia; North Africa from Algeria to Egypt, West and Southwest Asia eastward to Mongolia and China

Gonioma E. Mey. 1838. Apocynaceae. 2 (G. kamassii) South Africa (Cape), Madagascar

Goniorrhachis Taub. 1892. Fabaceae. 1 southeastern Brazil

Gonioscypha Baker. 1875. Convallariaceae. 2 Himalayas, Indochina

Goniostemma Wight. 1834. Asclepiadaceae. 2 eastern China

Goniothalamus (Blume) Hook. f. et Thomson. 1855. Annonaceae. 115 South and Southeast Asia, Malesia, Polynesia

Gonocalyx Planch. et Linden ex Lindl. 1856. Ericaceae. 8 Mexico, Costa Rica, Colombia, West Indies (Haiti, Puerto Rico, Lesser Antilles Is.)

Gonocarpus Thunb. 1783 (Haloragis J. R. Forst. et G. Forst.). Haloragaceae. c. 41 Himalayas, southern China, Taiwan, Japan, Malesia (5), Caroline Is., Australia and Tasmania (36), New Zealand

Gonocaryum Miq. 1861. Icacinaceae. 9–10 China, Taiwan, Indochina, Malesia

Gonocrypta Baill. 1889. Asclepiadaceae. 1 Madagascar

Gonocytisus Spach. 1845. Fabaceae. 3 Turkey, Syria, Lebanon, Israel, Jordan

Gonolobus Michx. 1803. Asclepiadaceae. 100 North and South America

Gonopyrum Fisch. et C. A. Mey. ex C. A. Mey. = Polygonella Michx. Polygonaceae

Gonospermum Less. 1832. Asteraceae. 4 Canary Is.

Gonostegia Turcz. 1846. Urticaceae. 12 Afghanistan, Himalayas, India, Sri Lanka, Burma, Tibet, southern China, southern Japan, Indochina, Malesia, Australia

Gontscharovia Borisova. 1953 (Satureja L.). Lamiaceae. 1 Middle Asia (Pamir-Alai)

Gonyanera Korth. = Acranthera Arn. ex Meisn. Rubiaceae

Gonypetalum Ule = Tapura Aubl. Dichapetalaceae

Gonystylus Teijsm. et Binn. 1862. Gonystylaceae. 20 Nicobar Is., Malesia (excl. Lesser Sunda Is.), Melanesia to Fiji

Gonzalagunia Ruiz et Pav. 1794. Rubiaceae. 25–40 trop. America West Indies

Goodallia Benth. 1845. Thymelaeaceae. 1 northeastern South America

Goodenia Sm. 1793. Goodeniaceae. More than 200, most endemic of Australia and Tasmania; a few in southern China, Thailand, Cambodia, Java, Lesser Sunda Is., Philippines, Moluccas, and New Guinea

Goodia Salisb. 1806. Fabaceae. 2 Australia, Tasmania

Goodmania Reveal et Ertter. 1977. Polygonaceae. 1 western U.S. (California, Nevada)

Goodyera R. Br. 1813. Orchidaceae. c. 100 temp. Eurasia, North America, Mascarene Is., trop. Asia from India, Sri Lanka, Himalayas through southern China to Taiwan and Ryukyu Is., Indochina, Malesia from Malay Peninsula to New Guinea, Solomon Is., Australia, Polynesia; Russia (3, all areas)

Gorceixia Baker. 1882. Asteraceae. 1 southeastern Brazil

Gordonia Ellis. 1771. Theaceae. c. 70 India, China, Taiwan, Indochina, Malesia from Malay Peninsula to New Guinea, New Britain, and New Ireland; southern North, Central and trop. South America

Gorgonidium Schott. 1864. Araceae. 1 Bolivia

Gormania Britton = Sedum L. Crassulaceae

Gorodkovia Botsch. et Karav. 1959. Brassicaceae. 1 Northeast Siberia

Gorteria L. 1759. Asteraceae. 3 South Africa

Gosela Choisy. 1848. Selaginaceae. 1 South Africa

Gossweilera S. Moore. 1908. Asteraceae. 2 Angola

Gossweilerochloa Renvoize. 1979 (Tridens Roem. et Schult.). Poaceae. 1 Angola

Gossweilerodendron Harms. 1925. Fabaceae. 2 trop. Africa (Ghana, Nigeria, Cameroun, Gabon, Zaire, Angola)

Gossypianthus Hook. = Guilleminea Kunth. Amaranthaceae

Gossypioides Skovst. ex J. B. Hutch. 1947. Malvaceae. 2 trop. Africa, Madagascar

Gossypiospermum (Griseb.) Urb. = Casearia Jacq. Flacourtiaceae

Gossypium L. 1753. Malvaceae. c. 40 tropics and subtropics; 7 Caucasus, Middle Asia

Gouania Jacq. 1763. Rhamnaceae. 26 trop. Africa, Madagascar, trop. and subtrop. Asia, Malesia, Indian Ocean islands, trop. America

Gouffeia Robill. et Castagne ex Lam. et DC. = Arenaria L. Caryophyllaceae

Gouinia E. Fourn. 1883. Poaceae. 13 Central and South America

Goulardia Husn. = Elymus L. Poaceae

Gouldia A. Gray = Hedyotis L. Rubiaceae

Gouldochloa J. Valdés, Morden et S. L. Hatch. 1986 (Chasmanthium Link). Poaceae. 1 Mexico

Goupia Aubl. 1775. Goupiaceae. 4 Guatemala, Guyana, northern Brazil

Gourliea Gillies ex Hook. et Arn. = Geoffroea Jacq. Fabaceae

Govenia Lindl. 1832. Orchidaceae. 20 Central and trop. South America, West Indies

Goyazia Taub. 1896. Gesneriaceae. 2 Brazil

Goyazianthus R. M. King et H. Rob. 1977. Asteraceae. 1 Brazil

Grabowskia Schltdl. 1832. Solanaceae. 6 Mexico (1), Galapagos Is. (1), temp. South America

Graciela Rzed. = Strotheria B. L. Turner. Asteraceae

Graderia Benth. 1846. Scrophulariaceae. 5 trop. and South Africa, Socotra.

Graellsia Boiss. 1841. Brassicaceae. 7 Morocco (1, G. hederifolia), Turkey, Iran, Middle Asia (2), Afghanistan, western Pakistan

Graffenrieda DC. 1828. Melastomataceae. c. 40 trop. South America, West Indies

Grafia A. D. Hawkes = Phalaenopsis Blume. Orchidaceae

Grafia Rchb. 1837. Apiaceae. 1 Italy, Yugoslavia

Grahamia Gillies ex Hook. et Arn. 1833. Portulacaceae. 2 temp. South America

Grajalesia Miranda. 1951. Nyctaginaceae. 1 Mexico

Grammadenia Benth. = Cybianthus Mart. Myrsinaceae

Grammangis Rchb. f. 1860. Orchidaceae. 1 Madagascar

Grammatophyllum Blume. 1825. Orchidaceae. 10 Burma, Indochina, Malesia from Malay Peninsula to New Guinea and New Britain, Admiralty Is., Bismarck Arch., Solomon Is. (2), Fiji

Grammatotheca C. Presl. 1836. Lobeliaceae. 2 South Africa (1), Australia (1)

Grammosciadium DC. 1829. Apiaceae. 7 West Asia: Caucasus (3), Turkey, Iraq, Iran

Grammosolen Haegi. 1981. Solanaceae. 2 arid and semiarid southern Australia

Grammosperma O. E. Schulz. 1929. Brassicaceae. 1 Patagonia

Grandidiera Jaub. 1866. Flacourtiaceae. 1 trop. East Africa (Kenya, Tanzania, Mozambique)

Grangea Adans. 1763. Asteraceae. 10 trop. and subtrop. Africa, Madagascar, trop. and subtrop. Asia

Grangeopsis Humbert. 1923. Asteraceae. 1 Madagascar

Grangeria Comm. ex Juss. 1789. Chrysobalanaceae. 2 Madagascar, Mascarene Is.

Grantia Boiss. = Iphiona Cass. Asteraceae

Graphandra J. B. Imlay. 1939. Acanthaceae. 1 Thailand

Graphardisia (Mez) Lundell = Ardisia Sw. Myrsinaceae

Graphephorum Desv. 1810. Poaceae. 3 North and Central America

Graphistemma (Benth.) Champ. ex Benth. 1876. Asclepiadaceae. 1 southern China

Graphistylis B. Nord. 1978. Asteraceae. 6 southern Brazil

Graphorkis Thouars. 1809. Orchidaceae. 5 trop. Africa (1, G. lurida), Madagascar, Mascarene Is.

Graptopetalum Rose. 1911 (Sedum L.). Crassulaceae. 10 America from southern U.S. to Paraguay

Graptophyllum Nees. 1832. Acanthaceae. 10 New Guinea (1), Australia, New Caledonia (4), Melanesia and Polynesia

Gratiola L. 1753. Scrophulariaceae. 25 temp. and warm–temp. regions and mountainous tropics

Gratwickia F. Muell. 1895. Asteraceae. 1 Australia

Grauanthus Fayed. 1979. Asteraceae. 2 trop. Africa

Gravesia Naudin. 1851. Melastomataceae. 110 trop. Africa (5), Madagascar

Gravesiella A. Fern. et R. Fern. = Cincinnobotrys Gilg. Melastoma taceae

Gravisia Mez = Aechmea Ruiz et Pav. Bromeliaceae

Grayia Hook. et Arn. = Eremosemium Greene. Chenopodiaceae

Grazielia R. M. King et H. Rob. 1972 (Eupatorium L.). Asteraceae. 10 Brazil, Uruguay, Argentina

Grazielodendron H. C. Lima. 1983. Fabaceae. 1 Brazil

Greenea Wight et Arn. 1834. Rubiaceae. 10 South and Southeast Asia

Greenella A. Gray = Gutierrezia Lag. Asteraceae

Greeneocharis Gürke et Harms = Cryptantha Lehm. ex G. Don. Boraginaceae

Greeniopsis Merr. 1909. Rubiaceae. 6 Philippines

Greenmania Hieron. = Unxia L. f. Asteraceae

Greenmaniella W. M. Sharp. 1935. Asteraceae. 1 Mexico

Greenovia Webb et Berthel. 1841. Crassulaceae. 4 Canary Is.

Greenwayodendron (Engl. et Diels) Verdc. 1969. Annonaceae. 2 trop. East Africa

Greenwoodia Burns-Bal. = Greenwoodia Burns-Bal. = Stenorrhynchos Rich. ex Spreng. Orchidaceae

Greigia Regel. 1865. Bromeliaceae. 26 Mexico, Guatemala, El Salvador, Costa Rica, Colombia, Venezuela, Ecuador, Peru, Chile, Juan Fernández Is.

Grenacheria Mez. 1902 (or = Embelia Burm. f.). Myrsinaceae. 12 Malesia

Greslania Balansa. 1873. Poaceae. 4 New Caledonia

Grevea Baill. 1884. Montiniaceae. 2 trop. East Africa, Madagascar

Grevillea R. Br. ex Knight. 1809. Proteaceae. c. 250 southern China (1, G. robusta), Sulawesi, New Guinea, Australia, New Caledonia, Vanuatu

Grewia L. 1753. Tiliaceae. c. 90 trop. and subtrop. Old World

Grewiopsis De Wild. et T. Durand = Desplatsia Bocquillon. Tiliaceae

Greyia Hook. et Harv. 1859. Greyiaceae. 3 southeastern South Africa

Grias L. 1759. Lecythidaceae. 6 Belize, Guatemala, Honduras, Costa Rica, Panama, western Colombia, Ecuador, Peru, Jamaica

Grielum L. 1764. Neuradaceae. 5 South Africa

Griffinia Ker-Gawl. 1820. Amaryllidaceae. 11 Brazil

Griffithella (Tul.) Warm. = Cladopus H. Moeller. Podostemaceae

Griffithia Maingay ex G. King = Enicosanthum Becc. Annonaceae

Griffithia Wight et Arn. = Benkara Adans. Rubiaceae

Griffithianthus Merr. = Enicosanthum Becc. Annonaceae

Griffithochloa G. J. Pierce. 1978. Poaceae. 1 Mexico

Griffonia Baill. 1865. Fabaceae. 4 trop. West Africa from Liberia to Angola, Zaire

Grimmeodendron Urb. 1908. Euphorbiaceae. 2 West Indies

Grindelia Willd. 1807. Asteraceae. c. 60 North (south to Veracruz and Oaxaca) and South America

Grisebachia Drude et H. A. Wendl. = Howea Bacc. Arecaceae

Grisebachia Klotzsch. 1838 (Eremia D. Don). Ericaceae. 8 South Africa (southwestern and southern Cape)

Grisebachianthus R. M. King et H. Rob. 1975. Asteraceae. 8 Cuba

Grisebachiella Lorentz = Astephanus R. Br. Ascelpiadaceae

Griselinia G. Forst. 1786. Griseliniaceae. 6 New Zealand, Chile, Paraguay, southeastern Brazil

Grisollea Baill. 1863–1864. Icacinaceae. 2 Seychelles, Madagascar

Grisseea Bakh. f. 1950. Apocynaceae. 1 Java

Grobya Lindl. 1835. Orchidaceae. 3 Brazil

Groenlandia J. Gay. 1854 (Potamogeton L.). Potamogetonaceae. 1 West Europe, North Africa, Southwest Asia

Gronophyllum Scheff. 1876. Arecaceae. 33 Malesia from Sulawesi through Moluccas to New Guinea, Bismarck Arch., northern Australia (Arnemland Peninsula)

Gronovia L. 1753. Loasaceae. 2 trop. America from Mexico to Ecuador and Venezuela

Grosourdya Rchb. f. 1864. Orchidaceae. 10 Andaman Is., Burma, Indochina, Malay Peninsula, Indonesia, Philippines

Grossera Pax. 1903. Euphorbiaceae. 11 trop. Africa, Madagascar

Grossheimia Sosn. et Takht. 1945 (Centaurea L.). Asteraceae. 2 Caucasus, Asia Minor

Grossularia Hill = Ribes L. Grossulariaceae

Grosvenoria R. M. King et H. Rob. 1975. Asteraceae. 4 Andes of Ecuador and Peru

Grubbia Bergius. 1767. Grubbiaceae. 3 South Africa (Cape)

Grumilea Gaertn. = Psychotria L. Rubiaceae

Grypocarpha Greenm. = Philactis Schrad. Asteraceae

Guacamaya Maguire. 1958. Rapateaceae. 1 Colombia, Venezuela

Guadua Kunth. 1822 (Bambusa Schreb.). Poaceae. c. 30 trop. America, Philippines

Guaduella Franch. 1887. Poaceae. 8 trop. Africa from Sierra Leone to Angola

Guaicaia Maguire = Glossarion Maguire et Wurdack. Asteraceae

Guajacum L. 1753. Zygophyllaceae. 6 trop. and subtrop. America, West Indies

Guamatela Donn. Sm. 1914. Rosaceae. 1 Central America

Guamia Merr. 1915. Annonaceae. 1 Mariana Is.

Guapeba Gomes = Pouteria Aubl. Sapotaceae

Guapira Aubl. 1775 (Pisonia L.). Nyctaginaceae. c. 70 Central and trop. South America, West Indies

Guardiola Cerv. ex Humb. et Bonpl. 1807. Asteraceae. 10 southwestern U.S., Mexico

Guarea Allamand ex L. 1771. Meliaceae. 40 trop. America (35), trop. Africa

Guatteria Ruiz et Pav. 1794. Annonaceae. 250 trop. America from southern Mexico to southern Brazil

Guatteriella R. E. Fr. 1939. Annonaceae. 2 Brazil

Guatteriopsis R. E. Fr. 1934. Annonaceae. 5 trop. South America

Guayania R. M. King et H. Rob. 1971. Asteraceae. 6 northern South America

Guaymasia Britton et Rose = Caesalpinia L. Fabaceae

Guazuma Mill. 1754. Sterculiaceae. 4 trop. America

Gueldenstaedtia Fisch. 1823. Fabaceae. 10 southern Siberia (2), China, India (Himalayas)

Gueldenstaedtia Neck. = Ceratoides Gagnebin. Chenopodiaceae

Guerkea K. Schum. = Baissea A. DC. Apocynaceae

Guerreroia Merr. = Glossocardia Cass. Asteraceae

Guettarda L. 1753. Rubiaceae. c. 80: 1 (G. speciosa) pantropics; trop. America, West Indies; 12 New Caledonia and Vanuatu

Guettardella Champ. ex Benth. 1852 (Antirhea Comm. ex Juss.). Rubiaceae. 20 southern China, Thailand, Malay Peninsula, Borneo, Sulawesi, Moluccas, Philippines, New Guinea, New Britain I., Solomon Is., trop. Australia, Fiji, Samoa Is.

Guevaria R. M. King et H. Rob. 1974. Asteraceae. 4 Ecuador, Peru

Guibourtia Benn. 1857. Fabaceae. 16–17 West Indies, trop. America, trop. Africa (13)

Guichenotia J. Gay. 1821. Sterculiaceae. 5 southwestern Australia

Guiera Adans. ex Juss. 1789. Combretaceae. 1 northern trop. Africa

Guihaia J. Dransf., S. K. Lee et F. N. Wei. 1985 (Maxburretia Furtado and Rhapis L. f. ex Aiton). Arecaceae. 2 southern China (Guangxi and Guangdong), northern Vietnam (1)

Guilfoylia F. Muell. 1873 (Cadellia F. Muell.). Surianaceae. 1 northeastern Australia

Guilielma Mart. = Bactris Jacq. ex Scop. Arecaceae

Guillainia Vieill. 1866. Zingiberaceae. 6 New Guinea, Vanuatu, New Caledonia

Guillauminia A. Bertrand = Aloe L. Asphodelaceae

Guilleminea Kunth. 1823. Amaranthaceae. 5 America from southern U.S. to Argentina

Guillenia Greene = Caulanthus S. Watson. Brassicaceae

Guillonea Coss. 1851 (Laserpitium L.). Apiaceae. 1 southern Spain

Guindilia Gillies ex Hook. et Arn. 1833. Sapindaceae. 2 Chile, Argentina

Guioa Cav. 1798. Sapindaceae. 64 Burma, Cambodia, Vietnam, Thailand, Malesia from Malay Peninsula to New Guinea and Solomon Is., Vanuatu, New Caledonia, Fiji, Tonga, Australia (Queensland and New South Wales), and Lord Howe I.

Guiraoa Coss. 1851. Brassicaceae. 1 Spain

Guizotia Cass. 1829. Asteraceae. 6 trop. and subtrop. Africa

Gularia Garay = Schiedeella Schltr. Orchidaceae

Gulubia Becc. 1885. Arecaceae. 9 eastern Malesia from Moluccas to New Guinea, Palau Is., Bismarck Arch., Solomon Is., northern Australia, Vanuatu, Fiji

Gulubiopsis Becc. = Gulubia Becc. Arecaceae

Gumillea Ruiz et Pav. 1794. Cunoniaceae. 1 Peru

Gundelia L. 1753. Asteraceae. 1 Asia Minor, Cyprus, Syria, Iraq, Iran, Transcaucasia, Middle Asia (Kopet Dagh), Afghanistan

Gundlachia A. Gray. 1880. Asteraceae. 9 Bahama Is., Cuba, Haiti, Curaçao I.

Gunillaea Thulin. 1974. Campanulaceae. 2 southern trop. Africa from Angola to Mozambique, Madagascar

Gunnarella Senghas. 1988 (Chamaeanthus Schltr. ex J. J. Sm.). Orchidaceae. 6 New Guinea, Solomon Is., Vanuatu, New Caledonia

Gunnarorchis Brieger. 1981. Orchidaceae. 1 Burma, Thailand, Vietnam

Gunnera L. 1767. Gunneraceae. 40 America from Mexico to Chile, trop. and South Africa, Madagascar, Malesia (excl. Malay Peninsula), Tasmania, New Zealand, Hawaiian Is., Juan Fernández Is.

Gunnessia P. I. Forst. 1990. Asclepiadaceae. 1 Australia (Queensland)

Gunniopsis Pax. 1889. Aizoaceae. 14 Australia

Gurania (Schltdl.) Cogn. 1875. Cucurbitaceae. 40 trop. America

Guraniopsis Cogn. 1908. Cucurbitaceae. 1 Peru

Gustavia L. 1775. Lecythidaceae. 41 trop. America from southern Costa Rica to northern Bolivia, Brazil (to state of Alagoas)

Gutenbergia Sch. Bip. 1843. Asteraceae. 20 trop. Africa

Guthriea Bolus. 1873. Achariaceae. 1 South Africa (Lesotho, Natal, Cape)

Gutierrezia Lag. 1816. Asteraceae. 20–30 from western North to subtrop. South America

Gutzlaffia Hance. 1849 (Strobilanthes Blume). Acanthaceae. 9 China, Southeast Asia

Guya Frapp. = Drypetes Vahl. Euphorbiaceae

Guyania Airy Shaw = Guayania R. M. King et H. Rob. Asteraceae

Guyonia Naudin. 1850. Melastomataceae. 2 trop. Africa from Guinea to Cameroun, Zaire, Angola, Uganda, Tanzania

Guzmania Ruiz et Pav. 1802. Bromeliaceae. c. 130 southern Florida, West Indies, trop. Central and South America

Gymapsis Bremek. 1957. Acanthaceae. 1 Java

Gyminda (Griseb.) Sarg. 1891. Celastraceae. 4 Florida, Mexico, Central America, West Indies

Gymnacanthus Nees = Ruellia L. Acanthaceae

Gymnachne Parodi. 1938 (Rhombolytrum Link). Poaceae. 1 Chile

Gymnacranthera Warb. 1896. Myristicaceae. 7 southern India (1, G. canarica), southern peninsular Thailand, Singapore, Malesia (Malay Peninsula, Sumatra, Borneo, Sulawesi, Philippines, Moluccas, New Guinea, excl. Java and Lesser Sunda Is.), Bismarck Arch.

Gymnadenia R. Br. 1813 (Habenaria Willd.). Orchidaceae. 10 northeastern Canada, Newfoundland I., Greenland, temp. Eurasia, incl. Russia and Caucasus (4)

Gymnadeniopsis Rydb. = Platanthera Rich. Orchidaceae

Gymnanthera R. Br. 1810. Asclepiadaceae. 4 Malesia

Gymnanthes Sw. 1788. Euphorbiaceae. 15 southern U.S., Mexico, Central America, Venezuela, northern Brazil; West Indies

Gymnarrhena Desf. 1818. Asteraceae. 1 Mediterranean, West and Middle Asia

Gymnartocarpus Boerl. = Parartocarpus Baill. Moraceae

Gymnaster Kitam. = Aster L. Asteraceae

Gymnelaea (Endl.) Spach = Olea L. Oleaceae

Gymnema R. Br. 1810. Asclepiadaceae. c. 20 South Africa (1), India, Sri Lanka, Himalayas, Burma, Bangladesh, southwestern China, continental Southeast Asia, northern Australia

Gymnemopsis Costantin. 1912. Asclepiadaceae. 2 Indochina

Gymnocactus Backeb. = Neolloydia Britton et Rose. Cactaceae

Gymnocalycium Pfeiff. 1845. Cactaceae. c. 50 Bolivia, southern Brazil, Paraguay, Argentina, Uruguay

Gymnocarpos Forssk. 1775. Caryophyllaceae. 2 Canary Is., North Africa, eastern Mediterranean, West Asia eastward to China, Mongolia, and Baluchistan

Gymnochilus Blume. 1859. Orchidaceae. 2 Mascarene Is.

Gymnocladus Lam. 1785. Fabaceae. 5 East and Southeast Asia (4), eastern North America (1)

Gymnocondylus R. M. King et H. Rob. 1972. Asteraceae. 1 Brazil

Gymnocoronis DC. 1836. Asteraceae. 5 Mexico (4), subtrop. South America (1)

Gymnodiscus Less. 1831. Asteraceae. 2 South Africa

Gymnogonum Parry = Goodmania Reveal et Ertter. Polygonaceae

Gymnolaena Rydb. 1915 (Dyssodia Cav.). Asteraceae. 3 Mexico

Gymnolomia Kunth. 1818. Asteraceae. 50 southwestern U.S. southward to trop. South America

Gymnopentzia Benth. 1876. Asteraceae. 1 South Africa

Gymnopetalum Arn. 1840. Cucurbitaceae. 6 Himalayas from Nepal to Assam, Burma, western and southern China (5), Taiwan, Indochina, Malesia

Gymnophragma Lindau. 1917. Acanthaceae. 1 New Guinea

Gymnophyton (Hook.) Gay. 1848. Apiaceae. 6 Andes of Chile and Argentina

Gymnopodium Rolfe. 1901. Polygonaceae. 3 southern Mexico (Yucatan), Guatemala, Belize

Gymnopogon P. Beauv. 1812. Poaceae. 15 South Asia, southern U.S., South America

Gymnopoma N. E. Br. = Skiatophyllum L. Bolus. Aizoaceae

Gymnoschoenus Nees. 1841. Cyperaceae. 6 Australia

Gymnosciadium Hochst. = Pimpinella L. Apiaceae

Gymnosiphon Blume. 1827. Burmanniaceae. 25 trop. America (15), trop. Africa (3), trop. Asia (7)

Gymnosperma Less. 1832. Asteraceae. 1 southern U.S., Mexico, Central America

Gymnospermium Spach. 1839 (Leontice L.). Berberidaceae. 8 Southeast Europe, West, Middle, and Central Asia

Gymnosporia (Wight et Arn.) Benth. et Hook. f. = Maytenus Molina. Celastraceae

Gymnostachys R. Br. 1810. Araceae. 1 eastern Australia

Gymnostachyum Nees. 1832. Acanthaceae. 30 India, Sri Lanka, southern China, Indochina, Java, Philippines

Gymnostemon Aubrév. et Pellegr. 1937. Simaroubaceae. 1 trop. West Africa

Gymnostephium Less. 1832. Asteraceae. 6 South Africa

Gymnosteris Greene. 1898. Polemoniaceae. 2 western U.S.

Gymnostoma L. A. S. Johnson. 1980 (Casuarina L.). Casuarinaceae. 20 Malesia, northeastern Australia, New Caledonia, Fiji

Gymnostyles Juss. = Soliva Ruiz et Pav. Asteraceae

Gymnotheca Decne. 1845. Saururaceae. 2 continental China

Gynandriris Parl. 1854 (Moraea Mill. ex L.). Iridaceae. 9: 2 Mediterranean, West Asia to Pakistan; 7 South Africa; G. maricoides—Middle Asia

Gynandropsis DC. = Cleome L. Capparaceae

Gynatrix Alef. 1862 (Plagianthus J. R. Forst. et G. Forst.). Malvaceae. 1 eastern Australia

Gynerium Humb. et Bonpl. (1813) = Gynerium P. Beauv. Poaceae

Gynerium P. Beauv. 1812. Poaceae. 1 trop. America from Mexico to Peru and Brazil, West Indies

Gynocardia R. Br. 1820. Flacourtiaceae. 1 Tibet, Assam, Sikkim, Burma

Gynochthodes Blume. 1827. Rubiaceae. 14 Andaman Is., Indochina, western Malesia, Caroline Is., Melanesia eastward to Tonga, Niue, and Samoa Is.

Gynocraterium Bremek. 1939. Scrophulariaceae. 1 trop. South America

Gynoglottis J. J. Sm. 1904. Orchidaceae. 1 Sumatra

Gynopachis Blume. 1825. Rubiaceae. 15 Indochina, Malesia

Gynophorea Gilli. 1955. Brassicaceae. 1 Afghanistan

Gynophyge Gilli. 1973 (Agrocharis Hochst.). Apiaceae. 1 Tanzania

Gynostemma Blume. 1825. Cucurbitaceae. 6 India, Sri Lanka, Himalayas, China, Japan, Malesia, New Guinea, New Britain I.

Gynotroches Blume. 1825. Rhizophoraceae. 1 Burma, Thailand, Malesia (all over, except Lesser Sunda Is.), Caroline Is., New Britain, and Solomon Is.

Gynoxys Cass. 1827. Asteraceae. 100 Central and trop. South America

Gynura Cass. 1825. Asteraceae. 40 trop. Africa, Madagascar, South and East Asia, Malesia, trop. Australia

Gypothamnium Phil. 1860 (Plazia Ruiz et Pav.). Asteraceae. 1 Chile

Gypsacanthus Lott, Jaramillo et Rzed. 1984 (Carlowrightia A. Gray). Acanthaceae. 1 southern Mexico

Gypsophila L. 1753. Caryophyllaceae. 125–150 temp. Eurasia, esp. (c. 65) Crimea, southern European part of Russia, Caucasus, and Middle Asia; eastern Mediterranean from Egypt to Iran, Somalia, Arabian Peninsula; 1 (G. australis) Australia and New Zealand; G. patrini is a copper indicator

Gyptidium R. M. King et H. Rob. 1972 (Eupatorium L.). Asteraceae. 2 southern Brazil, Argentina

Gyptis (Cass.) Cass. 1818 (Eupatorium L.). Asteraceae. 7 Brazil, Argentina

Gyranthera Pittier. 1914. Bombacaceae. 2 Panama, Venezuela

Gyrinops Gaertn. 1791. Thymelaeaceae. 9 Sri Lanka (1, G. walla), Vietnam, and Laos (1), Malesia (7 Sulawesi, Lesser Sunda Is., Moluccas, New Guinea)

Gyrinopsis Decne. = Aquilaria Lam. Thymelaeaceae

Gyrocarpus Jacq. 1763. Hernandiaceae. 4 tropics

Gyrocaryum B. Valdés. 1983. Boraginaceae. 1 Spain

Gyrocheilos W. T. Wang. 1981 (Didymocarpus Wall.). Gesneriaceae. 4 China

Gyrodoma Wild. 1974. Asteraceae. 1 Mozambique

Gyrogyne W. T. Wang. 1981 (Stauranthera Benth.). Gesneriaceae. 1 southern China

Gyroptera Botsch. 1967. Chenopodiaceae. 3 Somalia, Kenya, Abd-el-Kuri I. (area of Socotra)

Gyrostelma E. Fourn. = Matelea Aubl. Asclepiadaceae

Gyrostemon Desf. 1820. Gyrostemonaceae. 12 arid and southern Australia, Tasmania

Gyrostipula J.-F. Leroy. 1974 (1975). Rubiaceae. 2 Madagascar, Comoro Is.

Gyrotaenia Griseb. 1861. Urticaceae. 6 West Indies

Haageocereus Backeb. 1931. Cactaceae. 32 Peru, a few species extending to northern Chile and Bolivia

Haarera Hutch. et E. A. Bruce = Erlangea Sch. Bip. Asteraceae

Haastia Hook. f. 1864. Asteraceae. 3 New Zealand

Habenaria Willd. 1805. Orchidaceae. 500–600 tropics and subtropics, a few species in warm–temp. regions

Haberlea Friv. 1835. Gesneriaceae. 1 Greece, Balkan Peninsula

Hablitzia M. Bieb. 1817. Chenopodiaceae. 1 Turkey, Caucasus, Turkmenistan

Habracanthus Nees. 1847. Acanthaceae. c. 50 trop. America: Mexico (3), Guatemala (3), Colombia (37), Venezuela (5), Ecuador and Peru (1), Bolivia (2)

Habranthus Herb. 1824. Amaryllidaceae. 10 trop. America

Habrochloa C. E. Hubb. 1967. Poaceae. 1 trop. Africa: Tanzania, Zambia, Malawi, Zimbabwe

Habroneuron Stendl. 1927. Rubiaceae. 1 southern Mexico

Habropetalum Airy Shaw. 1952. Dioncophyllaceae. 1 trop. West Africa

Habrosia Fenzl. 1843. Caryophyllaceae. 1 Asia Minor, Syria, Lebanon, Iraq, western Iran

Hachettea Baill. 1880. Dactylanthaceae. 1 New Caledonia

Hackelia Opiz. 1838. Boraginaceae. 45 temp. Northern Hemisphere, incl. Russia (4, from European part to Far East) and Middle Asia; montane Central and South America

Hackelia Vasey = Gouinia E. Fourn. Poaceae

Hackelochloa Kuntze. 1891. Poaceae. 2 tropics

Hacquetia Neck. ex DC. 1830. Apiaceae. 1 (H. epipactis) Central, South and Southeast Europe

Hadrodemas H. E. Moore = Callisia Loefl. Commelinaceae

Haeckeria F. Muell. 1853. Asteraceae. 4 Australia

Haegiela P. S. Short et Paul G. Wilson. 1990. Asteraceae. 1 southwestern Australia, coastal southern Australia, and western Victoria

Haemacanthus S. Moore. 1899 (Satanocrater Schweinf.). Acanthaceae. 1 Somalia

Haemanthus L. 1753. Amaryllidaceae. 21 South Africa

Haemaria Lindl. = Ludisia A. Rich. Orchidaceae

Haematocarpus Miers. 1864. Menispermaceae. 3 eastern Himalayas, Assam, Bangladesh, Andaman Is., Thailand, Sumatra, Java, Borneo, Sulawesi, Philippines

Haematodendron Capuron. 1972. Myristicaceae. 1 Madagascar

Haematostaphis Hook. f. 1860. Anacardiaceae. 2 trop. West Africa

Haematostemon (Muell. Arg.) Pax et K. Hoffm. 1919. Euphorbiaceae. 2 Venezuela (Amazonas), Guyana

Haematoxylum L. 1753. Fabaceae. 3 trop. America and West Indies (2), Namibia (1, H. dinteri)

Haemodorum Sm. 1798. Haemodoraceae. 20 Australia and Tasmania, New Guinea (1, H. coccineum)

Haenianthus Griseb. 1861. Oleaceae. 2–3 West Indies

Hagenbachia Nees et Mart. 1823. Haemodoraceae. 5 Costa Rica (1), Panama (1), Colombia (1), Ecuador (1), Brazil (2), Bolivia (2), Paraguay (2)

Hagenia J. F. Gmel. 1791. Rosaceae. 1 Sudan, Ethiopia, Congo, Uganda, Kenya, Tanzania, Zambia, northern Zimbabwe

Hagsatera Tamayo. 1974. Orchidaceae. 2 Mexico

Hainania Merr. 1935. Tiliaceae. 1 southern China, Hainan

Hainardia Greuter = Monerma P. Beauv. Poaceae

Haitia Urb. 1919. Lythraceae. 2 Haiti

Haitiella L. H. Bailey = Coccothrinax Sargent. Arecaceae

Hakea Schrad. 1798. Proteaceae. 125 Australia

Hakoneaste F. Maek. 1935 (Ephippianthus Rchb. f.). Orchidaceae. 1 Japan

Hakonechloa Makino ex Honda. 1930. Poaceae. 1 Japan

Halacsya Doerfl. 1902. Boraginaceae. 1 Yugoslavia, Albania

Halacsyella Janchen = Edraianthus (A. DC.) DC. Campanulaceae

Halanthium K. Koch. 1844. Chenopodiaceae. 5 Caucasus (2), Asia Minor, Iran

Halarchon Bunge. 1862. Chenopodiaceae. 1 Afghanistan

Haldina Ridsdale. 1978. Rubiaceae. 1 India, Sri Lanka, southern China, Indochina

Halenbergia Dinter = Mesembryanthemum L. Aizoaceae

Halenia Borkh. 1796. Gentianaceae. c. 70 America, Europe (1), temp. Asia, Tibet, Himalayas; Russia (2, from European part to Far East)

Halerpestes Greene. 1900 (Ranunculus L.). Ranunculaceae. 10 North and South America; temp. Eurasia, incl. Russia (2 West and East Siberia, Far East)

Halesia Ellis ex L. 1759. Styracaceae. 5 eastern China (1, H. macgregorii), southeastern U.S.

Halfordia F. Muell. 1865. Rutaceae. 3 New Guinea (1), eastern Australia, New Caledonia

Halgania Gaudich. 1829. Ehretiaceae. 15 Australia

Halimione Aellen. 1938. Chenopodiaceae. 3 Central, South, and Southeast Europe; Caucasus; West Siberia; North Africa; West and Middle Asia eastward to northeastern China

Halimiphyllum (Engl.) A. Borissova = Zygophyllum L. Zygophyllaceae

Halimium (Dunal) Spach. 1836. Cistaceae. 11 Mediterranean

Halimocnemis C. A. Mey. 1829. Chenopodiaceae. 19: 13 southeastern European part of Russia, Caucasus, Middle Asia; Iran, Afghanistan, northwestern China

Halimodendron Fisch. ex DC. 1825. Fabaceae. 1 Europe, Caucasus, Asia Minor, Iran, Middle Asia, West Siberia, Afghanistan, Pakistan, northwestern China, Mongolia

Halimolobos Tausch. 1836. Brassicaceae.

19 western America from Rocky Mountains to Andes

Hallea J.-F. Leroy. 1975 (Mitrogyna Korth.). Rubiaceae. 3 trop. Africa

Halleria L. 1753. Scrophulariaceae. 10 trop. and South Africa, Madagascar

Hallia Thunb. 1799. Fabaceae. 9 South Africa

Hallianthus H. E. K. Hartmann. 1983 (Leipoldtia L. Bolus). Aizoaceae. 1 South Africa (western Cape)

Hallieracantha Stapf = Ptyssiglottis T. Anderson. Acanthaceae

Halmoorea J. Dransf. et N. W. Uhl. 1984. Arecaceae. 1 Madagascar

Halocarpus Quinn. 1982 (Dacrydium Sol. ex Lamb.). Podocarpaceae. 3 New Zealand

Halocharis Moq. 1849. Chenopodiaceae. 13 Syria, Iraq, Iran, Afghanistan, Pakistan, Middle Asia (4)

Halocnemum M. Bieb. 1819–1820. Chenopodiaceae. 1 Southeast Europe, Caucasus, West and East Siberia, Middle Asia; Mediterranean, Ethiopia, Arabian Peninsula, Iraq, Iran, Afghanistan, northwestern and western China, Mongolia

Halodule Endl. 1841. Cymodoceaceae. 8 tropical coasts

Halogeton C. A. Mey. 1829. Chenopodiaceae. 3–4 Spain, Morocco, Algeria, Middle Asia, West and East Siberia, Afghanistan, Pakistan, India, China, Mongolia

Halongia Jeanpl. 1971. Liliaceae. 1 Vietnam

Halopegia K. Schum. 1902. Maranthaceae. 6 trop. Africa, Madagascar, India, Indochina, Java

Halopeplis Bunge ex Ung.-Sternb. 1866. Chenopodiaceae. 3 Mediterranean, South Africa, Southwest and West Asia, Iran, northwestern China; H. pygmaea—Caucasus, Middle Asia

Halophila Thouars. 1806. Halophilaceae. 9 coastal Florida, West Indies, tropical coasts of the Indian and the Pacific Oceans

Halophytum Speg. 1902. Halophytaceae. 1 Patagonia

Halopyrum Stapf. 1896. Poaceae. 1 coasts of the Indian Ocean from Egypt, Somalia, Kenya, Tanzania, Mozambique, and Madagascar through Arabian Peninsula to Pakistan, India, and Sri Lanka

Haloragis J. R. Forst. et G. Forst. 1775. Haloragaceae. 28 East Asia and Malesia (5), temp. and eremaean Australia (23), Tasmania, Flinders Is., New Zealand, Chatham Is., Kermedec Is., New Caledonia, Rapa I., Juan Fernández Is.

Haloragodendron Orchard. 1975. Haloragaceae. 5 southwestern and southern Australia

Halosarcia Paul G. Wilson. 1980. Chenopodiaceae. 23: 1 (H. indica) trop. coasts of the Indian Ocean from East Africa to India, Sri Lanka, Malesia, Australia (23)

Halosciastrum Koidz. 1941. Apiaceae. 1 southern Soviet Far East and Korea

Halosicyos Mart. Crov. 1947. Cucurbitaceae. 1 northern Argentina

Halostachys C. A. Mey. 1843. Chenopodiaceae. 1 southeastern European part of Russia, Caucasus, Middle Asia, Iran, Afghanistan, northwestern China, Mongolia

Halothamnus Jaub. et Spach. 1845. Chenopodiaceae. 25 Northeast Africa, Arabian Peninsula, Asia Minor, Palestine, Iraq, Caucasus, Middle Asia, Iran, Afghanistan, Pakistan, northwestern China

Halotis Bunge. 1862 (Halimocnemis C. A. Mey.). Chenopodiaceae. 2 Syria, Palestine, Iraq, Iran, Afghanistan; H. pilifera—also Caucasus, Middle Asia

Haloxanthium Ulbr. = Atriplex L. Chenopodiaceae

Haloxylon Bunge. 1851. Chenopodiaceae. 3 Egypt, Palestine, Arabian Peninsula, Iraq, Middle Asia, Iran, Afghanistan, Pakistan, northwestern China, Mongolia

Halphophyllum Mansf. = Gasteranthus Benth. Gesneriaceae

Hamadryas Comm. ex Juss. 1789. Ranunculaceae. 6 Antarctic America

Hamamelis L. 1753. Hamamelidaceae. 5–6 East Asia, eastern North America

Hamatocactus Britton et Rose = Thelocactus (K. Schum.) Britton et Rose. Cactaceae

Hambergera Scop. = Combretum Loefl. Combretaceae

Hamelia Jacq. 1760. Rubiaceae. 16 America from Florida and Mexico to Paraguay, West Indies

Hamilcoa Prain. 1912. Euphorbiaceae. 1 trop. West Africa

Hamiltonia Roxb. = Spermadictyon Roxb. Rubiaceae

Hammada Iljin. 1948. Chenopodiaceae. 17 Spain, Mauritania, Morocco, Algeria, Tunisia, Libya, Egypt, Syria, Lebanon, Israel, Jordan, Arabian Peninsula, Iraq, Iran, Afghanistan, Pakistan, India, Burma, Middle Asia (3, Turkmenistan, Uzbekistan, Tajikistan)

Hammarbya Kuntze. 1891 (Malaxis Sol. ex Sw.). Orchidaceae. 1 temp. Northern Hemisphere

Hammatolobium Fenzl. 1842. (Tripodion Medik.). Fabaceae. 2 Mediterranean

Hampea Schltdl. 1837. Malvaceae. 21 trop. America from Mexico to Colombia

Hanabusaya Nakai. 1911. Campanulaceae. 2 Korea

Hanburia Seem. 1858. Cucurbitaceae. 2 Mexico, Guatemala

Hanceola Kudo. 1929. Lamiaceae. 8 China

Hancockia Rolfe. 1903. Orchidaceae. 2 southern China, Vietnam, Japan (H. japonica)

Hancornia Gomes. 1812. Apocynaceae. 1 Brazil, Paraguay

Handelia Heimerl. 1922. Asteraceae. 1 Middle Asia, western China

Handeliodendron Rehder. 1935. Sapindaceae. 1 China

Handroanthus Mattos = Tabebuia Gomes ex DC. Bignoniaceae

Hanghomia Gagnep. et Thénint. 1936. Apocynaceae. 1 Indochina

Hanguana Blume. 1827. Hanguanaceae. 2 Sri Lanka, Thailand, Malesia, Palau Is., northern Australia

Haniffia Holttum. 1950. Zingiberaceae. 2 Thailand, Malay Peninsula

Hannafordia F. Muell. 1860. Sterculiaceae. 3 Australia

Hannoa Planch. = Quassia L. Simaroubaceae

Hannonia Braun = Blanq. et Maire. 1931. Amaryllidaceae. 1 Northwest Africa

Hansemannia K. Schum. = Archidendron F. Muell. Fabaceae

Hansenia Turcz. 1844. Apiaceae. 1 Siberia, Mongolia

Hanseniella C. Cusset. 1992. Podostemaceae. 1 Thailand

Hanslia Schindl. 1924 (Desmodium Desv.). Fabaceae. 1 eastern Malesia, Bismarck Arch., Vanuatu

Hansteinia Oerst. 1854 (Habrachanthus Nees). Acanthaceae. 14 trop. America from Mexico to Bolivia

Hapaline Schott. 1858. Araceae. 5 Himalayas, Burma, China, Southeast Asia, Malesia

Hapalochilus (Schltr.) Senghas. 1978 (Bulbophyllum Thouars). Orchidaceae. 1 Southeast Asia, Malesia

Hapalorchis Schltr. 1919. Orchidaceae. 9 trop. South America, West Indies

Haplanthodes Kuntze = Andrographis Wall. ex Nees. Acanthaceae

Haplanthoides H. W. Li = Andrographis Wall. ex Nees. Acanthaceae

Haplanthus Nees. 1832 (Andrographis Wall. ex Nees). Acanthaceae. 2 Indo-Malesia

Haplocalymma S. F. Blake = Viguiera Kunth Asteraceae

Haplocarpha Less. 1831. Asteraceae. 10 Africa

Haplochorema K. Schum. 1899. Zingiberaceae. 7 Sumatra (1), Borneo (6)

Haploclathra Benth. 1861. Clusiaceae. 4 Brazil (Amazonia)

Haplocoelum Radlk. 1878. Sapindaceae. 7 trop. Africa

Haplodypsis Baill. = Neophloga Baill. Arecaceae

Haploesthes A. Gray. 1849. Asteraceae. 3 southern U.S., Mexico

Haplolobus H. J. Lam. 1931. Burseraceae. 22 Malesia (Borneo, Moluccas, New Guinea—c. 20), New Britain, Palau Is.,

Solomon and Santa Cruz Is., Vanuatu, Fiji, Samoa

Haplolophium Cham. 1832. Bignoniaceae. 1 Brazil

Haplopappus Cass. 1828. Asteraceae. 75 South America

Haplopetalon A. Gray = Crossostylis J. R. Forst. et G. Forst. Rhizophoraceae

Haplophandra Pichon. 1948 (Odontadenia Benth.). Apocynaceae. 1 Brazil

Haplophloga Baill. = Neophloga Baill. Arecaceae

Haplophragma Dop = Fernandoa Welw. ex Seem. Bignoniaceae

Haplophyllophorus (Brenan) A. Fern. et R. Fern. = Cincinnobotrys Gilg. Melastomataceae

Haplophyllum A. Juss. 1825 (Ruta L.). Rutaceae. c. 70 Mediterranean, Southeast Europe, West (incl. Caucasus) and Middle Asia (23), Siberia, Mongolia, China, and Afghanistan

Haplophyton A. DC. 1844. Apocynaceae. 3 southwestern U.S. (Arizona, New Mexico, Texas), Mexico, Cuba

Haplorhus Engl. 1881. Anacardiaceae. 1 Peru

Haplormosia Harms. 1915. Fabaceae. 1 trop. West Africa from Sierra Leone to Cameroun

Haplosciadium Hochst. 1844. Apiaceae. 1 Ethiopia, Uganda, Kenya, Tanzania

Haploseseli H. Wolff et Hand.-Mazz. = Physospermopsis H. Wolff. Apiaceae

Haplosphaera Hand. = Mazz. 1920. Apiaceae. 2–3 Himalayas, southwestern China

Haplostachys (A. Gray) W. F. Hillebr. 1888 (Phyllostegia Benth.). Lamiaceae. 5 Hawaiian Is.

Haplostephium Mart. ex DC. 1836. Asteraceae. 1 Brazil

Haplostichanthus F. Muell. 1891. Annonaceae. 1–2 Australia (Queensland)

Haplothismia Airy Shaw. 1952. Burmanniaceae. 1 southern India

Haptanthus Goldberg et Nelson. 1989. Uncertain familial position. Possibly related to Flacourtiaceae. 1 Honduras

Haptocarpum Ule. 1908. Capparaceae. 1 eastern Brazil

Haptotrichion Paul G. Wilson. 1992. Asteraceae. 2 western Australia

Haradjania Rech. f. 1950 (Myopordon Boiss.). Asteraceae. 1 Syria

Haraella Kudo = Gastrochilus D. Don. Orchidaceae

Harbouria J. M. Coult. et Rose. 1888. Apiaceae. 1 southwestern U.S. (Colorado, New Mexico)

Hardenbergia Benth. 1837. Fabaceae. 3 New Guinea, Australia

Hardwickia Roxb. 1811. Fabaceae. 1 western India

Harfordia Greene et Parry. 1886. Polygonaceae. 1–3 western Baja California

Harlanlewisia Epling = Scutellaria L. Lamiaceae

Harleye S. F. Blake. 1932. Asteraceae. 1 Mexico, Guatemala

Harleyodendron R. S. Cowan. 1979. Fabaceae. 1 eastern Brazil

Harmandia Baill. 1889. Olacaceae. 1 (H. mekongensis) Laos, Vietnam, Sumatra, Malay Peninsula, Borneo

Harmandiella Costantin. 1912. Asclepiadaceae. 1 Indochina

Harmsia K. Schum. 1898. Sterculiaceae. 3 trop. Africa

Harmsiella Briq. = Otostegia Benth. Lamiaceae

Harmsiodoxa O. E. Schulz. 1924. Brassicaceae. 3 Australia

Harmsiopanax Warb. 1897. Araliaceae. 3 Malesia

Harnackia Urb. 1925. Asteraceae. 1 Cuba

Haronga Thouars = Harungana Lam. Hypericaceae

Harpachne Hochst. ex A. Rich. 1847. Poaceae. 2 North and Northeast Africa

Harpagocarpus Hutch. et Dandy. 1926 (Fagopyrum Hill). Polygonaceae. 1 trop. Africa: Cameroun, Congo, Uganda, Kenya, Tanzania

Harpagonella A. Gray. 1876. Boraginaceae. 1 southwestern U.S., northwestern Mexico

Harpagophytum DC. ex Meisn. 1840. Pedaliaceae. 8 South Africa, Madagascar

Harpalyce Mociño et Sessé ex DC. 1825. Fabaceae. c. 20 Mexico, Central America, Cuba, Brazil

Harpanema Decne. 1844. Asclepiadaceae. 1 Madagascar

Harpephyllum Bernh. ex Krauss. 1844. Anacardiaceae. 1 South Africa

Harperia W. Fitzg. 1904. Restionaceae. 1 Australia

Harperocallis McDaniel. 1968. Melanthiaceae. 1 Florida Peninsula

Harpochilus Nees. 1847. Acanthaceae. 3 Brazil

Harpochloa Kunth. 1829. Poaceae. 1 South Africa

Harpullia Roxb. 1824. Sapindaceae. 26 India, Sri Lanka, Bangladesh, Burma, Indochina, southeastern China, Malesia (15), eastern Australia (8), Solomon Is., New Caledonia, Fiji, Tonga, and Samoa Is.

Harrimanella Coville = Cassiope D. Don. Ericaceae

Harrisella Fawc. et Rendle. 1909 (Campylocnetrum Benth.). Orchidaceae. 1 Florida, Mexico, Salvador, West Indies

Harrisia Britton. 1909. Cactaceae. 16 trop. America from Florida to Argentina, West Indies

Harrisonia R. Br. ex Adr. Juss. 1825. Simaroubaceae. 4 trop. Africa, South and Southeast Asia, Malesia, trop. Australia (1, H. brownii)

Harrysmithia H. Wolff. 1926. Apiaceae. 2 southwestern China

Harthamnus H. Rob. = Plazia Ruiz et Pav. Asteraceae

Hartia Dunn. 1902 (Stewartia L.). Theaceae. 13 China, Indochina

Hartleya Sleumer. 1969 (Gastrolepis Tiegh.). Icacinaceae. 1 New Guinea

Hartmannia Spach = Oenothera L. Onagraceae

Hartogia L. f. = Hartogiella Codd. Celastraceae

Hartogiella Codd. 1983. Celastraceae. 1 South Africa (Cape)

Hartwegiella O. E. Schulz = Mancoa Wedd. Brassicaceae

Hartwrightia A. Gray ex S. Watson. 1888. Asteraceae. 1 southeastern U.S. (Georgia and Florida)

Harungana Lam. 1796. Hypericaceae. 1 trop. Africa, Madagascar, Mauritius

Harveya Hook. 1837. Scrophulariaceae. 40 trop. and South Africa, Arabian Peninsula, Madagascar, Mascarene Is. (1)

Haselhoffia Lindau = Physacanthus Benth. Acanthaceae

Hasseanthus Rose = Dudleya Britton et Rose. Crassulaceae

Hasselquistia L. = Tordylium L. Apiaceae

Hasseltia Kunth. 1825. Flacourtiaceae. 3 trop. America from southern Mexico to southeastern Bolivia and Brazil

Hasseltiopsis Sleumer. 1938. Flacourtiaceae. 1 southern Mexico, Guatemala, Belize, Honduras, Costa Rica

Hasskarlia Baill. = Tetrorchidium Poepp. et Endl. Euphorbiaceae

Hasslerella Chodat. 1908. Scrophulariaceae. 1 Argentina

Hassleropsis Chodat = Basistemon Turcz. Scrophulariaceae

Hasteola Raf. 1838. Asteraceae. 1 North America

Hastingsia S. Watson = Schoenolirion Torr. ex E. M. Durand. Hyacinthaceae

Hatiora Britton et Rose. 1915 (Rhipsalis Gaertn.). Cactaceae. 5 eastern Brazil

Hatschbachiella R. M. King et H. Rob. 1972 (Eupatorium L.). Asteraceae. 2 Brazil, Argentina

Haumania J. Léonard. 1949. Maranthaceae. 3 trop. Africa

Haumaniastrum P. A. Duvign. et Plancke. 1959 (or = Acrocephalus Benth.). Lamiaceae. 23 trop. Africa

Haussknechtia Boiss. 1872. Apiaceae. 1 Iran

Haussmannia F. Muell. = Neosepicaea Diels. Bignoniaceae

Haussmannianthes Steenis = Neosepicaea Diels. Bignoniaceae

Hauya Mociño et Sessé ex DC. 1828. Onagraceae. 2 Central America

Havardia Small. 1901. Fabaceae. 20 India, Sri Lanka, Indochina, trop. and subtrop. America

Havetia Kunth. 1822. Clusiaceae. 1 Colombia

Havetiopsis Planch. et Triana. 1860. Clusiaceae. 5 Panama, trop. South America

Hawkesiophyton Hunz. 1977. Solanaceae. 4 Panama, Colombia, Peru, and Brazil

Haworthia Duval. 1809. Asphodelaceae. c. 70 trop. (1) and South Africa

Haya Balf. f. 1884. Caryophyllaceae. 1 Socotra

Hayataella Masam. 1934. Rubiaceae. 1 Taiwan

Haydonia R. Wilczek = Vigna Savi. Fabaceae

Haylockia Herb. 1930. Amaryllidaceae. 6 Andes

Haynaldia Schur = Dasypyrum (Coss. et Durieu) T. Durand. Poaceae

Haynea Schumach. et Thonn. = Laportea Gaudich. Urticaceae

Hazardia Greene. 1887. Asteraceae. 13 U.S. and western Mexico from Oregon to Baja California

Hazomalania Capuron. 1966 (Hernandia L.). Hernandiaceae. 1 western Madagascar

Hazunta Pichon = Tabernaemontana L. Apocynaceae

Hebe Comm. ex Juss. 1789 (Veronica L.). Scrophulariaceae. c. 75 New Guinea, Australia, Tasmania, New Zealand, Chatham Is., temp. South America, Falkland Is.

Hebea (Pers.) R. A. Hedw. = Tritoniopsis L. Bolus. Iridaceae

Hebecladus Miers. 1845 (Jaltomata Schltdl.). Solanaceae. 8 trop. South America (Andes)

Hebeclinium DC. 1836. Asteraceae. 20 trop. America from Mexico to Peru, West Indies

Hebecoccus Radlk. = Lepisanthes Blume. Sapindaceae

Hebenstretia L. 1753. Selaginaceae. 25 trop. and South Africa

Hebepetalum Benth. 1962. Hugoniaceae. 6 trop. South America

Heberdenia Banks ex A. DC. 1841. Myrsinaceae. 2 Mexico (1), Canary and Madeira Is.

Hebestigma Urb. 1900. Fabaceae. 1 Cuba

Hecastocleis A. Gray. 1882. Asteraceae. 1 southwestern U.S.

Hecatactis (F. Muell.) Mattf. = Keysseeria Lauterb. Asteraceae

Hecatostemon S. F. Blake. 1918. Flacourtiaceae. 1 (H. completus) Colombia, Venezuela, Guyana, Brazil

Hechtia Klotzsch. 1835. Bromeliaceae. 47 southern U.S. (3 Texas), Mexico, incl. Baja California, Guatemala, Honduras, El Salvador, Nicaragua

Heckeldora Pierre. 1897. Meliaceae. 2 trop. West Africa

Hectorella Hook. f. 1864 (Lyallia Hook. f.). Hectorellaceae. 1 New Zealand

Hecubaea DC. = Helenium L. Asteraceae

Hedbergia Molau. 1988. Scrophulariaceae. 1 Africa from Cameroun and Ethiopia to Uganda and Tanzania

Hedeoma Pers. 1806. Lamiaceae. 38 southwestern U.S., Mexico, Central and South America south to Argentina

Hedera L. 1753. Araliaceae. 5–15 Canary Is., West, Central, and Southeast (incl. Crimea) Europe; Mediterranean; Caucasus; West Asia; Himalayas; China; Korea; Japan; Australia (Queensland)

Hederella Stapf = Catanthera F. Muell. Melastomataceae

Hederopsis C. B. Clarke = Macropanax Miq. Araliaceae

Hederorkis Thouars. 1809 (Bulbophyllum Thouars). Orchidaceae. 2 Mascarene Is.

Hedinia Ostenf. 1922. Brassicaceae. 4 West and East (to Chukotka) Siberia, Middle Asia, Mongolia, western and southwestern China, Pakistan, Himalayas (India, Nepal)

Hediniopsis Botsch. et Petrovsky = Hedinia Ostenf.

Hedraianthera F. Muell. 1865. Celastraceae. 1 eastern Australia

Hedranthera (Stapf) Pichon = Callichilia Stapf. Apocynaceae

Hedstromia A. C. Sm. 1936. Rubiaceae. 1 Fiji

Hedyachras Radlk. = Glenniea Hook. f. Sapindaceae

Hedycarya J. R. Forst. et G. Forst. 1776.

Monimiaceae. 15 southeastern Australia, New Zealand, Solomon Is., New Caledonia (8), Vanuatu, Fiji, Samoa Is.

Hedycaryopsis Danguy = Ephippiandra Decne. Monimiaceae

Hedychium J. Koenig. 1783. Zingiberaceae. 55 Madagascar, India, Himalayas (Nepal, Bhutan), Burma, southwestern and southern China, Indochina, Malesia

Hedyosmos Mitch. = Cunila L. Lamiaceae

Hedyosmum Sw. 1788. Chloranthaceae. c. 40 trop. America from central Mexico to central Bolivia and Brazil, West Indies; a few species in southern China (Kwangsi–Tonkin, Hainan), southern Vietnam, Sumatra, Java, Borneo, Sulawesi

Hedyotis L. 1753. Rubiaceae. 350 tropics, esp. Asia and Malesia

Hedypnois Hill. 1753. Asteraceae. 3 Canary and Madeira Is., Mediterranean, Crimea, Caucasus, Asia Minor, Iran

Hedysarum L. 1753. Fabaceae. c. 170 temp. Northern Hemisphere, esp. (c. 120) European part of Russia, Caucasus, Middle Asia, Siberia, Far East; Mediterranean

Hedyscepe H. A. Wendl. et Drude. 1875. Arecaceae. 1 Lord Howe I.

Hedythyrsus Bremek. 1952. Rubiaceae. 2 trop. Africa (Zaire, Tanzania, Angola, Zambia, Malawi)

Heeria Meisn. 1837. Anacardiaceae. 1 South Africa

Hegemone Bunge ex Ledeb. = Trollius L. Ranunculaceae

Hegnera Schindl. = Desmodium Desv. Fabaceae

Heimerliodendron Skottsb. = Pisonia L. Nyctaginaceae

Heimia Link. 1822. Lythraceae. 3 America from southern U.S. to Argentina

Heimodendron Sillans = Entandrophragma C. DC. Meliaceae

Heinsenia K. Schum. 1897 (Aulacocalyx Hook. f.). Rubiaceae. 1 (H. diervilleoides) trop. Africa from Sudan south to Zimbabwe and Mozambique

Heinsia DC. 1830. Rubiaceae. 4–5 trop. and South (Transvaal) Africa

Heisteria Jacq. 1760. Olacaceae. 33: 30 Mexico, Guatemala, Belize, Honduras, El Salvador, Nicaragua, Costa Rica, Panama, Colombia, Venezuela, the Guianas, Ecuador, Peru, Bolivia, Brazil; Windward Is. (1, H. coccinea); 3 trop. West Africa

Hekistocarpa Hook. f. 1873. Rubiaceae. 1 trop. West Africa

Heladena A. Juss. 1840. Malpighiaceae. 6 trop. and subtrop. South America

Helanthium (Benth. et Hook. f.) Engelm. ex Britton. 1883 (Echinodorus Rich. ex Engelm.). Alismataceae. 1 from eastern Canada to South America

Helcia Lindl. 1845. Orchidaceae. 1 Colombia

Heldreichia Boiss. 1841. Brassicaceae. 3 Asia Minor, Iran (1)

Helenium L. 1753. Asteraceae. 40 West America from Wyoming, California, Utah, and Colorado, through Mexico and Central America to Andes

Heleochloa Host ex Roem. = Crypsis Aiton. Poaceae

Helia Mart. = Irlbachia Mart. Gentianaceae

Heliabravoa Backeb. = Myrtillocactus Console. Cactaceae

Heliamphora Benth. 1840. Sarraceniaceae. 6 northern South America (Guayana highlands)

Helianthella Torr. et A. Gray. 1842. Asteraceae. 8 western U.S., Mexico

Helianthemum Hill. 1753. Cistaceae. c. 110 Europe, Caucasus, Mediterranean, Cape Verde Is., Sakhara, Northeast Africa, West and Middle Asia eastward to western China (1, H. songaricum) and Iran

Helianthopsis H. Rob. = Pappobolus S. F. Blake. Asteraceae

Helianthostylis Baill. 1875. Moraceae. 1 Colombia, Brazil

Helianthus L. 1753. Asteraceae. 49 North America

Helicanthes Danser. 1933. Loranthaceae. 1 India

Helichrysopsis Kirp. 1950. Asteraceae. 1 trop. East Africa

Helichrysum Mill. 1754. Asteraceae. c. 600 South and Southeast Europe; trop. and South Africa; Madagascar; Socotra; West (incl. Caucasus), Southwest, and Middle Asia; southern India; Sri Lanka; Australia; South America

Helicia Lour. 1790. Proteaceae. c. 90 southern India, Sri Lanka, Himalayas, Tibet, southern and southeastern China, Taiwan, southern Japan, Indochina, Malesia, Australia, Melanesia

Helicilla Moq. = Suaeda Forssk. ex Scop. Chenopodiaceae

Heliciopsis Sleumer. 1955. Proteaceae. 7–10 northeastern India, Burma, southern China (incl. Hainan), Indochina, Malesia (Sumatra, Malay Peninsula, Java, Borneo, Philippines)

Helicodiceros Schott ex K. Koch = Dracunculus Hill. Araceae

Heliconia L. 1771. Heliconiaceae. c. 100 trop. America; 1 Melanesia

Helicostylis Trécul. 1847. Moraceae. 7 trop. America

Helicteres L. 1753. Sterculiaceae. 60 trop. Asia, Australia, and America

Helicteropsis Hochr. 1925. Malvaceae. 2 Madagascar

Helictonema Pierre = Hippocratea L. Celastraceae

Helictotrichon Besser ex Schult. et Schult. f. 1827. Poaceae. 50 temp. Northern Hemisphere, mountainous tropics, South Africa

Helietta Tul. 1847. Rutaceae. 9 trop. America, Cuba

Helinus E. Mey. ex Endl. 1840. Rhamnaceae. 5–6 trop. and South Africa, Madagascar, Arabian Peninsula, Himalayas from Kashmir to Nepal

Heliocarpus L. 1753. Tiliaceae. 22 trop. America

Heliocarya Bunge = Caccinia Savi. Boraginaceae

Heliocauta Humphries. 1977. Asteraceae. 1 North Africa

Heliocereus (A. Berger) Britton et Rose = Disocactus Lindl. Cactaceae

Heliomeris Nutt. = Viguiera Kunth. Asteraceae

Heliophila Burm. f. ex L. 1763. Brassicaceae. 75 South-West and South Africa

Heliopsis Pers. 1807. Asteraceae. 13 America

Heliosperma (Rchb.) Rchb. = Silene L. Caryophyllaceae

Heliostemma Woodson = Matelea Aubl. Asclepiadaceae

Heliotropium L. 1753. Boraginaceae. c. 280 temp., subtrop., and trop. regions, esp. more or less arid areas

Helipterum DC. ex Lindl. = Helichrysum Mill. Asteraceae

Helixanthera Lour. 1790. Loranthaceae. 25 trop. (south of the Sahara) and eastern South Africa (3), trop. Asia

Helixyra Salisb. ex N. E. Br. = Morea Mill. Iridaceae

Helladia M. Kral = Sedum L. Crassulaceae

Helleborus L. 1753. Ranunculaceae. 15 Europe from Britain and northern Germany to Mediterranean and eastward to Poland, western Ukraine, and Bulgaria; Balearic Is., Corsica, Sardinia, Sicily, northern and southern Turkey and adjacent northern Syria (H. vesicarius), western and central Caucasus (1, H. orientalis), China (1, H. thibetanus)

Hellenocarum H. Wolff. 1927 (or = Carum L.). Apiaceae. 3 South and Southeast Europe, Southwest Asia

Helleria Fourn. = Hellerochloa Rauschert. Poaceae

Helleriella A. D. Hawkes. 1966. Orchidaceae. 1 Nicaragua

Hellerochloa Rauschert. 1982 (Festuca L.). Poaceae. 2 Mexico, Venezuela

Hellerorchis A. D. Hawkes = Rodrigueziella Kuntze. Orchidaceae

Hellmuthia Steud. 1854. Cyperaceae. 1 South Africa

Helmholtzia F. Muell. 1866. Philydraceae. 3 Moluccas, New Guinea, eastern Australia (2)

Helminthia Juss. = Picris L. Asteraceae

Helminthocarpon A. Rich. = Lotus L. Fabaceae

Helminthotheca Zinn. 1757. Asteraceae.

1 Europe, Mediterranean, Asia Minor, Iran; Caucasus, Middle Asia

Helmiopsiella Arènes. 1956. Sterculiaceae. 4 Madagascar

Helmiopsis H. Perrier. 1944. Sterculiaceae. 8 Madagascar

Helmontia Cogn. 1875. Cucurbitaceae. 2 Venezuela, Guyana, Brazil

Helogyne Nutt. 1841. Asteraceae. 12 southern Andes

Helonema Suess. = Eleocharis R. Br. Cyperaceae

Helonias L. 1753. Melanthiaceae. 1 eastern U.S.

Heloniopsis A. Gray. 1858–1859. Melanthiaceae. 4 Sakhalin (1, H. orientalis), Korea, Japan, Taiwan

Helonoma Garay. 1980. Orchidaceae. 2 Venezuela, Guyana

Helosciadium W. D. J. Koch = Apium L. Apiaceae

Helosis Rich. 1822. Helosiaceae. 1 (H. cyanensis) trop. America, West Indies: Cuba, Dominica, Martinique, St. Lucia, Trinidad

Helwingia Willd. 1806. Helwingiaceae. 4–6 eastern Himalayas (Nepal, Bhutan), India (Assam, Manipur), northern Burma, continental China, northern Vietnam, Taiwan, Ryukyu Is.

Helxine Bubani = Soleirolia Gaudich. Urticaceae

Hemandradenia Stapf. 1908. Connaraceae. 2 West and Central Africa

Hemarthria R. Br. 1810. Poaceae. 12 trop. Old World, West and East Asia

Hemenatherum Cass. = Thymophylla Lagasca. Asteraceae

Hemerocallis L. 1753. Hemerocallidaceae. 15 Central Europe (1), Mediterranean, Caucasus, West and Southwest Asia eastward to China and Japan; Russia (West and East Siberia, Far East)

Hemiandra R. Br. 1810. Lamiaceae. 7 southwestern Australia

Hemiangium A. C. Sm. = Semialarium N. Hallé. Celastraceae

Hemianthus Nutt. 1817 (Micranthemum Michx.). Scrophulariaceae. 10 trop. America, West Indies

Hemiarrhena Benth. = Lindernia All. Scrophulariaceae

Hemiarthron (Eichler) Tiegh. = Psittacanthus Mart. Loranthaceae

Hemibaccharis S. F. Blake = Archibaccharis Heering. Asteraceae

Hemiboea C. B. Clarke. 1888. Gesneriaceae. 21 China, Taiwan, Japan, Indochina

Hemiboeopsis W. T. Wang. 1984. Gesneriaceae. 1 southern China (Yunnan)

Hemicarpha Nees. 1834 (Lipocarpha R. Br.). Cyperaceae. 6 trop., subtrop., and temp. regions

Hemichaena Benth. 1841. Scrophulariaceae. 1 Central America

Hemichlaena Schrad. = Ficinia Schrad. Cyperaceae

Hemichroa R. Br. 1810. Chenopodiaceae. 3 Australia

Hemicicca Baill. 1858 (Phyllanthus L.). Euphorbiaceae. 1 Japan

Hemicrambe Webb. 1851. Brassicaceae. 1 Morocco

Hemicrepidospermum Swart = Crepidospermum Hook. f. Burseraceae

Hemidesmus R. Br. 1810. Asclepiadaceae. 1 southern India, Southeast Asia, Malesia

Hemidiodia K. Schum. = Diodia L. Rubiaceae

Hemieva Raf. = Suksdorfia A. Gray. Saxifragaceae

Hemifuchsia Herrera = Fuchsia L. Onagraceae

Hemigenia R. Br. 1810. Lamiaceae. 37 Australia

Hemigraphis Nees. 1847. Acanthaceae. 100 South and Southeast Asia, Malesia, trop. Australia, New Caledonia, Melanesia, Polynesia

Hemilophia Franch. 1889. Brassicaceae. 2 southwestern China

Hemimeris L. f. 1782. Scrophulariaceae. 4 South Africa

Hemimunroa (Parodi) Parodi. 1937 (Munroa Torr.). Poaceae. 1 southern Andes

Hemiorchis Kurz. 1837. Zingiberaceae. 3 Himalayas (Nepal, Sikkim), India, Malesia

Hemipappus K. Koch = Tanacetum L. Asteraceae

Hemiperis Frapp. ex Cordem. = Habenaria Willd. Orchidaceae

Hemiphora F. Muell. 1882. Chloanthaceae. 1 western Australia

Hemiphragma Wall. 1822. Scrophulariaceae. 1 Himalayas, Assam, Burma, western and central China, Taiwan, Philippines

Hemiphylacus S. Watson. 1883. Hyacinthaceae. 1 northern Mexico

Hemipilia Lindl. 1835. Orchidaceae. 13 Himalayas, southwestern China, Thailand

Hemipogon Decne. 1844. Asclepiadaceae. 10 South America

Hemiptelea Planch. 1872 (Zelkova Spach). Ulmaceae. 1 China, Korea

Hemiscleria Lindl. 1853 (Diothonea Lindl.). Orchidaceae. 1 Ecuador, Peru

Hemiscolopia Slooten. 1925. Flacourtiaceae. 1 Indochina, Sumatra, Banka I., Java

Hemisiphonia Urb. 1909. Scrophulariaceae. 1 West Indies

Hemisorghum C. E. Hubb. ex Bor. 1960. Poaceae. 2 trop. Asia

Hemisphaera Kolak. = Campanula L. Campanulaceae

Hemisphaerocarya Brand = Oreocarya Greene. Boraginaceae

Hemistepta Bunge nom. non vite public. = Hemisteptia Bunge ex Fisch. et C. A. Mey. Asteraceae

Hemisteptia Bunge ex Fisch. et C. A. Mey. 1835 (Saussurea DC.). Asteraceae. 3 Himalayas, East Asia

Hemistylus Benth. 1843. Urticaceae. 4 trop. South America

Hemithrinax Hook. f. = Thrinax Sw. Arecaceae

Hemitomes A. Gray. 1857. Ericaceae. 1 western North America from British Columbia to California

Hemitria Raf. = Phthirusa Mart. Loranthaceae

Hemizonella (A. Gray) A. Gray = Madia Molina. Asteraceae

Hemizonia DC. 1836. Asteraceae. 31 southwestern U.S. and Baja California

Hemizygia (Benth.) Briq. 1897 (Syncolostemon E. Mey. ex Benth.). Lamiaceae. 35 trop. and South (28) Africa

Hemsleya Cogn. ex F. B. Forbes et Hemsl. 1889. Cucurbitaceae. 30 China, Himalayas

Hemsleyana Kuntze. 1891 (Thryallis Mart.). Malpighiaceae. 3 trop. South America

Henlea Karst. = Rustia Klotzsch. Rubiaceae

Henleophytum Karst. 1861. Malpighiaceae. 1 Cuba

Hennecartia Poiss. 1885. Monimiaceae. 1 southern Brazil, Paraguay, northeastern Argentina

Henonia Moq. 1849. Amaranthaceae. 1 Madagascar

Henoonia Griseb. 1866. Goetzeaceae. 3 Cuba

Henophyton Coss. et Durieu. 1856. Brassicaceae. 1 Morocco, Algeria, Tunisia, Libya

Henrardia C. E. Hubb. 1946. Poaceae. 2: H. persica—Asia Minor, Transcaucasia, Iraq, Middle Asia, Afghanistan, Pakistan, Himalayas; H. pubescens—Syria, Iraq, Iran, Turkmenistan

Henricia Cass. = Psiadia Jacq. Asteraceae

Henricksonia B. L. Turner. 1977. Asteraceae. 1 Mexico

Henriettea DC. 1828. Melastomataceae. 15 trop. America from Mexico to Amazonia

Henriettella Naudin. 1852. Melastomataceae. 40 trop. America from Guatemala to Bolivia, West Indies

Henriquezia Spruce ex Benth. 1854. Rubiaceae. 3 Colombia, Venezuela, Guyana, Brazil

Henrya Nees = Tetramerium Nees. Acanthaceae

Henryettana Brand = Antiotrema Hand.-Mazz. Boraginaceae

Hensmania W. Fitzg. 1903. Asphodelaceae. 3 southwestern Australia

Hepatica Hill. 1753 (Anemone L.). Ranunculaceae. 10 temp. Northern Hemisphere, northern Mediterranean from Spain to Balkan Peninsula; Middle Asia

Heppiella Regel. 1853 (Achimenes R. Br.). Gesneriaceae. 4 Andean Colombia, Venezuela, Ecuador, Peru

Heptacodium Rehder. 1916. Caprifoliaceae. 1 (H. miconioides) central and eastern China

Heptanthus Griseb. 1866. Asteraceae. 7 Cuba

Heptaptera Margot et Reut. 1839. Apiaceae. 6 South Europe, eastern Mediterranean, Iraq, Iran

Heracleum L. 1753. Apiaceae. c. 70 temp. Eurasia, North, Northeast, and trop. Africa, North America (1)

Herbabdiopsis Meisn. = Hernandia L. Hernandiaceae

Herbertia Sweet. 1827. Iridaceae. 6–8 temp. South America

Herbstia Sohmer. 1977. Amaranthaceae. 1 Peru, Bolivia, Paraguay, Argentina

Herderia Cass. 1829. Asteraceae. 1 trop. West Africa

Hereroa (Schwantes) Dinter et Schwantes. 1927. Aizoaceae. 34 southern Namibia, Lesotho, South Africa

Hericinia Fourr. = Ranunculus L. Ranunculaceae

Herissantia Medik. 1789 (Abutilon Hill). Malvaceae. 4: 1 (H. crispa) from southern U.S. and northern Mexico through Central America to Bolivia and Argentina; also in trop. Asia and Australia; H. dressleri—Mexico; Brazil (2),

Heritiera Dryand. ex Aiton. 1789. Sterculiaceae. c. 30 trop. Africa, Madagascar, trop. Asia, Malesia, trop. Australia, trop. Oceania

Hermannia L. 1753. Sterculiaccac. c. 300 tropics and subtropics, esp. South Africa

Hermas L. 1771. Apiaceae. 8–9 South Africa

Hermbstaedtia Rchb. 1828. Amaranthaceae. 15 Kenya (1, H. gregoryi), Namibia, and South Africa

Hermidium S. Watson = Mirabilis L. Nyctaginaceae

Herminiera Guill. et Perr. = Aeschynomene L. Fabaceae

Herminium L. 1758. Orchidaceae. 40 temp. Eurasia, Tibet, Himalayas, India, Burma, Indochina, Java, Philippines; H. monorchis—European part of Russia, Middle Asia, southern Siberia, and Far East

Hermodactylus Mill. 1754 (Iris L.). Iridaceae. 1 South Europe from France to Greece, Corsica, Sicily, and islands of Aegean Sea

Hernandia L. 1753. Hernandiaceae. 22 pantropics, but mostly in the Indo-Pacific region

Herniaria L. 1753. Caryophyllaceae. c. 45 Europe, Mediterranean, South Africa, West (incl. Caucasus) and Middle Asia, West Siberia, Afghanistan, Mongolia, China, northwestern India; Andean South America

Herodotia Urb. et Ekman. 1926. Asteraceae. 2 Haiti

Herpestis Gaertn. f. = Bacopa Aubl. Scrophulariaceae

Herpetacanthus Nees. 1847. Acanthaceae. 10 Costa Rica, Panama, trop. South America

Herpetospermum Wall. ex Hook. f. 1867. Cucurbitaceae. 2 Himalayas, southern China

Herpolirion Hook. f. 1853. Asphodelaceae. 1 Australia (New South Wales, Victoria, Tasmania), New Zealand

Herpysma Lindl. 1833. Orchidaceae. 2 Tibet, Nepal, northeastern India, northern Burma, Indochina, China, Sumatra, Philippines

Herpyza Sauvalle. 1868. Fabaceae. 1 Cuba, Puerto Rico

Herrania Goudot. 1844. Sterculiaceae. 20 trop. South America

Herrea Schwantes = Conicosia N. E. Br. Aizoaceae

Herreanthus Schwantes. 1928. Aizoaceae. 1 western South Africa

Herreria Ruiz et Pav. 1794. Herreriaceae. 7 subtrop. and trop. South America

Herreriopsis H. Perrier. 1934. Herreriaceae. 2 Madagascar

Herrickia Wooton et Standl. = Aster L. Asteraceae

Herschelia Lindl. 1838. Orchidaceae. 16 trop. and South Africa

Herschelianthe Rauschert = Herschelia Lindl. Orchidaceae

Herschellia T. E. Bowdich ex Rchb. = Physalis L. Solanaceae

Hersperethusa M. Roem. = Naringi Adans. Rutaceae

Hertia Less. = Othonna L. Asteraceae

Herya Cordem. 1895 (Pleurostylia Wight et Arn.). Celastraceae. 1 Réunion I.

Hesiodia Moench = Sideritis L. Lamiaceae

Hesperaloe Engelm. 1871. Agavaceae. 3 southern U.S., Mexico

Hesperantha Ker-Gawl. 1804. Iridaceae. 62 trop. (south from Sakhara) and South Africa

Hesperelaea A. Gray. 1876. Oleaceae. 1 northwestern Mexico

Hesperevax (A. Gray) A. Gray. 1868. Asteraceae. 3 southwestern U.S. (western California, southwestern Oregon)

Hesperhodos Cockerell = Rosa L. Rosaceae

Hesperidanthus (Rob.) Rydb. = Schoenocrambe Greene. Brassicaceae

Hesperis L. 1753. Brassicaceae. 30 Europe, Mediterranean, West (incl. Caucasus) and Middle Asia, West and East Siberia, western China, Mongolia

Hesperocallis A. Gray. 1868. Hesperocallidaceae. 1 deserts of California and Arizona

Hesperochiron S. Watson. 1871. Hydrophyllaceae. 2 southwestern North America from Washington to California, Nevada, Utah, and Arizona

Hesperochloa (Piper) Rydb. 1912 (Festuca L.). Poaceae. 1 western U.S.: Oregon and California to Montana, Nebraska, and Colorado

Hesperocnide Torr. 1857. Urticaceae. 2 Hawaiian Is. (H. sandwicensis), California (H. tenella)

Hesperodoria Greene. 1906. Asteraceae. 2 North America

Hesperogenia J. M. Coult. et Rose = Thauschia Schltdl. Apiaceae

Hesperogreigia Skottsb. = Greigia Regel. Bromeliaceae

Hesperolaburnum Maire. 1949. Fabaceae. 1 Morocco

Hesperolinon (A. Gray) Small. 1907. Linaceae. 13 western coastal U.S. from Oregon to California, Mexico (1)

Hesperomannia A. Gray. 1865. Asteraceae. 3 Hawaiian Is.

Hesperomecon Greene. 1903 (Meconella Nutt.). Papaveraceae. 1 western North America

Hesperomeles Lindl. 1837. Rosaceae. 11 Costa Rica, Colombia, Ecuador, Peru, Bolivia

Hesperonia Standl. = Mirabilis L. Nyctaginaceae

Hesperopeuce (Engelm.) Lemmon. 1890. Pinaceae. 1 western North America from southern Alaska to central California

Hesperoscordum Lindl. 1830 (Brodiaea Sm.). Alliaceae. 1–2 western North America

Hesperoseris Skottsb. = Dendroseris D. Don. Asteraceae

Hesperothamnus Brandegee = Millettia Wight et Arn. Fabaceae

Hesperoxiphion Baker. 1877. Iridaceae. 5 Andes

Hesperoyucca Baker = Yucca L. Agavaceae

Hesperozygis Epling. 1936. Lamiaceae. 8 trop. America from Mexico to Brazil

Hessea Herb. 1837. Amaryllidaceae. 16 South Africa

Hetaeria Blume. 1828. Orchidaceae. c. 70 trop. Africa (1, H. heterosepala—Liberia, Ivory Coast, Cameroun, Sao Tomé, Zaire, Tanzania); trop. Asia from India to Taiwan and Ryukyu Is., Malesia from Malay Peninsula to New Guinea, Melanesia to Fiji, Tonga, Niue, and Samoa; Brazil

Heterachne Benth. 1877. Poaceae. 2 northern Australia

Heteracia Fisch. et C. A. Mey. 1835. Asteraceae. 2 Crimea, Caucasus, Asia Minor, Middle Asia, Iran, China

Heteradelphia Lindau. 1893. Acanthaceae. 1 West Africa

Heteranthelium Hochst. ex Jaub. et Spach. 1851. Poaceae. 1 Northeast Africa, Caucasus, Asia Minor, Syria, Israel, Jordan, Iraq, Iran, Afghanistan, Pakistan

Heteranthemis Schott. 1816. Asteraceae. 1 Southwest Europe, North Africa

Heteranthera Ruiz et Pav. 1794. Pontederiaceae. 10 trop. and subtrop. America, Africa

Heteranthia Nees et Mart. 1823. Solanaceae. 1 Brazil

Heteranthoecia Stapf. 1911. Poaceae. 1 trop. Africa

Heteraspidia Rizzini = Justicia L. Acanthaceae

Heteroaridarum M. Hotta. 1976. Araceae. 1 Borneo

Heteroarisaema Nakai = Arisaema Mart. Araceae

Heterocalycium Rauschert = Xylophragma Sprague. Bignoniaceae

Heterocalymnantha Domin = Sauropus Blume. Euphorbiaceae

Heterocarpha Stapf et C. E. Hubb. 1929 (Drake-Brockmania Stapf). Poaceae. 1 trop. East Africa

Heterocarpus Phil. = Cardamine L. Brassicaceae

Heterocaryum A. DC. 1846 (Lappula Moench). Boraginaceae. 7 southeastern European part of Russia, Caucasus, Middle Asia, Iran, Afghanistan, China

Heterocentron Hook. et Arn. 1838. Melastomataceae. 27 Mexico, Central America

Heterochaenia A. DC. 1839. Campanulaceae. 3 Mascarene Is.

Heterochaeta DC. 1836. Asteraceae. 2 Middle Asia

Heterochiton Graebn. et Mattfeld = Herniaria L. Caryophyllaceae

Heterocodon Nutt. 1843. Campanulaceae. 1 western North America

Heterocoma DC. 1810. Asteraceae. 1 Brazil

Heterocondylus R. M. King et H. Rob. 1972. Asteraceae. 12 trop. America from Honduras to Bolivia and Paraguay, but chiefly Brazil

Heterocypsela H. Rob. 1979. Asteraceae. 1 Brazil

Heterodendrum Desf. = Alectryon Gaertn. Sapindaceae

Heteroderis (Bunge) Boiss. 1875. Asteraceae. 2–6 eastern Mediterranean, West Asia eastward to Middle Asia (2) and Beluchistan

Heterodon Meisn. = Brunia Lam. Bruniaceae

Heterodraba Greene. 1885. Brassicaceae. 2 western U.S., Baja California

Heterogaura Rothr. = Clarkia Pursh. Onagraceae

Heterolamium C. Y. Wu. 1965. Lamiaceae. 1 China

Heterolepis Cass. 1820. Asteraceae. 3 South Africa

Heterolobium Peter = Gonatopus Hook. f. ex Engl. Araceae

Heteromeles M. Roem. 1847 (Photinia Lindl.). Rosaceae. 1 California, Baja California

Heteromera Pomel. 1874. Asteraceae. 1 North Africa

Heteromma Benth. 1873. Asteraceae. 3 South Africa

Heteromorpha Cham. et Schltdl. 1826. Apiaceae. 8–10 Northeast, trop., and South Africa, Madagascar, Arabian Peninsula (Yemen)

Heteromyrtus Blume = Blepharocalyx O. Berg. Myrtaceae

Heteropanax Seem. 1866. Araliaceae. 5 Himalayas, India, Andaman Is., Indochina, southern China, Taiwan

Heteropappus Less. 1832. Asteraceae. 12 temp. and East Asia, esp. (8) Middle Asia, Siberia, and Far East

Heteropetalum Benth. 1861. Annonaceae. 2 Venezuela, Brazil

Heteropholis C. E. Hubb. 1956. Poaceae. 5 trop. Old World from Africa to Philippines and Australia

Heterophragma DC. 1838. Bignoniaceae. 2 India, Burma, Thailand, Cambodia, Laos

Heterophyllaea Hook. f. 1873. Rubiaceae. 4 Bolivia, Argentina

Heteroplexis C. C. Chang. 1937. Asteraceae. 2 eastern China

Heteropogon Pers. 1807. Poaceae. 8 tropics and subtropics

Heteropsis Kunth. 1841. Araceae. 4–5 Central and trop. South America

Heteropterys Kunth. 1822. Malpighiaceae. 120 trop. America from Mexico to Argentina, West Indies, trop. West Africa (1)

Heteroptilis E. Mey. ex Meisn. = Dasispermum Neck. ex Raf. Apiaceae

Heteropyxis Harv. 1836. Heteropyxidaceae. 3 Southeast Africa (Zimbabwe, Mozambique to Natal)

Heterorhachis Sch. Bip. ex Walp. 1846. Asteraceae. 1 South Africa

Heterosmilax Kunth. 1850. Smilacaceae. 11 India (Assam and Manipur), Bangladesh, Burma, continental China (7) and Hainan (1, H. micrandra), Thailand, Vietnam, Cambodia, Taiwan (2), southern Japan, incl. Ryukyu Is., western Malesia (1, H. borneensis): Sumatra, Malay Peninsula, Java, Borneo

Heterospathe Scheff. 1876. Arecaceae. 32 Philippines, Moluccas, New Guinea (16), Palau Is., Mariana and Solomon Is.

Heterosperma Cav. 1796. Asteraceae. 5 America from southwestern U.S. to Argentina

Heterostachys Ung.-Sternb. 1876. Chenopodiaceae. 2 Central and South America

Heterostemma Wight et Arn. 1834. Asclepiadaceae. 30 trop. Asia from India to China and Indochina, Malesia to New Guinea, Australia, New Caledonia, Fiji, Samoa, Tonga

Heterostemon Desf. 1818. Fabaceae. 7 trop. America, esp. Amazonia

Heterothalamus Less. 1830. Asteraceae. 8 South America

Heterotheca Cass. 1817. Asteraceae. 15 southern U.S., Mexico

Heterothrix (B. L. Rob.) Rydb. = Pennellia Nieuwl. Brassicaceae

Heterotis Benth. 1849 (Dissotis Benth.). Melastomataceae. 3 West Africa

Heterotoma Zucc. 1832. Lobeliaceae. 1 America from western Mexico to northern Costa Rica

Heterotrichum DC. 1828. Melastomataceae. 15 trop. America, West Indies

Heterotristicha Tobl. 1953. Podostemaceae. 1 Uruguay

Heterotropa C. Morren et Decne. = Asarum L. Aristololochiaceae

Heterozeuxine Hashim. = Zeuxine Lindl. Orchidaceae

Heterozostera (Setch.) Hartog. 1970. Zosteraceae. 1 coastal temp. Australia, Tasmania, Chile

Heuchera L. 1753. Saxifragaceae. 55 North America

Hevea Aubl. 1775. Euphorbiaceae. 12–17 Amazonia

Hewardia Hook. = Isophysis T. Moore. Iridaceae

Hewittia Wight et Arn. 1837. Convolvulaceae. 2 trop. Africa, trop. and East Asia, ? Jamaica

Hexachlamys O. Berg. 1856 (Eugenia L.). Myrtaceae. 3 temp. South America

Hexacyrtis Dinter. 1932. Colchicaceae. 1 Namibia

Hexadesmia Brongn. = Scaphyglottis Poepp. et Endl. Orchidaceae

Hexaglottis Vent. 1808. Iridaceae. 6 South Africa

Hexalectris Raf. 1825. Orchidaceae. 3 southern U.S., Mexico

Hexalobus A. DC. 1832. Annonaceae. 5 trop. and South Africa, Madagascar

Hexaneurocarpon Dop = Fernandoa Welw. ex Seem. Bignoniaceae

Hexapora Hook. f. 1886. Lauraceae. 1 Malay Peninsula

Hexaptera Hook. 1830. Brassicaceae. 13 temp. South America

Hexapterella Urb. 1903. Burmanniaceae. 1 Trinidad, northern South America

Hexaspermum Domin = Phyllanthus L. Euphorbiaceae

Hexaspora C. T. White. 1933. Celastraceae. 1 Australia (northern Queensland)

Hexastemon Klotzsch = Eremia D. Don. Ericaceae

Hexastylis Raf. = Asarum L. Aristolochiaceae

Hexatheca C. B. Clarke. 1883. Gesneriaceae. 3 Borneo

Hexisea Lindl. 1834. Orchidaceae. 5–6 trop. America from Mexico to Brazil, West Indies

Hexopetion Burret = Astrocaryum G. Mey. Arecaceae

Hexuris Miers = Peltophyllum Gardner. Triuridaceae

Heylandia DC. 1825 (or = Crotalaria L.). Fabaceae. 1 India, Sri Lanka

Heynea Roxb. ex Sims = Trichilia P. Browne. Meliaceae

Heynella Backer. 1950. Asclepiadaceae. 1 Java

Heywoodia T. R. Sim. 1907. Euphorbiaceae. 1 trop. East (Uganda, Kenya, Tanzania) and South (Natal, Transkei) Africa

Heywoodiella Svent. et Bramwell = Hypochaeris L. Asteraceae

Hibbertia Andrews. 1800. Dilleniaceae. 123 Madagascar, Aru Is., New Guinea, Australia, New Caledonia, Fiji

Hibiscadelphus Rock. 1911. Malvaceae. 6 Hawaiian Is.

Hibiscus L. 1753. Malvaceae. c. 200 warm–temp., subtrop., and trop. regions

Hickelia A. Camus. 1924. Poaceae. 1 Madagascar

Hickenia Lillo. 1919 (or Oxypetalum R. Br.). Asclepiadaceae. 1 Argentina

Hicksbeachia F. Muell. 1883. Proteaceae. 3 Australia (Queensland, New South Wales)

Hicoria Raf. = Carya Nutt. Juglandaceae

Hidalgoa La Llave ex Lex. 1824. Asteraceae. 5 Mexico, Central America, Colombia, Ecuador, Peru

Hieracium L. 1753. Asteraceae. c. 250 (groups of macrospecies) temp. regions and montane tropics (excl. Australasia)

Hieris Steenis. 1928. Bignoniaceae. 1 Malay Peninsula

Hiernia S. Moore. 1880. Scrophulariaceae. 1 Namibia

Hierobotana Briq. 1895. Verbenaceae. 1 Ecuador, Peru

Hierochloe R. Br. 1810. Poaceae. 30 temp. regions and mountainous tropics

Hieronima Allemao = Hyeronima Allemao. Euphorbiaceae

Hieronymiella Pax. 1890. Amaryllidaceae. 1 Argentina

Hieronymusia Engl. = Suksdorfia A. Gray. Saxifragaceae

Hilaria Kunth. 1816. Poaceae. 9 southern U.S., Mexico, Central America

Hilariophyton Pichon = Paragonia Bureau. Bignoniaceae

Hildebrandtia Vatke ex Braun. 1876. Convolvulaceae. 9 Africa

Hildegardia Schott et Endl. 1832. Sterculiaceae. 12 Cuba (1), trop. Africa (3), Madagascar (3), India (1), China (1), Philippines (1), Sumbawa I. (1), northern Australia (1)

Hildewintera F. Ritter = Cleistocactus Lem. Cactaceae

Hillebrandia Oliv. 1866. Begoniaceae. 1 Hawaiian Is.

Hilleria Vell. 1825. Petiveriaceae. 5 Central and trop. South America (4); H. latifolia—also in trop. Africa from Liberia to Ethiopia, Uganda and Kenya and south to Angola and Mozambique; Madagascar, Mascarene Is.

Hillia Jacq. 1760. Rubiaceae. 21 trop. America, West Indies

Hilliardia B. Nord. 1987. Asteraceae. 1 South Africa

Hilliella (O. E. Schulz) Y. H. Zhang et H. W. Li = Cochlearia L. Brassicaceae

Himalayacalamus Keng. f. = Thamnocalamus Munro. Poaceae

Himalrandia Yamaz. 1970. Rubiaceae. 1–3 South and Southeast Asia

Himantandra F. Muell. ex Diels = Galbulimima F. M. Bailey. Himantandraceae

Himantochilus T. Anderson ex Benth. = Anisotes Nees. Acanthaceae

Himantoglossum W. D. J. Koch. 1837. Orchidaceae. 4 Europe, North Africa, eastern Mediterranean; Crimea, Caucasus

Himantostemma A. Gray. 1885. Asclepiadaceae. 1 North America

Himatanthus Willd. ex Roem. et Schult. = Aspidosperma Mart. et Zucc. Apocynaceae

Hindsia Benth. ex Lindl. 1844. Rubiaceae. 9 trop. South America

Hinterhubera Sch. Bip. 1857. Asteraceae. 8 Andes of Colombia and Venezuela

Hintonella Ames. 1938. Orchidaceae. 1 Mexico

Hintonia Bullock. 1935. Rubiaceae. 4 Mexico, Central America

Hionanthera A. Fern. et Diniz. 1955. Lythraceae. 4 trop. East Africa

Hippeastrum Herb. 1821. Amaryllidaceae. 76 trop. and subtrop. America, West Africa (1)

Hippeophyllum Schltr. 1905. Orchidaceae. 5 southern China, Malesia

Hippia L. 1771. Asteraceae. 8 South Africa

Hippobroma G. Don. 1834 (Solenopsis C. Presl). Lobeliaceae. 1 Florida, Mexico, Central and South America, West Indies

Hippobromus Eckl. et Zeyh. 1836. Sapindaceae. 1 South Africa

Hippocastanum Hill = Aesculus L. Hippocastanaceae

Hippocratea L. 1753. Celastraceae. c. 100 tropics and subtropics

Hippocrepis L. 1753. Fabaceae. 21 Cape Verde Is., Mediterranean, Sudan, Asia Minor, Iran, Crimea, Caucasus

Hippodamia Decne. = Solenophora Benth. Gesneriaceae

Hippoglossum Breda = Bulbophyllum Thouars. Orchidaceae

Hippolytia Poljakov. 1957 (Tanacetum L.). Asteraceae. 20 Middle Asia (4), northwestern China and Tibet

Hippomane L. 1753. Euphorbiaceae. 5 Mexico, West Indies

Hippomarathrum Hoffmanns. et Link = Cachrys L. Apiaceae

Hippophae L. 1753. Elaeagnaceae. 4 temp. Eurasia; H. rhamnoides—southern European part of Russia, Caucasus, Middle Asia, southern Siberia

Hippotis Ruiz et Pav. 1794. Rubiaceae. 12 Panama, Colombia, Venezuela, Ecuador, Peru

Hippuris L. 1753. Hippuridaceae. 1 (H. vulgaris) cold and temp. Northern Hemisphere

Hiptage Gaertn. 1790. Malpighiaceae. 20–30 Mauritius, trop. Asia, Malesia, Melanesia to Fiji

Hiraea Jacq. 1760. Malpighiaceae. 45 trop. America from Mexico to Brazil and northern Argentina; West Indies

Hirpicium Cass. 1820. Asteraceae. 12 trop. East and South Africa

Hirschfeldia Moench. 1794. Brassicaceae. 2 West Europe, Mediterranean; H. incana—Crimea, Caucasus; Iraq, Iran, Socotra

Hirschia Baker = Iphiona Cass. Asteraceae

Hirtella L. 1753. Chrysobalanaceae. 103 Central and trop. South America, West Indies, trop. East Africa and Madagascar (1)

Hirtzia Dodson. 1981. Orchidaceae. 1 Ecuador

Hispaniella Braem. 1980 (Oncidium Sw.). Orchidaceae. 1 Haiti

Hispidella Barnadez ex Lam. 1789. Asteraceae. 1 Pyrenees

Hitchcockella A. Camus. 1925. Poaceae. 1 Madagascar

Hitchenia Wall. 1835. Zingiberaceae. 3 India, Burma, Malay Peninsula

Hitcheniopsis (Baker) Ridl. = Scaphochlamys Baker. Zingiberaceae

Hitoa Nadeaud = Ixora L. Rubiaceae

Hjaltalinia A. Löve et D. Löve = Sedum L. Crassulaceae

Hladnikia Rchb. 1831. Apiaceae. 1 Slovenia

Hochreutinera Krapov. 1970. Malvaceae. 2: H. amplexifolia—Mexico, Guatemala, El Salvador; H. hasslerana—temp. South America (Paraguay, Argentina)

Hochstetteria DC. = Dicoma Cass. Asteraceae

Hockinia Gardner. 1843. Gentianaceae. 1 eastern Brazil

Hodgkinsonia F. Muell. 1861. Rubiaceae. 2 eastern Australia

Hodgsonia Hook. f. et Thomson. 1854. Cucurbitaceae. 2 southwestern China, Indo-Malesia

Hodgsoniola F. Muell. 1861. Asphodelaceae. 1 southwestern Australia

Hoehnea Epling. 1939. Lamiaceae. 4 southern Brazil, northern Argentina

Hoehneella Ruschi. 1945. Orchidaceae. 3 Brazil

Hoehnelia Schweinf. = Ethulia L. f. Asteraceae

Hoehnephytum Cabrera. 1950. Asteraceae. 2 Brazil

Hoffmannia Sw. 1788. Rubiaceae. 125 America from Mexico to Argentina

Hoffmanniella Schltr. ex Lawalrée. 1943. Asteraceae. 1 trop. Africa

Hoffmannseggella H. Jones = Laelia Lindl. Orchidaceae

Hoffmannseggia Cav. 1798. Fabaceae. 28 America from southwestern U.S. to Chile and Argentina, southern Africa (3)

Hofmeisterella Rchb. f. 1852. Orchidaceae. 1 Ecuador

Hofmeisteria Walp. 1846. Asteraceae. 8 Mexico

Hohenackeria Fisch. et C. A. Mey. 1836. Apiaceae. 2 Spain, Northwest Africa, Caucasus, Turkey, Iran

Hohenbergia Schult. et Schult. f. 1830. Bromeliaceae. 40 West Indies and trop. South America

Hohenbergiopsis L. B. Sm. et Read. 1976. Bromeliaceae. 1 Mexico, Guatemala, West Indies

Hoheria A. Cunn. 1839. Malvaceae. 5 New Zealand

Hoita Rydb. 1919. Fabaceae. 3 California and Baja California

Holacantha A. Gray = Castela Turpin. Simaroubaceae

Holalafia Stapf = Alafia Thouars. Apocynaceae

Holarrhena R. Br. 1810. Apocynaceae. 4 trop. Africa, Madagascar, India, Sri Lanka, Nepal, Burma, southern China, Indochina, Malesia

Holboellia Wall. 1824 (Stauntonia Wall.). Lardizabalaceae. 13 Tibet, Nepal, Bhutan, Assam, Burma, China (Yunnan, Gansu), northern Vietnam

Holcoglossum Schltr. 1919. Orchidaceae. 8 Burma, southern China (incl. Hainan), Taiwan, Indochina

Holcolemma Stapf et C. E. Hubb. 1929. Poaceae. 4 East Africa, southern India, Sri Lanka, Australia

Holcus L. 1753. Poaceae. 8 Canary Is., Europe, North and South Africa, West (incl. Caucasus), Southwest, and East (southern Russian Far East) Asia

Holigarna Buch.-Ham. ex Roxb. 1820. Anacardiaceae. 8 India (6), Malesia

Hollandaea F. Muell. 1887. Proteaceae. 2 eastern Australia

Hollermayera O. E. Schulz. 1928. Brassicaceae. 1 Chile

Hollisteria S. Watson. 1879. Polygonaceae. 1 southwestern U.S.

Hollrungia K. Schum. 1887. Passifloraceae. 1 Moluccas, New Guinea, Solomon Is.

Holmbergia Hicken. 1909. Chenopodiaceae. 1 Argentina, Paraguay, Uruguay

Holmesia P. J. Cribb = Angraecopsis Kraenzl. Orchidaceae

Holmgrenanthe Elisens. 1985 (Asarina Hill). Scrophulariaceae. 1 U.S. (California)

Holmskioldia Retz. 1791. Verbenaceae. 11 trop. Africa, Madagascar, Mascarene Is., Himalayas, India, western Malesia

Holocalyx Micheli. 1883. Fabaceae. 1 northeastern Argentina, Paraguay, southeastern Brazil

Holocarpa Baker = Pentanisia Harv. Rubiaceae

Holocarpha Greene. 1897. Asteraceae. 4 California

Holocheila (Kudo) S. Chow. 1962 (Teucrium L.). Lamiaceae. 1 southwestern China

Holocheilus Cass. 1818. Asteraceae. 6 southern Brazil, Paraguay, northern and central Argentina, Uruguay

Holochlamys Engl. 1883. Araceae. 2 New Guinea

Holodiscus (K. Koch) Maxim. 1879. Rosaceae. 8 western North and Central America, Colombia

Holographis Nees. 1847. Acanthaceae. 15 Mexico

Hologyne Pfitzer = Coelogyne Lindl. Orchidaceae

Hololachna Ehrenb. = Reaumuria L. Tamaricaceae

Hololeion Kitam. 1941. Asteraceae. 2 northeastern China, Korea, Japan (Shikoku, Kyushu)

Holopogon Kom. et Nevski = Neottia Guett. Orchidaceae

Holoptelea Planch. 1848. Ulmaceae. 2 trop. Africa (1), India (1, H. integrifolia)

Holopyxidium Ducke = Lecythis Loefl. Lecythidaceae

Holoschoenus Link = Scirpoides Seguier. Cyperaceae

Holostemma R. Br. 1810. Asclepiadaceae. 1–2 western China, Himalayas, India, Sri Lanka, Burma

Holosteum L. 1753. Caryophyllaceae, 6 temp. Eurasia, esp. (5) southern European of Russia, Caucasus, and Middle Asia; North Africa, North America (1)

Holostyla Endl. = Caelospermum Blume. Rubiaceae

Holostylis Duch. 1854. Aristolochiaceae. 1 central and southeastern Brazil, Bolivia, Paraguay

Holostylon Robyns et Lebrun. 1929 (Plectranthus L'Hér.). Lamiaceae. 4 trop. and South (1, H. baumii) Africa

Holothrix Rich. ex Lindl. 1835. Orchidaceae. 45–55 trop. and South (32) Africa, Arabian Peninsula

Holozonia Greene. 1882. Asteraceae. 1 western U.S.

Holstia Pax = Neoholstia Rauschert. Euphorbiaceae

Holstianthus Steyerm. 1986. Rubiaceae. 1 Venezuela

Holtonia Standl. = Simira Aubl. Rubiaceae

Holtzea Schindl. = Dendrolobium Sw. Orchidaceae

Holubia A. Löve et D. Löve = Gentiana L. Gentianaceae

Holubia Oliv. 1884. Pedaliaceae. 1 South Africa

Holubogentia A. Löve et D. Löve = Gentiana L. Gentianaceae

Holzneria Speta. 1982 (Chaenorrhinum [Duby] Rchb.). Scrophulariaceae. 2 Southwest Asia

Homalachne (Benth.) Kuntze = Holcus L. Poaceae

Homalanthus A. Juss. = Omalanthus A. Juss. Euphorbiaceae

Homaliopsis S. Moore. 1920. Flacourtiaceae. 1 Madagascar

Homalium Jacq. 1760. Flacourtiaceae. c. 200 tropics and subtropics

Homalocalyx F. Muell. 1857. Myrtaceae. 10 northeastern Australia

Homalocarpus Hook. et Arn. 1833. Apiaceae. 6 Chile

Homalocenchrus Miègev. = Leersia Sw. Poaceae

Homalocheilos J. K. Morton = Rabdosia (Blume) Hassk. Lamiaceae

Homalocladium (F. Muell.) L. H. Bailey. 1929. Polygonaceae. 1 New Guinea, Solomon Is., New Caledonia

Homalodiscus Bunge ex Boiss. = Ochradenus Delile. Resedaceae

Homalomena Schott. 1832. Araceae. 140 trop. Asia and South America

Homalopetalum Rolfe. 1896. Orchidaceae. 1 Jamaica

Homalosciadium Domin. 1908. Apiaceae. 2 southwestern Australia

Homalospermum Schauer. 1843. Myrtaceae. 1 Australia

Homeria Vent. 1808. Iridaceae. 33 South-West and South Africa

Homochroma DC. = Mairia Nees. Asteraceae

Homocodon D. Y. Hong. 1980 (Heterocodon Nutt.). Campanulaceae. 1 southwestern China

Homoglossum Salisb. = Gladiolus L. Iridaceae

Homognaphalium Kirp. 1950 (Gnaphalium L.). Asteraceae. 1 North Africa

Homogyne Cass. 1816. Asteraceae. 3 mountainous Europe

Homoiachne Pilg. 1949 (Deschampsia P. Beauv.). Poaceae. 1 Spain

Homolepis Chase. 1911. Poaceae. 5 trop. America: Mexico, Belize, Guatemala, Honduras, Costa Rica, Panama, Colombia, Venezuela, Guyana, French Guiana,

Surinam, Ecuador, Peru, Bolivia, Brazil, Argentina, Paraguay; West Indies; 1 (H. glutinosa) Cuba, Haiti, Jamaica, Puerto Rico

Homollea Arènes. 1960. Rubiaceae. 3 Madagascar

Homolliella Arènes. 1960. Rubiaceae. 1 Madagascar

Homonoia Lour. 1790. Euphorbiaceae. 2 southern China, India, Southeast Asia, Malesia

Homopholis C. E. Hubb. 1934. Poaceae. 1 northeastern Australia

Homopogon Stapf. 1908 (Trachypogon Nees). Poaceae. 1 trop. Africa: Nigeria, Cameroun, Central African Rep., Zaire, Burundi, Zambia, Tanzania

Homoranthus A. Cunn. ex Schauer. 1836. Myrtaceae. 7 eastern Australia

Homozeugos Stapf. 1915. Poaceae. 5 trop. Africa

Honkenya Ehrh. 1788. Caryophyllaceae. 3 cold and temp. Northern Hemisphere, southern Patagonia; Russia (3, northern European part, Arctica, and Far East)

Hoodia Sweet ex Decne. 1844. Asclepiadaceae. 10 trop. South-West and South Africa

Hoodiopsis Luckhoff. 1933. Asclepiadaceae. 1 Namibia

Hookerochloa E. B. Alexeev. 1985. Poaceae. 1 Australia

Hopea Roxb. 1811. Dipterocarpaceae. 110 India, Sri Lanka, southern China (5), Indochina, Malesia

Hopkinsia W. Fitzg. 1904. Restionaceae. 2 western Australia

Hoplestigma Pierre. 1899. Hoplestigmataceae. 2 trop. West Africa from Cameroun to Gabon

Hoplophyllum DC. 1836. Asteraceae. 2 South Africa

Hoppea Willd. 1801. Gentianaceae. 2 India, Sri Lanka, Burma

Hoppia Nees = Bisboeckelera Kuntze. Cyperaceae

Horaninovia Fisch. et C. A. Mey. 1841. Chenopodiaceae. 7 Middle Asia (4), Iran, Afghanistan, northwestern China

Hordelymus (Jessen) Jessen. 1885. Poaceae. 1 Europe, North Africa; Russia—European part to Caucasus

Hordeum L. 1753. Poaceae. c. 40 temp. and subtrop. regions, mountainous tropics

Horichia Jenny. 1981 (Polycycnis Rchb. f.). Orchidaceae. 1 Panama

Horkelia Cham. et Schltdl. 1827. Rosaceae. 18 western North America

Horkeliella (Rydb.) Rydb. 1908 (Horkelia Cham. et Schltdl.). Rosaceae. 3 North America

Hormathophylla Cullen et T. R. Dudley = Alyssum L. Brassicaceae

Hormidium (Lindl.) Heynh. 1841. Orchidaceae. 7 North, Central, and trop. South America, West Indies

Horminum L. 1753. Lamiaceae. 1 mountainous South Europe

Hormocalyx Gleason = Myrmidone Mart. Melastomataceae

Hormuzakia Gusul. = Anchusa L. Boraginaceae

Hornea Baker. 1877. Sapindaceae. 1 Mauritius

Hornschuchia Nees. 1821. Annonaceae. 4 eastern Brazil

Hornstedtia Retz. 1791. Zingiberaceae. 24 northeastern India, southern China, Indochina, Malesia, northeastern Australia, Solomon Is., Vanuatu

Hornungia Rchb. 1842. Brassicaceae. 2: H. petraea—West and Central Europe, extending to Sweden, Estonia, and Ukraine; Mediterranean; Crimea; H. aragonensis—northeastern Spain

Horridocactus Backeb. = Neoporteria Britton et Rose. Cactaceae

Horsfieldia Willd. 1806. Myristicaceae. 30 India, Sri Lanka, Andaman Is., southern China, Indochina, Malesia, Solomon Is., northern Australia

Horsfordia A. Gray. 1887. Malvaceae. 4 southwestern U.S. (southern Califronia, Arizona), Mexico (Baja California, Sonora)

Horstrissea Egli, Gerstb., Greuter et Risse. 1990 (Scaliferia DC.). Apiaceae. 1 (H. dolinicola) Crete

Horta Vell. = Clavija Ruiz et Pav. Theophrastaceae

Hortensia Comm. = Hydrangea L. Hydrangeaceae

Hortia Vand. 1788. Rutaceae. 8 Panama, trop. South America, esp. Brazil

Hortonia Wight. 1838. Monimiaceae. 2 Sri Lanka

Horvatia Garay. 1977. Orchidaceae. 1 Ecuador

Horwoodia Turrill. 1939. Brassicaceae. 1 Arabian Peninsula, Iraq

Hosackia Benth. ex Lindl. = Lotus L. Fabaceae

Hosea Dennst. = Symplocos Jacq. Symplocaceae

Hosea Ridl. 1908. Verbenaceae. 1 Borneo

Hoseanthus Merr. = Hosea Ridl. Verbenaceae

Hoshiarpuria Hajra, P. Daniel et Philcox = Rotala L. Lythraceae

Hosiea Hemsl. et E. H. Wilson. 1906. Icacinaceae. 2 western and central China, Japan

Hoslundia Vahl. 1804. Lamiaceae. 1 (H. opposita) trop. and South Africa from Senegal, Sudan, and Ethiopia to Namibia and Natal

Hosta Tratt. 1814. Hostaceae. c. 40 China, Korea, Japan, Russian Far East (4)

Hottarum Bognor et Nicolson. 1978. Araceae. 3 Borneo

Hottea Urb. 1929. Myrtaceae. 6 Haiti

Hottonia L. 1753. Primulaceae. 2 Europe and Asia Minor (H. palustris), North America (H. inflata)

Houlletia Brongn. 1841. Orchidaceae. 12 trop. America from Guatemala to Peru, Bolivia, and Brazil

Houssayanthus Hunz. 1978. Sapindaceae. 3 Venezuela, Argentina

Houstonia L. = Hedyotes L. Rubiaceae

Houttuynia Thunb. 1784. Saururaceae. 1 Himalayas from Punjab to Assam, Thailand, southern China, Taiwan, Japan, Java

Hovea R. Br. ex W. T. Aiton. 1812. Fabaceae. 12 Australia, Tasmania

Hovenia Thunb. 1781. Rhamnaceae. 4–5 Himalayas from Punjab to Bhutan, northeastern India, China, Korea, Japan

Hoverdenia Nees. 1847. Acanthaceae. 1 Mexico

Howardia Klotzsch = Aristolochia L. Aristolochiaceae

Howea ("Howeia") Becc. 1877. Arecaceae. 2 Lord Howe I.

Howellia A. Gray. 1879. Lobeliaceae. 1 western North America

Howelliella Rothm. = Antirrhinum L. Scrophulariaceae

Howethoa Rauschert = Lepisanthes Blume. Sapindaceae

Howittia F. Muell. 1855. Malvaceae. 1 southeastern Australia

Hoya R. Br. 1810. Asclepiadaceae. c. 200 India, Sri Lanka, Nepal, southern China, Indochina, Malesia, Australia, Oceania to Tonga and Samoa Is.

Hoyella Ridl. 1917. Asclepiadaceae. 1 Sumatra

Hua Pierre ex De Wild. 1906. Huaceae. 1 trop. Africa

Huanaca Cav. 1800. Apiaceae. 4 Chile, Argentina

Huarpea Cabrera. 1951. Asteraceae. 1 southern Andes

Hubbardia Bor. 1951. Poaceae. 1 western India

Hubbardochloa Auquier. 1980. Poaceae. 1 trop. Africa (Ruanda, Burundi, Zambia)

Huberia DC. 1828. Melastomataceae. 10 Ecuador, Peru, Brazil

Huberodendron Ducke. 1935. Bombacaceae. 5 Central and trop. South America

Hubertia Bory. 1804 (Senecio L.). Asteraceae. 25 Madagascar, Comoro Is., Réunion I.

Hudsonia L. 1767. Cistaceae. 1 (H. ericoides) western North America

Hueblia Speta. 1982 (Chaenorrhinum [Duby] Rchb.). Scrophulariaceae. 2 Southwest Asia

Huernia R. Br. 1810. Asclepiadaceae. 68 trop. and South Africa, southern Arabian Peninsula

Huerniopsis N. E. Br. 1878. Asclepiadaceae. 2–4 South-West and South Africa

Huertea Ruiz et Pav. 1794. Tapisciaceae. 4 West Indies, Colombia, Peru

Huetia Boiss. = Geocaryum Cosson. Apiaceae

Hughesia R. M. King et H. Rob. 1980. Asteraceae. 1 Peru

Hugonia L. 1753. Hugoniaceae. 40 trop. Africa from Senegal to Kenya and Tanzania, Madagascar, Mauritius, India, Sri Lanka, Indochina, Malesia, Solomon Is., northeastern Australia, New Caledonia, Fiji

Hugueninia Rchb. 1832. Brassicaceae. 1 South Europe from Spain to Italy

Huilaea Wurdack. 1957. Melastomataceae. 3–4 Colombia

Hulemacanthus S. Moore. 1920. Acanthaceae. 1 New Guinea

Hullettia King ex Hook. f. 1888. Moraceae. 2 southern Burma, Thailand, Malay Peninsula, Sumatra

Hulsea Torr. et A. Gray. 1858. Asteraceae. 7 western U.S., Baja California

Hulteniella Tzvelev. 1987 (Dendranthema [DC] Des Moulins). Asteraceae. 1 Chukotka, Alaska, western Canada to British Columbia

Hulthemia Dumort. = Rosa L. Rosaceae

Humbertia Comm. ex Lam. 1786. Convolvulaceae. 1 Madagascar

Humbertianthus Hochr. 1948. Malvaceae. 1 Madagascar

Humbertiella Hochr. 1926. Malvaceae. 6 Madagascar

Humbertina Buchet = Arophyton Jum. Araceae

Humbertiodendron Leandri. 1949. Trigoniaceae. 1 Madagascar

Humbertioturraea J.-F. Leroy. 1969. Meliaceae. 3–4 Madagascar

Humbertochloa A. Camus et Stapf. 1934. Poaceae. 2 trop. East Africa, Madagascar

Humbertodendron Leandri = Humbertiodendron Leandri. Trigoniaceae

Humblotiodendron Engl. 1917 (Vepris Comm. ex A. Juss.). Rutaceae. 2 Comoro Is., Madagascar

Humboldtia Vahl. 1794. Fabaceae. 6 southern India, Sri Lanka

Humboldtiella Harms. 1923 (Coursetia DC.). Fabaceae. 2 trop. South America, West Indies

Humea Sm. = Calomeria Vent. Asteraceae

Humeocline Anderb. 1991. Asteraceae. 1 Madagascar

Humiria Aubl. 1775. Humiriaceae. 4 trop. South America

Humirianthera Huber = Casimirella Hassler. Icacinaceae

Humiriastrum (Urb.) Cuatrec. 1961. Humiriaceae. 13 Central and South America

Humularia P. A. Duvign. 1954. Fabaceae. 33 trop. Africa, mainly Zaire and Zambia

Humulopsis Grudz. = Humulus L. Cannabaceae

Humulus L. 1753. Cannabaceae. 2 temp. Northern Hemisphere south to southwestern U.S. and Indochina

Hunaniopanax C. J. Qi et T. R. Cao. 1988. Araliaceae. 1 southeastern China (Hunan)

Hunga Pancher ex Prance. 1979. Chrysobalanaceae. 11: New Guinea (3), New Caledonia and Loyalty Is. (8)

Hunnemannia Sweet. 1828. Papaveraceae. 2 southern U.S., Mexico

Hunteria Roxb. 1824. Apocynaceae. 9 trop. Africa, southern India, Sri Lanka, Andaman Is., southern China, Indochina, Malay Peninsula, Anambas Is.

Huntleya Bateman ex Lindl. 1837 (Zygopetalum Hook.). Orchidaceae. 10 trop. America from Costa Rica to Brazil and Trinidad

Hunzikeria D'Arcy. 1976. Solanaceae. 3 southern U.S. (Texas), Mexico, Venezuela

Huodendron Rehder. 1935. Styracaceae. 4 Tibet, Burma southern China, Indochina

Hura L. 1753. Euphorbiaceae. 2 trop. America from Costa Rica to northern Brazil and Bolivia, West Indies

Husnotia Fourn. = Ditassa R. Br. Asclepiadaceae

Hutchinsia R. Br. 1812 (Pritzelago Kuntze). Brassicaceae. 1 Europe

Hutchinsiella O. E. Schulz. 1933. Brassicaceae. 1 western Tibet

Hutchinsonia Robyns. 1928 (Rytigynia Blume). Rubiaceae. ? spp. trop. Africa

Hutera Porta et Rigo ex Porta = Coincya Porta et Rigo ex Rouy. Brassicaceae

Huthamnus Tsiang = Stephanotis Thouars. Asclepiadaceae

Huthia Brand. 1908. Polemoniaceae. 2 Peru

Huttonaea Harv. 1863. Orchidaceae. 5 South Africa

Huttonella Kirk = Carmichaelia R. Br. Fabaceae

Huxleya Ewart. 1912. Verbenaceae. 1 northern Australia

Huynhia Greuter. 1981 (Arnebia Forssk.). Boraginaceae. 1 Turkey

Hyacinthella Schur. 1856 (Hyacinthus L.). Hyacinthaceae. 16 Southeast Europe (4), Caucasus, Asia Minor, Syria, Lebanon, Israel, Jordan, Iraq, Iran

Hyacinthoides Heist. ex Fabr. 1759. Hyacinthaceae. 6–8 West Europe, North Africa

Hyacinthus L. 1753. Hyacinthaceae. 3 Asia Minor, Syria, Lebanon, Iraq, southern Turkmenistan and northeastern Iran

Hyaenanche Lambert. 1797. Euphorbiaceae. 1 South Africa

Hyalea (DC.) Jaub. et Spach. 1847. Asteraceae. 2 Caucasus, Turkey, Iran, Middle Asia, western China

Hyalis D. Don ex Hook. et Arn. 1835. Asteraceae. 2 southern Bolivia, Paraguay, Argentina

Hyalisma Champ. 1847 (Sciaphila Blume). Triuridaceae. 1 India

Hyalocalyx Rolfe. 1884. Turneraceae. 1 trop. East Africa, Madagascar

Hyalochaete Dittrich et Rech. f. 1979 (Jurinea Cass.). Asteraceae. 1 Afghanistan, Pakistan

Hyalochlamys A. Gray. 1851 (Angianthus J. C. Wendl.). Asteraceae. 1 southwestern Australia

Hyalocystis Hallier f. 1898. Convolvulaceae. 1 trop. Africa

Hyalolaena Bunge. 1854. Apiaceae. 10 Middle Asia (6), Iran, Afghanistan, China

Hyalopoa (Tzvelev) Tzvelev. 1965 (Poa L.). Poaceae. 5 Caucasus, Asia Minor, East Siberia, northwestern Himalayas

Hyalosepalum Troupin = Tinospora Miers ex Benth. Menispermaceae

Hyaloseris Griseb. 1879. Asteraceae. 5 Andes of Bolivia and Argentina

Hyalosperma Steetz. 1845. Asteraceae. 9 temp. western and southwestern Australia, Tasmania (1)

Hybanthus Jacq. 1760. Violaceae. 100–150 tropics and subtropics, esp. America

Hybochilus Schltr. 1920. Orchidaceae. 2 Costa Rica

Hybosema Harms = Gliricidia Kunth. Fabaceae

Hybosperma Urb. = Colubrina Rich. ex Brongn. Rhamnaceae

Hybridella Cass. 1817. Asteraceae. 3 Mexico

Hydatella Diels. 1904. Hydatellaceae. 5 southern Australia and Tasmania (4), New Zealand (1)

Hydnocarpus Gaertn. 1788. Flacourtiaceae. 40 India, Burma, southern China, incl. Hainan, Indochina, Malesia (Sumatra, Malay Peninsula, Java, Borneo, Philippines, Sulawesi)

Hydnophytum Jack. 1823. Rubiaceae. 90 Andaman Is., Indochina, Malesia from Malay Peninsula to New Guinea, Australia (Cape York), Solomon Is., Vanuatu, Fiji

Hydnora Thunb. 1775. Hydnoraceae. 4–5 Africa: Gabon, Congo, Zaire, Sudan, Ethiopia, Somalia, Kenya, Uganda, Tanzania, Angola, Namibia, Zambia, Zimbabwe, southern South Africa, Madagascar, Réunion I., southwestern Arabian Peninsula

Hydrangea L. 1753. Hydrangeaceae. 80 Himalayas, China, Russian Far East (2), Japan, Indochina, Java, Philippines, North and Central America, Andes to Chile

Hydranthelium Kunth = Bacopa Aubl. Scrophulariaceae

Hydrastis L. 1759. Hydrastidaceae. 2 Japan (1), North America (1)

Hydriastele H. A. Wendl. et Drude. 1875. Arecaceae. 8 New Guinea, Bismarck Arch., northern Australia

Hydrilla Rich. 1811. Hydrocharitaceae. 1 Eurasia, trop. Africa, Indian Ocean islands eastward to New Guinea and Australia

Hydrobryopsis Engl. 1930 (Hydrobrium Endl.). Podostemaceae. 1 southern India

Hydrobryum Endl. 1841. Podostemaceae. 10 Nepal, northeastern India, China, Indochina, southern Japan

Hydrocera Blume. 1825. Balsaminaceae. 1 India, Sri Lanka, Indochina, Hainan, Malay Peninsula, Java, Sulawesi

Hydrocharis L. 1753. Hydrocharitaceae. 2 Europe, Mediterranean, trop. Africa, Asia, temp. Australia

Hydrochloa P. Beauv. 1812 (Luziola Juss.). Poaceae. 1 southeastern U.S., Mexico

Hydrocleys Rich. 1815. Limnocharitaceae. 5 Mexico, Guatemala, Honduras, El Salvador, Nicaragua, Costa Rica, Colombia, Venezuela, Guyana, Surinam, Ecuador, Brazil, Bolivia, Paraguay, Argentina; West Indies (1, H. numphoides—Puerto Rico, Curaçao, Trinidad)

Hydrocotyle L. 1753. Apiaceae. c. 130 America; Europe; the Caucasus (3); North, trop., and South Africa; Madagascar; Southwest, South, East, and Southeast Asia; Malesia; Australia; Oceania

Hydrodea N. E. Br. = Mesembryanthemum L. Aizoaceae

Hydrodyssodia B. L. Turner. 1988. Asteraceae. 1 Mexico

Hydrogaster Kuhlm. 1935. Tiliaceae. 1 Brazil

Hydroidea Karis. 1990 (Atrichantha Hilliard et B. L. Burtt). Asteraceae. 1 South Africa (Cape)

Hydrolea L. 1762. Hydrophyllaceae. 20 trop. America, West Indies, trop. Africa, trop. Asia

Hydrolythrum Hook. f. 1867 (Rotala L.). Lythraceae. 1 Indo-Malesia

Hydromystria G. Mey. 1818 (Limnobium Rich.). Hydrocharitaceae. 1 trop. and subtrop. America, West Indies

Hydropectis Rydb. 1914. Asteraceae. 1 Mexico

Hydrophilus Linder. 1984 (Leptocarpus R. Br.). Restionaceae. 1 South Africa (southwestern Cape)

Hydrophylax L. f. 1782. Rubiaceae. 1 India, Sri Lanka, Thailand

Hydrophyllum L. 1753. Hydrophyllaceae. 8 North America

Hydrostachys Thouars. 1806. Hydrostachydaceae. 22 trop. and South (1, H. polymorpha) Africa, Madagascar

Hydrostemma Wall. = Barclaya Wall. . Barclayaceae

Hydrothauma C. E. Hubb. 1947. Poaceae. 1 Zaire, Zambia

Hydrothrix Hook. f. 1887. Pontederiaceae. 2 Brazil

Hydrotriche Zucc. 1832. Scrophulariaceae. 4 Madagascar

Hyeronima Allemao. 1848. Euphorbiaceae. 36 trop. America, West Indies

Hygea Hanst. 1853. Gesneriaceae. 1 Chile

Hygrochilus Pfitzer. 1897. Orchidaceae. 1 northeastern India, Burma, Thailand, Laos, Vietnam, ? southern China

Hygrochloa Lazarides. 1979. Poaceae. 2 Australia (Northern Territory)

Hygrophila R. Br. 1810. Acanthaceae. 100 pantropics

Hygroryza Nees. 1833. Poaceae. 1 South and Southeast Asia

Hylaeanthe Jonk.-Verh. et Jonk. 1955. Marantaceae. 4 trop. South America

Hylandia Airy Shaw. 1974. Euphorbiaceae. 1 Australia (Queensland)

Hylandra A. Löve = Arabidopsis Heynh. Brassicaceae

Hylebates Chippindall. 1945. Poaceae. 2 trop. Africa: Kenya, Tanzania, Zambia, Zimbabwe, Mozambique

Hylenaea Miers. 1872. Celastraceae. 3 Central America, West Indies, Venezuela, Guyana, Surinam

Hyline Herb. 1840. Amaryllidaceae. 2 Brazil

Hylocarpa Cuatrec. 1961. Humiriaceae. 1 Brazil (Amazonia)

Hylocereus (A. Berger) Britton et Rose. 1909. Cactaceae. 17 trop. America from Mexico to Peru, West Indies

Hylocharis Miq. 1861 (Oxyspora DC.). Melastomataceae. 5 Malay Peninsula, Sumatra

Hylodendron Taub. 1894. Fabaceae. 1 Nigeria, Cameroun, Gabon, Zaire

Hylomecon Maxim. 1859 (Chelidonium L.). Papaveraceae. 3 temp. East Asia; H. vernales—Russian Far East

Hylophila Lindl. 1833. Orchidaceae. 6 southern China, Malay Peninsula, New Guinea, Admiralty Is.

Hylotelephium H. Ohba. 1977 (Sedum L.). Crassulaceae. 30 China, Korea, Japan

Hymenachne P. Beauv. 1812. Poaceae. 8 tropics

Hymenaea L. 1753. Fabaceae. 15 trop. America, West Indies; 1 (H. verrucosa) Kenya, Tanzania, Mozambique, Madagascar, Mauritius, Seychelles

Hymenandra (A. DC.) Spach. 1840. Myrsinaceae. 3 eastern Himalayas, Assam

Hymenanthera R. Br. 1818 (Melicytus J. R. Forst. et G. Forst.). Violaceae. 10 southeastern Australia and Tasmania (1, H. dentata), New Zealand, Norfolk Is.

Hymenidium Lindl. 1835. Apiaceae. 1 Himalayas

Hymenocallis Salisb. 1812. Amaryllidaceae. 40 trop. and subtrop. America, West Indies

Hymenocardia Wall. ex Lindl. 1836. Euphorbiaceae. 5 trop. and South Africa (4), continental Southeast Asia, Malay Peninsula, Sumatra

Hymenocarpos Savi. 1798. Fabaceae. 1 Mediterranean, Arabian Peninsula, Iraq, Iran

Hymenocephalus Jaub. et Spach. 1847. Asteraceae. 1 Iran

Hymenochlaena Bremek. 1944 (Strobilanthes Blume). Acanthaceae. 3 northeastern India, Malay Peninsula, Philippines

Hymenoclea Torr. et A. Gray. 1849. Asteraceae. 2 southwestern U.S. (California, Utah, Nevada, Arizona), Mexico

Hymenocnemis Hook. f. 1873. Rubiaceae. 1 Madagascar

Hymenocoleus Robbr. 1975 (Geophila D. Don). Rubiaceae. 12 trop. Africa

Hymenocrater Fisch. et C. A. Mey. 1836. Lamiaceae. 12 eastern Asia Minor, Kurdistan, Iran, Caucasus, Middle Asia Afghanistan, western Pakistan

Hymenodictyon Wall. 1824. Rubiaceae. 20 trop. and South (1, Transvaal) Africa, Madagascar, India, Himalayas, southwestern China, Indochina, western Malesia, Sulawesi

Hymenogyne Haw. 1821. Aizoaceae. 2 South Africa

Hymenolaena DC. 1830. Apiaceae. 10–12 mountainous Middle (3) and Central Asia, Himalayas

Hymenolepis Cass. 1817. Asteraceae. 7 South Africa (Cape)

Hymenolobium Benth. 1837. Fabaceae. 10–15 trop. South America

Hymenolobus Nutt. 1838. Brassicaceae. 5 Europe, Mediterranean, Caucasus, West Siberia, Middle Asia, Mongolia, China, Australia, North America

Hymenolophus Boerl. 1900. Apocynaceae. 1 Sumatra

Hymenolyma Korovin = Hyalolaena Bunge. Apiaceae

Hymenonema Cass. 1817. Asteraceae. 2 Greece

Hymenopappus L'Hér. 1788. Asteraceae. 10 U.S., Mexico

Hymenophysa C. A. Mey. = Cardaria Desv. Brassicaceae

Hymenopogon Wall. = Neohymenopogon Bennet. Rubiaceae

Hymenopyramis Wall. ex Griff. 1842. Verbenaceae. 7 India, Burma, Indochina, Hainan

Hymenorchis Schltr. 1913. Orchidaceae. 6 New Guinea

Hymenosporum R. Br. ex F. Muell. 1860. Pittosporaceae. 1 New Guinea, eastern Australia

Hymenostegia (Benth.) Harms. 1897. Fabaceae. 17 trop. West Africa from Sierra Leone to Angola

Hymenostemma Kunze ex Willk. 1864. Asteraceae. 1 southwestern North America, Southwest Europe, Algeria

Hymenostephium Benth. = Viguiera Kunth. Asteraceae

Hymenothrix A. Gray. 1849. Asteraceae. 5 southwestern U.S., Mexico

Hymenoxys Cass. 1828. Asteraceae. 28 America from Saskatchewan and Alberta to Argentina

Hyobanche L. 1771. Scrophulariaceae. 8 South Africa

Hyophorbe Gaertn. 1791. Arecaceae. 5 Rodriguez (1), Réunion I. (1), Mauritius, and Round Is. (3)

Hyoscyamus L. 1753. Solanaceae. 16–20 Europe, North Africa (incl. Sakhara), Caucasus, Middle Asia, Southwest Asia to northwestern Pakistan, India, Himalayas, southwestern and central China, North America

Hyoseris L. 1753. Asteraceae. 3 Mediterranean

Hyospathe Mart. 1823. Arecaceae. 3 trop. America from Costa Rica to Peru

Hypacanthium Juz. 1936 (Cousinia Cass.). Asteraceae. 1 Middle Asia

Hypagophytum A. Berger. 1930. Crassulaceae. 1 Ethiopia

Hyparrhenia Fourn. 1886. Poaceae. c. 60 Mediterranean, Southwest Asia, trop. Old World, mainly Africa

Hypecoum L. 1753. Hypecoaceae. 20 Southeast Europe, Caucasus, Mediterranean, West and Middle Asia eastward to southern West and East Siberia, northern China, Mongolia, western Tibet and Himalayas (Nepal, Bhutan)

Hypelate P. Browne. 1756. Sapindaceae. 1 Florida, West Indies

Hypelichrysum Kirp. = Pseudognaphalium Kirp. Asteraceae

Hypenanthe (Blume) Blume. 1849. Melastomataceae. 4 Malesia (excl. Malay Peninsula, Java, and Borneo)

Hypenia (Mart. ex Benth.) Harley. 1988. Lamiaceae. 24 trop. South America

Hyperacanthus E. Mey. ex Bridson. 1985. Rubiaceae. 2 Mozambique, South Africa

Hyperaspis Briq. = Osmium L. Lamiaceae

Hyperbaena Miers ex Benth. 1861. Menispermaceae. 19 trop. America: Mexico, Guatemala, Belize, El Salvador, Honduras, Costa Rica, Panama, Colombia, Venezuela, the Guianas, Brazil, Bolivia, Paraguay, and northern Argentina; West Indies (from Cuba and Haiti through the Antilles to Trinidad and Tobago)

Hypericophyllum Steetz. 1864. Asteraceae. 7 trop. Africa

Hypericopsis Boiss. 1846. Frankeniaceae. 1 southern Iran

Hypericum L. 1753. Hypericaceae. c. 400 temp. regions and montane areas of tropics

Hypertelis E. Mey. ex Fenzl. 1840. Molluginaceae. 9 St. Helena, trop. and South Africa, Madagascar

Hyperthelia Clayton. 1967. Poaceae. 6 trop. Africa

Hyphaene Gaertn. 1788. Arecaceae. c. 10 arid and semiarid Africa southward to Natal, Madagascar, Red Sea, and Gulf of Elath coasts, coastal Arabian Peninsula, western coastal India, ? Sri Lanka

Hypobathrum Blume. 1827. Rubiaceae. 9 Burma, Indochina, Malay Peninsula, Sumatra, Java, Philippines

Hypocalymma (Endl.) Endl. 1840. Myrtaceae. 18 western Australia

Hypocalyptus Thunb. 1800. Fabaceae. 3 South Africa (Cape)

Hypochaeris L. 1753. Asteraceae. 50 Eurasia, North Africa, South America

Hypocylix Woloszczak = Salsola L. Chenopodiaceae

Hypocyrta Mart. = Nematanthus Schrader. Gesneriaceae

Hypodaphnis Stapf. 1909. Lauraceae. 1 trop. West Africa

Hypodiscus Nees. 1836. Restionaceae. 16 South Africa

Hypoestes Sol. ex R. Br. 1810. Acanthaceae. c. 40 trop. Old World

Hypogomphia Bunge. 1872. Lamiaceae. 4 Middle Asia (2), Afghanistan

Hypogon Raf. = Collinsonia L. Lamiaceae

Hypogynium Nees. 1829 (Andropogon L.). Poaceae. 2 South America, Africa

Hypolaena R. Br. 1810. Restionaceae. 4 Australia, Tasmania

Hypolobus Fourn. 1885. Asclepiadaceae. 1 Brazil

Hypolytrum Rich. ex Pers. 1805. Cyperaceae. 50 tropics and subtropics

Hypophyllanthus Regel = Helicteres L. Sterculiaceae

Hypopitys Hill = Monotropa L. Ericaceae

Hypoxidia Friedmann. 1984. Hypoxidaceae. 2 Seychelles (Mohé I.)

Hypoxis L. 1759. Hypoxidaceae. c. 100 tropics and subtropics, esp. Southern Hemisphere

Hypsela C. Presl. 1836. Lobeliaceae. 4–5 eastern Australia, New Zealand, Andes of the South America

Hypselandra Pax et K. Hoffm. 1936. Capparaceae. 1 Burma

Hypselodelphys (K. Schum.) Milne-Redh. 1950. Marantaceae. 4 trop. Africa

Hypseloderma Radlk. = Camptolepis Radlk. Sapindaceae

Hypseocharis J. Remy. 1847. Hypseocharitaceae. 8 Andes

Hypseochloa C. E. Hubb. 1936. Poaceae. 2 trop. Africa

Hypserpa Miers. 1851. Menispermaceae. 6 India, Sri Lanka, Assam, Lower Burma, China, Indochina, Malesia (excl. Java), trop. Australia, Caroline and Solomon Is., Vanuatu, Polynesia

Hypsophila F. Muell. 1887. Celastraceae. 2 northeastern Australia

Hyptianthera Wight et Arn. 1834. Rubiaceae. 2 northern and northeastern India, Nepal, Bangladesh, Burma, southern China, Indochina

Hyptidendron Harley. 1988. Lamiaceae. 16 trop. South America

Hyptis Jacq. 1787. Lamiaceae. 300 trop. and subtrop. America, West Indies, 6—trop. Africa

Hyrtanandra Miq. 1851 (Pouzolzia Gaudich.). Urticaceae. 15 South, East, and Southeast Asia

Hyssaria Kolak. = Campanula L. Campanulaceae

Hyssopus L. 1753. Lamiaceae. 5–10 Central and South Europe, Mediterranean, West (incl. Caucasus) and Middle Asia, Mongolia, China, western Himalayas

Hysterionica Willd. 1807. Asteraceae. 10 southern Brazil, Argentina, Uruguay

Hystrichophora Mattf. 1936. Asteraceae. 1 trop. East Africa

Hystrix Moench. 1794. Poaceae. 10 temp. Asia and North America; Russia (3, East Siberia, Far East)

Hytophrynium K. Schum. = Trachyphrynium Benth. Marantaceae

I

Iacranda Pers. = Jacaranda Juss. Bignoniaceae

Ianthe Salisb. = Spiloxena Salisb. Hypoxidaceae

Ibarraea Lundell = Ardisia Sw. Myrsinaceae

Ibatia Decne. 1844. Asclepiadaceae. 3 trop. America, West Indies

Iberidella Boiss. = Aethionema R. Br. Brassicaceae

Iberis L. 1753. Brassicaceae. 30 Eurasia, Mediterranean

Ibervillea Greene. 1895. Cucurbitaceae. 5 southwestern North America

Ibetralia Bremek. 1934 (Alibertia A. Rich. ex DC.). Rubiaceae. 1 Guyana

Ibicella Van Eselt. 1929. Martyniaceae. 2 trop. South America

Ibina Noronha = Saurous Blume. Euphorbiaceae

Iboza N. E. Br. = Tetradenia Benth. Lamiaceae

Icacina A. Juss. 1823. Icacinaceae. 6 trop. Africa

Icacinopsis Roberty = Dichapetalum Thouars. Dichapetalaceae

Icacorea Aubl. = Ardisia Sw. Myrsinaceae

Icaria J. F. Macbr. = Miconia Ruiz et Pav. Melastomataceae

Ichnanthus P. Beauv. 1812. Poaceae. 33 trop. America, 1 pantropics

Ichnocarpus R. Br. 1810. Apocynaceae. 10 India, Sri Lanka, Himalayas, southern China, Indochina, Malesia, Australia (Queensland)

Ichthyothere Mart. 1830. Asteraceae. 18 trop. South America

Icianthus Greene = Euklisia Rydb. Brassicaceae

Icma Phil. = Baccharis L. Asteraceae

Icomum Hua = Aeollanthus Mart. ex Spreng. Lamiaceae

Idahoa A. Nelson et J. F. Macbr. 1913. Brassicaceae. 1 western U.S.

Idanthisa Raf. = Anisacanthus Nees. Acanthaceae

Idertia Farron = Ouratea Aubl. Ochnaceae

Idesia Maxim. 1866. Flacourtiaceae. 1 China, Taiwan, Japan

Idiospermum S. T. Blake. 1972. Idiospermaceae. 1 Australia (northeastern Queensland)

Idiothamnus R. M. King et H. Rob. 1975. Asteraceae. 4 Venezuela, Peru, Brazil, Argentina

Idria Kellogg. 1863 (or = Fouquieria Kunth). Fouquieriaceae. 1 northwestern Mexico (Sonoran desert, Isla Angel de la Guarda)

Ifloga Cass. 1819. Asteraceae. 15 Canary Is., Mediterranean, South Africa, western India

Ighermia Wiklund. 1983 (Asteriscus Hill). Asteraceae. 1 southwestern Morocco

Iguanura Blume. 1838. Arecaceae. 18 southern Thailand, Malay Peninsula, Sumatra, Borneo

Ihlenfeldtia H. E. K. Hartmann. 1992. 2 northwestern and central South Africa

Ikonnikovia Lincz. 1952 (Limonium Mill.). Plumbaginaceae. 1 Middle Asia, northwestern China

Ildefonsia Gardner = Bacopa Aubl. Scrophulariaceae

Ileostylus Tiegh. 1894. Loranthaceae. 1 New Zealand

Ilex L. 1753. Aquifoliaceae. c. 400 trop., subtrop. and temp. regions, esp. South America, East Asia; absent in North America; 8 Caucasus and Russian Far East

Iliamna Greene. 1906 (Sphaeralcea A. St.-Hil.). Malvaceae. 7 western North America

Iljinia Korovin. 1936 (Haloxylon Bunge). Chenopodiaceae. 1 Middle Asia, northwestern China, Mongolia

Illecebrum L. 1753. Caryophyllaceae. 1 Canary Is., West Europe, western Mediterranean

Illicium L. 1759. Illiciaceae. 42 Bhutan, northeastern India, Burma, East and Southeast Asia, southeastern U.S., Mexico, West Indies

Illigera Blume. 1827. Hernandiaceae. c. 20: 1 (I. pentaphylla) trop. Africa from Ghana and Ivory Coast to Angola, Zaire, Uganda, Kenya; 1 (I. madagascariensis) Tanzania and Madagascar; c. 18 from southern China and Indochina through Malesia to New Guinea

Iltisia S. F. Blake. 1958. Asteraceae. 2 Costa Rica, Panama

Ilysanthes Raf. 1820 (Lindernia All.). Scrophulariaceae. 50 tropics and subtropics

Imantia Hook. f. = Morinda L. Rubiaceae

Imbralyx Geesink = Fordia Hemsl. Fabaceae

Imbricaria Comm. ex Juss. 1789 = Mimosops L. Sapotaceae

Imeria R. M. King et H. Rob. 1975. Asteraceae. 2 Venezuela

Imerinaea Schltr. 1925. Orchidaceae. 1 Madagascar

Imitaria N. E. Br. 1927 (or = Gibbaeum N. E. Br.). Aizoaceae. 1 South Africa (Cape)

Impatiens L. 1753. Balsaminaceae. c. 900 North and Central America, temp. Eurasia, tropics and subtropics of the Old

Impatientella H. Perrier = Impatiens L. Balsaminaceae

Imperata Cyr. 1792. Poaceae. 10 trop., subtrop. and temp. regions (absent in Russia)

Imperatoria L. 1753. Apiaceae. 3 West, Southwest, and Central Europe; Northwest and North Africa

Incarvillea Juss. 1789. Bignoniaceae. 12–14 Middle Asia, China (12), Himalayas from Punjab to Assam, East Asia, incl. Russian Far East

Indigofera L. 1753. Fabaceae. c. 700 tropics and subtropics

Indobanalia A. N. Henry et B. Roy. 1969. Amaranthaceae. 1 southern India

Indocalamus Nakai. 1925. Poaceae. 24 China, Taiwan, Indochina, Malesia

Indochloa Bor. 1954 (Euclasta Franch.). Poaceae. 2 India

Indocourtoisia Bennet et Raizada = Courtoisina Soják. Cyperaceae

Indofevillea Chatterjee. 1946. Cucurbitaceae. 1 Tibet, eastern Himalayas

Indokingia Hemsl. = Gastonia Comm. ex Lam. Araliaceae

Indoneesiella Sreem. 1968. Acanthaceae. 2 India

Indopiptadenia Brenan. 1955. Fabaceae. 1 India, Nepal

Indopoa Bor. 1958. Poaceae. 1 India

Indopolysolenia Bennet. 1981. Rubiaceae. 1 eastern Himalayas

Indorouchera Hallier f. 1923. Hugoniaceae. 2 Nicobar Is., Thailand, southern Vietnam, Cambodia, Malay Peninsula, Sumatra, Java, Borneo

Indoryza A. N. Henry et B. Roy = Porteresia Tateoka. Poaceae

Indosasa McClure. 1940. Poaceae. 12 South and Southeast Asia

Indosinia J. E. Vidal. 1965. Sauvagesiaceae. 1 Indochina

Indotristicha P. Royen. 1959 (Tristicha Thouars). Podostemaceae. 2 southern India, Malay Peninsula

Indovethia Boerl. 1897 (Sauvagesia L.). Sauvagesiaceae. 1 Sumatra, Borneo

Inezia E. Phillips. 1932. Asteraceae. 1 South Africa

Inga Mill. 1754. Fabaceae. c. 350 trop. and subtrop. America

Inhambanella (Engl.) Dubard. 1915. Sapotaceae. 2 trop. West (1) and Southeast (1) Africa

Inobulbon (Schltr.) Schltr. et Kranzlin. 1910 (Dendrobium Sw.). Orchidaceae. 2 New Caledonia

Inocarpus J. R. Forst. et G. Forst. 1775. Fabaceae. 3 Malesia to New Guinea, Oceania eastward to Marquesas Is., Society, and Austral Is.

Intsia Thouars. 1806. Fabaceae. 3 Tanzania (1, I. bijuga), trop. Asia, islands of the Indian and Pacific Oceans from Madagascar to Tonga and Samoa

Intybellia Cass. = Crepis L. Asteraceae

Inula L. 1753. Asteraceae. c. 100 Eurasia, Africa, Madagascar

Inulanthera Källersjö. 1985. Asteraceae. 10 South Africa, Madagascar

Inulopsis (DC.) O. Hoffm. = Podocoma Cass. Asteraceae

Inversodicraea Engl. = Ledermanniella Engl. Podostemaceae

Iocenes B. Nord. 1978 (Senecio L.). Asteraceae. 1 Patagonia

Iochroma Benth. 1845. Solanaceae. 15 Galapagos Is., Colombia, Ecuador, Peru, Bolivia, northwestern Argentina

Iodanthus Torr. et A. Gray ex Steud. 1848. Brassicaceae. 4 eastern U.S., Mexico (3)

Iodes Blume. 1825. Icacinaceae. 19 trop. Africa, Madagascar, India (Assam), southwestern and southern China, Taiwan, Indochina, Malesia

Iodina Miers = Jodina Hook. et Arn. ex Meisn. Santalaceae

Iodocephalus Thorel ex Gagnep. 1920. Asteraceae. 1 Indochina

Iogeton Strother. 1991. Asteraceae. 1 Panama

Ionacanthus Benoist. 1940. Acanthaceae. 1 Madagascar

Ionactis Greene = Aster L. Asteraceae

Ione Lindl. = Sunipia Buch.-Ham. ex Lindl. Orchidaceae

Ionidium Vent. = Hybanthus Jacq. Violaceae

Ionopsidium Rchb. = Jonopsidium Rchb. Brassicaceae

Ionopsis Kunth. 1816. Orchidaceae. 12 trop. America from Florida and Mexico to Bolivia and Paraguay, West Indies

Iostephane Benth. 1873. Asteraceae. 4 Mexico

Ipheion Raf. 1837. Alliaceae. c. 25 southern Brazil, Paraguay, Chile, Argentina, Uruguay

Iphigenia Kunth. 1843 (Aphoma Raf.). Colchicaceae. 13 trop. and South Africa (5), Madagascar (2), India, southern China, Philippines, New Guinea, Australia, New Zealand (South I.)

Iphigeniopsis F. Buxb. 1936 (Iphigenia Kunth). Colchicaceae. 4 trop. and South Africa

Iphiona Cass. 1817. Asteraceae. 11 Egypt, trop. Africa, Southwest Asia

Iphionopsis Anderb. 1985. Asteraceae. 2 northeastern trop. Africa, Madagascar

Ipomoea L. 1753. Convolvulaceae. c. 500 trop. and warm–temp. regions; Middle Asia (I. hispida), southern East Siberia and southern Far East of Russia (I. sibirica)

Ipomopsis Michx. 1803 (Gilia Ruiz et Pav.). Polemoniaceae. 24 western North America and Florida; 1 Chile and Argentina

Ipsea Lindl. 1831. Orchidaceae. 2 India, Sri Lanka

Iranecio B. Nord. 1989. Asteraceae. 16 Balkan Peninsula, Turkey (7), Caucasus, Iran, Iraq

Irania Hadac et Chrtek = Farsetia Turra. Brassicaceac

Irenella Suess. 1934. Amaranthaceae. 1 Ecuador

Irenepharsus Hewson. 1982. Brassicaceae. 3 Australia (South Australia, New South Wales, Victoria)

Iresine P. Browne. 1756. Amaranthaceae. 80 Australia, Galapagos Is., trop. and subtrop. America

Iriartea Ruiz et Pav. 1794. Arecaceae. 1 (I. deltoidea) trop. America from Nicaragua to Bolivia, east to western Amazonia of Venezuela and Brazil

Iriartella H. A. Wendl. 1860. Arecaceae. 2 Central–western Amazon basin in Colombia, Venezuela, Guyana, Ecuador, Peru, Brazil

Iridodictyum Rodion. 1961 (Iris L.). Iridaceae. 9 Asia Minor, Palestine; 5 Caucasus, Middle Asia

Iridosma Aubrév. et Pellegr. 1962. Simaroubaceae. 1 trop. West Africa

Iris L. 1753. Iridaceae. c. 300 temp. Northern Hemisphere

Irlbachia Mart. 1827. Gentianaceae. 17 Mexico, Central and trop. South America, West Indies

Irmischia Schltdl. 1847 (Metastelma R. Br.). Asclepiadaceae. 5 Mexico, West Indies, trop. South America

Irvingbaileya R. A. Howard. 1943. Icacinaceae. 1 Australia (Queensland)

Irvingia Hook. f. 1860, Irvingiaceae. 15 (or 4 ?) trop. Africa, India, Burma, Indochina, Malay Peninsula, Sumatra, Borneo

Irwinia Barroso. 1980. Asteraceae. 1 Brazil

Iryanthera Warb. 1896. Myristicaceae. 30 trop. South America

Isabelia Barb. Rodr. 1877. Orchidaceae. 1 Brazil

Isachne R. Br. 1810. Poaceae. 100 tropics and subtropics, esp. Asia

Isaloa Humbert = Barleria L. Acanthaceae

Isalus J. B. Phipps. 1966 (Tristachya Nees). Poaceae. 3 Madagascar

Isandra F. Muell. = Symonanthus Haegi. Solanaceae

Isandraea Rauschert = Symonanthus Haegi. Solanaceae

Isanthus Michx. = Trichostema L. Lamiaceae

Isatis L. 1753. Brassicaceae. c. 55 Europe, Mediterranean, West (incl. Caucasus), Central and East Asia; southern Siberia

Ischaemum L. 1753. Poaceae. 65 tropics and subtropics, esp. Asia

Ischnea F. Muell. 1889. Asteraceae. 3 New Guinea

Ischnocarpus O. E. Schulz. 1924. Brassicaceae. 1 New Zealand
Ischnocentrum Schltr. = Glossorhyncha Ridl. Orchidaceae
Ischnochloa Hook. f. 1896 (Microstegium Nees). Poaceae. 1 western Himalayas
Ischnolepis Jum. et H. Perrier. 1909. Asclepiadaceae. 1 Madagascar
Ischnosiphon Koern. 1859. Marantaceae. 39 trop. America, West Indies
Ischnostemma King et Gamble. 1908. Asclepiadaceae. 1 coasts of Malay Peninsula, Java, Philippines, New Guinea, trop. Australia
Ischnurus Balf. f. 1883 (Lepturus R. Br.). Poaceae. 1 Socotra
Ischynogyne Schltr. 1913. Orchidaceae. 1 China
Ischyrolepis Steud. 1855 (Restio Rottb.). Restionaceae. 48 South Africa
Iseia O'Don. 1953 (Ipomoea L.). Convolvulaceae. 1 trop. America from Honduras, Colombia, Venezuela and Trinidad southward to Argentina
Iseilema Andersson. 1856. Poaceae. 20 South and Southeast Asia, Australia
Isertia Schreb. 1789. Rubiaceae. 14 trop. America: Guatemala, Honduras, Nicaragua, Costa Rica, Panama, Colombia, Venezuela, French Guyana, Guyana, Surinam, Ecuador, Peru, Bolivia, Brazil; West Indies (Cuba, Martinique, Trinidad)
Isidorea A. Rich. ex DC. 1830. Rubiaceae. 11 West Indies
Isidrogalvia Ruiz et Pav. 1802 (Tofieldia Hudson). Melanthiaceae. 5 Colombia, Ecuador, Peru, Venezuela, Guyana
Isinia Rech. f. = Lavandula L. Lamiaceae
Iskandera N. Busch. 1939. Brassicaceae. 2 Middle Asia (Kirgizia and Tajikistan)
Islaya Backeb. = Neoporteria Britton et Rose. Cactaceae
Ismene Salisb. ex Herb. = Hymenocallis Salisb. Amaryllidaceae
Isoberlinia Craib et Stapf ex Holland. 1911. Fabaceae. 5 trop. Africa
Isocarpha R. Br. 1817. Asteraceae. 5 trop. America, West Indies
Isochilus R. Br. 1813 (Ponera Lindl.). Orchidaceae. 4 trop. America from Mexico to Argentina, West Indies
Isochoriste Miq. = Asystasia Blume. Acanthaceae
Isocoma Nutt. = Haplopappus Cass. Asteraceae
Isodendrion A. Gray. 1852. Violaceae. 4 Hawaiian Is.
Isodesmia Gardner = Chaetocalyx DC. Fabaceae
Isodictyophorus Briq. 1917 (Coleus Lour.). Lamiaceae. 1 West Africa
Isodon (Benth.) Spach = Rabdosia (Blume) Hassk. Lamiaceae
Isoetopsis Turcz. 1851. Asteraceae. 3 temp. Australia
Isoglossa Oerst. 1854. Acanthaceae. 8 trop. and South Africa (1), Southeast Asia (7)
Isolepis R. Br. = Scirpus L. Cyperaceae
Isoleucas O. Schwartz. 1939. Lamiaceae. 1 Arabian Peninsula
Isoloma Decne = Kohleria Regel. Gesneriaceae
Isolona Engl. 1897. Annonaceae. 20 trop. Africa, Madagascar
Isomacrolobium Aubrév. et Pellegr. = Anthonotha P. Beauv. Fabaceae
Isomeris Nutt. = Cleome L. Capparaceae
Isomerocarpa A. C. Sm. = Levieria Becc. Monimiaceae
Isometrum Craib. 1920. Gesneriaceae. 13 China, esp. Yunnan
Isonandra Wight. 1842 (Palaquium Blanco). Sapotaceae. 10 southern India, Sri Lanka, Malay Peninsula, Borneo
Isonema R. Br. 1810. Apocynaceae. 3 West Africa
Isopappus Torr. et A. Gray = Croptilon Raf. Asteraceae
Isophysis T. Moore ex Seem. 1853. Iridaceae. 1 Tasmania
Isoplexis (Lindl.) Benth. 1835 (Digitalis L.). Scrophulariaceae. 3 Madeira and Canary Is.
Isopogon R. Br. ex Knight. 1809. Proteaceae. 35 Australia
Isopyrum L. 1753. Ranunculaceae. 30 temp. Northern Hemisphere, montane areas of tropics; I. thalictroides—European part of Russia

Isostigma Less. 1831. Asteraceae. 11 subtrop. South America

Isotheca Turrill. 1922. Acanthaceae. 1 Trinidad

Isotoma (R. Br.) Lindl. = Solenopsis C. Presl. Lobeliaceae

Isotrema Raf. = Aristolochia L. Aristolochiaceae

Isotria Raf. 1808. Orchidaceae. 2 eastern U.S.

Isotropis Benth. 1837. Fabaceae. 10–11 Australia

Itaobimia Rizzini = Riedeliella Harms. Fabaceae

Itasina Raf. 1840 (Oenanthe L.). Apiaceae. 1 South Africa

Itatiaia Ule = Tibouchina Aubl. Melastomataceae

Itaya H. E. Moore. 1972. Arecaceae. 1 Amazonian Peru and Brazil

Itea L. 1753. Iteaceae. 10–15 Asia from Himalayas to Japan, western Malesia, eastern North America (1, I. virginica)

Iteiluma Baill. = Pouteria Aubl. Sapotaceae

Iti Garn.-Jones et P. N. John. 1987. Brassicaceae. 1 New Zealand (South I.)

Itoa Hemsl. 1901. Flacourtiaceae. 2 southern China, Southeast Asia (I. orientalis), Malesia (I. stapfii—Sulawesi, Moluccas, New Guinea)

Ituridendron De Wild. = Omphalocarpum P. Beauv. ex Vent. Sapotaceae

Itysa Ravenna = Calydorea Herb. Iridaceae

Itzaea Standl. et Steyerm. 1944. Convolvulaceae. 1 Central America

Iva L. 1753. Asteraceae. 15 North and Central America, West Indies

Ivania O. E. Schulz. 1933. Brassicaceae. 1 northern Chile

Ivanjohnstonia Kazmi. 1975 (Cynoglossum L.). Boraginaceae. 1 northwestern Himalayas

Ivesia Torr. et A. Gray. 1858 (Potentilla L.). Rosaceae. 23 western North America

Ivodea Capuron. 1961. Rutaceae. 6 Madagascar

Ixanthus Griseb. 1838. Gentianaceae. 1 Canary Is.

Ixerba A. Cunn. 1839. Ixerbaceae. 1 New Zealand

Ixeridium (A. Gray) Tzvelev. 1964. Asteraceae. 20–25 temp. and trop. Asia, Malesia; Russia (5 East Siberia, Far East)

Ixeris Cass. 1822. Asteraceae. 25 East and Southeast Asia to New Guinea

Ixia L. 1762. Iridaceae. 49 trop. (1) and South Africa

Ixianthes Benth. 1836. Scrophulariaceae. 1 South Africa

Ixiochlamys F. Muell. et Sond. 1853. Asteraceae. 4 Australia

Ixiolaena Benth. 1837. Asteraceae. 6 Australia

Ixiolirion Fisch. ex Herb. 1821. Ixioliriaceae. 4–6 eastern Anatolia, western Syria, Lebanon, Israel, Jordan, southern Iran, Baluchistan, Middle Asia, Afghanistan, western China, Kashmir

Ixoca Raf. = Silene L. Caryophyllaceae

Ixocactus Rizzini. 1952. Loranthaceae. 6 west–central Mexico, Colombia, Ecuador, Peru, Venezuela, Brazil

Ixodia R. Br. 1812. Asteraceae. 2 southeastern Australia

Ixodonerium Pit. 1933. Apocynaceae. 1 Southeast Asia

Ixonanthes Jack. 1822. Ixonanthaceae. 3 India (Assam), southern China (incl. Hainan), Indochina, Malesia (excl. Java, Lesser Sunda, and Moluccas)

Ixophorus Schltdl. 1861–1862. Poaceae. 3 Mexico

Ixora L. 1753. Rubiaceae. c. 400 trop. Africa, trop. Asia, Malesia to northern Australia, Oceania, rarely in trop. America

Ixorhea Fenzl. 1886. Boraginaceae. 1 Andes of Argentina

Ixtlania M. E. Jones = Justicia L. Acanthaceae

Izabalaea Lundell. 1971. Nyctaginaceae. 1 Guatemala

J

Jablonskia G. L. Webster. 1984. Euphorbiaceae. 1 Venezuela, Guyana, Surinam, Peru, Brazil (Amazonia)

Jaborosa Juss. 1789. Solanaceae. 23 southern Peru, Bolivia, Brazil, Paraguay, Chile, Argentina (21 [11 endemic, J. magellanica] to Tierra del Fuego), Uruguay

Jacaima Rendle. 1936. Asclepiadaceae. 1 Jamaica

Jacaranda Juss. 1789. Bignoniaceae. 49 Belize, Honduras, Costa Rica, Panama, Colombia, Venezuela, Guyana, Surinam, French Guiana, Ecuador, Peru, Brazil, Bolivia, Paraguay, Argentina; West Indies: Bahamas, Cuba, Haiti

Jacaratia A. DC. 1864. Caricaceae. 6 trop. America, Africa

Jackia Wall. = Jackiopsis Ridsdale. Rubiaceae

Jackiopsis Ridsdale. 1979. Rubiaceae. 1 Malay Peninsula, Sumatra, Banka I., Borneo

Jacksonia R. Br. ex Sm. 1811. Fabaceae. c. 45 Australia

Jacmaia B. Nord. 1978. Asteraceae. 1 Jamaica

Jacobinia Nees ex Moricand = Justicia L. Acanthaceae

Jacobsenia L. Bolus et Schwantes. 1954 (or = Drosanthemum Schwantes). Aizoaceae. 2 western South Africa

Jacquemontia Choisy. 1834. Convolvulaceae. 120 tropics and subtropics, esp. America

Jacquesfelixia J. B. Phipps. 1964 (Danthoniopsis Stapf). Poaceae. 1 Angola, Namibia, Zimbabwe, Mozambique, Botswana, South Africa (Transvaal)

Jacqueshuberia Ducke. 1922. Fabaceae. 3 Colombia, Brazil, Peru

Jacquinia L. 1759. Theophrastaceae. 50 trop. America, West Indies

Jacquiniella Schltr. 1920. Orchidaceae. 3 trop. America from Mexico to Brazil, ? West Indies

Jadunia Lindau. 1913. Acanthaceae. 1 New Guinea

Jaegeria Kunth. 1820. Asteraceae. 17 America from Mexico to Argentina, Galapagos Is.

Jaeschkea Kurz. 1870. Gentianaceae. 3 Kashmir, western Himalayas, Tibet, China

Jagera Blume. 1849. Sapindaceae. 4 Moluccas, New Guinea, Australia

Jahnia Pittier et S. F. Blake = Turpinia Vent. Staphyleaceae

Jaimenostia Guinea et Gómez Moreno = Sauromatum Schott. Araceae

Jainia Balakr. = Coptophyllum Korth. Rubiaceae

Jalcophila Dillon et Sagast. 1986. Asteraceae. 3 Ecuador, Peru, Bolivia

Jaliscoa S. Watson. 1890. Asteraceae. 3 Mexico

Jaltomata Schltdl. 1838. Solanaceae. c. 25 trop. and subtrop. America from southwestern U.S. to Bolivia, Galapagos Is. (3), West Indies (1)

Jamaiciella Braem. 1980 (Oncidium Sw.). Orchidaceae. 2 West Indies

Jambosa Adans. = Syzygium Gaertn. Myrtaceae

Jamesbrittenia Kuntze. 1891 (Satura Roth). Scrophulariaceae. 1 Egypt, Southwest Asia from Sinai and Arabian Peninsula eastward to northwestern India

Jamesia Torr. et A. Gray. 1840. Hydrangeaceae. 1 western North America

Jamesianthus S. F. Blake et Sherff. 1940.

Asteraceae. 1 southern U.S.
Janakia Joseph et Chandras. 1978. Asclepiadaceae. 1 India
Jancaea Boiss. 1875. Gesneriaceae. 1 Greece (Olympic Mts.)
Janotia J.-F. Leroy. 1975. Rubiaceae. 1 Madagascar
Jansenella Bor. 1955. Poaceae. 1 India
Jansonia Kippist. 1847. Fabaceae. 1 western and southwestern Australia
Janusia A. Juss. 1840. Malpighiaceae. 12 America from California to Argentina
Japonolirion Nakai. 1930. Japonoliriaceae. 1 Japan (Honshu and Hokkaido Is.)
Jaramilloa R. M. King et H. Rob. 1980. Asteraceae. 2 Colombia
Jarandersonia Kosterm. 1960. Tiliaceae. 1 Borneo
Jardinea Steud. 1854 (Phacelurus Griseb.). Poaceae. 3 trop. Africa
Jarilla Rusby. 1921. Caricaceae. 1 Mexico
Jasarum Bunting. 1975. Araceae. 1 Venezuela
Jasione L. 1753. Campanulaceae. 20 Europe, Mediterranean, Asia Minor
Jasionella Stoj. et Stef. = Jasione L. Campanulaceae
Jasminocereus Britton et Rose. 1920. Cactaceae. 1–2 Galapagos Is.
Jasminochyla (Stapf) Pichon = Landolphia P. Beauv. Apocynaceae
Jasminum L. 1753. Oleaceae. c. 450 trop. and subtrop. Old World, a few species in temp. Eurasia
Jasonia (Cass.) Cass. 1815. Asteraceae. 1 southeastern Spain, southern France
Jateorhiza Miers. 1849. Menispermaceae. 2 trop. East Africa
Jatropha L. 1753. Euphorbiaceae. 175 North America, South Africa, tropics and subtropics
Jaubertia Guill. 1841. Rubiaceae. 16 North Africa, Socotra, West and Southwest Asia
Jaumea Pers. 1807. Asteraceae. 2 southwestern U.S., Mexico, temp. South America
Jaundea Gilg. 1894 (Rourea Aubl.). Connaraceae. 9 trop. Africa

Javorkaea Borhidi et Koml. = Rondeletia L. Rubiaceae
Jedda J. R. Clarkson. 1986. Thymelaeaceae. 1 Australia (Queensland)
Jefea Strother. 1991. Asteraceae. 5 U.S. (1, J. brevifolia—New Mexico, Texas), Mexico (4), Guatemala (1, J. phyllocephala)
Jeffersonia Barton. 1793. Berberidaceae. 2 East Asia, North America; Russia (1, J. dubia: Far East; sometimes J. dubia as a separate genus—Plagiorhegma Maxim.)
Jeffreya Wild. 1974. Asteraceae. 1 trop. East and South Africa
Jehlia Rose = Lopezia Cav. Onagraceae
Jejosephia A. N. Rao et Mani. 1986. Orchidaceae. 1 India
Jenkinsia Griff. = Miquilia Meisn. Icacinaceae
Jenmaniella Engl. 1927. Podostemaceae. 7 Guyana, Brazil
Jensenobotrya A. G. J. Herre. 1951. Aizoaceae. 1 western coastal Namibia
Jepsonia Small. 1896. Saxifragaceae. 1–4 southern California
Jepsonisedum Kral = Sedum L. Crassulaceae
Jerdonia Wight. 1848. Gesneriaceae. 1 southern India
Jessenia Karst. 1857. Arecaceae. 1 Panama, Colombia, Venezuela, Guyana, Surinam, Trinidad, Peru, Brazil (Amazonia)
Joannesia Vell. 1798. Euphorbiaceae. 2 Venezuela, northern Brazil
Jobinia Fourn. 1885. Asclepiadaceae. 3–4 trop. South America
Jodina Hook. et Arn. ex Meisn. 1837. Santalaceae. 1 southern Brazil, Uruguay, Argentina
Jodrellia Baijnath. 1978. Asphodelaceae. 3 trop. Africa from Ethiopia to Zimbabwe
Johannesteijsmannia H. E. Moore. 1961. Arecaceae. 4 southern Thailand, Malay Peninsula, Sumatra, Borneo
Johnsonia R. Br. 1810. Asphodelaceae. 5 southwestern Australia
Johnstonella Brand = Cryptantha Lehm. ex G. Don. Boraginaceae
Johrenia DC. 1829. Apiaceae. 15 South Eu-

rope, southern Caucasus; 1 (J. pauci-juga) West Asia eastward to Afghanistan

Johreniopsis Pimenov. 1987. Apiaceae. 4 Eastern Mediterranean, West Asia; J. seseloides—Caucasus, Middle Asia (Turkmenistan)

Joinvillea Gaudich. ex Brongn. et Gris. 1861. Joinvilleaceae. 2 Malay Peninsula, Sumatra, Borneo, Philippines, Solomon Is., New Caledonia, Fiji, Samoa, Hawaiian Is., Caroline atoll

Jollydora Pierre ex Gilg. 1896. Connaraceae. 6 trop. West Africa

Jonopsidium Rchb. 1829. Brassicaceae. 6 Mediterranean (incl. Europe)

Joosia Karst. 1859 (Ladenbergia Klotzsch). Rubiaceae. 4 Andes of Panama, Colombia, and Peru

Jordaaniella H. E. K. Hartmann. 1983 (Fenestraria N. E. Br.). Aizoaceae. 4 western Namibia and coastal western South Africa

Joseanthus H. Rob. 1989 (Vernonia Schreb.). Asteraceae. 5 trop. South America

Josephinia Vent. 1804. Pedaliaceae. 5 trop. Northeast Africa (1), Java, Sulawesi, Lesser Sunda Is., New Guinea, Australia

Jossinia Comm. ex DC. = Eugenia L. Myrtaceae

Jouvea Fourn. 1876. Poaceae. 2 Mexico, Central America

Jovellana Ruiz et Pav. 1798. Scrophulariaceae. 6 New Zealand

Jovetia Guédès. 1975. Rubiaceae. 2 Madagascar

Jovibarba Opez. 1852 (Sempervivum L.). Crassulaceae. 6 Europe, eastward to southwestern Ukraine

Juania Drude. 1878. Arecaceae. 1 Juan Fernández Is.

Juanulloa Ruiz et Pav. 1794. Solanaceae. 9–10 southern Mexico through Andes to Bolivia, Brazil (Amazonia, 1)

Jubaea Kunth. 1816. Arecaceae. 1 central Chile

Jubaeopsis Becc. 1913. Arecaceae. 1 South Africa (Transkei)

Jubelina A. Juss. 1838 (Diplopterys A. Juss.). Malpighiaceae. 6 trop. America

Juelia Aspl. = Ombrophytum Poepp. ex Endl. Lophophytaceae

Jugastrum Miers = Eschweilera Mart. ex DC. Lecythidaceae

Juglans L. 1753. Juglandaceae. 21 Southeast Europe (1), Turkey, Caucasus, Middle Asia, Iran, Himalayas, China, Japan, Russian Far East, North America, Andes to northern Argentina

Julbernardia Pellegr. 1943. Fabaceae. 11 trop. Africa

Julibaria Mez = Loheria Merr. Myrsinaceae

Julocroton Mart. 1837. Euphorbiaceae. 45 trop. America

Julostylis Thwaites. 1858. Malvaceae. 1 Sri Lanka

Jumellea Schltr. 1914. Orchidaceae. 57 trop. East and South Africa (2), Madagascar, Mascarene Is.

Jumelleanthus Hochr. 1924. Malvaceae. 1 Madagascar

Juncago Hill = Triglochin L. Juncaginaceae

Juncellus (Griseb.) C. B. Clarke. 1893 (Cyperus L.). Cyperaceae. 18 temp. and subtrop. regions; Russia (3, southern European part, southern West Siberia, Far East),

Juncus L. 1753. Juncaceae. 250 cosmopolitan

Junellia Moldenke. 1940. Verbenaceae. 57 South America

Jungia L. f. 1782. Asteraceae. 30 Mexico, Central America through Andes to northern Argentina, Uruguay

Juniperus L. 1753. Cupressaceae. c. 50 Northern Hemisphere, esp. (c. 30) Russia, Crimea, Caucasus, Middle Asia; southward to trop. Africa, India, Taiwan, and West Indies

Juno Tratt. = Iris L. Iridaceae

Junopsis Wern. Schulze = Iris L. Iridaceae

Jurinea Cass. 1821. Asteraceae. c. 250 Eurasia from Atlantic Europe to East Siberia and China

Jurinella Jaub. et Spach = Jurinea Cass. Asteraceae

Jurtsevia A. Löve et D. Löve. 1976 (Anemone L.). Ranunculaceae. 1 arctic East Siberia, Far East, northeastern North America (Hudson Bay)

Juruasia Lindau. 1904. Acanthaceae. 2 Brazil

Jussiaea L. = Ludwigia L. Onagraceae

Justago Kuntze = Cleome L. Capparaceae

Justenia Hiern = Bertiera Aubl. Rubiaceae

Justicia L. 1753. Acanthaceae. 300 tropics and subtropics

Juttadinteria Schwantes. 1926 [Dracophilus (Schwantes) Dinter et Schwantes]. Aizoaceae. 5 southwestern Namibia, western South Africa

Juzepczukia Khrzhanovsky = Rosa L. Rosaceae

Kabulia Bor et C. Fisch. 1939. Caryophyllaceae. 1 Afghanistan

Kadenia Lavrova et V. N. Tichom. 1986 (Selinum L.). Apiaceae. 2 Europe, Kazakhstan, Siberia, Mongolia, northern China

Kadsura Juss. 1810. Schisandraceae. 22 India, China, Japan, Indochina, western Malesia, Moluccas

Kadua Cham. et Schltdl. = Hedyotis L. Rubiaceae

Kaempferia L. 1753. Zingiberaceae. c. 50 South and Southeast Asia from India to Philippines and Indonesia

Kafirnigania Kamelin et Kinzik. 1984. Apiaceae. 1 Middle Asia (Gissaro-Darvaz)

Kageneckia Ruiz et Pav. 1794. Rosaceae. 1–2 Chile, Peru

Kaieteuria Dwyer = Ouratea Aubl. Ochnaceae

Kailarsenia Tirveng. 1983. Rubiaceae. 6 Burma, Thailand, southern China, Indochina, Malesia (Malay Peninsula, Sumatra, Borneo)

Kairoa Philipson. 1980. Monimiaceae. 1 New Guinea

Kairothamnus Airy Shaw. 1980. Euphorbiaceae. 1 New Guinea

Kajewskia Guillaumin = Veitchia H. A. Wendl. Arecaceae

Kajewskiella Merr. et Perry. 1947. Rubiaceae. 2 Solomon Is.

Kalaharia Baill. 1891 (Clerodendrum L.). Verbenaceae. 1 trop. and South Africa

Kalakia Alava. 1975. Apiaceae. 1 Iran

Kalanchoe Adans. 1763. Crassulaceae. c. 200 (or c. 60 ? trop. and South Africa (13), Madagascar, Arabian Peninsula, Socotra, trop. and East Asia, Moluccas, New Guinea, trop. Australia, trop. South America (1)

Kalappia Kosterm. 1952. Fabaceae. 1 Sulawesi

Kalbfussia Sch. Bip. = Leontodon L. Asteraceae

Kalbreyera Burret = Geonoma Willd. Arecaceae

Kalbreyeracanthus Wassh. = Habracanthus Nees. Acanthaceae

Kalbreyeriella Lindau. 1922. Acanthaceae. 3 Panama, Colombia

Kalidiopsis Aellen = Kalidium Moq. Chenopodiaceae

Kalidium Moq. 1849. Chenopodiaceae. 5 Spain, Turkey, southeastern European part of Russia, Caucasus, West and East Siberia, Middle Asia, Iran, Afghanistan, China, Mongolia

Kalimeris (Cass.) Cass. 1825 (Aster L.). Asteraceae. 10 East and Southeast Asia; 2 extending to southern East Siberia and Russian Far East

Kalimpongia Pradhan = Dickasonia L. O. Williams. Orchidaceae

Kaliphora Hook. f. 1867. Kaliphoraceae. 1 Madagascar

Kallstroemia Scop. 1777. Zygophyllaceae. 23 northern and northeastern Australia (7), America from southern U.S. to Argentina, West Indies

Kalmia L. 1753. Ericaceae. 10 North America, Cuba

Kalmiella Small = Kalmia L. Ericaceae

Kalmiopsis Rehder. 1932. Ericaceae. 1 northwestern U.S.

Kalopanax Miq. 1863 (Acanthopanax Decne.). Araliaceae. 1 Soviet Far East, China, Korea, Japan

Kalopternix Garay et Dunst. 1976 (Epidendrum L.). Orchidaceae. 3 trop. America

Kamettia Kostel. 1834. Apocynaceae. 4 Madagascar, Seychelles, India, Philippines

Kamiella Vassilcz. = Medicago L. or Trigonella L. Fabaceae

Kamiesbergia Snijman. 1991 (Namaquanula D. Müll.-Doblies et U. Müll.-Doblies). Asphodelaceae. 1 South Africa (northwestern Cape)

Kampmannia Steud. = Cortaderia Stapf. Poaceae

Kampochloa Clayton. 1967. Poaceae. 1 Angola, Zambia

Kanahia R. Br. 1810. Asclepiadaceae. 5 trop. East Africa, Arabian Peninsula

Kandaharia Alava. 1976. Apiaceae. 1 Afghanistan

Kandelia (DC.) Wight et Arn. 1834. Rhizophoraceae. 1 India, Banglasesh, Burma, Indochina, China, Taiwan, Kyukyu and Kyushu Is., Malesia (Malay Peninsula, Sumatra, Borneo)

Kania Schltr. 1914 (Metrosideros Banks ex Gaertn.). Myrtaceae. 4 Philippines, New Guinea

Kanimia Gardner = Mikania Willd. Asteraceae

Kanjarum Ramam. = Strobilanthes Blume. Acanthaceae

Kantou Aubrév. et Pellegr. = Inhambanella (Engl.) Dubard. Sapotaceae

Kaokochloa De Winter. 1961. Poaceae. 1 Namibia

Kaoue Pellegr. 1933 (Stachyothyrsus Harms). Fabaceae. 2 trop. America

Karamyschewia Fisch. et C. A. Mey. = Oldenlandia L. Rubiaceae

Karatas Mill. = Bromelia L. Bromeliaceae

Karatavia Pimenov et Lavrova. 1987. Apiaceae. 1 Middle Asia

Kardanoglyphos Schltdl. = Rorippa Scop. Brassicaceae

Karelinia Less. 1834 (Pluchea Cass.). Asteraceae. 1 (K. caspia) southeastern European part of Russia, Middle Asia, northeastern Iran, northwestern China, western Mongolia

Karimbolea Descoings. 1960. Asclepiadaceae. 1 Madagascar

Karina Boutique. 1971. Gentianaceae. 1 Zaire

Karnataka P. K. Mukh. et Constance. 1986. Apiaceae. 1 India

Karomia Dop. 1932. Verbenaceae. 9 East, Southeast and South Africa, Madagascar, Vietnam

Karroochloa Conert et Turpe. 1969 (Notodanthonia Zotov). Poaceae. 4 South-West and South Africa

Karvandarina Rech. f. 1950. Asteraceae. 1 Iran

Karwinskia Zucc. 1832. Rhamnaceae. 17 America from southwestern U.S. to Bolivia, West Indies

Kaschgaria Poljakov. 1957. Asteraceae. 2 Middle Asia (eastern Kazakhstan), western China, western Mongolia

Kashmiria D. Y. Hong. 1980. Scrophulariaceae. 1 India (Kashmir)

Katafa Costantin et J. Poiss. = Cedrelopsis Baill. Ptaeroxylaceae

Kaufmannia Regel. 1875. Primulaceae. 1 Middle Asia

Kaunia R. M. King et H. Rob. 1980. Asteraceae. 14 South America

Kayea Wall. 1831. Clusiaceae. 35 South and Southeast Asia

Kearnemalvastrum D. M. Bates. 1967. Malvaceae. 2 Mexico, Guatemala, Honduras, Costa Rica, Colombia

Keayodendron Leandri. 1959. Euphorbiaceae. 1 trop. Africa

Kechiella Straw. 1967. Scrophulariaceae. 7 North America

Kedarnatha P. K. Mukh. et Constance. 1986. Apiaceae. 1 India (Himalayas)

Kedrostis Medik. 1791. Cucurbitaceae. 20 trop. and subtrop. Africa, Madagascar, trop. Asia and Malesia

Keenania Hook. f. 1880. Rubiaceae. 5 northeastern India, Indochina

Keerlia A. Gray et Engelm. = Chaetopappa DC. Asteraceae

Keetia E. Phillips. 1926. Rubiaceae. c. 40 trop. and South Africa

Kefersteinia Rchb. f. 1852. Orchidaceae. 20 Central and trop. South America (Colombia, Venezuela, Ecuador, Peru, and Bolivia)

Kegeliella Mansf. 1934. Orchidaceae. 2 Costa Rica, Panama, Venezuela, Surinam, Jamaica, and Trinidad

Keiria S. Bowdich = ? Cissus L. Vitaceae

Keiskea Miq. 1865. Lamiaceae. 6 southern China, Japan

Keithia Benth. = Hoehnea Epling. Lamiaceae

Kelissa Ravenna. 1981 (Cypella Herb.). Iridaceae. 1 southeastern Brazil (Rio Grande do Sul)

Kelleria Endl. 1848 (Drapetes Banks ex Lamarck). Thymelaeaceae. 11 Borneo, New Guinea, Australia and Tasmania, New Zealand (9)

Kelleronia Schinz. 1895. Zygophyllaceae. 10 Somalia, Arabian Peninsula

Kelloggia Torr. ex Benth. et Hook. f. 1873. Rubiaceae. 2 China (1), western U.S. (1, K. galioides—Washington, Oregon, Wyoming, Utah, California, Arizona)

Kelseya (S. Watson) Rydb. 1900 (Luetkea Bong.). Rosaceae. 1 western U.S.

Kemulariella Tamamsch. = Aster L. Asteraceae

Kendrickia Hook. f. 1867. Melastomataceae. 1 southern India, Sri Lanka

Kengia Packer = Cleistogenes Keng. Poaceae

Kengyilia C. Yen et J.-L. Yang. 1990 (Agropyron Gaertn.). Poaceae. 1 western China and Mongolia (the Gobi Desert)

Keniochloa Meld. = Colpodium Trin. Poaceae

Kennedia Vent. 1805. Fabaceae. 15 Australia, Tasmania

Kenopleurum Candargy. 1897 (Thapsia L.). Apiaceae. 1 Lesbos

Kensitia Fedde. 1940. Aizoaceae. 1 South Africa

Kentia Blume = Gronophyllum Scheff. Arecaceae

Kentiopsis Brongn. 1873. Arecaceae. 1 New Caledonia

Kentranthus Raf. = Centranthus DC. Valerianaceae

Kentrosiphon N. E. Br. = Gladiolus L. Iridaceae

Kentrothamnus Suess. et Overk. 1941. Rhamnaceae. 2 Bolivia, Argentina

Keracia (Coss.) Calest. = Hohenackeria Fisch. et C. A. Mey. Apiaceae

Keramanthus Hook. f. = Adenia Forssk. Passifloraceae

Keraudrenia J. Gay. 1821. Sterculiaceae. 9 Madagascar (1), Australia (8)

Keraymonia Farille. 1985. Apiaceae. 3 Nepal

Kerbera E. Fourn. 1885 (Melinia Decne.). Asclepiadaceae. 1 trop. South America

Kerianthera J. H. Kirkbr. 1985. Rubiaceae. 1 Brazil

Kerigomnia P. Royen = Chitonanthera Schltr. Orchidaceae

Kermadecia Brongn. et Gris. 1863. Proteaceae. 4 New Caledonia

Kernera Medik. 1792 (Cochlearia L.). Brassicaceae. 2–3 mountainous Central and South Europe

Kerria DC. 1818. Rosaceae. 1 temp. East Asia (China, Japan)

Kerriochloa C. E. Hubb. 1950. Poaceae. 1 Thailand, Indochina

Kerriodoxa J. Dransf. 1983. Arecaceae. 1 peninsular Thailand

Kerriothyrsus C. Hansen. 1988. Melastomataceae. 1 Laos

Kerstania Rech. f. 1957. Fabaceae. 1 Afghanistan

Kerstingiella Harms = Macrotyloma (Wight et Arn.) Verdc. Fabaceae

Keteleeria Carriere. 1866. Pinaceae. 2 (or 11) continental and southern China, Taiwan, Vietnam, Laos

Keyserlingia Bunge ex Boiss. = Sophora L. Fabaceae

Keysseria Lauterb. 1914. Asteraceae. 15 Malesia, New Guinea, Fiji, Hawaiian Is.

Khadia N. E. Br. 1930 (Acrodon N. E. Br.). Aizoaceae. 5 South Africa (Transvaal)

Khasiaclunea Ridsdale. 1978. Rubiaceae. 1 India, Burma, China

Khaya A. Juss. 1830. Meliaceae. 7 trop. Africa, Madagascar, Comoro Is.

Kibara Endl. 1837. Monimiaceae. 43 Nicobar Is., peninsular Thailand, Malesia (39), chiefly New Guinea (37); K. coriacea—throughout Malesia, Australia (Queensland)

Kibaropsis Vieill. ex Jérémie. 1977 (Hedycarya J. R. Forst. et G. Forst.). Monimiaceae. 1 New Caledonia

Kibatalia G. Don. 1837. Apocynaceae. 26 trop. Africa, western Malesia

Kibessia DC. = Pternandra Jack. Melastomataceae

Kichxia Dumort. 1827 (Linaria L.). Scrophulariaceae. 27 Europe (5), Mediterranean, West (incl. Caucasus) and Middle Asia, western India, eastern Himalayas (Bhutan)

Kielmeyera Mart. 1826. Clusiaceae. 35 Brazil

Kigelia DC. 1838. Bignoniaceae. 1 trop. Africa

Kigelianthe Baill. = Fernandoa Welw. ex Seem. Bignoniaceae

Kiggelaria L. 1753. Flacourtiaceae. 1 trop. (Tanzania, Malawi, Mozambique, Zimbabwe), South-West and South Africa

Kiharopyrum A. Löve = Aegilops L. Poaceae

Killipia Gleason. 1925. Melastomataceae. 2 Colombia

Killipiella A. C. Sm. 1943 [Disterigma (Klotzsch) Nied.]. Ericaceae. 2 Colombia

Killipiodendron Kobuski. 1942. Theaceae. 1 Colombia

Kinepetalum Schltr. = Brachystelma R. Br. Asclepiadaceae

Kinetochilus (Schltr.) Brieger = Dendrobium Sw. Orchidaceae

Kingdon-Wardia Marq. = Swertia L. Gentianaceae

Kingdonia Balf. f. et W. W. Sm. 1914. Kingdoniaceae. 1 western and northern China

Kingella Tiegh. = Trithecanthera Tiegh. Loranthaceae

Kinghamia C. Jeffrey. 1988. Asteraceae. 5 trop. West and Central Africa

Kingia R. Br. 1826. Dasypogonaceae. 1 southwestern Australia

Kingianthus H. Rob. 1978. Asteraceae. 1 South America

Kingidium P. F. Hunt. 1970. Orchidaceae. 5 India, Sri Lanka, Himalayas, Burma, southern China (Yunnan), Indochina, Malay Peninsula, Sumatra, Java, Ambon, Lesser Sunda Is., Sulawesi, Borneo, Philippines

Kingiodendron Harms. 1897. Fabaceae. 6 India, Malesia to New Guinea, Solomon Is., Fiji

Kingstonia Hook. f. et Thomson = Dendrokingstonia Rauschert. Annonaceae

Kinia Raf. 1814. Hyacinthaceae. 1 Comoro Is.

Kinostemon Kudo. 1929. Lamiaceae. 3 southwestern and eastern China, Taiwan, Japan

Kinugasa Tatew. et Suto. 1935. Trilliaceae. 1 Japan (Honshu)

Kionophyton Garay = Stenorrhynchos Rich. ex Spreng. Orchidaceae

Kippistia F. Muell. 1858. Asteraceae. 1 central and southern Australia

Kirengeshoma Yatabe. 1890. Hydrangeaceae. 1 continental China, Korea, Japan

Kirilowia Bunge. 1843. Chenopodiaceae. 1 Middle Asia, Afghanistan, northwestern China

Kirkbridea Wurdack. 1976. Melastomataceae. 2 Colombia (Sierra Nevada de Santa Marta)

Kirkia Oliv. 1868. Kirkiaceae. 5 trop. (Congo, Malawi, Zambia, Zimbabwe, Botswana, Angola, Namibia, Ethiopia, Somalia, Kenya, and Mozambique) and South Africa (Transvaal)

Kirkianella Allan. 1961. Asteraceae. 1 New Zealand

Kirkophytum (Harms) Allan = Stilbocarpa (Hook. f.) Decne. et Planch. Araliaceae

Kirpicznikovia A. Löve et D. Löve. 1976 (Sedum L.). Crassulaceae. 1 Arctica, West and East Siberia, northern China, Mongolia

Kissenia T. Anderson. 1860. Loasaceae. 2 southwestern Arabian Peninsula, northeastern Ethiopia, Somalia, Namibia (1)

Kissodendron Seem. = Polyscias J. R. Forst. et G. Forst. Araliaceae

Kita A. Chev. = Hygrophila R. Br. Acanthaceae

Kitagawia Pimenov. 1986. Apiaceae. 5 temp. Asia from eastern Kazakhstan to Russian Far East, northern and northeastern China, Mongolia, Korea, Japan, ? Taiwan

Kitaibelia Willd. 1799. Malvaceae. 2 Southeast Europe (lower Danube, Yugoslavia—K. vitifolia), Asia Minor (K. balansae)

Kitamuraea Rauschert = Aster L. Asteraceae

Kitchingia Baker = Kalanchoe Adans. Crassulaceae

Kitigorchis Maekawa. 1971. Orchidaceae. 4 China, Japan

Kjellbergia Bremek. 1948. Acanthaceae. 1 Sulawesi

Kjellbergiodendron Burret. 1936. Myrtaceae. 3 Sulawesi

Klaineanthus Pierre ex Prain. 1912. Euphorbiaceae. 1 trop. West Africa

Klaineastrum Pierre ex A. Chev. = Memecylon L. Melastomataceae

Klainedoxa Pierre ex Engl. 1896. Irvingiaceae. 3 trop. Africa

Klaprothia Kunth. 1823. Loasaceae. 2 trop. America from southern Mexico to Peru, northern Bolivia, northern Venezuela, Haiti

Klattia Baker. 1877. Iridaceae. 3 South Africa (Cape)

Kleinhovia L. 1763. Sterculiaceae. 1 trop. Asia, Australia

Kleinia Hill. 1753 (Senecio L.). Asteraceae. 50 Canary Is.; North, trop., and South Africa; Arabian Peninsula; India

Kleinodendron L. B. Sm. et Downs = Savia Willd. Euphorbiaceae

Klemachloa R. Parker. 1932 (Dendrocalamus Nees). Poaceae. 1 Burma

Klingia Schoenl. = Gethyllis L. Amaryllidaceae

Klopstockia Karst. = Ceroxylon Bonpl. ex DC. Arecaceae

Klossia Ridl. 1909. Rubiaceae. 1 Malay Peninsula

Klotzschia Cham. 1833. Apiaceae. 3 southern Brazil

Klugia Schltr. = Rhinchoglossum Blume. Gesneriaceae

Klugiodendron Britton et Killip. 1936 (Pithecellobium Mart.). Fabaceae. 1–9 trop. South America

Kmeria (Pierre) Dandy. 1927. Magnoliaceae. 2 southern China, Indochina

Knautia L. 1753 (Scabiosa L.). Dipsacaceae. 60 temp. Europe, Mediterranean, West Asia; Caucasus, West Siberia, Middle Asia

Knema Lour. 1790. Myristicaceae. c. 90 India, southern China, Indochina, Malesia

Knightia R. Br. 1810. Proteaceae. 3 New Zealand, New Caledonia

Kniphofia Moench. 1794. Asphodelaceae. 70 trop. East and South Africa, Madagascar

Knorringia (Czukav.) S.-P. Hong. 1989 = Knorringia (Czukav.) Tzvelev. Polygonaceae

Knorringia (Czukav.) Tsvelev. 1987. Polygonaceae. 2 Middle Asia (K. pamirica—Pamir-Alai; K. sibirica—eastern Kazakhstan, Tien Shan), West and East Siberia, Russian Far East, China, Mongolia

Knowltonia Salisb. 1796. Ranunculaceae. 8 trop. and South Africa

Knoxia L. 1753. Rubiaceae. 9 (or 2 ?) trop. Africa ?, India, Himalayas, southern China, Indochina, Malesia, trop. Australia

Koanophyllon Arruda. 1810. Asteraceae. 114 trop. America from southern U.S. to Paraguay, West Indies

Kobresia Willd. 1805. Cyperaceae. 40 cold and temp. regions, mountainous trop. and subtrop. Northern Hemisphere

Kochia Roth. 1801. Chenopodiaceae. 16 Central, South, and East Europe; North and South Africa; Asia; North America

Kochummenia K. M. Wong. 1984. Rubiaceae. 2 Malaya

Kodalyodendron Borhidi et Acuña. 1973. Rutaceae. 1 Cuba

Koeberlinia Zucc. 1832. Capparaceae. 1 southern U.S., Mexico, Bolivia

Koechlea Endl. = Ptilostemon Cass. Asteraceae

Koehneola Urb. 1901. Asteraceae. 1 Cuba

Koehneria S. A. Graham, Tobe et Baas. 1986 (Lagerstroemia L.). Lythraceae. 2 southern Madagascar

Koeiea Rech. f. = Prionotrichon Botsch. et Vved. Brassicaceae

Koeleria Pers. 1805. Poaceae. 60 temp. and subtrop. regions and mountainous tropics

Koellensteinia Rchb. f. 1854. Orchidaceae. 16 trop. America from Panama to Peru, Bolivia, and Brazil; West Indies

Koellikeria Regel. 1847. Gesneriaceae. 3 Costa Rica, Panama, Venezuela, Bolivia, and Argentina

Koelpinia Pall. 1776. Asteraceae. 5 from North Africa through southeastern European part of Russia and Middle Asia eastward to East Asia

Koelreuteria Laxm. 1772. Sapindaceae. 4 China, Taiwan, Fiji

Koelzella Hiroe = Prangos Lindl. Apiaceae

Koenigia L. 1767. Polygonaceae. 8 arctic and temp. Asia, temp. South America (1); K. islandica—Arctica, Siberia, Russian Far East, mountainous Middle Asia

Koernickanthe L. Andersson. 1981 (Ischnosiphon Koern.). Marantaceae. 1 trop. America

Kohautia Cham. et Schltdl. 1829. Rubiaceae. 60 Cape Verde Is., trop., South-West and northern South Africa, Madagascar, Comoro Is., Arabian Peninsula, trop. Asia, Malesia, trop. Australia

Kohleria Regel. 1847. Gesneriaceae. 17 trop. America: Mexico, Guatemala, Belize, El Salvador, Honduras, Nicaragua, Costa Rica, Panama, Colombia (14), Venezuela, Trinidad, Guyana, Surinam, Ecuador, Peru, northern Brazil

Kohlerianthus Fritsch. 1897 (Columnea L.). Gesneriaceae. 1 Bolivia

Kohlrauschia Kunth = Petrorhagia (Ser. ex DC.) Link. Caryophyllaceae

Koilodepas Hassk. 1856. Euphorbiaceae. 11 southern India, Indochina, Hainan, western Malesia, New Guinea

Kokia Lewton. 1912. Malvaceae. 4 Hawaiian Is.

Kokoona Thwaites. 1853. Celastraceae. 8 southern India and Sri Lanka (1), Burma (1), Malesia (6—Sumatra, Malay Peninsula, Borneo, Philippines)

Kolkwitzia Graebn. 1901. Caprifoliaceae. 1 central China

Kolobochilus Lindau. 1900. Acanthaceae. 2 Costa Rica

Kolobopetalum Engl. 1899. Menispermaceae. 4 trop. Africa

Kolowratia C. Presl. 1827. Zingiberaceae. 90 Philippines, Moluccas, New Guinea, northern Australia, Micronesia, Polynesia

Kolpakowskia Regel = Ixiolirion Fisch. ex Herb. Ixioliriaceae

Komaroffia Kuntze. 1887 (Nigella L.). Ranunculaceae. 2 Central Asia and Iran

Komarovia Korovin. 1939. Apiaceae. 1 Middle Asia

Kompitsia Costantin et Gallaud. 1906. Asclepiadaceae. 1 Madagascar

Konantzia Dodson et N. H. Williams. 1980. Orchidaceae. 1 western Ecuador

Koniga Adans. = Lobularia Desv. Brassicaceae

Koompassia Maingay. 1873. Fabaceae. 3 Malay Peninsula, Sumatra, Philippines, New Guinea

Koordersiodendron Engl. 1898. Anacardiaceae. 1 Sulawesi, Philippines, New Guinea

Kopsia Blume. 1823. Apocynaceae. 30 Southeast Asia, western Malesia, Caroline Is.

Kopsiopsis (Beck) Beck. 1930. Orobanchaceae. 2 western North America

Korolkowia Regel. 1873 (Fritillaria L.). Liliaceae. 1 Middle Asia

Korovinia Nevski et Vved. = Galagania Lipsky. Apiaceae

Korshinskya Lipsky. 1901. Apiaceae. 4 Middle Asia (3), Afghanistan

Korthalsella Tiegh. 1896. Viscaceae. 25–30 Northeast Africa (Ethiopia), Madagascar, Mascarene Is., India, Himalayas, southwestern China, Japan, Indochina, Malesia, Australia (6), Lord Howe I., New Zealand, Oceania, West Indies

Korthalsia Blume. 1843. Arecaceae. 26 Andaman and Nicobar Is., Burma, Indochina, Malesia

Kosmosiphon Lindau. 1913. Acanthaceae. 1 trop. West Africa

Kosopoljanskia Korovin. 1923. Apiaceae. 2 Middle Asia (Tien Shan and Pamir-Alai)

Kosteletzkya C. Presl. 1835. Malvaceae. 17–30 North, Central, and South America; South Europe (1 Spain, Italy); Balearic Is.; Corsica; trop. and South Africa; Madagascar; Malesia; K. pentacarpos—southern European part of Russia, Caucasus

Kostermansia Soegeng. 1959. Bombacaceae. 1 Malay Peninsula

Kostermanthus Prance. 1979. Chrysobalanaceae. 2 Malay Peninsula, Sumatra, Borneo, Sulawesi, Philippines (Mindanao)

Kotchubaea Regel ex Hook. f. = Kutchubaea Fisch. ex DC. Rubiaceae

Kotschya Endl. 1839. Fabaceae. 30 trop. Africa, Madagascar

Kotschyella F. K. Mey. = Thlaspi L. Brassicaceae

Koyamacalia H. Rob. et Brettell = Parasenecio W. W. Sm. et Small. Asteraceae

Koyamaea W. W. Thomas et Davidse. 1989. Cyperaceae. 1 Amazonas (Neblina Massif), Venezuela, Brazil

Kozlovia Lipsky. 1904. Apiaceae. 1 Middle Asia (Turkmenistan, Tien Shan, Pamir-Alai), Afghanistan

Kraenzlinella Kuntze. 1903. Orchidaceae. 5 trop. America

Krameria L. ex Loefl. 1758. Krameriaceae. 17 America from southwestern U.S. to Chile and Argentina

Krankofa Raf. = Ceratoides Gagnebin. Chenopodiaceae

Krankovia Raf. = Ceratoides Gagnebin. Chenopodiaceae

Krapfia DC. 1817 (Ranunculus L.) Ranunculaceae. 8 Andes

Krapovickasia Fryxell. 1978. Malvaceae. 4 Mexico (1, K. physaloides), Peru ?, Bolivia, Paraguay, Brazil, Argentina, Uruguay

Krascheninnikovia Gueldenst. = Ceratoides Gagnebin. Chenopodiaceae

Krascheninnikowia Turcz. ex Fenzl = Pseudostellaria Pax. Caryophyllaceae

Krasnovia Popov ex Schischk. 1950. Apiaceae. 1 Middle Asia, China

Krassera O. Schwartz = Anerincleistus Korth. Melastomataceae

Krausella H. J. Lam = Pouteria Aubl. Sapotaceae

Krauseola Pax et K. Hoffm. 1934. Caryophyllaceae. 2 northern Kenya (1), Mozambique (1)

Kraussia Harv. 1842. Rubiaceae. 3 trop. and South (1) Africa

Kremeria Durieu = Coleostephus Cass. Asteraceae

Kremeriella Maire. 1932. Brassicaceae. 1 Morocco, Algeria

Kreodanthus Garay. 1977. Orchidaceae. 6 trop. America

Kreysigia Rchb. = Tripladenia D. Don. Melanthiaceae

Krigia Schreb. 1791. Asteraceae. 7 North America

Krokia Urb. = Pimenta Lindl. Myrtaceae

Krubera Hoffm. 1814. Apiaceae. 1 Southwest, South, and Southeast Europe; North Africa

Krugia Urb. = Marlierea Cambess. Myrtaceae

Krugiodendron Urb. 1902. Rhamnaceae. 1 Central America, Venezuela, West Indies

Krukoviella A. C. Sm. 1939. Sauvagesiaceae. 1 Peru, Brazil

Krylovia Schischk. = Rhinactinidia Novopokr. Asteraceae

Kryptostoma (Summerh.) Geerinck. 1982 (Habenaria Willd.). Orchidaceae. 2 trop. Africa

Kubitzkia van der Werff = Systemonodaphne Mez. Lauraceae

Kudoacanthus Hosok. 1933. Acanthaceae. 1 Taiwan

Kudrjaschevia Pojark. 1953 (or = Nepeta L.). Lamiaceae. 5 Middle Asia (5), Iran, northern Afghanistan

Kuepferella M. Lainz = Gentiana L. Gentianaceae

Kuhitangia Ovcz. 1967. Caryophyllaceae. 2 Middle Asia

Kuhlhasseltia J. J. Sm. 1910. Orchidaceae. 5 Malesia

Kuhlmannia J. C. Gomes = Pleonotoma Miers. Bignoniaceae

Kuhlmanniella Barroso = Dicranostyles Benth. Convolvulaceae

Kuhnia L. = Brickellia Elliott. Asteraceae

Kumlienia Greene. 1886 (Ranunculus L.). Ranunculaceae. 1 western North America

Kummerowia Schindl. 1912. Fabaceae. 2 Asia, North America

Kundmannia Scop. 1777. Apiaceae. 2 South Europe, Mediterranean, Asia Minor

Kunhardtia Maguire. 1958. Rapateaceae. 2 Venezuela

Kunkeliella W. T. Stearn. 1972. Santalaceae. 3 Canary Is.

Kunstleria Prain. 1897. Fabaceae. 9 Malesia, New Guinea, northern Australia

Kuntheria Conran et Clifford. 1987 (Schelhammera R. Br.). Melanthiaceae. 1 Australia (northern Queensland)

Kunzea Rchb. 1828. Myrtaceae. 25–30 Australia, Tasmania

Kurrimia Wall. ex Meisn. = Itea L. Iteaceae

Kurrimia Wall. ex Thwaites = Bhesa Buch.-Ham. ex Arn. Celastraceae

Kurzamra Kuntze. 1891. Lamiaceae. 1 temp. South America

Kutchubaea Fisch. ex DC. 1830. Rubiaceae. 11 trop. South America

Kydia Roxb. 1811. Malvaceae. 4 (or 1 ?) Himalayas from Kashmir to Bhutan, India, Burma, Thailand, southern China (Yunnan)

Kyllinga Rottb. 1773. Cyperaceae. 70 trop., subtrop., and warm–temp. regions

Kyllingiella R. W. Haines et Lye. 1978. Cyperaceae. 3–4 trop. East Africa

Kyrsteniopsis R. M. King et H. Rob. 1971 (Eupatorium L.). Asteraceae. 4 Mexico

L

Labatia Sw. = Pouteria Aubl. Sapotaceae

Labichea Gaudich. ex DC. 1825. Fabaceae. 9 Australia

Labidostelma Schltr. 1906. Asclepiadaceae. 1 Central America

Labisia Lindl. 1845. Myrsinaceae. 9 Malesia

Lablab Adans. 1763. Fabaceae. 1 Africa; very largely cultivated

Labordia Gaudich. 1828 (Geniostoma J. R. Forst. et G. Forst.). Loganiaceae. 15 Hawaiian Is.

Labourdonnaisia Bojer. 1841. Sapotaceae. 4 South Africa (Natal), Madagascar, Mauritius

Labramia A. DC. 1844 (Mimusops L.). Sapotaceae. 8 Madagascar

Laburnum Fabr. 1759. Fabaceae. 2 France, Switzerland, Italy, Yugoslavia, Albania, Greece, Austria, Czechoslovakia, Hungary, Romania

Lacaena Lindl. 1843. Orchidaceae. 2 Mexico, Central America

Lacaitaea Brand. 1914 (Trichodesma R. Br.). Boraginaceae. 1 eastern Himalayas

Lacandonia E. Martinez et Ramos. 1989 (Triurus L.). Triuridaceae. 1 Mexico

Laccodiscus Radlk. 1879. Sapindaceae. 4 trop. West Africa

Laccopetalum Ulbr. 1906 (Ranunculus L.). Ranunculaceae. 1 Peru

Laccospadix Drude et H. A. Wend. 1875 (Calyptrocalyx Blume). Arecaceae. 1 Australia (northern Queensland)

Laccosperma (G. Mann et H. A. Wendl.) Drude. 1877. Arecaceae. 7 trop. West Africa and Congo Basin

Lachanodes DC. 1833. Asteraceae. 1 St. Helena

Lachemilla (Kocke) Rydb. = Alchemilla L. Rosaceae

Lachenalia J. F. Jacq. ex Murray. 1780–1784. Hyacinthaceae. 80 South Africa

Lachnaea L. 1753. Thymelaeaceae. 29 South Africa

Lachnagrostis Trin. 1822 (Agrostis L.). Poaceae. 5 New Zealand, Auckland Is., Campbell I., Antipodes Is.

Lachnanthes Elliott. 1816. Haemodoraceae. 1 North America

Lachnocapsa Balf. f. 1882. Brassicaceae. 1 Socotra

Lachnocaulon Kunth. 1841. Eriocaulaceae. 7 southeastern U.S., Cuba

Lachnochloa Steud. 1855. Poaceae. 1 trop. West Africa

Lachnoloma Bunge. 1843. Brassicaceae. 1 Middle Asia, northeastern Iran, China

Lachnophyllum Bunge. 1852. Asteraceae. 2 West Asia from Israel (1, L. noëanum) to Middle Asia (1, L. gossypinum) and Afghanistan

Lachnopylis Hochst. = Nuxia Comm. ex Lam. Buddlejaceae

Lachnorhiza A. Rich. 1850 (Vernonia Schreb.). Asteraceae. 2 Cuba

Lachnosiphonium Hochst. = Catunaregam Wolf. Rubiaceae

Lachnospermum Willd. 1803. Asteraceae. 3 South Africa

Lachnostachys Hook. 1841. Chloanthaceae. 6 western and southern Australia

Lachnostoma Kunth = Matelea Aubl. Asclepiadaceae

Lachnostylis Turcz. 1846. Euphorbiaceae. 2 South Africa

Lacistema Sw. 1788. Lacistemataceae. 11 trop. America from southern Mexico to northern Argentina, Paraguay, southeastern Brazil and Uruguay, West Indies

Lacmellea Karst. 1857. Apocynaceae. 20 trop. South America

Lactoris Phil. 1865. Lactoridaceae. 1 Juan Fernández Is. (Más a Tierra I.)

Lactuca L. 1753. Asteraceae. c. 100 cosmopolitan, esp. temp. Northern Hemisphere

Lactucella Nazarova. 1990 (Lactuca L.). Asteraceae. 1 Transcaucasia, eastern Asia Minor, Jordan, Iraq, Iran, Afghanistan, Pakistan, Middle Asia, West Siberia

Lactucosonchus (Sch. Bip.) Svent. 1969. Asteraceae. 1 Canary Is.

Lacunaria Ducke. 1925. Quiinaceae. 12 Costa Rica, Panama, and trop. South America, esp. Brazil and Argentina

Ladenbergia Klotzsch. 1846. Rubiaceae. 31 Costa Rica, Panama, trop. South America

Ladyginia Lipsky. 1904 (Ferula L.). Apiaceae. 3 Middle Asia, Afghanistan

Laelia Lindl. 1831. Orchidaceae. 50 trop. America from Mexico to Brazil, West Indies

Laeliopsis Lindl. ex Paxton. 1853. Orchidaceae. 1 West Indies

Laennecia Cass. 1822. Asteraceae. 15 trop. America from Mexico to Bolivia

Laestadia Kunth ex Less. 1832. Asteraceae. 6 trop. Andes, West Indies

Laetia Loefl. ex L. 1759. Flacourtiaceae. 10 trop. America from southern Mexico to Bolivia, Brazil, Paraguay, West Indies

Lafoensia Vand. 1788. Lythraceae. 11 trop. America

Lafuentia Lag. 1816. Scrophulariaceae. 2 southern Spain (1), Morocco

Lagarinthus E. Mey. = Schizoglossum E. Mey. Asclepiadaceae

Lagarosiphon Harv. 1841. Hydrocharitaceae. 9–16 trop. and South Africa, Madagascar

Lagarosolen W. T. Wang. 1984. Gesneriaceae. 1 southern China (Yunnan)

Lagarostrobos Quinn. 1982 (or = Dacrydium Sol. ex Lamb.). Podocarpaceae. 2 Tasmania (1), New Zealand (1)

Lagascea Cav. 1803. Asteraceae. 8 trop. America, West Indies

Lagedium Soják = Mulgedium Cass. Asteraceae

Lagenandra Dalzell. 1852 (Cryptocoryne Fisch. ex Wydl.). Araceae. 12 southern (3) and northeastern (1) India, Sri Lanka (8)

Lagenantha Chiov. 1929. Chenopodiaceae. 1 Somalia

Lagenanthus Gilg = Lehmanniella Gilg. Gentianaceae

Lagenaria Ser. 1825. Cucurbitaceae. 6 trop. and subtrop. Africa, Madagascar; L. siceraria—pantropics

Lagenias E. Mey. = Sebaea Sol. ex R. Br. Gentianaceae

Lagenifera Cass. = Lagenophora Cass. Asteraceae

Lagenocarpus Nees. 1834. Cyperaceae. 70 West Indies, trop. South America

Lagenophora Cass. 1818. Asteraceae. 15 Australia, New Zealand, Central and South America

Lagerstroemia L. 1759. Lythraceae. 55 trop. and subtrop. Asia, Malesia, and trop. Australia

Lagetta Juss. 1789. Thymelaeaceae. 4 West Indies

Laggera Sch. Bip. ex Oliv. 1847. Asteraceae. 15 trop. Africa, Arabian Peninsula, India, southern China, Taiwan

Lagoa T. Durand. 1888. Asclepiadaceae. 1 Brazil

Lagochilopsis Knorring = Lagochilus Bunge ex Benth. Lamiaceae

Lagochilus Bunge ex Benth. 1834. Lamiaceae. 35 Caucasus (1), Middle Asia (c. 25—endemic), West Siberia, Iran, Afghanistan, northwestern India, China, Mongolia

Lagoecia L. 1753. Apiaceae. 1 Mediterranean, eastward to western and southern Iran, the Crimea

Lagonychium M. Bieb. = Prosopis L. Fabaceae

Lagophylla Nutt. 1841. Asteraceae. 5 western North America

Lagopsis (Benth.) Bunge. 1835 (Marrubium L.). Lamiaceae. 4 Middle Asia, West and East Siberia, Mongolia, China, Japan

Lagoseriopsis Kirp. = Launaea Cass. Asteraceae

Lagoseris M. Bieb. = Crepis L. Asteraceae

Lagotis Gaertn. 1770. Scrophulariaceae. c. 30 northeastern European part of Russia, Caucasus, Middle Asia, West and East Siberia, Far East to Aleutian Is.); Asia Minor, Iran, Pakistan, Himalayas from Kashmir to Bhutan, China (17), Mongolia

Lagrezia Moq. 1849 (Celosia L.). Amaranthaceae. 17 trop. East Africa, Madagascar, Chagos Arch., ? Mexico (1)

Laguna Cav. = Abelmoschus Medik. Malvaceae

Lagunaria (DC.) Rchb. 1828. Malvaceae. 1 eastern Australia, Lord Howe I., Norfolk I.

Laguncularia C. F. Gaertn. 1807. Combretaceae. 2 trop. America, trop. West Africa

Lagurus L. 1753. Poaceae. 1 Mediterranean, Crimea, Caucasus

Lagynias E. Mey. ex Robyns. 1928. Rubiaceae. 4 trop. East (2), Southeast, and South Africa

Lahia Hassk. = Durio Adans. Bombacaceae

Lalldhwojia Farille. 1984. Apiaceae. 2 Nepal, Sikkim

Lallemantia Fisch. et C. A. Mey. = Dracocephalum L. Lamiaceae

Lamanonia Vell. 1825. Cunoniaceae. 6–8 Paraguay, southeastern Brazil

Lamarchea Gaudich. = Meleleuca L. Myrtaceae

Lamarckia Moench. 1794. Poaceae. 1 Mediterranean, Southwest Asia, Iran, Afghanistan, Pakistan

Lambertia Sm. 1798. Proteaceae. 9 Australia

Lamechites Markgr. = Micrechites Miq. Apocynaceae

Lamellisepalum Engl. 1897. Rhamnaceae. 1 trop. Africa

Lamiacanthus Kuntze = Strobilanthus Blume. Acanthaceae

Lamiastrum Heist. ex Fabr. = Lamium L. Lamiaceae

Lamiodendron Steenis. 1957. Bignoniaceae. 1 New Guinea

Lamiophlomis Kudo. 1929 (or = Phlomis L.). Lamiaceae. 1 Himalayas, Tibet

Lamium L. 1753. Lamiaceae. c. 40 temp. Eurasia, North, trop. (a few sp.), and South Africa (1)

Lamourouxia Kunth. 1818. Scrophulariaceae. 26 trop. America from Mexico to Peru

Lampas Danser. 1929. Loranthaceae. 1 Borneo

Lampaya Phil. = Lampayo Phil. ex Murillo. Verbenaceae

Lampayo Phil. ex Murillo. 1889. Verbenaceae. 3 Bolivia, Chile, and Argentina

Lamprachaenium Benth. 1873. Asteraceae. 1 India

Lampranthus N. E. Br. 1930. Aizoaceae. c. 180 southern Namibia, South Africa

Lamprocaulos Mast. = Elegia L. Restionaceae

Lamprocephalus B. Nord. 1976. Asteraceae. 1 South Africa (southwestern Cape)

Lamprococcus Beer. 1857 (Aechmea Ruiz et Pav.). Bromeliaceae. 9 trop. America

Lamproconus Lem. = Pitcairnia L'Her. Bromeliaceae

Lamprolobium Benth. 1864. Fabaceae. 1 Australia (Queensland)

Lamprophragma O. E. Schulz = Pennellia Nieuwl. Brassicaceae

Lamprothamnus Hiern. 1877 (Polysphaeria Hook. f.) Rubiaceae. 1 (L. zanguebaricus) East Africa (Somalia, Kenya, Tanzania)

Lamprothyrsus Pilg. 1906. Poaceae. 3 South America: Bolivia to Argentina

Lamyra (Cass.) Cass. = Ptilostemon Cass. Asteraceae

Lamyropappus Knorring et Tamamsch. 1954. Asteraceae. 1 Middle Asia

Lamyropsis (Charadze) Dittrich = Cirsium Hill. Asteraceae

Lanaria Aiton. 1789. Lanariaceae. 1 South Africa

Lancea Hook. f. et Thomson. 1857. Scrophulariaceae. 2 western China, Tibet, Himalayas from Kashmir to Bhutan

Lancisia Farb. = Cotula L. Asteraceae

Landolphia P. Beauv. 1806. Apocynaceae. 60 trop. and South Africa, Madagascar, Mascarene Is.

Landtia Less. = Haplocarpha Less. Asteraceae

Lanessania Baill. = Trymatococcus Poepp. et Endl. Moraceae

Langebergia Anderb. 1991. Asteraceae. 1 South Africa

Langlassea H. Wolff = Prionosciadium S. Watson. Apiaceae

Langloisia Greene. 1896 (Gilia Ruiz et Pav.). Polemoniaceae. 4 western U.S.: Idaho, Nevada, California, Utah, Arizona; northwestern Mexico

Langsdorffia Mart. 1818. Balanophoraceae. 3 Madagascar (1), New Guinea (1), trop. America (1 from Mexico to trop. South America)

Lanium (Lindl.) Benth. 1881. Orchidaceae. 6 Colombia, Venezuela, the Guianas, Peru, and Brazil

Lankesterella Ames. 1923 (Stenorrhynchos Rich. ex Spreng.). Orchidaceae. 8 trop. America from Costa Rica and Venezuela to Ecuador and Brazil

Lankesteria Lindl. 1845. Acanthaceae. 7 trop. Africa, Madagascar

Lannea A. Rich. 1831. Anacardiaceae. 40 trop. and South Africa, southern Arabian Peninsula, Socotra, South and Southeast Asia

Lansium Corrêa. 1807. Meliaceae. 3 Thailand, Malesia

Lantana L. 1753. Verbenaceae. 150–160 trop. America, West Indies, trop. and

South Africa, trop. Asia (2)

Lantanopsis C. Wright ex Griseb. 1862. Asteraceae. 2 West Indies

Lanugia N. E. Br. = Mascarhenasia A. DC. Apocynaceae

Lapageria Ruiz et Pav. 1802. Philesiaceae. 1 Chile and western Argentina

Lapeirousia Pourr. 1788. Iridaceae. 38 trop. and South Africa from northern Ethiopia to Angola, northern and southwestern Cape

Laphamia A. Gray = Perityle Benth. Asteraceae

Lapidaria (Dinter et Schwantes) Schwantes ex N. E. Br. 1928 (Dintheranthus Schwantes). Aizoaceae. 2 southern Namibia, western South Africa

Lapiedra Lag. 1816. Amaryllidaceae. 1 Spain, ? Morocco

Laplacea Kunth = Gordonia Ellis. Theaceae

Laportea Gaudich. 1830. Urticaceae. c. 50 pantropics, eastern North America, South Africa, temp. East Asia; L. bulbifera—extending to southern Far East of Russia

Lappula Moench. 1794. Boraginaceae. c. 60 North America (5) temp. Eurasia, Mediterranean, Australia

Lapsana L. 1753. Asteraceae. 9 temp. Eurasia; Russia from European part and northern Caucasus to East Siberia

Lardizabala Ruiz et Pav. 1794. Lardizabalaceae. 2 central and southern Chile and Juan Fernández Is.

Larentia Klatt = Cypella Herb. Iridaceae

Laretia Gillies et Hook. 1830. Apiaceae. 2 Chilean Andes

Larix Hill. 1753. Pinaceae. 9 North America, cold and temp. Eurasia, esp. Russia (5 West and East Siberia, Far East)

Larnax Miers = Deprea Raf. Solanaceae

Larrea Cav. 1800. Zygophyllaceae. 5 southwestern North and temp. South America

Larsenaikia Tirveng. 1993. Rubiaceae. 3 trop. northeastern Australia

Larsenia Bremek. 1965. Acanthaceae. 1 Thailand

Lasallea Greene = Aster L. Asteraceae

Laseguea A. DC. = Mandevilla Lindl. Apocynaceae

Laser Borkh. ex P. Gaertn., B. Mey. et Schreb. 1799. Apiaceae. 1 Central and South Europe, West Asia, incl. Caucasus

Laserpitium L. 1753. Apiaceae. c. 25 Canary Is., Mediterranean, West (incl. Caucasus, 6) and Southwest Asia

Lasersisia Liben. 1991. Sapotaceae. 1 Gabon, Congo, Zaire

Lasia Lour. 1790. Araceae. 2 India, Sri Lanka, Nepal, Burma, southern China (incl. Hainan), Indochina, Malay Peninsula, Sumatra, Java, Borneo, New Guinea

Lasiacis (Griseb.) Hitchc. 1910. Poaceae. c. 30 trop. and subtrop. America, Madagascar

Lasiadenia Benth. 1845. Thymelaeaceae. 2 Colombia, Venezuela, Brazil

Lasiagrostis Link = Achnatherum P. Beauv. Poaceae

Lasianthaea DC. 1836. Asteraceae. 11 America from Arizona and Mexico to Panama

Lasianthera P. Beauv. 1807. Icacinaceae. 1 trop. West Africa

Lasianthus Jack. 1823. Rubiaceae. 150 trop. Africa (20), trop. (esp. Southeast) Asia, Malesia south to Australia, West Indies (1)

Lasiarrhenum I. M. Johnst. 1924. Boraginaceae. 1 Mexico

Lasimorpha Schott. 1857. Araceae. 1 trop. Africa: Senegal, Gambia, Guinea, Sierra Leone, Liberia, Ivory Coast, Ghana, Benin, Nigeria, Chad, Cameroun, Central African Rep., Gabon, Congo, Zaire, Angola

Lasiobema (Korth.) Miq. = Bauhinia L. Fabaceae

Lasiocarpus Liebm. 1853. Malpighiaceae. 4 Mexico

Lasiocaryum I. M. Johnst. 1925. Boraginaceae. 4–5 Himalayas from Kashmir to Bhutan, southern Tibet

Lasiocephalus Schltdl. 1818. Asteraceae. 21 Andes

Lasiochlamys Pax et K. Hoffm. 1922. Flacourtiaceae. 13 New Caledonia

Lasiochloa Kunth = Tribolium Desv. Poaceae

Lasiocladus Bojer ex Nees. 1847. Acanthaceae. 5 Madagascar

Lasiococca Hook. f. 1887. Euphorbiaceae. 3 eastern Himalayas, India, Hainan, Malay Peninsula

Lasiocoma Bolus = Euryops (Cass.) Cass. Asteraceae

Lasiocorys Benth. = Leucas R. Br. Lamiaceae

Lasiocroton Griseb. 1859. Euphorbiaceae. 6 West Indies

Lasiodiscus Hook. f. 1862 (Colubrina Rich. ex Brongn.). Rhamnaceae. 9 trop. Africa, Madagascar

Lasiolaena R. M. King et H. Rob. 1972. Asteraceae. 5 Brazil

Lasiopetalum Sm. 1789. Sterculiaceae. 37 Australia

Lasiopogon Cass. 1818. Asteraceae. 3 Mediterranean, South Africa, West Asia eastward to India; L. muscoides—also in Caucasus

Lasiorrhachis (Hack.) Stapf. 1927 (Saccharum L.). Poaceae. 3 Madagascar

Lasiosiphon Fres. = Gnidia L. Thymelaeaceae

Lasiospermum Lag. 1816. Asteraceae. 1 South Africa

Lasiostelma Benth. = Brachystelma R. Br. Asclepiadaceae

Lasiurus Boiss. 1859. Poaceae. 3 trop. East Africa, Southwest and South Asia

Lastarriaea J. Remy. 1851–1852. Polygonaceae. 1 North America (coastal southwestern California to central Baja California) and South America (deserts of Chile)

Lasthenia Cass. 1834. Asteraceae. 16 southwestern U.S., Chile

Latace Phil. = Leucocoryne Lindl. Alliaceae

Latania Comm. ex Juss. 1789. Arecaceae. 3 Mascarene Is.

Lateropora A. C. Sm. 1932. Ericaceae. 2 Panama

Lathraea L. 1753. Orobanchaceae. 5 temp. Eurasia; L. squamaria—European part of Russia and Caucasus

Lathraeocarpa Bremek. 1957. Rubiaceae. 2 Madagascar

Lathriogyna Eckl. et Zeyh. = Amphithalea Eckl. et Zeyh. Fabaceae

Lathrophytum Eichler. 1868. Lophophytaceae. 1 southeastern Brazil

Lathyrus L. 1753. Fabaceae. 150 temp. Northern Hemisphere, Mediterranean, montane areas of trop. East Africa, temp. South America

Latipes Kunth = Leptothrium Kunth Poaceae

Latouchea Franch. 1899. Gentianaceae. 1 eastern China

Latourea Blume = Dendrobium Sw. Orchidaceae

Latourorchis Brieger = Dendrobium Sw. Orchidaceae

Latrobea Meisn. 1848. Fabaceae. 6 western and southwestern Australia

Latua Phil. 1858. Solanaceae. 1 southern Chile

Laubertia A. DC. 1844. Apocynaceae. 6 Central and trop. South America

Laumoniera Nooteb. 1987 (Brucea J. F. Mill.). Simaroubaceae. 1 Sumatra

Launaea Cass. 1822. Asteraceae. 40 Canary Is.; Mediterranean; trop. and South Africa; West, South, East, and Southeast Asia; L. korovinii—Middle Asia

Lauradia Vand. 1788. Sauvagesiaceae. 7 Brazil

Laurelia Juss. 1809. Monimiaceae. 2 New Zealand (1), Peru and Chile (1)

Laureliopsis Schodde. 1983 (Atherosperma Labill.). Monimiaceae. 1 Chile and Patagonia

Laurembergia Bergius. 1767. Haloragaceae. 4 tropics and subtropics

Laurentia Adans. = Solenopsis C. Presl. Lobeliaceae

Laurocerasus Hill. 1753 (Prunus L.). Rosaceae. c. 80 Canary Is., Portugal, Balkan Peninsula, Caucasus (1, L. officinalis), trop. Africa, Madagascar, temp. and trop. Asia, America

Laurophyllus Thunb. 1792. Anacardiaceae. 1 South Africa

Laurus L. 1753. Lauraceae. 2 Azores, Madeira and Canary Is., Morocco (L. azorica), Mediterranean (L. nobilis)

Lautembergia Baill. 1858. Euphorbiaceae. 4 Madagascar, Mauritius

Lauterbachia Perkins. 1900. Monimiaceae. 1 New Guinea

Lavandula L. 1753. Lamiaceae. 28 Madeira, Canary and Cape Verde Is., Europe (7), Mediterranean (15) southward to Somalia, West and South Asia

Lavatera L. 1753. Malvaceae. 25 Canary Is., Mediterranean (18), West (incl. Caucasus) and Middle Asia, West and East Siberia south to northwestern Himalayas, Australia, western North America (southern and Baja California and adjacent islands)

Lavauxia Spach = Oenothera L. Onagraceae

Lavigeria Pierre. 1892. Icacinaceae. 1 trop. West Africa

Lavoisiera DC. 1828. Melastomataceae. 60 Brazil

Lavoixia H. E. Moore. 1978. Arecaceae. 1 New Caledonia

Lavradia Roem. = Lauradia Vand. Sauvagesiaceae

Lavrania Plowes. 1986. Asclepiadaceae. 1 South-West Africa (Namibia)

Lawia Griff. ex Tul. = Dalzellia Wight. Podostemaceae

Lawrencella Lindl. 1839 (Helichrysum L.). Asteraceae. 35 Australia, New Zealand

Lawrencia Hook. 1840 (Plagianthus J. R. Forst. et G. Forst.). Malvaceae. 7 Australia

Lawsonia L. 1753. Lythraceae. 1 North Africa, trop. Asia and Australia

Laxmannia R. Br. 1810. Asphodelaceae. 13 Australia, esp. southwestern part, Tasmania (1)

Laxoplumeria Markgr. 1926. Apocynaceae. 3 eastern Peru, Brazil

Layia Hook. et Arn. ex DC. 1838. Asteraceae. 15 western North America from British Columbia and Idaho to Baja California

Lazarum A. Hay. 1992. Araceae. 1 Australia (Melville Is.)

Leandra Raddi. 1820. Melastomataceae. 200 trop. America from Mexico to Paraguay and Argentina, West Indies

Leandriella Benoist. 1939. Acanthaceae. 1 Madagascar

Leaoa Schltr. et C. Porto. 1922 (or = Scaphyglottis Poepp. et Endl.). Orchidaceae. 2 Venezuela, Brazil

Leavenworthia Torr. 1837. Brassicaceae. 7 southeastern U.S.

Lebeckia Thunb. 1800. Fabaceae. 44 South Africa

Lebetanthus Endl. 1841 (Prionotes R. Br.). Epacridaceae. 1 temp. South America (Patagonia, Tierra del Fuego)

Lebetina Cass. = Adenophyllum Pers. Asteraceae

Lebronnecia Fosberg. 1966. Malvaceae. 1 Marquesas Is.

Lebrunia Staner. 1934. Clusiaceae. 1 trop. Africa

Lebruniodendron J. Léonard. 1951. Fabaceae. 1 Cameroun, Zaire

Lecananthus Jack. 1822. Rubiaceae. 2 western Malesia

Lecaniodiscus Planch. ex Benth. 1849. Sapindaceae. 3 trop. Africa

Lecanorchis Blume. 1856. Orchidaceae. 5 East and Southeast Asia, Malesia (Malay Peninsula, Indonesia, Philippines, New Guinea)

Lecanosperma Rusby. 1893 (Heterophyllaea Hook. f.). Rubiaceae. 1 Bolivia

Lecanthus Wedd. 1854. Urticaceae. 5 trop. Africa (1), India, Burma, southern China, Taiwan, Indochina, Malesia, Fiji

Lecardia Poiss. et Guillaumin. 1927 (Salaciopsis Baker f.). Celastraceae. 1 New Caledonia

Lecariocalyx Bremek. 1940. Rubiaceae. 1 Borneo

Lechea L. 1753. Cistaceae. 21 North and Central America, West Indies

Lechenaultia R. Br. 1810. Goodeniaceae. 25 New Guinea (1), Australia

Lecocarpus Decne. 1846. Asteraceae. 3 Galapagos Is.

Lecointea Ducke. 1922. Fabaceae. 4 trop. South America

Lecokia DC. 1829. Apiaceae. 1 Crete, Cyprus, Anatolia, Syria, Lebanon, Israel, northern Iran

Lecomtea Koidz. = Cladopus H. Moeller. Podostemaceae

Lecomtedoxa Dubard. 1914. Sapotaceae. 5 trop. West Africa (Gabon)

Lecomtella A. Camus. 1925. Poaceae. 1 Madagascar

Lecontea A. Rich. ex DC. = Paederia L. Rubiaceae

Lecostemon Mociño et Sessé ex DC. = Rhabdodendron Gilg et Pilg. Rhabdodendraceae

Lecythis Loefl. 1758. Lecythidaceae. 26 trop. America from Nicaragua to southern Brazil (Sao Paulo)

Leda C. B. Clarke = Leptostachya Nees. Acanthaceae

Ledebouria Roth. 1821 (Scilla L.). Hyacinthaceae. 30 trop. and South Africa, India

Ledebouriella H. Wolff. 1910. Apiaceae. 2 eastern Kazakhstan, Altai

Ledenbergia Klotzsch ex Moq. = Flueckigera Kuntze. Petiveriaceae

Ledermannia Mildbr. et Burret = Desplatsia Bocquillon. Tiliaceae

Ledermanniella Engl. 1909. Podostemaceae. 23 trop. Africa, Madagascar

Ledothamnus Meisn. 1863. Ericaceae. 9 Venezuela, Guyana

Ledum L. 1753 (Rhododendron L.). Ericaceae. 10 arctic and temp. Northern Hemisphere; Russia (4 northern European part, Siberia, Far East)

Leea Royen ex L. 1767. Leeaceae. 34: 2 trop. Africa and Madagascar, 32—South and Southeast Asia, Australia eastward to Fiji

Leersia Sw. 1788. Poaceae. 18 trop. and warm–temp. regions; L. oryzoides—European part of Russia, Caucasus, Altai, Middle Asia, southern Russian Far East

Leeuwenbergia Letouzey et N. Hallé. 1974 (Joannesia Vell.). Euphorbiaceae. 2 Cameroun, Gabon, Congo

Lefebvrea A. Rich. 1840. Apiaceae. 6 trop. and South Africa

Legazpia Blanco. 1845. Scrophulariaceae. 1 trop. Asia

Legenere McVaugh. 1943. Lobeliaceae. 2 California (1), Chile (1)

Legnephora Miers. 1867. Menispermaceae. 5 Philippines (1), Moluccas, Timor I., New Guinea, New Ireland I., Solomon Is. (Santa Cruz Is.), eastern Australia

Legousia Durande. 1782. Campanulaceae. 15 temp. Northern Hemisphere, South America; 4—coasts of the Black Sea, Crimea, Caucasus, Middle Asia

Legrandia Kausel. 1944. Myrtaceae. 1 Chile

Lehmanniella Gilg. 1895. Gentianaceae. 4 Panama, Colombia, Peru

Leiandra Raf. = Callisia Loefl. Commelinaceae

Leibergia J. M. Coult. et Rose = Tauschia Schltr. Apiaceae

Leibnitzia Cass. 1822. Asteraceae. 5 Middle Asia, East Siberia, Russian Far East, Himalayas, China, Mongolia, Korea, Japan, Central America (1)

Leiboldia Schltdl. ex Gleason. 1847. Asteraceae. 1 Mexico, Central America

Leichardtia R. Br. = Leichhardtia R. Br. Asclepiadaceae

Leichhardtia F. Muell. (1876) = ? Phyllanthus L. Euphorbiaceae

Leichhardtia R. Br. 1849. Asclepiadaceae. 6 Australia, New Caledonia, Fiji (1)

Leichhardtia Shepherd (1851) = Callitris Vent. Cupressaceae

Leidesia Muell. Arg. 1866 (Seidelia Baill.). Euphorbiaceae. 3 South Africa

Leiophaca Lindau = Whitfieldia Hook. Acanthaceae

Leiophyllum (Pers.) R. G. Hedw. 1806. Ericaceae. 1 eastern North America

Leiopoa Ohwi = Leucopoa Griseb. Poaceae

Leiospora (C. A. Mey.) Dvorak. 1968. Brassicaceae. 6 Middle Asia, West and East Siberia, Pakistan, India, western China, Mongolia

Leiostemon Raf. = Penstemon Schmidel. Scrophulariaceae

Leiothrix Ruhland. 1903. Eriocaulaceae. 37 South America, esp. Brazil (35)

Leiothylax Warm. 1899. Podostemaceae. 3 trop. Africa

Leiphaimos Cham. et Schltdl. = Voyria Aubl. Gentianaceae

Leipoldtia L. Bolus. 1927 (Hallianthus H. E. K. Hartmann). Aizoaceae. 10 southwestern Namibia, western and southern South Africa

Leitgebia Eichler. 1871 (Sauvagesia L.). Sauvagesiaceae. 4 northern South America

Leitneria Chapm. 1860. Leitneriaceae. 1 southeastern U.S. from southern Missouri to Texas and Florida

Leleba (Kurz) Nakai. 1933 (Bambusa Schreb.). Poaceae. 6 Southeast Asia, western Oceania

Lelya Bremek. 1952. Rubiaceae. 1 trop. Africa (Nigeria, Zaire, Angola, Tanzania, Zambia, Malawi)

Lemaireocereus Britton et Rose = Pachycereus (A. Berger) Britton et Rose. Cactaceae

Lembertia Greene. 1897. Asteraceae. 1 California

Lembocarpus Leeuwenb. 1958. Gesneriaceae. 1 northern South America

Lemboglossum Halbinger. 1984. Orchidaceae. 14 trop. America

Lembotropis Griseb. (or = Cytisus L.). Fabaceae. 2–3 Central, South, and East Europe

Lemmonia A. Gray. 1876. Hydrophyllaceae. 1 California

Lemna L. 1753. Lemnaceae. 9–10 cosmopolitan, esp. cold and temp. regions

Lemooria P. S. Short. 1989. Asteraceae. 1 southwestern, southern, and southeastern Australia

Lemphoria O. E. Schulz. 1924 (Arabidella [F. Muell.] O. E. Schulz). Brassicaceae. 1 southern Australia

Lemurella Schltr. 1925. Orchidaceae. 7 Madagascar

Lemurodendron Villiers et Guinet. 1990. Fabaceae. 1 Madagascar

Lemurophoenix J. Dransf. 1991. Arecaceae. 1 Madagascar

Lemuropisum H. Perrier. 1939. Fabaceae. 1 Madagascar

Lemurorchis Kraenzl. 1893. Orchidaceae. 1 Madagascar and neighboring islands

Lemurosicyos Keraudren. 1964. Cucurbitaceae. 1 Madagascar

Lemyrea (A. Chev.) A. Chev. et Beille. 1939 (Coffea L.). Rubiaceae. 3 Madagascar

Lenbrassia G. W. Gillett. 1974. Gesneriaceae. 1 Australia (northern Queensland)

Lencymmoea C. Presl. 1851. Myrtaceae. 1 Burma

Lennea Klotzsch. 1842. Fabaceae. 6 trop. America from Mexico to Uruguay

Lennoa Llave ex Lex. 1824. Lennoaceae. 1 Mexico, Guatemala, Colombia, Venezuela

Lenophyllum Rose. 1904. Crassulaceae. 6 southern U.S., Mexico

Lens Mill. 1754. Fabaceae. 4–7 Mediterranean; L. ervoide—extending to Ethiopia, Uganda, and Zaire; Arabian Peninsula; Iraq; Iran; Afghanistan; ? China; India

Lentibularia Séguier = Utricularia L. Lentibulariaceae

Lenzia Phil. 1863. Portulacaceae. 1 Chile

Leocereus Britton et Rose. 1920. Cactaceae. 1 eastern Brazil

Leochilus Knowles et Westc. 1838. Orchidaceae. 9 trop. America: Mexico (6), Guatemala, El Salvador, Honduras, Nicaragua, Costa Rica, Panama, Colombia, Venezuela, Ecuador; West Indies (Cuba, Jamaica, Haiti, Puerto Rico, Martinique, Trinidad)

Leocus A. Chev. 1909. Lamiaceae. 1 trop. West Africa

Leonardendron Aubrév. = Anthonotha P. Beauv. Fabaceae

Leonardoxa Aubrév. 1968. Fabaceae. 3 Ghana, Nigeria, Cameroun, Gabon, Zaire, Angola

Leonia Ruiz et Pav. 1799. Violaceae. 6 trop. South America

Leonohebe Heads = Hebe Comm. ex Juss. Scrophulariaceae

Leonotis (Pers.) R. Br. 1810. Lamiaceae. 15 trop. Africa (south of the Sahara) and South Africa (3); L. nepetifolia—pantropics

Leontice L. 1753. Berberidaceae. 3–5 from Mediterranean through Caucasus, Middle Asia, and southern West Siberia to East Asia

Leontochir Phil. 1873. Alstroemeriaceae. 1 Chile

Leontodon L. 1753. Asteraceae. 40 temp. Eurasia from tundra zone to Mediterranean and Iran

Leontonyx Cass. = Helichrysum Mill. Asteraceae

Leontopodium (Pers.) R. Br. 1810. Asteraceae. 35 mountainous Eurasia (esp. Southeast Asia), Andes

Leonuroides Rauschert = Panzerina Soják. Lamiaceae

Leonurus L. 1753. Lamiaceae. 4 (or c. 20 ?) temp. Eurasia

Leopoldia Parl. = Muscari Hill. Hyacinthaceae

Leopoldinia Mart. 1824. Arecaceae. 3–4 Amazonian Colombia, southern Venezuela, western Brazil

Lepanthes Sw. 1799. Orchidaceae. 100 trop. America from Mexico to Bolivia, Chile, and Brazil, esp. Costa Rica and Colombia; West Indies

Lepanthopsis Ames. 1933. Orchidaceae. 23 trop. America: Mexico, Guatemala, Honduras, El Salvador, Nicaragua, Costa Rica, Panama, Colombia, Venezuela, Ecuador, Peru, Bolivia, Brazil; West Indies

Lepargochloa Launert. 1963 (Loxodera Launert). Poaceae. 1 southern trop. Africa

Lepechinella Airy Shaw = Lepechiniella Popov. Boraginaceae

Lepechinia Willd. 1804. Lamiaceae. 38 western America from California to Argentina, Hawaiian Is.

Lepechiniella Popov. 1953. Boraginaceae. 14 Middle Asia, China (1)

Lepeostegeres Blume. 1830. Loranthaceae. 13 western Malesia to Sulawesi

Lepianthes Raf. = Piper L. Piperaceae

Lepidacanthus C. Presl. 1845 (Aphelandra R. Br.). Acanthaceae. 2 Brazil

Lepidagathis Willd. 1800. Acanthaceae. 100 tropics and subtropics

Lepidaria Tiegh. 1895. Loranthaceae. 12 western Malesia

Lepiderema Radlk. 1879. Sapindaceae. 8 New Guinea (2), northeastern Australia (6)

Lepidesmia Klatt. 1896. Asteraceae. 1 Cuba, Colombia, Venezuela

Lepidium L. 1753. Brassicaceae. 150 cosmopolitan

Lepidobolus Nees. 1846. Restionaceae. 5 southern Australia

Lepidobotrys Engl. 1902. Lepidobotryaceae. 1 trop. Africa

Lepidocaryum Mart. 1824. Arecaceae. 3 Colombia, Peru, Venezuela, Guyana, Brazil

Lepidoceras Hook. f. 1846. Eremolepidaceae. 2 southern Peru, Chile

Lepidocordia Ducke. 1925. Ehretiaceae. 1 Brazil

Lepidogyne Blume. 1859. Orchidaceae. 3 Malaysia

Lepidolopha C. Winkl. 1894. Asteraceae. 6 Middle Asia

Lepidolopsis Poljakov. 1956. Asteraceae. 1 Middle Asia, Iran, Afghanistan

Lepidonia S. F. Blake. 1936. Asteraceae. 7 Mexico, Guatemala, Costa Rica (1, L. lankesteri)

Lepidopetalum Blume. 1849. Sapindaceae. 6 Andaman and Nicobar Is., Sumatra, Philippines, Tanimbar Is., Key I., New Guinea, Bismarck Arch., trop. Australia

Lepidophorum Neck. ex DC. 1838. Asteraceae. 1 Portugal

Lepidophyllum Cass. 1816. Asteraceae. 1 Patagonia

Lepidorrhachis (H. A. Wendl. et Drude) Becc. ex Martellii. 1934. Arecaceae. 1 Lord Howe I.

Lepidospartum (A. Gray) A. Gray. 1883. Asteraceae. 3 southwestern U.S.

Lepidosperma Labill. 1805. Cyperaceae. 50 southern China, Malay Peninsula, Sumatra, Borneo, New Guinea, Australia, New Caledonia, New Zealand

Lepidostemon Hook. f. et Thomson. 1861. Brassicaceae. 1 eastern Himalayas

Lepidostephium Oliv. 1868. Asteraceae. 2 South Africa

Lepidostoma Bremek. 1940. Rubiaceae. 1 Sumatra

Lepidothamnus Phil. 1860 (or = Dacrydium Sol. ex Lamb.). Podocarpaceae. 3 New Zealand (2), southern Chile (1)

Lepidotrichilia (Harms) J.-F. Leroy. 1958. Meliaceae. 4 trop. Africa (1, L. volkensii—Ethiopia, Sudan, Zaire, Rwanda, Burundi, Malawi, Uganda, Kenya, Tanzania), Madagascar (3)

Lepidotrichum Velen. et Bornm. = Alyssum L. Brassicaceae

Lepidozamia Regel. 1857. Zamiaceae. 1 northeastern and eastern Australia

Lepilaena J. L. Drumm. ex Harv. 1855. Zannichelliaceae. 4 Australia, New Zealand

Lepinia Decne. 1849. Apocynaceae. 3 Caroline and Solomon Is., Tahiti

Lepiniopsis Valeton. 1895. Apocynaceae. 2 Philippines, Moluccas, Palau Is., Micronesia

Lepionurus Blume. 1872. Opiliaceae. 1 (L. sylvestris) Nepal, Sikkim, Assam, Burma, southern China, Thailand, southern Vietnam, Sumatra, Malay Peninsula, Java, Borneo, New Guinea

Lepipogon G. Bertol. = Catunaregam Adans. Boraginaceae

Lepironia Pers. 1805. Cyperaceae. 1 Madagascar, trop. and subtrop. Asia, Australia, Polynesia

Lepisanthes Blume. 1825. Sapindaceae. 24 West Africa, China, India, Southeast Asia, Philippines, New Guinea, Australia (1, L. rubiginosa)

Lepismium Pfeiff. 1835. Cactaceae. 16 eastern Bolivia, Brazil, Argentina

Lepistemon Blume. 1826. Convolvulaceae. 10 trop. Old World from Africa to Asia and Australia

Lepistemonopsis Dammer. 1895. Convolvulaceae. 1 Ethiopia, Kenya, Tanzania

Leporella A. S. George. 1971. Orchidaceae. 1 southwestern Australia, Victoria, and Tasmania

Leptacanthus Nees. 1832 (Strobilanthes Blume). Acanthaceae. 5 India, Sri Lanka

Leptactina Hook. f. 1871. Rubiaceae. 25 (or 4 ?) trop. and South Africa

Leptadenia R. Br. 1810. Asclepiadaceae. 10 North and trop. Africa, Madagascar, Mascarene Is., Arabian Peninsula, South Asia to Burma

Leptagrostis C. E. Hubb. 1937. Poaceae. 1 Ethiopia

Leptaleum DC. 1821. Brassicaceae. 2 Southeast Europe, eastern Mediterranean, Iraq, Iran, Afghanistan, Pakistan, China; 1, L. filifolium—southeastern European part of Russia, Caucasus, and Middle Asia

Leptaloe Stapf = Aloe L. Asphodelaceae

Leptandra Nutt. = Veronicastrum Heist. ex Fabr. Scrophulariaceae

Leptarrhena R. Br. 1823. Saxifragaceae. 1 western North America

Leptaspis R. Br. 1810. Poaceae. 5 trop. Africa from Guinea to Madagascar, Mascarene Is., South and Southeast Asia eastward to Taiwan and Bougainville Is., Australia, New Caledonia, Fiji

Leptaulus Benth. 1862. Icacinaceae. 6 West and Central Africa, Madagascar

Leptinella Cass. 1822. Asteraceae. 33 New Guinea, Australia, New Zealand, subantarctic islands, South America

Leptobaea Benth. 1876. Gesneriaceae. 3 eastern Himalayas from Nepal to Assam, China (1), Thailand, Borneo (1)

Leptocallisia (Benth. et Hook. f.) Pichon = Callisia Loefl. Commelinaceae

Leptocanna L. C. Chia et H. L. Fung = Schizostachyum Nees. Poaceae

Leptocarpha DC. 1836. Asteraceae. 1 Andes

Leptocarpus R. Br. 1810. Restionaceae. 16 Southeast Asia, Malesia (3, Malay Peninsula, Aru Is., New Guinea), Australia, Tasmania, New Zealand, Chile, and southern Argentina (1)

Leptocarydion Stapf. 1898. Poaceae. 1 trop. Africa from Ethiopia and Sudan through East Africa (Uganda, Kenya, Tanzania) and Zimbabwe to Angola; South Africa (Transvaal, Natal), Madagascar

Leptoceras Fitzg. = Leporella A. S. George. Orchidaceae

Leptocereus (A. Berger) Britton et Rose. 1909. Cactaceae. 10 West Indies

Leptochiton Sealy. 1937 (Hymenocallis Salisb.). Amaryllidaceae. 1 Andes

Leptochloa P. Beauv. 1812. Poaceae. c. 40 tropics and subtropics

Leptochloopsis H. O. Yates = Uniola L. Poaceae

Leptoclinium (Nutt.) Benth. 1873. Asteraceae. 1 Brazil

Leptocodon Hook. f. ex Lem. 1856. Campanulaceae. 2 eastern Himalayas (Nepal, Sikkim), western China, Burma

Leptocoryphium Nees. 1829. Poaceae. 1 America from Mexico to Argentina, West Indies

Leptodactylon Hook. et Arn. 1839. Polemoniaceae. 5 western North America from British Columbia and Montana to Baja California and New Mexico

Leptodermis Wall. 1824. Rubiaceae. 30 Himalayas, northeastern India, China, Japan

Leptoderris Dunn. 1910. Fabaceae. 26 trop. Africa from Senegal to Tanzania, south to Angola and Zaire

Leptodesmia Benth. 1865. Fabaceae. 6 Madagascar (5), India (1)

Leptofeddea Diels = Leptoglossis Benth. Solanaceae

Leptoglossis Benth. 1845. Solanaceae. 7 Peru, Argentina

Leptoglottis DC. = Schrankia Willd. Fabaceae

Leptogonum Benth. 1880. Polygonaceae. 1 (L. domingense) Haiti (Hispaniola)

Leptolaena Thouars. 1805. Sarcolaenaceae. 12 Madagascar

Leptoloma Chase. 1906 (Digitaria Haller). Poaceae. 5 China (1, L. fujianensis), Australia, U.S.

Leptomeria R. Br. 1810. Santalaceae. 18 southern Australia, Tasmania

Leptomischus Drake. 1895. Rubiaceae. 1 Southeast Asia

Leptonema A. Juss. 1824. Euphorbiaceae. 2 Madagascar

Leptonychia Turcz. 1858. Sterculiaceae. 45 trop. Africa, southern India, Burma, western Malesia (excl. Java), New Guinea

Leptonychiopsis Ridl. = Leptonychia Turcz. Sterculiaceae

Leptopharyngia (Stapf) Boiteau = Tabernaemontana L. Apocynaceae

Leptopharynx Rydb. = Perityle Benth. Asteraceae

Leptophoenix Becc. = Gronophyllum Scheff. Arecaceae

Leptophyllochloa Calderon ex E. G. Nicora. 1978 (Koeleria Pers.). Poaceae. 1 southern South America

Leptoplax O. E. Schulz = Peltaria Jacq. Brassicaceae

Leptopus Decne. 1836 (Andrachne L.). Euphorbiaceae. 20 western Himalayas to northern and southwestern China, continental Southeast Asia, Malesia (excl. Sumatra, Borneo, Sulawesi), Australia; L. colchicus—Caucasus

Leptopyrum Rchb. 1832. Ranunculaceae. 1 West and East Siberia, Russian Far East, Mongolia, western and southwestern China, Korea, Japan

Leptorhabdos Schrenk. 1841. Scrophulariaceae. 1 Caucasus, Middle Asia, Iran, northwestern and western China, Himalayas from Kashmir to Nepal

Leptorhoeo C. B. Clarke ex Hemsl. = Callisia Loefl. Commelinaceae

Leptorhynchos Less. 1832. Asteraceae. 9 temp. Australia

Leptosaccharum (Hack.) A. Camus. 1923 (Eriochrysis P. Beauv.) Poaceae. 1 Brazil, Paraguay

Leptoscela Hook. f. 1873. Rubiaceae. 1 eastern Brazil

Leptosema Benth. 1837 (Brachysema R. Br.). Fabaceae. 8 Australia

Leptosiphon Benth. = Gilia Ruiz et Pav. Polemoniaceae

Leptosiphonium F. Muell. 1886. Acanthaceae. 10 China (1), New Guinea, Solomon Is.

Leptosolena C. Presl. 1827 (Alpinia Roxb.). Zingiberaceae. 1 Philippines

Leptospermopsis S. Moore = Leptospermum J. R. Forst. et G. Forst. Myrtaceae

Leptospermum J. R. Forst. et G. Forst.

1775. Myrtaceae. 79: 1 (L. javanicum) from southern Burma and Thailand through Malesia to Philippines, Moluccas, and Lesser Sunda Is.; New Guinea; northern, southern, and eastern Australia; Tasmania; New Zealand

Leptostachya Nees. 1832 (Justicia L.). Acanthaceae. 1 southern India, Southeast Asia

Leptostigma Arn. = Nertera Banks et Sol. ex Gaertn. Rubiaceae

Leptostylis Benth. 1876. Sapotaceae. 8 New Caledonia

Leptotaenia Nutt. ex Torr. et A. Gray = Lomatium Raf. Apiaceae

Leptoterantha Louis ex Troupin. 1949. Menispermaceae. 1 Gabon, Angola, Congo

Leptotes Lindl. 1833. Orchidaceae. 6 trop. South America

Leptothrium Kunth. 1829. Poaceae. 2: L. rigidum—West Indies; L. senegalense (Kunth) W. Clayton—trop. Africa from Senegal to Ethiopia, Uganda, Kenya, and Tanzania; Southwest Asia eastward to Pakistan

Leptothyrsa Hook. f. 1862. Rutaceae. 1 Brazil (Amazonia)

Leptunis Steven. 1856. Rubiaceae. 2–3 Iran; L. trichoides—Caucasus and Middle Asia

Lepturella Stapf. 1912 (Oropetium Trin.). Poaceae. 2 trop. West and South Africa

Lepturidium Hitchc. et Ekman. 1936. Poaceae. 1 Cuba

Lepturopetium Morat. 1981. Poaceae. 1 New Caledonia

Lepturus R. Br. 1810. Poaceae. 15 coasts of the Indian and the western Pacific Oceans, Hawaiian Is. (1, L. repens)

Lepuropetalon Elliott. 1817. Lepuropetalaceae. 1 southeastern U.S., Mexico, central Chile, Uruguay

Lepyrodia Juss. 1827. Restionaceae. 17 Australia, Tasmania, New Zealand, Chatham Is.

Lepyrodiclis Fenzl. 1840. Caryophyllaceae. 5 West (incl. Caucasus) and Middle Asia, Afghanistan, northwestern China, Mongolia, Tibet, Himalayas from Kashmir to Nepal

Lerchea L. 1771. Rubiaceae. 8 Sumatra, Java, Lesser Sunda Is.

Lerchenfeldia Schur. 1866 (Deschampsia P. Beauv.). Poaceae. 5 temp. and subtrop. regions and mountainous tropics; L. flexuosa—European part of Russia, Caucasus, Siberia, Far East

Leretia Vell. 1825 (Mappia Jacq.). Icacinaceae. 1 Panama, Venezuela, Peru, Guyana, Brazil

Leroya Cavaco = Pyrostria Comm. ex Juss. Rubiaceae

Lescaillea Griseb. 1866. Asteraceae. 1 Cuba

Leschenaultia Benth. = Lechenaultia R. Br. Goodeniaceae

Lesliea Seidenf. 1988. Orchidaceae. 1 Thailand

Lespedeza Michx. 1803. Fabaceae. 40 trop. and East Asia, Australia, temp. North America; Russia (5 southern East Siberia and Far East)

Lesquerella S. Watson. 1888 (Discovium Raf.). Brassicaceae. 40 North America; L. arctica—in Russian arctic areas

Lesquereuxia Boiss. et Reut. 1853. Scrophulariaceae. 1 Greece, Asia Minor (southern Anatolia), northern Syria

Lessertia DC. 1802. Fabaceae. 52 mainly South Africa (47), a few species in Angola, Namibia, Botswana, Zambia, Zimbabwe; L. pauciflora—extending to Kenya and Tanzania

Lessingia Cham. 1829. Asteraceae. 14 southwestern U.S. (California, Nevada, Arizona), northern Baja California

Lessingianthus H. Rob. = Vernonia Schreb. Asteraceae

Letestua Lecomte. 1920. Sapotaceae. 2 trop. West Africa

Letestudoxa Pellegr. 1920. Annonaceae. 2 trop. West Africa

Letestuella G. Taylor. 1953. Podostemaceae. 1 trop. West Africa

Lethedon Spreng. 1807. Thymelaeaceae. 12 northeastern Australia (1, L. setosa), New Caledonia, Vanuatu

Lethia Ravenna. 1986 (Herbertia Sweet). Iridaceae. 1 eastern Brazil

Lettowianthus Diels. 1936. Annonaceae. 1 trop. East Africa

Leucactinia Rydb. 1915. Asteraceae. 1 Mexico

Leucadendron R. Br. 1810. Proteaceae. c. 80 South Africa

Leucaena Benth. 1842. Fabaceae. 40 trop. America

Leucampyx A. Gray ex Benth. et Hook. f. = Hymenopappus L'Her. Asteraceae

Leucanthemella Tzvelev. 1961 (Chrysanthemum L.). Asteraceae. 2 Central and Southeast Europe, East Asia; Russia (2, southeastern Europe [L. serotina], Far East [L. linearis])

Leucanthemopsis (Giroux) Heywood. 1975. Asteraceae. 6 Europe, North Africa

Leucanthemum Hill. 1753. Asteraceae. c. 25 temp. Eurasia

Leucas R. Br. 1810. Lamiaceae. 160 trop. America, trop. and South Africa, Arabian Peninsula, South Asia, southern China, Taiwan (2), Indochina, Malesia, trop. Australia

Leucaster Choisy. 1849. Nyctaginaceae. 1 southeastern Brazil

Leucelene Greene = Chaetopappa DC. Asteraceae

Leucheria Lag. 1811. Asteraceae. 46 southern Andes, Patagonia, Falkland Is.

Leuchtenbergia Hook. 1848. Cactaceae. 1 northern Mexico

Leucobarleria Lindau = Neuracanthus Nees. Acanthaceae

Leucoblepharis Arn. 1838. Asteraceae. 1 India

Leucocalantha Barb. Rodr. 1891. Bignoniaceae. 1 Brazil (Amazonia)

Leucocarpus D. Don. 1831. Scrophulariaceae. 1 trop. America

Leucochrysum (DC.) Paul G. Wilson. 1992. Asteraceae. 5 temp. Australia

Leucocodon Gardner. 1846. Rubiaceae. 1 Sri Lanka

Leucocorema Ridl. = Trichadenia Thwaites. Flacourtiaceae

Leucocoryne Lindl. 1830. Alliaceae. 12 Chile

Leucocrinum Nutt. ex A. Gray. 1837. Hemerocallidaceae. 1 western U.S. from Oregon and northern California to Colorado and New Mexico

Leucocroton Griseb. 1861. Euphorbiaceae. 20 Cuba, Haiti

Leucocyclus Boiss. 1849 (Anacyclus L.). Asteraceae. 1 Southwest Asia

Leucogenes Beauverd. 1910. Asteraceae. 3 New Zealand

Leucoglossum Wilcox, Bremer et Humphries. 1993. Asteraceae. 3 sp. Distr. Spain, Morocco, Algeria, Tunisia, Libya

Leucohyle Klotzsch. 1855. Orchidaceae. 7 trop. America, West Indies

Leucojum L. 1753. Amaryllidaceae. 9 Europe (7), Morocco, Southwest Asia to Caucasus and Iran

Leucolophus Bremek. 1940. Rubiaceae. 3 Malay Peninsula, Sumatra

Leucomphalos Benth. ex Planch. 1848. Fabaceae. 1 Ghana, Nigeria, Cameroun, Gabon

Leuconotis Jack. 1823. Apocynaceae. 10 western Malesia

Leucopholis Gardner = Chionolaena DC. Asteraceae

Leucophrys Rendle. 1899 (Brachiaria [Trin.] Griseb.). Poaceae. 3 South-West and South Africa

Leucophyllum Bonpl. 1812. Scrophulariaceae. 12 southern U.S., Mexico

Leucophysalis Rydb. 1896 (Physaliastrum Makino). Solanaceae. 9 from eastern Himalayas to East Asia, North America (2)

Leucopoa Griseb. 1852 (Festuca L.). Poaceae. 10 montane temp. Eurasia

Leucopogon R. Br. 1810 (Styphelia J. E. Sm.). Epacridaceae. 150 Malesia, Australia (143), New Caledonia

Leucopsis (DC.) Baker = Noticastrum DC. Asteraceae

Leucoptera B. Nord. 1976. Asteraceae. 3 South Africa

Leucorchis E. Mey. = Pseudorchis Séguier. Orchidaceae

Leucosalpa Scott Elliott. 1891. Scrophulariaceae. 3 Madagascar

Leucosceptrum Sm. 1805. Lamiaceae. 3 Himalayas, Assam, Burma, southern China (Yunnan), Taiwan (1), Japan

Leucosidea Eckl. et Zeyh. 1836. Rosaceae. 1 Zimbabwe, South Africa

Leucospermum R. Br. 1810. Proteaceae. 46 southern trop. and South Africa

Leucosphaera Gilg. 1897. Amaranthaceae. 1 South-West Africa

Leucospora Nutt. 1834. Scrophulariaceae. 1 eastern North America

Leucostegane Prain. 1901. Fabaceae. 2 Malay Peninsula, Borneo

Leucosyke Zoll. et Moritzi. 1845–1846. Urticaceae. 35 China, Taiwan, Indochina, Malesia, Oceania

Leucothoe D. Don. 1834. Ericaceae. 45 America (40), East Asia (5)

Leunisia Phil. 1864. Asteraceae. 1 Chile

Leurocline S. Moore = Echiochilon Desf. Boraginaceae

Leutea Pimenov. 1987. Apiaceae. 6 Iraq, Iran, Turkmenistan (1, L. petiolaris)

Leuzea DC. 1805. Asteraceae. c. 25 temp. Eurasia, incl. Russia (c. 10 southern European part), 1 Siberia and Far East, 1 Caucasus; Middle Asia; eastern Australia

Levenhookia R. Br. 1810. Stylidiaceae. 10 southern Australia

Levieria Becc. 1877. Monimiaceae. 7 Sulawesi, Moluccas, New Guinea, Bismarck Arch., Australia (northern Queensland)

Levisticum Hill. 1756. Apiaceae. 1 (L. officinale) North America, Europe, Caucasus, Middle Asia

Levya Bureau ex Baill. = Cydista Miers. Bignoniaceae

Lewisia Pursh. 1814. Portulacaceae. c. 22 western North America

Leycephallum Piper = Rhynchosia Lour. Fabaceae

Leycesteria Wall. 1824. Caprifoliaceae. 8 Himalayas from Kashmir to Bhutan, Assam, Burma, southwestern China

Leymus Hochst. 1848. Poaceae. 50 temp. and subtrop. Northern Hemisphere, mountainous South America

Leysera L. 1763. Asteraceae. 4 North and South Africa, Southwest Asia

Lhotzkya Schauer = Calytrix Lab. Myrtaceae

Lhotzkyella Rauschert. 1982. Asclepiadaceae. 1 Brazil

Liabellum Cabrera (not Rydb.) = Microliabum Cabrera. Asteraceae

Liabellum Rydb. 1927 (Sinclairia Hook. et Arn.). Asteraceae. 4 southeastern Mexico

Liabum Adans. 1763. Asteraceae. 37 America, West Indies

Liatris Gaertn. ex Schreb. 1791. Asteraceae. 33 eastern North America

Libanothamnus Ernst. 1870. Asteraceae. 18 Colombia, Venezuela

Libanotis Haller ex Zinn. = Seseli L. Apiaceae

Libanotis Hill = Seseli L. Apiaceae

Liberatia Rizzini. 1947 (Lophostachys Pohl). Acanthaceae. 1 Brazil

Liberbaileya Furtado = Maxburretia Furtado. Arecaceae

Libertia Spreng. 1824. Iridaceae. 12 New Guinea, eastern Australia, New Zealand, Andes of the South America

Libocedrum Endl. 1847. Cupressaceae. 5 New Caledonia (3), New Zealand (2)

Librevillea Hoyle. 1956. Fabaceae. 1 Cameroun, Gabon, Angola

Libyella Pamp. 1925. Poaceae. 1 Libya

Licania Aubl. 1775. Chrysobalanaceae. 193 trop. America (188), West Africa from Guinea to Gabon (1), Malesia (3, except Sulawesi and Lesser Sunda Is.)

Licaria Aubl. 1775. Lauraceae. 45 trop. America from southern Florida and Mexico to southern Brazil and Bolivia

Lichtensteinia Cham. et Schltdl. 1826. Apiaceae. 7 St. Helena, South Africa

Licuala Thunb. 1782. Arecaceae. 110 northeastern India, southern China, Indochina, Malesia, Solomon Is., Vanuatu, Australia (Queensland)

Lidbeckia Bergius. 1767. Asteraceae. 2 South Africa

Lidia A. Löve et D. Löve = Minuartia L. Caryophyllaceae

Lietzia Regel. 1880. Gesneriaceae. 1 Brazil

Lifago Schweinf. et Muschl. 1911. Asteraceae. 1 Morocco, Algeria

Ligaria Tiegh. 1895 (Loranthus Jacq.). Loranthaceae. 2 arid South America from central Peru and eastern Bolivia to central Chile, Uruguay, northern Argentina, southern Brazil

Ligeophila Garay. 1977. Orchidaceae. 8 trop. America

Lightfootia L'Her. = Wahlenbergia Schrad. ex Roth. Campanulaceae

Lightia R. H. Schomb. = Euphronia Mart. et Zucc. Euphroniaceae

Lightiodendron Rauschert = Euphronia Mart. et Zucc. Euphroniaceae

Ligia Fasano = Thymelaea Mill. Thymelaeaceae

Lignariella Baehni. 1955. Brassicaceae. 2 Tibet, Himalayas from Kashmir to Bhutan

Lignocarpa J. W. Dawson. 1967. Apiaceae. 2 New Zealand

Ligularia Cass. 1816. Asteraceae. 120 temp. Eurasia, mostly China (111)

Ligusticella J. M. Coult. et Rose. 1909. Apiaceae. 2 North America (Alaska, Colorado)

Ligusticopsis Leute. 1960 (or = Ligusticum L.). Apiaceae. 14 Mongolia, China

Ligusticum L. 1753. Apiaceae. 1 (L. scoticum) coastal North Europe, coastal North and East Asia, North America

Ligustrina Rupr. = Syringa L. Oleaceae

Ligustrum L. 1753. Oleaceae. c. 50 Europe; North Africa; West, East and Southeast Asia; Malesia; trop. Australia; Melanesia to Vanuatu; L. vulgare—southern European part of Russia and Caucasus; L. tschonoskii—Russian Far East

Lijndenia Zoll. et Moritzi. 1846. Melastomataceae. c. 10 trop. Africa (2), Madagascar, Sri Lanka, Malesia

Lilaea Humb. et Bonpl. 1808. Juncaginaceae. 1 coastal Pacific America from Canada to Chile and Argentina

Lilaeopsis Greene. 1891. Apiaceae. 25 America: Canada, U.S., Mexico, Colombia, Ecuador, Peru, Bolivia, Chile, southeastern Brazil, Paraguay, Argentina (south to Tierra del Fuelgo), Uruguay, Falkland Is.; ? Madagascar; Kerguelen; Australia and Tasmania; New Zealand; Stewart, Chatham, and Auckland Is.

Lilium L. 1753. Liliaceae. c. 100 temp. Northern Hemisphere, esp. western and southwestern China (c. 40), Russia (16 European part, Siberia, Far East) and Caucasus

Lilloa Speg. = Synandrospadix Engl. Araceae

Limacia Lour. 1790. Menispermaceae. 3 Lower Burma, peninsular Thailand, Indochina, Malesia (excl. Moluccas and New Guinea)

Limaciopsis Engl. 1899. Menispermaceae. 1 trop. Africa, China

Limbarda Adans. 1763 (Inula L.). Asteraceae. 1 Mediterranean

Limeum L. 1759. Molluginaceae. 26 trop. and South Africa, Arabian Peninsula, Pakistan, northwestern India (Punjab)

Limlia Masamune et Tomiza = Lithocarpus Blume. Fagaceae

Limnalsine Rydb. 1932 (Montia L.). Portulacaceae. 1 western North America from Vancouver Is. to San Francisco

Limnanthemum S. G. Gmel. = Nymphoides Hill. Menyanthaceae

Limnanthes R. Br. 1833. Limnanthaceae. 9 coastal western North America

Limnas Trin. 1822. Poaceae. 3 Altai, East Siberia, Okhotsk coast

Limniboza R. E. Fr. 1916. Lamiaceae. 2 Zaire, Zambia

Limnobium Rich. 1814. Hydrocharitaceae. 1 (L. spongia) southeastern North America

Limnocharis Bonpl. 1807. Limnocharitaceae. 2 trop. America: Mexico, Guatemala, Honduras, El Salvador, Nicaragua, Costa Rica, Panama, Colombia, Venezuela, Ecuador, Peru, Brazil, Bolivia, Paraguay, Argentina; West Indies: Haiti, Windward Is., Santo Domingo

Limnocitrus Swingle = Pleiospermium (Engl.) Swingle. Rutaceae

Limnodea L. H. Dewey. 1894. Poaceae. 1 U.S.

Limnophila R. Br. 1810. Scrophulariaceae. 36 trop. Old World

Limnophyton Miq. 1856. Alismataceae. 1–3 trop. Africa, Madagascar, India, Indochina, Java, Timor I.

Limnopoa C. E. Hubb. 1943. Poaceae. 1 southern India

Limnorchis Rydb. = Platanthera Rich. Orchidaceae

Limnosciadium Mathias et Constance. 1941. Apiaceae. 2 southern U.S.

Limnosipanea Hook. f. 1868. Rubiaceae. 7 Panama, trop. South America

Limodorum Boehm. 1760. Orchidaceae. 2 South Europe (1), Mediterranean, West Asia; L. abortivum also—Crimea, Caucasus

Limonia L. 1762 (Feroniella Swingle). Rutaceae. 1 India, Sri Lanka, Burma, Indochina, Malay Peninsula, Sumatra, Java

Limoniastrum Fabr. 1759. Plumbaginaceae. 5–9 Iberian Peninsula, southern France, Sardinia and Sicily, Italy, Crete, North Africa from Morocco to Egypt, Sinai Peninsula

Limoniopsis Lincz. 1952. Plumbaginaceae. 2 Caucasus (1, L. owerinii), eastern Asia Minor

Limonium Mill. 1754. Plumbaginaceae. c. 150 cosmopolitan, mainly Mediterranean and arid areas of the Northern Hemisphere

Limosella L. 1753. Scrophulariaceae. 15 cosmopolitan; L. aquatica—from Arctica to Caucasus, Middle Asia, and Russian Far East

Linanthastrum Ewan = Linanthus Benth. Polemoniaceae

Linanthus Benth. 1833. Polemoniaceae. 37 western U.S. from Washington and Idaho to southern California and Texas; Baja California and northern Mexico; Chile

Linaria Hill. 1753. Scrophulariaceae. c. 100 temp. Northern Hemisphere

Lineriantha B. L. Burtt et Ros. M. Sm. 1965. Acanthaceae. 1 Borneo

Linariopsis Welw. 1869. Pedaliaceae. 2 trop. West and South-West Africa

Linconia L. 1771. Bruniaceae. 2 South Africa

Lindackeria C. Presl. 1835. Flacourtiaceae. 13 trop. America (6) from Mexico to Bolivia and Brazil (Amazonia); trop. Africa (7) from Cameroun to Somalia and south to Zambia and Malawi

Lindauea Rendle. 1896. Acanthaceae. 1 Somalia

Lindbergella Bor. 1969. Poaceae. 1 Cyprus

Lindbergia Bor, non Kindb. = Lindbergella Bor. Poaceae

Lindelofia Lehm. 1850 (Cynoglossum L.). Boraginaceae. 10–12 Middle Asia (6), Afghanistan, China, Himalayas from Kashmir to Nepal, India

Lindenbergia Lehm. 1828. Scrophulariaceae. 20 trop. Africa; L. sinaica—through Arabian Peninsula extending to Israel (Negev); trop. and East Asia

Lindenia Benth. 1857 (Augusta Pohl). Rubiaceae. 3 New Caledonia (1), Fiji (1), southern Mexico and Central America (1)

Lindeniopiper Trel. 1929. Piperaceae. 2 Mexico

Lindera Thunb. 1783. Lauraceae. c. 80 trop. and temp. East Asia, eastern North America

Lindernia All. 1766. Scrophulariaceae. c. 50 tropics and subtropics, esp. Africa and Asia; L. procumbens—European part of Russia, Caucasus, Middle Asia, West Siberia, Far East

Lindheimera A. Gray et Engelm. 1846–1847. Asteraceae. 1 southern U.S.

Lindleya Kunth. 1824. Rosaceae. 2 Mexico

Lindmannia Mez = Cottendorfia Schult. f. Bromeliaceae

Lindsayella Ames et C. Schweinf. 1937 (or = Sobralia Ruiz et Pav.). Orchidaceae. 1 Panama

Lindsayomyrtus B. Hyland et Steenis. 1973. Myrtaceae. 1 eastern Malesia, Australia (Queensland)

Lingelsheimia Pax. 1909 (or = Drypetes Vahl). Euphorbiaceae. 2 trop. Africa

Lingnania McClure. 1940 (Bambusa

Schreb.). Poaceae. 10 India, southern China

Linkagrostis Rom. Garcia, Blanca et C. Morales. 1987 (Agrostis L.). Poaceae. 1 Pyrenees, Northwest Africa

Linnaea L. 1753. Caprifoliaceae. 1 circumpolar Northern Hemisphere

Linnaeobreynia Hutch. 1967 (Capparis L.). Capparaceae. 10 trop. America

Linnaeopsis Engl. 1900. Gesneriaceae. 3 trop. East Africa

Linocalix Lindau. 1913 (Justicia L.). Acanthaceae. 1 trop. West Africa

Linociera Sw. ex Schreb. = Chionanthus L. Oleaceae

Linodendron Griseb. 1861. Thymelaeaceae. 3 Cuba

Linospadix H. A. Wendl. 1875 (non Becc. ex Hook. f.). Arecaceae. 11 New Guinea (5), Australia

Linostoma Wall. ex Endl. 1837. Thymelaeaceae. 6 northeastern India, Burma, Indochina, Malesia (Sumatra, Malay Peninsula, Borneo), trop. Australia

Lintonia Stapf. 1911. Poaceae. 2 trop. East Africa

Linum L. 1753. Linaceae. c. 230 temp. and subtrop. regions, esp. Mediterranean (c. 80)

Liodendron H. Keng. 1951 (Drypetes Vahl). Euphorbiaceae. 2 Taiwan, Ryukyu Is.

Liparia L. 1771. Fabaceae. 2 South Africa

Liparis Rich. 1817 (Malaxis Sol. ex Sw.). Orchidaceae. c. 250 cosmopolitan, excl. New Zealand

Liparophyllum Hook. f. 1847. Menyanthaceae. 1 Tasmania, New Zealand

Lipocarpha R. Br. 1818. Cyperaceae. 15 trop. and South Africa, Madagascar, trop. and subtrop. Asia, Australia, trop. America

Lipochaeta DC. 1836. Asteraceae. 20 Hawaiian Is.

Lipostoma D. Don = Coccocypselum P. Browne. Rubiaceae

Lippia L. 1753. Verbenaceae. c. 200 trop. America, trop. Africa, India, Sri Lanka, Indochina, Malesia; L. nodiflora—Caucasus, Middle Asia

Lipskya Nevski. 1937 (Schrenkia Fisch. et C. A. Mey.). Apiaceae. 1 (L. insignis) Middle Asia (Pamir-Alai)

Lipskyella Juz. 1937. Asteraceae. 1 Middle Asia

Liquidambar L. 1953. Altingiaceae. 5–6 Cyprus and Rhodes, southwestern Asia Minor, southeastern China, Taiwan, Indochina, eastern North and Central America

Liriodendron L. 1753. Magnoliaceae. 2 southern China and northern Vietnam (1), southeastern North America (1)

Liriope Lour. 1790. Convallariaceae. 8 continental China, Korea, Japan, Taiwan, Vietnam, Philippines

Liriosma Poepp. et Endl. = Dulacia Vell. Olacaceae

Liriothamnus Schltr. = Trachyandra Kunth. Asphodelaceae

Lisaea Boiss. 1844. Apiaceae. 3 Turkey, Transcaucasia (2), Syria, Lebanon, Israel, Iraq, Iran

Lisianthius P. Browne. 1756. Gentianaceae. 27 trop. America, West Indies

Lisianthus L. = Lisianthius P. Browne. Gentianaceae

Lissanthe R. Br. 1810 (Styphelia Sm.). Epacridaceae. 2 Australia

Lissocarpa Benth. 1876. Lissocarpaceae. 3 trop. South America from Guyana to Bolivia

Lissochilus R. Br. = Eulophia R. Br. ex Lindl. Orchidaceae

Lissospermum Bremek. 1944. Acanthaceae. 1 Sumatra, Java

Listera R. Br. 1813. Orchidaceae. 40 cold and temp. Northern Hemisphere

Listia E. Mey. = Lotononis (DC.) Eckl. et Zeyh. Fabaceae

Listrobanthes Bremek. 1944. Acanthaceae. 1 eastern Himalayas, Assam

Listrostachys Rchb. f. 1852. Orchidaceae. 2 southern trop. Africa

Litanthus Harv. 1844. Hyacinthaceae. 1 South Africa

Litchi Sonn. 1782. Sapindaceae. 1 southern China, Philippines

Lithachne P. Beauv. 1812. Poaceae. 4 Central and trop. South America, West Indies

Lithobium Bong. 1838. Melastomataceae. 1 Brazil

Lithocarpus Blume. 1826. Fagaceae. 300 eastern Himalayas, China, Korea, Japan, Taiwan, Indochina, Philippines, New Guinea, western U.S. (1, L. densiflora)

Lithococca Small ex Rydb. = Heliotropium L. Boraginaceae

Lithodora Griseb. 1844. Boraginaceae. 9 Europe, Mediterranean

Lithodraba Boelcke. 1951 (Xerodraba Skottsb.). Brassicaceae. 1 Argentina

Lithophila Sw. 1788. Amaranthaceae. 15 Galapagos Is., West Indies

Lithophragma (Nutt.) Torr. et A. Gray. 1840. Saxifragaceae. 10 western North America from British Columbia and Alberta to California, Arizona and New Mexico

Lithophytum Brandegee. 1911 (Plocosperma Benth.). Plocospermataceae. 1 Mexico

Lithops N. E. Br. 1922. Aizoaceae. 37 Namibia, southeastern Botswana, South Africa

Lithosciadium Turcz. 1844 (Cnidium Cuss. ex Juss.). Apiaceae. 2 East Siberia, northern Mongolia

Lithospermum L. 1753. Boraginaceae. 50–60 temp. regions and mountainous tropics and subtropics, excl. Australia

Lithrea Miers ex Hook. et Arn. 1833. Anacardiaceae. 3 South America

Litogyne Harv. 1863. Asteraceae. 1 trop. and South Africa

Litosanthes Blume. 1823. Rubiaceae. 5 Taiwan (1), Java (1), New Guinea (3)

Litothamus R. M. King et H. Rob. 1979. Asteraceae. 1 Brazil

Litrisa Small. 1924. Asteraceae. 1 Florida

Litsea Lam. 1792. Lauraceae. c. 400 tropics and subtropics, esp. Asia, Australia, and the Pacific Islands

Littledalea Hemsl. 1896. Poaceae. 3 Middle Asia (1, L. alaica), western China

Littonia Hook. 1853. Colchicaceae. 7 trop. and South Africa, Arabian Peninsula

Littorella Bergius. 1768. Plantaginaceae. 3 North America (1) and South America (1), Azores, Europe, Morocco

Litwinowia Woronow. 1931. Brassicaceae. 1 (L. tenuissima) southeastern European part of Russia, Caucasus, West Siberia, Middle Asia, Iran, Afghanistan, Pakistan, western China, India

Livistona R. Br. 1810. Arecaceae. 28 Horn of Africa, Arabian Peninsula eastward to Himalayas through China to Ryukyu Is., south through Indochina and Malesia to Solomon Is. and northern and eastern Australia

Llagunoa Ruiz et Pav. 1794. Sapindaceae. 3 western trop. South America

Llerasia Triana. 1858 (Vernonia Schreb.). Asteraceae. 14 Andes from Colombia to Bolivia

Llewelynia Pittier. 1939. Melastomataceae. 1 Venezuela

Lloydia Rchb. 1830. Liliaceae. c. 20 temp. Eurasia, western North America; Russia (2, arctic areas, northern Caucasus, Ural, Siberia, Far East), Middle Asia

Loasa Adans. 1763. Loasaceae. 105 Mexico, Central and South America

Lobanilia Radcl.-Sm. 1989. Euphorbiaceae. 7 Madagascar

Lobelia L. 1753. Lobeliaceae. 365 tropics and subtropics, esp. America, a few species in temp. regions; Russia (3, European part, East Siberia, Far East)

Lobivia Britton et Rose = Echinopsis Zucc. Cactaceae

Lobostemon Lehm. 1830. Boraginaceae. 28 South Africa

Lobostephanus N. E. Br. 1901 (Emicocarpus K. Schum. et Schltr.). Asclepiadaceae. 1 South Africa

Lobularia Desv. 1815. Brassicaceae. 4–5 Azores, Canary Is., Cape Verde Is., South Europe from Portugal and Spain to Greece and Romania, Crimea; North Africa (Morocco to Libya and Egypt),

Aegean islands, Cyprus, Turkey, Caucasus, Lebanon, Israel, Sinai and Arabian peninsulas, southern Iran, China

Lochia Balf. f. 1884. Caryophyllaceae. 2 Socotra, Abdul Kuri, and Oman (1), Somalia (1)

Lockhartia Hook. 1827. Orchidaceae. 20 trop. America from Mexico to Peru and Brazil

Lodoicea Comm. ex DC. 1800. Arecaceae. 1 Seychelles

Loefgrenianthus Hoehne. 1927. Orchidaceae. 1 Brazil

Loeflingia L. 1753. Caryophyllaceae. 5 North America (1), Southwest Europe, North Africa, Asia Minor, Israel, Jordan, Sinai Peninsula

Loerzingia Airy Shaw. 1963. Euphorbiaceae. 1 Sumatra

Loeselia L. 1753. Polemoniaceae. 10 America from California to Venezuela

Loeseliastrum (Brand) Timbrook. 1986. Polemoniaceae. 2 southwestern U.S. (California, Utah, Nevada, Arizona), northern Mexico

Loesenera Harms. 1897. Fabaceae. 4 Liberia, Nigeria, Cameroun, Gabon

Loeseneriella A. C. Sm. 1941 (Hippocratea L.). Celastraceae. 20 southern China, Indochina, Malesia

Loewia Urb. 1897. Turneraceae. 3 trop. East Africa

Logania R. Br. 1810. Loganiaceae. 15 Australia (13), New Zealand (1), New Caledonia (1)

Logfia Cass. 1819 (Filago L.). Asteraceae. 9 Europe, North Africa, Southwest Asia

Loheria Merr. 1910. Myrsinaceae. 6 Philippines, New Guinea

Loiseleuria Desv. 1813. Ericaceae. 1 circumpolar Northern Hemisphere

Lojaconoa Bobrov = Trifolium L. Fabaceae

Loliolum Krecz. et Bobrov. 1934. Poaceae. 1 eastern Mediterranean; West (incl. Caucasus), Southwest, and Middle Asia

Lolium L. 1753. Poaceae. 12 temp. Eurasia, North Africa

Lomandra Labill. 1805. Dasypogonaceae. 50 northern, eastern, and southeastern Australia and Tasmania (50), New Guinea (2), New Caledonia (1, L. banksii)

Lomanodia Raf. = Astronidium A. Gray. Melastomanaceae

Lomatia R. Br. 1810. Proteaceae. 12 eastern Australia, Tasmania, South America

Lomatium Raf. 1819. Apiaceae. c. 80 western North America (U.S., Canada)

Lomatocarpa Pimenov. 1982. Apiaceae. 3 Middle Asia (2), Afghanistan, western China

Lomatogoniopsis T. N. Ho et S. W. Liu. 1980. Gentianaceae. 3 southwestern China, Tibet

Lomatogonium A. Braun. 1830. Gentianaceae. 18 Europe (2), temp. Asia: Caucasus, Middle Asia, China (esp. Yunnan), Mongolia, Afghanistan, Pakistan, Tibet, Kashmir, Himalayas (India, Nepal, Bhutan, Sikkim); Russia (2 European part from Arctica south to Caucasus, West and East Siberia, Far East)

Lomatophyllum Willd. 1811 (Aloe L.). Asphodelaceae. 12 Madagascar, Mascarene Is.

Lomatozona Baker. 1876. Asteraceae. 4 Brazil

Lombardochloa Roseng. et B. R. Arill. = Briza L. Poaceae

Lomelosia Raf. 1838. Dipsacaceae. c. 40 Europe, Mediterranean, West Asia

Lonas Adans. 1763. Asteraceae. 1 southwestern Mediterranean

Lonchocarpus Kunth. 1824. Fabaceae. c. 150 trop. America, West Indies, trop. Africa and Madagascar (17), ? Australia

Lonchophora Durieu = Matthiola R. Br. Brassicaceae

Lonchostephus Tul. 1852. Podostemaceae. 1 Brazil (Amazonia)

Lonchostigma P. & K. = Jaborosa Juss. Solanaceae

Lonchostoma Wikstr. 1818. Bruniaceae. 4 South Africa

Londesia Fisch. et C. A. Mey. 1836. Chenopodiaceae. 1 Middle Asia, Iran, Afghanistan, Pakistan, northwestern China, Mongolia

Longetia Baill. 1866 (Austrobuxus Miq.).

Euphorbiaceae. 1 New Caledonia

Lonicera L. 1753. Caprifoliaceae. c. 200 Northern Hemisphere southward to Himalayas, Philippines, Mexico

Loniceroides Bullock = Marsdenia R. Br. Asclepiadaceae

Lopezia Cav. 1791. Onagraceae. 21 Mexico, Central America

Lophachme Stapf. 1898. Poaceae. 2 trop. Africa

Lophanthera A. Juss. 1840. Malpighiaceae. 5 Costa Rica (1, L. hammelii), Amazonian Colombia, and Brazil

Lophanthus Adans. 1763. Lamiaceae. 22 Iran, Afghanistan, Middle Asia, western Pakistan, western China, Mongolia, Russia (2 West and East Siberia)

Lophatherum Brongn. 1831. Poaceae. 2 trop. Asia, New Guinea

Lophiaris Raf. = Oncidium Sw. Orchidaceae

Lophiocarpus Turcz. 1843 (Microtea Sw.). Petiveriaceae. 4: 3 Namibia, Zimbabwe, South Africa; 1 Mozambique

Lophiola Ker-Gawl. 1813. Melanthiaceae. 2 eastern North America

Lophira Banks ex Gaertn. f. 1805. Lophiraceae. 2 trop. Africa

Lophocarpinia Burkart. 1957. Fabaceae. 1 Paraguay, Argentina

Lophocarpus Boeck. = Neolophocarpus E. G. Camus. Cyperaceae

Lophocereus (A. Berger) Britton et Rose = Pachycereus (A. Berger) Britton et Rose. Cactaceae

Lophochlaena Nees. 1838 (Pleuropogon R. Br.). Poaceae. 5 western North America

Lophochloa Rchb. = Rostraria Trin. Poaceae

Lophogyne Tul. 1849. Podostemaceae. 2 eastern Brazil

Lopholaena DC. 1838. Asteraceae. 19 trop. and South Africa

Lopholepis Decne. 1839. Poaceae. 1 southern India, Sri Lanka

Lophomyrtus Burret. 1941. Myrtaceae. 2 New Zealand

Lophopappus Rusby. 1894. Asteraceae. 5 Andes

Lophopetalum Wight ex Arn. 1839. Celastraceae. 20 India, Burma, Indochina, Malesia (except Lesser Sunda and east Java), Australia (1, L. arnhemicum)

Lophophora J. M. Coult. 1894. Cactaceae. 2 southern U.S. (Texas), northern and eastern Mexico

Lophophytum Schott et Endl. 1832. Lophophytaceae. 3 trop. and subtrop. South America

Lophopogon Hack. 1887. Poaceae. 2 India

Lophopterys A. Juss. 1838. Malpighiaceae. 3 Venezuela, Guyana, Surinam

Lophopyrum A. Löve = Elytrigia Desv. Poaceae

Lophopyxis Hook. f. 1887. Lophopyxidaceae. 2 Malay Peninsula, Borneo, eastern Malesia, Palau Is., Solomon Is.

Lophoschoenus Stapf = Costularia C. B. Clarke. Cyperaceae

Lophosciadium DC. = Ferulago Koch. Apiaceae

Lophospatha Burret = Salacca Reinw. Arecaceae

Lophospermum D. Don. 1827 (Asarina Hill). Scrophulariaceae. 10 Mexico, Guatemala (2)

Lophostachys Pohl. 1830–1831. Acanthaceae. 15 Central and trop. South America

Lophostemon Schott. 1830. Myrtaceae. 4 New Guinea (1), Australia (3)

Lophostigma Radlk. 1897. Sapindaceae. 1 Bolivia

Lophostoma (Meisn.) Meisn. 1857. Thymelaeaceae. 4 northern South America

Lophothecium Rizzini = Justicia L. Acanthaceae

Lophotocarpus T. Durand = Sagittaria L. Alismataceae

Lopimia Mart. 1823 (Pavonia Cav.). Malvaceae. 2 southern Mexico, Central and trop. South America

Lopriorea Schinz. 1911. Amaranthaceae. 1 East Africa: southern Ethiopia, southern Somalia, and Kenya

Loranthus Jacq. 1762. Loranthaceae. 2: L. europaeus—Europe (except southwestern part), Asia Minor, Syria, Lebanon; L.

acaciae—Israel, Jordan, Sinai Peninsula (the remaining species assigned to other genera); parasitic on Fagaceae

Lordhowea B. Nord. 1978. Asteraceae. 1 Lord Howe I.

Lorentzia Griseb. = Wedelia Jacq. Asteraceae

Lorentzianthus R. M. King et H. Rob. 1975. Asteraceae. 1 Bolivia, Argentina

Lorenzochloa J. R. Reeder et C. G. Reeder = Ortachne Nees ex Steud. Poaceae

Loretoa Standl. = Capirona Spruce. Rubiaceae

Loreya DC. 1828. Melastomataceae. 14 Nicaragua, Costa Rica, Panama, Colombia, Venezuela, the Guianas, Ecuador, Peru, Bolivia, Brazil

Loricalepis Brade. 1938. Melastomataceae. 1 Brazil

Loricaria Wedd. 1856. Asteraceae. 17 Andes

Loropetalum R. Br. ex Rchb. 1828. Hamamelidaceae. 4 northeastern India, Thailand, southern China, Japan

Lorostelma Fourn. = Stenomeria Turcz. Asclepiadaceae

Lorostemon Ducke. 1935. Clusiaceae. 3 Brazil

Lotononis (DC.) Eckl. et Zeyh. 1836. Fabaceae. c. 150: 144 South Africa (Cape, Transvaal, Orange Free State, Natal, Drakensberg Range); southern Spain (1), Morocco (4), southeastern Bulgaria and Turkey (1, L. genistoides); trop. Africa (Cameroun, Uganda, Kenya, Tanzania, Burundi, Zaire, Angola, Zambia, Malawi, Namibia, Zimbabwe, Botswana, Mozambique); L. platycarpa—Cape Verde Is., Mauritania, Morocco, Algeria, Libya, Egypt, Chad, Ethiopia, Kenya, Tanzania, southern Angola, Namibia, South Africa, Arabian Peninsula, Israel, Jordan, southern Iran, Pakistan, northwestern India

Lotus L. 1753. Fabaceae. Over 100 Macaronesia, temp. Eurasia, Africa (70), Australia, South America

Loudetia Hochst. ex Steud. 1854. Poaceae. c. 40 South America, trop. and South Africa, Madagascar

Loudetiopsis Conert. 1957. Poaceae. 10 trop. Africa

Loudonia Lindl. = Glischrocaryon Endl. Haloragaceae

Louiseania Carriere = Amygdalus L. Rosaceae

Louisiella C. E. Hubb. et Leonard. 1952. Poaceae. 1 Sudan, Zaire

Lourteigia R. M. King et H. Rob. 1971 (Eupatorium L.). Asteraceae. 9 Colombia, Venezuela

Lourtella S. A. Graham, Baas et Tobe. 1987. Lythraceae. 1 Peru

Louteridium S. Watson. 1888. Acanthaceae. 5 Mexico, Central America

Louvelia Jum. et H. Perrier. 1912. Arecaceae. 3 coastal eastern Madagascar

Lovanafia M. Peltier = Dicraeopetalum Harms. Fabaceae

Lovoa Harms. 1896. Meliaceae. 2 trop. Africa from Sierra Leone to Uganda, Kenya and Tanzania, Zaire, Zimbabwe, Mozambique

Lowia Scort. = Orchidantha N. E. Br. Lowiaceae

Loxanthera (Blume) Blume. 1830. Loranthaceae. 2 western Malesia

Loxocalyx Hemsl. 1890. Lamiaceae. 2 China

Loxocarpus Benn. et R. Br. 1840. Gesneriaceae. 15 Malay Peninsula, Java

Loxocarya R. Br. 1810. Restionaceae. 8 southwestern Australia

Loxococcus H. A. Wendl. et Drude. 1875. Arecaceae. 1 Sri Lanka

Loxodera Launert. 1963. Poaceae. 4 trop. Africa

Loxodiscus Hook. f. 1857. Sapindaceae. 1 New Caledonia

Loxoma Garay = Smithsonia C. J. Saldanha. Orchidaceae

Loxomorchis Rauschert = Smithsonia C. J. Saldanha. Orchidaceae

Loxonia Jack. 1823. Gesneriaceae. 3 Japan, Malay Peninsula, Sumatra, Java, Borneo

Loxoptera O. E. Schulz. 1933. Brassicaceae. 1 Peru

Loxopterygium Hook. f. 1862. Anacardiaceae. 5 trop. South America

Loxostemon Hook. f. et Thomson. 1861. Brassicaceae. 10 western China, southeastern Tibet, Himalayas from Nepal to Bhutan

Loxostigma C. B. Clarke. 1883. Gesneriaceae. 7–8 eastern Himalayas from Nepal to Bhutan, India, Burma, southwestern China (7), northern Indochina

Loxostylis Spreng. ex Rchb. 1830. Anacardiaceae. 1 South Africa

Loxothysanus B. L. Rob. 1907. Asteraceae. 2 Mexico

Lozanella Greenm. 1905. Ulmaceae. 2 trop. America from Mexico to Peru and Bolivia

Lozania S. Mutis. 1810. Lacistemataceae. 3 trop. America: Costa Rica, Panama, Colombia, Venezuela, Ecuador, Peru, Bolivia and Brazil

Lubaria Pittier. 1929. Rutaceae. 2 Venezuela

Lucilia Cass. 1817. Asteraceae. 8 southern Brazil, Paraguay, Uruguay, northern and central Argentina and along the Andes in Bolivia, Argentina, and Chile

Luciliocline Anderb. et Freire. 1991. Asteraceae. 5 Andean region of Peru, Bolivia, and Argentina; S. santanica—Mexico

Luciliopsis Wedd. = Chactanthera Ruiz et Pav. Asteraceae

Lucinaea DC. 1830. Rubiaceae. 25 Malesia, New Caledonia

Luckhoffia A. C. White et B. Sloane. 1935. Asclepiadaceae. 1 South Africa

Luculia Sweet. 1832. Rubiaceae. 5 Himalayas, southwestern and southern China, Indochina

Lucuma Molina = Pouteria Aubl. Sapotaceae

Lucya DC. 1830. Rubiaceae. 1 West Indies

Ludekia Ridsdale. 1978. Rubiaceae. 2 Borneo, Philippines

Ludia Comm. ex Juss. 1789. Flacourtiaceae. 23 East Africa (Kenya, Tanzania, Mozambique), Madagascar, Seychelles, Aldabra I., Mascarene Is.

Ludisia A. Rich. 1825. Orchidaceae. 1 Burma, southern China, Indochina, Malay Peninsula, Sumatra I., Java, Philippines

Ludovia Brongn. 1861. Cyclanthaceae. 3 Panama, Colombia, Peru, northern Brazil, Guyana, Surinam

Ludwigia L. 1753. Onagraceae. 82 cosmopolitan; 3 spp.—Caucasus, Middle Asia, southern Far East of Russia

Lueddemannia Rchb. f. 1854. Orchidaceae. 5 Colombia, Venezuela, Ecuador, Peru

Luehea Willd. 1803. Tiliaceae. 25 trop. America from southern Mexico to Argentina and Uruguay, West Indies

Lueheopsis Burret. 1926. Tiliaceae. 9 trop. South America

Luerella Braas = Masdevallia Ruiz et Pav. Orchidaceae

Luerssenidendron Domin = Acradenia Kippist. Rutaceae

Luetkea Bong. 1833. Rosaceae. 1 western North America

Luetzelburgia Harms. 1922. Fabaceae. 6 Brazil

Luffa Mill. 1754. Cucurbitaceae. 8 trop. Old World (7), trop. America (1)

Lugoa DC. 1838. Asteraceae. 1 Canary Is.

Lugonia Wedd. 1859. Asclepiadaceae. 3 Peru, Bolivia, Argentina

Luina Benth. 1873. Asteraceae. 2 northwestern North America

Luisia Gaudich. 1827. Orchidaceae. 47 Himalayas, India, Sri Lanka, Andaman Is., Burma, Indochina, southern China, Korea, Japan; Malesia from Malay Peninsula to New Guinea, Mariana Is., Solomon Is., New Caledonia, Vanuatu, Fiji (1, L. teretifolia) and Samoa

Lulia Zardini. 1980. Asteraceae. 1 Brazil

Luma A. Gray. 1854. Myrtaceae. 4 Chile, Argentina

Lumnitzera Willd. 1803. Combretaceae. 2 East Africa, Madagascar, trop. Asia, Malesia, northern Australia, Oceania

Lunania Hook. 1844. Flacourtiaceae. 14 trop. America from southern Mexico to eastern Bolivia and Brazil, West Indies

Lunaria L. 1753. Brassicaceae. 3 Central, South, and East Europe; Russia (1, L. rediviva, European part)

Lunasia Blanco. 1837. Rutaceae. 10 Java, Lesser Sunda Is., Borneo, Philippines, Sulawesi, Moluccas, New Guinea, Australia (Cape York Peninsula)

Lundellia Leonard = Holographis Nees. Acanthaceae

Lundellianthus H. Rob. 1979. Asteraceae. 8 Mexico, Central America

Lundia DC. 1838. Bignoniaceae. 12 trop. America from Mexico, Guatemala and Belize to Brazil and Bolivia, Trinidad

Lupinaster Adans. = Trifolium L. Fabaceae

Lupinophyllum Gillett ex Hutch. = Tephrosia Pers. Fabaceae

Lupinus L. 1753. Fabaceae. c. 200 North and South America (esp. Rocky Mts. and Andes), Mediterranean, montane trop. East Africa

Luronium Raf. 1840. Alismataceae. 1 Europe

Luteidiscus J. St. John = Tetramolopium Nees. Asteraceae

Lutzia Gand. 1923. Brassicaceae. 1 Greece, Crete

Luvunga Buch.-Ham. ex Wight et Arn. 1834. Rutaceae. 12 India, Sri Lanka, Indochina, Malesia

Luxemburgia A. St.-Hil. 1822. Sauvagesiaceae. 23 Venezuela, Brazil

Luziola Juss. 1789. Poaceae. 10 southern U.S., Central and South America

Luzonia Elmer. 1907. Fabaceae. 1 Philippines

Luzula DC. 1805. Juncaceae. 80 cosmopolitan; esp. Russia (c. 30) chiefly tundra, forest zones, and mountainous areas

Luzuriaga Ruiz et Pav. 1802. Luzuriagaceae. 4: 1 New Zealand and Stewart I.; 3 Peru, Chile, Argentina to Tierra del Fuego, Falkland Is.

Lyallia Hook. f. 1847. Hectorellaceae. 1 Kerguelen

Lycapsus Phil. 1870. Asteraceae. 1 Chile

Lycaste Lindl. 1842. Orchidaceae. 40–45 trop. America from southern Mexico to Peru and Brazil, West Indies

Lychniothyrsus Lindau. 1914. Acanthaceae. 5 Brazil

Lychnis L. 1753. Caryophyllaceae. 35 arctic and temp. Northern Hemisphere; Russia (c. 10, northern European part, Siberia, Far East)

Lychnodiscus Radlk. 1878. Sapindaceae. 8 trop. Africa

Lychnophora Mart. 1822. Asteraceae. 18 southern Brazil

Lychnophoriopsis Sch. Bip. = Lychnophora Mart. Asteraceae

Lycianthes (Dunal) Hassl. 1917 (Solanum L.). Solanaceae. 180–200 Himalayas, India, East and Southeast Asia, Hawaiian Is., trop. America

Lycium L. 1753. Solanaceae. c. 100 North and temp. South America (45), temp. and subtrop. Eurasia (incl. 7—Caucasus and Middle Asia), South Africa, Australia (1), Pacific islands

Lycocarpus O. E. Schulz. 1924. Brassicaceae. 1 southern Spain

Lycochloa Sam. 1933. Poaceae. 1 Syria

Lycomormium Rchb. f. 1852. Orchidaceae. 2 trop. South America

Lycopersicon Hill. 1753 (Solanum L.). Solanaceae. 10 Galapagos Is., western South America from Colombia to central Chile

Lycopsis L. 1753. (or = Anchusa L.). Boraginaceae. 3 temp. Eurasia

Lycopus L. 1753. Lamiaceae. 10 temp. Northern Hemisphere, esp. Russia (7: 2, European part, Caucasus, Siberia; 5, Far East); Australia (1, L. australis R. Br.)

Lycoris Herb. 1821. Amaryllidaceae. 11 eastern Himalayas, Burma, China, Taiwan, Japan

Lycoseris Cass. 1824. Asteraceae. 11 trop. America: Guatemala (1, L. crocata), Costa Rica (1, L. grandis), Panama (1, L. triplinervia), Colombia (5), Venezuela (2), Ecuador (2), Peru (3), Bolivia (2), Brazil (1, L. boliviana)

Lycurus Kunth. 1816. Poaceae. 3 southern U.S., Mexico, Central and South America

Lydenburgia N. Robson. 1965 (Maytenus Molina). Celastraceae. 1 South Africa

Lygeum Loefl. ex L. 1754. Poaceae. 1 Mediterranean

Lyginia R. Br. 1810. Restionaceae. 1 southwestern Australia

Lygisma Hook. f. 1883. Asclepiadaceae. 3 Southeast Asia

Lygodesmia D. Don. 1829. Asteraceae. 8 North America from British Columbia to California, Texas, and Mexico

Lygodisodea Ruiz et Pav. = Paederia L. Rubiaceae

Lygos Adans. = Retama Raf. Fabaceae

Lymanbensonia Kimnach = Lepismium Pfeiff. Cactaceae

Lymania Read. 1984. Bromeliaceae. 4 trop. South America

Lyonia Nutt. 1818. Ericaceae. 35 Himalayas, northeastern India, China, Japan, Indochina, Malay Peninsula, eastern U.S., Mexico, West Indies

Lyonothamnus A. Gray. 1885. Rosaceae. 1 islands off southern California (Santa Catalina, Santa Cruz, San Clemente)

Lyonsia R. Br. = Parsonsia R. Br. Apocynaceae

Lyperanthus R. Br. 1810. Orchidaceae. 8 Australia, Tasmania, New Caledonia, New Zealand

Lyperia Benth. = Sutera Roth. Scrophulariaceae

Lyrocarpa Hook. et Harv. 1845. Brassicaceae. 4 California and Baja California, Arizona

Lyroglossa Schltr. 1920 (Stenorrhynchos Rich. ex Spreng.). Orchidaceae. 3 trop. South America

Lyrolepis Rech. f. = Carlina L. Asteraceae

Lysiana Tiegh. 1894. Loranthaceae. 8 Australia, Vanuatu

Lysicarpus F. Muell. 1858. Myrtaceae. 1 northeastern Australia

Lysichiton Schott. 1857. Araceae. 2: L. camtschatcense—Kamchatka, Sakhalin, Kuril Is., Japan; L. americanum—western North America

Lysichlamys Compton = Euryops (Cass.) Cass. Asteraceae

Lysiclesia A. C. Sm. = Orthaea Klotzsch. Ericaceae

Lysidice Hance. 1867. Fabaceae. 1 southern China, Vietnam

Lysiella Rydb. 1900 (Platanthera Rich.). Orchidaceae. 1 East Siberia, Russian Far East

Lysiloma Benth. 1844. Fabaceae. 38 trop. America northwestern to southwestern U.S., West Indies

Lysimachia L. 1753. Primulaceae. c. 180 cosmopolitan, mostly East Asia (esp. China—134)

Lysinema R. Br. 1810. Epacridaceae. 5 western Australia

Lysionotus D. Don. 1822. Gesneriaceae. 31 Asia from eastern Himalayas through China (28) to Japan, Taiwan, and Indochina

Lysiosepalum F. Muell. 1859. Sterculiaceae. 3 western Australia

Lysiostyles Benth. 1846. Convolvulaceae. 1–4 Central and trop. South America

Lysiphyllum (Benth.) de Wit = Bauhinia L. Fabaceae

Lysipomia Kunth. 1819. Lobeliaceae. 27 Andes

Lythrum L. 1753. Lythraceae. 38 cosmopolitan, excl. South America

Lytocaryum Toledo. 1944. Arecaceae. 3 southeastern Brazil

Maackia Rupr. et Maxim. 1856. Fabaceae. 8 China, Korea, Japan (Honshu, Shikoku, Kyushu, Ryuku Is.), M. floribunda—extending to Taiwan; Russia (1 [M. amurensis] Far East—Amur, Ussuri)

Maba J. R. Forst. et G. Forst. = Diospyros L. Ebenaceae

Mabea Aubl. 1775. Euphorbiaceae. 50 Central and trop. South America, Trinidad

Mabrya Elisens. 1985 (Asarina Hill). Scrophulariaceae. 5 U.S. (M. acerifolia—Arizona), Mexico (4)

Macadamia F. Muell. 1858. Proteaceae. 10–14: Madagascar (1), southern China (2), Sulawesi (1), eastern Australia (5), New Caledonia (3)

Macairea DC. 1828. Melastomataceae. 22 Colombia, Venezuela, Guyana, Surinam, Peru, Brazil, Bolivia

Macaranga Thouars. 1806. Euphorbiaceae. c. 300 trop. Africa, Madagascar, South and Southeast Asia, Malesia, Australia, Oceania

Macarenia P. Royen. 1951. Podostemaceae. 1 Colombia

Macarisia Thouars. 1806. Rhizophoraceae. 7 Madagascar

Macarthuria Huegel ex Endl. 1837. Molluginaceae. 3 southwestern and southeastern Australia

Macbridea Elliott ex Nutt. 1818. Lamiaceae. 2 southeastern U.S.

Macbrideina Standl. 1929. Rubiaceae. 1 Peru

Macdougalia A. Heller. 1898. Asteraceae. 1 U.S.

Macfadyena A. DC. 1845. Bignoniaceae. 4 trop. America from Mexico to northern Argentina, Uruguay, West Indies

Macgregoria F. Muell. 1874. Stackhousiaceae. 1 arid Australia

Machadoa Welw. ex Hook. f. = Adenia Forssk. Passifloraceae

Machaeranthera Nees. 1832. Asteraceae. c. 35 western North America

Machaerina Vahl. 1805–1806. Cyperaceae. 25 trop. Africa, trop. Asia, Japan, Australia, New Zealand, Oceania, trop. America

Machaerium Pers. 1807. Fabaceae. 120 trop. America from Mexico to Argentina; 1 (M. lunatum) trop. Africa from Senegal to Zaire and Angola

Machaerocarpus Small = Damasonium Hill. Alismataceae

Machaerocereus Britton et Rose = Stenocereus (A. Berger) Riccob. Cactaceae

Machairophyllum Schwantes. 1927. Aizoaceae. 10 South Africa

Machaonia Humb. et Bonpl. 1806. Rubiaceae. 25 trop. America, West Indies

Machilus Nees. 1831 (Persea Mill.). Lauraceae. 100 tropics and subtropics

Mackaya Harv. 1859. Acanthaceae. 1 South Africa

Mackeea H. E. Moore. 1978. Arecaceae. 1 northeastern New Caledonia

Mackenziea Nees. 1847 (Strobilanthes Blume). Acanthaceae. 9 India, Sri Lanka

Mackinlaya F. Muell. 1864. Araliaceae. 12 Sulawesi, Philippines, New Guinea, Solomon Is., northeastern Australia

Macleania Hook. 1837. Ericaceae. 40 central and western trop. South America

Macleaya R. Br. 1826. Papaveraceae. 2 China, Taiwan, Japan

Maclura Nutt. 1818. Moraceae. 1 (M. pomifera) southern U.S. (Arkansas, Texas)

Maclurodendron T. G. Hartley. 1982. Rutaceae. 6 northern Vietnam, Hainan, Malay Peninsula, Sumatra, Philippines

Maclurolyra Calderón et Soderstr. 1973. Poaceae. 1 Panama

Macnabia Benth. ex Endl. = Nabea Lehm. ex Klotzsch. Ericaceae

Macodes (Blume) Lindl. 1840. Orchidaceae. 10 Malesia from Malay Peninsula to New Guinea, Solomon Is., Vanuatu

Macoubea Aubl. 1775. Apocynaceae. 2 trop. South America

Macowania Oliv. 1870. Asteraceae. 11: Ethiopia and southern Arabian Peninsula (2), South Africa (9)

Macphersonia Blume. 1849. Sapindaceae. 8 trop. East Africa, Madagascar, Comoro Is.

Macrachaenium Hook. f. 1846. Asteraceae. 1 Patagonia and Tierra del Fuego

Macradenia R. Br. 1822. Orchidaceae. 8 trop. America from Florida and Mexico to Peru and Brazil, West Indies

Macraea Hook. f. 1847. Asteraceae. 1 Galapagos Is.

Macranthera Nutt. ex Benth. 1835. Scrophulariaceae. 1 southeastern U.S.

Macranthisiphon Bureau ex K. Schum.

1894. Bignoniaceae. 1 coastal Ecuador and Peru

Macroberlinia Hauman = Berlinia Sol. ex Hook. f. Fabaceae

Macrobia (Webb et Berth.) G. Kunkel = Aichryson Webb et Berth. Crassulaceae

Macrobriza (Tzvelev) Tzvelev. 1987 (Briza L.). Poaceae. 1 East Europe, Caucasus, southern Far East of Russia, Mediterranean, South Africa

Macrocarpaea (Griseb.) Gilg. 1895. Gentianaceae. 35 trop. America, West Indies

Macrocarpium (Spach) Nakai = Cornus L. Cornaceae

Macrocaulon N. E. Br. = Carpanthea N. E. Br. Aizoaceae

Macrocentrum Hook. f. 1867. Melastomataceae. 15 northern South America

Macrochaetium Steud. 1855 (Tetraria P. Beauv.) Cyperaceae. 6 South America, South Africa

Macrochlaena Hand.-Mazz. = Nothosmyrnium Miq. Apiaceae

Macrochloa Kunth. 1829 (Stipa L.). Poaceae. 1 western Mediterranean

Macrochordion De Vriese = Aechmea Ruiz et Pav. Bromeliaceae

Macroclinidium Maxim. = Pertya Sch. Bip. Asteraceae

Macroclinium Barb. Rodr. = Notylia Lindl. Orchidaceae

Macrocnemum P. Browne. 1756. Rubiaceae. 20 Central America, Colombia, West Indies

Macrococculus Becc. 1877. Menispermaceae. 1 New Guinea, New Britain, and New Ireland Is.

Macrodiervilla Nakai = Weigela Thunb. Caprifoliaceae

Macroditassa Malme = Ditassa R. Br. Asclepiadaceae

Macrohasseltia L. O. Williams. 1961. Flacourtiaceae. 1 Central America: Honduras, Nicaragua, Costa Rica, Panama

Macrolenes Naudin ex Miq. 1856. Melastomataceae. 15 Thailand, Malesia

Macrolobium Schreb. 1789. Fabaceae. 60 trop. America

Macromeria D. Don. 1832. Boraginaceae. 8 southern U.S., Mexico, Central and northern South America

Macronema Nutt. = Ericameria Nutt. Asteraceae

Macropanax Miq. 1856. Araliaceae. 8–14 Himalayas; Assam; Burma; southwestern, central, and southern China (incl. Hainan); Indochina; Malesia

Macropelma K. Schum. 1895. Asclepiadaceae. 1 East Africa

Macropeplus Perkins. 1898. Monimiaceae. 1 eastern Brazil

Macropetalum Burch. ex Decne. 1844. Asclepiadaceae. 1 South Africa

Macropharynx Rusby. 1927. Apocynaceae. 4 trop. South America

Macropidia J. L. Drumm. ex Harv. 1855 (Anigozanthos Labill.). Conostylidaceae. 1 southwestern Australia

Macropiper Miq. 1839 (Piper L.). Piperaceae. 9 Ogasawara (Bonin) Is., Mariana and Caroline Is., Santa Cruz Is., Vanuatu, Australia, Lord Howe I., New Zealand, Chatham Is., eastward to Society and Marquesas Is.

Macroplectrum Pfitzer = Angraecum Bory. Orchidaceae

Macropodanthus L. O. Williams. 1938. Orchidaceae. 6 Andaman and Nicobar Is., Thailand, Malay Peninsula, Indonesia, Philippines

Macropodiella Engl. 1926. Podostemaceae. 6 trop. West Africa

Macropodina R. M. King et H. Rob. 1972. Asteraceae. 14 trop. America from Mexico to Argentina

Macropodium R. Br. 1812. Brassicaceae. 2 West and East Siberia, Russian Far East, Mongolia, China

Macropsychanthus Harms ex K. Schum. et Lauterb. 1900. Fabaceae. 4 Philippines, New Guinea, Caroline and Solomon Is., Fiji (1, M. lauterbachii)

Macropteranthes F. Muell. ex Benth. 1864. Combretaceae. 4 northeastern Australia

Macroptilium (Benth.) Urb. 1928. Fabaceae. 20 trop. and subtrop. America

Macrorhamnus Baill. = Colubrina Rich. ex Brongn. Rhamnaceae

Macrorungia C. B. Clarke = Anisotes Nees. Acanthaceae

Macrosamanea Britton et Rose = Albizia Durazz. Fabaceae

Macroscepis Kunth. 1819. Asclepiadaceae. 20 Central and trop. South America, West Indies

Macrosciadium V. N. Tikhom. et Lavrova. 1988 (Ligusticum L.). Apiaceae. 2 Caucasus, Southwest Asia

Macroselinum Schur. 1853. Apiaceae. 1 South and Southeast Europe, northern Caucasus

Macrosepalum Regel et Schmalh. 1882 (Sedum L.). Crassulaceae. 2 Mediterranean, northern coast of the Black Sea, Crimea, Caucasus, Asia Minor, Iran

Macrosiphonia Muell. Arg. 1860. Apocynaceae. 10 trop. America

Macrosolen (Blume) Rchb. 1830. Loranthaceae. 35–40 India, Sri Lanka, southern China, Indochina, Malesia

Macrosphyra Hook. f. 1873. Rubiaceae. 3—5 trop. Africa from Senegal to Ethiopia, south to Angola and Zaire

Macrostegia Nees = Vitex L. Verbenaceae

Macrostelia Hochr. 1952. Malvaceae. 4 Madagascar (3), northeastern Australia (1)

Macrostemon Boriss. = Veronica L. Scrophulariaceae

Macrostigmatella Rauschert = Eigia Soják. Brassicaceae

Macrostylis Bartl. et H. L. Wendl. 1824 (Euchaetis Bartl. et H. L. Wendl.). Rutaceae. 10 South Africa

Macrosyringion Rothm. = Odontites Ludwig. Scrophulariaceae

Macrotomia DC. ex Meisn. = Arnebia Forssk. Boraginaceae

Macrotorus Perkins. 1898. Monimiaceae. 1 southeastern Brazil

Macrotyloma (Wight et Arn.) Verdc. 1970. (Dolichos L.). Fabaceae. 30 trop. Old World, esp. Africa (23)

Macroule Pierce. 1942 (Ormosia G. Jacks.). Fabaceae. 4: Southeast Asia (2), eastern South America (2)

Macrozamia Miq. 1842. Zamiaceae. 14: southwestern (1), central (1), and eastern (12) Australia

Macvaughiella R. M. King et H. Rob. 1968 (Eupatorium L.). Asteraceae. 1 Mexico, Central America

Madarosperma Benth. = Tassadia Decne. Asclepiadaceae

Maddenia Hook. f. et Thomson. 1854. Rosaceae. 5–6 China, Tibet, Himalayas

Madhuca Ham. ex J. F. Gmel. 1791. Sapotaceae. 85 India, China, Indochina, Malesia, Australia

Madia Molina. 1781. Asteraceae. 21 coastal Pacific America

Maerua Forssk. 1775. Capparaceae. 100 trop. and South Africa, West Asia eastward to India

Maesa Forssk. 1775. Myrsinaceae. c. 230 Africa, Madagascar, India, Sri Lanka, Burma, China, Japan, Indochina, Malesia, Australia, Melanesia to Fiji, Samoa, and Tonga

Maesobotrya Benth. 1879. Euphorbiaceae. 20 trop. Africa

Maesopsis Engl. 1895. Rhamnaceae. 1 trop. Africa from Liberia to Uganda, Kenya, and Tanzania, south to Angola and Zambia

Maeviella Rossow. 1985. Scrophulariaceae. 1 Brazil

Maferria C. Cusset. 1992. Podostemaceae. 1 India

Maga Urb. = Thespesia Sol. ex Corrêa. Malvaceae

Magadania Pimenov et Lavrova. 1985 (Ochotia Khokhr.). Apiaceae. 2 Russian Far East

Magallana Cav. 1798. Tropaeolaceae. 1 temp. South America

Magdalenaea Brade. 1935. Scrophulariaceae. 1 southeastern Brazil

Magnistipula Engl. 1905. Chrysobalanaceae. 11 trop. Africa (9), Madagascar (2)

Magnolia L. 1753. Magnoliaceae. c. 80 Himalayas, China, Russia (1, M. hypoleuca, southern Sakhalin and southern Kuril Is.), Korea, Japan, Taiwan, Indochina, Western and Central Malesia, southeastern U.S., Central and northern

South America, Bahama Is., West Indies

Magodendron Vink. 1957. Sapotaceae. 1 New Guinea

Magonia A. St. = Hil.1824. Sapindaceae. 2 Brazil, Bolivia, Paraguay

Maguirea A. D. Hawkes = Dieffenbachia Schott. Araceae

Maguireanthus Wurdack. 1964. Melastomataceae. 1 northeastern South America

Maguireocharis Steyerm. 1972. Rubiaceae. 1 northern South America

Maguireothamnus Steyerm. 1964. Rubiaceae. 2 Venezuela

Magydaris W. D. J. Koch ex DC. 1829. Apiaceae. 2 western Mediterranean (Iberian Peninsula, Balearic Is., Sardinia and Sicily, southern Italy)

Mahafalia Jum. et H. Perrier. 1911. Asclepiadaceae. 1 Madagascar

Maharanga A. DC. 1846 (Onosma L.). Boraginaceae. 9 Himalayas, southern Tibet

Mahawoa Schltr. 1916. Asclepiadaceae. 1 Sulawesi

Mahonia Nutt. 1818. Berberidaceae. c. 110 Himalayas, Tibet eastward to Japan, south to Sumatra, North and Central America

Mahurea Aubl. 1775. Clusiaceae. 7 Colombia, Venezuela, Brazil

Mahya Cordem. 1895 (Lepechinia Willd.). Lamiaceae. 1 Réunion I.

Maianthemum Wiggers. 1780. Convallariaceae. 3 temp. Northern Hemisphere; Russia (3, European part, West and East Siberia, Far East)

Maidenia Domin = Uldinia J. M. Black. Apiaceae

Maidenia Rendle. 1916. Hydrocharitaceae. 1 northwestern Australia

Maieta Aubl. 1775. Melastomataceae. 3 Colombia, Venezuela, Brazil, Bolivia

Maihuenia Phil. ex K. Schum. 1898. Cactaceae. 4 Chile, Argentina

Maihueniopsis Speg. = Opuntia Hill. Cactaceae

Maillarida Frapp. et Duchartre. 1862 (Trophis P. Browne). Moraceae. 5 Madagascar, Réunion I.

Maillea Parl. 1842 (Phelum L.). Poaceae. 1 Mediterranean islands

Maingaya Oliv. 1873. Hamamelidaceae. 1 Malay Peninsula

Maireana Moq. 1840. Chenopodiaceae. 58 Australia

Mairetis I. M. Johnst. 1953. Boraginaceae. 1 Canary Is., Morocco

Mairia Nees. 1932. Asteraceae. 15 South Africa

Majidea J. Kirk ex Oliv. 1871. Sapindaceae. 5 trop. Africa, Madagascar

Majorana Mill. = Origanum L. Lamiaceae

Malabaila Hoffm. 1814 (Pastinaca L.). Apiaceae. 10 Southeast Europe, eastern Mediterranean, Northeast and trop. Africa, West (incl. Caucasus) and Middle Asia

Malacantha Pierre = Pouteria Aubl. Sapotaceae

Malaccotristicha C. Cusset et G. Cusset. 1988. Podostemaceae. 1 Malay Peninsula

Malachium Fires = Myosoton Moench. Caryophyllaceae

Malachra L. 1767. Malvaceae. 8–10 trop. and subtrop. America, West Indies

Malacocarpus Fisch. et C. A. Mey. 1843 (Peganum L.). Peganaceae. 1 Middle Asia, northern Iran

Malacocera R. H. Anderson. 1926. Chenopodiaceae. 4 arid areas of southern Australia

Malacomeles (Decne.) Engl. 1945. Rosaceae. 3 Texas, Mexico, Guatemala

Malacothamnus Greene. 1906. Malvaceae. 12 southwestern U.S., Mexico, Chile (1)

Malacothrix DC. 1838. Asteraceae. 15 western North America

Malacurus Nevski = Leymus Hochst. Poaceae

Malagasia L. A. S. Johnson et B. G. Briggs. 1975. Proteaceae. 1 Madagascar

Malaisia Blanco = Trophis P. Browne. Moraceae

Malanea Aubl. 1775. Rubiaceae. 21 trop. America, West Indies

Malania Chun et S. K. Lee ex S. K. Lee. 1980. Olacaceae. 1 China

Malaxis Sol. ex Sw. 1788. Orchidaceae. c. 300 temp. and trop. regions, esp. trop. Asia; Russia (1, M. monophyllos: northern and central European part, Siberia)

Malcolmia R. Br. 1812. Brassicaceae. 35 Mediterranean, Southwest and West Asia eastward to Afghanistan, Kashmir, and western Tibet

Malea Lundell. 1943. Ericaceae. 1 Mexico

Malephora N. E. Br. 1927. Aizoaceae. 8–9 South Africa

Malesherbia Ruiz et Pav. 1794. Malesherbiaceae. c. 30 Andes from southern Peru to northern Chile and western Argentina

Malinvaudia Fourn. = Matelea Aubl. Asclepiadaceae

Malleastrum (Baill.) J.-F. Leroy. 1964. Meliaceae. 13 Comoro Is., Madagascar, Aldabra I. I.

Malleola J. J. Sm. et Schltr. ex Schltr. 1913. Orchidaceae. 20 China, Indochina, Malesia

Malleostemon J. W. Green. 1983. Myrtaceae. 6 southwestern Australia

Mallinoa J. M. Coult. = Ageratina Spach. Asteraceae

Mallophora Endl. 1838. Chloanthaceae. 2 western Australia

Mallophyton Wurdack. 1964. Melastomataceae. 1 Venezuela

Mallotonia (Griseb.) Britton = Tournefortia L. Boraginaceae

Mallotopus Franch. et Sav. = Arnica L. Asteraceae

Mallotus Lour. 1790. Euphorbiaceae. c. 114 trop. Africa and Madagascar (2), South, East, and Southeast Asia; Malesia; northern and eastern Australia; Melanesia to Fiji

Malmea R. E. Fr. 1905. Annonaceae. 14 trop. America from Mexico to Brazil

Malmeanthus R. M. King et H. Rob. 1980. Asteraceae. 3 Brazil, Argentina, Uruguay

Malope L. 1753. Malvaceae. 4 Mediterranean, Asia Minor, Caucasus (1, M. malacoides)

Malortiea H. A. Wendl. = Reinhardtia Liebm. Arecaceae

Malosma (Nutt.) Raf. = Rhus L. Anacardiaceae

Malouetia A. DC. 1844. Apocynaceae. 4 Central and trop. South America, West Indies, trop. Africa

Malouetiella Pichon = Malouetia DC. Apocynaceae

Malperia S. Watson. 1889. Asteraceae. 1 southwestern U.S., California Peninsula

Malpighia L. 1753. Malpighiaceae. 40 trop. America, West Indies

Malpighiantha Rojas Acosta. 1897. Malpighiaceae. 2 Argentina

Malpighioides Nied. = Mascagnia (Bertero ex DC.) Colla. Malpighiaceae

Maltebrunia Kunth. 1829. Poaceae. 5 trop. East Africa, Madagascar

Malus Hill. 1753. Rosaceae. 55 temp. Northern Hemisphere

Malva L. 1753. Malvaceae. 30–40 temp. and subtrop. regions of the Old World

Malvalthaea Iljin. 1924. Malvaceae. 1 Caucasus

Malvastrum A. Gray. 1849. Malvaceae. 14 trop. and subtrop. America, 1 (M. coromandelianum) Egypt, trop. Asia, Malesia, Australia, and Pacific Islands

Malvaviscus Fabr. 1759. Malvaceae. 6 North (Texas, Mexico), Central and South America, West Indies

Malvella Jaub. et Spach. 1855 (Sida L.). Malvaceae. 4: America (3, southern U.S., Mexico); 1 (M. sherardiana) Mediterranean, Crimea, Caucasus, Southwest Asia

Malveopsis C. Presl = Anisodontea C. Presl. Malvaceae

Mamillopsis E. Morren ex Britton et Rose = Mammillaria Haw. Cactaceae

Mammea L. 1753. Clusiaceae. c. 50 pantropics

Mammillaria Haw. 1812. Cactaceae. c. 150–200 trop. America from southwestern U.S. and Mexico to Colombia and Venezuela; West Indies

Mammilloydia Buxb. 1951 (Mammillaria Haw.). Cactaceae. 1 northeastern Mexico

Mamorea Sota. 1960 (Thismia Griff.). Burmanniaceae. 1 Bolivia

Mananthes Bremek. 1948 (Justicia L.). Acanthaceae. 5 southern China, Indochina, western Malesia

Manaosella J. C. Gomes. 1949. Bignoniaceae. 1 Brazil

Mancoa Wedd. 1859. Brassicaceae. 11 Mexico, Andes of Peru, Bolivia, and Argentina

Mandenovia Alava. 1973 (Tordylium L.). Apiaceae. 1 Caucasus

Mandevilla Lindl. 1840. Apocynaceae. c. 116 Central and trop. South America

Mandragora L. 1753. Solanaceae. 6 Eurasia from Iberian Peninsula to southwestern China, Himalayas and Burma, North Africa; M. turcomanica—Middle Asia

Manekia Trel. 1927. Peperomiaceae. 1 West Indies

Manettia Mutis ex L. 1771. Rubiaceae. 80 trop. America, West Indies

Manfreda Salisb. 1866 (Agave L.). Agavaceae. 18 southeastern U.S., Mexico

Mangenotia Pichon. 1954. Asclepiadaceae. 1 trop. West Africa

Mangifera L. 1753. Anacardiaceae. 62 Himalayas, India, Sri Lanka, southern China, Indochina, Malesia, New Britain, Solomon Is.

Manglietia Blume. 1823. Magnoliaceae. 27 Burma, southern China, Vietnam, Laos, Sumatra, Java, Sulawesi

Manglietiastrum Law = Magnolia L. Magnoliaceae

Mangonia Schott. 1857. Araceae. 2 southern Brazil, Uruguay

Manicaria Gaertn. 1791. Arecaceae. 1 (M. saccifera) Panama, Colombia, Venezuela, Trinidad, Guyana, Brazil

Manihot Hill. 1753. Euphorbiaceae. c. 100 trop. and subtrop. America

Manihotoides D. J. Rogers et Appan. 1973. Euphorbiaceae. 1 Mexico

Manilkara Adans. 1763. Sapotaceae. 70: 30—trop. America (Florida, Bahamas, Mexico, Guatemala, Belize, El Salvador, Nicaragua, Costa Rica, Panama, Colombia, Venezuela, the Guianas, Ecuador, Peru, Brazil—18, Bolivia) and West Indies; c. 20 trop. and South (3) Africa, Mascarene Is.; c. 20 trop. Asia, Malesia to Pacific Islands

Manilkariopsis (Gilly) Lundell = Manilkara Adans. Sapotaceae

Maniltoa Scheff. 1876. Fabaceae. 20–25 South and Southeast Asia, Malesia (esp. New Guinea), trop. Australia, Oceania eastward to Fiji (4) and Tongo

Manisuris L. 1771. Poaceae. 5 India

Mannagettaea Harry Sm. 1933. Orobanchaceae. 3 East Siberia (1, M. hummelii), western China (2)

Manniella Rchb. f. 1881. Orchidaceae. 1 trop. Africa from Sierra Leone to Zaire, Uganda, and Tanzania; ? Hong Kong

Manniophyton Muell. Arg. 1864. Euphorbiaceae. 1 trop. Africa

Manochlamys Aellen. 1939. Chenopodiaceae. 1 South-West and South Africa

Manoelia S. Bowdich = Withania Pauquy. Solanaceae

Manongarivea Choux = Lepisanthes Blume Sapindaceae

Manostachya Bremek. 1952. Rubiaceae. 3 trop. Africa

Manotes Sol. ex Planch. 1850. Connaraceae. 11 trop. Africa

Manothrix Miers. 1878. Apocynaceae. 2 Brazil

Mansoa DC. 1838. Bignoniaceae. 15 trop. America from Mexico to Argentina

Mansonia J. R. Drumm. ex Prain. 1905. Sterculiaceae. 5 West, Central, and East Africa; northeastern India, Burma

Mantalania Capuron ex J.-F. Leroy. 1973. Rubiaceae. 2–3 Madagascar

Mantisalca Cass. 1818. Asteraceae. 4 South Europe, North Africa, Asia Minor

Mantisia Sims. 1810. Zingiberaceae. 3 South and Southeast Asia

Manulea L. 1767. Scrophulariaceae. 50–60 South Africa, 3 extending to Botswana, Zimbabwe, and Mozambique

Manuleopsis Thell. ex Schinz. 1915. Scrophulariaceae. 2 Namibia

Maoutia Wedd. 1854. Urticaceae. 15

northern India, Burma, Thailand, southern China (2), Indochina, Malesia, and Polynesia

Mapania Aubl. 1775. Cyperaceae. 80 tropics, excl. Madagascar

Mapaniopsis C. B. Clarke. 1908. Cyperaceae. 1 northern Brazil

Mapouria Aubl. = Psychotria L. Rubiaceae

Mappia Jacq. 1797. Icacinaceae. 5 Mexico, Guatemala, Belize, Panama, the Greater Antilles

Mappianthus Hand.-Mazz. 1921. Icacinaceae. 2 southern China, Borneo

Maprounea Aubl. 1775. Euphorbiaceae. 4 trop. America (2), trop. Africa (2 from Nigeria to Tanzania and south to Angola and Mozambique)

Maquira Aubl. 1775. Moraceae. 5 trop. America

Maracanthus Kuijt = Oryctina Tiegh. Loranthaceae

Marah Kellogg. 1854. Cucurbitaceae. 6–7 western North America from British Columbia to Baja California

Marahuacaea Maguire. 1984. Rapateaceae. 1 northern South America

Maranta L. 1753. Marantaceae. 23 trop. America, West Indies

Maranthes Blume. 1825. Chrysobalanaceae. 12 trop. Africa (10); 1 (M. corymbosa) southern Thailand, Malesia to Caroline and Solomon Is., Australia (Northern Territory and Queensland); Central America (1) from Nicaragua to Panama

Marantochloa Brongn. et Cris. 1860. Marantaceae. 15 trop. Africa, Mascarene Is.

Marasmodes DC. 1838. Asteraceae. 4 South Africa

Marathrum Bonpl. 1806. Podostemaceae. 25 trop. America

Marcania J. B. Imlay. 1939. Acanthaceae. 1 Thailand

Marcelliopsis Schinz. 1934. Amaranthaceae. 9 trop. Africa

Marcetella Svent. 1948 (Bencomia Webb et Berthel.). Rosaceae. 2 Macaronesia

Marcetia DC. 1828. Melastomataceae. 10–15 Brazil, Uruguay

Marcgravia L. 1753. Marcgraviaceae. 60 trop. America, West Indies

Maresia Pomel. 1874. Brassicaceae. 6 Mediterranean, West Asia to Caucasus (2), Caspian Sea, and southern Iran

Mareya Baill. 1860. Euphorbiaceae. 3 trop. West Africa

Mareyopsis Pax et K. Hoffm. 1919. Euphorbiaceae. 1 trop. West Africa

Margaranthus Schltdl. 1838. Solanaceae. 1 southwestern U.S. (Arizona, New Mexico, Texas), Mexico, Guatemala, Honduras, West Indies

Margaretta Oliv. 1875. Asclepiadaceae. 1 trop. Africa

Margaritaria L. f. 1782. Euphorbiaceae. 14 tropics

Margaritolobium Harms. 1923. Fabaceae. 1 Venezuela

Margaritopsis Sauvalle. 1869. Rubiaceae. 4 West Indies

Margelliantha P. J. Cribb. 1979. Orchidaceae. 4 northeastern Zaire, Kenya, and Tanzania

Margotia Boiss. 1838 (Elaeoselinum W. D. J. Koch ex DC.). Apiaceae. 1 western Mediterranean

Margyricarpus Ruiz et Pav. 1794. Rosaceae. 1–3 Andes south to Chile

Marianthus Huegel ex Endl. 1837 (Billardiera Sm.). Pittosporaceae. 15 Australia

Marila Sw. 1788. Clusiaceae. 20 trop. America, West Indies (2)

Marina Liebm. 1854. Fabaceae. 38 southwestern U.S., Mexico, Central America

Maripa Aubl. 1775. Convolvulaceae. 20 trop. America: Guatemala, Belize, Honduras, Nicaragua, Costa Rica, Panama, Colombia, Venezuela, Guyana, Surinam, French Guiana, Ecuador, Peru, Brazil

Mariposa (Wood) Hoover = Calochortus Pursh. Calochortaceae

Mariscopsis Cherm. = Queenslandiella Domin. Cyperaceae

Marisculus Goetgh. = Cyperus L. Cyperaceae

Mariscus Vahl. 1805–1806 (Cyperus L.). Cyperaceae. c. 200 tropics and subtropics

Markea Rich. 1792. Solanaceae. 9 central and trop. South America, Trinidad

Markhamia Seem. ex Baill. 1888. Bignoniaceae. 13 trop. Africa and trop. Asia from Burma and Andaman Is. through Indochina to southeastern China and Malesia

Markleya Bondar = Maximiliana Mart. Arecaceae

Marlierea Cambess. 1833. Myrtaceae. c. 50 trop. America

Marlieriopsis Kiaersk. = Blepharoxalyx O. Berg. Myrtaceae

Marlothia Engl. = Helinus E. Mey. ex Endl. Rhamnaceae

Marlothiella H. Wolff. 1912. Apiaceae. 1 Namibia

Marlothistella Schwantes = Ruschia Schwantes. Aizoaceae

Marmaroxylon Killip. 1940 (Zygia P. Browne). Fabaceae. 8 Colombia, Peru, Venezuela, Guyana, Surinam, Brazil

Marmoritis Benth. 1833. Lamiaceae. 4 Tibet, Himalayas, southwestern China

Marojejya Humbert. 1955. Arecaceae. 2 northeastern Madagascar

Marquesia Gilg. 1908. Monotaceae. 4 trop. Africa

Marrubium L. 1753. Lamiaceae. 40 Macaronesia, Mediterranean, temp. Eurasia

Marsdenia R. Br. 1811. Asclepiadaceae. c. 100 trop. America, trop. Africa, trop. Asia, trop. Australia

Marshallfieldia J. F. Macbr. = Adelobotrys DC. Melastomataceae

Marshallia Schreb. 1791. Asteraceae. 10 southern U.S.

Marshalljohnstonia Henrickson. 1976. Asteraceae. 1 Mexico

Marsippospermum Desv. 1809. Juncaceae. 4 New Zealand, Auckland, and Campbell Is.; southern South America; Tierra del Fuego; Falkland Is.

Marssonia Karst. = Napeanthus Gardner. Gesneriaceae

Marsypianthes Mart. ex Benth. 1833. Lamiaceae. 5 trop. America from southern U.S. to Paraguay, West Indies

Marsypopetalum Scheff. 1870. Annonaceae. 1 western Malesia: Malay Peninsula, Java, Borneo

Marthella Urb. 1903. Burmanniaceae. 1 Trinidad

Marticorenia Crisci. 1974. Asteraceae. 1 Chile

Martinella Baill. 1888. Bignoniaceae. 2 trop. America from southern Mexico to Brazil and Bolivia, West Indies

Martinezia auct., non Ruiz et Pav. = Aiphanes Willd. Arecaceae

Martinezia Ruiz et Pav. = Prestoea Hook. f. Arecaceae

Martiodendron Gleason. 1935. Fabaceae. 4 trop. South America

Martretia Beille. 1907. Euphorbiaceae. 1 trop. Africa: Sierra Leone, Ivory Coast, Ghana, Nigeria, Cameroun, Gabon, Congo, Zaire

Martynia L. 1753. Martyniaceae. 1 Mexico, Central America, West Indies

Mascagnia (Bertero ex DC.) Colla. 1824. Malpighiaceae. 55 trop. America from Mexico to Argentina; West Indies

Mascarena L. H. Bailey = Hyophorbe Gaertn. Arecaceae

Mascarenhasia A. DC. 1844. Apocynaceae. 12 trop. East Africa (1), Madagascar

Maschalocephalus Gilg et K. Schum. 1900. Rapateaceae. 1 trop. West Africa

Maschalocorymbus Bremek. 1940. Rubiaceae. 4 western Malesia

Maschalodesme K. Schum. et Lauterb. 1900. Rubiaceae. 2 New Guinea

Masdevallia Ruiz et Pav. 1794. Orchidaceae. c. 300 mountainous trop. America from Mexico to Ecuador, Peru, Brazil

Masmenia F. K. Mey. = Thlaspi L. Brassicaceae

Masoala Jum. 1933. Arecaceae. 1 northeastern Madagascar

Massangea E. Morren = Guzmania Ruiz et Pav. Bromeliaceae

Massartina Maire = Elizaldia Willk. Boraginaceae

Massia Balansa = Eriachne R. Br. Poaceae

Massonia Thunb. ex L. f. 1782. Hyacinthaceae. 8 South Africa

Massularia (K. Schum.) Hoyle. 1973. Rubiaceae. 1 trop. Africa

Mastersia Benth. 1865. Fabaceae. 2 South and Southeast Asia

Mastersiella Gilg-Ben. 1930 (Hypolaena R. Br.). Restionaceae. 3 South Africa (Cape)

Mastichodendron (Engl.) H. J. Lam. = Sideroxylon L. Sapotaceae

Mastigosciadium Rech. f. et Kuber. 1964. Apiaceae. 1 Afghanistan

Mastigostyla I. M. Johnst. 1928. Iridaceae. 17 Peru, Argentina

Mastixia Blume. 1826. Mastixiaceae. 13 India, Sri Lanka, Andaman Is., eastern Himalayas, southwestern and southern China, Malesia eastward to New Britain and Solomon Is.

Mastixiodendron Melch. 1925. Rubiaceae. 7 Moluccas, New Guinea, Solomon Is., Fiji (3)

Matayba Aubl. 1775. Sapindaceae. 50 trop. and subtrop. America

Matelea Aubl. 1775. Asclepiadaceae. c. 150 from central U.S. to trop. South America, West Indies

Mathewsia Hook. et Arn. 1833. Brassicaceae. 6 Peru, Chile

Mathiasella Constance et C. Hitchc. 1954. Apiaceae. 1 Mexico

Mathieua Klotzsch. 1853 (Stenomesson Herb.). Amaryllidaceae. 1 Peru

Mathurina Balf. f. 1876. Turneraceae. 1 Mascarene Is. (Rodriguez)

Matisia Humb. et Bonpl. 1805. Bombacaceae. 35 trop. South America

Matricaria L. 1753. Asteraceae. 5 Eurasia, North America

Mattfeldanthus H. Rob. et R. M. King. 1979. Asteraceae. 2 Brazil

Mattfeldia Urb. 1931. Asteraceae. 1 Haiti

Matthaea Blume. 1856 (Steganthera Perkins). Monimiaceae. 6–7 Malesia (Sumatra, Malay Peninsula, Anambas Is., Borneo, Philippines, Taluad Is., Sulawesi)

Matthiola R. Br. 1812. Brassicaceae. c. 60 Macaronesia, West and Southeast Europe, Mediterranean, South Africa, West (incl. Caucasus) and Middle Asia, southern Siberia, China

Mattiastrum (Boiss.) Brand = Cynoglossum L. Boraginaceae

Matucana Britton et Rose. 1922. Cactaceae. 6–7 (or. 19 ?) Peru

Matudacalamus F. Maek. = Aulonemia Goudot. Poaceae

Matudaea Lundell. 1940. Hamamelidaceae. 1 Mexico

Matudanthus D. R. Hunt. 1978 (Tradescantia L.). Commelinaceae. 1 Mexico

Matudina R. M. King et H. Rob. 1973. Asteraceae. 1 Mexico

Maughaniella L. Bolus = Dilposoma Schwantes. Aizoaceae

Mauloutchia Warb. 1895 (Brochoneura Warb.). Myristicaceae. 6 Madagascar

Maundia F. Muell. 1858. Juncaginaceae. 1 Australia, Tasmania

Maurandella (A. Gray) Rothm. 1943 (Asarina Hill). Scrophulariaceae. 2–3 southwestern U.S., Mexico, West Indies

Maurandya Ortega. 1797 (Asarina Hill). Scrophulariaceae. 4 U.S. (California, Arizona, New Mexico, Texas), Mexico (4), Bermuda, Bahamas, Cuba, Jamaica

Mauranthe O. F. Cook = Chamaedorea Willd. Arecaceae

Mauria Kunth. 1824. Anacardiaceae. 10 trop. America from Honduras to Ecuador, Peru, and southern Brazil

Mauritia L. f. 1782. Arecaceae. 3 Trinidad, Colombia, Ecuador, Peru, Venezuela, Guyana, Surinam, French Guiana, Brazil

Mauritiella Burret. 1935 (Mauritia L. f.). Arecaceae. 5 northern South America

Maurocenia Hill. 1753. Celastraceae. 1 South Africa

Mausolea Bunge ex Podlech. 1986 (Artemisia L.). Asteraceae. 1 Middle Asia, Iran

Maxburretia Furtado. 1941. Arecaceae. 3 southern Thailand, Malay Peninsula

Maxia O. E. G. Nilsson. 1967 (Montia L.). Portulacaceae. 1 western North America from Vancouver to northern California

Maxillaria Ruiz et Pav. 1794. Orchidaceae.

c. 400 trop. America from Mexico to Peru and Brazil, West Indies

Maximiliana Mart. 1824. Arecaceae. 1 (M. regia) Colombia, Peru, Trinidad, Venezuela, Guyana, Surinam, Brazil

Maximoviczia A. P. Khokhr. = Scirpus L. Cyperaceae

Maximowicziella A. P. Khokhr. = Scirpus L. Cyperaceae

Maxwellia Baill. 1871. Sterculiaceae. 1 New Caledonia

Mayaca Aubl. 1775. Mayacaceae. 4 America from southeastern U.S. to Paraguay, West Indies; 1 (M. baumii) Zaire, Angola, Zambia

Mayanaea Lundell. 1974 (Orthion Standl. et Steyerm.). Violaceae. 1 Guatemala

Mayna Aubl. 1775. Flacourtiaceae. 6 trop. America from Honduras to eastern Peru, eastern Bolivia, and Amazonian Brazil

Mayodendron Kurz. 1875 (Radermachera Zoll. et Moritzi). Bignoniaceae. 1 Burma, southern China, Indochina

Maytenus Molina. 1781. Celastraceae. c. 225 tropics and subtropics

Mazaea Krug et Urb. 1897. Rubiaceae. 4 West Indies

Mazus Lour. 1790. Scrophulariaceae. c. 50 Middle Asia and Russian Far East (2), Himalayas from Kashmir to Bhutan, India, Burma, western and southeastern China, Korea, Japan, Taiwan, Malesia, Australia, Tasmania, New Zealand, Melanesia to Fiji

Mazzettia Iljin = Dolomiaea DC. Asteraceae

Mcvaughia W. R. Anderson. 1979. Malpighiaceae. 1 Brazil

Mearnsia Merr = Metrosideros Gaertn. Myrtaceae

Mecardonia Ruiz et Pav. 1794. Scrophulariaceae. 10 trop. and subtrop. America from southern U.S. to Argentina, West Indies

Mechowia Schinz. 1893. Amaranthaceae. 1 trop. Namibia

Mecomischus Coss. ex Benth. 1873. Asteraceae. 1 Algeria

Meconella Nutt. ex Torr. et A. Gray. 1838. Papaveraceae. 3–4 western North America

Meconopsis Vig. 1814. Papaveraceae. 50 West Europe (1, M. cambrica); 49 Himalayas, India, Burma, China (c. 40)

Mecopus Benn. 1840. Fabaceae. 1 South and Southeast Asia

Mecranium Hook. f. 1867. Melastomataceae. 15 Greater Antilles (Cuba, Haiti, Jamaica, Puerto Rico)

Medemia Wuerttemb. ex H. A. Wendl. 1881. Arecaceae. 1 southern Egypt, northeastern Sudan (Nubian desert)

Medeola L. 1753. Medeolaceae. 2 North America from Ontario and Quebec to Louisiana, Alabama, and Florida

Mediasia Pimenov. 1974. Apiaceae. 1 (M. macrophylla) Middle Asia and Afghanistan

Medicago L. 1753. Fabaceae. 83 Europe, Mediterranean, trop. (3) and South (1) Africa, West and Central Asia eastward to China and Russian Far East, south to India, Australasia

Medicosma Hook. f. 1862. Rutaceae. 22 New Guinea (1), eastern Australia (6), New Caledonia (15)

Medinilla Gaudich. 1826. Melastomataceae. 350–400 (or 150) trop. Africa, Madagascar, Himalayas, India, Sri Lanka, Burma, southern China, Indochina, Taiwan, Malesia (esp. New Guinea—c. 70), northern Australia (1, M. balls-headleyi), Micronesia, Admiralty and Solomon Is., Fiji to Polynesia

Mediocalcar J. J. Sm. 1900. Orchidaceae. 37 Amboina and Key Is. (1), New Guinea

Mediusella (Cavaco) Dorr. 1987. Sarcolaenaceae. 1 Madagascar

Medusagyne Baker. 1877. Medusagynaceae. 1 Seychelles

Medusandra Brenan. 1952. Medusandraceae. 2 trop. West Africa

Medusanthera Seem. 1864. Icacinaceae. 4–5 Andaman and Nicobar Is., southern Thailand, Malay Peninsula, Sumatra, Philippines, Sulawesi, Moluccas, New

Guinea, Bismarck Arch., Caroline and Solomon Is., Fiji, Samoa Is.

Meeboldia H. Wolff. 1924. Apiaceae. 2 northwestern Himalayas, Nepal

Meeboldina Suess. 1943. Restionaceae. 1 southwestern Australia

Meehania Britton. 1894. Lamiaceae. 7 East and Southeast Asia (6), eastern U.S. (1, M. cordata); Russia (1, M. urticifolia: Far East)

Meehaniopsis Kudo = Glechoma L. Lamiaceae

Megacarpaea DC. 1821. Brassicaceae. 8: 7 southeastern European part of Russia, West Siberia, Middle Asia; Mongolia, western Tibet, Himalayas from Kashmir to Bhutan; 1 eastern China

Megacaryon Boiss. = Echium L. Boraginaceae

Megacodon (Hemsl.) Harry Sm. 1936. Gentianaceae. 1–2 Tibet, Himalayas from Nepal to Assam

Megadenia Maxim. 1889. Brassicaceae. 3 southern East Siberia and Russian Far East (2), China

Megalachne Steud. 1854. Poaceae. 1 Juan Fernández Is.

Megaleranthis Ohwi. 1935 (Eranthis Salisb.). Ranunculaceae. 1 Korea

Megalochlamys Lindau. 1899. Acanthaceae. 10 Ethiopia, Somalia, Kenya, Tanzania, Mozambique, Angola, Zimbabwe, Botswana, northern South Africa, southern Arabian Peninsula (2)

Megalodonta Greene. 1901. Asteraceae. 1 North America

Megalopanax Ekman. 1924. Araliaceae. 1 Cuba

Megaloprotachne C. E. Hubb. 1929. Poaceae. 1 Angola, Zambia, Zimbabwe, Botswana, South Africa

Megalorchis H. Perrier. 1937. Orchidaceae. 1 Madagascar

Megalostoma Leonard. 1940. Acanthaceae. 1 Central America

Megalostylis S. Moore = Dalechampia L. Euphorbiaceae

Megalotheca F. Muell. = Restio Rottb. Restionaceae

Megalotus Garay. 1972. Orchidaceae. 1 Philippines

Megaphrynium Milne-Redh. 1952. Marantaceae. 3 trop. Africa from Sierra Leone to Sudan and Uganda

Megaphyllaea Hemsl. = Chisocheton Blume. Meliaceae

Megarrhena Schrad. ex Nees = Androtrichum (Brongn.) Brongn. Cyperaceae

Megasea Haw. = Bergenia Moench. Saxifragaceae

Megaskepasma Lindau. 1897. Acanthaceae. 1 Panama, Venezuela

Megastachya P. Beauv. 1812. Poaceae. 1 trop. Africa, Madagascar

Megastigma Hook. f. 1862. Rutaceae. 2 Mexico, Central America

Megastoma (Benth.) Bonnet et Baratte = Ogastemma Brummitt. Boraginaceae

Megastylis (Schltr.) Schltr. 1911. Orchidaceae. 8 eastern Australia (1, M. elliptica), New Caledonia, and Vanuatu (1)

Megatritheca Cristóbal. 1965. Sterculiaceae. 2 trop. Africa

Megistostegium Hochr. 1916. Malvaceae. 3 Madagascar

Megistostigma Hook. f. 1887. Euphorbiaceae. 3–4 Southeast Asia, western Malesia

Meiandra Markgr. 1927. Melastomataceae. 2 eastern Peru

Meineckia Baill. 1858. Euphorbiaceae. 20 Southwest and Northeast trop. Africa, Madagascar, southern Arabian Peninsula, Socotra, southern and northeastern (Assam) India, Sri Lanka, Central and trop. South America

Meiocarpidium Engl. et Diels. 1900. Annonaceae. 2 trop. West Africa

Meiogyne Miq. 1865. Annonaceae. 9–10 western India, southern China, Indochina, Malesia: Malay Peninsula, Philippines, and Sunda Is.

Meiomeria Standl. = Chenopodium L. Chenopodiaceae

Meionandra Gauba = Valantia L. Rubiaceae

Meiostemon Exell et Stace. 1966. Combretaceae. 2 trop. Southeast Africa (1), Madagascar (1)

Meiracyllium Rchb. f. 1854. Orchidaceae. 2 Mexico and Central America

Melachone Gilli = Amaracarpus Blume. Rubiaceae

Meladenia Turcz. = Cullen Medik. Fabaceae

Meladerma Kerr. 1938. Asclepiadaceae. 3 Thailand

Melaleuca L. 1767. Myrtaceae. c. 150 Southeast Asia, Malesia, New Guinea, Australia (c. 100), New Caledonia (1)

Melampodium L. 1753. Asteraceae. 37 trop. and subtrop. America

Melampyrum L. 1753. Scrophulariaceae. 35 temp. Northern Hemisphere

Melananthus Walp. 1850. Solanaceae. 5 West India, Brazil, northeastern Argentina

Melancium Naudin. 1862. Cucurbitaceae. 1 eastern and southern Brazil

Melandrium Roehl. 1812 (Vaccaria Wolf). Caryophyllaceae. c. 100 Northern Hemisphere, trop. and South Africa, Andes of South America

Melanocenchris Nees. 1841. Poaceae. 3 trop. Northeast Africa, Arabian Peninsula, southern Iran, Pakistan, India, Sri Lanka

Melanochyla Hook. f. 1876. Anacardiaceae. 17 western Malesia

Melanocommia Ridl. = Semecarpus L. f. Anacardiaceae

Melanodendron DC. 1836. Asteraceae. 1 St. Helena

Melanodiscus Radlk. = Glenniea Hook. f. Sapindaceae

Melanolepis Rchb. f et Zoll. 1856. Euphorbiaceae. 1 Indochina, Taiwan, Malesia, Melanesia

Melanophylla Baker. 1884. Melanophyllaceae. 8 Madagascar

Melanopsidium Colla. 1824. Rubiaceae. 1 Brazil

Melanorrhoea Wall. = Gluta L. Anacardiaceae

Melanosciadium Boissieu. 1902. Apiaceae. 1 western China

Melanoselinum Hoffm. 1814. Apiaceae. 4 Macaronesia (Madeira, Cape Verde Is., Canary Is.)

Melanospermum Hilliard. 1988–89 (Phyllopodium Benth.). Scrophulariaceae. 6 South Africa

Melanoxylon Schott = Melanoxylum Schott. Fabaceae

Melanoxylum Schott. 1822. Fabaceae. 3 trop. South America

Melanthera Rohr. 1792. Asteraceae. 20 trop. and subtrop. America

Melanthium L. 1753 (Zygadenus Michx.). Melanthiaceae. 5 North America

Melanthium Medik. = Nigella L. Ranunculaceae

Melasma Bergius. 1767. Scrophulariaceae. 5 Central and South America, trop. Africa

Melasphaerula Ker-Gawl. 1803. Iridaceae. 1 Namibia, South Africa

Melastoma L. 1753. Melastomataceae. 70–100 South and Southeast Asia from India to southern China, Indochina, Malesia, coastal northern Australia (1, M. affine), trop. Oceania

Melastomastrum Naudin. 1850. Melastomataceae. 6 trop. Africa

Melchiora Kubuski = Balthasaria Verdc. Theaceae

Melhania Forssk. 1775. Sterculiaceae. 60 trop. Old World

Melia L. 1753. Meliaceae. 3 trop. Old World

Melianthus L. 1753. Melianthaceae. 7 South Africa

Melica L. 1753. Poaceae. c. 80 temp. regions and montane areas of tropics, excl. Australia

Melichrus R. Br. 1810 (Styphelia Sm.). Epacridaceae. 4 Australia

Melicocca L. = Melicoccus P. Browne. Sapindaceae

Melicoccus P. Browne. 1756. Sapindaceae. 2 trop. America, West Indies

Melicope J. R. Forst. et G. Forst. 1775. Rutaceae. c. 70 Madagascar, India, southern China, Malesia, trop. Australia, New Zealand, Oceania eastward to Society Is.

Melicytus J. R. Forst. et G. Forst. 1775. Violaceae. 5 Solomon Is., New Zealand, Norfolk Is., Fiji

Melientha Pierre. 1888. Opiliaceae. 1 (M. suavis) Thailand, Laos, Cambodia, Vietnam, Malay Peninsula, Philippines

Melilotoides Heist. ex Fabr. 1763. Fabaceae. 28 West (incl. Caucasus) and Middle Asia, southern Siberia, Russian Far East, Mongolia, China, India (Himalayas)

Melilotus Hill. 1753. Fabaceae. 20 temp. and subtrop. Eurasia, esp. (12) Crimea and chiefly Caucasus; North and Northeast Africa

Melinia Decne. 1844. Asclepiadaceae. 10 South America

Melinis P. Beauv. 1812. Poaceae. 15 trop. and South Africa, Madagascar

Meliosma Blume. 1823. Sabiaceae. c. 50 Himalayas from Kashmir to Bhutan, India, Sri Lanka, Burma, China (c. 30), Taiwan, southern Korea, Japan, Indochina, Malesia (8), trop. America (10) from Mexico to Brazil, West Indies

Melissa L. 1753. Lamiaceae. 4 Europe, Mediterranean, West (incl. Caucasus) and Middle Asia, Himalayas, northeastern India, Burma, southern China, Taiwan, Malesia

Melissitus Medik. = Melilotoides Heist. ex Fabr. Fabaceae

Melittacanthus S. Moore. 1906. Acanthaceae. 1 Madagascar

Melittis L. 1753. Lamiaceae. 2 Central, South and Southeast Europe, Asia Minor

Mellera S. Moore. 1879. Acanthaceae. 4–5 trop. and subtrop. Africa

Mellichampia A. Gray ex S. Watson = Cynanchum L. Asclepiadaceae

Melliniella Harms. 1914 (Alysicarpus Desv.). Fabaceae. 1 trop. West Africa

Melliodendron Hand.-Mazz. 1922. Styracaceae. 1 (M. xylocarpum) southwestern and southern China

Mellissia Hook. f. 1867. Solanaceae. 1 St. Helena

Melloa Bureau. 1868. Bignoniaceae. 1 trop. America from Mexico to Brazil and Argentina

Melo Hill = Cucumis L. Cucurbitaceae

Melocactus Link et Otto. 1872. Cactaceae. 20 trop. America, West Indies

Melocalamus Benth. 1883. Poaceae. 2 South and Southeast Asia

Melocanna Trin. 1820. Poaceae. 2 South and Southeast Asia

Melochia L. 1753. Sterculiaceae. 54 tropics, esp. America

Melodinus J. R. Forst. et G. Forst. 1776. Apocynaceae. 75 India, southern China, Indochina, Malesia to New Guinea, trop. Australia, New Caledonia, Vanuatu, Norfolk I., Fiji, and Tonga

Melodorum Lour. 1790. Annonaceae. 3–4 trop. Africa (1), Southeast Asia, and western Malesia

Melolobium Eckl. et Zeyh. 1836. Fabaceae. 28 South Africa (25), Namibia, Botswana

Melosperma Benth. 1846. Scrophulariaceae. 1 Chile

Melothria L. 1753. Cucurbitaceae. 12 trop. America

Melothrianthus Mart. Crov. 1954. Cucurbitaceae. 1 Brazil, Paraguay

Memecylanthus Gilg et Schltr. = Wittsteinia F. Muell. Alseuosmiaceae

Memecylon L. 1753. Melastomataceae. 200–300 (or c. 150 ?) trop. Old World from Africa and Asia through Malesia to Australia and the Pacific Islands

Memora Miers. 1863. Bignoniaceae. 30 trop. South America

Memorialis Buch.-Ham. ex Wedd. = Gonostegia Turcz. Urticaceae

Menabea Baill. 1889. Asclepiadaceae. 1 Madagascar

Menais Loefl. 1758. Boraginaceae. 1 South Africa

Mendoncella A. D. Hawkes = Galeottia A. Rich. Orchidaceae

Mendoncia Vell. ex Vand. 1788. Mendonciaceae. 60 Central and trop. South America, trop. Africa, Madagascar

Mendoravia Capuron. 1968. Fabaceae. 1 Madagascar

Menendezia Britton. 1925 (Tetrazygia Rich.). Melastomataceae. 3 Puerto Rico

Menepetalum Loes. 1906. Celastraceae. 6 New Caledonia

Meniocus Desv. = Alyssum L. Brassicaceae

Meniscogyne Gagnep. 1928. Urticaceae. 2 Indochina

Menispermum L. 1753. Menispermaceae. 2 temp. East Asia (M. dauricum), eastern North America (M. canadense)

Menkea Lehm. 1843. Brassicaceae. 6 Australia

Menodora Bonpl. 1812. Oleaceae. 25 trop. and subtrop. America, South Africa (3)

Menonvillea R. Br. ex DC. 1821. Brassicaceae. c. 30 Peru, Chile, Argentina

Menstruocalamus Yi. 1992. Poaceae. 1 China (Sichuan Prov.)

Mentha L. 1753. Lamiaceae. 25–30 temp. Northern Hemisphere, South Africa, Australia

Mentocalyx N. E. Br. = Gibbaeum Haw. Aizoaceae

Mentodendron Lundell = Pimentia Lindl. Myrtaceae

Mentzelia L. 1753. Loasaceae. 60 trop. and subtrop. America, West Indies

Menyanthes L. 1753. Menyanthaceae. 1 (M. trifoliata) temp. Northern Hemisphere, south to northern Mediterranean and Morocco

Menziesia Sm. 1791. Ericaceae. 7 East Asia, North America; Russia (1, M. pentandra—Sakhalin, Kuril Is.)

Meopsis (Calest.) Kozo = Polj. = Daucus L. Apiaceae

Merciera A. DC. 1830. Campanulaceae. 5 South Africa

Merckia Fisch. ex Cham. et Schltr. = Wilhelmsia Rchb. Caryophyllaceae

Mercurialis L. 1753. Euphorbiaceae. 8 Mediterranean, temp. Eurasia south to northern Thailand, southern China, and Taiwan

Merendera Ramond. 1801 (Colchicum L.). Colchicaceae. 10–15 Mediterranean to Ethiopia, West (incl. Caucasus) and Middle Asia eastward to Punjab

Meresaldia Bullock. 1965. Asclepiadaceae. 1 Venezuela

Meriandra Benth. 1829. Lamiaceae. 2 Ethiopia (1, M. bengalensis), India (1)

Merania Sw. 1798. Melastomataceae. 52 West Indies, trop. South America

Merianthera Kuhlm. 1935. Melastomataceae. 1 Brazil

Mericarpaea Boiss. 1843. Rubiaceae. 1 West Asia

Mericocalyx Bamps = Otiophora Zuch. Rubiaceae

Meringogyne H. Wolff. 1927 (or = Angoseseli Chiov.). Apiaceae. 1 trop. Namibia

Meringurus Murb. 1900 (Gaudinia P. Beauv.). Poaceae. 1 Tunisia

Merinthopodium Donn. Sm. = Markea Rich. Solanaceae

Meriolix Raf. ex Endl. = Calylophus Spach. Onagraceae

Merismia Tiegh. = Psittacanthus Mart. Loranthaceae

Merismostigma S. Moore = Coelospermum Blume. Rubiaceae

Meristotropis Fisch. et C. A. Mey. 1843 (Glycyrrhiza L.). Fabaceae. 2 Kazakhstan, Middle Asia, Iran

Merope M. Roem. 1846 (Atalantia Corrêa). Rutaceae. 1 trop. Asia

Merostachys Spreng. 1824. Poaceae. c. 40 Central and trop. South America

Merremia Dennst. ex Endl. 1841. Convolvulaceae. c. 80 pantropics

Merrillanthus Chun et Tsiang. 1941. Asclepiadaceae. 1 Hainan

Merrillia Swingle. 1919. Rutaceae. 1 Burma, Thailand, Malay Peninsula

Merrilliodendron Kaneh. 1934. Icacinaceae. 1 Philippines, Sulawesi, northern New Guinea, Mariana and Caroline Is., New Ireland, New Britain, Solomon Is., Santa Cruz Is.

Merrilliopanax H. L. Li. 1942. Araliaceae. 1 eastern Nepal, Bhutan, China (Yunnan)

Merrittia Merr. = Blumea DC. Asteraceae

Mertensia Roth. 1797. Boraginaceae. c. 50 Arctica and temp. Northern Hemisphere south to mountainous Middle Asia, Afghanistan, Mexico; Russia (c. 10, arctic

European part, West and East Siberia, Far East)

Merumea Steyerm. 1972. Rubiaceae. 2 northern South America

Merwia Fedtsch. = Ferula L. Apiaceae

Merwiopsis Safina = Pilopleura Schischk. Apiaceae

Merxmuellera Conert. 1970 (Notodanthonia Zotov). Poaceae. 14 trop. and South Africa, Madagascar

Meryta J. R. Forst. et G. Forst. 1775. Araliaceae. 30 Micronesia (Yap), New Guinea, Solomon Is., Vanuatu, New Caledonia (11), New Zealand, Norfolk I. eastward to Fiji, Samoa, Tonga, Marquesas Is.

Mesadenella Pabst et Garay. 1952 (Stenorrhynchos Rich. ex Spreng.). Orchidaceae. 7 trop. America

Mesadenus Schltr. 1920 (Brachystele Schltr.). Orchidaceae. 8 trop. America

Mesanthemum Koern. 1856. Eriocaulaceae. 12 trop. Africa, Madagascar

Mesechites Muell. Arg. 1860. Apocynaceae. 10 Central and trop. South America, West Indies

Mesembryanthemum L. 1753. Aizoaceae. c. 25 (or c. 75) South Europe, arid areas of Africa (esp. South), Middle East, Arabian Peninsula, western North America (California) and western coast of South America, ? Australia

Mesoglossum Halbinger. 1982 (Odontoglossum Kunth). Orchidaceae. 1 trop. America

Mesogyne Engl. 1894. Moraceae. 1 trop. Africa (Sao Tomé, Tanzania)

Mesomelaena Nees. 1846. Cyperaceae. 5 southwestern Australia

Mesona Blume. 1826. Lamiaceae. 2–3 northeastern India, southern China, Taiwan, Indochina, Malesia

Mesoptera Hook. f. = Psydrax Gaertn. Rubiaceae

Mesosetum Steud. 1854. Poaceae. 25: 20 Brazil; 5 Mexico, Central America, Colombia, Venezuela, Guyana, Surinam, Bolivia, Paraguay, Argentina, Cuba, Jamaica, Trinidad

Mesospinidium Rchb. f. 1852. Orchidaceae. 7 trop. America

Mesostemma Vved. 1941. Caryophyllaceae. 6 Central Asia from Karatau and Altai to Himalayas

Mespilus L. 1753. Rosaceae. 2: M. germanica—South and Southeast Europe, Asia Minor, northern Iran, Caucasus, and Middle Asia; M. canescens—U.S. (Arkansas)

Messerschmidia Hebenstr. = Argusia Boehm. Boraginaceae

Messerschmidia Roem. et Schult. = Tournefortia L. Boraginaceae

Mestoklema N. E. Br. ex Glen. 1981. Aizoaceae. 6 South Africa (Cape)

Mesua L. 1754. Clusiaceae. 48 India, Sri Lanka, Burma, southern China (1, M. ferrea), Indochina, Malesia to New Guinea

Metabolos Blume = Hedyotis L. Rubiaceae

Metabriggsia W. T. Wang. 1983. Gesneriaceae. 2 southern China

Metadina Bakh. f. 1970. Rubiaceae. 1 India (Meghalaya, Assam, Manipur), Burma, Thailand, Cambodia, Vietnam, Malay Peninsula, Sumatra, Java, Borneo, Philippines

Metaeritrichium W. T. Wang. 1980 (Eritrichium Schrad. ex Gaudin). Boraginaceae. 1 Tibet

Metalasia R. Br. 1817. Asteraceae. 52 South Africa

Metalepis Griseb. = Cynanchum L. Asclepiadaceae

Metalonicera M. Wang et A. G. Gu. 1988 (Lonicera L.). Caprifoliaceae. 1 China

Metanarthecium Maxim. 1867. Melanthiaceae. 5 Kuril Is. (1, M. luteo-viride), Japan, Taiwan

Metanemone W. T. Wang. 1980. Ranunculaceae. 1 China (Yunnan)

Metapetrocosmea W. T. Wang. 1981 (Petrocosmea Oliv.). Gesneriaceae. 1 China

Metaplexis R. Br. 1810. Asclepiadaceae. 6 China, Korea, Japan; M. japonica—extending to Russian Far East

Metaporana N. E. Br. 1914 (Bonamia

Thouars). Convolvulaceae. 4 trop. East Africa, Madagascar

Metarungia Baden. 1984. Acanthaceae. 3 Africa from Ethiopia to Cape

Metasasa Lin. 1988 (Acidosasa C. D. Chu et C. S. Chao). Poaceae. 1 China (Guangdong)

Metasequoia Miki ex Hu et Cheng. 1948. Taxodiaceae. 1 central China

Metasocratea Dugand = Socratea Karst. Arecaceae

Metastachydium Airy Shaw ex C. Y. Wu et H. W. Li. 1975. Lamiaceae. 1 Middle Asia, northwestern China

Metastachys Knorring = Metastachydium Airy Shaw ex C. Y. Wu et H. W. Li. Lamiaceae

Metastelma R. Br. 1810 (Cynanchum L.). Asclepiadaceae. c. 150 trop. and subtrop. America

Metastevia Grashoff. 1975. Asteraceae. 1 Mexico

Metatrophis F. B. H. Br. 1935. Moraceae. 1 Polynesia

Metcalfia Conert. 1960. Poaceae. 1 Mexico

Meteoromyrtus Gamble. 1918. Myrtaceae. 1 southern India

Meterostachys Nakai. 1935 (Orostachys Fisch.). Crassulaceae. 1 Korea, Japan

Metharme Phil. ex Engl. 1890. Zygophyllaceae. 1 northern Chile

Methysticodendron R. E. Schult. = Brugmansia Pers. Solanaceae

Metopium P. Browne. 1756. Anacardiaceae. 3 Florida, Mexico, West Indies

Metrodorea A. St.-Hil. 1825. Rutaceae. 5 Brazil (5); M. flavida also in Surinam and Bolivia

Metrosideros Banks ex Gaertn. 1788. Myrtaceae. 60 South Africa, Philippines, Moluccas, New Guinea, Ogasawara (Bonin) Is., Solomon Is., Australia, Lord Howe I., New Zealand (10), New Caledonia, Vanuatu, Fiji, eastward to Hawaii, Marquesas and Tuamotu, and Austral Is., Chile

Metroxylon Rottb. 1783. Arecaceae. 8 Thailand, Malesia, Caroline and Solomon Is., Vanuatu, Samoa and Fiji

Mettenia Griseb. 1859. Euphorbiaceae. 6 West India

Metteniusa Karst. 1860. Metteniusaceae. 7 Colombia, Venezuela, Peru

Metternichia J. C. Mikan. 1823. Solanaceae. 1 (M. principis) eastern Brazil

Meum Hill. 1753. Apiaceae. 3 Europe to Carpathians, North Africa

Mexacanthus T. F. Daniel. 1981. Acanthaceae. 1 western Mexico

Mexerion Nesom. 1990. Asteraceae. 2 Mexico

Mexianthus B. L. Rob. 1928. Asteraceae. 1 Mexico

Mexicoa Garay. 1974. Orchidaceae. 1 Mexico

Meximalva Fryxell. 1975. Malvaceae. 2 Mexico and adjacent Texas

Mexipedium V. A. Albert et M. W. Chase. 1993 (Phragmipedium Rolfe). Orchidaceae. 1 Mexico (Oaxaca)

Meyenia Nees. 1832 (Thunbergia Retz.). Acanthaceae. 1 India, Sri Lanka

Meyerophytum Schwantes. 1927. Aizoaceae. 1 western South Africa

Meyna Roxb. ex Link. 1820. Rubiaceae. 11 trop. Africa, Comoro Is., South and Southeast Asia

Mezia Schwacke ex Nied. 1890. Malpighiaceae. 4 Panama, trop. South America

Meziella Schindl. 1905 (Haloragis J. R. Forst. et G. Forst.). Haloragaceae. 1 southwestern Australia

Mezilaurus Kuntze ex Taub. 1892. Lauraceae. 16 trop. South America (Amazonian Basin, incl. Colombia, Venezuela, Peru, Bolivia, and the Guianas)

Mezleria C. Presl. 1836 (Lobelia L.). Lobeliaceae. 8: South Africa (7), New Zealand (1)

Mezobromelia L. B. Sm. 1935. Bromeliaceae. 2 Colombia, Ecuador

Mezochloa Butzin. 1966 (Alloteropsis C. Presl). Poaceae. 1 Madagascar

Mezoneuron Desf. = Caesalpinia L. Fabaceae

Mezzettia Becc. 1871. Annonaceae. 4 Thailand, Sumatra, Malay Peninsula, Borneo, Moluccas

Mezzettiopsis Ridl. 1912. Annonaceae. 1–3 Andaman Is., Thailand, Malay Peninsula, Java, Borneo, Philippines, southern China

Mibora Adans. 1763. Poaceae. 2 West and East Europe, North Africa

Michauxia L'Hér. 1788. Campanulaceae. 7 eastern Mediterranean, West Asia, incl. southern Caucasus (1, M. laevigata)

Michelaria Dumort. = Bromus L. Poaceae

Michelia L. 1753. Magnoliaceae. c. 60 southern and northeastern India, Sri Lanka, Burma, China, Taiwan, Japan (Honshu), Indochina, western Malesia, southern Borneo

Micheliopsis H. Keng = Magnolia L. Magnoliaceae

Michelsonia Hauman. 1952. Fabaceae. 2 Cameroun and Gabon (1), Zaire (1)

Micholitzia N. E. Br. 1909. Asclepiadaceae. 1 India

Miconia Ruiz et Pav. 1794. Melastomataceae. c. 1,000 trop. America, West Indies, trop. Africa (1)

Micractis DC. 1836 (Sigesbeckia L.). Asteraceae. 3 trop. Africa, Madagascar

Micraeschynanthus Ridl. 1925. Gesneriaceae. 1 Malay Peninsula

Micraira F. Muell. 1866. Poaceae. 13 Australia

Micrandra Benth. 1854. Euphorbiaceae. 14 trop. South America

Micrandropsis Rodrigues. 1973. Euphorbiaceae. 1 Amazonia

Micrantha Dvorak. 1968 (or = Hesperis L.). Brassicaceae. 1 Iran

Micranthemum Michx. 1803. Scrophulariaceae. 4 America from eastern U.S. to Brazil and Argentina; Cuba, Haiti, Lesser Antilles Is., Trinidad

Micrantheum Desf. 1818. Euphorbiaceae. 3 Australia

Micranthocereus Backeb. 1938 (Arrojadoa Britton et Rose). Cactaceae. 9 central and eastern Brazil

Micranthus (Pers.) Eckl. 1827. Iridaceae. 3 South Africa (Cape)

Micrargeria Benth. 1846. Scrophulariaceae. 4–5 trop. Africa, India

Micrargeriella R. E. Fr. 1916. Scrophulariaceae. 1 Zambia

Micrasepalum Urb. 1913. Rubiaceae. 2 West Indies

Micrechites Miq. 1857. Apocynaceae. 20 South and Southeast Asia, Malesia

Microberlinia A. Chev. 1946. Fabaceae. 2 Cameroun, Gabon

Microbiota Kom. 1923 (Thuja L.). Cupressaceae. 1 southern Russian Far East (Sikhote-Alin)

Microbriza Parodi ex Nicora et Rugolo. 1981. Poaceae. 2 Brazil, Argentina

Microcachrys Hook. f. 1845. Podocarpaceae. 1 Tasmania

Microcala Hoffmanns. et Link = Cicendia Adans. Gentianaceae

Microcalamus Franch. 1889. Poaceae. 4 trop. West Africa

Microcardamum O. E. Schulz = Hymenolobus Nutt. Brassicaceae

Microcarpaea R. Br. 1810. Scrophulariaceae. 1 India, Sri Lanka, Nepal, southern China, Japan, Malesia, Australia

Microcaryum I. M. Johnst. 1924. Boraginaceae. 3–5 Himalayas, western and southwestern China

Microcasia Becc. = Bucephalandra Schott. Araceae

Microcephala Pobed. 1961. Asteraceae. 4 Middle Asia, Iran

Microcharis Benth. 1865 (Indigofera L.). Fabaceae. 34 trop. and South Africa, Madagascar, Arabian Peninsula (Yemen)

Microchloa R. Br. 1810. Poaceae. 3 tropics

Microcitrus Swingle. 1915. Rutaceae. 5–6 New Guinea (1), eastern Australia

Microcnemum Ung.-Sternb. 1876. Chenopodiaceae. 1 Spain, Asia Minor, Armenia

Micrococca Benth. 1849. Euphorbiaceae. 14 trop. Africa, Madagascar, trop. Asia, northern Australia

Microcodon A. DC. 1830. Campanulaceae. 4 South Africa

Microcoelia Lindl. 1830. Orchidaceae. 27 trop. and South Africa, Madagascar

Microcoelum Burret et Potztal = Lytocaryum Toledo. Arecaceae

Microconomorpha (Mez) Lundell = Cybianthus Mart. Myrsinaceae

Microcorys R. Br. 1810. Lamiaceae. 17 southwestern Australia

Microcos L. 1753. Tiliaceae. c. 60 trop. Asia from India to southern China (3) and Indochina, Malesia, Fiji

Microculcas Peter = Gonatopus Hook. f. ex Engl. Araceae

Microcybe Turcz. 1852. Rutaceae. 3 Australia

Microcycas (Miq.) A. DC. 1868. Zamiaceae. 1 Cuba

Microdactylon Brandegee. 1908. Asclepiadaceae. 1 Mexico

Microderis A. DC. = Leontodon L. Asteraceae

Microdesmis Hook. f. ex Hook. 1848. Pandaceae. 11 trop. Africa, Southeast Asia, western Malesia

Microdon Choisy. 1824. Selaginaceae. 5 South Africa (Cape)

Microdracoides Hua. 1906. Cyperaceae. 1 trop. West Africa

Microglossa DC. 1836. Asteraceae. 8 trop. Africa, trop. Asia

Microgyne Less. = Microgynella Grau. Asteraceae

Microgynella Grau. 1975. Asteraceae. 1 Argentina

Microgynoecium Hook. f. 1880. Chenopodiaceae. 1 Middle Asia, China, Tibet, India

Microholmesia P. J. Cribb = Angraecopsis Kraenzl. Orchidaceae

Microlaena R. Br. 1810. Poaceae. 10 Southeast Asia, Malesia, Australia, New Zealand

Microlagenaria (C. Jeffrey) A. M. Lu et J. Q. Li. 1993. Cucurbitaceae. 1 trop. Africa (Nigeria and Tanzania)

Microlecane Sch. Bip. = Bidens L. Asteraceae

Microlepidium F. Muell. 1853 (Capsella Medik.). Brassicaceae. 2 Australia

Microlepis (DC.) Miq. 1840. Melastomataceae. 4 southern Brazil

Microliabum Cabrera. 1955. Asteraceae. 6 Andes from central Bolivia to central Argentina

Microlicia D. Don. 1823. Melastomataceae. c. 100 trop. South America

Microloma R. Br. 1810. Asclepiadaceae. 12 South Africa to Namibia

Microlonchoides Candargy = Jurinea Cass. Asteraceae

Microlonchus Cass. = Mantisalca Cass. Asteraceae

Microlophopsis Czerep. = Schumeria Iljin and Serratula L. Asteraceae

Micromeles Decne. = Aria (Pers.) Host. Rosaceae

Micromelum Blume. 1825. Rutaceae. 11 western Pakistan, India, Sri Lanka, southern China, Indochina, Malesia, Australia, New Caledonia, Melanesia to Fiji, Tonga, and Samoa Is.

Micromeria Benth. = Satureja L. Lamiaceae

Micromonolepis Ulbr. = Monolepis Schrader. Chenopodiaceae

Micromyrtus Benth. 1865. Myrtaceae. 16 Australia

Micromystria O. E. Schulz. 1924 [Arabidella (F. Muell.) O. E. Schulz]. Brassicaceae. 2 Australia

Micronoma H. A. Wendl. ex Hook. f. 1883. Arecaceae. 1 eastern Peru

Micronychia Oliv. 1881. Anacardiaceae. 5 Madagascar

Micropapyrus Suess. = Rhynchospora Vahl. Cyperaceae

Microparacaryum (Popov ex Riedl) Hilger et Podlech. 1985. Boraginaceae. 2 Asia from Sinai Peninsula to Middle Asia (2) and Kashmir

Micropeplis Bunge = Halogeton C. A. Mey. Chenopodiaceae

Micropera Lindl. 1832. Orchidaceae. 14–15 trop. Asia from Sikkim, northeastern India and Burma to Indochina, Malesia from Malay Peninsula eastward to Bougainville and Solomon Is., New Caledonia, and northern Australia (1, M. fasciculata)

Micropholis (Griseb.) Pierre. 1891 (Sideroxylon L.). Sapotaceae. 38 trop. America (Mexico, Guatemala, Belize, Costa Rica, Panama, Colombia, Venezuela, Guyana, Surinam, French Guiana, Ecuador, Peru, Brazil, Bolivia), West Indies

Microphyes Phil. 1860. Caryophyllaceae. 3 Chile

Microphysa Schrenk. 1844. Rubiaceae. 1 Middle Asia

Microphysca Naudin = Tococa Aubl. Melastomataceae

Microphytanthe (Schltr.) Brieger = Dendrobium Sw. Orchidaceae

Micropleura Lag. 1825. Apiaceae. 2 Colombia, Chile

Microplumeria Baill. 1889. Apocynaceae. 1 Brazil (Amazonia)

Micropsis DC. 1836. Asteraceae. 5 Chile, Argentina, Uruguay

Micropterum Schwantes = Cleretum N. E. Br. Aizoaceae

Micropus L. 1753. Asteraceae. 1 Mediterranean, West Asia, incl. Caucasus

Micropyropsis Rom. Zarco et Cabezudo. 1983. Poaceae. 1 Spain

Micropyrum (Gaudin) Link. 1844. Poaceae. 3 Central Europe, Mediterranean

Microrphium C. B. Clarke. 1906. Gentianaceae. 2 peninsular Thailand, Malay Peninsula, Philippines

Microsaccus Blume. 1825. Orchidaceae. 12 Burma, Thailand, Cambodia, Vietnam, Malay Peninsula, Sumatra, Java (4), Borneo, Philippines

Microschoenus C. B. Clarke. 1894. Cyperaceae. 1 western Himalayas

Microsciadium Boiss. 1844. Apiaceae. 1 Southeast Europe, Asia Minor

Microsechium Naudin. 1866. Cucurbitaceae. 1 Mexico

Microsemia Greene = Streptanthus Nutt. Brassicaceae

Microseris D. Don. 1832. Asteraceae. 14 western North America, Chile (1), Australia and New Zealand (1)

Microsisymbrium O. E. Schulz. 1924. Brassicaceae. 7 Northwest Africa (1), Middle Asia (2), Iran, Afghanistan, Pakistan, Himalayas, China

Microspermum Lag. 1816. Asteraceae. 7 Mexico

Microstegium Nees. 1836. Poaceae. 20 trop., subtrop. and temp. Africa and Asia

Microsteira Baker. 1883. Malpighiaceae. 25 Madagascar

Microstelma Baill. 1890. Asclepiadaceae. 1–2 Mexico

Microstemma R. Br. = Brachystelma Sims. Asclepiadaceae

Microsteris Greene. 1898 (Phlox L.). Polemoniaceae. 1 western North America (from Alaska to Montana and south to New Mexico, Arizona, and Baja California) and South America

Microstigma Trautv. 1845 (Matthiola R. Br.). Brassicaceae. 4 West Siberia, Mongolia, China

Microstrobilus Bremek. = Strobilanthes Blume. Acanthaceae

Microstrobos J. Garden et L. A. S. Johnson. 1950. Podocarpaceae. 2 Australia (New South Wales), Tasmania

Microstylis (Nutt.) Eaton = Malaxis Sol. ex Sw. Orchidaceae

Microtatorchis Schltr. 1905. Orchidaceae. 47 southern China (2), Java (3), Sulawesi (1), Moluccas, Philippines (2), New Guinea, Micronesia, Solomon Is., New Caledonia (2), Fiji (2), Samoa, Tahiti

Microtea Sw. 1788. Petiveriaceae. 10 trop. America, West Indies

Microterangis Senghas. 1985. Orchidaceae. 7 Madagascar and adjacent islands

Microthelys Garay. 1982 (Brachystele Schltr.). Orchidaceae. 3 Central America

Microthlaspi F. K. Mey. = Thlaspi L. Brassicaceae

Microtis R. Br. 1810. Orchidaceae. 12 trop Asia from Sri Lanka north to China, eastward through Indochina to Taiwan, Ryukyu Is., Japan; Malesia, Australia, western Oceania southward to New Caledonia and New Zealand

Microtoena Prain. 1889. Lamiaceae. 21–25 southwestern China, Himalayas, northeastern India, Java

Microtrichia DC. = Grangea Adans. Asteraceae

Microtropis Wall. ex Meisn. 1837. Celastraceae. c. 80 India, Sri Lanka, Burma, China (Yunnan, Szechuan, Kwantung,

Fukien, Hainan), Taiwan, Ryukyu Is., Indochina, Malesia (Sumatra, Malay Peninsula, Java, Borneo, Philippines); Mexico and Central America

Microula Benth. 1876. Boraginaceae. c. 30 western China, Pakistan, Himalayas, Tibet

Mida A. Cunn. ex Endl. 1837. Santalaceae. 2 New Zealand (1), Juan Fernández Is. (1)

Middendorfia Trautv. = Lythrum L. Lythraceae

Miersia Lindl. 1826. Alliaceae. 5 Bolivia, Chile

Miersiella Urb. 1903. Burmanniaceae. 1 trop. South America

Miersiophyton Engl. = Rhigiocarya Miers. Menispermaceae

Migandra O. F. Cook = Chamaedorea Willd. Arecaceae

Mikania Willd. 1803. Asteraceae. 415 pantropics, esp. America (c. 400 from southern U.S. to Argentina)

Mikaniopsis Milne-Redh. 1956. Asteraceae. 12 trop. Africa

Mila Britton et Rose. 1922. Cactaceae. 1 Peru

Mildbraedia Pax. 1909. Euphorbiaceae. 3–4 trop. Africa

Mildbraediochloa Butzin. 1971 (Melinis P. Beauv.). Poaceae. 1 trop. West Africa

Mildbraediodendron Harms. 1911. Fabaceae. 1 trop. Africa from Ghana to Sudan south to Zaire and Uganda

Milicia T. R. Sim. 1909. Moraceae. 2 trop. Africa from Guinea Bissau to Ethiopia and south to Angola, Zimbabwe, and Mozambique

Milium L. 1753. Poaceae. 7 temp. Northern Hemisphere

Miliusa Lesch. ex A. DC. 1832. Annonaceae. 40 India, Sri Lanka, Burma, China, Indochina, Philippines, Australia (Queensland)

Milla Cav. 1794. Alliaceae. 6 southern U.S., Mexico, Central America

Milleria L. 1753. Asteraceae. 1 trop. America from Mexico to Ecuador

Millettia Wight et Arn. 1834. Fabaceae. c. 100 trop. Africa and trop. Asia

Milligania Hook. f. 1853. Asteliaceae. 5 Tasmania

Millingtonia L. f. 1782. Bignoniaceae. 1 Burma, southern China, Thailand, Cambodia, Laos, Vietnam, Malesia

Millotia Cass. 1829. Asteraceae. 5 temp. Australia

Miltianthus Bunge. 1847. Zygophyllaceae. 1 Middle Asia, Afghanistan

Miltitzia A. DC. = Phacelia Juss. Hydrophyllaceae

Miltonia Lindl. 1837. Orchidaceae. 25 Central and trop. South America

Miltonioides Brieger et Lückel = Oncidium Sw. Orchidaceae

Miltoniopsis God.-Leb. 1889. Orchidaceae. 1 Central and trop. South America

Milula Prain. 1896. Alliaceae. 1 northwestern India (Kashmir), Tibet, eastern Himalayas

Mimetanthe Greene = Mimulus L. Scrophulariaceae

Mimetes Salisb. 1807. Proteaceae. 12 South Africa (Cape)

Mimetophytum L. Bolus = Mirtophyllum Schwantes. Aizoaceae

Mimophytum Greenm. 1905. Boraginaceae. 1 Mexico

Mimosa L. 1753. Fabaceae. 400–450 tropics and subtropics, esp. America

Mimosopsis Britton et Rose = Mimosa L. Fabaceae

Mimozyganthus Burkart. 1939. Fabaceae. 1 Paraguay, Argentina

Mimulicalyx Tsoong. 1979. Scrophulariaceae. 2 China (Sichuan and Yunnan)

Mimulopsis Schweinf. 1868. Acanthaceae. 30 trop. Africa, Madagascar

Mimulus L. 1753. Scrophulariaceae. c. 150 America, South Africa, Asia; 3–5 spp. extending to Russian Far East

Mimusops L. 1753. Sapotaceae. c. 40: 20 trop. and South (3) Africa, 15 Madagascar, 4 Mascarene Is., 1 Seychelles; 1 (M. elengi) Malesia, northern Australia, Melanesia, New Caledonia, Polynesia to Hawaiian Is.

Mina La Llave ex Lex. 1824 (Ipomoea L.). Convolvulaceae. 1 trop. America

Minquartia Aubl. 1775. Olacaceae. 1 Nicaragua, Costa Rica, Panama, Colombia, Venezuela, the Guianas, Ecuador, Peru, Bolivia, Brazil

Minthostachys (Benth.) Spach. 1840 (Bystropogon L'Hér.). Lamiaceae. 12 Andes

Minuartia L. 1753. Caryophyllaceae. c. 120 arctic and temp. regions southward to Mexico, Ethiopia, and Himalayas; 1 (M. acutiflora) Chile

Minuopsis W. A. Weber = Minuartia L. Caryophyllaceae

Minuria DC. 1836. Asteraceae. 9 central and southeastern Australia

Minuriella Tate = Minuria DC. Asteraceae

Minurothamnus DC. = Heterolepis Cass. Asteraceae

Mionandra Griseb. 1874. Malpighiaceae. 1 Bolivia, Paraguay, Argentina

Miquelia Meisn. 1838. Icacinaceae. 8 southwestern and northeastern India, Burma, Indochina, Malesia

Miqueliopuntia Fric ex F. Ritter = Opuntia Hill. Cactaceae

Mirabella F. Ritter = Acanthocereus (Engelm. ex A. Berger) Britton et Rose. Cactaceae

Mirabilis L. 1753. Nyctaginaceae. c. 55 trop. and subtrop. America from western and central U.S. to northern South America, West Indies; 1 western Himalayas; 1 Europe

Miraglossum Kupicha. 1984. Asclepiadaceae. 7 South Africa

Mirandaceltis A. J. Sharp = Gironniera Gaudich. Ulmaceae

Mirandea Rzed. 1959. Acanthaceae. 2 Mexico

Mirbelia Sm. 1805. Fabaceae. 27 western and southwestern Australia

Miricacalia Kitam. 1936. Asteraceae. 1 Japan

Miscanthidium Stapf. 1917 (Miscanthus Andersson). Poaceae. 7 trop. and South Africa

Miscanthus Andersson. 1855. Poaceae. 15 East and Southeast Asia; Russia (2, southern Far East)

Mischobulbum Schltr. 1911. Orchidaceae. 8 Himalayas, northeastern India, Burma, southern China, Taiwan, Indochina, Malesia from Malay Peninsula to Solomon Is. (1, M. papuanum)

Mischocarpus Blume. 1825. Sapindaceae. 15 northeastern India, China, Indochina, Malay Peninsula, Sumatra, Borneo, New Guinea, Solomon Is., eastern Australia

Mischocodon Radlk. = Mischocarpus Blume. Sapindaceae

Mischodon Thwaites. 1854. Euphorbiaceae. 1 southern India, Sri Lanka

Mischogyne Exell. 1932. Annonaceae. 2 trop. Africa

Mischophloeus Scheff. = Areca L. Arecaceae

Mischopleura Wernham ex Ridl. = Sericolea Schltr. Elaeocarpaceae

Misodendron G. Don = Misodendrum Banks ex DC. Misodendraceae

Misodendrum Banks ex DC. 1830. Misodendraceae. 8 South America (esp. Andes)

Misopates Raf. 1840 (Antirrhinum L.). Scrophulariaceae. 3 Cape Verde Is., Mediterranean through Asia Minor to northwestern India, Ethiopia, Malawi

Mitchella L. 1753. Rubiaceae. 2 Japan (1), North America (1)

Mitella L. 1753. Saxifragaceae. 15–20 West and East Siberia (1, M. nuda), China, Japan, North America

Mitolepis Balf. f. 1883. Asclepiadaceae. 2 Socotra

Mitophyllum Greene = Streptanthus Nutt. Brassicaceae

Mitostemma Mast. 1883. Passifloraceae. 3 trop. South America

Mitostigma Decne. 1844. Asclepiadaceae. 20 South America

Mitracarpus Zucc. 1827. Rubiaceae. 30–40 trop. America, West Indies; 1 (M. villosus) trop. Africa, Cape Verde and Seychelles, India, Burma, New Guinea, Mariana Is., Fiji (probably naturalized)

Mitragyna Korth. 1839. Rubiaceae. 10 trop. Africa (4), India, Sri Lanka, Bangladesh, Burma, Indochina, Malay Peninsula, Sumatra, Java, Borneo, Philippines, New Guinea

Mitranthes O. Berg. 1856. Myrtaceae. 11 trop. America, West Indies

Mitrantia Peter G. Wilson et B. Hyland. 1988. Myrtaceae. 1 Australia (Queensland)

Mitraria Cav. 1801. Gesneriaceae. 1 Chile

Mitrasacme Labill. 1805. Spigeliaceae. 40 India, eastern Himalayas, Burma, southern China, Taiwan, Korea, Japan, Indochina, Malesia, Australia, Tasmania, New Zealand, New Caledonia

Mitrasacmopsis Jovet. 1935. Rubiaceae. 1 trop. Africa (Zaire, Angola, Burundi, Zambia, Tanzania), Madagascar

Mitrastemma Makino. 1909. Mitrastemmataceae. 2 (M. yamamotoi) India (Assam), China, Taiwan, Japan, Thailand, Cambodia, Melesia (Sumatra, Borneo, New Guinea); (M. matudai) Mexico, Central America to Colombia

Mitrastemon Makino = Mitrastemma Makino. Mitrastemmataceae

Mitrastylus Alm et T. C. E. Fr. 1927 (or = Erica L.). Ericaceae. 2 Madagascar

Mitratheca K. Schum. = Oldenlandia L. Rubiaceae

Mitrella Miq. = Fissistigma Griffith. Annonaceae

Mitreola L. 1758. Spigeliaceae. 7 pantropics

Mitrephora (Blume) Hook. f. et Thomson. 1855. Annonaceae. c. 40 South and Southeast Asia, western Malesia to Philippines

Mitriostigma Hochst. 1842. Rubiaceae. 5 trop. West (2), East (2, Kenya, Tanzania) and South (1 [M. axillare] extends to Mozambique) Africa

Mitrocereus (Backeb.) Backeb. = Pachycereus (A. Berger) Britton et Rose. Cactaceae

Mitrophyllum Schwantes. 1926. Aizoaceae. 6 South Africa (Cape)

Mitropsidium Burret = Psidium L. Myrtaceae

Mitwabachloa J. B. Phipps. 1967. [Zonotriche (C. E. Hubb.) J. B. Phipps]. Poaceae. 2 trop. Africa

Miyakea Miyabe et Tatew. 1935 (Pulsatilla Hill). Ranunculaceae. 1 Sakhalin I.

Miyamayomena Kitam. 1982. Asteraceae. 3 East Asia

Mizonia A. Chev. = Pancratium L. Amaryllidaceae

Mkilua Verdc. 1970. Annonaceae. 1 trop. East Africa

Mnesithea Kunth. 1829. Poaceae. 5 South and Southeast Asia, Malesia

Mniochloa Chase. 1908. Poaceae. 2 Cuba

Mniodes (A. Gray) Benth. 1873. Asteraceae. 5 Peru

Mniopsis Mart. et Zucc. = Cladopus H. Moeller. Podostemaceae

Mniothamnea (Oliv.) Nied. 1891. Bruniaceae. 2 South Africa

Moacroton Croizat. 1945. Euphorbiaceae. 6 Cuba

Mobilabium Rupp. 1946. Orchidaceae. 1 Australia (Queensland)

Mocquerysia Hua. 1893. Flacourtiaceae. 2 trop. Africa

Modestia Charadze et Tamamsch. = Anacantha (Iljin) Soják. Asteraceae

Modiola Moench. 1794. Malvaceae. 1 warm–temp. and trop. America from U.S. to Argentina and Uruguay

Modiolastrum K. Schum. 1891. Malvaceae. 7 South America

Moehringia L. 1753 (Arenaria L.). Caryophyllaceae. 25 temp. Northern Hemisphere

Moenchia Ehrh. 1788. Caryophyllaceae. 6 West and Central Europe, Mediterranean

Moerenhoutia Blume = Malaxis Sol. ex Sw. Orchidaceae

Moghamia Steud. = Flemingia Roxb. ex W. T. Aiton. Fabaceae

Mogoltavia Korovin. 1947. Apiaceae. 2 Middle Asia (Tien Shan, Pamir-Alai)

Mohavea A. Gray. 1856. Scrophulariaceae. 2 southwestern U.S. (California, Nevada, Arizona), Mexico (Baja California)

Moldenhauera Spreng. = Pyrenacantha Wight. Icacinaceae

Moldenhawera Schrad. 1802. Fabaceae. 6 Brazil

Moldenkea Traub = Hippeastrum Herb. Amaryllidaceae

Moldenkeanthus Morat = *Paepalanthus* Kunth. Eriocaulaceae

Molinadendron P. K. Endress. 1969. Hamamelidaceae. 3 Mexico, Central America

Molinaea Comm. ex Juss. 1789. Sapindaceae. 10 Madagascar, Mascarene Is.

Molineria Colla. 1826 (*Curculigo* Gaertn.). Hypoxidaceae. 7 South and Southeast Asia, Malesia, northeastern Australia (1, M. capitulata)

Molineria Parl., non Colla = *Molineriella* Rouy. Poaceae

Molineriella Rouy. 1913 (*Periballia* Trin.). Poaceae. 2 Mediterranean

Molinia Schrank. 1789. Poaceae. 2–3 temp. Eurasia (incl. European part of Russia, Caucasus, Kazakhstan, southern West Siberia); North America, Ethiopia

Moliniopsis Hayata. 1925 (*Molinia* Schrank). Poaceae. 2 East Asia; Russia (1, M. japonica: Sakhalin I., Kuril Is.)

Mollera O. Hoffm. = *Calostephane* Benth. Asteraceae

Mollia Mart. 1826. Tiliaceae. 18 trop. South America

Mollinedia Ruiz et Pav. 1794. Monimiaceae. 90 Central and trop. South America

Mollugo L. 1753. Molluginaceae. 15 tropics and subtropics; M. cerviana—extending to southern European part of Russia, West Siberia, and Middle Asia

Molongum Pichon. 1948. Apocynaceae. 3 trop. South America

Molopanthera Turcz. 1848. Rubiaceae. 1 eastern Brazil

Molopospermum W. D. J. Koch. 1824. Apiaceae. 1 (M. cicutarium) Central and South Europe, western Mediterranean

Molpadia Cass. 1824. Asteraceae. 1 Hungary, ? Caucasus

Moltkia Lehm. 1817. Boraginaceae. 6 South Europe from northern Italy to northern Greece, Asia Minor (2), Syria, Transcaucasia (1, M. caerulea), northwestern Iran, Iraq

Moltkiopsis I. M. Johnst. 1953. Boraginaceae. 1 North Africa, West Asia

Moluccella L. 1753. Lamiaceae. 4 Mediterranean, Southeast Europe, West Asia to northwestern India; M. laevis—Caucasus, Middle Asia

Mommsenia Urb. et Ekman. 1926. Melastomataceae. 1 Haiti

Momordica L. 1753. Cucurbitaceae. c. 80 trop. and subtrop. Old World

Mona O. E. G. Nilsson. 1966 (*Montia* L.). Portulacaceae. 1 Colombia, Venezuela

Monachather Steud. 1854. Poaceae. 1 Australia

Monachyron Parl. = *Rhynchelytrum* Nees. Poaceae

Monactis Kunth. 1820. Asteraceae. 8 trop. South America

Monadenia Lindl. 1838. Orchidaceae. 16 trop. and South Africa, Madagascar

Monadenium Pax. 1894. Euphorbiaceae. 54 Central, trop. Southwest, and (chiefly) trop. East Africa

Monandriella Engl. = *Ledermanniella* Engl. Podostemaceae

Monanthes Haw. 1821. Crassulaceae. 12 Canary Is.

Monanthochloe Engelm. 1859. Poaceae. 2 southern U.S. from California to Texas, Mexico, Cuba, Argentina

Monanthocitrus Tanaka. 1928. Rutaceae. 4 New Guinea

Monanthotaxis Baill. 1890. Annonaceae. 56 trop. Africa, Madagascar

Monanthus (Schltr.) Brieger = *Dendrobium* Sw. Orchidaceae

Monarda L. 1753. Lamiaceae. 12–16 North America

Monardella Benth. 1834. Lamiaceae. 23 western North America

Monarrhenus Cass. 1817. Asteraceae. 2 Mascarene Is.

Monarthrocarpus Merr. = *Desmodium* Desv. Fabaceae

Mondia Skeels. 1911. Asclepiadaceae. 2 trop. Africa

Monechma Hochst. 1841. Acanthaceae. 60 trop. Africa, India (1, M. debilis)

Monelytrum Hack. ex Schinz. 1888. Poaceae. 2 Namibia

Monenteles Labill. 1825. Asteraceae. 6 from East and Southeast Asia, and Malesia to Australia

Monerma P. Beauv. 1812 (Lepturus R. Br.). Poaceae. 2 Mediterranean, West and Southwest Asia; M. cylindrica—the Crimea, eastern Transcaucasia

Moneses Salisb. ex Gray. 1821. Ericaceae. 1 Arctic and temp. regions south to northern Mediterranean

Moniera Loefl. = Monnieria Loefl. Rutaceae

Monilaria (Schwantes) Schwantes. 1929. Aizoaceae. 5 western South Africa

Monimia Thouars. 1804. Monimiaceae. 3 Madagascar (?), Mascarene Is. (Réunion—3, Mauritius—1, M. ovalifolia)

Monimiastrum Guého et A. J. Scott. 1980. Myrtaceae. 5 Mauritius

Monimopetalum Rehder. 1926. Celastraceae. 1 China

Monium Stapf. 1919 (Anadelphia Hack.). Poaceae. 7 West Africa

Monizia Lowe. 1856. Apiaceae. 1 Madeira Is.

Monnieria Loefl. 1758 (Ertela Adans.). Rutaceae. 2 trop. South America

Monnina Ruiz et Pav. 1798. Polygalaceae. c. 120–125 America from southwestern U.S. and Mexico to southern Chile

Monocardia Pennell = Bacopa Aubl. Scrophulariaceae

Monocarpia Miq. 1865. Annonaceae. 1 Thailand, Malay Peninsula, Sumatra, Borneo

Monocelastrus F. T. Wang et T. Tang. 1951 (Celastrus L.). Celastraceae. 2 western China, Himalayas

Monochaetum (DC.) Naudin. 1845. Melastomataceae. 40 trop. America

Monochasma Maxim. ex Franch. et Sav. 1878. Scrophulariaceae. 4 China, Korea, Japan

Monochilus Fisch. et C. A. Mey. 1835. Verbenaceae. 1 Brazil

Monochoria C. Presl. 1827. Pontederiaceae. 6 Northeast Africa, Southwest Asia, India, Sri Lanka throughout China eastward to Russian Far East (2) and Japan, Philippines and New Guinea, trop. Australia (4)

Monocladus L. C. Chia, H. L. Fung et Y. L. Yang. 1988 (Bambusa Schreb.). Poaceae. 4 southern China (incl. Hainan)

Monococcus F. Muell. 1858. Petiveriaceae. 1 Australia, New Caledonia, Vanuatu

Monocosmia Fenzl. 1839 (Calyptridium Nutt. ex Torr. et A. Gray). Portulacaceae. 1 Chile, Argentina

Monocostus K. Schum. 1904. Costaceae. 1 eastern Peru

Monocyclanthus Keay. 1953. Annonaceae. 1 trop. West Africa

Monocymbium Stapf. 1919. Poaceae. 4 trop. and South Africa

Monodia S. W. L. Jacobs. 1985. Poaceae. 1 western Australia

Monodiella Maire. 1943. Gentianaceae. 1 North Africa (Sakhara)

Monodora Dunal. 1817. Annonaceae. 20 trop. Africa, Madagascar

Monogereion G. M. Barroso et R. M. King. 1971. Asteraceae. 1 northeastern Brazil

Monolena Triana. 1867. Melastomataceae. 8–10 trop. South America

Monolepis Schrad. 1831. Chenopodiaceae. 5 Arctica and East Siberia (1, M. asiatica), North (U.S.) and South (Argentina) America

Monolophus Wall. = Kaempferia L. Zingiberaceae

Monolopia DC. 1838. Asteraceae. 4 California

Monomeria Lindl. 1830. Orchidaceae. 4 eastern Himalayas (Nepal, Sikkim, Bhutan), Burma, northeastern India, southern China, Indochina

Monopera Barringer. 1983. Scrophulariaceae. 2 Brazil and Paraguay

Monopetalanthus Harms. 1897. Fabaceae. 20 trop. West Africa

Monopholis S. F. Blake = Monactis Kunth. Asteraceae

Monophrynium K. Schum. 1902. Marantaceae. 3 Philippines, Moluccas

Monophyllaea Bennett et R. Br. 1840. Gesneriaceae. 20 southern Thailand, Malesia

Monophyllanthe K. Schum. 1902. Marantaceae. 1 French Guiana, Surinam

Monophyllorchis Schltr. 1920. Orchidaceae. 2 Colombia, Ecuador

Monoplegma Piper = Oxyrhynchus Brandegee. Fabaceae

Monoporus A. DC. 1841. Myrsinaceae. 8 Madagascar

Monopsis Salisb. 1817. Lobeliaceae. 20 trop. and South Africa

Monopteryx Spruce ex Benth. 1862. Fabaceae. 3 Venezuela, Brazil

Monoptilon Torr. et A. Gray. 1844. Asteraceae. 2 arid southwestern U.S. and northern Mexico

Monopyle Moritz ex Benth. 1876. Gesneriaceae. 8 trop. America

Monopyrena Speg. = Junellia Moldenke. Verbenaceae

Monosalpinx N. Hallé. 1968. Rubiaceae. 1 trop. West Africa

Monoschisma Brenan = Pseudopiptadenia Rauschert. Fabaceae

Monosepalum Schltr. = Bulbophyllum Thouars. Fabaceae

Monostachya Merr. ex Merr. et Merritt. = Notodanthonia Zotov. Poaceae

Monostemon Henrard = Microbriza Parodi ex Nicora et Rugolo. Poaceae

Monostylis Tul. = Apinagia Tul. Podostemaceae

Monotagma K. Schum. 1902. Marantaceae. 21 trop. South America

Monotaxis Brongn. 1829. Euphorbiaceae. 10 Australia

Monotes A. DC. 1868. Monotaceae. 32–36 trop. Africa, Madagascar

Monotheca A. DC. = Sideroxylon L. Sapotaceae

Monothecium Hochst. 1842. Acanthaceae. 3 trop. Africa, Madagascar, southern India, Sri Lanka

Monotoca R. Br. 1810. Epacridaceae. 11 Australia

Monotrema Koern. 1872. Rapateaceae. 3–5 Colombia, Venezuela, Brazil

Monotropa L. 1753. Ericaceae. 2 Europe, Russia (2, forest areas from European part to Far East), Algeria, Asia Minor, Cyprus, Caucasus, Afghanistan, Himalayas from Kashmir to Bhutan, northeastern India, Thailand, China, Korea, Japan, North and Central America, extending to northern Colombia

Monotropanthum Andres = Cheilotheca Hook. f. Ericaceae

Monotropastrum Andres = Cheilotheca Hook. f. Ericaceae

Monotropsis Schwein. ex Elliott. 1817. Ericaceae. 1 North America

Monroa Torr. 1857. Poaceae. 4 America from southern Canada to Mexico and from Peru to Argentina

Monrosia Grondona. 1949 (Polygala L.). Polygalaceae. 1 Argentina

Monsonia L. 1767. Geraniaceae. 25 Africa, Madagascar, Southwest Asia eastward to northwestern India

Monstera Adans. 1763. Araceae. 22 trop. America from Mexico to Brazil, Lesser Antille Is.

Montagueia Baker f. 1921. Anacardiaceae. 1 New Caledonia

Montamans Dwyer. 1980. Rubiaceae. 1 Panama

Montanoa Llave et Lex. 1825. Asteraceae. 23 trop. America: Mexico, Guatemala, Belize, El Salvador, Honduras, Nicaragua, Costa Rica, Panama, Colombia, Venezuela, Ecuador, Peru

Montbretia DC. = Tritonia Ker-Gawl. Iridaceae

Montbretiopsis L. Bolus = Tritonia Ker-Gawl. Iridaceae

Monteiroa Krapov. 1951. Malvaceae. 5 Brazil, Argentina, Uruguay

Montezuma Moçino et Sessé ex DC. 1824. Bombacaceae. 1 Puerto Rico

Montia L. 1753. Portulacaceae. c. 15 temp. Eurasia, Mediterranean, montane trop. Africa and Asia, southeastern Australia, North and South America; Russia (2, Arctic and northern European part, Northeast Siberia, Far East)

Montiastrum (A. Gray) Rydb. 1917 (Montia L.). Portulacaceae. 4 northeastern Siberia and western North America

Monticalia C. Jeffrey. 1992. Asteraceae. c. 60 Costa Rica, Andes of Colombia, Venezuela, Peru, Ecuador

Montinia Thunb. 1776. Montiniaceae. 1 southern Africa from southern Angola to Cape

Montiopsis Kuntze. 1898 (Calandrinia Kunth). Portulacaceae. 18 Peru, Bolivia, Chile, Argentina

Montrichardia Grueg. 1854. Araceae. 1–2 trop. America, West Indies

Montrouziera Panch. ex Planch. et Triana. 1860. Clusiaceae. 5 New Caledonia

Monttea C. Gay. 1849. Scrophulariaceae. 3 Chile

Monvillea Britton et Rose = Acanthocereus (Engelm. ex A. Berger) Britton et Rose. Cactaceae

Moonia Arn. 1836. Asteraceae. 1 southern India, Sri Lanka

Moorcroftia Choisy = Argyreia Lour. Convolvulaceae

Mooria Montrouz. = Cloezia Brongn. et Gris. Myrtaceae

Mopania Lundell = Manilkara Adans. Sapotaceae

Moquinia DC. 1838. Asteraceae. 1 Brazil

Moquiniella Balle. 1954. Loranthaceae. 1 South Africa (Cape)

Mora Schomb. ex Benth. 1839. Fabaceae. 6 Central and trop. South America, West Indies

Moraea Mill. 1758. Iridaceae. 120 trop. and South (c. 100) Africa, Mascarene Is.

Morangaya G. D. Rowley = Echinocereus Engelm. Cactaceae

Moratia H. E. Moore. 1980 (Cyphokentia Brongn.). Arecaceae. 1 New Caledonia

Morelia A. Rich. ex DC. 1830. Rubiaceae. 1 trop. Africa

Morelotia Gaudich. 1829. Cyperaceae. 2 New Zealand, Hawaiian Is.

Morenia Ruiz et Pav. = Chamaedorea Willd. Arecaceae

Morettia DC. 1821. Brassicaceae. 3 North Africa to Somalia and Arabian Peninsula and northward to Israel

Morgania R. Br. = Stemodia L. Scrophulariaceae

Moricandia DC. 1821. Brassicaceae. 8 Europe, Mediterranean, West Asia to Baluchistan

Moriera Boiss. 1841. Brassicaceae. 2 Turkmenistan, Iran, Afghanistan

Morierina Vieill. 1865. Rubiaceae. 2 New Caledonia

Morina L. 1753. Morinaceae. 4 Southeast Europe (Balkan), West Asia eastward to Middle Asia (4), Himalayas, and southwestern China

Morinda L. 1753. Rubiaceae. 85 tropics, esp. Old World

Morindopsis Hook. f. 1873. Rubiaceae. 2 Southeast Asia

Moringa Adans. 1763. Moringaceae. 14: North, Northeast (9), Southwest (Angola), and Southeast (1) Africa; Madagascar (2), Israel, Sinai and Arabian peninsulas, southern Iran, Pakistan, India

Morisia J. Gay. 1832. Brassicaceae. 1 Corsica and Sardinia

Morisonia L. 1753. Capparaceae. 4 North and trop. South America, West Indies

Morithamnus R. M. King, H. Rob. et G. M. Barroso. 1979. Asteraceae. 2 Brazil

Moritzia DC. ex Meisn. 1840. Boraginaceae. 5 Central and trop. South America

Morkillia Rose et Painter. 1907. Zygophyllaceae. 2 Mexico

Mormodes Lindl. 1836. Orchidaceae. 50 trop. America from Mexico to Peru, Bolivia, and Brazil

Mormolyca Fenzl. 1850. Orchidaceae. 6 Central and South America

Morolobium Kosterm. = Archidendron F. Muell. Fabaceae

Morongia Britton = Schrankia Willd. Fabaceae

Moronobea Aubl. 1775. Clusiaceae. 7 trop. South America

Morrenia Lindl. 1838. Asclepiadaceae. 10 trop. and temp. South America

Morrisiella Aellen = Atriplex L. Chenopodiaceae

Morsacanthus Rizzini. 1952. Acanthaceae. 1 Brazil

Mortonia A. Gray. 1852. Celastraceae. 8 southern U.S., Mexico

Mortoniella Woodson. 1939. Apocynaceae. 1 Central America

Mortoniodendron Standl. et Steyerm. 1938. Tiliaceae. 5 Central America

Morus L. 1753. Moraceae. 10–15 Africa, West and Southwest Asia eastward to China (9), Japan and Java, warm–temp. North America; Russia (1 [M. bombycis] Far East)

Moscharia Ruiz et Pav. 1794. Asteraceae. 2 Chile

Moschopsis Phil. 1865. Calyceraceae. 8 Chile, Argentina, incl. Patagonia

Mosdenia Stent. 1922. Poaceae. 1 South Africa (Transvaal)

Mosheovia Eig. 1938. Scrophulariaceae. 1 Israel (Upper Galilee), southern Syria

Mosiera Small. 1933 (Eugenia L.). Myrtaceae. 20 Florida, Bahamas, Great Antilles except Jamaica

Mosla (Benth.) Buch.-Ham. ex Maxim. 1875. Lamiaceae. 22 Himalayas, northern India, Malesia, China, Taiwan, Russian Far East; 1 (M. dianthera) Korea, Japan

Mosquitoxylum Krug et Urb. 1895. Anacardiaceae. 1 Mexico, Central America, West Indies

Mossia N. E. Br. 1930. Aizoaceae. 1 South Africa (Transvaal), Lesotho

Mostacillastrum O. E. Schulz = Sisymbrium L. Brassicaceae

Mostuea Didr. 1853. Loganiaceae. 8: trop. South America (1), trop. Africa and Madagascar (7)

Motandra A. DC. 1844. Apocynaceae. 3 trop. West and Central Africa

Motherwellia F. Muell. 1870. Araliaceae. 1 northeastern Australia

Motleyia Johansson. 1987 (Prismatomeris Thwaites). Rubiaceae. 1 Borneo

Moullava Adans. 1763. Fabaceae. 1 southern India

Moultonia Balf. f. et W. W. Sm. = Monophyllaea R. Br. Gesneriaceae

Moultonianthus Merr. 1916. Euphorbiaceae. 1 Sumatra, Borneo

Mourera Aubl. 1775. Podostemaceae. 6 northern South America

Mouretia Pit. 1922. Rubiaceae. 1 Indochina

Mouriri Aubl. 1775. Melastomataceae. 78 Central and trop. South America, West Indies

Moussonia Regel. 1847. Gesneriaceae. 10 Mexico, Central America

Moutabea Aubl. 1775. Polygalaceae. 10 trop. America

Moya Griseb. 1874. Celastraceae. 4 Bolivia, Argentina

Mozartia Urb. 1923. Myrtaceae. 7 Cuba

Msuata O. Hoffm. 1894. Asteraceae. 1 trop. Africa

Muantijamvella J. B. Phipps. 1964 (or = Tristachya Nees). Poaceae. 1 Zaire, Angola, Zambia, Malawi

Muantum Pichon. 1948 (Beaumontia Wall.). Apocynaceae. 1 southern Burma, southern Thailand

Mucizonia (DC.) Batt. et Trabut. 1905. Crassulaceae. 2 Canary Is., Iberian Peninsula, southern France, Morocco, Algeria

Mucoa Zarucchi. 1988. Apocynaceae. 2 trop. South America

Mucronea Benth. 1836. Polygonaceae. 2 southwestern California

Mucuna Adans. 1763. Fabaceae. c. 110 tropics and subtropics

Muehlenbeckia Meisn. 1841. Polygonaceae. 25 New Guinea, Australia, Tasmania, New Zealand, southern Pacific Islands, western South America

Muellera L. f. 1782. Fabaceae. 2 trop. America

Muelleranthus Hutch. 1964. Fabaceae. 3 Australia

Muellerargia Cogn. 1881. Cucurbitaceae. 2 Madagascar (1); M. timorensis—Malesia and Australia (Queensland)

Muellerina Tiegh. 1895 (Notanthera [DC.] G. Don). Loranthaceae. 4 eastern Australia, ? New Zealand

Muellerolimon Lincz. 1982. Plumbaginaceae. 1 western Australia

Muhlenbergia Schreb. 1789. Poaceae. c. 150 temp. and trop. Asia and America

Muilla S. Watson ex Benth. 1883. Alliaceae. 4–5 southwestern U.S., Mexico

Muiria N. E. Br. 1927. Aizoaceae. 1 South Africa

Muiriantha C. A. Gardner. 1942. Rutaceae. 1 southwestern Australia

Mukdenia Koidz. 1935. Saxifragaceae. 2 northeastern China, Korea

Mukia Arn. 1841. Cucurbitaceae. 4–7 trop. Africa, trop. Asia, Malesia, Australia

Mulgedium Cass. 1824 (Lactuca L.). Asteraceae. 5 temp. Eurasia

Mulinum Pers. 1805. Apiaceae. 20 southern Andes

Multidentia Gilli. 1973. Rubiaceae. 11 trop. Africa: Guinea, Ghana, Cameroun, Zaire, Sudan, Uganda, Malawi, Tanzania, Zambia, Zimbabwe, Mozambique

Muluorchis J. J. Wood = Tropidia Lindl. Orchidaceae

Munbya Pomel = Psoralea L. Fabaceae

Mundulea (DC.) Benth. 1852. Fabaceae. 15 Madagascar; M. sericea—trop. and South Africa, India, Sri Lanka

Munnozia Ruiz et Pav. 1794. Asteraceae. 40 Andes from Costa Rica and Panama to Argentina

Munroa Hack. = Monroa Torr. Poaceae

Munroidendron Sherff. 1952. Araliaceae. 1 Hawaiian Is.

Munronia Wight. 1838. Meliaceae. 10 western and northeastern India, Sri Lanka, Nepal, southern China, Indochina, Malesia

Muntafara Pichon = Tabernaemontana L. Apocynaceae

Muntingia L. 1753. Tiliaceae. 1 (M. calabura) West Indies and trop. South America

Munzothamnus Raven. 1963. Asteraceae. 1 California

Muraltia DC. 1824. Polygalaceae. c. 115 Africa from Tanzania to southern Cape (c. 100)

Murbeckiella Rohtm. 1939. Brassicaceae. 4 Southwest Europe, Morocco, Algeria, northeastern Asia Minor; M. huetii—Caucasus

Murchisonia Brittan. 1972 (Thysanotus R. Br.). Asphodelaceae. 2 western and central Australia

Murdannia Royle. 1839. Commelinaceae. c. 50 trop., subtrop. and warm–temp. regions

Muretia Boiss. = Elaeosticta Fenzl. Apiaceae

Muricaria Desv. 1815. Brassicaceae. 1 North Africa from Morocco to Libya

Muriea Hartog = Manilkara Adans. Sapotaceae

Murieanthe (Baill.) Aubrév. = Manilkara Adans. Sapotaceae

Murraya Koenig ex L. 1771. Rutaceae. 11 trop. Asia from India and Sri Lanka through Indochina to southern China and southern Ryuku Is., Malesia, Mariana Is., northeastern Australia, New Caledonia, Vanuatu, Fiji (2)

Murtonia Craib = Desmodium Desv. Fabaceae

Musa L. 1753. Musaceae. c. 40 trop. Asia, trop. Australia, Oceania

Musanga R. Br. 1818. Cecropiaceae. 2 trop. Africa from Senegal to northern Angola, Zaire, Burundi, Uganda

Muscadinia (Planch.) Small = Vitis L. Vitaceae

Muscari Hill. 1753. Hyacinthaceae. c. 30: Europe (13), Mediterranean, Caucasus, Iraq, Arabian Peninsula, Iran, Afghanistan, Pakistan, Middle Asia

Muscarimia Kostel. = Muscari Hill. Hyacinthaceae

Muschleria S. Moore. 1914. Asteraceae. 1 Angola

Musella (Franch.) C. Y. Wu ex H. W. Li. 1978 (Musa L.). Musaceae. 1 southern China (Yunnan)

Museniopsis (A. Gray) J. M. Coult. et Rose = Tauschia Schltr. Apiaceae

Musgravea F. Muell. 1891. Proteaceae. 2 Australia (Queensland)

Musilia Velen. = Rhanterium Desf. Asteraceae

Musineon Raf. 1820. Apiaceae. 4–6 U.S., Mexico, Central America

Mussaenda L. 1753. Rubiaceae. c. 190 trop. Africa, Madagascar, trop. Asia, Malesia, Australia, islands of the southern Pacific to Society Is.

Mussaendopsis Baill. 1879. Rubiaceae. 2 western Malesia

Mussatia Bureau ex Baill. 1888. Bignoniaceae. 3 trop. America from Mexico to Amazonia

Musschia Dumort. 1822. Campanulaceae. 2 Madeira Is.

Mutellina Wolf. 1776. Apiaceae. 3 Central, South, and East Europe; the Caucasus

Mutisia L. f. 1782. Asteraceae. c. 60 Andes from Colombia to Chile, Argentina, Paraguay, southeastern Brazil, Uruguay

Myagrum L. 1753. Brassicaceae. 1 Central Europe, Mediterranean, Caucasus, Turkey, northern Iraq, western Iran

Mycaranthes Blume = Eria Lindl. Orchidaceae

Mycelis Cass. 1824. Asteraceae. 1 (M. muralis) temp. Eurasia

Mycerinus A. C. Sm. 1931. Ericaceae. 5 Venezuela, Guyana

Mycetia Reinw. 1825. Rubiaceae. c. 25 India, Nepal (1), southern China, Indochina, Malesia (1)

Myginda Jacq. 1760. Celastraceae. 15 trop. America, West Indies

Myladenia Airy Shaw. 1977. Euphorbiaceae. 1 Thailand

Myllanthus R. S. Cowan = Raputia Aubl. Rutaceae

Myodocarpus Brongn. et Gris. 1861. Araliaceae. 12 New Caledonia

Myonima Comm. ex Juss. 1789 (Ixora L.). Rubiaceae. 5 Mascarene Is.

Myopordon Boiss. 1846. Asteraceae. 3 Iran

Myoporum Banks et Sol. ex G. Forst. 1786. Myoporaceae. 32 Mauritius, East Asia, New Guinea, Australia, New Zealand, Oceania

Myoschilos Ruiz et Pav. 1794. Santalaceae. 1 Chile

Myosotidium Hook. 1859. Boraginaceae. 1 Chatham Is.

Myosotis L. 1753. Boraginaceae. c. 50 temp. Eurasia, mountainous trop. and South Africa, New Guinea, Australia, New Zealand

Myosoton Moench. 1794. Caryophyllaceae. 1 temp. Eurasia

Myosurus L. 1753. Ranunculaceae. 15 temp. regions of both hemispheres, Mediterranean; M. minimus—European part of Russia, Caucasus, West Siberia, Middle Asia

Myoxanthus Poepp. et Endl. 1835 (Pleurothallis R. Br.). Orchidaceae. ? trop. America

Myrceugenella Kausel = Luma A. Gray. Myrtaceae

Myrceugenia O. Berg. 1855. Myrtaceae. 38 Juan Fernández Is., Chile, Argentina, southeastern Brazil

Myrcia DC. ex Guill. 1827. Myrtaceae. c. 250 West Indies and trop. South America

Myrcialeucus Rojas Acosta = Eugenia L. Myrtaceae

Myrcianthes O. Berg. 1856. Myrtaceae. 50 trop. America, esp. Andes and West Indies

Myrciaria O. Berg. 1856. Myrtaceae. 40 trop. America from Mexico to Brazil, West Indies

Myrciariopsis Kausel = Siphoneugenia O. Berg. Myrtaceae

Myriactis Less. 1831. Asteraceae. 9 temp. and trop. Asia; M. wallichii—Caucasus, Middle Asia

Myrialepis Becc. 1893. Arecaceae. 1 Burma, Indochina, Malay Peninsula, Sumatra

Myrianthemum Gilg = Medinilla Gaudich. Melastomataceae

Myrianthus P. Beauv. 1805. Cecropiaceae. 7 trop. Africa

Myriaspora DC. 1828. Melastomataceae. 1–2 Colombia, Venezuela, Guyana, Brazil

Myrica L. 1753. Myricaceae. c. 50 almost cosmopolitan, excl. Central and Southeast Europe, North Africa, Southwest Asia, and Australia; Russia (2 [M. gale] northwestern European part; M. tomentosa—Far East)

Myricanthe Airy Shaw. 1980. Euphorbiaceae. 1 New Caledonia

Myricaria Desv. 1825. Tamaricaceae. 13 Europe (M. germanica—from Scandinavia to northern Mediterranean), temp. Asia from Asia Minor, Caucasus to West and East Siberia, Mongolia, Himalayas, and southern China

Myriocarpa Benth. 1846. Urticaceae. 18 Central and trop. South America south to Peru, Bolivia, and Brazil

Myriocephalus Benth. 1837. Asteraceae. 11 temp. Australia

Myriocladus Swallen. 1951. Poaceae. c. 20 Venezuela

Myrioneuron R. Br. ex Kurz. 1880. Rubiaceae. 15 eastern Himalayas, northeastern India, China, Indochina, Malesia

Myriophyllum L. 1753. Haloragaceae. c. 60 almost cosmopolitan, but chiefly Australia and Tasmania (36, 31 endemic), North America (13, 7 endemic), trop. Asia (10, 7 endemic)

Myriopteron Griff. 1844. Asclepiadaceae. 1 northeastern India, southwestern China, Indochina, Malay Peninsula

Myriostachya (Benth.) Hook. f. 1897. Poaceae. 1 southern India, Sri Lanka, Bangladesh, southern Burma, Malay Peninsula

Myripnois Bunge. 1833. Asteraceae. 1 northern China

Myristica Gronov. 1755. Myristicaceae. c. 80 trop. Asia and Australia

Myrmechis (Lindl.) Blume. 1858. Orchidaceae. 4 India, China, Japan, Sumatra, Malay Peninsula, Java (2), Philippines (3)

Myrmecodia Jack. 1823. Rubiaceae. 26 Vietnam, Malay Peninsula, Sumatra, Java, Borneo, Lesser Sunda Is., Philippines, Sulawesi, Moluccas, New Guinea (c. 25), trop. Australia, Solomon Is.

Myrmeconauclea Merr. 1920. Rubiaceae. 3 Borneo, Anambas I., Philippines

Myrmecophila Rolfe = Schomburgkia Lindl. Orchidaceae

Myrmecosicyos C. Jeffrey. 1962. Cucurbitaceae. 1 Kenya

Myrmedoma Becc. = Myrmephytum Becc. Rubiaceae

Myrmephytum Becc. 1884. Rubiaceae. 5–9 Philippines, Sulawesi, western New Guinea (Irian Jaya)

Myrmidone Mart. 1832. Melastomataceae. 4 trop. South America

Myrocarpus Allemao. 1847. Fabaceae. 4 Venezuela, Brazil, Paraguay, Argentina

Myrosma L. f. 1882. Marantaceae. 15 Central and trop. South America

Myrosmodes Rchb. f. 1854. Orchidaceae. 12 Andes of South America

Myrospermum Jacq. 1760. Fabaceae. 1–2 southern Mexico, Central and northern South America, West Indies

Myrothamnus Welw. 1859. Myrothamnaceae. 2 trop. and South Africa, Madagascar

Myroxylon L. f. 1882. Fabaceae. 2–3 Mexico, Central and trop. South America

Myrrhidendron J. M. Coult. et Rose. 1894. Apiaceae. 5 Central America, Colombia

Myrrhinium Schott. 1827. Myrtaceae. 1 (M. atropurpureum) trop. South America

Myrrhis Hill. 1753. Apiaceae. 1 (M. odorata) mountainous Europe eastward to Caucasus

Myrrhoides Heist. ex Fabr. = Physocaulis (DC.) Tausch. Apiaceae

Myrsine L. 1753. Myrsinaceae. 5 Azores, Africa, West Asia eastward to China

Myrsiphyllum (L.) Willd. 1808. Asparagaceae. 12 South Africa

Myrtama Ovcz. et Kinzik. = Myricaria Desv. Tamaricaceae

Myrtastrum Burret. 1941. Myrtaceae. 1 New Caledonia

Myrtekmania Urb. = Pimenta Lindl. Myrtaceae

Myrtella F. Muell. 1877. Myrtaceae. 10 Micronesia, New Guinea, Australia

Myrteola O. Berg. 1856. Myrtaceae. 3 Juan Fernández Is., Colombia, Venezuela, Ecuador, Peru, Bolivia, Brazil, Chile to Tierra del Fuego, Falkland Is.

Myrtillocactus Console. 1897. Cactaceae. 4 Mexico, Guatemala

Myrtopsis Engl. 1896. Rutaceae. 8 New Caledonia

Myrtus L. 1753. Myrtaceae. 2 (s. str.): M. communis—Mediterranean (excl. Bulgaria, Crimea, Egypt, and Sinai Peninsula); M. nivellei—North Africa (Algeria); or c. 100 (S. lato) tropics and subtropics

Mystacidium Lindl. 1836. Orchidaceae. 13 East (Tanzania, 3), south-central, and South Africa

Mystropetalon Harv. 1818. Mystropetalaceae. 2 South Africa (Cape)

Mystroxylon Eckl. et Zeyh. = Cassine L. Celastraceae

Mytilaria Lecomte. 1924. Hamamelidaceae. 1 southern China, Indochina

Myxopappus Källersjö. 1988. Asteraceae. 2 South Africa

Myxopyrum Blume. 1826. Oleaceae. 15 India, Burma, southern China, Indochina, Sumatra, Java, Borneo, New Guinea, Admiralty Is.

Myzodendron Banks et Sol. ex R. Br. = Misodendrum Banks ex DC. Misodendraceae

Myzorrhiza Phil. 1858 (or = Orobanche L.). Orobanchaceae. 10 America

Mzymtella Kolak. 1981 (Campanula L.). Campanulaceae. 1 Caucasus (Colchis)

N

Nabaluia Ames. 1920. Orchidaceae. 3 Borneo

Nabalus Cass. 1825. Asteraceae. 15 temp. East Asia, North America

Nabea Lehm. ex Klotzsch = Erica L. Ericaceae

Nabelekia Roshev. 1937 (Festuca L.). Poaceae. 1 Turkey

Nablonium Cass. = Ammobium R. Br. Asteraceae

Nacrea A. Nelson = Anaphalis DC. Asteraceae

Nageia Gaertn. 1788. Nageiaceae. 5 northeastern India, Assam, Bangladesh, Burma, southern China (Zhejiang, Fujian, Guangdong, Hainan), Taiwan, Ryukyu Is., southern Japan, Thailand, Cambodia, Laos, Vietnam, Malesia (Sumatra, Malay Peninsula, Borneo, Philippines, Moluccas, Sabah, New Guinea, New Britain)

Nagelia Lindl. = Malacomeles (Decne.) Engl. Rosaceae

Nageliella L. O. Williams. 1940. Orchidaceae. 3 Central America, Venezuela

Nagelocarpus Bullock. 1954. Ericaceae. 1 (N. serratus) South Africa (southwestern Cape)

Naiocrene (Torr. et A. Gray) Rydb. 1906 (Montia L.). Portulacaceae. 2 western North America from Alaska to northern California

Najas L. 1753. Najadaceae. 40–50 cosmopolitan, excl. Arctica and taiga of Eurasia

Naletonia Bremek. = Psychotria L. Rubiaceae

Nama L. 1759. Hydrophyllaceae. 35–55 America from California and Nevada, through Mexico and Central Ametica to Brazil, West Indies and (1, N. sandwicensis) Hawaiian Is.

Namacodon Thulin. 1974. Campanulaceae. 1 South-West Africa

Namaquanthus L. Bolus. 1954. Aizoaceae. 2 South Africa (Cape)

Namaquanula D. Müll.-Doblies et U. Müll.-Doblies. 1985. Amaryllidaceae. 1 South Africa (Cape)

Namation Brand. 1912. Scrophulariaceae. 1 Mexico

Namibia (Schwantes) Dinter et Schwantes. 1927 (Juttadintera Schwantes). Aizoaceae. 1 southwestern Namibia

Nananthea DC. 1837. Asteraceae. 1 Corsica and Sardinia

Nananthus N. E. Br. 1925 (Rabiea N. E. Br.). Aizoaceae. 7 central South Africa

Nanarepenta Matuda = Dioscorea L. Dioscoreaceae

Nandina Thunb. 1781. Nandinaceae. 1 continental China, Japan

Nannoglottis Maxim. 1881. Asteraceae. 8–9 western China

Nannorrhops H. A. Wendl. 1879. Arecaceae. 1 Arabian Peninsula, semiarid Iran, Afghanistan, Pakistan

Nannoseris Hedb. = Dianthoseris A. Rich. Asteraceae

Nanochilus K. Schum. 1899. Zingiberaceae. 2 Sumatra (1), New Guinea (1)

Nanocnide Blume. 1856. Urticaceae. 4 East Asia to Taiwan and Ryukyu Is.

Nanodea Banks ex C. F. Gaertn. 1807. Santalaceae. 1 southern temp. South America

Nanolirion Benth. = Caesia R. Br. Asphodelaceae

Nanophyton Less. 1834–1835. Chenopodiaceae. 8 southeastern European part of Russia, West and East Siberia, Middle Asia, China, Mongolia

Nanorrhinum Betsche = Kichxia Dumort. Scrophulariaceae

Nanostelma Baill. = Tylophora R. Br. Asclepiadaceae

Nanothamnus Thomson. 1867. Asteraceae. 1 western India

Nanuza L. B. Sm. et Ayensu. 1976. Velloziaceae. 1 Brazil

Napaea L. 1753. Malvaceae. 1 central and eastern U.S.

Napeanthus Gardner. 1843. Gesneriaceae. 16 trop. America from southern Mexico and Guatemala to Bolivia and southeastern Brazil

Napoleonaea P. Beauv. 1804. Lecythidaceae. 15 trop. Africa

Naravelia Adans. 1763. Ranunculaceae. 7 trop. Asia from India to southern China, Indochina, Malesia

Narcissus L. 1753. Amaryllidaceae. 27 Europe (26), Mediterranean, West Asia (incl. Caucasus)

Nardophyllum (Hook. et Arn.) Hook. et Arn. 1836. Asteraceae. 7 southern Andes

Nardosmia Cass. = Petasites Hill. Asteraceae

Nardostachys DC. 1830. Valerianaceae. 1 (N. jatamansi) Himalayas from north-western India to Bhutan, western and southwestern China

Narduroides Rouy. 1913. Poaceae. 1 Mediterranean

Nardurus (Bluff, Nees et Schauer) Rchb. 1841 (Vulpia C. C. Gmel.). Poaceae. 2 Mediterranean, West, South-West Asia; M. krausei also in Crimea, Caucasus, Middle Asia

Nardus L. 1753. Poaceae. 1 Europe, South-West Asia, southern Siberia, northeastern North America

Naregamia Wight et Arn. 1834. Meliaceae. 2 Angola (1), India (1)

Narenga Bor. 1940 (Saccharum L.). Poaceae. 2 South Asia, Malesia

Nargedia Bedd. ex Hook. f. 1880. Rubiaceae. 1 Sri Lanka

Naringi Adans. 1763. Rutaceae. 1 Pakistan, India, Sri Lanka, Bangladesh, Burma, southwestern China, Indochina

Narthecium Huds. 1762. Melanthiaceae. 7 Europe (3), Caucasus and Asia Minor (1, N. balansae), Japan (1, N. asiaticum), North America (2)

Narvalina Cass. 1825. Asteraceae. 4 trop. America, West Indies

Nashia Millsp. 1906. Verbenaceae. 7 West Indies

Nassauvia Comm. ex Juss. 1789. Asteraceae. 37 southern Andes, Patagonia

Nassella (Trin.) E. Desv. 1854 (Piptochaetium J. Presl and Stipa L.). Poaceae. 79 Central and South America, esp. Andes

Nastanthus Miers = Acarpha Griseb. Calyceraceae

Nasturtiicarpa Gilli = Calymmatium O. E. Schulz. Brassicaceae

Nasturtiopsis Boiss. 1867. Brassicaceae. 2 North Africa, Arabian Peninsula north to Israel

Nasturtium R. Br. 1812 (Rorippa Scop.). Brassicaceae. 6 Europe, West (incl. Caucasus) and Middle Asia eastward to China and Pakistan

Nastus Juss. 1789. Poaceae. 18 Madagascar, Mascarene Is., Bourbon I., Malesia from Sumatra eastward to New Guinea (7) and Solomon Is.

Nathaliella Fedtsch. 1932. Scrophulariaceae. 1 Middle Asia

Natsiatopsis Kurz. 1876. Icacinaceae. 1 Burma, southern China

Natsiatum Buch.-Ham. ex Arn. 1834. Icacinaceae. 1 eastern Himalayas (Nepal, Sikkim), eastern India, Assam, Bangladesh, Burma, southern China, northern Thailand, northern Vietnam

Nauclea L. 1762. Rubiaceae. 6 trop. Africa (4), India, Sri Lanka, Burma, southern China, Thailand, Indochina, Malesia to New Guinea, trop. Australia

Naucleopsis Miq. 1853. Moraceae. 18 trop. America

Naudinia Planch. et Linden. 1853. Rutaceae. 1 Colombia

Naufraga Constance et Cannon. 1967. Apiaceae. 1 (N. balearica) Balearic Is.

Naumburgia Moench = Lysimachia L. Primulaceae

Nauplius (Cass.) Cass. 1822. Asteraceae. 8 Canary Is., Cape Verde Is., Mediterranean

Nautilocalyx Linden ex Hanst. 1854. Gesneriaceae. c. 60 trop. America, West Indies

Nautochilus Bremek. = Orthosiphon Benth. Lamiaceae

Nautonia Decne. 1844. Asclepiadaceae. 1 southern Brazil

Nautophylla Guillaumin = Logania R. Br. Loganiaceae

Navarretia Ruiz et Pav. 1794. Polemoniaceae. c. 30 western North America, Chile, Argentina

Navia Mart. ex Schult. f. 1830. Bromeliaceae. c. 80 northern South America

Nayariophyton T. K. Paul. 1988. Malvaceae. 1 eastern Himalayas, southern China (Yunnan)

Nealchornea Huber. 1913. Euphorbiaceae. 1 Colombia (Upper Amazon)

Neamyza Tiegh. = Peraxilla Tiegh. Loranthaceae

Neanotis W. H. Lewis. 1966. Rubiaceae. 28 trop. Asia and Australia

Neatostema I. M. Johnst. 1953. Boraginaceae. 1 Canary Is., Mediterranean, Crimea, West Asia

Nebelia Neck. ex Sweet = Brunia Lam. Bruniaceae

Neblinaea Maguire et Wurdack. 1957. Asteraceae. 1 Venezuela, Guyana

Neblinantha Maguire. 1985. Gentianaceae. 2 southern Venezuela, northern Brazil

Neblinanthera Wurdack. 1964. Melastomataceae. 1 Venezuela, Brazil

Neblinaria Maguire = Bonnetia Mart. Bonnetiaceae

Neblinathamnus Steyerm. 1964. Rubiaceae. 2 Venezuela

Necepsia Prain. 1910. Euphorbiaceae. 3 forests of trop. Africa and Madagascar

Nechamandra Planch. 1849 (Lagarosiphon Harv.). Hydrocharitaceae. 1 trop. Asia

Neckia Korth. 1848. Sauvagesiaceae. 1 Malesia: Sumatra, Malay Peninsula, Borneo, Philippines

Necramium Britton. 1924. Melastomataceae. 1 Trinidad

Necranthus Gilli. 1968. Orobanchaceae. 1 Asia Minor

Nectandra Rol. ex Rottb. 1778 (Ocotea Aubl.). Lauraceae. c. 110 trop. and subtrop. America

Nectaropetalum Engl. 1902. Erythroxylaceae. 6 trop. and South Africa from northern Kenya to coastal Cape

Nectaroscordum Lindl. 1836 (Allium L.). Alliaceae. 6 Mediterranean, Crimea (1, N. dioscoridis), Asia Minor, Iran, Caucasus (1, N. tripedale)

Nectouxia Kunth. 1818. Solanaceae. 1 Mexico, ? Texas

Neea Ruiz et Pav. 1794. Nyctaginaceae. c. 80 trop. America from southern Florida to Bolivia; West Indies

Needhamiella L. Watson. 1965. Epacridaceae. 1 southwestern Australia

Neeopsis Lundell. 1976. Nyctaginaceae. 1 Guatemala

Neeragrostis Bush = Eragrostis Wolf. Poaceae

Neesenbeckia Levyns. 1947. Cyperaceae. 1 South Africa

Neesia Blume. 1835. Bombacaceae. 8 Burma, Thailand, western Malesia

Neesiochloa Pilg. 1940. Poaceae. 1 northeastern Brazil

Negria F. Muell. 1871. Gesneriaceae. 1 Lord Howe I.

Neillia D. Don. 1825. Rosaceae. 13 Himalayas from Nepal to Bhutan, northeastern India, Bangladesh, Burma, China (10), Indochina, western Malesia

Neisosperma Raf. 1838. Apocynaceae. 20 Seychelles, Maldive and Andaman Is., Sri Lanka, Thailand, Vietnam, Malesia, Ryukyu and Ogasawara (Bonin) Is., Micronesia, Australia (Queensland), Solomon Is., Vanuatu, New Caledonia, Fiji eastward to Marquesas Is.

Nelia Schwantes. 1928. Aizoaceae. 1–4 South Africa (Cape)

Nelmesia Veken. 1955. Cyperaceae. 1 trop. Africa

Nelsia Schinz. 1912. Amaranthaceae. 2 southern trop. and South Africa

Nelsonia R. Br. 1810. Acanthaceae. 2 trop. Africa, South and Southeast Asia, Malesia, trop. Australia

Nelsonianthus H. Rob. et Brettell. 1973. Asteraceae. 1 Mexico, Guatemala

Nelumbo Hill. 1753. Nelumbonaceae. 2 Southeast Europe (?); West (?), South, East, and Southeast Asia; Malesia; northern Australia; America from eastern U.S. through Mexico and Central America to Colombia and northern Brazil, West Indies; N. nucifera—also the delta of the Volga and Kura rivers, Far East: Primorski and Khabarovski kraies

Nemacaulis Nutt. 1848. Polygonaceae. 1 southwestern North America: California, Baja California, western Arizona, northwestern Sonora

Nemacladus Nutt. 1843. Nemacladaceae. 10 southwestern U.S. and Mexico

Nemaluma Baill. = Pouteria Aubl. Sapotaceae

Nemastylis Nutt. 1835. Iridaceae. 25 America

Nematanthus Schrad. 1821. Gesneriaceae. 30 South America

Nematolepis Turcz. 1852. Rutaceae. 1 southwestern Australia

Nematopoa C. E. Hubb. 1957. Poaceae. 1 Zimbabwe

Nematosciadium H. Wolff. 1911 (Arracacia Bancr.). Apiaceae. 1 Mexico

Nematostemma Choux. 1921. Asclepiadaceae. 1 Madagascar

Nematostylis Hook. f. 1873. Rubiaceae. 1 Madagascar

Nematuris Turcz. = Enslenia Nutt. Asclepiadaceae

Nemcia Domin = Oxylobium Andrews. Fabaceae

Nemesia Vent. 1804. Scrophulariaceae. 70 trop. and South Africa

Nemopanthus Raf. 1819. Aquifoliaceae. 2 northeastern North America from Newfoundland to North Carolina

Nemophila Nutt. 1822. Hydrophyllaceae. 11 western and southeastern North America

Nemosenecio (Kitam.) B. Nord. 1978. Asteraceae. 7 China, Japan

Nemuaron Baill. 1873. Monimiaceae. 1 (N. vieillardii) New Caledonia

Nemum Desv. ex Ham. 1825. Cyperaceae. 6 trop. Africa

Nenax Gaertn. 1788. Rubiaceae. 9 South Africa

Nenga H. A. Wendl. et Drude. 1875. Arecaceae. 5 Burma, Indochina, Malay Peninsula, Sumatra, Java, Borneo

Nengella Becc. = Gronophyllum Scheff. Arecaceae

Neoabbottia Britton et Rose = Leptocereus (A. Berger) Britton et Rose. Cactaceae

Neoacanthophora Bennet = Aralia L. Araliaceae

Neoalsomitra Hutch. 1942. Cucurbitaceae. 22 India, eastern Himalayas, Burma, southern China, Malesia, Australia (Queensland), Polynesia

Neoancistrophyllum Rauschert = Laccosperma (G. Mann et H. A. Wendl.) Drude. Arecaceae

Neoapaloxylon Rauschert. 1982. Fabaceae. 2 Madagascar

Neoastelia J. B. Williams. 1987. Asteliaceae. 1 Australia (New South Wales)

Neoaulacolepis Rauschert = Aniselytron Merr. Poaceae

Neobaclea Hochr. 1930. Malvaceae. 2 temp. South America

Neobakeria Schltr. = Massonia Houtt. Hyacinthaceae

Neobalanocarpus P. S. Ashton. 1978. Dipterocarpaceae. 1 Thailand, Malay Peninsula

Neobartlettia R. M. King et H. Rob. = Bartlettia A. Gray. Asteraceae

Neobartlettia Schltr. 1920 (or = Palmorchis Barb. Rodr.). Orchidaceae. 6 trop. South America

Neobassia A. J. Scott. 1978. Chenopodiaceae. 2 northwestern (1) and southeastern (1) Australia

Neobathiea Schltr. 1925. Orchidaceae. 7 Madagascar

Neobaumannia Hutch. et Dalziel = Knoxia L. Rubiaceae

Neobeguea J.-F. Leroy. 1958. Meliaceae. 3 Madagascar

Neobenthamia Rolfe. 1891 (Polystachya Hook.). Orchidaceae. 1 East Africa (Tanzania)

Neobertiera Wernham. 1917. Rubiaceae. 1 Guyana

Neobesseya Britton et Rose = Escobaria Britton et Rose. Cactaceae

Neoblakea Standl. 1930. Rubiaceae. 1 Venezuela

Neobolusia Schltr. 1895. Orchidaceae. 4 trop. East and South Africa

Neobouteloua Gould. 1968. Poaceae. 1 Chile, Argentina

Neoboutonia Muell. Arg. 1864. Euphorbiaceae. 3 trop. Africa from Nigeria eastward to the southern Sudan, Kenya, Tanzania, and south to Angola and Mozambique

Neobracea Britton. 1920. Apocynaceae. 4 Bahama Is., Cuba

Neobreonia Ridsdale. 1975. Rubiaceae. 3 Madagascar

Neobrittonia Hochr. 1905. Malvaceae. 1 Mexico, Guatemala, Honduras, Nicaragua, Costa Rica, Panama

Neobuchia Urb. 1902. Bombacaceae. 1 West Indies

Neobuxbaumia Backeb. 1938 (Carnegiea Britton et Rose). Cactaceae. 7 Mexico

Neobyrnesia J. A. Amrstr. 1980. Rutaceae. 1 northern Australia

Neocabreria R. M. King et H. Rob. 1972 (Eupatorium L.). Asteraceae. 5 Brazil, Argentina

Neocaldasia Guatrec. 1944. Asteraceae. 1 Colombia

Neocallitropsis Florin. 1944. Cupressaceae. 1 New Caledonia

Neocalyptrocalyx Hutch. 1967 (Capparis L.). Capparaceae. 2 trop. South America

Neocarya (DC.) Prance ex F. White. 1976. Chrysobalanaceae. 1 trop. West Africa from Senegal to Liberia

Neocentema Schinz. 1911. Amaranthaceae. 2 Somalia (1), Tanzania (1)

Neochamaelea (Engl.) Erdtman = Cneorum L. Cneoraceae

Neochevalierodendron J. Léonard. 1951. Fabaceae. 1 Gabon

Neocinnomomum H. Liou. 1934 (Cinnamomum L.). Lauraceae. 7 Himalayas from Nepal to Bhutan, India, Burma, southwestern China, Indochina

Neoclemensia C. E. Carr. 1935. Orchidaceae. 1 Borneo

Neocodon Kolak. et Serdyuk. = Campanula L. Campanulaceae

Neocogniauxia Schltr. 1913. Orchidaceae. 2 West Indies

Neocollettia Hemsl. 1890. Fabaceae. 1 Burma, Java

Neoconopodium (Kozo-Polj.) Pimenov et Kljuykov. 1987. Apiaceae. 1 Afghanistan, Himalayas

Neocouma Pierre. 1898. Apocynaceae. 2 Brazil

Neocracca Kuntze. 1898. Fabaceae. 1 Andes of Bolivia, Argentina

Neocryptodiscus Hedge et Lamond = Prangos Lindl. Apiaceae

Neocuatrecasia R. M. King et H. Rob. 1970 (Eupatorium L.). Asteraceae. 8 Andes of Peru and Bolivia

Neocussonia Hutch. = Cussonia Thunb. Araliaceae

Neodielsia Harms = Astragalus L. Fabaceae

Neodissochaeta Bakh. f. = Dissochaeta Blume. Melastomataceae

Neodistemon Babu et A. N. Henry. 1970. Urticaceae. 1 South and Southeast Asia from India to Philippines

Neodonnellia Rose = Tripogandra Raf. Commelinaceae

Neodregea C. H. Wright. 1909. Colchicaceae. 1 South Africa

Neodriessenia M. P. Nayar. 1977. Melastomataceae. 6 northern Vietnam, Borneo

Neodryas Rchb. f. 1852. Orchidaceae. 5 trop. South America

Neodunnia R. Vig. 1950. Fabaceae. 5 Madagascar

Neodypsis Baill. 1894. Arecaceae. 15 Madagascar

Neoeplingia Ramam., Hiriart et Medran. 1982. Lamiaceae. 1 Mexico

Neoescobaria Garay = Helcia Lindl. Orchidaceae

Neofabricia Joy Thompson. 1983. Myrtaceae. 3 Australia (Queensland)

Neofinetia Hu. 1925 (Holcoglossum Schltr.). Orchidaceae. 1 China, Japan

Neofranciella Guillaumin. 1925. Rubiaceae. 1 New Caledonia

Neogaerrhinum Rothm. = Antirrhinum L. Scrophulariaceae

Neogaillonia Lincz. 1973. Rubiaceae. 10 West (incl. Caucasus) and Middle Asia, Northeast Africa

Neogardneria Schltr. ex Garay. 1973. Orchidaceae. 3 Central and northeastern South America

Neoglaziovia Mez. 1894. Bromeliaceae. 2 eastern Brazil

Neogleasonia Maguire = Bonnetia Mart. Bonnetiaceae

Neogoezea Hemsl. 1894. Apiaceae. 5 North America, mainly Mexico

Neogontscharovia Lincz. 1971. Plumbaginaceae. 3 Middle Asia and Afghanistan

Neogoodenia C. A. Gardner et A. S. George = Goodenia Sm. Goodeniaceae

Neoguillauminia Croizat. 1938. Euphorbiaceae. 1 New Caledonia

Neogunnia Pax et K. Hoffm. = Gunniopsis Pax. Aizoaceae

Neogyne Rchb. f. 1852. Orchidaceae. 1 Nepal, Bhutan, northeastern India, Burma, southern China (Yunnan), Thailand, Laos

Neohallia Hemsl. 1882. Acanthaceae. 1 southern Mexico

Neoharmsia R. Vig. 1951. Fabaceae. 2 Madagascar

Neohemsleya T. D. Penn. 1991. Sapotaceae. 1 Tanzania

Neohenricia L. Bolus. 1938. Aizoaceae. 1 central South Africa

Neohenrya Hemsl. = Tylophora R. Br. Asclepiadaceae

Neohintonia R. M. King et H. Rob. 1971 (Eupatorium L.). Asteraceae. 1 Mexico

Neoholstia Rauschert. 1982. Euphorbiaceae. 1 (N. tenuifolia) Kenya, Tanzania, Zambia, Malawi, Zimbabwe, Mozambique

Neohouzeaua A. Camus. 1922 (Schizostachyum Nees). Poaceae. 5 South and Southeast Asia, Malesia

Neohuberia Ledoux = Eschweilera Mart. ex DC. Lecythidaceae

Neohumbertiella Hochr. = Humbertiella Hochr. Malvaceae

Neohusnotia A. Camus. 1921 (Acroceras Stapf). Poaceae. 1 Southeast Asia

Neohymenopogon Bennet. 1981. Rubiaceae. 3 Himalayas from Kumaon to Khasi Mts., Assam, northern Burma, southwestern China, Indochina

Neohyptis J. K. Morton. 1962. Lamiaceae. 1 trop. West and East Africa to Angola, Zambia, and northern Botswana

Neojatropha Pax = Mildbraedia Pax. Euphorbiaceae

Neojeffreya Cabrera. 1978. Asteraceae. 1 Africa, Madagascar

Neojobertia Baill. 1888. Bignoniaceae. 1 northeastern Brazil

Neokoehleria Schltr = Scelochilus Klotzsch. Orchidaceae

Neolabatia Aubrév. = Pouteria Aubl. Sapotaceae

Neolamarckia Bosser. 1984. Rubiaceae. 2 South and Southeast Asia, Malesia, Australia, Oceania

Neolauchea Kraenzl. = Isabelia Barb. Rodr. Orchidaceae

Neolaugeria Nicolson. 1979. Rubiaceae. 3 Bahama Is., Cuba, Haiti, Puerto Rico, Guadeloupe, St. Vincent, Montserrat, Dominica and Martinique Is.

Neolchmannia Kraenzl. = Epidendrum L. Orchidaceae

Neolemonniera Heine. 1960. Sapotaceae. 3 trop. Africa

Neoleptopyrum Hutch. = Leptopyrum Rchb. Ranunculaceae

Neoleroya Cavaco = Pyrostria Comm. ex Juss. Rubiaceae

Neolindenia Baill. 1890. Acanthaceae. 1 Mexico

Neolindleya Kraenzl. = Platanthera Rich. Orchidaceae

Neolitsea (Benth.) Merr. 1906. Lauraceae. 85 eastern Himalayas, India, China, Taiwan, Japan, Indochina, Malesia, Australia

Neolloydia Britton et Rose. 1922. Cactaceae. c. 15 southern U.S. (Texas), northeastern and eastern Mexico, Cuba

Neolophocarpus E. G. Camus. 1912. Cyperaceae. 1 Indochina

Neolourya L. Rodrig. = Peliosanthes Andrews. Convallariaceae

Neoluederitzia Schinz. 1894. Zygophyllaceae. 1 South-West Africa

Neoluffa Chakrav. = Siraitia Merr. Cucurbitaceae

Neomacfadya Baill. = Arrabidaea DC. Bignoniaceae

Neomammillaria Britton et Rose = Mammillaria Haw. Cactaceae

Neomandonia Hutch. = Tradescantia L. Commelinaceae

Neomangenotia J.-F. Leroy = Commiphora Jacq. Burseraceae

Neomarica Sprague. 1928. Iridaceae. 13 trop. America, West Africa

Neomartinella Pilg. 1906. Brassicaceae. 1 China

Neomazaea Krug et Urb. = Mazaea Krug et Urb. Rubiaceae

Neomezia Votsch. 1904 (or = Deherainia Decne.). Theophrastaceae. 1 Cuba

Neomicrocalamus Keng f. = Racemobambos Holttum. Poaceae

Neomillspaughia S. F. Blake. 1921. Polygonaceae. 2 Mexico (Yucatan), Central America

Neomirandea R. M. King et H. Rob. 1970 (Eupatorium L.). Asteraceae. 27 trop. America from Mexico to Ecuador

Neomitranthes Legrand. 1977. Myrtaceae. 3 Brazil

Neomolinia Honda et Sakisaka. 1930 (Diarrhena P. Beauv.). Poaceae. 5 East Asia, incl. (3) southern Far East of Russia

Neomoorea Rolfe. 1904. Orchidaceae. 1 Panama, Colombia, Ecuador

Neomortonia Wiehler. 1975. Gesneriaceae. 1 Costa Rica, Panama, Colombia

Neomuellera Briq. = Plectranthus L'Her. Lamiaceae

Neomyrtus Burret. 1941. Myrtaceae. 1 New Zealand

Neonauclea Merr. 1915. Rubiaceae. 64 Bhutan, Burma, southern China, Andaman Is., peninsular Thailand, Indochina, Taiwan, Malesia from Sumatra to New Ireland and New Britain, Australia (northern Queensland), Bougainville and Solomon Is., Vanuatu, Fiji, Tonga, Samoas, Society Is.

Neonelsonia J. M. Coult. et Rose. 1895. Apiaceae. 2 Mexico, Central and western trop. South America

Neonicholsonia Dammer. 1901. Arecaceae. 1 Nicaragua, Panama

Neonotonia Lackey. 1977 (Glycine Willd.) Fabaceae. 2 Tanzania (1, N. verdcourtii), trop. Asia (N. wightii)

Neopalissya Pax = Necepsia Prain. Euphorbiaceae

Neopallasia Poljakov. 1955. Asteraceae. 1 Middle Asia, East Siberia, Mongolia, China

Neopanax Allan = Pseudopanax K. Koch. Araliaceae

Neopatersonia Schoenl. 1912. Hyacinthaceae. 3 South-West and South Africa

Neopaulia Pimenov et Kljukov = Paulita Soják. Apiaceae

Neopaxia O. E. G. Nilsson. 1966 (Montia L.). Portulacaceae. 1 southeastern Australia, Tasmania, New Zealand

Neopentanisia Verdc. 1953. Rubiaceae. 2 southern trop. Africa

Neopetalonema Brenan = Gravesia Naudin. Melastomataceae

Neophloga Baill. 1894. Arecaceae. c. 30 Madagascar

Neopicrorhiza D. Y. Hong. 1984. Scrophulariaceae. 1 China

Neopilea Leandri = Pilea Lindl. Urticaceae

Neoplatytaenia Geld. et Nikitin = Semenovia Regel et Herd. Apiaceae

Neopometia Aubrév. et Pellegr. = Pradosia Liais. Sapotaceae

Neoporteria Britton et Rose. 1922. Cactaceae. c. 25 southern Peru, Chile, western Argentina

Neopreissia Ulbr. = Atriplex L. Chenopodiaceae

Neopringlea S. Watson. 1891. Flacourtiaceae. 3 Mexico, Guatemala

Neoptychocarpus Buchheim. 1959. Flacourtiaceae. 2 trop. South America: Surinam, French Guiana, Amazonian Colombia, Peru, and Brazil

Neoraimondia Britton et Rose. 1920. Cactaceae. 2 Andes from northern Peru to northern Chile and Bolivia

Neorapinia Moldenke. 1955. Verbenaceae. 1 New Caledonia

Neoraputia Emmerich. 1978 (Raputia Aubl.). Rutaceae. 5 trop. South America

Neorautanenia Schinz. 1899. Fabaceae. 3 Africa from Togo, Chad, Sudan, Ethiopia to South Africa

Neoregelia L. B. Sm. 1934. Bromeliaceae. c. 80 Colombia, Venezuela, Ecuador, Peru, Brazil

Neoregnellia Urb. 1924. Sterculiaceae. 1 Cuba

Neorhine Schwantes = Rhinephyllum N. E. Br. Aizoaceae

Neorites L. S. Sm. 1969. Proteaceae. 1 northeastern Australia

Neoroepera Muell. Arg. ex F. Muell. 1866. Euphorbiaceae. 2 northeastern Australia

Neorosea N. Hallé = Tricalysia A. Rich. ex DC. Rubiaceae

Neorudolphia Britton. 1924. Fabaceae. 1 West Indies

Neosabicea Wernham = Manettia Mutis ex L. Rubiaceae

Neosasamorpha Tatew. = Sasa Makino et Shib. Poaceae

Neoschimpera Hemsl. = Psychotria L. Rubiaceae

Neoschischkinia Tzvelev. 1968 (Agrostis L.). Poaceae. 4 Southwest Europe, North Africa

Neoschroetera Briq. = Larrea Cav. Zygophyllaceae

Neoschumannia Schltr. 1905. Asclepiadaceae. 1 trop. West Africa

Neosciadium Domin. 1908. Apiaceae. 1 southwestern Australia

Neoscortechinia Pax. 1897. Euphorbiaceae. 4 Burma, Nicobar Is., Thailand, Malesia, Solomon Is.

Neosepicaea Diels. 1922. Bignoniaceae. 4 New Guinea, northeastern Australia (1)

Neosieversia Bolle = Novosieversia Bolle. Rosaceae

Neosinocalamus Keng f. = Dendrocalamus Nees. Poaceae

Neosloetiopsis Engl. = Sloetiopsis Engl. Moraceae

Neosparton Griseb. 1874. Verbenaceae. 4 temp. South America

Neosprucea Sleumer. 1938. Flacourtiaceae. 5 trop. America: Panama, Colombia, Ecuador, Peru, and Amazonian Brazil

Neostapfia Burtt Davy. 1899. Poaceae. 1 California

Neostapfiella A. Camus. 1926. Poaceae. 1 Madagascar

Neostenanthera Exell. 1935. Annonaceae. 10 trop. Africa

Neostrearia L. S. Sm. 1958. Hamamelidaceae. 1 northeastern Australia

Neostricklandia Rauschert = Eucrosia Ker-Gawl. Amaryllidaceae

Neotainiopsis Bennet et Raizada = Eriodes Rolfe. Orchidaceae

Neotatea Maguire. 1972. Clusiaceae. 3 Colombia, Venezuela

Neotessmannia Burret. 1924. Tiliaceae. 1 Peru

Neothorelia Gagnep. 1908. Capparaceae. 1 Indochina

Neothymopsis Britton et Millsp. = Thymopsis Benth. Asteraceae

Neotina Capuron. 1969. Sapindaceae. 2 Madagascar

Neotinea Rchb. f. 1852. Orchidaceae. 1 Canary Is., West Europe, Mediterranean

Neotorularia Hedge et J. Léonard. 1987. Brassicaceae. 21 Europe, North Africa, West (incl. Caucasus) and Middle Asia, eastward to Siberia, Mongolia, China, and India

Neotreleasia Rose = Tradescantia L. Commelinaceae

Neotrewia Pax et K. Hoffm. 1914. Euphorbiaceae. 1 Sulawesi, Philippines

Neotschihatchewia Rauschert. 1982. Brassicaceae. 1 Armenia

Neottia Guett. 1754. Orchidaceae. 11 temp. Eurasia

Neottianthe (Rchb.) Schltr. 1919. Orchidaceae. 5–6 temp. Eurasia, incl. Russia (1, N. cucullata—central European part, Siberia, Far East), Himalayas

Neotuerckheimia Donn. Sm. = Amphitecna Miers. Bignoniaceae

Neoturczaninowia Kozo-Polj. 1924. Apiaceae. ? sp. South America (nomen dubium)

Neotysonia Dalla Torre et Harms. 1905. Asteraceae. 1 southwestern Australia

Neo-urbania Fawcett et Rendle. 1909 (or = Maxilaria Ruiz et Pav.). Orchidaceae. 2 Cuba, Jamaica

Neo-uvaria Airy Shaw. 1939. Annonaceae. 2 western Malesia

Neoveitchia Becc. 1920. Arecaceae. 1 Fiji (Viti Levu I.)

Neowawraea Rock = Flueggea Willd. Euphorbiaceae

Neowerdermannia Fric. 1930 (Neoporteria Britton et Rose). Cactaceae. 2 Peru, southern Bolivia, northern Chile, northern Argentina

Neowilliamsia Garay. 1977. Orchidaceae. 6 Costa Rica, Panama

Neowimmeria O. Deg. et I. Deg. = Lobelia L. Loberliaceae

Neowollastonia Wernham ex Ridl. = Melodinus J. R. Forst. et G. Forst. Apocynaceae

Neowormia Hutch. et Summerh. 1928 (Dillenia L.). Dilleniaceae. 1 Seychelles

Neoxythece Aubrév. et Pellegr. = Pouteria Aubl. Sapotaceae

Neozenkerina Mildbr. = Staurogyne Wall. Acanthaceae

Nepa Webb = Stauracanthus Link. Fabaceae

Nepenthes L. 1753. Nepenthaceae. 72 Madagascar, Seychelles, Sri Lanka, India (Assam), southern China, Indochina, Malesia, northern Australia (1, N. mirabilis), New Caledonia

Nepeta L. 1753. Lamiaceae. c. 215 temp. Eurasia (esp. Mediterranean, Caucasus, Middle Asia, Iran, Afghanistan, and China), mountainous North and East Africa

Nephelaphyllum Blume. 1825. Orchidaceae. 16–17 trop. Asia from Himalayas and India eastward to Indochina and Malesia

Nephelium L. 1767. Sapindaceae. 38 (or 22) India (Assam), Burma, southern China, Indochina, Malesia (except Lesser Sunda Is.)

Nephelochloa Boiss. 1844. Poaceae. 1 Southwest Asia

Nephradenia Decne. 1844. Asclepiadaceae. 10 trop. America from Mexico to Brazil

Nephrangis (Schltr.) Summerh. 1948. Orchidaceae. 1 trop. Africa: Liberia, Zaire, Zambia, Uganda, Kenya, Tanzania

Nephrocarpus Dammer = Basselinia Vieill. Arecaceae

Nephrocarya Candargy = Nonea Medik. Boraginaceae

Nephrodesmus Schindl. 1916 (Arthroclianthus Baill.). Fabaceae. 10 New Caledonia

Nephromeria (Benth.) Schindl. = Desmodium Desv. Fabaceae

Nephropetalum B. L. Rob. et Greenm. = Ayenia L. Sterculiaceae

Nephrophyllidium Gilg. 1895. Menyanthaceae. 1 northern Japan, northwestern North America

Nephrophyllum A. Rich. 1850. Convolvulaceae. 1 Ethiopia

Nephrosperma Balf. f. 1877. Arecaceae. 1 Seychelles

Nephthytis Schott. 1857. Araceae. 7 trop. Africa

Nepsera Naudin. 1849. Melastomataceae. 1 trop. America, West Indies

Neptunia Lour. 1790. Fabaceae. 12 tropics and subtropics, esp. America

Neraudia Gaudich. 1830. Urticaceae. 5 Hawaiian Is.

Neriacanthus Benth. 1876. Acanthaceae. 3 trop. America, West Indies

Nerine Herb. 1820. Amaryllidaceae. c. 30 South Africa

Nerisyrenia Greene. 1900. Brassicaceae. 9 western U.S., Mexico

Nerium L. 1753. Apocynaceae. 3–4 Cape Verde Is., Mediterranean, subtrop. Asia; N. oleander cultur. in Crimea and Caucasus

Nernstia Urb. 1923. Rubiaceae. 1 Mexico

Nerophila Naudin. 1850. Melastomataceae. 1 trop. West Africa

Nertera Banks et Sol. ex Gaertn. 1788. Rubiaceae. 12–15 Madagascar, southern China, Taiwan, Malesia, Australia, New Zealand, Hawaiian Is., Society Is., temp. South America, Is. of Tristan da Cunha

Nervilia Comm. ex Gaudich. 1829. Orchidaceae. c. 80 trop. and South Africa, Madagascar, Mascarene Is., southern Arabian Peninsula, India, Burma, southern China, Taiwan, Ryukyu Is., Indochina, Malesia from Malay Peninsula to New Guinea, Solomon Is. (1, N. aragoana), Micronesia, Australia, Vanuatu, New Caledonia, Horn Is., Fiji (3), Samoa, Niue Is.

Nesaea Kunth. 1823. Lythraceae. 56 trop. and South Africa, Madagascar, trop. Asia, Australia, trop. America

Nesiota Hook. f. 1862. Rhamnaceae. 1 St. Helena

Neslia Desv. 1814. Brassicaceae. 2 Mediterranean, Southeast Europe, West and Middle Asia eastward to Mongolia, China, and northwestern India

Nesocaryum I. M. Johnst. 1927. Boraginaceae. 1 Chile

Nesocodon Thulin. 1980. Campanulaceae. 1 Mascarene Is.

Nesogenes A. DC. 1847. Nesogenaceae. 7 Tanzania, Madagascar, Seychelles and Mascarene Is., Malesia, Oceania (Tuamotu Arch.)

Nesogordonia Baill. 1886. Sterculiaceae. 18 trop. Africa, Madagascar (14)

Nesohedyotis (Hook. f.) Bremek. 1952. Rubiaceae. 1 St. Helena

Nesoluma Baill. 1891. Sapotaceae. 3 Hawaiian Is., Henderson I., Rapa, and Tahiti

Nesomia B. Turner. 1991. Asteraceae. ??

Nesothamnus Rydb. = Perityle Benth. Asteraceae

Nesphostylis Verdc. 1970. Fabaceae. 2 trop. Africa (1, N. holose ricea), Burma (1)

Nestegis Raf. 1838. Oleaceae. 5 New Zealand, Norfolk I., Hawaiian Is. (1, N. sandwicensis)

Nestlera Apreng = Relhania L'Her. Asteraceae

Nestoria Urb. = Pleonotoma Miers. Bignoniaceae

Nestronia Raf. 1837. Santalaceae. 1 southeastern U.S.

Nettoa Baill. 1866. Tiliaceae. 1 Australia

Neuburgia Blume. 1850. Loganiaceae. 10–12 Philippines, Sulawesi, New Guinea, Caroline and Solomon Is., Vanuatu, New Caledonia, Fiji

Neuontobotrys O. E. Schulz. 1924. Brassicaceae. 2 Chile, Argentina

Neuracanthus Nees. 1832. Acanthaceae. 21 trop. Africa, Madagacar, Arabian Peninsula, Socotra, India

Neurachne R. Br. 1810. Poaceae. 6 Australia

Neuractis Cass. 1825. Asteraceae. 6 Tanzania, India, Sri Lanka, Thailand, China, Taiwan, Japan, Vietnam, Malay Peninsula, Philippines, northern Marianas, Micronesia, Caroline Is., New Guinea, eastern Australia, New Caledonia, Fiji

Neurada L. 1753. Neuradaceae. 1 North Africa, Cyprus, Syria, Lebanon, Israel, Sinai, Arabian Peninsula, Iraq, northern Iran, Afghanistan, Pakistan, India

Neuradopsis Bremek. et Oberm. 1935. Neuradaceae. 3 southwestern Africa

Neurocalyx Hook. 1837. Rubiaceae. 4 southern India, Sri Lanka

Neurolaena R. Br. 1817. Asteraceae. 10 trop. America, West Indies

Neurolakis Mattf. 1924. Asteraceae. 1 trop. West Africa

Neurolepis Meisn. 1843. Poaceae. 12 Andes from Venezuela and Colombia to Peru, Trinidad

Neurolobium Baill. 1888. Apocynaceae. 1 Brazil

Neuroloma Andrz. = Achoriphragma Soják. Brassicaceae

Neuropeltis Wall. 1824. Convolvulaceae. 13 trop. Africa (9), trop. Asia (4)

Neuropeltopsis Ooststr. 1964. Convolvulaceae. 1 Borneo

Neurophyllodes (A. Gray) O. Deg. = Geranium L. Geraniaceae

Neuropoa Clayton. 1985. Poaceae. 1 Australia

Neurotheca Salisb. ex Benth. 1876. Gentianaceae. 1 (N. loeselioides) northeastern South America, trop. Africa, Madagascar

Neurotropis (DC.) F. K. Mey. = Thlaspi L. Brassicaceae

Neustruevia Juz. = Pseudomarrubium Popov. Lamiaceae

Neuwiedia Blume. 1833. Orchidaceae. 8 Indochina, Malesia from Malay Peninsula to New Guinea, Solomon Is., Southwest Pacific Is.

Neves-Armondia K. Schum. = Pithecoctenium Mart. ex Meisn. Bignoniaceae

Nevillea Esterh. et P. H. Linder. 1984 (Restio Rottb.). Restionaceae. 2 South Africa (southwestern Cape)

Neviusia A. Gray. 1858. Rosaceae. 2 California (N. cliftonii), southeastern U.S. (Alabama)

Nevskiella Krecz. et Vved. 1934 (Bromus L.). Poaceae. 1 Southwest and Middle Asia

Newbouldia Seem. ex Bureau. 1864. Bignoniaceae. 1 trop. West Africa

Newcastelia F. Muell. 1857. Chloanthaceae. 12 trop. Australia

Newtonia Baill. 1888. Fabaceae. 11 trop. Africa from Sierra Leone to Somalia south to Angola and Zimbabwe

Neyraudia Hook. f. 1896. Poaceae. 6 trop. Old World

Nicandra Adans. 1763. Solanaceae. 1 (N. physalodes) Peru

Nichallea Bridson. 1978. Rubiaceae. 1 trop. Africa

Nicobariodendron Vasudeva Rao et Chakrab. 1985. Celastraceae. 1 Nicobar Is.

Nicodemia Ten. = Buddleja L. Buddlejaceae

Nicolaia Horan. = Etlingera Giseke. Zingiberaceae

Nicolasia S. Moore. 1900. Asteraceae. 6 trop. Africa

Nicolletia A. Gray. 1845. Asteraceae. 3 southwestern U.S. (California, New Mexico, Texas), Mexico

Nicolsonia DC. = Desmodium Desv. Fabaceae

Nicotiana L. 1753. Solanaceae. c. 70 mostly trop. and subtrop. America (46), South-West Africa (1, N. africana, Namibia), Australia (16), Lord Howe I., New Caledonia and the other islands of the southern Pacific; widely naturalized

Nidema Britton et Millsp. 1920. Orchidaceae. 2 Mexico, Central and northern South America, West Indies

Nidorella Cass. 1825. Asteraceae. 13 Cape Verde Is., trop. and South Africa

Nidularium Lem. 1854. Bromeliaceae. 26 eastern Brazil

Niederleinia Hieron = Frankenia L. Frankeniaceae

Niedzwedzkia Fedtsch. 1915 (Incarvillea Juss.). Bignoniaceae. 1 Middle Asia

Niemeyera F. Muell. 1870. Sapotaceae. c. 20 Australia (Queensland) and New Caledonia

Nierembergia Ruiz et Pav. 1794. Solanaceae. c. 36 South America, mostly Argentina; N. angustifolia—Mexico; N. repens—from Colombia, Ecuador, Peru to Uruguay and eastern Argentina

Nietneria Klotzsch ex Benth. 1883. Melanthiaceae. 1 Venezuela, Guyana

Nigella L. 1753. Ranunculaceae. 14–20 Europa (12), Mediterranean, West (incl. Caucasus) and Middle Asia

Nigritella Rich. 1818. Orchidaceae. 2 Europe

Nigromnia Carolin = Scaevola L. Goodeniaceae

Nikitinia Iljin. 1960. Asteraceae. 1 Middle Asia, Iran

Nilgirianthus Bremek. 1944. Acanthaceae. 20 South Asia

Nimiria Prain ex Craib = Acacia Hill. Fabaceae

Niphaea Lindl. 1841. Gesneriaceae. 5 trop. America, West Indies

Niphogeton Schltdl. 1856 (Apium L.). Apiaceae. 18 Andes from Costa Rica to Peru

Nipponanthemum (Kitam.) Kitam. 1978. Asteraceae. 1 Japan

Nipponobambusa Muroi et Shib. = Sasa Makino et Shib. Poaceae

Nipponocalamus Nakai = Arundinaria Michx. Poaceae

Nirarathamnos Balf. f. 1882. Apiaceae. 1 Socotra

Nispero Aubrév. = Manilkara Adans. Sapotaceae

Nissolia Jacq. 1760. Fabaceae. 13 trop. and subtrop. America from Arizona and Texas to Argentina and Paraguay, but chiefly Mexico (10–12)

Nitraria L. 1759. Nitrariaceae. 10 Southeast Europe (1), North Africa, Crimea, lower Volga, West and East Siberia, Middle Asia, Afghanistan, Mongolia, China, Indochina, southwestern Australia

Nitrophila S. Watson. 1871. Chenopodiaceae. 8 U.S., Mexico, Chile, Argentina

Nivellea Wilcox, Bremer et Humphries. 1991–1992. Asteraceae. 1 Morocco

Nivenia Vent. 1808 (Aristea Sol. ex Aiton). Iridaceae. 9 South Africa (Cape)

Niveophyllum Matuda = Hechtia Klotzsch. Bromeliaceae

Noaea Moq. 1849. Chenopodiaceae. 6 Greece, Morocco, Algeria, Libya, Egypt, Turkey, Syria, Lebanon, Israel, Jordan, Arabian Peninsula, Iraq, Caucasus, Turkmenistan, Iran, Afghanistan,

Noahdendron Endress, B. Hyland et Tracey. 1985 (Ostrearia Baill.). Hamamelidaceae. 1 Australia (northern Queensland)

Nocca Cav. = Lagascea Cav. Asteraceae

Noccaea Moench. 1802 (Thlaspi L.). Brassicaceae. 13 Southeast Europe, Caucasus, Middle Asia, West and East Siberia, Russian Far East

Noccidium F. K. Mey. = Thlaspi L. Brassicaceae

Nodocarpaea A. Gray. 1883. Rubiaceae. 1 Cuba

Nodonema B. L. Burtt. 1982 (Acanthonema Hook. f. and Trachystigma C. B. Clarke). Gesneriaceae. 1 southeastern Nigeria, southwestern Cameroun

Noeparrya Mathias. 1929. Apiaceae. 2 southwestern U.S.

Nogalia Verdc. 1988. Boraginaceae. 1 Africa, Arabian Peninsula

Nogo Baehni = Lecomtedoxa Dubard. Sapotaceae

Nogra Merr. 1935. Fabaceae. 3 trop. Asia from India to Thailand

Noisettia Kunth. 1823. Violaceae. 1 Guyana, Peru, Brazil

Nolana L. f. 1762. Nolanaceae. 18 Galapagos Is., western South America from southern Peru to northern Chile

Noldeanthus Knobl. 1935 (or = Jasminum L.). Oleaceae. 1 Angola

Nolina Michx. 1803. Nolinaceae. 30 southwestern U.S., Mexico

Nolletia Cass. 1825. Asteraceae. 4 Morocco, South Africa

Noltea Rchb. 1828. Rhamnaceae. 1 (N. africana) South Africa from Natal to southwestern Cape

Nomaphila Blume = Hydrophila R. Br. Acanthaceae

Nomocharis Franch. 1889. Liliaceae. 16 western China, southeastern Tibet, India (Assam), northern Burma

Nomosa I. M. Johnst. 1954. Boraginaceae. 1 Mexico

Nonatelia Aubl. 1775 (Palicourea Aubl.). Rubiaceae. 1 trop. America

Nonea Medik. 1789. Boraginaceae. 35 Europe, Mediterranean, West (incl. Caucasus) and Middle Asia to Mongolia,

China, northwestern and western India
Nopalea Salm-Dyck = Opuntia Hill. Cactaceae
Nopalxochia Britton et Rose = Disocactus Lindl. Cactaceae
Norantea Aubl. 1775. Marcgraviaceae. 35 trop. America, West Indies
Normanbokea Klad. et Buxb. = Neolloydia Britton et Rose. Cactaceae
Normanboria Butzin. 1978 (Acrachne Chiov.). Poaceae. 1 India
Normanbya F. Muell. ex Becc. 1885. Arecaceae. 1 Australia (northern Queensland)
Normandia Hook. f. 1872. Rubiaceae. 1 New Caledonia
Normania Lowe. 1872 (Solanum L.). Solanaceae. 2 Macaronesia: Madeira, Tenerife, and Grand Canary
Noronhia Stadmam ex Thouars. 1806. Oleaceae. c. 40 Comoro Is., Madagascar, Mauritius
Norrisia Gardner. 1849. Loganiaceae. 2 Malay Peninsula, Sumatra, Borneo, Philippines
Northea Hook. f. 1884. Sapotaceae. 1 (N. seychellana) Seychelles
Northiopsis Kaneh. = Manilkara Adans. Sapotaceae
Nosema Prain. 1904. Lamiaceae. 6 China, Indochina, Malesia
Nostolachma T. Durand. 1888. Rubiaceae. 10 Indochina, Malesia
Notanthera (DC.) G. Don. 1834 (Loranthus Jacq.). Loranthaceae. 1 Andes of Peru, Bolivia, and northern Chile
Notechidnopsis Lavranos et Bleck. 1985. Asclepiadaceae. 2 South Africa
Notelaea Vent. 1804. Oleaceae. 9 eastern Australia
Nothaphoebe Blume. 1851. Lauraceae. 30–40 southern China, Indochina, Malesia
Nothapodytes Blume. 1850. Icacinaceae. 5 southern India, Sri Lanka, Assam, Burma, Central and southern China (incl. Hainan), Taiwan, Ryukyu Is., Indochina, western Malesia: Sumatra, Java, Lesser Sunda Is., Philippines

Nothoalsomitra Telford. 1982. Cucurbitaceae. 1 Australia (southeastern Queensland)
Nothobaccharis R. M. King et H. Rob. 1979. Asteraceae. 1 Peru
Nothocalais Greene = Microseris D. Don. Asteraceae
Nothocestrum A. Gray. 1862. Solanaceae. 4–6 Hawaiian Is.
Nothochelone (A. Gray) Straw. 1966. Scrophulariaceae. 1 western North America from British Columbia to California
Nothochilus Radlk. 1889. Scrophulariaceae. 1 Brazil
Nothocissus (Miq.) Latiff. 1982. Vitaceae. ? sp.
Nothocnide Blume ex Chew. 1856. Urticaceae. 5 Malesia, Bismarck Arch., Solomon Is., Australia (1, N. repanda)
Nothodoritis Tsi. 1989. Orchidaceae. 1 China
Nothofagus Blume. 1850. Nothofagaceae. 37 New Guinea, New Britain I., temp. Australia, Tasmania, New Zealand, New Caledonia, subantarctic Pacific islands, temp. South America
Notholirion Wall. ex Boiss. 1945. Liliaceae. 6 western Iraq, Iran, Afghanistan, Pakistan, Himalayas from Kashmir to Bhutan, southeastern Tibet, Central and southern China
Nothomyrcia Kausel = Myrceugenia O. Berg
Nothopanax Miq. = Polyscias J. R. Forst. et G. Forst. Araliaceae
Nothopegia Blume. 1850. Anacardiaceae. 6 India, Sir Lanka, Borneo
Nothopegiopsis Lauterb. = Semecarpus L. f. Anacardiaceae
Nothophlebia Standl. = Pentagonia Benth. Rubiaceae
Nothoruellia Bremek. et Nannenga-Bremek. = Ruellia L. Acanthaceae
Nothosaerva Wight. 1853. Amaranthaceae. 1 trop. Africa from Senegal through Chad to Somalia and Ethiopia, south to Angola and Zimbabwe, Mauritius, India from Punjab to Madras, Sri Lanka, Burma, ? Borneo

Nothoscordum Kunth. 1843. Alliaceae. 20–25 America, esp. South America; naturalized in Europe, Asia, and Australia

Nothosmyrnium Miq. 1867. Apiaceae. 2 China, Japan

Nothospondias Engl. 1905. Simaroubaceae. 1 trop. West Africa

Nothostele Garay. 1982. Orchidaceae. 1 Brazil

Nothotaxus Florin = Pseudotaxus Cheng. Taxaceae

Nothotsuga H.-H. Hu ex C. N. Page. 1989. Pinaceae. 1 southern China (Hainan, Guishou, Guangdong, Guangxi)

Noticastrum DC. 1836. Asteraceae. 12 trop. South America

Notiosciadium Speg. 1924. Apiaceae. 1 Argentina

Notobasis Cass. 1825. Asteraceae. 1 (N. syriaca) Mediterranean, Asia Minor, eastern Caucasus, Iran

Notobuxus Oliv. 1882. Buxaceae. 8 trop. and South Africa, Madagascar

Notocactus (K. Schum.) Backeb. et F. M. Knuth = Parodia Speg. Cactaceae

Notoceras R. Br. 1812. Brassicaceae. 2 Canary Is., Europe, Mediterranean, Arabian Peninsula, Iraq, Kuwait, Iran, Afghanistan, Pakistan, and northwestern India

Notochaete Benth. 1829. Lamiaceae. 2 Himalayas from Nepal to Bhutan, India, Burma, southwestern China

Notochloe Domin. 1911. Poaceae. 1 Australia (New South Wales)

Notodanthonia Zotov. 1963. Poaceae. c. 60 Malesia, Australia, New Zealand, Argentina

Notodon Urb. 1899. Fabaceae. 4 Cuba

Notodontia Pierre ex Pitard = Ophiorrhiza L. Rubiaceae

Notonerium Benth. 1876. Apocynaceae. 1 southern Australia

Notonia DC. = Kleinia Hill. Asteraceae

Notoniopsis R. Nord. = Kleinia Hill. Asteraceae

Notopleura (Hook. f.) Bremek. = Psychotria L. Rubiaceae

Notopora Hook. f. 1873. Ericaceae. 5 eastern Venezuela, Guyana

Notoptera Urb. = Otopappus Benth. Asteraceae

Notopterygium Boissieu. 1903. Apiaceae. 2 southwestern China

Notosceptrum Benth. = Kniphofia Moench. Asphodelaceae

Notoseris C. Shih. 1987. Asteraceae. 14 trop. and subtrop. China

Notospartium Hook. f. 1857. Fabaceae. 3 New Zealand

Notothixos Oliv. 1863. Viscaceae. 8 Sri Lanka, Andaman Is., Burma, Malay Peninsula, Java, Philippines, Moluccas, New Guinea, Bismarck Arch., Solomon Is., Australia (4)

Notothlaspi Hook. f. 1862. Brassicaceae. 2 New Zealand

Nototriche Turcz. 1863. Malvaceae. 100 South America

Nototrichium (A. Gray) W. F. Hillebr. 1888 (Achyranthes L.). Amaranthaceae. 2 Hawaiian Is.

Notoxylinon Lewton. 1915 (Cienfuegosia Cav.). Malvaceae. 8 Australia

Notylia Lindl. 1825. Orchidaceae. c. 60 trop. America from Mexico to Brazil

Nouelia Franch. 1888. Asteraceae. 1 southwestern China

Nouettea Pierre. 1898. Apocynaceae. 1 Indochina

Novenia Freire. 1986. Asteraceae. 2 Andes of Bolivia and Argentina

Novosieversia F. Bolle. 1933. Rosaceae. 1 arctic and temp. West and East Siberia, Kamchatka, northwestern North America

Nowickea J. Martinez et J. A. McDonald. 1989. Phytolaccaceae. 2 Mexico

Nowodworskya C. Presl = Polypogon Desf. Poaceae

Noyera Trécul = Perebea Aubl. Moraceae

Nucularia Batt. 1903. Chenopodiaceae. 1 western Sakhara, Mauritania, Morocco, Algeria, Libya

Nuihonia Dop = Craibiodendron W. W. Sm. Ericaceae

Numaeacampa Gagnep. 1948 (Codonopsis Wall.). Campanulaceae. 1 Indochina

Nuphar Sm. 1809. Nymphaeaceae. 20 cold and temp. Northern Hemisphere, Mediterranean (2)
Nurmonia Harms = Turraea L. Meliaceae
Nuttallanthus D. A. Sutton. 1988 (Antirrhinum L.). Scrophulariaceae. 4 North America
Nuxia Comm. ex Lam. 1791. Buddlejaceae. 15 trop. (from Guinea and Sierra Leone to Ethiopia, Kenya, Tanzania and southward to Angola, Zimbabwe and Mozambique) and South Africa (Transvaal, Natal, northeastern Cape), Madagascar, Comoro and Mascarene Is., southern Arabian Peninsula
Nuytsia R. Br. ex G. Don. 1831. Loranthaceae. 1 southwestern Australia
Nyctaginia Choisy. 1849. Nyctaginaceae. 1 southern U.S., Mexico
Nyctanthes L. 1753. Oleaceae. 1–2 India, Sri Lanka, Thailand, Sumatra, Java
Nycticalanthus Ducke. 1932. Rutaceae. 1 Amazonia
Nyctocalos Teijsm. et Binn. 1861. Bignoniaceae. 5 northeastern India, southwestern China, Indochina, western Malesia
Nyctocereus (A. Berger) Britton et Rose = Peniocereus (A. Berger) Britton et Rose. Cactaceae
Nylandtia Dumort. 1822. Polygalaceae. 1 South Africa
Nymania Lindb. 1868. Meliaceae. 1 South Africa
Nymphaea L. 1753. Nymphaeaceae. 35 cosmopolitan in forest and tundra South America and Cape
Nymphoides Hill. 1753. Menyanthaceae. 21 cosmopolitan
Nypa Steck. 1757. Arecaceae. 1 Sri Lanka, the Ganges Delta, Malay Peninsula eastward to Ryukyu and Solomon Is., northeastern Australia
Nyssa L. 1753. Nyssaceae. 6 northeastern India, China, western Malesia, southeastern North (3) and Central (1, Costa Rica, Panama) America
Nyssanthes R. Br. 1810. Amaranthaceae. 2 eastern Australia
Nyssopsis Kuntze = Camptotheca Decne. Nyssaceae

Oakes-Amesia C. Schweinf. et P. H. Allen. 1948 (Sphyrastylis Schltr.). Orchidaceae. 1 Panama
Oakesiella Small = Uvularia L. Melanthiaceae
Oaxacania B. L. Rob. et Greenm. 1895 (Hofmeisteria Walp.). Asteraceae. 1 Mexico
Obbea Hook. f. = Bobea Gaudich. Rubiaceae
Oberna Adans. = Silene L. Caryophyllaceae
Oberonia Lindl. 1830. Orchidaceae. 100–160 trop. Old World from Africa amd western Indian Ocean islands through trop. Asia to Japan, Malesia, trop. Australia, Pacific Islands to Tahiti and Samoa Is.
Obetia Gaudich. 1844. Urticaceae. 8 trop. and South Africa, Aldabra I., Madagascar, Mascarene Is.
Obione Gaertn. = Atriplex L. Chenopodiaceae
Oblivia Strother. 1989. Asteraceae. 1 Panama, Colombia, Venezuela, Ecuador, Peru, Bolivia
Obolaria L. 1753. Gentianaceae. 1 eastern North America
Obolinga Barneby. 1989. Fabaceae. 1 Haiti
Obregonia Fric. 1928. Cactaceae. 1 northeastern Mexico

Oceanopapaver Guillaumin. 1932. Capparaceae. 1 New Caledonia

Ochagavia Phil. 1856. Bromeliaceae. 3 Juan Fernández Is. (O. elegans), northern and central Chile

Ochanostachys Mast. 1875. Olacaceae. 1 Malay Peninsula, Sumatra, Banka I., Borneo

Ochlandra Thwaites. 1864. Poaceae. 12 Madagascar, South Asia

Ochna L. 1753. Ochnaceae. 86 trop. and South (12) Africa, Madagascar, Mascarene Is., Yemen, trop. Asia from India and Sri Lanka (3) through Nicobar Is. and Indochina to Hainan, Malesia (1, O. integerrima—Malay Peninsula)

Ochoterenaea F. A. Barkley. 1942. Anacardiaceae. 1 Colombia

Ochotia A. P. Khokhr. = Magadanis Pimenov et Lavrova. Apiaceae

Ochotonophila Gilli. 1956. Caryophyllaceae. 1 Afghanistan

Ochradenus Delile. 1813. Resedaceae. 6 Africa (Libya, Egypt, Ethiopia, Somalia, Sudan, Chad), Israel, Sinai and Arabian peninsulas, Socotra, Jordan, Iran to Pakistan, southern Turkmenistan (O. ochradeni)

Ochreata (Lojac.) Bobrov = Trifolium L. Fabaceae

Ochreinauclea Ridsdale et Bakh. f. 1978. Rubiaceae. 2 India, Thailand, Malay Peninsula, Sumatra, Borneo

Ochrocarpus A. Juss. = Mammea L. Clusiaceae

Ochrocephala Dittrich. 1983. Asteraceae. 1 Sudan, Ethiopia

Ochroma Sw. 1788. Bombacaceae. 1 trop. America from southern Mexico to Bolivia

Ochrosia Juss. 1789. Apocynaceae. 30 Réunion I. (1), Sri Lanka, Malesia, northern Australia, Ogasawara (Bonin) Is., Mariana Is., Nauru I., New Caledonia, Vanuatu, Hawaiian Is., Fiji, Tonga, Marquesas Is.

Ochrosperma Trudgen. 1987. Myrtaceae. 3 Australia (Queensland and New South Wales)

Ochrothallus Pierre ex Baill. = Niemeyera F. Muell. Sapotaceae

Ochrus Hill = Lathyrus L. Fabaceae

Ochthephilus Wurdack. 1972. Melastomataceae. 1 Guyana

Ochthocharis Blume. 1831. Melastomataceae. 7 trop. Africa (2), trop. Asia

Ochthochloa Edgew. 1842. Poaceae. 1 Northeast Africa, Southwest and South Asia

Ochthocosmus Benth. 1843. Ixonanthaceae. 14 trop. South America (6—southeastern Colombia, Venezuela, Guyana, Amazonian Brazil), trop. Africa (8)

Ochthodium DC. 1821. Brassicaceae. 1 Asia Minor, western Syria, Israel

Ocimum L. 1753. Lamiaceae. c. 150 (or 6 ?!) warm–temp. and trop. regions, esp. Africa; 1 (O. basilicum) Ukraine, Crimea, southern European part of Russia, Caucasus, Middle Asia, Far East of Russia

Oclorosis Raf. = Iodanthus Torr. et A. Gray. Brassicaceae

Ocotea Aubl. 1775. Lauraceae. c. 400 trop. and subtrop. America (c. 300, from Mexico and southern Florida to Argentina), trop. and South Africa (6), Madagascar (50), Mascarene Is.

Octamyrtus Diels. 1922. Myrtaceae. 3 Moluccas, New Guinea

Octarrhena Thwaites. 1861. Orchidaceae. 20–30 Sri Lanka, Malesia from Malay Peninsula to New Guinea, and eastward to New Caledonia, Vanuatu, and Fiji

Octoceras Bunge. 1848. Brassicaceae. 1 Middle Asia, Iran, Afghanistan, western Pakistan

Octoknema Pierre. 1897. Octoknemaceae. 6 trop. Africa

Octolepis Oliv. 1865. Thymelaeaceae. 6 trop. Africa

Octolobus Welw. 1869. Sterculiaceae. 5 trop. Africa

Octomeles Miq. 1861. Datiscaceae. 1 Malesia (excl. Malay Peninsula, Java, and Lesser Sunda Is.), New Britain I., Solomon Is.

Octomeria R. Br. 1813. Orchidaceae. c. 130 trop. America from Honduras and

Costa Rica to Peru, Bolivia, and Uruguay; West Indies

Octomeron Robyns. 1943. Lamiaceae. 1 trop. Africa

Octopoma N. E. Br. 1930. Aizoaceae. 8 western and southern South Africa

Octospermum Airy Shaw. 1965. Euphorbiaceae. 1 New Guinea

Octotheca R. Vig. = Schefflera J. R. Forst. et G. Forst. Araliaceae

Octotropis Bedd. 1873. Rubiaceae. 2 southern and northeasterm India, Burma

Ocyroe Phil. = Nardophyllum (Hook. et Arn.) Hook. et Arn. Asteraceae

Oddoniodendron De Wild. 1925. Fabaceae. 3 Cameroun, Gabon, Zaire, Angola

Odicardis Raf. = Veronica L. Scrophulariaceae

Odixia Orchard. 1982 (Ixodia R. Br.). Asteraceae. 2 Tasmania

Odoniellia K. R. Robertson. 1982 (Jacquemontia Choisy). Convolvulaceae. 2 southern Mexico, Guatemala, Belize, Costa Rica, Panama, Colombia, Venezuela, Ecuador, Peru, and Brazil (1, O. eriocephala)

Odontadenia Benth. 1841. Apocynaceae. 31 Central and trop. South America, West Indies

Odontanthera Wight. 1838. Asclepiadaceae. 1 (O. reniformis) eastern Mediterranean, Northeast Africa, Arabian Peninsula

Odontella Tiegh. 1895. Loranthaceae. 20 arid trop. and South (1, O. welwitschii) Africa

Odontelytrum Hack. 1898. Poaceae. 1 East Africa, southern Arabian Peninsula

Odontitella Rothm. = Odontites Ludw. Scrophulariaceae

Odontites Ludw. 1757. Scrophulariaceae. c. 30 West and South Europe, Mediterranean, West Asia eastward to northeastern China; Russia (5–7; O. vulgaris—from European part to East Siberia and Far East)

Odontocarya Miers. 1851. Menispermaceae. 25–30 trop. and subtrop. South America, 4 spp. extending to Central America and Lesser Antilles

Odontochilus Blume. 1858 (or = Anoectochilus Blume). Orchidaceae. 21 China, Indochina, Malesia, Melanesia to Fiji

Odontocline B. Nord. 1978 (Senecio L.). Asteraceae. 6 Jamaica

Odontoglossum Kunth. 1816. Orchidaceae. c. 250 trop. America from Mexico to Bolivia and Brazil, Jamaica

Odontonema Nees. 1842. Acanthaceae. c. 35 southern U.S., Mexico, Central and trop. South America, West Indies

Odontophorus N. E. Br. 1927. Aizoaceae. 3 western South Africa

Odontophyllum Sreem. = Steemadhavana Rauschert. Acanthaceae

Odontorrhynchus M. N. Corrêa. 1953. Orchidaceae. 5 South America

Odontospermum Neck. ex Sch. Bip. = Nauplius (Cass.) Cass. Asteraceae

Odontostelma Rendle. 1894. Asclepiadaceae. 1 southern trop. Africa

Odontostomum Torr. 1856. Tecophilaeaceae. 1 California

Odontotrichum Zucc. = Psacalium Cass. Asteraceae

Odosicyos Keraudren. 1980 (Ampelosicyos Thouars). Cucurbitaceae. 1 Madagascar

Odyendea Pierre ex Endl. = Quassia L. Simaroubaceae

Odyssea Stapf. 1922. Poaceae. 2 trop. and South Africa, Ethiopia, southwestern Arabian Peninsula

Oeceoclades Lindl. 1832 (Angraecum Bory). Orchidaceae. 33 Florida Peninsula, Bahama Is., West Indies, trop. South America, Africa, Madagascar, Seychelles, Comoro and Mascarene Is., Southeast Asia, Malesia, Australia, Mariana Is., Solomon Is., Vanuatu, Fiji, Tonga, Niue Is.

Oecopetalum Greenm. et C. H. Thomps. 1915. Icacinaceae. 3 Mexico, Central America

Oedematopus Planch. et Triana. 1860. Clusiaceae. 10 trop. America

Oedera L. 1771. Asteraceae. 6 South Africa

Oedibasis Kozo-Polj. 1916. Apiaceae. 4 Middle Asia (3), Afghanistan

Oedina Tiegh. = Dendrophthoe Mart. Loranthaceae

Oemleria Rchb. 1841. Rosaceae. 1 western North America

Oenanthe L. 1753. Apiaceae. c. 40 North America, Eurasia (incl. Russia, Caucasus and Middle Asia), Northwest, Northeast and trop. Africa, Malesia, Australia

Oenocarpus Mart. 1823. Arecaceae. 8 Costa Rica, Panama, Colombia, Ecuador, Peru, Venezuela, Guyana, Surinam, French Guiana, Bolivia, Brazil

Oenosciadium Pomel = Oenanthe L. Apiaceae

Oenostachys Bullock = Gladiolus L. Iridaceae

Oenothera L. 1753. Onagraceae. 125 temp. and subtrop. North and South America

Oenotheridium Reiche = Clarkia Pursh. Onagraceae

Oeonia Lindl. 1824. Orchidaceae. 4 Madagascar, Mascarene Is. (Réunion I.)

Oeoniella Schltr. 1918. Orchidaceae. 3 Mascarene Is.

Oerstedella Rchb. f. = Epidendrum L. Orchidaceae

Oerstedianthus Lundell = Ardisia Sw. Myrsinaceae

Oerstedina Wiehler. 1977. Gesneriaceae. 3 Mexico, Costa Rica, Panama

Ofaiston Raf. 1837. Chenopodiaceae. 1 southeastern European part of Russia, West Siberia, Kazakhstan, Uzbekistan

Oftia Adans. 1763. Oftiaceae. 3 South Africa

Ogastemma Brummitt. 1982. Boraginaceae. 1 Canary Is., North Africa, Sinai Peninsula, Israel, Jordan

Ogcodeia Bureau = Naucleopsis Miq. Moraceae

Oianthus Benth. = Heterostemma Wight et Arn. Asclepiadaceae

Oiospermum Less. 1829. Asteraceae. 1 northeastern Brazil

Oistanthera Markgr. = Tabernaemontana L. Apocynaceae

Oistonema Schltr. = Dischidia R. Br. Asclepiadaceae

Okenia Schltdl. et Cham. 1830. Nyctaginaceae. 2 southeastern U.S., Mexico, Central America south to Nicaragua

Okoubaka Pellegr. et Normand. 1946. Santalaceae. 2 trop. Africa

Olax L. 1753. Olacaceae. 65 trop. and South Africa, Madagascar, India, Sri Lanka, Andaman and Nicobar Is., Himalayas, Burma, southwestern and southern China (incl. Hainan), Taiwan, Indochina, Malesia, Australia, Micronesia, Melanesia, Polynesia

Oldenburgia Less. 1830. Asteraceae. 4 South Africa

Oldenlandia L. 1753. Rubiaceae. c. 300 tropics, esp. Africa

Oldenlandiopsis Terrell et W. H. Lewis. 1990 (Oldenlandia L.). Rubiaceae. 1 U.S. (Florida), Mexico (Yucatan), Nicaragua, Panama, West Indies (Cuba, Cayman Is., Haiti, Guadeloupe, Martinique, Trinidad, Puerto Rico, Virgin Is., Bahama Is.), Guyana

Oldfieldia Benth. et Hook. 1850. Euphorbiaceae. 4 Africa from Sierra Leone to Somalia, from Angola and Zambia to Tanzania

Olea L. 1753. Oleaceae. 20–30 Mediterranean, Africa, Madagascar, trop. and subtrop. Asia, Malesia, eastern Australia, New Caledonia; O. europaea—also possibly Crimea, Caucasus, Middle Asia

Olearia Moench. 1802. Asteraceae. c. 140 New Guinea, Australia, New Zealand

Oleiocarpon Dwyer = Dipteryx Schreb. Fabaceae

Olgaea Iljin. 1922. Asteraceae. 12 Middle Asia (9), northern China (6)

Oligactis (Kunth) Cass. 1825. Asteraceae. 12 Andes from Costa Rica to Peru

Oligandra Less. = Lucilia Cass. Asteraceae

Oliganthemum F. Muell. = Allopterigeron Dunlop. Asteraceae

Oliganthes Cass. 1817. Asteraceae. 9 Madagascar

Oligarrhena R. Br. 1810. Epacridaceae. 1 southwestern Australia

Oligobotrya Baker = Smilacina Desf. Convallariaceae

Oligocarpus Less. = Osteospermum L. Asteraceae

Oligoceras Gagnep. 1925. Euphorbiaceae. 1 Southeast Asia

Oligochaeta (DC.) K. Koch. 1843. Asteraceae. 4 Caucasus, Middle Asia, eastern Turkey, Iran, China (1), India

Oligocladus Chodat et Wilczek. 1902. Apiaceae. 2 Argentina

Oligocodon Keay. 1958. Rubiaceae. 1 trop. West Africa

Oligomeris Cambess. 1844. Resedaceae. 10 Macaronesia (Canary Is.), North (Morocco, Algeria, Central Sahara, Egypt), trop. (the Sudan) and South (3) Africa; Syria; Israel; Jordan; Arabian Peninsula; Iraq; Iran; Afghanistan; Pakistan; India (Punjab); O. linifolia—also in arid areas of the southwestern U.S. and Mexico

Oligophyton Linder. 1986. Orchidaceae. 1 Zimbabwe

Oligospermum D. Y. Hong = Veronica L. Scrophulariaceae

Oligostachyum Z. P. Wang et G. H. Ye = Arundinaria Michx. Poaceae

Oligothrix DC. 1838. Asteraceae. 1 (O. gracilis) South Africa

Olinia Thunb. 1799. Oliniaceae. 8 St. Helena (1, O. ventosa), trop. (from eastern Zaire to Ethiopia and south to southern Angola and Tanzania) and South Africa

Olivaea Sch. Bip. ex Benth. 1872. Asteraceae. 2 Mexico

Oliveranthus Rose = Echeveria DC. Crassulaceae

Oliverella Tiegh. = Tapinanthus (Blume) Rchb. Loranthaceae

Oliveria Vent. 1801. Apiaceae. 1 Turkey, Syria, Iraq, Iran

Oliveriana Rchb. f. 1876. Orchidaceae. 1 Colombia

Olmeca Soderstr. 1982. Poaceae. 2 southern Mexico

Olmedia Ruiz et Pav. = Trophis P. Browne. Moraceae

Olmediella Baill. 1880. Flacourtiaceae. 1 southern Mexico, Guatemala, Honduras, Nicaragua, El Salvador

Olmedioperebea Ducke = Maquira Aubl. Moraceae

Olmediophaena Karst. = Maquira Aubl. Moraceae

Olneya A. Gray. 1855. Fabaceae. 1 southwestern U.S., northwestern Mexico

Olsynium Raf. 1837. Iridaceae. 11 North and South (Andes) America

Olymposciadium H. Wolff. 1922 (Aegokeras Raf.). Apiaceae. 1 Turkey

Olympusa Klotzsch. 1848. Asclepiadaceae. 1 Guyana

Olyra L. 1759. Poaceae. 25 trop. America, trop. Africa, Madagascar

Omalanthus A. Juss. 1824. Euphorbiaceae. c. 40 Southeast Asia, Malesia, Australia, Kermadec Is., New Caledonia (3), Fiji, Tonga, Samoa

Omalocarpus Choux. 1926 (Deinbollia Schumach. et Thonn.). Sapindaceae. 1 Madagascar

Omalotheca Cass. 1828 (Gnaphalium L.). Asteraceae. 8 eastern North America, Eurasia

Omania S. Moore. 1901. Scrophulariaceae. 1 Arabian Peninsula

Ombrocharis Hand.-Mazz. 1936. Lamiaceae. 1 China (western Hunan)

Ombrophytum Poepp. ex Endl. 1836. Lophophytaceae. 4 Galapagos Is. (Santa Cruz I.), Peru, western Brazil, northern Argentina

Omeiocalamus Keng f. = Arundinaria Michx. Poaceae

Omiltemia Standl. 1918. Rubiaceae. 4 Mexico

Omoea Blume. 1825. Orchidaceae. 2 Java (O. micranta), Philippines (O. philippinensis)

Omphacomeria (Engl.) A. DC. 1857. Santalaceae. 1 southeastern Australia

Omphalea L. 1759. Euphorbiaceae. 20 pantropics

Omphalocarpum P. Beauv. ex Vent. 1800. Sapotaceae. 6 trop. West and Central Africa

Omphalodes Hill. 1753. Boraginaceae. 28 temp. Eurasia, incl. European part of Russia (7) and Caucasus; Mexico

Omphalogramma (Franch.) Franch. 1898. Primulaceae. 13–15 western China (10),

Himalayas from Nepal to Bhutan, northeastern India

Omphalolappula Brand. 1931. Boraginaceae. 1 temp. Australia

Omphalopappus O. Hoffm. 1891. Asteraceae. 1 Angola

Omphalophthalmum Karst. = Matelea Aubl. Asclepiadaceae

Omphalopus Naudin. 1851. Melastomataceae. 1 Sumatra, Java, New Guinea

Omphalothrix Maxim. 1859. Scrophulariaceae. 1 Russian Far East, China (Manchuria), Korea

Omphalotrigonotis W. T. Wang. 1984 (Trigonotis W. T. Wang). Boraginaceae. 1 China

Ona Ravenna = Olsynium Raf. Iridaceae

Onagra Hill = Oenothera L. Onagraceae

Oncella Tiegh. 1895. Loranthaceae. 4 trop. Africa

Oncidium Sw. 1800. Orchidaceae. c. 350 trop. America from Florida and Mexico to Peru, Bolivia, Brazil, West Indies

Oncinema Arn. 1834. Asclepiadaceae. 1 South Africa

Oncinocalyx F. Muell. 1883. Verbenaceae. 1 eastern Australia

Oncinotis Benth. 1849. Apocynaceae. 7 trop. and South Africa (6), Madagascar (1)

Oncoba Forssk. 1775. Flacourtiaceae. 4–5 trop. and South (1) Africa, Arabian Peninsula

Oncocalamus (G. Mann et H. A. Wendl.) G. Mann et H. A. Wendl. ex Hook. f. 1883. Arecaceae. 2–5 equatorial West Africa and Congo Basin

Oncocalyx Tiegh. 1985. Loranthaceae. 4 trop. Africa

Oncocarpus A. Gray = Semecarpus L. f. Anacardiaceae

Oncodostigma Diels. 1912. Annonaceae. 4 southern China, western Malesia, New Guinea, Vanuatu

Oncosiphon Källersjö. 1988. Asteraceae. 8 South Africa

Oncosperma Blume. 1838. Arecaceae. 5 Sri Lanka (1), Indochina, western Malesia eastward to Philippines, Sulawesi and western Moluccas

Oncostemma K. Schum. 1893. Asclepiadaceae. 1 Sao Tomé I. (Gulf of Guinea)

Oncostemum A. Juss. 1830. Myrsinaceae. c. 100 Madagascar, Mascarene Is.

Oncostylus (Schltdl.) Bolle. 1933 (Geum L.). Rosaceae. 9 Tasmania, New Zealand, Snares Is., temp. South America

Oncotheca Baill. 1891. Oncothecaceae. 2 New Caledonia

Ondetia Benth. 1872. Asteraceae. 1 Namibia

Ondinea Hartog. 1970. Nymphaeaceae. 1 northwestern Australia

Ongokea Pierre. 1897. Olacaceae. 2 trop. West Africa

Onira Ravenna. 1983. Iridaceae. 1 Brazil, Uruguay

Onixotis Raf. 1837. Melanthiaceae. 2 South Africa

Onobrychis Hill. 1754. Fabaceae. c. 130 Eurasia, esp. (c. 75) Russia, Caucasus, Middle Asia, southern Siberia; Mediterranean, Sudan, Ethiopia, Somalia, Djibouti

Onochualcoa Lundell = Mansoa DC. Bignoniaceae

Ononis L. 1753. Fabaceae. c. 75 Europe (c. 50), Canary Is., Mediterranean (mainly), Ethiopia and Kenya (1, O. reclinata), Djibouti (1), Arabian Peninsula, Iraq, Iran, Afghanistan, Pakistan; 6 Russia (6, European part, except Arctica, southern Siberia), Caucasus, Middle Asia

Onopordum L. 1753. Asteraceae. 40 Europe, North Africa, West (incl. Caucasus) and Middle Asia; O. acanthium—from European part of Russia to East Siberia; China (2)

Onoseris Willd. 1803. Asteraceae. 29 Mexico and Andes from Central America to Argentina

Onosma L. 1762. Boraginaceae. c. 150 Mediterranean, Southeast Europe, West (incl. Caucasus) and Middle Asia eastward to Mongolia, China, and Himalayas

Onosmodium Michx. 1803. Boraginaceae. 15 North America

Onuris Phil. 1872. Brassicaceae. 5 Chile, Argentina

Onus Gilli = Mellera S. Moore. Acanthaceae

Onychopetalum R. E. Fr. 1931. Annonaceae. 4 Brazil

Onychosepalum Steud. 1855. Restionaceae. 2 southwestern Australia

Oonopsis Greene. 1896. Asteraceae. 3 North America

Oophytum N. E. Br. 1925 (or = Conophytum N. E. Br.). Aizoaceae. 2 western South Africa

Oosterdyckia Boehm. = Cunonia L. Cunoniaceae

Oparanthus Sherff. 1937 (Bidens L.). Asteraceae. 2 Polynesia (Rapa I.)

Opercularia Gaertn. 1788. Rubiaceae. 15 Australia

Operculicarya H. Perrier. 1944. Anacardiaceae. 3 Madagascar

Operculina Silva Manso. 1836 (Merremia Dennst. ex Endl.). Convolvulaceae. 15–20 pantropics

Ophelia D. Don. 1837 (Swertia L.). Gentianaceae. 5 West and East Siberia, Far East

Ophellantha Standl. 1924. Euphorbiaceae. 2 Central America

Ophidion Luer. 1982. Orchidaceae. 4 trop. America

Ophiobortys Gilg. 1908 (Osmelia Thwaites). Flacourtiaceae. 1 trop. West Africa

Ophiocaryon Endl. 1841. Sabiaceae. 7 trop. South America

Ophiocephalus Wiggins. 1933. Scrophulariaceae. 1 southern California

Ophiocolea H. Perrier. 1938. Bignoniaceae. 5 Comoro Is., Madagascar

Ophiomeris Miers. 1847 (Thismia Griff.). Burmanniaceae. 10 trop. South America

Ophionella Bruyns. 1981 (Pectinaria Haw.). Asclepiadaceae. 1 South Africa (Cape)

Ophiopogon Ker-Gawl. 1807. Convallariaceae. c. 50 Himalayas, India, continental China (c. 40), Hainan, Japan, northern Vietnam, Thailand, Malesia

Ophiorrhiza L. 1753. Rubiaceae. c. 150 India, Sri Lanka, China, Southeast Asia, Malesia, Micronesia, Melanesia, Polynesia to Society Is.

Ophiorrhiziphyllon Kurz. 1871. Acanthaceae. 5 China, Southeast Asia

Ophiuros C. F. Gaertn. 1805. Poaceae. 4 Northeast and trop. Africa, South Asia, Malesia, Australia

Ophrestia H. M. L. Forbes. 1948. Fabaceae. 13 trop. Africa (8) and trop. Asia

Ophryococcus Oerst. 1852. Rubiaceae. 1 Central America

Ophryosporus Meyen. 1834. Asteraceae. 37 trop. and subtrop. South America

Ophrypetalum Diels. 1936. Annonaceae. 1 trop. East Africa

Ophrys L. 1753. Orchidaceae. 35 Europe, Mediterranean, North Africa, West (incl. Caucasus) and Middle Asia

Ophthalmoblapton Allemao. 1849. Euphorbiaceae. 3 Brazil

Ophthalmophyllum Dinter et Schwantes. 1927 (or = Conophytum N. E. Br.). Aizoaceae. 15 southern Namibia, western South Africa

Opilia Roxb. 1802. Opiliaceae. 2 trop. Africa; O. amentacea also southern and eastern India, Sri Lanka, Burma, southern China, Indochina, Malesia (excl. Sumatra and Malay Peninsula) eastward to New Britain, Solomon Is., southward to northern Australia

Opisthiolepis L. S. Sm. 1952. Proteaceae. 1 Australia (Queensland)

Opisthocentra Hook. f. 1867. Melastomataceae. 1 Colombia, Venezuela, northern Brazil

Opisthopappus C. Shih. 1979. Asteraceae. 2 China

Opithandra B. L. Burtt. 1956. Gesneriaceae. 9 East Asia

Opizia C. Presl. 1830. Poaceae. 1 southern Mexico, Central America, West Indies

Oplismenopsis Parodi. 1937. Poaceae. 1 Uruguay, Argentina

Oplismenus P. Beauv. 1810. Poaceae. 10

tropics and subtropics; 3 spp. extending to Caucasus

Oplonia Raf. 1838. Acanthaceae. 19 West Indies, Peru (1), Madagascar (5)

Oplopanax (Torr. et A. Gray) Miq. 1863. Araliaceae. 3 China, Korea, Japan, northwestern North America

Opoidia Lindl. 1839 (Peucedanum L.). Apiaceae. 1 Iran

Opopanax W. D. J. Koch. 1824. Apiaceae. 3 West and South Europe, West Asia, incl. Caucasus (2)

Opophytum N. E. Br. 1925 (Mesembryanthemum L.). Aizoaceae. 5 Morocco, Algeria, trop. and South Africa, Arabian Peninsula

Opsiandra O. F. Cook = Gaussia H. A. Wendl. Arecaceae

Opuntia Hill. 1753. Cactaceae. c. 200 America from British Columbia and Massachusetts to Strait of Magellan, Galapagos Is.

Orania Zipp. 1829. Arecaceae. 17 Madagascar (1), southern Thailand, Malay Peninsula, Sumatra, Java, Borneo, Philippines, Sulawesi, Moluccas, New Guinea

Oraniopsis (Becc.) J. Dransf., A. K. Irvine et N. Uhl. 1985. Arecaceae. 1 Australia (Queensland)

Orbea Haw. 1812 (Stapelia L.). Asclepiadaceae. 10 trop. and South Africa

Orbeanthus L. C. Leach. 1978. Asclepiadaceae. 2 South Africa

Orbeopsis L. C. Leach. 1978 (Caralluma R. Br.). Asclepiadaceae. 10 trop. and South Africa

Orbexilum Raf. 1832. Fabaceae. 8 U.S. from Virginia to Florida west into the Great Plains of Kansas, Oklahoma, and Texas south in mountainous Mexico to Chiapas

Orbignya Mart. ex Endl. 1837. Arecaceae. 12 trop. America from Mexico to Bolivia and Brazil

Orchadocarpa Ridl. 1905. Gesneriaceae. 1 Malay Peninsula

Orchidantha N. E. Br. 1886. Lowiaceae. 9 southern China (incl. Hainan), Vietnam, Malay Peninsula, Borneo, Oceania

Orchipedum Breda. 1829. Orchidaceae. 1 Malay Peninsula, Java, Philippines

Orchis L. 1753. Orchidaceae. c. 80 temp. Northern Hemisphere, Mediterranean, Tibet, Himalayas

Orcuttia Vasey. 1886. Poaceae. 5 California and western Mexico

Oreacanthus Benth. 1876. Acanthaceae. 4 Cameroun, Sudan, Tanzania, Zambia

Oreanthes Benth. 1844. Ericaceae. 4 Andes of Ecuador and Peru

Orectanthe Maguire et Wurdack. 1958. Xyridaceae. 1 Venezuela

Oregandra Standl. 1929. Rubiaceae. 1 Panama

Oreiostachys Gamble. 1908 (Nastus Juss.). Poaceae. 5 Southeast Asia

Oreithales Schltdl. 1856. Ranunculaceae. 1 Andes of Ecuador, Peru, and Bolivia

Oreobambos K. Schum. 1896. Poaceae. 1 trop. East Africa

Oreoblastus Susl. 1972 (or = Christolea Cambess. ex Jacquem.). Brassicaceae. 9 Middle Asia (5, Kirgizia, Tajikistan), Pakistan, India, western China, Tibet

Oreobliton Durieu et Moq. 1847. Chenopodiaceae. 1 Algeria, Tunisia

Oreobolopsis Koyama et Guagl. 1987. Cyperaceae. 1 Bolivia

Oreobolus R. Br. 1810. Cyperaceae. 14 Malesia, Australia, Tasmania, New Zealand, Stewart I., Campbell and Auckland Is., Hawaiian Is., Juan Fernández Is., trop. and temp. (Chile, Tierra del Fuego) America, Falkland Is.

Oreocalamus Keng = Chimonobambusa Makino. Poaceae

Oreocallis R. Br. 1810. Proteaceae. 2 Ecuador, Peru

Oreocarya Greene = Cryptantha Lehm. ex G. Don. Boraginaceae

Oreocereus (A. Berger) Riccob. 1909. Cactaceae. 6–7 Andes of southern Peru, southern Bolivia, northern Chile, and northern Argentina

Oreocharis Benth. 1876. Gesneriaceae. 27 southern China (26) (incl. Hainan), Thailand, Vietnam

Oreochloa Link. 1827 (Sesleria Scop.). Poaceae. 4 mountainous Europe eastward to Carpathians (1, O. disticha)

Oreochorte Kozo-Polj. = Anthriscus Pers. Apiaceae

Oreochrysum Rudb. 1906. Asteraceae. 1 U.S., Mexico

Oreocnide Miq. 1851. Urticaceae. 20 Himalayas from Punjab to Bhutan, India, Burma, western and southern China, southern Japan, Indochina, Malesia

Oreocome Edgew. 1845. Apiaceae. 1 Himalayas

Oreodendron C. T. White. 1933. Thymelaeaceae. 1 Australia (Queensland)

Oreodoxa auct., non Willd. = Roystonea O. F. Cook. Arecaceae

Oreodoxa Willd. = Prestoea Hook. f. Arecaceae

Oreograstis K. Schum. 1895 (Carpha Banks et Sol. ex R. Br.). Cyperaceae. 1 trop. East Africa

Oreoherzogia W. Vent. = Rhamnus L. Rhamnaceae

Oreoleysera Bremer. 1978. Asteraceae. 1 South Africa (Cape)

Oreoloma Botsch. 1980. Brassicaceae. 4 northwestern China, Mongolia

Oreomitra Diels. 1912. Annonaceae. 1 New Guinea

Oreomunnea Oerst. 1856. Juglandaceae. 2 Mexico, Central America

Oreomyrrhis Endl. 1839. Apiaceae. 25 Taiwan, Borneo, Mindoro I., New Guinea, southeastern Australia, New Zealand, Mexico, Central America, Andes, Tierra del Fuego, Falkland Is.

Oreonana Jeps. 1925. Apiaceae. 3 California

Oreonesion A. Raynal. 1965. Gentianaceae. 1 trop. West Africa

Oreopanax Decne. et Planch. 1854. Araliaceae. c. 120 trop. America, West Indies

Oreophysa (Bunge ex Boiss.) Bornm. 1905. Fabaceae. 1 northern Iran

Oreophyton O. E. Schulz. 1924. Brassicaceae. 1 mountainous Northeast and East Africa

Oreopolus Schltdl. 1857 (Cruckshanksia Hook. et Arn.). Rubiaceae. 2 Andes, Patagonia

Oreoporanthera Hutch. 1969. Euphorbiaceae. 1 New Zealand

Oreorchis Lindl. 1858. Orchidaceae. 14 Himalayas, Tibet, northeastern India, Burma, China, Japan; Far East of Russia (1, O. patens)

Oreoschimperella Rauschert. 1982. Apiaceae. 3 Ethiopia, Kenya, Yemen

Oreosedum Grulich = Sedum L. Crassulaceae

Oreoselinum Hill. 1753. Apiaceae. 1 Europe

Oreosolen Hook. f. 1884. Scrophulariaceae. 3 Tibet, Nepal, Bhutan

Oreosparte Schltr. 1916. Asclepiadaceae. 1 Sulawesi

Oreosphacus Leyb. 1873. Lamiaceae. 1 Chile

Oreostemma Greene. 1900 (Aster L.). Asteraceae. 3 western U.S. (Oregon, Nevada, California)

Oreostylidium Berggr. 1878. Stylidaceae. 1 New Zealand

Oreosyce Hook. f. 1871. Cucurbitaceae. 2 trop. Africa, Madagascar

Oreothyrsus Lindau = Ptyssiglottis T. Anderson. Acanthaceae

Oreoxis Raf. 1830. Apiaceae. 4 southwestern U.S.

Oresitrophe Bunge. 1837. Saxifragaceae. 1 northeastern China

Orestias Ridl. 1887. Orchidaceae. 3 trop. Africa

Orias Dode = Lagerstroemia L. Lythraceae

Oricia Pierre. 1897. Rutaceae. 8 trop. and South Africa

Oriciopsis Engl. 1931. Rutaceae. 1 trop. West Africa

Origanum L. 1753. Lamiaceae. 42 Europe (13), Mediterranean, West (incl. Caucasus) and Middle Asia, West Siberia; O. vulgare—extending to Mongolia, China, and Himalayas

Orinus Hitchc. 1933. Poaceae. 3 western China, Himalayas

Orites R. Br. 1810. Proteaceae. 9 temp. eastern Australia (6), Andes (3)

Oritrephes Ridl. 1908. Melastomataceae. 6 Burma, Malay Peninsula

Oritrophium (Kunth) Cuatrec. 1961. Asteraceae. 15 Andes from Colombia and Venezuela to Peru and Bolivia

Orixa Thunb. 1783. Rutaceae. 1 China, Japan

Orlaya Hoffm. 1814. Apiaceae. 3 Central and Southeast Europe, Mediterranean, West (incl. Caucasus) and Middle Asia

Orleanesia Barb. Rodr. 1877. Orchidaceae. 3 Venezuela, Brazil

Ormenis (Cass.) Cass. = Chamaemelum Hill. Asteraceae

Ormocarpopsis R. Vig. 1951. Fabaceae. 5 Madagascar

Ormocarpum P. Beauv. 1810. Fabaceae. 22 trop. Africa (17), Madagascar, Arabian Peninsula, South and Southeast Asia, Malesia, Caroline Is., northern Australia, Fiji, ? trop. America

Ormopterum Schischk. 1950. Apiaceae. 2 Middle Asia (1, O. turkomanicum), Afghanistan, Pakistan

Ormosciadium Boiss. 1844. Apiaceae. 1 Asia Minor, Iran

Ormosia G. Jacks. 1811. Fabaceae. c. 100 South, East, and Southeast Asia; Malesia to northeastern Australia; trop. America from Mexico to southern Brazil; West Indies

Ormosiopsis Ducke = Ormosia G. Jacks. Fabaceae

Ormosolenia Tausch. 1834. Apiaceae. 1 Crete, Southwest Asia

Ornichia Klack. 1986. Gentianaceae. 3 Madagascar

Ornithidium R. Br. = Maxillaria Ruiz et Pav. Orchidaceae

Ornithoboea Parish ex C. B. Clarke. 1883. Gesneriaceae. 11 India, China, Indochina, Malesia

Ornithocarpa Rose. 1905. Brassicaceae. 2 Mexico

Ornithocephalus Hook. 1875. Orchidaceae. 50 trop. America from Mexico to Brazil

Ornithochilus (Lindl.) Wall. ex Benth. 1883. Orchidaceae. 3 Himalayas, north-eastern India, Burma, China (Yunnan, Kwangtung, Kwangsi, Guangxi, and Sichuan), Vietnam, Thailand, Malay Peninsula, Sumatra, Borneo

Ornithogalum L. 1753. Hyacinthaceae. c. 120 Europe (34), Africa (esp. South—54 spp.), West and Southwest Asia eastward to Afghanistan, Caucasus, and Middle Asia

Ornithoglossum Salisb. 1806. Colchicaceae. 8 trop., South-West and South Africa

Ornithophora Barb. Rodr. 1882 (Sigmatostalix Rchb. f.). Orchidaceae. 1 Brazil

Ornithopodium Hill = Ornithopus L. Fabaceae

Ornithopus L. 1753. Fabaceae. 6–15 islands of the eastern Atlantic Ocean, Europe, Mediterranean, Caucasus, Iran, ? temp. South America (1)

Ornithostaphylos Small. 1914. Ericaceae. 1 North America

Orobanche L. 1753. Orobanchaceae. 100–150 temp. warm–temp. and subtrop. Northern Hemisphere

Orobanchia Vandelli = Nematanthus Schrad. Gesneriaceae

Orobus L. = Lathyrus L. Fabaceae

Orochaenactis Coville. 1893. Asteraceae. 1 California

Orogenia S. Watson. 1871. Apiaceae. 2 western North America

Orontium L. 1753. Araceae. 1 eastern North America

Oropetium Trin. 1822. Poaceae. 5 Africa, India

Orophea Blume. 1825. Annonaceae. 37 southern India, Sri Lanka (1, O. polycarpa), Burma, Andaman Is., Thailand, southern China, Indochina, Malesia (Malay Peninsula, Sumatra, Java, Borneo, Sulawesi, Philippines, Moluccas)

Orophochilus Lindau. 1897. Acanthaceae. 1 Peru

Orostachys Fisch. ex A. Berger. 1930 (Sedum L.). Crassulaceae. 13 Europe (2), temp. Asia; Russia (8, from eastern European part [Ural Mountains] to Far East) and Middle Asia

Orothamnus Pappe ex Hook. 1848. Proteaceae. 1 South Africa

Oroxylum Vent. 1808. Bignoniaceae. 2 India, Sri Lanka, trop. Himalayas, Burma, western and southern China, Indochina, Malesia from Malay Peninsula to Philippines, Sulawesi, and Timor

Oroya Britton et Rose. 1922 (Oreocereus [A. Berger] Britton et Rose). Cactaceae. 1–2 Peru

Orphanidesia Boiss. et Balansa = Epigaea L. Ericaceae

Orphanodendron Barneby et J. W. Grimes. 1990. Fabaceae. 1 northwestern Colombia

Orphium E. Mey. 1838. Gentianaceae. 1 South Africa

Orrhopygium A. Löve = Aegilops L. Poaceae

Ortachne Nees ex Steud. 1854. Poaceae. 3 Andes from Costa Rica to Peru; Patagonia

Ortegia L. 1753. Caryophyllaceae. 1 (O. hispanica) Portugal, Spain

Ortegocactus Alexander. 1961 (Mammilaria Haw.). Cactaceae. 1 southeastern Mexico

Ortgiesia Regel. 1867. Bromeliaceae. 21 trop. America

Orthaea Klotzsch. 1851. Ericaceae. 31 trop. America

Orthandra Burret = Mortoniodendron Standl. et Steyerm. Tiliaceae

Orthantha (Benth.) Kerner = Odontites Ludw. Scrophulariaceae

Orthanthella Rauschert. 1983 (Odontites Ludw.). Scrophulariaceae. 3 Southeast Europe, Asia Minor

Orthanthera Wight. 1834. Asclepiadaceae. 3 trop. Old World

Orthechites Urb. = Secondatia A. DC. Apocynaceae

Orthilia Raf. 1840 (Pyrola L.). Ericaceae. 1–2 circumboreal regions, northern Mediterranean (O. secunda)

Orthion Standl. et Steyerm. 1940. Violaceae. 3 Central America

Orthocarpus Nutt. 1818. Scrophulariaceae. 27 western North America (25; O. luteus—from California to Minnesota and Ontario, Canada) and Andean South America (1), China (1)

Orthoceras R. Br. 1810. Orchidaceae. 2 southeastern Australia, Tasmania, New Zealand, New Caledonia

Orthoclada P. Beauv. 1812. Poaceae. 2 trop. America and Africa

Orthodon Benth. = Mosla (Benth.) Buch.-Ham. ex Maxim. Lamiaceae

Orthogoneuron Gilg = Gravesia Naudin. Melastomataceae

Orthogynium Baill. 1885. Menispermaceae. 1 Madagascar

Orthomene Barneby et Krukoff. 1971. Menispermaceae. 4 trop. South America

Orthopappus Gleason = Elephantopus L. Asteraceae

Orthopenthea Rolfe = Disa Bergius. Orchidaceae

Orthophytum Beer. 1854. Bromeliaceae. 17 eastern Brazil

Orthopichonia H. Huber. 1962. Apocynaceae. 9 trop. West Africa

Orthopterum L. Bolus. 1927. Aizoaceae. 2 South Africa (Cape)

Orthopterygium Hemsl. 1907. Anacardiaceae. 1 Peru

Orthoraphium Nees. 1841 (Stipa L.). Poaceae. 2 Himalayas, Southeast Asia

Orthosia Decne. 1844. Asclepiadaceae. 20 trop. South America, West Indies

Orthosiphon Benth. 1830. Lamiaceae. c. 50 trop. and South Africa, Himalayas, India, Burma, Andaman Is., East and Southeast Asia, Malesia, Australia

Orthosphenia Standl. 1923. Celastraceae. 1 Mexico

Orthotactus Nees = Justicia L. Acanthaceae

Orthotheca Pichon = Xylophragma Sprague. Bignoniaceae

Orthothylax (Hook. f.) Skottsb. = Helmholtzia F. Muell. Philydraceae

Orthrosanthus Sweet. 1829. Iridaceae. 9 southwestern Australia (4), trop. America

Orthurus Juz. = Geum L. Rosaceae

Orumbella J. M. Coult. et Rose = Podistera S. Watson. Apiaceae

Orvala L. = Lamium L. Lamiaceae

Orychophragmus Bunge. 1835. Brassicaceae. 1 China

Oryctanthus (Griseb.) Eichler. 1868. Loranthaceae. 9–10 trop. America from southern Mexico to Bolivia and northern Brazil

Oryctes S. Watson. 1871. Solanaceae. 1 southwestern U.S. (California and Nevada)

Oryctina Tiegh. 1895 (Oryctanthus [Griseb.] Eichler). Loranthaceae. 6 Colombia, Venezuela, the Guianas, Brazil

Oryctina Tiegh. = Oryctanthus (Griseb.) Eichler. Loranthaceae

Orygia Forssk. 1775 (Corbichonia Scop.). Molluginaceae. 2 North, trop. and Namibia, South Asia from Arabian Peninsula to India

Oryza L. 1753. Poaceae. 25 trop., subtrop. and warm–temp. regions

Oryzidium C. E. Hubb. et Schweick. 1936. Poaceae. 1 Namibia, Botswana, Zambia, Zimbabwe

Oryzopsis Michx. 1803. Poaceae. 1 warm–temp. North America

Osa Aiello. 1979 (Hintonia Bullock). Rubiaceae. 1 Costa Rica

Osbeckia L. 1753. Melastomataceae. c. 60 trop. Africa, Madagascar, trop. Asia, Malesia, northern Australia

Osbertia Greene. 1895. Asteraceae. 2 Mexico, Guatemala

Osbornia F. Muell. 1862. Myrtaceae. 1 Bali, Borneo, Philippines, New Guinea, northeastern Australia

Oschatzia Walp. 1848 (Azorella Lam.). Apiaceae. 2 Australia

Oscularia Schwantes. 1927 (Lampranthus N. E. Br.). Aizoaceae. 1–3 southwestern South Africa

Oserya Tul. et Wedd. 1849. Podostemaceae. 6 trop. America

Osmadenia Nutt. 1841. Asteraceae. 1 California and Baja California

Osmanthus Lour. 1790. Oleaceae. 40: 1 (O. decorus) Asia Minor and western Transcaucasus (Adzharia), East and Southeast Asia, Polynesia, southern U.S.

Osmaronia Greene = Oemleria Rchb. Rosaceae

Osmelia Thwaites. 1858. Flacourtiaceae. 4 Sri Lanka (1), Malesia (except Java)

Osmiopsis R. M. King et H. Rob. 1975. Asteraceae. 1 Haiti

Osmites L. = Relhania L'Hér. Asteraceae

Osmitopsis Cass. 1817. Asteraceae. 9 South Africa (Cape)

Osmoglossum (Schltr.) Schltr. 1922. Orchidaceae. 5 trop. America

Osmorhiza Raf. 1819. Apiaceae. 11: 1 (O. aristata) Caucasus, southern Siberia, Russian Far East; Himalayas from Kashmir to Bhutan; East and Southeast Asia; North, Central, and South (Andes) America

Osmoxylon Miq. 1863. Araliaceae. 55 Taiwan, Micronesia, Malesia (Malay Arch., Borneo, Philippines, Sulawesi, New Guinea), Solomon Is., Vanuatu

Ossaea DC. 1828. Melastomataceae. c. 80 trop. America from Mexico to Brazil, West Indies

Ossiculum P. J. Cribb et Laan. 1986. Orchidaceae. 1 Cameroun

Ostenia Buchenau = Hydrocleys Rich. Limnocharitaceae

Osteocarpum F. Muell. 1854 (Threlkeldia R. Br. and Sclerolaena R. Br.). Chenopodiaceae. 5 Australia

Osteomeles Lindl. 1821. Rosaceae. 3–10 East and Southeast Asia eastward to Hawaiian Is. and New Zealand

Osteophloeum Warb. 1897. Myristicaceae. 1 Brazil (Amazonia)

Osteospermum L. 1753. Asteraceae. 67 St. Helena (1), trop. and South Africa, Jordan, Arabian Peninsula

Ostericum Hoffm. = Angelica L. Apiaceae

Ostodes Blume. 1826. Euphorbiaceae. 10 eastern Himalayas, Burma, southwestern China, Indochina, Sumatra, Java, Borneo

Ostrearia Baill. 1871. Hamamelidaceae. 1 Australia (Queensland)

Ostrowskia Regel. 1884. Campanulaceae. 1 Middle Asia (Tien Shan, Pamir-Alai)

Ostrya Scop. 1760. Betulaceae. 7 temp. Northern Hemisphere; Caucasus (1, O. carpinifolia)

Ostryocarpus Hook. f. 1849. Fabaceae. 2

trop. West Africa from Guinea to Gabon and Zaire

Ostryoderris Dunn. 1911 (Aganope Miq.). Fabaceae. 7 trop. and South Africa

Ostryopsis Decne. 1873. Betulaceae. 2 eastern Mongolia, southwestern China

Osyridicarpos A. DC. 1857. Santalaceae. 6 trop. and South Africa

Osyris L. 1753. Santalaceae. 5–6 Mediterranean, Africa, South and Southeast Asia north to China (1, O. wightiana)

Otacanthus Lindl. 1862. Scrophulariaceae. 4 Brazil

Otachyrium Nees. 1829. Poaceae. 7 trop. South America

Otanthera Blume. 1831. Melastomataceae. 8 Nicobar Is., southern China, Taiwan, Malesia, trop. Australia

Otanthus Hoffmanns. et Link. 1809. Asteraceae. 1 coastal West Europe, Mediterranean, Caucasus (Abkhazia)

Otatea (McClure et E. W. Sm.) Calderón et Soderstr. = Sinarundinaria Nakai. Poaceae

Oteiza La Llave. 1832. Asteraceae. 3 Mexico, Guatemala

Othake Raf. = Palafoxia Lagasca. Asteraceae

Otherodendron Makino = Microtropis Wall. ex Meisn. Celastraceae

Otholobium C. H. Stirt. 1981. Fabaceae. 53 trop. America (8, from Mexico through Andes of Colombia and Venezuela to Chile and Argentina, South Africa (mostly), 1 (O. foliosum) also in Kenya, Tanzania, Malawi, Zambia, Zimbabwe

Othonna L. 1753. Asteraceae. 140–150 North and South Africa, Southeast Asia, Australia (1)

Othonnopsis Jaub. et Spach = Othonna L. Asteraceae

Otiophora Zucc. 1832. Rubiaceae. 15 trop. and South Africa, Madagascar

Otites Adans. 1763. Caryophyllaceae. 26 Europe, Asia Minor, Caucasus, Middle Asia, Siberia, Far East of Russia; O. parviflorus–extending to Mongolia and China

Otoba (DC.) Karst. 1880–1883. Myristicaceae. 9 trop. America

Otocalyx Brandegee. 1914. Rubiaceae. 1 Mexico

Otocarpus Durieu. 1847. Brassicaceae. 1 Algeria

Otocephalus Chiov. = Calanda K. Schum. Rubiaceae

Otochilus Lindl. 1830. Orchidaceae. 5 Nepal, Sikkim, Bhutan, Tibet, northwestern (Kumaon) and northeastern India, Burma, China (Yunnan), Thailand, Laos, Cambodia, Vietnam

Otoglossum (Schltr.) Garay et Dunst. 1976 (Odontoglossum Kunth). Orchidaceae. 8 northern South America

Otomeria Benth. 1849. Rubiaceae. 8 trop. Africa, Madagascar

Otonephelium Radlk. 1890. Sapindaceae. 1 India

Otopappus Benth. 1873. Asteraceae. 15 Mexico (14), Guatemala, Belize, El Salvador, Honduras, Nicaragua, Costa Rica, Panama, Jamaica (1, O. hirsutus)

Otophora Blume. 1849 (Lepisanthes Blume). Sapindaceae. 30 China, Indochina, western Malesia

Otoptera DC. 1825. Fabaceae. 1 (O. burchellii) Namibia, Botswana, Zimbabwe, South Africa

Otospermum Willk. 1864. Asteraceae. 1 Southwest Europe

Otostegia Benth. 1834. Lamiaceae. 21 eastern Mediterranean, Northeast trop. Africa, West and Southwest Asia to Middle Asia (8) and northwestern India

Otostylis Schltr. 1918. Orchidaceae. 4 Colombia, Venezuela, Trinidad, Guyana, French Guiana, Surinam, northern Brazil

Ottelia Pers. 1805. Hydrocharitaceae. 21 Africa (13), trop. and East Asia, Borneo (1), trop. Australia and New Caledonia, South America (1, O. brasiliensis); O. alismoides—Europe, West Asia, North America

Ottoa Kunth. 1821. Apiaceae. 1 Mexico, Guatemala, south to Venezuela, Colombia and Ecuador

Ottochloa Dandy. 1931. Poaceae. 6 South

and Southeast Asia, Malesia, northern Australia

Ottonia Spreng. 1820 (Piper L.). Piperaceae. c. 70 trop. South America

Ottoschmidtia Urb. 1924. Rubiaceae. 3 West Indies

Ottoschulzia Urb. 1912. Icacinaceae. 3 Guatemala, West Indies

Ottosonderia L. Bolus. 1958 (or = Ruschia Schwantes). Aizoaceae. 1 western South Africa

Oubanguia Baill. 1890. Scytopetalaceae. 3 trop. Africa

Oudneya R. Br. 1826. Brassicaceae. 1 Algeria

Ougeinia Benth. = Desmodium Desv. Fabaceae

Ouratea Aubl. 1775. Ochnaceae. c. 120 pantropics

Ourisia Comm. ex Juss. 1789. Scrophulariaceae. 25 Tasmania, New Zealand, southern Oceania, Andes of South America

Ourisianthus Bonati. 1925 (Artanema D. Don). Scrophulariaceae. 1 Indochina

Outreya Jaub. et Spach. 1843. Asteraceae. 1 Southwest Asia

Ovidia Meisn. 1857. Thymelaeaceae. 4 temp. South America

Owenia F. Muell. 1857. Meliaceae. 6 eastern Australia (Queensland and New South Wales)

Oxalis L. 1753. Oxalidaceae. c. 600 cosmopolitan, chiefly trop. America and South Africa

Oxandra A. Rich. 1845. Annonaceae. 23 Central and trop. South America, West Indies

Oxanthera Montrouz. 1860 (Citrus L.). Rutaceae. 5 New Caledonia

Oxera Labill. 1824. Verbenaceae. 20 New Caledonia

Oxyanthera Brongn. = Thelasis Blume. Orchidaceae

Oxyanthus DC. 1807. Rubiaceae. c. 40 trop. and South Africa

Oxybaphus L'Hér. ex Willd. = Mirabilis L. Nyctaginaceae

Oxycarpha S. F. Blake. 1918. Asteraceae. 1 Venezuela

Oxycaryum Nees. 1842. Cyperaceae. 1 trop. Africa, Central and South America

Oxyceros Lour. 1790. Rubiaceae. c. 30 Southeast Asia

Oxychlamys Schltr. 1923. Gesneriaceae. 1 New Guinea

Oxychloe Phil. 1860. Juncaceae. 7 Andes of South America

Oxychloris Lazarides. 1984. Poaceae. 1 Australia

Oxycoccus Hill = Vaccinium L. Ericaceae

Oxydendrum DC. 1839. Ericaceae. 1 eastern U.S.

Oxygonum Burch. ex Campdera. 1819. Polygonaceae. c. 30 trop. and South Africa, Madagascar (1, O. tristachyum)

Oxygraphis Bunge. 1836. Ranunculaceae. 6 temp. Asia; O. glacialis—from northern Ural to Middle Asia and Far East

Oxygyne Schltr. 1906. Burmanniaceae. 1 trop. West Africa

Oxylaena Benth. ex Anderb. 1991. Asteraceae. 1 South Africa

Oxylobium Andrews. 1807. Fabaceae. 15 Australia, Tasmania

Oxylobus (Moq. ex DC.) A. Gray. 1880. Asteraceae. 4 trop. America from Mexico to Venezuela

Oxyosmyles Speg. 1901. Boraginaceae. 1 Argentina

Oxypappus Benth. 1845. Asteraceae. 2 Mexico

Oxypetalum R. Br. 1811. Asclepiadaceae. c. 150 trop. America, West Indies

Oxyphyllum Phil. 1860. Asteraceae. 1 Chile

Oxypolis Raf. 1825. Apiaceae. 7 North America

Oxyrhachis Pilg. 1932. Poaceae. 1 trop. Africa: Sierra Leone, Cameroun, Zambia, Tanzania, Madagascar

Oxyrhynchus Brandegee. 1912. Fabaceae. 4 New Guinea (1), Mexico and Central America (3)

Oxyria Hill. 1765. Polygonaceae. 2 arctic, subarctic and montane temp. Eurasia, California; Russia (2: almost all areas, except steppes)

Oxyspora DC. 1828 (Allomorphia Blume). Melastomataceae. 24 India, southern China, Indochina, Malesia

Oxystelma R. Br. 1810. Asclepiadaceae. 2 trop. West Africa, Egypt, West Asia, trop. Himalayas, India, Sri Lanka, southern China, Indochina, Malesia

Oxystigma Harms. 1897. Fabaceae. 7 trop. Africa from Cameroun to Kenya, Angola to Tanzania

Oxystylis Torr. et Frém. 1845. Capparaceae. 1 southwestern U.S.

Oxytenanthera Munro. 1868. Poaceae. 2 trop. Africa

Oxytenia Nutt. = Iva L. Asteraceae

Oxytheca Nutt. 1848. Polygonaceae. 7 western North America (Great Basin and the Mojave Desert) and central eastern Chile and adjacent Argentina

Oxytropis DC. 1802. Fabaceae. 300–360 Eurasia (chiefly Middle and Central Asia), North America

Oyedaea DC. 1836. Asteraceae. 13 Central and trop. South America

Ozodia Wight et Arn. = Foeniculum Hill. Apiaceae

Ozoroa Delile. 1843 (Heeria Meisn.). Anacardiaceae. 40 Africa south of the equator extending to Senegal and southern Ethiopia, ? Madagascar, southern Arabian Peninsula (Yemen)

Ozothamnus R. Br. 1818 (Helichrysum Mill.). Asteraceae. c. 40 Australasia

P

Pabellonia Quezada et Martic. 1976. Alliaceae. 2 Chile

Pabstia Garay. 1973. Orchidaceae. 5 Brazil

Pabstiella Brieger et Senghas. 1976 (Pleurothallis R. Br.). Orchidaceae. 1 Brazil

Pachecoa Standl. et Steyerm. 1943. Fabaceae. 1 Mexico, Guatemala

Pachira Aubl. 1775 (Bombax L.). Bombacaceae. 24 trop. America

Pachites Lindl. 1835. Orchidaceae. 2 South Africa (southwestern Cape)

Pachyacris Schltr. ex Bullock = Xysmalobium R. Br. Asclepiadaceae

Pachyanthus A. Rich. 1846. Melastomataceae. 20 West Indies, Colombia

Pachycarpus E. Mey. 1838. Asclepiadaceae. c. 50 trop. and South Africa

Pachycentria Blume. 1831. Melastomataceae. 8 Burma, China, Taiwan (1), Malesia

Pachycereus (A. Berger) Britton et Rose. 1909. Cactaceae. 12 Mexico

Pachycladon Hook. f. 1864. Brassicaceae. 1 New Zealand

Pachycormus Coville ex Standl. 1923. Anacardiaceae. 1 southern California

Pachycornia Hook. f. 1880. Chenopodiaceae. 1 (P. triandra) central and southern Australia

Pachyctenium Maire et Pamp. ex Pamp. 1936. Apiaceae. 1 northern Ethiopia

Pachycymbium L. C. Leach. 1978 (Caralluma R. Br.). Asclepiadaceae. 2 South Africa

Pachydesmia Gleason = Miconia Ruiz et Pav. Melastomataceae

Pachydiscus Gilg et Schltr. = Wittsteinia F. Muell. Alseuosmiacea

Pachyelasma Harms. 1913. Fabaceae. 1 Nigeria, Cameroun, Gabon, Zaire

Pachygone Miers. 1851. Menispermaceae. 10 southern India, Sri Lanka, China, Indochina, Malesia (1, P. ovata—Java, Lesser Sunda Is., Borneo, Sulawesi, Moluccas, New Guinea), northeastern Australia, Oceania

Pachylaena D. Don ex Hook. et Arn. 1835. Asteraceae. 2 Andes of Chile and Argentina

Pachylarnax Dandy. 1927. Magnoliaceae. 4 northeastern India (Assam), Thailand, Malay Peninsula, Sumatra

Pachylecythis Ledoux = Lecythis Loefl. Lecythidaceae

Pachyloma DC. 1828. Melastomataceae. 4 Colombia, Venezuela, Brazil

Pachymitus O. E. Schulz. 1924. Brassicaceae. 1 southeastern Australia

Pachynema R. Br. ex DC. 1817. Dilleniaceae. 5 northern Australia

Pachyneurum Bunge. 1840. Brassicaceae. 1 West and East Siberia, Mongolia

Pachypharynx Aellen = Atriplex L. Chenopodiaceae

Pachyphragma (DC.) Rchb. 1841. Brassicaceae. 1 Armenia, Asia Minor

Pachyphyllum Kunth. 1816. Orchidaceae. 12 western trop. South America

Pachyphytum Link, Klotzsch et Otto. 1841. Crassulaceae. 12 Mexico

Pachyplectron Schltr. 1906. Orchidaceae. 2 New Caledonia

Pachypleurum Ledeb. 1829 (Ligusticum L.). Apiaceae. 2 Arctica, montane Eurasia; P. alpinum—from West Europe through Russia to Kamchatka, south to Middle Asia and China

Pachypodanthium Engl. et Diels. 1900. Annonaceae. 4 trop. West Africa

Pachypodium Lindl. 1830. Apocynaceae. 13 South-West and South Africa, Madagascar

Pachyptera DC. = Mansoa DC. Bignoniaceae

Pachypteris Kar. et Kir. = Pachypterygium Bunge. Brassicaceae

Pachypterygium Bunge. 1843. Brassicaceae. 2–6 Kazakhstan, Middle Asia, Iran, Afghanistan, Pakistan, northwestern China

Pachyrhizanthe (Schltr.) Nakai = Cymbidium Sw. Orchidaceae

Pachyrhizus A. Rich. ex DC. 1825. Fabaceae. 5 trop. America: Mexico, Guatemala, El Salvador, Honduras, Nicaragua, Costa Rica, Panama, Colombia, Venezuela, Guyana, Ecuador, Peru, Brazil, Bolivia; West Indies

Pachyrhynchus DC. = Lucilia Cass. Asteraceae

Pachysandra Michx. 1803. Buxaceae. 5 East Asia (4), eastern U.S.; Russia (1, P. terminalis: Far East)

Pachystachys Nees. 1847. Acanthaceae. 12 trop. America, West Indies

Pachystegia Cheeseman = Olearia Moench. Asteraceae

Pachystela Pierre ex Radlk. = Synsepalum (A. DC.) Daniell. Sapotaceae

Pachystele Schltr. 1923. Orchidaceae. 6 Central America

Pachystelis Rauschert = Pachystele Schltr. Orchidaceae

Pachystelma Brandegee = Matelea Aubl. Asclepiadaceae

Pachystigma Hochst. 1842. Rubiaceae. 14 trop. and subtrop. Africa

Pachystoma Blume. 1825. Orchidaceae. 11 India, Andaman Is., Nepal, Bhutan, southern China (incl. Hainan), Taiwan, Indochina, Malesia from Malay Peninsula to New Guinea, northern Australia, New Caledonia, western Oceania

Pachystrobilus Bremek. = Strobilanthes Blume. Acanthaceae

Pachystroma Muell. Arg. 1865. Euphorbiaceae. 1 southern Brazil

Pachystylidium Pax et K. Hoffm. 1919. Euphorbiaceae. 1 India, Indochina, Java, Philippines

Pachystylus K. Schum. 1889. Rubiaceae. 1 New Guinea

Pachythamnus (R. M. King et H. Rob.) R. M. King et H. Rob. 1972 (Eupatorium L.). Asteraceae. 1 Mexico, Central America

Pachytrophe Bureau = Streblus Lour. Moraceae

Packera A. Löve et D. Löve. 1976 (Senecio L.). Asteraceae. 67 North America south to Mexico, Arctic Siberia

Pacouria Aubl. = Landolphia P. Beauv. Apocynaceae

Pacourina Aubl. 1775. Asteraceae. 2 trop. South America

Padbruggea Miq. 1855. Fabaceae. 10 Southeast Asia

Padellus Vassilcz. 1973 (Cerasus L.). Rosaceae. 1 Southeast Europe, Mediterranean, Asia Minor, Caucasus, Middle Asia

Padus Hill. 1753 (Prunus L.). Rosaceae. c. 20 temp. Eurasia

Paederia L. 1767. Rubiaceae. 20 tropics

Paederota Hill. 1753 (Veronica L.). Scrophulariaceae. 2 South Europe

Paederotella (Wulf) Kem.-Nath. 1952 (Veronica L.). Scrophulariaceae. 3 Caucasus, Asia Minor

Paedicalyx Pierre ex Pit. = Xanthophytum Reinw. ex Blume. Rubiaceae

Paeonia L. 1753. Paeoniaceae. 35 temp. and subtrop. Eurasia (esp. c. 20—Caucasus, Middle Asia, southern Siberia, Far East); western North America (2)

Paepalanthus Kunth. 1841. Eriocaulaceae. 485 trop. South America, West Indies

Paepalanthus Mart. = Paepalanthus Kunth. Eriocaulaceae

Pagaea Griseb. = Iribachia Mart. Gentianaceae

Pagamea Aubl. 1775. Rubiaceae. 22 trop. South America

Pagameopsis Steyerm. 1965. Rubiaceae. 2 Venezuela

Pagella Schoenl. = Crassula L. Crassulaceae

Pagesia Raf. = Mecardonia Ruiz et Pav. Scrophulariaceae

Pagetia F. Muell. = Bosistoa F. Muell. Rutaceae

Pagiantha Markgr. = Tabernaemontana L. Apocynaceae

Painteria Britton et Rose = Havardia Small. Fabaceae

Paivaea O. Berg = Campomanesia Ruiz et Pav. Myrtaceae

Pajanelia DC. 1838. Bignoniaceae. 1 India, Andaman Is., Malesia

Pakaraimaea Maguire et P. S. Ashton. 1977. Monotaceae. 1 Venezuela, Guyana

Paladelpha Pichon = Alstonia R. Br. Apocynaceae

Palaeocyanus Dostál. 1976. Asteraceae. 1 Malta and Gozo

Palafoxia Lag. 1816. Asteraceae. 12 southwestern U.S., northern Mexico

Palandra O. F. Cock. 1927 (Phytelephas Ruiz et Pav.). Arecaceae. 1 western Colombia, Ecuador

Palaquium Blanco. 1837. Sapotaceae. c. 120 India, southern China, Taiwan, Indochina, Malesia, Solomon Is.

Palaua Cav. 1785. Malvaceae. 15 Andes

Paleaepappus Cabrera. 1969. Asteraceae. 1 Patagonia

Palenia Phil. = Baccharis L. Asteraceae

Paleodicraeia C. Cusset. 1973. Podostemaceae. 1 Madagascar

Paliavana Vell. ex Vand. 1788. Gesneriaceae. 3 Brazil

Palicourea Aubl. 1775. Rubiaceae. c. 250 trop. America, West Indies

Palimbia Besser ex DC. 1830. Apiaceae. 3 Russia from eastern European part to eastern Kazakhstan

Palisota Rchb. ex Endl. 1836. Commelinaceae. 18 trop. Africa

Paliurus Hill. 1754. Rhamnaceae. 6–8 Eurasia eastward to Japan; 1 (P. spina-christi) from Iberian Peninsula and Algeria through northeastern Mediterranean, Crimea, Caucasus, and Middle Asia to northern China, also in Syria, Lebanon, Israel, Iraq, Iran, Afghanistan

Pallasia Klotzsch = Wittmackanthus Kuntze. Rubiaceae

Pallenis (Cass.) Cass. = Asteriscus Hill. Asteraceaes

Palmerella A. Gray. 1876 (Solenopsis C. Presl). Lobeliaceae. 2 southwestern North America

Palmeria F. Muell. 1864. Monimiaceae. 15 Malesia (12: Sulawesi, Manus I., Japen I., New Guinea, New Britain), Australia (Queensland)

Palmervandenbroeckia Gibbs = Polyscias J. R. Forst. et G. Forst. Araliaceae

Palmolmedia Ducke = Naucleopsis Miq. Moraceae

Palmorchis Barb. Rodr. 1887. Orchidaceae. 10 Central and South (Colombia, Venezuela, Trinidad, Guyana, Peru, Brazil) America

Paloue Aubl. 1775. Fabaceae. 4 trop. America

Paloveopsis R. S. Cowan. 1957. Fabaceae. 1 northeastern South America

Palumbina Rchb. f. 1863. Orchidaceae. 1 Guatemala

Pamburus Swingle. 1916 (Atalantia Corrêa). Rutaceae. 1 southern India, Sri Lanka

Pamianthe Stapf. 1933. Amaryllidaceae. 2–3 northern Andes

Pamphalea Lag. 1811. Asteraceae. 9 trop. and subtrop. South America

Pamphilia Mart. ex A. DC. 1844. Styracaceae. 3 Brazil

Pamplethantha Bremek. = Pauridiantha Hook. f. Rubiaceae

Panamanthus Kuijt. 1991 (Gaiadendron G. Don). Loranthaceae. 1 Panama, Costa Rica

Panax L. 1753. Araliaceae. 7 eastern Himalayas, northeastern India, northern Burma, China, Indochina

Pancheria Brongn. et Gris. 1862. Cunoniaceae. 26 New Caledonia

Pancicia Vis. et Schltdl. 1858. Apiaceae. 1 Southeast Europe

Pancovia Willd. 1799. Sapindaceae. 12 trop. Africa

Pancratium L. 1753. Amaryllidaceae. 20 Mediterranean, trop. Africa, West (incl. Caucasus—P. maritimum) and South Asia

Panda Pierre. 1896. Pandaceae. 1 trop. West Africa

Pandaca Noronha ex Thouars = Tabernaemontana L. Apocynaceae

Pandacastrum Pichon = Tabernaemontana L. Apocynaceae

Pandanus Parkinson. 1773. Pandanaceae. c. 600 trop. Old World, esp. Madagascar (90 endemic), Malesia and Melanesia, few species in temp. China and Japan

Panderia Fisch. et C. A. Mey. 1836. Chenopodiaceae. 2 Turkey, Syria, Israel, Jordan, Iraq, Caucasus, Kazakhstan, Kirgizia, Iran, Afghanistan, northwestern China, Mongolia

Pandiaka (Moq.) Hook. f. 1880. Amaranthaceae. 12 trop. and South Africa

Pandorea Endl. ex Spach. 1838. Bignoniaceae. 6 Sulawesi, Lesser Sunda Is. and Moluccas, New Guinea, Solomon Is., Australia, Lord Howe I., New Caledonia

Pangium Reinw. 1823. Flacourtiaceae. 1 India, Malesia, Palau Is., Bismarck Arch.

Panicum L. 1753. Poaceae. c. 600 warm–temp., subtrop. and trop. regions

Panisea (Lindl.) Lindl. 1854. Orchidaceae. 7 southern and northeastern India, Sri Lanka, Nepal, Bhutan, Burma, southern China, Thailand, Laos, Cambodia, Vietnam

Panopsis Salisb. ex Knight. 1809. Proteaceae. 20 trop. America

Pantacantha Speg. 1902 (Lycium L.). Solanaceae. 1 Patagonia

Pantadenia Gagnep. 1925. Euphorbiaceae. 1 Indochina

Pantathera Phil. = Megalachne Steud. Poaceae

Pantlingia Prain. 1896. Orchidaceae. 5 East, Southeast Asia, Malesia

Panulia (Baill.) Kozo-Polj. = Apium L. Apiaceae

Panurea Spruce ex Benth. 1865. Fabaceae. 1 Colombia, Brazil

Panzeria J. F. Gmel. = Lycium L. Solanaceae

Panzeria Moench = Panzerina Soják. Lamiaceae

Panzerina Soják. 1982. Lamiaceae. 6 southern Siberia, Mongolia, northern China

Paolia Chiov. = Coffea L. Rubiaceae

Papaver L. 1753. Papaveraceae. c. 100 Eurasia, esp. (c. 75) Caucasus, Middle Asia, and Arctica; North and South Africa, Australia, western North America

Paphia Seem. = Agapetes D. Don ex G. Don. Ericaceae

Paphinia Lindl. 1843. Orchidaceae. 7 trop. America from Guatemala to Peru and Brazil

Paphiopedilum Pfitzer. 1886. Orchidaceae. 65 trop. Asia from India to southern China and Indochina, Malesia from Malay Peninsula to Philippines (13) and New Guinea, Solomon Is., western Oceania

Papilionanthe Schltr. 1915. Orchidaceae. 11 trop. Asia, Malesia

Papilionopsis Steenis = Desmodium Desv. Fabaceae

Papillilabium Dockrill. 1967. Orchidaceae. 1 East Asia, Australia

Pappagrostis Roshev. 1934 (Stephanachne

Keng). Poaceae. 1 Central Asia, incl. central Tien Shan

Pappea Eckl. et Zeyh. 1835. Sapindaceae. 1 (P. capensis) trop. East and South Africa

Papperitzia Rchb. f. 1852. Orchidaceae. 1 Mexico

Pappobolus S. F. Blake. 1916. Asteraceae. 38 Andes of Colombia, Ecuador, Peru

Pappophorum Schreb. 1791. Poaceae. c. 10 southwestern U.S., Mexico, Central and South America southward to Argentina, Uruguay, and Paraguay

Pappothrix (A. Gray) Rydb. = Perityle Benth. Asteraceae

Papuacalia Veldk. 1991. Asteraceae. 14 New Guinea

Papuacedrus Li. 1953 (Libocedrus Endl.). Cupressaceae. 1 (P. papuana) Moluccas, New Guinea

Papuaea Schltr. 1919. Orchidaceae. 1 New Guinea

Papualthia Diels. 1912. Annonaceae. 20 Philippines, New Guinea

Papuanthes Danser. 1931. Loranthaceae. 1 New Guinea

Papuastelma Bullock. 1965. Asclepiadaceae. 1 New Guinea

Papuechties Markgr. 1927. Apocynaceae. 3 New Guinea

Papuodendron C. T. White = Hibiscus L. Malvaceae

Papuzilla Ridl. 1916. Brassicaceae. 2 New Guinea

Parabaena Miers ex Hook. f. et Thomson. 1851. Menispermaceae. 6: P. sagittata—Nepal, Bhutan, northeastern India, Burma, southern China (Yunnan), Andaman Is., Indochina; Borneo (P. megalocarpa), Philippines (3), New Guinea, and Solomon Is. (P. tuberculata)

Parabarium Pierre ex C. Spire et A. Spire. 1906 (Ecdysanthera Hook. et Arn.). Apocynaceae. 20 eastern Himalayas, Burma, southern China, Indochina

Parabeaumontia (Baill.) Pichon = Vallaris Burm. f. Apocynaceae

Paraberlinia Pellegr. = Julbernardia Pellegr. Fabaceae

Parabignonia Bureau ex K. Schum. 1894. Bignoniaceae. 1 northeastern Brazil

Paraboea (C. B. Clarke) Ridl. 1905. Gesneriaceae. c. 85 northeastern India, China (13), Thailand, Malesia

Parabouchetia Baill. 1887. Solanaceae. 1 Brazil

Paracaleana Blaxell. 1972 (Caleana R. Br.). Orchidaceae. 3 Australia, Tasmania, New Zealand

Paracalia Cuatrec. 1960. Asteraceae. 2 Peru, Bolivia

Paracalyx Ali. 1968. Fabaceae. 5–6 Ethiopia (1), Somalia (3), southern Arabian Peninsula

Paracarpaea (K. Schum.) Pichon. 1946 (Arrabidaea DC.). Bignoniaceae. 1 Brazil

Paracaryopsis (H. Riedl) R. R. Mill. 1991. Boraginaceae. 3 Oman (1), India (3)

Paracaryum Boiss. = Cynoglossum L. Boraginaceae

Paracautleya Ros. M. Sm. 1977. Zingiberaceae. 1 southern India

Paracephaelis Baill. 1879. Rubiaceae. 1 Madagascar

Parachampionella Bremek. 1944 (Strobilanthes Blume). Acanthaceae. 2 Taiwan

Parachimarrhis Ducke. 1922. Rubiaceae. 1 Brazil (Amazonia)

Parachionolaena Dillon et Sagastegui. 1990. Asteraceae. 1 Colombia

Paracoffea (Miq.) J.-F. Leroy = Psilanthus Hook. f. Rubiaceae

Paracolpodium (Tsvelev) Tsvelev. 1965 (Colpodium Trin.). Poaceae. 3 Caucasus, Middle and Central Asia, southern Siberia

Paracorynanthe Capuron. 1978. Rubiaceae. 2 Madagascar

Paracryphia Baker f. 1921. Paracryphiaceae. 1–2 New Caledonia

Paractaenum P. Beauv. 1812. Poaceae. 1 Australia

Paracyclea Kudo et Yamamoto. 1932 (Cissampelos L.). Menispermaceae. 3 Ryukyu Is. (Lanyu and Lutao), Taiwan

Paracynoglossum Popov. 1953 (Cynoglossum L.). Boraginaceae. c. 50 tropics and subtropics

Paraderris (Miq.) R. Geesink. 1984 (Derris

Lour.). Fabaceae. 6 South and Southeast Asia

Paradina Pierre ex Pit. = Mitragyna Korth. Rubiaceae

Paradisanthus Rchb. f. 1852. Orchidaceae. 5 trop. South America

Paradisea Mazzuc. 1811. Asphodelaceae. 2 mountainous South Europe

Paradolichandra Hassl. = Parabignonia Bureau ex K. Schum. Bignoniaceae

Paradombeya Stapf. 1902. Bombacaceae. 5 Burma, southwestern China

Paradrymonia Hanst. 1854. Gesneriaceae. c. 40 trop. America

Paradrypetes Kuhlm. 1935. Euphorbiaceae. 2 Brazil: southwestern Amazonas (P. subintegrifolia), Espirito Santo and Minas Gerais (P. ilicifolia)

Paraeremostachys Adylov, Kamelin et Machm. 1986 (Eremostachys Bunge.). Lamiaceae. 15 Middle Asia

Parafaujasia C. Jeffrey. 1992. Asteraceae. 2 Mascarene Is.

Parafestuca E. B. Alexeev. 1985. Poaceae. 1 Madeira Is.

Paragelonium Leandri = Aristogeitonia Prain. Euphorbiaceae

Paragenipa Baill. 1879. Rubiaceae. 1 Madagascar

Parageum Nakai et H. Hara = Geum L. Rosaceae

Paraglycine F. J. Herm. = Ophrestia H. M. L. Forbes. Fabaceae

Paragoldfussia Bremek. 1944 (Strobilanthes Blume). Acanthaceae. 2 Sumatra

Paragonia Bureau. 1872. Bignoniaceae. 2 trop. America from Mexico to Argentina, West Indies

Paragophyton K. Schum. = Spermacoce L. Rubiaceae

Paragrewia Gagnep. ex R. S. Rao = Leptonychia Turcz. Sterculiaceae

Paragulubia Burret = Gulubia Becc. Arecaceae

Paragutzlaffia H. P. Tsui = Strobilanthes Blume. Acanthaceae

Paragynoxys (Cuatrec.) Cuatrec. 1955. Asteraceae. 14 northwestern South America

Parahancornia Ducke. 1922. Apocynaceae. 7 eastern Colombia (2), eastern Venezuela, the Guianas, Peru, and Brazil (northwestern Amazonas)

Parahebe W. R. B. Oliv. 1944. Scrophulariaceae. 30 New Guinea, Australia, Tasmania, New Zealand, Polynesia

Parahyparrhenia A. Camus. 1950. Poaceae. 5 trop. Africa, South Asia

Paraia Rohwer, H. G. Richt. et van der Werff. 1991. Lauraceae. 1 Brazil

Paraixeris Nakai = Crepidiastrum Nakai. Asteraceae

Parajaeschkea Burkill. 1911 (Gentianella Moench). Gentianaceae. 1 eastern Himalayas (Sikkim)

Parajubaea Burret. 1930. Arecaceae. 2 Ecuador, central and southern Bolivia

Parajusticia Benoist = Gymnostachyum Nees. Acanthaceae

Parakaempferia A. S. Rao et D. M. Verma. 1971. Zingiberaceae. 1 India (Assam)

Parakibara Philipson. 1985. Monimiaceae. 1 Moluccas (Halmahera)

Parakmeria Hu et Cheng. 1957 (Magnolia L.). Magnoliaceae. 2–5 central, southwestern, and southern (incl. Hainan) China, Taiwan

Paraknoxia Bremek. 1952. Rubiaceae. 1 trop. Africa from Central African Rep., Zaire, Uganda, Kenya south to Angola and Zimbabwe

Parakohleria Wiehler. 1978. Gesneriaceae. 20 Colombia, Andes of South America

Paralabatia Pierre = Pouteria Aubl. Sapotaceae

Paralamium Dunn. 1913. Lamiaceae. 1 Burma, southern China (Yunnan), Vietnam

Paralbizzia Kosterm. = Archidendron F. Muell. Fabaceae

Paralepistemon Lejoly et Lisowski. 1986. Convolvulaceae. 1 trop. Africa

Paraligusticum V. N. Tikhom. 1973. Apiaceae. 1 (P. discolor) Middle Asia, southern West Siberia

Paralinospadix Burret = Calytprocalyx Blume. Arecaceae

Paralstonia Baill. = Alyxia R. Br. Apocynaceae

Paralychnophora MacLeish = Eremanthus Less. Asteraceae

Paralyxia Baill. 1888 (Aspidosperma Mart. et Zucc.). Apocynaceae. 1 northeastern South America

Paramachaerium Ducke. 1925. Fabaceae. 5 trop. America: Costa Rica, Panama, Guyana, French Guiana, Brazil, Peru

Paramacrolobium J. Léonard. 1954. Fabaceae. 1 trop. Africa from Guinea to Kenya south to Angola and Tanzania

Paramammea J.-F. Leroy = Mammea L. Clusiaceae

Paramanglietia Hu et Cheng = Manglietia Blume. Magnoliaceae

Paramansoa Baill. = Arrabidaea DC. Bignoniaceae

Paramapania Uittien. 1935. Cyperaceae. 7 Malesia, Caroline and Solomon Is., Fiji, Samoa Is.

Paramelhania Arènes. 1949. Sterculiaceae. 1 Madagascar

Parameria Benth. 1876. Apocynaceae. 6 India, Andaman Is., China, Indochina, Malesia

Parameriopsis Pichon = Parameria Benth. Apocynaceae

Paramichelia Hu. 1940 (Michelia L.). Magnoliaceae. 3 India (Assam), southern China, Indochina, Malay Peninsula, Sumatra

Paramicropholis Aubrév. et Pellegr. = Micropholis (Griseb.) Pierre. Sapotaceae

Paramicrorhynchus Kirp. = Launaea Cass. Asteraceae

Paramignya Wight. 1831 (Atalantia Corrêa). Rutaceae. 12 India, Sri Lanka, Burma, southern China, Indochina, Malesia from Malay Peninsula to Philippines and Timor I., northeastern Australia

Paramitranthes Burret = Siphoneugenia O. Berg. Myrtaceae

Paramogaia Velarde. 1948. Amaryllidaceae. 1 Peru

Paramoltkia Greuter. 1981. Boraginaceae. 1 Yugoslavia, Albania

Paramomum S. Q. Tong. 1985. Zingiberaceae. 1 southern China (Yunnan)

Paramyrciaria Kausel. 1967. Myrtaceae. 1 Argentina

Paranecepsia Radcl.-Sm. 1976. Euphorbiaceae. 1 Tanzania, northeastern Mozambique

Paranephelium Miq. 1861. Sapindaceae. 4–8 Burma, southern China, Indochina, Sumatra, Borneo, Philippines

Paranephelius Poepp. et Endl. 1843. Asteraceae. 7 Peru, Bolivia, Argentina

Paraneurachne S. T. Blake. 1972. Poaceae. 1 Australia

Paranneslea Gagnep. 1948. Theaceae. 1 Indochina

Paranomus Salisb. 1807. Proteaceae. 18 South Africa

Parantennaria Beauverd. 1911. Asteraceae. 1 eastern Australia

Parapachygone Forman = Pachygone Miers. Menispermaceae

Parapantadenia Capuron. 1972 (Pantadenia Gagnep.). Euphorbiaceae. 1 Madagascar

Parapentapanax Hutch. = Pentapanax Seem. Araliaceae

Parapentas Bremek. 1952. Rubiaceae. 3–4 trop. Africa, ? Madagascar

Paraphalaenopsis A. D. Hawkes. 1964. Orchidaceae. 3 Borneo

Paraphlomis Prain. 1908. Lamiaceae. 24 eastern Himalayas, northeastern India, southern China, Indochina, Malesia

Parapholis C. E. Hubb. 1946. Poaceae. 6 West and Southeast (Crimea) Europe, Mediterranean, West (incl. Caucasus) and Middle Asia to northwestern India

Paraphyadanthe Mildbr. = Caloncoba Gilg. Flacourtiaceae

Parapiptadenia Brenan. 1963. Fabaceae. 3 trop. South America

Parapiqueria R. M. King et H. Rob. 1980. Asteraceae. 1 Brazil

Parapodium E. Mey. 1838. Asclepiadaceae. 1 South Africa

Paraprenanthes Chang et C. Shih. 1988. Asteraceae. 11 South and East Asia

Paraprotium Cuatrec. = Protium Burman f. Burseraceae

Parapteroceras Averyanov. 1990 (Tuberolabium Yamam.). Orchidaceae. 5–6

Thailand, Vietnam, Indonesia, Philippines
Parapteropyrum A. J. Li. 1981. Polygonaceae. 1 Tibet
Parapyrenaria H. T. Chang. 1963 (Pyrenaria Blume). Theaceae. 1 southern China (incl. Hainan)
Paraquilegia J. R. Drumm. et Hutch. 1920. Ranunculaceae. 5 Middle Asia, southern Siberia, Mongolia, southwestern China, eastern Iran, Afghanistan, Himalayas from Kashmir to Bhutan
Pararchidendron I. C. Nielsen. 1983. Fabaceae. 1 Java, Lesser Sunda Is., Timor I., New Guinea, Australia (Queensland and New South Wales)
Parardisia M. P. Nayar et Giri = Ardisia Sw. Myrsinaceae
Pararistolochia (Hutch. et Dalziel) Hutch. et Dalziel. 1928 (Aristolochia L.). Aristolochiaceae. 14 West and Central Africa (9), Borneo and New Guinea (5)
Parartocarpus Baill. 1875. Moraceae. 4 Thailand, Malesia (excl. Lesser Sunda Is.). New Guinea, Solomon Is.
Pararuellia Bremek. et Nannenga-Bremek. 1948. Acanthaceae. 6 China, Indochina, western Malesia
Parasamanea Kosterm. = Albizia Durazz. Fabaceae
Parasarcochilus Dockrill = Pteroceras Hasselt ex Hassk. Orchidaceae
Parasassafras D. C. Long = Actinodaphne Nees. Lauraceae
Parascheelea Dugand = Orbignya Mart. ex Endl. Arecaceae
Parascopolia Baill. = Lycianthes (Dunel) Hassl. Solanaceae
Paraselinum H. Wolff. 1921. Apiaceae. 1 Peru
Parasenecio W. W. Sm. et Small. 1922. Asteraceae. 50 temp. Eurasia, esp. East Asia, North America
Paraserianthes I. C. Nielsen. 1983. Fabaceae. 4 Malesia, Solomon Is., Australia
Parashorea Kurz. 1870. Dipterocarpaceae. 15 Burma, southern China (1, P. chinensis), Indochina, Malesia
Parasicyos Dieterle. 1975. Cucurbitaceae. 1 Guatemala
Parasilaus Leute. 1972. Apiaceae. 1 (P. asiaticus) southern Tajikistan, Afghanistan
Parasitaxus de Laub. 1972. Podocarpaceae. 1 (P. ustus) New Caledonia (parasitic on the roots of other members of the Podocarpaceae)
Paraskevia W. Sauer et G. Sauer = Nonea Medik. Boraginaceae
Parasponia Miq. 1851. Ulmaceae. 5 Malesia, Solomon Is., Vanuatu, Fiji, Tahiti
Parastemon A. DC. 1842. Chrysobalanaceae. 3–4 Nicobar Is., Malay Peninsula, Sumatra, Borneo, Moluccas, New Guinea, Admiralty Is.
Parastrephia Nutt. 1841. Asteraceae. 5 Andes
Parastriga Mildbr. 1930. Scrophulariaceae. 1 trop. Africa
Parastrobilanthes Bremek. 1944. Acanthaceae. 4 Sumatra, Java
Parastyrax W. W. Sm. 1920. Styracaceae. 2 Burma, China
Parasympagis Bremek. 1944. Acanthaceae. 3 Burma, Thailand
Parasyringa W. W. Sm. = Ligustrum L. Oleaceae
Paratecoma Kuhlm. 1931. Bignoniaceae. 1 Brazil
Paratephrosia Domin. 1912. Fabaceae. 1 central Australia
Paratheria Griseb. 1866. Poaceae. 2 West Indies, South America, West Africa, Madagascar
Parathesis (A. DC.) Hook. f. 1876. Myrsinaceae. c. 130 (or c. 75) trop. America from southern Mexico to Panama and through the Andes from Venezuela to Peru; West Indies
Paratriaina Bremek. 1956. Rubiaceae. 1 Madagascar
Paravallaris Pierre ex Hua = Kinatalia G. Don. Apocynaceae
Paravitex H. R. Fletcher. 1937. Verbenaceae. 1 Thailand
Pardanthopsis (Hance) Lenz. 1972. Iridaceae. 2 temp. Asia; Russia (1, P. dichotoma: southern East Siberia, Far East)
Pardoglossum Barbier et Mathez = Cynoglossum L. Boraginaceae

Parduyna Salisb. = Schelhammera R. Br. Melanthiaceae

Parenterolobium Kosterm. = Albizia Durazz. Fabaceae

Parentucellia Viv. 1824. Scrophulariaceae. 4 West Europe, Mediterranean, West (incl. Caucasus) and Middle Asia, Iran

Parepigynum Tsiang et P. T. Li. 1973. Apocynaceae. 1 southern China

Parhabenaria Gagnep. 1932. Orchidaceae. 2 Indochina

Pariana Aubl. 1775. Poaceae. c. 30 Costa Rica, trop. South America

Parietaria L. 1753. Urticaceae. 10–20 temp. and subtrop. regions, mountainous tropics

Parinari Aubl. 1775. Chrysobalanaceae. 44 trop. Africa and Madagascar (6), trop. Asia (P. anamensis), Malesia (13), Australia (northern Queensland), Vanuatu, Fiji, Tonga, Samoa, trop. South America, Trinidad

Paris L. 1753. Trilliaceae. 5–6 Europe, Mediterranean, Caucasus, West and East Siberia, East Asia, incl. Far East of Russia

Parishella A. Gray. 1882. Nemacladaceae. 1 California

Parishia Hook. f. 1860. Anacardiaceae. 12 Burma, Andaman Is., western Malesia

Pariti Adans. = Thespesia Sol. ex Corrêa. Malvaceae

Parkia R. Br. 1826. Fabaceae. c. 40 trop. South America, Africa, Madagascar, trop. Asia, Malesia to Caroline and Solomon Is., Fiji (1)

Parkinsonia L. 1753. Fabaceae. 19 arid America (15), northeastern (3, Ethiopia, Somalia, Kenya) and southern (1) Africa

Parlatorea Barb. Rodr. = Sanderella Kuntze. Orchidaceae

Parlatoria Boiss. 1842. Brassicaceae. 2 southeastern Asia Minor, northern Iraq, western Iran

Parmentiera DC. 1838. Bignoniaceae. 9 Mexico, Guatemala, Belize, El Salvador, Honduras, Costa Rica, Panama, Colombia

Parnassia L. 1753. Parnassiaceae. c. 70 temp. Northern Hemisphere, esp. East Asia and northwestern North America

Parochetus Buch.-Ham. ex D. Don. 1825. Fabaceae. 1 montane areas of Ethiopia, Kenya, Tanzania, Uganda, Malawi, Mozambique, and Zaire; trop. Asia

Parodia Speg. 1923. Cactaceae. c. 60 Bolivia, southern Brazil, Paraguay, northern Argentina, Uruguay

Parodianthus Tronc. 1973. Verbenaceae. 1 Argentina

Parodiella J. R. Reeder et C. G. Reeder = Ortachne Nees ex Steud. Poaceae

Parodiochloa A. M. Molina, non C. E. Hubb. = Koeleria Pers. Poaceae

Parodiochloa C. E. Hubb. = Poa L. Poaceae

Parodiodendron Hunz. 1969. Euphorbiaceae. 1 Argentina

Parodiodoxa O. E. Schulz. 1929. Brassicaceae. 1 high mountains of northern Argentina

Parodiolyra Soderstr. et Zuolaga. 1989. Poaceae. 3 trop. America

Parolinia Webb. 1840. Brassicaceae. 3 Canary Is.

Paronychia Hill. 1753. Caryophyllaceae. c. 100 America, Europe, Mediterranean, Africa, West Asia

Paropsia Noronha ex Thouars. 1805. Passifloraceae. 11 trop. and South Africa (4), Madagascar (6), Malay Peninsula, and Sumatra (1)

Paropsiopsis Engl. 1891. Passifloraceae. 7 trop. West Africa

Paropyrum Ulbr. 1925 (Isopyrum L.). Ranunculaceae. 1 Middle Asia

Paroxygraphis W. W. Sm. 1913. Ranunculaceae. 1 eastern Himalayas (Nepal, Bhutan)

Parquetina Baill. 1889. Asclepiadaceae. 1 trop. West Africa

Parrotia C. A. Mey. 1831. Hamamelidaceae. 1 southeastern Caucasus, Iran

Parrotiopsis (Nied.) C. K. Schneid. 1905. Hamamelidaceae. 1 northwestern Himalayas

Parrya R. Br. 1823. Brassicaceae. 1 arctic Canada

Parryella Torr. et A. Gray. 1868. Fabaceae. 1 southwestern U.S., Mexico

Parryodes Jafri. 1957. Brassicaceae. 1 southern Tibet, eastern Himalayas

Parryopsis Botsch. 1955. Brassicaceae. 1 Tibet

Parsana Parsa et Maleki. 1952 (or = Laportea Gaudich.). Urticaceae. 1 Iran

Parsonsia R. Br. 1810. Apocynaceae. 80–100 India, Sri Lanka, southern China, Taiwan, Indochina, Malesia, Australia, New Zealand, New Caledonia (16), Vanuatu, Fiji (2)

Parthenice A. Gray. 1853. Asteraceae. 1 southwestern U.S., Mexico

Parthenium L. 1753. Asteraceae. 18 America, esp. Mexico, West Indies

Parthenocissus Planch. 1887. Vitaceae. 10 temp. Asia, North America; Russia (1, P. tricuspidata: Far East)

Parvatia Decne. 1837 (Stauntonia Wall.). Lardizabalaceae. 3 northeastern India

Parvisedum R. T. Clausen = Sedum L. Crassulaceae

Parvotrisetum Chrtek. 1965 (Trisetaria Forssk.). Poaceae. 1 South Europe from Italy to Greece

Pasania Oerst. = Lithocarpus Blume. Fagaceae

Pasaniopsis Kudo = Castanopsis (D. Don) Spach. Fagaceae

Pascalia Ortega = Wedelia Jacq. Asteraceae

Paschalococos J. Dransf. 1992. Arecaceae. ? sp.

Pascopyrum A. Löve. 1980 (Elytrigia Desv.). Poaceae. 1 temp. North America

Pasithea D. Don. 1832. Asphodelaceae. 1 Peru, Chile

Paspalidium Stapf. 1920. Poaceae. c. 40 tropics

Paspalum L. 1759. Poaceae. c. 330 temp. and trop. regions; 4 spp. Crimea, Caucasus, Middle Asia

Passacardoa Kuntze. 1891. Asteraceae. 3 trop. and South Africa

Passacardoa Wild = Pasaccardoa Kuntze. Asteraceae

Passaea Adans. = Ononis L. Fabaceae

Passerina L. 1753. Thymelaeaceae. 18 South Africa (southwestern Cape)

Passiflora L. 1753. Passifloraceae. c. 430: 410 trop. and subtrop. America and West Indies; 1 Madagascar, Mascarene Is.; c. 20 trop. East and Southeast Asia, Malesia; 3 Australia, Tasmania, Melanesia, Hawaiian Is.

Pastinaca L. 1753. Apiaceae. 14 temp. Eurasia; P. sativa cultivated

Pastinacopsis Golosk. 1950. Apiaceae. 1 (P. glacialis) Middle Asia (Tien Shan)

Pastorea Tod. = Ionopsidium Rchb. Brassicaceae

Patagonula L. 1753. Cordiaceae. 2 northern (P. babiensis) and southern Brazil, northern Argentina

Patascoya Urb. = Freziera Willd. Theaceae

Patellaria J. T. Williams, A. S. Scott et Ford-Lloyd = Patellifolia A. J. Scott, Ford-Lloyd et J. T. Williams. Chenopodiaceae

Patellifolia A. J. Scott, Ford-Lloyd et J. T. Williams. 1977. Chenopodiaceae. 3 Madeira, Salvage, Canary and Cape Verde Is., Morocco

Patersonia R. Br. 1807. Iridaceae. 19 Sumatra, Borneo, Philippines, New Guinea, Australia, and Tasmania

Patima Aubl. = Sabicea Aubl. Rubiaceae

Patinoa Cuatrec. 1953. Bombacaceae. 4 trop. South America

Patis Ohwi = Stipa L. Poaceae

Patosia Buchenau. 1890 (Oxychloe Phil.). Juncaceae. 2 Andes of Chile and Argentina

Patrinia Juss. 1807. Valerianaceae. 20: 5 eastern European part of Russia, Middle Asia, Siberia, Far East; Mongolia, China (15), Korea, Japan, eastern Himalayas, eastern India, northern Burma

Patropyrum A. Löve = Aegilops L. Poaceae

Pattalias S. Watson. 1889. Asclepiadaceae. 2 southwestern U.S., Mexico

Paua Caball. = Andryala L. Asteraceae

Pauella Ramam. et Sebastine = Theriophonum Blume. Araceae

Pauia Deb et Dutta. 1965 (Atropa L.). Solanaceae. 1 northeastern India

Pauldopia Steenis. 1969. Bignoniaceae. 1 (P. ghorta) northeastern India, Nepal, northern Burma, southwestern China (Yunnan), Thailand, Laos, Vietnam

Paulia Korovin = Paulita Soják. Apiaceae

Paulita Soják. 1982. Apiaceae. 3 montane Middle Asia

Paullinia L. 1753. Sapindaceae. 194 trop. and subtrop. America, 1 (P. pinnata) trop. Africa

Paulownia Siebold et Zucc. 1836. Bignoniaceae. 17 East and Southeast Asia

Paulseniella Briq. = Elsholzia Willd. Lamiaceae

Pauridia Harv. 1838. Hypoxidaceae. 2 South Africa

Pauridiantha Hook. f. 1873. Rubiaceae. 25 trop. Africa, Madagascar

Paurolepis S. Moore = Gutenbergia Sch. Bip. Asteraceae

Pausandra Radlk. 1870. Euphorbiaceae. 12 trop. America

Pausinystalia Pierre ex Beille. 1906. Rubiaceae. 13 trop. West Africa

Pavetta L. 1753. Rubiaceae. c. 350 trop. Old World

Pavieasia Pierre. 1894. Sapindaceae. 3 China, Indochina

Pavonia Cav. 1787. Malvaceae. c. 200 tropics and subtropics, esp. South America (over 100) and Africa

Paxia Gilg = Rourea Aubl. Connaraceae

Paxia O. E. G. Nilsson = Neopaxia O. E. G. Nilsson. Portulacaceae

Paxistima Raf. 1838. Celastraceae. 2 North America

Paxiuscula Herter = Argythamnia P. Browne. Euphorbiaceae

Payena A. DC. 1844. Sapotaceae. 16 Burma, Andaman Is., Indochina, Malay Peninsula, Sumatra, Java, Borneo, Philippines

Payera Baill. 1878. Rubiaceae. 1 Madagascar

Paypayrola Aubl. 1775. Violaceae. 7 trop. South America

Pearcea Regel. 1867. Gesneriaceae. 2 Ecuador

Pearsonia Duemmer. 1912. Fabaceae. 12 Zaire, Zambia, Tanzania, Malawi, Zimbabwe, Mozambique, South Africa, Madagascar (1)

Pechuel-Loeschea O. Hoffm. 1888 (Pluchea Cass.). Asteraceae. 1 Namibia

Peckelia Harms = Oxyrhynchus Brandegee. Fabaceae

Peckoltia Fourn. 1885. Asclepiadaceae. 1 Brazil

Pecteilis Raf. 1837. Orchidaceae. 8–10 trop. and East Asia, Malesia

Pectinaria Haw. 1819. Asclepiadaceae. 3 South Africa

Pectis L. 1759. Asteraceae. c. 75 trop. and subtrop. America from southern U.S. to Brazil, Galapagos Is., West Indies

Pectocarya DC. ex Meisn. 1840. Boraginaceae. 15 Pacific coastal America from British Columbia to Chile

Pedaliodiscus Ihlenf. 1968. Pedaliaceae. 1 trop. East Africa

Pedalium Royen ex L. 1759. Pedaliaceae. 1 trop. Africa, Madagascar, Socotra, trop. Asia

Peddiea Harv. ex Hook. 1840. Thymelaeaceae. 10 trop. and Southeast Africa, Madagascar (1)

Pedicellarum M. Hotta. 1976 (Pothos L.). Araceae. 1 Borneo

Pediculariopsis A. Löve et D. Löve = Pedicularis L. Scrophulariaceae

Pedicularis L. 1753. Scrophulariaceae. c. 500 Northern Hemisphere, esp. China (c. 330), 1 sp. reaches to Andes

Pedilanthus Neck. ex Poit. 1812. Euphorbiaceae. 14 Florida, Mexico, Central and trop. South America, West Indies

Pedilochilus Schltr. 1905. Orchidaceae. 25 Sulawesi (1), New Guinea, Vanuatu (1)

Pedinogyne Brand = Trigonotis Stev. Boraginaceae

Pedinopetalum Urb. et H. Wolff ex Urb. 1929. Apiaceae. 1 Hispaniola

Pediocactus Britton et Rose. 1913. Cactaceae. 6 western U.S.

Pediomelum Rydb. 1919 (Orbexilum Raf.). Fabaceae. 21 North America from southern central Canada through the U.S. to central Mexico

Pedistylis Wiens. 1978 (Emelianthe Danser). Loranthaceae. 1 South Africa (Mozambique, Transvaal, Swasiland)

Peekeliopanax Harms = Gastonia Comm. ex Lam. Araliaceae

Peersia L. Bolus = Rhinephyllum N. E. Br. Aizoaceae

Pegaeophyton Hayek et Hand.-Mazz. 1922. Brassicaceae. 2–3 western China, Tibet, Himalayas from Kashmir to Bhutan

Peganum L. 1753. Peganaceae. 5–6 southern U.S. (Texas), Mexico, Mediterranean, Southeast Europe, Caucasus, Middle Asia, southern Siberia; Iran, Afghanistan, Mongolia, China, India

Pegia Colebr. 1827. Anacardiaceae. 3 eastern Himalayas, northeastern India, southern China, Philippines

Pegolettia Cass. 1825. Asteraceae. 9 North, trop., and South Africa, Madagascar, Arabian Peninsula, South Asia southward to Java

Pehria Sprague. 1923. Lythraceae. 1 Colombia, Venezuela

Peixotoa A. Juss. 1833. Malpighiaceae. 11 Brazil

Pelagatia O. E. Schulz = Weberbauera Gilg et Muschler. Brassicaceae

Pelagodendron Seem. 1866. Rubiaceae. 1 Fiji

Pelagodoxa Becc. 1917. Arecaceae. 1 (P. henryana) Marquesas Is.

Pelargonium L'Hér. ex Aiton. 1789. Geraniaceae. c. 280 trop. and South (esp. Cape) Africa, St. Helena (1, P. cotyledonis), Tristan da Cunha (1), Caucasus (1, P. endlicheranum), Asia Minor, Iraq, southern Arabian Peninsula eastward to southern India, Australia, Lord Howe I., New Zealand

Pelatantheria Ridl. 1896. Orchidaceae. 8–10 trop. Asia from Himalayas and India through Burma, Indochina, and southern China to Taiwan; Malesia: Malay Peninsula, Sumatra

Pelea A. Gray = Melicope J. R. Forst. et G. Forst. Rutaceae

Pelecostemon Leonard. 1958. Acanthaceae. 1 Colombia

Pelecyphora Ehrenb. 1843. Cactaceae. 2 northeastern Mexico

Pelexia Poit. ex Lindl. 1826. Orchidaceae. 67 trop. and subtrop. America

Peliosanthes Andrews. 1808. Convallariaceae. 10 India, eastern Himalayas, Burma, southern and southeastern China, Hainan and Taiwan, Indochina, Malesia

Peliostomum E. Mey. ex Benth. 1836. Scrophulariaceae. 7 trop. and South Africa

Pellacalyx Korth. 1836. Rhizophoraceae. 8 Burma, Thailand, southern China, Malesia (Malay Peninsula, Borneo, Sulawesi, Philippines)

Pellegrinia Sleumer. 1935. Ericaceae. 5 Andes

Pellegriniodendron J. Léonard. 1955. (Macrolobium Schreb.). Fabaceae. 1 Ivory Coast, Ghana, Cameroun, Gabon

Pelletiera A. St.-Hil. 1822. Primulaceae. 2 subtrop. South America (1), Macaronesia (1)

Pelliceria Planch. et Triana = Pelliciera Planch. et Triana ex Benth. Pellicieraceae

Pelliciera Planch. et Triana ex Benth. 1862. Pellicieraceae. 1 from Costa Rica to Ecuador, West Indies

Pellionia Gaudich. 1830 (or = Elatostema J. R. Forst. et G. Forst.). Urticaceae. c. 50 trop. and East Asia, Oceania

Pelozia Rose = Lopezia Cav. Onagraceae

Peltaea (C. Presl) Standl. 1916 (Pavonia Cav.). Malvaceae. 18 trop. America, West Indies (2)

Peltandra Raf. 1819. Araceae. 3 North America

Peltanthera Benth. 1876. Buddlejaceae. 1 montane trop. America

Peltaria Jacq. 1762. Brassicaceae. 7 Europe (2), eastern Mediterranean, West Asia; 1–2 southern Transcaucasia, Middle Asia

Peltariopsis (Boiss.) N. Busch. 1927. Brassicaceae. 3 Caucasus and Turkmenistan (2), Turkey, northern Iran

Peltastes Woodson. 1932. Apocynaceae. 7 Central and trop. South America

Pelticalyx Griff. 1854. Annonaceae. 1 (P. argentea) South Asia

Peltiphyllum (Engl.) Engl. = Darmera Voss. Saxifragaceae

Peltoboykinia (Engl.) Hara. 1937 (Boykinia Nutt.). Saxifragaceae. 2 Japan

Peltobractea Rusby = Peltaea (C. Presl) Standl. Malvaceae

Peltodon Pohl. 1827. Lamiaceae. 6 Brazil, Paraguay

Peltogyne Vogel. 1837. Fabaceae. 23 trop. America, West Indies

Peltophoropsis Chiov. = Parkinsonia L. Fabaceae

Peltophorum (Vogel) Benth. 1840. Fabaceae. 8–9 tropics

Peltophyllum Gardner. 1843. Triuridaceae. 2 northern Argentina, Paraguay, southeastern Brazil

Peltostigma Walp. 1846. Rutaceae. 3 Mexico, Central America, West Indies

Pelucha S. Watson. 1889. Asteraceae. 1 Baja California

Pemphis J. R. Forst. et G. Forst. 1775. Lythraceae. 1 (P. acidula) trop. Old World from coasts of East Africa and Indian Ocean islands to southern China, coastal trop. Australia, and eastward to Polynesia

Penaea L. 1753. Penaeaceae. 3 South Africa (Cape)

Penelopeia Urb. 1921. Cucurbitaceae. 1 Haiti

Penianthus Miers. 1864. Menispermaceae. 4 trop. West and Central Africa

Peniocereus (A. Berger) Britton et Rose. 1909. Cactaceae. 20 southwestern U.S., Mexico, Central America

Peniophyllum Pennell = Oenothera L. Onagraceae

Pennantia J. R. Forst. et G. Forst. 1775. Icacinaceae. 4 eastern Australia, New Zealand, Norfolk I.

Pennellia Nieuwl. 1918. Brassicaceae. 9 southern U.S. (2, Colorado, Arizona, New Mexico, Texas), Mexico, Guatemala (1), Bolivia and northern Argentina (1)

Pennellianthus Crosswh. et Kawano. 1970 (Penstemon Schmidel). Scrophulariaceae. 1 Russian Far East (Kamchatka, Okhotski Krai, Sakhalin I., Kuril Is.), Japan

Pennilabium J. J. Sm. 1914. Orchidaceae. 4–6 northeastern India, Thailand, Malay Peninsula, Sumatra, Java, Philippines (2)

Pennisetum Rich. ex Pers. 1805. Poaceae. c. 100 trop. and temp. regions

Penstemon Schmidel. 1763. Scrophulariaceae. 265 North and Central (1) America

Pentabothra Hook. f. 1883. Asclepiadaceae. 1 northeastern India

Pentabrachion Muell. Arg. = Microdesmis Hook. f. ex Hook. Pandaceae

Pentacalia Cass. 1827. Asteraceae. c. 200 Central and South America

Pentacarpaea Hiern = Pentanisia Harv. Rubiaceae

Pentace Hassk. 1858. Tiliaceae. 25 Southeast Asia, western Malesia

Pentaceras Hook. f. 1862. Rutaceae. 1 eastern Australia

Pentachaeta Nutt. 1840. Asteraceae. 6 California and Baja California

Pentachlaena H. Perrier. 1920. Sarcolaenaceae. 2 Madagascar

Pentachondra R. Br. 1810. Epacridaceae. 4 southeastern Australia, Tasmania, New Zealand

Pentaclethra Benth. 1840. Fabaceae. 2 Central and northern South America; West Indies (1, P. macroloba); trop. Africa (2, from Nigeria to Angola and Congo)

Pentacme A. DC. = Shorea Roxb. ex C. F. Gaertn. Dipterocarpaceae

Pentacoilanthus Rappa et Cammarone. 1954. Aizoaceae. 7 South Africa (nomen dubium ?)

Pentacrostigma K. Afzel. 1929. Convolvulaceae. 1 Madagascar

Pentactina Nakai. 1917. Rosaceae. 1 Korea

Pentacyphus Schltr. 1906. Asclepiadaceae. 1 Peru

Pentadesma Sabine. 1824. Clusiaceae. 5 trop. Africa, Seychelles, ? China

Pentadiplandra Baill. 1886. Capparaceae. 1–2 trop. West Africa

Pentadynamis R. Br. = Crotalaria L. Fabaceae

Pentaglottis Tausch. 1829 (Anchusa L.). Boraginaceae. 1 Southwest Europe

Pentagonanthus Bullock. 1962. Asclepiadaceae. 2 trop. Africa

Pentagonia Benth. 1844. Rubiaceae. 25 trop. America

Pentalepis F. Muell. 1863 (Chrysogonum L.). Asteraceae. 2–3 Australia

Pentalinon Voigt. 1845. Apocynaceae. 2 Florida, Central America, West Indies

Pentaloncha Hook. f. 1873. Rubiaceae. 3 trop. West Africa

Pentameris P. Beauv. 1812. Poaceae. 9 South Africa

Pentamerista Maguire. 1972. Tetrameristaceae. 1 northern South America

Pentanema Cass. 1818. Asteraceae. 10–12 Mediterranean, Africa, West Asia eastward to Middle Asia (8), China, and India

Pentanisia Harv. 1842. Rubiaceae. 18 trop. and northeastern South Africa, Madagascar

Pentanopsis Rendle. 1898. Rubiaceae. 1 Ethiopia, Somalia, Kenya

Pentanura Blume. 1850. Asclepiadaceae. 2 Burma, Sumatra

Pentapanax Seem. 1864. Araliaceae. 18 India, Sri Lanka, Himalayas, southwestern and southern China, Taiwan (1), Indochina, Malesia to Australia

Pentapeltis Bunge. 1845. Apiaceae. 1 Australia

Pentapera Klotzsch = Erica L. Ericaceae

Pentapetes L. 1753. Sterculiaceae. 1 India, China, Indochina, Malesia

Pentaphalangium Warb. = Garcinia L. Clusiaceae

Pentaphragma Wall. ex G. Don. 1834. Pentaphragmataceae. 30 southern China, Indochina, Malesia

Pentaphylax Gardner et Champ. 1849. Pentaphylacaceae. 1–2 southern China, Indochina, Malay Peninsula, Sumatra

Pentaphylloides Hill = Potentilla L. Rosaceae

Pentaplaris L. O. Williams et Standl. 1952. Tiliaceae. 1 Central America

Pentapleura Hand.-Mazz. 1913. Lamiaceae. 1 Asia Minor

Pentapogon R. Br. 1810. Poaceae. 1 southeastern Australia, Tasmania

Pentaptilon E. Pritz. 1904. Goodeniaceae. 1 southwestern Australia

Pentaraphia Lindl. = Gesneria L. Gesneriaceae

Pentarhaphia Decne. = Gesneria L. Gesneriaceae

Pentarhopalopilia (Engl.) Hiepko. 1987. Opiliaceae. 4 Central Africa, Madagascar (2)

Pentarrhaphis Kunth. 1816. Poaceae. 3 trop. America from Mexico to Colombia

Pentarrhinum E. Mey. 1838. Asclepiadaceae. 3 trop. and South Africa

Pentas Benth. 1844. Rubiaceae. 40–50 Africa from West Africa to Somalia and south to Angola and Natal, Comoro Is., Madagascar, Arabian Peninsula

Pentasacme Wall. ex Wight. 1834. Asclepiadaceae. 4 Himalayas from Kumaun through Nepal to Bhutan, northeastern India (Khasia, Assam), Bangladesh, Burma, Thailand, southern China, incl. Hainan, Malay Peninsula

Pentaschistis (Nees) Spach. 1841. Poaceae. c. 66 trop. (6) and South (c. 60) Africa, Madagascar (3), southwestern Arabian Peninsula

Pentascyphus Radlk. 1879. Sapindaceae. 1 Guyana

Pentaspadon Hook. f. 1860. Anacardiaceae. 5 Southeast Asia, Malesia, Solomon Is.

Pentaspatella Gleason = Sauvagesia L. Sauvagesiaceae

Pentastelma Tsiang et P. T. Li. 1974. Asclepiadaceae. 1 southern China (incl. Hainan)

Pentastemon Batsch = Penstemon Schmidel. Scrophulariaceae

Pentastemona Steenis. 1982. Pentastemonaceae. 2 Sumatra

Pentastemonodiscus Rech. f. 1965. Caryophyllaceae. 1 Afghanistan
Pentasticha Turcz. 1862 (Fuirena Rottb.). Cyperaceae. 1 trop. Africa, Madagascar
Pentataenium Tamamsch. = Stenotaenia Boiss. Apiaceae
Pentatherum Nábêlek. 1929 (Agrostis L.). Poaceae. 7 West and Middle Asia
Pentathymelaea Lecomte = Wikstroemia Endl. Thymelaeaceae
Pentatrichia Klatt. 1895. Asteraceae. 4 Namibia
Pentatropis Wight et Arn. 1834. Asclepiadaceae. 6 Africa, Mascarene Is., Arabian Peninsula, India, Sri Lanka, Southeast Asia, Australia (4)
Penthea Lindl. = Disa Bergius. Orchidaceae
Pentheriella O. Hoffm. et Muschl. = Heteromma Benth. Asteraceae
Penthorum L. 1753. Penthoraceae. 3 East Asia, Indochina, eastern North America; Russia (1, P. chinense—Far East)
Pentodon Hochst. 1844. Rubiaceae. 2–3: P. pentandrus [? incl. P. halei (Torr. et A. Gray) A. Gray—U.S. from Texas to Florida, Nicaragua, Cuba, ? Brazil] trop. Africa from Cape Verde Is. to Somalia (2) south to Swaziland, Seychelles, Madagascar, Arabian Peninsula
Pentopetia Decne. 1844. Asclepiadaceae. 10 Madagascar
Pentopetiopsis Costantin et Gallaud. 1906. Asclepiadaceae. 1 Madagascar
Pentossaea Judd = Ossaea DC. Melastomataceae
Pentstemonacanthus Nees. 1847. Acanthaceae. 1 Brazil
Pentzia Thunb. 1800. Asteraceae. 25 North, trop., and South Africa
Peperomia Ruiz et Pav. 1794. Peperomiaceae. Over 1,000 tropics and subtropics, chiefly America
Pepinia Brongn. ex André. 1870. Bromeliaceae. 42 trop. America from southern Mexico to Ecuador and Brazil
Peplidium Delile. 1813. Scrophulariaceae. 10 North and trop. Africa, trop. Asia, Australia
Peplis L. 1753 (Lythrum L.). Lythraceae. 3 temp. Northern Hemisphere in wet places
Peplonia Decne. 1844. Asclepiadaceae. 3 Brazil
Peponia Naudin = Peponium Engl. Cucurbitaceae
Peponidium (Baill.) Arènes. 1960. Rubiaceae. 20 Comoro Is., Madagascar
Peponiella Kuntze = Peponium Engl. Cucurbitaceae
Peponium Engl. 1897. Cucurbitaceae. 20 trop. and South Africa, Madagascar, Aldabra I., and Seychelles
Peponopsis Naudin. 1859. Cucurbitaceae. 1 Mexico
Pera Mutis. 1784. Euphorbiaceae. 40 trop. America, West Indies
Peracarpa Hook. f. et Thomson. 1858. Campanulaceae. 2: P. carnosa—Himalayas (India, Nepal, Bhutan), northern Burma, northern Thailand, southern China (Yunnan, Kweichow), Taiwan, Japan, Malesia (Philippines, New Guinea); P. circaeoides—southern Kamchatka, Sakhalin)
Perakanthus Robyns. 1925. Rubiaceae. 2 Malay Peninsula
Perama Aubl. 1775. Rubiaceae. 12 trop. America: Colombia, Peru, Venezuela, Guyana, Surinam, French Guiana, Brazil; West Indies: Trinidad, Martinique
Perantha Craib = Oreocharis Benth. Gesneriaceae
Perapentacoilanthus Rappa et Camarrone. 1956. Aizoaceae. 1 South Africa (nomen dubium ?)
Peraphyllum Nutt. ex Torr. et A. Gray. 1840. Rosaceae. 1 western U.S. from Oregon and California to Colorado and New Mexico
Peratanthe Urb. 1921. Rubiaceae. 2 Cuba, Haiti
Peratetracoilanthus Rappa et Camarrone = Mesembryanthemum L. Aizoaceae
Peraxilla Tiegh. 1894. Loranthaceae. 2 New Zealand
Perdicium L. 1760. Asteraceae. 2 South Africa (southwestern Cape)
Perebea Aubl. 1775. Moraceae. 8 trop. America

Peregrina W. R. Anderson. 1985. Malpighiaceae. 1 Brazil, Paraguay

Pereilema C. Presl. 1830. Poaceae. 4 trop. America from Mexico to Brazil

Perella Tiegh. = Peraxilla Tiegh. Loranthaceae

Perenideboles Ram. Goyena. 1911. Acanthaceae. 1 Nicaragua

Pereskia Mill. 1754. Cactaceae. 16 Trop. America from southern Mexico to northern Argentina; West Indies

Pereskiopsis Britton et Rose. 1907. Cactaceae. 12 Mexico, Central America

Perezia Lag. 1811. Asteraceae. 30 trop. and subtrop. America from southern U.S. to Patagonia

Pereziopsis Coulter = Onoseris Willd. Asteraceae

Pergularia L. 1767. Asclepiadaceae. 4 trop. and North Africa, Madagascar, Southwest Asia to India, Sri Lanka, Nepal, Burma, Malay Peninsula

Periandra Mart. ex Benth. 1837. Fabaceae. 6 Haiti (1), Brazil (5)

Perianthomega Bureau ex Baill. 1888. Bignoniaceae. 1 central Brazil, Bolivia, Paraguay

Periarrabidaea A. Samp. 1934. Bignoniaceae. 2 Brazil (Amazonia)

Periballia Trin. 1822. Poaceae. 3 Mediterranean

Pericalia Cass. = Roldana La Llave et Lex. Asteraceae

Pericallis Webb. 1839. Asteraceae. 14 Macaronesia

Pericalymma Endl. 1840. Myrtaceae. 3 southwestern Australia

Pericalypta Benoist. 1962. Acanthaceae. 1 Madagascar

Pericampylus Miers. 1851. Menispermaceae. 2–3 India, eastern Himalayas, Burma, southern China, Taiwan, southern Japan, Indochina, western Malesia, Moluccas

Perichasma Miers = Stephania Lour. Menispermaceae

Perichlaena Baill. 1888 (Fernandoa Welw. ex Seem.). Bignoniaceae. 1 Madagascar

Pericome A. Gray. 1853. Asteraceae. 2 southwestern U.S., Mexico

Pericopsis Thwaites. 1864. Fabaceae. 4 trop. Africa (3), Sri Lanka, Malesia, Micronesia

Perictenia Miers = Odontadenia Benth. Apocynaceae

Perideridia Rchb. 1837. Apiaceae. 13 U.S., esp. western areas

Peridiscus Benth. 1862. Peridiscaceae. 1 Venezuela, Brazil (Amazonia)

Periestes Baill. 1890 (Hypoestes Sol. ex R. Br.). Acanthaceae. 2 Comoro Is., Madagascar

Periglossum Decne. 1844. Asclepiadaceae. 4 South Africa

Perilepta Bremek. 1944. Acanthaceae. 8 India, Himalayas, western and southwestern China, Indochina

Perilimnastes Ridl. = Anerincleistus Korth. Melastomataceae

Perilla L. 1764. Lamiaceae. 6 Himalayas from Kashmir to Burma, China, Malesia; naturalized in Russian Far East and Japan

Perillula Maxim. 1875. Lamiaceae. 1 Japan

Perilomia Kunth. 1818. Lamiaceae. 8 China

Periomphale Baill. = Wittsteinia F. Muell. Alseuousmiaceae

Peripentadenia L. S. Sm. 1957. Elaeocarpaceae. 2 Australia (Queensland)

Peripeplus Pierre. 1898. Rubiaceae. 1 trop. West Africa

Periphanes Salish. = Hessea Herb. Amaryllidaceae

Periploca L. 1753. Asclepiadaceae. 13 Europe, Mediterranean, trop. Africa, temp. and subtrop. Asia; P. graeca—Middle Asia; P. sepium—Russian Far East

Periptera DC. 1824 (Anoda Cav.). Malvaceae. 5 Mexico from Sonora to Chiapas; P. punicea—extending to Guatemala

Peripterygia (Baill.) Loes. 1906. Celastraceae. 1 New Caledonia

Peripterygium Hassk. = Cardiopteris Blume. Cardiopteridaceae

Perispermum O. Deg. = Bonamia Thouars. Convolvulaceae

Perissandra Gagnep. 1948. Violaceae. 1 Laos

Perissocarpa Steyerm. et Maguire. 1984. Ochnaceae. 2 Venezuela

Perissocoeleum Mathias et Constance. 1952. Apiaceae. 4 Colombia

Perissolobus N. E. Br. = Machaerophyllum Schwantes. Aizoaceae

Peristeranthus T. F. Hunt. 1954. Orchidaceae. 1 northeastern Australia

Peristeria Hook. 1831. Orchidaceae. 10 trop. America from Costa Rica to Peru, Brazil, and Surinam

Peristrophe Nees. 1832. Acanthaceae. c. 40 North (Egypt), trop. and South Africa, Southwest Asia eastward to China, Himalayas and eastern Malesia

Peristylus Blume. 1825 (Herminium Guett.). Orchidaceae. 70–80 Himalayas, India, Sri Lanka, Andaman Is., Burma, China, Hong Kong, Taiwan, Korea, Japan, Indochina, Malesia from Malay Peninsula to New Guinea, Australia, western Oceania to Fiji (4), Samoa, and Society Is.

Peritassa Miers. 1872. Celastraceae. 14 trop. South America, Tobago

Perithrix Pierre = Batesanthus N. E. Br. Asclepiadaceae

Peritoma DC. = Cleome L. Capparaceae

Perittostema I. M. Johnst. 1954. Boraginaceae. 1 Mexico

Perityle Benth. 1844. Asteraceae. 55 southwestern U.S., Mexico

Pernettya Gaudich. 1825 (Gaultheria Kalm ex L.). Ericaceae. 14 Tasmania, New Zealand, Galapagos Is., America from Mexico to Chile and southern Argentina, Falkland Is.

Pernettyopsis King et Gamble. 1906. Ericaceae. 1 Malay Peninsula

Peronema Jack. 1822. Verbenaceae. 1 southern Burma, Thailand, Malay Peninsula, Sumatra, Borneo

Perotis Aiton. 1789. Poaceae. 10 trop. Africa, trop. Asia and Australia

Perotriche Cass. = Stoebe L. Asteraceae

Perovskia Kar. 1841. Lamiaceae. 8 Middle Asia (5) northern and southeastern Iran, Afghanistan, Pakistan, northwestern India, western Tibet, continental China

Perplexia Iljin = Jurinella Jaub. ex Spach. Asteraceae

Perralderia Coss. 1859. Asteraceae. 3 Morocco, Algeria, Libya

Perralderiopsis Rauschert = Iphiona Cass. Asteraceae

Perriera Courchet. 1905. Simaroubaceae. 1 Madagascar

Perrieranthus Hochr. = Perrierophytum Hochr. Malvaceae

Perrierastrum Guillaumin. 1931. Lamiaceae. 1 Madagascar

Perrierbambus A. Camus. 1924. Poaceae. 2 Madagascar

Perrieriella Schltr. = Oeonia Lindl. Orchidaceae

Perrierodendron Cavaco. 1951. Sarcolaenaceae. 1 Madagascar

Perrierophytum Hochr. 1916. Malvaceae. 9 Madagascar

Perrierosedum (Berger) H. Ohba. 1978. Crassulaceae. 1 Madagascar

Perrottetia Kunth. 1824. Celastraceae. 25–30 Southeast Asia from central China and Taiwan (1) to Malesia (1, P. alpestris Sumatra, Malay Peninsula, Java, Borneo, Philippines, Sulawesi, Moluccas, New Guinea), Solomon Is. (Bougainville, Isabel, and Guadalcanal Is.), Australia (Queensland), Hawaiian Is., trop. America from Mexico to Colombia and Peru

Persea Mill. 1754 (Machilus Nees). Lauraceae. c. 150 tropics and subtropics

Persica Hill = Amygdalus L. Rosaceae

Persicaria Hill. 1753. Polygonaceae. 35 (or 150 ?) temp. Northern Hemisphere south to Mediterranean, India, southern China, and Malay Peninsula

Persoonia Sm. 1798. Proteaceae. c. 60 Australia, New Zealand

Pertusadina Ridsdale. 1978 (Adina Salisb.). Rubiaceae. 4 China, Hainan, Thailand, Malay Peninsula, Sumatra, Borneo, Philippines, Moluccas, New Guinea

Pertya Sch. Bip. 1862. Asteraceae. 18 Afghanistan, China, Japan

Perularia Lindl. = Tulotis Raf. Orchidaceae

Perulifera A. Camus = Pseudoechinolaena Stapf. Poaceae

Perymeniopsis H. Rob. 1978. Asteraceae. 1 Mexico

Perymenium Schrad. 1830. Asteraceae. 40 trop. America from Mexico to Peru

Pescatorea Rchb. f. 1852. Orchidaceae. 17 Costa Rica, Panama, Colombia

Peschiera A. DC. = Tabernaemontana L. Apocynaceae

Petagnaea Caruel. 1894. Apiaceae. 1 Sicily

Petagnia Guss. = Petagnaea Caruel. Apiaceae

Petagomoa Bremek. 1934 (Psychotria L.). Rubiaceae. 7 trop. South America

Petalacte D. Don. 1826. Asteraceae. 1 South Africa

Petaladenium Ducke. 1938. Fabaceae. 1 Brazil (Basin of Rio Negro)

Petalidium Nees. 1832. Acanthaceae. 35 trop. and South Africa 1 (P. barlerioides) Himalayas and western India

Petalocaryum Pierre ex A. Chev. 1917. Olacaceae. 1 trop. West Africa

Petalocentrum Schltr. = Sigmatostalix Rchb. f. Orchidaceae

Petalochilus R. S. Rogers = Caladenia R. Br. Orchidaceae

Petalolophus K. Schum. 1905. Annonaceae. 1 (P. megalopus) New Guinea

Petalonyx A. Gray. 1855. Loasaceae. 5 southwestern U.S., Mexico

Petalostelma E. Fourn. 1885 (Metastelma R. Br.). Asclepiadaceae. 2 Brazil

Petalostemon Michx. = Dalea L. Fabaceae

Petalostigma F. Muell. 1857. Euphorbiaceae. 7 New Guinea (1), Australia

Petalostylis R. Br. 1849. Fabaceae. 3 Australia

Petamenes Salisb. ex J. W. Loudon = Gladiolus L. Iridaceae

Petasites Hill. 1753. Asteraceae. 15 North America, Europe, temp. Asia; Russia (c. 10, from European part to Far East)

Petastoma Miers = Arrabidaea DC. Bignoniaceae

Petchia Livera. 1926. Apocynaceae. 1 Sri Lanka

Petelotiella Gagnep. 1929. Urticaceae. 1 northeastern Vietnam

Petenaea Lundell. 1962. Elaeocarpaceae. 1 Guatemala

Peteniodendron Lundell = Pouteria Aubl. Sapotaceae

Peteravenia R. M. King et H. Rob. 1971 (Eupatorium L.). Asteraceae. 5 Mexico, Central America

Peteria A. Gray. 1852. Fabaceae. 4 southern and southeastern U.S., northern Mexico

Petermannia F. Muell. 1860. Pettermanniaceae. 1 Australia (Queensland and New South Wales)

Peterodendron Sleumer. 1936. Flacourtiaceae. 1 Tanzania

Petersianthus Merr. 1916. Lecythidaceae. 2 trop. West Africa

Petiniotia J. Léonard. 1980 (Sterigma DC.). Brassicaceae. 1 Iran, Afghanistan

Petitia Jacq. 1760. Verbenaceae. 2 West Indies

Petitiocodon Robbr. 1988. Rubiaceae. 1 Nigeria

Petitmenginia Bonati. 1911. Scrophulariaceae. 1–2 southern China and Indochina

Petiveria L. 1753. Petiveriaceae. 2 trop. and subtrop. America, West Indies

Petkovia Stef. 1936 (Campanula L.). Campanulaceae. 1 Balkan Peninsula

Petopentia Bullock = Tacazzea Decne. Asclepiadaceae

Petradoria Greene. 1895 (Solidago L.). Asteraceae. 2 southwestern U.S.

Petraeovitex Oliv. 1883. Verbenaceae. 7 Malesia (excl. Java), New Zealand, Oceania

Petrea L. 1753. Verbenaceae. c. 40 trop. America, West Indies

Petriella Zotov. 1943. Poaceae. 2 New Zealand

Petrina J. B. Phipps. 1964 (Danthoniopsis Stapf). Poaceae. 3 trop. and South Africa

Petrobium R. Br. 1818. Asteraceae. 1 St. Helena

Petrocallis R. Br. 1812. Brassicaceae. 2 mountains of South Europe (1), northern Iran (1)

Petrocodon Hance. 1883. Gesneriaceae. 1 (P. dealbatus) China

Petrocoma Rupr. 1869. Caryophyllaceae. 1 Caucasus

Petrocoptis A. Braun ex Endl. 1842. Caryophyllaceae. 12 Iberian Peninsula, France (2)

Petrocosmea Oliv. 1887. Gesneriaceae. 27 East and Southeast Asia, esp. Yunnan (20 endemic)

Petrodavisia Holub = Centaurea L. Asteraceae

Petrodoxa Anthony = Beccarinda Kuntze. Gesneriaceae

Petroedmondia Tamamsch. 1987. Apiaceae. 1 (P. syriaca) Turkey (Anatolia), Syria, Israel, Jordan, Iraq, western Iran

Petrogenia I. M. Johnst. = Bonamia Thouars. Convolvulaceae

Petrollinia Chiov. = Inula L. Asteraceae

Petromarula Vent. ex Hedw. f. 1806. Campanulaceae. 1 Crete

Petronymphe H. E. Moore. 1951. Alliaceae. 1 Mexico

Petrophile R. Br. ex Knight. 1809. Proteaceae. 42 Australia

Petrophyton Rydb. 1908. Rosaceae. 3 western North America

Petrophytum (Nutt. ex Torr. et A. Gray) Rydb. = Petrophyton Rydb. Rosaceae

Petrorhagia Link. 1829 (Gypsophila L.). Caryophyllaceae. 26 Canary Is., Europe, Mediterranean, West (incl. Caucasus) and Middle Asia, West and East Siberia

Petrosavia Becc. 1871. Melanthiaceae. 2 East and Southeast Asia, western Malesia

Petrosciadium Edgew. = Eriocycla Lindl. Apiaceae

Petrosedum Grulich = Sedum L. Crassulaceae

Petroselinum Hill. 1756. Apiaceae. 2–3 Southwest and South Europe

Petrosimonia Bunge. 1862. Chenopodiaceae. 12 Southeast Europe (Greece, Romania, Bulgaria, Ukraine), Caucasus, Middle Asia, Turkey, Syria, Iraq, Iran, Mongolia, northwestern China

Petteria C. Presl. 1845. Fabaceae. 1 Balkans and Dalmatia

Petunga DC. = Hypobathrum Blume. Rubiaceae

Petunia Juss. 1803 (Nicotiana L.). Solanaceae. 3 trop. and subtrop. South America, Cuba

Peucedanum L. 1753. Apiaceae. 6–7 warm–temp. Eurasia from Mediterranean to Siberia

Peucephyllum A. Gray. 1859. Asteraceae. 1 southwestern U.S.

Peumus Molina. 1782. Monimiaceae. 1 Chile

Peyritschia Fourn. 1886. Poaceae. 2 Mexico, Costa Rica

Peyrousea DC. 1838. Asteraceae. 1 South Africa

Pezisicarpus Vernet. 1904. Apocynaceae. 1 Indochina

Pfaffia Mart. 1826. Amaranthaceae. 53 Central and South America

Pfeiffera Salm-Dyck = Lipismium Pfeiff. Cactaceae

Phacelia Juss. 1789. Hydrophyllaceae. c. 200 North and South (Andes) America

Phacellanthus Siebold et Zucc. 1846. Orobanchaceae. 1 Russian Far East (Ussuriyski Krai and Sakhalin I.), China, Japan

Phacellaria Benth. 1880. Santalaceae. 8 northeastern India, China (5), Indochina

Phacellothrix F. Muell. 1878. Asteraceae. 1 eastern Malesia, trop. eastern Australia

Phacelophrynium K. Schum. 1902. Maranbtaceae. 9 Nicobar Is., Malesia

Phacelurus Griseb. 1846. Poaceae. 9 East Africa, Southwest and trop. Asia

Phacocapnos Bernh. = Cysticapnos Mill. Fumariaceae

Phaeanthus Hook. f. et Thomson. 1855. Annonaceae. c. 20 southern India, southern China, Lower Burma, Thailand, Indochina, Malesia to Philippines

Phaedranassa Herb. 1845. Amaryllidaceae. 8 Andes from Costa Rica to Peru

Phaedranthus Miers = Distictis Mart. ex Meisn. Bignoniaceae

Phaenanthoecium C. E. Hubb. 1936. Poaceae. 1 mountains of trop. Northeast Africa, southwestern Arabian Peninsula

Phaenocoma D. Don. 1829. Asteraceae. 1 South Africa (Cape)

Phaenohoffmannia Kuntze = Pearsonia Dümmer. Fabaceae

Phaenosperma Munro ex Benth. 1881. Poaceae. 1 East and Southeast Asia

Phaeocephalus S. Moore = Hymenolepis Cass. Asteraceae

Phaeoneuron Gilg = Ochthocharis Blume. Melostomataceae

Phaeonychium O. E. Schulz. 1927. Brassicaceae. 4 Middle Asia (2, Tajikistan), western China, Afghanistan, Pakistan, India, Nepal, Bhutan

Phaeopappus Boiss. = Centaurea L. Asteraceae

Phaeoptilum Radlk. 1883. Nyctaginaceae. 1 Namibia

Phaeosphaerion Hassk. = Commelina L. Commelinaceae

Phaeostemma Fourn. = Matelea Aubl. Asclepiadaceae

Phaeostigma Muldashev. 1981. Asteraceae. 3 China

Phagnalon Cass. 1819. Asteraceae. 30 Canary Is., Mediterranean, West Asia eastward to Middle Asia (2) and Himalayas

Phainantha Gleason. 1948. Melastomataceae. 1 trop. South America

Phaiophleps Raf. = Olsynium Raf. Iridaceae

Phaius Lour. 1790. Orchidaceae. 40–50 trop. Africa (2), Madagascar, Mascarene Is., Himalayas, India, Sri Lanka, Burma, southern China, Taiwan, Japan, Indochina, Malesia from Malay Peninsula to New Guinea, Solomon Is., trop. Australia, Vanuatu, New Caledonia, Fiji, Samoa, and several Pacific islands

Phalacrachena Iljin. 1937 (Centaurea L.). Asteraceae. 2 southeastern European part of Russia, Caucasus, Middle Asia

Phalacraea DC. 1836. Asteraceae. 4 Colombia, Ecuador, Peru

Phalacrocarpum (DC.) Willk. 1864. Asteraceae. 2 Portugal, Spain

Phalacroseris A. Gray. 1868. Asteraceae. 1 California

Phalaenopsis Blume. 1825. Orchidaceae. c. 40 Himalayas, Tibet, India, Burma, China, Taiwan, Indochina, Malesia, esp. Borneo and Philippines (c. 20), eastward to New Guinea, New Britain, northern Australia, Oceania

Phalaris L. 1753. Poaceae. c. 20 Southeast Europe, Mediterranean, mountains of trop. Africa and South Africa, West (incl. Caucasus) and Middle Asia

Phalaroides Wolf. 1781 (Phalaris L.). Poaceae. 1 extratrop. regions of the Northern Hemisphere, South Africa

Phaleria Jack. 1822. Thymelaeaceae. 20 Sri Lanka (1, P. capitata), continental Southeast Asia, Malesia from Sumatra to Admiralty Is., New Britain and Louisiade Arch., Australia, Caroline Is., the Pacific eastward to Samoa and Tonga

Phalocallis Herb. = Cypella Herb. Iridaceae

Phanera Lour. = Bauhinia L. Fabaceae

Phanerodiscus Cavaco. 1954. Olacaceae. 1 Madagascar

Phaneroglossa B. Nord. 1978 (Senecio L.). Asteraceae. 1 South Africa (Cape)

Phanerogonocarpus Cavaco = Tambourissa Sonn. Monimiaceae

Phanerostylis (A. Gray) R. M. King et H. Rob. 1972. Asteraceae. 5 Mexico

Phania DC. 1836. Asteraceae. 5 West Indies

Phanopyrum (Raf.) Nash. 1903 (Panicum L.). Poaceae. 1 U.S. from Georgia to eastern Texas

Pharbitis Choisy. 1833 (Ipomoea L.). Convolvulaceae. 24 warm–temp. and trop. regions

Pharnaceum L. 1753. Molluginaceae. 25 South Africa

Pharus P. Browne. 1756. Poaceae. 8 trop. and subtrop. America from Florida and Mexico to Argentina and Uruguay; West Indies

Phaseolus L. 1753. Fabaceae. c. 50 trop. and subtrop. America

Phaulanthus Ridl. = Anerincleistus Korth. Melastomataceae

Phaulopsis Willd. 1800. Acanthaceae. 20 trop. Africa, Mascarene Is., Himalayas, India, western China, Indochina

Phaulothamnus A. Gray. 1885. Achatocarpaceae. 1 Texas, northern Mexico

Phebalium Vent. 1804. Rutaceae. 45 Australia, New Zealand (1)

Pheidochloa S. T. Blake. 1945. Poaceae. 2 New Guinea, Australia (Queensland)

Pheidonocarpa L. E. Skog. 1976. Gesneriaceae. 1 Cuba, Jamaica

Phelipaea Desf. = Phelypaea L. Orobanchaceae

Phelline Labill. 1824. Phellinaceae. 10 New Caledonia

Phellocalyx Bridson. 1980. Rubiaceae. 1 Tanzania, Malawi, Mozambique

Phellodendron Rupr. 1857. Rutaceae. 10 East Asia (incl. 2 Russian Far East)

Phellolophium Baker. 1894. Apiaceae. 1 Madagascar

Phellopterus Benth. = Glehnia F. Schmidt ex Miq. Apiaceae

Phelpsiella Maguire. 1958. Rapateaceae. 1 Venezuela

Phelypaea Hill. 1753. Orobanchaceae. 4 Yugoslavia, Crimea, Caucasus, Asia Minor, ? Syria, Iran

Phenakospermum Endl. 1833. Strelitziaceae. 1 eastern South America

Phenax Wedd. 1854. Urticaceae. 12 trop. America south to northern Argentina, West Indies

Pherolobus N. E. Br. = Dorotheanthus Schwantes. Aizoaceae

Pherosphaera Archer = Microstrobos J. Garden et L. A. S. Johnson. Podocarpaceae

Pherotrichis Decne. 1838. Asclepiadaceae. 2 southern Arizona, Mexico

Phialacanthus Benth. 1876. Acanthaceae. 5 northeastern India, Southeast Asia

Phialanthus Griseb. 1861. Rubiaceae. 10 West Indies

Phialodiscus Radlk. = Blighia Koenig. Sapindaceae

Phidiasia Urb. 1923 (Odontonema Nees). Acanthaceae. 1 Cuba

Philacra Dwyer. 1944. Sauvagesiaceae. 3 trop. South America

Philactis Schrad. 1833. Asteraceae. 3 southern Mexico, Guatemala

Philadelphus L. 1753. Hydrangeaceae. 75 temp. Northern Hemisphere, esp. East Asia

Philbornea Hallier f. 1912. Hugoniaceae. 1 Sumatra, Borneo, Philippines

Philenoptera Hochst. ex A. Rich. 1847 (Lonchocarpus Kunth). Fabaceae. 19 South America (4), trop. Africa (15)

Philesia Comm. ex Juss. 1789. Philesiaceae. 1 southern Chile

Philgamia Baill. 1894. Malpighiaceae. 4 Madagascar

Philibertia Kunth = Sarcostemma R. Br. Asclepiadaceae

Philippia Klotzsch. 1834 (Erica L.). Ericaceae. 40 trop. and South Africa, Madagascar, Mascarene Is.

Philippiamra Kuntze = Silvaea Phil. Portulacaceae

Philippiella Speg. 1897. Caryophyllaceae. 1 Patagonia

Philippinaea Schltr. et Ames = Orchipedum Breda. Orchidaceae

Philippodendrum Poit. = Plagianthus J. R. Forst. et G. Forst. Malvaceae

Phillyrea L. 1753. Oleaceae. 4 Madeira, Mediterranean (excl. Egypt) eastward to northern Iran

Philocrena Bong. = Tristicha Thouars. Podostemaceae

Philodendron Schott. 1829. Araceae. c. 350–400 trop. America, West Indies

Philodice Mart. 1835. Eriocaulaceae. 2 Brazil

Philoglossa DC. 1836. Asteraceae. 5 Colombia, Ecuador, Peru, Bolivia

Philonotion Schott = Schismatoglottis Zoll. et Moritzi. Araceae

Philotheca Rudge. 1816. Rutaceae. 2 Australia

Philoxerus R. Br. 1810. Amaranthaceae. 15 West Africa, East Asia, coastal trop. Australia, trop. and subtrop. America, West Indies

Philydrella Caruel. 1878. Philydraceae. 2 southwestern Australia

Philydrum Banks et Sol. ex Gaertn. 1788. Philydraceae. 1 Burma, southern China, Taiwan, southern Japan, Indochina, Malay Peninsula, New Guinea, northern and eastern Australia from Kimberley region to Victoria

Philyra Klotzsch. 1841. Euphorbiaceae. 1

southern Brazil, Paraguay

Philyrophyllum O. Hoffm. 1890. Asteraceae. 1 South Africa

Phinaea Benth. 1876. Gesneriaceae. 9 trop. America from Mexico to Colombia

Phippsia (Trin.) R. Br. 1823. Poaceae. 4 Arctica, mountainous areas of Siberia, North and temp. South America

Phitopis Hook. f. 1871. Rubiaceae. 2 Peru

Phlebiophragmus O. E. Schulz. 1924. Brassicaceae. 1 Peru

Phlebocarya R. Br. 1810. Conostylidaceae. 3 southwestern Australia

Phlebolobium O. E. Schulz. 1933. Brassicaceae. 1 Falkland Is.

Phlebophyllum Nees. 1832 (Strobilanthes Blume). Acanthaceae. 8 India

Phlebotaenia Griseb. 1861 (Polygala L.). Polygalaceae. 3 Cuba, Puerto Rico

Phlegmatospermum O. E. Schulz. 1933. Brassicaceae. 4 Australia

Phleum L. 1753. Poaceae. 15 North and temp. South America, temp. Eurasia, North Africa

Phloeophila Hoehne et Schltr. = Pleurothallis R. Br. Orchidaceae

Phloga Noronha ex Hook. f. 1883. Arecaceae. 2 Madagascar

Phlogacanthus Nees. 1832. Acanthaceae. 17 Himalayas from Kashmir to Bhutan, Burma, China, Indochina, Malesia

Phlogella Baill. = Chrysalidocarpus H. A. Wendl. Arecaceae

Phlojodicarpus Turcz. ex Ledeb. 1844. Apiaceae. 4 northern Ural, Siberia, Russian Far East, Middle Asia

Phlomidoschema (Benth.) Vved. 1941 (Stachys L.). Lamiaceae. 1 Middle Asia, Iran, Afghanistan, Pakistan, northwestern India

Phlomis L. 1753. Lamiaceae. Over 100 temp. Eurasia, Mediterranean; Russia (European part, Siberia, Far East), Caucasus, Middle Asia

Phlomoides Moench. 1794. Lamiaceae. c. 115 Russia (from southeastern European part to Far East), Caucasus, Middle Asia, Iran, Afghanistan, Mongolia, China, India

Phlox L. 1753. Polemoniaceae. 67 North America (66); P. sibirica—eastern European part of Russia, West and East Siberia, Far East

Phlyctidocarpa Cannon et Theobald. 1967. Apiaceae. 1 Namibia

Phoebanthus S. F. Blake = Helianthella Torr. et A. Gray. Asteraceae

Phoebe Nees. 1836. Lauraceae. 94 eastern Pakistan, India, eastern Himalayas, Burma, China, Indochina, Malesia

Phoenicanthus Alston. 1931. Annonaceae. 2 Sri Lanka

Phoenicaulis Nutt. 1838. Brassicaceae. 1 western North America

Phoenicophorium H. A. Wendl. 1865. Arecaceae. 1 Seychelles

Phoenicoseris (Skottsb.) Skottsb. = Dendroseris D. Don. Asteraceae

Phoenix L. 1753. Arecaceae. 17 Canary and Cape Verde Is., trop. and subtrop. Africa (excl. South-West), northern Madagascar, Comoro Is., Crete, southern Turkey, eastern Mediterranean and Arabian Peninsula through India, Himalayas, Sri Lanka eastward to southern China, Hong Kong, Taiwan, Philippines, Sumatra, and Malay Peninsula

Pholidia R. Br. = Eremophila R. Br. Myoporaceae

Pholidocarpus Blume. 1830. Arecaceae. 6 Thailand, Malay Peninsula, Sumatra, Borneo, Sulawesi, and Moluccas

Pholidostachys H. A. Wendl. ex Hook. f. 1883. Arecaceae. 4 trop. America from Costa Rica to Peru

Pholidota Lindl. ex Hook. 1825. Orchidaceae. 29 Himalayas, India, Sri Lanka, Andaman Is., Burma, southern China, Indochina, Malesia from Malay Peninsula to New Guinea, trop. Australia, Solomon Is., Vanuatu, Loyalty Is., New Caledonia, Fiji

Pholisma Nutt. ex Hook. 1844. Lennoaceae. 3 southwestern U.S., northwestern Mexico

Pholistoma Lilja. 1839 (Nemophila Nutt.). Hydrophyllaceae. 3 southwestern U.S. (California, Nevada), Baja California

Pholiurus Trin. 1820. Poaceae. 1 Southeast Europe, West Asia; Caucasus, West Siberia, Middle Asia

Phoradendron Nutt. 1848. Viscaceae. 25 (or c. 150 ?) America from U.S. to Bolivia and Argentina, West Indies

Phormium J. R. Forst. et G. Forst. 1775. Phormiaceae. 2 New Zealand, Stewart I., Auckland Is., Norfolk I., Chatham Is.

Phornothamnus Baker = Gravgesia Naudin. Melastomataceae

Photinia Lindl. 1821. Rosaceae. 54 India, Sri Lanka, Himalayas, Burma, southwestern and southern China, Taiwan, Korea, Japan, Indochina, eastern Malesia to Borneo, Mexico, Central America

Phragmanthera Tiegh. = Tapinanthus (Blume) Rchb.

Phragmipedium Rolfe. 1896. Orchidaceae. 11 trop. America from Mexico to Peru, Bolivia, and Brazil

Phragmites Adans. 1763. Poaceae. 5 almost cosmopolitan, excl. Arctica

Phragmocarpidium Krapov. 1969. Malvaceae. 1 Brazil

Phragmorchis L. O. Williams. 1938. Orchidaceae. 1 Philippines

Phragmotheca Cuatrec. 1946. Bombacaceae. 2 trop. South America

Phreatia Lindl. 1830. Orchidaceae. c. 150–200 trop. Asia from Himalayas, India, and Sri Lanka to southern China, Taiwan, Indochina, Malesia from Malay Peninsula to New Ireland, Micronesia, Bougainville and Solomon Is., Australia, New Zealand, Norfolk Is., Vanuatu, the Horn Is., New Caledonia, Fiji, Society Is., Samoa

Phrissocarpus Miers = Tabernaemontana L. Apocynaceae

Phrodus Miers. 1849. Solanaceae. 1 (P. microphyllus) northern Chile

Phryganocydia Mart. ex Bur. 1872. Bignoniaceae. 3 trop. America from Costa Rica to Bolivia and Brazil, West Indies

Phrygilanthus Eichler = Notanthera (DC.) G. Don. Loranthaceae

Phryma L. 1753. Verbenaceae. 1–2 India, Himalayas, Indochina, China, Korea, Japan, Russian Far East (Ussuriyski Krai)

Phryne Bubani = Lycocarpus O. Schulz (p. p.) and Murbeckiella Rothm. (p. p.). Brassicaceae

Phrynella Pax et K. Hoffm. 1934. Caryophyllaceae. 1 Asia Minor

Phrynium Willd. 1797. Marantaceae. c. 30 trop. Africa, trop. Asia, Malesia

Phtheirospermum Bunge ex Fisch. et C. A. Mey. 1835. Scrophulariaceae. 7 Himalayas, China, Korea, Japan, Russian Far East (1, P. chinense—Ussuriyski Krai)

Phthirusa Mart. 1830 (Hemitria Raf.). Loranthaceae. c. 70 trop. America

Phuodendron Graebn. ex Dalla Torre et Harms = Valeriana L. Valerianaceae

Phuopsis (Griseb.) Hook. f. 1873. Rubiaceae. 1 Caucasus, eastern Asia Minor, northwestern Iran

Phycella Lindl. = Hippeastrum Herb. Amaryllidaceae

Phygelius E. Mey. ex Benth. 1836. Scrophulariaceae. 2 South Africa

Phyla Lour. 1790 (Lippia L.). Verbenaceae. 15 tropics and subtropics

Phylacium Benn. 1840. Fabaceae. 2 Burma, Thailand, Laos, ? northern Australia

Phylibertella Vail = Sarcostemma R. Br. Asclepiadaceae

Phylica L. 1753. Rhamnaceae. c. 150 South Africa (140 endemic to Cape), Madagascar, Mascarene Is., Amsterdam I., Tristan da Cunha Is. (1, P. arborea)

Phyllacantha Hook. f. = Phyllacanthus Hook. f. Rubiaceae

Phyllacanthus Hook. f. 1871. Rubiaceae. 1 Cuba

Phyllachne J. R. Forst. et G. Forst. 1775. Stylidaceae. 4 Tasmania, New Zealand, temp. South America

Phyllactis Pers. = Valeriana L. Valerianaceae

Phyllagathis Blume. 1831. Melastomataceae. c. 60 southern China, Indochina (14), western Malesia, Borneo

Phyllanoa Croizat. 1943. Euphorbiaceae. 1 Colombia

Phyllanthera Blume. 1827. Asclepiadaceae. 2 Malay Peninsula, Java

Phyllanthodendron Hemsl. 1898 (Phyllanthus L.). Euphorbiaceae. 16 southeastern China, Indochina, Malay Peninsula

Phyllanthus L. 1753. Euphorbiaceae. c. 700–750 tropics and subtropics, few temp. North America and temp. Pacific regions

Phyllapophysis Mansf. = Catanthera F. Muell. Melastomataceae

Phyllarthron DC. 1839. Bignoniaceae. 14 Comoro Is., Madagascar

Phyllis L. 1753. Rubiaceae. 2 Madeira Is., Canary Is.

Phyllobaea Benth. 1876. Gesneriaceae. 2 Burma, Malay Peninsula

Phylloboea C. B. Clarke = Phyllobaea Benth. Gesneriaceae

Phyllobolus N. E. Br. 1925 (Sphalmanthus N. E. Br.). Aizoaceae. 35 southern Namibia, western and central South Africa

Phyllobotryon Muell. Arg. 1864. Flacourtiaceae. 4 trop. Africa

Phyllocara Guzul. 1927 (Anchusa L.). Boraginaceae. 1 southern Transcaucasia, northern Iraq

Phyllocarpus Riedel ex Endl. 1842. Fabaceae. 2 trop. America

Phyllocarpus Riedel ex Tul. (1843) = Phyllocarpus Riedel ex Endl. Fabaceae

Phyllocephalum Blume. 1826. Asteraceae. 3 India, Java

Phyllocharis Diels = Ruthiella Steenis. Lobeliaceae

Phyllocladus Rich. ex Mirb. 1825. Phyllocladaceae. 5 Philippines, Borneo, Sulawesi, New Guinea, Tasmania (1), New Zealand (3)

Phylloclinium Baill. 1890. Flacourtiaceae. 4 trop. Africa

Phyllocomos Mast. = Anthochortus Nees. Restionaceae

Phyllocosmus Klotzsch. 1857 (Ochthocosmus Benth.). Ixonanthaceae. 6 trop. Africa

Phyllocrater Wernham. 1914 (Oldenlandia L.). Rubiaceae. 1 Borneo

Phylloctenium Baill. 1887. Bignoniaceae. 2 Madagascar

Phyllodium Desv. 1813. Fabaceae. 6 South and Southeast Asia, northern Australia

Phyllodoce Salisb. 1806. Ericaceae. 6–7 circumpolar and temp. Northern Hemisphere; Russia (2, arctic and northern areas of the European part, East Siberia, Far East)

Phyllogeiton (Weberb.) Herzog = Berchemia Neck. ex DC. Rhamnaceae

Phyllomelia Griseb. 1866. Rubiaceae. 1 Cuba

Phyllonoma Willd. ex Schult. 1820. Dulongiaceae. 4 trop. America: Mexico, Guatemala, El Salvador, Honduras, Costa Rica, Panama, Colombia, Peru, northwestern Bolivia

Phyllophyton Kudo = Marmoritis Benth. Lamiaceae

Phyllopodium Benth. 1836 (or = Polycarena Benth.). Scrophulariaceae. 15 South Africa

Phyllorachis Trimen. 1879. Poaceae. 1 trop. Africa: Angola, Zambia, Malawi, Tanzania, Mozambique

Phylloscirpus C. B. Clarke. 1908. Cyperaceae. 1 Argentina

Phyllosma Bolus. 1898. Rutaceae. 2 South Africa

Phyllospadix Hook. 1838. Zosteraceae. 7 northern part of Pacific coasts of Asia and North America

Phyllostachys Siebold et Zucc. 1843. Poaceae. c. 70 Caucasus and Middle Asia (7), Himalayas, northeastern India, East Asia

Phyllostegia Benth. 1830. Lamiaceae. 27 Hawaiian Is., Tahiti (1, P. tahitensis)

Phyllostelidium Beauverd = Baccharis L. Asteraceae

Phyllostemonodaphne Kosterm. 1936. Lauraceae. 1 southern Brazil

Phyllostylon Capan. ex Benth. et Hook. f. 1880. Ulmaceae. 2: P. rhamnoides—Mexico, Guatemala, Belize, El Salvador, Honduras, Nicaragua, Colombia, Venezuela, Brazil, Bolivia, Paraguay, Argentina, West Indies (Cuba, Haiti); P. brasiliense—Brazil

Phyllota (DC.) Benth. 1837. Fabaceae. 10 Australia

Phyllotrichum Thorel ex Lecomte. 1911. Sapindaceae. 1 Indochina

Phylloxylon Baill. 1861. Fabaceae. 5 Madagascar, Mauritius

Phylogyne Salisb. ex Haw. = Narcissus L. Amaryllidaceae

Phylohydrax Puff. 1986. Rubiaceae. 2 East Africa, Madagascar

Phymaspermum Less. 1832. Asteraceae. 17 South Africa

Phymatarum M. Hotta. 1965. Araceae. 1 Borneo

Phymatidium Lindl. 1833. Orchidaceae. 5 Brazil

Phymatocarpus F. Muell. 1862. Myrtaceae. 2 western Australia

Phymosia Desv. 1825. Malvaceae. 8 Mexico, Guatemala, El Salvador; P. abutiloides—Bahama Is. and Haiti (Hispaniola)

Phyodina Raf. = Callisia Loefl. Commelinaceae

Physacanthus Benth. 1876. Acanthaceae. 5 trop. Africa

Physalastrum Monteiro = Krapovickasia Fryxell. Malvaceae

Physaliastrum Makino. 1914 (Leucophysalis Rydb.). Solanaceae. 7 East Asia; Russia (1, P. echinatum: Russian Far East)

Physalidium Fenzl = Graellsia Boiss. Brassicaceae

Physalis L. 1753. Solanaceae. c. 75 America, warm–temp. and subtrop. Eurasia, Mediterranean

Physandra Botsch. 1956 (Salsola L.). Chenopodiaceae. 1 Kazakhstan

Physaria (Nutt. ex Torr. et A. Gray) A. Gray. 1848. Brassicaceae. 22 western North America

Physena Noronha ex Thouars. 1806. Physenaceae. 2 Madagascar

Physetobasis Hassk. = Holarrhena R. Br. Apocynaceae

Physinga Lindl. = Epidendrum L. Orchidaceae

Physocalymma Pohl. 1827. Lythraceae. 1 trop. South America

Physocalyx Pohl. 1827. Scrophulariaceae. 2 Brazil

Physocardamum Hedge. 1968. Brassicaceae. 1 eastern Turkey

Physocarpus (Cambess.) Maxim. 1879. Rosaceae. 7–10 northern China, Korea, Japan, Russian Far East (2), North America

Physocaulis (DC.) Tausch. 1834. Apiaceae. 1 (P. nodosus) Mediterranean from Portugal and Algeria to the Crimea, Caucasus, Syria, Israel, Iran, Middle Asia

Physoceras Schltr. 1925. Orchidaceae. 7 Madagascar

Physochlaina G. Don. 1838. Solanaceae. 9–12 Caucasus, Middle Asia, southern Siberia, Russian Far East, China southward to Himalayas, eastward to Japan

Physodium C. Presl = Melochia L. Sterculiaceae

Physogyne Garay. 1982 (Pseudogoodyera Schltr.). Orchidaceae. 2 Mexico

Physokentia Becc. 1934. Arecaceae. 7 New Britain I., Solomon Is., Vanuatu (New Hebrides), Fiji

Physoleucas (Benth.) Jaub. et Spach. 1855 (Leucas R. Br.). Lamiaceae. 1 Arabian Peninsula

Physoplexis (Endl.) Schur. 1853 (Phyteuma L.). Campanulaceae. 1 montane Italy and Yugoslavia

Physopsis Turcz. 1849. Chloanthaceae. 2 western and southern Australia

Physoptychis Boiss. 1867. Brassicaceae. 2 Transcaucasia (1, P. caspica), Turkey, Iraq, northwestern Iran

Physopyrum Popov = Atraphaxis L. Polygonaceae

Physorhynchus Hook. 1851. Brassicaceae. 2 Iran, Afghanistan, Pakistan, northwestern India

Physosiphon Lindl. = Pleurothallis R. Br. Orchidaceae

Physospermopsis H. Wolff. 1925. Apiaceae. 18 western China, Tibet, eastern Himalayas (Nepal to Bhutan), northeastern India (Assam)

Physospermum Cusson ex Juss. 1787. Apiaceae. 2 West, Central, and South and Southeast Europe; North Africa; West (incl. Caucasus) and Southwest Asia

Physostegia Benth. 1829. Lamiaceae. 12 North America

Physostelma Wight = Hoya R. Br. Asclepiadaceae

Physostemon Mart. 1824 (Cleome L.). Capparaceae. 7 trop. America

Physostigma Balf. 1861. Fabaceae. 5 trop. Africa

Physothallis Garay = Pleurothallis R. Br. Orchidaceae

Physotrichia Hiern. 1873 (Dilpolophium Turcz.). Apiaceae. 10 trop. and South Africa

Phytelephas Ruiz et Pav. 1789. Arecaceae. 6 Panama, Colombia, Venezuela, Amazonian Basin in Bolivia, Ecuador, Peru, and ? Brazil

Phyteuma L. 1753. Campanulaceae. c. 40 Eurasia, Mediterranean

Phytocrene Wall. 1831. Icacinaceae. 11 Southeast Asia, Malesia

Phytolacca L. 1753. Phytolaccaceae. 35 tropics and subtropics, esp. America

Piaranthus R. Br. 1811. Asclepiadaceae. 16 South-West and South Africa

Picardaea Urb. 1903. Rubiaceae. 2 Cuba, Haiti

Picconia A. DC. 1844. Oleaceae. 2 Azores, Madeira Is., and Canary Is.

Picea A. Dietr. 1824. Pinaceae. 34 cold and temp. Northern Hemisphere; Russia (11, all over, except steppes and arid areas)

Pichisermollia Monteiro-Neto = Areca L. Arecaceae

Pichleria Stapf et Wettst. = Zosimia Hoffm. Apiaceae

Pichonia Pierre. 1890 (Pouteria Aubl.). Sapotaceae. 5 New Guinea, Solomon Is., New Caledonia

Pickeringia Nutt. ex Torr. et A. Gray. 1840. Fabaceae. 1 western and southwestern U.S., Mexico (Baja California)

Picnomon Adans. 1763. Asteraceae. 1 South Europe, eastern Mediterranean, Iran; Crimea, Caucasus, Middle Asia

Picradeniopsis Rydb. = Bahia Lag. Asteraceae

Picralima Pierre. 1896. Apocynaceae. 1 trop. West Africa

Picramnia Sw. 1788. Simaroubaceae. 41 trop. America from Florida and Mexico south to Paraguay and northern Argentina, West Indies

Picrasma Blume. 1825. Simaroubaceae. 8 Himalayas from Kashmir to Bhutan, northeastern India, Burma, China, Taiwan, Korea, Japan, Indochina, Malesia, Solomon Is., Fiji, trop. America (6)

Picria Lour. 1790. Scrophulariaceae. 2 India, China, Indochina

Picris L. 1753. Asteraceae. 40–50 temp. Eurasia, Mediterranean, Ethiopia; Russia (c. 10, from European part to Far East)

Picrodendron Griseb. 1860. Euphorbiaceae. 3 Bahama Is., Haiti, Jamaica

Picrolemma Hook. f. 1862. Simaroubaceae. 3 eastern Peru, Brazil (Amazonia)

Picrorhiza Royle ex Benth. 1835. Scrophulariaceae. 1 (P. kurrooa) northern Burma, western and southwestern China (P. scrophula riiflora = Neopicrorhiza scrophulariiflora)

Picrosia D. Don. 1830. Asteraceae. 2 subtrop. South America

Picrothamnus Nutt. 1841. Asteraceae. 1 North America

Pictetia DC. 1825. Fabaceae. 6 West Indies

Pieris D. Don. 1834. Ericaceae. 7 Himalayas, northeastern India, Burma, China (6), North America

Pierranthus Bonati. 1912. Scrophulariaceae. 1 Southeast Asia

Pierrea F. Heim = Hopea Roxb. Dipterocarpaceae

Pierreodendron Engl. = Quassia L. Simaroubaceae

Pierrina Engl. 1909. Scytopetalaceae. 1 trop. West Africa

Pietrosia Nyárády = Andryala L. Asteraceae

Pigafetta (Blume) Becc. 1877. Arecaceae. 1 Sulawesi, Moluccas, New Guinea

Pilea Lindl. 1821. Urticaceae. c. 250 tropics and subtropics, few spp. in New Zealand; Russia (2, southern Siberia, Far East)

Pileanthus Labill. 1806. Myrtaceae. 3 western Australia

Pileostegia Hook. f. et Thomson. 1857

(Schizophragma Siebold et Zucc.). Hydrangeaceae. 3–4 eastern Himalayas, northeastern India, China, Taiwan

Pileus Ramírez = Jacaratia Endl. Caricaceae

Pilgerochloa Eig. 1929 (Ventenata Koeler). Poaceae. 2 Southwest Asia

Pilgerodendron Florin. 1930 (Libocedrus Endl.). Cupressaceae. 1 Chilean Andes, Chiloé I., southern Argentina

Pilidiostigma Burret. 1941. Myrtaceae. 4 New Guinea (1), northeastern Australia

Piliocalyx Brongn. et Gris. 1865. Myrtaceae. 10 New Caledonia, Vanuatu, Fiji (1)

Piliostigma Hochst. 1846 = Bauhinia L.. Fabaceae. 3 trop. and South Africa, trop. Asia, Malesia

Pillansia L. Bolus. 1914. Iridaceae. 1 South Africa

Pilocarpus Vahl. 1796. Rutaceae. 13 Mexico, El Salvador, Costa Rica, Venezuela, Guyana, Surinam, French Guiana, Colombia, Peru, Brazil, Argentina, Paraguay; West Indies from Cuba to Martinique

Pilocopiapoa F. Ritter = Pilosocereus Byles et Rowley. Cactaceae

Pilocosta Almeda et Whiffin. 1981. Melastomataceae. 3 trop. America

Pilophyllum Schltr. 1914. Orchidaceae. 1–2 Malay Peninsula, Java, Borneo, Philippines, New Guinea, Solomon Is.

Pilopleura Schischk. 1951. Apiaceae. 2 Middle Asia (Dzungarian Alatau, western Tien Shan)

Pilosella Hill = Hieracium L. Asteraceae

Piloselloides (Less.) C. Jeffrey = Gerbera L. ex Cass. Asteraceae

Pilosocereus Bybles et Rowley. 1957. Cactaceae. 35 Mexico, trop. South America, esp. Brazil; West Indies

Pilosperma Planch. et Triana. 1860. Clusiaceae. 2 Colombia

Pilostemon Iljin. 1961. Asteraceae. 2 Middle Asia, Afghanistan, western China

Pilostigma Tiegh. = Amyema Tiegh. Loranthaceae

Pilostyles Guill. 1834. Apodanthaceae. c. 25 America from southern U.S. to the Strait of Magellan; 1 (P. haussknechtii) Asia Minor, Iraq, Iran; 2 western Australia

Pilothecium (Kiaersk.) Kausel = Eugenia L. Myrtaceae

Pimelea Banks ex Sol. 1788. Thymelaeaceae. 108 northern Philippines, Lesser Sunda Is., New Guinea, New Britain Is., Louisiade Arch., Australia and Tasmania (90), King Is., Lord Howe I., New Zealand, Chatam Is.

Pimelodendron Hassk. 1856. Euphorbiaceae. 6–8 Malesia

Pimenta Lindl. 1821. Myrtaceae. 15 trop. America from Mexico to Brazil, West Indies

Pimentelia Wedd. 1849. Rubiaceae. 1 Peru

Pimia Seem. 1862. Sterculiaceae. 1 Fiji

Pimpinella L. 1753. Apiaceae. c. 150 Eurasia, esp. China (over 40), Africa from Mediterranean to Cape, Madagascar

Pinacantha Gilli. 1959. Apiaceae. 1 Afghanistan

Pinacopodium Exell et Mendonça. 1951. Erythroxylaceae. 2 trop. Africa

Pinanga Blume. 1838. Arecaceae. c. 120 Sri Lanka, Himalayas, northeastern India, southern China, Indochina, Malesia

Pinaropappus Less. 1832. Asteraceae. 7 southern U.S. (Arizona, New Mexico, Texas), Mexico, Guatemala

Pinarophyllon Brandegee. 1914. Rubiaceae. 2 Mexico, Central America

Pinckneya Michx. 1803. Rubiaceae. 1 southeastern U.S.

Pinda P. K. Mukh. et Constance. 1986 (Heracleum L.). Apiaceae. 1 India

Pineda Ruiz et Pav. 1794. Flacourtiaceae. 1 Andes of Ecuador and Peru

Pinelia Lindl. = Pinelianthe Rauschert. Orchidaceae

Pinelianthe Rauschert. 1983. Orchidaceae. 3 Brazil

Pinellia Ten. 1853. Araceae. 8 China, Japan

Pinguicula L. 1753. Lentibulariaceae. 47 America from southeastern U.S. to northern Andes, Chile, Tierra del Fuego, West Indies, Mediterranean, temp. Eurasia (incl. 9 Arctica, northern European part of Russia, Caucasus, Siberia, Far East), Himalayas, Tibet, southwestern and southern China

Pinillosia Ossa ex DC. 1836. Asteraceae. 2 Cuba, Haiti

Pinosia Urb. 1930. Caryophyllaceae. 1 Cuba

Pintoa Gay. 1846. Zygophyllaceae. 1 Chile

Pinus L. 1753. Pinaceae. Over 100 Northern Hemisphere and southward into Central America, Cuba, Hispaniola; Southeast Asia and Philippines; (P. merkusii—across the equator to Sumatra)

Pinzona Mart. et Zucc. 1832. Dilleniaceae. 1–2 trop. America, West Indies

Piofontia Cuatrec. = Diplostephium Kunth. Asteraceae

Pionocarpus S. F. Blake = Iostephane Benth. Asteraceae

Piora J. Kost. 1966. Asteraceae. 1 New Guinea

Piper L. 1753. Piperaceae. c. 2,000 tropics, chiefly Central and South America

Piperanthera C. DC. 1923 (Peperomia Ruiz et Pav.). Peperomiaceae. 1 West Indies

Piperia Rydb. 1901 (Habenaria Willd.). Orchidaceae. 4 North America

Pippenalia McVaugh. 1972. Asteraceae. 1 Mexico

Piptadenia Benth. 1840. Fabaceae. c. 25 trop. America, West Indies

Piptadeniastrum Brenan. 1955. Fabaceae. 1 trop. Africa from Senegal to Angola and the Congo Basin

Piptadeniopsis Burkart. 1944 (Piptadenia Benth.). Fabaceae. 1 Paraguay

Piptanthus Sweet. 1828. Fabaceae. 8–9 Himalayas, Tibet

Piptatherum P. Beauv. 1812. Poaceae. c. 50 warm–temp. regions and montane areas of tropics

Piptocalyx Oliv. ex Benth. = Trimenia Seem. Trimeniaceae

Piptocarpha R. Br. 1817. Asteraceae. c. 50 Central and trop. South America, West Indies

Piptochaetium J. Presl. 1830. Poaceae. c. 30 temp. North, Central, and South America

Piptocoma Cass. 1817. Asteraceae. 3 Haiti, Puerto Rico

Piptolepis Sch. Bip. 1863. Asteraceae. 8 Brazil

Piptophyllum C. E. Hubb. 1957. Poaceae. 1 Angola

Piptoptera Bunge. 1877. Chenopodiaceae. 1 Kazakhstan, Uzbekistan, Turkmenistan

Piptospatha N. E. Br. 1879. Araceae. 10 western Malesia, Borneo

Piptostachya (C. E. Hubb.) J. B. Phipps. 1964 (Zonotriche [C. E. Hubb.] J. B. Phipps). Poaceae. 1 Angola, Zaire, Zambia, Zimbabwe, Malawi, Tanzania, Mozambique

Piptostigma Oliv. 1865. Annonaceae. 15 trop. Africa (Ivory Coast, Cameroun, Gabon)

Piptothrix A. Gray. 1886. Asteraceae. 7 Mexico, Guatemala

Pipturus Wedd. 1854. Urticaceae. 40 Madagascar, Comoro Is., Mascarene Is., South and Southeast Asia, Australia, Oceania eastward to Hawaiian Is. and Fiji

Piqueria Cav. 1795. Asteraceae. 7 Mexico (7), P. trinervia also in Central America and West Indies

Piqueriella R. M. King et H. Rob. 1974 (Eupatorium L.). Asteraceae. 1 Brazil

Piqueriopsis R. M. King. 1965. Asteraceae. 1 Mexico

Piquetia (Pierre) Hallier f. = Camellia L. Theaceae

Piranhea Baill. 1866. Euphorbiaceae. 1–2 Venezuela, Guyana, Brazil

Piratinera Aubl. = Brosimum Sw. Moraceae

Piresia Swallen. 1964. Poaceae. 2 trop. South America, Trinidad

Piresodendron Auvrév. ex A. Thomas = Pouteria Aubl. Sapotaceae

Piriadacus Pichon = Arrabideae DC. Bignoniaceae

Pirinia M. Kral. 1984. Caryophyllaceae. 1 Bulgaria

Piriqueta Aubl. 1775. Turneraceae. 20 mostly trop. America from Georgia and

Florida to Paraguay; trop. and South Africa (1), Madagascar

Pironneava Gaudich. ex Regel = Hohenbergia Schult. et Schult. f. Bromeliaceae

Piscaria Piper = Eremocarpus Benth. Euphorbiaceae

Piscidia L. 1759. Fabaceae. 8 Florida, Mexico, Central America, northern Peru, Venezuela, West Indies

Pisonia L. 1753. Nyctaginaceae. 35 tropics and subtropics, chiefly Southeast Asia and America

Pisoniella (Heimerl) Standl. 1911. Nyctaginaceae. 1 trop. and subtrop. America from Mexico to Bolivia and Argentina

Pistacia L. 1753. Anacardiaceae. 12 Macaronesia, Mediterranean, Northeast and East Africa, Crimea, Caucasus, West and Middle Asia eastward to northwestern India, East and Southeast Asia, Malesia, southern U.S., Mexico, Guatemala

Pistaciovitex Kuntze = Vitex L. Verbenaceae

Pistia L. 1753. Araceae. 1 tropics and subtropics

Pistorina DC. 1828. Crassulaceae. 4 Iberian Peninsula, Morocco, Algeria, Tunisia

Pisum L. 1753. Fabaceae. 2 Mediterranean, Kenya, Iraq, Iran, Afghanistan

Pitardia Batt. ex Pit. 1918. Lamiaceae. 2 Morocco

Pitavia Molina. 1810. Rutaceae. 1 Chile

Pitcairnia L'Hér. 1789. Bromeliaceae. c. 280 Mexico, Central and trop. South America, West Indies; 1 (P. feliciana) West Africa (Guinea)

Pithecellobium Mart. 1837. Fabaceae. c. 20 Central and South America, West Indies

Pithecoctenium Mart. ex Meisn. 1840. Bignoniaceae. 4 trop. America from Mexico to Argentina and Brazil, West Indies

Pithecoseris Mart. ex DC. 1836. Asteraceae. 1 northern Brazil

Pithocarpa Lindl. 1839. Asteraceae. 4 southwestern Australia

Pitraea Turcz. 1836. Verbenaceae. 1 South America

Pittierothamnus Steyerm. = Amphidasya Standl. Rubiaceae

Pittocaulon H. Rob. et Brettell. 1973. Asteraceae. 5 Mexico, Central America

Pittoniotis Griseb. 1858. Rubiaceae. 1 Panama, Colombia, Venezuela

Pittosporopsis Craib. 1911. Icacinaceae. 1 Burma, southern China (Yunnan), Thailand, Laos, Vietnam

Pittosporum Banks ex Sol. 1788. Pittosporaceae. c. 200 trop. and South Africa, Madagascar, trop. Asia eastward to China and Japan, Malesia, Australia, New Zealand, Oceania

Pituranthos Viv. = Deverra DC. Apiaceae

Pitygentias Gilg = Gentianella Moench. Gentianaceae

Pityopsis Nutt. 1840 (Heterotheca Cass.). Asteraceae. 7–8 eastern U.S., Mexico, Central America

Pityopus Small. 1914. Ericaceae. 1 western U.S.

Pityphyllum Schltr. 1920. Orchidaceae. 1 Colombia

Pityranthe Thwaites = Diplodiscus Turcz. Tiliaceae

Pityrodia R. Br. 1810. Chloanthaceae. 41 Australia

Placea Miers. 1841. Amaryllidaceae. 5 Chile

Placocarpa Hook. f. 1873. Rubiaceae. 1 Mexico

Placodiscus Radlk. 1878. Sapindaceae. 16 trop. Africa

Placolobium Miq. = Ormosia Jacks. Fabaceae

Placopoda Balf. f. 1882. Rubiaceae. 1 Socotra

Placospermum C. T. White et W. D. Francis. 1924. Proteaceae. 1 Australia (Queensland)

Pladaroxylon (Endl.) Hook. f. 1870. Asteraceae. 1 St. Helena

Plaesianthera (C. B. Clarke) Livera = Brillantaisia P. Beauv. Acanthaceae

Plagiantha Renvoize. 1982. Poaceae. 1 Brazil

Plagianthus J. R. Forst. et G. Forst. 1775. Malvaceae. 2 New Zealand

Plagiarthron P. A. Duvign. = Loxodera Launert. Poaceae

Plagiobasis Schrenk. 1845. Asteraceae. 1 (P. centauroides) Middle Asia, western China

Plagiobothrys Fisch. et C. A. Mey. 1836. Boraginaceae. c. 50 Australia (4), Pacific coastal America

Plagiocarpus Benth. 1873. Fabaceae. 1 northern Australia

Plagioceltis Mildbr. ex Baehni = Ampelocera Klotzsch. Ulmaceae

Plagiocheilus Arn. ex DC. 1838. Asteraceae. 7 South America

Plagiochloa Adamson et Sprague. 1941 (Tribolium Desv.). Poaceae. 7 South Africa

Plagiolirion Baker. 1883 (Hymenocallis Salisb.). Amaryllidaceae. 1 western Colombia

Plagiolophus Greenm. 1904. Asteraceae. 1 Mexico (Yucatán Peninsula)

Plagiopetalum Rehder. 1917. Melastomataceae. 2 India, Burma, southern continental China, Vietnam

Plagiopteron Griff. 1843. Plagiopteraceae. 1–2 India, Bangladesh, southern Burma, southern China, Thailand

Plagiorhegma Maxim. 1859 (or = Jeffersonia Barton). Berberidaceae. 1 (P. dubium) Russian Far East (Ussuriiski Krai), China, Japan

Plagioscyphus Radlk. 1878. Sapindaceae. 2 Madagascar

Plagiosetum Benth. 1877. Poaceae. 1 Australia

Plagiosiphon Harms. 1897. Fabaceae. 5 trop. West Africa from Sierra Leone to Gabon

Plagiospermum Oliv. = Prinsepia Royle. Rosaceae

Plagiostachys Ridl. 1899. Zingiberaceae. 16 southern China, Malesia (excl. New Guinea)

Plagiostyles Pierre. 1897. Euphorbiaceae. 1 trop. Africa

Plagiotheca Chiov. = Issoglossa Oerst. Acanthaceae

Plagius L'Hér. ex DC. 1838. Asteraceae. 1 Europe

Plakothira Florence. 1985. Loasaceae. 1 Marquesas Is.

Planaltoa Taub. 1895. Asteraceae. 2 central Brazil

Planchonella Pierre = Pouteria Aubl. Sapotaceae

Planchonia Blume. 1851–1852. Lecythidaceae. 14 Andaman Is., Indochina, Malesia to Philippines and New Guinea, northern and northeastern Australia (1, P. careya)

Planea Karis. 1990 (Metalasia L. Bolus). Asteraceae. 1 South Africa (Cape)

Planera J. F. Gmel. 1791. Ulmaceae. 1 southeastern North America

Planichloa B. K. Simon = Ectrosia R. Br. Poaceae

Planodes Greene = Sibara Greene. Brassicaceae

Planotia Munro = Neurolepis Meisn. Poaceae

Plantago L. 1753. Plantaginaceae. 256 cosmopolitan

Plarodrigoa Looser = Cristaria Cav. Malvaceae

Platanocephalus Crantz = Anthocephalus A. Rich. Rubiaceae

Platanthera Rich. 1817. Orchidaceae. 85 boreal and temp. Eurasia and North America, tropics and subtropics; Russia (European part, West and East Siberia, Far East), Caucasus

Platanus L. 1753. Platanaceae. 6–7 southwestern and southeastern U.S., Mexico and Guatemala (1, P. mexicana); P. orientalis—Sicily, Italy, Balkan Peninsula, Crete, Rhodes and Cyprus, Asia Minor, Syria, Lebanon, Israel, Jordan, Caucasus, Middle Asia, northern Iran, Afghanistan, Kashmir; P. kerii—Indochina

Platea Blume. 1826. Icacinaceae. 5 northeastern India, Burma, southern China, Indochina, Malesia, New Britain I.

Plateilema (A. Gray) Cockerell. 1904. Asteraceae. 1 Mexico

Plathymenia Benth. 1840. Fabaceae. 4 trop. South America

Platonia Mart. 1832. Clusiaceae. 1–2 Brazil

Platostoma P. Beauv. 1808. Lamiaceae. 4 trop. Africa, western India

Platyadenia B. L. Burtt. 1971. Gesneriaceae. 1 Borneo

Platyaechmea (Baker) L. B. Sm. et Kress. 1989 (Aechmea Ruiz et Pav.). Bormeliaceae. 21 trop. America

Platycalyx N. E. Br. 1905. Ericaceae. 1 South Africa (southern Cape)

Platycapnos (DC.) Bernh. 1833. Fumariaceae. 3 Macaronesia, Iberian Peninsula, southern France, Sicily, Sardinia, Italy, North Africa

Platycarpha Less. 1831. Asteraceae. 3 South Africa

Platycarpum Bonpl. 1811. Rubiaceae. 14 Colombia, Venezuela, Guyana, Brazil

Platycarya Siebold et Zucc. 1843. Juglandaceae. 3 China, Korea, Japan, Vietnam

Platycaulos Linder. 1984 (Restio Rottb). Restionaceae. 8 South Africa (Cape)

Platycelyphium Harms. 1905. Fabaceae. 1 Ethiopia, Somalia, Kenya, Tanzania

Platycentrum Naudin = Leandra Raddi. Melastomataceae

Platychaete Boiss. = Pulicaria Gaertn. Asteraceae

Platycladus Spach. 1841 (Thuja L.). Cupressaceae. 1 (P. orientalis) northern and northeastern (Manchuria) China, Korea

Platycodon A. DC. 1830. Campanulaceae. 1 East Siberia, Russian Far East, northeastern China, Korea, Japan

Platycoryne Rchb. f. 1855. Orchidaceae. 18 trop. Africa, Madagascar

Platycraspedum O. E. Schulz. 1922. Brassicaceae. 1 eastern Tibet

Platycrater Siebold et Zucc. 1838. Hydrangeaceae. 1 Japan

Platycyamus Benth. 1862. Fabaceae. 2 trop. South America

Platydesma H. Mann. 1866. Rutaceae. 4 Hawaiian Is.

Platyglottis L. O. Williams. 1942. Orchidaceae. 1 Panama

Platygyna Mercier = Tragia L. Euphorbiaceae

Platykeleba N. E. Br. 1895. Asclepiadaceae. 1 Madagascar

Platylepis A. Rich. 1828. Orchidaceae. 10 trop. and South Africa, Madagascar, Seychelles, Mascarene Is.

Platylobium Sm. 1793. Fabaceae. 4 Australia, Tasmania

Platylophus D. Don. 1830. Cunoniaceae. 1 South Africa (Cape)

Platymischium Vogel. 1837. Fabaceae. c. 20 Mexico, Central and South America, West Indies

Platymitra Boerl. 1899. Annonaceae. 2 Thailand, Malay Peninsula, Sumatra, Java, Philippines

Platyopuntia (Engelm.) Ritter = Opuntia Hill. Cactaceae

Platyosprion (Maxim.) Maxim. = Cladrastis Raf. Fabaceae

Platypholis Maxim. 1887. Orobanchaceae. 1 Ogasawara (Bonin) Is.

Platypodanthera R. M. King et H. Rob. 1972. Asteraceae. 1 Brazil

Platypodium Vogel. 1837. Fabaceae. 1–2 Central and South America

Platypterocarpus Dunkley et Brenan. 1948. Celastraceae. 1 trop. East Africa

Platyrhiza Barb. Rodr. 1881. Orchidaceae. 1 Brazil

Platyrhodon Decne. ex Hurst = Rosa L. Rosaceae

Platysace Bunge. 1845. Apiaceae. 25 Australia

Platyschkuhria (A. Gray) Rudb. 1906 (Bahia Lag.). Asteraceae. 1 southwestern U.S.

Platysepalum Welw. ex Baker. 1871. Fabaceae. 12 trop. Africa, mainly Zaire (10)

Platyspermation Guillaumin. 1950. Rutaceae. 1 New Caledonia

Platystele Schltr. 1910. Orchidaceae. c. 10 trop. America

Platystemma Wall. 1831. Gesneriaceae. 1 Himalayas from Simla to Bhutan, Tibet

Platystemon Benth. 1835. Papaveraceae. 1 (P. californicus) western North America

Platystigma Benth. = Meconella Nutt. Papaveraceae

Platytaenia Nevski et Vved. = Semenovia Regel et Herd. Apiaceae

Platytheca Steetz. 1845. Tremandraceae. 2 southwestern Australia

Platythelys Garay. 1977. Orchidaceae. 8 trop. and subtrop. America

Platythyra N. E. Br. 1925 (Aptenia N. E. Br.). Aizoaceae. 1 South Africa (Cape)

Platytinospora (Engl.) Diels. 1910. Menispermaceae. 1 trop. West Africa

Plazia Ruiz et Pav. 1794. Asteraceae. 2 southern Andes, Argentina

Pleconax Raf. = Sylene L. Caryophyllaceae

Plecospermum Trécul. 1847 (Maclura Nutt.). Moraceae. 3 India, Sri Lanka, Andaman Is., Burma, Indochina

Plecostachys Hilliard et B. L. Burtt. 1981. Asteraceae. 2 South Africa from Natal to Cape

Plectaneia Thouars. 1806. Apocynaceae. 12 Madagascar

Plectis O. F. Cook = Euterpe Mart. Arecaceae

Plectocephalus D. Don. 1830. Asteraceae. 5 southern U.S., Mexico, temp. South America, Ethiopia

Plectocomia Mart. ex Blume. 1830. Arecaceae. 16 Himalayas, northeastern India, Burma, southern China (incl. Hainan), Indochina, Malay Peninsula, Sumatra, Java, Borneo, Philippines

Plectocomiopsis Becc. 1893. Arecaceae. 5 southern Thailand (1), Malay Peninsula, Sumatra, Borneo

Plectomirtha W. R. B. Oliv. 1948. Anacardiaceae. 1 Three Kings Is.

Plectorhiza Dockrill. 1967. Orchidaceae. 3 eastern Australia (2), Lord Howe I. (1)

Plectrachne Henrard. 1929. Poaceae. 16 Australia

Plectranthastrum T. C. E. Fr. 1924 (Alvesia Welw.). Lamiaceae. 1 trop. East Africa

Plectranthus L'Hér. 1788. Lamiaceae. c. 350 trop. and subtrop. Old World from Africa to Japan and Oceania

Plectrelminthus Raf. 1838. Orchidaceae. 10 trop. West Africa, Comoro Is.

Plectritis (Lindl.) DC. 1830. Valerianaceae. 5 western North America (4), Chile (1)

Plectrocarpa Gillies ex Hook. et Arn. 1833. Zygophyllaceae. 3 temp. South America

Plectroniella Robyns. 1928. Rubiaceae. 2 trop. Africa

Plectrophora H. Focke. 1848. Orchidaceae. 5 northeastern South America, Trinidad

Pleea Michx. = Tofieldia Hudson. Melanthiaceae

Plegmatolemma Bremek. = Justicia L. Acanthaceae

Pleiacanthus (Nutt.) Rydb. = Lygodesmia D. Don. Asteraceae

Pleiadelphia Stapf. 1927 (Elymandra Stapf). Poaceae. 1 trop. Africa

Pleioblastus Nakai = Arundinaria Michx. Poaceae

Pleiocardia Greene = Streptanthus Nutt. Brassicaceae

Pleiocarpa Benth. 1876. Apocynaceae. 5 trop. Africa

Pleiocarpidia K. Schum. 1897. Rubiaceae. 27 Andaman Is., western Malesia

Pleioceras Baill. 1888. Apocynaceae. 3 trop. West Africa

Pleiochiton Naudin ex A. Gray. 1854. Melastomataceae. 9 southern Brazil

Pleiococca F. Muell. = Acronychia J. R. Forst. et G. Forst. Rutaceae

Pleiocoryne Rauschert. 1982. Rubiaceae. 1 trop. West Africa

Pleiocraterium Bremek. 1939. Rubiaceae. 4 southern India, Sri Lanka, Sumatra

Pleiogynium Engl. 1883. Anacardiaceae. 2–3 Sulawesi, Borneo, Philippines, Moluccas, Lesser Sunda Is., New Guinea, Solomon Is., trop. Australia, Pacific islands eastward to Tonga

Pleiokirkia Capuron. 1961. Simaroubaceae. 1 Madagascar

Pleiomeris A. DC. 1841. Myrsinaceae. 1 Madeira and Canary Is.

Pleione D. Don. 1825. Orchidaceae. 16 Nepal, Sikkim, Bhutan, India, Burma, Thailand, China, Taiwan

Pleioneura (C. E. Hubb.) J. B. Phipps, non Rech. = Rattraya J. B. Phipps. Poaceae

Pleioneura Rech. f. 1951. Caryophyllaceae. 2 Middle Asia (1, P. griffithiana), Iran, Afghanistan, western Himalayas

Pleiosepalum Hand.-Mazz. = Aruncus L. Rosaceae

Pleiospermium (Engl.) Swingle. 1916. Rutaceae. 5 Sri Lanka, Sumatra (1), Java (1), Borneo (2)

Pleiospilos N. E. Br. 1925. Aizoaceae. 4 South Africa (Cape)

Pleiostachya K. Schum. 1902. Marantaceae. 2 Central America, Ecuador

Pleiostachyopiper Trel. = Piper L. Piperaceae

Pleiostemon Sond. = Flueggea Willd. Euphorbiaceae

Pleiotaenia J. M. Coult. et Rose = Polytaenia DC. Apiaceae

Pleiotaxis Steetz. 1864. Asteraceae. 26 trop. Africa

Plenckia Reisseck. 1861. Celastraceae. 4 South America

Pleocarphus D. Don. 1830 (Jungia L. f.). Asteraceae. 1 northern Chile

Pleocaulus Bremek. 1944 (Strobilanthes Blume). Acanthaceae. 3 India

Pleodendron Tiegh. 1899. Canellaceae. 2 West Indies

Pleogyne Miers. 1851. Menispermaceae. 1 trop. eastern Australia

Pleomele Salisb. 1796 (Dracaena Vand. ex L.). Dracaenaceae. c. 140 trop. and subtrop. Old World from Africa to Hawaiian Is. and northern Australia

Pleonotoma Miers. 1863. Bignoniaceae. 14 trop. America from Costa Rica to Argentina, Trinidad

Plerandra A. Gray = Schefflera J. R. Forst. et G. Forst. Araliaceae

Plesmonium Schott. 1856 (Amorphophallus Blume ex Decne.). Araceae. 2 northern India, Indochina

Plethadenia Urb. 1912. Rutaceae. 2 West Indies

Plethiandra Hook. f. 1867. Melastomataceae. 7 western Malesia

Plettkea Marrf. 1934. Caryophyllaceae. 4 Andes of Peru

Pleurandropsis Baill. = Asterolasia F. Muell. Rutaceae

Pleuranthemum (Pichon) Pichon = Hunteria Roxb. Apocynaceae

Pleuranthodendron L. O. Williams. 1961. Flacourtiaceae. 2 trop. America from southern Mexico to Ecuador, Amazonian Peru and Brazil

Pleuranthodes Weberb. = Gouania Jacq. Rhamnaceae

Pleuranthodium (K. Schum.) Ros. M. Sm. 1991. Zingiberaceae. c. 30 New Guinea, Bismarck Arch., Australia

Pleuraphis Torr. = Hilaria Kunth. Poaceae

Pleuriarum Nakai = Arisaema Mart. Araceae

Pleuricospora A. Gray. 1868. Ericaceae. 1 western North America

Pleurisanthes Baill. 1874. Icacinaceae. 5 trop. South America

Pleuroblepharis Baill. 1890 (Crossandra Salisb.). Acanthaceae. 1 Madagascar

Pleurobotryum Barb. Rodr. = Pleurothallis R. Br. Orchidaceae

Pleurocalyptus Brong. et Gris. 1868. Myrtaceae. 1 New Caledonia

Pleurocarpaea Benth. 1867. Asteraceae. 1 trop. Australia

Pleurocitrus Tanaka = Microcitrus Swingle. Rutaceae

Pleurocoffea Baill. 1880 (Coffea L.). Rubiaceae. 1 Madagascar

Pleurocoronis R. M. King et H. Rob. 1966 (Eupatorium L.). Asteraceae. 3 southwestern U.S., northwestern Mexico

Pleurogyna Eschsch. ex Cham. et Schltdl. = Lomatogonium A. Braun. Gentianaceae

Pleurogyne Eschsch. ex Griseb. = Swertia L. Gentianaceae

Pleurogynella Ikonn. = Lomatogonium A. Braun. Gentianaceae

Pleuropappus F. Muell. 1855. Asteraceae. 1 southern Australia

Pleuropetalum Hook. f. 1846. Amaranthaceae. 5 Galapagos Is., Central and trop. South America

Pleurophora D. Don. 1837. Lythraceae. 11 South America

Pleurophragma Rydb. = Thelypodium Endl. Brassicaceae

Pleurophyllum Hook. f. 1844. Asteraceae. 3 islands south of New Zealand

Pleuropogon R. Br. 1823. Poaceae. 1 (P. sabinii) Russian Arctica and Altai

Pleuropterantha Franch. 1882. Amaranthaceae. 2 Somalia

Pleuropteropytum Gross = Polygonum L. Polygonaceae

Pleurospa Raf. = Montrichardia Grüger. Araceae

Pleurospermopsis C. Norman. 1938. Apiaceae. 1 eastern Himalayas, Sikkim

Pleurospermum Hoffm. 1814. Apiaceae. c. 40 temp. Eurasia; Russia (2, all over in forest zones)

Pleurostachys Brongn. 1829. Cyperaceae. 50 South America

Pleurostelma Baill. 1890. Asclepiadaceae. 1 trop. East Africa, Madagascar, Aldabra I.

Pleurostima Raf. 1836 (Barbacenia Vend.). Velloziaceae. 12 trop. South America

Pleurostylia Wight et Arn. 1834. Celastraceae. 6 Africa, Madagascar, Mascarene Is., India, Sri Lanka, Indochina, Hainan, Malesia (1, P. opposita—Malay Peninsula, Philippines, New Guinea), Australia, New Caledonia

Pleurothallis R. Br. 1813. Orchidaceae. c. 2,500 trop. America from Mexico and Florida to Argentina, mainly in Mexico, Costa Rica, Colombia, Venezuela, Peru, and Brazil

Pleurothallopsis Porto et Brade = Octomeria R. Br. Orchidaceae

Pleurothyrium Nees. 1836 (Ocotea Aubl.). Lauraceae. 40 trop. America from Guatemala south to Amazonian Basin of Peru, Bolivia, and Brazil

Plexipus Raf. = Chascanum E. Mey. Verbenaceae

Plicosepalus Tiegh. 1894. Loranthaceae. 10 arid trop. and South (3) Africa

Plinia L. 1753. Myrtaceae. 30 West Indies, trop. South America

Plinthanthesis Steud. 1853. Poaceae. 5 Australia

Plinthus Fenzl. 1839. Aizoaceae. 5–6 Namibia, South Africa

Plocama Aiton. 1789. Rubiaceae. 1 Canary Is.

Plocaniophyllon Brandegee = Deppea Cham. et Schltdl. Rubiaceae

Plocoglottis Blume. 1825. Orchidaceae. 35–40 Andaman Is., peninsular Thailand, Cambodia, Malesia from Malay Peninsula to New Guinea, Solomon Is., Oceania

Plocosperma Benth. 1876. Plocospermataceae. 1 southern Mexico, Guatemala

Ploiarium Korth. 1842. Bonnetiaceae. 3 Indochina, western Malesia, Moluccas, New Guinea

Plowmania Hunz. et Subils. 1986. Solanaceae. 1 Guatemala and South America

Pluchea Cass. 1817. Asteraceae. 40 tropics and subtropics

Plukenetia L. 1753. Euphorbiaceae. 10 trop. America from Mexico to Brazil, and the Lesser Antilles; 1 Madagascar

Plumbagella Spach. 1841 (Plumbago L.). Plumbaginaceae. 1 Middle Asia, West and East Siberia, western China, Mongolia

Plumbago L. 1753. Plumbaginaceae. 12 trop. and warm–temp. regions; P. europaea–extending to Caucasus and Middle Asia

Plumeria L. 1753. Apocynaceae. 8 trop. and subtrop. America, West Indies

Plumeriopsis Rusby et Woodson = Thevetia L. Apocynaceae

Plummera A. Gray. 1882. Asteraceae. 2 southwestern U.S. (Arizona)

Plumosipappus Czerep. = Centaurea L. Asteraceae

Plutarchia A. C. Sm. 1936. Ericaceae. 12 northern Andes

Poa L. 1753. Poaceae. c. 500 extratrop. regions and montane areas of tropics

Poacynum Baill. 1888 (Apocynum L.). Apocynaceae. 2 Middle Asia (2), western China, Mongolia

Poaephyllum Ridl. 1907. Orchidaceae. 3 Thailand, Malay Peninsula, Sumatra, Java (1, P. pauciflorum), Philippines (2), New Guinea (2)

Poagrostis Stapf. 1899. Poaceae. 1 South Africa

Poarium Desv. 1825. Scrophulariaceae. 20 trop. America

Pobeguinea (Stapf) Jacq.-Fèl. = Anadelphia Hack. Poaceae

Pochota Ram. Goyena = Bombacopsis Pittier. Bombacaceae

Poculodiscus Danguy et Choux = Plagioscyphus Radkl. Sapindaceae

Podachaenium Benth. ex Oerst. 1852. Asteraceae. 2 Mexico, Central America

Podadenia Thwaites. 1861 (Ptychopyxis Miq.). Euphorbiaceae. 1 Sri Lanka

Podaechmea (Mez) L. B. Sm. et Kress. 1989 (Aechmea Ruiz et Pav.). Bromeliaceae. 5 trop. America

Podagrostis (Griseb.) Scribn. et Merr. = Agrostis L. Poaceae

Podalyria Willd. 1799. Fabaceae. 28 South Africa, esp. Cape; P. calyptrata—extending to Asia (? introduce)

Podandra Baill. 1890. Asclepiadaceae. 1 Bolivia

Podandrogyne Ducke. 1930. Capparaceae. 11 Central and South (Andes) America

Podangis Schltr. 1918. Orchidaceae. 1 trop. Africa (Guinea, Sierra Leone, Ghana, Ivory Coast, Nigeria, Cameroun, Zaire, Angola, Uganda, Tanzania), ? Madagascar

Podanthus Lag. 1816. Asteraceae. 2 Chile, Argentina

Podistera S. Watson. 1887. Apiaceae. 4 western North America, Chukotka (1, P. macounii)

Podocaelia (Benth.) A. Fern. et R. Fern. 1962. Melastomataceae. 1 trop. West Africa

Podocalyx Klotzsch. 1841. Euphorbiaceae. 1 Venezuela, Brazil (Amazonia)

Podocarpium (Benth.) Y. C. Yang et S. H. Huang = Desmodium Desv. Fabaceae

Podocarpus L'Hér. ex Pers. 1807. Podocarpaceae. 95–110 Cameroun, Angola, East, Southeast and South Africa, Madagascar; Asia: northern Iran and from eastern Nepal through Central China to southern Japan, Indochina; Malesia, eastern and southeastern Australia, Tasmania, New Zealand, Oceania from New Britain and Solomon Is. to New Caledonia and Tonga; Mexico, Central and South America, West Indies

Podochilopsis Guillaumin = Adenocos Blume. Orchidaceae

Podochilus Blume. 1825. Orchidaceae. c. 100 trop. Asia from Himalayas, India, and Sri Lanka to Burma, Indochina, Malesia from Malay Peninsula to New Guinea, western Oceania

Podochrosia Baill. = Rauvolfia L. Apocynaceae

Podococcus G. Mann et H. A. Wendl. 1864. Arecaceae. 1 trop. West Africa from Nigeria (the Niger delta) to Gabon

Podocoma Cass. 1817. Asteraceae. 12 South America

Podocytisus Boiss. et Heldr. 1849. Fabaceae. 1 Yugoslavia, Albania, Greece, Asia Minor

Podogynium Taub. = Zenkerella Taub. Fabaceae

Podolasia N. E. Br. 1882. Araceae. 1 Malay Peninsula, Batu Is., Sumatra, Borneo

Podolepis Labill. 1806. Asteraceae. 18 Australia

Podolotus Royly. 1835. (Astragalus L.). Fabaceae. 1 Iran, Afghanistan, Pakistan, India

Podonephelium Baill. 1874. Sapindaceae. 1 New Caledonia

Podopetalum F. Muell. = Ormosia Jacks. Fabaceae

Podophania Baill. = Hofmeisteria Walp. Asteraceae

Podophorus Phil. 1856. Poaceae. 1 Juan Fernández Is.

Podophyllum L. 1753. Berberidaceae. 10 Asia from western Himalayas to Japan, eastern North America (1, P. peltatum)

Podopterus Bonpl. 1812. Polygonaceae. 4 Mexico, Central America

Podorungia Baill. 1891. Acanthaceae. 1 Madagascar

Podosperma Labill. = Podotheca Cass. Asteraceae

Podospermum DC. = Scorzonera L. Asteraceae

Podostelma K. Schum. 1893. Asclepiadaceae. 1 trop. Northeast Africa

Podostemum Michx. 1803. Podostemaceae. 18 trop. Africa, Madagascar, trop. Asia, North (1), Central, and trop. South America

Podostigma Elliott = Asclepias L. Asclepiadaceae

Podotheca Cass. 1822. Asteraceae. 6 western Australia

Podranea Sprague. 1904. Bignoniaceae. 1–2 trop. and South Africa

Poecilandra Tul. 1847. Sauvagesiaceae. 3 northern South America

Poecilanthe Benth. 1859. Fabaceae. 7 trop. South America

Poecilocalyx Bremek. 1940. Rubiaceae. 2 trop. Africa

Poecilochroma Miers = Saracha Ruiz et Pav. Solanaceae

Poecilodermis Schott et Endl. = Brachychiton Schott et Endl. Sterculiaceae

Poecilolepis Grau. 1977 (Aster L.). Asteraceae. 2 South Africa (Cape)

Poeciloneuron Bedd. 1865. Clusiaceae. 2 southern India

Poecilostachys Hack. 1884. Poaceae. c. 20 Madagascar; P. oplismenoides—trop. Africa from Nigeria to Ethiopia and south to Zimbabwe and Mozambique

Poellnitzia Uitew. 1940 (Haworthia Duval.). Asphodelaceae. 1 South Africa (Cape)

Poenosedum Holub = Sedum L. Crassulaceae

Poeppigia C. Presl. 1830. Fabaceae. 1 Central and South America, West Indies

Poga Pierre. 1896. Anisophylleaceae. 1 trop. West Africa

Poggea Gürke. 1893. Flacourtiaceae. 5 Guinea, Congo, Zaire, Angola

Pogogyne Benth. 1834. Lamiaceae. 7 western U.S. (Oregon, California), Baja California

Pogonachne Bor. 1949. Poaceae. 1 India

Pogonanthera Blume. 1831. Melastomataceae. 1–4 Malesia

Pogonarthria Stapf. 1898. Poaceae. 4 trop. and South Africa

Pogonatherum P. Beauv. 1812. Poaceae. 3 trop. Asia

Pogonia Juss. 1789. Orchidaceae. c. 50 temp. Asia, North America; Russia (1, P. japonica: Far East)

Pogoniopsis Rchb. f. 1881. Orchidaceae. 2 Brazil

Pogonochloa C. E. Hubb. 1940. Poaceae. 1 Zambia and Zimbabwe

Pogonolepis Steetz. 1845. Asteraceae. 6 Australia

Pogononeura Napper. 1963. Poaceae. 1 trop. East Africa

Pogonophora Miers ex Benth. 1854. Euphorbiaceae. 2–3 trop. South America (1–2), trop. Africa (1)

Pogonopus Klotzsch. 1853. Rubiaceae. 3 trop. America

Pogonorrhinum Betsche = Kichxia Dumort. Scrophulariaceae

Pogonotium J. Dransf. 1980. Arecaceae. 3 Malay Peninsula (1), Borneo

Pogostemon Desf. 1815. Lamiaceae. 71 India, Sri Lanka, eastern Himalayas, Burma, southern China, Taiwan, Japan, Indochina, Malesia, Australia

Pohlidium Davidse, Soderstr. et R. P. Ellis. 1986. Poaceae. 1 Panama

Pohliella Engl. = Saxicolella Engl. Podostemaceae

Poicilanthe Schltr. = ? Cymbidium Sw. Orchidaceae

Poicilla Griseb. = Matelea Aubl. Asclepiadaceae

Poicillopsis Schltr. ex Rendle = Matelea Aubl. Asclepiadaceae

Poikilacanthus Lindau. 1895. Acanthaceae. 9 Central and South America

Poikilogyne Baker f. 1917. Melastomataceae. c. 20 Borneo, New Guinea

Poikilospermum Zipp. ex Miq. 1864. Cecropiaceae. c. 20 India, eastern Himalayas, Burma, China, Thailand, Malesia, southward to Australia

Poilanedora Gagnep. 1948. Capparaceae. 1 Indochina

Poilaniella Gagnep. 1925. Euphorbiaceae. 1 Indochina

Poilannammia C. Hansen. 1987. Melastomataceae. 4 Vietnam

Poinciana L. = Caesalpinia L. Fabaceae

Poinsettia Graham. 1836 (Euphorbia L.). Euphorbiaceae. 11–12 America from eastern U.S. to northern Argentina

Poiretia Vent. 1807. Fabaceae. 6 trop. America

Poissonia Baill. = Coursetia DC. Fabaceae

Poitea Vent. 1808. Fabaceae. 5–6 West Indies

Pojarkovia Askerova. 1984. Asteraceae. 4 West Asia, incl. Caucasus (4)

Polakia Stapf = Salvia L. Lamiaceae

Polakowskia Pittier = Sechium P. Browne. Cucurbitaceae

Polanisia Raf. 1819 (Cleome L.). Capparaceae. 7 temp. North America; Russia (1, P. dodecandra: Far East—Primorski Krai)

Polaskia Backeb. 1949 (Myrtillocactus Console). Cactaceae. 2 southern Mexico

Polemannia Eckl. et Zeyh. 1837. Apiaceae. 3–4 South Africa

Polemanniopsis B. L. Burtt. 1988. Apiaceae. 1 Namibia, South Africa

Polemonium L. 1753. Polemoniaceae. c. 50 temp. Eurasia, North America, Chile (2)

Polevansia De Winter. 1966. Poaceae. 1 South Africa

Polhillia C. H. Stirt. 1986 (Argyrolobium Eckl. et Zeyh.). Fabaceae. 7 South Africa (southwestern Cape)

Polianthes L. 1753. Agavaceae. 13 Mexico, Trinidad

Poliomintha A. Gray. 1870. Lamiaceae. 4 southwestern U.S. eastward to Colorado and Texas, northern Mexico

Poliophyton O. E. Schulz = Mancoa Wedd. Brassicaceae

Poliothyrsis Oliv. 1889. Flacourtiaceae. 1 China

Pollalesta Kunth. 1818. Asteraceae. 16 trop. America from Costa Rica to Ecuador, Peru, Venezuela, and northern Brazil

Pollia Thunb. 1781. Commelinaceae. 17 trop. Old World, Panama (1)

Pollichia Aiton. 1789. Caryophyllaceae. 1 eastern Africa from Ethiopia to eastern Cape, Arabian Peninsula

Polliniopsis Hayata. 1918 (Microstegium Nees). Poaceae. 1 Taiwan

Polpoda C. Presl. 1829. Molluginaceae. 2 South Africa (Cape)

Polyachyrus Lag. 1811. Asteraceae. 7 Peru, Chile

Polyadoa Stapf = Hunteria Roxb. Apocynaceae

Polyalthia Blume. 1830. Annonaceae. c. 120 trop. East Africa (3), Madagascar (18), trop. Asia (c. 100)

Polyandra Leal. 1951. Euphorbiaceae. 1 Brazil

Polyandrococos Barb. Rodr. 1901. Arecaceae. 2 Brazil

Polyanthina R. M. King et H. Rob. 1970 (Eupatorium L.). Asteraceae. 1 Central America, Colombia, Ecuador, Peru

Polyanthus C. H. Hu et Y. C. Hu. 1991 (Arundinaria Michx.). Poaceae. 1 China (Hunan)

Polyarrhena Cass. 1828 (Felicia Cass.). Asteraceae. 4 South Africa (Cape)

Polyaster Hook. f. 1862. Rutaceae. 1–2 Mexico

Polyaulax Backer. 1945. Annonaceae. 1 Sumatra, Java, Borneo, Lesser Sunda Is., Moluccas, New Guinea

Polybactrum Salisb. = Pseudorchis Séguier. Orchidaceae

Polycalymma F. Muell. et Sond. 1853 (Myriocephalus Benth.). Asteraceae. ? sp. Australia

Polycardia Juss. 1789. Celastraceae. 9 Madagascar

Polycarena Benth. 1836. Scrophulariaceae. c. 50 South Africa; 1 (P. transvaalensis) extending to Botswana and Zimbabwe

Polycarpaea Lam. 1792. Caryophyllaceae. 50 tropics and subtropics

Polycarpon L. 1759. Caryophyllaceae. 16 Europe and Mediterranean, 2 South America; P. tetraphyllum—cosmopolitan, except Russia and Middle Asia

Polycephalium Engl. 1897. Icacinaceae. 2 trop. Africa

Polyceratocarpus Engl. et Diels. 1900. Annonaceae. 7 trop. Africa

Polychrysum (Tzvelev) Kovalevsk. 1962. Asteraceae. 1 Middle Asia

Polyclathra Bertol. 1840. Cucurbitaceae. 1 Central America

Polyclita A. C. Sm. 1936. Ericaceae. 1 Bolivia

Polycnemum L. 1753. Chenopodiaceae. 6–7 Central, South, and Southeast Europe, Morocco, Algeria; Caucasus, Middle Asia, West Siberia

Polycodium Raf. ex Greene = Vaccinium L. Ericaceae

Polycoryne Keay = Pleiocoryne Rauschert. Rubiaceae

Polyctenium Greene. 1912 (Smelowskia C. A. Mey.). Brassicaceae. 2 western U.S.

Polycycliska Ridl. = Lerchea L. Rubiaceae

Polycycnis Rchb. f. 1855. Orchidaceae. 12 trop. America from Costa Rica to Peru and Brazil

Polygala L. 1753. Polygalaceae. c. 500 trop., subtrop. and temp. regions, montane areas (excl. New Zealand, Polynesia, and Arctica), esp. trop. America

Polygaloides Haller = Polygala L. Polygalaceae

Polygonanthus Ducke. 1932. Anisophylleaceae. 2 Brazil (Amazonia)

Polygonataceae Salisb. = Convallariaceae Horan.

Polygonatum Hill. 1753. Convallariaceae. 55 North America, temp. Eurasia, Himalayas from Kashmir to Sikkim, montane areas of trop. and subtrop. China (c. 30) and Indochina

Polygonella Michx. 1803 (Polygonum L.). Polygonaceae. 9 eastern North America

Polygonum L. 1753. Polygonaceae. 200–250 (or 600, or 20 ?) cosmopolitan

Polylepis Ruiz et Pav. 1794. Rosaceae. 15 South America (Andes)

Polylophium Boiss. 1844 (Laserpitium L.). Apiaceae. 2 Caucasus (1, P. panjutinii), Iran

Polylychnis Bremek. 1938. Acanthaceae. 2 northeastern South America

Polymeria R. Br. 1810. Convolvulaceae. 10 Timor I., Australia, New Caledonia

Polymita N. E. Br. 1930 (Ruschia Schwantes). Aizoaceae. 2 northwestern South Africa

Polymnia L. 1753. Asteraceae. 2 eastern North America

Polyosma Blume. 1826. Escalloniaceae. c. 60 eastern Himalayas, northeastern India, Andaman Is., China, Indochina, Malesia, trop. Australia, New Caledonia

Polyotidium Garay. 1958. Orchidaceae. 1 Colombia

Polypleurella Engl. = Polypleurum (Taylor ex Tul.) Warm. Podostemaceae

Polypleurum (Taylor ex Tul.) Warm. 1901. Podostemaceae. 7 trop. West Africa (1), India, Sri Lanka, Burma, Thailand (1), India, Sri Lanka

Polypogon Desf. 1798. Poaceae. 15 Europe, North Africa, West (incl. Caucasus) and Central Asia to West Siberia, montane areas of tropics

Polypompholyx Lehm. 1844 (Utricularia L.). Lentibulariaceae. 3 Australia

Polyporandra Becc. 1877. Icacinaceae. 1 Moluccas, New Guinea, Melanesia

Polypremum L. 1753. Buddlejaceae. 1 southern U.S., Mexico, Central and South America, West Indies

Polypsecadium O. E. Schulz. 1924. Brassicaceae. 3 Central Andes, northern Argentina

Polypteris Nutt. = Palafoxia Lag. Asteraceae

Polyradicion Garay. 1969. Orchidaceae. 2–4 Florida Peninsula, West Indies

Polyrhabda C. C. Towns. 1984. Amaranthaceae. 1 Somalia

Polyscias J. R. Forst. et G. Forst. 1776. Araliaceae. c. 150 trop. Old World, esp. Madagascar, Malesia, extending into Pacific islands

Polysolen Rauschert = Indopolysolenia Bennet. Rubiaceae

Polysolenia Hook. f. = Indopolysolenia Bennet. Rubiaceae

Polyspatha Benth. 1849. Commelinaceae. 3 trop. West Africa

Polysphaeria Hook. f. 1873. Rubiaceae. 24 trop. Africa, Madagascar, Comoro Is., Aldabra I.

Polystachya Hook. 1824. Orchidaceae. c. 210 tropics and subtropics: Central and South America, West Indies, trop. and South Africa, Madagascar, Mauritius, trop. Asia from India, Sri Lanka through Indochina to the Philippines (1, P. luteola), Indonesia, New Guinea, and Australia

Polystemma Decne. = Matelea Aubl. Asclepiadaceae

Polystemonanthus Harms. 1897. Fabaceae. 1 Liberia, Ivory Coast

Polytaenia DC. 1829. Apiaceae. 2 North America

Polytaxis Bunge. 1843 (Jurinea Cass.). Asteraceae. 2 Middle Asia

Polytepalum Suess. et Beyerle. 1838. Caryophyllaceae. 1 Angola

Polytoca R. Br. 1838. Poaceae. 2 India, Southeast Asia

Polytrema C. B. Clarke = Ptyssiglottis T. Anderson. Acanthaceae

Polytrias Hack. 1887. Poaceae. 1 Southeast Asia

Polyura Hook. f. 1868. Rubiaceae. 1 India (Assam)

Polyxena Kunth. 1843. Hyacinthaceae. 2 South Africa (Cape)

Polyzygus Dalzell. 1850. Apiaceae. 1 southern India

Pomaderris Labill. 1805. Rhamnaceae. 47 Australia, Tasmania, New Zealand

Pomatocalpa Breda, Kuhl et Hasselt. 1829. Orchidaceae. c. 30 Sikkim, northeastern India, Sri Lanka, Andaman Is., Burma, southern China, Taiwan, Indochina, Malesia from Malay Peninsula to New Guinea, Solomon Is., northern Australia, Oceania, incl. Fiji and Samoa

Pomatosace Maxim. 1881. Primulaceae. 1 northwestern China

Pomatostoma Stapf = Anerincleistus Korth. Melastomataceae

Pomax Sol. ex Gaertn. 1788. Rubiaceae. 1 eastern Australia

Pomazota Ridl. = Coptophyllum Korth. Rubiaceae

Pometia J. R. Forst. et G. Forst. 1775. Sapindaceae. 8 India, Sri Lanka, Andaman Is., China, Taiwan, Indochina, Malesia and eastward to Tonga, Niue, and Samoa

Pommereschea Wittm. 1895. Zingiberaceae. 2 Burma, China

Pommereulla L. f. 1780. Poaceae. 1 southern India, Sri Lanka

Ponapea Becc. = Ptychosperma Labill. Arecaceae

Poncirus Raf. 1838 (Citrus L.). Rutaceae. 1 northern and central China

Ponera Lindl. 1831. Orchidaceae. 8 trop. America from Mexico to Peru, Bolivia, and Brazil; West Indies

Ponerorchis Rchb. f. = Habenaria Willd. Orchidaceae

Pongamia Vent. 1803. Fabaceae. 1 Mascarene Is., coastal Southeast Asia eastward to Taiwan and Ryukyu Is., south to northern Australia, Oceania eastward to Samoa Is.

Pongamiopsis R. Vig. 1950. Fabaceae. 2 Madagascar

Pontederia L. 1753. Pontederiaceae. 5 America from southeastern Canada to northeastern Argentina, West Indies

Ponthieva R. Br. 1813. Orchidaceae. 60 America from southern U.S. to Chile and northern Argentina

Pontya A. Chev. = Trilepisium Thouars. Moraceae

Popoviocodonia Fedorov. 1957. Campanulaceae. 2 Russian Far East: Okhotski Krai, Sakhalin I.

Popoviolimon Lincz. 1971. Plumbaginaceae. 1 Middle Asia

Popowia Endl. 1839. Annonaceae. c. 100 trop. Asia, Malesia, trop. Australia

Populina Baill. 1891. Acanthaceae. 2 Madagascar

Populus L. 1753. Salicaceae. 35 temp. and warm–temp. regions of the Northern Hemisphere; 1 (P. ilicifolia) trop. East Africa

Porana Burm. f. 1768. Convolvulaceae. c. 20 trop. Africa (3), India, Nepal, northern Burma, China, Indochina, Malesia, Australia (1)

Porandra D. Y. Hong. 1974 (Amischotolype Hassk.). Commelinaceae. 3 southern China

Poranopsis Roberty. 1952 (Parana Burm. f.). Convolvulaceae. 3 Pakistan, Tibet, China, northeastern India

Poranthera Rudge. 1811. Euphorbiaceae. 10 Australia, New Zealand

Poraqueiba Aubl. 1775. Icacinaceae. 3 Panama, Venezuela, Peru, Surinam, Brazil

Porcelia Ruiz et Pav. 1794. Annonaceae. 5 Central and trop. South America

Porlieria Ruiz et Pav. 1794. Zygophyllaceae. 6 Mexico, Andes

Porocystis Radlk. 1878. Sapindaceae. 2 trop. South America

Porodittia G. Don. 1838. Scrophulariaceae. 1 Andes of Peru

Porolabium R. Tang et F. T. Wang. 1940. Orchidaceae. 1 China, Mongolia

Porophyllum Adans. 1763. Asteraceae. 28 trop. and subtrop. America

Porospermum F. Muell. = Delarbrea Vieill. Araliaceae

Porpax Lindl. 1845. Orchidaceae. 11 trop. Asia from Deccan and Himalayas eastward to China, Indochina, and Malay Peninsula

Porphyrocoma Scheidw. ex Hook. = Justicia L. Acanthaceae

Porphyrodesme Schltr. 1913. Orchidaceae. 2 Philippines (P. elongata), New Guinea (P. papuana)

Porphyroglottis Ridl. 1896. Orchidaceae. 1 west Malesia

Porphyroscias Miq. = Angelica L. Apiaceae

Porphyrospatha Engl. = Syngonium Schott. Araceae

Porphyrostachys Rchb. f. 1854. Orchidaceae. 1 Ecuador, Peru

Porphyrostemma Benth. ex Oliv. 1873. Asteraceae. 2 trop. Africa

Porroglossum Schltr. 1920. Orchidaceae. 6 trop. South America

Porrorhachis Garay. 1972. Orchidaceae. 2 Malesia

Portea Brongn. et K. Koch. 1856. Bromeliaceae. 7 east Brazil

Portenschlagiella Tutin. 1967. Apiaceae. 1 southern Italy, Yugoslavia, Albania

Porterandia Ridl. 1940. Rubiaceae. 9 Malesia from Malay Peninsula to Borneo, and a few species from Borneo eastward to Fiji and Tonga

Porteranthus Britton ex Small. 1894. Rosaceae. 2 North America

Porterella Torr. 1872 (Solenopsis C. Presl). Lobeliaceae. 1 western North America

Porteresia Tateoka. 1965. Poaceae. 1 India, Burma

Portlandia P. Browne. 1756. Rubiaceae. 18 Mexico, Central America, West Indies

Portulaca L. 1753. Portulacaceae. 40 (or 100–200) trop., subtrop., and warm–temp. regions; P. oleracea—southern European part of Russia, Caucasus, Middle Asia, Far East

Portulacaria Jacq. 1787. Portulacaceae. 3 Namibia, South Africa

Posadaea Cogn. 1890. Cucurbitaceae. 1 Central and trop. South America, West Indies

Posidonia K. Koenig. 1805. Posidoniaceae. 3–5 coastal Mediterranean (1), southern Australia and Tasmania

Poskea Vatke. 1882. Globulariaceae. 2 Somalia, Socotra

Posoqueria Aubl. 1775. Rubiaceae. 12 trop. America, West Indies

Postia Boiss. et Blanche = Rhanteriopsis Rauschert. Asteraceae

Postiella Kljuykov. 1985. Apiaceae. 1 southern Turkey

Potalia Aubl. 1775. Gentianaceae. 1 trop. South America from Colombia to northeastern Brazil (Amapa)

Potameia Thouars. 1806. Lauraceae. c. 20 Madagascar

Potamoganos Sandwith. 1937. Bignoniaceae. 1 northeastern South America

Potamogeton L. 1753. Potamogetonaceae. c. 100 cosmopolitan, but chiefly extratrop. regions—fresh, rarely brackish waters

Potamophila R. Br. 1810. Poaceae. 1 Australia (New South Wales)

Potaninia Maxim. 1881. Rosaceae. 1 Mongolia, China

Potarophytum Sandwith. 1939. Rapataceae. 1 northeastern South America

Potentilla L. 1753. Rosaceae. c. 500 cosmopolitan, chiefly temp. and arctic regions of the Northern Hemisphere; a few species in montane trop. areas, Tasmania, New Zealand

Poteranthera Bong. 1838. Melastomataceae. 1 trop. South America

Poterium L. = Sanguisorba L. Rosaceae

Pothoidium Schott. 1856–1857. Araceae. 1 China, Java, Philippines, Sulawesi, Moluccas

Pothomorphe Miq. = Piper L. Piperaceae

Pothos L. 1753. Araceae. 50–75 Comoro Is., Madagascar, trop. Asia, Malesia, Bismarck Arch., eastern Australia

Pothuava Gaudich. 1844–1852 (Aechmea Ruit et Pav.). Bromeliaceae. 21 trop. America

Potoxylon Kosterm. 1978. Lauraceae. 1 Borneo

Pottingeria Prain. 1898. Pottingeriaceae. 1 northeastern India, northern Burma, northern Thailand

Pottsia Hook. et Arn. 1837. Apocynaceae. 5 India, continental Southeast Asia, Java

Pouchetia A. Rich. ex DC. 1830. Rubiaceae. 6 trop. Africa

Poulsenia Eggers. 1898. Moraceae. 1 Ecuador

Pounguia Benoist = Whitfieldia Hook. Acanthaceae

Poupartia Comm. ex Juss. 1789. Anacardiaceae. 12 tropics

Pourouma Aubl. 1775. Cecropiaceae. 25 trop. America from Guatemala to Peru, Bolivia, and Brazil; not in West Indies

Pouteria Aubl. 1775. Sapotaceae. c. 320 trop. America (c. 200 Mexico, Guatemala, Belize, Honduras, Nicaragua, Costa Rica, Panama, Colombia, Venezuela, Guyana, Surinam, French Guiana, Ecuador, Peru, Brazil, Bolivia, Paraguay) and West Indies; trop. Africa (5), Seychelles, trop. Asia, Malesia, Australia, New Zealand, New Caledonia, eastward to Polynesia

Pouzolzia Gaudich. 1830. Urticaceae. 52 trop. America, trop. and South Africa, trop. Asia

Povedadaphne W. C. Burger. 1988 (Ocotea Aubl.). Lauraceae. 1 Costa Rica

Pozoa Lag. 1816. Apiaceae. 2 Andes of Chile and Argentina

Pradosia Liais. 1872. Sapotaceae. 23 Colombia, Venezuela, Trinidad (3), Guyana, Surinam, French Guiana, Ecuador, Peru (1, P. montana), Brazil (15), Paraguay; P. atroviolacea—extending to Panama and Costa Rica

Praecereus F. Buxb. = Cereus Mill. Cactaceae

Praecitrullus Pang. 1966. Cucurbitaceae. 1 India

Prainea King ex Hook. f. 1888. Moraceae. 7 Malesia

Pranceacanthus Wassh. 1984. Acanthaceae. 1 Brazil (Amazonia)

Prangos Lindl. 1825. Apiaceae. c. 40 arid areas from Mediterranean through Caucasus to Central Asia

Praravinia Korth. 1842. Rubiaceae. c. 50 Philippines, Borneo, Sulawesi

Prasium L. 1753. Lamiaceae. 1 Mediterranean

Prasophyllum R. Br. 1810. Orchidaceae. c. 80–90 Australia, Tasmania, New Zealand, New Caledonia (1, P. calopterum)

Pratia Gaudich. = Lobelia L. Lobeliaceae

Pravinaria Bremek. 1940. Rubiaceae. 2 Borneo

Praxeliopsis G. M. Barroso. 1949. Asteraceae. 1 Brazil

Praxelis Cass. 1826 (Eupatorium L.). Asteraceae. 14 trop. America

Premna L. 1771. Verbenaceae. c. 200 trop. and subtrop. Africa, Asia, Australia, and Oceania

Prenanthella Rydb. 1906 (Prenanthes L.). Asteraceae. 1 northwestern North America

Prenanthes L. 1753. Asteraceae. 25 temp. Eurasia, Canary Is., montane areas of Africa

Prenia N. E. Br. 1925 (Phyllobolus N. E. Br.). Aizoaceae. 5 South Africa

Prepodesma N. E. Br. = Nananthus N. E. Br. Aizoaceae

Prepusa Mart. 1827. Gentianaceae. 5 Brazil

Prescotia Lindl. 1824. Orchidaceae. 22 trop. America

Prescottia Lindl. = Prescotia Lindl. Orchidaceae

Preslia Opiz = Mentha L. Lamiaceae

Prestelia Sch. Bip. 1865 (?). Asteraceae. 1 Brazil

Prestoea Hook. f. 1883. Arecaceae. 35 trop. America from Nicaragua to Peru and Brazil, West Indies

Prestonia R. Br. 1810. Apocynaceae. 66 Central and trop. South America, West Indies

Preussiella Gilg. 1897. Melastomataceae. 4 trop. West Africa

Preussiodora Keay. 1958. Rubiaceae. 1 trop. West Africa

Priamosia Urb. 1919 (Xylosma G. Forst.). Flacourtiaceae. 1 Haiti

Pridania Gagnep. = Pycnarrhena Miers ex Hook. f. et Thomsom. Menispermaceae

Priestleya DC. 1825. Fabaceae. 22 South Africa (Cape)

Prieurella Pierre = Chrysophyllum L. Sapotaceae

Primula L. 1753. Primulaceae. c. 400–500 Northern Hemisphere (esp. China [c. 300] and Russia [c. 80]) south to Ethiopia, trop. Asia, few species in the Southern Hemisphere—montane Java, New Guinea; P. magellanica—in temp. South America

Primularia Brenen = Cincinnobotrys Gilg. Melastomataceae

Primulina Hance. 1883. Gesneriaceae. 1 China

Princea Dubard et Dop = Triainolepis Hook. f. Rubiaceae

Principina Uittien. 1935. Cyperaceae. 1 trop. West Africa

Pringlea T. Anderson ex Hook. f. 1845. Brassicaceae. 1 Kerguelen and Crozet Is. (southern Indian Ocean)

Pringleochloa Scribn. 1896. Poaceae. 1 Mexico

Prinsepia Royle. 1835. Rosaceae. 5 Himalayas, northeastern India, China (4), Taiwan; Russia (1, P. sinensis: Russian Far East—Primorski Krai)

Printzia Cass. 1825. Asteraceae. 6 South Africa

Prionanthium Desv. 1831. Poaceae. 3 South Africa

Prionium E. Mey. 1832. Juncaceae. 1 South Africa

Prionocarpus S. F. Blake = Iostephane Benth. Asteraceae

Prionophyllum K. Koch = Dyckia Schultes f. Bromeliaceae

Prionopsis Nutt. 1841. Asteraceae. 1 North America

Prionosciadium S. Watson. 1888. Apiaceae. c. 20 America from Mexico ro Ecuador

Prionostemma Miers. 1872. Celastraceae. 5 Panama, trop. South America, Trinidad, trop. Africa (3), southwestern India (1)

Prionotes R. Br. 1810. Epacridaceae. 1 Tasmania

Prionotrichon Botsch. et Vved. 1848. Brassicaceae. 7 Middle Asia (3), Afghanistan, Mongolia

Prioria Griseb. 1860. Fabaceae. 1 Mexico, Central America, Colombia, Jamaica

Prismatocarpus L'Hér. 1789. Campanulaceae. 31 trop. (1), Southwest and South Africa

Prismatomeris Thwaites. 1856. Rubiaceae. 25 eastern Himalayas, southern China, Sri Lanka, Andaman Is., Borneo

Pristiglottis Cretz. et J. J. Sm. 1934. Orchidaceae. 10–20 India, southern China, Indochina, Malesia from Malay Peninsula, Sumatra, and Java to Philippines and New Guinea, Solomon Is., New Caledonia (1), Vanuatu eastward to Samoa Is.

Pristimera Miers. 1872. Celastraceae. c. 30 South, Southeast and East Asia, Central and trop. South America, West Indies

Pritchardia Seem. et H. A. Wendl. 1862. Arecaceae. 37 Hawaiian Is., Fiji (2), Tonga (1) and Danger Is.

Pritchardiopsis Becc. 1910. Arecaceae. 1 New Caledonia

Pritzelago Kuntze. 1891. Brassicaceae. 1 western Mediterranean

Priva Adans. 1763. Verbenaceae. c. 35

trop. and subtrop. America, Africa and Asia

Proboscidea Schmidel. 1763. Martyniaceae. 14 trop. and subtrop. America

Prochnyanthes S. Watson. 1887. Agavaceae. 1 (P. mexicana) Mexico

Prockia P. Browne ex L. 1759. Flacourtiaceae. 2: P. flava—Venezuela; P. crucis—trop. America from Mexico to Uruguay, northern Argentina and southern Brazil, West Indies

Prockiopsis Baill. 1886. Flacourtiaceae. 1 Madagascar

Procopiania Gusul. = Symphytum L. Boraginaceae

Procris Comm. ex Juss. 1789. Urticaceae. 30 trop. Old World

Proiphys Herb. 1821. Amaryllidaceae. 3 Malesia (P. amboinensis), New Guinea (P. alba), northern and eastern Australia (3)

Prolobus R. M. King et H. Rob. 1982. Asteraceae. 1 Brazil

Prolongoa Boiss. 1839 (Chrysanthemum L.). Asteraceae. 1 Spain

Promenaea Lindl. 1843. Orchidaceae. 15 Brazil

Prometheum (A. Berger) H. Ohba. 1978. Crassulaceae. 2 eastern Turkey, Iran, ? Caucasus

Pronaya Huegel. 1837. Pittosporaceae. 1 southwestern Australia

Prosanerpis S. F. Blake = Clidemia D. Don. Melastomataceae

Prosartes D. Don = Disporum Salisb. Convallariaceae

Proscephaleium Korth. 1851 (Psychotria L.). Rubiaceae. 1 Java

Proserpinaca L. 1753. Haloragaceae. 2–3 eastern North America, West Indies

Prosopanche Bary. 1868. Hydnoraceae. 2 Paraguay, Argentina

Prosopidastrum Burkart. 1964. Fabaceae. 2 Mexico (1), Argentina (1)

Prosopis L. 1767. Fabaceae. 44 trop. and subtrop. America, Africa (4), Arabian Peninsula, Turkey, Syria, Lebanon, Isbrael, Jordan, Iraq, Iran, Afghanistan, Pakbistan, India

Prosopostelma Baill. 1890. Asclepiadaceae. 3 trop. Africa, Madagascar

Prospero Salisb. 1866 (Scilla L.). Hyacinthaceae. 3 Europe, North Africa, West (incl. Caucasus—P. autumnalis) Asia

Prosphytochloa Schweick. 1961. Poaceae. 1 South Africa

Prostanthera Labill. 1806. Lamiaceae. 50 Australia

Prosthecidiscus Donn. Sm. = Marsdenia R. Br. Asclepiadaceae

Protarum Engl. 1901. Araceae. 1 Seychelles (Mahé I.)

Protasparagus Oberm. 1983 (or = Asparagus L.). Asparagaceae. c. 100 trop. and South Africa, trop. Asia, Australia (1)

Protea L. 1771. Proteaceae. 115 trop. and South Africa (85 endemic to Cape)

Proteopsis Mart. et Zucc. ex Sch. Bip. 1863. Asteraceae. 5 southern Brazil

Protium Burm. f. 1768. Burseraceae. 91 mainly trop. America; also in Madagascar, Mascarene Is., trop. Asia from India to Malesia, eastward to New Guinea

Protoceras J. Joseph et Vajr. = Pteroceras Hasselt ex Hassk. Orchidaceae

Protocyrtandra Hosok. 1934 (Cyrtandra J. R. Forst et G. Forst.). Gesneriaceae. 1 Taiwan

Protogabunia Boiteau = Tabernaemontana L. Apocynaceae

Protolirion Ridl. = Petrosavia Becc. Melanthiaceae

Protomegabaria Hutch. 1911. Euphorbiaceae. 2 trop. West Africa

Protorhus Engl. 1881. Anacardiaceae. 21 Namibia (1), Madagascar (20)

Protoschwenckia Soler. 1898. Solanaceae. 1 (P. mandonii) Bolivia and Brazil

Proustia Lag. 1811. Asteraceae. 4 West Indies, Chile, Argentina

Provancheria B. Boivin = Cerastium L. Caryophyllaceae

Prumnopitys Phil. 1860. Podocarpaceae. 8 trop. America from Costa Rica and Venezuela to southern Chile and Bolivia, New Caledonia, New Zealand

Prunella L. 1753. Lamiaceae. 8–9 North America, Mediterranean, temp. Eurasia

Prunus L. 1753. Rosaceae. c. 430 cosmopolitan, but chiefly temp. Northern Hemisphere

Przewalskia Maxim. 1882. Solanaceae. 2 China, Tibet

Psacadocalymma Bremek. = Justicia L. Acanthaceae

Psacadopaepale Bremek. 1944 (Strobilanthes Blume). Acanthaceae. 2 Sumatra

Psacaliopsis H. Rob. et Brettell. 1974 (Senecio L.). Asteraceae. 2 Mexico

Psacalium Cass. 1826. Asteraceae. 38 Central America

Psammagrostis C. A. Gardner et C. E. Hubb. 1938. Poaceae. 1 western Australia

Psammetes Hepper. 1962. Scrophulariaceae. 1 trop. West Africa

Psammisia Klotzsch. 1851. Ericaceae. 55 Central (Costa Rica, Panama) and trop. South America south to Bolivia, eastward to the Guianas

Psammochloa Hitchc. 1927. Poaceae. 1 Mongolia

Psammogeton Edgew. 1845. Apiaceae. 7 arid areas or Iraq, Iran, Afghanistan, Middle Asia (1, P. canesum), Pakistan, and India (1, P. biternatus)

Psammomoya Diels et Loes. 1904. Celastraceae. 2 western Australia

Psammophila Fourr. = Gypsophila L. Caryophyllaceae

Psammophiliella Iconn. = Gypsophila L. Caryophyllaceae

Psammophora Dinter et Schwantes. 1926. Aizoaceae. 2–3 southwestern Namibia, western South Africa

Psammopyrum A. Löve = Elytrigia Desv. Poaceae

Psammosilene W. C. Wu et C. Y. Wu. 1945. Caryophyllaceae. 1 Tibet

Psammotropha Eckl. et Zeyh. 1836. Molluginaceae. 11 trop. (2) and Southeast Africa

Psathura Comm. ex Juss. 1789. Rubiaceae. 8 Madagascar, Mascarene Is.

Psathyranthus Ule = Psittacanthus Mart. Loranthaceae

Psathyrastachys Nevski. 1934. Poaceae. 8 European part of Russia, Caucasus, West and East Siberia, Altai, Middle Asia; Turkey, Transcaucasia, Kurdistan, Iraq, Iran, Afghanistan, Pakistan, China (Shaanxi, Xinjiang, Gansu, Qinghai), Mongolia

Psathyrotes A. Gray. 1853. Asteraceae. 5 western U.S., northern Mexico

Psathyrotopsis Rydb. 1927 (Psathyrotes A. Gray). Asteraceae. 1 Mexico

Psednotrichia Hiern. 1898. Asteraceae. 1 South Africa

Pseudabutilon R. E. Fr. 1908. Malvaceae. 18 trop. and subtrop. America

Pseudacanthopale Benoist. 1950 (Strobilanthopsis S. Moore). Acanthaceae. 1 trop. Southeast Africa

Pseudacoridium Ames. 1922. Orchidaceae. 2 Philippines

Pseudactis S. Moore = Emilia (Cass.) Cass. Asteraceae

Pseudaechmanthera Bremek. 1944 (Strobilanthes Blume). Acanthaceae. 1 Himalayas from Kashmir to Nepal

Pseudaechmea L. B. Sm. et Read. 1982. Bromeliaceae. 1 Colombia

Pseudaegiphila Rusby = Aegiphila Jacq. Verbenaceae

Pseudagrostistachys Pax et K. Hoffm. 1912. Euphorbiaceae. 2 trop. Africa

Pseudaidia Tirveng. 1986. Rubiaceae. 1 India

Pseudais Decne. = Phaleria Jack. Thymelaeaceae

Pseudammi H. Wolff = Seseli L. Apiaceae

Pseudanamomis Kausel. 1956. Myrtaceae. 1 Puerto Rico

Pseudananas Hassl. ex Harms. 1930. Bromeliaceae. 1 Ecuador, Brazil, Paraguay, northern Argentina

Pseudanastatica (Boiss.) Grossh. 1930. Brassicaceae. 1 Caucasus, Iran, Afghanistan

Pseudannona (Baill.) Saff. = Xylopia L. Annonaceae

Pseudanthistiria (Hack.) Hook. f. 1896. Poaceae. 4 South and Southeast Asia

Pseudanthus Sieber ex A. Spreng. 1827. Euphorbiaceae. 7 Australia

Pseudarabidella O. E. Schulz. 1924 (Ara-

bidella (F. Muell.) O. E. Schulz). Brassicaceae. 1 southern Australia

Pseudarrhenatherum Rouy. 1921 (Arrhenatherum P. Beauv.). Poaceae. 1 western France, northwestern Spain, Portugal

Pseudartabotrys Pellegr. 1920. Annonaceae. 1 (P. letestui) trop. West Africa

Pseudarthria Wight et Arn. 1834. Fabaceae. 6 trop. Old World, mainly Africa (5); P. hookeri—extending to South Africa

Pseudechinolaena Stapf. 1919. Poaceae. 6 pantropics (1, P. polystachya), Madagascar

Pseudelephantopus Rohr = Elephantopus L. Asteraceae

Pseudelleanthus Brieger = Elleanthus C. Presl. Orchidaceae

Pseudellipanthus Schellenb. 1922 (Ellipanthus Hook. f.). Connaraceae. 2 Borneo

Pseudeminia Verdc. 1970. Fabaceae. 4 Angola (3 endemic); P. comosa—Malawi, Tanzania, Zambia, Zimbabwe, Mozambique

Pseudephedranthus Aristeg. 1969. Annonaceae. 1 (P. fragrans) Venezuela, Brazil

Pseuderanthemum Radlk. 1893. Acanthaceae. 60–100 pantropics

Pseuderemostachys Popov. 1940. Lamiaceae. 1 Middle Asia

Pseuderia Schltr. 1912. Orchidaceae. 20 Malesia, Micronesia, Melanesia

Pseuderucaria (Boiss.) O. E. Schulz. 1916. Brassicaceae. 2 southern Mediterranean from Morocco to Israel

Pseudeugenia Legrand et Mattos. 1966. Myrtaceae. 1 Brazil

Pseudibatia Malme = Matelea Aubl. Asclepiadaceae

Pseudima Radlk. 1878. Sapindaceae. 3 Central and trop. South America

Pseudiosma DC. 1824. Rutaceae. 1 Southeast Asia

Pseudoacanthocereus F. Ritter = Acanthocereus (Engelm. ex A. Berger) Britton et Rose. Cactaceae

Pseudoanastatica (Boiss.) Grossh. = Clypeola L. Brassicaceae

Pseudobaccharis Cabrera = Baccharis L. Asteraceae

Pseudobaeckea Nied. 1891. Bruniaceae. 4 South Africa

Pseudobahia (A. Gray) Rydb. 1915. Asteraceae. 3 California

Pseudobartlettia Rydb. = Psathyrotopsis Rydb. Asteraceae

Pseudobartsia D. Y. Hong. 1979. Scrophulariaceae. 1 southern China (Yunnan)

Pseudoberlinia P. A. Duvign. = Julbernardia Pellegr. Fabaceae

Pseudobersama Verdc. 1956 (Trichilia P. Browne). Meliaceae. 1 eastern coastal Africa from Kenya to Natal

Pseudobetckea (Hock) Lincz. 1958. Valerianaceae. 1 Caucasus

Pseudoblepharis Baill. = Sclerochiton Harv. + Crossandrella C. B. Clarke. Acanthaceae

Pseudoblepharispermum J.-P. Lebrun et Stork. 1982. Asteraceae. 1 Ethiopia

Pseudoboivinella Aubrév. et Pellegr. = Englerophytum Krause. Sapotaceae

Pseudobombax Dugand. 1943. Bombacaceae. 20 trop. America

Pseudobotrys Moeser. 1912. Icacinaceae. 2 New Guinea

Pseudobrachiaria Launert = Brachiaria (Trin.) Griseb. Poaceae

Pseudobrassaiopsis R. N. Banerjee = Brassaiopsis Decne. et Planch. Araliaceae

Pseudobravoa Rose. 1899. Agavaceae. 1 Mexico

Pseudobrickellia R. M. King et H. Rob. 1972 (Eupatorium L.). Asteraceae. 2 Brazil

Pseudobromus K. Schum. 1895 (Festuca L.). Poaceae. 8 montane trop. Africa, Madagascar (4)

Pseudobrownanthus Ihlenf. et Bittrich. 1985 (Brownanthus Schwantes). Aizoaceae. 1 South Africa (Orange River Basin), Namibia

Pseudocadiscus Lisowski. 1987. Asteraceae. 1 Zaire

Pseudocalymma A. Samp. et Kuhlm. = Mansoa DC. Bignoniaceae

Pseudocalyx Radlk. 1883. Acanthaceae. 5

trop. Africa, Madagascar
Pseudocamelina (Boiss.) N. Busch. 1928. Brassicaceae. 3 Iran
Pseudocampanula Kolak. = Campanula L. Campanulaceae
Pseudocarapa Hemsl. 1884. Meliaceae. 1 Sri Lanka
Pseudocarpidium Millsp. 1906. Verbenaceae. 8 West Indies
Pseudocarum C. Norman. 1924. Apiaceae. 1–2 Ethiopia, Uganda, Kenya
Pseudocaryophyllus O. Berg = Pimenta L. Myrtaceae
Pseudocatalpa A. H. Gentry. 1973. Bignoniaceae. 1 southern Mexico, Belize
Pseudocedrela Harms. 1895. Meliaceae. 1 (P. kotschyi) trop. Africa
Pseudocentrum Lindl. 1858. Orchidaceae. 6 trop. America
Pseudochaetochloa Hitchc. 1924. Poaceae. 1 Australia
Pseudochamaespacos Parsa. 1946. Lamiaceae. 1 Iran
Pseudochimarrhis Ducke = Chimarrhis Jacq. Rubiaceae
Pseudochirita W. T. Wang. 1983 (Chirita Buch.-Ham. ex D. Don). Gesneriaceae. 1 southern China
Pseudocinchona A. Chev. ex E. Perrot = Pausinystalia Pierre ex Beille. Rubiaceae
Pseudoclappia Rydb. 1923. Asteraceae. 1 southwestern U.S.
Pseudoclausia Popov. 1955. Brassicaceae. 10 Kazakhstan, Middle Asia, Iran, Afghanistan, China
Pseudocoix A. Camus. 1924. Poaceae. 1 Tanzania, Madagascar
Pseudoconnarus Radlk. 1886. Connaraceae. 6 Colombia, Guyana, Surinam, Brazil
Pseudoconyza Cuatrec. 1961 (or = Laggera Sch. Bip. ex Oliv.). Asteraceae. 1 pantropics
Pseudocorchorus Capuron. 1963. Tiliaceae. 6 Madagascar
Pseudocranichis Garay. 1982. Orchidaceae. 1 Mexico
Pseudocroton Muell. Arg. 1872. Euphorbiaceae. 1 Central America
Pseudocrupina Velen. = Leysera L. Asteraceae
Pseudoctomeria Kraenzl. = Pleurothallis R. Br. Orchidaceae
Pseudocunila Brade = Hedeoma Pers. Lamiaceae
Pseudocyclanthera Mart. Crov. 1954. Cucurbitaceae. 1 South America
Pseudocydonia (C. K. Schneid.) C. K. Schneid. 1906. Rosaceae. 1 eastern China
Pseudocymopterus J. M. Coult. et Rose = Cymopterus Raf. Apiaceae
Pseudodanthonia Bor et C. E. Hubb. 1958. Poaceae. 1 Himalayas
Pseudodichanthium Bor. 1940. Poaceae. 1 India
Pseudodicliptera Benoist. 1939. Acanthaceae. 2 Madagascar
Pseudodigera Chiov. = Digera Forssk. Amaranthaceae
Pseudodiphryllum Nevski = Platanthera Rich. Orchidaceae
Pseudodissochaeta M. P. Nayar. 1969. Melastomataceae. 5 trop. Asia from northern India to Indochina and Hainan
Pseudodovouapa Britton et Killip = Macrolobium Schreb. Fabaceae
Pseudodracontium N. E. Br. 1882. Araceae. 7 Indochina
Pseudoentada Britton et Rose = Adenopodia C. Presl. Fabaceae
Pseudoeriosema Hauman. 1955. Fabaceae. 5 trop. Africa, mainly Zaire (4)
Pseudoernestia (Cogn.) Krasser = Ernestia DC. Melastomataceae
Pseudoeurya Yamam. = Eurya Thunb. Theaceae
Pseudoeurystyles Hoehne = Eurystyles Wawra. Orchidaceae
Pseudoeverardia Gilly = Everardia Ridl. Cyperaceae
Pseudofortuynia Hedge. 1968. Brassicaceae. 1 Iran
Pseudofumaria Medik. 1789 (or = Corydalis Vent.). Fumariaceae. 2 Italy (Alps), Yugoslavia, Albania, Greece
Pseudogaillonia Lincz. 1973. Rubiaceae. 1 Southwest Asia from Arabian Peninsula to Beluchistan

Pseudogaltonia Kuntze. 1886. Hyacinthaceae. 1 South Africa

Pseudogardenia Keay. 1958. Rubiaceae. 1 trop. Africa

Pseudoglycine F. J. Herm. 1962 (Ophrestia H. M. L. Forbes). Fabaceae. 1 Madagascar

Pseudognaphalium Kirp. 1950. Asteraceae. c. 90 tropics and subtropics (America, Africa, Asia)

Pseudognidia E. Phillips = Gnidia L. Thymelaeaceae

Pseudogomphrena R. E. Fr. 1920. Amaranthaceae. 1 Brazil

Pseudogonocalyx Bisse et Berazain = Schoepfia Schreb. Olacaceae

Pseudogoodyera Schltr. 1920. Orchidaceae. 1 Central America

Pseudogynoxys (Greenm.) Cabrera. 1950. Asteraceae. 13 trop. South America

Pseudohamelia Wernham. 1912. Rubiaceae. 1 Andes

Pseudohandelia Tzvelev. 1961 (Tanacetum L.). Asteraceae. 1 Middle Asia, Iran, Afghanistan

Pseudohexadesmia Brieger = Hexadesmia Brongn. Orchidaceae

Pseudohomalomena A. D. Hawkes = Zantedeschia Spreng. Araceae

Pseudohydrosme Engl. 1892. Araceae. 2 trop. West Africa

Pseudojacobaea (Hook. f.) Mathur = Senecio L. Asteraceae

Pseudokyrsteniopsis R. M. King et H. Rob. 1973. Asteraceae. 1 Mexico, Guatemala

Pseudolabatia Aubrév. et Pellegr. = Pouteria Aubl. Sapotaceae

Pseudolachnostylis Pax. 1899. Euphorbiaceae. 1 trop. (Zaire, Burundi, Tanzania, Zambia, Zimbabwe, Angola, Namibia, Botswana, Mozambique) and South (Transvaal) Africa

Pseudolaelia Porto et Brade. 1935. Orchidaceae. 6 southeastern Brazil

Pseudolarix Gordon. 1858. Pinaceae. 1 Central and northeastern China (Zheziang, Jiangsu, Guangxi, and Anhui)

Pseudolasiacis (A. Camus) A. Camus. 1945 (Lasiacis [Griseb.] Hitchc.). Poaceae. 4 Madagascar

Pseudoligandra Dillon et Sagast. = Chionolaena DC. Asteraceae

Pseudolinosyris Novopokr. = Aster L. Asteraceae

Pseudolitchia Danguy et Choux = Stadmannia Lam. Sapindaceae

Pseudolithos Bally. 1965. Asclepiadaceae. 4 trop. Northeast Africa

Pseudolmedia Trécul. 1847. Moraceae. 9 Central and trop. South America, West Indies

Pseudolopezia Rose = Lopezia Cav. Onagraceae

Pseudolophanthus Levin = Marmoritis Benth. Lamiaceae

Pseudolotus Rech. f. 1958. (Lotus L.). Fabaceae. 1 Iran, Afghanistan, Pakistan, Oman

Pseudoloxocarya Linder. 1991 ?. Restionaceae. ? sp. Distr. ?

Pseudoludovia Harling = Sphaeradenia Harling. Cyclanthaceae

Pseudolysimachion Opiz = Veronica L. Scrophulariaceae

Pseudomachaerium Hassl. = Nissolia Jacq. Fabaceae

Pseudomacrolobium Hauman. 1952. Fabaceae. 1 Zaire

Pseudomalachra Monteiro = Sida L. Malvaceae

Pseudomantalania J.-F. Leroy. 1973. Rubiaceae. 1–2 Madagascar

Pseudomariscus Rauschert = Courtoisina Soják. Cyperaceae

Pseudomarrubium Popov. 1940. Lamiaceae. 1 Middle Asia

Pseudomarsdenia Baill. = Marsdenia R. Br. Asclepiadaceae

Pseudomaxillaria Hoehne = Maxillaria Ruiz et Pav. Orchidaceae

Pseudomelissitus Ovcz., Rasulova et Kinzik. = Medicago L. Fabaceae

Pseudomertensia Riedl. 1967. Boraginaceae. 8 Iran, Afghanistan, Tajikistan (1, P. rosulata), Himalayas

Pseudomuscari Garbari et Greuter = Muscari Hill. Hyacinthaceae

Pseudomussaenda Wernham. 1916 (Mussaenda L.). Rubiaceae. 4–5 trop. Africa

Pseudomyrcianthes Kausel = Eugenia L. Myrtaceae

Pseudonemacladus McVaugh. 1943. Nemacladaceae. 1 Mexico

Pseudonesohedyotis Tennant. 1965. Rubiaceae. 1 trop. East Africa (Tanzania)

Pseudonoseris H. Rob. et Brettell. 1974. Asteraceae. 3 Peru

Pseudoorleanesia Rauschert = Orleanesia Barb. Rodr. Orchidaceae

Pseudopachystela Aubrév. et Pellegr. = Synsepalum (A. DC.) Daniell. Sapotaceae

Pseudopaegma Urb. = Anemopaegma Mart. ex Meisn. Bignoniaceae

Pseudopanax K. Koch. 1859. Araliaceae. 6 New Zealand, Chile

Pseudopancovia Pellegr. 1955. Sapindaceae. 1 trop. West Africa

Pseudoparis H. Perrier. 1936. Commelinaceae. 2 Madagascar

Pseudopavonia Hassl. = Pavonia Cav. Malvaceae

Pseudopectinaria Lavranos (1971) = Echidnopsis Hook. f. Asclepiadaceae

Pseudopentameris Conert. 1971. Poaceae. 2 South Africa

Pseudopentatropis Costantin = Gymnema R. Br. Asclepiadaceae

Pseudopeponidium Homolle ex Arènes = Pyrostria Commers. ex Juss. Rubiaceae

Pseudophleum Dogan. 1982 (Phleum L.). Poaceae. 1 Turkey

Pseudophoenix H. A. Wendl. ex Sarg. 1886. Arecaceae. 4 southern Florida, Bahama Is., southern Mexico, Belize, Cuba, Haiti, Mona and Dominica Is.

Pseudopilocereus Buxb. = Pilosocereus Byles et Rowley. Cactaceae

Pseudopinanga Burret = Pinanga Blume. Arecaceae

Pseudopiptadenia Rauschert. 1982. Fabaceae. 3–4 trop. South America

Pseudoplantago Suess. 1934. Amaranthaceae. 2 Venezuela and Argentina

Pseudopogonatherum A. Camus. 1921 (Eulalia Kunth). Poaceae. 2 trop. Asia

Pseudoprimula (Pax) O. Schwarz = Primula L. Primulaceae

Pseudoprosopis Harms. 1902. Fabaceae. 7 trop. Africa (Guinea, Liberia, Sierra Leone, Congo, Gabon, Zaire, Tanzania, Zimbabwe)

Pseudoprotorhus H. Perrier. 1944. Anacardiaceae. 1 Madagascar

Pseudopteris Baill. 1874. Sapindaceae. 1 Madagascar

Pseudopyxis Miq. 1867. Rubiaceae. 2 Japan

Pseudoraphis Griff. 1851. Poaceae. 5 South Asia, Japan, Malesia, Australia

Pseudorchis Séguier. 1754. Orchidaceae. 3 eastern North America, Europe, West Siberia (1, P. albida)

Pseudorhipsalis Britton et Rose. 1923. Cactaceae. 5 Central and trop. South America, West Indies

Pseudorlaya (Murb.) Murb. 1897. Apiaceae. 3 Southwest, South, and Southeast Europe, Northwest and North Africa; P. pumila—extending to Israel

Pseudorleanesia Rauschert = Orleanesia Barb. Rodr. Orchidaceae

Pseudorobanche Rouy = Alectra Thunb. Scrophulariaceae

Pseudoroegneria (Nevski) A. Löve = Elytrigia Desv. Poaceae

Pseudorontium (A. Gray) Rothm. 1943 (Antirrhinum L.). Scrophulariaceae. 1 southwestern U.S. (California, Arizona), Mexico (Baja California, Sonora)

Pseudorosularia Gurgen. = Sedum L. Crassulaceae

Pseudoruellia Benoist. 1962. Acanthaceae. 1 Madagascar

Pseudosabicea N. Hallé. 1963. Rubiaceae. 12 trop. Africa

Pseudosagotia Secco. 1985. Euphorbiaceae. 1 southern Venezuela

Pseudosalacia Codd. 1972. Celastraceae. 1 South America (Natal)

Pseudosamanea Harms = Albizia Durazz. Fabaceae

Pseudosaponaria (F. Williams) Ikonn. = Gypsophila L. Caryophyllaceae

Pseudosarcolobus Costantin = Gymnema R. Br. Asclepiadaceae

Pseudosasa Nakai. 1925. Poaceae. c. 20: 4 Caucasus, Middle Asia; East Asia

Pseudosassafras Lecomte = Sassafras Nees et Eberm. Lauraceae

Pseudosbeckia A. Fern. et R. Fern. 1956. Melastomataceae. 1 trop. East Africa

Pseudoscabiosa Devesa. 1984 (Scabiosa L.). Dipsacaceae. 4 Spain, Morocco, Sicily

Pseudoschoenus (C. B. Clarke) Oteng-Yeb. 1974. Cyperaceae. 1 South Africa

Pseudosciadium Baill. 1878. Araliaceae. 1 New Caledonia

Pseudoscolopia Gilg. 1917. Flacourtiaceae. 1 South Africa (Natal, Cape)

Pseudosedum (Boiss.) A. Berger. 1930. Crassulaceae. 10: 9 Middle Asia, West Siberia; Iran (1), China (1)

Pseudoselinum C. Norman. 1929. Apiaceae. 1 Angola

Pseudosempervivum (Boiss.) Grossh. = Cochlearia L. Brassicaceae

Pseudosericocoma Cavaco. 1962. Amaranthaceae. 1 South-West and South Africa

Pseudosicydium Harms. 1927. Cucurbitaceae. 1 Colombia, Peru, Bolivia

Pseudosindora Symington = Copaifera L. Fabaceae

Pseudosmelia Sleumer. 1954. Flacourtiaceae. 1 Moluccas

Pseudosmilax Hayata = Heterosmilax Kunth. Smilacaceae

Pseudosmodingium Engl. 1881. Anacardiaceae. 7 Mexico

Pseudosopubia Engl. 1897. Scrophulariaceae. 7 trop. Africa

Pseudosorghum A. Camus. 1921. Poaceae. 2 trop. Asia

Pseudospigelia Klett = Spigelia L. Spigeliaceae

Pseudospondias Engl. 1883. Anacardiaceae. 2 trop Africa from Senegal to Sudan, Angola, Zaire, Zambia and Tanzania; fruits edible and sweet

Pseudostachyum Munro. 1868 (Schizostachyum Nees). Poaceae. 2 South and Southeast Asia

Pseudostelis Schltr. = Pleurothallis R. Br. Orchidaceae

Pseudostellaria Pax. 1934. Caryophyllaceae. 16 temp. Eurasia, North America (1); Russia (8, West and East Siberia, Far East)

Pseudostenomesson Velarde. 1949. Amaryllidaceae. 2 Andes

Pseudostenosiphonium Lindau. 1893 (Strobilanthes Blume). Acanthaceae. 9 Sri Lanka

Pseudostifftia H. Rob. 1979. Asteraceae. 1 Brazil

Pseudostreptogyne A. Camus = Streblochaete Pilg. Poaceae

Pseudostriga Bonati. 1911. Scrophulariaceae. 1 Indochina

Pseudotaenidia Mack. 1903 (or = Taenidia [Torr. et A. Gray] Drude). Apiaceae. 1 North America

Pseudotaxus W. C. Cheng. 1948. Taxaceae. 1 continental China

Pseudotrimezia R. C. Foster. 1945. Iridaceae. 7 Brazil

Pseudotsuga Carrière. 1867. Pinaceae. 4 central and southern China, Taiwan, southern Japan, western North America from Canada to Baja California

Pseudourceolina Vargas = Urceolina Rchb. Amaryllidaceae

Pseudovanilla Garay. 1986 (Vanilla Mill.). Orchidaceae. 8 Java, Philippines, Moluccas, Ponape and Ternate Is., New Guinea, Australia, Fiji

Pseudovesicaria (Boiss.) Rupr. 1869. Brassicaceae. 1 Caucasus

Pseudovigna (Harms) Verdc. 1970. Fabaceae. 2 trop. Africa (Ghana, Togo, Nigeria, Kenya, Tanzania, Mozambique)

Pseudovossia A. Camus. 1921 (Phacelurus Griseb.). Poaceae. 1 Indochina

Pseudoweinmannia Engl. 1930. Cunoniaceae. 2 Australia (Queensland)

Pseudowillughbeia Markgr. = Melodinus J. R. Forst. et G. Forst. Apocynaceae

Pseudowintera Dandy. 1933. Winteraceae. 3 New Zealand, Stewart I.

Pseudowolffia Hartog et Van der Plas. 1970 (Wolffiella (Hegelm.) Hegelm.). Lemnaceae. 3 North and Central Africa

Pseudoxandra R. E. Fr. 1937. Annonaceae. 10 trop. South America

Pseudoxytenanthera Soderstr. et Ellis = Schizostachyum Nees. Poaceae

Pseudoxythece Aubrév. = Pouteria Aubl. Sapotaceae

Pseudozoysia Chiov. 1928. Poaceae. 1 Somalia

Psueduvaria Miq. 1858. Annonaceae. c. 20 Southeast Asia, western Malesia, New Guinea (1), Australia (1)

Psiadia Jacq. 1797. Asteraceae. c. 60 St. Helena, trop. Africa, Madagascar, Mascarene Is., Socotra

Psiadiella Humbert. 1923. Asteraceae. 1 Madagascar

Psidiopsis O. Berg = Calycolpus O. Berg. Myrtaceae

Psidium L. 1753. Myrtaceae. c. 100 trop. America, West Indies

Psiguria Neck. ex Arn. 1841. Cucurbitaceae. 12 trop. America

Psila Phil. = Baccharis L. Asteraceae

Psilactis A. Gray = Machaeranthera Nees. Asteraceae

Psilantha (K. Koch) Tzvelev = Boriskellera Terekhov. Poaceae

Psilanthele Lindau. 1897. Acanthaceae. 1 Ecuador

Psilanthopsis A. Chev. 1939 (Coffea L.). Rubiaceae. 1 Angola

Psilanthus Hook. f. 1873. Rubiaceae. 5 trop. West Africa

Psilathera Link. 1827 (Sesleria Scop.). Poaceae. 1 southern Germany, northern Italy, Austria

Psilocarphus Nutt. 1841. Asteraceae. 6 western U.S. (5), western temp. South America (1)

Psilocaulon N. E. Br. 1925. Aizoaceae. 15 southern Angola, southern Namibia, western and southern South Africa

Psilochilus Barb. Rodr. 1882 (Pogonia Juss.). Orchidaceae. 1 Venezuela, Brazil

Psilochloa Launert = Panicum L. Poaceae

Psiloesthes Benoist. 1936 (Peristrophe Nees). Acanthaceae. 1 Indochina

Psilolaemus I. M. Johnst. 1954. Boraginaceae. 1 Mexico

Psilolemma S. M. Phillips. 1974. Poaceae. 1 East Africa

Psilopeganum Hemsl. 1886. Rutaceae. 1 China

Psilostrophe DC. 1838. Asteraceae. 7 southern U.S., northern Mexico

Psilothonna E. Mey. ex DC. = Steirodiscus Less. Asteraceae

Psilotrichopsis C. C. Towns. 1974. Amaranthaceae. 3 Thailand (2), Vietnam (1, P. cochinchinensis), Malay Peninsula (1, P. curtisii), Hainan (1, P. hainanensis)

Psilotrichum Blume. 1826. Amaranthaceae. 15 Northeast, trop. East and Southeast Africa, Madagascar, Socotra, trop. Arabian Peninsula, India, Sri Lanka, Nepal, China, Indochina, Malesia

Psiloxylon Thouars ex Tul. 1856. Psiloxylaceae. 1 Mauritius and Réunion Is.

Psilurus Trin. 1822. Poaceae. 1 southern Europe, Mediterranean, Crimea, Caucasus, Syria, Iraq, Iran, Middle Asia, Afghanistan, Pakistan

Psithyrisma Herb. = Olsynium Raf. Iridaceae

Psittacanthus Mart. 1830. Loranthaceae. c. 60 trop. America from northwestern Mexico to northern Argentina

Psophocarpus DC. 1825. Fabaceae. 9 trop. Africa and Madagascar (9); P. tetragonolobus—extending to trop. Asia and Australia

Psoralea L. 1753. Fabaceae. 34 South Africa, P. pinnata—extending to Mozambique

Psoralidium Rydb. 1919 (Orbexilum Raf.). Fabaceae. 3 North America from Saskatchewan and Alberta to northern Mexico, eastward to Minnesota and Indiana

Psorospermum Spach. 1836. Hypericaceae. 40–45 trop. Africa, Madagascar

Psorothamnus Rydb. 1919. Fabaceae. 9 southwestern U.S., northwestern Mexico

Psychanthus (K. Schum.) Ridl. = Pleuranthodium (K. Schum.) Ros. M. Sm. Zingiberaceae

Psychilus Raf. = Epidendrum L. Orchidaceae

Psychine Desf. 1798. Brassicaceae. 1 North Africa

Psychopsiella Lueckel et Braem. 1982. Orchidaceae. 1 Venezuela

Psychopsis Raf. = Oncidium Sw. Orchidaceae

Psychotria L. 1795. Rubiaceae. c.1400 tropics and subtropics

Psychrogeton Boiss. = Erigeron L. Asteraceae

Psychrophila (DC.) Bercht. et J. Presl = Caltha L. Ranunculaceae

Psychrophyton Beauverd. 1910 (Raoulia Hook. f.). Asteraceae. 10 New Zealand

Psydrax Gaertn. 1788. Rubiaceae. c. 110: c. 35 trop. and South Africa; c. 75 Arabian Peninsula (Yemen), India, Sri Lanka, southern China, Indochina, Malesia, Oceania to Hawaiian Is., and Tuamotu Is.

Psygmorchis Dodson et Dressler. 1972. Orchidaceae. 4 trop. America

Psylliostachys (Jaub. et Spach) Nevski. 1927. Plumbaginaceae. 6 Syria, Palestine, Iraq, Iran, Middle Asia (Kazakhstan, Uzbekistan, Turkmenistan, southern Tajikistan), Afghanistan, Pakistan; P. spicata—extending to Caucasus and southeastern Russia

Psyllocarpus Mart. et Zucc. 1824. Rubiaceae. 8 Brazil

Ptaeroxylon Eckl. et Zeyh. 1835. Ptaeroxylaceae. 1 trop. East and South Africa

Ptelea L. 1753. Rutaceae. 3 southwestern U.S. (California, Arizona, New Mexico), Mexico

Pteleocarpa Oliv. 1873. Ehretiaceae. 2 western Malesia

Pteleopsis Engl. 1894. Combretaceae. 9 trop. and South Africa

Ptelidium Thouars. 1804. Celastraceae. 2 Madagascar

Pteracanthus (Nees) Bremek. 1944. Acanthaceae. 20 Himalayas from Kashmir to Bhutan, northeastern India, China (10)

Pterachaenia (Benth.) Lipschitz. 1939. Asteraceae. 2 Afghanistan, Pakistan

Pteralyxia K. Schum. 1895. Apocynaceae. 2 Hawaiian Is.

Pterandra A. Juss. 1833. Malpighiaceae. 7 Panama, Colombia, Gorgona I., Venezuela, Brazil

Pteranthus Forssk. 1775. Caryophyllaceae. 2 Spain, Malta, Cyprus, North Africa, Syria, Israel, Sinai Peninsula, Iraq, Iran, Caucasus (1, P. dichotomus)

Pterichis Lindl. 1840. Orchidaceae. 15 Costa Rica (1), Jamaica (1), Andes of Colombia, Venezuela, Ecuador, Peru, Bolivia, and Brazil

Pteridiphyllum Siebold et Zucc. 1843. Pteridophyllaceae. 1 Japan

Pteridocalyx Wernham. 1911. Rubiaceae. 2 Guyana

Pterigeron (DC.) Benth. = Streptoglossa Steetz ex F. Muell. Asteraceae

Pterisanthes Blume. 1825. Vitaceae. 20 Burma, western Malesia

Pternandra Jack. 1822. Melastomataceae. 15 (or 2) China, Indochina, Malesia to New Guinea, Australia (1)

Pternopetalum Franch. 1885. Apiaceae. 27 China, ? Japan (P. tanakae), South Asia

Pterobesleria C. Morton = Besleria L. Gesneriaceae

Pterocactus K. Schum. 1897. Cactaceae. 5 Argentina

Pterocarpus Jacq. 1763. Fabaceae. 22 pantropics, esp. Africa (22, 2 spp. in South Africa)

Pterocarya Kunth. 1824. Juglandaceae. 8–10 Asia from Caucasus (1, P. pterocarpa), Turkey, and Iran to Japan and Indochina

Pterocaulon Elliott. 1824. Asteraceae. 18 Southeast Asia, Malesia, Australia, trop. and subtrop. America

Pterocelastrus Meisn. 1837. Celastraceae. 4–5 trop. and South Africa

Pteroceltis Maxim. 1873. Ulmaceae. 1 northern China, Mongolia

Pterocephalidium Lopez Gonz. 1987 (Pterocephalus Adans.). Dipsacaceae. 1 Iberian Peninsula

Pterocephalus Adans. 1763. Dipsacaceae. 25 Europe, Mediterranean, trop. Africa, Crimea, West (incl. Caucasus) and Middle Asia to western China, eastern Himalayas, northeastern India

Pteroceras Hasselt ex Hassk. 1842 (Sarcochilus R. Br.). Orchidaceae. c. 20 India, Nepal, Bangladesh, Andaman Is., Burma, southern China (Yunnan—*P. asperatum*), Thailand, Laos, Cambodia, Vietnam, Malay Peninsula, Sumatra, Java, Bali, Borneo (10), Sulawesi, Philippines (5), Moluccas

Pterocereus MacDougall et Miranda = Pachycereus (A. Berger) Britton et Rose. Cactaceae

Pterochaeta Steetz. 1845 (Waitzia Wendl.). Asteraceae. 1 southwestern Australia

Pterochloris (A. Camus) A. Camus. 1957 (Chloris Sw.). Poaceae. 1 Madagascar

Pterocissus Urb. et Ekman. 1926. Vitaceae. 1 Haiti

Pterocladon Hook. f. = Miconia Ruiz et Pav. Melastomataceae

Pterococcus Hassk. 1842 (Plukenetia L.). Euphorbiaceae. 1 eastern Himalayas, northeastern India, Indochina, western Malesia, Moluccas

Pterocyclus Klotzsch. 1862 (or = Pleurospermum Hoffm.). Apiaceae. 3 western China, western Himalayas

Pterocymbium R. Br. 1844. Sterculiaceae. 15 Southeast Asia, Malesia, Oceania to Fiji

Pterocypsela C. Shih. 1988 (Lactuca L.). Asteraceae. 7 South Africa, Madagascar, trop. and East Asia

Pterodiscus Hook. 1844. Pedaliaceae. 18 trop. and South Africa

Pterodon Vogel. 1837. Fabaceae. 6 Bolivia, Brazil

Pterogaillonia Lincz. 1973. Rubiaceae. 3 Iran

Pterogastra Naudin. 1849. Melastomataceae. 4 northern South America

Pteroglossa Schltr. 1920. Orchidaceae. 8 South America

Pteroglossaspis Rchb. f. 1878. Orchidaceae. 7 southeastern U.S. and Cuba (1), Argentina and Brazil (1), trop. and subtrop. Africa (5)

Pterogonum Gross = Eriogonum Michx. Polygonaceae

Pterogyne Tul. 1843. Fabaceae. 1 Brazil, Paraguay, Argentina

Pterolepis (DC.) Miq. 1839–1840 (Tibouchina Aubl.). Melastomataceae. 25–30 trop. America from Mexico to Brazil and Paraguay, West Indies

Pterolobium R. Br. ex Wight et Arn. 1834. Fabaceae. 11 trop. and South Africa (1, *P. stellatum*), South and Southeast Asia

Pteroloma Desv. ex Benth. = Tadehagi H. Ohashi. Fabaceae

Pteromonnina Eriksen. 1993. Polygalaceae. 26 trop. America

Pteronia L. 1763. Asteraceae. 79 trop. and South Africa, Madagascar, western Australia

Pteropentacoilanthus Rappa et Camarrone = Mesembryanthemum L. Aizoaceae

Pteropepon (Cogn.) Cogn. 1916. Cucurbitaceae. 3 South America

Pteropogon DC. 1838. Asteraceae. 10 Australia

Pteroptychia Bremek. 1944 (Strobilanthes Blume). Acanthaceae. 5 China, Indochina, western Malesia

Pteropyrum Jaub. et Spach. 1844. Polygonaceae. 3–5 Asia Minor, Iran, ? Afghanistan; *P. aucheri*—extending to Middle Asia

Pterorhachis Harms. 1895. Meliaceae. 1–2 Cameroun, Gabon

Pteroscleria Nees = Diplacrum R. Br. Cyperaceae

Pterosicyos Brandegee. 1914. Cucurbitaceae. 1 Mexico, Guatemala

Pterospermum Schreb. 1791. Sterculiaceae. 25 eastern Himalayas, China, Indochina, western Malesia

Pterospora Nutt. 1818. Ericaceae. 1 North America

Pterostegia Fisch. et C. A. Mey. 1835. Polygonaceae. 1 western U.S.: Oregon, Nevada, Utah, California, Arizona; northwestern Mexico, incl. Baja California

Pterostemma Kraenzl. 1899. Orchidaceae. 1 Colombia

Pterostemon Schauer. 1847. Pterostemonaceae. 2 Mexico

Pterostylis R. Br. 1810. Orchidaceae. 106 New Guinea (3), Australia, Tasmania, New Zealand, New Caledonia

Pterostyrax Siebold et Zucc. 1839. Styracaceae. 4 Burma, China, Japan

Pterotaberna Stapf = Tabernaemontana L. Apocynaceae

Pterotetracoilanthus Rappa et Camarrone = Mesembryanthemum L. Aizoaceae

Pterothrix DC. 1838. Asteraceae. 4 South Africa

Pteroxygonum Dammer et Diels. 1905 (Fagopyrum Hill). Polygonaceae. 1 China

Pterygiella Oliv. 1896. Scrophulariaceae. 4 China

Pterygiosperma O. E. Schulz. 1924. Brassicaceae. 1 Argentina (Patagonia)

Pterygocalyx Maxim. 1859. Gentianaceae. 1 China, Japan, Soviet Far East—Primorski Krai

Pterygodium Sw. 1800. Orchidaceae. 18 Tanzania (1, P. ukingense), South Africa (17, mainly Cape)

Pterygopappus Hook. f. 1874. Asteraceae. 1 Tasmania

Pterygopleurum Kitag. 1937. Apiaceae. 1 China, southern Korea, Japan

Pterygopodium Harms = Oxystigma Harms Fabaceae

Pterygostemon V. V. Boczantzeva. 1977. Brassicaceae. 1 Kazakhstan

Pterygota Schott et Endl. 1832. Sterculiaceae. c. 20 tropics, esp. Old World

Pteryxia Nutt. ex Torr. = Cymopterus Raf. Apiaceae

Ptilagrostis Griseb. 1852 (Stipa L.). Poaceae. 9 Middle Asia, Siberia, Russian Far East, western North America

Ptilanthelium Steud. 1855. Cyperaceae. 2 eastern Australia

Ptilanthus Gleason = Graffenrieda DC. Melastomataceae

Ptilimnium Raf. 1825 (Discopleura DC.). Apiaceae. 5–10 North America

Ptilochaeta Turcz. 1843. Malpighiaceae. 5 subtrop. South America

Ptilostemon Cass. 1816 (Cirsium Hill). Asteraceae. 14 Europe, Mediterranean; P. echinocephalus—Crimea, Caucasus

Ptilotrichum C. A. Mey. 1831 (Alyssum L.). Brassicaceae. 12 Europe, Mediterranean, West Asia eastward to China and Mongolia; Middle Asia, West and East Siberia

Ptilotus R. Br. 1810. Amaranthaceae. c. 110 Australia, Tasmania; P. conicus R. Br.—extending to Lesser Sunda Is. and Moluccas

Ptycanthera Decne. 1844. Asclepiadaceae. 1 Haiti

Ptychandra Scheff. = Heterospathe Scheff. Arecaceae

Ptychococcus Becc. 1885. Arecaceae. 7 New Guinea (6), Solomon Is. (1)

Ptychogyne Pfitzer = Coelogyne Lindl. Orchidaceae

Ptycholobium Harms. 1915. Fabaceae. 3 arid areas of trop. and South Africa, southern Arabian Peninsula

Ptychomeria Benth. = Gymnosiphon Blume. Burmanniaceae

Ptychopetalum Benth. 1843. Olacaceae. 4: 2 Surinam, French Guiana, Brazil; 2 trop. West and Central Africa

Ptychopyxis Miq. 1861. Euphorbiaceae. 13 Indochina, western Malesia, eastern New Guinea

Ptychosema Benth. 1839. Fabaceae. 2 Australia

Ptychosperma Labill. 1809. Arecaceae. 28 Moluccas, New Guinea, New Britain I., D'Entrecasteaux and Louisiade Archs., Caroline and Solomon Is., Australia (Queensland)

Ptychotis W. D. J. Koch. 1824. Apiaceae. 1–2 West and South Europe, North Africa

Ptyssiglottis T. Anderson. 1860. Acanthaceae. c. 30 Sri Lanka, Indochina, western Malesia

Pubistylus Thoth. 1966. Rubiaceae. 1 Andaman Is.

Pucara Ravenna. 1972. Amaryllidaceae. 1 northern Peru

Puccinellia Parl. 1848. Poaceae. c. 150 extratrop. regions and montane tropics

Puccionia Chiov. 1929. Brassicaceae. 1 Somalia

Puelia Franch. 1887. Poaceae. 5 trop. West Africa; P. olyriformis—extending to Congo and Tanzania

Pueraria DC. 1825. Fabaceae. 20 trop. and East Asia, Malesia, western Oceania; P. lobata—Russian Far East

Pugionium Gaertn. 1791. Brassicaceae. 5 China, Mongolia

Pulchranthus V. M. Baum, Reveal et Nowicke. 1983. Acanthaceae. 4 Colombia, eastern Peru, Surinam, French Guiana, Brazil

Pulicaria Gaertn. 1791. Asteraceae. 50–60 temp. and warm–temp. Eurasia, trop. and South Africa

Pullea Schltr. 1914. Cunoniaceae. 3 Moluccas, New Guinea, Australia (Queensland), Fiji

Pulmonaria L. 1753. Boraginaceae. 15 Europe; P. mollissima—from Europe and West Asia to China and Mongolia; Russia (6, from European part and Caucasus to East Siberia)

Pulsatilla Hill. 1753 (Anemone L.). Ranunculaceae. 43 temp. Eurasia

Pultenaea Sm. 1794. Fabaceae. c. 150 Australia, Tasmania

Pulvinaria Fourn = Lhotzkyella Rauschert. Asclepiadaceae

Puna Kiesling = Opuntia Hill. Cactaceae

Punica L. 1753. Punicaceae. 2 Balkan Peninsula, West Asia to western Himalayas (P. granatum), Socotra (P. protopunica)

Punjuba Britton et Rose. 1928. Fabaceae. 3 Central and South America

Puntia Hedge. 1983. Lamiaceae. 1 Somalia

Pupalia Juss. 1803. Amaranthaceae. 4 trop. and South (1, P. lappacea) Africa, Madagascar, Arabian Peninsula, India, Sri Lanka, Malay Peninsula, Java, Sulawesi, Philippines, New Guinea

Pupilla Rizzini. 1949 (Justicia L.). Acanthaceae. 2 trop. South America

Purdiaea Planch. 1846. Cyrillaceae. 12 trop. America, West Indies, esp. Cuba (10)

Purdieanthus Gilg = Lehmanniella Gilg. Gentianaceae

Purpureostemon Gugerli. 1939. Myrtaceae. 1 New Caledonia

Purpusia Brandegee = Ivesia Torr. et A. Gray. Rosaceae

Purshia DC. ex Poir. 1816. Rosaceae. 7 western North America

Puschkinia M. F. Adams. 1805. Hyacinthaceae. 1 (P. scilloides) Caucasus, Transcaucasia, Asia Minor, Lebanon, Iraq, Iran

Putoria Pers. 1805. Rubiaceae. 3 Mediterranean

Putranjiva Wall. = Drypetes Vahl. Euphorbiaceae

Putterlickia Endl. 1840. Celastraceae. 3 Namibia, South Africa (Transkei, Cape)

Puya Molina. 1781. Bromeliaceae. ca. 170 Central and South America, esp. Andes

Pycnandra Benth. 1876. Sapotaceae. 12 New Caledonia

Pycnanthemum Michx. 1803. Lamiaceae. 17 North America

Pycnanthus Warb. 1895. Myristicaceae. 8 trop. Africa

Pycnarrhena Miers ex Hook. f. et Thomson. 1855. Menispermaceae. 9 northeastern India, Andaman and Nicobar Is., Thailand, Cambodia, southern China (incl. Hainan), Malesia (7, Malay Peninsula, Sumatra, Java, Borneo, Lesser Sunda Is., Sulawesi, Philippines, Moluccas, Kangean Is., New Guinea, New Ireland, New Britain), Australia (Queensland), Solomon Is., Vanuatu

Pycnobotrya Benth. 1876. Apocynaceae. 2 trop. Africa

Pycnobregma Baill. 1890. Asclepiadaceae. 1 Colombia

Pycnocephalum MacLeish = Chresta Vell. Asteraceae

Pycnocoma Benth. 1849. Euphorbiaceae. 15 trop. Africa, Madagascar, Mascarene Is.

Pycnocomon Hoffmanns. et Link. 1820 (Scabiosa L.). Dipsacaceae. 2 Portugal, Spain, Corsica, Sardinia, Sicily, Italy, Morocco, Algeria, Tunisia

Pycnocycla Lindl. 1835. Apiaceae. 12 West and Central Africa (1), Egypt, Ethiopia, Arabian Peninsula, Iran, Pakistan, India

Pycnoneurum Decne. 1838. Asclepiadaceae. 2 Madagascar

Pycnonia L. A. S. Johnson et B. G. Briggs. 1975. Proteaceae. 1 Australia

Pycnophyllopsis Skottsb. 1916. Caryophyllaceae. 2 Andes of Bolivia, Paraguay, montane regions of Patagonia

Pycnophyllum J. Remy. 1846. Caryophyllaceae. 17 Andes

Pycnoplinthopsis Jafri. 1972 (? = Pycnoplinthus O. E. Schulz). Brassicaceae. 1 Nepal, Bhutan

Pycnoplinthus O. E. Schulz. 1924. Brassicaceae. 1 China

Pycnorhachis Benth. 1876. Asclepiadaceae. 1 Malay Peninsula

Pycnospatha Thorel ex Gagnep. 1941. Araceae. 2 Indochina

Pycnosphaera Gilg. 1903. Gentianaceae. 5 trop. Africa

Pycnospora R. Br. ex Wight et Arn. 1834. Fabaceae. 1 trop. Africa, trop. Asia, Malesia, northeastern Australia

Pycnostachys Hook. 1825. Lamiaceae. 40 trop. and South (3) Africa, Madagascar (1, P. coerulea)

Pycnostelma Bunge ex Decne. = Cynanchum L. Asclepiadaceae

Pycnostylis Pierre = Triclisia Benth. Menispermaceae

Pycreus P. Beauv. 1816. Cyperaceae. c. 100 warm–temp., trop. and subtrop. regions

Pygeum Gaertn. 1788 (Prunus L.). Rosaceae. 40 India, Himalayas, China, Korea, Japan, Southeast Asia

Pygmaea B. D. Jackson = Chionohebe B. G. Briggs et Ehrend. Scrophulariaceae

Pygmaeocereus J. H. Johnson et Backeb. = Haageocereus Backeb. Cactaceae

Pygmaeopremna Merr. 1910 (Premna L.). Verbenaceae. 2–3 India, Sri Lanka, Nepal, Bhutan, southern China (incl. Hainan), Indochina, Malesia, North Australia

Pygmaeorchis Brade. 1939. Orchidaceae. 2 Brazil

Pygmaeothamnus Robyns. 1928. Rubiaceae. 4 trop. and South Africa

Pygmea Hook. f. = Chionohebe B. G. Briggs et Ehrend. Scrophulariaceae

Pynaertiodendron De Wild = Cryptosepalum Benth. Fabaceae

Pyracantha M. Roem. 1847. Rosaceae. 9: 1 (P. coccinea) Southeast Europe (incl. Crimea), Caucasus, Turkey and Iran; 8: Himalayas (1) from Kashmir to Nepal, China (8), Taiwan, Indochina (Vietnam, Laos), Philippines (1)

Pyragra Bremek. 1958. Rubiaceae. 2 Madagascar

Pyramia Cham. = Cambessedesia DC. Melastomataceae

Pyramidanthe Miq. = Fissistigma Griff. Annonaceae

Pyramidium Boiss. = Vaselskya Opiz. Brassicaceae

Pyramidoptera Boiss. 1856. Apiaceae. 1 Afghanistan

Pyrenacantha Hook. 1830. Icacinaceae. c. 30 trop. and South Africa, Madagascar, southern India, Sri Lanka, Thailand, Vietnam, Hainan (1), Philippines (1)

Pyrenaria Blume. 1827. Theaceae. 20 northeastern India, southern China, Taiwan, Ryukyu Is., Indochina, western Malesia

Pyrenocarpa H. T. Chang et R. H. Miau. 1975 (Decaspermum J. R. Forst. et G. Forst.). Myrtaceae. 2 Hainan

Pyrenoglyphis Karst. = Bactris Jacq. ex Scop. Arecaceae

Pyrethropsis (Giroux) Wilcox, Bremer et Humphries. 1992. Asteraceae. ? sp. Distr. ?

Pyrethrum Zinn = Tanacetum L. Asteraceae

Pyrgophyllum (Gagnep.) T. L. Wu et Z. Y. Chen. 1989 (Kaempferia L.). Zingiberaceae. 1 China (Sichuan, Yunnan)

Pyriluma (Baill.) Aubrév. = Pouteria Aubl. Sapotaceae

Pyrogennema Lunell = Epilobium L. Onagraceae

Pyrola L. 1753. Ericaceae. 35 circumpolar and temp. regions South to northern Mediterranean, Himalayas, Sumatra and Mexico, temp. South America

Pyrolirion Herb. 1821 (Zephyranthes Herb.). Amaryllidaceae. 4 Andes of Peru, Bolivia, Chile

Pyrostegia C. Presl. 1845. Bignoniaceae. 3–4 trop. South America from Colombia to Paraguay and Brazil

Pyrostria Comm. ex Juss. 1789. Rubiaceae. 18 trop. (Guinea, Liberia, Ghana, Cameroun, Zaire, Angola, Ethiopia, Somalia, Kenya, Uganda, Tanzania, Rwanda, Zambia, Mozambique, Zimbabwe) and South Africa (Natal, Transvaal, Swaziland), Arabian Peninsula, Socotra, Madagascar, Seychelles, and Aldabra I.

Pyrrhanthera Zotov. 1963. Poaceae. 1 New Zealand

Pyrrhocactus (A. Berger) Backeb. et F. M. Knuth = Neoporteria Britton et Rose. Cactaceae

Pyrrhopappus DC. 1838. Asteraceae. 3–5 southern U.S., Mexico

Pyrrocoma Hook. 1833. Asteraceae. 10 North America

Pyrrorhiza Maguire et Wurdack. 1957. Haemodoraceae. 1 Venezuela

Pyrrothrix Bremek. 1944 (Strobilanthes Blume). Acanthaceae. 10 northeastern India, Indochina to Sumatra

Pyrularia Michx. 1803. Santalaceae. 5 Himalayas (Nepal, Bhutan), India, Burma, China (4), southeastern U.S.

Pyrus L. 1753. Rosaceae. 76 Eurasia, North Africa

Pyxidanthera Michx. 1803. Diapensiaceae. 1 southeastern U.S.

Qaisera Omer. 1989 (Gentiana L.). Gentianaceae. 3 northern Pakistan, Himalayas from Kashmir to Nepal, Tibet

Qiongzhuea C. J. Hsueh et T. P. Yi. 1980 (Chimonobamnusa Makino). Poaceae. 3 southwestern China

Quadrangula Baum.-Bod. = Gymnostoma L. A. S. Johnson. Casuarinaceae

Quadricasia Woodson = Tabernaemontana L. Apocynaceae

Quadripterygium Tardieu. 1948 (Euonymus L.). Celastraceae. 1 Indochina

Qualea Aubl. 1775. Vochysiaceae. 63 trop. America

Quamoclidion Choisy = Mirabilis L. Nyctaginaceae

Quamoclit Hill. 1753 (Ipomoea L.). Convolvulaceae. 10 tropics

Quapoya Aubl. 1775. Clusiaceae. 3 Peru, Guyana

Quaqua N. E. Br. 1879 (Caralluma R. Br.). Asclepiadaceae. 5 Southwest Africa

Quararibea Aubl. 1775. Bombacaceae. 25–30 trop. America, West Indies

Quassia L. 1762. Simaroubaceae. 35 tropics and subtropics, esp. America

Quaternella Pedersen. 1990. Amaranthaceae. 1 (Q. confusa) South America (Chapada de Contagem)

Queenslandiella Domin. 1915. Cyperaceae. 1 trop. East Africa, Madagascar, Mauritius, India, Sri Lanka, Java, Timor, Moluccas, Australia (Queensland)

Quekettia Lindl. 1839. Orchidaceae. 6 trop. South America, Trinidad

Quelchia N. E. Br. 1901. Asteraceae. 5 Venezuela, Guyana

Quercus L. 1753. Fagaceae. c. 600 America from Canada to Colombia; Europe from Scandinavia to Mediterranean; North Africa; West (incl. Caucasus, South, East, and Southeast Asia; Malesia; Russia (European part and Russian Far East)

Queria L. = Minuartia L. Caryophyllaceae

Quesnelia Gaudich. 1842. Bromeliaceae. 14 southeastern Brazil

Quetzalia Lundell. 1970. Celastraceae. 10 trop. America

Quezelia H. Scholz = Quezelianthe H. Scholz. Brassicaceae

Quezelianthe H. Scholz ex Rauschert. 1982. Brassicaceae. 1 Sakhara

Quiabentia Britton et Rose = Pereskiopsis Britton et Rose. Cactaceae

Quidproquo Greuter et Burdet. 1983. Brassicaceae. 1 Syria, Lebanon, Israel, Jordan

Quiducia Gagnep. = Silvianthus Hook. f. Carlemanniaceae

Quiina Aubl. 1775. Quiinaceae. 25 trop. South America

Quillaja Molina. 1781. Rosaceae. 4 temp. South America

Quinchamalium Molina. 1782. Santalaceae. 25 Andes

Quincula Raf. 1832 (Physalis L.). Solanaceae. 1 North America

Quinetia Cass. 1830. Asteraceae. 1 western Australia

Quinqueremulus Paul G. Wilson. 1987. Asteraceae. 1 western Australia

Quintinia A. DC. 1830. Escalloniaceae. 25 Philippines, Sulawesi, Moluccas, New Guinea, Solomon Is., Australia, New Zealand, New Caledonia

Quiotania Zarucchi. 1991. Apocynaceae. 1 Colombia

Quisqualis L. 1762. Combretaceae. 17 trop. and South Africa, trop. Asia; Q. indica—extending eastward to New Guinea and New Britain

Quisqueya Dod. 1979. Orchidaceae. 5 Haiti

Quisumbingia Merr. 1936. Asclepiadaceae. 1 Philippines

Quivisianthe Baill. 1893. Meliaceae. 1–2 Madagascar

Quoya Gaudich. = Pityrodia R. Br. Chloanthaceae

Rabdosia (Blume) Hassk. 1842 (Plectranthus L'Hér.). Lamiaceae. c. 100 trop. and subtrop. Old World, esp. Asia; Russia (2, Far East)

Rabdosiella Codd. 1984. Lamiaceae. 2 northern and eastern South Africa (1, R. calycina), India (1)

Rabiea N. E. Br. 1930 (Nananthus N. E. Br.). Aizoaceae. 4 central South Africa

Racemobambos Holttum. 1956. Poaceae. 18 South and Southeast Asia, New Guinea, Solomon Is.

Racinaea Spencer et L. B. Sm. 1993 (Tillandsia L.). Bromeliaceae. c. 60 Mexico, Nicaragua, Costa Rica, Colombia, Venezuela, Trinidad, Guyana, Ecuador, Peru, Brazil, Bolivia; West Indies (Cuba, Jamaica)

Racosperma (DC.) Mart. = Acacia Hill. Fabaceae

Radamaea Benth. 1846. Scrophulariaceae. 5 Madagascar

Raddia Bertol. 1819. Poaceae. 7 trop. South America

Raddiella Swallen. 1948. Poaceae. 8 Panama, trop. South America

Radermachera Zoll. et Moritzi. 1855. Bignoniaceae. c. 60 India, China, Taiwan, Ryukyu Is., Indochina, Java, Sulawesi, Philippines, Moluccas, New Guinea

Radinosiphon N. E. Br. 1932. Iridaceae. 1–2 trop. and eastern South Africa

Radiola Hill. 1756. Linaceae. 1 Europe (incl. European part of Russia), Mediterranean, temp. Asia

Radlkofera Gilg. 1897. Sapindaceae. 1 trop. Africa

Radlkoferotoma Kuntze. 1891. Asteraceae. 3 southern Brazil, Uruguay

Radyera Bullock. 1957. Malvaceae. 2 South Africa (1, R. urens), Australia (1, R. farragei)

Raffenaldia Godr. 1853. Brassicaceae. 2 Morocco, Algeria

Rafflesia R. Br. 1821. Rafflesiaceae. 14 Malay Peninsula, Sumatra, Java, Borneo, Philippines

Rafinesquia Nutt. 1841. Asteraceae. 2 southwestern U.S. (California, Utah, Arizona, western Texas), Mexico (northern Baja California)

Rafnia Thunb. 1800. Fabaceae. 25 South Africa

Ragala Pierre = Chrysophyllum L. Sapotaceae

Rahowardiana D'Arcy. 1974 (Markea Rich.). Solanaceae. 1 Panama

Raillardella (A. Gray) Benth. 1873. Asteraceae. 5 western U.S.

Raillardiopsis Rydb. 1927. Asteraceae. 2 California

Railliardia Gaudich. = Dubautia Gaudich. Asteraceae

Raimondia Saff. 1913. Annonaceae. 2 Colombia, Ecuador

Raimondianthus Harms = Chaetocalyx DC. Fabaceae

Raimundochloa Molina. 1986 (Koeleria Pers.). Poaceae. 1 Chile

Rainiera Greene. 1898. Asteraceae. 1 U.S. (Washington and Oregon)

Rajania L. 1753. Dioscoreaceae. 20 Bahama Is., West Indies

Ramatuela Kunth = Terminalia L. Combretaceae

Ramelia Baill. = Bocquillonia Baill. Euphorbiaceae

Rameya Baill. 1870 (Triclisia Benth.). Menispermaceae. 2 Comoro Is., Madagascar

Ramirezella Rose. 1903. Fabaceae. 8 Mexico, Central America

Ramischia Opiz ex Garcke = Orthilia Raf. Ericaceae

Ramisia Glaz. ex Baill. 1887. Nyctaginaceae. 1 southeastern Brazil

Ramonda Rich. 1805. Gesneriaceae. 3 West (Spain, France) and Southeast (Yugoslavia, Greece, Bulgaria) Europe

Ramorinoa Speg. 1924. Fabaceae. 1 Argentina

Ramosia Merr. 1916 (Centotheca Desv.). Poaceae. 1 Philippines

Ramosmania Tirving. et Verdc. 1982 (Randia L.). Rubiaceae. 2 Rodriguez I.

Ranalisma Stapf. 1900 (Echinodorus Rich.). Alismataceae. 2 trop. Africa from Senegal to the Sudan and through Congo and Zaire to Zambia and Tanzania (1, R. humile); continental Southeast Asia and Malesia (1)

Randia L. 1753. Rubiaceae. c. 200 tropics

Randonia Coss. 1859. Resedaceae. 3 Africa (Morocco, Mauritania, Algeria, Libya, Egypt, Ethiopia, Somalia), Arabian Peninsula

Ranevea L. H. Bailey = Ravenea C. D. Bouché. Arecaceae

Rangaeris (Schltr.) Summerh. 1936. Orchidaceae. 6 trop. and South Africa

Ranopisoa J.-F. Leroy. 1977 (Oftia Adans.). Oftiaceae. 1 Madagascar

Ranunculus L. 1753. Ranunculaceae. 250–400 cosmopolitan, chiefly cold, temp. regions and montane areas of the tropics

Ranzania T. Itô. 1888. Ranzaniaceae. 1 Japan (Honshu)

Raoulia Hook. f. ex Raoul. 1846. Asteraceae. 20 Tasmania, New Zealand

Raouliopsis S. F. Blake. 1938. Asteraceae. 2 Andes

Rapanea Aubl. 1775 (Myrsine L.). Myrsinaceae. 140–200 Africa, India, Sri Lanka, China, Korea, Japan, Indochina, Malesia, Bonin Is., Palau Is., Caroline and Solomon islands, Australia, Lord Howe I., New Zealand, New Caledonia, Fiji, Samoa Is., Tonga, Hawaiian Is.

Raparia F. K. Mey. = Thlaspi L. Brassicaceae

Rapatea Aubl. 1775. Rapateaceae. 20 trop. South America

Raphanorhyncha Rollins. 1976 Brassicaceae. 1 Mexico

Raphanus L. 1753. Brassicaceae. 8 West, Central, and East Europe, Mediterranean, temp. Asia; Russia (3 European part, Siberia, Far East)

Raphia P. Beauv. 1806. Arecaceae. 28 trop. America (1 [R. taedigera] from Nicaragua to northern Colombia and northwestern Venezuela, eastern Brazil [adjacent areas of the Amazon delta]), trop. equatorial and East Africa, northern Madagascar (1)

Raphidiocystis Hook. f. 1867. Cucurbitaceae. 5 trop. Africa, Madagascar

Raphiocarpus Chun = Didissandra C. B. Clarke. Gesneriaceae

Raphionacme Harv. 1842. Asclepiadaceae. 36 trop. and South Africa, southern Arabian Peninsula

Raphistemma Wall. 1831. Asclepiadaceae. 3 western and southwestern China, Nepal, northeastern India, Burma, Thailand

Rapicactus Buxb. et Oehme = Neolloydia Britton et Rose. Cactaceae

Rapistrum Crantz. 1769. Brassicaceae. 3 Central and East Europe, Mediterranean, West (incl. Caucasus) and Middle Asia

Rapona Baill. 1890. Convolvulaceae. 1 Madagascar

Raputia Aubl. 1775. Rutaceae. 4 Venezuela, Brazil (Amazonia)

Raputiarana Emmerich. 1978 (Raputia Aubl.). Rutaceae. 1 trop. America

Raritebe Wernham. 1917. Rubiaceae. 6 Costa Rica, Panama, Colombia, Ecuador, Peru

Raspalia Brongn. 1826. Bruniaceae. 16 South Africa

Rastrophyllum Wild et G. V. Pope. 1977. Asteraceae. 2 Zambia

Rathbunia Britton et Rose. 1909. Cactaceae. 2 Mexico

Ratibida Raf. 1817. Asteraceae. 6 North America

Rattraya J. B. Phipps. 1964 (Danthoniopsis Stapf). Poaceae. 1 Zambia, Zimbabwe

Ratzeburgia Kunth. 1831. Poaceae. 1 Burma

Rauhia Traub. 1957. Amaryllidaceae. 3 Peru

Rauhiella Pabst et Braga. 1978 (Ornithocephalus Hook.). Orchidaceae. 1 Brazil

Rauhocereus Backeb. = Weberbauerocereus Backeb. Cactaceae

Rauia Nees et Mart. 1823. Rutaceae. 3 trop. South Africa

Raulinoa R. S. Cowan. 1960. Rutaceae. 1 Brazil

Raulinoreitzia R. M. King et H. Rob. 1971 (Eupatorium L.). Asteraceae. 2 Brazil, Bolivia, Argentina

Rautanenia Buchenau = Burnatia Micheli. Alismataceae

Rauvolfia L. 1753. Apocynaceae. c. 100 trop. America, trop. and South Africa, trop. Asia, trop. Pacific Is.

Rauwenhoffia Scheff. 1885 (Melodorum Lour.). Annonaceae. 5 Thailand, Indochina, Malay Peninsula, New Guinea, eastern Australia (1)

Ravenala Adans. 1763. Strelitziaceae. 1 Madagascar

Ravenea C. D. Bouché ex H. A. Wendl. 1884. Arecaceae. 10 Comoro Is. (2), Madagascar (8)

Ravenia Vell. 1825. Rutaceae. 18 Central and trop. South America, West Indies

Raveniopsis Gleason. 1939. Rutaceae. 16 Venezuela, Brazil

Ravensara Sonn. 1782 (Cryptocarya R. Br.). Lauraceae. c. 30 Madagascar

Ravnia Oerst. = Hillia Jacq. Rubiaceae

Rawsonia Harv. et Sond. 1860. Flacourtiaceae. 2 trop. (from Sudan and Somalia to Angola, Zimbabwe, and Mozambique) and South (1) Africa

Raycadenco Dodson. 1989. Orchidaceae. 1 Ecuador

Rayleya Cristóbal. 1981. Sterculiaceae. 1 Brazil

Raynalia Soják = Alinula J. Raynal. Cyperaceae

Razisea Oerst. 1854. Acanthaceae. 3 Central America, Colombia

Rea Bertero ex Decne. = Dendroseris D. Don. Asteraceae

Readea Gillespie. 1930. Rubiaceae. 1 Fiji

Reaumuria L. 1759. Tamaricaceae. 13 eastern Mediterranean, West (incl. Caucasus) and Middle Asia eastward to West Siberia, Mongolia, continental China, and Pakistan

Reboudia Coss. et Durieu = Eruca Hill. Brassicaceae

Rebutia K. Schum. 1895. Cactaceae. c. 40 Andes from Bolivia to northwestern Argentina

Recchia Moçiño et Sessé ex DC. 1817. Simaroubaceae. 3 Mexico

Rechsteineria Regel = Sinningia Nees. Gesneriaceae

Recordia Moldenke. 1934. Verbenaceae. 1 Bolivia

Recordoxylon Ducke. 1934. Fabaceae. 2 Brazil (Amazonia)

Rectanthera O. Deg. = Callisia L. Commelinaceae

Redfieldia Vasey. 1887. Poaceae. 2 U.S.

Redowskia Cham. et Schltdl. 1826. Brassicaceae. 1 northeastern Siberia

Reederochloa Soderstr. et H. F. Decker. 1964. Poaceae. 1 Mexico

Reedia F. Muell. 1859. Cyperaceae. 1 southwestern Australia

Reedrollinsia J. W. Walker. 1971 (Stenanona Standl.). Annonaceae. 1 (R. cauliflora) Mexico

Reevesia Lindl. 1827. Sterculiaceae. 18 eastern Himalayas, northeastern India, China, Taiwan

Regelia Schauer. 1843. Myrtaceae. 4 southwestern Australia

Registaniella Rech. f. 1987. Apiaceae. 1 Afghanistan (Kandagar)

Rehdera Moldenke. 1935. Verbenaceae. 3 Central America

Rehderodendron Hu. 1932. Styracaceae. 4 China, Indochina

Rehderophoenix Burret = Drymophloeus Zipp. Arecaceae

Rehia Fijten. 1975. Poaceae. 1 Brazil

Rehmannia Libosch. ex Fisch. et C. A. Mey. 1835. Scrophulariaceae. 6 East Asia

Rehsonia Stritch = Wisteria Nutt. Fabaceae

Reichardia Roth. 1789. Asteraceae. 8 Canary Is., Mediterranean, Ethiopia, West Asia eastward to northwestern India; R. glauca—Caucasus

Reichea Kausel = Myrcianthes O. Berg. Myrtaceae

Reicheëlla Pax. 1900. Caryophyllaceae. 1 Chile

Reicheia Kausel = Reichea Kausel. Myrtaceae

Reichenbachanthus Barb. Rodr. 1882. Orchidaceae. 3 trop. America

Reichenbachia Spreng. 1823. Nyctaginaceae. 2 Bolivia, Brazil, Paraguay, Argentina

Reimaria Humb. et Bonpl. ex Fluegge = Paspalum L. Poaceae

Reimarochloa Hitchc. 1909. Poaceae. 4 America from southern U.S. to Argentina

Reineckea Kunth. 1844. Convallariaceae. 1 (R. carnea) China, Japan

Reinhardtia Liebm. 1849. Arecaceae. 6 Mexico, Central America, northwestern Colombia (1), Dominican Republic (1)

Reinwardtia Dumort. 1822. Linaceae. 2 Himalayas from Kashmir to Bhutan, India, Burma, China, Indochina

Reinwardtiodendron Koord. 1898. Meliaceae. 7 southern India, Indochina, Hainan, Malesia

Reissantia N. Hallé. 1958. Celastraceae. 7 trop. Africa, trop. Asia

Reissekia Endl. 1840. Rhamnaceae. 1 Brazil

Reitzia Swallen. 1956. Poaceae. 1 Brazil

Rejoua Gaudich. = Tabernaemontana L. Apocynaceae

Relbunium (Endl.) Benth. et Hook. f. = Galium L. Rubiaceae

Relchela Steud. 1854. Poaceae. 1 Chile, Argentina

Reldia Wiehler. 1977. Gesneriaceae. 5 trop. America from Panama to northern Peru

Relhania L'Hér. 1789. Asteraceae. 29 South Africa

Remijia DC. 1829. Rubiaceae. 35 trop. South America

Remirea Aubl. 1775. Cyperaceae. 1 pantropics

Remirema Kerr. 1943. Convolvulaceae. 1 Indochina

Remusatia Schott. 1832. Araceae. 2–4: R. vivipara (Roxb.) Schott—trop. West, Central, and East (Ethiopia, Uganda) Africa; Madagascar; Oman; India; Sri Lanka; Thailand, southern China (Yunnan); Taiwan; Indochina; Java; and northern Australia; R. hookeriana Schott—Hialayas from Simla to Sikkim

Remya W. F. Hillebr. ex Benth. 1873 (Olearia Moench). Asteraceae. 3 Hawaiian Is.

Renanthera Lour. 1790. Orchidaceae. 10–15 Himalayas, China, Indochina, Malesia from Malay Peninsula to New Guinea, Solomon Is.

Renantherella Ridl. 1896. Orchidaceae. 1 Thailand, Malay Peninsula

Renata Ruschi. 1946. Orchidaceae. 2 Venezuela, Brazil

Rendlia Chiov. 1914 (Microchloa R. Br.). Poaceae. 3 trop. Southeast and South Africa

Renealmia L. f. 1782. Zingiberaceae. c. 80 trop. America, West Indies, trop. Africa

Renggeria Meisn. 1837 (or = Clusia L.). Clusiaceae. 3 Brazil

Rengifa Poepp. et Endl. = Quapoya Aubl. Clusiaceae

Rennellia Korth. 1851. Rubiaceae. 4 southern Burma, peninsular Thailand, Malay Peninsula, Sumatra, Borneo

Rennera Merxm. 1957. Asteraceae. 1 Namibia

Renschia Vatke. 1881. Lamiaceae. 2 Somalia

Rensonia S. F. Blake. 1923. Asteraceae. 1 Mexico (Chiapas), Central America

Reptonia A. DC. = Monotheca A. DC. Sapotaceae

Requienia DC. 1825 (Tephrosia Pers.). Fabaceae. 3 Africa: 1 (R. obcordata) Senegal, Mali, Niger, Nigeria, Chad, Sudan; 2 Zambia (1), Namibia, Botswana, South Africa

Reseda L. 1753. Resedaceae. c. 60 Macronesia (Azores, Canary Is.), Europe (20), Mediterranean, trop. East Africa (Ethiopia, Somalia, Kenya); Southwest, West (incl. Caucasus), and Middle Asia eastward to southwestern West Siberia, China, and northwestern India

Resia H. E. Moore. 1962. Gesneriaceae. 1 Colombia

Resnova Van der Merwe = Drimiopsis Lindl. Hyacinthaceae

Restella Pobed. 1941 (Wikstroemia Endl.). Thymelaeaceae. 1 Middle Asia

Restio Rottb. 1772. Restionaceae. 89 trop. and South (c. 40 endemics of Cape) Africa, Madagascar

Restrepia Kunth. 1816. Orchidaceae. c. 60 Mexico, Central and trop. South America south to Argentina

Restrepiella Garay et Dunst. 1966. Orchidaceae. 8 trop. America

Restrepiopsis Luer. 1978. Orchidaceae. 15 trop. America: Guatemala, Nicaragua, Costa Rica, Panama, Colombia, Venezuela, Ecuador, Peru, Bolivia

Retama Raf. 1838. Fabaceae. 4 Canary Is., Iberian Peninsula, Sicily, North Africa from Morocco to Egypt, Syria, Palestine, Sinai, and Arabian Peninsula

Retanilla (DC.) Brongn. 1827. Rhamnaceae. 2 Peru, Chile

Retiniphyllum Humb. et Bonpl. 1806. Rubiaceae. 22 trop. South America

Retispatha J. Dransf. 1979. Arecaceae. 1 Borneo

Retrophyllum C. N. Page. 1988. Podocarpaceae. 5: 1 eastern Malesia from Moluccas to New Britain, Melanesia eastward to New Caledonia (2), Fiji, South America (2, Colombia, Venezuela, Peru, western Brazil)

Retzia Thunb. 1776. Retziaceae. 1 South Africa (Cape of Good Hope)

Reussia Endl. = Pontederia L. Pontederiaceae

Reutealis Airy Shaw. 1967. Euphorbiaceae. 1 Philippines

Reutera Boiss. = Pimpinella L. Apiaceae

Revealia R. M. King et H. Rob. 1976. Asteraceae. 1 Mexico

Reverchonia A. Gray. 1880. Euphorbiaceae. 1 southern U.S., Mexico

Reyesia Gay. 1848–1849 (Salpiglossis Ruiz et Pav.). Solanaceae. 4–5 Chile, Argentina

Reynaudia Kunth. 1830. Poaceae. 1 West Indies

Reynoldsia A. Gray. 1854. Araliaceae. 14 western Oceania

Reynosia Griseb. 1866. Rhamnaceae. 16 Florida peninsula, West Indies

Reynoutria Houtt. = Fallopia Adans. Polygonaceae

Rhabdadenia Muell. Arg. 1860. Apocynaceae. 4 Florida, Mexico, Central and trop. South America

Rhabdocaulon (Benth.) Epling. 1936. Lamiaceae. 7 trop. South America

Rhabdodendron Gilg et Pilg. 1905. Rhabdodendraceae. 3 trop. South America

Rhabdophyllum Tiegh. 1902 (Ouratea Aubl.). Ochnaceae. 25 trop. Africa

Rhabdosciadium Boiss. 1844. Apiaceae. 4–5 Turkey, Iran, Afghanistan

Rhabdothamnopsis Hemsl. 1903. Gesneriaceae. 1 China

Rhabdothamnus A. Cunn. 1838. Gesneriaceae. 1 New Zealand

Rhabdotheca Cass. = Launaea Cass. Asteraceae

Rhabdotosperma Hartl. 1977 (Verbascum L.). Scrophulariaceae. 7 trop. Africa

Rhachicallis DC. = Arcytophyllum Willd. ex Schult. et Schult. f. Rubiaceae

Rhacodiscus Lindau = Justicia L. Acanthaceae

Rhacoma P. Browne ex L. = Crossopetalum P. Browne. Celastraceae

Rhadamanthus Salisb. 1866. Hyacinthaceae. 10 South Africa

Rhadinopus S. Moore. 1930. Rubiaceae. 2 New Guinea

Rhadinothamnus Paul G. Wilson. 1971 (Nematolepis Turcz.). Rutaceae. 1 western Australia

Rhaesteria Summerh. 1966. Orchidaceae. 1 trop. East Africa (Uganda)

Rhagadiolus Hill. 1753. Asteraceae. 7 Mediterranean northeast to Crimea, West (incl. Caucasus) Asia eastward to Afghanistan and northeastern India (Kashmir)

Rhagodia R. Br. 1810. Chenopodiaceae. 11 Australia (excl. northern part), New Zealand (1, R. triandra)

Rhammatophyllum O. E. Schulz = Erysimum L. Brassicaceae

Rhamnella Miq. 1867. Rhamnaceae. 10 Asia from western Himalayas to China and Japan, Malesia, trop. Australia, Melanesia to Fiji

Rhamnidium Reissek. 1861. Rhamnaceae. 12 trop. America, West Indies

Rhamnoluma Baill. = Pichonia Pierre. Sapotaceae

Rhamnoneuron Gilg. 1894. Thymelaeaceae. 1 China, Southeast Asia

Rhamnus L. 1753. Rhamnaceae. 125 (or c. 200) temp. Northern Hemisphere southward to Brazil and South Africa; c. 25 European part of Russia, Caucasus, Middle Asia, Siberia, Far East

Rhamphicarpa Benth. 1836. Scrophulariaceae. 6 trop. Africa, Madagascar, India, Australia; R. medwedewii—Caucasus, Asia Minor

Rhamphocarya Kuang = Carya Nutt. Juglandaceae

Rhamphogyne S. Moore. 1914. Asteraceae. 2 Rodriguez I. (1), New Guinea (1)

Rhamphorhynchus Garay. 1977. Orchidaceae. 1 Brazil

Rhanteriopsis Rauschert. 1982. Asteraceae. 4 West Asia

Rhanterium Desf. 1799. Asteraceae. 3 Northwest and North Africa, West Asia to Pakistan

Rhaphidanthe Hiern ex Guerke = Diospyros L. Ebenaceae

Rhaphidophora Hassk. 1842. Araceae. 60–100 West Africa, India, Sri Lanka, Nepal, Burma, southwestern China, Indochina, Malesia, New Caledonia

Rhaphidophyton Iljin. 1936 (Noaea Moq.). Chenopodiaceae. 1 Kazakhstan, Kirgizstan

Rhaphidospora Nees. 1832. Acanthaceae. 12 trop. Africa, China, Himalayas, India, Sri Lanka, western Malesia

Rhaphidura Bremek. 1940. Rubiaceae. 1 Borneo

Rhaphiodon Schauer = Hyptis Jacq. Lamiaceae

Rhaphiolepis Lindl. 1820. Rosaceae. 9 India (1), southern China, Hainan, Taiwan, Japan, Indochina, Indonesia, Borneo, Philippines

Rhaphiostylis Planch. ex Benth. 1849. Icacinaceae. 6 trop. West and East (R. beninensis—Uganda, Tanzania) Africa

Rhaphis Lour. = Chrysopogon Trin. Poaceae

Rhaphispermum Benth. 1846. Scrophulariaceae. 1 Madagascar

Rhaphithamnus Miers. 1870. Verbenaceae. 2 Juan Fernández Is., Chile, Argentina

Rhapidophyllum H. A. Wendl. et Drude. 1876. Arecaceae. 1 southeastern U.S. from Beaufort County, South Carolina, south to Florida (Highlands and Hardee counties and west to Simpson County)

Rhapis L. f. ex Aiton. 1789. Arecaceae. 12 southern China, Indochina, Sumatra (1)

Rhaponticum Ludw. = Centaurea L. Asteraceae

Rhaptonema Miers. 1867. Menispermaceae. 7 Madagascar

Rhaptopetalum Oliv. 1864. Scytopetalaceae. 10 trop. Africa

Rhazya Decne. 1835. Apocynaceae. 2 Egypt, Arabian Peninsula, West Asia to northwestern India

Rheedia L. 1753 (Garcinia L.). Clusiaceae. 25 trop. America, Africa, Madagascar

Rhektophyllum N. E. Br. 1882. Araceae. 2 trop. West, Central, and East (R. mirablie—Uganda) Africa

Rheome Goldblatt. 1980. Iridaceae. 3 South Africa

Rhetinodendron Meisn. = Robinsonia DC. Asteraceae

Rhetinosperma Radlk. = Chisocheton Blume. Meliaceae

Rheum L. 1753. Polygonaceae. c. 60 temp. and subtrop. Asia, esp. China (40); Bulgaria (1, R. rhaponticum); Caucasus, Middle Asia, southern Siberia, Russian Far East

Rhexia L. 1753. Melastomataceae. 11 North America

Rhigiocarya Miers. 1864. Menispermaceae. 3 trop. West Africa

Rhigiophyllum Hochst. 1842. Campanulaceae. 1 South Africa

Rhigospira Miers. 1878. Apocynaceae. 1 trop. South America

Rhigozum Burch. 1822. Bignoniaceae. 6 trop. and South Africa, Madagascar

Rhinacanthus Nees. 1832. Acanthaceae. 15 trop. Africa, Madagascar, Socotra, western India, Malesia, East Asia

Rhinactinidia Novopokr. 1948. Asteraceae. 4 Middle Asia, West and East Siberia

Rhinanthus L. 1753. Scrophulariaceae. c. 50 North America, temp. Eurasia

Rhinchoglossum Blume = Rhynchoglossum Blume. Gesneriaceae

Rhinephyllum N. E. Br. 1927. Aizoaceae. 14 South Africa

Rhinerrhiza Rupp. 1951. Orchidaceae. 2 New Guinea, New Britain, Solomon Is., northeastern Australia

Rhinopetalum Fisch. ex Alexand. 1830 (Fritillaria L.). Liliaceae. 4 Caucasus, Middle Asia, Iran

Rhinopterys Nied. = Acridocarpus Guill., Perr. et A. Rich. Malpighiaceae

Rhipidantha Bremek. 1940. Rubiaceae. 1 trop. Africa (Tanzania)

Rhipidia Markgr. = Condylocarpon Desf. Apocynaceae

Rhipidocladum McClure. 1973. Poaceae. 14 trop. America from Mexico and Trinidad to northern Argentina

Rhipidoglossum Schltr. = Diaphananthe Schltr. Orchidaceae

Rhipsalidopsis Britton et Rose = Hatiora Britton et Rose. Cactaceae

Rhipsalis Gaertn. 1788. Cactaceae. 35 trop. America from Mexico to Brazil and Argentina, West Indies; 1 (R. baccifera) forests of West and eastern South Africa, Madagascar, Mascarene Is., Sri Lanka

Rhizanthella R. S. Rogers. 1928. Orchidaceae. 2 southwestern Australia

Rhizanthes Dumort. 1829. Rafflesiaceae. 2 Malay Peninsula, Sumatra, Java, Borneo

Rhizobotrya Tausch. 1836. Brassicaceae. 1 southern Germany, northern Italy

Rhizocephalum Wedd. = Lysipomia Kunth. Lobeliaceae

Rhizocephalus Boiss. 1844. Poaceae. 1 (R. orientalis) Transcaucasia, Jordan, Iran, Middle Asia, Afghanistan

Rhizomonanthes Danser. 1933. Loranthaceae. 3 New Guinea

Rhizophora L. 1753. Rhizophoraceae. 8–9 trop. coasts of India, Pacific and Atlantic Oceans from East Africa eastward to western Pacific and northern Australia (north to Taiwan), and from mid-Pacific through trop. America to West Africa

Rhodalsine J. Gay. 1845 (Minuartia L.). Caryophyllaceae. 2 Mediterranean

Rhodamnia Jack. 1822. Myrtaceae. 23 China, Indochina, Malesia, eastern Australia, New Caledonia

Rhodanthe Lindl. 1834. Asteraceae. 43 Australia

Rhodiola L. 1753 (Sedum L.). Crassulaceae. c. 90 temp. Northern Hemisphere, esp. China (75) and Mediterranean; Russia (c. 25, Arctica, European part, Siberia, Far East) and Middle Asia

Rhodocactus (A. Berger) F. Knuth = Pereskia Hill. Cactaceae

Rhodocalyx Muell. Arg. 1860. Apocynaceae. 1 Brazil

Rhodochiton Zucc. ex Otto et Dietr. 1834. Scrophulariaceae. 1 Mexico

Rhodocodon Baker. 1881. Hyacinthaceae. 8 Madagascar

Rhodocolea Baill. 1887. Bignoniaceae. 6 Madagascar

Rhodocoma Nees. 1836 (Restio Rottb.). Restionaceae. 6 South Africa (from Natal to Cape)

Rhododendron L. 1753. Ericaceae. c. 850 Europe (9), Asia, esp. China (c. 650), Himalayas and montane Malesia, Australia (1), North America (c. 25); Russia (southern Siberia, Far East) and Caucasus (c. 20)

Rhododon Epling. 1939. Lamiaceae. 1 southern U.S. (Texas)

Rhodogeron Griseb. = Sachsia Griseb. Asteraceae

Rhodognaphalon (Ulbr.) Roberty = Bombax L. Bombacaceae

Rhodognaphalopsis A. Robyns. 1963 (Bombacopsis Pittier). Bombacaceae. 9 trop. America

Rhodohypoxis Nel. 1914. Hypoxidaceae. 6 South and Southeast Africa

Rhodolaena Thouars. 1805. Sarcolaenaceae. 5 Madagascar

Rhodoleia Champ. ex Hook. f. 1850. Rhodoleiaceae. 1–9 Burma, southern China, Indochina, Malay Peninsula, Sumatra

Rhodomyrtus (DC.) Rchb. 1841. Myrtaceae. c. 20 southern India, Sri Lanka, Thailand, China, Taiwan, Philippines, Moluccas, New Guinea, Solomon Is., Australia, New Caledonia

Rhodophiala C. Presl = Hippeastrum Herb. Amaryllidaceae

Rhodopis Urb. 1900. Fabaceae. 1 West Indies

Rhodosciadium S. Watson. 1889. Apiaceae. 12 Mexico

Rhodosepala Baker = Dissotis Benth. Melastomataceae

Rhodospatha Poepp. 1845. Araceae. 15 Central and trop. South America

Rhodosphaera Engl. 1881. Anacardiaceae. 1 Australia (Queensland and New South Wales)

Rhodostachys Phil. = Ochagavia Phil. Bromeliaceae

Rhodostemonodaphne Rohwer et Kubitzki. 1985. Lauraceae. 12 trop. America from Costa Rica to Peru and Brazil

Rhodothamnus Rchb. 1827. Ericaceae. 2 northern Italy, montane Yugoslavia (1), northeastern Turkey (1)

Rhodotypos Siebold et Zucc. 1841. Rosaceae. 1 China, Japan

Rhodusia Vassilcz. = Medicago L. Fabaceae

Rhoeo Hance ex Walp. = Tradescantia L. Commelinaceae

Rhoiacarpos A. DC. 1857. Santalaceae. 1 South Africa

Rhoicissus Planch. 1887. Vitaceae. 10 trop. and South Africa

Rhoiptelea Diels et Hand.-Mazz. 1932. Rhoipteleaceae. 1 southwestern China, northern Vietnam

Rhombochlamys Lindau. 1897. Acanthaceae. 2 Colombia

Rhombolytrum Link. 1883. Poaceae. 1 Chile, Uruguay, southern Brazil

Rhombonema Schltr. = Parapodium E. Mey. Asclepiadaceae

Rhombophyllum (Schwantes) Schwantes. 1927. Aizoaceae. 3 South Africa (Cape)

Rhoogeton Leeuwenb. 1958. Gesneriaceae. 3 northeastern South America

Rhopalephora Hassk. 1864 (Aneilema R. Br.). Commelinaceae. 4 Madagascar, South and Southeast Asia, Malesia, Melanesia eastward to Fiji

Rhopaloblaste Scheff. 1876. Arecaceae. 6 Nicobar Is., Singapore, Moluccas, New Guinea, Solomon Is.

Rhopalobrachium Schltr. et K. Krause. 1908. Rubiaceae. 2 Caroline Is. (? 1), New Caledonia

Rhopalocarpus Bojer. 1846. Sphaerosepalaceae. 14 Madagascar

Rhopalocnemis Jungh. 1841. Helosiaceae. 2 Madagascar (1), Himalayas, northern and northeastern India, southern China, Indochina, Malesia from Sumatra to Moluccas

Rhopalocyclus Schwantes = Leipoldtia L. Bolus. Aizoaceae

Rhopalopilia Pierre. 1896. Opiliaceae. 3 trop. Africa (R. halei—southern Gabon; R. pallens—Cameroun, Gabon, Congo, Zaire; R. altescandens—Cameroun, southern Central African Rep., and northern Zaire)

Rhopalopodium Ulbr. = Krapfia DC. Ranunculaceae

Rhopalosciadium Rech. f. 1952. Apiaceae. 1 Iran

Rhopalostylis H. A. Wendl. et Drude. 1875 (Hedyscepe H. A. Wendl. et Drude). Arecaceae. 2–3 New Zealand, Chatham I., Norfolk I., Raoul (Sunday) I.

Rhopalota N. E. Br. = Crassula L. Crassulaceae

Rhuacophila Blume. 1827. Phormiaceae. 1 Southwest, South, and Southeast Asia; Malesia; Solomon Is.; New Caledonia

Rhus L. 1753. Anacardiaceae. c. 200 warm–temp., subtrop., and trop. regions; R. coriaria—extending to Crimea, Caucasus, and Middle Asia

Rhynchanthera DC. 1828. Melastomataceae. 15 trop. America from southern Mexico to Bolivia and Paraguay

Rhynchanthus Hook. f. 1886. Zingiberaceae. 5–6 Burma, southwestern China, western Malesia

Rhyncharrhena F. Muell. = Pentatropis R. Br. Asclepiadaceae

Rhynchelytrum Nees. 1836. Poaceae. c. 15 Africa, Madagascar, Arabian Peninsula; naturalized in tropics

Rhynchocalyx Oliv. 1894. Rhynchocalycaceae. 1 South Africa (Natal, Transkei)

Rhynchocarpa Becc., non Schrad. ex Endl. = Burretiokentia Pichi-Serm. Arecaceae

Rhynchocladium T. Koyama. 1972. Cyperaceae. 1 Venezuela

Rhynchocorys Griseb. 1844. Scrophulariaceae. 6 South Europe, West Asia to Caucasus (4–5) and Iran

Rhynchodia Benth. 1876. Apocynaceae. 8 India, southern China, Indochina, western Malesia

Rhynchoglossum Blume. 1826. Gesneriaceae. 11–13 southern and northeastern India, Sri Lanka, southern China, Taiwan, Indochina, Malesia, trop. South America (1)

Rhynchogyna Seidenf. et Garay. 1973. Orchidaceae. 3 Thailand, Vietnam, Malay Peninsula

Rhyncholacis Tul. 1849. Podostemaceae. 25 northern South America

Rhyncholaelia Schltr. 1918 (Brassavola R. Br.). Orchidaceae. 2 Mexico, Central America

Rhynchophora Arènes. 1946. Malpighiaceae. 1 Madagascar

Rhynchophreatia Schltr. 1921 (Phreatia Lindl.). Orchidaceae. 7 Micronesia, New Guinea, New Caledonia (1, R. micrantha), Vanuatu, Fiji, Samoa Is.

Rhynchoryza Baill. 1893. Poaceae. 1 South America (southern Brazil, Paraguay and northern Argentina)

Rhynchosia Lour. 1790. Fabaceae. c. 200 pantropics, esp. Africa (150)

Rhynchosida Fryxell. 1978. Malvaceae. 2: R. physocalyx—U.S. (Arizona, New Mexico, Oklahoma, Texas), Mexico (from Sonora south to Nuevo Leon, San Luis Potosi, Puebla), Bolivia, Brazil, Argentina; R. kearneyi Fryxell—Bolivia (Santa Cruz)

Rhynchosinapis Hayek = Coincya Porta

et Rigo ex Rouy. Brassicaceae

Rhynchospermum Reinw. 1825. Asteraceae. 2 China, Southeast Asia

Rhynchospora Vahl. 1806. Cyperaceae. c. 200 cosmopolitan

Rhynchostigma Benth. 1876 (Toxocarpus Wight et Arn.). Asclepiadaceae. 1 trop. West Africa

Rhynchostylis Blume. 1825. Orchidaceae. 4 trop. Asia from India and Sri Lanka to Burma, Indochina, Malesia (2): Malay Peninsula, Java and Borneo (1, R. retusa), Philippines (2, R. retusa and R. violacea)

Rhynchotechum auctt. = Rynchotoechum Blume. Gesneriaceae

Rhynchotheca Ruiz et Pav. 1794. Rhynchothecaceae. 1 Andes

Rhynchotoechum Blume. 1829. Gesneriaceae. 14 India, Sri Lanka, eastern Himalayas, Burma, southern China, Thailand, Taiwan, Malesia

Rhynchotropis Harms. 1901. Fabaceae. 2 Zaire, Angola, Zambia

Rhynea DC. = Tenrhynea Hilliard et B. L. Burtt. Asteraceae

Rhysolepis S. F. Blake. 1917. Asteraceae. 3 Mexico

Rhysopterus J. M. Coult. et Rose. 1900. Apiaceae. 3 North America

Rhysotoechia Radlk. 1879. Sapindaceae. 14 Philippines, Borneo, Aru Is., New Guinea, Australia (Queensland)

Rhyssolobium E. Mey. 1838. Asclepiadaceae. 1 South Africa

Rhyssopteris Blume ex A. Juss. 1837. Malpighiaceae. 6 southern China, Taiwan, Malesia (Java, Lesser Sunda Is., Sulawesi, Philippines, Moluccas, New Guinea), Micronesia (Caroline Is.), Solomon Is., trop. Australia, New Caledonia

Rhyssostelma Decne. 1844. Asclepiadaceae. 1 Argentina

Rhytachne Desv. 1825. Poaceae. 12 trop. America and Africa

Rhyticalymma Bremek. = Justicia L. Acanthaceae

Rhyticarpus Sond. = Anginon Raf. Apiaceae

Rhyticaryum Becc. 1877. Icacinaceae. 12 Moluccas (1, R. oleraceum), New Guinea (12); R. longifolium—also in northeastern Australia and Melanesia (New Britain, New Ireland, Admiralty Is., Solomon Is.)

Rhyticocos Becc. = Syagrus Mart. Arecaceae

Rhytidanthera (Planch.) Tiegh. 1904. Sauvagesiaceae. 6 Colombia

Rhytidocaulon Bally. 1962. Asclepiadaceae. 6 trop. Northeast Africa, Arabian Peninsula

Rhytidophyllum Mart. 1829 (Gesneria L.). Gesneriaceae. 20 West Indies

Ribes L. 1753 (incl. Grossularia Hill). Grossulariaceae. 150 North, Central and western South America from Canada throughout Andes to Tierra del Fuego, temp. Eurasia (incl. Middle Asia, Siberia, and Far East), Northwest Africa

Richardia L. 1753. Rubiaceae. 15 trop. and subtrop. America from Texas to Argentina, West Indies; now widely naturalized

Richardsiella Elffers et Kenn.-O'Byrne. 1957. Poaceae. 1 Zambia

Richea R. Br. 1810. Epacridaceae. 11 southeastern Australia, Tasmania

Richella A. Gray. 1852. Annonaceae. 3 southern China (1), Borneo (1), New Caledonia, and Fiji (1)

Richeria Vahl. 1797. Euphorbiaceae. 8 trop. South America

Richeriella Pax et K. Hoffm. 1922. Euphorbiaceae. 2 Hainan, Philippines, Borneo, Malay Peninsula

Richteria Kar. et Kir. 1842. Asteraceae. 25–30 montane Central and Middle Asia

Ricinocarpodendron Boehm = Dysoxylum Blume. Meliaceae

Ricinocarpos Desf. 1817. Euphorbiaceae. 16 Australia, Tasmania, New Caledonia (1)

Ricinodendron Muell. Arg. 1864. Euphorbiaceae. 2: R. rautanenii—from southern Angola and northern Namibia eastward to Tanzania and Mozambique; R. heudelotii—from Nigeria eastward to Sudan and Uganda, south to Angola and Mozambique

Ricinus L. 1753. Euphorbiaceae. 1 (R. communis) Mediterranean, Northeast, and trop. East Africa

Ricotia L. 1763. Brassicaceae. 9 Europe (2), Mediterranean

Riddellia Nutt. = Psilostrophe DC. Asteraceae

Ridleyella Schltr. 1913. Orchidaceae. 1 New Guinea

Ridolfia Moris. 1842. Apiaceae. 1 West and South Europe, northwest and north Africa, Asia Minor

Riedelia Oliv. 1883. Zingiberaceae. c. 70 Malesia from eastern Moluccas to Solomon Is., esp. New Guinea

Riedeliella Harms. 1903. Fabaceae. 2 Brazil, Paraguay

Riencourtia Cass. 1818. Asteraceae. 8 northern South America

Riesenbachia C. Presl = Lopezia Cav. Onagraceae

Rigidella Lindl. 1840. Iridaceae. 4 Mexico, Guatemala, Peru

Rigiolepis Hook. f. = Vaccinium L. Ericaceae

Rigiopappus A. Gray. 1865. Asteraceae. 1 southwestern U.S.

Rikliella J. Raynal. 1973 (Lipocarpha R. Br.). Cyperaceae. 4 trop. Africa, Madagascar, South and Southeast Asia, northern and eastern Australia

Rimacola Rupp. 1942. Orchidaceae. 1 Australia (New South Wales)

Rimaria N. E. Br. = Gibbaeum N. E. Br. Aizoaceae

Rindera Pall. 1771 (Cynoglossum L.). Boraginaceae. 14 Asia Minor, Transcaucasia, Iraq, Iran, Middle Asia, China (1, R. tetraspis); (European species = Cynoglossum L.)

Ringentiarum Nakai = Arisaema Mart. Araceae

Rinorea Aubl. 1775. Violaceae. c. 300 pantropics

Rinoreocarpus Ducke. 1925. Violaceae. 1 Brazil (Amazonia)

Rinzia Schauer. 1843. Myrtaceae. 12 western Australia

Riocreuxia Decne. 1844. Asclepiadaceae. 9 trop. and South Africa (8), Nepal (1)

Ripogonum J. R. Forst. et G. Forst. 1775. Ripogonaceae. 6 New Guinea, northern and eastern Australia, New Zealand, Chatham Is., Stewart I.

Riqueuria Ruiz et Pav. 1794. Rubiaceae. 1 Peru

Risleya King et Pantl. 1898. Orchidaceae. 1 western China, eastern Himalayas

Ristantia Peter G. Wilson et J. T. Waterh. 1982 (Tristania R. Br.). Myrtaceae. 3 Australia (Queensland)

Ritchiea R. Br. ex G. Don. 1831. Capparaceae. c. 30 trop. Africa

Ritonia Benoist. 1962. Acanthaceae. 3 Madagascar

Rivasgodaya E. Chueca = Genista L. Fabaceae

Rivea Choisy. 1833. Convolvulaceae. 3 India, Nepal, Sri Lanka, Southeast Asia

Rivina L. 1753. Petiveriaceae. 1 (P. humilis) trop. America; naturalized in Madagascar, Sri Lanka, Southeast Asia, Malesia, Australia

Robbairea Boiss. = Polycarpaea Lam. Caryophyllaceae

Robeschia Hochst. ex O. E. Schulz. 1865. Brassicaceae. 1 Syria, Lebanon, Israel, Sinai Peninsula eastward to Iran, Afghanistan, Pakistan

Robinia L. 1753. Fabaceae. 10 U.S., Mexico

Robinsonella Rose et Baker f. 1897. Malvaceae. 16 Mexico (14), Guatemala, El Salvador, Honduras, Nicaragua, Costa Rica, Panama

Robinsonia DC. 1833. Asteraceae. 6 Juan Fernández Is.

Robinsoniodendron Merr. = Maoutia Wedd. Urticaceae

Robiquetia Gaudich. 1829. Orchidaceae. 20–25 trop. Asia, Malesia from Malay Peninsula to New Guinea, Solomon Is., Australia (Queensland), Micronesia, Melanesia to Fiji and Tonga

Roborowskia Batalin. 1893 (Corydalis DC.). Fumariaceae. 1 (R. mira) Middle Asia

Robynsia Hutch. 1931. Rubiaceae. 1 trop. West Africa

Robynsiella Suess. = Centemopsis Schinz. Amaranthaceae

Robynsiochloa Jacq. = Fél.1960 (Rottboellia L. f.). Poaceae. 1 trop. Africa

Robynsiophyton R. Wilczek. 1953. Fabaceae. 1 Zaire, Angola, Zambia

Rochea DC. = Crassula L. Crassulaceae

Rochefortia Sw. 1788. Boraginaceae. 6–7 southern Mexico, Central America, Colombia, West Indies

Rochelia Rchb. 1824. Boraginaceae. 20 Eurasia, Mediterranean, China

Rochonia DC. 1836. Asteraceae. 4 Madagascar

Rockia Heimerl = Pisonia L. Nyctaginaceae

Rockinghamia Airy Shaw. 1966. Euphorbiaceae. 2 Australia (northeastern Queensland)

Rodgersia A. Gray. 1858. Saxifragaceae. 6 Tibet, eastern Himalayas, East Asia

Rodriguezia Ruiz et Pav. 1794. Orchidaceae. 30 trop. America from Costa Rica to Peru and Brazil

Rodrigueziella Kuntze. 1891. Orchidaceae. 2 Brazil

Rodriqueziopsis Schltr. 1920. Orchidaceae. 2 Brazil

Roebelia Engel. 1865. Arecaceae. 1 Colombia

Roegneria K. Koch = Elymus L. Poaceae

Roella L. 1753. Campanulaceae. 25 South Africa

Roemeria Medik. 1792. Papaveraceae. 8 Mediterranean (excl. Sardinia, Sicily, and Malta), Crimea, Caucasus, West and Middle Asia to Afghanistan

Roentgenia Urb. 1916. Bignoniaceae. 2 northern South America

Roeperocharis Rchb. f. 1881. Orchidaceae. 5 trop. Africa

Roezliella Schltr. 1918. Orchidaceae. 5 Colombia

Rogeria J. Gay ex Delile. 1826. Pedaliaceae. 6 Brazil, trop. and South Africa

Rogersonanthus Maguire et B. M. Boom. 1989. Gentianaceae. 3 Venezuela, Guyana, Brazil

Roggeveldia Goldblatt. 1980. Iridaceae. 1 South Africa (Cape)

Rogiera Planch. = Rondeletia L. Rubiaceae

Rohdea Roth. 1821. Convallariaceae. 2–3 continental China, Korea, Japan

Roigella Borhidi et Fernández. 1981 (Rondeletia L.). Rubiaceae. 1 Cuba

Rojasia Malme. 1905. Asclepiadaceae. 1 Brazil

Rojasianthe Standl. et Steyerm. 1940. Asteraceae. 1 Central America

Rojasimalva Fryxell. 1984. Malvaceae. 1 Venezuela

Rolandra Rottb. 1775. Asteraceae. 1 Honduras, Costa Rica, Panama, trop. South America, West Indies from Puerto Rico to the Lesser Antilles

Roldana La Llave ex Lex. 1825 (Senecio L.). Asteraceae. 48 Mexico, Central America

Rolfeella Schltr. = Benthamia A. Rich. Orchidaceae

Rollandia Gaudich. 1829. Lobeliaceae. 8 Hawaiian Is.

Rollinia A. St.-Hil. 1825. Annonaceae. c. 80 Central and trop. South America south to Paraguay and northern Argentina; West Indies

Rolliniopsis Saff. = Rollinia A. St.-Hil. Annonaceae

Rollinsia Al = Shehbaz. 1982 (Thelypodium Endl.). Brassicaceae. 1 southern U.S., Mexico

Romanoa Trevis. 1848. Euphorbiaceae. 1 eastern Brazil

Romanschulzia O. E. Schulz. 1933. Brassicaceae. 12 Mexico, Central America south to Panama

Romanzoffia Cham. 1820. Hydrophyllaceae. 4 western North America from Aleutian Is. and Alaska to California

Romeroa Dugand. 1952. Bignoniaceae. 1 Colombia

Romnalda P. F. Stevens. 1978. Dasypogonaceae. 3 Japen I., New Guinea, New Britain I., Australia (Queensland)

Romneya Harv. 1845. Papaveraceae. 2 California, northwestern Mexico

Romulea Maratti. 1772. Iridaceae. 95 Southwest Europe, Canary Is., Mediterranean, trop. and South (c. 70) Africa

Rondeletia L. 1753. Rubiaceae. c. 125 trop. America, West Indies

Rondonanthus Herz. 1931 (Paepalanthus Kunth). Eriocaulaceae. 6 Venezuela (6), Guyana (1), Brazil (1)

Ronnbergia E. Morren et André. 1874. Bromeliaceae. 9 Costa Rica, Panama, Colombia, Peru

Roodia N. E. Br. = Argyroderma N. E. Br. Aizoaceae

Rooseveltia O. F. Cook = Euterpe Mart. Arecaceae

Roridula Burm. f. ex L. 1764. Roridulaceae. 2 South Africa (southwestern Cape)

Roripella (Maire) Greuter et Burdet. 1984. Brassicaceae. 1 Morocco

Rorippa Scop. 1760. Brassicaceae. c. 70 cosmopolitan

Rosa L. 1753. Rosaceae. c. 250 (or 100, or c. 400) temp. Northern Hemisphere and montane areas of tropics

Roscheria H. A. Wendl. ex Balf. f. 1877. Arecaceae. 1 Seychelles (Mahé and Silhouette Is.)

Roscoea Sm. 1805. Zingiberaceae. 15–17 Himalayas from Kashmir to Bhutan, India, southwestern and southern China

Rosenbergiodendron Fagerl. = Randia L. Rubiaceae

Rosenia Thunb. 1800. Asteraceae. 4 South Africa

Roseodendron Miranda = Tabebuia Gomes ex DC. Bignoniaceae

Roshevitzia Tzvelev = Diandrochloa de Winter. Poaceae

Rosifax C. C. Towns. 1991. Amaranthaceae. 1 Somalia

Rosilla Less. = Dyssodia Cav. Asteraceae

Rosmarinus L. 1753. Lamiaceae. 2 Mediterranean; R. officinalis largely cultivated

Rossioglossum (Schltr.) Garay et G. C. Kenn. 1976. Orchidaceae. 6 Mexico, Guatemala, Costa Rica, Panama

Rostellularia Rchb. = Justicia L. Acanthaceae

Rostkovia Desv. 1809. Juncaceae. 2 New Zealand, Campbell Is., Auckland Is., temp. South America to Tierra del Fuego, Falkland Is., South Georgia I., Tristan da Cunha

Rostraria Trin. 1822 (Koeleria Pers.). Poaceae. 15 Mediterranean, North Africa, West Asia, montane South America; R. cristata—Crimea, Caucasus, and Middle Asia

Rostrinucula Kudo. 1929 (Elsholtzia Willd.). Lamiaceae. 2 East Asia

Rosularia (DC.) Stapf. 1923. Crassulaceae. 37 Europe, North Africa, West Asia to northwestern China, Himalayas from Kashmir to Nepal; Caucasus and Middle Asia (c. 20)

Rotala L. 1771. Lythraceae. 40–50 trop., subtrop., and warm–temp. regions (incl. 2 southern Caucasus, Middle Asia)

Rothia Pers. 1807. Fabaceae. 2 arid areas of Africa (1, R. hirsuta), South Asia (from Baluchistan to Indochina), Malesia, and Australia

Rothmaleria Font Quer. 1940. Asteraceae. 1 southern Spain

Rothmannia Thunb. 1776. Rubiaceae. 25–30 trop. and South Africa, Seychelles, Thailand, Cambodia, Vietnam

Rothrockia A. Gray. 1885. Asclepiadaceae. 3 southwestern U.S.

Rottboellia L. f. 1782. Poaceae. 3 trop. Old World

Rotula Lour. 1790. Ehretiaceae. 1 (R. aquatica) eastern Brazil and trop. Old World

Roubieva Moq. = Chenopodium L. Chenopodiaceae

Roucelea (L.) Dum. = Campanula L. Campanulaceae

Roucheria Planch. 1847. Hugoniaceae. 8 trop. South America

Roulinia Decne. = Cynanchum L. Asclepiadaceae

Rouliniella Vail = Telminostelma Fourn. Asclepiadaceae

Roupala Aubl. 1775. Proteaceae. 50 trop. America

Roupellina (Baill.) Pichon = Strophanthus DC. Apocynaceae

Rourea Aubl. 1775. Connaraceae. c. 90 trop. America (42, Mexico, Guatemala, Honduras, El Salvador, Nicaragua,

Costa Rica, Panama, Colombia, Venezuela, Ecuador, Peru, Bolivia, Brazil), West Indies, trop. Africa, Madagascar, trop. Asia from India and Sri Lanka through Andaman and Nicobar Is., Burma, Indochina to southern China, Taiwan, Malesia (7, from Malay Peninsula and Sumatra to New Caledonia), northeastern Australia, New Caledonia, New Hebrides (Vanuatu), Fiji, Samoa

Roureopsis Planch. 1850 (Rourea Aubl.). Connaraceae. 11 trop. West Africa (2), Burma, southern China, Indochina, Malesia (4, Malay Peninsula, Sumatra, Java, Borneo)

Roussea Sm. 1789. Rousseaceae. 1 Mauritius

Rousseauxia DC. 1828. Melastomataceae. 13 Madagascar

Rousselia Gaudich. 1830. Urticaceae. 3 Central America, Colombia, West Indies

Rouya Coincy. 1901 (Thapsia L.). Apiaceae. 1 Corsica, Sardinia, North Africa

Roycea C. A. Gardner. 1948. Chenopodiaceae. 3 southwestern Australia

Royena L. = Diospyros L. Ebenaceae

Roylea Wall. ex Benth. 1829. Lamiaceae. 1 Himalayas from Kashmir to Nepal

Roystonea O. F. Cook. 1900. Arecaceae. 10–12 southern Florida peninsula, Greater Antilles, southeastern Mexico, Honduras, northeastern Venezuela, Trinidad, Guyana

Ruagea Karst. 1863. Meliaceae. 6 Guatemala (1, R. insignis), Costa Rica, Panama, Colombia, Venezuela, Ecuador, Peru, Bolivia

Rubacer Rydb. = Rubus L. Rosaceae

Rubia L. 1753. Rubiaceae. c. 40 West, Central, and Southeast Europe; Mediterranean; trop. and South Africa; temp. Asia (incl. c. 20, Caucasus, Middle Asia, southern Siberia, and Russian Far East); Himalayas; trop. America

Rubiteucris Kudo. 1929. Lamiaceae. 1 southeastern Tibet, Nepal, Bhutan, India (Sikkim), southwestern China, Taiwan

Rubrivena M. Král = Polygonum L. Polygonaceae

Rubus L. 1753. Rosaceae. c. 250 (or c. 700) cosmopolitan, but chiefly Northern Hemisphere (c. 300, North America; 150, China)

Ruckeria DC. = Euryops (Cass.) Cass. Asteraceae

Rudbeckia L. 1753. Asteraceae. 15 North America

Ruddia Yakovlev. 1971 (Ormosia G. Jacks.). Fabaceae. 1 southern China, Vietnam

Rudgea Salisb. 1807. Rubiaceae. 100–150 trop. America, West Indies

Rudolfiella Hoehne. 1944 (Bifrenaria Lindl.). Orchidaceae. 9 West Indies, trop. South America

Rudua F. Maek. = Vigna Savi. Fabaceae

Ruehssia Karst. ex Schltdl. = Marsdenia R. Br. Asclepiadaceae

Ruellia L. 1753. Acanthaceae. c. 150 tropics and a few species in temp. North America

Ruelliola Baill. = Brillantaisia P. Beauv. Acanthaceae

Ruelliopsis C. B. Clarke. 1899. Acanthaceae. 2–3 trop. and South Africa

Rufodorsia Wiehler. 1975. Gesneriaceae. 3 Costa Rica, Panama

Rugelia Shuttlew. ex Chapman. 1860 (Senecio L.). Asteraceae. 1 southeastern U.S.

Ruilopezia Cuatrec. 1976. Asteraceae. 25 Venezuela

Ruizia Cav. 1786. Sterculiaceae. 3 Réunion I.

Ruizodendron R. E. Fr. 1936. Annonaceae. 1 (R. ovale) Peru, Bolivia

Ruizterania Marc.-Berti. 1969 (Qualea Aubl.). Vochisiaceae. 19 trop. South America

Rulingia R. Br. 1820. Sterculiaceae. 20 Madagascar, Australia

Rumex L. 1753. Polygonaceae. c. 200 cosmopolitan

Rumfordia DC. 1836. Asteraceae. 6–10 Mexico, Central America

Rumia Hoffm. 1816 (Trinia Hoffm.). Apiaceae. 1 (R. orithmifolia) Crimea

Rumicastrum Ulbr. 1934. Portulacaceae. c. 50 Australia

Rumicicarpus Chiov. = Triumfetta L. Tiliaceae

Rungia Nees. 1832. Acanthaceae. c. 50 trop. Africa, India, Nepal, China, Indochina, western Malesia, Sulawesi

Runyonia Rose = Manfreda Salisb. Agavaceae

Rupertia Grimes. 1990 (Psoralea L.). Fabaceae. 3 western North America from Vancouver I. to north–central Baja California

Rupicapnos Pomel. 1860. Fumariaceae. 7 southern Spain (1, R. africana), Morocco, Algeria, Tunisia

Rupicola J. H. Maiden et E. Betche. 1899. Epacridaceae. 3 southeastern Australia

Rupiphila Pimenov et Lavrova. 1986. Apiaceae. 1 (R. tachiroei) southern Russian Far East, northeastern China, Korea, Japan

Ruppia L. 1753. Ruppiaceae. 2–10 temp. and subtrop. regions and montane tropics; Russia (5, brackish lakes and sea coasts)

Ruprechtia C. A. Mey. 1840. Polygonaceae. 19 trop. America from Mexico to northern Argentina and Uruguay, Trinidad

Rusbya Britton. 1893. Ericaceae. 1 Bolivia

Rusbyanthus Gilg = Macrocarpaea Gilg. Gentianaceae

Rusbyella Rolfe ex Rusby. 1896. Orchidaceae. 1 Bolivia

Ruschia Schwantes. 1926. Aizoaceae. c. 350 arid areas in Namibia, South Africa, Lesotho

Ruschianthemum Friedrich. 1960. Aizoaceae. 2 southwestern Namibia, western South Africa

Ruschianthus L. Bolus. 1960. Aizoaceae. 1 southern Namibia

Ruscus L. 1753. Ruscaceae. 7 Azores; Mediterranean; West, Central, and Southeast Europe; West Asia; 4 Crimea and western Transcaucasia

Ruspolia Lindau. 1896. Acanthaceae. 4 trop. Africa

Russelia Jacq. 1760. Scrophulariaceae. 52 trop. America from Mexico to Colombia, Cuba

Russowia C. Winkl. 1892. Asteraceae. 1 Middle Asia, western China

Rustia Klotzsch. 1846. Rubiaceae. 15 trop. America, West Indies

Ruta L. 1753. Rutaceae. 7 Macaronesia, Europe, Mediterranean, West Asia

Rutaneblina Steyerm. et Luteyn. 1984. Rutaceae. 1 Venezuela

Ruthalicia C. Jeffrey. 1962. Cucurbitaceae. 2 trop. West Africa

Ruthea Bolle = Rutheopsis A. Hansen et Kunkel. Apiaceae

Rutheopsis A. Hansen et Kunkel. 1976. Apiaceae. 1 Canary Is.

Ruthiella Steenis. 1965. Lobeliaceae. 4 New Guinea

Rutidea DC. 1807. Rubiaceae. 22 trop. Africa, Madagascar

Rutidosis DC. 1838. Asteraceae. 7 Australia

Rutilia Vell. 1825. Sapindaceae. 1 Brazil

Ruttya Harv. 1842. Acanthaceae. 3 trop. and South Africa

Ruyschia Jacq. 1760. Marcgraviaceae. 6 trop. America

Ryania Vahl. 1796. Flacourtiaceae. 8 trop. America from Nicaragua to Peru and Brazil, Trinidad

Rylstonea R. T. Baker. 1899 (Homoranthus Cunn. ex Schauer). Myrtaceae. 1 eastern Australia

Ryncholeucaena Britton et Rose = Leucaena Benth. Fabaceae

Ryparosa Blume. 1826. Flacourtiaceae. 18 Andaman and Nicobar Is., Malesia (Sumatra, Malay Peninsula, Java, Borneo, Philippines, and New Guinea), Australia (Queensland)

Ryssopterys Blume ex A. Juss. = Rhyssopteris Blume ex A. Juss.

Rytidocarpus Coss. 1889. Brassicaceae. 1 Morocco

Rytidosperma Steud. = Notodanthonia Zotov. Poaceae

Rytidostylis Hook. et Arn. 1840. Cucurbitaceae. 9 trop. America, West Indies

Rytigynia Blume. 1850. Rubiaceae. c. 70 trop. and South Africa, Madagascar

Rytilix Hitchc. = Hackelochloa Kuntze. Poaceae

Rzedowskia Medrano. 1981. Celastraceae. 1 Mexico

S

Saba (Pichon) Pichon. 1953. Apocynaceae. 3 trop. Africa: Senegal, Gambia, Guinea, Sierra Leone, Ivory Coast, Mali, Ghana, Togo, Nigeria, Cameroun, Chad, Central African Rep., Gabon, Congo, Zaire, Burundi, Angola, Ethiopia, Sudan, Somalia, Kenya, Uganda, Tanzania, Malawi, Mozambique; Comoro Is., Madagascar

Sabal Adans. 1763. Arecaceae. 15 southeastern U.S., southwestern and southeastern Mexico, Guatemala, Nicaragua, Panama, northern Colombia, northern Venezuela, West Indies to Bermuda Is.

Sabatia Adans. 1763. Gentianaceae. 21 eastern and southeastern U.S., extending to Kansas, Oklahoma, and central Texas; Mexico, West Indies: Bahamas, Cuba, Hispaniola

Sabaudia Buscal. et Muschler. 1913. Lamiaceae. 2 Northeast and trop. Southeast Africa

Sabaudiella Chiov. 1929. Convolvulaceae. 1 trop. Northeast Africa

Sabazia Cass. 1827. Asteraceae. 13 Mexico, Central America to Colombia (2)

Sabbata Vell. 1825. Asteraceae. 2 Brazil

Sabia Colebr. 1819. Sabiaceae. 30 Himalayas from Kashmir to Bhutan, India, Burma, China (c. 25), Indochina, Malesia, Solomon Is.

Sabicea Aubl. 1775. Rubiaceae. 135 trop. America, trop. Africa, Madagascar (4)

Sabina Mill. = Juniperus L. Cupressaceae

Sabinea DC. 1825. Fabaceae. 3 Central America, West Indies

Saccardophytum Speg. = Benthamiella Speg. Solanaceae

Saccellium Humb. et Bonpl. 1806. Ehretiaceae. 3 trop. South America

Saccharum L. 1753. Poaceae. 5 tropics and subtropics; 2 sp. extending to Middle Asia

Saccia Naudin. 1889. Convolvulaceae. 1 Bolivia

Saccifolium Maguire et Pires. 1978. Saccifoliaceae. 1 Venezuela

Sacciolepis Nash. 1901. Poaceae. c. 30 tropics and subtropics, esp. Africa

Saccocalyx Coss. et Durieu. 1853 (Satureja L.). Lamiaceae. 1 Morocco, Algeria

Saccoglossum Schltr. 1912. Orchidaceae. 2 New Guinea

Saccolabiopsis J. J. Sm. 1918. Orchidaceae. 12 India, Indochina, Malesia (incl. Java—1, S. bakhuizenii) eastward to New Guinea, south to northern Australia, Fiji (1)

Saccolabium Blume. 1825. Orchidaceae. 4 Java (3 endemics), Sumatra (1)

Saccolena Gleason = Salpinga Mart. ex DC. Melastomataceae

Saccopetalum Benn. = Miliusa Lesch. ex A. DC. Annonaceae

Saccularia Kellogg = Gambelia Nutt. Scrophulariaceae

Sachokiella Kolak. = Campanula L. Campanulaceae

Sachsia Griseb. 1866. Asteraceae. 3–4 Florida peninsula, Bahama Is., Cuba

Sacleuxia Baill. 1890. Asclepiadaceae. 2 trop. East Africa

Sacoglottis Mart. 1827. Humiriaceae. 10 Central (Costa Rica, Panama) and trop. South America (8), trop. West Africa (1)

Sacoila Raf. 1837 (Stenorhynchos Rich. ex Spreng.). Orchidaceae. 10 subtrop. America

Sacosperma G. Taylor. 1944. Rubiaceae. 2 trop. Africa

Sadiria Mez. 1902. Myrsinaceae. 5 India

Saffordiella Merr. = Myrtella F. Muell. Myrtaceae

Sageraea Dalzell. 1851. Annonaceae. 9 India, Sri Lanka, Indochina, Philippines

Sageretia Brongn. 1827. Rhamnaceae. c. 40 trop. and subtrop. America; Somalia; West, Southwest, South, and Southeast Asia (from Asia Minor, Palestine, and Sinai Peninsula to Taiwan); S. thea— Middle Asia

Sagina L. 1753. Caryophyllaceae. 20–30 temp. Northern Hemisphere, montane trop. East Africa, Tibet, Himalayas, New Guinea, Australia, New Zealand, Andes

Sagittaria L. 1753. Alismataceae. 25 cosmopolitan, but chiefly America

Saglorithys Rizzini = Justicia L. Acanthaceae

Sagotia Baill. 1860. Euphorbiaceae. 2 Costa Rica, Panama, Venezuela, the Guianas, northern Brazil

Sahagunis Liebm. 1851. Moraceae. 3 trop. America

Saintpaulia H. A. Wendl. 1893. Gesneriaceae. 20 trop. East Africa

Saintpauliopsis Staner. 1934 (Staurogyne Wall.). Acanthaceae. 1 Gabon, Zaire, Rwanda, Burundi, Tanzania

Sairocarpus D. A. Sutton. 1988. Scrophulariaceae. 13, southwestern North America

Sajanella Soják. 1980. Apiaceae. 1 (S. momstrosa) southern West and East Siberia, northern Mongolia

Sajania Pimenov = Sajanella Soják. Apiaceae

Sakoanala R. Vig. 1951. Fabaceae. 3 Madagascar

Salacca Reinw. 1826. Arecaceae. 15 India (1, Assam—S. secunda), Burma, Indochina, Java, Sumatra, Borneo, Philippines

Salacia L. 1771. Celastraceae. c. 150 pantropics

Salacicratea Loes. 1910 (Salacia L.). Celastraceae. 1 Australia (Queensland)

Salacighia Loes. 1940. Celastraceae. 1 trop. West Africa

Salaciopsis Baker f. 1921. Celastraceae. 5 New Caledonia

Salaxis Salisb. 1802. Ericaceae. 8 South Africa (southwestern and southern Cape)

Salazaria Torr. = Scutellaria L. Lamiaceae

Saldanhaea Bureau = Cuspidaria DC. Bignoniaceae

Saldinia A. Rich. ex DC. 1830. Rubiaceae. 2 Madagascar

Salicornia L. 1753. Chenopodiaceae. 14 cosmopolitan, excl. Australia

Salix L. 1753. Salicaceae. c. 300 Arctic, temp. and warm–temp. regions of the Northern Hemisphere, few species in tropics, temp. South America (1), South Africa (1)

Salmea DC. 1813. Asteraceae. 10 Mexico, Guatemala, Belize, El Salvador, Honduras, Nicaragua, Costa Rica, Panama, Colombia, Ecuador, Peru, Venezuela, Brazil, Bolivia, Paraguay, Argentina, West Indies (esp. Cuba)

Salmeopsis Benth. = Salmea DC. Asteraceae

Salomonia Lour. 1790. Polygalaceae. 8–14 southern and eastern India, Sri Lanka, Nepal, Bhutan, Bangladesh, Burma, China, southern Korea, Japan, Indochina, Taiwan, Ryukyu Is., Mariana and Caroline Is., Malesia (2), northern and northeastern Australia

Salpianthus Humb. et Bonpl. 1807. Nyctaginaceae. 1–4 Mexico, Central America, Colombia, Venezuela, Ecuador

Salpichroa Miers. 1845. Solanaceae. 15 trop. and subtrop. America from southwestern U.S. to northern Argentina and Uruguay; widely cultivated and naturalized

Salpiglossis Ruiz et Pav. 1794. Solanaceae. 2 Chile

Salpinctes Woodson. 1931. Apocynaceae. 2 Venezuela

Salpinctium F. J. Edwards. 1989 (or = Asystasia Blume). Acanthaceae. 3 South Africa

Salpinga Mart. ex DC. 1828. Melastomataceae. 10 trop. South America

Salpingantha Hort. ex Lem. = Salpixantha Hook. Acanthaceae

Salpingostylis Small = Calydorea Herb. Iridaceae

Salpistele Dressler. 1979. Orchidaceae. 6 Costa Rica, Panama, Ecuador (2)

Salpixantha Hook. 1845. Acanthaceae. 1–2 Jamaica

Salsola L. 1753. Chenopodiaceae. c. 240 Europe, Africa, Asia, esp. Middle Asia; Russia—from European part to Far East

Saltera Bullock. 1958. Penaeaceae. 1 South Africa

Saltia R. Br. ex Moq. 1849. Amaranthaceae. 1 Sinai and Arabian Peninsula

Salvadora L. 1753. Salvadoraceae. 5 trop. and South (2) Africa, Madagascar, Mascarene Is., Arabian Peninsula, trop. Asia eastward to China

Salvadoropsis H. Perrier. 1944. Celastraceae. 1 Madagascar

Salvertia A. St.-Hil. 1820. Vochysiaceae. 1 campos of southern Brazil

Salvia L. 1753. Lamiaceae. 800–900 temp. and trop. regions, esp. Mediterranean (c. 130), Southwest Asia, China (c. 80), Himalayas; c. 85 Crimea, Caucasus, Middle Asia, and adjacent areas

Salviastrum Scheele = Salvia L. Lamiaceae

Salweenia Baker f. 1935. Fabaceae. 1 Tibet

Salzmannia DC. 1830. Rubiaceae. 1 eastern Brazil

Samadera Gaertn. = Quassia L. Simaroubaceae

Samaipaticereus Cárdenas. 1952. Cactaceae. 1 Bolivia

Samanea (DC.) Merr. = Albizia Durazz. Fabaceae

Sambucus L. 1753. Sambucaceae. c. 40 temp. and subtrop. regions, chiefly East Asia and eastern North America, trop. East Africa (1), Australia and Tasmania (2); Russia (European part, West and East Siberia, Far East), Caucasus, Middle Asia

Sameraria Desv. 1814. Brassicaceae. 12 Caucasus, Middle Asia, Turkey, Iraq, Iran, Afghanistan, Pakistan

Samolus L. 1753. Primulaceae. 10 cosmopolitan; S. valerandii—European part of Russia, Caucasus, Middle Asia

Sampantaea Airy Shaw. 1972. Euphorbiaceae. 1 Southeast Asia

Samuela Trel. 1902. Agavaceae. 2 southern U.S., Mexico

Samuelssonia Urb. et Ekman. 1929. Acanthaceae. 1 Haiti

Samyda Jacq. 1760. Flacourtiaceae. 9 Mexico (2), West Indies

Sanango Bunting et Duke. 1961. Buddlejaceae. 1 Peru

Sanblasia L. Andersson. 1984. Marantaceae. 1 eastern Panama

Sanchezia Ruiz et Pav. 1974. Acanthaceae. 30 trop. South America

Sanctambrosia Skottsb. ex Kuschel. 1962. Caryophyllaceae. 1 Chile (San Ambrosio I.)

Sandbergia Greene = Halimolobos Tausch. Brassicaceae

Sandemania Gleason. 1939. Melastomataceae. 1 Venezuela, Peru, Brazil

Sanderella Kuntze. 1891. Orchidaceae. 1 Brazil

Sandersonia Hook. 1853. Colchicaceae. 1 South Africa (Natal)

Sandoricum Cav. 1789. Meliaceae. 5 Malesia (3, Bornean endemic)

Sandwithia Lanj. 1933 (Sagotia Baill.). Euphorbiaceae. 1 Venezuela, Guyana, French Guiana, northern Brazil

Sandwithiodoxa Aubrév. et Pellegr. = Pouteria Aubl. Sapotaceae

Sanguinaria L. 1753. Papaveraceae. 1 eastern North America

Sanguisorba L. 1753. Rosaceae. 30 cold and temp. Northern Hemisphere; Russia (European part, Siberia, Far East)

Sanhilaria Baill. = Paragonia Bur. Bignoniaceae

Sanicula L. 1753. Apiaceae. c. 40 North and Central America; trop. western and temp. South America; North, trop., and South Africa; Madagascar; Eurasia; Oceania; absent in Australia and New Zealand

Saniella Hilliard et B. L. Burtt. 1978. Hypoxidaceae. 1 South Africa (Natal, Draco Hills), Lesotho

Sanmartinia Buchinger = Eriogonum Michx. Polygonaceae

Sanseverinia Petagna = Sansevieria Thunb. Dracaenaceae

Sansevieria Thunb. 1794. Dracaenaceae. 12 trop. and South Africa, Madagascar, Arabian Peninsula, South Asia

Santaloides Schellenb. = Rourea Aubl. Connaraceae

Santalum L. 1753. Santalaceae. 8–9 Java, Sulawesi, Lesser Sunda Is., New Guinea, Australia, Oceania eastward to Hawaiian Is., eastern Polynesia and Juan Fernández Is.

Santapaua N. P. Balakr. et K. Subramanyam = Hygrophila R. Br. Acanthaceae

Santiria Blume. 1850. Burseraceae. 24 trop. Old World

Santiriopsis Engl. = Santiria Blume. Burseraceae

Santisukia Brummitt. 1992. Bignoniaceae. 2 Thailand

Santolina L. 1753. Asteraceae. 18 Mediterranean

Santomasia N. Robson. 1981 (Hypericum L.). Hypericaceae. 1 Mexico and Guatemala

Santosia R. M. King et H. Rob. 1980. Asteraceae. 1 Brazil

Sanvitalia Gualt. ex Lam. 1792. Asteraceae. 7 southwestern U.S., Mexico, Central America

Saphesia N. E. Br. 1932. Aizoaceae. 1 South Africa (Cape)

Sapindopsis F. C. How et C. N. Ho = Aphania Blume. Sapindaceae

Sapindus L. 1753. Sapindaceae. 13 trop. and subtrop. Asia, Oceania, America (absent in Africa and Australia)

Sapium P. Browne. 1756. Euphorbiaceae. c. 125. tropics and subtropics, temp. South America

Saponaria L. 1753. Caryophyllaceae. 30 temp. Eurasia, esp. Mediterranean; 8 European part of Russia, Caucasus, West Siberia, and Middle Asia

Saposhnikovia Schischk. 1951. Apiaceae. 1 East Siberia, Russian Far East, East Asia

Sapota Mill. = Manilkara Adans. Sapotaceae

Sapphoa Urb. 1922. Acanthaceae. 2 Cuba

Sapranthus Seem. 1866. Annonaceae. 9 Mexico, Central America

Sapria Griff. 1849. Rafflesiaceae. 2 Bhutan, northeastern India, Burma, southern China, Thailand, Indochina

Saprosma Blume. 1827. Rubiaceae. 25 eastern Himalayas, India, Sri Lanka, Andaman Is., Burma, southwestern and southern China, Indochina, Malesia

Sapucaya Knuth = Lecythis Loefl. Lecythidaceae

Saraca L. 1767. Fabaceae. 8 South and Southeast Asia

Saracha Ruiz et Pav. 1794. Solanaceae. 3 Andes from Venezuela to Bolivia

Saranthe (Regel et Koern.) Eichler. 1884. Marantaceae. 10 Brazil

Sararanga Hemsl. 1894. Pandanaceae. 2 Philippines, New Guinea, Admiralty, and Solomon Is.

Sarawakodendron Ding Hou. 1967. Celastraceae. 1 Borneo

Sarcandra Gardner. 1845. Chloranthaceae. 2 India, Sri Lanka, Burma, China, Korean peninsula, Japan, Indochina, Malay Peninsula, Sumatra, Borneo, Lesser Sunda Is., Philippines, New Guinea

Sarcanthidion Baill. = Citronella D. Don. Icacinaceae

Sarcanthopsis Garay. 1972. Orchidaceae. 5–7 trop. Asia, Malesia, Caroline Is. (Palau), Bougainville and Solomon Is., Vanuatu, Fiji

Sarcanthus Lindl. = Cleisostoma Blume. Orchidaceae

Sarcathria Raf. = Halocnemum M. Bieb. Chenopodiaceae

Sarcaulus Radlk. 1882. Sapotaceae. 5 Costa Rica, Panama, Colombia, Venezuela, Guyana, Surinam, French Guiana, Ecuador, Peru, Bolivia, Brazil

Sarcobatus Nees. 1841. Chenopodiaceae. 2 western U.S.

Sarcobodium Beer = Bulbophyllum Thouars. Orchidaceae

Sarcoca Raf. = Phytolacca L. Phytolaccaceae

Sarcocapnos DC. 1821. Fumariaceae. 3 Spain, southern France (Pyrenees) Morocco (Atlas Mts.), Algeria

Sarcocaulon (DC.) Sweet. 1827. Geraniaceae. 14 South-West and South Africa

Sarcocephalus Afzel. ex Sabine. 1824 (Nauclea Merr.). Rubiaceae. 2 trop. Africa

Sarcochilus R. Br. 1810. Orchidaceae. 16 trop. Asia, Malesia eastward to Solomon Is., northern and eastern Australia, Tasmania (1), New Caledonia, Fiji (1, S. williamsianus)

Sarcochlamys Gaudich. 1844. Urticaceae. 1 Tibet, Bhutan, northeastern India (Assam), China (Yunnan), Thailand, Indochina, ? Malesia

Sarcococca Lindl. 1826. Buxaceae. 11 Afghanistan, Himalayas, India, Sri Lanka, Burma, China, Hainan and Taiwan, Indochina, Sumatra, Java, Philippines

Sarcocornia A. J. Scott. 1978. Chenopodiaceae. 16 Eurasia (excl. Russia, Caucasus, and Middle Asia), Africa, Australia, Tasmania, New Zealand, North and South America

Sarcodes Torr. 1853. Ericaceae. 1 western U.S. (Oregon, California), Baja California

Sarcodraba Gilg et Muschl. 1909. Brassicaceae. 4 Andes

Sarcodum Lour. 1790. Fabaceae. 3 Japan, Southeast Asia, ? Solomon Is. (1)

Sarcoglottis C. Presl. 1827. Orchidaceae. 52 trop. America

Sarcoglyphis Garay. 1972. Orchidaceae. 11 northeastern India, Burma, southern China (Yunnan), Thailand, Vietnam, Malay Peninsula, Sumatra, Java, Borneo

Sarcolaena Thouars. 1805. Sarcolaenaceae. 10 Madagascar

Sarcolobus R. Br. 1810. Asclepiadaceae. 4 northeastern India, Indochina, ? Hainan, Malesia, Caroline and Solomon Is., New Caledonia, ? northeastern Australia

Sarcolophium Troupin. 1960. Menispermaceae. 1 Cameroun, Gabon

Sarcomelicope Engl. 1896. Rutaceae. 9: New Caledonia (8), 1 (S. simplicifolia) from eastern Australia eastward to Vanuatu and Fiji

Sarcopetalum F. Muell. 1862. Menispermaceae. 1 (S. harveyanum) New Guinea, eastern Australia

Sarcophagophilus Dinter = Quaqua N. E. Br. Asclepiadaceae

Sarcopharyngia (Stapf) Boiteau = Tabernaemontana L. Apocynaceae

Sarcophrynium K. Schum. 1902. Maranthaceae. 3 trop. Africa

Sarcophyte Sparrm. 1776. Sarcophytaceae. 2 trop. East and South (Transvaal) Africa

Sarcophyton Garay. 1972. Orchidaceae. 3 Southeast Asia, Philippines (2)

Sarcopilea Urb. 1912. Urticaceae. 1 Haiti

Sarcopoterium Spach. 1846. Rosaceae. 1 Italy, eastern Mediterranean

Sarcopteryx Radlk. 1879. Sapindaceae. 11 Moluccas, New Guinea, Australia (5)

Sarcopygme Setch. et Christoph. 1935. Rubiaceae. 4 Samoa Is.

Sarcopyramis Wall. 1824. Melastomataceae. 7 Himalayas, northeastern India, central and southern China, Indochina (1), western Malesia

Sarcorhachis Trel. 1927. (Piper L.). Piperaceae. 6 trop. America, West Indies

Sarcorhynchus Schltr. = Diaphananthe Schltr. Orchidaceae

Sarcorrhiza Bullock. 1962. Asclepiadaceae. 1 trop. Africa

Sarcosperma Hook. f. 1876. Sapotaceae. 8–9 eastern Himalayas, northeastern India, southern and eastern China, Indochina, western Malesia, and Moluccas

Sarcostemma R. Br. 1810. Asclepiadaceae. 10 trop. and subtrop. Old World

Sarcostigma Wight et Arn. 1833. Icacinaceae. 6 India, Burma, Andaman Is., Indochina, western Malesia

Sarcostoma Blume. 1825. Orchidaceae. 3 Malay Peninsula, Java, Sulawesi

Sarcotheca Blume. 1850–1851. Oxalidaceae. 11 western Malesia and Sulawesi

Sarcotoechia Radlk. 1879. Sapindaceae. 6 New Guinea, eastern Australia

Sarcoyucca (Trel.) Lindinger = Yucca L. Agavaceae

Sarcozona J. M. Black. 1934 (Carpobrotus N. E. Br.). Aizoaceae. 2 southern and southeastern Australia

Sarcozygium Bunge = Zygophyllum L. Zygophyllaceae

Sargentia S. Watson = Casimiroa La Llave et Lex. Rutaceae

Sargentodoxa Rehder et E. H. Wilson. 1913. Sargentodoxaceae. 2 central China, northern Laos, northern Vietnam

Sarinia O. F. Cook = Attalea Kunth. Arecaceae

Saritaea Dugand. 1945. Bignoniaceae. 1 Colombia, Ecuador

Sarmienta Ruiz et Pav. 1794. Gesneriaceae. 1 southern Chile

Sarojusticia Bremek. 1962 (or = Justicia L.). Acanthaceae. 1 northern and central Australia

Sarothamnus Wimm. = Cytisus Desf. Fabaceae

Sarracenella Luer = Pleurothallis R. Br. Orchidaceae

Sarracenia L. 1753. Sarraceniaceae. 8 eastern North America

Sartidia De Winter. 1963. Poaceae. 4 trop. and South Africa, Madagascar

Sartoria Boiss. et Heldr. 1849. Fabaceae. 1 southern Asia Minor

Sartorina R. M. King et H. Rob. 1974. Asteraceae. 1 Mexico

Sartwellia A. Gray. 1852. Asteraceae. 4 southern U.S., Mexico

Saruma Oliv. 1889. Aristolochiaceae. 1 northwestern and southwestern China

Sarx St. John = Sicyos L. Cucurbitaceae

Sasa Makino et Shib. 1901. Poaceae. c. 60 East Asia, incl. southern Russian Far East (17)

Sasaella Makino. 1929 (Sasa Makino et Shib.). Poaceae. 12 East Asia

Sasamorpha Nakai. 1931 (Sasa Makino et Shib.). Poaceae. 15 East Asia

Sassafras Nees et Eberm. 1833. Lauraceae. 3 China, Taiwan (1), eastern North America from Canada to Florida (1)

Sassafridium Meisn. = Cinnamomum Schaeffer. Lauraceae

Satakentia H. E. Moore. 1969. Arecaceae. 1 Ryukyu Is. (Ishigaki, Iriomote)

Satanocrater Schweinf. 1868. Acanthaceae. 4 trop. Africa

Sattadia Fourn. = Metastelma R. Br. Asclepiadaceae

Satureja L. 1753. Lamiaceae. c. 200 temp. and warm–temp. regions, esp. Mediterranean (c. 120); c. 20 southern European part of Russia, Crimea, Caucasus, Middle Asia

Satyria Klotzsch. 1851. Ericaceae. 23 Central and trop. South America

Satyridium Lindl. 1838. Orchidaceae. 1 South Africa (Cape)

Satyrium Sw. 1800. Orchidaceae. c. 170 trop. and South (68) Africa, Madagascar, Arabian Peninsula, trop. Asia

Saugetia Hitchc. et Chase. 1917 (Enteropogon Nees). Poaceae. 2 West Indies

Saundersia Rchb. f. 1866. Orchidaceae. 1 Brazil

Saurauia Willd. 1801. Actinidiaceae. c. 300 Himalayas, East and Southeast Asia, Australia (1, Queensland), montane trop. and subtrop. America

Sauroglossum Lindl. 1833. Orchidaceae. 7 South America

Sauromatum Schott. 1832. Araceae. 2: S. venosum (Aiton) Kunth—Cameroun, Central African Rep., Ethiopia, Malawi, Uganda, Kenya, Tanzania, Zambia, Angola; Arabian Peninsula (Yemen), India, Himalayas from Simla to Nepal, southeastern Tibet; S. brevipes (Hook. f.) N. E. Br.—Nepal, Sikkim

Sauropus Blume. 1826. Euphorbiaceae. c. 40 Himalayas, India, Burma, southwestern China, Indochina, Malesia, Australia

Saururus L. 1753. Saururaceae. 3 southern China, Taiwan, Indochina, Philippines, eastern U.S. (1, S. cernuus)

Saussurea DC. 1810. Asteraceae. c. 400 Europe (9), temp. Asia (c. 395: c. 300, China; c. 120, Middle Asia, Siberia, Far East of Russia); western North America (4)

Sautiera Decne. 1834. Acanthaceae. 1 Timor I.
Sauvagesia L. 1753. Sauvagesiaceae. 34 tropics, esp. South America
Sauvallea W. Wright. 1871 (Commelina L.). Commelinaceae. 1 Cuba
Sauvallella Rydb. = Poitea Vent. Fabaceae
Savannosiphon Goldblatt et Marais. 1980. Iridaceae. 1 southern trop. Africa
Savia Willd. 1806. Euphorbiaceae. 30 trop. America from southern U.S. (2) to southern Brazil (3), West Indies (12), trop. East and South Africa, Madagascar
Savignya DC. 1821. Brassicaceae. 1 North Africa, West Asia to Pakistan
Saxegothaea Lindl. 1851. Podocarpaceae. 1 Chilean Andes and western Argentina
Saxicolella Engl. 1926. Podostemaceae. 3 trop. West Africa
Saxifraga L. 1753. Saxifragaceae. c. 300 subartic and temp. regions of the Northern Hemisphere to Andes of South America, North Africa and Himalayas; Russia (c. 130, chiefly Arctica and alpine areas)
Saxifragella Engl. 1891. Saxifragaceae. 2 antarctic South America
Saxifragodes D. M. Moore. 1969. Saxifragaceae. 1 Tierra del Fuego
Saxifragopsis Small. 1896 (Saxifraga L.). Saxifragaceae. 1 western U.S. (Oregon and California)
Saxofridericia R. H. Schomb. 1845. Rapateaceae. 9 trop. South America
Scabiosa L. 1753. Dipsacaceae. c. 80 temp. Eurasia, Mediterranean, montane areas of East and South Africa, West and East Asia; c. 35 Russia (European part, Siberia, Far East), Caucasus, Middle Asia
Scabiosella Tiegh. = Scabiosa L. Dipsacaceae
Scabiosiopsis Rech. f. 1989. Dipsacaceae. 1 southwestern Iran
Scadoxus Raf. 1838 (Haemanthus L.). Amaryllidaceae. 9 trop., South-West and South Africa, Arabian Peninsula (Yemen)
Scaevola L. 1771. Goodeniaceae. c. 140 tropics and subtropics, most endemics of Australia; S. frutescent—Indo-Pacific area; S. plumieri—Atlantic and southwest Indian oceans north to Socotra and Hainan
Scagea McPherson. 1985. Euphorbiaceae. 2 New Caledonia
Scalesia Arn. ex Lindl. 1836. Asteraceae. 14 Galapagos Is.
Scaligeria DC. 1829. Apiaceae. 3 Mediterranean
Scambopus O. E. Schulz. 1924. Brassicaceae. 1 southern and eastern Australia
Scandia J. W. Dawson. 1967. Apiaceae. 2 New Zealand
Scandicium Thell = Scandix L. Apiaceae
Scandivepres Loes. = Acanthothamnus T. S. Brandegee. Celastraceae
Scandix L. 1753. Apiaceae. 15–20 Europe; Mediterranean; Southwest and South Asia; Northwest, North, and South Africa; 8 southern European part, Crimea, Caucasus, Middle Asia
Scaphiophora Schltr. = Thismia Griff. Burmanniaceae
Scaphispatha Brongn. ex Schott. 1860. Araceae. 1 southern Peru and Brazil
Scaphium Schott et Endl. 1832. Sterculiaceae. 6 Burma, Indochina, western Malesia: Malay Peninsula, Borneo
Scaphocalyx Ridl. 1920. Flacourtiaceae. 2 Malay Peninsula
Scaphochlamys Baker. 1892. Zingiberaceae. 30 Southeast Asia, Malesia
Scaphopetalum Mast. 1867. Sterculiaceae. 15 trop. Africa, Malesia
Scaphosepalum Pfitzer. 1888. Orchidaceae. 25 trop. America from southern Mexico to central Bolivia
Scaphospermum Korovin = Parasilaus Leute. Apiaceae
Scaphyglottis Poepp. et Endl. 1836. Orchidaceae. 35 trop. America from Mexico to Peru, Bolivia, and Brazil; West Indies
Scapicephalus Ovcz. et Chukav. = Pseudomertensia H. Riedl. Boraginaceae
Scariola F. W. Schmidt. 1795. Asteraceae. 5 Mediterranean, West (incl. Caucasus) and Middle Asia to Tibet
Scaryomyrtus F. Muell. 1854. Myrtaceae. 1 ? Distr.

Scassellatia Chiov. 1932. Anacardiaceae. 1 Somalia

Sceletium N. E. Br. = Phyllobolus N. E. Br. Aizoaceae

Scelochiloides Dodson et M. W. Chase. 1989. Orchidaceae. 1 Bolivia

Scelochilus Klotzsch. 1841 (Neokoehleria Schltr.). Orchidaceae. 27 Colombia, Venezuela, Ecuador, Peru

Sceptrocnide Maxim. 1877 (Laportea Gaudich.). Urticaceae. 1 China, Japan

Schaefferia Jacq. 1760. Celastraceae. 16 trop. and subtrop. America, West Indies

Schaenomorphus Thorel ex Gagnep. 1933 (or = Tropidia Lindl.). Orchidaceae. 1 Laos

Schaetzellia Sch. Bip. = Macvaughiella R. M. King et H. Rob. Asteraceae

Schaffnera Benth. = Schaffnerella Nash. Poaceae

Schaffnerella Nash. 1912. Poaceae. 1 Mexico

Schaueria Nees. 1838. Acanthaceae. 9 Brazil

Schedonnardus Steud. 1854. Poaceae. 1 southern U.S., Mexico

Scheelea Karst. 1857. Arecaceae. 32 Central and South America south to Peru, Bolivia, and Brazil

Schefferomitra Diels. 1912. Annonaceae. 1 New Guinea

Schefflera J. R. Forst. et G. Forst. 1775. Araliaceae. c. 600–650 (750 ?) Africa, Madagascar, India, Sri Lanka, southern China, Taiwan, Indochina, Malesia, Caroline Is., Solomon Is., Australia, New Zealand, New Caledonia, Vanuatu, Fiji, Samoa, Hawaiian Is., Central and South America, West Indies

Schefflerodendron Harms. 1901. Fabaceae. 4 Cameroun, Gabon, Angola, Zaire, Tanzania

Scheffleropsis Ridl. = Schefflera J. R. Forst. et G. Forst. Araliaceae

Schelhammera R. Br. 1810. Melanthiaceae. 2 New Guinea (1, S. multiflora), eastern Australia (2, from Queensland to Victoria)

Schellenbergia C. E. Parkinson. 1936 (Vismianthus Mildbr.). Connaraceae. 1 southern Burma

Schenckia K. Schum. = Deppea Cham. et Schltdl. Rubiaceae

Scherya R. M. King et H. Rob. 1977. Asteraceae. 1 Brazil

Scheuchzeria L. 1753. Scheuchzeriaceae. 1 Arctic and temp. regions of the Northern Hemisphere

Schickendantzia Pax. 1890. Alstroemeriaceae. 2 Argentina

Schickendantziella Speg. 1903. Alliaceae. 1 Argentina

Schiedea Cham. et Schltdl. 1826. Caryophyllaceae. 23 Hawaiian Is.

Schiedeella Schltr. 1920. Orchidaceae. 11 southern U.S. (New Mexico, Arizona, Texas), Mexico, Guatemala, Honduras, El Salvador, Nicaragua, Costa Rica; Haiti

Schiekia Meisn. 1842. Haemodoraceae. 1 trop. South America

Schima Reinw. ex Blume. 1823. Theaceae. c. 30 eastern Himalayas, northeastern India, China (c. 20), Taiwan, Ryukyu and Bonin Is., Indochina, western Malesia

Schimpera Steud. et Hochst. ex Endl. 1839. Brassicaceae. 1 Egypt, Sinai Peninsula, Syria, Israel, Jordan eastward to southern Iran

Schimperella H. Wolff = Oreoschimperella Rauschert. Apiaceae

Schindleria H. Walter. 1906. Petiveriaceae. 6 Peru, Bolivia

Schinopsis Engl. 1876. Anacardiaceae. 7 South America

Schinus L. 1753. Anacardiaceae. 28–30 trop. America from Mexico to Argentina

Schinziella Gilg. 1895. Gentianaceae. 2 trop. Africa

Schinziophyton Hutch. ex Radcl.-Sm. 1990 (Ricinodendron Muell. Arg.). Euphorbiaceae. 1 Zaire, Tanzania, Mozambique, Malawi, Zambia, Zimbabwe, Botswana, Angola, Namibia

Schippia Burret. 1933. Arecaceae. 1 Belize

Schisandra Michx. 1803. Schisandraceae.

27 trop. and subtrop. Asia, southeastern North America (1); Russia (1, S. chinensis—Far East, Ussuriyski Krai)

Schischkinia Iljin. 1935. Asteraceae. 1 Middle Asia, Iran, Afghanistan, Pakistan, western China

Schischkiniella Steenis = Silene L. Caryophyllaceae

Schismatoclada Baker. 1883. Rubiaceae. 20 Madagascar

Schismatoglottis Zoll. et Moritzi. 1846. Araceae. c. 100 trop. Asia, Malesia, trop. South America (3)

Schismocarpus S. F. Blake. 1918. Loasaceae. 2 Mexico

Schismus P. Beauv. 1812. Poaceae. 5 South Africa, Mediterranean, West (incl. Caucasus) and Middle Asia, southern West Siberia, to northwestern India

Schistocarpaea F. Muell. 1891. Rhamnaceae. 1 Australia (Queensland)

Schistocarpha Less. 1831. Asteraceae. 16 trop. America from Mexico to Bolivia

Schistocaryum Franch. = Microula Benth. Boraginaceae

Schistogyne Hook. et Arn. 1834. Asclepiadaceae. 12 South America

Schistolobos W. T. Wang. 1983. Gesneriaceae. 1 southern China

Schistonema Schltr. 1906. Asclepiadaceae. 1 Peru

Schistophragma Benth. ex Endl. 1839. Scrophulariaceae. 2–5 America

Schistophyllidium (Juz. ex Fedorov) Ikonn. = Potentilla L. Rosaceae

Schistostemon (Urb.) Cuatrec. 1961. Humiriaceae. 8 trop. South America

Schistostephium Less. 1832. Asteraceae. 12 trop. and South Africa

Schistotylus Dockrill. 1967. Orchidaceae. 1 Australia (New South Wales)

Schivereckia Andrz. ex DC. 1821. Brassicaceae. 5 from northern European part of Russia to Balkan Peninsula and Asia Minor

Schizachne Hack. 1909. Poaceae. 2 eastern European part of Russia, Siberia, Russian Far East, temp. East Asia, temp. North America

Schizachyrium Nees. 1829. Poaceae. c. 60 tropics

Schizanthus Ruiz et Pav. 1794. Solanaceae. 12 Chile, Argentina (1)

Schizeilema (Hook. f.) Domin. 1908. Apiaceae. 13 Australia (1), New Zealand (11), South America (1)

Schizenterospermum Homolle ex Arènes. 1960. Rubiaceae. 4 Madagascar

Schizobasis Baker. 1873. Hyacinthaceae. 2 South Africa

Schizoboea (Fritsch) B. L. Burtt. 1974 (Saintpaulia H. A. Wendl.). Gesneriaceae. 1 Cameroun

Schizocalomyrtus Kausel = Calycorectes O. Berg. Myrtaceae

Schizocalyx Wedd. 1854. Rubiaceae. 2 Colombia

Schizocapsa Hance. 1881. Taccaceae. 2 southeastern China

Schizocardia A. C. Sm. et Standl. = Purdiaea Planch. Cyrillaceae

Schizocarphus Van der Merwe = Scilla L. Hyacinthaceae

Schizocarpum Schrad. 1830. Cucurbitaceae. 6 Central America

Schizocasia Schott ex Engl. = Xenophya Schott. Araceae

Schizochilus Sond. 1846. Orchidaceae. 10 trop. and South (10) Africa

Schizococcus Eastw. = Arctostaphylos Adans. Ericaceae

Schizocodon Siebold et Zucc. = Shortia Torr. et A. Gray. Diapensiaceae

Schizocolea Bremek. 1950. Rubiaceae. 1 trop. West Africa

Schizocorona F. Muell. 1853. Asclepiadaceae. 1 ? Dist.

Schizodium Lindl. 1838. Orchidaceae. 6 South Africa (Cape)

Schizoglossum E. Mey. 1838. Asclepiadaceae. 14 trop. and South Africa

Schizogyne Cass. 1828 (Inula L.). Asteraceae. 2 Canary Is.

Schizolaena Thouars. 1805. Sarcolaenaceae. 12 Madagascar

Schizolobium Vogel. 1837. Fabaceae. 1–2 trop. America

Schizomeria D. Don. 1830. Cunoniaceae.

15 Moluccas, New Guinea, Solomon Is., Australia (Queensland)

Schizomeryta R. Vig. 1906. Araliaceae. 1 New Caledonia

Schizomussaenda H. L. Li. 1943 (Mussaenda L.). Rubiaceae. 1 Southeast Asia

Schizonepeta (Benth.) Briq. 1896. Lamiaceae. 3 West and East Siberia, Russian Far East, Mongolia, China, Tibet, India (Kashmir)

Schizopepon Maxim. 1859. Cucurbitaceae. 8 from northeastern India to East Asia; Russia (1, S. bryoniifolius—Far East, Sakhalin I., and Kuril Is.)

Schizopetalon Sims. 1823. Brassicaceae. 10 north–central Chile and adjacent Argentina

Schizophragma Siebold et Zucc. 1838. Hydrangeaceae. 8 Himalayas, northeastern India, China, Taiwan, Korea, Japan, Kunashiri (1, S. hydrangeoides)

Schizopsera Turcz. 1851. Asteraceae. 1 Ecuador

Schizoptera Benth. = Schizopsera Turcz. Asteraceae

Schizoscyphus K. Schum. ex Taub. = Maniltoa Scheffer. Fabaceae

Schizosepala G. M. Barroso. 1955. Scrophulariaceae. 1 Brazil

Schizosiphon K. Schum. = Maniltoa Scheff. Fabaceae

Schizospatha Furtado = Calamus L. Arecaceae

Schizostachyum Nees. 1829. Poaceae. 35 Madagascar, South and Southeast Asia, New Guinea

Schizostigma Arn. ex Meisn. 1838. Rubiaceae. 1 Sri Lanka

Schizostylis Backh. et Harv. ex Hook. f. 1864. Iridaceae. 1 (S. coccinea) South Africa

Schizotorenia Yamaz. 1978 (or = Lindernia All.). Scrophulariaceae. 2 Southeast Asia, western Malesia

Schizotrichia Benth. 1873. Asteraceae. 5 Peru

Schizozygia Baill. 1888. Apocynaceae. 1 trop. East Africa

Schkuhria Roth. 1797. Asteraceae. 6 trop. and subtrop. America

Schlagintweitiella Ulbr. = Thalictrum L. Ranunculaceae

Schlechtendalia Less. 1830. Asteraceae. 1 Brazil, Argentina, Uruguay

Schlechteranthus Schwantes. 1929. Aizoaceae. 2 South Africa (Cape)

Schlechterella K. Schum. 1899. Asclepiadaceae. 1 East Africa

Schlechteria Bolus. 1897. Brassicaceae. 1 South Africa (western Cape)

Schlechterina Harms. 1902 (Crossostemma Planch. ex Hook.). Passifloraceae. 1 Kenya, Tanzania, Zanzibar, Mozambique, and northern Natal

Schlechterosciadium H. Wolff = Chamarea Eckl. et Zeyh. Apiaceae

Schlegelia Miq. 1844. Bignoniaceae. 15 trop. America from Mexico and Guatemala to Brazil, West Indies

Schleichera Willd. 1806. Sapindaceae. 1 South and Southeast Asia, Malesia (Java, Lesser Sunda Is., Sulawesi, Moluccas)

Schleinitzia Warb. ex Guinet. 1891. Fabaceae. 4 Philippines, New Guinea, Mariana Is., Solomon Is., New Caledonia, Vanuatu, Fiji, and eastward to Society and Austral Is.

Schliebenia Mildbr. = Isoglossa Oerst. Acanthaceae

Schlimmia Planch. et Linden. 1852. Orchidaceae. 4 Andes of Colombia, Venezuela, Ecuador

Schlumbergera C. J. Morren = Guzmania Ruiz et Pav. Bromeliaceae

Schlumbergera Lem. 1858. Cactaceae. 6 southeastern Brazil

Schmalhausenia C. Winkl. 1892. Asteraceae. 1 (S. nidulans) Middle Asia, western China

Schmalzia Desv. = Rhus L. Anacardiaceae

Schmardaea Karst. 1861. Meliaceae. 1–2 Andes from Venezuela to Peru

Schmidtia Steud. ex J. A. Schmidt. 1852. Poaceae. 2 trop. and South Africa, Pakistan

Schmidtottia Urb. 1923. Rubiaceae. 10 Cuba

Schnabelia Hand.-Mazz. 1924. Verbenaceae. 2 southwestern China

Schnella Raddi = Bauhinia L. Fabaceae

Schoenefeldia Kunth. 1829. Poaceae. 2: S. transiens—Sudan, Uganda, Kenya, Tanzania, Zimbabwe, Mozambique; S. gracilis—trop. Africa, Madagascar, Arabian Peninsula, Pakistan, southern India

Schoenia Steetz. 1845. Asteraceae. 5 temp. Australia

Schoenobiblus Mart. et Zucc. 1824. Thymelaeaceae. 10 Andes of South America, West Indies

Schoenocaulon A. Gray. 1837. Melanthiaceae. 10 America from Florida peninsula to Peru

Schoenocephalium Seub. 1847. Rapateaceae. 5 Colombia, Venezuela

Schoenocrambe Greene. 1896 (Sisymbrium L.). Brassicaceae. 4, western North America, northern Mexico

Schoenoides Seberg. 1986. Cyperaceae. 1 Tasmania

Schoenolaena Bunge. 1845 (Xanthosia Rudge). Apiaceae. 2 western Australia

Schoenolirion Torr. ex E. M. Durand. 1855. Hyacinthaceae. 4 western (1, Oregon, California) and southeastern (3) U.S.

Schoenoplectus (Rchb.) Palla = Scirpus L. Cyperaceae

Schoenorchis Blume. 1825. Orchidaceae. 10–15 Himalayas, Tibet, India, Sri Lanka, Burma, Indochina, southern China (Yunnan, Hainan); Malesia: Malay Peninsula (3), Sumatra (4), Java (4), Borneo, Philippines (2), New Guinea; trop. Australia, Bougainville and Solomon Is., Vanuatu, New Caledonia, Fiji

Schoenoxiphium Nees. 1832. Cyperaceae. 12 West and South Africa, Madagascar

Schoenus L. 1753. Cyperaceae. c. 100 extra-trop. South America, Eurasia, North Africa, Malesia, Micronesia, Australia, New Zealand, New Caledonia; 2 European part of Russia, except northern areas, Caucasus, Middle Asia

Schoepfia Schreb. 1789. Olacaceae. 23: 19 U.S. (Florida), Mexico, Guatemala, El Salvador, Belize, Honduras, Nicaragua, Costa Rica, Panama, Colombia, Venezuela, Ecuador, Peru, Bolivia, Brazil, Argentina; West Indies from Bahamas to Trinidad; 4 trop. Asia and Malesia (1, S. fragrans)

Scholleropsis H. Perrier. 1936. Pontederiaceae. 1 Madagascar

Scholtzia Schauer. 1843. Myrtaceae. 13 southwestern Australia

Schomburgkia Lindl. 1838. Orchidaceae. 17 trop. America from Mexico to Peru and Brazil, West Indies

Schotia Jacq. 1787. Fabaceae. 4–5 trop. southern and South Africa

Schoutenia Korth. 1848. Tiliaceae. 8 Indochina, Java, Borneo

Schouwia DC. 1821. Brassicaceae. 1 deserts of North Africa (except Tunisia), Sinai and Arabian peninsulas

Schradera Vahl. 1796. Rubiaceae. 25 Panama, West Indies, trop. South America

Schraderia Medik. = Salvia L. Lamiaceae

Schrameckia Danguy = Tambourissa Sonn. Monimiaceae

Schranckia Willd. = Schrankia Willd. Fabaceae

Schranckiastrum Hassl. 1919. Fabaceae. 1 Paraguay

Schrankia Willd. 1806. Fabaceae. 19 America from southern U.S. to Argentina

Schrebera Roxb. 1799. Oleaceae. 6 trop. and South (2) Africa, Madagascar; 1 (S. swietenioides) trop. Himalayas, India, and Burma; 1 Malesia; 1 Peru

Schreiteria Carolin. 1985 (Calandrinia Kunth). Portulacaceae. 1 (T. macrocarpa) northern Argentina

Schrenkia Fisch. et C. A. Mey. 1841. Apiaceae. 12 Kazakhstan, Middle Asia, western China (1, S. vaginata)

Schtschurowskia Regel et Schmalh. 1882. Apiaceae. 2 montane areas of the Middle Asia

Schubertia Mart. 1824. Asclepiadaceae. 6 South America

Schultesia Mart. 1827. Gentianaceae. 20 trop. America, trop. Africa

Schultesianthus Hunz. 1977. Solanaceae. 5 Mexico, Central America, Colombia, Venezuela

Schultesiophytum Harling. 1958. Cyclanthaceae. 1 southern Colombia
Schultzia Spreng. = Schulzia Spreng. Apiaceae
Schulzia Spreng. 1813. Apiaceae. 6: 1 (S. crinata) Middle Asia, southern Siberia; China, northwestern India, Nepal
Schumacheria Vahl. 1810. Dilleniaceae. 3 Sri Lanka
Schumannia Kuntze = Ferula L. Apiaceae
Schumannianthus Gagnep. 1904. Marantaceae. 2 India, Sri Lanka, Malesia
Schumanniophyton Harms. 1897. Rubiaceae. 5 trop. Africa
Schumeria Iljin. 1960 (Serratula L.). Asteraceae. 6 eastern Mediterranean, West Asia to Middle Asia (2) and Iran
Schuurmansia Blume. 1850. Sauvagesiaceae. 3 Borneo, Philippines, Sulawesi, Moluccas, New Guinea, Bismarck Arch., and Solomon Is.
Schuurmansiella Hallier f. 1913. Sauvagesiaceae. 1 Borneo
Schwabea Endl. et Fenzl. 1839 (Monechma Hochst.). Acanthaceae. 5 trop. Africa
Schwackaea Cogn. 1891. Melastomataceae. 1 Mexico, Central America
Schwalbea L. 1753. Scrophulariaceae. 1 (S. americana) eastern North America
Schwannia Endl. 1840 (Janusia A. Juss.). Malpighiaceae. 6 trop. and subtrop. South America
Schwantesia Dinter. 1927 (Lithops N. E. Br.). Aizoaceae. 3–5 southeastern Namibia, western South Africa
Schwartzkopffia Kraenzl. 1900. Orchidaceae. 2 trop. Africa: Malawi, Zimbabwe, Mozambique; E. lastii—Angola, Congo, Zaire, Zambia, Tanzania
Schweiggeria Spreng. 1820. Violaceae. 2 Mexico, Brazil
Schweinfurthia A. Braun. 1866. Scrophulariaceae. 8 Northeast Africa, Southwest Asia eastward to northwestern India
Schwenckia L. 1764. Solanaceae. 22 trop. America, West Indies
Schwenckiopsis Dammer = Protoschwenckia Soler. Solanaceae
Schwendenera K. Schum. 1886. Rubiaceae. 1 southeastern Brazil
Schwenkia Vahl = Schwenckia L. Solanaceae
Sciadocephala Mattf. 1838. Asteraceae. 5 Panama, Colombia, Ecuador, Haiti
Sciadodendron Griseb. 1858. Araliaceae. 1 trop. America and West Indies
Sciadopanax Seem. = Schefflera J. R. Forst. et G. Forst. Araliaceae
Sciadophyllum P. Browne = Schefflera J. R. Forst. et G. Forst. Araliaceae
Sciadopitys Siebold et Zucc. 1842. Sciadopityaceae. 1 central and southern Japan
Sciadotenia Miers. 1851. Menispermaceae. 18 Panama, Colombia, Peru, Venezuela, Guyana, Brazil
Sciaphila Blume. 1826. Triuridaceae. c. 50 West (Ivory Coast, Nigeria, Cameroun), Central and trop. South Africa, Sri Lanka, Assam, Thailand, Hainan and Botel; Tobago Is. northward to southern Japan and Bonin Is., Malesia, Micronesia, Melanesia, western Polynesia, Australia (northern Queensland)
Sciaphyllum Bremek. 1940. Acanthaceae. 1 original growth is unknown, cult.
Sciaplea Rauschert. 1982. Sapindaceae. 1 Somalia
Scilla L. 1753. Hyacinthaceae. c. 50 Canary Is., Mediterranean, Africa, temp. Eurasia, incl. (14) southern European part of Russia, Caucasus, Middle Asia, and Russian Far East
Scindapsus Schott. 1832. Araceae. c. 40 India, southeastern China, Indochina, Malesia, Solomon Is., Brazil (1)
Sciodaphyllum P. Browne = Schefflera J. R. Forst. et G. Forst. Araliaceae
Sciothamnus Endl. = Peucedanum L. Apiaceae
Scirpidiella Rauschert. 1983. Cyperaceae. 6 West Europe, trop. East Africa, Malesia (Sumatra, Java, New Guinea), Australia, Tasmania, New Zealand
Scirpodendron Zipp. ex Kurz. 1869. Cyperaceae. 1 Indo-Malesia, New Guinea, Australia, Polynesia

Scirpoides Séguier. 1754. Cyperaceae. 2 temp. and subtrop. Eurasia (incl. European part of Russia, Caucasus, southern West Siberia, Middle Asia), North and South Africa

Scirpus L. 1753. Cyperaceae. c. 300 cosmopolitan

Sclerachne R. Br. 1838. Poaceae. 1 Thailand, Indochina, Malesia

Sclerandrium Stapf et C. E. Hubb. 1935 (Germainia Bal. et Poitr.). Poaceae. 3 New Guinea, northeastern Australia

Scleranthera Pichon = Wrightia R. Br. Apocynaceae

Scleranthopsis Rech. f. 1967. Caryophyllaceae. 1 Afghanistan

Scleranthus L. 1753. Caryophyllaceae. 15 Europe, Caucasus, temp. and subtrop. Asia, North Africa, New Guinea, Australia, New Zealand

Scleria Bergius. 1765. Cyperaceae. c. 200 trop., subtrop., and warm–temp. (Japan, North America) regions

Sclerobassia Ulbr. = Sclerolaena R. Br. Chenopodiaceae

Scleroblitum Ulbr. 1934 (Chenopodium L.). Chenopodiaceae. 1 southeastern Australia

Sclerocactus Britton et Rose. 1922. Cactaceae. c. 20 southwestern U.S., northern Mexico

Sclerocarpus Jacq. 1780–1784. Asteraceae. 8 trop. and subtrop. America, trop. Africa (1)

Sclerocarya Hochst. 1844 (Poupartia Comm. ex Juss.). Anacardiaceae. 2: S. birrea—trop. (from Senegal to Ethiopia south to Namibia, Zimbabwe, and Tanzania) and South (Transvaal, Natal) Africa, Madagascar; S. gillettii Kokwaro—Kenya

Sclerocaryopsis Brand = Lappula Moench. Boraginaceae

Sclerocephalus Boiss. 1843. Caryophyllaceae. 1 Canary Is., Cape Verde Is., North Africa to Sinai Peninsula, Syria, Lebanon, Israel, Jordan, Iraq, Iran

Sclerochiton Harv. 1842. Acanthaceae. 19 trop. Africa from Sierra Leone to Kenya and Mozambique, southward to Angola and Zimbabwe, South Africa

Sclerochlamys F. Muell. 1857. Chenopodiaceae. 1 temp. eastern Australia

Sclerochloa P. Beauv. 1812. Poaceae. 3 West, Central, South, and Southeast Europe; North Africa; West (incl. Caucasus) and Middle Asia, Himalayas

Sclerochorton Boiss. 1872. Apiaceae. 1 Iran

Sclerodactylon Stapf. 1911. Poaceae. 1 trop. East Africa, Madagascar, islands of the Indian ocean

Sclerodeyeuxia (Stapf) Pilg. = Calamagrostis Adans. Poaceae

Sclerolaena R. Br. 1810. Chenopodiaceae. 77 Australia and Tasmania (esp. semi-arid areas)

Sclerolepis Cass. 1816. Asteraceae. 1 eastern U.S. from New Hampshire to Alabama and Florida

Sclerolinon C. M. Rogers. 1966. Linaceae. 1 western U.S. from Washington and Idaho to California

Sclerolobium Vogel. 1837. Fabaceae. c. 40 Costa Rica (1, S. costaricense Zamora et Poveda), trop. South America, esp. Amazonian Brazil

Scleronema Benth. 1862. Bombacaceae. 5 trop. South America

Sclerophylax Miers. 1848. Sclerophylacaceae. 12 Argentina, Paraguay (1), Uruguay (1)

Scleropoa Griseb. 1846 (Catapodium Link). Poaceae. 3 Mediterranean, West Asia; S. rigida—extending to Crimea and Caucasus

Scleropogon Phil. 1870. Poaceae. 1 southwestern U.S., Mexico, Chile, Argentina

Scleropyrum Arn. 1838. Santalaceae. 6 India, southern China, Indochina, Malesia

Sclerorhachis (Rech. f.) Rech. f. 1969. Asteraceae. 4 Iran, Afghanistan, Turkmenistan (1, S. platyrachis)

Sclerosperma G. Mann et H. A. Wendl. 1864. Arecaceae. 3 trop. West Africa (Nigeria, Cameroun, Gabon)

Sclerostachya (Hack.) A. Camus. 1922

(Miscanthus Andersson). Poaceae. 3 South and Southeast Asia

Sclerostegia Paul G. Wilson. 1980. Chenopodiaceae. 5 Tasmania

Sclerostephane Chiov. 1929. Asteraceae. 5 Somalia

Sclerotheca A. DC. 1839. Lobeliaceae. 4 Cook and Society Is.

Sclerothrix C. Presl = Klaprothia Kunth. Loasaceae

Sclerotiaria Korovin. 1962. Apiaceae. 1 Middle Asia

Scobinaria Seibert. 1940 (Arrabidaea DC.). Bignoniaceae. 1 trop. America from southern Mexico to Amazonia

Scoliaxon Payson. 1924. Brassicaceae. 1 northeastern Mexico

Scoliopus Torr. 1856. Melanthiaceae. 2 western North America

Scoliotheca Baill. = Monopyle Benth. Gesneriaceae

Scolochloa Link. 1827. Poaceae. 1 (S. festucacea) Eurasia and North America; Russia (European part), West and East Siberia; Caucasus, montane Middle Asia,

Scolophyllum Yamaz. 1978. Scrophulariaceae. 1 Southeast Asia

Scolopia Schreb. 1789. Flacourtiaceae. c. 40 trop. and South (5) Africa, Madagascar, Comoro and Mascarene Is., South and Southeast Asia, Malesia to New Ireland I., Australia (Queensland, New South Wales)

Scolosanthus Vahl. 1796. Rubiaceae. 16 West India

Scolymus L. 1753. Asteraceae. 3 Canary Is., Mediterranean, Southeast Europe, West Asia (incl. Caucasus)

Scoparia L. 1753. Scrophulariaceae. 20 trop. America, West Indies

Scopelogena L. Bolus. 1962. Aizoaceae. 1 southwestern South Africa

Scopolia Jacq. 1764. Solanaceae. 5 Central and Southeast Europe, West (incl. Caucasus) Asia eastward to India, Himalayas, southwestern and central China, Korea, Japan

Scopulophila M. E. Jones. 1908. Caryophyllaceae. 2 southwestern U.S. (California, Nevada), Mexico

Scorodocarpus Becc. 1877. Olacaceae. 1 (S. borneensis) peninsular Thailand, Sumatra, Lingga Arch., Malay Peninsula, Borneo

Scorodophloeus Harms. 1901. Fabaceae. 2: (S. zenkeri) Cameroun, Gabon, Zaire, Angola; (S. fischeri) Kenya, Tanzania

Scorpiothyrsus H. L. Li. 1944. Melastomataceae. 6 Indochina, Hainan

Scorpiurus L. 1753. Fabaceae. 4 Macaronesia, Mediterranean, Southeast Europe, Caucasus, Arabian Peninsula, Iraq, Iran

Scorzonella Nutt. = Microseris D. Don. Asteraceae

Scorzonera L. 1753. Asteraceae. c. 170 Central Europe, Mediterranean, Asia; c. 30 Russia, Caucasus, and Middle Asia

Scottellia Oliv. 1893. Flacourtiaceae. 3 trop. Africa

Scribneria Hack. 1886. Poaceae. 1 western U.S. from Washington to California

Scrithacola Alava. 1980 (Semenovia Regel bet Herder). Apiaceae. 1 Afghanistan, Pakistan

Scrobicaria Cass. 1827. Asteraceae. 2 Colombia, Venezuela

Scrobicularia Mansf. = Poikilogyne Baker f. Melastomataceae

Scrofella Maxim. 1888. Scrophulariaceae. 1 northwestern China

Scrophularia L. 1753. Scrophulariaceae. c. 200 America (from temp. to trop. regions), temp. Eurasia, esp. Mediterranean, Caucasus, and Middle Asia

Scrotochloa Judz. 1984 (or = Leptaspis R. Br.). Poaceae. 2 southern India, Sri Lanka, Indochina, Malesia eastward to Solomon Is., Australia

Scurrula L. 1753. Loranthaceae. c. 20 trop. Asia: India, Sri Lanka, eastern Himalayas, Bangladesh, Burma, Indochina, southern China (11), Taiwan; Malesia: Malay Peninsula, Java, Lesser Sunda Is., Philippines, Sulawesi, Moluccas

Scutachne Hitchc. et Chase. 1911. Poaceae. 2 West Indies

Scutellaria L. 1753. Lamiaceae. 360–425 nearly cosmopolitan, except Amazonian, lowland trop. and South Africa, Pacific Is., desert Central Asia, north of the Arctic circle; c. 130 Russia, Caucasus, and Middle Asia

Scutia (Comm. ex DC.) Brongn. 1827. Rhamnaceae. 3 trop. America; trop. and South Africa, Madagascar, Seychelles, Mascarene Is., India, Sri Lanka, Burma, southern China, Indochina

Scuticaria Lindl. 1843. Orchidaceae. 5 Colombia, Venezuela, Guyana, French Guiana, Surinam, northern Brazil

Scutinanthe Thwaites. 1856. Burseraceae. 2 Sri Lanka, southern Burma, Malay Peninsula, Sumatra, Borneo, Sulawesi

Scybalium Schott et Endl. 1832. Helosiaceae. 4 trop. South America, West Indies

Scyphanthus Sweet. 1828. Loasaceae. 2 Chile

Scyphellandra Thwaites. 1858 (Rinorea Aubl.). Violaceae. 4: Sri Lanka (1), southern China and Indochina (3)

Scyphiphora C. F. Gaertn. 1806. Rubiaceae. 1 coastal trop. Asia, Malesia and Australia, Caroline Is., Melanesia

Scyphocephalium Warb. 1896. Myristicaceae. 3 trop. West Africa

Scyphochlamys Balf. f. 1879. Rubiaceae. 1 Rodriguez I.

Scyphocoronis A. Gray. 1852. Asteraceae. 2 southwestern Australia

Scyphogyne Decne. ex Brongn. 1834. Ericaceae. 7 South Africa (southwestern and southern Cape)

Scyphonychium Radlk. 1879. Sapindaceae. 1 northeastern Brazil

Scyphopappus B. Nord. = Argyranthemum Sch. Bip. Asteraceae

Scyphostachys Thwaites. 1859. Rubiaceae. 2 Sri Lanka

Scyphostegia Stapf. 1894. Scyphostegiaceae. 1 Borneo

Scyphostelma Baill. 1890. Asclepiadaceae. 1 Colombia

Scyphostrychnos S. Moore = Strychnos L. Loganiaceae

Scyphosyce Baill. 1875. Moraceae. 2 trop. Africa

Scytopetalum Pierre ex Engl. 1897. Scytopetalaceae. 3 trop. West Africa

Sebaea Sol. ex R. Br. 1810. Gentianaceae. c. 100 trop. and South (over 40) Africa, Madagascar; S. microphylla—extending to Himalayas (Garhwal, Nepal), India (Khasi Hills), southern China, Indochina, and Malesia; 2 Australia, New Zealand

Sebastiania Spreng. 1820. Euphorbiaceae. c. 100 eastern coast of the U.S., trop. America, trop. Africa, Madagascar, India, southern China, Malesia, Australia

Sebastiano-schaueria Nees. 1847. Acanthaceae. 1 Brazil

Sebertia Pierre ex Engl. = Niemeyera F. Muell. Sapotaceae

Sebestena Boehm. = Cordia L. Cordiaceae

Secale L. 1753. Poaceae. 8 Mediterranean, Southeast Europe, South Africa (1), Southwest, West, and Middle Asia

Secamone R. Br. 1810. Asclepiadaceae. c. 40 trop. and South Africa, Madagascar, trop. Asia, and Australia

Secamonopsis Jum. 1908. Asclepiadaceae. 1 Madagascar

Sechiopsis Naudin. 1866. Cucurbitaceae. 5 Mexico, Guatemala

Sechium P. Browne. 1756. Cucurbitaceae. 7–8 trop. America

Secondatia A. DC. 1844. Apocynaceae. 7 trop. South America, Jamaica

Securidaca Hill. 1754, non L. = Securigera DC. Fabaceae

Securidaca L. 1759. Polygalaceae. 80 Central and trop. South America, trop. Africa, South and Southeast Asia, Malesia, absent in Australia

Securigera DC. 1805 (Coronilla L.). Fabaceae. 12 northern–northeastern Mediterranean from Spain to Israel, Morocco, Tunisia, Algeria (1), Somalia (1), Iraq, Iran

Securinega Comm. ex Juss. 1789. Euphorbiaceae. 20 temp., subtrop. and trop. regions; Russia (1, S. suffruticosa: southern East Siberia and Far East)

Seddera Hochst. 1844. Convolvulaceae. 15–20 Egypt, trop. and South Africa, Madagascar, Arabian Peninsula

Sedella Britton et Rose = Sedum L. Crassulaceae

Sedirea Garay et H. R. Sweet. 1974 (Aerides Lour.). Orchidaceae. 1 East Asia

Sedopsis (Engl.) Exell et Mendonça = Portulaca L. Portulacaceae

Sedum L. 1753. Crassulaceae. c. 470 temp. and subtrop. Northern Hemisphere, montane areas of tropics, Peru (1)

Seemannaralia R. Vig. 1906. Araliaceae. 1 South Africa

Seemannia Regel = Gloxinia L'Hér. Gesneriaceae

Seetzenia R. Br. ex Decne. 1826. Zygophyllaceae. 2 North, Northeast, and South Africa; Arabian Peninsula; Israel; Afghanistan; northwestern India

Segetella Desv. = Spergularia J. Presl et C. Presl. Caryophyllaceae

Seguieria Loefl. 1758. Petiveriaceae. 6 trop. South America

Sehima Forssk. 1775. Poaceae. 7 trop. Old World

Seidelia Baill. 1858. Euphorbiaceae. 2 South Africa

Seidenfadenia Garay. 1972. Orchidaceae. 1 Burma, Thailand

Seidlitzia Bunge ex Boiss. 1879. Chenopodiaceae. 3 Caucasus and Middle Asia (2), Turkey, Syria, Lebanon, Israel, Egypt, Arabian Peninsula, Iraq, Iran

Selago L. 1753. Selaginaceae. c. 150 trop. and South Africa

Selenia Nutt. 1825. Brassicaceae. 6 southern U.S., Mexico

Selenicereus (A. Berger) Britton et Rose. 1909. Cactaceae. 25 southern U.S., Mexico, Central, and northern coastal South America; 1 Argentina and Uruguay

Selenipedium Rchb. f. 1854. Orchidaceae. 4 trop. America

Selenothamnus Melville = Lawrencia Hook. Malvaceae

Selera Ulbr. = Gossypium L. Malvaceae

Selinocarpus A. Gray. 1853. Nyctaginaceae. 9 southwestern U.S., Mexico, Somalia (1)

Selinopsis Coss. et Durieu ex Batt. et Trab. 1859. Apiaceae. 2 Spain, Algeria, Tunisia

Selinum L. 1762 (Cnidium Cusson ex Juss.). Apiaceae. 8 Europe, West Siberia, China, Himalayas from Kashmir to Sikkim, Assam

Selkirkia Hemsl. 1884. Boraginaceae. 1 Juan Fernández Is.

Selleola Urb. = Minuartia L. Caryophyllaceae

Selleophytum Urb. = Coreopsis L. Asteraceae

Selliera Cav. 1799. Goodeniaceae. 2 Australia, New Zealand, temp. South America

Selloa Kunth. 1820. Asteraceae. 3 Mexico, Central America

Sellocharis Taub. 1889. Fabaceae. 1 southeastern Brazil

Selysia Cogn. 1881. Cucurbitaceae. 3 trop. South America

Semecarpus L. f. 1782. Anacardiaceae. c. 60 India, Sri Lanka, Indochina, southern China, Taiwan, Malesia eastward to New Britain I., Micronesia, Solomon Is., northern Australia (1, S. australiensis)

Semeiandra Hook. et Arn. = Lopezia Cav. Onagraceae

Semeiocardium Zoll. = Impatiens L. Balsaminaceae

Semele Kunth. 1844. Ruscaceae. 5 Madeira Is., Canary Is.

Semenovia Regel et Herder. 1866. Apiaceae. 18 Middle Asia (10), Iran, Afghanistan

Semialarium N. Hallé. 1983 (Hippocratea L.). Celastraceae. 2: Mexico (1), Brazil (1)

Semiaquilegia Makino. 1902. Ranunculaceae. 7 China, Korea, Japan, Russian Far East (1, S. manshurica)

Semiarundinaria Makino ex Nakai. 1925. Poaceae. 5 East Asia

Semibegoniella C. DC. = Begonia L. Begoniaceae

Semiliquidambar H. T. Chang. 1962 (Altingia Noronha). Altingiaceae. 3 eastern China, Hainan

Semiramisia Klotzsch. 1851. Ericaceae. 4 Colombia, Venezuela, Ecuador

Semnanthe N. E. Br. 1927 (Erepsia N. E. Br.). Aizoaceae. 1 South Africa

Semnostachya Bremek. 1944 (Strobilanthes Blume). Acanthaceae. 9 China, western Malesia

Semnothyrsus Bremek. 1944 (or = Strobilanthes Blume). Acanthaceae. 1 Sumatra

Semonvillea J. Gay = Limeum L. Aizoaceae

Sempervivella Stapf. 1923 (Rosularia [DC.] Stapf). Crassulaceae. 4 China, India

Sempervivum L. 1753. Crassulaceae. 42 South and Southeast Europe, Morocco, West Asia (incl. Caucasus), ? China (1)

Senaea Taub. 1893. Gentianaceae. 2 Brazil

Senecio L. 1753. Asteraceae. c. 1,000 cosmopolitan, excl. Antarctica

Senefeldera Mart. 1841. Euphorbiaceae. 10 trop. South America

Senefelderopsis Steyerm. 1951. Euphorbiaceae. 2 northwestern South America

Senisetum Honda. 1932 (Agrostis L.). Poaceae. 1 Japan

Senna Mill. 1754. Fabaceae. c. 240–260 pantropics, esp. South America; few species extending into subtropics and warm temperate (rarely into cool temperate) areas of both hemispheres

Sennia Chiov. = Sciaplea Rauschert. Sapindaceae

Senniella Aellen = Atriplex L. Chenopodiaceae

Senra Cav. 1786. Malvaceae. 3 Arabian Peninsula, eastern Australia

Sepalosaccus Schltr. 1923. Orchidaceae. 1 trop. America

Sepalosiphon Schltr. 1912. Orchidaceae. 1 New Guinea

Separotheca Waterf. = Tradescantia L. Commelinaceae

Sepikea Schltr. 1923. Gesneriaceae. 1 New Guinea

Septogarcinia Kosterm. = Garcinia L. Clusiaceae

Septotheca Ulbr. 1924. Bombacaceae. 1 Peru

Septulina Tiegh. 1895 (Taxillus Tiegh.). Loranthaceae. 2 South-West (southern Namibia) and South (Cape) Africa

Sequoia Endl. 1847. Taxodiaceae. 1 (S. sempervirens) western North America from southwestern Oregon to California

Sequoiadendron Buchholz. 1939. Taxodiaceae. 1 (S. giganteum) southwestern U.S.

Seraphyta Fisch. et C. A. Mey. = Epidendrum L. Orchidaceae

Serapias L. 1753. Orchidaceae. 10 Azores, Europe, Mediterranean, Caucasus (1, S. vomeracea)

Serenoa Hook. f. 1883 (Acoelorraphe H. A. Wendl.). Arecaceae. 1 southeastern U.S. from Mississippi to South Carolina and Florida

Seretoberlinia P. A. Duvign. = Julbernardia Pellegr. Fabaceae

Sergia Fedorov. 1957. Campanulaceae. 2 Middle Asia

Serialbizzia Kosterm. = Albizia Durazz. Fabaceae

Serianthes Benth. 1844. Fabaceae. 13 Malay Peninsula, Philippines, Sulawesi, Sumbawa, Moluccas, New Guinea, Caroline Is., Admiralty Is., Bismarck Arch., Solomon Is., New Caledonia, Vanuatu, Fiji, and eastward to Marquesas Is., Society and Austral Is.

Sericanthe Robbr. 1978 (Tricalysia A. Rich. ex DC.). Rubiaceae. 17 trop. and South Africa

Sericocalyx Bremek. 1944 (Strobilanthes Blume). Acanthaceae. 15 northeastern India, southern China, Indochina, Malesia

Sericocarpus Nees = Aster L. Asteraceae

Sericocoma Fenzl. 1842. Amaranthaceae. 6 trop. and South Africa

Sericocomopsis Schinz ex Gilg. 1895. Amaranthaceae. 2 trop. East Africa (Ethiopia, Somalia, Uganda, Kenya, Tanzania)

Sericodes A. Gray. 1852. Zygophyllaceae. 1 northern Mexico

Sericographis Nees = Justicia L. Acanthaceae

Sericolea Schltr. 1916. Elaeocarpaceae. 15 montane New Guinea

Sericorema (Hook. f.) Lopr. 1899. Amaranthaceae. 3 trop. Africa, Madagascar

Sericospora Nees. 1847. Acanthaceae. 1 West Indies

Sericostachys Gilg et Lopr. 1899. Amaranthaceae. 1 trop. Africa: Fernando Po, Ivory Coast, Nigeria, Cameroun, Zaire, Sudan, Ethiopia, Uganda, Kenya, Tanzania, Angola, Malawi

Sericostoma Stocks. 1848. Boraginaceae. 8 Northeast and trop. East Africa, Arabian Peninsula, West Asia to northwestern India

Seringia J. Gay. 1821. Sterculiaceae. 1 New Guinea, eastern Australia

Seriphidium (Cass.) Poljak. 1961. Asteraceae. c. 130 North America, Europe, Mediterranean, West (incl. Caucasus) and Middle Asia (c. 60), Iran eastward to Monogolia, China (esp. Tibet), Afghanistan, Pakistan, and India (Kashmir); Russia (European part, West Siberia)

Serissa Comm. ex Juss. 1789. Rubiaceae. 1–3 Southeast Asia

Serjania Mill. 1754. Sapindaceae. c. 230 southern U.S., trop. America, West Indies

Serratula L. 1753. Asteraceae. c. 70 Europe, North Africa, Asia from Mediterranean to Japan

Serruria Burm. ex Salisb. 1807. Proteaceae. 55–60 South Africa

Sertifera Lindl. et Rchb. f. 1876 (Elleanthus C. Presl). Orchidaceae. 4 Andes of Colombia, Ecuador, and Venezuela

Sesamoides Hill. 1753. Resedaceae. 1 (S. canescens) Portugal, Spain, France, southern Germany, Corsica and Sardinia, Italy, Morocco, Algeria, Tunisia

Sesamothamnus Welw. 1896. Pedaliaceae. 7 trop. Africa

Sesamum L. 1753. Pedaliaceae. 18 Morocco and Egypt (1, S. alatum), trop. and South Africa, trop. Asia and Australia

Sesbania Scop. 1777. Fabaceae. c. 50 tropics and subtropics, esp. Africa (35)

Seseli L. 1753. Apiaceae. c. 120 Eurasia (esp. Russia, from European part to Far East), Northwest, North and trop. Africa, South America (1)

Seselopsis Schischk. 1950. Apiaceae. 2 Middle Asia, western China

Seshagiria Ansari et Hemadri. 1971. Asclepiadaceae. 1 western India

Sesleria Scop. 1760. Poaceae. c. 30 Europe, West Asia

Sesleriella Deyl = Sesleria Scop. Poaceae

Sessea Ruiz et Pav. 1794. Solanaceae. 5 Andes from Colombia to Bolivia, Haiti

Sesseopsis Hassl. = Sessea Ruiz et Pav. Solanaceae

Sessilanthera Molseed et Cruden. 1969. Iridaceae. 3 Mexico, Guatemala

Sessilistigma Goldblatt = Homeria Vent. Iridaceae

Sesuvium L. 1759. Aizoaceae. 6: 1 (S. portulacastrum) pantropical littoral, trop. Africa (4), Galapagos Is. (1)

Setaria P. Beauv. 1812. Poaceae. c. 140 warm–temp. and trop. regions

Setariopsis Scribn. et Millsp. 1896. Poaceae. 2 Mexico, Central America, Colombia

Setchellanthus Brandegee. 1909. Capparaceae. 1 Mexico

Setcreasea K. Schum. et Sidow = Tradescantia L. Commelinaceae

Setiacis S. L. Chen et Jin. 1988 (or = Panicum L.). Poaceae. 1 Hainan

Seticleistocactus Backeb. = Cleisotcactus Lem. Cactaceae

Setilobus Baill. 1888. Bignoniaceae. 3 Brazil

Seutera Rchb. 1828 (Vincetoxicum Moench). Asclepiadaceae. 2 Russian Far East (1 [S. wilfordii] Ussuriyski Krai), Korea, Japan, western North America

Sevada Moq. 1849. Chenopodiaceae. 1 Egypt, the Sudan, Ethiopia, Somalia, Arabian Peninsula

Severinia Ten. 1842 (Atalantia Corrêa). Rutaceae. 7 East and Southeast Asia

Seychellaria Hemsl. 1907. Triuridaceae. 3 Tanzania (1), Seychelles (1), Madagascar (1)

Seymeria Pursh. 1814. Scrophulariaceae. 25 southern U.S., Mexico

Seymeriopsis Tzvelev. 1987. Scrophulariaceae. 1 Cuba

Seyrigia Keraudren. 1961. Cucurbitaceae. 5 Madagascar

Shafera Greenm. 1912. Asteraceae. 1 Cuba

Shaferocharis Urb. 1912. Rubiaceae. 1 Cuba

Shaferodendron Gilly = Manilkara Adans. Sapotaceae

Shaniodendron M. B. Deng, H. T. Wei et X. Q. Wang. 1992. Hamamelidaceae. 1 southern China (Jiangsu)

Sheareria S. Moore. 1875. Asteraceae. 2 China

Sheilanthera I. Williams. 1981. Rutaceae. 1 South Africa (southwestern Cape)

Shepherdia Nutt. 1818. Elaeagnaceae. 3 North America

Sherardia L. 1753. Rubiaceae. 1 Europe, North Africa, West (incl. Caucasus) and ? Middle Asia

Sherbournia G. Don. 1855. Rubiaceae. 10 trop. Africa

Sherwoodia (Torr. et A. Gray) House = Shortia Torr. et A. Gray. Diapensiaceae

Shibataea Makino ex Nakai. 1933. Poaceae. 5 China (4), Japan

Shibateranthis Nakai = Eranthis Salisb. Ranunculaceae

Shinnersia R. M. King et H. Rob. 1970. Asteraceae. 1 southern U.S., Mexico

Shinnersoseris Tomb. 1973. Asteraceae. 1 North America

Shiuyinghua Paclt. 1962 (= ? Paulownia Siebold et Zucc.). Scrophulariaceae. 1 central China

Shorea Roxb. ex C. F. Gaertn. 1805. Dipterocarpaceae. c. 200 (or 375 ?) India, Sri Lanka eastward to southern China (1, S. assamica), Indochina, western Malesia, Moluccas and Lesser Sunda Is., Philippines

Shortia Torr. et A. Gray. 1842. Diapensiaceae. 6 southwestern and southern China, Taiwan, Japan (5), southeastern U.S. (1, S. galacifolia)

Shoshonea Evert et Constance. 1982. Apiaceae. 1 U.S. (Wyoming)

Shuteria Wight et Arn. 1834. Fabaceae. 5 India, Sri Lanka, Himalayas, China, Indochina, Malesia

Siamosia K. Larsen et Pedersen. 1987 (Herbstia Sohmer and Chamissoa Kunth). Amaranthaceae. 1 Thailand

Sibangea Oliv. 1883 (Drypetes Vahl). Euphorbiaceae. 3 trop. Africa

Sibara Greene. 1896. Brassicaceae. 11 southeastern North America

Sibbaldia L. 1753 (Potentilla L.). Rosaceae. 20 temp. Eurasia south to Himalayas and Taiwan; 5 Russia (Arctica, Urals, Siberia, Far East), Caucasus, and montane Middle Asia

Sibbaldianthe Juz. = Sibbaldia L. Rosaceae

Sibiraea Maxim. 1879 (Eleiosine Raf.). Rosaceae. 6 Balkan Peninsula (1, S. croatica), Middle Asia and West Siberia (2), western and southwestern China (3)

Sibthorpia L. 1753. Scrophulariaceae. 5 Central and South America, Azores, Madeira Is., West and South Europe, montane areas of trop. Africa

Sicana Naudin. 1862. Cucurbitaceae. 3 trop. America, West Indies

Siccobaccatus P. J. Braun et E. Est. Pereira = Micranthocereus Backeb. Cactaceae

Sicrea (Pierre) Hallier f. 1923. Tiliaceae. 1 Indochina

Sicydium Schltdl. 1832. Cucurbitaceae. 7 trop. America from southern Mexico to Bolivia, West Indies

Sicyocarya (A. Gray) St. John = Sicyos L. Cucurbitaceae

Sicyocaulis Wiggins = Sicyos L. Cucurbitaceae

Sicyos L. 1753. Cucurbitaceae. c. 50 Australia, Tasmania, New Zealand, southwestern Oceania, Hawaiian Is., North and South America

Sicyosperma A. Gray. 1853. Cucurbitaceae. 1 southern U.S.

Sida L. 1753. Malvaceae. c. 150 tropics and subtropics, esp. America, a few species in warm–temp. regions; S. spinosa—extending to Caucasus

Sidalcea A. Gray. 1849. Malvaceae. 20 western U.S. and Mexico (2)

Sidastrum Baker f. 1892 (Sida L.). Malvaceae. 7 southern U.S. (Texas), Mexico (4, 3 endemic), Guatemala, Honduras, Costa Rica, Panama, Colombia, Venezuela, the Guianas, Peru, Bolivia, Brazil,

Argentina, and Paraguay; West Indies (Bahamas, Cuba, Hispaniola, Lesser Antilles); S. paniculatum—extending to Africa

Siderasis Raf. 1837. Commelinaceae. 2 trop. South America

Sideria Ewart et A. H. K. Petrie = Melhania Forssk. Sterculiaceae

Sideritis L. 1753. Lamiaceae. c. 100 Macaronesia, Mediterranean (c. 80), temp. Eurasia; 10 Crimea, Caucasus, and adjacent areas

Siderobombyx Bremek. 1947. Rubiaceae. 1 Borneo

Sideropogon Pichon. 1946 (Cuspidaria DC.). Bignoniaceae. 1 Brazil

Sideroxylon L. 1753. Sapotaceae. c. 80: 49 trop. America from southern U.S. to Argentina and West Indies; 6 Africa; 6 Madagascar; 8 Mascarenes Is.; 4 Asia; S. mascatense (A. DC.) Pennington—1 southern Ethiopia, northern Somalia, Oman, Afghanistan, northwestern Pakistan

Sidopsis Rydb. = Malvastrum A. Gray. Malvaceae

Siebera J. Gay. 1827. Asteraceae. 1 (S. pungens) eastern Mediterranean to Middle Asia and Iran

Siegfriedia C. A. Gardner. 1933. Rhamnaceae. 1 western Australia

Sieglingia Bernh. 1800 (Danthonia DC.). Poaceae. 1 Madeira Is., Europe, Algeria, Asia Minor, Caucasus

Siella Pimenov = Berula Koch. Apiaceae

Siemensia Urb. 1923. Rubiaceae. 1 Cuba

Sievekingia Rchb. f. 1871. Orchidaceae. 15 trop. America from Costa Rica to Colombia and Guyana, Ecuador, Peru, Bolivia

Sieversia Willd. 1811. Rosaceae. 2 Northeast Asia (Anadyr), Russian Far East

Sigesbeckia L. 1753. Asteraceae. 9 warm–temp. and trop. regions;

Sigmatanthus Huber ex Emmerich = Raputia Aubl. Rutaceae

Sigmatogyne Pfitzer. 1907 (Panisea [Lindl.] Lindl.). Orchidaceae. 2–3 northeastern India, Indochina

Sigmatostalix Rchb. f. 1852. Orchidaceae. 16 trop. America from Mexico to Brazil

Silaum Mill. 1754. Apiaceae. 1 (S. silaus) Europe, Caucasus, Kazakhstan, West Siberia, China

Silaus Bernh. = Silaum Mill. Apiaceae

Silene L. 1753. Caryophyllaceae. c. 500 temp. Northern Hemisphere, esp. Mediterranean (c. 350), few in South Africa

Silentvalleya V. J. Nair, Sreek., Vajr. et Barghavan. 1982–1983. Poaceae. 1 southern India

Siler Crantz = Laser Borkh. ex Gaertn., B. Mey. et Schreb. Apiaceae

Silicularia Compton. 1953. Brassicaceae. 1 South Africa (Cape)

Siliquamomum Baill. 1895. Zingiberaceae. 1 southern China, Indochina

Siloxerus Labill. 1806. Asteraceae. 3 southwestern Australia

Silphium L. 1753. Asteraceae. 23 eastern U.S.

Silvaea Phil. = Cistanthe Spach. Portulacaceae

Silvianthus Hook. f. 1868. Carlemanniaceae. 2 northeastern India, southern China, Indochina

Silviella Pennell. 1928. Scrophulariaceae. 2 Mexico

Silvorchis J. J. Sm. 1907. Orchidaceae. 1 Java

Silybum Adans. 1763. Asteraceae. 2 Central and Southeast Europe, Mediterranean; Asia from Caucasus and Asia Minor eastward to West Siberia, Middle Asia, China, Iran, Afghanistan, and northwestern India

Simaba Aubl. = Quassia L. Simaroubaceae

Simarouba Aubl. = Quassia L. Simaroubaceae

Simenia Szabó. 1940 (Cephalaria Schrad.). Dipsacaceae. 1 Ethiopia

Simethis Kunth. 1843. Asphodelaceae. 1 West and South Europe, North Africa

Simicratea N. Hallé. 1983 (Simirestis N. Hallé). Celastraceae. 1 trop. West Africa

Similisinocarum Cauwet et Farille = Pimpinella L. Apiaceae

Simira Aubl. 1775. Rubiaceae. 35 trop. America

Simirestis N. Hallé. 1958 (or = Hippocratea L.). Celastraceae. 8 trop. Africa from Sierra Leone to Ethiopia, south to Angola, Tanzania,

Simmondsia Nutt. 1844. Simmondsiaceae. 1 southwestern U.S. and northwestern Mexico

Simocheilus Klotzsch. 1838 (Eremia D. Don). Ericaceae. 11 South Africa (southwestern and southern Cape)

Simplicia T. Kirk. 1897. Poaceae. 2 New Zealand

Simsia Pers. 1807. Asteraceae. 18 southwestern U.S. (S. calva—New Mexico, Texas; S. lagascaeformis—Arizona), Mexico (12), Guatemala, El Salvador, Honduras, Nicaragua, Costa Rica, Panama (1), Venezuela, Colombia, Ecuador, Peru, Bolivia, Brazil, Argentina, Jamaica (1, S. foedita)

Sinacalia H. Rob. et Brettell. 1973 (Ligularia Cass.). Asteraceae. 4 China

Sinadoxa C. Y. Wu, Z. L. Wu et R. F. Huang. 1982 (Adoxa L.). Adoxaceae. 1 southwestern China (Qinghai, Yushu)

Sinapidendron Lowe. 1831. Brassicaceae. 5–6 Madeira Is., Canary Is.

Sinapis L. 1753 (Brassica L.). Brassicaceae. 10 Europe, Mediterranean; West (incl. Caucasus), Southwest, and Middle Asia; Russia (all areas)

Sinarundinaria Nakai. 1935 (Fargesia Franch.). Poaceae. c. 50 Asia from Pakistan through Himalayas to China and Japan, southward to southern India, Sri Lanka, Burma, and Indochina; 2 Mexico, El Salvador, Nicaragua; S. alpina—western Cameroun, Zaire, Sudan, Ethiopia, Uganda, Kenya, Tanzania; 2 Madagascar

Sinclairia Hook. et Arn. 1841. Asteraceae. 25 Mexico, Central America, Colombia

Sincorea Ule = Orthophytum Beer. Bromeliaceae

Sindechites Oliv. 1888. Apocynaceae. 3 central and southwestern China, Thailand

Sindora Miq. 1861. Fabaceae. 18–20 Gabon (1, S. klaineana), Southeast Asia, Malesia

Sindoropsis J. Léonard. 1957. Fabaceae. 1 Gabon

Sindroa Jum. = Orania Zipp. Arecaceae

Sineoperculum Jaarsveld = Dorotheanthus Schwantes. Aizoaceae

Singana Aubl. 1775. ? Fabaceae. 1 Guyana

Sinia Diels. 1930. Sauvagesiaceae. 1 southeastern China

Sinningia Nees. 1825. Gesneriaceae. 75 trop. America from Mexico to northern Argentina and Uruguay

Sinoadina Ridsdale. 1978. Rubiaceae. 1 Burma, Thailand, southern and southeastern China, Taiwan, Japan (Shikoku, Kyushu, Ryukyu)

Sinobacopa D. Y. Hong = Bacopa Aubl. Scrophulariaceae

Sinobambusa Makino ex Nakai. 1925. Poaceae. 17 East and Southeast Asia

Sinocalamus McClure. 1940 (Dendrocalamus Nees). Poaceae. 10 South and Southeast Asia

Sinocalycanthus (C. C. Cheng et S. Y. Chang) C. C. Cheng et S. Y. Chang = Calycanthus L. Calycanthaceae

Sinocarum H. Wolff ex Shan et Pu. 1927. Apiaceae. 11 China, Sikkim (3)

Sinochasea Keng. 1958 (Pseudodanthonia Bor et C. E. Hubb.). Poaceae. 1 China

Sinocrassula A. Berger. 1930. Crassulaceae. 8–9 Himalayas, southwestern China

Sinodielsia H. Wolff. 1925. Apiaceae. 4 China

Sinodolichos Verdc. 1970. Fabaceae. 2 southern China

Sinofranchetia Hemsl. 1907. Lardizabalaceae. 1 central and eastern China from Yunnan to Honan and Kiangsu to Kwangsi and Hunan

Sinoga S. T. Blake = Asteromyrtus Schauer. Myrtaceae

Sinojackia Hu. 1928. Styracaceae. 4 southern China

Sinojohnstonia Hu. 1936. Boraginaceae. 3 western China

Sinoleontopodium Y. L. Chen. 1985. Asteraceae. 1 Tibet

Sinolimprichtia H. Wolff. 1922. Apiaceae. 1 eastern Tibet

Sinomenium Diels. 1910. Menispermaceae. 1 East Asia

Sinomerrillia Hu. 1937. Celastraceae. 1 southwestern China

Sinopanax Li. 1949. Araliaceae. 1 Taiwan

Sinopimelodendron Tsiang = Cleidiocarpon Airy Shaw. Euphorbiaceae

Sinoplagiospermum Rauschert = Prinsepia Royle. Rosaceae

Sinopodophyllum Ying. 1979. Berberidaceae. 1 southwestern China

Sinopyrenaria Hu = Pyrenaria Blume. Theaceae

Sinoradlkofera F. G. Mey. = Boniodendron Gagnep. Sapindaceae

Sinorchis S. C. Chen. 1978 (Aphyllorchis Blume). Orchidaceae. 1 China

Sinosassafras H. W. Li. 1985 (or = Lindera Thunb.). Lauraceae. 1 China

Sinosenecio B. Nord. 1978. Asteraceae. 31 East and Southeast Asia

Sinosideroxylon (Engl.) Aubrév. = Sideroxylon L. Sapotaceae

Sinowilsonia Hemsl. 1906. Hamamelidaceae. 1 western and central China

Sinthroblastes Bremek. 1957 (Strobilanthes Blume). Acanthaceae. 1 Timor I.

Siolmatra Baill. 1885. Cucurbitaceae. 3 South America

Sipanea Aubl. 1775. Rubiaceae. 18 trop. South America

Sipaneopsis Steyerm. 1967. Rubiaceae. 7 trop. South America

Sipapoa Maguire = Diacidia Griseb. Malpighiaceae

Sipapoantha Maguire et B. M. Boom. 1989. Gentianaceae. 1 Venezuela

Siparuna Aubl. 1775. Monimiaceae. c. 150 trop. America from Mexico to Peru, Bolivia, and Brazil; West Indies

Siphanthera Pohl ex DC. 1828. Melastomataceae. 20 Guyana, Brazil

Siphantheropsis Brade = Macairea DC. Melastomataceae

Siphocampylus Pohl. 1830–1831. Lobeliaceae. 215 trop. America, West Indies

Siphocodon Turcz. 1852. Campanulaceae. 2–3 South Africa (Cape)

Siphocranion Kudo. 1929. Lamiaceae. 3 Tibet, northern India, northern Burma, southern China, northern Vietnam

Siphokentia Burret. 1927. Arecaceae. 2 Moluccas (Obi I., Kahatola I.)

Siphonandra Klotzsch. 1951. Ericaceae. 3 Andes

Siphonandrium K. Schum. 1905. Rubiaceae. 1 New Guinea

Siphonella (Torr. et A. Gray) Small = Valerianella Hill. Valerianaceae

Siphoneugenia O. Berg. 1856 (Eugenia L.). Myrtaceae. 8 Puerto Rico, Lesser Antilles, Venezuela, the Guianas, Brazil, southern Bolivia

Siphonochilus J. M. Wood et Franks. 1911. Zingiberaceae. 16 trop. and South Africa

Siphonodon Griff. 1844. Celastraceae. 7 Southeast Asia, Malesia, northeastern Australia

Siphonoglossa Oerst. 1854. Acanthaceae. 8 trop. America, trop. and South Africa (1)

Siphonosmanthus Stapf = Osmanthus Lour. Oleaceae

Siphonostegia Benth. 1835. Scrophulariaceae. 3 Southeast Europe and Asia Minor (1), East Asia (2); Russia (1, S. chinensis: Far East)

Siphonostelma Schltr. = Brachystelma R. Br. Asclepiadaceae

Siphonostylis Wern. Schulze. 1965 (Iris L.). Iridaceae. 3 Mediterranean, Asia Minor, Caucasus (1, S. lazica)

Siphonychia Torr. et A. Gray = Paronychia Hill. Caryophyllaceae

Sipolisia Glaz. ex Oliv. 1894. Asteraceae. 1 Brazil

Siraitia Merr. 1934. Cucurbitaceae. 4 eastern Himalayas, southwestern and southern China, Thailand, Indochina, Indonesia

Sirhookera Kuntze. 1891. Orchidaceae. 2 southern India, Sri Lanka

Sison L. 1753. Apiaceae. 2 West and South Europe, West (incl. Caucasus) and Southwest Asia

Sisymbrella Spach. 1838. Brassicaceae. 5 Southwest Europe, western and central Mediterranean

Sisymbriopsis Botsch. et Tzvelev. 1961. Brassicaceae. 2 Kirgizia, Tajikistan, western China, Tibet

Sisymbrium L. 1753. Brassicaceae. c. 100 North and South (esp. Andes) America, temp. Eurasia, Mediterranean, South Africa

Sisyndite E. Mey. ex Sond. 1860. Zygophyllaceae. 1 South Africa

Sisyranthus E. Mey. 1838. Asclepiadaceae. 12 southern trop. and South Africa

Sisyrinchium L. 1753. Iridaceae. c. 100 North, Central, and South America, West Indies; S. montanum—Europe to eastern European part of Russia; S. iridifolium—naturalized in New Guinea, Australia, Tasmania, New Zealand; Hawaiian Is. (1, S. acre)

Sisyrolepis Radlk. 1905. Sapindaceae. 1 Thailand, Cambodia

Sitanion Raf. 1819. Poaceae. 4 western North America from Canada to Mexico

Sitella L. H. Bailey = Waltheria L. Sterculiaceae

Sium L. 1753. Apiaceae. 16 North America, Eurasia, North, trop. and South Africa

Sixalix Raf. 1838 (Scabiosa L.). Dipsacaceae. 10 Mediterranean

Skapanthus C. Y. Wu et H. W. Li. 1975 (Plectranthus L'Hér.). Lamiaceae. 1 southwestern China

Skeptrostachys Garay = Stenorrhynchos Rich. ex Spreng. Orchidaceae

Skiatophytum L. Bolus. 1927. Aizoaceae. 1 South Africa (Cape)

Skimmia Thunb. 1783. Rutaceae. 7–8 Afghanistan, Pakistan, Himalayas from Kashmir to Bhutan, northeastern India, China, Japan, Indochina, Philippines; Russia (1, S. repens—Sakhalin I., Kuril Is.)

Skoliopteris Cuatrec. = Clonodia Griseb. Malpighiaceae

Skottsbergianthus Boelcke = Xerodraba Skottsb. Brassicaceae

Skottsbergiella Boelcke = Xerodraba Skottsb. Brassicaceae

Skottsbergiella Epling = Cuminia Colla. Lamiaceae

Skottsbergiliana St. John = Sicyos L. Cucurbitaceae

Skytanthus Meyen. 1834. Apocynaceae. 3 Brazil, Chile

Sladenia Kurz. 1873. Theaceae. 1 Burma, southern China (Yunnan), northern Thailand

Sleumerodendron Virot. 1868. Proteaceae. 1 New Caledonia

Sloanea L. 1753. Elaeocarpaceae. c. 100–120 Madagascar (1), trop. Asia, Malesia, Australia, New Caledonia, trop. America

Sloetiopsis Engl. 1907 (Streblus Lour.). Moraceae. 1 trop. Africa from Guinea to Zaire, coastal Kenya, Tanzania, Mozambique

Smallanthus Mack. 1933. Asteraceae. 20 trop. and subtrop. America from eastern U.S. to Argentina

Smeathmannia Sol. ex R. Br. 1821. Flacourtiaceae. 2 trop. West Africa

Smelophyllum Radlk. 1878. Sapindaceae. 1 South Africa

Smelowskia C. A. Mey. 1830. Brassicaceae. 14: 8 Arctica, West and East Siberia, Kazakhstan, Middle Asia; Afghanistan, Pakistan, China, Mongolia, Canada from Yukon to Alberta, U.S. from Alaska to Colorado

Smicrostigma N. E. Br. 1930. Aizoaceae. 1 South Africa

Smilacina Desf. 1807. Convallariaceae. 25 continental China, Taiwan, Himalayas, India, Burma, Vietnam, Malesia, America from Canada to Guatemala; Russia (3, East Siberia, Far East)

Smilax L. 1753. Smilacaceae. 300–350 tropics and subtropics, few species in warm–temp. Europe, Asia, and North America; Caucasus (2), Russian Far East (1)

Smirnovia Bunge. 1876 (Sphaerophysa DC.). Fabaceae. 1 Middle Asia, Iran, Afghanistan

Smithia Aiton. 1789. Fabaceae. c. 30 trop. Old World, chiefly Madagascar and Asia

Smithiantha Kuntze. 1891. Gesneriaceae. 4 Mexico

Smithiella Dunn. = Aboriella Bennet. Urticaceae

Smithiodendron Hu = Broussonetia L'Hér. ex Vent. Moraceae

Smithorchis T. Tang et F. T. Wang. 1936. Orchidaceae. 1 China

Smithsonia C. J. Saldanha. 1974. Orchidaceae. 3 western India

Smitinandia Holttum. 1969. Orchidaceae. 2 India, Nepal, Sikkim, Bhutan, Burma, Andaman Is., Indochina, Malay Peninsula

Smodingium E. Mey. ex Sond. 1860. Anacardiaceae. 1 South Africa

Smyrniopsis Boiss. 1844. Apiaceae. 1 southern Transcaucasia, Asia Minor, Iraq, Iran

Smyrnium L. 1753. Apiaceae. 7 Europe, Northwest and North Africa, Southwest Asia eastward to Afghanistan; 3 Crimea, Caucasus, and Middle Asia

Smythea Seem. ex A. Gray. 1862. Rhamnaceae. 7 northeastern India, Burma, southern China, Andaman Is., Malesia, Oceania

Snowdenia C. E. Hubb. 1929. Poaceae. 4 trop. Northeast (Sudan, Ethiopia) and East (Uganda, Kenya, Tanzania) Africa, southern Arabian Peninsula (Yemen)

Soaresia Sch. Bip. 1863. Asteraceae. 1 southern Brazil

Sobennikoffia Schltr. 1925. Orchidaceae. 4 Mascarene Is.

Sobolewskia M. Bieb. 1832. Brassicaceae. 4 Crimea, Caucasus, eastern Turkey

Sobralia Ruiz et Pav. 1794. Orchidaceae. 40 trop. America from Mexico to Peru, Bolivia, and Brazil

Socotora Balf. f. 1883 (Periploca L.). Asclepiadaceae. 1 trop. Northeast Africa, Socotra

Socotranthus Kuntze. 1903. Asclepiadaceae. 1 Socotra

Socotria Levin = Punica L. Punicaceae

Socratea Karst. 1857. Arecaceae. 5 Nicaragua, Costa Rica, Colombia, Ecuador, Peru, Venezuela, Guyana, Surinam, French Guiana, Brazil, Bolivia

Socratina Balle. 1964. Loranthaceae. 2 Madagascar

Soderstromia C. Morton. 1966. Poaceae. 1 Mexico, Central America

Sodiroa E. F. André = Guzmania Ruiz et Pav. Bromeliaceae

Sodiroella Schltr. = Stellilabium Schltr. Orchidaceae

Soehrensia (Backeb.) Backeb. = Echinopsis Zucc. Cactaceae

Soemmerignia Mart. 1828. Fabaceae. 1 trop. South America

Sogerianthe Danser. 1933. Loranthaceae. 4 New Guinea, Solomon Is.

Sohnsia Airy Shaw. 1965. Poaceae. 1 Mexico

Soja Moench = Glycine Willd. Fabaceae

Solandra Sw. 1787. Solanaceae. 10 trop. America from Mexico to Peru, Bolivia, and southeastern Brazil; West Indies

Solanecio Sch. Bip. ex Welp. 1846. Asteraceae. 15 trop. Africa, Madagascar, southern Arabian Peninsula

Solanoa Greene = Asclepias L. Asclepiadaceae

Solanum L. 1753. Solanaceae. c. 1,500 (or 1,400–2,000) almost cosmopolitan, except Arctica and Antarctica, chiefly trop. and subtrop. America (1,000–1,100), secondary centers in Africa and Australia (c. 110)

Solaria Phil. 1858. Alliaceae. 3 Chile

Soldanella L. 1753. Primulaceae. 11 montane South and Central Europe, incl. the Carpathian Mts. (2)

Soleirolia Gaudich. 1830. Urticaceae. 1 Corsica, Sardinia, Italy

Solena Lour. 1790. Cucurbitaceae. 2 Afghanistan, Himalayas, India, Sri Lanka, Burma, southwestern China, Taiwan, Indochina, Malesia, Australia

Solenachne Steud. 1853 (Spartina Schreb.). Poaceae. 1 temp. South America

Solenangis Schltr. 1918. Orchidaceae. 5 trop. Africa, Madagascar, Mascarene Is.

Solenanthus Ledeb. 1829 (Cynoglossum L.). Boraginaceae. 16 Mediterranean, Crimea, West (incl. Caucasus) and Middle Asia, southern West Siberia, western China, and western Himalayas

Solenidiopsis Senghas. 1986. Orchidaceae. 1 Andes of Peru

Solenidium Lindl. 1846. Orchidaceae. 4 Costa Rica, trop. South America

Solenixora Baill. = Coffea L. Rubiaceae

Solenocarpus Wight et Arn. 1834 (or = Spondias L.). Anacardiaceae. 1 India

Solenocentrum Schltr. 1911. Orchidaceae. 1 Costa Rica

Solenogyne Cass. 1827. Asteraceae. 2 Australia

Solenomelus Miers. 1842. Iridaceae. 3 Chile

Solenophora Benth. 1840. Gesneriaceae. 16 Mexico, Central America

Solenopsis C. Presl. 1836. Lobeliaceae. 25 Mediterranean, South Africa, Australia, Tasmania, New Zealand, Society Is., western trop. America south to Bolivia, West Indies

Solenoruellia Baill. = Henrya Nees ex Benth. Acanthaceae

Solenospermum Zoll. = Lophopetalum Wight ex Arn. Celastraceae

Solenostemma Hayne. 1825. Asclepiadaceae. 1 Algeria, Libya, Egypt, Mali, Niger, Chad; Israel, Jordan, Arabian Peninsula

Solenostemon Thonn. 1827 (Plectranthus L'Hér.). Lamiaceae. 60 trop. and South (2) Africa, trop. Asia, Malesia

Solidago L. 1753. Asteraceae. c. 100 America, Macaronesia, Eurasia

Soliva Ruiz et Pav. 1794. Asteraceae. 9 South America

Sollya Lindl. 1832. Pittosporaceae. 3 southwestern Australia

Solms-Laubachia Muschler. 1912. Brassicaceae. 13 China (Tibet, Szechwan, Yunnan), eastern Himalayas

Solmsia Baill. 1871. Thymelaeaceae. 2 New Caledonia

Solonia Urb. 1922. Myrsinaceae. 1 Cuba

Sommera Schltdl. 1835. Rubiaceae. 12 trop. America

Sommerfeltia Less. 1832. Asteraceae. 2 southern Brazil, Argentina, Uruguay

Sommieria Becc. 1877. Arecaceae. 3 New Guinea

Somphoxylon Eichler = Odontocarya Miers. Menispermaceae

Sonchus L. 1753. Asteraceae. c. 70 islands of the Atlantic ocean, Mediterranean, trop. Africa, Eurasia

Sonderina H. Wolff. 1927. Apiaceae. 5 South Africa

Sonderothamnus R. Dahlgren. 1968. Penaeaceae. 2 South Africa

Sondottia P. S. Short. 1989. Asteraceae. 2 western Australia

Sonerila Roxb. 1820. Melastomataceae. c. 170 trop. and subtrop. Asia, Malesia

Sonnea Greene = Plagiobothrys Fisch. et C. A. Mey. Boraginaceae

Sonneratia L. f. 1782. Sonneratiaceae. 6–7 coastal trop. East Africa and Madagascar, South and Southeast Asia east to Hainan and Ryukyu Is., Malesia, Micronesia, Solomon Is., coastal northern Australia (3), New Caledonia, Vanuatu (New Hebrides)

Sooia Pócs. 1973 (Epiclastopelma Lindau). Acanthaceae. 1 Tanzania

Sophiopsis O. E. Schulz. 1924. Brassicaceae. 4 montane Kazakhstan, Middle Asia, Pakistan, western China

Sophora L. 1753. Fabaceae. 52 almost cosmopolitan; S. flavescens—southern European part of Russia, Caucasus, West Siberia, Middle Asia, Russian Far East

Sophronanthe Benth ex Lindl. = Gratiola L. Scrophulariaceae

Sophronitella Schltr. 1925. Orchidaceae. 1 Brazil

Sophronitis Lindl. 1828. Orchidaceae. 8 Paraguay, eastern Brazil

Sopubia Buch.-Ham. ex D. Don. 1825. Scrophulariaceae. c. 60 trop. and South Africa, Madagascar, western and central China, Himalayas, India, Indochina, Taiwan, Malesia, Australia (1, Queensland)

Soranthus Ledeb. = Ferula L. Apiaceae

Sorbaria (Ser. ex DC.) A. Braun. 1860. Rosaceae. 9 Afghanistan, northwestern India, Nepal, Mongolia, China, Korea, Japan, North America; 6 Middle Asia, Siberia, Russian Far East

Sorbus L. 1753. Rosaceae. c. 260 temp.,

subtrop., and trop. regions of the Northern Hemisphere, esp. China (c. 90)

Sorghastrum Nash. 1901. Poaceae. 16 trop. America, Africa

Sorghum Moench. 1794. Poaceae. c. 30 tropics and few in temp. regions; 8 European part of Russia, Caucasus, Middle Asia, southern Siberia and southern Russian Far East

Soridium Miers. 1850. Triuridaceae. 1 Central and northeastern South America

Sorindeia Thouars. 1806. Anacardiaceae. 40 trop. Africa, Madagascar, Mascarene Is.

Sorocea A. St.-Hil. 1821. Moraceae. 16 trop. America

Sorocephalus R. Br. 1810. Proteaceae. 11 South Africa (Cape)

Soroseris Stebb. 1940. Asteraceae. 8 India, Himalayas, Tibet, western China

Sorostachys Steud. 1850 (Cyperus L.). Cyperaceae. 2 trop. Africa, Madagascar, trop. Asia, Australia

Sosnovskya Takht. = Centaurea L. Asteraceae

Soterosanthus Lehm. ex Jenny. 1986. Orchidaceae. 1 Central America

Soulamea Lam. 1785. Simaroubaceae. 8 Seychelles (1, Mahé I.), Borneo, Moluccas, New Guinea, Melanesia to New Caledonia (6)

Souliea Franch. 1898. Ranunculaceae. 1 China

Souroubea Aubl. 1775. Marcgraviaceae. 20 trop. America, West Indies

Sowerbaea Sm. 1798–1799. Asphodelaceae. 5 northern (1), southwestern (2), and eastern Australia; Tasmania (1)

Soyauxia Oliv. 1882. Flacourtiaceae. 5 trop. West Africa

Soymida A. Juss. 1830. Meliaceae. 1 India, Sri Lanka

Spachea A. Juss. 1838. Malpighiaceae. 6 trop. America, West Indies

Spananthe Jacq. 1791. Apiaceae. 1 Mexico, Central and trop. South America

Spaniopappus B. L. Rob. 1926. Asteraceae. 5 Cuba

Spanizium Griseb. 1843 (or = Saponaria L.). Caryophyllaceae. 1 southwestern Georgia, Asia Minor, northern Syria, Iran

Sparattanthelium Mart. 1841. Hernandiaceae. 13 trop. South America; S. amazonum—also in Mexico, Guatemala, and Honduras

Sparattosperma Mart. ex Meisn. 1840. Bignoniaceae. 2 Venezuela, Peru, Brazil, Bolivia, Paraguay

Sparattosyce Bureau. 1869. Moraceae. 2 New Caledonia

Sparaxis Ker-Gawl. 1804. Iridaceae. 13 South Africa (northern and southwestern Cape)

Sparganium L. 1753. Sparganiaceae. 20 cold, temp. and warm regions of the Northern Hemisphere, esp. Russia (c. 10, all areas); Sumatra, New Guinea, southeastern Australia, New Zealand

Sparganophorus Boehm. 1760. Asteraceae. 1 pantropics

Sparmannia L. f. = Sparrmannia L. f. Tiliaceae

Sparrea Hunz. et Dottori = Celtis L. Ulmaceae

Sparrmannia L. f. 1782. Tiliaceae. 3 trop. and South Africa, Madagascar

Spartidium Pomel. 1874 (Lebeckia Thunb.). Fabaceae. 1 Morocco, Algeria, Tunisia, Libya

Spartina Schreb. 1789. Poaceae. 20 America, Atlantic coastal West Europe and West Africa

Spartium L. 1753. Fabaceae. 1 Canary Is., Mediterranean (excl. Cyprus and Egypt), Crimea

Spartochloa C. E. Hubb. 1952. Poaceae. 1 southwestern Australia

Spartothamnella Briq. 1895. Chloanthaceae. 3 Australia

Spatalla Salisb. 1807. Proteaceae. 21 South Africa (Cape)

Spatallopsis E. Phillips = Spatalla Salisb. Proteaceae

Spathacanthus Baill. 1891. Acanthaceae. 5 Mexico, Central America

Spathandra Guill. et Perr. 1833. Melastomataceae. 6 trop. Africa

Spathantheum Schott. 1859. Araceae. 2

Andes of Bolivia and northern Argentina

Spathanthus Desv. 1828. Rapateaceae. 2 northern South America

Spathelia L. 1762. Rutaceae. 20 West Indies, northern South America

Spathia Ewart. 1917. Poaceae. 1 trop. Australia

Spathicalyx J. C. Gómes. 1956. Bignoniaceae. 2 Brazil

Spathicarpa Hook. 1831. Araceae. 7 southern Brazil, Paraguay, northern Argentina

Spathichlamys R. Parker. 1931. Rubiaceae. 1 Burma

Spathidolepis Schltr. = Dischidia R. Br. Asclepiadaceae

Spathionema Taub. 1895. Fabaceae. 1 Kenya, Tanzania

Spathiostemon Blume. 1826. Euphorbiaceae. 3 Thailand, western Malesia, New Guinea

Spathipappus Tzvelev = Tanacetum L. Asteraceae

Spathiphyllum Schott. 1832. Araceae. 45 Philippines, Palau Is., Moluccas, New Guinea, Bismarck Arch., Solomon Is., Central and trop. South America

Spathodea P. Beauv. 1805. Bignoniaceae. 1 (S. campanulata) trop. Africa; cult. tree in tropics

Spathodeopsis Dop = Fernandoa Welw. ex Seem. Bignoniaceae

Spathoglottis Blume. 1825. Orchidaceae. c. 40–50 trop. Asia: Himalayas, India, Sri Lanka, Burma, Andaman Is., Indochina, China, Taiwan, Ryukyu Is.; Malesia from Malay Peninsula to New Guinea, northern Australia, Micronesia, New Britain I., Bougainville and Solomon Is., Vanuatu, Horn Is., New Caledonia, Fiji, Samoa, Tonga, Niue, and Tahiti Is.

Spatholirion Ridl. 1896. Commelinaceae. 3 southern China, Indochina

Spatholobus Hassk. 1842. Fabaceae. 15 trop. Asia

Spathulata (Borisova) A. Löve et D. Löve = Sedum L. Crassulaceae

Speea Loes. 1927. Alliaceae. 1–2 Chile

Speirantha Baker. 1875. Convallariaceae. 1 eastern China

Spenceria Trimen. 1879. Rosaceae. 2 western China

Speranskia Baill. 1858. Euphorbiaceae. 2 China

Spergula L. 1753. Caryophyllaceae. 8 temp. Eurasia, esp. Russia (6, European part, Caucasus, Siberia, Far East) and Mediterranean; northern Patagonia (1)

Spergularia (Pers.) J. Presl et C. Presl. 1819. Caryophyllaceae. 25–40 cosmopolitan

Spermacoce L. 1753. Rubiaceae. 150–200 trop. America, few sp. in trop. and subtrop. Africa

Spermadictyon Roxb. 1815. Rubiaceae. 1–6 Himalayas from Kashmir to Bhutan, India, southern China

Spermolepis Raf. 1825. Apiaceae. 6 Hawaiian Is. (1), North America (4), Argentina (1)

Sphacanthus Benoist. 1939. Acanthaceae. 2 Madagascar

Sphacele Benth. 1829 (Lepechinia Willd.). Lamiaceae. 26 Mexico, Venezuela, Andes, southeastern Brazil (1)

Sphacophyllum Benth. = Anisopappus Hook. et Arn. Asteraceae

Sphaenolobium Pimenov. 1975. Apiaceae. 3 Middle Asia (Tien Shan)

Sphaeradenia Harling. 1954. Cyclanthaceae. Over 40 Nicaragua, Costa Rica, Panama, Colombia, western Venezuela, Ecuador, Peru, Bolivia

Sphaeralcea A. St.-Hil. 1825. Malvaceae. c. 60 arid areas of western U.S. and northern Mexico, subtrop. and temp. South America, esp. Argentina; South Africa

Sphaeranthus L. 1753. Asteraceae. 38 Africa (except northwestern), Madagascar, Iraq, Iran, India, Southeast Asia, western Malesia, Sulawesi, northeastern Australia

Sphaerantia Peter G. Wilson et B. Hyland. 1988. Myrtaceae. 2 Australia (Queensland)

Sphaereupatorium (O. Hoffm.) Kuntze ex

B. L. Rob. 1920. Asteraceae. 1 Bolivia, Brazil

Sphaerobambos S. Dransf. 1989. Poaceae. 3 Borneo, Philippines, Sulawesi

Sphaerocardamum Schauer. 1847. Brassicaceae. 1 Mexico

Sphaerocaryum Nees ex Hook. f. 1896. Poaceae. 1 trop. Asia from India to China and Taiwan, Malesia

Sphaeroclinium (DC.) Sch. Bip. 1844. Asteraceae. 1 South Africa

Sphaerocodon Benth. 1876. Asclepiadaceae. 2 Africa

Sphaerocoma T. Anderson. 1861. Caryophyllaceae. 2 Egypt, northeastern Sudan, eastern Somalia, southern Arabian Peninsula, Iran

Sphaerocoryne Scheff. ex Ridl. 1917. Annonaceae. 2 East Africa: Kenya, Tanzania, Mozambique

Sphaerocyperus Lye. 1972 (Cyperus L.). Cyperaceae. 1 trop. Africa

Sphaerolobium Sm. 1805. Fabaceae. 15 Australia

Sphaeromariscus E. G. Camus = Cyperus L. Cyperaceae

Sphaeromeria Nutt. 1841. Asteraceae. 6 U.S. (Utah, Nevada, Oregon, California, Montana, Wyoming, Colorado), Mexico (Baja California)

Sphaeromorphaea DC. 1838 (Epaltes Cass.). Asteraceae. 1 East and Southeast Asia, Malesia, Australia

Sphaerophysa DC. 1825. Fabaceae. 2 Turkey, Syria, Iraq, Caucasus, Middle Asia, Iran, Afghanistan, southern West and East Siberia, Mongolia, China, western Tibet

Sphaerosacme Wall. ex Royle. 1835 (Lansium Corrêa). Meliaceae. 1 Himalayas from Nepal to Bhutan

Sphaerosciadium Pimenov et Kljuykov. 1981. Apiaceae. 1 Middle Asia (Gissar Mts.)

Sphaerostylis Baill. 1858. Euphorbiaceae. 5 trop. Africa, Madagascar, western Malesia

Sphaerothylax Bisch. ex Krauss. 1844. Podostemaceae. 10 trop. and South Africa, Madagascar

Sphaerotylos C. J. Chen = Sarcochlamys Gaudich. Urticaceae

Sphagneticola O. Hoffm. 1900. Asteraceae. 1 southeastern Brazil

Sphallerocarpus Besser ex DC. 1929. Apiaceae. 1 East Siberia, Russian Far East, Mongolia, East Asia

Sphalmanthus N. E. Br. 1925 (Phyllobolus N. E. Br.). Aizoaceae. c. 100 South Africa

Sphalmium B. G. Briggs, B. Hyland et L. A. S. Johnson. 1975. Proteaceae. 1 Australia (northern Queensland)

Sphedamnocarpus Planch. ex Benth. 1862. Malpighiaceae. 18 trop. and South Africa, Madagascar

Sphenandra Benth. = Sutera Roth. Scrophulariaceae

Spheneria Kuhlm. 1922. Poaceae. 1 trop. South America

Sphenocarpus Korovin = Seseli L. Apiaceae

Sphenocentrum Pierre. 1898. Menispermaceae. 1 West Africa: Ivory Coast, Ghana, Nigeria, Cameroun

Sphenoclea Gaertn. 1788. Sphenocleaceae. 2 West Africa (1), trop. Old World (1); S. zeylanica—extending to southern Middle Asia

Sphenodesme Jack. 1820. Symphorematanceae. 23 India, Sri Lanka, Bangladesh, Burma, Indochina, Hainan, Borneo

Sphenopholis Scribn. 1906. Poaceae. 5 U.S., Mexico, West Indies

Sphenopus Trin. 1822. Poaceae. 2 Mediterranean, West Asia; S. divaricatus—extending to Caucasus and Middle Asia

Sphenosciadium A. Gray. 1865. Apiaceae. 1–2 western U.S.

Sphenostemon Baill. 1875. Sphenostemonaceae. 8: 3 Sulawesi, Moluccas, New Guinea; 1 Australia (Queensland); 4 New Caledonia

Sphenostigma Baker = Gelasine Herb. Iridaceae

Sphenostylis E. Mey. 1836. Fabaceae. 7: 6 trop. and South (2) Africa; 1 (S. bracteata) India

Sphenotoma (R. Br.) Sweet. 1828. Epacridaceae. 6 southwestern Australia

Sphinctacanthus Benth. 1876. Acanthaceae. 1 India, Bangladesh, Burma, China

Sphinctanthus Benth. 1841. Rubiaceae. 6 South America

Sphinctospermum Rose. 1906. Fabaceae. 1 southwestern U.S., northwestern Mexico

Sphingiphila A. H. Gentry. 1990. Bignoniaceae. 1 Paraguay

Sphyranthera Hook. f. 1887. Euphorbiaceae. 2 Andaman Is.

Sphyrarhynchus Mansf. 1935. Orchidaceae. 1 East Africa (Tanzania)

Sphyrastylis Schltr. 1920. Orchidaceae. 1 Colombia

Sphyrospermum Poepp. et Endl. 1835. Ericaceae. 20 trop. America from Mexico to Bolivia, West Indies

Spiculaea Lindl. 1840. Orchidaceae. 1 western Australia

Spigelia L. 1753. Spigeliaceae. 50 trop. and subtrop. America; S. anthelmia—naturalized in trop. Africa and trop. Asia

Spilanthes Jacq. 1760. Asteraceae. 6 trop. America (4, Mexico, Costa Rica, Colombia, Venezuela, Peru, Brazil, Bolivia, northern Chile, Paraguay, trop. Africa (S. costata—from Cameroun and Angola to Ethiopia, Uganda, and Tanzania); S. anactina—southern India, Sri Lanka, Malay Peninsula, Sumatra, Java, Lesser Sunda Is., Borneo, New Guinea, northern Australia

Spiloxene Salisb. 1866. Hypoxidaceae. 30 South Africa

Spinacia L. 1753. Chenopodiaceae. 3 Transcaucasia, Middle Asia, Turkey, Syria, Iraq, Iran, Afghanistan

Spinifex L. 1771. Poaceae. 4 South, Southeast, and East Asia; Malesia; Australia; Oceania

Spiniluma (Baill.) Aubrév. = Sideroxylon L. Sapotaceae

Spiracantha Kunth. 1818. Asteraceae. 1 Central America, Colombia

Spiradiclis Blume. 1827. Rubiaceae. 10 India, southern China, Indochina, western Malesia

Spiraea L. 1753. Rosaceae. 100–120 temp. Northern Hemisphere south to Himalayas, southern China, and Mexico

Spiraeanthemum A. Gray. 1854. Cunoniaceae. 6 New Britain I., Bougainville I., Solomon Is., Vanuatu (New Hebrides Is.), Fiji, Samoa Is.

Spiraeanthus (Fisch. et C. A. Mey.) Maxim. 1879. Rosaceae. 1 Middle Asia

Spiraeopsis Miq. = Caldcluvia D. Don. Cunoniaceae

Spiranthera A. St.-Hil. 1823. Rutaceae. 4 Venezuela, Guyana, Brazil, Bolivia

Spiranthes Rich. 1817. Orchidaceae. c. 30 trop., subtrop. and temp. regions; Russia (3, eastern European part, West and East Siberia, Far East), Caucasus

Spirella Costantin. 1912. Asclepiadaceae. 2 Indochina

Spiroceratium H. Wolff = Pimpinella L. Apiaceae

Spirodela Schleid. 1839. Lemnaceae. 4 temp. and trop. regions; S. polyrhiza—European part of Russia, Caucasus, Middle Asia, Siberia, Russian Far East

Spirogardnera Stauffer. 1968 (Choretrum R. Br.). Santalaceae. 1 southwestern Australia

Spirolobium Baill. 1889. Apocynaceae. 1 Southeast Asia

Spiropetalum Gilg. 1891. Connaraceae. 5 trop. West Africa

Spirorhynchus Kar. et Kir. 1842. Brassicaceae. 1 (S. sabulosus) Kazakhstan, Middle Asia, Iran, Pakistan, western China

Spiroseris Rech. f. 1977. Asteraceae. 1 Pakistan

Spirospermum Thouars. 1806. Menispermaceae. 1 Madagascar

Spirostachys Sond. 1850. Euphorbiaceae. 2: S. africana—Kenya, Tanzania, Zimbabwe, Angola, Namibia, Botswana, Swaziland, and South Africa (Transvaal, Natal, Cape); S. venenifera (Pax) Pax—Somalia, Kenya, Tanzania

Spirostegia Ivanina. 1955. Scrophulariaceae. 1 Middle Asia

Spirostigma Nees. 1847. Acanthaceae. 1 Brazil

Spirotecoma Baill. ex Dalla Torre et

Harms. 1891. Bignoniaceae. 4 West Indies (Cuba, Haiti)
Spirotheca Ulbr. = Ceiba Mill. Bombacaceae
Spirotropis Tul. 1844. Fabaceae. 1 northeastern South America
Spodiopogon Trin. 1822. Poaceae. 10 temp. and subtrop. Asia; S. sibiricus—southern East Siberia, Far East
Spondianthus Engl. 1905. Euphorbiaceae. 1 (S. preussii) trop. Africa from Liberia to southern Egypt, south to southern Angola and Tanzania
Spondias L. 1753. Anacardiaceae. 10–12 trop. Himalayas, southern China, Indochina, trop. America
Spondogona Raf. = Dipholis A. DC. Sapotaceae
Spondylantha C. Presl. 1831 (or = Vitis L.). Vitaceae. 1 Mexico
Spongiocarpella Yakovlev et N. Ulziykh. 1987. Fabaceae. 8 Mongolia, China, Tibet, Himalayas
Spongiola J. J. Wood et A. L. Lamb. 1993. Orchidaceae. 1 Borneo
Spongiosperma Zarucchi. 1988. Apocynaceae. 6 trop. South America
Spongiosyndesmus Gilli = Ladyginia Lipsky. Apiaceae
Sporadanthus F. Muell. 1874. Restionaceae. 1 New Zealand, Chatham Is.
Sporobolus R. Br. 1810. Poaceae. c. 160 warm–temp. and trop. regions
Sporoxeia W. W. Sm. 1917. Melastomataceae. 6 China, Indochina
Spraguea Torr. = Calyptridium Nutt. Portulacaceae
Spragueanella Balle. 1954. Loranthaceae. 1 trop. East Africa
Sprekelia Heist. 1755. Amaryllidaceae. 1 Mexico, Guatemala
Sprengelia Sm. 1794. Epacridaceae. 4 southeastern Australia, Tasmania
Spruceanthus Sleumer = Hasseltia Kunth. Flacourtiaceae
Spryginia Popov. 1923. Brassicaceae. 7 Turkmenistan, Uzbekistan, Afghanistan
Spuriodaucus C. Norman. 1930. Apiaceae. 3 trop. Africa
Spuriopimpinella (H. Boissieu) Kitag. 1941. (Pimpinella L.). Apiaceae. 4 East Asia
Spyridium Fenzl. 1837. Rhamnaceae. c. 30 temp. Australia
Squamellaria Becc. 1886. Rubiaceae. 3 Fiji
Squamopappus Jansen, Harriman et Urbatsch. 1982. Asteraceae. 1 Mexico, Guatemala, El Slavador, Honduras, Nicaragua, Costa Rica
Sredinskya (Stein) Fedorov. 1950 (Primula L.). Primulaceae. 1 Caucasus
Sreemadhavana Rauschert = Aphelandra R. Br. Acanthaceae
Staavia Dahl. 1787. Bruniaceae. 9 South Africa
Staberoha Kunth. 1841. Restionaceae. 9 South Africa (Cape)
Stachyacanthus Nees. 1847. Acanthaceae. 1 Brazil
Stachyandra J.-F. Leroy ex Radcl.-Sm. 1990 (Androstachys Prain). Euphorbiaceae. 4 Madagascar
Stachyanthus Engl. 1897. Icacinaceae. 6 trop. Africa
Stachyarrhena Hook. f. 1870. Rubiaceae. 10 trop. America
Stachycarpus Tiegh. = Prumnopitys Phil. Podocarpaceae
Stachycephalum Sch. Bip. ex Benth. 1872. Asteraceae. 2 Mexico (1), Argentina (1)
Stachydeoma Small = Hedeoma Pers. Lamiaceae
Stachyococcus Standl. 1936. Rubiaceae. 1 Peru
Stachyophorbe (Liebm. ex Mart.) Liebm. ex Klotzsch = Chamaedorea Willd. Arecaceae
Stachyopsis Popov et Vved. 1923. Lamiaceae. 4 Middle Asia, western China
Stachyothyrsus Harms. 1897. Fabaceae. 3 Sierra Leone, Liberia, Ivory Coast, Cameroun, Gabon, Zaire
Stachyphrynium K. Schum. 1902. Marantaceae. 14 India, Sri Lanka, southern China, Indochina, western Malesia
Stachys L. 1753. Lamiaceae. c. 450 temp., subtrop. and trop. regions, except Malesia, Australia, and New Zealand
Stachystemon Planch. 1845. Euphorbiaceae. 3 southwestern Australia

Stachytarpheta Vahl. 1804. Verbenaceae. 65 trop. and subtrop. America, few species in trop. Old World

Stachyurus Siebold et Zucc. 1836. Stachyuraceae. 16 western and central China, southern Tibet, Nepal, Bhutan, northeastern India, northern Burma, northern Indochina, Taiwan, Japan

Stackhousia Sm. 1798. Stackhousiaceae. 14 Australia and Tasmania (13); S. intermedia—extending to Malesia and Micronesia (Caroline Is.), New Zealand (1)

Stadiochilus Ros. M. Sm. 1980. Zingiberaceae. 1 Burma

Stadmannia Lam. 1793. Sapindaceae. 1 trop. East Africa, Madagascar, Mascarene Is.

Staehelina L. 1753. Asteraceae. 8 Mediterranean

Staëlia Cham. et Schltdl. 1828. Rubiaceae. 12 trop. South America

Stahelia Jonk. = Tapeinostemon Benth. Gentianaceae

Stahlia Bello. 1881. Fabaceae. 1 Puerto Rico

Stahlianthus Kuntze. 1891. Zingiberaceae. 6–7 eastern Himalayas, northeastern India, northern Indochina, Hainan, ? Philippines

Staintoniella Hara. 1974. Brassicaceae. 2 Tibet, Nepal, southern China (Yunnan)

Stalagmitis Murray = Garcinia L. Clusiaceae

Stalkya Garay. 1980. Orchidaceae. 1 Andes of Venezuela

Standleya Brade. 1932. Rubiaceae. 4 Brazil

Standleyacanthus Leonard. 1952. Acanthaceae. 1 Costa Rica

Standleyanthus R. M. King et H. Rob. 1971 (Eupatorium L.). Asteraceae. 1 Costa Rica

Stanfieldia Small. 1903. Asteraceae. 1 North America

Stanfieldiella Brenan. 1960. Commelinaceae. 4 trop. Africa

Stanfordia S. Watson = Caulanthus S. Watson. Brassicaceae

Stangea Graebn. = Valeriana L. Valerianaceae

Stangeria T. Moore. 1853. Stangeriaceae. 1 South Africa (Natal and Cape)

Stanhopea Frost ex Hook. 1829. Orchidaceae. 45 trop. America from southern Mexico to Peru and Brazil, Trinidad

Stanleya Nutt. 1818. Brassicaceae. 6 western U.S.

Stapelia L. 1753. Asclepiadaceae. c. 50 trop. and South Africa

Stapelianthus Choux ex A. C. Wight et B. Sloane. 1933. Asclepiadaceae. 10 Madagascar

Stapeliopsis Pillans. 1928 (Orbea Haw.). Asclepiadaceae. 6 South Africa

Stapfia Burtt Davy = Neostapfia Burtt Davy. Poaceae

Stapfiella Gilg. 1913. Turneraceae. 5 trop. Africa

Stapfiola Kuntze = Desmostachya (Hook. f.) Stapf. Poaceae

Stapfiophyton H. L. Li. 1944. Melastomataceae. 3 southern China

Staphylea L. 1753. Staphyleaceae. 11 temp. Northern Hemisphere

Stathmostelma K. Schum. 1893. Asclepiadaceae. 12 trop. Africa

Statice L. = Armeria Willd. + Limonium Mill. Plumbaginaceae

Staudtia Warb. 1897. Myristicaceae. 2–3 West Africa

Stauntonia DC. 1817. Lardizabalaceae. 15–25 Burma, southern China, Taiwan, Japan

Stauracanthus Link. 1808 (Ulex L.). Fabaceae. 2 Iberian Peninsula, Morocco, Algeria

Stauranthera Benth. 1835. Gesneriaceae. 11 India (Assam), China, Indochina, western Malesia, New Guinea

Stauranthus Liebm. 1853. Rutaceae. 1 southern Mexico

Staurochilus Ridl. ex Pfitzer. 1900. Orchidaceae. 6 Sikkim, northeastern India, Burma, Thailand, Laos, Cambodia, Vietnam, Malay Peninsula, Sumatra, Batoe Is., Anambas, Borneo, Philippines

Staurochlamys Baker. 1889. Asteraceae. 1 northern Brazil

Staurogyne Wall. 1831. Acanthaceae. c. 85 tropics, esp. western Malesia

Staurogynopsis Mangenot et Ake Assi = Staurogyne Wall. Acanthaceae

Staurophragma Fisch. et C. A. Mey. 1842 (Verbascum L.). Scrophulariaceae. 1 Asia Minor, Caucasus

Stauropsis Rchb. f. 1860 (Trichoglottis Blume). Orchidaceae. 1 China

Stawellia F. Muell. 1870 (Hensmania W. Fitz). Asphodelaceae. 2 southwestern Australia

Stayneria L. Bolus. 1961. Aizoaceae. 1 southwestern South Africa

Stebbinsia Lipsch. 1956. Asteraceae. 1 Middle Asia

Stebbinsoseris K. L. Chambers. 1991. Asteraceae. 2 California, Arizona, Baja California

Steenisia Bakh. f. 1952 (Neurocalyx Hook.). Rubiaceae. 5 Borneo, Natuna Is.

Stefanoffia H. Wolff. 1925. Apiaceae. 2–3 Balkan Peninsula, western Asia Minor

Steganotaenia Hochst. 1844. Apiaceae. 2 Northeast, trop., and northern South Africa

Steganotropis Lehm. = Centrosema (DC.) Benth. Fabaceae

Steganthera Perkins. 1898. Monimiaceae. 16 chiefly New Guinea (16); S. hirsuta—Sulawesi, Moluccas, Aru Is., Manus I., New Ireland and New Britain Is., Solomon Is. from Bougainville to San Cristobal and Rennell, Australia (northern Queensland)

Stegia DC. = Lavatera L. Malvaceae

Stegnosperma Benth. 1844. Stegnospermataceae. 4 Mexico, Central America south to Nicaragua, West Indies

Stegolepis Klotzsch ex Koern. 1872. Rapateaceae. 23 northern South America

Steinbachiella Harms = Diphysa Jacq. Fabaceae

Steinchisma Raf. 1830. Poaceae. 4 America from southern U.S. to Argentina

Steinheilia Decne. = Odontanthera Wight. Asclepiadaceae

Steirachne Ekman. 1911. Poaceae. 2 northeastern Brazil, Guyana, Venezuela

Steiractinia S. F. Blake. 1915. Asteraceae. 12 Colombia, Ecuador, Venezuela

Steirodiscus Less. 1832. Asteraceae. 6 South Africa

Steirosanchezia Lindau. 1904 (Sanchezia Ruiz et Pav.). Acanthaceae. 1 Peru

Steirotis Raf. = Struthanthus Mart. Loranthaceae

Stelechantha Bremek. 1940. Rubiaceae. 1 Angola

Stelechocarpus (Blume) Hook. f. et Thomson. 1855. Annonaceae. 5 Thailand, Malesia eastward to New Guinea

Steleocodon Gilli = Phalacraea DC. Asteraceae

Steleostemma Schltr. 1906. Asclepiadaceae. 1 Bolivia

Stelestylis Drude. 1881. Cyclanthaceae. 4 northern Venezuela, Guyana, Surinam, eastern Brazil

Stelis Sw. 1800. Orchidaceae. c. 300 trop. America from Mexico to Peru, Bolivia, and Brazil; West Indies

Stellaria L. 1753. Caryophyllaceae. c. 120 cosmopolitan

Stellariopsis (Baill.) Rydb. = Ivesia Torr. et A. Gray. Rosaceae

Stellera L. 1753 (Wikstroemia Endl.). Thymelaeaceae. 3 Russia (1, S. chamaejasme—East Siberia), Mongolia, northern and southwestern China, northwestern India, Bhutan, Taiwan

Stelleropsis Pobed. = Diarthron Turcz. Thymelaeaceae

Stelligera A. J. Scott. 1978 (Maireana Moq.). Chenopodiaceae. 1 temp. eastern Australia

Stellilabium Schltr. 1914. Orchidaceae. 2 Colombia, Peru

Stellularia Benth. = Benthamistella Kuntze. Scrophulariaceae

Stelmacrypton Baill. 1889 (Pentanura Blume). Asclepiadaceae. 1 northeastern India, southern China

Stelmagonum Baill. 1890. Asclepiadaceae. 2 trop. America

Stelmation Fourn. = Metastelma R. Br. Asclepiadaceae

Stelmatocodon Schltr. 1906. Asclepiadaceae. 1 Bolivia

Stemmacantha Cass. = Cirsium Hill. Asteraceae

Stemmadenia Benth. 1845. Apocynaceae. 20 trop. America from Mexico to Ecuador

Stemmatella Wedd. ex Benth. = Galinsoga Ruiz et Pav. Asteraceae

Stemmatium Phil. = Tristagma Poepp. et Endl. Alliaceae

Stemmatodaphne Gamble. 1910 (Alseodaphne Nees). Lauraceae. 1 Malay Peninsula, Borneo

Stemodia L. 1759. Scrophulariaceae. 37 trop. America, West Indies

Stemodiopsis Engl. 1897. Scrophulariaceae. 10 trop. and South Africa, Madagascar

Stemona Lour. 1791. Stemonaceae. c. 30 eastern India, Sri Lanka, Bangladesh, Burma, central, southwestern and southern China, Japan, Thailand, Malesia, northern and northeastern Australia

Stemonocoleus Harms. 1905. Fabaceae. 1 trop. West and Central Africa

Stemonoporus Thwaites. 1854. Dipterocarpaceae. 26 Sri Lanka

Stemonurus Blume. 1826. Icacinaceae. 12 India, Sri Lanka, Burma, Thailand, southern Vietnam, Malesia east to Solomon Is.

Stemotria Wettst. et Harms ex Engl. = Porodittia G. Don. Scrophulariaceae

Stenachaenium Benth. 1873. Asteraceae. 4 Argentina, southern Brazil

Stenadenium Pax = Monadenium Pax. Euphorbiaceae

Stenandriopsis S. Moore = Crossandra Salisb. Acanthaceae

Stenandrium Nees. 1836. Acanthaceae. c. 60 America from southern U.S. (New Mexico, Texas, Florida) to central Chile, Argentina and Brazil, West Indies

Stenanona Standl. 1929. Annonaceae. 2 Costa Rica, Panama

Stenanthella (A. Gray) Rydb. = Stenanthium (A. Gray) Kunth. Melanthiaceae

Stenanthium (A. Gray) Kunth. 1843. Melanthiaceae. 4 Sakhalin I. (1, S. sachalinense), Canada, U.S., Mexico

Stenia Lindl. 1837. Orchidaceae. 2 Venezuela, Trinidad, Guyana, Brazil, Peru

Stenocactus (K. Schum.) A. W. Hill. 1933. Cactaceae. c. 10 Mexico

Stenocarpha S. F. Blake = Galinsoga Ruiz et Pav. Asteraceae

Stenocarpus R. Br. 1810. Proteaceae. 25 Aru Is., New Guinea, northern and eastern Australia, New Caledonia

Stenocereus (A. Berger) Riccob. 1909. Cactaceae. 25 Mexico, Central America, Venezuela, Colombia; West Indies

Stenochasma Miq. 1851 (Broussonetia L'Hér. ex Vent.). Moraceae. 1 Sumatra

Stenochilus R. Br. = Eremophila R. Br. Myoporaceae

Stenocline DC. 1838. Asteraceae. 6 Brazil, Madagascar, Mauritius

Stenocoelium Ledeb. 1829. Apiaceae. 3–4 southeastern Kazakhstan and southern West Siberia (3), northwestern China, Mongolia

Stenocoryne Lindl. 1843 (Bifrenaria Lindl.). Orchidaceae. 10 northeastern South America

Stenodon Naudin. 1844. Melastomataceae. 2 southern Brazil

Stenodraba O. E. Schulz = Weberbauera Gilg et Muschl. Brassicaceae

Stenodrepanum Harms. 1921. Fabaceae. 1 Argentina

Stenofestuca (Honda) Nakai. 1950 (Bromus L.). Poaceae. 1 Japan

Stenoglossum Kunth. 1816 (Epidendrum L.). Orchidaceae. 1 Colombia, Venezuela, Ecuador

Stenoglottis Lindl. 1837. Orchidaceae. 3 trop. East and South (3) Africa

Stenogonum Nutt. 1848. Polygonaceae. 2 western U.S.: Wyoming, Utah, New Mexico, Colorado

Stenogyne Benth. 1830. Lamiaceae. 20 Hawaiian Is.

Stenolirion Baker = Ammocharis Herb. Amaryllidaceae

Stenolobium D. Don = Tecoma Juss. Boraginaceae

Stenomeria Turcz. 1852. Asclepiadaceae. 3 Colombia

Stenomeris Planch. 1852. Stenomeridaceae. 5 Malesia

Stenomesson Herb. 1821. Amaryllidaceae. 35 Andes

Stenopadus S. F. Blake. 1931. Asteraceae. 16 northwestern South Africa

Stenopetalum R. Br. ex DC. 1821. Brassicaceae. 9 western and southern Australia

Stenophalium Anderb. 1991. Asteraceae. 3 Brazil

Stenops B. Nord. 1978. Asteraceae. 1 Tanzania

Stenoptera C. Presl. 1827. Orchidaceae. 5 trop. South America

Stenorrhynchos Rich. ex Spreng. 1826. Orchidaceae. c. 80 trop. America

Stenoschista Bremek. = Ruelia L. Acanthaceae

Stenosemis E. Mey. ex Harv. et Sond. 1862 (Annesorhiza Cham. et Schltdl.). Apiaceae. 2 South Africa

Stenoseris C. Shih. 1991. Asteraceae. 5 Tibet, southern China (Yunnan)

Stenosiphon Spach. 1835. Onagraceae. 1 central and southern U.S.

Stenosiphonium Nees. 1832. Acanthaceae. 6 India, Sri Lanka

Stenosolen (Muell. Arg.) Markgr. = Tabernaemontana L. Apocynaceae

Stenosolenium Turcz. 1840. Boraginaceae. 1 East Siberia, Mongolia, China

Stenospermation Schott. 1858. Araceae. c. 300 trop. America from Nicaragua to Bolivia

Stenostelma Schltr. 1894. Asclepiadaceae. 5 southern trop. and South Africa

Stenostephanus Nees. 1847. Acanthaceae. 7 trop. South America

Stenotaenia Boiss. 1844 (Heracleum L.). Apiaceae. 5–6 Caucasus (1, S. daralaghezica), Asia Minor, Iran

Stenotaphrum Trin. 1822. Poaceae. 7 tropics and subtropics

Stenothyrsus C. B. Clarke. 1908. Acanthaceae. 1 Malay Peninsula

Stenotopsis Rydb. = Ericameria Nutt. Asteraceae

Stenotus Nutt. 1841. Asteraceae. 5 North America

Stephanachne Keng. 1934. Poaceae. 1 China

Stephanandra Siebold et Zucc. 1843. Rosaceae. 4 China, Korea, Japan

Stephania Lour. 1790. Menispermaceae. c. 30 trop. Africa (5), trop. Asia from India to continental and southern China, Indochina, and Malesia (12, eastward to New Guinea)

Stephanocaryum Popov. 1951. Boraginaceae. 2 Middle Asia

Stephanocereus A. Berger. 1926. Cactaceae. 2 Brazil (Bahia)

Stephanochilus Coss. et Durieu ex Hook. f. 1873. Asteraceae. 1 North Africa

Stephanococcus Bremek. 1952. Rubiaceae. 1 trop. Africa

Stephanodaphne Baill. 1875. Thymelaeaceae. 8–9 Comoro Is., Madagascar

Stephanodoria Greene. 1895 (Xanthocephalum Willd.). Asteraceae. 1 Mexico

Stephanolepis S. Moore = Erlangea Sch. Bip. Asteraceae

Stephanomeria Nutt. 1841. Asteraceae. 24 western North America

Stephanopholis S. F. Blake = Chromolepis Benth. Asteraceae

Stephanophysum Pohl. 1830–1831 (Ruellia L.). Acanthaceae. 5 trop. America

Stephanopodium Poepp. 1843. Dichapetalaceae. 9 trop. South America

Stephanorossia Chiov. = Oenanthe L. Apiaceae

Stephanostegia Baill. 1888. Apocynaceae. 3 Madagascar

Stephanostema K. Schum. 1904. Apocynaceae. 1 Tanzania

Stephanotella E. Fourn. = Marsdenia R. Br. Asclepiadaceae

Stephanothelys Garay. 1977. Orchidaceae. 4 Andes

Stephanotis Thouars. 1806. Asclepiadaceae. 5–15 Madagascar (5), trop. Old World

Steptorhamphus Bunge. 1852. Asteraceae. c. 10 eastern Mediterranean, Crimea (1), Caucasus, Middle Asia, Iran, Afghanistan, Pakistan

Sterculia L. 1753. Sterculiaceae. c. 200 tropics

Stereocaryum Burret. 1941. Myrtaceae. 5 Malay Peninsula (2), New Caledonia (3)

Stereochilus Lindl. 1858. Orchidaceae. 6 trop. Asia from Sikkim to Burma, Thailand, Vietnam, Philippines

Stereochlaena Hack. 1908. Poaceae. 5 trop. and South Africa

Stereosandra Blume. 1856 (Epipogium J. F. Gmel. ex Borkh.). Orchidaceae. 1 Thailand, Taiwan, Ryukyu Is., Malay Peninsula, Sumatra, Java, Borneo, Philippines, New Guinea, and Solomon Is.

Stereospermum Cham. 1833. Bignoniaceae. 24 trop. Africa, Madagascar, trop. Asia from Himalayas, India, and Sri Lanka through Burma to southern China and Indochina, Malesia

Sterigma DC. = Sterigmostemum M. Bieb. Brassicaceae

Sterigmapetalum Kuhlm. 1925. Rhizophoraceae. 7 trop. South America

Sterigmostemum M. Bieb. 1819. Brassicaceae. 8 Europe (1), Asia Minor, Iran, Caucasus, West Siberia, Middle Asia, China

Steriphoma Spreng. 1827. Capparaceae. 8 Central and trop. South America, Trinidad

Sternbergia Waldst. et Kit. 1804. Amaryllidaceae. 7 South and Southeast Europe (2), Southwest, West, and Middle Asia eastward to Kashmir; S. colchiciflora—extending to northern coast of Black Sea, Crimea, Caucasus

Sterropetalum N. E. Br. = Nelia Schwantes. Aizoaceae

Stethoma Raf. = Justicia L. Acanthaceae

Stetsonia Britton et Rose. 1920. Cactaceae. 1 southern Bolivia, northwestern Argentina, Paraguay

Steudnera K. Koch. 1862. Araceae. 8 India, southern China, Indochina, Malay Peninsula

Stevenia Adams et Fisch. 1817. Brassicaceae. 3 West and East Siberia, Russian Far East, Mongolia, China, Korea

Steveniella Schltr. 1918. Orchidaceae. 1 Crimea, Caucasus, Asia Minor, Iran

Stevensia Poit. 1804. Rubiaceae. 6 West Indies

Stevia Cav. 1797. Asteraceae. c. 230 southwestern U.S., Mexico, Guatemala, Honduras, Nicaragua, Costa Rica, Panama, Colombia, Venezuela, Ecuador, Peru, Brazil, Bolivia, Chile, Argentina, Paraguay, Uruguay

Steviopsis R. M. King et H. Rob. 1971 (Eupatorium L.). Asteraceae. 5 Mexico

Stewartia L. 1753. Theaceae. 11 East Asia, eastern U.S.

Stewartiella Nasir. 1972. Apiaceae. 1 Afghanistan, Pakistan

Steyerbromelia L. B. Sm. 1984. Bromeliaceae. 3 Venezuela (Amazonias)

Steyermarkia Standl. 1940. Rubiaceae. 1 Central America

Steyermarkina R. M. King et H. Rob. 1971 (Eupatorium L.). Asteraceae. 4 Venezuela, Brazil

Steyermarkochloa Davidse et R. P. Ellis. 1985. Poaceae. 1 Colombia, Venezuela

Stiburus Stapf = Eragrostis Wolf. Poaceae

Stichianthus Valeton. 1920. Rubiaceae. 3 Borneo

Stichoneuron Hook. f. 1883. Stemonaceae. 2 eastern India (Assam, Manipur), Bangladesh, Thailand, Malay Peninsula

Stictocardia Hallier f. 1893. Convolvulaceae. 12 tropics, esp. Malesia

Stictophyllum Dodson et M. W. Chase. 1989. Orchidaceae. ? sp. Distr. ?

Stifftia J. C. Mikan. 1820. Asteraceae. 7 northeastern South America

Stigmaphyllon A. Juss. 1833. Malpighiaceae. c. 80 trop. America, West Indies (1)

Stigmatella Eig = Eigia Soják. Brassicaceae

Stigmatodactylus Maxim. ex Makino. 1905. Orchidaceae. 4 India (1, Sikkim, Assam), western Malesia, Japan

Stigmatorhynchus Schltr. 1913. Asclepiadaceae. 3 trop. East and South Africa

Stigmatorthos M. W. Chase and D. E. Bennett. 1993. Orchidaceae. 1 Peru

Stigmatosema Garay. 1980. Orchidaceae. 2 trop. South America

Stilbanthus Hook. f. 1879. Amaranthaceae. 1 northeastern India (Sikkim, Naga Hills)

Stilbe Bergius. 1767. Stilbaceae. 6 South Africa (Cape)

Stilbocarpa (Hook. f.) Decne. et Planch. 1854. Araliaceae. 2–3 New Zealand and neighboring islands

Stillingia Garden ex L. 1767. Euphorbiaceae. c. 30 Madagascar (4), Mascarene Is. (1–2), eastern Malesia, Melanesia to Fiji, trop. and subtrop. America

Stilpnogyne DC. 1838. Asteraceae. 1 South Africa

Stilpnolepis I. M. Kraschen. 1946 (Artemisia L.). Asteraceae. 1 Mongolia, China

Stilpnopappus Mart. ex DC. 1836. Asteraceae. 20 trop. South America

Stilpnophleum Nevski = Calamagrostis Adans. Poaceae

Stilpnophyllum Hook. f. 1873. Rubiaceae. 1 Peru

Stilpnophytum Less. = Athanasia L. Asteraceae

Stimpsonia Wright ex A. Gray. 1858. Primulaceae. 1 China, Taiwan, Korea, Japan

Stipa L. 1753. Poaceae. c. 300 warm–temp. Northern Hemisphere, montane areas of tropics, southern South America

Stipagrostis Nees. 1832. Poaceae. c. 50 arid and semiarid Southeast Europe, Africa, West (incl. Caucasus), Southwest, and Middle Asia eastward to Afghanistan and Pakistan

Stipecoma Muell. Arg. 1860. Apocynaceae. 1 Brazil

Stiptanthus (Benth.) Briq. 1897. Lamiaceae. 1 eastern Himalayas, northeastern India

Stipularia P. Beauv. = Sabicea Aubl. Rubiaceae

Stipulicida Michx. 1803. Caryophyllaceae. 2 southeastern U.S.

Stirlingia Endl. 1837. Proteaceae. 6 Australia

Stironeurum Radlk. = Synsepalum (A. DC.) Daniell. Sapotaceae

Stixis Lour. 1790. Capparaceae. 12–15 Nepal, Sikkim, northeastern India (Assam), Burma, southern China (Yunnan, Hainan), Indochina, Malesia (Sumatra, Malay Peninsula, Borneo, Philippines)

Stizolobium P. Browne = Mucuna Adans. Fabaceae

Stizolophus Cass. = Centaurea L. Asteraceae

Stizophyllum Miers. 1863. Bignoniaceae. 3–4 trop. America from Mexico to Brazil

Stoberia Dinter et Schwantes. 1927. Aizoaceae. 3 southern Namibia, western South Africa

Stocksia Benth. 1853. Sapindaceae. 1 eastern Iran, Afghanistan, Pakistan

Stoebe L. 1753. Asteraceae. 34 trop. Africa, Madagascar, Mascarene Is.

Stoeberia Dinter et Schwantes = Ruschia Schwantes. Aizoaceae

Stoibrax Raf. 1840. Apiaceae. 5 Southwest Europe, North and South Africa

Stokesia L'Hér. 1789. Asteraceae. 1 southeastern U.S.

Stokoeanthus E. G. H. Oliv. 1976. Ericaceae. 1 South Africa (southwestern Cape)

Stolzia Schltr. 1915. Orchidaceae. 15 trop. Africa

Stomandra Standl. 1947. Rubiaceae. 1 Central America

Stomatanthes R. M. King et H. Rob. 1970 (Eupatorium L.). Asteraceae. 15 Brazil and Uruguay (12), trop. Africa (3)

Stomatium Schwantes. 1926. Aizoaceae. 44 South Africa

Stomatochaeta (S. F. Blake) Maguire et Wurdack. 1957. Asteraceae. 4 Venezuela, Guyana

Stomatostemma N. E. Br. = Cryptolepis R. Br. Asclepiadaceae

Stonesia G. Taylor. 1953. Podostemaceae. 4 trop. West Africa

Storckiella Seem. 1861. Fabaceae. 5 northeastern Australia (1), New Caledonia (3), Fiji (1)

Storthocalyx Radlk. 1879. Sapindaceae. 4 New Caledonia

Stracheya Benth. 1853. Fabaceae. 1 Himalayas, Tibet

Strailia T. Durand Lecythis Loefl. Lecythidaceae

Stramentopappus H. Rob. et V. A. Funk. 1987. Asteraceae. 1 Mexico

Strangea Meisn. 1855. Proteaceae. 3 Australia

Strangweja Bertol. = Bellevalia Lapeyr. Hyacinthaceae

Stranvaesia Lindl. 1837 (Photinia Lindl.). Rosaceae. 8 Himalayas; northeastern India; Burma; central, eastern, and southern China; Taiwan; Indochina; Sumatra; Philippines

Strasburgeria Baill. 1876. Strasburgeriaceae. 1 New Caledonia

Stratiotes L. 1753. Hydrocharitaceae. 1 Europe

Straussiella Hausskn. 1897. Brassicaceae. 1 Iran

Streblacanthus Kuntze. 1891. Acanthaceae. 7 Central America (6), Bolivia (1)

Streblochaete Hochst. ex Pilg. 1906. Poaceae. 1 trop. (mountains of Cameroun, Ethiopia, Malawi, Uganda, Kenya, Tanzania) and South (Natal) Africa, Réunion I., Malesia (Java, Lombok I., Philippines)

Streblorrhiza Endl. 1833. Fabaceae. 1 Phillip Is. (near Norfolk I.)

Streblosa Korth. 1851 (Psychotria L.). Rubiaceae. 25 western Malesia

Streblosiopsis Valeton. 1910. Rubiaceae. 1 Borneo

Streblus Lour. 1790. Moraceae. 24 trop. East Africa, Madagascar, India, Sri Lanka, Andaman Is., Burma, southern China, Hainan, Indochina, Malesia (Malay Peninsula, Sumatra, Java, Bali, Lesser Sunda Is., Philippines, Sulawesi, New Guinea), eastern Australia, New Zealand, Society Is., Fiji, and Hawaiian Is.

Strelitzia Banks. 1789. Strelitziaceae. 5 South Africa

Strempelia A. Rich. ex DC. 1830 (Rudgea Salisb.). Rubiaceae. 5 trop. South America

Strempeliopsis Benth. 1876. Apocynaceae. 2 Cuba, Jamaica

Strephium Schrad. ex Nees. 1829 (Raddia Bertel.). Poaceae. 5 trop. South America

Strephonema Hook. f. 1867. Combretaceae. 6 trop. West Africa

Streptachne R. Br. = Aristida L. Poaceae

Streptanthella Rydb. 1918. Brassicaceae. 1 western U.S.

Streptanthera Sweet = Sparaxis Ker-Gawl. Iridaceae

Streptanthus Nutt. 1825. Brassicaceae. c. 40 western and southern U.S., northen Mexico

Streptocalyx Beer. 1854. Bromeliaceae. 14 Colombia, Ecuador, Peru, Venezuela, Guyana, Surinam, Amazonian and eastern Brazil, Bolivia

Streptocarpus Lindl. 1828. Gesneriaceae. c. 120 trop. and South Africa, Madagascar

Streptocaulon Wight et Arn. 1834. Asclepiadaceae. 5 northeastern India (2), China, Indochina, Malesia

Streptochaeta Schrad. ex Nees. 1829. Poaceae. 3 Central and South America

Streptoglossa Steetz ex F. Muell. 1863. Asteraceae. 8 northern, western, and central Australia

Streptogyna P. Beauv. 1812. Poaceae. 2 trop. America (1, S. americana C. E. Hubbard) S. crinata—trop. West Africa, Congo, Sudan, Uganda, Tanzania, and South Asia (India, Sri Lanka)

Streptolirion Edgew. 1845. Commelinaceae. 2 Tibet, eastern Himalayas, Burma, Indochina, Korea, Japan

Streptoloma Bunge. 1848. Brassicaceae. 2 Kazakhstan, Middle Asia, Iran, Afghanistan

Streptolophus D. K. Hughes. 1923. Poaceae. 1 Angola

Streptomanes K. Schum. ex Schltr. = Cryptolepis R. Br. Asclepiadaceae

Streptopetalum Hochst. 1841. Turneraceae. 5 Central, trop. East, and South (1, S. serratum) Africa

Streptopus Michx. 1803. Convallariaceae. 10 Central Europe, Mediterranean, Himalayas, Tibet, northern Burma, China, northern Korea, Japan, North America; 2 East Siberia and Russian Far East

Streptosiphon Mildbr. 1935. Acanthaceae. 1 trop. East Africa

Streptosolen Miers. 1850 (Browallia L.).

Solanaceae. 1 (S. jamessnii) Ecuador, Peru

Streptostachys Desv. 1810. Poaceae. 1 trop. South America, Trinidad

Streptothamnus F. Muell. 1862. Berberidopsidaceae. 1 (S. moorei) Australia (Queensland and New South Wales)

Streptotrachelus Greenm. 1897 (Laubertia A. DC.). Apocynaceae. 1 Mexico

Stricklandia Baker = Eucrosia Ker-Gawl. Amaryllidaceae

Striga Lour. 1790. Scrophulariaceae. 44 trop. and South Africa, trop. Asia, Australia

Strigina Engl. = Lindernia All. Scrophulariaceae

Strigosella Boiss. 1854 (Malcolmia R. Br.). Brassicaceae. 23 Mediterranean, Southeast Europe, West (incl. Caucasus), Southwest and Middle Asia, West Siberia, east to Mongolia, central China, and Pakistan

Striolaria Ducke. 1945. Rubiaceae. 1 Brazil (Amazonia)

Strobilacanthus Griseb. 1858. Acanthaceae. 1 Panama

Strobilanthes Blume. 1826. Acanthaceae. c. 250 trop. Asia

Strobilanthopsis S. Moore. 1900. Acanthaceae. 5 trop. Africa

Strobilocarpus Klotzsch = Grubbia Bergius. Grubbiaceae

Strobilopanax R. Vig. = Meryta J. R. Forst. et G. Forst. Araliaceae

Strobilopsis Hilliard et B. L. Burtt. 1977 (Glumicalyx Hiern). Scrophulariaceae. 1 South Africa (Natal), Lesotho

Strobopetalum N. E. Br. 1894. Asclepiadaceae. 1 trop. Northeast Africa, Arabian Peninsula

Stroganowia Kar. et Kir. 1841. Brassicaceae. 21 Middle Asia (incl. Kazakhstan), Iran, Afghanistan, China (1), southwestern U.S. (Nevada)

Stromanthe Sond. 1849. Maranthaceae. 13 trop. South America

Strombocactus Britton et Rose. 1922. Cactaceae. 1 central Mexico

Strombosia Blume. 1827. Olacaceae. 12 trop. Africa (9), southwestern India, Sri Lanka, Burma, Thailand, Malesia (Sumatra, Malay Peninsula, Java, Borneo, Philippines, Moluccas)

Strombosiopsis Engl. 1897. Olacaceae. 1 trop. Africa

Strongylocaryum Burret = Ptychosperma Labill. Arecaceae

Strongylodon Vogel. 1836. Fabaceae. 20 trop. Old World from Madagascar and Mascarene Is. through trop. Asia, Malesia, and trop. Australia eastward to Hawaii and Society Is.

Strophacanthus Lindau = Ptyssiglottis T. Anderson. Acanthaceae

Strophanthus DC. 1802. Apocynaceae. 38 trop. and South Africa, Madagascar, trop. Asia, Malesia

Strophioblachia Boerl. 1900. Euphorbiaceae. 2 Indochina, Hainan, Philippines, Sulawesi

Strophiodiscus Choux. 1926 (or = Plagioscyphus Radlk.). Sapindaceae. 1 Madagascar

Strophocactus Britton et Rose = Selenicereus Britton et Rose. Cactaceae

Stropholirion Torr. = Dichelostemma Kunth. Alliaceae

Strophostyles Elliott. 1823. Fabaceae. 3 North America

Strotheria B. L. Turner. 1972. Asteraceae. 1 Mexico

Struchium P. Browne = Sparganophorus Boehm. Asteraceae

Strumaria Jacq. ex Willd. 1797. Amaryllidaceae. 9 Namibia, South Africa

Strumpfia Jacq. 1760. Rubiaceae. 1 Mexico, West Indies, Curaçao

Struthanthus Mart. 1830. Loranthaceae. c. 50 trop. America from northern Mexico to northern Argentina

Struthiola L. 1767. Thymelaeaceae. 30 trop. (2) and South Africa

Struthiolopsis E. Phillips = Gnidia L. Thymelaeaceae

Strychnopsis Baill. 1885. Menispermaceae. 1 Madagascar

Strychnos L. 1753. Loganiaceae. c. 200 tropics and subtropics

Stryphnodendron Mart. 1837. Fabaceae. c. 20 trop. America; 1 (S. obovatum) Tanzania

Stuartina Sond. 1853. Asteraceae. 2 southern and eastern Australia

Stubendorffia Schrenk ex Fisch., C. A. Mey. et Avé-Lall. 1844. Brassicaceae. 9 Transcaucasia (1), eastern Turkey, Middle Asia (8), Afghanistan

Stuckertia Kuntze. 1903. Asclepiadaceae. 1 South America

Stuckertiella Beauverd. 1913. Asteraceae. 2 Argentina

Stuessya B. L. Turner et F. G. Davies. 1980. Asteraceae. 3 central and southern Mexico

Stuhlmannia Taub. 1895. Fabaceae. 1 Tanzania

Stultitia E. Phillips = Orbea Haw. Asclepiadaceae

Stussenia C. Hansen. 1985 (Blastus Lour.). Melastomataceae. 1 Southeast Asia

Styasasia S. Moore. 1905 (or = Asystasia Blume). Acanthaceae. ? sp. Distr. ?

Stylapterus A. Juss. 1846 (Penaea L.). Penaeaceae. 8 South Africa (southwestern Cape)

Stylidium Sw. ex Willd. 1805. Stylidiaceae. c. 150 Sri Lanka, northeastern India, southern China, Indochina, Malesia, Australia, New Zealand

Stylisma Raf. 1818. Convolvulaceae. 6 southern and eastern U.S.

Stylobasium Desf. 1819. Surianaceae. 2 southwestern Australia

Styloceras Kunth ex A. Juss. 1824. Styloceratanceae. 5 trop. Andes from Colombia and Venezuela to Bolivia

Stylochaeton Lepr. 1834. Araceae. 22 Central, East, and Southeast trop. Africa

Stylochiton Lepr. = Stylochaeton Lepr. Araceae

Stylocline Nutt. 1840. Asteraceae. 6 southwestern U.S. (California, Utah, Arizone, New Mexico), northern Mexico

Styloconus Baill. = Blancoa Lindl. Conostylidaceae

Stylodon Raf. 1825 (Verbena L.). Verbenaceae. 1 southeastern U.S.

Stylogyne A. DC. 1841. Myrsinaceae. 50–60 trop. South America, West Indies

Stylolepis Lehm. = Podolepis Labill. Asteraceae

Stylomecon G. Taylor. 1930 (Meconopsis Vig.). Papaveraceae. 1 California and northern Baja California

Stylophorum Nutt. 1818. Papaveraceae. 3 East Asia (2), eastern North America (1)

Stylophyllum Britton et Rose = Dudleya Britton et Rose. Crassulaceae

Stylosanthes Sw. 1788. Fabaceae. 25 tropics and subtropics

Stylosiphonia Brandegee. 1914. Rubiaceae. 2 Mexico, Central America

Stylotrichium Mattf. 1923. Asteraceae. 4 northeastern Brazil

Stylurus Salisb. ex J. Knight = Grevillea R. Br. Proteaceae

Stypandra R. Br. 1810. Phormiaceae. 1 (S. glauca R. Br., very variable) southwestern and southeastern temp. Australia, ? New Caledonia

Styphelia Sm. 1795. Epacridaceae. 12 continental Southeast Asia, eastern Malesia (incl. New Guinea, 8), Australia and Tasmania, New Zealand, Mariana Is., Melanesia to New Caledonia, ? Hawaiian Is.

Styphnolobium Schott ex Endl. 1831. Fabaceae. 2 China, Japan (1), North America (1)

Styppeiochloa De Winter. 1966. Poaceae. 2 Namibia, Madagascar

Styrax L. 1753. Styracaceae. 120–130 Mediterranean (incl. South Europe); 1, (S. officinale), South, East, and Southeast Asia; western Malesia; America from southern U.S. to northern Argentina; West Indies

Styrophyton S. Y. Hu. 1952 (or = Allomorphia Blume). Melastomataceae. 1 southwestern China

Suaeda Forssk. ex Scop. 1777. Chenopodiaceae. Over 100 cosmopolitan

Suarezia Dodson. 1989. Orchidaceae. 1 Ecuador

Suberanthus Borhidi et Fernández. 1981 (Rondeletia L.). Rubiaceae. 5 Cuba, Haiti

Subularia L. 1753. Brassicaceae. 2 temp. Eurasia, North America, montane areas of trop. East Africa (1); Russia (1, S. aquatica, European part, West Siberia, Far East)

Succisa Haller. 1768. Dipsacaceae. 3 Europe, Mediterranean, montane Cameroun (1), Asia Minor; Russia (S. pratensis—European part, Caucasus, West and East Siberia)

Succisella Beck. 1893 (Succisa Haller). Dipsacaceae. 4 South, Southeast, and East Europe; S. inflexa—European part of Russia, Caucasus

Succowia Medik. 1792. Brassicaceae. 2: 1 Canary Is.; 1 (S. balearica) Spain, Balearic Is., Corsica, Sardinia, Sicily, Italy, Morocco, Tunisia, Algeria

Suchtelenia Kar. ex Meisn. 1840. Boraginaceae. 2 Caucasus, Middle Asia

Suckleya A. Gray. 1876. Chenopodiaceae. 1 U.S.: Rocky Mts. (Montana to Colorado)

Sucrea Soderstr. 1981. Poaceae. 3 Brazil

Suddia Renvoize. 1984. Poaceae. 1 the Sudan

Suessenguthia Merxm. 1953. Acanthaceae. 1 Bolivia

Suessenguthiella Friedrich. 1955 (Pharnaceum L.). Molluginaceae. 2 South-West and South Africa

Suksdorfia A. Gray. 1880. Saxifragaceae. 3 western North America from British Columbia and Alberta to Montana, Oregon, and Idaho; southern Andes

Sukunia A. C. Sm. 1936. Rubiaceae. 2 Fiji, ? Solomon Is., Vanuatu

Sulaimania Hedge et Rech. f. 1982. Lamiaceae. 1 Pakistan

Sulcorebutia Backeb. = Rebutia K. Schum. Cactaceae

Sulitia Merr. 1926. Rubiaceae. 1 Philippines

Sullivania F. Muell. = Caleana R. Br. Orchidaceae

Sullivantia Torr. et A. Gray. 1842. Saxifragaceae. 3 U.S. (Washington, Oregon, Montana, Wisconsin, Wyoming, Colorado, Missouri, Illinois, Indiana, Ohio, Indiana, Virginia)

Sumatroscirpus Oteng-Yeb. 1974. Cyperaceae. 1 Sumatra

Sumbaviopsis J. J. Sm. 1910. Euphorbiaceae. 2 northeastern India, southern China, Indochina, western Malesia

Summerhayesia P. J. Cribb. 1977. Orchidaceae. 2 Liberia, Ivory Coast, Ghana, Zaire, Zambia, Zimbabwe

Sundacarpus (Buchh. et E. Gray) C. N. Page. 1988. Podocarpaceae. 1 Malesia from Sumatra to Philippines, New Ireland I., Australia (northern Queensland)

Sunipia Buch.-Ham. ex Lindl. 1826. Orchidaceae. 20 eastern Himalayas, northeastern India, Tibet, Burma, southern China, Taiwan, Indochina

Supushpa Suryanarayana. 1970. Acanthaceae. 1 western India

Suregada Roxb. ex Rottler. 1803. Euphorbiaceae. c. 40 trop. and South Africa, Madagascar, South and Southeast Asia eastward to Taiwan, Malesia

Surfacea Moldenke = Premna L. Verbenaceae

Suriana L. 1753. Surianaceae. 1 coastal eastern trop. America, trop. East Africa, Madagascar, Mascarene Is., Sri Lanka, Malay Peninsula, Taiwan, Philippines, eastern Malesia, northeastern Australia, Polynesia

Susanna E. Phillips = Amellus L. Asteraceae

Susilkumara Bennet = Alajja Ikonn. Lamiaceae

Sutera Roth. 1807. Scrophulariaceae. c. 130 Canary Is., trop. and South Africa, 1 from Southwest Asia to India

Sutherlandia R. Br. ex W. T. Aiton. 1812. Fabaceae. 6 South Africa (6); S. frutescens and S. microphylla—extending to Namibia and Botswana

Sutrina Lindl. 1842. Orchidaceae. 1 Peru

Suttonia A. Rich. = Rapanea Aubl. Myrsinaceae

Suzukia Kudo. 1930. Lamiaceae. 2 Taiwan, Ryukyu Is.

Svenkoeltzia Burns-Bal. = Spiranthes Rich. Orchidaceae

Svensonia Moldenke = Chascanum E. Mey. Verbenaceae

Sventenia Font Quer. 1949. Asteraceae. 1 Canary Is.

Svitramia Cham. 1835. Melastomataceae. 2 southern Brazil

Swainsonia Salisb. 1806. Fabaceae. c. 50 arid Australia (7), New Zealand (montane areas of South I.)

Swallenia Soderstr. et H. F. Decker. 1963. Poaceae. 1 California

Swallenochloa McClure. 1973 (Chusquea Kunth). Poaceae. 10 Costa Rica, Panama, Venezuela, Colombia, Ecuador, Peru, Bolivia

Swartzia Schreb. 1791. Fabaceae. c. 135 trop. America, trop. Africa (2), Madagascar (1)

Sweetia Spreng. 1825. Fabaceae. 2 Brazil, Bolivia, Paraguay

Swertia L. 1753. Gentianaceae. c. 150 temp. regions and montane areas of the Northern Hemisphere (esp. China, c. 80), Africa, Madagascar (1, S. rosulata)

Swida Opiz. 1838 (or = Cornus L.). Cornaceae. c. 40 temp. Northern Hemisphere, Himalayas from Kashmir to Bhutan, northeastern India, northern Burma, Indochina; Mexico (3), northern Andes (1)

Swietenia Jacq. 1760. Meliaceae. 4 U.S. (Florida), Mexico, Guatemala, Belize, Honduras, El Salvador, Nicaragua, Costa Rica, Panama, Colombia, Venezuela, French Guiana, Peru, Bolivia, Brazil, and West Indies (2)

Swinglea Merr. 1927. Rutaceae. 1 Philippines

Swintonia Griff. 1846. Anacardiaceae. 14 Andaman Is., Indochina, western Malesia

Swynnertonia S. Moore. 1908. Asclepiadaceae. 1 Zimbabwe

Syagrus Mart. 1824. Arecaceae. 29 trop. South America from Colombia and Venezuela to Argentina and Uruguay (esp. Brazil), Lesser Antilles (1)

Sycopsis Oliv. 1860. Hamamelidaceae. 9 central and southern China, northeastern India, Philippines, Sulawesi, New Guinea

Symbegonia Warb. 1894. Begoniaceae. 13 New Guinea

Symbolanthus G. Don. 1837. Gentianaceae. 15 trop. America: Costa Rica, Panama, Colombia, Peru, and Bolivia

Symingtonia Steenis = Exbucklandia R. W. Br. Hamamelidaceae

Symmeria Benth. 1845. Polygonaceae. 1 northern South America, trop. West Africa

Symonanthus Haegi. 1981 (Isandra F. Muell.). Solanaceae. 2 southwestern Australia

Sympa Ravenna. 1981. Iridaceae. 1 Brazil

Sympagis (Nees) Bremek. 1944 (Strobilanthes Blume). Acanthaceae. 5 Nepal, Bhutan, northeastern India

Sympegma Bunge. 1879. Chenopodiaceae. 1 Kazakhstan, Kirgizia, Tajikistan, northwestern and central China, Mongolia

Sympetalandra Stapf. 1891. Fabaceae. 5 Malesia

Sympetaleia A. Gray = Eucnide Zucc. Loasaceae

Symphionema R. Br. 1810. Proteaceae. 2 Australia (New South Wales)

Symphonia L. f. 1782. Clusiaceae. 17: S. globulifera L. f.—Central and trop. South America, West Indies (Santo Domingo, Jamaica, Dominica, Guadeloupe, Trinidad), and trop. Africa (from Sierra Leone to Cameroun, Sao Tomé, Gabon, Angola, Zambia, Uganda, Tanzania); 16 Madagascar

Symphorema Roxb. 1805. Symphoremataceae. 3 India, Sri Lanka, Burma, Thailand, China, Philippines

Symphoricarpos Duhamel. 1755. Caprifoliaceae. 17 North America (16, incl. Mexico), Guatemala; central China (1)

Symphostemon Hiern. 1900. Lamiaceae. 2 Angola

Symphyandra A. DC. 1830. Campanulaceae. 12 eastern Mediterranean, West (incl. Caucasus) and Middle Asia

Symphyglossum Schltr. 1919. Orchidaceae. 2 trop. South America

Symphyllarion Gagnep. = Hedyotis L. Rubiaceae

Symphyllocarpus Maxim. 1859. Asteraceae. 1 northeastern China, Soviet Far East

Symphyllophyton Gilg. 1897. Gentianaceae. 2 Brazil

Symphyobasis K. Krause = Goodenia Sm. Goodeniaceae

Symphyochaeta (DC.) Skottsb. = Robinsonia DC. Asteraceae

Symphyochlamys Gürke. 1903. Malvaceae. 1 trop. Northeast Africa

Symphyoloma C. A. Mey. 1831. Apiaceae. 1 high mountainous Caucasus

Symphyonema Spreng. = Symphionema R. Br. Proteaceae

Symphyopappus Turcz. 1848. Asteraceae. 11 southern Brazil

Symphyosepalum Hand.-Mazz. 1936. Orchidaceae. 1 China

Symphysia C. Presl = Vaccinium L. Ericaceae

Symphytonema Schltr. 1895 (Tanulepis Balf. f.). Asclepiadaceae. 3 Madagascar

Symphytum L. 1753. Boraginaceae. 35 Europe (11), Mediterranean, West (incl. Caucasus) Asia; rarely in West Siberia and Middle Asia

Sympieza Lichtenst. ex Roem. et Schult. 1818. Ericaceae. 4 South Africa (southwestern Cape)

Symplectochilus Lindau = Anisotes Nees. Acanthaceae

Symplectrodia Lazarides. 1985. Poaceae. 2 Australia

Symplocarpus Salisb. ex W. Barton. 1817. Araceae. 1–3 China, Japan, Russian Far East, North America

Symplococarpon Airy Shaw. 1937. Theaceae. 9 trop. America from Mexico to Colombia

Symplocos Jacq. 1760. Symplocaceae. c. 250 South, Southeast and East Asia, New Guinea, Solomon Is., Australia (Queensland, New South Wales), Lord Howe I., New Caledonia, Fiji, America from northwestern U.S. to southern Brazil

Synadenium Boiss. 1862. Euphorbiaceae. 13 trop. Africa, Madagascar, Mascarene Is.

Synandra Nutt. 1818. Lamiaceae. 1 eastern U.S.

Synandrina Standl. et L. O. Williams = Casearia Jacq. Flacourtiaceae

Synandrodaphne Gilg. 1915. Thymelaeaceae. 1 trop. West Africa

Synandrogyne Buchet = Arophyton Jum. Araceae

Synandropus A. C. Sm. 1931. Menispermaceae. 1 northeastern Brazil

Synandrospadix Engl. 1883. Araceae. 1 Bolivia, northern Argentina

Synanthes Burns-Bal., H. Rob. et M. S. Foster. 1985. Orchidaceae. 2 Nicaragua (1, S. borealis), Paraguay (1)

Synaphea R. Br. 1810. Proteaceae. 10 southwestern Australia

Synapsis Griseb. 1866. Bignoniaceae. 1 Cuba

Synaptantha Hook. f. = Hedyotis L. Rubiaceae

Synaptolepis Oliv. 1870. Thymelaeaceae. 4–5 trop. and South (Natal) Africa, Madagascar (1)

Synaptophyllum N. E. Br. 1925. Aizoaceae. 1 southwestern Namibia

Synardisia (Mez) Lundell = Ardisia Sw. Myrsinaceae

Synassa Lindl. = Sauroglossum Lindl. Orchidaceae

Syncalathium Lipsch. 1956. Asteraceae. 6 Tibet, southern China

Syncarpha DC. 1810. Asteraceae. 25 South Africa (Cape)

Syncarpia Ten. 1839. Myrtaceae. 5 Moluccas, trop. Australia

Syncephalantha Bartl. = Dyssodia Cav. Asteraceae

Syncephalum DC. 1838. Asteraceae. 5 Madagascar

Synchaeta Kirp. = Gnaphalium L. Asteraceae

Synchoriste Baill. 1891. Acanthaceae. 1 Madagascar

Synclisia Benth. 1862. Menispermaceae. 3 trop. Africa

Syncolostemon E. Mey. ex Benth. 1837 (Hemizygia [Benth.] Briq.). Lamiaceae. 9 South Africa

Syncretocarpus S. F. Blake. 1916. Asteraceae. 2 Peru

Syndesmanthus Klotzsch. 1838. Ericaceae. 12 South Africa (southwestern and southern Cape)

Syndiclis Hook. f. 1886 (Potameia Thouars). Lauraceae. 10 eastern Himalayas from Bhutan to Assam, Burma, southern China

Syndyophyllum K. Schum. et Lauterb. 1900. Euphorbiaceae. 1 Sumatra, Borneo, New Guinea

Synechanthus H. A. Wendl. 1858. Arecaceae. 2–3 southern Mexico, Central America, northwestern South America

Synedrella Gaertn. 1791. Asteraceae. 2 trop. and subtrop. America, trop. Africa, Madagascar; S. nodiflora—introduced in India, Sri Lanka, Indochina, southern China, Taiwan, Ryukyu Is., Malesia

Synedrellopsis Hieron. et Kuntze. 1898. Asteraceae. 1 Argentina

Syneilesis Maxim. 1859. Asteraceae. 5 northeastern China, Korea, Japan, Taiwan; S. aconitifolia—Russian Far East

Synelcosciadium Boiss. = Tordylium L. Apiaceae

Synepilaena Baill. = Kohleria Regel. Gesneriaceae

Syngonanthus Ruhland. 1900. Eriocaulaceae. c. 200 Central and trop. South America, West Indies, trop. Africa, Madagascar

Syngonium Schott. 1829. Araceae. 33 Central and trop. South America, West Indies

Synima Radlk. 1879. Sapindaceae. 2 New Guinea, northeastern Australia

Synisoon Baill. = Retiniphyllum Humb. et Bonpl. Rubiaceae

Synnema Benth. = Hygrophila R. Br. Acanthaceae

Synnotia Sweet = Sparaxis Ker-Gawl. Iridaceae

Synosma Raf. ex Britton et A. Br. = Hasteola Raf. Asteraceae

Synostemon F. Muell. = Sauropus Blume. Euphorbiaceae

Synotis (C. B. Clarke) C. Jeffrey et Y. L. Chen. 1984. Asteraceae. c. 50 China, Tibet, Himalayas, Burma (1)

Synotoma (G. Don) R. Schulz = Physoplexis (Endl.) Schur. Campanulaceae

Synoum A. Juss. 1830. Meliaceae. 2 eastern Australia

Synsepalum (A. DC.) Daniell. 1852. Sapotaceae. c. 20 trop. and South Africa

Synstemon Botsch. = Synstemonanthus Botsch. Brassicaceae

Synstemonanthus Botsch. 1980. Brassicaceae. 2 China

Synthlipsis A. Gray. 1849. Brassicaceae. 3 southern U.S., Mexico

Synthyris Benth. 1846. Scrophulariaceae. 15 montane areas of the western North America

Syntriandrum Engl. 1899. Menispermaceae. 1–4 trop. Africa: Nigeria, Cameroun, Gabon, Congo

Syntrichopappus A. Gray. 1857. Asteraceae. 2 southwestern U.S.

Syntrinema H. Pfeiff. 1925 (Rhynchospora Vahl). Cyperaceae. 1 Brazil

Synurus Iljin. 1926. Asteraceae. 6 Russia (1, S. deltoides—East Siberia, Far East), China, Mongolia, Korea, Japan

Sypharissa Salisb. = Urginea Steinh. Hyacinthaceae

Syreitschikovia Pavlov. 1933. Asteraceae. 2 Middle Asia, northwestern China

Syrenia Andrz. ex Besser. 1822. Brassicaceae. 10 temp. Eurasia from Central Europe through Caucasus, Middle Asia, and West Siberia to China and Mongolia

Syrenopsis Jaub. et Spach = Thlaspi L. Brassicaceae

Syringa L. 1753. Oleaceae. c. 40 Eurasia from Southeast Europe to East Asia

Syringantha Standl. 1930. Rubiaceae. 1 Mexico

Syringidium Lindau = Habracanthus Nees. Acanthaceae

Syringodea Hook. f. 1873. Iridaceae. 8 South Africa

Syringodium Kütz. 1860. Cymodoceaceae. 2 coasts of Caribbean Sea (1), coasts of the Indian and western Pacific oceans (1)

Syrrheonema Miers. 1864. Menispermaceae. 3 trop. West Africa

Systeloglossum Schltr. 1923. Orchidaceae. 1 Costa Rica

Systemonodaphne Mez. 1889. Lauraceae. 1–2 Venezuela, Guyana, Surinam, Brazil

Systenotheca Reveal et Hardham. 1989 (Chorizanthe R. Br.). Polygonaceae. 1 western America

Syzygiopsis Ducke = Pouteria Aubl. Sapotaceae

Syzygium Gaertn. 1788. Myrtaceae. c. 500 tropics and subtropics of the Old World

Szovitsia Fisch. et C. A. Mey. 1835. Apiaceae. 1 southern Transcaucasia, Turkey, Iran

T

Tabascina Baill. = Justicia L. Acanthaceae

Tabebuia Gomes ex DC. 1838. Bignoniaceae. c. 100 trop. America: Mexico, Guatemala, Belize, El Salvador, Honduras, Nicaragua, Costa Rica, Panama, Colombia, Venezuela, Guyana, Surinam, French Guiana, Ecuador, Peru, Brazil, Bolivia, Paraguay, northern Argentina; West Indies: Bahamas, Cuba, Cayman Is., Haiti, Jamaica, Puerto Rico, Virgin Is., Leeward Is., Windward Is., southern Dutch Antilles, Trinidad, and Tobago

Taberna (A. DC.) Miers = Tabernaemontana L. Apocynaceae

Tabernaemontana L. 1753. Apocynaceae. c. 110 trop. America, trop. and South Africa, Madagascar, Seychelles and Comoro Is., trop. Asia, Malesia, trop. Australia, Oceania

Tabernanthe Baill. 1888. Apocynaceae. 7 trop. Africa

Tacareuna Hulf. 1989. Euphorbiaceae. 3 Panama, Colombia, Venezuela, Peru

Tacazzea Decne. 1844. Asclepiadaceae. 4 trop. (from Senegal to Ethiopia and Kenya and south to Namibia and Mozambique), and South Africa; T. apiculata—very polymorphic

Tacca J. R. Forst. et G. Forst. 1776. Taccaceae. 11 eastern part of the trop. Old World (9), T. leontopetaloides—from trop. West Africa and Madagascar to Easter I.; T. parkeri—northern and northeastern South America

Taccarum Brongn. ex Schott. 1857. Araceae. 5 southern Brazil, northern Argentina, Paraguay

Tachia Aubl. 1775. Gentianaceae. 9 trop. South America

Tachiadenus Griseb. 1838. Gentianaceae. 11 Madagascar

Tachigali Aubl. 1775. Fabaceae. 24 trop. America, esp. Amazonia

Tachigalia Juss. = Tachigali Aubl. Fabaceae

Tacinga Britton et Rose. 1919. Cactaceae. 2 northeastern Brazil

Tacitus Moran = Graptopetalum Rose. Crassulaceae

Tacoanthus Baill. 1890. Acanthaceae. 1 Bolivia

Tadehagi H. Ohashi. 1973. Fabaceae. 3 trop. Asia, Australia, Oceania

Taeckholmia Boulos = Sonchus L. Asteraceae

Taeniandra Bremek. 1944 (Strobilanthes Blume). Acanthaceae. 1 India

Taenianthera Burret = Geonoma Willd. Arecaceae

Taeniatherum Nevski. 1934. Poaceae. 3 Mediterranean, Southeast Europe, West (incl. Caucasus), Southwest, and Middle Asia eastward to Himalayas

Taenidia (Torr. et A. Gray) Drude. 1898 (Zizia W. D. J. Koch). Apiaceae. 2 eastern U.S.

Taeniochlaena Hook. f. = Roureopsis Planch. Connaraceae

Taeniopetalum Vis. 1850 (Peucedanum L.). Apiaceae. 2–3 Europe; Russia (1, T. arenarium—southern European part)

Taeniophyllum Blume. 1825. Orchidaceae. c. 120 trop. Africa (1), trop. and East Asia, Micronesia, Malesia, northern Australia, Oceania eastward to Tahiti

Taeniopleurum J. M. Coult. et Rose = Perideridia Rchb. Apiaceae

Taeniorhachis Cope. 1993 (Digitaria Haller). Poaceae. 1 Somalia

Taeniorrhiza Summerh. 1943. Orchidaceae. 1 Gabon

Tagetes L. 1753. Asteraceae. c. 50 trop. and subtrop. America, trop. Africa (1)

Tahitia Burret. 1926 (Berrya Roxb.). Tiliaceae. 1 Society Is.

Taihangia T. T. Yü et C. L. Li. 1980 (Geum L.). Rosaceae. 1 northern China

Tainia Blume. 1825. Orchidaceae. c. 25–30 trop. Asia from India and Burma through southern China (Yunnan, Hainan) and Indochina eastward to Taiwan, Malesia from Malay Peninsula to New Guinea, Solomon Is., Australia (1, T. parviflora)

Tainionema Schltr. 1899. Asclepiadaceae. 1 Haiti

Tainiopsis Schltr. 1915 (Eriodes Rolfe). Orchidaceae. 1 northeastern India, southern China, Burma, Thailand

Taitonia Yamam. = Gomphostemma Wall. ex Benth. Lamiaceae

Taiwania Hayata. 1906. Taxodiaceae. 2: T. floussiana—eastern Burma, southern China (Yunnan and Hupeh); T. cryptomerioides—Taiwan

Takasagoya Y. Kimura = Hypericum L. Hypericaceae

Takeikadzuchia Kitag. et Kitam. = Olgaea Iljin. Asteraceae

Takhtajania Baranova et J.-F. Leroy. 1978. Winteraceae. 1 Madagascar

Takhtajaniantha Nazarova. 1990 (Scorzonera L.). Asteraceae. 1 Transcaucasia, eastern Asia Minor, Jordan, Iran, Afghanistan, Pakistan, Middle Asia, southwestern Siberia

Takhtajanianthus De = Rahnteriopsis Rauschert. Asteraceae

Takhtajaniella V. Avet. 1980. Brassicaceae. 1 West Asia

Talassia Korovin = Ferula L. Apiaceae

Talauma Juss. 1789. Magnoliaceae. 66 trop. America from Mexico to northern Brazil, West Indies, eastern Himalayas, southern China, Indochina, Malesia (excl. Tanimbar and Aru Is., and most parts of New Guinea)

Talbotia Balf. = Xerophyta Juss. Velloziaceae

Talbotia S. Moore. 1913 (Afrofittonia Lindau). Acanthaceae. 1 South Africa

Talbotiella Baker f. 1914. Fabaceae. 3 Ghana (1), southern Nigeria (1), Cameroun (1)

Talbotiopsis L. B. Sm. = Xerophyta Juss. Velloziaceae

Talguenea Miers ex Endl. 1840. Rhamnaceae. 1 Chile

Talinaria Brandegee. 1908 (Grahamia Gillies ex Hook. et Arn.). Portulacaceae. 3 Mexico (1), Argentina (2)

Talinella Baill. 1886. Portulacaceae. 2 Madagascar

Talinopsis A. Gray. 1852. Portulacaceae. 1 southern U.S., Mexico

Talinum Adans. 1763. Portulacaceae. c. 40 North and South America, trop. and subtrop. Africa

Talisia Aubl. 1775. Sapindaceae. c. 40 trop. America, Trinidad

Tamamschjania Pimenov et Kljuykov. 1981. Apiaceae. 2 South Europe, Caucasus, and Asia Minor

Tamananthus Badillo. 1985. Asteraceae. 1 Venezuela

Tamania Cuatrec. 1976. Asteraceae. 1 Colombia, Venezuela

Tamaricaria Qaiser et Ali = Myricaria Desv. Tamaricaceae

Tamarindus L. 1753. Fabaceae. 1 (T. indica) probably native to southern Madagascar, trop. Africa, or India; very widely introduced in tropics

Tamarix L. 1753. Tamaricaceae. 60–90 Canary Is., West and Southeast Europe, Mediterranean; Southwest, South (1), and trop. East Africa; West (incl. Caucasus), Southwest, and Middle Asia; West Siberia, eastward to northeastern China and Burma

Tamaulipa R. M. King et H. Rob. 1971 (Eupatorium L.). Asteraceae. 1 southern U.S., northeastern Mexico

Tamayoa Badillo = Lepidesmia Klatt. Asteraceae

Tambourissa Sonn. 1782. Monimiaceae. 43 Madagascar (26), Comoro Is. (5) and Mascarene Is. (12)

Tamilnadia Tirveng. et Sastre. 1979. Rubiaceae. 1 India

Tammsia Karst. 1861. Rubiaceae. 1 Colombia, Venezuela

Tamonea Aubl. 1775. Verbenaceae. 7 trop. America, West Indies

Tamus L. 1753. Dioscoreaceae. 4–5 Madeira and Canary Is., Europe, Mediterranean eastward to Iran; T. communis—also in Caucasus

Tanacetopsis (Tzvelev) Kovalevsk. 1971. Asteraceae. 16 Uzbekistan, Tajikistan

Tanacetum L. 1753. Asteraceae. c. 70 temp. Northern Hemisphere

Tanaecium Sw. 1788. Bignoniaceae. 6 trop. America from Costa Rica to Brazil, West Indies

Tanakaea Franch. et Sav. 1878. Saxifragaceae. 2 China, Japan

Tanaosolen N. E. Br. = Tritoniopsis L. Bolus. Iridaceae

Tanghinia Thouars = Cerbera L. Apocynaceae

Tangtsinia S. C. Chen. 1965 (Cephalanthera Rich.). Orchidaceae. 1 southwestern China

Tannodia Baill. 1861. Euphorbiaceae. 3 trop. Africa, Comoro Is.

Tanquana N. E. K. Hartmann et Liede. 1986. Aizoaceae. 3 South Africa

Tanulepis Balf. f. 1877. Asclepiadaceae. 5 Madagascar, Rodriguez I.

Tanzaniochloa Rauschert. 1982 (Setaria P. Beauv.). Poaceae. 1 Tanzania

Tapeinanthus Herb. = Narcissus L. Amaryllidaceae

Tapeinia Comm. ex Juss. 1789. Iridaceae. 1 southern Chile and Argentina

Tapeinocheilos Miq. 1869. Costaceae. 12 Moluccas, New Guinea, Bismarck Arch., Australia (1, T. ananassae—Queensland)

Tapeinoglossum Schltr. = Bulbophyllum Thouars. Orchidaceae

Tapeinosperma Hook. f. 1876. Myrsinaceae. c. 40 Philippines and Moluccas (?), New Guinea, New Britain, New Ireland, Solomon Is., Australia (Queensland), New Caledonia, Vanuatu, Fiji

Tapeinostemon Benth. 1854. Gentianaceae. 6 northeastern South America

Taphrospermum C. A. Mey. 1831. Brassicaceae. 2 West Siberia, Middle Asia, Mongolia, China

Tapinanthus (Blume) Rchb. 1841. Loranthaceae. c. 200 trop. and South (c. 20) Africa

Tapinopentas Bremek. = Otomeria Benth. Rubiaceae

Tapinostemma (Benth.) Tiegh. = Plicosepalum Tiegh. Loranthaceae

Tapiphyllum Robyns. 1928. Rubiaceae. 12 trop. Africa

Tapirira Aubl. 1775. Anacardiaceae. 10–15 trop. America

Tapirocarpus Sagot. 1882. Burseraceae. 1 Guyana

Tapiscia Oliv. 1890. Tapisciaceae. 3 continental and southern (incl. Hainan) China

Taplinia Lander. 1989. Asteraceae. 1 western Australia

Tapoides Airy Shaw. 1960. Euphorbiaceae. 1 Borneo

Tapura Aubl. 1775. Dichapetalaceae. 28 West Indies and trop. South America, trop. Africa (7)

Taraktogenos Hassk. = Hydnocarpus Gaertn. Flacourtiaceae

Taralea Aubl. 1775. Fabaceae. 5 trop. South America

Tarasa Phil. 1891. Malvaceae. c. 30 Mexico

(2), Andes from Peru to Chile and Argentina

Taravalia Greene = Ptelea L. Rutaceae

Taraxacum G. H. Weber ex Wigg. 1780. Asteraceae. 60 (or incl. apomictic races c. 2,000) temp. Northern Hemisphere and temp. South America (2)

Taraxia (Nutt.) Raimann = Camissonia Link. Onagraceae

Tarchonanthus L. 1753. Asteraceae. 2 South Africa

Tarenna Gaertn. 1788. Rubiaceae. c. 180 trop. Africa (c. 50), Madagascar, Seychelles, Indian Ocean islands, trop. Asia, Malesia, Australia, Oceania eastward to Tuamotu Arch.

Tarennoidea Tirveng. et Sastre. 1979. Rubiaceae. 2 Himalayas, India, Burma

Tarigidia Stent. 1932. Poaceae. 1 South Africa

Tarphochlamys Bremek. 1944 (Strobilanthes Blume). Acanthaceae. 1 Nepal, northeastern India, southern China

Tashiroea Matsum. = Bredia Blume. Melastomataceae

Tasmannia R. Br. ex DC. 1817 (Drimys J. R. Forst. et G. Forst.). Winteraceae. 5 Philippines, Borneo, Sulawesi, New Guinea, Solomon Is., Australia (Queensland, Victoria, Tasmania)

Tassadia Decne. 1844. Asclepiadaceae. 17 trop. America

Tateanthus Gleason. 1931. Melastomataceae. 1 Venezuela, Brazil

Tatianyx Zuloaga et Soderstr. 1985. Poaceae. 1 Brazil

Taubertia K. Schum. 1893. Menispermaceae. 1 Brazil

Tauscheria Fisch. ex DC. 1821. Brassicaceae. 1 (T. lasiocarpa) southeastern European part of Russia, West Siberia, Middle Asia, Iran, Afghanistan, Pakistan, northwestern China, Mongolia

Tauschia Schltdl. 1835. Apiaceae. 31 western U.S., Mexico, Central and trop. western South America

Tavaresia Welw. ex N. E. Br. 1903. Asclepiadaceae. 3 trop. and South Africa

Taverniera DC. 1825. Fabaceae. 15 Egypt (1), Arabian Peninsula, Socotra, Iraq, Iran, Pakistan, India

Taveunia Burret = Cyphosperma H. A. Wendl. ex Hook. f. Arecaceae

Taxanthema R. Br. 1810. Plumbaginaceae. 1 Australia (New South Wales)

Taxillus Tiegh. 1895. Loranthaceae. 25–30 ? Madagascar, Mascarene Is., India, Sri Lanka, southern China (15), Taiwan, Indochina, western Malesia (1, T. chinensis—Malay Peninsula, Borneo, Philippines)

Taxodium Rich. 1810. Taxodiaceae. 2 southeastern U.S., Mexico, Guatemala

Taxus L. 1753. Taxaceae. 7–8 Europe, Morocco, Algeria, Tunisia, Turkey, northwestern Iran, Himalayas, southern and eastern China, Taiwan, Korea, Japan, southern Vietnam, Sumatra, Philippines, Sulawesi, western and eastern North America, Mexico, northern Central America

Tayloriophyton M. P. Nayar. 1968. Melastomataceae. 2 Malay Peninsula, Borneo

Tchihatchewia Boiss. = Neotchihatchewia Rauschert. Brassicaceae

Teagueia (Luer) Luer. 1991. Orchidaceae. 6 Colombia, Ecuador

Teclea Delile. 1843. Rutaceae. 22 trop. and South Africa, Comoro Is., Madagascar

Tecleopsis Hoyle et Leakey = Vepris Comm. ex A. Juss. Rutaceae

Tecoma Juss. 1789. Bignoniaceae. 12 trop. America: U.S. (Arizona, New Mexico, Texas, Florida), Mexico, Guatemala, El Salvador, Honduras, Nicaragua, Costa Rica, Panama, Colombia, Venezuela, Ecuador, Peru, Bolivia, Chile, Argentina; West Indies: Bahamas, Cuba, Cayman Is., Jamaica, Haiti, Puerto Rico, Virgin Is., Leeward and Windward Is., southern Dutch Antilles, Trinidad

Tecomanthe Baill. 1888 (Campsis Lour.). Bignoniaceae. 5 Moluccas, New Guinea, Australia (Queensland), New Zealand (Three Kings Is.)

Tecomaria (Endl.) Spach. 1840 (Tecoma Juss.). Bignoniaceae. 1 (T. capensis) trop. East and South Africa

Tecomella Seem. 1863. Bignoniaceae. 1 Arabian Peninsula, western India

Tecophilaea Bertero ex Colla. 1836 (7EZephyra D. Don). Tecophilaeaceae. 7 Chile

Tecticornia Hook. f. 1880. Chenopodiaceae. 3: T. australasica—Java, New Guinea, northern Australia, 2—endemic in western Australia

Tectiphiala H. E. Moore. 1978 (Acanthophoenix H. A. Wendl.). Arecaceae. 1 (T. ferox) Mauritius

Tectona L. f. 1782. Chloanthaceae. 4 India, Sri Lanka, Burma, Thailand, Laos, Malesia

Tecunumania Standl. et Steyerm. 1944. Cucurbitaceae. 1 southern Mexico, Guatemala

Tedingea D. Müll.-Doblies et U. Müll.-Doblies. 1985. Amaryllidaceae. 1 South Africa

Teedia Rudolphi. 1799. Scrophulariaceae. 4 South Africa; T. lucida—extending to Zimbabwe

Teesdalia R. Br. ex Aiton et Aiton f. 1812. Brassicaceae. 2 Europe, Mediterranean

Teesdaliopsis (Willk.) Rothm. 1940 (Iberis L.). Brassicaceae. 1 (T. conferta) Iberian Peninsula

Tegicornia Paul G. Wilson. 1980. Chenopodiaceae. 1 western Australia

Teijsmanniodendron Koord. 1904. Verbenaceae. 14 southern Indochina, Malesia (except Java and Lesser Sunda Is.)

Teinosolen Hook. f. = Heterophyllaea Hook. f. Rubiaceae

Teinostachyum Munro. 1868 (Schizostachyum Nees). Poaceae. 6 India, Sri Lanka, Burma

Teixeiranthus R. M. King et H. Rob. 1980. Asteraceae. 2 Brazil

Telanthophora H. Rob. et Brettell. 1974. Asteraceae. 14 Mexico, Central America

Telectadium Baill. 1889. Asclepiadaceae. 3 Indochina

Telekia Baumg. 1816. Asteraceae. 2 Eurasia from Central Europe to Asia Minor; T. speciosa—also in European part of Russia and Caucasus

Telemachia Urb. 1916 (Cassine L.). Celastraceae. 1 Trinidad

Telephium L. 1753. Caryophyllaceae. 6 Mediterranean, Somalia, sothernmost Arabian Peninsula and eastern part facing Gulf of Oman, West Asia to Caucasus (2), Iran and western Pakistan, Madagascar (1)

Telesilla Klotzsch. 1848. Asclepiadaceae. 1 Guyana

Telfairia Hook. 1827. Cucurbitaceae. 3 trop. Africa

Teline Webb. = Genista L. Fabaceae

Teliostachya Nees. 1847. Acanthaceae. 10 Central and trop. South America, West Indies

Telipogon Mutis ex Kunth. 1816. Orchidaceae. c. 80 trop. America from Costa Rica to Peru and Brazil

Telitoxicum Moldenke. 1938. Menispermaceae. 8 Colombia, Guyana, Peru, Brazil

Tellima R. Br. 1823. Saxifragaceae. 1 western North America (Alaska, British Columbia, Washington, Oregon, California)

Telmatophila Mart. ex Baker. 1873. Asteraceae. 1 Brazil

Telminostelma Fourn. 1885. Asclepiadaceae. 1 Brazil

Telmissa Fenzl. 1842. Crassulaceae. 1 eastern Mediterranean from Turkey to Sinai Peninsula

Telopea R. Br. 1810. Proteaceae. 4 eastern Australia, Tasmania

Telosma Coville. 1905. Asclepiadaceae. 10 trop. Old World

Teloxys Moq. = Chenopodium L. Chenopodiaceae

Temmodaphne Kosterm. 1973. Lauraceae. 1 Thailand

Temnadenia Miers. 1878. Apocynaceae. 5 trop. South America

Temnocalyx Robyns. 1928. Rubiaceae. 1 Tanzania

Temnopteryx Hook. f. 1873. Rubiaceae. 1 trop. West Africa

Templetonia R. Br. ex W. T. Aiton. 1812. Fabaceae. 11 Australia

Temu O. Berg = Blepharocalyx O. Berg. Myrtaceae

Tenagocharis Hochst. = Butomopsis Kunth. Limnocharitaceae

Tenaris E. Mey. 1838. Asclepiadaceae. 7 trop. and South Africa

Tengia Chun. 1946. Gesneriaceae. 1 southwestern China

Tenicroa Raf. 1837 (Drimia Jacq.). Hyacinthaceae. 5 South Africa

Tennantia Verdc. 1981. Rubiaceae. 1 Somalia, Kenya

Tenrhynea Hilliard et B. L. Burtt. 1981. Asteraceae. 1 South Africa

Tephrocactus Lem. = Opuntia Hill. Cactaceae

Tephroseris Rchb. 1842. Asteraceae. c. 50 cold and temp. Northern Hemisphere, chiefly Eurasia

Tephrosia Pers. 1807. Fabaceae. c. 400 pantropics, esp. Africa (166), a few species in subtropics

Tepualia Griseb. 1854. Myrtaceae. 1 Chile

Tepuia Camp. 1939. Ericaceae. 8 Venezuela, Guyana

Tepuianthus Maguire et Steyerm. 1981. Tepuianthaceae. 6 Colombia, Venezuela, Brazil (Amazonia)

Teramnus P. Browne. 1756. Fabaceae. 8 pantropics; T. labialis—extending to South Africa

Terana La Llave. 1884. Asteraceae. 1 Mexico (nomen dubium)

Terauchia Nakai. 1913. Asphodelaceae. 1 Korea

Terebraria Sessé ex Kuntze = Neolaugeria Nicolson. Rubiaceae

Terminalia L. 1767. Combretaceae. c. 200: Africa (30), Madagascar (c. 35), trop. Asia and Malesia (c. 70), trop. Australia, Oceania (30), trop. America (c. 30)

Terminaliopsis Danguy = Terminalia L. Combretaceae

Terminthia Bernh. = Rhus L. Anacardiaceae

Terminthodia Ridl. = Tetractomia Hook. f. Rutaceae

Terniola Tul. = Dalzellia Wight. Podostemaceae

Terniopsis Chao. 1948 (Dalzellia Wight). Podostemaceae. 1 southern China (Yunnan)

Ternstroemia Mutis ex L. f. 1782. Theaceae. c. 85 Central and trop. South America, trop. (esp. South-West) Africa, trop. Asia from Sri Lanka through Indochina to China, Taiwan, Japan, Malesia to Bismarck Arch.; T. cherryi—extending into Queensland

Ternstroemiopsis Urb. = Eurya Thunb. Theaceae

Terrellia Lunell = Elymus L. Poaceae

Tersonia Moq. 1849. Gyrostemonaceae. 2 southwestern Australia

Terua Standl. et F. J. Herm. = Lonchocarpus Kunth. Fabaceae

Tessarandra Miers = Chionanthus L. Oleaceae

Tessaria Ruiz et Pav. 1794. Asteraceae. 1 trop. America from southwestern U.S. to Argentina

Tessmannia Harms. 1910. Fabaceae. 12 trop. Africa, esp. Zaire (7)

Tessmanniacanthus Mildbr. 1926. Acanthaceae. 1 eastern Peru

Tessmannianthus Markgr. 1927. Melastomataceae. 7 Panama, Colombia, Ecuador, eastern Peru

Tessmanniodoxa Burret = Chelyocarpus Dammer. Arecaceae

Testudipes Markgr. = Tabernaemontana L. Apocynaceae

Testulea Pellegr. 1924. Sauvagesiaceae. 1 trop. Africa

Tetilla DC. 1930. Francoaceae. 1 Chile

Tetraberlinia (Harms) Hauman. 1952. Fabaceae. 3 trop. Africa from Liberia to Zaire and Angola

Tetracanthus A. Rich = Pectis L. Asteraceae

Tetracarpaea Hook. 1840. Tetracarpaeaceae. 1 Tasmania

Tetracarpidium Pax. 1899. Euphorbiaceae. 1 trop. West Africa

Tetracentron Oliv. 1889. Tetracentraceae. 1 eastern Nepal, northeastern India, northern Burma, central and southwestern China

Tetracera L. 1753. Dilleniaceae. c. 40 pantropics

Tetrachaete Chiov. 1902. Poaceae. 1 Ethiopia, Kenya, Tanzania, Arabian Peninsula

Tetrachne Nees. 1841. Poaceae. 1 South Africa, Pakistan

Tetrachondra Petrie ex Oliv. 1892. Tetrachondraceae. 2 New Zealand (1), temp. South America (1, Patagonia and Tierra del Fuego)

Tetrachyron Schltdl. 1847. Asteraceae. 5 Mexico, Guatemala

Tetraclea A. Gray. 1853. Verbenaceae. 2 southern U.S., Mexico

Tetraclinis Mast. 1892. Cupressaceae. 1 southern Spain, Malta, Morocco, Algeria, Tunisia

Tetraclis Hiern. 1873 (or = Diospyros L.). Ebenaceae. 3 Madagascar

Tetracme Bunge. 1836. Brassicaceae. 5 eastern European part of Russia and Middle Asia (4), Iran, Afghanistan, Pakistan, northwestern China

Tetracmidion Korsh. 1898 (Tetracme Bunge). Brassicaceae. 2 Middle Asia

Tetracoccus Engelm. ex Parry. 1885. Euphorbiaceae. 5 southwestern North America

Tetracoilanthus Rappa et Camarrone = Aptenia N. E. Br. Aizoaceae

Tetractomia Hook. f. 1875 (Melicope J. R. Forst. et G. Forst.). Rutaceae. 6 Malesia from Malay Peninsula and Sumatra to New Guinea, Solomon Is.

Tetractys Spreng. 1822. ? Rutaceae. 1 South Africa

Tetracustelma Baill. 1890 (Matelea Aubl.). Asclepiadaceae. 2 Mexico

Tetradapa Osbeck = Erythrina L. Fabaceae

Tetradenia Benth. 1830. Lamiaceae. 6 trop. (1, T. riparia—from Ethiopia to Angola and Natal) and South Africa, Madagascar (3)

Tetradia Benn. ex R. Br. 1844 (Pterygota Schott et Endl.). Sterculiaceae. 1 Java

Tetradiclis Steven ex M. Bieb. 1819. Tetradiclidaceae. 1 Crimea, southern European part of Russia (Caspian lowland), Caucasus, Middle Asia, southeastern Mediterranean eastward to Iran

Tetradium Lour. 1790. Rutaceae. 9 Himalayas, northeastern India, China, Japan, Indochina, Sumatra, Java

Tetradoa Pichon. 1947 (Hunteria Roxb.). Apocynaceae. 2 trop. West Africa

Tetradoxa C. Y. Wu. 1981 (Adoxa L.). Adoxaceae. 1 southwestern China

Tetradyas Danser. 1931. Loranthaceae. 1 New Guinea

Tetradymia DC. 1838. Asteraceae. 10 western North America (British Columbia, Washington, Idaho, Montana, Wyoming, California, Nevada, Utah, Colorado, Arizona, New Mexico, and northern Baja California)

Tetraedrocarpus O. Schwarz = Echiochilon Desf. Boraginaceae

Tetraena Maxim. 1889. Zygophyllaceae. 1 China (Inner Mongolia, Gansu, Shensy)

Tetragamestus Rchb. f. = Scaphyglottis Poepp. et Endl. Orchidaceae

Tetragastris Gaertn. 1790. Burseraceae. 9 Central and northern South America, West Indies

Tetraglochidium Bremek. 1944 (Strobilanthes Blume). Acanthaceae. 8 western Malesia

Tetraglochin Kunze ex Poepp. 1833 (Margyricarpus Ruiz et Pav.). Rosaceae. 8 Andes and temp. South America

Tetragoga Bremek. 1944 (Strobilanthes Blume). Acanthaceae. 4 northeastern India, China, Indochina, Sumatra, Malay Peninsula

Tetragompha Bremek. 1944 (Strobilanthes Blume). Acanthaceae. 2 Sumatra

Tetragonia L. 1753. Aizoaceae. 50–60 North and South Africa, Southeast and East Asia, Australia, Tasmania, New Zealand, Polynesia, South America (arid parts of Peru and Chile)

Tetragonocalamus Nakai = Bambusa Schreb. Poaceae

Tetragonolobus Scop. = Lotus L. Fabaceae

Tetragonotheca L. 1753. Asteraceae. 4 southern U.S., Mexico

Tetralix Griseb. 1866. Tiliaceae. 2 Cuba

Tetralocularia O'Don. 1960. Convolvulaceae. 1 Colombia

Tetralopha Hook. f. = Gynochtodes Blume. Rubiaceae

Tetrameles R. Br. 1826. Datiscaceae. 1 southwestern India, Sri Lanka, Bangladesh, Andaman Is., Malay Peninsula, Indonesia, New Guinea, northeastern Australia

Tetrameranthus R. E. Fr. 1939. Annonaceae. 6 Colombia, Venezuela, Peru, Brazil

Tetramerista Miq. 1861. Tetrameristaceae. 1 Sumatra, Malay Peninsula, Borneo

Tetramerium Nees. 1846. Acanthaceae. 27 southern U.S. (Arizona, New Mexico, Texas), Mexico (25), Guatemala, Belize, El Salvador, Honduras, Nicaragua, Costa Rica, Panama, Colombia, Venezuela, Ecuador, Peru, Bolivia

Tetramicra Lindl. 1831. Orchidaceae. 10 West Indies

Tetramolopium Nees. 1832. Asteraceae. 36 New Guinea, Cook Is., Hawaiian Is. (11)

Tetranema Benth. 1843. Scrophulariaceae. 5 Mexico, Central America

Tetraneuris Greene. 1898. Asteraceae. 12 U.S., Mexico

Tetranthus Sw. 1788. Asteraceae. 4 West Indies

Tetrapanax (K. Koch) K. Koch. 1859 (Didymopanax Decne. ex Planch.). Araliaceae. 2 southern China, northern Vietnam, Taiwan

Tetrapathaea (DC.) Rchb. 1828. Passifloraceae. 1 New Zealand

Tetraperone Urb. 1901. Asteraceae. 1 Cuba

Tetrapetalum Miq. 1865. Annonaceae. 2 Borneo

Tetraphyllaster Gilg. 1897. Melastomataceae. 1 trop. West Africa

Tetraphyllum Griff. ex C. B. Clarke. 1883. Gesneriaceae. 2 northeastern India, Thailand

Tetraphysa Schltr. 1906. Asclepiadaceae. 1 Colombia

Tetrapilus Lour. 1790. Oleaceae. 10–15 Southeast Asia

Tetraplandra Baill. 1858. Euphorbiaceae. 5 Brazil

Tetraplasandra A. Gray. 1854 (Gastonia Comm. ex Lam.). Araliaceae. c. 20 Philippines, Sulawesi, New Guinea, Solomon Is., Hawaiian Is.

Tetraplasia Rehder = Damnacanthus C. F. Gaertn. Rubiaceae

Tetrapleura Benth. 1841. Fabaceae. 2 trop. Africa from Senegal to Sudan and Kenya and south to Angola and Zaire

Tetrapodenia Gleason = Burdachia Mart. ex A. Juss. Malpighiaceae

Tetrapogon Desf. 1799. Poaceae. 5 Canary Is., Cape Verde Is., Africa from Mauritania to Somalia and southward to Angola and Zimbabwe; West Asia to southwestern and southern India; T. villosus—extending to southern Middle Asia

Tetrapollinia Maguire et B. M. Boom. 1989. Gentianaceae. 1 Venezuela, Guyana, Surinam, Brazil

Tetrapteris Cav. 1790. Malpighiaceae. c. 90 Mexico, Central and trop. South America, West Indies

Tetrapterocarpon Humbert. 1939. Fabaceae. 1 Madagascar

Tetrardisia Mez. 1902. Myrsinaceae. 2 Java, Australia (Queensland)

Tetraria P. Beauv. 1816. Cyperaceae. 40 trop. and South Africa, Borneo, Australia

Tetrariopsis C. B. Clarke. 1908 (Tetraria P. Beauv.). Cyperaceae. 1 Australia

Tetrarrhena R. Br. 1810 (Ehrharta Thunb.). Poaceae. 4 Australia

Tetraselago Junell. 1961. Selaginaceae. 4 South Africa

Tetrasida Ulbr. 1916. Malvaceae. 1 Peru

Tetrasiphon Urb. 1904. Celastraceae. 1 Jamaica

Tetraspidium Baker. 1884. Scrophulariaceae. 1 Madagascar

Tetrastigma (Miq.) Planch. 1887. Vitaceae. c. 90 India, Himalayas, China (35), Taiwan, Indochina, Malesia, trop. Australia, and eastward to Fiji

Tetrastylidium Engl. 1872. Olacaceae. 2 Peru, southern Brazil

Tetrastylis Barb. Rodr. 1882 (Passiflora L.). Passifloraceae. 1 Brazil

Tetrasynandra Perkins. 1898. Monimiaceae. 3 northeastern Australia

Tetrataenium (DC.) Manden. 1959 (Heracleum L.). Apiaceae. 7–8 Middle Asia (1, T. olgae—Pamir-Alai), Iran, Afghanistan, Pakistan

Tetrataxis Hook. f. 1867. Lythraceae. 1 Mauritius

Tetratelia Sond. 1860 (Cleome L.). Capparaceae. 2 trop. Southeast Africa

Tetrathalamus Lauterb. = Zygogynum Baill. Winteraceae

Tetratheca Sm. 1793. Tremandraceae. 39 Australia

Tetrathylacium Poepp. 1841 (1843). Flacourtiaceae. 2 trop. America from Costa Rica to Peru and Amazonian Brazil

Tetrathyrium Benth. 1861. Hamamelidaceae. 1 southern China, Hong Kong

Tetraulacium Turcz. 1843. Scrophulariaceae. 1 Brazil

Tetrazygia Rich. ex DC. 1828. Melastomataceae. 25 Florida peninsula (1), West Indies

Tetrazygiopsis Borhidi = Tetrazygia Rich. ex DC. Melastomataceae

Tetroncium Willd. 1808. Juncaginaceae. 1 temp. South America (Strait of Magellan)

Tetrorchidiopsis Rauschert = Tetrorchidium Poepp. et Endl. Euphorbiaceae

Tetrorchidium Poepp. et Endl. 1841. Euphorbiaceae. c. 20 Central and trop. South America, West Indies, trop. Africa (5, from Guinea-Bissau eastward to Uganda and Tanzania, south to Angola)

Teucridium Hook. f. 1853. Verbenaceae. 1 New Zealand

Teucrium L. 1753. Lamiaceae. c. 200 cosmopolitan, chiefly Northern Hemisphere, esp. Mediterranean (c. 140), a few species in South America, mountainous trop. Northeast and South Africa, Australia

Teuscheria Garay. 1958. Orchidaceae. 3 Mexico, Central America, Colombia, Venezuela, Ecuador

Teyleria Backer. 1939. Fabaceae. 1 Java, Hainan

Thacla Spach = Caltha L. Ranunculaceae

Thaia Seidenf. 1975. Orchidaceae. 1 Thailand

Thailentadopsis Kosterm. = ? Havardia Small. Fabaceae

Thalassia Banks et C. Koenig. 1805. Thalassiaceae. 2 islands of the Indian and Pacific oceans (1), Caribbean coasts (1)

Thalassodendron Hartog. 1970. Cymodoceaceae. 2 coasts of the Red Sea, western Indian Ocean, eastern Malesia, southwestern and northeastern Australia

Thalestris Rizzini = Justicia L. Acanthaceae

Thalia L. 1753. Marantaceae. 11 trop. America, Africa (1)

Thalictrum L. 1753. Ranunculaceae. c. 330 North and South America; North, trop., and South Africa; Eurasia south to Mediterranean, Indochina, and New Guinea; Russia (all areas)

Thaminophyllum Harv. 1865. Asteraceae. 3 South Africa

Thamnea Sol. ex Brongn. 1826. Bruniaceae. 7 South Africa

Thamnocalamus Munro. 1868. Poaceae. 6 South Africa (1, T. tessellatus), Himalayas (3, India, Nepal), China (2)

Thamnocharis W. T. Wang. 1981. Gesneriaceae. 1 southwestern China

Thamnochortus Bergius. 1767. Restionaceae. 33 South Africa

Thamnojusticia Mildbr. = Justicia L. Acanthaceae

Thamnosciadium Hartvig. 1984. Apiaceae. 1 Greece

Thamnoseris F. Phil. 1875. Asteraceae. 1 Chile (San Ambrosio I.)

Thamnosma Torr. et Frém. 1845. Rutaceae. 8–9 U.S. (Nevada, southeastern California, Utah, Colorado, Arizona, New Mexico, Texas) and northern Mexico (incl. Baja California); South-West and South Africa, Arabian Peninsula, Socotra

Thamnus Klotzsch. 1838. Ericaceae. 1 South Africa (southern Cape)

Thapsia L. 1753. Apiaceae. 3 Southwest and South Europe, North Africa

Thaspium Nutt. 1818. Apiaceae. 3 North America

Thaumasianthes Danser. 1933. Loranthaceae. 2 Philippines

Thaumastochloa C. E. Hubb. 1936. Poaceae. 7 Southeast Asia, New Guinea, Australia

Thaumatocaryon Baill. 1890. Boraginaceae. 4 Brazil

Thaumatococcus Benth. 1883. Marantaceae. 1 trop. West Africa

Thaumatophyllum Schott = Philodendron Schott. Araceae

Thea L. = Camellia L. Theaceae

Thecacoris A. Juss. 1824. Euphorbiaceae. 20 trop. Africa, Madagascar

Thecagonum Babu = Oldenlandia L. Rubiaceae

Thecanthes Wikstr. 1818. Thymelaeaceae. 5 Malesia (Lesser Sunda Is., Philippines, New Guinea, New Ireland, New Britain, and Louisiade Arch.), western and northern Australia

Thecocarpus Boiss. 1844 (Echinophora L.). Apiaceae. 2 Turkey, Iran

Thecophyllum E. F. André = Vriesea Lindl. = Guzmania Ruiz et Pav. Bromeliaceae

Thecopus Seidenf. 1983. Orchidaceae. 2–3 Southeast Asia, Borneo

Thecorchus Bremek. 1952. Rubiaceae. 1 trop. Africa

Thecostele Rchb. f. 1857. Orchidaceae. 1 (T. alata) Himalayas, northeastern India, Burma, Thailand, Cambodia, Laos, Vietnam, Malay Peninsula, Sumatra, Java, Borneo, Philippines

Theileamea Baill. 1890. Acanthaceae. 4 trop. East Africa, Madagascar

Theilera E. Phillips. 1926. Campanulaceae. 1 (T. guthriei) South Africa (Cape)

Thelasis Blume. 1825. Orchidaceae. 20–25 Himalayas, India, Burma, southern China, Indochina, Malesia from Malay Peninsula to New Guinea and Solomon Is.; 1 (T. pygmaea) Admiralty Is.

Thelechitonia Cuatrec. = Wedelia Jacq. Asteraceae

Theleophyton (Hook. f.) Moq. = Atriplex L. Chenopodiaceae

Thelepaepale Bremek. 1944 (Strobilanthes Blume). Acanthaceae. 1 India, ? Sri Lanka

Thelepogon Roth ex Roem. et Schult. 1817. Poaceae. 1 trop. Africa and Asia, Malesia

Thelesperma Less. 1831. Asteraceae. 12 western North America

Thelethylax C. Cusset. 1973. Podostemaceae. 2 Madagascar

Theligonum L. 1753. Theligonaceae. 3 Canary Is., Mediterranean, Crimea (1, T. cynocrambe), West Asia, southwestern China, Japan

Thelionema R. J. F. Hend. 1985. Phormiaceae. 3 southeastern Australia, Tasmania

Thellungia Stapf = Eragrostis Wolf. Poaceae

Thellungiella O. E. Schulz. 1924. Brassicaceae. 3 Southeast Europe, Asia Minor, Caucasus, Middle Asia, West and East Siberia, Iran, northern China, Mongolia, western North America

Thelocactus (K. Schum.) Britton et Rose. 1922. Cactaceae. 11 southern U.S. (Texas), northern to central Mexico

Thelymitra J. R. Forst. et G. Forst. 1776. Orchidaceae. 45–50 Malesia, Australia, Tasmania, New Caledonia, New Zealand

Thelypodiopsis Rydb. = Thelypodium Endl. Brassicaceae

Thelypodium Endl. 1839. Brassicaceae. 19 western U.S., Mexico

Thelyschista Garay. 1982. Orchidaceae. 1 Brazil

Themeda Forssk. 1775. Poaceae. c. 20 trop. Old World

Themistoclesia Klotzsch. 1851. Ericaceae. 25 Andes from Costa Rica and Panama to Peru and Venezuela

Thenardia Kunth. 1819. Apocynaceae. 6 Mexico, Central America

Theobroma L. 1753. Sterculiaceae. 20 trop. America

Theodorovia Kolak. = Campanula L. Campanulaceae

Theophrasta L. 1753. Theophrastaceae. 2 Haiti

Thereianthus G. L. Lewis. 1941 (Micranthus [Pers.] Eckl.) Iridaceae. 7 South Africa (Cape)

Theriophonum Blume. 1837. Araceae. 5 India, Sri Lanka

Thermopsis R. Br. 1811. Fabaceae. 23 North America (10), East Europe, temp. Asia; Russia (13, eastern and southeastern European part, southern Siberia, Far East) and Middle Asia

Therocistus Holub. 1986 (Cistus L.). Cistaceae. 11 West Europe, Mediterranean

Theropogon Maxim. 1871. Convallariaceae. 1 Himalayas, Tibet, northeastern India

Therorhodion (Maxim.) Small. 1914 (Rhododendron L.). Ericaceae. 2 East Asia, Northwest America

Thesidium Sond. 1857. Santalaceae. 8 South Africa (Cape)

Thesium L. 1753. Santalaceae. c. 300 Europe, Africa, Asia, southeastern Australia (1), South America; c. 25 Russia (European part, Siberia, Far East), Caucasus, and Middle Asia

Thespesia Sol. ex Corrêa. 1807. Malvaceae. 18: 1 in the littoral zone of the pantropics; 17 West Indies, trop. Africa, trop. Asia to New Guinea

Thespesiopsis Exell et Hillc. = Thespesia Sol. ex Corrêa. Malvaceae

Thespidium F. Muell. ex Benth. 1867. Asteraceae. 1 trop. Australia

Thespis DC. 1833. Asteraceae. 1 Southeast Asia

Thevenotia DC. 1833. Asteraceae. 2 Middle Asia (1, T. scabra), Iran

Thevetia L. 1758. Apocynaceae. 9 trop. America, West Indies

Thibaudia Ruiz et Pav. 1805. Ericaceae. c. 60 trop. America from Costa Rica and Panama (1, T. costaricensis) through Colombia to Peru and eastward to Mt. Roraima on the Venezuela–Guyana border

Thieleodoxa Cham. = Alibertia A. Rich. Rubiaceae

Thilachium Lour. 1790. Capparaceae. 15 trop. East and South (1) Africa, Madagascar, Mascarene Is.

Thiloa Eichler. 1866. Combretaceae. 3 trop. South America

Thinopyrum A. Löve = Elytrigia Desv. Poaceae

Thinouia Triana et Planch. 1862. Sapindaceae. 12 trop. and subtrop. South America

Thiseltonia Hemsl. 1905. Asteraceae. 1 western Australia

Thismia Griff. 1845. Burmanniaceae. 25 trop. Asia, Australia, New Zealand, trop. America from Panama to Peru, Bolivia, and southeastern Brazil

Thladiantha Bunge. 1833. Cucurbitaceae. 24 eastern Himalayas, northeastern India, China, Korea, Japan, Indochina, Malesia; T. dubia—southern Far East of Russia

Thlaspeocarpa C. A. Sm. 1931. Brassicaceae. 2 South Africa

Thlaspi L. 1753. Brassicaceae. c. 60 temp. regions and mountainous trop. areas of the Northern Hemisphere, a few species in South America

Thlaspiceras F. K. Mey. = Thlaspi L. Brassicaceae

Thodaya Compton = Euryops (Cass.) Cass. Asteraceae

Thogsennia Aiello. 1979. Rubiaceae. 1 Cuba

Thomandersia Baill. 1891. Acanthaceae. 6 trop. Africa

Thomandersiaceae Sreemadh. = Acanthaceae Juss.

Thomasia J. Gay. 1821. Sterculiaceae. 31 Australia

Thompsonella Britton et Rose. 1909 (Echeveria DC.). Crassulaceae. 2 Mexico

Thomsonia Wall. 1830 (Amorphophallus Blume ex Decne.). Araceae. 2 Nepal, northeastern India, Thailand, Taiwan

Thonningia Vahl. 1810. Balanophoraceae. 1 trop. Africa from Senegal eastward to southwestern Ethiopia and south to Zambia

Thoracocarpus Harling. 1958. Cyclanthaceae. 1 Panama, trop. South America, Trinidad

Thoracosperma Klotzsch. 1834. Ericaceae. 5 South Africa (southern Cape)

Thoracostachyum Kurz. 1869 (Mapania Aubl.). Cyperaceae. 8 Seychelles, southern China, Indochina, Malesia, Caroline Is., Australia (Queensland), Fiji, Polynesia

Thoreldora Pierre = Glycosmis Corrêa. Rutaceae

Thorelia Gagnep. = Camchaya Gagnep. Asteraceae

Thorella Briq. = Caropsis (Rouy et Camus) Rauschert. Apiaceae

Thoreochloa Holub = Pseudarrhenatherum Rouy. Poaceae

Thornbera Rydb. = Dalea L. Fabaceae

Thorncroftia N. E. Br. 1912 (Plectranthus L'Hér.). Lamiaceae. 3 South Africa (Transvaal)

Thornea Breedlove et E. McClintock. 1976 (Triadenum Raf.). Hypericaceae. 2 Mexico, Guatemala

Thorntonia Rchb. = Kosteletzkya C. Presl. Malvaceae

Thottea Rottb. 1783. Aristolochiaceae. 26 India (4), Sri Lanka (1), Andaman Is., Bangladesh, Burma, southern China (1, Hainan), Thailand, Vietnam, Malesia (22, Sumatra, Malay Peninsula, Java, Borneo, Philippines, Sulawesi)

Thouarsiora Homolle ex Arènes = Ixora L. Rubiaceae

Thouinia Poit. 1804. Sapindaceae. 28 Mexico, Central America, West Indies

Thouinidium Radlk. 1878. Sapindaceae. 7 Mexico, Central America, West Indies

Thozetia F. Muell. ex Benth. 1868. Asclepiadaceae. 1 Australia (eastern Queensland)

Thrasya Kunth. 1816. Poaceae. c. 20 Central and trop. South America

Thrasyopsis Parodi. 1946. Poaceae. 2 southern Brazil

Thraulococcus Radlk. = Lepisanthes Blume. Sapindaceae

Threlkeldia R. Br. 1810. Chenopodiaceae. 4 Australia and Tasmania

Thrinax Sw. 1788. Arecaceae. 7 Caribbean coasts from Florida to Belize, Bahama Is., Cuba (3), Jamaica (3), Puerto Rico, Virgin Is., Anguilla

Thrixgyne Keng = Duthiea Hack. Poaceae

Thrixspermum Lour. 1790. Orchidaceae. c. 160 trop. Asia from India and Sri Lanka through Indochina to Taiwan and Ryukyu Is., Malesia from Malay Peninsula to New Guinea and New Ireland, trop. Australia, Caroline Is., Bougainville and Solomon Is., Vanuatu, New Caledonia, Fiji, Samoa

Thryallis Mart. 1829. Malpighiaceae. 12 trop. and subtrop. America

Thryothamnus Phil. = Junellia Moldenke. Verbenaceae

Thryptomene Endl. 1839. Myrtaceae. 25 Australia

Thuarea Pers. 1805. Poaceae. 2 Madagascar, South Asia, Malesia, northern Australia, Polynesia

Thuja L. 1753. Cupressaceae. 5 China, northern Korea, Japan; northwestern and eastern (from Hudson Bay to Virginia) North America

Thujopsis Siebold et Zucc. ex Endl. 1842. Cupressaceae. 1 Japan, southern China (Yunnan)

Thulinia P. J. Cribb. 1985. Orchidaceae. 1 Tanzania

Thunbergia Retz. 1780. Thunbergiaceae. c. 100 trop. Old World

Thunbergianthus Engl. 1897. Scrophulariaceae. 2 Sao Tomé I. (1), trop. East Africa (1)

Thunbergiella H. Wolff = Itasina Raf. Apiaceae

Thunia Rchb. f. 1852. Orchidaceae. 4–6 Himalayas (Nepal, Sikkim, Bhutan), India, Burma, southern China (Yunnan), Andaman Is., Indochina

Thuranthos C. H. Wright. 1916. Hyacinthaceae. 2 South Africa

Thurnia Hook. f. 1883. Thurniaceae. 2 Venezuela, Guyana, Brazil (Amazonia)

Thurovia Rose = Gutierrezia Lag. Asteraceae

Thurya Boiss. et Balansa. 1856. Caryophyllaceae. 1 Asia Minor

Thuspeinanta T. Durand. 1888. Lamiaceae. 2 Middle Asia (1, T. persica), Iraq, Iran, Afghanistan

Thylacanthus Tul. 1844. Fabaceae. 1 Brazil (Amazonia)

Thylacodraba (Nábêlek) O. E. Schulz = Draba L. Brassicaceae

Thylacoglossum (Schltr.) Brieger = Glossorhyncha Ridl. Orchidaceae

Thylacophora Ridl. 1816 (Riedelia Oliv.). Zingiberaceae. 1 western New Guinea

Thylacospermum Fenzl. 1840. Caryophyllaceae. 1 Middle Asia, western China, Tibet, Himalayas

Thymbra L. 1753. Lamiaceae. 4 Mediterranean, Turkey, Iraq, Iran

Thymelaea Mill. 1754. Thymelaeaceae. c. 30 temp. Eurasia, Mediterranean; T. passerina—southern European part of Russia, Caucasus, Middle Asia, West Siberia

Thymocarpus Nicolson, Steyerm. et Sivad. 1981 (Calathea G. Mey.). Marantaceae. 1 Venezuela, Brazil

Thymophylla Lag. 1816. Asteraceae. 11 southern U.S., Mexico, West Indies, Argentina

Thymopsis Benth. 1873. Asteraceae. 2 West Indies

Thymus L. 1753. Lamiaceae. c. 350 Canary Is., Mediterranean (c. 120), Northeast Africa to southern Ethiopia, temp. and subtrop. Eurasia

Thyridachne C. E. Hubb. 1949. Poaceae. 1 trop. Africa

Thyridocalyx Bremek. 1956. Rubiaceae. 1 Madagascar

Thyridolepis S. T. Blake. 1972. Poaceae. 3 Australia

Thyrocarpus Hance. 1862. Boraginaceae. 3 China, Taiwan

Thyrsacanthus Nees = Odontonema Nees. Acanthaceae

Thyrsanthella (Baill.) Pichon = Trachelospermum Lem. Apocynaceae

Thyrsanthemum Pichon. 1946. Commelinaceae. 3 Mexico

Thyrsanthera Pierre ex Gagnep. 1925. Euphorbiaceae. 1 Southeast Asia

Thyrsia Stapf. 1917 (Phacelurus Griseb.). Poaceae. 3–4 trop. Africa, South and Southeast Asia

Thyrsodium Salzm. ex Benth. 1852. Anacardiaceae. 7–8 trop. South America, West Africa

Thyrsosalacia Loes. 1940. Celastraceae. 1–2 trop. West Africa

Thyrsostachys Gamble. 1896. Poaceae. 2 Burma and Thailand

Thysanella A. Gray = Polygonella Michx. Polygonaceae

Thysanocarpus Hook. 1830. Brassicaceae. 4–5 western and southwestern U.S.

Thysanoglossa Porto et Brade. 1940. Orchidaceae. 2 Brazil

Thysanolaena Nees. 1835. Poaceae. 1 trop. Asia

Thysanostemon Maguire. 1964. Clusiaceae. 2 Guyana

Thysanostigma J. B. Imlay. 1939. Acanthaceae. 2 Thailand, Malay Peninsula

Thysanotus R. Br. 1810. Asphodelaceae. 49 Australia (49) and Tasmania (1, T. patersonii); T. chinensis—southern China, Hong Kong, northern Vietnam, southern Thailand, northern Malay Peninsula, Philippines, Sulawesi, Flores I., Aru Is., New Guinea (2)

Thysanurus O. Hoffm. = Geigeria Griessel. Asteraceae

Thyselium Raf. 1840. Apiaceae. 1 Europe, North Asia

Tianschaniella Fedtsch. et Popov. 1951. Boraginaceae. 1 Middle Asia

Tiarella L. 1753. Saxifragaceae. 3–7 Nepal; Bhutan; western, central, and southwestern China; Japan; North America

Tiarocarpus Rech. f. 1972. Asteraceae. 3 Afghanistan

Tibestina Maire = Dicoma Cass. Asteraceae

Tibetia (Ali) H. P. Tsui. 1979 (Gueldenstaedtia Fisch.). Fabaceae. 3–4 Himalayas, Tibet

Tibouchina Aubl. 1775. Melastomataceae. c. 350 trop. America from Mexico to Argentina, West Indies

Tibouchinopsis Markgr. 1927. Melastomataceae. 1 northeastern Brazil

Ticodendron Gómez-Laur. et L. D. Gómez. 1989. Ticodenraceae. 1 southern Mexico, Guatemala, Nicaragua, Costa Rica, Panama

Ticoglossum Luc. Rodr. et Halbinger. 1983 (Odontoglossum Kunth). Orchidaceae. 2 trop. America

Ticorea Aubl. 1775. Rutaceae. 3 Panama, Colombia, Venezuela, Guyana, Brazil

Tidestromia Standl. 1916. Amaranthaceae. 7 southwestern U.S., Mexico

Tieghemella Pierre. 1890. Sapotaceae. 2 trop. West Africa

Tieghemia Balle. 1956 (Oncocalyx Tiegh.). Loranthaceae. 3 Zimbabwe, Mozambique, South Africa from Natal to eastern Cape

Tieghemopanax R. Vig. = Polyscias J. R. Forst. et G. Forst. Araliaceae

Tienmuia Hu = Phacellanthus Siebold et Zucc. Orobanchaceae

Tietkensia P. S. Short. 1990. Asteraceae. 1 western and central Australia

Tigridia Juss. 1789. Iridaceae. 35 Mexico, Guatemala, Andes from Peru to Chile

Tigridiopalma C. Chen. 1979. Melastomataceae. 1 southern China

Tikalia Lundell = Blomia Miranda. Sapindaceae

Tilia L. 1753. Tiliaceae. 60–80 temp. Northern Hemisphere (esp. China—32), southward to Mexico and Indochina; Russia (16, European part, West Siberia, Far East) and Caucasus

Tiliacora Colebr. 1821. Menispermaceae. 22 trop. Africa (19), northeastern India, Burma, Indochina, Malay Peninsula, Penang and Langkawi Is., northern Australia (1)

Tilingia Regel. 1859 (Ligusticum L.). Apiaceae. 3 East Asia (incl. Russian Far East [1, T. ajanensis]), Alaska

Tillaea L. 1753 (Crassula L.). Crassulaceae. c. 60 cosmopolitan; 3 Russia (European part, Far East), Caucasus, Middle Asia

Tillandsia L. 1753. Bromeliaceae. c. 440 trop. and subtrop. America, West Indies

Timonius DC. 1830. Rubiaceae. c. 180 Seychelles, Mauritius, Sri Lanka, Andaman Is., southern China, Malesia, Micronesia, Australia, New Caledonia, Fiji, east to Tuamotu Arch.

Timouria Roshev. = Achnatherum P. Beauv. Poaceae

Tina Schult. 1819–1820. Sapindaceae. 16 Madagascar

Tinantia Scheidw. 1839. Commelinaceae. 13 trop. America, West Indies

Tinguarra Parl. 1843. Apiaceae. 1 Canary Is.

Tinna Kotschy ex Hook. f. 1867. Lamiaceae. 19 trop. and South (4) Africa

Tinomiscium Miers ex Hook. f. et Thomson. 1855. Menispermaceae. 1 northeastern India, Nicobar Is., Burma, southern China (Yunnan), Thailand, Vietnam, Sumatra, Malay Peninsula, Java, Borneo, Natuna Is., Philippines, New Guinea

Tinopsis Radlk. 1887 (Tina Schult.). Sapindaceae. ? sp. Madagascar

Tinospora Miers. 1851. Menispermaceae. 33 trop. Africa (7), Madagascar (2), trop. Asia, Malesia (14), Australia, Pacific Is.

Tintinnabularia Woodson. 1936. Apocynaceae. 1 Central America

Tipuana (Benth.) Benth. 1859. Fabaceae. 1 Bolivia, northwestern Argentina

Tipularia Nutt. 1818. Orchidaceae. 4 eastern Himalayas, Tibet, China, Japan, eastern U.S.

Tiquilia Pers. 1805 (Coldenia L.). Boraginaceae. 27 arid North and South America

Tiquiliopsis A. Heller = Tiquilia Pers. Boraginaceae

Tirania Pierre. 1887. Capparaceae. 1 Indochina

Tirpitzia Hallier f. 1923. Linaceae. 2 southwestern China, northern Vietnam

Tischleria Schwantes = Carruanthus (Schwantes) Schwantes. Aizoaceae

Tisonia Baill. 1886. Flacourtiaceae. 15 Madagascar

Tisserantia Humbert = Sphaeranthus L. Asteraceae

Tisserantiodoxa Aubrév. et Pellegr. = Englerophytum Krause. Sapotaceae

Titania Endl. = Oberonia Lindl. Orchidaceae

Titanopsis Schwantes. 1926. Aizoaceae. 6 southern Namibia, central South Africa

Titanotrichum Soler. 1909. Gesneriaceae. 1 China, Taiwan

Tithonia Desf. ex Juss. 1789. Asteraceae. 11 southwestern U.S., Mexico, Central America (1), West Indies

Tithymalus Gaertn. = Euphorbia L. Euphorbiaceae

Tittmannia Brongn. 1826. Bruniaceae. 3–4 South Africa (Cape)

Tobagoa Urb. 1916. Rubiaceae. 1 West Indies, Panama, northern coastal Venezuela, Tobago

Tococa Aubl. 1775. Melastomataceae. c. 50 trop. America from Mexico to Bolivia and Brazil

Tocoyena Aubl. 1775. Rubiaceae. 20 trop. America, West Indies

Todaroa Parl. 1843. Apiaceae. 2 Canary Is.

Toddalia Juss. 1789. Rutaceae. 1 trop. (Sudan, Zaire, Ethiopia, Uganda, Kenya, Tanzania, Zambia, Zimbabwe) and South Africa, Madagascar, Mauritius, India, Sri Lanka, southern China, Indochina, Indonesia, and Philippines

Toddaliopsis Engl. 1895 (Vepris A. Juss). Rutaceae. 2–3 trop. and South Africa

Toechima Radlk. 1879. Sapindaceae. 8 Lesser Sunda Is. (Flores I.), New Guinea, eastern Australia (6)

Tofieldia Huds. 1778. Melanthiaceae. 10–17 cold and temp. Northern Hemisphere, Andes of South America; Russia (4–5, arctic and central European part, East Siberia, Far East)

Tolbonia Kuntze. 1891 (Calotis R. Br.). Asteraceae. 1 Southeast Asia

Tolmachevia A. Löve et D. Löve = Rhodiola L. Crassulaceae

Tolmiea Torr. et A. Gray. 1840. Saxifragaceae. 1 western North America (Alaska, British Columbia, Washington, Oregon, northern California)

Tolpis Adans. 1763. Asteraceae. 25 Asores, Canary Is., Cape Verde Is., Mediterranean, Ethiopia, Somalia, 1 sp. extending to Cape

Tolypanthus (Blume) Blume. 1830. Loranthaceae. 5 India, Sri Lanka eastward to southeastern China, Indochina, and Malesia

Tomanthea DC. = Centaurea L. Asteraceae

Tomanthera Raf. = Agalinis Raf. Scrophulariaceae

Tomentaurum G. L. Nesom. 1991. Asteraceae. 1 Mexico

Tommasinia Bertol. 1837 (Angelica L.). Apiaceae. 1 Central and South Europe

Tonduzia Pittier = Alstonia R. Br. Apocynaceae

Tonella Nutt. ex A. Gray. 1868. Scrophulariaceae. 2 western U.S.

Tonestus A. Nelson. 1904. Asteraceae. 8 North America

Tongoloa H. Wolff. 1925. Apiaceae. 10–14 western China, Tibet, Himalayas

Tonina Aubl. 1775. Eriocaulaceae. 1 Central and trop. South America, West Indies

Tontelea Aubl. 1775. Celastraceae. 30 Central and trop. South America

Toona (Endl.) M. Roem. 1846 (Cedrela P. Browne). Meliaceae. 6 Afghanistan, Himalayas from Kashmir to Bhutan, India, Sri Lanka, southern China, Indochina, Malesia, northern Australia

Topobea Aubl. 1775. Melastomataceae. c. 50 trop. America

Tordyliopsis DC. 1830. Apiaceae. 1 Himalayas from Kumaon to Sikkim

Tordylium L. 1753. Apiaceae. 18 Europe, North Africa, West and Southwest Asia; T. maximum—Caucasus, Middle Asia

Torenia L. 1753. Scrophulariaceae. 53 trop. Old World, naturalized in America

Toricellia DC. = Torricellia DC. Torricelliaceae

Torilis Adans. 1763. Apiaceae. 15 Canary Is., Eurasia, incl. (8) southern European part of Russia, Caucasus, Middle Asia, southern Russian Far East), North Africa

Torminalis Medik. 1789 (Crataegus L.). Rosaceae. 2: T. clusii—southern Europe, North Africa, Asia Minor; T. orientalis—Iran

Tornabenea Parl. ex Webb. 1850. Apiaceae. 3 Cape Vedre Is.

Toronia L. A. S. Johnson et B. G. Briggs. 1975. Proteaceae. 1 New Zealand

Torralbasia Krug et Urb. 1900. Celastraceae. 2 West Indies

Torrenticola Domin ex Steenis. 1947. Podostemaceae. 1 New Guinea, Australia (Queensland)

Torresea Allemao = Amburana Schwacke et Taub. Fabaceae

Torreya Arn. 1838. Taxaceae. 6 Burma, China, Japan, U.S. (California—Sierra Nevada Mts, Georgia, northwestern Florida)

Torreyochloa Church. 1949. Poaceae. 8 East Asia, North America; 2 extending to southern Russian Far East

Torricellia DC. 1830. Torricelliaceae. 3 continental China, eastern Himalayas (Nepal, Bhutan), northern Burma

Tortuella Urb. 1927. Rubiaceae. 1 Haiti

Torularia (Cass.) O. E. Schulz (non Bonnemaison—Algae) = Neotorularia Hedge et J. Léonard. Brassicaceae

Torulinium Desv. ex Ham. 1825 (Cyperus L.). Cyperaceae. 10 tropics and subtropics; T. odoratum—Caucasus

Toubaouate Aubrév. et Pellegr. = Didelotia Baill. Fabaceae

Touchardia Gaudich. 1847–1848. Urticaceae. 1 Hawaiian Is.

Toulicia Aubl. 1775. Sapindaceae. 14 northern South America

Tournaya A. Schmitz = Bauhinia L. Fabaceae

Tournefortia L. 1753. Boraginaceae. c. 150 tropics, subtropics, and warm–temp. regions

Tournefortiopsis Rusby = Guettarda L. Rubiaceae

Tourneuxia Coss. 1859. Asteraceae. 1 Algeria

Tournonia Moq. 1849. Basellaceae. 1 Colombia

Touroulia Aubl. 1775. Quiinaceae. 4 trop. South America

Tourrettia Foug. 1787. Bignoniaceae. 1 Mexico, Guatemala, Honduras, Nicaragua, Costa Rica, Panama, Colombia, Venezuela, Ecuador, Peru, Bolivia, Argentina

Toussaintia Boutique. 1951. Annonaceae. 3 trop. Africa (Congo, Gabon, Tanzania)

Tovara Adans. = Persicaria Hill. Polygonaceae

Tovaria Ruiz et Pav. 1794. Tovariaceae. 2 trop. America from Mexico to Peru and Venezuela, West Indies (1, T. diffusa)

Tovarochloa Macfarlane et But. 1982. Poaceae. 1 Andean Peru

Tovomita Aubl. 1775. Clusiaceae. 12–25 trop. America

Tovomitidium Ducke. 1935. Clusiaceae. 2 Brazil

Tovomitopsis Planch. et Triana. 1860. Clusiaceae. 50 trop. America

Townsendia Hook. 1834. Asteraceae. 25 western and central U.S. and adjacent Mexico

Townsonia Cheeseman. 1906 (Acianthus R. Br.). Orchidaceae. 2 Tasmania, New Zealand

Toxanthes Turcz. 1851. Asteraceae. 2 western and southern Australia

Toxicodendron Hill. 1753 (Rhus L.). Anacardiaceae. c. 40 East Asia, North and South America

Toxicoscordion Rydb. = Zigadenus Michx. Melanthiaceae

Toxocarpus Wight et Arn. 1834. Asclepiadaceae. 70 trop. Africa, Madagascar, Mascarene Is., India, Sri Lanka, Indochina, southern China, Malesia

Tozzia L. 1753. Scrophulariaceae. 1 (T. carpathica) Central Europe and Carpathian Mts.

Trachelanthus Kunze. 1850. Boraginaceae. 2 Asia Minor, Iran, Middle Asia

Tracheliopsis Buser = Campanula L. Campanulaceae

Trachelium L. 1753. Campanulaceae. 1 western Mediterranean

Trachelospermum Lem. 1851. Apocynaceae. c. 30 Himalayas, India, southern China, Vietnam, Japan, southeastern U.S.

Trachoma Garay = Tuberolabium Yamam. Orchidaceae

Trachomitum Woodson. 1930 (Apocynum L.). Apocynaceae. 1–6 South and Southeast Europe; West (incl. Caucasus),

Southwest, and Middle Asia eastward to southern Siberia and China

Trachyandra Kunth. 1843. Asphodelaceae. 65 trop. and South Africa, Madagascar (1)

Trachycalymma (K. Schum.) Bullock = Asclepias L. Asclepiadaceae

Trachycarpus H. A. Wendl. 1863. Arecaceae. 8 Himalayas, northeastern India, northern Burma, northern Thailand, China

Trachydium Lindl. 1835. Apiaceae. 1–2 Himalayas, Tibet

Trachylobim Hayne. 1827 (Hymenaea L.). Fabaceae. 1 trop. East Africa, Madagascar, Mauritius

Trachymene Rudge. 1811. Apiaceae. 45 Philippines, Borneo, Australia, New Caledonia, Fiji

Trachynia Link. 1827 (Brachypodium P. Beauv.). Poaceae. 2 Canary Is.; Mediterranean; Ethiopia; South Africa; Southeast Europe; West (incl. Caucasus), Southwest, and Middle Asia eastward to Pakistan and northern India

Trachyphrynium Benth. 1883. Marantaceae. 1 trop. Africa from Sierra Leone to Sudan and Uganda

Trachyphrynium K. Schum. = Hypselodelphys (K. Schum.) Milne-Redh.

Trachypogon Nees. 1829. Poaceae. 5 trop. America, Africa, Madagascar

Trachyrhizum (Schltr.) Brieger = Dendrobium Sw. Orchidaceae

Trachys Pers. 1805. Poaceae. 1 southern India, Sri Lanka, Burma

Trachyspermum Link. 1821. Apiaceae. 15 North, Northeast, and trop. Africa; West and Southwest Asia eastward to western China and India; T. ammi—Middle Asia

Trachystemon D. Don. 1832. Boraginaceae. 1 (T. orientalis) Bulgaria, Turkey, Caucasus

Trachystigma C. B. Clarke. 1883. Gesneriaceae. 1 trop. Africa

Trachystoma O. E. Schulz. 1916. Brassicaceae. 3 Morocco

Trachystylis S. T. Blake. 1937. Cyperaceae. 2 Australia (Queensland)

Tractocopevodia Raizada et Naray. 1946. Rutaceae. 1 Burma

Tracyina S. T. Blake. 1937. Asteraceae. 1 California

Tradescantia L. 1753. Commelinaceae. 65 warm–temp. and trop. America

Traganopsis Maire et Wilczek. 1936. Chenopodiaceae. 1 Morocco

Traganum Delile. 1813–1814. Chenopodiaceae. 2 Canary Is., western Sahara, Mauritania, Morocco, Algeria, Tunisia, Libya, Egypt, Syria, Lebanon, Israel, Jordan, Iraq, Arabian Peninsula

Tragia L. 1753. Euphorbiaceae. 125 tropics and subtropics

Tragiella Pax et K. Hoffm. 1919 (Sphaerostylis Baill.). Euphorbiaceae. 5 trop. and South Africa

Tragiola Small et Pennell = Gratiola L. Scrophulariaceae

Tragiopsis Pomel = Brachyapium (Baill.) Maire. Apiaceae

Tragoceros Kunth = Zinnia L. Asteraceae

Tragopogon L. 1753. Asteraceae. c. 100 temp. Eurasia, esp. Caucasus and Middle Asia; Mediterranean

Tragus Haller. 1768. Poaceae. 8 trop. Old World; T. racemosus—extending to South Europe, southern European part of Russia, Caucasus, and Middle Asia

Trailliaedoxa W. W. Sm. et Forrest. 1917. Rubiaceae. 1 southwestern China

Transcaucasia M. Hiroe = Astrantia L. Apiaceae

Trapa L. 1753. Trapaceae. 1–5 (or 15–40) trop., subtrop., and temp. regions of the Old World, except Australia

Trapella Oliv. 1887. Trapellaceae. 1–2 China, Korea, Japan, southern Russian Far East (1, T. sinensis)

Trattinnickia Willd. 1806. Burseraceae. 11 northern South America

Traubia Moldenke. 1963. Amaryllidaceae. 1 Chile

Traunia K. Schum. = Toxocarpus Wight et Arn. Asclepiadaceae

Traunsteinera Rchb. 1842. Orchidaceae. 2 Europe, Mediterranean, Caucasus

Trautvetteria Fisch. et C. A. Mey. 1835. Ranunculaceae. 2: T. japonica—Sakhalin I., Kuril Is., Japan; T. caroliniensis—western and southern North America

Traversia Hook. f. 1864. Asteraceae. 1 New Zealand

Trechonaetes Miers = Jaborosa Juss. Solanaceae

Treculia Decne. ex Trécul. 1847. Moraceae. 3 trop. Africa, Madagascar

Treichelia Vatke. 1874. Campanulaceae. 1 South Africa (Cape)

Trema Lour. 1790. Ulmaceae. 10–15 tropics and subtropics

Tremacanthus S. Moore. 1904. Acanthaceae. 1 Brazil

Tremacron Craib. 1916. Gesneriaceae. 7 southwestern China

Tremandra R. Br. ex DC. 1824. Tremandraceae. 2 southwestern Australia

Tremastelma Raf. = Lomelosia Raf. Dipsacaceae

Trematocarpus Zahlbr. = Trematolobelia Zahlbr. ex Rock. Lobeliaceae

Trematolobelia Zahlbr. ex Rock. 1913 (Lobelia L.). Lobeliaceae. 4 Hawaiian Is.

Trembleya DC. 1828 (Microlicia D. Don). Melastomataceae. 15 southern Brazil

Trepocarpus Nutt. ex DC. 1829. Apiaceae. 1 southern U.S.

Tresanthera Karst. 1859. Rubiaceae. 2 Venezuela, Tobago

Treutlera Hook. f. 1883. Asclepiadaceae. 1 eastern Himalayas (Nepal, Bhutan)

Trevesia Vis. 1842. Araliaceae. 13 India, Nepal, southern and southeastern China, Indochina, Malesia, Oceania

Trevoa Miers ex Hook. 1829. Rhamnaceae. 6 Andes

Trevoria F. Lehm. 1897. Orchidaceae. 2 Colombia, Ecuador

Trewia L. 1753. Euphorbiaceae. 2 Himalayas from Kashmir to Bhutan, India, Sri Lanka, Burma, Indochina, Hainan, western Malesia

Triadenum Raf. 1837. Hypericaceae. 6–10 temp. East Asia, eastern North America; T. japonicum—extending to Russia Far East

Triadodaphne Kosterm. = Endiandra R. Br. Lauraceae

Triaena Kunth = Bouteloua Kunth. Poaceae

Triaenacanthus Nees. 1847 (Strobilanthus Blume). Acanthaceae. 1 India (Assam)

Triaenanthus Nees = Triaenacanthus Nees. Acanthaceae

Triaenophora (Hook. f.) Soler. 1909. Scrophulariaceae. 2 China

Triainolepis Hook. f. 1873. Rubiaceae. 2 East Africa (Kenya, Tanzania, Zambia, Malawi, Mozambique), Comoro Is., Madagascar, Aldabra I.

Trianaea Planch. et Linden. 1853. Solanaceae. 4 Colombia, Ecuador

Trianaeopiper Trel. 1928. Piperaceae. 18 northern Andes

Trianoptiles Fenzl ex Endl. 1836. Cyperaceae. 3 South Africa

Triantha Baker = Tofieldia Huds. Melanthiaceae

Trianthema L. 1753. Aizoaceae. 20 tropics and subtropics, esp. northern and arid Australia (c. 12—endemics)

Triaristella (Rchb. f.) Brieger ex Luer = Trisetella Luer. Orchidaceae

Triaristellina Rauschert = Trisetella Luer. Orchidaceae

Trias Lindl. 1830. Orchidaceae. 6–11 western coastal and northeastern India, Sri Lanka, Bangladesh, southern Burma, Thailand, Laos, Vietnam

Triaspis Burch. 1824. Malpighiaceae. 18 trop. and South (12) Africa

Triavenopsis Candargy = Duthiea Hack. Poaceae

Tribeles Phil. 1864. Tribelaceae. 1 southern Chile and Tierra del Fuego

Tribolium Desv. 1831. Poaceae. 5 South Africa

Tribonanthes Endl. 1839. Conostylidaceae. 5 southwestern Australia

Tribroma O. F. Cook = Theobroma L. Sterculiaceae

Tribulocarpus S. Moore. 1921. Aizoaceae. 1 Northeast (Ethiopia) and South-West (Namibia) Africa

Tribulopis R. Br. 1849. Zygophyllaceae. 5 trop. Australia

Tribulus L. 1753. Zygophyllaceae. 25 trop., subtrop., and warm–temp. regions of the Old World from Africa to Polynesia; 2 sp. extending to southern European part of Russia, Caucasus, Middle Asia, and southern Siberia

Tricalistra Ridl. 1909. Hyacinthaceae. 1 Malay Peninsula

Tricalysia A. Rich. ex DC. 1830. Rubiaceae. More than 100 trop. Africa (95), Madagascar (7), Comoro Is., Aldabra I., Assumption I., and several small Indian Ocean islands

Tricardia Torr. 1871. Hydrophyllaceae. 1 southwestern U.S.

Tricarpelema J. K. Morton. 1966. Commelinaceae. 7 trop. Africa, eastern Himalayas, Tibet

Tricarpha Longpre = Sabazia Cass. Asteraceae

Triceratella Brenan. 1961. Commelinaceae. 1 southern trop. Africa (Zimbabwe)

Triceratorhynchus Summerh. 1951. Orchidaceae. 1 trop. East Africa (Uganda, Kenya)

Tricerma Liebm. 1854 (Maytenus Molina). Celastraceae. 7 trop. America

Trichacanthus Zoll. et Moritzi = Blepharis Juss. Acanthaceae

Trichachne Nees. 1829 (Digitaria Haller). Poaceae. 15 Australia, trop. and subtrop. America

Trichadenia Thwaites. 1855. Flacourtiaceae. 2 Sri Lanka (1), Philippines and New Guinea (1)

Trichantha Hook. = Columnea L. Gesneriaceae

Trichanthemis Regel et Schmalh. 1877. Asteraceae. 7 montane Middle Asia

Trichanthera Kunth. 1818. Acanthaceae. 2 Central and trop. South America

Trichanthodium Sond. et F. Mueller ex Sond. 1853. Asteraceae. 4 Australia from western Australia to central part and Victoria

Trichapium Gilli = Clibadium Allem. ex L. Asteraceae

Trichaulax Vollesen. 1992. Acanthaceae. 1 Kenya and Tanzania (coastal areas)

Trichilia P. Browne. 1756. Meliaceae. c. 90 trop. America (c. 70), trop. Africa and Madagascar (c. 20), trop. Asia (2)

Trichlora Baker. 1877. Alliaceae. 1 Peru

Trichloris E. Fourn. ex Benth. 1881. Poaceae. 2 southwestern U.S., Central and South America

Trichocalyx Balf. f. 1884. Acanthaceae. 2 Socotra

Trichocaulon N. E. Br. 1878. Asclepiadaceae. 15 South Africa, Madagascar

Trichocentrum Poepp. et Endl. 1836. Orchidaceae. 30 trop. America from Mexico to Peru and Brazil

Trichocereus (Berger) Riccob. = Echinopsis Zucc. Cactaceae

Trichoceros Kunth. 1816. Orchidaceae. 4 northern South America

Trichochiton Kom. 1896 (Cryptospora Kar. et Kir.). Brassicaceae. 2 Kazakhstan, Middle Asia, Afghanistan

Trichocladus Pers. 1807. Hamamelidaceae. 5 trop. (Congo, Ethiopia, Uganda, Kenya, Malawi, Tanzania, Angola, Zambia, Mozambique) and South Africa

Trichocline Cass. 1817. Asteraceae. 22 South America (21), western Australia (1, T. scapigera)

Trichocoronis A. Gray. 1849. Asteraceae. 2 southwestern U.S., Mexico

Trichocoryne S. F. Blake. 1924. Asteraceae. 1 Mexico

Trichocyamos Yakovlev. 1972 (Ormosia Jacks.). Fabaceae. 4 southern China, Vietnam

Trichodesma R. Br. 1810. Boraginaceae. 45 trop. and subtrop. Old World; T. incanum—extending to Middle Asia

Trichodiadema Schwantes. 1926. Aizoaceae. 30 southern Namibia, Lesotho, South Africa

Trichodypsis Baill. = Dypsis Noronha ex Mart. Arecaceae

Trichoglottis Blume. 1825. Orchidaceae. c. 60 Himalayas, India, Sri Lanka, Burma, Nicobar Is., Indochina, Taiwan, Ryukyu Is., Malesia from Malay Peninsula to New Guinea, Solomon Is., Australia (1), Pacific Is.

Trichogonia (DC.) Gardner. 1846. Asteraceae. 30 trop. South America

Trichogoniopsis R. M. King et H. Rob. 1972 (Eupatorium L.). Asteraceae. 4 Brazil

Trichogyne Less. 1831. Asteraceae. 8 South Africa

Tricholaena Schrad. ex Schult. et Schult. f. 1824. Poaceae. 8 Canary Is., Mediterranean, Africa, Madagascar

Tricholaser Gilli. 1959. Apiaceae. 2 Afghanistan, Pakistan, India (Himalayas)

Tricholepis DC. 1833. Asteraceae. 15 Himalayas from Kashmir to Bhutan, Tibet, India (11), Burma; T. trichocephala—montane Middle Asia

Trichoneura Andersson. 1855. Poaceae. 5 southern U.S., South America, Galapagos Is.

Trichopetalum Lindl. 1832. Asphodelaceae. 2 Chile

Trichophorum Pers. = Scirpus L. Cyperaceae

Trichopilia Lindl. 1836. Orchidaceae. c. 30 trop. America from southern Mexico to Peru and Brazil, West Indies

Trichopteryx Nees. 1836. Poaceae. 7 trop. and South Africa, Madagascar

Trichoptilium A. Gray. 1859. Asteraceae. 1 southwestern U.S. (California, Nevada, Arizona), Mexico (Baja California)

Trichopus Gaertn. 1788. Trichopodaceae. 1 southern India, Sri Lanka, Malay Peninsula

Trichopyrum A. Löve = Elytrigia Desv. Poaceae

Trichosacme Zucc. 1846. Asclepiadaceae. 1 Mexico

Trichosalpinx Luer. 1983. Orchidaceae. c. 90 trop. America

Trichosanchezia Mildbr. 1926. Acanthaceae. 1 eastern Peru

Trichosandra Decne. 1844. Asclepiadaceae. 1 Mauritius

Trichosanthes L. 1753. Cucurbitaceae. c. 50 South, East, and Southeast Asia, Malesia, trop. Australia, Melanesia to Fiji

Trichoschoenus J. Raynal. 1968. Cyperaceae. 1 Madagascar

Trichoscypha Hook. f. 1862. Anacardiaceae. c. 70 forests of West and Central Africa; T. ulugurensis—also in Tanzania, Uganda, and Mozambique

Trichosiphum Schott et Endl. = Brachychiton Schott et Endl. Sterculiaceae

Trichosma Lindl. = Eria Lindl. Orchidaceae

Trichospermum Blume. 1825. Tiliaceae. 36 Nicobar Is., Malesia, western Pacific Is., trop. America (3)

Trichospira Kunth. 1818. Asteraceae. 1 trop. and subtrop. America

Trichostachys Hook. f. 1873. Rubiaceae. 10 trop. Africa

Trichostelma Baill. 1890. Asclepiadaceae. 3 Mexico, Guatemala

Trichostema L. 1753. Lamiaceae. 17 North America

Trichostephania Tardieu. 1949. Sterculiaceae. 1 Indochina

Trichostephanus Gilg. 1908. Flacourtiaceae. 1 trop. West Africa

Trichostigma A. Rich. 1845. Petiveriaceae. 3 Florida, Central and trop. South America, West Indies

Trichostomanthemum Domin = Melodinus J. R. Forst. et G. Forst. Apocynaceae

Trichotaenia Yamazaki = Lindernia All. Scrophulariaceae

Trichotolinum O. E. Schulz. 1933. Brassicaceae. 1 Argentina

Trichotosia Blume. 1825. Orchidaceae. c. 50 northeastern India, Indochina, Malesia, western Oceania

Trichovaselia Tiegh. = Elvasia DC. Ochnaceae

Trichuriella Bannet. 1985. Amaranthaceae. 1 India, Sri Lanka, Burma, southern China, Thailand, Cambodia, Vietnam

Trichurus C. C. Towns. (non Clements—Fungae) = Trichuriella Bannet. Amaranthaceae

Tricliceras DC. = Wormskioldia Schumacher et Thonn. Turneraceae

Triclisia Benth. 1862. Menispermaceae. 25 trop. Africa, Madagascar

Tricomaria Hook. et Arn. 1833. Malpighiaceae. 1 Argentina

Tricoryne R. Br. 1810. Asphodelaceae.

7 New Guinea (1), Australia, Tasmania (1)

Tricostularia Nees ex Lehm. 1844. Cyperaceae. 5 Sri Lanka, Indochina, Malesia, Australia

Tricycla Cav. = Bougainvillea Comm. ex Juss. Nyctaginaceae

Tricyclandra Keraudren. 1966. Cucurbitaceae. 1 Madagascar

Tricyrtis Wall. 1826. Melanthiaceae. 10–15 eastern Himalayas, northeastern India, northern Burma, continental China, Taiwan, Japan

Tridactyle Schltr. 1914. Orchidaceae. 42 trop. and South Africa

Tridactylina (DC.) Sch. Bip. 1844. Asteraceae. 1 East Siberia

Tridax L. 1753. Asteraceae. 26 trop. America, mostly Mexico and West Indies

Tridens Roem. et Schult. 1817. Poaceae. 16 eastern U.S., Mexico, Cuba, and South America from Guyana to Argentina

Tridentea Haw. 1812 (Stapelia L.). Asclepiadaceae. 17 South Africa

Tridesmostemon Engl. 1905. Sapotaceae. 2–3 trop. Africa (Cameroun, Gabon, Zaire)

Tridianisia Baill. 1879. Icacinaceae. 1 Madagascar

Tridimeris Baill. 1869. Annonaceae. 1 Mexico

Tridynamia Gagnepain. 1950 (Porana Burm. f.). Convolvulaceae. 4 northeastern India, Andaman Is., Thailand, southern China (incl. Hainan), Indochina, Malay Peninsula

Trieenea Hilliard. 1989 (Polycarene Benth.). Scrophulariaceae. 9 South Africa

Trientalis L. 1753. Primulaceae. 2–3 temp. Northern Hemisphere; Russia (2, Arctica, European part, Siberia, Far East)

Trifidacanthus Merr. 1917. Fabaceae. 1–2 Vietnam, Hainan, Philippines, Lesser Sunda Is.

Trifoliada Rojas Acosta. 1897. Oxalidaceae. 2 Argentina

Trifolium L. 1753. Fabaceae. c. 250 temp. and subtrop. regions, except Australia; mainly temp. Northern Hemisphere, but also in trop. upland and montane regions

Trifurcia Herb. = Herbertia Sweet. Iridaceae

Triglochin L. 1753. Juncaginaceae. 15 cosmopolitan; Russia (2, European part, Siberia, Far East)

Trigonachras Radlk. 1879. Sapindaceae. 9 Malay Peninsula (1, T. acuta), Sumatra, Borneo, Philippines (8), Sulawesi, New Guinea

Trigonanthe (Schltr.) Brieger = Dryadella Luer. Orchidaceae

Trigonella L. 1753. Fabaceae. c. 80 temp. Eurasia, esp. (25) Russia (European part, Caucasus, southern Siberia, Far East) and Caucasus; Afghanistan (31), southward to Himalayas; Macaronesia, Mediterranean, Djibouti (1), Sudan (2), Zambia (1), South Africa, Arabian Peninsula, Australia

Trigonia Aubl. 1775. Trigoniaceae. 24 trop. America

Trigoniastrum Miq. 1861. Trigoniaceae. 1 western Malesia

Trigonidium Lindl. 1837. Orchidaceae. 14 trop. America from Mexico to Brazil

Trigoniodendron E. F. Guim. et Miguel. 1987. Trigoniaceae. 1 eastern Brazil (Espíritu Santo)

Trigonobalanus Forman. 1962. Fagaceae. 3 southern China, northern Thailand, Malay Peninsula, Borneo, Sulawesi, South America (Colombia)

Trigonocapnos Schltr. 1899. Fumariaceae. 1 (T. lichtensteinii) South Africa

Trigonocaryum Trautv. 1875. Boraginaceae. 1 Caucasus

Trigonopleura Hook. f. 1887. Euphorbiaceae. 2 western Malesia (except Java), Borneo, New Caledonia

Trigonopyren Bremek. 1963 (Psychortia L.). Rubiaceae. 9 Comoro Is., Madagascar

Trigonosciadium Boiss. 1844. Apiaceae. 4 Turkey, Iraq, Iran

Trigonospermum Less. 1832. Asteraceae. 5 southern Mexico, Guatemala, Nicaragua

Trigonostemon Blume. 1825. Euphorbiaceae. c. 50 Sri Lanka, eastern Himalayas, northeastern India, China, Indochina, western Malesia, Fiji

Trigonotis Steven. 1851. Boraginaceae. 57: 3 Russia (southeastern European part, Siberia, Far East), Caucasus, Middle Asia; Himalayas, India, China, Korea, Japan, Indochina, Malesia

Triguera Cav. 1786. Solanaceae. 2–3 southern Spain, North Africa

Trigynaea Schltdl. 1834. Annonaceae. 5 northern South America

Trigynia Jacq.-Fél. = Tryginia Jacq.-Fél. Melastomataceae

Trikeraia Bor. 1955. Poaceae. 2 Pakistan, western Himalayas, Tibet

Trilepidea Tiegh. 1895. Loranthaceae. 1 New Zealand (North I.)

Trilepis Nees. 1834. Cyperaceae. 5 Guyana, eastern Brazil

Trilepisium Thouars. 1806. Moraceae. 1 Africa from Guinea to southern Ethiopia and south to Angola and South Africa (Natal), Madagascar, Seychelles, and ? Mascarene Is.

Trilisia (Cass.) Cass. 1820. Asteraceae. 2 southeastern U.S.

Trillidium Kunth. 1850 (Trillium L.). Trilliaceae. 1 Himalayas from Kashmir to Bhutan

Trillium L. 1753. Trilliaceae. c. 40 Asia from Afghanistan and Pakistan through Himalayas to Russian Far East (5, to Kamchatka) and Japan, North America (35)

Trilobachne Schenck ex Henrard. 1931. Poaceae. 1 western India

Trilocularia Schltr. 1906 (Balanops Baill.). Balanopaceae. 1 New Caledonia, Fiji

Trimenia Seem. 1873. Trimeniaceae. 5 Sulawesi, Moluccas, New Guinea, Bougainville I., Solomon Is., eastern Australia, New Caledonia, Fiji, Samoa Is., Marquesas Is.

Trimeria Harv. 1838. Flacourtiaceae. 3 trop. (Cameroun, Zaire, Uganda, Kenya, Tanzania, Burundi, Zimbabwe) and South Africa

Trimeris C. Presl. 1836 (Lobelia L.). Lobeliaceae. 1 St. Helena

Trimerocalyx (Murb.) Murb. = Linaria Hill. Scrophulariaceae

Trimezia Salisb. ex Herb. 1944. Iridaceae. c. 20 trop. America, West Indies

Trimorpha Cass. 1825. Asteraceae. c. 45 North America, Eurasia

Trimorphopetalum Baker = Impatiens L. Balsaminaceae

Trinacte Gaertn. = Jungia L. f. Asteraceae

Trinia Hoffm. 1814. Apiaceae. 12 Europe, Mediterranean, West (incl. Caucasus) and Middle Asia, southern West Siberia

Triniochloa Hitchc. 1913. Poaceae. 4 trop. America from Mexico to Bolivia

Triodanis Raf. 1836. Campanulaceae. 8 North America through Mexico to Guatemala, ? Mediterranean (1), China (2)

Triodia R. Br. 1810. Poaceae. c. 35 Australia

Triodoglossum Bullock. 1962. Asclepiadaceae. 1 trop. Africa

Triolena Naudin. 1851. Melastomataceae. 20–25 trop. America

Triomma Hook. f. 1860. Burseraceae. 1 Malesia: Malay Peninsula, Sumatra, Banka I., Borneo

Trioncinia (F. Muell.) Veldkamp. 1991. Asteraceae. ? sp. Distr. ?

Triopteris L. 1753. Malpighiaceae. 3 trop. America, West Indies

Triosteum L. 1753. Caprifoliaceae. 7–8 eastern Himalayas, Tibet, China (3), Japan, eastern North America; Russia (1, T. sinuatum—Far East)

Tripetaleia Siebold et Zucc. = Elliottia Muhlenb. ex Elliott. Ericaceae

Tripetalum K. Schum. = Garcinia L. Clusiaceae

Triphasia Lour. 1790. Rutaceae. 3 trop. Asia, Philippines

Triphora Nutt. 1818. Orchidaceae. 13 North, Central and trop. South America, West Indies

Triphylleion Suess. = Niphogeton Schltdl. Apiaceae

Triphyophyllum Airy Shaw. 1952. Dioncophyllaceae. 1 Sierra Leone, Liberia, Ivory Coast

Triplachne Link. 1833. Poaceae. 1 Mediterranean

Tripladenia D. Don. 1839. Melanthiaceae. 1 (T. cunninghamii) Australia (Queensland, New South Wales)

Triplaris Loefl. ex L. 1759. Polygonaceae. 18 trop. America from Mexico to southeastern Brazil

Triplasis P. Beauv. 1812. Poaceae. 2 southeastern U.S., Mexico, Central America

Tripleurospermum Sch. Bip. 1844. Asteraceae. 30 temp. Northern Hemisphere

Triplisomeris Aubrév. et Pellegr. = Anthonotha P. Beauv. Fabaceae

Triplocephalum O. Hoffm. 1894. Asteraceae. 1 trop. East Africa

Triplochiton K. Schum. 1900. Sterculiaceae. 3 trop. Africa

Triplochlamys Ulbr. = Pavonia Cav. Malvaceae

Triplopogon Bor. 1954. Poaceae. 1 western India

Triplostegia Wall. ex DC. 1830. Triplostegiaceae. 2 Himalayas, Tibet, northeastern India, Burma, continental China, Taiwan, Malesia

Triplotaxis Hutch. = Vernonia Schreb. Asteraceae

Tripodandra Baill. = Rhaptonema Miers. Menispermaceae

Tripodanthus (Eichler) Tiegh. 1895. Loranthaceae. 2 Venezuela, Colombia, Ecuador, Peru, Bolivia, north-central Argentina, southern Brazil, Uruguay

Tripodion Medik. 1787 (Anthyllis L.). Fabaceae. 3 Algeria, Morocco, Tunisia, Libya, Egypt,

Tripogandra Raf. 1837. Commelinaceae. 22 trop. America

Tripogon Roem. et Schult. 1817. Poaceae. c. 30 tropics and subtropics, esp. Africa and India; Russia (1, T. chinensis southern East Siberia and Far East)

Tripolium Nees = Aster L. Asteraceae

Tripsacum L. 1759. Poaceae. 13 trop. and subtrop. America from southern U.S. to Paraguay

Tripteris Less. = Osteospermum L. Asteraceae

Tripterocalyx Hook. 1909. Nyctaginaceae. 3 southwestern U.S. (Nevada, Arizona, California)

Tripterococcus Endl. 1837. Stackhousiaceae. 2 southwestern Australia

Tripterodendron Radlk. 1891. Sapindaceae. 1 Brazil

Tripterospermum Blume. 1826. Gentianaceae. 17 Himalayas, Burma, southwestern and southern China (15), Taiwan (3), Korea, Japan, Thailand (1, T. trinerve), Malay Peninsula, Philippines; T. japonicum—extending to Russian Far East (Sakhalin)

Tripterygium Hook. f. 1862. Celastraceae. 4 China, Taiwan (1, T. wilfordii), Korea, Japan (2)

Triptilion Ruiz et Pav. 1794. Asteraceae. 12 Chile, Argentina (1)

Triptilodiscus Turcz. 1851. Asteraceae. 1 Australia

Triraphis R. Br. 1810. Poaceae. 7 trop. and South Africa, Arabian Peninsula, Australia (1)

Trirostellum Z. P. Wang et Xie = Gynostemma Blume. Cucurbitaceae

Triscenia Griseb. 1863. Poaceae. 1 Cuba

Triscyphus Taub. ex Warm. = Thismia Griff. Burmanniaceae

Trisepalum C. B. Clarke. 1883. Gesneriaceae. 13 Burma, Thailand, southern China (1, Yunnan), Langkawi Is.

Trisetaria Forssk. 1775. Poaceae. 15 Southeast Europe, Mediterranean, West (incl. Caucasus), Southwest and Middle Asia, western Himalayas

Trisetella Luer. 1980. Orchidaceae. 12 trop. America

Trisetobromus Nevski. 1934 (Bromus L.). Poaceae. 1 Chile

Trisetum Pers. 1805. Poaceae. c. 70 cold and temp. regions and montane areas of tropics

Tristachya Nees. 1829. Poaceae. 10 trop. America, Africa, Madagascar

Tristagma Poepp. et Endl. 1833. Alliaceae. 3–5 Chile, southern Argentina, Uruguay

Tristania R. Br. 1812. Myrtaceae. 1 (T.

neriifolia) Australia (New South Wales)

Tristaniopsis Brongn. et Gris. 1863 (Tristania R. Br.). Myrtaceae. 44 Burma, Indochina, Malesia, eastern Australia, New Caledonia

Tristellateia Thouars. 1806. Malpighiaceae. 22 trop. East Africa (1, T. africana—Kenya, Tanzania, Mozambique), Madagascar (20); 1 (T. australasiae) Indochina, Taiwan, Malesia, New Ireland, Micronesia (Caroline Is.), Australia (Queensland), New Caledonia

Tristemma Juss. 1789. Melastomataceae. 15 trop. Africa, Madagascar, Mascarene Is.

Tristemonanthus Loes. 1940 (Campylostemon Welw.). Celastraceae. 2 trop. West Africa

Tristerix Mart. 1830 (Peraxilla Tiegh.). Loranthaceae. 11 Andes from north-central Colombia through Ecuador, Peru, western Bolivia to northwestern Argentina, central Chile (incl. Chiloé I.)

Tristicha Thouars. 1806. Podostemaceae. 2 Central and trop. South America, West Indies, Africa, Madagascar, Mascarene Is., India, Sri Lanka, northwestern Australia

Tristira Radlk. 1879. Sapindaceae. 4 Philippines, Sulawesi, Moluccas

Tristiropsis Radlk. 1887. Sapindaceae. 13–14 Malay Peninsula, Borneo, Philippines, New Guinea, Mariana and Solomon Is., Australia (1, T. canarioides), Polynesia (Christmas I.)

Trisyngyne Baill. = Nothofagus Blume. Nothofagaceae

Tritaenicum Turcz. = Gymnophyton Clos. Apiaceae

Triteleia Douglas ex Lindl. 1830 (Bordiaea Sm.). Alliaceae. 16 western America

Triteleiopsis Hoover. 1941 (Brodiaea Sm.). Alliaceae. 1 western U.S.

Trithecanthera Tiegh. 1894. Loranthaceae. 4 Malay Peninsula, Borneo

Trithrinax Mart. 1837. Arecaceae. 5 Bolivia, western trop. and southern Brazil, Paraguay, Uruguay, and Argentina

Trithuria Hook. f. 1858. Hydatellaceae. 3 Australia, Tasmania

Triticum L. 1753. Poaceae. c. 20 South and Southeast Europe, West (incl. Caucasus) and Middle Asia, Russia (all areas except Arctica and forest zone)

Tritonia Ker-Gawl. 1802. Iridaceae. 28 southern trop. and South Africa

Tritoniopsis L. Bolus. 1929. Iridaceae. 22 South Africa (Cape)

Triumfetta L. 1753. Tiliaceae. c. 100 (or 60–150 ?) tropics and subtropics

Triumfettoides Rauschert. 1982 (Triumfetta L.). Tiliaceae. 1 trop. East Africa

Triunia L. A. S. Johnson et B. G. Briggs. 1975 (Helicia Lour.). Proteaceae. 4 eastern Australia

Triuranthera Backer = Driessenia Korth. Melastomataceae

Triuris Miers. 1841. Triuridaceae. 3 Guatemala, Colombia, Guyana, Surinam, Brazil

Triurocodon Schltr. = Thismia Griff. Burmanniaceae

Trivalvaria Miq. 1865. Annonaceae. 5 northeastern India, Burma, Thailand, western Malesia

Trixis P. Browne. 1756. Asteraceae. 60 America from southwestern U.S. to Chile, West Indies

Trizeuxis Lindl. 1821. Orchidaceae. 2–3 trop. America from Costa Rica to Peru, Bolivia and Brazil, Trinidad

Trochetia DC. 1823. Sterculiaceae. 6 Madagascar ?, Mascarene Is.

Trochetiopsis Marais. 1981. Sterculiaceae. 2 St. Helena

Trochiscanthes W. D. J. Koch. 1824. Apiaceae. 1 France, Switzerland, Italy

Trochiscus O. E. Schulz. 1933. Brassicaceae. 1 northeastern India

Trochocarpa R. Br. 1810. Epacridaceae. 12 Borneo, Sulawesi, Kinabalu M., New Guinea, eastern Australia, Tasmania

Trochocodon Candargy = Campanula L. Campanulaceae

Trochodendron Siebold et Zucc. 1839. Trochodendraceae. 1 Korea, Japan south to Ryukyu Is., Taiwan

Trochomeria Hook. f. 1867. Cucurbitaceae. 8 Africa

Trochomeriopsis Cogn. 1881. Cucurbitaceae. 1 Madagascar

Troglophyton Hilliard et B. L. Burtt. 1981 (Gnaphalium L.). Asteraceae. 6 South Africa

Trollius L. 1753. Ranunculaceae. 31 cold and temp. Northern Hemisphere; Russia (25, all areas, but chiefly Siberia and Far East)

Tromotriche Haw. 1812 (Stapelia L.). Asclepiadaceae. 3 South Africa (southwestern Cape)

Tropaeastrum Sparr. = Tropaeolum L. Tropaeolaceae

Tropaeolum L. 1753. Tropaeolaceae. c. 90 montane regions from Mexico to central Chile and Argentina

Trophis P. Browne. 1756. Moraceae. 11 Madagascar, East and Southeast Asia, Malesia, Australia, Oceania, trop. America, West Indies

Tropidia Lindl. 1833. Orchidaceae. 12–20 trop. Asia from India, Burma and Sri Lanka eastward through China to Japan and Taiwan, Malesia from Malay Peninsula to New Guinea, Australia, Melanesia to Fiji, and Samoa; trop. America (1, T. polystachya Ames—southern Florida, Mexico, Central America, Galapagos Is., West Indies)

Tropidocarpum Hook. 1836. Brassicaceae. 2 California

Tropilis Raf. 1837 (Dendrobium Sw.). Orchidaceae. 15 Australia

Trouettia Pierre ex Baill. = Niemeyera F. Muell. Sapotaceae

Trudelia Garay. 1986. Orchidaceae. 3 Bhutan

Trukia Kaneh. 1935. Rubiaceae. 5 Aru Is., New Guinea, New Britain I., eastern Australia, Caroline and Solomon Is., Hawaiian Is., Tahiti

Trungbao Rauschert. 1982. Scrophulariaceae. 1 Southeast Asia

Trybliocalyx Lindau. 1904 (Chileranthemum Oerst.). Acanthaceae. 1 Central America

Trychinolepis B. L. Rob. = Ophryosporus Meyen. Asteraceae

Tryginia Jacq.-Fél. 1936. Melastomataceae. 1 trop. West Africa

Trymalium Fenzl. 1837. Rhamnaceae. 12–13 Australia

Trymatococcus Poepp. et Endl. 1838. Moraceae. 3 trop. America

Tryphostemma Harv. = Basananthe Peyr. Passifloraceae

Tryssophyton Wurdack. 1964. Melastomataceae. 1 northeastern South America

Tsaiorchis T. Tang et F. T. Wang. 1936. Orchidaceae. 1 China

Tsavo Jarmol. = Polulus L. Salicaceae

Tschulaktavia Bajt. 1983. Apiaceae. 1 Middle Asia

Tsebona Capuron. 1962. Sapotaceae. 1 Madagascar

Tsiangia But, H. H. Hsue et P. T. Li. 1986. Rubiaceae. 1 Hong Kong

Tsimatimia Jum. et H. Perrier = Rheedia L. Clusiaceae

Tsingya Capuron. 1969. Sapindaceae. 1 Madagascar

Tsoala Bosser et D'Arcy. 1992. Solanaceae. 1 Madagascar

Tsoongia Merr. 1923. Verbenaceae. 1 southern China, Indochina

Tsoongiodendron Chun. 1963 (Michelia L.). Magnoliaceae. 1 southeastern China

Tsuga Carrière. 1855. Pinaceae. 10–14 western, central, and southern China; Himalayas; northern Burma; Japan; Taiwan; western coastal Canada; western and eastern U.S.

Tsusiophyllum Maxim. 1870. Ericaceae. 1 Japan

Tuberaria (Dunal) Spach. 1836. Cistaceae. 14 West and Central Europe, Mediterranean

Tuberolabium Yamam. 1924. Orchidaceae. 11 northeastern India, Thailand, Taiwan, Malesia (Sumatra, Java, Sulawesi, Philippines, Moluccas, New Guinea), Australia, Mariana Is., Vanuatu, New Caledonia, Fiji, Tubuai Is., Society Is.

Tuberostylis Steetz. 1853. Asteraceae. 2 Panama, Colombia, Ecuador

Tubilabium J. J. Sm. 1928. Orchidaceae. 2 Indonesia

Tubocapsicum (Wettst.) Makino. 1908. Solanaceae. 2 China, Korea, Japan, Philippines, Borneo

Tucma Ravenna. 1973. Iridaceae. 1 Andean Argentina

Tuctoria J. R. Reeder. 1982. Poaceae. 3 southwestern U.S., Mexico

Tuerckheimocharis Urb. 1912. Scrophulariaceae. 1 Haiti

Tugarinovia Iljin. 1928. Asteraceae. 1 Mongolia, northern China

Tulasnea Naudin = Siphanthera Pohl ex DC. Melastomataceae

Tulasneantha P. Royen. 1951. Podostemaceae. 1 western Brazil

Tulbaghia L. 1771. Alliaceae. 22 trop. and South Africa

Tulestea Aubrév. et Pellegr. = Synsepalum (A. DC.) Daniell. Sapotaceae

Tulipa L. 1753. Liliaceae. c. 100 temp. Eurasia (esp. Russia central and southern European part, northern Caucasus, southern Siberia); North Africa, eastern Mediterranean, Arabian Peninsula, eastward to Iran, Middle Asia

Tuloclinia Raf. = Metalasia R. Br. Asteraceae

Tulotis Raf. 1833 (Platanthera Rich.). Orchidaceae. 3 East Asia, North America; 2 extending to Russian Far East

Tumamoca Rose. 1912. Cucurbitaceae. 1 southwestern U.S., northern Mexico

Tumidinodus H. W. Li = Anna Pellegr. Gesneriaceae

Tunaria Kuntze = Cantua Lam. Polemoniaceae

Tunica (Haller) Scop. = Petrorhagia (Ser. ex DC.) Link. Caryophyllaceae

Tupeia Cham. et Schltdl. 1828. Loranthaceae. 1 New Zealand

Tupeianthus Maguire et Steyerm. = Tepuianthus Maguire et Steyerm. Tepuianthaceae

Tupidanthus Hook. f. et Thomson = Schefflera J. R. Forst. et G. Forst. Araliaceae

Tupistra Ker-Gawl. 1814. Convallariaceae. c. 20 Himalayas, India, Burma, southwestern and southern China, Indochina, Malay Peninsula, Sumatra

Turanga (Bunge) Kimura = Populus L. Salicaceae

Turaniphytum Poljakov. 1961. Asteraceae. 2 Middle Asia

Turbina Raf. 1838. Convolvulaceae. 14 trop. America, trop. and South Africa, New Caledonia (1, T. inopinata)

Turbinicarpus (Backeb.) Buxb. = Neolloydia Britton et Rose. Cactaceae

Turczaninovia DC. = Aster L. Asteraceae

Turczaninoviella Kozo-Polj. 1924. Apiaceae. ? (nomen nudum ?)

Turgenia Hoffm. 1814. Apiaceae. 2: T. latifolia—Central, South, and Southeast Europe; Mediterranean; Caucasus; West, Southwest, and Middle Asia; Afghanistan; Pakistan, Kashmir; T. lisaeoides—Iraq

Turgeniopsis Boiss. = Glochidotheca Fenzl. Apiaceae

Turnera L. 1753. Turneraceae. c. 100 mostly trop. and subtrop. America from Mexico to Argentina, trop. and South-West (1) Africa

Turpinia Vent. 1807. Staphyleaceae. 30–40 (or 10 ?) India, Sri Lanka, Burma, southern China, Taiwan, southern Japan (Kyushu), Indochina, Malesia; Mexico, Central, and trop. South America; West Indies

Turraea L. 1771. Meliaceae. 66: trop. Africa (24); 36 Madagascar, Mascarene and Comoro Is.; 6 trop. Asia and Australia

Turraeanthus Baill. 1874. Meliaceae. 2 trop. West Africa

Turricula J. F. Macbr. 1917. Hydrophyllaceae. 1 southwestern North America

Turrigera Decne. = Tweedia Hook. et Arn. Asclepiadaceae

Turrillia A. C. Sm. 1985 (Bleasdalea F. Muell. ex Domin). Proteaceae. 5 New Guinea, Australia (Queensland), Vanuatu, Fiji

Turritis L. 1753 (Arabis L.). Brassicaceae. 1 (T. glabra) Eurasia, incl. European part of Russia, Caucasus, Middle Asia, Siberia, Russian Far East

Turukhania Vassilcz. = Medicago L. or Trigonella L. Fabaceae

Tussilago L. 1753. Asteraceae. 1 (T. farfara) temp. Eurasia, except Japan, North Africa

Tutcheria Dunn. 1908 (Pyrenaria Blume). Theaceae. 20 East Asia

Tuxtla Villaseñor et Strother. 1989. Asteraceae. 1 southern Mexico, Costa Rica

Tuyamaea Yamaz. = Legazpia Blanco. Scrophulariaceae

Tweedia Hook. et Arn. 1834. Asclepiadaceae. 6 Bolivia, Uruguay, Chile, Argentina

Tylanthera C. Hansen. 1990. Melastomataceae. 2 Thailand

Tylecodon Tölken. 1978 (Cotyledon L.). Crassulaceae. 27 South-West (Namibia) and South (Cape) Africa

Tyleria Gleason. 1931. Sauvagesiaceae. 17 trop. South America

Tyleropappus Greenm. 1931. Asteraceae. 1 Venezuela

Tylocarya Nelmes. 1949 (Fimbristylis Vahl). Cyperaceae. 1 Thailand

Tylodontia Griseb. = Astephanus R. Br. Asclepiadaceae

Tylopetalum Barneby et Krukoff = Sciadotenia Miers. Menispermaceae

Tylophora R. Br. 1810. Asclepiadaceae. c. 60 trop. and South Africa, Mascarene Is., trop. Asia eastward to Japan, Australia, Melanesia to Tongo and Samoa Is.

Tylophoropsis N. E. Br. 1894. Asclepiadaceae. 2 East Africa

Tylopsacas Leeuwenb. 1960. Gesneriaceae. 1 trop. America

Tylosema (Schweinf.) Torre et Hillcoat. 1955 (Bauhinia L.). Fabaceae. 4 trop. Central and East (from Sudan to Angola and Namibia) and South (2, Transvaal, Natal, Cape) Africa

Tylosperma Botsch. 1952 (Potentilla L.). Rosaceae. 1 Middle Asia, northern Iran

Tylostigma Schltr. 1916. Orchidaceae. 3 Madagascar

Tynnanthus Miers. 1863. Bignoniaceae. 14 trop. America from southern Mexico to Brazil and Bolivia, West Indies

Typha L. 1753. Typhaceae. 15 cosmopolitan, esp. (12) European part of Russia, Caucasus, Middle Asia, Siberia, Far East

Typhonium Schott. 1829. Araceae. c. 40 Himalayas, India, Sri Lanka, Burma, Thailand, Indochina, China, Japan, Malesia from Malay Peninsula and Sumatra to New Guinea, northern and eastern Australia; T. roxburghii—naturalized in Venezuela, Brazil, Ghana, Comoro Is.

Typhonodorum Schott. 1857. Araceae. 1 trop. East Africa, Madagascar, Mascarene Is., Zanzibar and Pemba Is.

Tyrimnus Cass. 1826. Asteraceae. 1 South Europe, Cyprus, Turkey, western Syria

Tysonia Bolus = Afrotysonia Rauschert. Boraginaceae

Tytthostemma Nevski. 1937 (or = Stellaria L.). Caryophyllaceae. 1 Middle Asia, eastern Iran

Tzellemtinia Chiov. = Bridelia Willd. Euphorbiaceae

Tzvelevia E. B. Alexeev. 1985 (Festuca L.). Poaceae. 1 Kerguelen

Uapaca Baill. 1858. Euphorbiaceae. c. 50 trop. Africa, Madagascar

Ubochea Baill. 1891. Verbenaceae. 1 Cape Verde Is.

Uebelinia Hochst. 1841. Caryophyllaceae. 10 trop. Africa

Uebelmannia Buining. 1967. Cactaceae. 5 eastern Brazil (Minas Gerais)

Uechtritzia Freyn. 1892. Asteraceae. 2 Asia Minor, Caucasus, Middle Asia

Ugamia Pavlov. 1950. Asteraceae. 1 Middle Asia

Ugni Turcz. 1848. Myrtaceae. 5–15 trop. and subtrop. America from Mexico through Andes to Chile, Juan Fernández Is.

Uladendron Marc.-Berti. 1971. Malvaceae. 1 Venezuela

Ulbrichia Urb. = Thespesia Sol. ex Corrêa. Malvaceae

Uldinia J. M. Black. 1922. Apiaceae. 1 central Australia

Uleanthus Harms. 1905. Fabaceae. 1 South America (Amazonia)

Ulearum Engl. 1905. Araceae. 1 South America (Amazonia)

Uleiorchis Hoehne. 1944. Orchidaceae. 1 Brazil

Uleodendron Rauschert = Naucleopsis Miq. Moraceae

Uleophytum Hieron. 1907. Asteraceae. 1 Peru

Ulex L. 1753. Fabaceae. c. 20 West Europe, Morocco, Algeria

Ulleria Bremek. 1969 (Ruellia L.). Acanthaceae. 4 Central and South America

Ullucus Caldas. 1809. Basellaceae. 1 Andes

Ulmus L. 1753. Ulmaceae. c. 50 temp. Northern Hemisphere south to northern Mexico, Himalayas, and Indochina; Russia (c. 10, European part, southern East Siberia, Far East), Caucasus, Middle Asia

Ultragossypium Roberty = Gossypium L. Malvaceae

Ulugbekia Zak. 1961 (Arnebia Forssk.). Boraginaceae. 1 Middle Asia, northwestern China

Umbellularia (Nees) Nutt. 1842. Lauraceae. 1 U.S. (California and southern Oregon)

Umbilicus DC. 1801. Crassulaceae. 18 Europe, Mediterranean, montane trop. Africa from Cameroun to Ethiopia, Somalia and Tanzania, West Asia to Caucasus (1, U. oppositifolius) and Iran

Umtiza T. R. Sim. 1907. Fabaceae. 1 South Africa (Cape)

Unanuea Ruiz et Pav. ex Pennell. = Stemodia L. Scrophulariaceae

Uncaria Schreb. 1789. Rubiaceae. c. 60 pantropics, but mostly trop. Asia, Malesia, Pacific Is.; trop. America (2), trop. Africa (3), Madagascar (1)

Uncarina (Baill.) Stapf. 1895. Pedaliaceae. 9 Madagascar

Uncariopsis Karst. = Schradera Vahl. Rubiaceae

Uncifera Lindl. 1858. Orchidaceae. 5 Himalayas, northeastern India, Indochina, western Malesia

Uncinia Pers. 1807. Cyperaceae. c. 50 Borneo, Philippines, Sulawesi, New Guinea, Australia, Tasmania, New Zealand, Campbell Is., Lord Auckland Is., Hawaiian Is., Mexico, Central and South America, West Indies, Amsterdam I., Kerguelen, Tristan da Cunha

Ungeria Schott et Endl. 1832. Sterculiaceae. 1 eastern Australia, Norfolk I.

Ungernia Bunge. 1875. Amaryllidaceae. 6–8 Middle Asia, Iran (1)

Ungnadia Endl. 1835. Sapindaceae. 1 southern U.S., Mexico

Ungula Barlow = Amyema Tiegh. Loranthaceae

Ungulipetalum Moldenke. 1938. Menispermaceae. 1 Brazil

Unigenes E. Wimm. 1948. Lobeliaceae. 1 South Africa

Uniola L. 1753. Poaceae. 4 America from southern U.S. to Argentina, and West Indies

Unonopsis R. E. Fr. 1900. Annonaceae. 29 trop. America, West Indies

Unxia L. f. 1782. Asteraceae. 2 Panama, northern South America

Upudalia Raf. = Eranthemum L. Acanthaceae

Upuna Symington. 1941. Dipterocarpaceae. 1 Borneo

Urandra Thwaites. 1855 (Stemonurus Blume). Icacinaceae. 17 South and Southeast Asia, Malesia

Uranodactylus Gilli = Winklera Regel. Brassicaceae

Uranthoecium Stapf. 1916. Poaceae. 1 Australia

Uraria Desv. 1813. Fabaceae. c. 20 trop. Old World

Urariopsis Schindl. 1916. Fabaceae. 1 Burma, Indochina

Urbananthus R. M. King et H. Rob. 1971 (Eupatorium L.). Asteraceae. 2 Cuba, Jamaica

Urbania Phil. 1891. Verbenaceae. 2 Chile, Argentina

Urbanodendron Mez. 1889. Lauraceae. 3 southeastern Brazil

Urbanodoxa Muschl. = Cremolobus DC. Brassicaceae

Urbanoguarea Harms = Guarea Allemand ex L. Meliaceae

Urbanolophium Melch. 1927. Bignoniaceae. 2 Brazil

Urbanosciadium H. Wolff. = Niphogeton Schltdl. Apiaceae

Urbinella Greenm. 1903. Asteraceae. 1 Mexico

Urceola Roxb. 1799. Apocynaceae. 15 Burma, western Malesia

Urceolaria Herb. = Urceolina Rchb. Amaryllidaceae

Urceolina Rchb. 1828. Amaryllidaceae. 8 Andes

Urechites Muell. Arg. = Pentalinon Voigt. Apocynaceae

Urelytrum Hack. 1887. Poaceae. 10 trop. and South Africa, Madagascar

Urena L. 1753. Malvaceae. 6 tropics and subtropics

Urera Gaudich. 1830. Urticaceae. 35–40 trop. and South Africa; Madagascar; Mascarene Is.; Pacific Is., incl. Hawaiian Is. (2); trop. and subtrop. America; West Indies

Urginea Steinh. 1834. Hyacinthaceae. c. 40 Macaronesia, South Europe, Africa, Asia Minor, Iraq, Iran, India

Urgineopsis Compton = Urginea Steinh. Hyacinthaceae

Uribea Dugand et Romero. 1962. Fabaceae. 1 Central and South America (Colombia)

Urmenetea Phil. 1860. Asteraceae. 1 northern Chile, northwestern Argentina

Urnularia Stapf. 1901. Apocynaceae. 7 western Malesia

Urobotrya Stapf. 1905 (Opilia Roxb.). Opiliaceae. 7: 2 trop. Africa (Nigeria, Cameroun, Guinea, Gabon, Congo, Sierra Leone, Liberia to Zaire and northeastern Angola); 5 Southeast Asia: southern China (Yunnan), Thailand, southern Burma, Laos, Vietnam, Borneo, Flores I.

Urocarpidium Ulbr. 1916. Malvaceae. 15 Galapagos Is., Mexico (2), Guatemala, South America from Venezuela to Chile and Argentina

Urocarpus J. L. Drumm. ex Harv. = Asterolasia F. Muell. Rutaceae

Urochlaena Nees. 1841. Poaceae. 2 South Africa

Urochloa P. Beauv. 1812. Poaceae. 15 trop. Old World

Urochondra C. E. Hubb. 1947. Poaceae. 1 trop. Northeast Africa, Arabian Peninsula, Socotra, Pakistan

Urodon Turcz. = Pultenaea Sm. Fabaceae

Urogentias Gilg et Gilg-Ben. 1933. Gentianaceae. 1 Tanzania

Urolepis (A. DC.) R. M. King et H. Rob. 1971 (Eupatorium L.). Asteraceae. 1 Brazil, Bolivia, Argentina

Uromyrtus Burret. 1941. Myrtaceae. 13 Borneo, New Guinea (3), northern Australia (1), New Caledonia (9)

Uropappus Nutt. 1841 (Microseris D. Don). Asteraceae. 1 southwestern U.S., Baja California

Urophyllum Jack. ex Wall. 1824. Rubiaceae. c. 150 trop. Africa, trop. Asia eastward to Japan and New Guinea

Urophysa Ulbr. 1929. Ranunculaceae. 2 China

Uroskinnera Lindl. 1857. Scrophulariaceae. 4 Mexico, Central America

Urospatha Schott. 1853. Araceae. 20 Central and trop. South America

Urospathella Bunting. 1988 (Cyrtosperma Griff.). Araceae. 1 Colombia, Venezuela

Urospermum Scop. 1777. Asteraceae. 2 Mediterranean, West Asia to Caucasus (1, U. picroides) and Iran

Urostemon B. Nord. = Brachyglottis J. R. Forst. et G. Forst. Asteraceae

Urostephanus B. L. Rob. et Greenm. 1895. Asclepiadaceae. 1 Mexico

Urotheca Gilg = Gravesia Naudin. Melastomataceae

Ursia Vassilcz. = Trifolium L. Fabaceae

Ursinia Gaertn. 1791. Asteraceae. 40 Ethiopia (1), South Africa

Ursiniopsis E. Phillips = Ursinia Gaertn. Asteraceae

Urtica L. 1753. Urticaceae. c. 70 almost cosmopolitan, but chiefly in temp. regions, a few in the tropics; Russia (c. 10, all areas)

Urvillea Kunth. 1821. Sapindaceae. 13 trop. and subtrop. America

Usteria Willd. 1790. Loganiaceae. 1 Africa from Senegal to Angola and east to eastern Zaire

Utleria Bedd. ex Benth. 1876. Asclepiadaceae. 1 southern India

Utleya Wilbur et Luteyn. 1977. Ericaceae. 1 Costa Rica

Utricularia L. 1753. Lentibulariaceae. c. 215 cosmopolitan, chiefly in tropics

Utsetela Pellegr. 1928. Moraceae. 1 trop. West Africa

Uvaria L. 1753. Annonaceae. c. 150 trop. Africa, Madagascar, trop. Asia, Malesia, Australia, New Caledonia

Uvariastrum Engl. et Diels. 1901. Annonaceae. 8 trop. Africa from Cameroun and Gabon to Angola and Zimbabwe

Uvariodendron (Engl. et Diels) R. E. Fr. 1931. Annonaceae. 12–16 trop. Africa

Uvariopsis Engl. ex Engl. et Diels. 1899. Annonaceae. 13 trop. Africa

Uvularia L. 1753. Melanthiaceae. 5 eastern North America from New Scotland Peninsula to Florida

Vaccaria Wolf. 1781. Caryophyllaceae. 1 (V. hispanica) Central and East Europe, Mediterranean, temp. Asia; Russia—all areas, except polar

Vacciniopsis Rusbey = Disterigma (Klotzsch) Nied. Ericaceae

Vaccinium L. 1753. Ericaceae. c. 450 Europe (8), Northeast Africa and Madagascar (5), temp. and montane trop. Asia, Malesia (c. 240); a few species in Melanesia, Polynesia: Rapa I. (1, V. rapae), Samoa Is. (1, V. whitmeei), Hawaii (3), Society Is., and Marquesas Is. (1, V. cereum); North (65) and trop. (30) America; Russia (c. 10, Arctica, European part, Caucasus, Siberia, Far East)

Vagaria Herb. 1837. Amaryllidaceae. 1 Asia Minor

Vahadenia Stapf. 1902. Apocynaceae. 2 trop. West Africa

Vahlia Thunb. 1782. Vahliaceae. 5 trop. Northeast and South Africa, Madagascar, Southwest Asia to northwestern India

Vahlodea Fr. 1842 (Deschampsia P. Beauv.). Poaceae. 3 Arctica, Northeast Asia, northern North and extratrop. South America; Russia (2, northern European part, Far East)

Vailia Rusby. 1898. Asclepiadaceae. 1 Bolivia

Valantia L. 1753. Rubiaceae. 3–4 Canary Is., Mediterranean, West Asia to Caucasus (1, V. muralis) and northern Iran

Valdivia C. Gay ex J. Remy. 1848. Escalloniaceae. 1 Chile

Valentiana Raf. = Thunbergia Retz. Acanthaceae

Valentiniella Speg. 1903 (Heliotropium L.). Boraginaceae. 1 temp. South America

Valenzuelia Bertero ex Cambess. = Guindilia Gillies ex Hook. et Arn. Sapindaceae

Valeria Minod. 1918. Scrophulariaceae. 1 eastern Brazil

Valeriana L. 1753. Valerianaceae. c. 250 Eurasia, South Africa, temp. North and South (Andes) America; Russia (c. 20, almost all areas, except polar and arid areas)

Valerianela Hill. 1753. Valerianaceae. c. 80 temp. Eurasia from West Europe and Mediterranean to Caucasus, Middle Asia, Afghanistan and southwestern Pakistan, North and South Africa, North America

Valerioa Stand. et Steyerm. = Peltanthera Benth. Buddlejaceae

Valerioanthus Lundell = Ardisia Sw. Myrsinaceae

Vallariopsis Woodson. 1936. Apocynaceae. 1 western Malesia

Vallaris Burm. f. 1768. Apocynaceae. 3–4 Himalayas, India, Sri Lanka, Burma, southern China, Indochina, Malay Peninsula, Philippines

Vallea Mutis ex L. f. 1782. Elaeocarpaceae. 2 Andes of Colombia, Ecuador, Peru, Bolivia

Vallesia Ruiz et Pav. 1794. Apocynaceae. 8 trop. America from Florida peninsula to Argentina, West Indies

Vallisneria L. 1753. Hydrocharitaceae. 2–8 tropics and subtropics; V. spiralis—extending to southern European part of Russia, Middle Asia, and Russian Far East

Vallota Salisb. ex Herb. = Cyrtanthus Aiton. Amaryllidaceae

Valovaea Chiov. = ? Vigna Savi. Fabaceae

Valvanthera C. T. White = Hernandia L. Hernandiaceae

Van-Royena Aubrév. = Pouteria Aubl. Sapotaceae

Vanasushava P. K. Murh. et Constance. 1974. Apiaceae. 1 southern India

Vanclevea Greene. 1899 (Grindelia Willd.). Asteraceae. 1 U.S.: southeastern Utah and northeastern Arizona

Vancouveria C. Morren et Decne. 1834 (Epimedium L.). Berberidaceae. 3 western North America from Washington state to California

Vanda Jones ex R. Br. 1820. Orchidaceae. c. 60 trop. Asia from India and Sri Lanka to China and Indochina, Malesia from Malay Peninsula to Bougainville and Solomon Is. (1, V. hindsii), northern Australia

Vandasia Domin = Vandasina Rauschert. Fabaceae

Vandasina Rauschert. 1982. Fabaceae. 1 New Guinea, Australia (Queensland)

Vandellia P. Browne ex L. = Lindernia All. Scrophulariaceae

Vandopsis Pfitzer. 1889. Orchidaceae. 21 trop. Asia, Malesia from Malay Peninsula to New Guinea, western Oceania

Vangueria Comm. ex Juss. 1789. Rubiaceae. 8–15 trop. and South (Transvaal) Africa, Madagascar

Vangueriella Verdc. 1987. Rubiaceae. 17 trop. (mainly West) Africa eastward to Uganda (1, V. rhamnoides)

Vangueriopsis Robyns. 1928. Rubiaceae. 18 trop. Africa

Vanheerdia L. Bolus. 1938. Aizoaceae. 7 central South Africa, Botswana

Vanhouttea Lem. 1845. Gesneriaceae. 4 Brazil

Vania F. K. Mey. = Thlaspi L. Brassicaceae

Vanilla Mill. 1754. Orchidaceae. c. 100 trop. equatorial Africa, Madagascar, southern India, Sri Lanka, Indochina, southern China, Taiwan, Malesia from Malay Peninsula to New Guinea and New Britain I., Bougainville and Solomon Is.; North, Central, and trop. South America

Vanillosmopsis Sch. Bip. = Eremanthus Less. Asteraceae

Vanoverberghia Merr. 1912. Zingiberaceae. 1 Philippines

Vanroyenella Novelo et Philbrick. 1993. Podostemaceae. 1 Mexico (Jalisco)

Vantanea Aubl. 1775. Humiriaceae. 16 Central and trop. South America

Vanwykia Wiens. 1978 (Taxillus Tiegh.). Loranthaceae. 1–2 Tanzania, Zambia, Malawi, Mozambique, northeastern South Africa

Vanzijlia L. Bolus. 1927. Aizoaceae. 1–2 coastal western South Africa

Vargasiella C. Schweinf. 1952. Orchidaceae. 2 Venezuela, Peru

Varilla A. Gray. 1849. Asteraceae. 2 southern U.S. (Texas), Mexico

Varronia P. Browne = Cordia L. Cordiaceae

Varthemia DC. 1836. Asteraceae. 1 (V. persica) Middle Asia, Iran

Vaseyanthus Cogn. 1891. Cucurbitaceae. 1 southwestern North America

Vaseyochloa Hitchc. 1933. Poaceae. 1 U.S. (Texas)

Vasivaea Baill. 1872. Tiliaceae. 2 Peru, Brazil, Bolivia

Vasqueziella Dodson. 1982. Orchidaceae. 1 Bolivia

Vassilczenkoa Lincz. 1979 (Chaetolimon [Bunge] Lincz.). Plumbaginaceae. 1 Middle Asia, Afghanistan

Vassobia Rusby. 1907. Solanaceae. 4 Bolivia, southern Brazil, Paraguay, Uruguay, northern Argentina

Vatairea Aubl. 1775. Fabaceae. 7 trop. America from Mexico to Brazil

Vataireopsis Ducke = Vatairea Aubl. Fabaceae

Vateria L. 1753. Dipterocarpaceae. 2 southern India (1, V. indica), Sri Lanka

Vateriopsis F. Heim. 1892. Dipterocarpaceae. 1 Seychelles

Vatica L. 1771. Dipterocarpaceae. 65–80 southern India, Sri Lanka, Burma, southern China (incl. Hainan), Indochina, Malesia

Vatovaea Chiov. 1951. Fabaceae. 1 Sudan, Ethiopia, Somalia, Kenya, Uganda, Tanzania, Oman

Vauanthes Haw. = Crassula L. Crassulaceae

Vaughania S. Moore = Indigofera L. Fabaceae

Vaupelia Brand = Cystostemon Balf. f. Boraginaceae

Vaupesia R. E. Schult. 1955. Euphorbiaceae. 1 Colombia, western Brazil

Vauquelinia Corrêa ex Humb. et Bonpl. 1807. Rosaceae. 3 U.S. (California, Arizona, New Mexico, Texas), Mexico

Vausagesia Baill. 1890. Sauvagesiaceae. 2 trop. South Africa

Vavaea Benth. 1843. Meliaceae. 4 Malesia from Sumatra to Solomon Is., northern Australia, Melanesia, Polynesia

Vavara Benoist. 1962. Acanthaceae. 1 Madagascar

Vavilovia Fedorov. 1939. Fabaceae. 1 Caucasus, Asia Minor, Syria, Lebanon, Iraq, Iran

Veconcibea (Muell. Arg.) Pax et K. Hoffm. 1914. Euphorbiaceae. 2 trop. America

Veeresia Monach. et Moldenke = Reevesia Lindl. Sterculiaceae

Vegaea Urb. 1913. Myrsinaceae. 1 West Indies

Veillonia H. E. Moore. 1978 (Burretiokentia Pic. Serm.). Arecaceae. 1 New Caledonia

Veitchia H. A. Wendl. 1868. Arecaceae. 18 Philippines, Vanuatu, Fiji (10)

Velezia L. 1753. Caryophyllaceae. 6 Mediterranean, West Asia to Afghanistan; V. rigida—Crimea, Caucasus, Middle Asia

Vella L. 1753. Brassicaceae. 5 Europe (Spain), Morocco, Algeria

Velleia Sm. 1798. Goodeniaceae. 20 New Guinea (1), Australia, Tasmania

Vellereophyton Hilliard et B. C. Burtt. 1981. Asteraceae. 7 South Africa

Vellosiella Baill. 1887. Scrophulariaceae. 3 Brazil

Vellozia Vand. 1788. Velloziaceae. 125 trop. America

Velophylla Benj. Clarke ex Durand (nomen dubium). Podostemaceae. 1 Brazil

Veltheimia Gled. 1771. Hyacinthaceae. 2 South Africa

Velvitsia Hiern = Melasma Bergius. Scrophulariaceae

Venegasia DC. 1838. Asteraceae. 1 U.S. (southwestern California), Mexico (coastal Baja California and Channel Is.)

Venidium Less. = Arctotis L. Asteraceae

Ventenata Koeler. 1802. Poaceae. 3 South and Southeast Europe, Mediterranean, West Asia, incl. Caucasus (1, V. dubia)

Ventilago Gaertn. 1788. Rhamnaceae. 37 trop. Africa (1), Madagascar (1), Himalayas, India, Burma, China, Indochina, Malesia, Australia, Pacific Is.

Ventricularia Garay. 1972. Orchidaceae. 1 peninsular Thailand and Malaya

Veprecella Naudin = Gravesia Naudin. Melastomataceae

Vepris Comm. ex A. Juss. 1825. Rutaceae. 15 trop. and South Africa, Madagascar, Mascarene Is., southern India

Veratrilla Baill. et Franch. 1899. Gentianaceae. 2 China, eastern Himalayas

Veratrum L. 1753. Melanthiaceae. 25 temp. Northern Hemisphere; Russia (from European part and Caucasus to Far East)

Verbascum L. 1753. Scrophulariaceae. c. 360 temp. Eurasia, a few species in trop. and subtrop. regions; Caucasus—over 40, Middle Asia, southern Siberia

Verbena L. 1753. Verbenaceae. c. 250 temp. and trop. America, a few species in the Old World

Verbenoxylum Tronc. 1971. Verbenaceae. 1 southern Brazil (Santa Catarina I.)

Verbesina L. 1753. Asteraceae. c. 150 trop. and subtrop. America, West Indies

Verdcourtia R. Wilczek = Dipogon Liebm. Fabaceae

Verdickia De Wild = Chlorophytum Ker-Gawl. Asphodelaceae

Verena Minod. 1918. Scrophulariaceae. 1 Paraguay

Verheullia Miq. 1844. Peperomiaceae. 3 West Indies

Verlotia E. Fourn. = Marsdenia R. Br. Asclepiadaceae

Vermeulenia A. Löve et D. Löve = Orchis L. Orchidaceae

Vermifrux J. B. Gillett. 1966 (Lotus L.). Fabaceae. 1 Sudan, Ethiopia, Somalia, Arabian Peninsula

Vernicia Lour. 1790 (Aleurites J. R. Forst. et G. Forst.). Euphorbiaceae. 3 East and Southeast Asia

Vernonanthura H. Rob. 1992. Asteraceae. 62 America from the eastern U.S. and Mexico to Argentina, West Indies

Vernonia Schreb. 1791. Asteraceae. c. 1,000 tropics (mostly) and subtropics

Vernoniopsis Humbert. 1955. Asteraceae. 1 Madagascar

Veronica L. 1753. Scrophulariaceae. c. 250 temp. Northern Hemisphere, a few species in montane tropics and temp. Southern Hemisphere

Veronicastrum Heist. ex Fabr. 1759 (Veronica L.). Scrophulariaceae. 2 temp. East Asia (1), northeastern North America (1, V. virginicum)

Verreauxia Benth. 1868. Goodeniaceae. 3 southwestern Australia

Verrucifera N. E. Br. = Titanopsis Schwantes. Aizoaceae

Verrucularia A. Juss. = Verrucularina Rauschert. Malpighiaceae

Verrucularina Rauschert. 1982. Malpighiaceae. 2 eastern Brazil

Verschaffeltia H. A. Wendl. 1865. Arecaceae. 1 Seychelles (Mahé, Silhouette, and Praslin Is.)

Versteegia Valeton. 1911. Rubiaceae. 5–6 New Guinea

Verticordia DC. 1828. Myrtaceae. 40 Australia

Veselskya Opiz. 1856. Brassicaceae. 1 Afghanistan

Veseyochloa J. B. Phipps. 1964 (Tristachya Nees). Poaceae. 1 Zambia, Burundi, Tanzania

Vesicarex Steyerm. = Carex L. Cyperaceae

Vesicaria Adans. = Alyssoides Hill. Brassicaceae

Vesselowskya Pamp. 1905. Cunoniaceae. 1 Australia (New South Wales)

Vestia Willd. 1809. Solanaceae. 1 (V. foetida) Chile

Vetiveria Lam. ex Cass. 1822. Poaceae. 10 trop. Old World

Vexatorella Rourke. 1984 (Leucospermum R. Br.). Proteaceae. 4 South Africa

Vexibia Raf. 1825 (Sophora L.). Fabaceae. 5 North and South (Argentina) America, Southeast Europe, West Asia; 2 Crimea, Caucasus, Middle Asia, southern West Siberia

Vexillabium F. Maek. 1935. Orchidaceae. 4 China, Korea, Japan, Taiwan, Philippines (1, V. yakushimense)

Viburnum L. 1753. Viburnaceae. c. 225 temp. and subtrop. regions, mostly East Asia (China—75) and eastern North America, Malesia (16, excl. New Guinea); Russia (c. 10, European part, Siberia, Far East), Caucasus, Middle Asia

Vicatia DC. 1830. Apiaceae. 4–5 Russia (1, V. coniifolia—Middle Asia, southern Siberia), Afghanistan, Pakistan, western China, India

Vicia L. 1753. Fabaceae. c. 140 temp. Northern Hemisphere, esp. Mediterranean, a few species in trop. Africa (6), South America, and Hawaiian Is.

Vicoa Cass. = Pentanema Cass. Asteraceae

Victoria Lindl. 1837. Nymphaeaceae. 2 trop. South America from Guyana to Paraguay, Jamaica

Victorinia Léon. 1941 (Cnidoscolus Pohl). Euphorbiaceae. 2 Cuba

Vieillardorchis Kraenzl. = Goodyera R. Br. Orchidaceae

Vieraea Sch. Bip. = Vieria Webb. et Berth. Asteraceae

Viereckia R. M. King et H. Rob. 1975. Asteraceae. 1 Mexico

Vieria Webb et Berth. 1839. Asteraceae. 1 Canary Is.

Vietnamosasa T. Q. Nguyen. 1990. Poaceae. 3 Vietnam

Vietsenia C. Hansen. 1984. Melastomataceae. 4 Vietnam

Vigethia W. A. Weber. 1943. Asteraceae. 1 Mexico

Vigna Savi. 1824. Fabaceae. c. 150 pantropics, esp. Africa (70), a few species extending to South Africa

Viguiera Kunth. 1820. Asteraceae. c. 150 trop. and subtrop. America, West Indies

Viguierella A. Camus. 1926. Poaceae. 1 Madagascar

Villadia Rose. 1903. Crassulaceae. 25–30 trop. and subtrop. America from Texas to Peru

Villanova Lag. 1816. Asteraceae. 10 America from Mexico to Chile

Villaresia Ruiz et Pav. 1793. ? Celastraceae. 1 Peru

Villaresiopsis Sleumer. 1940. Icacinaceae. 1 Peru

Villaria Rolfe. 1884. Rubiaceae. 5 Philippines

Villarsia Vent. 1803. Menyanthaceae. 16 South Africa (1, V. capensis), Southeast Asia, southwestern Australia (9)

Villebrunea Gaudich. ex Wedd. 1854. Urticaceae. 8 Sri Lanka, Indochina, China, Taiwan, Japan

Villocuspis (A. DC.) Aubrév. et Pellegr. = Chrysophyllum L. Sapotaceae

Vilobia Strother. 1968. Asteraceae. 1 Bolivia

Viminaria Sm. 1805. Fabaceae. 1 southeastern Australia

Vinca L. 1753. Apocynaceae. 10 Europe, Mediterranean, West (incl. Caucasus) and Middle Asia to China

Vincentella Pierre. 1891. Sapotaceae. 4 trop. and South Africa

Vincentia Gaudich. 1829. Cyperaceae. 10 trop. Old World from Mascarene Is. to Polynesia

Vincetoxicopsis Costantin. 1912. Asclepiadaceae. 1 Indochina

Vincetoxicum Wolf. 1776. Asclepiadaceae. c. 30 temp. Eurasia, incl. European part of Russia, Caucasus, Middle Asia

Vindasia Benoist. 1962. Acanthaceae. 1 Madagascar

Vinkia Meijden = Myriophyllum L. Haloragaceae

Vinticena Steud. 1841 (Crewia L.). Tiliaceae. 30–35 trop. and South Africa, Madagascar, Mascarene Is., Arabian Peninsula, Socotra, India

Viola L. 1753. Violaceae. c. 450 cosmopolitan, but chiefly temp. regions

Viposia Lundell. 1939 (Plenckia Reisseck). Celastraceae. 1 Argentina

Virecta Afzel ex Sm. = Virectaria Bremek. Rubiaceae

Virectaria Bremek. 1952. Rubiaceae. 8 trop. Africa

Virga Hill. = Dipsacus L. Dipsacaceae

Virgilia Poir. 1808. Fabaceae. 2 South Africa (Cape)

Virginea (DC.) Nicoli = Helichrysum Mill. Asteraceae

Virgulaster Semple = Aster L. Asteraceae

Virgulus Raf. = Aster L. Asteraceae

Viridivia J. H. Hemsl. et Verdc. 1956. Passifloraceae. 1 southwestern Tanzania and Zambia

Virola Aubl. 1775. Myristicaceae. 48 trop. America

Virotia L. A. S. Johnson et B. G. Briggs. 1975 (Kermadecia Brongn. et Gris). Proteaceae. 6 Australia (1), New Caledonia

Viscainoa Greene. 1888. Zygophyllaceae. 1–2 Baja California

Viscaria Roehl. = Lychnis L. Caryophyllaceae

Viscum L. 1753. Viscaceae. 70–100 temp. Europe, West (incl. Caucasus) and East Asia, Africa, Madagascar, trop. Asia, Malesia, Australia (14)

Vismia Vand. 1788. Hypericaceae. 35 trop. America, trop. West (3) and East (2) Africa

Vismianthus Mildbr. 1935. Connaraceae. 1 Tanzania

Visnaga Gaertn. 1788. Apiaceae. 1 Mediterranean, Northwest Africa, West Asia to Caucasus and Iran

Visnea L. f. 1782. Theaceae. 1 Canary Is.

Vitaliana Sesl. = Androsace L. Primulaceae

Vitellaria C. F. Gaertn. 1807. Sapotaceae. 1 northern trop. Africa, southern China

Vitellariopsis (Baill.) Dubard. 1915. Sapotaceae. 5 trop. East Africa (1), Zimbabwe (1), and South Africa

Vitex L. 1753. Verbenaceae. c. 250 tropics and subtropics, a few species in temp. regions

Viticipremna Lam. 1919. Verbenaceae. 5 Java, Philippines, Moluccas, New Guinea, Admiralty Is., Bismarck Arch., Australia (northern Queensland), Fiji (1, V. vitilevuensis)

Vitiphoenix Becc. = Veitchia H. A. Wendl. Arecaceae

Vitis L. 1753. Vitaceae. 60–70 temp. Northern Hemisphere; Russia (southern European part, Far East), Caucasus, Middle Asia

Vittadinia A. Rich. 1832. Asteraceae. 29 New Guinea, Australia, New Zealand, New Caledonia

Vittetia R. M. King et H. Rob. 1974 (Eupatorium L.). Asteraceae. 2 Brazil

Viviania Cav. 1804. Vivianaceae. 6 Chile, southern Brazil

Vladimiria Iljin = Dolomiaea DC. Asteraceae

Vleisia Toml. et Posl. 1976. Zannichelliaceae. 1 South Africa (Cape)

Voacanga Thouars. 1806. Apocynaceae. 12: 7 trop. and South (1) Africa, Madagascar, Mascarene Is.; 5 Malesia

Voandzeia Thouars = Vigna Savi. Fabaceae

Voanioala J. Dransf. 1989. Arecaceae. 1 Madagascar

Voatamalo Capuron ex Bosser. 1976. Euphorbiaceae. 2 Madagascar

Vochysia Aubl. 1775. Vochysiaceae. c. 110 trop. America

Voharanga Costantin et Bois. 1908 (Cynanchum L.). Asclepiadaceae. 1 Madagascar

Vohemaria Buchenau. 1889. Asclepiadaceae. 2 Madagascar

Voladeria Benoist = Oreobolus R. Br. Cyperaceae

Volkensia O. Hoffm. = Bothriocline Oliv. ex Benth. Asteraceae

Volkensiella H. Wolff. 1912 (or = Oenanthe L.). Apiaceae. 1 (V. procumbens) trop. East Africa

Volkensinia Schinz. 1912. Amaranthaceae. 1 (V. prostrata) East Africa: southeastern Sudan, southern Ethiopia, Kenya, Tanzania

Volkensiophyton Lindau = Lepidagathis Willd. Acanthaceae

Volkiella Merxm. et Czech. 1953. Cyperaceae. 1 South-West (Namibia) and central (Zambia) Africa

Volutaria Cass. 1816. Asteraceae. 10 Mediterranean, trop. Northeast Africa, Southwest Asia

Vonitra Becc. 1906. Arecaceae. 4 Madagascar

Vossia Wall. et Griff. 1836. Poaceae. 1 trop. Africa, South Asia

Votomita Aubl. 1775. Melastomataceae. 7 trop. America

Vouacapoua Aubl. 1775. Fabaceae. 3 trop. South America

Vouarana Aubl. 1775. Sapindaceae. 1 Guyana, northern Brazil

Voyria Aubl. 1775. Gentianaceae. 18 trop. America, West Indies, trop. West Africa (1)

Voyriella Miq. 1851. Gentianaceae. 1 Panama, trop. South America

Vriesea Lindl. 1843. Bromeliaceae. c. 260 Mexico, Central and trop. South America, West Indies

Vrydagzynea Blume. 1858. Orchidaceae. 20–40 trop. Asia from Himalayas to Taiwan, Malesia from Malay Peninsula to New Guinea, Micronesia, Pacific Is. to Vanuatu, Fiji, Tonga, and Samoa; ? Australia (1)

Vulpia C. C. Gmel. 1805. Poaceae. 25 temp. North and temp. South America; West, South, and Southeast Europe; North Africa; West (incl. Caucasus), Southwest, and Middle Asia

Vulpiella (Batt. et Trab.) Burollet. 1934. Poaceae. 3 western Mediterranean

Vvedenskia Korovin. 1947. Apiaceae. 1 Middle Asia (Gessar Mts.)

Vvedenskyella Botsch. 1955. Brassicaceae. 3 Pakistan, western China, Tibet, India (Kashmir)

Wachendorfia Burm. 1757. Haemodoraceae. 5 South Africa (Cape)

Wagatea Dalzell = Moullava Adans. Fabaceae

Wagenitzia Dostál. 1973 (Centaurea L.). Asteraceae. 1 Crete

Wahlenbergia Schrad. ex Roth. 1821. Campanulaceae. c. 200 temp. regions, mainly Southern Hemisphere, esp. South Africa

Waitzia J. C. Wendl. 1808. Asteraceae. 6 temp. western and southern Australia

Wajira Thulin. 1982. Fabaceae. 1 eastern Kenya

Wakilia Gilli = Phaeonychium O. E. Schulz. Brassicaceae

Walafrida E. Mey. 1838. Selaginaceae. c. 40 trop. and South Africa, Madagascar

Waldheimia Kar. et Kir. = Allardia Decne. Asteraceae

Waldsteinia Willd. 1799. Rosaceae. 6 temp. Northern Hemisphere, incl. (2) southern European part of Russia and Caucasus (W. geoides), East Siberia, and Far East (V. ternata)

Walidda (A. DC.) Pichon. 1951 (Wrightia R. Br.). Apocynaceae. 1 Sri Lanka

Wallacea Spruce ex Benth. et Hook. f. 1862. Sauvagesiaceae. 3 Brazil (Amazonia)

Wallaceodendron Koord. 1898. Fabaceae. 1 Philippines, Sulawesi

Wallenia Sw. 1788 (Cybianthus Mart.). Myrsinaceae. 26 West Indies

Walleniella P. Wilson = Solonia Urb. Myrsinaceae

Walleria J. Kirk. 1864. Tecophilaeaceae. 3 trop. and South Africa, Madagascar

Wallichia Roxb. 1820. Arecaceae. 7 trop. Asia from Indian Himalayas and northern Burma to southern China and peninsular Thailand

Walpersia Harv. 1862 (or = Phyllota [DC.] Benth.). Fabaceae. 1 Australia

Walsura Roxb. 1832. Meliaceae. c. 10 India, Sri Lanka, Andaman Is., southern China, Indochina, western Malesia to Sulawesi

Walteranthus Keighery. 1985. Gyrostemonaceae. 1 western Australia

Waltheria L. 1753. Sterculiaceae. 67 trop.

America, West Indies, trop. Africa (1), Madagascar (1), Malay Peninsula (1), Taiwan (1)

Wamalchitamia Strother. 1991. Asteraceae. 5 Mexico, Honduras, Nicaragua, Costa Rica

Wangenheimia Moench. 1794. Poaceae. 2 Iberian Peninsula, North Africa

Wangerinia E. Franz. 1908. Caryophyllaceae. 1 Chile

Warburgia Engl. 1895. Canellaceae. 3 trop. East and South Africa from Kenya and Uganda to Transvaal and Natal

Warburgina Eig. 1927. Rubiaceae. 1 Syria, Israel

Wardaster Small = Aster L. Asteraceae

Wardenia King = Brassaiopsis Decne. et Planch. Araliaceae

Warea Nutt. 1834. Brassicaceae. 4 southeastern U.S.

Warionia Benth. et Coss. 1872. Asteraceae. 1 northwestern Sahara

Warmingia Rchb. f. 1881. Orchidaceae. 2 Brazil

Warneckea Gilg. 1904. Melastomataceae. 9 trop. Africa, Madagascar, Mauritius (1)

Warpuria Stapf. 1908 (Podorungia Baill.). Acanthaceae. 2 Madagascar

Warrea Lindl. 1843. Orchidaceae. 6 trop. America from Costa Rica to Peru and Brazil

Warreella Schltr. 1914. Orchidaceae. 1 Colombia, Venezuela

Warreopsis Garay. 1973. Orchidaceae. 3 trop. America

Warszewiczia Klotzsch. 1853. Rubiaceae. 4 trop. America, West Indies

Wasabia Matsum. 1899 (Eutrema R. Br.). Brassicaceae. 2 East Asia

Washingtonia H. A. Wendl. 1879. Arecaceae. 2: W. filifera—southeastern California, western Arizona, and Baja California; W. robusta—Baja California and Sonora

Waterhousea B. Hyland. 1983 (Syzygium Gaertn.). Myrtaceae. 4 trop. Australia

Watsonia Mill. 1758. Iridaceae. 52 South Africa

Wattakaka (Decne.) Hassk. = Dregea E. Mey. Asclepiadaceae

Weberaster A. Löve et D. Löve = Aster L. Asteraceae

Weberbauera Gilg et Muschler. 1909. Brassicaceae. 16 Andes of Peru, Bolivia, Chile, Argentina

Weberbauerella Ulbr. 1906. Fabaceae. 2 Peru

Weberbauerocereus Backeb. = Haageocereus Backeb. Cactaceae

Weberocereus Britton et Rose. 1909. Cactaceae. 9 southern Mexico, Central America (mainly Costa Rica), Ecuador

Websteria S. H. Wright. 1887. Cyperaceae. 1 tropics and subtropics

Weddellina Tul. 1849. Podostemaceae. 1 northern South America

Wedelia Jacq. 1760. Asteraceae. c. 70 U.S. (Texas), Mexico (c. 30), Guatemala, Belize, Honduras, El Salvador, Nicaragua, Costa Rica, Panama; trop. and subtrop. South America, West Indies

Wehlia F. Muell. = Homalocalyx F. Muell. Myrtaceae

Weigela Thunb. 1780 (Diervilla Adans.). Caprifoliaceae. 10 East Asia; 3 extending to Russian Far East

Weigeltia A. DC. = Cybianthus Mart. Myrsinaceae

Weihea Spreng. = Cassipourea Aubl. Rhizophoraceae

Weinmannia L. 1759. Cunoniaceae. c. 130 Madagascar, Comoro Is., Malesia to New Guinea, Solomon Is., New Zealand, New Caledonia, Vanuatu, Fiji, Tahiti, Mexico, Central America, Chilean Andes

Welchiodendron Peter G. Wilson et J. T. Waterh. 1982. Myrtaceae. 1 northeastern Australia, New Guinea

Weldenia Schult. f. 1829. Commelinaceae. 1 Mexico, Guatemala

Welfia H. A. Wendl. 1869. Arecaceae. 1 (W. regia H. A. Wendl.) trop. America from Honduras to western and eastern Colombia

Wellstedia Balf. f. 1884. Wellstediaceae. 3 Ethiopia, Somalia, Socotra, and Namibia

Welwitschia Hook. f. 1862. Welwitschiaceae. 1 deserts of Angola and coastal Namibia

Welwitschiella O. Hoffm. 1894. Asteraceae. 1 Angola, Zambia

Wenchengia C. Y. Wu et S. Chow. 1965. Lamiaceae. 1 Hainan

Wendelboa Soest = Taraxacum G. H. Weber ex Wigg. Asteraceae

Wendlandia Bartl. ex DC. 1830. Rubiaceae. 44 Himalayas, India, Burma, China, Taiwan, Indochina, Malesia, Australia (Queensland)

Wendlandiella Dammer. 1905. Arecaceae. 3 Amazonian Peru

Wendtia Meyen. 1834. Ledocarpaceae. 3 Chile, Argentina

Wenzelia Merr. 1915. Rutaceae. 9 Philippines, New Guinea, Solomon Is., Vanuatu, Fiji, and Hawaiian Is.

Wercklea Pittier et Standl. 1916. Malvaceae. 12 trop. America

Werdermannia O. E. Schulz. 1928. Brassicaceae. 3–4 northern China

Werneria Kunth. 1818. Asteraceae. 40 Andes

Wernhamia S. Moore. 1922. Rubiaceae. 1 Bolivia

Westia Vahl = Berlinia Sol. ex Hook. f. Fabaceae

Westoniella Cuatrec. 1977. Asteraceae. 5 Costa Rica

Westphalina A. Robyns et Bamps. 1977. Tiliaceae. 1 Guatemala

Westringia Sm. 1797. Lamiaceae. 25 Australia, Tasmania

Wetria Baill. 1858. Euphorbiaceae. 1 Burma, Thailand, Malesia

Wettinella O. F. Cook et Doyle = Wettinia Poepp. Arecaceae

Wettinia Poepp. 1837. Arecaceae. 9 Panama, Colombia, Ecuador, Peru, and western Brazil

Wettiniicarpus Burret = Wettinia Poepp. Arecaceae

Wettsteiniola Suess. 1935. Podostemaceae. 3 Brazil, Argentina

Whipplea Torr. 1857. Hydrangeaceae. 1 western coastal U.S.

Whiteheadia Harv. 1868. Hyacinthaceae. 1 South Africa

Whiteochloa C. E. Hubb. 1952. Poaceae. 5 trop. Australia

Whiteodendron Steenis. 1952. Myrtaceae. 1 Borneo

Whitesloanea Chiov. 1937. Asclepiadaceae. 1 Somalia

Whitfieldia Hook. 1845. Acanthaceae. 10 trop. Africa

Whitfordiodendron Elmer = Callerya Endl. Fabaceae

Whitleya Sweet = Anisodus Link et Spreng. Solanaceae

Whitmorea Sleumer. 1969. Icacinaceae. 1 Solomon Is.

Whitneya A. Gray. 1865. Asteraceae. 1 California

Whittonia Sandwith. 1962. Peridiscaceae. 1 northeastern South America

Whyanbeelia Airy Shaw et B. Hyland. 1976. Euphorbiaceae. 1 Australia (Queensland)

Whytockia W. W. Sm. 1919 (Monophyllaea R. Br.). Gesneriaceae. 3 southwestern China (2), Taiwan (1)

Wiborgia Thunb. 1800. Fabaceae. 10 South Africa (Cape)

Widdringtonia Endl. 1842. Cupressaceae. 3 trop. and South Africa

Widgrenia Malme. 1900. Asclepiadaceae. 1 Brazil

Wiedemannia Fisch. et C. A. Mey. 1838. Lamiaceae. 1–3 eastern Mediterranean to Caucasus (1, W. multifida)

Wielandia Baill. 1858. Euphorbiaceae. 1 Seychelles

Wiesneria Micheli. 1818. Alismataceae. 4 trop. Africa, Madagascar, India

Wigandia Kunth. 1819. Hydrophyllaceae. 2–3 trop. America from Mexico to Peru, West Indies

Wigginsia D. M. Porter = Parodia Speg. Cactaceae

Wightia Wall. 1830. Bignoniaceae. 2–3 eastern Himalayas, northeastern India, northern Burma, western China, Indochina, Malesia

Wikstroemia Endl. 1833. Thymelaeaceae.

75 Afghanistan, Himalayas, India, Sri Lanka, Burma, southern China, Indochina, Malesia, Australia, Melanesia eastward to Fiji, Hawaiian Is. (12), Polynesia

Wilbrandia Silva Manso. 1836. Cucurbitaceae. 5 trop. South America

Wilcoxia Britton et Rose = Peniocereus (A. Berger) Britton et Rose. Cactaceae

Wildemaniodoxa Aubrév. et Pellegr. = Englerophytum Krause. Sapotaceae

Wilhelminia Hochr. = Hibiscus L. Malvaceae

Wilhelmsia Rchb. 1828. Caryophyllaceae. 1 arctic Northeast Asia, Northwest America

Wilkesia A. Gray. 1852. Asteraceae. 2 Hawaiian Is.

Wilkiea F. Muell. 1858. Monimiaceae. 7 New Guinea (1), Australia (Queensland, New South Wales)

Willardia Rose = Lonchocarpus Kunth. Fabaceae

Willbleibia Herter. 1953 (Willkommia Hach.). Poaceae. 2 southern U.S., Argentina

Willdenowia Thunb. 1788. Restionaceae. 11 South Africa

Williamodendron Kubitzki et H. G. Richt. 1987. Lauraceae. 3 Costa Rica, northern Colombia, Amazonia, and southern Brazil

Williamsia Merr. = Praravinia Korth. Rubiaceae

Willisia Warm. 1901. Podostemaceae. 1 southern India

Willkommia Hack. 1888. Poaceae. 3 trop. and Namibia

Willughbeia Roxb. 1820. Apocynaceae. 25 India, Sri Lanka, Andaman and Nicobar Is., Malesia

Willughbeiopsis Rauschert = Urnularia Stapf. Apocynaceae

Willwebera A. Löve et D. Löve = Arenaria L. Caryophyllaceae

Wilsonia R. Br. 1810. Convolvulaceae. 4 Australia

Wimmeria Schltdl. et Cham. 1831. Celastraceae. 14 Mexico, Central America

Winchia A. DC. 1844 (Alstonia R. Br.). Apocynaceae. 2 Southeast Asia

Windsorina Gleason. 1923. Rapateaceae. 1 northeastern South America

Winifredia L. A. S. Johnson et B. G. Briggs. 1986. Restionaceae. 1 Tasmania

Winklera Regel. 1886. Brassicaceae. 2 Tajikistan, Afghanistan, Pakistan

Winklerella Engl. 1905. Podostemaceae. 1 trop. West Africa

Wislizenia Engelm. 1848. Capparaceae. 1 southwestern U.S. from southern California to western Texas, Mexico

Wissadula Medik. 1787. Malvaceae. 25 pantropics, chiefly America; W. amplissima—reaches southern Texas

Wissmannia Burret = Livistona R. Br. Arecaceae

Wisteria Nutt. 1818. Fabaceae. 6 China, Japan, North America

Withania Pauquy. 1825. Solanaceae. 11 Canary Is., Cape Verde Is., Mediterranean, Ethiopia, Somalia, South Africa, West and Southwest Asia eastward to southern China

Witheringia L'Hér. 1789. Solanaceae. 15 Mexico, Central and trop. South America, West Indies

Witsenia Thunb. 1782. Iridaceae. 1 South Africa (Cape)

Wittia K. Schum. = Disocactus Lindl. Cactaceae

Wittiocactus Rauschert = Disocactus Lindl. Cactaceae

Wittmackanthus Kuntze. 1891. Rubiaceae. 1 trop. America from Panama to Peru

Wittmackia Mez = Aechmea Ruiz et Pav. Bromeliaceae

Wittrochia Lindm. 1891 (Canistrum E. Morren). Bromeliaceae. 7 eastern Brazil

Wittsteinia F. Muell. 1861. Alseuosmiaceae. 3–4 New Guinea, southeastern Australia (Victoria), New Caledonia

Wodyetia Irvine. 1983 (Normanbya F. Muell. ex Becc.). Arecaceae. 1 Australia (northeastern Queensland)

Woehleria Griseb. 1861. Amaranthaceae. 1 Cuba

Wokoia Baehni = Pouteria Aubl. Sapotaceae

Wolffia Horkel ex Schleid. 1844. Lemnaceae. 7 temp. and trop. regions; W. arrhiza—also in European part of Russia and Caucasus

Wolffiella Hegelm. 1895. Lemnaceae. 6 warm–temp., subtrop., and trop. America, South Africa (1)

Wolffiopsis Hartog et Van der Plas. 1970 (Wolffiella Hegelm.). Lemnaceae. 1–2 trop. America and West Indies, trop. Africa from Senegal and Angola to Sudan and Tanzania

Wollastonia DC. ex Decne. 1834. Asteraceae. 15 trop. Old World from Africa to Polynesia

Woodburnia Prain. 1904. Araliaceae. 1 Burma

Woodfordia Salisb. 1806. Lythraceae. 2 Ethiopia (1); W. floribunda—Madagascar, Southwest Asia, subtrop. Himalayas, India, Sri Lanka, Burma, China

Woodia Schltr. 1894. Asclepiadaceae. 3 South Africa

Woodiella Merr. = Woodiellantha Rauschert. Annonaceae

Woodiellantha Rauschert. 1982. Annonaceae. 1 Borneo

Woodrowia Stapf. 1896 (Dimeria R. Br.). Poaceae. 1 southern India

Woodsonia L. H. Bailey = Neonicholsonia Dammer. Arecaceae

Wooleya L. Bolus. 1960. Aizoaceae. 1 Namibia, western South Africa

Woollsia F. Muell. 1873 (Lysinema R. Br.). Epacridaceae. 1 northeastern Australia

Wootonella Standl. = Werbesina L. Asteraceae

Wootonia Greene = Dicranocarpus A. Gray. Asteraceae

Wormskioldia Schumach. et Thonn. 1827. Turneraceae. 11 trop. and South (7) Africa

Woronowia Juzep. 1941 (Sieversia Willd.). Rosaceae. 1 Caucasus

Worsleya (Traub.) Traub. 1944 (Hippeastrum Herb.). Amaryllidaceae. 1 Brazil

Woytkowskia Woodson. 1960. Apocynaceae. 2 Peru

Wrightia R. Br. 1810. Apocynaceae. 23 trop. and South (1, W. natalensis) Africa, trop. Asia from India and Sri Lanka (4) to Burma, southern China, and Indochina; Malesia from Malay Peninsula eastward to Solomon Is. and northern Australia

Wrixonia F. Muell. 1876. Lamiaceae. 2 western Australia

Wulfenia Jacq. 1781–1782. Scrophulariaceae. 2 Southeast Europe

Wulfeniopsis D. Y. Hong. 1980 (Wulfenia Jacq.). Scrophulariaceae. 2 Afghanistan, Himalayas from Kashmir to Nepal

Wulffia Neck. ex Cass. 1823. Asteraceae. 4 West India, Panama, South America

Wullschlaegelia Rchb. f. 1863. Orchidaceae. 2 trop. America from Guatemala to Brazil and Paraguay; West Indies

Wunderlichia Riedel ex Benth. 1873. Asteraceae. 6 Brazil

Wunschmannia Urb. = Distictic Mart ex Meisn. Bignoniaceae

Wurdackanthus Maguire. 1985. Gentianaceae. 2 Venezuela, northern Brazil; West Indies (Guadeloupe, Dominica, St. Vincent)

Wurdackia Moldenke = Rondonanthus Herzog. Eriocaulaceae

Wurmbea Thunb. 1781. Colchicaceae. 40 trop. and South Africa (c. 20), western Australia (19), and Tasmania (3)

Wythia Nutt. 1834. Asteraceae. 14 western North America

Xantheranthemum Lindau. 1895. Acanthaceae. 1 Andes of Peru

Xanthisma DC. 1836. Asteraceae. 1 southern U.S. (Texas)

Xanthium L. 1753. Asteraceae. 2 cosmopolitan, incl. European part of Russia, Caucasus, Middle Asia, West Siberia, Far East

Xanthobrychis Galushko = Onobrychis Hill. Fabaceae

Xanthocephalum Willd. 1807. Asteraceae. 5 southern U.S., Mexico

Xanthoceras Bunge. 1833. Sapindaceae. 1 northern China

Xanthocercis Baill. 1870. Fabaceae. 2: 1 (X. zambesiaca) Malawi, Zambia, Zimbabwe, Mozambique, South Africa; 1 Madagascar

Xanthogalum Avé-Lall. = Angelica L. Apiaceae

Xanthomyrtus Diels. 1922. Myrtaceae. 23 Philippines, Borneo, Sulawesi, Moluccas, New Guinea, Bismarck Arch., New Caledonia

Xanthopappus C. Winkl. 1894. Asteraceae. 1 (X. subacaulis) northwestern China

Xanthophyllum Roxb. 1820. Polygalaceae. 94 trop. Southeast Asia from Bangladesh to southern China and Hainan, through Indochina to Malesia (76, mostly Borneo, not in Lesser Sunda Is.), Solomon Is., Australia (northern Queensland)

Xanthophytopsis Pit. = Xanthophytum Reinw. ex Blume. Rubiaceae

Xanthophytum Reinw. ex Blume. 1827. Rubiaceae. 30 southern China (incl. Hainan), Vietnam, Laos, Malesia (Malay Peninsula, Sumatra, Java, Banguran I., Borneo; 18, Philippines, New Guinea), Vanuatu and Fiji (X. calycinum)

Xanthorhiza Marshall. 1785. Ranunculaceae. 1 eastern North America

Xanthorrhoea Sm. 1798. Xanthorrhoeaceae. 28 Australia and Tasmania

Xanthoselinum Schur. 1866. Apiaceae. 1 (X. alsaticum) Eurasia from West Europe to central Kazakhstan

Xanthosia Rudge. 1811. Apiaceae. 25 Australia

Xanthosoma Schott. 1832. Araceae. c. 45 trop. America, West Indies

Xanthostachya Bremek. 1944. Acanthaceae. 2 Lesser Sunda Is.

Xanthostemon F. Muell. 1857. Myrtaceae. 45 Philippines; eastern Indonesia; New Guinea; Solomon Is.; western, northern, and eastern Australia (13); New Zealand; New Caledonia (32)

Xantolis Raf. 1838. Sapotaceae. 14 southern India, southern China, Indochina, northern Philippines

Xantonnea Pierre ex Pit. 1923. Rubiaceae. 3 Indochina

Xantonneopsis Pit. 1923. Rubiaceae. 1 Indochina

Xatardia Meisn. ex Zeyh. 1838. Apiaceae. 1 southern France, Spain

Xenacanthus Bremek. 1944 (Strobilanthes Blume). Acanthaceae. 4 India

Xenikophyton Garay. 1974. Orchidaceae. 1 western Oceania

Xenophya Schott = Alocasia (Schott) G. Don. Araceae

Xenostegia D. F. Austin et Staples. 1981. Convolvulaceae. 2 Africa, India, Sri Lanka

Xeranthemum L. 1753. Asteraceae. 25 Southeast Europe, Mediterranean, West (incl. Caucasus) and Middle Asia to southwestern China

Xeroaloysia Tronc. 1963. Verbenaceae. 1 Argentina

Xerocarpa H. J. Lam. 1919 (Teijsmanniodendron Koord.). Verbenaceae. 1 New Guinea

Xerochlamys Baker. 1882 (Leptolaena Thouars). Sarcolaenaceae. 16 Madagascar

Xerochloa R. Br. 1810. Poaceae. 4 Java, Australia

Xerochrysum Tzvelev. 1990 (Helichrysum Mill.). Asteraceae. 1 Australia

Xerocladia Harv. 1862. Fabaceae. 1 Namibia, South Africa

Xerococcus Oerst. 1852 (Hoffmannia Sw.). Rubiaceae. 2 Costa Rica, Panama

Xerodanthia J. B. Phipps. 1966 (Danthoniopsis Stapf). Poaceae. 2 Northeast Africa, Southwest Asia

Xeroderris Roberty. 1954. Fabaceae. 1 savannas of trop. and South Africa

Xerodraba Skottsb. 1916. Brassicaceae. 6 montane areas of southern Argentina

Xerolakia Anderb. 1991. Asteraceae. 1 Europe

Xerolirion A. S. George. 1986 (Lomandra Labill.). Dasypogonaceae. 1 southwestern Australia

Xeromphis Raf. 1838 (Randia L.). Rubiaceae. 10 Himalayas, India, Burma, southern China, Indochina, Malesia

Xeronema Brongn. et Gris. 1865. Phormiaceae. 2 New Caledonia, New Zealand

Xerophyllum Michx. 1803. Xerophyllaceae. 2 North America

Xerophyta Juss. 1789 (Barbacenia Vand.). Velloziaceae. 31 trop. Africa, Madagascar, Arabian Peninsula

Xeroplana Briq. 1895. Stilbaceae. 2 South Africa (Cape)

Xerorchis Schltr. 1912. Orchidaceae. 2 northern South America

Xerosicyos Humbert. 1939. Cucurbitaceae. 4 Madagascar

Xerosiphon Turcz. 1843 (Gomphrena L.). Amaranthaceae. 2 Brazil

Xerospermum Blume. 1849. Sapindaceae. 2 (or 20 ?) India (Assam), Bangladesh, Burma, southern China, Indochina, Malay Peninsula, Sumatra, Java, Borneo

Xerosphaera Soják = Trifolium L. Fabaceae

Xerospiraea Henrickson. 1986 (Spiraea L.). Rosaceae. 1 Mexico

Xerotecoma J. C. Gomes = Godmania Hemsl. Bignoniaceae

Xerothamnella C. T. White. 1944. Acanthaceae. 1 Australia (Queensland)

Xerotia Oliv. 1895. Caryophyllaceae. 1 Arabian Peninsula

Ximenia L. 1753. Olacaceae. 8–9: 7 U.S. (Florida), Mexico, Guatemala, Belize, Honduras, El Salvador, Nicaragua, Costa Rica, Panama, Colombia, Venezuela, the Guianas, Brazil, Bolivia, Argentina, Paraguay; West Indies; trop. and South Africa, trop. Asia, Australia (1, Queensland), New Caledonia (1)

Ximeniopsis Alain = Ximenia L. Olacaceae

Xiphidium Loefl. ex Aubl. 1775. Haemodoraceae. 1–2 trop. America, West Indies

Xiphion Hill = Iris L. Iridaceae

Xiphium Mill. = Iris L. Iridaceae

Xiphochaeta Poepp. 1843. Asteraceae. 1 the Guianas

Xizangia D. Y. Hong. 1986. Scrophulariaceae. 1 Tibet

Xolocotzia Miranda. 1965. Verbenaceae. 1 Mexico

Xylanche Beck = Boschniakia C. A. Mey. ex Bongard. Orobanchaceae

Xylanthemum Tzvelev. 1961. Asteraceae. 5 Middle Asia (3), northeastern Iran, Afghanistan

Xylia Benth. 1842. Fabaceae. 13 trop. (Sierra Leone to Ghana, Zaire, Tanzania, Malawi, Mozambique) and South Africa, Madagascar, India, Indochina

Xylinabaria Pierre. 1898. Apocynaceae. 4 Indochina (2), Java (2)

Xylinabariopsis Pit. 1933. Apocynaceae. 2–3 Indochina, ? Hainan

Xylobium Lindl. 1825. Orchidaceae. 33 trop. America from southern Mexico to Peru and Brazil, West Indies

Xylocalyx Balf. f. 1883. Scrophulariaceae. 5 Somalia, Socotra

Xylocarpus J. Koenig. 1784. Meliaceae. 3 coastal trop. East Africa from Somalia to Mozambique, Aldabra I., Madagascar, Mascarene Is., Sri Lanka, Andaman Is., Malesia, northern Austraia, trop. Pacific to Fiji and Tonga

Xylococcus Nutt. = Arctostaphylos Adans. Ericaceae

Xylomelum Sm. 1798. Proteaceae. 5 Australia

Xylonagra Donn. Sm. et Rose. 1913. Onagraceae. 1 Baja California

Xylonymus Kalkman ex Ding Hou. 1963. Celastraceae. 1 western New Guinea

Xyloölaena Baill. 1886. Sarcolaenaceae. 1 Madagascar

Xylophragma Sprague. 1903. Bignoniaceae. 5 trop. America from Mexico to Bolivia, Paraguay, Brazil, Trinidad

Xylopia L. 1759. Annonaceae. c. 160 pantropics, esp. Africa (c. 60) and America (c. 50)

Xylorhiza Nutt. 1840. Asteraceae. 8 western U.S. (Wyoming, Montana, South Dakota, Colorado, California, southern Nevada, Utah, Arizona, Texas), Mexico (Baja California and Chihuahua)

Xylosma G. Forst. 1786. Flacourtiaceae. c. 100 trop. America and West Indies (c. 50), Himalayas, India, Sri Lanka, southern China, Indochina, Malesia, Australia (Queensland, New South Wales), Lord Howe I., New Caledonia (19), Polynesia

Xylosterculia Kosterm. = Sterculia L. Sterculiaceae

Xylothamia G. L. Nesom, Y. B. Suh, D. R. Morgan et B. B. Simpson. 1990. Asteraceae. 8 Mexico, southern Texas

Xylotheca Hochst. 1843. Flacourtiaceae. 3 trop. East (Kenya, Tanzania, Malawi, Mozambique) and South (1, Transvaal, Natal) Africa, ? Madagascar

Xymalos Baill. ex Warb. 1893. Monimiaceae. 3 trop. and South Africa

Xyridopsis Welw. ex B. Nord. = Emilia (Cass.) Cass. Asteraceae

Xyris Gronov. ex L. 1753. Xyridaceae. c. 250 tropics and subtropics

Xysmalobium R. Br. 1810. Asclepiadaceae. 40 trop. and South Africa

Yabea Kozo-Polj. 1914 (Caucalis L.). Apiaceae. 1 western North America

Yadakeya Makino = Pseudosasa Nakai. Poaceae

Yakirra Lazarides et R. d. Webster. 1985 (Panicum L.). Poaceae. 6 Burma, trop. Australia

Yarina O. F. Cook = Phytelephas Ruiz et Pav. Arecaceae

Yeatesia Small. 1896. Acanthaceae. 3 southwestern U.S. from Texas to western Florida, northwestern Mexico

Yermo Dorn. 1991. Asteraceae. 1 U.S. (Wyoming)

Yinquania Z. Y. Zhu. 1984 (or = Corns L.). Cornaceae. 2 China

Yinshania Y. C. Ma et Y. Z. Zhao. 1979 (Cochlearia L.). Brassicaceae. 7–9 China from Inner Mongolia to eastern Tibet, Yunnan, Szechwan

Ynesa O. F. Cook = Attalea Kunth. Arecaceae

Yoania Maxim. 1872. Orchidaceae. 2 Himalayas, Japan, New Zealand

Yolanda Hoehne. 1919 (Brachionidium Lindl.). Orchidaceae. 1 Brazil

Youngia Cass. 1831. Asteraceae. c. 40 temp. and trop. Asia, North America (Alaska); 6 Middle Asia, East and West Siberia, Russian Far East

Ypsilandra Franch. 1888. Melanthiaceae. 5 western and southwestern China, Nepal, Bhutan, northern Burma

Ypsilopus Summerh. 1949 (Tridactyle

Schltr.). Orchidaceae. 4 trop. Africa: Kenya, Tanzania, Zambia, Malawi, Zimbabwe, and Swaziland

Ystia Compere = Schizachyrium Nees. Poaceae

Yua C. L. Li. 1990 (Parthenocissus L.). Vitaceae. 2 subtrop. China, northern India, Nepal

Yucaratonia Burkart. 1969. Fabaceae. 1 Ecuador, Peru

Yucca L. 1753. Agavaceae. c. 40 southern U.S., Mexico, West Indies

Yunckeria Lundell = Ardisia Sw. Myrsinaceae

Yungasocereus F. Ritter = Haageocereus Backeb. Cactaceae

Yunnanea Hu = Camellia L. Theaceae

Yunquea Skottsb. = Centaurodendron Johow. Asteraceae

Yushania Keng f. = Sinarundinaria Nakai. Poaceae

Yutajea Steyerm. 1987. Rubiaceae. 1 Venezuela (Guiana Highlands)

Yuyba (Barb. Rodr.) L. Bailey = Bactris Jacq. ex Scop. Arecaceae

Yvesia A. Camus. 1927. Poaceae. 1 Madagascar

Z

Zabelia (Rehder) Makino. 1948 (Abelia R. Br.). Caprifoliaceae. 15 East Asia

Zacateza Bullock. 1954. Asclepiadaceae. 1 trop. Africa

Zacintha Hill = Crepis L. Asteraceae

Zaczatea Baill. = Raphionacme Harv. Asclepiadaceae

Zahlbrucknera Rchb. = Saxifraga L. Saxifragaceae

Zalaccella Becc. = Calamus L. Arecaceae

Zaleya Burm. f. 1768. Aizoaceae. 6 southeastern Mediterranean, trop. Africa, India, Sri Lanka, trop. Australia

Zaluzania Pers. 1807. Asteraceae. 9 southern U.S. (Arizona), Mexico (8), Ecuador (1)

Zaluzianskya F. W. Schmidt. 1793. Scrophulariaceae. 40 South Africa; 1 extending to Zimbabwe, Uganda, and Mozambique

Zamia L. 1763. Zamiaceae. c. 50 America: southeastern U.S. (Z. integrifolia L. f.—Georgia, Florida), Mexico, Guatemala, Belize, El Salvador, Honduras, Nicaragua, Costa Rica, Panama, Colombia, Venezuela, Ecuador, Peru, Bolivia, and Brazil; West Indies: Cuba, Hispaniola, Jamaica, Puerto Rico

Zamioculcas Schott. 1856. Araceae. 1 Kenya, Tanzania, Malawi, Zimbabwe, Mozambique, South Africa (Natal)

Zandera D. L. Schulz. 1988. Asteraceae. 3 Mexico

Zanha Hiern. 1896. Sapindaceae. 1–2 trop. Africa

Zannichellia L. 1753. Zannichelliaceae. 3–8: Z. palustris—cosmopolitan; Z. aschersoniana—South Africa; Z. andina—Peru; Russia (6, from European part to Kamchatka)

Zanonia L. 1753. Cucurbitaceae. 1 South and Southeast Asia, Malesia

Zantedeschia Spreng. 1826. Araceae. 8–9 trop. and South (6) Africa

Zanthorhiza L'Hér. = Xanthorhiza Marshall. Ranunculaceae

Zanthoxylum L. 1753. Rutaceae. c. 250 tropics and (few species) subtropics

Zapoteca H. M. Hern. 1986 (Calliandra Benth.). Fabaceae. 17 trop. America from southwestern U.S. and northern Mexico to northern Argentina, West Indies

Zataria Boiss. 1844. Lamiaceae. 1 southern Iran, eastern Afghanistan, western Pakistan

Zauschneria C. Presl = Epilobium L. Onagraceae

Zea L. 1753. Poaceae. 4 Central and South America

Zebrina Schnizl. = Tradescantia L. Commelinaceae

Zederbauera H. P. Fuchs = Erysimum L. Brassicaceae

Zehnderia C. Cusset. 1987. Podostemaceae. 1 Cameroun

Zehneria Endl. 1833. Cucurbitaceae. 35 trop. Old World

Zehntnerella Britton et Rose = Facheiroa Britton et Rose. Cactaceae

Zelkova Spach. 1841. Ulmaceae. 5 eastern Mediterranean, West and East Asia; Z. carpinifolia—Caucasus

Zemisne O. Deg. et Sherff = Scalesia Arn. ex Lindl. Asteraceae

Zenia Chun. 1946. Fabaceae. 1 southern China, Thailand, Vietnam

Zenkerella Taub. 1894. Fabaceae. 6 (Z. citrina) Ghana, Nigeria, Cameroun, Gabon; 5 Tanzania

Zenkeria Trin. 1837. Poaceae. 4 India, Sri Lanka

Zenkerophytum Engl. ex Diels. 1910 (Syrrheonema Miers). Menispermaceae. 1 trop. West Africa

Zenobia D. Don. 1834. Ericaceae. 1 southeastern U.S.

Zephyra D. Don. 1832. Tecophilaeaceae. 1 (?) Chile

Zephyranthella Pax. 1930 (Habranthus Herb.). Amaryllidaceae. 1 Argentina

Zephyranthes Herb. 1821. Amaryllidaceae. 71 trop. and subtrop. America, West Indies

Zeravschania Korovin. 1948. Apiaceae. 6–7 Caucasus and Middle Asia (3), Iran, Afghanistan

Zerdana Boiss. 1842. Brassicaceae. 1 montane Iran

Zerna Panz. = Vulpia C. C. Gmel. Poaceae

Zerumbet J. C. Wendl. 1798 (Alpinia Roxb.). Zingiberaceae. 5 Malay Peninsula

Zetagyne Ridl. 1921. Orchidaceae. 1 Southeast Asia

Zeugandra P. H. Davis. 1950. Campanulaceae. 2 Iran

Zeugites P. Browne. 1756. Poaceae. 10 America from Mexico through Andes to Peru, West Indies

Zeuktophyllum N. E. Br. 1927. Aizoaceae. 1 South Africa

Zeuxanthe Ridl. = Prismatomeris Thwaites. Rubiaceae

Zeuxine Lindl. 1826. Orchidaceae. 76 Europe, Africa, Madagascar, trop. and subtrop. Asia eastward to Ryukyu Is., Malesia, Australia, New Caledonia, Vanuatu, Fiji, Tonga, Niue, and Samoa; Z. strateumatica—extending to Middle Asia

Zexmenia Llave ex Lex. 1824. Asteraceae. 2 Mexico, Guatemala, Belize, Honduras, Costa Rica

Zeyheria Mart. 1826. Bignoniaceae. 2 Brazil, Bolivia

Zeylanidium (Tul.) Engl. 1930 (Hydrobryum Endl.). Podostemaceae. 4 southern and northeastern (Assam) India, Sri Lanka, Burma

Zhumeria Rech. f. et Wendelbo. 1967. Lamiaceae. 1 southern Iran

Zieria Sm. 1798. Rutaceae. 25 eastern Australia

Zieridium Baill. 1872. Rutaceae. 3 New Caledonia

Zigadenus Michx. 1803. Melanthiaceae. 15 Russia (1, Z. sibiricus—from southern Ural to Far East), northeastern China, Japan (Risiri and Hokkaido Is.), North America

Zilla Forssk. 1775. Brassicaceae. 1 (Z. spinosa) Morocco, Algeria, Tunisia, Libya, Egypt, Lebanon, Syria, Israel, Jordan, Sinai and Arabian peninsulas

Zimmermannia Pax. 1910. Euphorbiaceae. 6: Kenya (1), Tanzania (5)

Zimmermanniopsis Radcl.-Sm. 1990. Euphorbiaceae. 1 Tanzania

Zingeria P. A. Smirn. 1946. Poaceae. 5 Southeast Europe, West (incl. Caucasus) and Southwest Asia

Zingeriopsis Probat. 1977 (Zingeria P. A. Smirn.). Poaceae. 1 Turkey

Zingiber Boehm. 1760. Zingiberaceae. c. 100 Himalayas, India, East and Southeast Asia, Malesia

Zinnia L. 1759. Asteraceae. 33 America from southern U.S. to Chile and Brazil

Zinowiewia Turcz. 1859. Celastraceae. 11 trop. America from Mexico to Venezuela

Zippelia Blume ex Schult. et Schult. f. 1830 (Piper L.). Piperaceae. 1 southwestern China, Indochina, Java

Zizania L. 1753. Poaceae. 4 South and East Asia, North America; southeastern East Siberia, southern Russian Far East

Zizaniopsis Doell et Asch. 1871. Poaceae. 5 southeastern U.S., trop. South America

Zizia W. D. J. Koch. 1824. Apiaceae. 4 North America

Ziziphora L. 1753. Lamiaceae. 25 Mediterranean, Southeast Europe, West (incl. Caucasus), Southwest and Middle Asia eastward to southern Siberia, Mongolia, western China, and Afghanistan

Ziziphus Mill. 1754. Rhamnaceae. c. 90 tropics and subtropics; Z. jujuba—from Southeast Europe to China; 28 Caucasus and Middle Asia

Zizyphus Adans. = Ziziphus Mill. Rhamnaceae

Zoegea L. 1767. Asteraceae. 6 West Asia from Sinai Peninsula to Middle Asia (2), Afghanistan, and northwestern India

Zoellnerallium Crosa. 1975. Alliaceae. 1 Chile

Zollernia Wied-Neuw. et Nees. 1826. Fabaceae. 10–12 trop. America

Zollingeria Kurz. 1872. Sapindaceae. 4 Burma, Thailand, Laos, Borneo (1, Z. borneensis)

Zombia L. H. Bailey. 1939. Arecaceae. 1 Haiti

Zombitsia Keraudren. 1963. Cucurbitaceae. 1 Madagascar

Zomicarpa Schott. 1856. Araceae. 3 southern Brazil

Zomicarpella N. E. Br. 1881. Araceae. 1 Colombia

Zonanthus Griseb. 1862. Gentianaceae. 1 Cuba

Zonotriche (C. E. Hubb.) J. B. Phipps. 1964. Poaceae. 3 trop. Africa

Zootrophion Luer. 1982. Orchidaceae. 12 trop. America

Zornia J. F. Gmel. 1792. Fabaceae. c. 80 tropics and subtropics

Zosima Hoffm. 1814 ("Zozima," "Zozimia," "Zosimia"). Apiaceae. 4 Southwest Asia eastward to Pakistan; 2 Caucasus, Middle Asia

Zostera L. 1753. Zosteraceae. 15 subtrop. and extratrop. seas of the Northern Hemisphere (subgenus Zostera); temp. and subtrop. (a few species in trop. Africa, Southeast Asia, and Australia) seas (subgenus Zosterella)

Zosterella Small. 1913 (Heteranthera Ruiz et Pav.). Pontederiaceae. 2 temp. and subtrop. North America

Zoutpansbergia Hutch. = Callilepis DC. Asteraceae

Zoysia Willd. 1801. Poaceae. 10 Mascarene Is., South and East Asia, Australia; Z. japonica—extending Far East

Zschokkea Muell. Arg. = Lacmellea Karst. Apocynaceae

Zuccagnia Cav. 1799. Fabaceae. 1 Chile, Argentina

Zuccarinia Blume. 1827. Rubiaceae. 2 Sumatra and Java

Zuckia Standl. 1915. Chenopodiaceae. 1 U.S. (Arizona)

Zuelania A. Rich. 1841 (1843). Flacourtiaceae. 1 (Z. guidonia) trop. America from southern Mexico to northern Colombia and northern Venezuela, West Indies

Zunilia Lundell = Ardisia Sw. Myrsinaceae

Zuvanda Askerova. 1985. Brassicaceae. 3 southern Caucasus, Asia Minor, Syria, Lebanon, Israel, Iraq, Iran

Zygia P. Browne. 1756. Fabaceae. c. 20 Mexico, Guatemala, Belize, Nicaragua, Costa Rica, Panama, Colombia, Ecuador, Peru, Venezuela, Surinam, Guyana, Brazil, Bolivia, Paraguay, Argentina

Zygochloa S. T. Blake. 1941. Poaceae. 1 central Australia

Zygodia Benth. 1876. Apocynaceae. 5 trop. Africa

Zygogynum Baill. 1867. Winteraceae. 41 Madagascar, Malesia, Australia, Lord Howe I., Solomon Is., New Caledonia

Zygonerion Baill. = Strophanthus DC. Apocynaceae

Zygoon Hiern = Tarenna Gaertn. Rubiaceae

Zygopetalon Hook. 1827. Orchidaceae. 35 trop. America from southern Mexico to Peru, Bolivia, Paraguay, and Brazil

Zygopetalum Lindl. = Zygopetalon Hook. Orchidaceae

Zygophyllum L. 1753. Zygophyllaceae. c. 100 South and Southeast Europe, Mediterranean, West (incl. Caucasus) and Middle Asia, West Siberia eastward to Mongolia, western China and Afghanistan; South Africa, Australia

Zygoruellia Baill. 1890. Acanthaceae. 1 Madagascar

Zygosepalum Rchb. f. 1859. Orchidaceae. 5 trop. South America: Colombia, Venezuela, the Guianas, Peru, and Brazil

Zygosicyos Humbert. 1945. Cucurbitaceae. 2 Madagascar

Zygostates Lindl. 1837. Orchidaceae. 4 Brazil

Zygostelma Benth. 1876. Asclepiadaceae. 1 Thailand

Zygostigma Griseb. 1838. Gentianaceae. 2 Brazil, Argentina

Zygotritonia Mildbr. 1923. Iridaceae. 4 trop. Africa from Senegal to southern Sudan, Zaire, Zambia, Tanzania

Zyzyxia Strother. 1991. Asteraceae. 1 Guatemala, Belize

8905217656?

b89052176567a

NON-CIRCULATING